Separations Division 2014

Core Programming Area at the 2014 AIChE Annual Meeting

Atlanta, Georgia, USA
16-21 November 2014

Volume 1 of 2

ISBN: 978-1-5108-1272-7

Printed from e-media with permission by:

Curran Associates, Inc.
57 Morehouse Lane
Red Hook, NY 12571

proceedings.com

Some format issues inherent in the e-media version may also appear in this print version.

Printed by Curran Associates, Inc. (2015)

For permission requests, please contact AIChE
at the address below.

AIChE
120 Wall Street, FL 23
New York, NY 10005-4020

Phone: (800) 242-4363
Fax: (203) 775-5177

www.aiche.org

Additional copies of this publication are available from:

Curran Associates, Inc.
57 Morehouse Lane
Red Hook, NY 12571 USA
Phone: 845-758-0400
Fax: 845-758-2634
Email: curran@proceedings.com
Web: www.proceedings.com

TABLE OF CONTENTS

VOLUME 1

VOLUME 2

The Combination of Forward Osmosis (FO)/Membrane Distillation (MD) Membrane Process with Thermophilic Anaerobic Bioprocess for Sustainable Energy Recovery and Wastewater Reclamation

Zhang Jinsong,* Kwan Zheng Xin Joseph, Goh Shuwen, Anthony Gordon Fane*

Singapore Membrane Technology Centre(SMTC),Nanyang Environment and Water Research Institute(NEWRI),Nanyang Technological University,1 Cleantech Loop, CleanTech One, #06-08, Singapore 637141

*Corresponding author: jszhang@ntu.edu.sg, *Tel: 65-65921835, Fax: 65- 67910756*

Abstract

The organic loading and temperature of wastewater from industries such as chemical, petroleum and pharmaceutical are typically high. These characteristics make them potential candidates for anaerobic bioprocess, which is able to process high strength wastewater and generate energy in the form of methane. Due to the high temperature of wastewater, applying thermophilic anaerobic digestion has additional benefits over mesophilic digestion, including a higher COD removal efficiency, faster degradation rate, and lower methane solubility in wastewater. However, due to the incomplete degradation of organics, the effluent from the anaerobic bioprocess usually contains a high concentration of organics, which makes it difficult to reuse or dispose according to the discharge standard.

In this study, a thermophilic upflow anaerobic sludge blanket (UASB) combined with the forward osmosis (FO)/membrane distillation (MD) integrated process has been demonstrated for the first time. The schematic diagram of anaerobic bioprocess with FO/MD module is shown in Figure 1. The MD process uses microporous hydrophobic membranes that allow water vapour to pass through the dry pores, while retaining the non-volatile organics and inorganics. If the MD membrane was directly used to treat the UASB effluent, serious biofouling, scaling and wetting will occur. The FO process is a membrane separation process which uses semi-permeable membranes that allows water molecules to pass through while retaining bigger size solutes. The driving force for water transport is an osmotic pressure gradient achieved by using an osmotic 'draw solution' (DS). The FO/MD integrated process captures the complementary advantages provided by FO and MD in a single step. By applying the FO membrane process at the anaerobic bioprocess effluent side, nearly all the organics and inorganics can be rejected at a low fouling rate. This makes the FO membrane a perfect barrier to prevent MD fouling and wetting. High temperature feed water will also mitigate the concentration polarization effects usually observed in FO processes, resulting in enhanced FO membrane flux. The draw solution is simultaneously regenerated in-situ via the MD process, which completely retains the solute (such as NaCl) in DS while allowing water to diffuse across the membrane into the permeate reservoir. The FO/MD process allows us to achieve a product that is of distilled water quality and ready for reuse in a single step.

Our experiments have shown that the anaerobic bioreactor with FO/MD module had a high rejection for TOC, with a calculated average of 99.55 ± 0.24 %. TOC levels in the permeate remained fairly constant below 0.5 mg/L. Permeate conductivity was stable at 5 μS cm^{-1} , which is comparable to the RO permeate of wastewater reclamation. The average composition of the biogas generated from the UASB process was found to be 67.4 % methane, with the highest percentage being 83.6 %. Optimally, the bioprocess is able to produce a maximum of 0.35 L methane for every gram of chemical oxygen demand (COD) that is degraded.

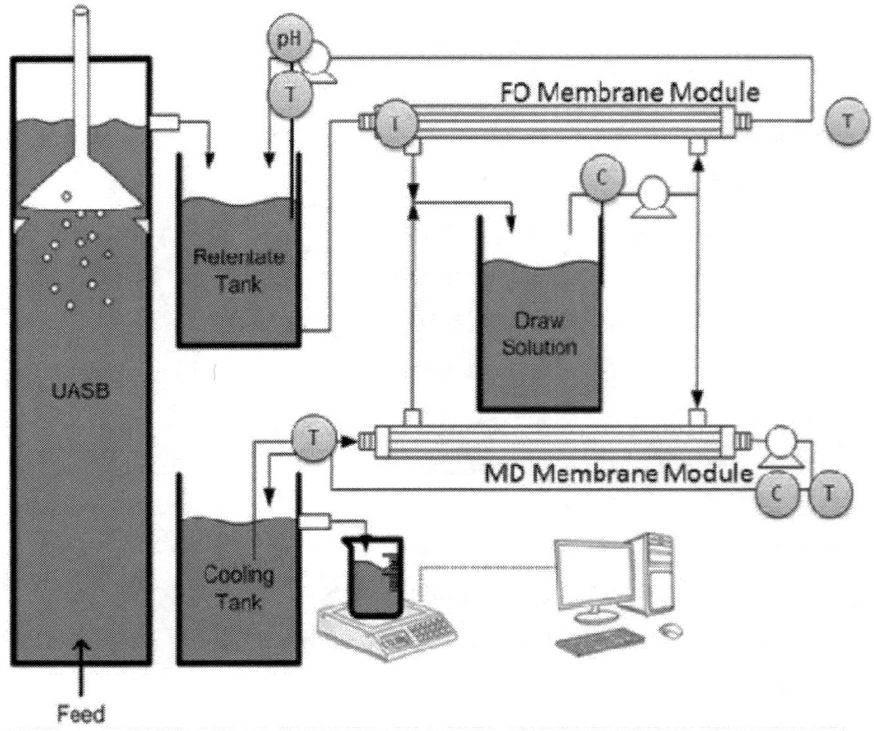

Figure 1 Schematic diagram of side-stream UASB-FO/MD system

In conclusion, the integrated process of combining the anaerobic bioprocess with the FO/MD process is a state-of-art technology and a viable alternative to wastewater reclamation. This technology appears to be particularly well suited for high-strength industrial waste streams reclamation due to the potential to achieve high quality water product with lower energy consumption and higher methane production.

Keywords: Thermophilic, Upflow Anaerobic Sludge Blanket (UASB), Forward Osmosis (FO), Membrane Distillation (MD), FO/MD integrated module

387586 Developing Advanced Antifouling Membranes Using Complementary Experimental and Theoretical Methods

Abstract:

Membrane fouling is very complex and affected by concentration polarization, the hydrophobicity, charge and polarity of the foulants and the properties of the membrane surface. Designing polymer surfaces that are non-fouling has been a central issue in membrane research. Developing advanced antifouling membranes requires both fundamental understanding of the fouling mechanisms as well as implementing novel approaches to membrane processes. Here both modeling and experimental techniques are used to understand and develop advanced anti-fouling membranes.

Surface modification with zwitterionic materials has been found to be effective in resisting protein attachment. Atomistic molecular dynamics simulations are used to understand charge and charge separation distance on the hydration of zwitterionic materials and their role in resisting fouling. In addition, membrane surface modification of hydrophilic polymers with superparamagnetic nanoparticles attached at the polymer chain ends has been implemented to develop advanced antifouling membranes. Under an appropriate external magnetic field, these functionalized nanoparticles can act as localized nanoheaters or as micro-mixers, both of which are effective for reducing surface fouling. Current progress on these three areas will be discussed.

389935 Bioinspired Hybrid Membranes for Efficient CO2 Capture

Abstract:

Bioinspired hybrid membranes for efficient CO_2 capture

Authors: Yifan Li, Shaofei Wang, Xueqin Li, Qingping Xin, Hong Wu, Zhongyi Jiang*

Collaborative Innovation Center of Chemical Science and Engineering (Tianjin); Key Laboratory for Green Chemical Technology of Ministry of Education School of Chemical Engineering and Technology, Tianjin University, 300072, P. R. China

[*] Corresponding author and presenter: zhyjiang@tju.edu.cn

Abstract Hybrid membranes composed of continuous polymer phase and filler phase have been extensively explored for gas separation, including CO_2 capture. The multiphase characteristics of composite membranes render more degree of freedom to manipulate multiple interactions, tailor multiscale structures and integrate multiple functionalities, compared to pristine polymer membranes. However, the tradeoff effect between selectivity and permeability is a daunting challenge. Learning from nature is a shortcut for dealing with this challenge and designing new generation hybrid membrane. In this talk, several kinds of bioinspired hybrid membranes were designed and fabricated for CO_2 capture. On one hand, the ways living organisms employ to manipulate multiple interactions were borrowed to tailor hierarchical structures of hybrid membranes, especially for free volume characteristics. On the other hand, the selective transport mechanisms of cell membranes based on channel proteins and carrier proteins were imitated to confer high CO_2capture performance on membrane. We hope our efforts will offer new design alternatives for hybrid membranes, and foster the development of hybrid membranes for energy-related applications.

Key words Bioinspiration; hybrid membranes; CO_2 capture.

387799 The Use of Green Nanoparticles As a Biofouling-Resistant Agent in Membrane Separations

Abstract:

With the rapid growth and development of countries around the world, and the inherent need of clean water, it is imperative to treat brackish- and sea-water sources. The two leading desalination technologies are thermal and membrane technologies, specifically reverse osmosis (RO). While nanofiltration (NF) is used for the removal of other substances from water source, it is also commonly used for the desalination of water. While both RO and NF can produce potable water, RO removes a large majority of solutes, which leaves the permeate water void of any essential nutrients (calcium, magnesium ions, etc.), placing the nutrient levels below that of the required World Health Organization standards. This means nutrients must be added back (i.e. reminerailization) to bring the water back to the standard levels for drinking water. NF methods, on the other hand, remove fewer solutes, and thus may avoid the need to remineralize the water. Thus, it is a more desirable technology for developing nations.

This project seeks to address a major technical challenge in membrane technology, mitigation of membrane biofouling, due to rejected chemicals and microbes, by impregnating the solutions with cost effective nanoparticles. Most of the research and development in the area of biofouling prevention has focused on pretreatment of the feed water, improved cleaning solutions and cleaning procedures. In this project, biofouling-resistant (BRN) nanocomposite membranes, loaded with green silver nanoparticles were studied from synthesis to macro-scale production.

Filtration experiments were performed with synthetic brackish water containing biofoulant *Pseudomonas fluorescens Migula* to test regular cellulose acetate (CA) membranes and CA membranes loaded with silver nanoparticles (CA-AgNP). CA-AgNP membranes resulted in lower flux declines when compared to CA membranes. CA membranes showed a flux decline of approximately 26% as compared to 20% for CA-AgNP membranes during 7 hours of filtration. Though the differences are small in the decline of flux, the flux recovered after backwashing was higher for the CA-AgNP membranes (93%) than for the CA membrane (84%). Therefore, filtration experiments using *p. fluorscens* showed that CA-AgNP membranes were more effective at mitigating biofouling formation.

391676 Pressure Recovery Systems for Pressure Driven Membrane Processes

Abstract:

The majority of gas and liquid membrane separation processes are pressure driven. For some separations, such as nitrogen production from air, the desired product is the high pressure retentate stream and the product is desired to be delivered at pressure.

For many other separations, such as filtration or oxygen production from air, the desired product is the low pressure permeate and the high pressure retentate commonly is depressurized without attempt to capture the energy released upon depressurization. Pressure recovery devices for use in membrane desalination plants are increasingly commonplace and can recover in excess of 90% of the energy release.

We describe alternative devices for gas separation processes in which the permeate is the desired product and may require recompression for a second stage to achieve desired purities. The devices relay upon direct contacting between high and low pressure streams in devices that closely resemble the modules used in the separation process. The use of such devices could have a significant impact on energy consumption in pressure driven membrane gas separation processes.

Analytical Models for Osmotically Driven Membrane Processes at Osmotic Equilibrium

Sherwood Benavides, Alex Oloriz, William A. Phillip
182 Fitzpatrick Hall
Department of Chemical and Biomolecular Engineering
University of Notre Dame
Notre Dame, IN 46556

Forward osmosis (FO) is an emerging water treatment technology that has demonstrated a unique ability to handle difficult and complex feed streams (*e.g.*, those found in the desalination of high-salinity brines and wastewater reclamation). In the FO process, an engineered draw solution is used to generate an osmotic pressure difference that drives the flux of water through a semipermeable membrane from the feed solution into the draw solution. Simultaneously, solutes present in the feed solution and the draw solution move down their respective concentration gradients. Understanding the evolution of the water and solute fluxes along the length of the membrane module is vital to determine the optimal operating conditions for an FO process. This study seeks to quantify the effects of water and solute fluxes on the performance of novel FO systems using simple mathematical models. Specifically, because the presence of a water flux renders the governing equations non-linear, it is useful to find solutions that describe the system in appropriate and physically relevant limits. Here, a binary system (*i.e.*, one feed solute and one draw solute) is studied, and the performance of the system is described in terms of the recovery rate, which quantifies the amount of water that moves from the feed solution to the draw solution, and the separation factors, which quantify the amount of the solutes that permeate through the membrane. The influence of membrane properties (*e.g.*, the relative magnitude of the permeability of water to the permeability of solutes) and operating conditions (*e.g.*, the ratio of the initial flow rate of the feed solution to the initial flow rate of the draw solution) on the percent recovery and the separation factors will be discussed. These models and their solutions will ultimately allow for the prediction of system level performance based on lab scale membrane characterization.

Thin and Durable Thin Film Composite (TFC) Membranes with Enhanced FO Performance

Young-Hye Na,[1*] Daniel Stillman,[2] Leslie Krupp,[1] Joe Sly,[1] and Victoria Piunova[1]

[1]*IBM Almaden Research Center, 650 Harry Road, San Jose, CA 95120*
[2]*Department of Chemical Engineering, San Jose Sate University, San Jose, CA95192*

[*]Young-Hye Na: 408-927-1256 (phone), 408-927-3310 (fax)
yna@us.ibm.com (e-mail)

Forward Osmosis (FO), an osmotically driven membrane process, has become a robust technology for various water treatments including seawater desalination and waste water treatment. Compared to pressure-driven membrane processes such as reverse osmosis (RO), FO operates at near zero hydraulic pressure, resulting in lower operating cost and lower membrane fouling tendency. Despite the advantages and applications, the lack of a suitable FO membrane hinders FO application on a commercial level, and additional research is required to develop a high-performance FO membrane. One major concern in using traditional thin film composite (TFC) polyamide RO membranes for FO applications is internal concentration polarization (ICP), which is induced by salt build up within thick, hydrophobic support layers (a porous PSF mid-layer on a non-woven PE backing). To reduce ICP, FO membranes should be designed to have small thicknesses with low tortuosity and hydrophilic surface nature in their support layers.

Here, we demonstrated thin and durable TFC polyamide membranes comprised of dense aromatic polyamide active layers and mesh-embedded porous PSF supports. The strategy to embed a polyester (PE) mesh into a PSF support without using an additional thick backing layer (ex. non-woven PE fabric) enabled a reduction in the overall thickness of a TFC membrane while retaining good mechanical durability (easy handling). For the mesh-embedded PSF support prepared with a relatively low concentration casting solution (9 wt% PSF in NMP), the PE mesh was oriented towards the bottom of the PSF support, creating macroscopic pores between the PSF matrix and the mesh lines in the backside surface, and the PSF matrix showed a well-developed finger-like cross-sectional morphology. These structural characteristics effectively mitigated the tendency of ICP, resulting in outstanding FO performance of the corresponding TFC membrane; almost 2.5 times higher water flux (49 LMH) with about 6 times lower reverse salt flux (7.0 GMH) compared to a commercial CTA membrane in PRO mode evaluation. In the presentation, the influence of a PE mesh on the formation of PSF pore structures and consequently on the final FO performance of corresponding TFC membranes will be thoroughly discussed. Furthermore, we will also introduce novel hydrophilic functional polymers which we have recently developed with the aim of improving wetting behavior of hydrophobic PSF supports. The hydrophilic polymer coatings formed on the surface of PSF pores resulted in better transport of water and salts within the support layers (efficient reduction of IC)) during FO operation and consequently the modified membranes showed enhanced FO performance with higher water flux.

390041 Ion Exchange in Forward Osmosis: A New Membrane and Draw Solution Design Problem

Abstract:

For the past decade, most of the high profile research on osmotic membrane processes has focused on making new membranes with low structural parameter support layers. As such, the active layer has been by comparison ignored in its design. Conventional wisdom in the field has suggested that a typical reverse osmosis-type selective layer (aromatic polyamide) would suffice as a selective layer for FO as well, so long as the supporting structure were designed appropriately. Unfortunately, this view may have been misguided. Polyamide layers contain carboxylic acid groups which deprotenate and become negatively charged in most aqueous environments. Many draw solutions contain cations which can exchange with cations in the feed solution across this negatively charged membrane. Such transport will negatively impact permeate water quality in FO. This study examines this phenomenon both from the practical perspective of desalination performance as well as a more systematic look at the transport. We believe that understanding this problem will help redirect some attention of osmotic membrane design back toward the selective layer properties where adjustments can be made to mitigate cation transport and improve solute rejection in FO.

383013 Outer-Selective Thin-Film Composite (TFC) Hollow Fiber Membranes for Osmotic Power Generation

Abstract:

An outer-selective thin-film composite (TFC) hollow fiber membrane using Ultem® as the support material has been developed for osmotic power generation with the pressure-retarded osmosis (PRO) process. Firstly, a defect-free TFC membrane has been achieved on the shell side of the fiber by a simple pre-wetting step with alcohols. The PRO performance has further been improved by optimizing the support structure in terms of morphology, pore size and mechanical strength. The study may provide insightful guidelines for developing a new TFC membrane generation for PRO power production.

367772 Forward Osmosis Polysaccharide Fouling: Effect of Solution Viscosity and Foulant Deposit Resistance

Abstract:

In recent years, forward osmosis (FO) has been attracting an intensive research interest due to its potential for low-energy water desalination and wastewater treatment. Membrane fouling, especially that caused by polysaccharides, has been regarded as one of the significant challenges retarding the development of FO process in various applications. In order to efficiently alleviate polysaccharide fouling in FO separation, it is necessary to strengthen the fundamental understanding on its mechanisms. This study systematically investigated the fouling behaviors of different polysaccharides, namely alginate, xanthan and pullulan, in FO. The three selected model foulants behaved differently; specifically the polysaccharide solution of high viscosity in the presence of calcium ions are likely to lead to severe membrane fouling, due to adsorption and accumulation of the polysaccharide on the membrane surface.

383695 Separation of Binary Mixtures Containing CO2: From Adsorption Equilibrium to Fixed Bed Dynamics

Abstract:

The increasing emissions of carbon dioxide have been claimed to lead to global warming, as a consequence of the intensification of the greenhouse effect in our planet. Therefore, technologies under the concept of Carbon Capture and Storage (CCS) have been developed with the purpose of reducing, or at least stabilizing, the concentration of CO_2 in the atmosphere. Adsorption-based methods are a promising cost-efficient technology, particularly in continuous mode, such as Pressure Swing Adsorption (PSA). The knowledge of adsorption fundamental equilibrium relationships, together with a comprehensive understanding of fixed bed dynamics are essential tools for the design of industrial units like PSA. The objective of the present work is to evaluate the suitability of commercially available activated carbon (AC) and MOF samples for CO_2 capture from CO2/N2 (typical of flue gas scenarios) and CO2/CH4 (typical of biogas and natural gas upgrading) mixtures. Single component adsorption data and binary breakthrough curves were obtained experimentally and modelled by pertinent thermodynamic and transport phenomena equations. The CO2/CH4 and CO2/N2 selectivity of the samples was also analyzed and compared under static and dynamic conditions. Finally, different PSA configurations were examined and compared for the two binary separations under study.

383770 Development of Adsorption Equilibrium Relations for Mixtures from Pure Component Isotherms and Henry's Law Behavior with Components in Excess

Abstract:

A new method for developing adsorption equilibrium relations for gas mixtures is presented. First, pure component isotherms are measured and described by appropriate isotherm equations. Then, binary measurements are performed with one component at a trace level and the other at a substantial loading, a condition we refer to as Henry's law behavior with a second component in excess. The virial excess mixture coefficient method is used to obtain correction terms to the ideal adsorbed solution theory, with the coefficients determined solely from the Henry's law data. Then, for data measured over the full composition space, the theory provides simple and accurate improvements to predictions of the ideal adsorbed solution theory. As an example of implementation, the method is applied to the system of nitrogen and oxygen adsorbed on LiLSX zeolite.

1

Purification of Hydrogen Using Formulated Metal-Organic Framework

Carlos A. Grande [1], Richard Blom [1], Aud I. Spjelkavik [1], Vicente I. Águeda [2] and José A. Delgado-Dobladez [2]

1- SINTEF Materials and Chemistry. Forskningsveien 1, 0373, Oslo, Norway
2. Dept. Chem. Engineering, Universidad Complutense Madrid. 28040, Madrid, Spain.

2

Outline

- Introduction
- Synthesis and formulation of UTSA-16
- Measurement of adsorption equilibrium
- Measurement of adsorption kinetics
- Breakthrough curve measurements
- PSA simulations
- Conclusions
- Acknowledgments

Pre-combustion capture of CO_2

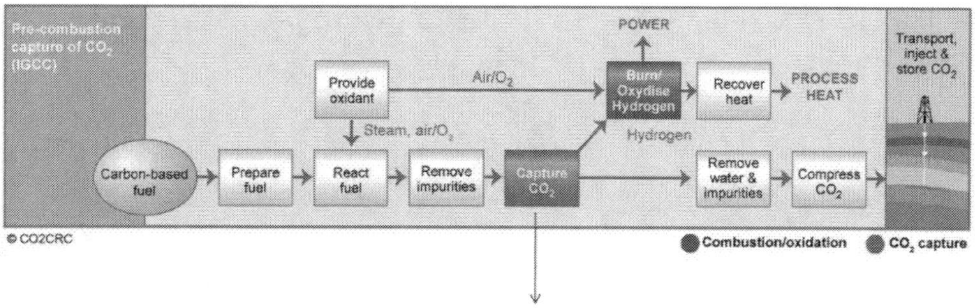

- If we use a pressure swing adsorption (PSA) process some alternatives are possible
  - Single-stage PSA (complex?)
  - Dual-stage (example in Port Arthur, Texas)

Pre-combustion capture of CO_2

- PSA technological alternatives
  - Single-stage PSA (complex if possible)
  - Dual-stage (H_2 purification + CO_2 capture)
- The dual PSA alternative has good flexibility and give us the possibility to use the knowledge developed in last 40 years of H_2-PSA.
- New materials and new operation modes should be studied to optimize the process in terms of energy requirements and environmental costs.

Materials for H_2-PSA

- Contaminants in the production of hydrogen are diverse

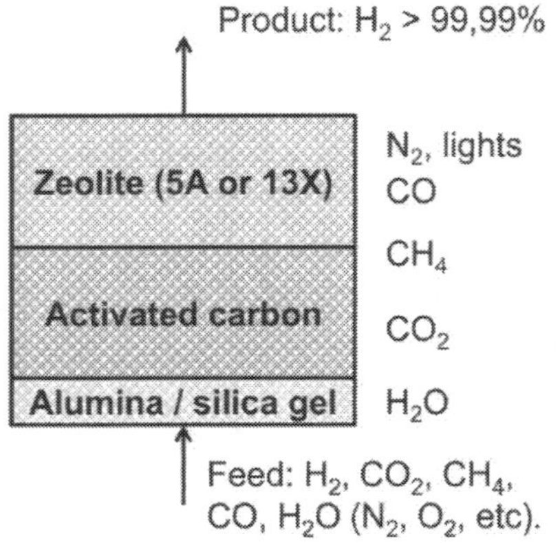

These materials are hard-to-beat in price... But, can we beat them in performance with something not so expensive?

BIGCCS International CCS Research Centre

MOF Material: UTSA-16

- Main selection criteria
  - Relatively cheap starting materials
  - High adsorption of CO_2 at reasonable pressures
  - High volumetric density

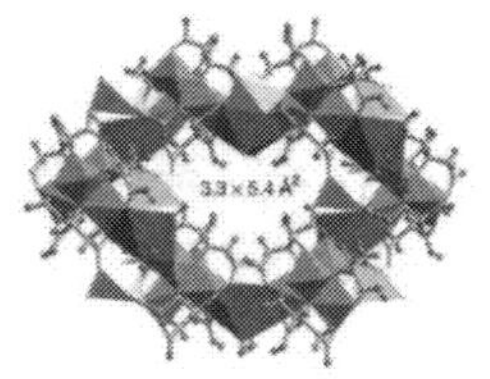

Can host 2 molecules of CO_2

Synthesis:
Citric acid
Ni-acetate 4*H_2O
Potassium hydroxide
Water
Ethanol

Heat in autoclave at 383K.

Xiang, S., He, Y., Zhang, Z., Wu, H., Zhou, W., Krishna, R., Chen, B., 2012. Microporous Metal-Organic Framework with Potential for Carbon Dioxide Capture at Ambient Conditions. Nature Commun.3, 954.

BIGCCS International CCS Research Centre

MOF formulation

7

- Extrusion is the most important technique of forming
- Extrusion of MOFs should be different
  - We cannot burn-off the plastiziser.

- Mostly due to specific properties of UTSA-16, we have used poly-vinyl alcohol (PVA).
  - We wanted to study the "compounding" properties and thus we have produced extrudates with variable amounts of PVA.

BIGCCS International CCS Research Centre

MOF formulation: results

8

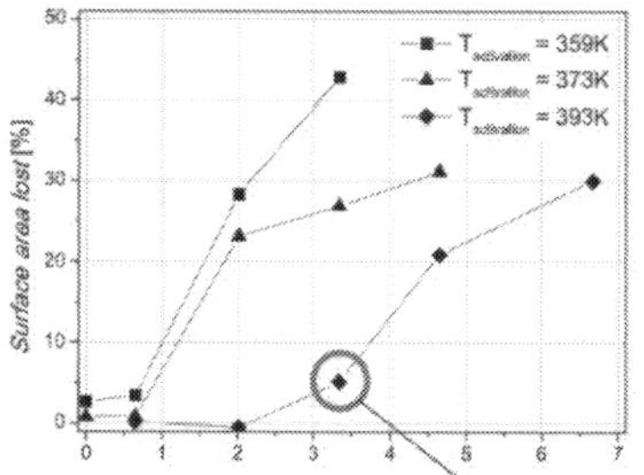

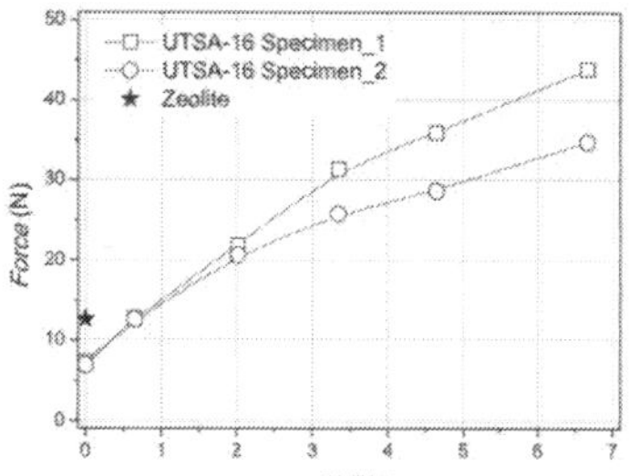

Small surface area loss.
Particles are porous.

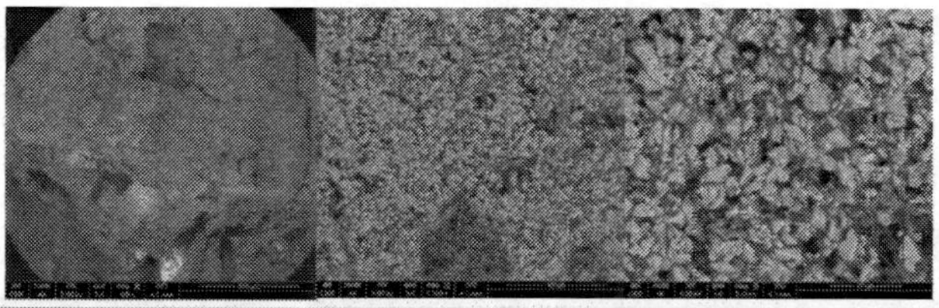

BIGCCS International CCS Research Centre

9

MOF formulation: results

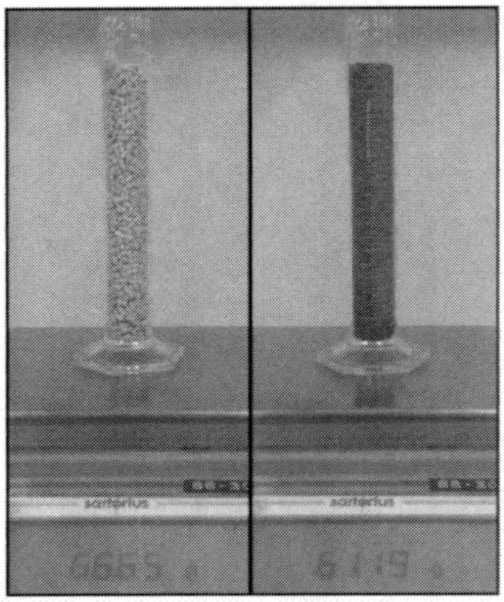

Indeed high density (~600 kg/m^3 bulk density)
We have produced around 200 grams of extrudates

BIGCCS International CCS Research Centre

10

Adsorption equilibrium results

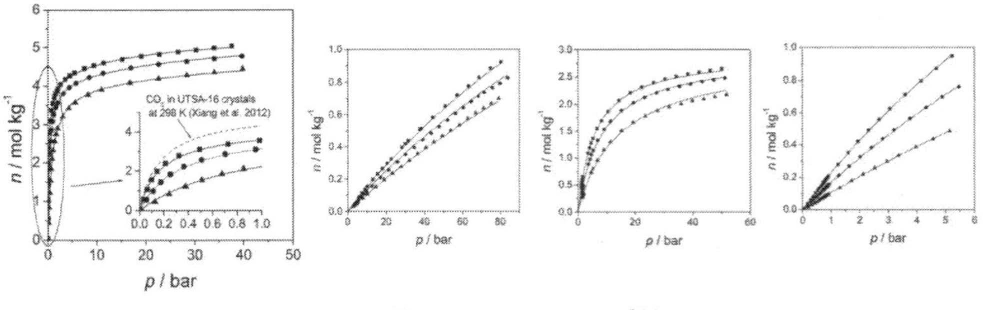

CO_2 H_2 CH_4 CO

- Isotherms at 298, 313, and 348 K.
- Carbon dioxide more adsorbed than other gases
- Steepness of CO_2 is intermediate between AC and zeolite
- Mathematical model: dual Langmuir

BIGCCS International CCS Research Centre

Adsorption kinetic results

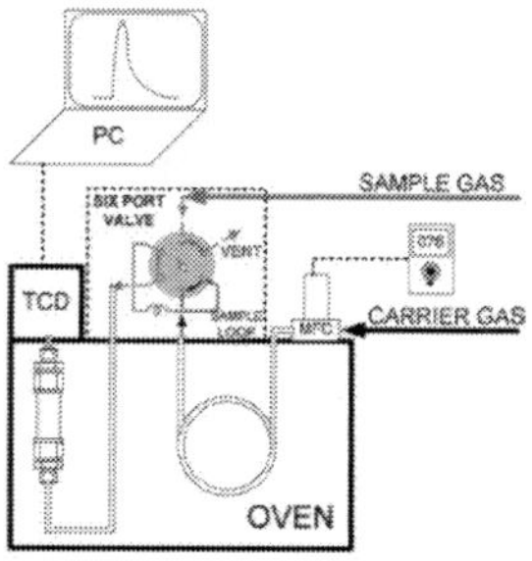

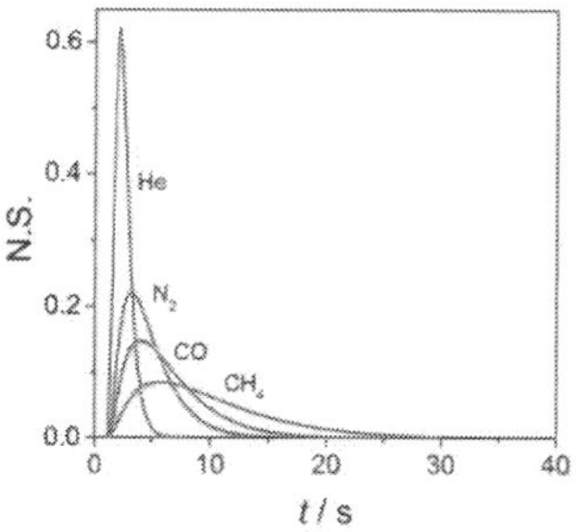

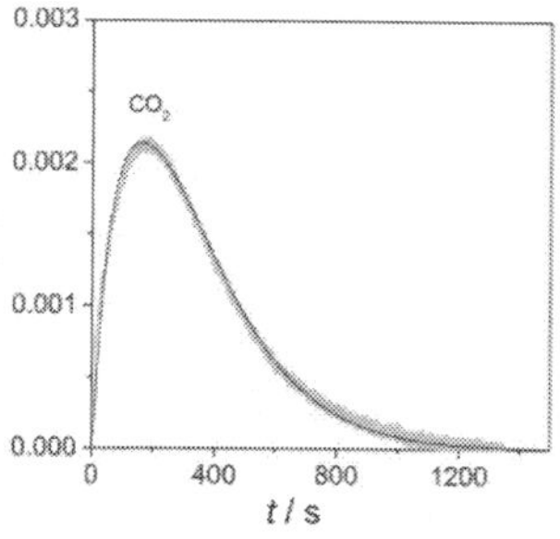

- T = 298, 313, and 348 K.
- Diffusion of hydrogen is very fast

Gas	Q, 10^{-6} m^3 s^{-1}	T, K	K_H	^aK_H manometric	$^b(D_c/r_c^2)'$, s^{-1}
H_2	0.5	298	0.22	0.882	∞
N_2	0.5	298	7.45	8.56	0.096
CO	0.5	298	13.0	13.6	0.070
CH_4	0.5	298	24.3	27.9	0.043
CO_2	0.5	298	1108	1416	0.0012

Breakthrough curves

- Very recent results. Pressure = 16 bar. Temp = 30 °C
- 76% H_2, 17% CO_2, 3% CH_4, 4% CO
- Column: 1" diameter x 0.56 m length

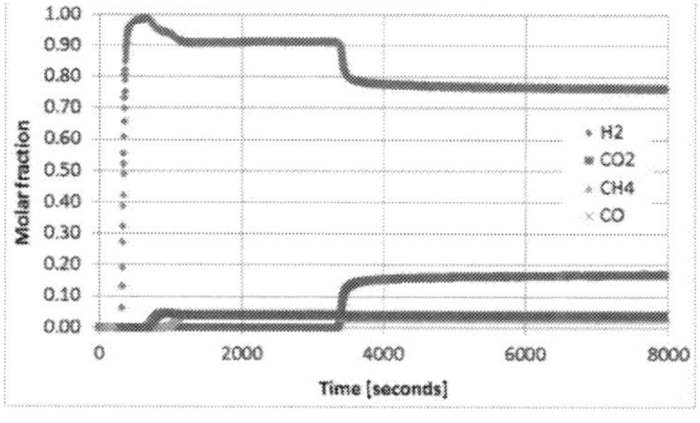

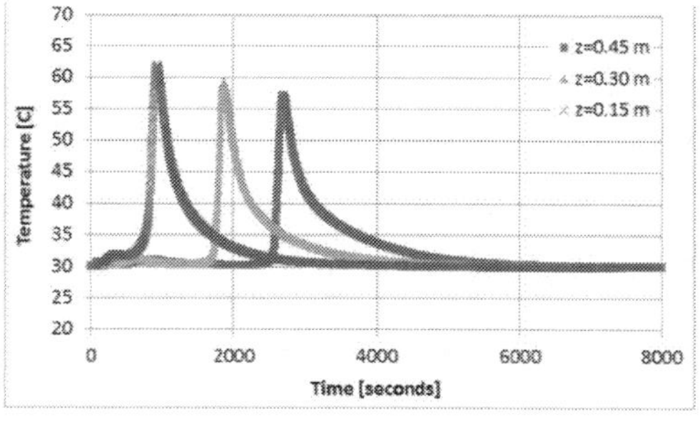

- Results confirm the very high selectivity towards CO_2

H_2-PSA results

- The 10-step cycle has two-pressure equalizations, a provide purge step and a rinse with part of the heavy product aiming to increase the cycle performance

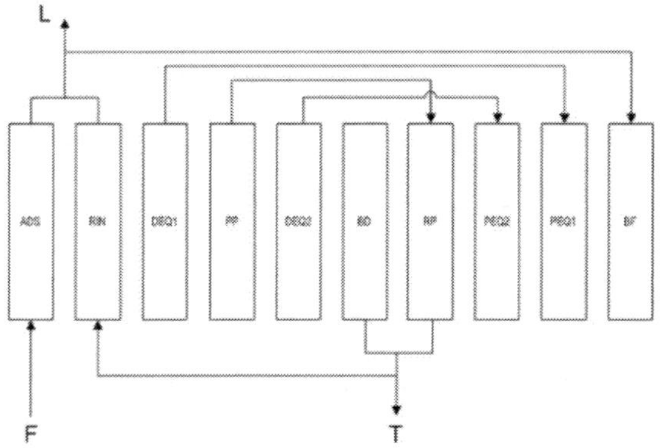

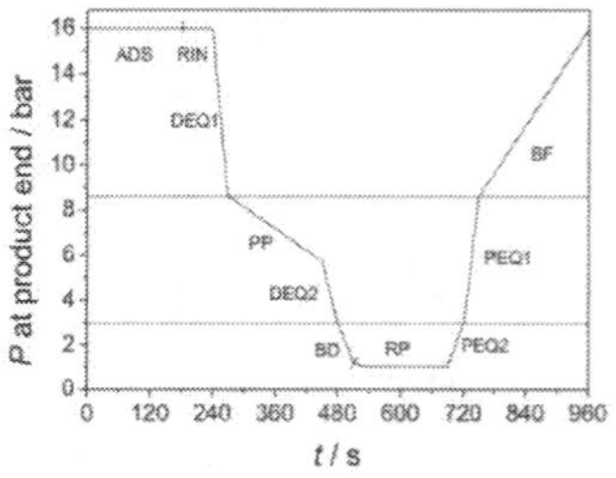

H_2-PSA results

- The 10-step cycle can be accommodated in a 4-column PSA unit
- More work has to be done to put it in "continuous feed".

Duration, s	30	180	30	30	180	30	30	180	30	30	210	
STEP	1		2	3	4	5	6	7	8	9	10	
COLUMN 1	ADS		RIN	DEQ1	PP	DEQ2	BD	RP	PEQ2	PEQ1	BF	
COLUMN 2	PEQ1	BF		ADS		RIN	DEQ1	PP	DEQ2	BD	RP	PEQ2
COLUMN 3	BD	RP	PEQ2	PEQ1	BF		ADS		RIN	DEQ1	PP	DEQ2
COLUMN 4	DEQ1	PP	DEQ2	BD	RP	PEQ2	PEQ1	BF		ADS		RIN

H_2-PSA results

15

- The 10-step cycle can be accommodated in a 4-column PSA unit
- More work has to be done to put it in "continuous feed".

Simulation	u_F, m s^{-1}	t_{RIN}, s	$^{a}P_{DEQ1}$, bar	$^{a}P_{DEQ2}$, bar	Pur., %	Rec., %	Productivity, mol$_{H2}$ kg^{-1} h^{-1}
1	0.0529	0	9.16	3.21	99.9902	80.3	3.71
2	0.0462	65	8.50	2.89	99.9947	94.8	2.78
3	0.0398	95	8.37	2.80	99.9980	96.7	2.03
4	0.0460	60	8.61	2.96	99.9991	93.0	2.80

- Quite good results so far...

BIGCCS International CCS Research Centre

H2-PSA results

16

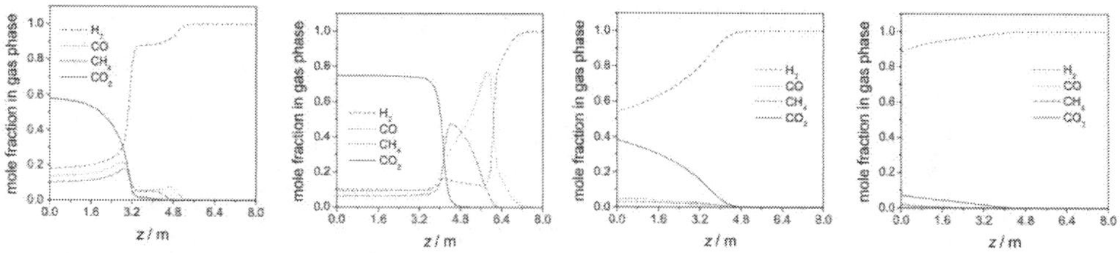

Feed Deq2 Purge Peq1

- The diffusion of all gases is relatively fast.
- Temperature variations are contained (±25K)
- Better cyclic capacity than AC, less steepness to regenerate than zeolites.

BIGCCS International CCS Research Centre

Conclusions

- We have successfully synthesized, scaled-up, formulated and tested UTSA-16 for hydrogen purification.
- The volumetric density of the extruded UTSA-16 is similar to a zeolite material.
- Diffusion is relatively fast compared to other commercial materials.
- The PSA performance is good: H_2 purity > 99.99% with H_2 recovery ranging 93-96% and unit productivity higher than 2 mol H_2. kg^{-1}.h^{-1}.

In summary

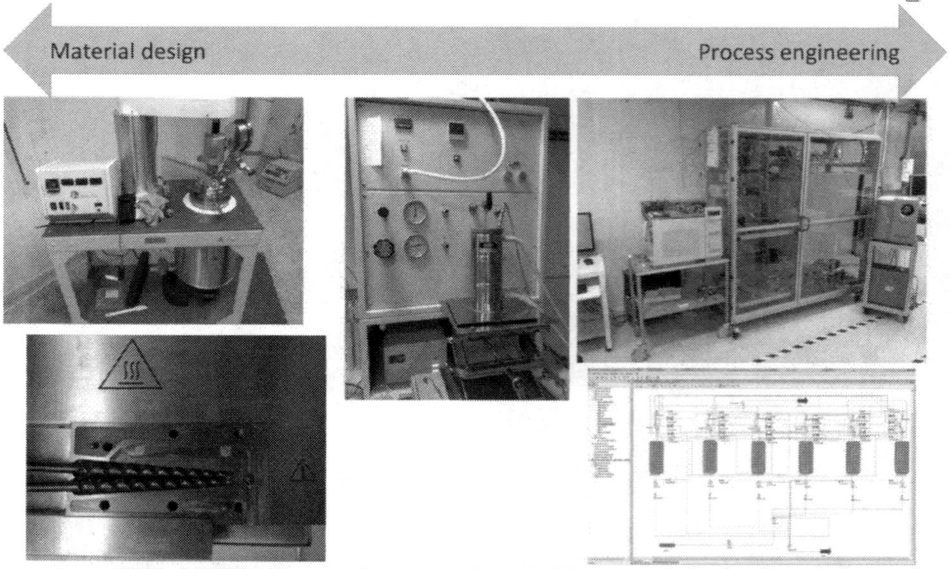

19

Acknowledgments

- The support of the Research Council of Norway through the CLIMIT program by the SINTERCAP project (233818) and through the RENERGI program (190980/S60).

- This publication has been produced with support from the BIGCCS Centre, performed under the Norwegian research program *Centres for Environment-friendly Energy Research (FME)*. The author acknowledge the following partners for their contributions: ConocoPhillips, Gassco, Shell, Statoil, TOTAL, GDF SUEZ and the Research Council of Norway (193816/S60).

BIGCCS International CCS Research Centre

20

Acknowledgments

- To Prof. Rodrigues for introducing me to the challenges of using honeycomb monoliths in adsorption

BIGCCS International CCS Research Centre

378631 Integration of Two-Column, Open-Loop, Size-Exclusion, Simulated Countercurrent Chromatography into the Downstream Processing Train of a Viral Vector

Abstract:

The purification of complex biopharmaceuticals, such as recombinant proteins, antibodies or vaccine, is gaining increasing importance to fulfill the requirements of a reliable downstream purification process with high purity and yield, and, most importantly, cost efficiency. Viruses, in particular, play an important role in the vaccine and gene therapy fields.

The downstream biopurification train of viral vectors has been extensively developed in the past years by combining different chromatography steps, namely ion-exchange (IEX)) and size-exclusion chromatography (SEC), and, less frequently, affinity chromatography, intermingled with concentration and ultra/diafiltration steps.

To be more specific, the classical approach for adenovirus purification consists of three major steps—clarification, concentration/purification, and polishing—applied sequentially: (i) clarification of the harvested bioreaction bulk to remove cell and cell debris; (ii) ultra/diafiltration; (iii) anion exchange; (iv) ultra/diafiltration to concentrate the product; and (v) SEC as a final polishing step. The SEC step is usually carried out last, mainly because of its low productivity and low product titer. Usually in this final step the amount processed is reduced by 50–100 fold.

The current paradigm in the downstream processing of viral vectors is to operate each chromatographic step using classical single-column batch chromatography. However, the cost of goods and overall processing time can be reduced by switching to multicolumn, (semi-)continuous chromatography, which gives higher throughput, has lower buffer consumption, provides for higher capacity of utilization of the stationary phase (or smaller column volume), and hence gives increased productivity. Unfortunately, unlike other industrial fields, biotechnology is late to embrace continuous processing or multicolumn preparative chromatography.

The present work reports on the design and experimental validation of a simple quasi-continuous, open-loop, two-column countercurrent chromatographic process for size-exclusion purification of adenovirus serotype 5 (Ad5). As mentioned above, SEC has been used in the past for both polishing and intermediate purification of adenoviruses. However, it is often claimed that the main drawbacks of SEC, namely low productivity and high product dilution, make it a costly purification step. Here, it is shown that these drawbacks can be eliminated to a large extent by switching from single-column batch operation to two-column SMB-type operation.

Because of the anticipated boost in performance obtained by switching to multi-column (quasi-)continuous operation, the Ad5 purification train employed for assessing the effectiveness of the enhanced SEC step was streamlined: the SMB-SEC step was applied after the first ultra/diafiltration step and replaced the last three steps of the standard purification train. The main reason for working with the streamlined purification train was to challenge the SMB-SEC step with a less purified bulk in order to prove that, once the classical limitations of SEC are alleviated, the standard purification train can be changed to meet specific needs in terms of cost reduction or purity requirements. Moreover, running the process with a less clean bulk gave more confidence on the performance assessment of the SMB-SEC step.

In a two-column SEC process the cycle is divided into two equal-length time intervals, which we shall henceforth refer to as switching intervals. To design the cycle it suffices to define the sequence of flow-line configurations for one of the switching intervals; the sequence for the other twin switching interval is similar, but the positions of the two columns are exchanged. In practice, it is easier to keep the columns fixed in space and to move the positions of the inlet and outlet lines by means of an automated arrangement of two- or three-way valves. Over a full cycle, each column undergoes the same sequence of steps but phased out in time by one switching interval.

In the developed process, each half-cycle is divided into the following sequence of steps:

(1) Elution of the upstream column while the effluent of the downstream column is diverted to the waste line.

(2) Elution of the upstream column but its effluent is diverted to the waste line instead of being directed to the downstream column; the latter column is fed with the clarified bioreaction bulk while is effluent is collected as purified product;

(3) Operation of the system as in step 1, but the effluent of the downstream column is collected as product.

(4) Elution of the upstream column but its effluent is again diverted to the waste line instead of being directed to the downstream column; the latter column is kept frozen, that is, the flow through it is temporarily halted.

At the end of every half-cycle the positions of the two columns are exchanged to implement the simulated moving-bed contact between the solid and the fluid.

Clearance of impurities, namely DNA and host cell protein (HCP), were experimentally assessed. The obtained performance compares very favorably against single-column batch chromatography for the same volume of size-exclusion resin. In particular, our two-column SEC process achieves about the same DNA and HCP clearances as the standard batch process. However, and most importantly, the virus yield is increased from 57% for the batch process to 86% for our two-column because of internal recycling of the mixed fractions of contaminated Ad5 even though the process is operated strictly in an open-loop configuration. And last, but not least, the productivity is boosted by 6-fold with the two-column process. This is a clear demonstration that with this increased productivity, size exclusion can do much more than just final polishing and buffer exchange in the downstream train of a biopurification.

In conclusion, the main drawbacks of size-exclusion chromatography, namely low productivity and low product titer, were easily overcome by an innovative two-column configuration that keeps the mixed fractions inside the system at all times.

386290 Ultrafiltration Behavior of Bacterial Polysaccharides

Abstract:

Bacterial polysaccharides are an important component of newly developed vaccines against pneumococci, and meningococci. These vaccines induce a strong immune response against the capsular polysaccharides that are unique to specific microorganisms. Ultrafiltration can be used for concentration and purification of polysaccharides; and the study aims to further understand the behavior of different polysaccharides during ultrafiltration.

Experiments were performed using purified bacterial polysaccharides provided by Pfizer Inc. Data were obtained in a stirred ultrafiltration cell using Ultracel composite regenerated cellulose and Biomax polyethersulfone membranes with nominal molecular weight cutoffs of 300, 500, and 1000 kDa. Polysaccharide concentrations were determined using high performance liquid chromatography (HPLC) with refractive index detection.

The filtrate flux and polysaccharide transmission were both strong functions of the solution ionic strength due to inter- and intra-molecular electrostatic interactions. The filtrate flux attained a pressure-independent value at relatively low pressures (around 1 psi = 6.9 kPa) even in dilute polysaccharide solutions (< 1 g/L). There was clear evidence of polysaccharide fouling, particularly in the high ionic strength solutions. Polysaccharide transmission was also a function of the stirring rate, providing clear evidence for the effects of concentration polarization. Much higher transmission could be obtained by using the membranes in the skin-side down orientation due to the strong degree of internal polarization under these conditions. These results provided important insights into the factors controlling the ultrafiltration behavior of bacterial polysaccharides of interest in bioprocessing applications.

391944 Development of Platform Host Cell Protein Enrichment Strategies for Use in Impurity Spike Challenge Studies

Abstract:

Controlling process related impurities is essential for minimizing risk-to-patients, however, for platform processes producing monoclonal antibodies the risk associated with host impurities may be low owing to process refinement over time. As an alternative to establishing measurement controls with acceptance criteria, ICH Q6B authorities' state "clearance studies, which could include spiking experiments at the laboratory scale" can be used to demonstrate removal of such impurities. This paper presents the development and characterization of a platform methodology to enable the enrichment of highly representative and product-specific host cell protein impurities. These impurities can then be used in high resolution spike challenge studies to monitor host cell protein clearance by successive chromatography steps. The results of these spike challenge studies provide greater understanding of purity identity, process step redundancy and, correspondingly, the necessity for in-process measurement controls.

377229 Purification of Plasminogen from Blood Sources Using Affinity Membranes

Abstract:

Plasmin is protein of the fibrinolityc system, obtained by activation of plasminogen, which is of interest for ophthalmology applications. It can be used as a treatment for diabetic rethinopaty, macular pukers, but also as a facilitator for vitreoctomy since it has the properties to hydrolize a variety of glycoproteins, including laminin and fibronectin, by degrading the links between these components of the vitreoretinal interface and the inner limiting membrane.

The purification of plasminogen from blood is conventionally performed with bead-based affinity chromatography, by exploiting the affinity between plasminogen and L-lysine. However, due to its low concentration, which is about 0.2 g/L in human blood, with a single step affinity purification from serum or plasma it is not possible to obtain high levels of purity and a sufficient amount of protein.

In this work we propose an alternative two stage purification process which includes a precipitation stage before the affinity chromatography which is performed with lysine affinity membranes. Different precipitation methods have been investigated and a simple procedure inspired by the Cohn method has been optimized and successfully implemented.

The affinity membranes were obtained by immobilization of L-lysine on regenerated cellulose membrane supports and characterized in terms of ligand density, binding capacity and selectivity. The effect of operating parameters like flow rate and feed concentration have been investigated in detail. The comparison with a chromatography column packed with a commercial lysine affinity beads demonstrated the superior performance of the affinity membranes and the feasibility of the proposed process.

Acknowledgements

This work was financially supported by MIUR, Italian Ministry of Education, University and Research (PRIN 2008) and by the University of Bologna, Italy.

380931 Combination of Mixed-Mode Chromatography and Expanded-Bed Adsorption for Antibody Recovery

Abstract:

Mixed-mode chromatography (MMC) is a novel technology for bioproduct separation, especially antibodies, which combines multiple binding modes like hydrophobic interaction, electrostatic forces, hydrogen bonding, etc.. High capacity, salt-tolerance, good selectivity and relatively low cost are the major advantages of MMC for direct capture process. Expanded-bed adsorption (EBA) was developed in 1990s as an innovative technology that allows capturing target proteins directly from unclarified feedstock, e.g., culture suspensions and cell homogenates. EBA technology combines solid-liquid separation with an adsorption step in a single-unit operation, aiming at increased overall yield, reduced operational time, and less requirements for capital investment and consumables. Two chromatographic techniques, MMC and EBA, would be integrated in the present work to develop new separation technology, mixed-mode EBA, improving the capture efficiency and reducing the pretreatments on the feedstock, such as clarification, dilution and salt-adjustment.

MMC ligands (such as 4-Pyridylethylmercaptan (MEP), 2-mercapto-1-methylimidazole (MMI) and 5-amine-1,3-benzimidazol (ABM)) were coupled onto the matrix for EBA (3% crosslinked agarose containing tungsten carbide). The density of matrix is up to 3.0 mg/ml with a suitable size distribution, which is very beneficial to operate at high velocity in an expanded bed. The matrix activation and the ligand coupling were optimized, and the ligand density could be controlled at the range of 40~100 μmol/ml resin. The saturated adsorption capacities of IgG could reach more than 60 mg IgG/ml resin. The effects of pH and salt addition on the adsorption of IgG were investigated, and the adsorption showed high pH-sensitivity and low salt-dependence. In addition, high dynamic binding capacity was found even at high operation velocity (>400cm/h), and the efficient elution could be achieved at a mild acidic condition. The adsorption in expanded-bed mode could match that in packed bed. Finally, new technology was challenged to capture IgG directly from the crude cell-containing feedstock with relative high conductivity, and good performance and high separation efficiency was obtained.

For biopharmaceuticals such as monoclonal antibodies (mAbs), downstream processes require highly productive and robust technologies to improve the process efficiency. The combination of MMC and EBA has been demonstrated to be a potential new platform for antibody capture with reduced feedstock pretreatments, high efficiency and relative low cost, which can be expanded to other bioproduct separation.

*This work was supported by National Natural Science Foundation of China and the Zhejiang Provincial Natural Science Foundation of China.

382187 Gigaporous Chromatography of Virus-like Particles

Abstract:

Virus-like particles (VLPs) represent a class of modern vaccines that are designed specifically for prevention of virus infections. These particles are produced through cultivation of recombinant microorganisms or mammalian cells. The cultivation suspension contains very low concentration of the product and large quantities of impurities. Hence it is necessary to purify the VLPs by chromatographic methods. Unfortunately, common chromatographic packing beads have pore sizes in the range of 20-30nm which close to the size of VLPs and may cause blockage of the pores. It is a prerequisite to develop suitable microsphere beads with much greater size, i.e. >100 nm to allow the VLPs to diffuse in and out of the beads freely. This study introduces the preparations of gigaporous microsphere beads which are made of polystyrene, poly-glycidylmethacrylate (PGMA) and agarose respectively. A simple method was developed for creating the super large pores, which was based on reverse micelles swelling. We found that surfactant reverse micelles could absorb water into the oil droplets, forming a bi-continuous W/S/O (water/surfactant/oil) emulsion. The aqueous channels became large pores after solidification of the oil droplets by polymerization. At the end, integral microspheres with giga-pores were obtained.

The effects of pore size on adsorption equilibrium, adsorption kinetics, dynamic binding capacity, and recovery were investigated. According to the confocal laser scanning microscopy, adsorption of VLPs in commercial chromatographic beads was mostly confined to a thin shell on the outer surface, leaving the underlying pore space and the binding sites inaccessibly, while the large pores in gigaporous beads enabled the VLPs to access to the interior pore spaces by diffusion transport efficiently. Ion exchange chromatography with the gigaporous beads gained about 12 times increase in static adsorption capacity, 8 times increase in dynamic binding capacity, and 11 times increase in effective pore diffusivity. The gigaporous structure also significantly improve the stability of the VLPs during intensive adsorption-desorption process by lowing the multi-point interaction between the VLPs and binding sites in the pores.

New Approaches in Separation Process Development

Tim Frank

Dow Fellow and Assoc. R&D Director
Process Separations
Engineering & Process Science Laboratory
The Dow Chemical Company

Gerhold and Kunesh Award Plenary Session on Separations

AIChE Annual Meeting, Atlanta

November 17, 2014

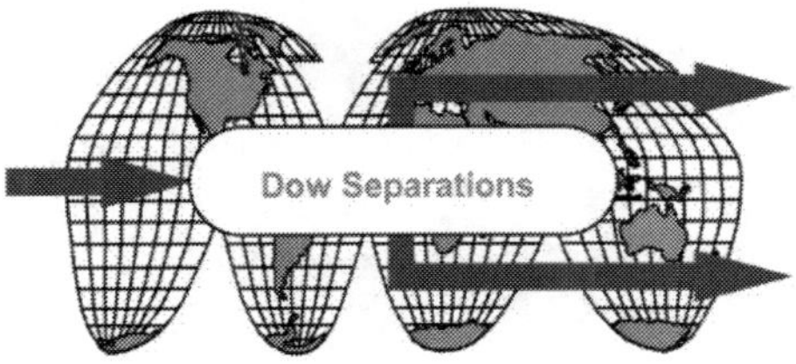

- Introduction
  - Clarence G. (Larry) Gerhold
  - Thanks to all
- Separation Process Development at Dow
  - Brief overview
  - Tools of the trade
- Some New Approaches
  - Rapid screening methods – γ_{ij}^{∞} temperature effects
  - New uses for glycol ethers
    - Recovery of recombinant proteins
    - Chiral glycol ethers – via reactive distillation

tcfrank@dow.com The Dow Chemical Company Page 2

Clarence G. Gerhold, 1907 - 2001

- BS in Chem E (1928), MS in Chem (1929) – University of Illinois
- Universal Oil Products, 1929 – 1974 (45 years)
- Ground-breaking process innovations
  - 78 US patents
  - Thermal cracking and thermal reforming
  - Catalytic reforming
  - Catalytic polymerization
  - Aromatic/aliphatic liquid-liquid extraction using a heavy polar solvent with reflux
  - Simulated Moving Bed (SMB) chromatography
    "*Continuous sorption with a fixed bed and moving inlets and outlets*"

- Thank you!
  - UOP for supporting this award
  - AIChE Awards Committee and nominators

tcfrank@dow,com The Dow Chemical Company Page 3

Thank you!

Many generous mentors and collaborators

- *John Falconer*
- *Lloyd Berg*
- *Lanny Robbins*
- *Charles Eckert*
- *Ed Cussler*
- *Hank Kohlbrand*
- *Bruce Holden*
- *Felipe Donate*
- *James Ringer*
- *Pat Au-Yeung*
- *Sol Resnick*
- *Chris Christenson*
- *John Pendergast*
- *Herb Johnson*
- *Nancy Duty*
- *Victor Atiemo-Obeng*
- *Yujun Liu*
- *Andrei Merenov*
- *Anant Patkar*
- *Ray Rooks*
- *Scott Christensen*
- *Craig Schubert*
- *Jefry Shields*
- *Chuck Squires*
- *Paul Gillis*
- *Pramod Patil*
- *George Frycek*

And many others

- *Kathy Lichtenwald*
- *Lise Dahuron*
- *Shawn Feist*
- *Dennis Jewell*
- *Chris Jones*
- *Kishore Kar*
- *"J" Moore*
- *Paul Larsen*
- *Megan Donaldson*
- *Vishesh Shah*
- *Tom Thyne*
- *Steve Arturo*
- *Yoshiaki Kawajiri*
- *Jacinto Lopez-Toledo*

tcfrank@dow.com The Dow Chemical Company Page 4

The Dow Chemical Company

- Polymers	polyethylene, polypropylene, polystyrene-divinylbenzene, Saran, PEG, PPG, P(EP)G, epoxies, urethanes (TDI, MDI, and polyols)
- Acrylic Latex	adhesives, paint additives, sealants, ...
- Cellulosics	hydroxymethylcellulose, hydroxyethylcellulose, thickeners, excipients, ...
- Ethylene Oxide	ethylene glycol, polyethylene glycol, nonionic surfactants, ...
- Propylene Oxide	propylene glycol, polypropylene glycol, polyetherpolyols, ...
- Specialty Monomers	divinyl benzene, acrylic acid and acrylates, vinyl acetate, ...
- Alkanolamines	acid gas scrubbing fluids, formulation additives, ...
- Chlorine and Caustic	Cl_2, H_2, HCl, NaOH, ethylene dichloride, vinyl chloride, ...
- Organic Solvents	*n*-butanol, 2-ethylhexanol, ketones, glycol ethers, $MeCl_2$, ...
- Formulated Products	surfactants, lubricants, thickeners, chelants, biocides, ...
- Industrial Products	heat-transfer fluids, nitroalkanes, industrial coatings, lubricants, ...
- Consumer Products	cosmetics, personal care, foam insulation, appliance mat'ls, flame retardants, ...
- Building Materials	insulation batting, insulation board, solar roofing mat'ls, ...
- Coating Materials	acrylic latex binder, pigment extender, co-solvents, ...
- Water Treatment	RO and UF membranes, ion-exchange beads, arsenic adsorbents, ...
- Oil & Gas Processing	acid gas scrubbing fluids, natural gas treating, LNG, LPG, absorbers, liquid treaters, ...
- Agricultural Products	herbicides, insecticides, healthy oils, corn and soybean seeds with custom traits,...
- Biotechnology	fermentation-derived products (spinosad), herbicide-resistant seed crops, ...
- Electronic Materials	photoresist, planarization (CMP), displays (oLEDs), metal deposition, ...
- Automotive Materials	exterior/interior plastics, sealants, adhesives, seat cushions, foams, ...
- Energy	solar roof shingles, wind turbine materials, battery materials, ...

www.dow.com

Process Separations: A Core Capability for Dow

- Focus on fundamentals
- Process and equipment
- Experiments and modeling
- Commercial-scale vision

→ Commercial Success

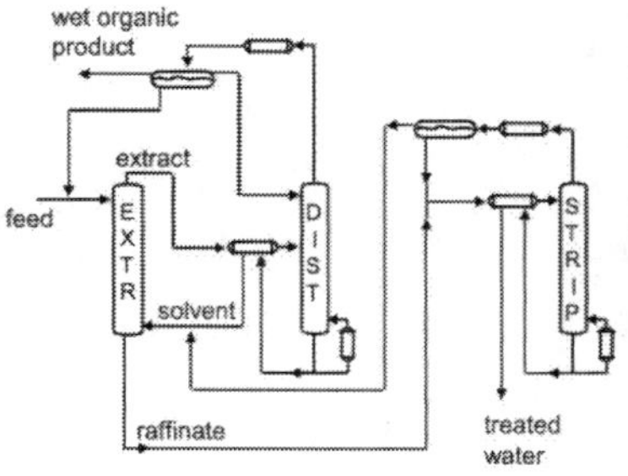

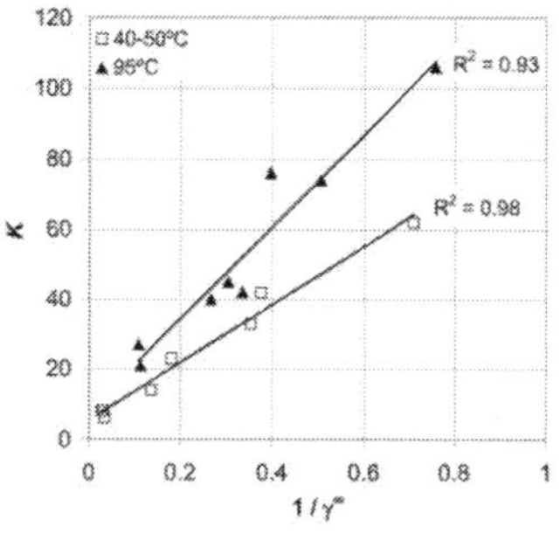

$$\ln \gamma_{ij}^{\infty} = \frac{V_i^L (T_{ref}/T)^{\theta_{ij}}}{RT} \left[(\lambda_j - \lambda_i)^2 + \frac{q_i^2 q_j^2 (\tau_j - \tau_i)^2}{\psi_j} + \frac{(\alpha_j - \alpha_i)(\beta_j - \beta_i)}{\xi_j} \right]$$

Key Technologies

Distillation*
- fractional distillation trains
- batch rectification (and other configurations)

Evaporation
- forced-circulation, falling-film
- wiped-film or short path (deep vacuum)
- multiple effect, vapor recompression

Stripping (to ppm and ppb levels)
- organics from aq. soln., brine
- water from organics

Absorption
- organics into oil
- acid-gas scrubbing (reactive absorption)

Enhanced Distillations (highly non-ideal systems)
- azeotropic
- extractive**
- pressure-swing***
- reactive distillation (equilibrium-limited rxn)

Distillation Tower

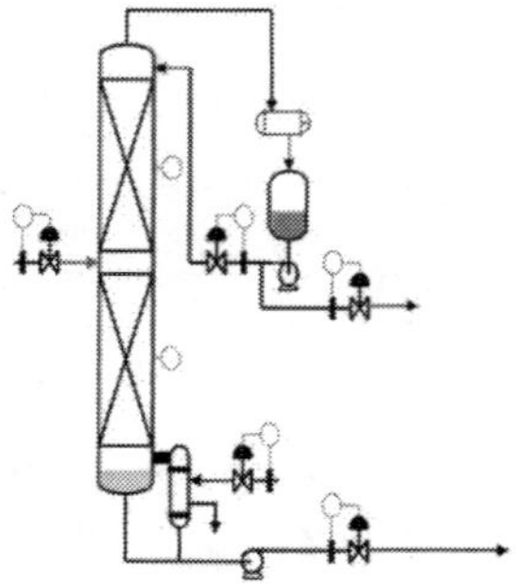

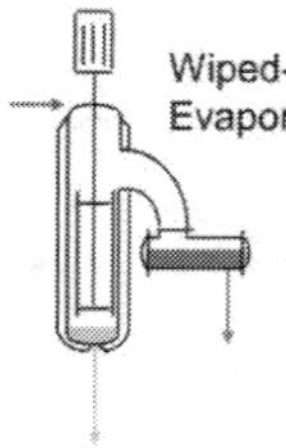

Wiped-Film Evaporator

*Pilling, Holden, "Choosing Trays and Packings," *Chem. Eng. Progress* **2009**, *105*(9), 44-50.
Pendergast, Jain, Martinez, Brayden, Hasabnis, "Evaluation of Reactive Absorption," *Ind. Eng. Chem. Res.* **2014, *53*(6), 2403-2411.
***Frank, "Pressure Sensitive Distillation," *Chem. Eng. Progress* **1997**, *93*(4), 52-63.

Key Technologies

Liquid-Liquid Extraction*
- standard extraction for solute recovery or removal
- fractional extraction for product purification
- extractive reaction (biodiesel processing, aromatic nitration, ...)
- reaction-enhanced extraction (carboxylic acid/trioctyl amine, ...)

*Frank, Dahuron, Holden, Prince, Seibert, Wilson, Section 15, "Liquid-Liquid Extraction," in Perry's Chem. Engineers' Handbook, 8th Ed. (McGraw-Hill, New York, 2007).

Karr Extraction Column

Crystallization from Solution**
- cooling
- evaporative (forced-circulation, multiple effect)
- drowning out (anti-solvent, including polymer precip./fractionation)
- CO_2 expanded-liquid (anti-solvent/reactive gas)
- reactive, adductive

Solution Crystallizer

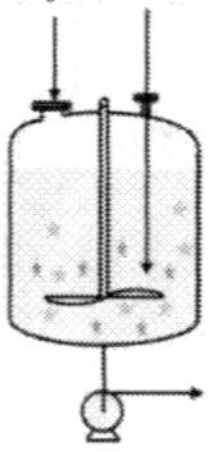

Hamer, Ressler, Briggs, Hitt, Larsen, Frank, "Use of a Fiber-Optic Turbidity Probe," *Organic Process Res. Dev.* **2009, *13*(1), 114-124;
Frank, Fort, Jones, "Treat Precipitation Reactions as Reactive Crystallizations," *Chemical Processing* **2005**, *68*(12), 37-41.
Frank, Gupta, Downey, "Quickly Screen Solvents for Organic Solids," *Chemical Engineering Progress* **1999**, *95*(12), 41-61.

Crystallization from the Melt
- batchwise (static or forced-circulation)
- suspension crystallization with wash column
- multistage processing (Brodie crystallizer)

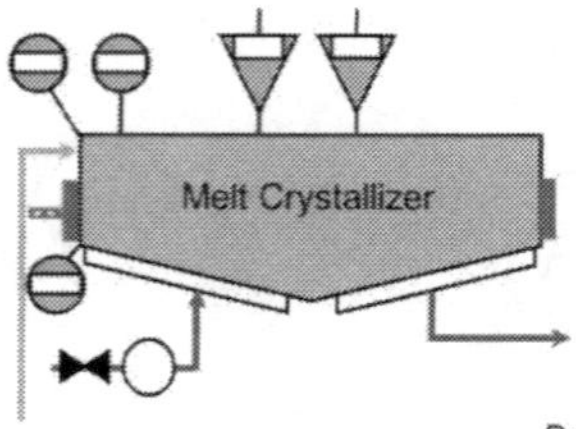

Key Technologies

Gas-Solid Adsorption (zeolites, silicas, aluminas, carbons, polymers)
- temperature-swing
- pressure-swing[a]
- vacuum swing (SORBATHENE®)[b]
- reactive adsorption (chemisorption, specialty media)

[a] Liu, Feist, Jones, Armstrong, "Alcohol Drying by Hot Gas PSA," *Ind. Eng. Chem. Res.* **2014**, *53*(20), 8599.
[b] Pezolt, Collick, Johnson, Robbins, "PSA for VOC Recovery," *Environ. Prog.* **1997**, 16(1), 16-19.
[b] Robbins, Frank, US 4,857,084, Aug. 15, 1989.

Pressure Swing Adsorption

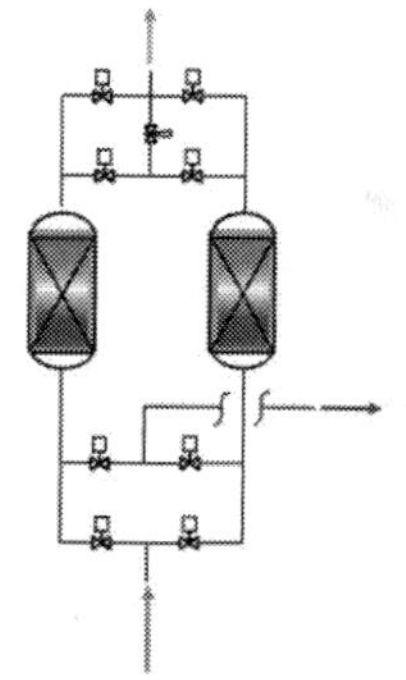

Liquid-Solid Adsorption (zeolites, silicas, aluminas, carbons, polymers)
- wastewater treatment (carbon beds)
- removal of polar impurities from monomer feeds

Process-Scale Liquid Chromatography
- bioseparations, ion exchange, …
- product purification, product recovery
- Simulated Moving Bed (SMB)[c]
- reactive chromatography[d]

[c] Feist, Hasabnis, Pynnonen, Frank, "SMB Chromatography Design Using Profile Advancement Factors," *AIChE J.* **2009**, *55*(11), 2848-2860;
[c] Pynnonen, Feist, Hasabnis, Albers, Frank, "Quickly Evaluate SMB Chromatography for Your Application," *Chem. Proc.* **2010**, 73(5), 29-33;
[c,d] Oh, Agrawal, Sreedhar, Donaldson, Schultz, Frank, Bommarius, Kawajiri, "Conversion Improvement for Catalytic Synthesis of Propylene Glycol Methyl Ether Acetate," *Chem. Eng. J.* **2015**, *259*, 397-409.

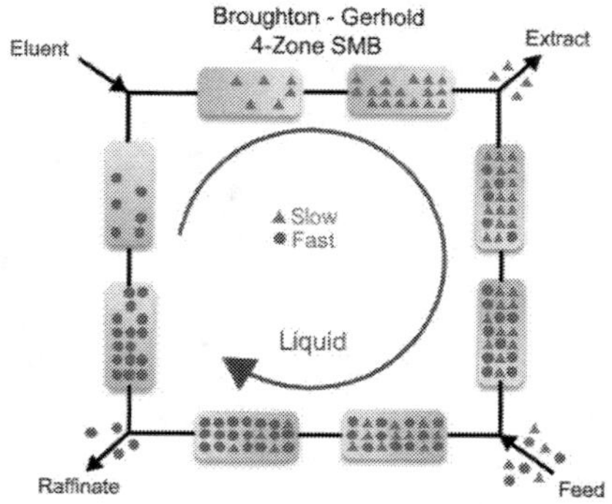

tcfrank@dow.com

Key Technologies

Liquid-Liquid Phase Separation*
- gravity decanters (settlers)
- hydrocyclones
- packed-bed coalescers
- fiber-bed or filter-type coalescers
- centrifuges

*Frank, Dahuron, Holden, Prince, Seibert, Wilson, "Liquid-Liquid Extraction," Perry's Chem. Engineers' Handbook, 8th Ed. (McGraw-Hill, 2007).

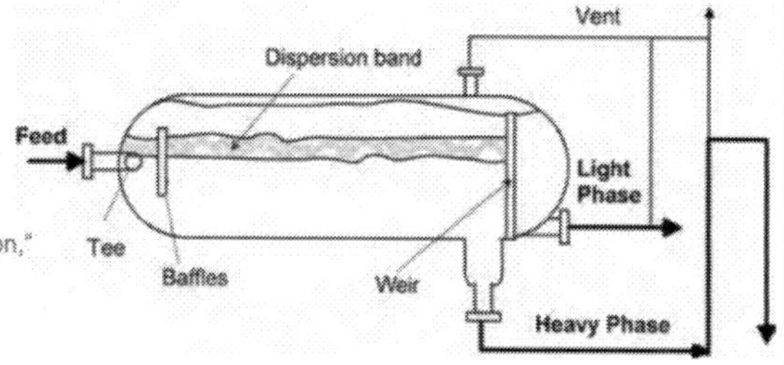

Membranes (polymers, zeolites, ceramics, …)
- reverse osmosis/nanofiltration (water purification, conc'n)
- bioseparations (microfiltration, ultrafiltration, diafiltration)
- latex stripping**
- gas-phase separations (ethylene, hydrogen, …)

Ulrich, Frank, McCormick, Cussler, *J. Membrane Sci.*, **2014, *452*, 426-432;
Zeng, McCabe, Frank, Carpenter, Arnold, Cussler, *Ind. Eng. Chem. Res.* **2014**, *53*(31), 12420-12427.

Hybrid Processing
- distillation-adsorption, distillation-membranes
- distillation-melt crystallization, distillation-reaction
- distillation-extraction, extraction-crystallization

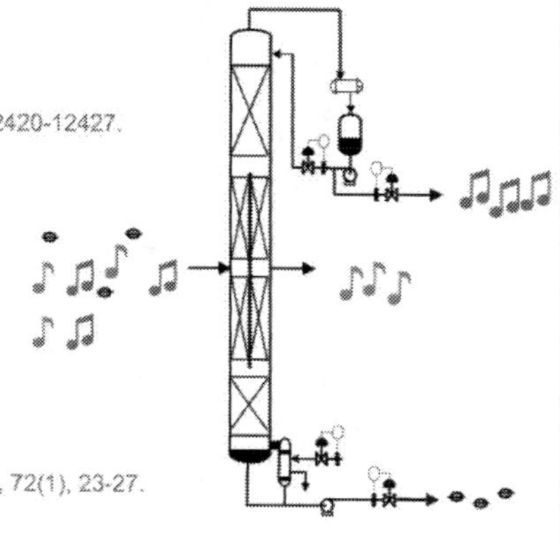

Process Intensification/Simplification
- combined reactor-separator
  (reactive distillation, reaction-enhanced extraction, …)
- dividing-wall distillation column***, microwaves, ultrasonics, …

***Pendergast, Vickery, Au-Yeung, Anderson, "Consider DWCs", *Chem. Processing* **2009**, 72(1), 23-27.

tcfrank@dow.com

Typical Pilot-Plant Equipment

tcfrank@dow.com The Dow Chemical Company Page 11

Shipping a Unit for a Test at the Plant Site

tcfrank@dow.com The Dow Chemical Company Page 12

More Pilot-Plant Equipment – Installed on a Slip Stream at the Plant

tcfrank@dow.com The Dow Chemical Company Page 13

How to Quickly Identify the Appropriate Technology?

One way is to focus on γ_{ij}^{∞}

Solution Crystallization?

Gas-phase Adsorption?

Absorption?

Distillation?

Extraction?

Extractive Distillation?

Liquid-phase Adsorption?

Melt Crystallization?

Anti-solvent Crystallization?

Stripping?

Pressure-Swing Distillation?

Fractional Extraction?

tcfrank@dow.com The Dow Chemical Company Page 14

Screening Process Separation Options

Distillation Relative volatility	$\alpha_{ij}^{\infty} = \frac{\gamma_i^{\infty}(T)}{\gamma_j^{\infty}(T)} \frac{p_i^{SAT}(T)}{p_j^{SAT}(T)}$ $\quad \alpha_{ik}^{\infty}\Big\vert_{\text{in a solvent or entrainer}} = \frac{\gamma_{i,solvent}^{\infty}(T)}{\gamma_{k,solvent}^{\infty}(T)} \frac{p_i^{SAT}(T)}{p_k^{SAT}(T)}$
Liquid-liquid extraction Partition ratio	$K_i^{\infty} = \frac{x_i^{extract}}{x_i^{raffinate}} = \frac{\gamma_{i,raffinate}^{\infty}(T)}{\gamma_{i,extract}^{\infty}(T)}$
Henry's Law constants for organics in water	$H_i = \gamma_{i,water}^{\infty}(T) \frac{p_i^{SAT}(T)}{\rho_{water}(T)}$
Octanol-water partition ratio	$K_{ow,i}^{\infty} \equiv \frac{x_i^{\text{octanol-rich phase}} \rho_{\text{octanol-rich phase}}}{x_i^{\text{aqueous phase}} \rho_{\text{aqueous phase}}} \approx \frac{\gamma_{i,water}^{\infty}(T)}{\gamma_{i,organic\ phase}^{\infty}(T)} \frac{\rho_{organic\ phase}(T)}{\rho_{water}(T)}$
Crystallization Solid solubility	$\ln x_i^{SAT} = \frac{\Delta_{fus} S}{R}\left(1 - \frac{T_m}{T}\right) - \ln \gamma_i^{SAT}$, for $T < T_m$

$$RT \ln \gamma_{ij}^{\infty} = \overline{g}_{ij}^{E,\infty} = \overline{h}_{ij}^{E,\infty} - T\overline{s}_{ij}^{E,\infty}$$

Numerous models can be used to estimate $\gamma = f(\text{composition})$
UNIFAC, Hansen, MOSCED, COSMO-RS, COSMO-SAC, ...

Also need to consider $\gamma = f(T)$

Well Known Approach
$$\left[\frac{\partial \ln \gamma_{ij}^{\infty}}{\partial(1/T)}\right]_P = \frac{\overline{h}_{ij}^{E,\infty}}{R}, \quad \gamma_{ij}^{\infty}\Big\vert_{\text{at } T_2} \approx \gamma_{ij}^{\infty}\Big\vert_{\text{at } T_1} \exp\left[\frac{\overline{h}_{ij}^{E,\infty}}{R}\left(\frac{1}{T_2} - \frac{1}{T_1}\right)\right]$$

New Approach* $\quad \ln \gamma_{ij}^{\infty}(\text{at } T) = a_{ij}(T_{ref}/T)^{\theta_{ij}}$

$$a_{ij} = \ln \gamma_{ij}^{\infty} \text{ at } T = T_{ref}, \quad \theta_{ij} = \frac{1}{1 - (T\overline{s}_{ij}^{E,\infty} / \overline{h}_{ij}^{E,\infty})}$$

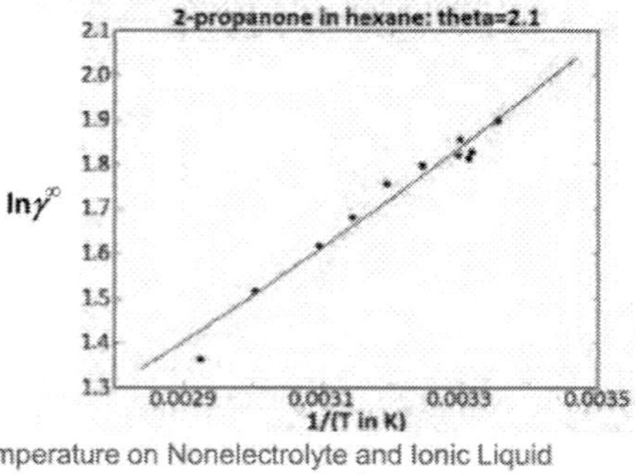

*Frank, Arturo, Holden, "Framework for Correlating the Effect of Temperature on Nonelectrolyte and Ionic Liquid Activity Coefficients," *AIChE J.* **2014**, 60(10), 3675-3690.

$$RT \ln \gamma_{ij}^{\infty} = \overline{g}_{ij}^{E,\infty} = \overline{h}_{ij}^{E,\infty} - T\overline{s}_{ij}^{E,\infty}$$

Numerous models can be used to estimate $\gamma = f(\text{composition})$
UNIFAC, Hansen, MOSCED, COSMO-RS, COSMO-SAC, ...

Also need to consider $\gamma = f(T)$

Well Known Approach
$$\left[\frac{\partial \ln \gamma_{ij}^{\infty}}{\partial (1/T)}\right]_P = \frac{\overline{h}_{ij}^{E,\infty}}{R} \quad , \quad \gamma_{ij}^{\infty}\Big|_{\text{at } T_2} \approx \gamma_{ij}^{\infty}\Big|_{\text{at } T_1} \exp\left[\frac{\overline{h}_{ij}^{E,\infty}}{R}\left(\frac{1}{T_2} - \frac{1}{T_1}\right)\right]$$

New Approach* $\ln \gamma_{ij}^{\infty}(\text{at } T) = a_{ij}(T_{ref}/T)^{\theta_{ij}}$

$$a_{ij} = \ln \gamma_{ij}^{\infty} \text{ at } T = T_{ref}, \quad \theta_{ij} = \frac{1}{1 - (T\overline{s}_{ij}^{E,\infty} / \overline{h}_{ij}^{E,\infty})}$$

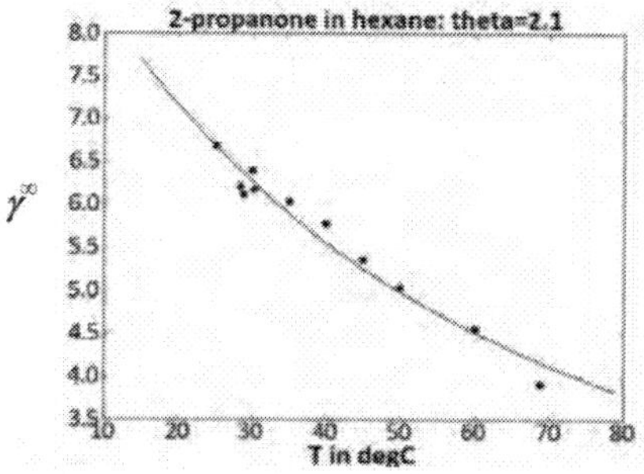

*Frank, Arturo, Holden, "Framework for Correlating the Effect of Temperature on Nonelectrolyte and Ionic Liquid Activity Coefficients," *AIChE J.* **2014**, 60(10), 3675-3690.

van Laar

$$\ln \gamma_i = \frac{a_{ij}(T_{ref}/T)^{\theta_{ij}}}{\left(1 + \dfrac{a_{ij}x_i}{a_{ji}x_j}\right)^2} \qquad \ln \gamma_j = \frac{a_{ji}(T_{ref}/T)^{\theta_{ji}}}{\left(1 + \dfrac{a_{ji}x_j}{a_{ij}x_i}\right)^2}$$

$$a_{ij} = \ln \gamma_{ij}^{\infty} \text{ at } T = T_{ref}, \qquad \theta_{ij} = \frac{1}{1 - (T\overline{s}_{ij}^{E,\infty} / \overline{h}_{ij}^{E,\infty})}$$

Extended Hansen

$$\ln \gamma_{ij}^{\infty} = a_{ij}(T_{ref}/T)^{\theta_{ij}} = \frac{V_i^L (T_{ref}/T)^{\theta_{ij}}}{RT_{ref}} \left\{ \left(\delta_i^d - \delta_j^d\right)^2 + 0.25\left[\left(\delta_i^p - \delta_j^p\right)^2 + \left(\delta_i^h - \delta_j^h\right)^2\right] \right\}$$

MOSCED

$$\ln \gamma_{ij}^{\infty} = \frac{V_i^L (T_{ref}/T)^{\theta_{ij}}}{RT}\left[\left(\lambda_j - \lambda_i\right)^2 + \frac{q_i^2 q_j^2 \left(\tau_j - \tau_i\right)^2}{\psi_j} + \frac{\left(\alpha_j - \alpha_i\right)\left(\beta_j - \beta_i\right)}{\xi_j}\right]$$

Wilson

$$\ln \gamma_{ij}^{\infty} = a_{ij}(T_{ref}/T)^{\theta_{ij}} = 1 - \ln \Lambda_{ij} - \Lambda_{ji}$$
$$\Lambda_{ij} = 1 - a_{ji}(T_{ref}/T)^{\theta_{ji}} - \ln \Lambda_{ji}$$
$$\Lambda_{ji} = 1 - a_{ij}(T_{ref}/T)^{\theta_{ij}} - \ln \Lambda_{ij}$$

NRTL

$$\ln \gamma_{ij}^{\infty} = a_{ij}(T_{ref}/T)^{\theta_{ij}} = \tau_{ij}\exp\left(-\alpha_{ij}\tau_{ij}\right) + \tau_{ji}$$
$$\tau_{ij} = a_{ji}(T_{ref}/T)^{\theta_{ji}} - \tau_{ji}\exp(-\alpha_{ji}\tau_{ji})$$
$$\tau_{ji} = a_{ij}(T_{ref}/T)^{\theta_{ij}} - \tau_{ij}\exp(-\alpha_{ij}\tau_{ij})$$

An alternative to "ln***A*** or ***A*** = ***a*** + ***b***/***T*** + **c**ln***T***" where ***A*** is a model parameter and ***a***, ***b***, and ***c*** are correlation constants.

New Approaches

Involving
Glycol Ethers

E-Series Glycol Ethers

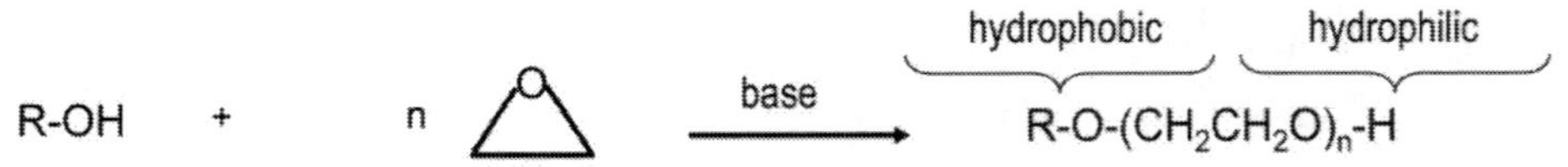

R = n-propyl, n-butyl, n-hexyl, phenyl

n = 1	Ethylene glycol alkyl (or aryl) ether	R-O-CH_2CH_2-OH
n = 2	Diethylene glycol alkyl ether	R-O-(CH_2CH_2-O)$_2$ H
n = 3	Triethylene glycol alkyl ether	R-O-(CH_2CH_2-O)$_3$ H

Compounds with n > 3 are polyethylene glycols

P-Series Glycol Ethers

R-OH + n  $\xrightarrow{\text{base}}$ R-O-($CH_2CHO)_n$-H (with CH_3 on CH)

n = 1	Propylene glycol alkyl (or aryl) ether	R-O-CH_2CH-OH (with CH_3 on CH)
n = 2	Dipropylene glycol alkyl ether	R-O-(CH_2CH-O$)_2$ H (with CH_3 on CH)
n = 3	Tripropylene glycol alkyl ether	R-O-(CH_2CH-O$)_3$ H (with CH_3 on CH)

Compounds with n > 3 are polypropylene glycols

Inverse Solubility in Water

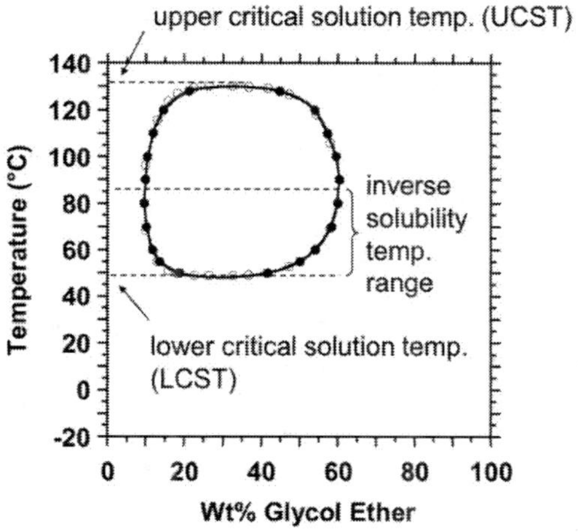

- Complete miscibility below LCST
- Above LCST, solubility decreases with increasing temperature
- Christensen, Donate, Frank, LaTulip, and Wilson, "Mutual Solubility and LCST for Water + Glycol Ether Systems," *J. Chem. Eng. Data* **2005**, *50*(3), 869-877.

Pfenex Expression Technology™

for protein biopharmaceutical production

- Therapeutic proteins via high productivity fermentation
- *Pseudomonas fluorescens* **bacteria**
- Primary recovery using glycol ethers **(one of several options)**
- See *Pfenex.com*

Release of Periplasmic Proteins Using Glycol Ethers

Pfenex Expression Technology™

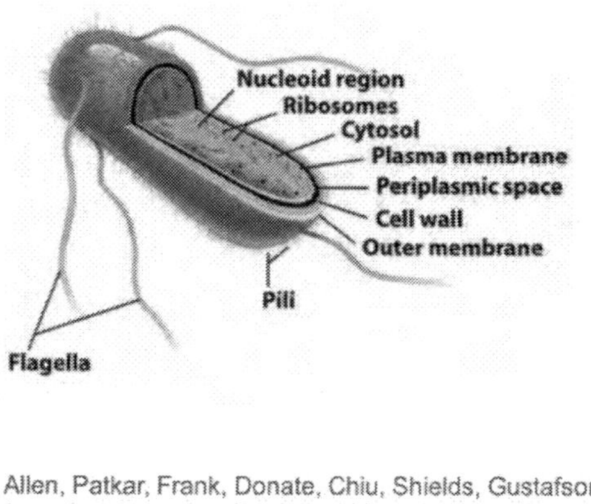

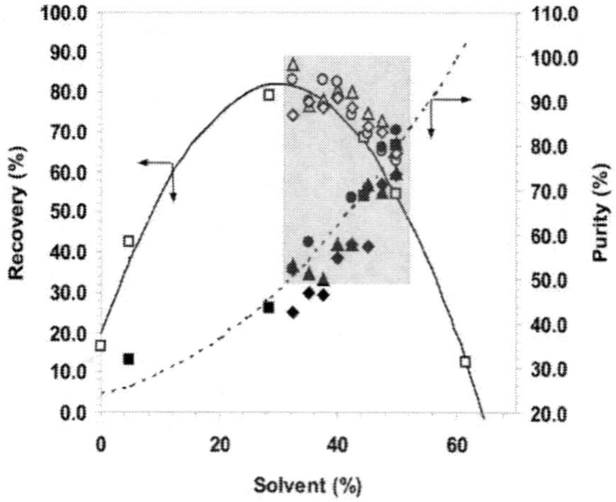

Allen, Patkar, Frank, Donate, Chiu, Shields, Gustafson, "Use of Glycol Ethers for Selective Release of Periplasmic Proteins from Gram-Negative Bacteria," *Biotechnology Progress* **2007**, *23*(5), 1163-1170.

Retallack, Schneider, Mitchell, Chew, Liu, "Transport of Heterologous Proteins to the Periplasmic Space of Pseudomonas fluorescens Using a Variety of Native Signal Sequences," *Biotechnology Letters* **2007**, *29*(10), 1483-1491.

Temperature-Change-Induced Phase Separation

Pfēnex
Expression Technology™

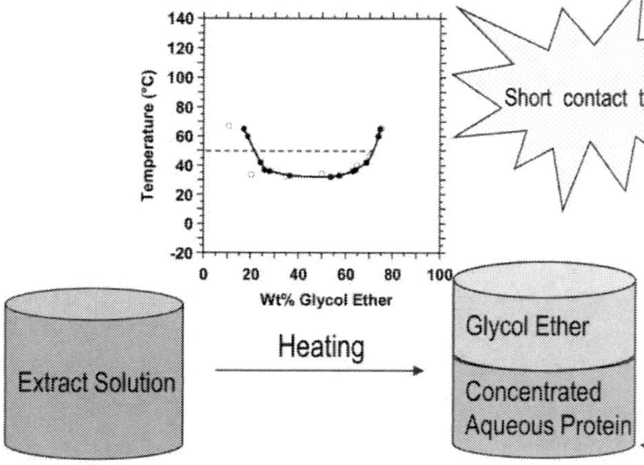

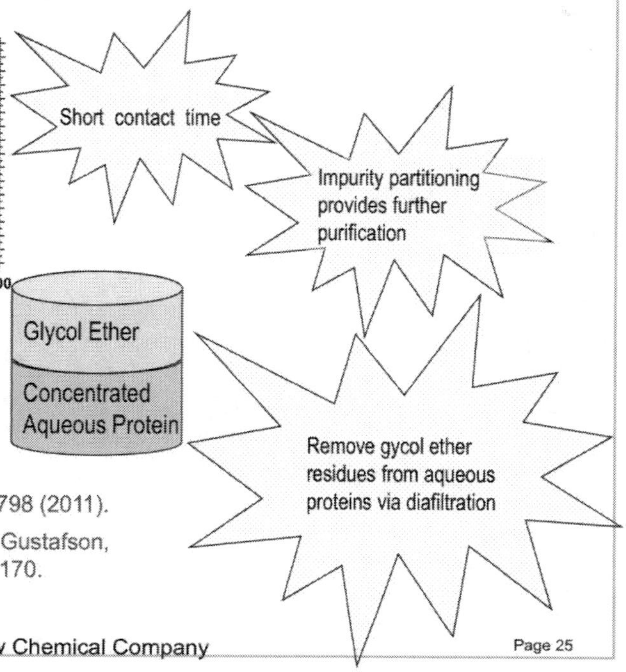

- Frank, Donate, Shields, Li, Allen, US 7,935,798 (2011).
- Allen, Patkar, Frank, Donate, Chiu, Shields, Gustafson, *Biotechnology Progress* **2007**, *23*(5), 1163-1170.

Reactive Distillation for Chiral Resolution

Chiral Synthons – Synthetic Building Blocks (pharma, fragrances, flavors, …)

Chiral Propylene Glycol Ethers

(*R*)-1-methoxy-2-propanol (propylene glycol methyl ether)

(*S*)-1-methoxy-2-propanol

Isolation of Enantiomers

- Well-known tactic: enzymatic enantioselective acylation + distillation
- New approach uses 2 chiral reagents (racemates), giving 4 enantiomers

$$(R,S)\text{-A} + (R,S)\text{-B} \xleftrightarrow{(R)\text{-Selective Catalyst}} (R)\text{-C} + (R)\text{-D} + (S)\text{-A} + (S)\text{-B}$$

- Reaction + *in-situ* distillation drives conversion (reactive distillation)
- Novel horizontal apparatus provides long residence time (hours) with very low ΔP_{vapor} for operation at P < 10 Torr and T < 30 °C
- The new process allows simultaneous production of 4 enantiomers.

tcfrank@dow.com The Dow Chemical Company Page 27

Overall Process for Multicomponent Chiral Resolution*

(*R,S*)-PnB + (*R,S*)-PMA →

(*R*)-PnBA + (*R*)-PM + (*S*)-PnB + (*S*)-PMA

volatility ↑ PM, PMA, PnB, PnBA

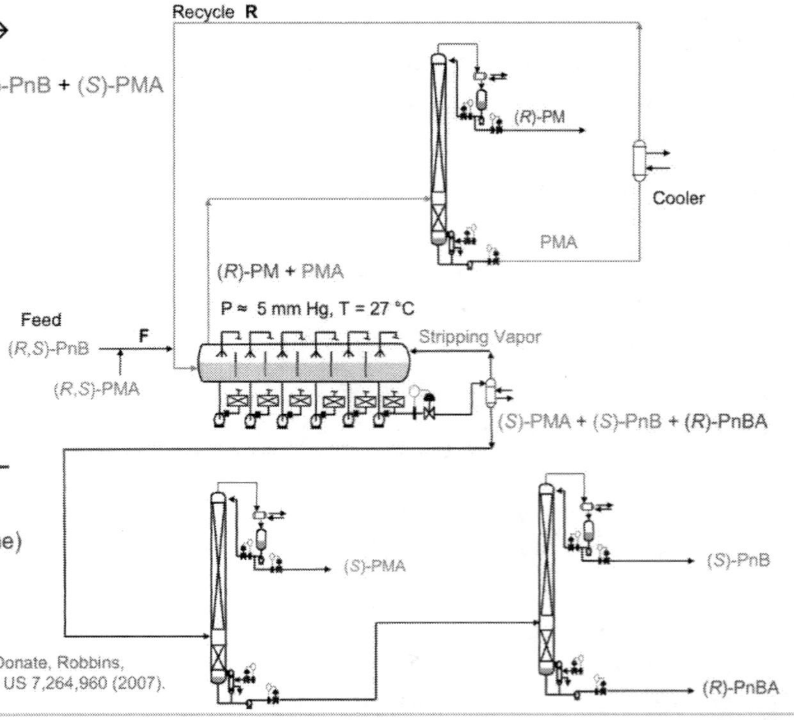

Key Process Variables

R/F (stripping power)

τ_{liquid} (liquid residence time)

*Au-Yeung, Resnick, Witt, Frank, Donate, Robbins, *AIChE J.* **2013**, *59*(7), 2603-2620; US 7,264,960 (2007).

tcfrank@dow.com The Dow Chemical Company Page 28

Optimizing Recycle Vs. Equipment Size

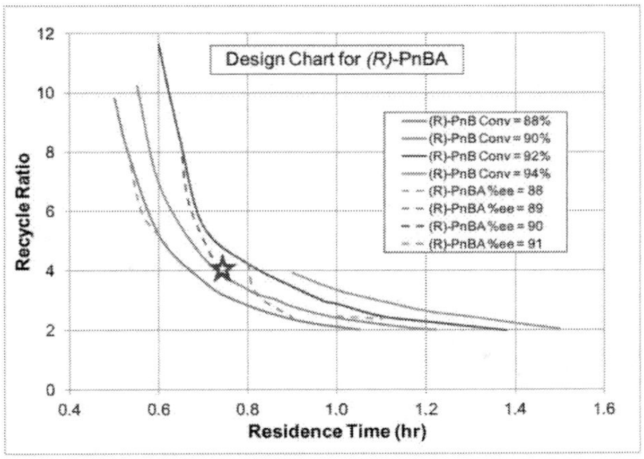

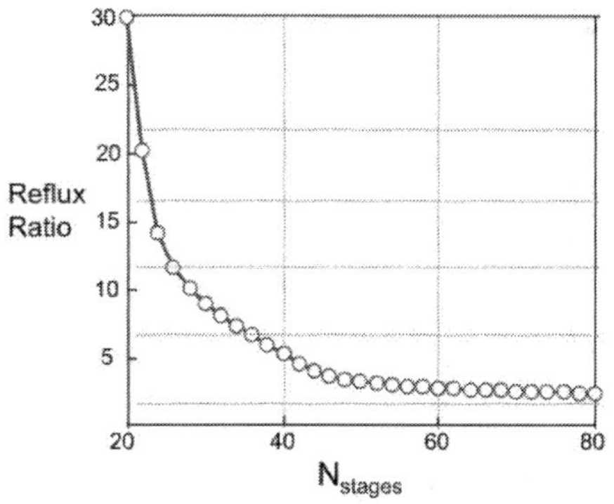

- At R/F = 4, τ_{liquid} = 0.77 hr → 90% conversion, 130 g/L/h productivity
- Familiar tradeoff: R/F *Vs.* τ_{liquid} is analogous to Reflux *Vs.* N_{stages}
- Resnick, Donate, Frank, Thyne, Foley, US 7,264,960 (2007).
- Au-Yeung, Resnick, Witt, Frank, Donate, Robbins, *AIChE J.* **2013**, *59*(7), 2603-2620.

tcfrank@dow.com The Dow Chemical Company Page 29

Thank you!

- Introduction
  - Clarence G. (Larry) Gerhold
  - Thanks to all
- Separation Process Development at Dow
  - Brief overview
  - Tools of the trade
- Some New Approaches
  - Rapid screening methods – γ_{ij}^{∞} temperature effects
  - New uses for glycol ethers
    - Recovery of recombinant proteins
    - Chiral glycol ethers – via reactive distillation

Visit the Dow Lab Safety Academy at http://safety.dow.com

- Understand hazards
- Think about what you are doing – ahead of time
- Protect yourself and others around you.

tcfrank@dow.com The Dow Chemical Company Page 30

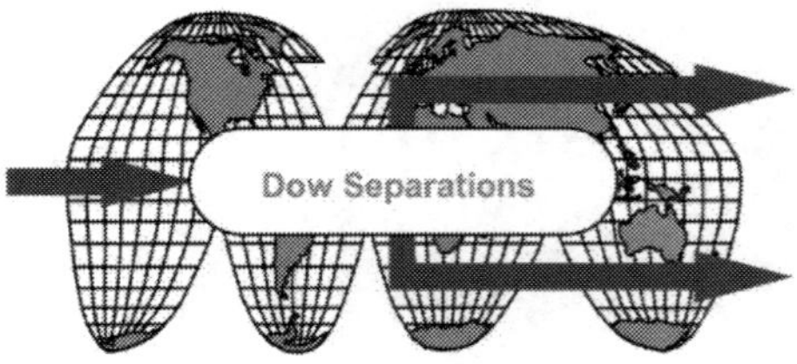
Dow Separations

393758 Nanofiber Thin Film Composite Membranes for Engineered Osmosis

Abstract:

Engineered osmosis (EO) is a rapidly emerging technology platform that harnesses osmotic pressure for desalination, water reuse, dewatering, and power generation. However, a vast majority of EO development has been hindered by the lack of membranes designed for use with this unusual driving force. At the University of Connecticut, we have developed a number of new polymeric thin film composite membranes that function well in a number of EO applications. One of our most unique is our electrospun nanofiber supported TFC membrane. The high porosity and low tortuosity of the nanofiber nonwoven support facilitates mass transport during osmosis. These membranes demonstrate substantially better flux performance than commercial membranes and prove that radical departures in TFC membrane design are not only possible but also can yield high performance osmotic membranes.

393000 Breaking the Rules for Selecting Protein-Resistant Surfaces

Abstract:

Both rational and trial-and-error methods have been used to search for low protein adhesive (fouling) surfaces for marine, medical, separations technology and other applications. Although these approaches to-date have produced some successes, we still do not fully understand how to rationally choose a protein-resistant surface. Unfortunately, surface science has not yet developed to the point that allows prediction of the surface or functional characteristics needed to minimize undesirable interactions with solution components, and thus to control fouling.

About 14 years ago, George Whitesides and his research group listed a set of rules that, if followed, constitute protein-resistant surfaces. The chemistry of the surface needs to (i) be hydrophilic (wettable), (ii) contain hydrogen bond acceptors, (iii) lack hydrogen bond donors, and (iv) be electrically neutral (Ostuni et al. (2001) Langmuir, 17 (18) 5605-5620).

Over the past five years we have screened monomer libraries and discovered a series of protein-resistant monomers using an inexpensive, fast, reproducible and scalable high throughput method. Both photo-oxidation with vacuum and atmospheric plasma without vacuum are used to graft-polymerize commercially available and combinatorially produced monomers onto light sensitive poly(ether sulfone) synthetic membranes. In addition to protein-resistant surfaces identified by the Whitesides group and others such as polyethylene glycol and zwitterionic monomers, the high throughput method selected a series of new monomers such as amines, mixed polar/apolar and chiral compounds. Using a membrane filtration assay as a surrogate measure for protein adhesion, and analyzing these new protein-resistant materials in light of the Whitesides et al's rules, we propose new modified rules.

360622 Ammonia Conversion Enhanced By Magnesium Chloride Absorption

Abstract:

Ammonia conversion with Haber-Bosch catalysts can be increased to above 95% by simultaneously and selectively adsorbing the ammonia with magnesium chloride. The maximum conversion possible depends both the reaction and the absorption equilibria. The conversion rate is a weighted harmonic average of the rates of reaction and of absorption. Under the conditions used, the chemical kinetics control at very low conversion, but the absorption rate is controlling at high conversion. This means that the overall reaction rate is as much as fifty times slower than the chemical kinetically controlled limit. The absorption rates could be increased by using a larger mass of much smaller particles, possibly supported on an inert substrate. The results presented provide a basis for designing small plants for distribute ammonia production

393757 Multimodal Membrane Adsorbers for Protein Purifications at High Ionic Strength

Abstract:

Development of efficient bioseparation materials and processes has been highlighted as one of the most pressing challenges facing the pharmaceutical and biotechnology industries. Without intervention to improve the downstream manufacturing capacity, there will be increasing shortages of biologic products, especially for those used in high doses to treat chronic diseases. In this talk, I will describe our latest research efforts to develop adsorptive membranes for chromatographic bioseparations. Specifically, I will describe the formation, characterization, and performance of a new class of multimodal membranes that can be used for protein purifications at high ionic strength. The additional flexibility in designing purification strategies that employ these multimodal membranes makes them attractive candidates to face emerging challenges, particularly tolerance of feedstock conditions. Using this membrane platform is further appealing because volumetric productivity can be high compared to more traditional chromatography media. Our work suggests that multimodal membranes have great potential to compete with more traditional ion-exchange materials following the affinity purification step in the downstream processing of biologics.

369567 Comparison of Thin Versus Thick Glassy Polymer Films for Membrane Applications

Abstract:

Gas permeation properties of thick laboratory films have long been used as a guide to the performance of asymmetric membranes with a very thin selective skin (~ 100 nm). However, much research has shown that very thin films of glassy polymers undergo more rapid physical aging than such thick films, leading to different permeation characteristics. This presentation will summarize recent results from collaborations with the National University of Singapore. The focus will be on 1) a first effort to explore comparisons of mixed gas permeation, where one component is CO_2, in thin films versus thick films made of Matrimid and 2) how film thickness affects the thermal rearrangement (TR) process of an *ortho*-functional polymide (6FDA-HAB) and its subsequent gas permeation characteristics.

385254 Designing Membranes with Ion-Conducting Nanochannels

Abstract:

Polymer electrolyte membranes have applications in energy storage and transformation devices, such as fuel cells, flow batteries, water electrolysis, or in water treatment processes such as desalination. The presence of ion-conducting nanochannels, which enable passage of hydrated ions, are important for efficient ion transport in these membranes. Some approaches to induce nanochannel formation using polymer architecture will be discussed.

386824 Polymer Hydroxide Exchange Membranes for Electrochemical Energy Conversion and Storage

Abstract:

Energy has been central to human civilization. The availability of cheap and abundant fossil energy is primarily responsible for the fastest ever human progress for the past 100+ years. With the depletion of fossil energy, the increase of world population, the improvement of living standard in large population centers, and the concern of the air quality and climate change, one of the grand challenges facing humanity today is the development of an alternative energy system that is clean, safe, and sustainable and where combustion of fossil fuels no longer dominates. Fuel cells, electrolyzers, solar hydrogen, and redox flow batteries when combined could represent such as system.[1] For our fuel cell efforts, we have been focusing on developing catalyst and membrane materials that will help to solve the cost and durability problems. Our fuel cell work began with the exploration of carbon nanotubes (CNTs) as durable catalyst supports[12]. We later explored the concept of platinum nanotubes (PtNTs) catalysts that were shown to have significantly improved durability and activity for the oxygen reduction reaction in proton exchange membrane fuel cells (PEMFCs)[3]. This catalyst platform has the potential to meet the mass and specific activity targets for vehicle applications specified by the US Department of Energy. We have also studied hydrogen generation by photocatalytic water splitting.[4] In this presentation, I will focus on our recent work on hydroxide exchange membrane (HEMs)[5]. HEMs, when used in fuel cells (HEMFCs), allow the platinum catalysts to be replaced by non-platinum-group-metals such as nickel and silver while the expensive fluorinated polymer membrane substituted by a hydrocarbon membrane, thus drastically reducing the cost of fuel cells and making them potentially economically viable. HEMs can also be used in electrolyzers and solar hydrogen generators or redox flow batteries for hydrogen generation and large scale solar/wind electricity storage.

References:

1. S. Gu, B. Xu, Y. Yan, **2014.** Electrochemical Energy Engineering: A New Frontier of Chemical Engineering Innovation, *Annu. Rev. Chem. Biomol. Eng.*, http://www.annualreviews.org/doi/abs/10.1146/annurev-chembioeng-060713-040114.
2. G. Gu, B.J. Xu, Y. Yan C. Wang, M. Waje, X. Wang, J. M. Tang, R. C. Haddon, Y. Yan **2004**. Proton exchange membrane fuel cells with carbon nanotube based electrodes, *Nano Lett.* 4:345-348.
3. Z. Chen, W. Li, M. Waje, Y. S. Yan **2007**. Supportless Pt and PtPd nanotubes as electrocatalysts for oxygen reduction reaction, *Angew. Chem. Int. Ed.* 46:4060-4063.
4. H. Luo, T. Takata, Y. Lee, K. Domen, J. Zhao, Y. Yan **2004.** Photocatalytic activity enhancing for titanium dioxide by co-doping with bromine and chlorine, *Chem. Mater.* 16:846-849.
5. S. Gu, R. Cai, T. Luo, Z. Chen, M. Sun, Y. Liu, G. He, Y. Yan **2009**. A Soluble and Highly Conductive Ionomer for High Performance Hydroxide Exchange Membrane Fuel Cells, *Angew. Chem. Int. Ed.*, 48: **6499-6502.**

373901 Development of New, Low-Cost and Improved Membranes and Membrane-Based Systems for Water, Energy and Environmental Applications in Singapore

Abstract:

Membrane technology is an emerging technique that has become increasingly important in our life. With nearly 50 years of rapid development, today, various membrane processes have found numerous applications in industries from biotechnology, pharmaceuticals, agro-food, chemical processing, as well as in water, energy and the environment. While membrane technology has greatly enhanced our capability of tackling the challenges of increased population growth and scarcity of freshwater resources, as well as more stringent environment regulations, it is recognized that the current membrane processes still suffer from high energy consumption and generation of a considerable amount of waste stream. In addition, their benefits are also constrained by the less than ideal separations (low permeability & selectivity) of synthetic membranes due to the lack of functionality and controlled architecture.

Singapore membrane Technology centre (SMTC) was set up to spearhead Singapore's R&D efforts in fundamental and applied research in membrane science and technology. This presentation will review our efforts in developing new, low-cost and improved membranes and membrane-based systems for water, energy and environmental applications. Our activities and specialisation involve following seven core areas: (1) water production – water treatment, desalination; (2) water reclamation – pre-treatment and reversed reverse osmosis; (3) membrane bioreactors – conventional and novel; (4) energy Issues – improving energy efficiency and energy production from chemical potential gradients; (5) special Needs – chronic and acute, decentralised; (6) sensors and process control – membrane monitoring and performance enhancement; and (7) cleaner production and environment – modification of production process to reduce its environmental impact and improve its profitability.

382025 Integrated Membrane Material and Process Development for Gas Separation

Abstract:

The popularization of membrane systems for gas separation has been realized by effective process designs and high performance membrane materials, achieving low cost and high energy efficiency. This presentation will highlight three areas of the research, based on our experience. First, the importance of membrane process designs will be illustrated by examples of sweep/countercurrent designs to seamlessly integrate membrane units into process trains to enhance energy efficiency. This approach is particularly useful for separations with low feed to permeate pressure ratios such as CO_2 capture from flue gas. Second, polymeric membranes with high permeability and selectivity are preferred. However, there exists a tradeoff, i.e., materials with higher permeability often have lower selectivity and vice versa. The selection of membrane materials will be dictated by process designs with judicious balance of permeability and selectivity. The presentation will give examples of optimal membranes identified for O_2/N_2 separation and CO_2/CH_4 separation. Finally, high performance polymers need to be made into thin film composite membranes to achieve high flux and establish industrial relevancy. Thin films of many hydrocarbon polymers often show decreasing permeability with time. Interestingly, thin film perfluoropolymers show stable permeability over time, which may shed some light on the molecular level understanding of thin film polymer aging.

378647 Beyond Upper Bound: Toward High Separation Performance Hybrid Membranes

Abstract:

Beyond Upper Bound: Toward High Separation Performance Hybrid Membranes

Authors: Yifan Li, Guangwei He, Shaofei Wang, Shengnan Yu, Fusheng Pan, Hong Wu, Dong Yang and Zhongyi Jiang*

Key Laboratory for Green Chemical Technology of Ministry of Education School of Chemical Engineering and Technology, Tianjin University, 300072, P. R. China

Collaborative Innovation Center of Chemical Science and Engineering (Tianjin), Tianjin 300072, China

[*] Corresponding author and presenter: zhyjiang@tju.edu.cn

Abstract Hybrid membranes comprising continuous polymer phase and dispersed inorganic filler phase have been extensively studied in diverse separation processes including gas separation and pervaporation. The multiphase characteristics of composite membranes render more degree of freedom to manipulate multiple interactions, tailor multiscale structures, and integrate multiple functionalities, thus win the much more chance in solving the tradeoff relation between permeability and selectivity. In this presentation, we will describe the various methods for the fabrication of hybrid membranes including physical blending method, sol–gel method, biomimetic mineralization method and biomimetic adhesion method. For each method, the major advances and challenges have been summarized. The application of hybrid membranes in carbon capture, ethanol dehydration and gasoline desulfurization will be illustrated. Future directions toward designing advanced hybrid membranes with tunable structures and high separation performances are proposed tentatively.

Key words Hybrid membranes, Fabrication, Pervaporation, Gas separation, Tradeoff relation.

394663 Engineering New Ion-Conducting Membranes using Tapered Block Polymers

Abstract:

Ion-conducting block polymers have enormous potential for membrane applications such as lithium battery separator/electrolytes due to their ability to form ordered and co-continuous nanostructures. These self-assembling macromolecules can allow the simultaneous control over ionic transport, processing, and mechanical strength, thus providing an enticing opportunity toward fabricating designer polymer electrolytes for flexible, lightweight, and efficient devices. From the standpoint of increasing processability in polymer electrolytes, tapered block polymers provide an efficient way to decouple processing temperatures from the polymer molecular weight by reducing interfacial interactions between pure blocks through adjustments in the interfacial composition; in other words, the order–disorder transition temperatures (T_{ODT}) and glass transition temperatures can be tuned independent of molecular weight and overall block composition using tapered interfaces. Recently, we have successfully synthesized a tapered block polymer electrolyte system that shows significantly improved room temperature lithium ion conductivity, co-continuous nanoscale network structures, lower temperature processability, robust cycling behavior, and lower energetics associated with lithium ion motion. This behavior can be correlated directly to the composition of the tapered interfaces in our modified block polymers, and these tapered architectures may demonstrate similar advantages in other nanostructured polymer systems.

385210 Gas Separation in Cross-Linked Liquid Crystal Polymer Membranes

Abstract:

We have synthesized stable, cross-linked, liquid crystalline polymer (LCP) films for membrane separation applications using a combination of acrylate mesogenic and non-mesogenic monomers and a dimethacrylate cross-linker via an *in situ* free radical polymerization technique. The phase behavior of the LCP membranes was characterized using DSC and X-ray scattering, and indicated the formation of a nematic liquid crystalline (LC) phase above the glass transition temperature. Single gas transport of CO_2, H_2, CH_4, propane and propylene in the cross-linked LCP membranes was investigated over a range of temperatures in the LC mesophase and the isotropic phase. The solubility of the gases was dependent not only on the condensability in the LC mesophase, but also on favorable molecular interactions of certain penetrant gas molecules with the ordered polar mesogenic side chains of the LCP. Selectivities for various gas pairs generally decreased with increasing temperature and were discontinuous across the nematic/isotropic transition. Higher cross-link densities in the membrane generally led to decreased selectivity at low temperatures when the main chain motion was limited by the lack of mesogen mobility in the ordered nematic phase. However, at higher temperatures increasing the cross-link density increased selectivity as the cross-links acted to limit chain mobility.

379501 Fouling Resistant, High Flux Membranes with ~1nm Effective Pore Size By Zwitterionic Copolymer Self-Assembly

Abstract:

Membranes with permeate size cut-offs in the nanometer range have numerous applications in the biochemical, pharmaceutical and food industries, as well as in specialized wastewater treatment processes. To be commercially viable, membranes for molecular-size-based separations need to exhibit high flux and fouling resistance, and be readily synthesized and fabricated into large-area membranes. Zwitterionic groups strongly resist biomacromolecular fouling due to their high degree of hydration, which makes them promising materials for membrane applications. Zwitterions are also documented to self-assemble into channel-type clusters 0.6-2 nm in size. Within certain composition ranges, random copolymers of zwitterionic and hydrophobic monomers self-assemble to form bicontinuous networks of nanochannels. We introduce a new class of high flux, fouling resistant, size-selective nanofiltration membranes whose selective layers are made of this family of copolymers. We have synthesized such copolymers and formed thin film composite membranes by coating them onto commercial ultrafiltration membrane supports. These membranes exhibit fluxes as high as 21 L/m^2.h.bar, which can be further improved by better coating methods. Based on the rejection of anionic dyes of varying sizes, they show size-based selectivity with a cut-off around 1 nm. This pore size closely matches the size of the zwitterionic nanochannels, measured to be ~1.1-1.4 nm in diameter by transmission electron microscopy (TEM). These are the first examples of membranes that gain their selectivity from the microphase separation of zwitterionic groups, in addition to exploiting this functionality for fouling resistance. We expect these membranes to be promising candidates for various applications including the purification of pharmaceuticals, antioxidants, and textile wastewater treatment.

Both surface chemistry and surface structure are believed to affect membrane fouling. Researchers have screened a wide range of surface chemistries for antifouling properties, finding that neutral, hydrophilic, and low-hydrogen bond forming surfaces have a lower fouling propensity. Zwitterions, neutral molecules with positive and negative charges on individual atoms within the molecule, meet each of these criteria and are promising materials for antifouling surfaces. With regards to surface structure, researchers have demonstrated that amphiphilic surfaces with compositional heterogeneities on the length scale of the foulant will discourage adsorption by disallowing favorable thermodynamic interactions between the foulant and the surface. While the nanopatterned surfaces have received a great deal of attention, no studies have systematically isolated the effects of size scale and chemical structure in a single set of experiments.

This work characterizes the fundamental mechanisms of fouling resistance by novel amphiphilic polymeric membrane surfaces formed through the self-assembly of zwitterionic copolymers. The first stage is to synthesize a library of amphiphilic, random copolymers that self-assemble into hydrophilic and hydrophobic domains and can be cast into membrane films. These films will vary both in the chemical composition and in the hydrophobic/hydrophilic domain size, but each will contain A) zwitterionic domains that allow water permeation but prevent salt passage, and B) rigid hydrophobic domains that resist membrane swelling. We have selected two commercially available zwitterionic monomers and four commercially available hydrophobic monomers. Thus far, we have successfully synthesized four copolymer films exhibiting different domain sizes and chemical structures, with rejection rates in the nanofiltration to reverse osmosis range.

Using QCM-D, we report a quantitative assessment fouling resistance for the four copolymer films. We find effects of both chemistry and polymer structure on foulant deposition, the structure of the foulant on the surface, and on cleaning rates. Additional experiments with other pore chemistries, including ionic liquids, could further expand this work.

This work will facilitate the creation of fouling resistant membranes through a fundamental understanding of the direct correlation between membrane structure, membrane chemistry, and membrane fouling. These improved membranes have the potential to dramatically increase the efficiency of existing membrane-based separation technologies by reducing energy losses associated with fouling.

Chemical crosslinking of LBL polyelectrolytes NF membranes for performance improvement

Rui Jia[1], Baowei Su[1,*]

[1] Key Laboratory of Marine Chemistry Theory and Technology (Ocean University of China), Ministry of Education; College of Chemistry & Chemical Engineering, Ocean University of China, 238 Songling Road, Qingdao 266100, China

*Corresponding author: Tel/Fax: +86−532−66786371; Email address: subaowei@ouc.edu.cn

ABSTRACT

A novel kind of nanofiltration (NF) membrane with merely two bilayers of poly (ethylene imine acid) (PEI) / poly (acrylic acid) (PAA) / glutaraldehyde (GA), abbreviated as $(PEI/PAA/GA)_2$, was self-assembled using layer-by-layer (LBL) and covalent crosslinking method on hydrolysis-modified polyacrylonitrile (PAN) ultrafiltration (UF) substrate. The effect of GA concentration and crosslinking time on the membrane separation performance was investigated. The formation mechanism, surface morphology, separation performance, solvent resistance, as well as the stability of the chemical crosslinked $(PEI/PAA/GA)_2$ NF membrane were also extensively investigated. The results showed that this novel preparation method could increase the membrane separation performance considerably. Under optimal crosslinking condition, the prepared two bilayer $(PEI/PAA/GA)_2$ membrane could reach a rejection of 98.13% for Na_2SO_4 with a permeate flux of 12.64 L m^{-2} h^{-1}; and a rejection of 80.89% for

NaCl with a permeate flux of 16.78 L m^{-2} h^{-1}. The crosslinked $(PEI/PAA/GA)_2$ NF membrane is superior and more stable compared with the electrostatic membrane with even more bilayers, and shows excellent anti-solvent and anti-bacterial performance.

Keywords: layer-by-layer, self-assembly; nanofiltration; crosslinking; solvent resistance; anti-bacterial property

378643 Fabrication of Gelatin/Graphene Oxide Ultrathin Hybrid Membrane Via Layer-By-Layer Self-Assembly for Pervaporation Dehydration Process

Abstract:

Fabricating composite membrane with ultrathin, defect-free and stable active layer is an effective method to achieve separation membrane with high permeability, selectivity and stability simultaneously. Layer-by-layer (LBL) self-assembly method provides an attractive option for the fabrication of ultrathin, defect-free membranes with tailored composition and tunable properties. Inspired by the distinctive mechanical properties of nacre from the ordered brick-and-mortar arrangement of organic and inorganic layers, ultrathin hybrid membranes were fabricated via LBL self-assembly of graphene oxide and gelatin for pervaporation dehydration process. Graphene oxide possesses amphiphilic structure with oxygenated polar groups on the basal plane and edge, and the hydrophobic sp^2 hybridized carbon atoms on backbone. Gelatin is a natural protein composed of various hydrophobic and hydrophilic amino acids. The matching chemical structures of graphene oxide and gelatin endows the formation of multiple interactions between them (including electrostatic attraction, hydrogen bond, and hydrophobic interaction), which differs from the existing researches. The dense interface regions between graphene oxide and gelatin achieved efficient separation of water from ethanol solutions. Moreover, the micrometer-scale size of graphene oxide in horizontal direction favored the coverage of nanopores on substrate, which greatly reduced the required deposition cycles, and thus simplified the membrane-fabrication procedure.

Pros and Cons of Absolute, Excess and Net Adsorption Thermodynamics

Orhan Talu
Cleveland State University
Cleveland, Ohio 44115, USA
(o.talu@csuohio.edu)

The *net* adsorption thermodynamic framework was recently introduced as an alternative to *excess* and *absolute* adsorption (Gumma and Talu, *Langmuir* 2010, 26(22), 17013-17023). It is a fundamentally different thermodynamic description of adsorption since the placement of Gibbs dividing surface, necessary for any treatment of adsorption, is different than the known/used alternatives. Net adsorption is a natural choice for microporous solids. It removes any ambiguity in theoretical treatment of adsorption and circumvents all questionable measurements/quantities in experiments.

Initial development of net adsorption was prompted by the interest in adsorptive gas storage and by the need to unambiguously compare experiments to molecular simulation results. Previously, our research group and others have shown that: 1) non-adsorbing helium assumption commonly used to place the Gibbs dividing surface in experiments does not hold (e.g. Gumma and Talu, *Adsorption J.* 2003, 9, 17-28), 2) molecular simulation results can only be compared to excess data by performing more simulations with helium (e.g. Talu and Myers, *AIChE J.* 2001, 47(5), 1160-1168), and 3) volume occupied and thermal mass of solid adsorbent play a significant role in equilibrium and dynamics of adsorptive storage systems (e.g. Chang and Talu, *Applied Thermal Engineering* 1996, 16(5), 359-374). As such, our previous work has concentrated on pure component adsorption equilibrium and dynamics.

This contribution involves extension of net adsorption framework to gas mixtures. Theoretical developments on equilibrium formulations will be included as well as its implications on adsorption processes such as column dynamics in separations and in storage. The theory will be supplement with data examples and implementation methods. The advantages/disadvantages and limitations of net adsorption will be discussed.

382137 Robust Algorithms for the Solution of the Ideal Adsorbed Solution Theory Equations

Abstract:

The Ideal Adsorbed Solution Theory is used extensively for the prediction of multicomponent adsorption equilibria, yet based on recent publications there appears to be a lack of understanding of when solution algorithms are guaranteed to converge.

In its original formulation the IAST is used to predict the behaviour of mixtures starting from the pure component isotherms. The solution of the IAST equations is also needed when non-ideality of the adsorbed phase is considered, since the IAST is used in the definition of the reference state for the mixture in the definition of the activity coefficients. The availabitily of robust methods to solve the IAST equations is therefore at the heart of thermodynamically consistent approaches to describe multicomponent adsorption.

We investigate in detail the robustness and convergence criteria for the nested loop and the Fast-IAS algorithms. A formal proof of guaranteed convergence is presented for the nested loop method of Valenzuela and Myers. Based on the analysis of this algorithm it is also possible to understand some partial instabilities present in the Fast-IAS procedure.

Detailed comparisons of the different algorithms is presented to show also that it is possible to extend the Fast-IAS to cases where there are no analytical expressions available for the spreading pressure and to non-type I isotherms.

373106 Modeling the Thermodynamics and Dynamics of Fluids Confined in Three-Dimensionally Ordered Mesoporous (3DOm) Carbon Materials

Abstract:

Three-dimensionally ordered mesoporous (3DOm) carbons recently emerged as a step in the synthesis of mesoporous zeolite-based materials [1]. 3DOm carbons are templated by colloidal crystals formed from silica nanoparticles. A range of pore sizes of the 3DOm carbon can be obtained by using colloidal crystals with various primary particle sizes (e.g., 10, 20, 30, 40 nm). The resulting materials consist of spherical pores interconnected by windows created by sintering in the silica sphere assembly. They can be used to template the assembly of zeolites within the pores created by the pores. These materials represent an important approach in the search for carbon materials with a high degree of order.

In this work we use a lattice gas model to study adsorption equilibrium and dynamics in 3DOm carbons. The 3DOm structure is superimposed onto a lattice model and this is done by the coarse graining of configurations from Monte Carlo simulations of hard sphere solids. This approach allows us to study several important effects upon the adsorption properties, including spatial disorder in the initial sphere assembly, as well the degree of sintering and size polydispersity in the silica spheres. Using mean field density functional theory we consider the effects of window size upon the adsorption hysteresis including the transition from cavitation to pore blocking. DMFT calculations provide visualizations of the evolution of the three-dimensional density distribution and reveal mechanisms in the dynamics of pore filling in these systems.

[1] Fan, W., M.A. Snyder, S. Kumar, P.-S. Lee, W.C. Yoo, A.V. McCormick, R.L. Penn, A. Stein, and M. Tsapatsis, *Hierarchical nanofabrication of microporous crystals with ordered mesoporosity*. Nat Mater, 2008. **7**(12): p. 984-991.

A General Method to Model Column Separations with Breathing MOF Adsorbents

Tom R.C. Van Assche[a], Gino V. Baron[a], Joeri F.M. Denayer[a]

[a]Department of Chemical Engineering, Vrije Universiteit Brussel, Pleinlaan 2, 1050 Brussels, Belgium

Metal-organic frameworks (MOF) have a distinctive chemistry, often allowing a large structural flexibility. Upon adsorption, such highly flexible materials can display remarkable guest induced transitions as gate-opening, swelling and 'breathing'. Significant progress has been made in understanding the adsorptive behavior of such complex materials using a suitable thermodynamic framework [1] as developed by Fuchs and coworkers, which also allows understanding and predicting the adsorption of mixtures using the osmotic framework adsorbed solution theory (OFAST [2]). These thermodynamic models are considering the phase transformations of the material, e.g. from a large pore form (LP) to a narrow pore form (NP) and back for the MIL-53(Al) MOF, depending on the adsorbate and temperature. Such models can be used to predict the adsorption behavior in equilibrium conditions, but must be implemented into a dynamic model to provide information regarding the separation in dynamic conditions, i.e. in a packed bed of adsorbent.

In this work, the dynamic adsorption behavior of such flexible adsorbents was modeled, by combining general equilibrium models for flexible structures with the balance equations of the adsorption column. Modeling was performed using a partial differential equations approach (PDE) as well as a robust Stop&Go based (SG) [3] modeling method for equilibrium profiles. We have studied the general dynamic adsorption behavior on flexible adsorbents for pure adsorbates (single component) as well as binary mixtures. This is performed for diluted (1 or 2 adsorbates in an excess of inert carrier) and non-diluted (1 or 2 adsorbates) cases. The model solutions are compared with their characteristics using the hodograph technique presented by Glueckauf [4], allowing the general classification of diluted breakthrough profiles on phase transforming adsorbents.

During dynamic adsorption (breakthrough experiments) the behavior of flexible adsorbents as breathing MOFs is quite complex. Elution of the initial desorbed state by the feed state can occur by a succession of phase transformations. For the diluted binary mixture case, a variety of unusual breakthrough profiles can be observed. For example, intermediate steps and roll-up of the most strongly adsorbed component can be observed. In general, several cases can be distinguished, based on the relative position of the feed concentration to the phase-coexistence line (e.g. NP and LP form phase coexistence). For non-diluted cases, these effects are less pronounced, yet intermediate steps or kinks in the breakthrough profiles can often still be observed.

Based on the general breakthrough modeling methodology developed in this work, we have performed 2 case studies, one in diluted form and one in non-diluted form on MIL-53. This approach leads to very good predictions of the experimental breakthrough profiles shapes. For example, the remarkable double roll-up effect for CH_4 in CH_4/CO_2 separation on MIL-53, as observed by Hamon et al.[5], is predicted as shown in figure 1.

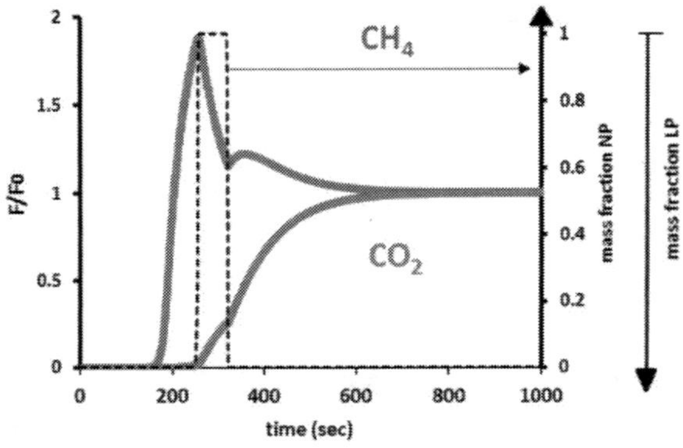

Figure 1:
Predicted breakthrough curve for a 75-25 CO_2-CH_4 mixture at 10 bar using MIL-53(Al) parameters. The dashed line shows the occurence of the NP and LP form.

Acknowledgements

T. Van Assche gratefully acknowledges the Agency for Innovation by Science and Technology (IWT) in Flanders for financial support. J. Denayer is grateful to FWO Vlaanderen for financial support.

[1] F.X. Coudert, M. Jeffroy, A.H. Fuchs, A. Boutin, C. Mellot-Draznieks, JACS 130, 14294, 2008
[2] F.X. Coudert, Phys. Chem. Chem. Phys. 12, 10904, 2012
[3] K. Chihara, Y. Yoneda, S. Morishita, , Stud. Surf. Sci. Cat. 28, 563, 1986
[4] E. Glueckauf, Discuss. Faraday Soc. 7, 12, 1949
[5] L. Hamon, P.L. Lewellyn, T. Devic, A. Ghoufi, G. Clet, V. Guillerm, G. D. Pirngruber, G. Maurin, C. Serre, G. Driver, W. van Beek, E. Jolimaître, A. Vimont, M. Daturi, G. Férey, JACS 131, 17490, 2009

384586 Computational Design of Metal-Organic Frameworks for Gas Storage

Abstract:

Fuels such as natural gas (NG) and hydrogen produce less CO_2emissions than gasoline. They are therefore attractive to power next-generation vehicles. However, a challenge that must be addressed first is the storage of these gases in compact volumes such that the vehicle tank is of a reasonable size but allows for a practical driving range. Vehicles powered by natural gas currently store the gas compressed to 250 bar (CNG) in cumbersome tanks. An attractive alternative is to use an adsorbent material that enables densification of the fuel at safer, cheaper, and lower pressures.

Metal-organic frameworks (MOF) are porous, crystalline materials resulting from the assembly of inorganic "nodes" and organic "linkers" into a three-dimensional network. The essentially limitless possible combinations of inorganic and organic building blocks make MOFs highly tunable and promising for gas storage. Nonetheless, a major concern in the applications of MOFs is the stability of this class of materials. Recent synthesis of remarkably stable MOFs, however, suggests that the identity of the inorganic nodes may be the key to stable structures. In this contribution, we first use a high-throughput computational screening approach, along with data analysis techniques, to investigate thousands of hypothetical MOF structures and explore the performance limits of nanoporous materials based on their methane (the major component of natural gas) "deliverable capacity" at various operation scenarios. Then, we continue our computational work using a hypothesis-driven approach where we focus on building and investigating the adsorption characteristics of hypothetical MOFs based on stable $Zr_6O_4(OH)_4(CO_2)_n$ inorganic nodes.

After computational characterization and simulation of methane adsorption in over 200 hypothetical zirconium MOFs, some relationships became apparent, which constitute important guidelines for material design. One example is the link between methane deliverable capacity and the presence *and* location of triple bonds in the structure of the screened MOFs. Following these emerging guidelines, we constructed and identified *in silico* a material with a predicted deliverable methane capacity of 197 cc(STP)/cc between 65 and 5.8 bar. The identified hypothetical zirconium MOF, based on 1,4-benzenedipropynoic acid linker, was synthesized and named **NU-800**. Experimentally, it was possible to activate around 90% of the porosity of **NU-800**, which allowed us to confirm the promising methane adsorption properties of this material. The measured deliverable capacity between 65 and 5.8 bar was 167 cc(STP)/cc, with the discrepancies between simulation and experiment probably due to the somewhat incomplete activation of the material. Our screening also allowed us to identify zirconium MOFs with large gravimetric surface areas, which is a key MOF property to achieve high gravimetric hydrogen adsorption. Some of these materials were synthesized, for instance MOF **NU-1100**, and their promising hydrogen adsorption properties were confirmed by experimental measurements. All of the synthesized structures were confirmed to be stable over a dozen adsorption/desorption cycles.

Computational Screening of Mixed Cation Zeolites for CO_2 Capture

Hanjun Fang[1§], Ambarish Kulkarni[1§], Preeti Kamakoti[2*], Rohan Awati[1], Peter I. Ravikovitch[2], and David S. Sholl[1*]

[1]School of Chemical & Biomolecular Engineering, Georgia Institute of Technology, Atlanta, GA 30332-0100, USA

[2]ExxonMobil Research and Engineering, 1545 Route 22 East, Annandale, NJ 08801, USA

Abstract

Zeolites are a valuable class of porous materials and widely used commercially as adsorbents due to the high adsorption capacities and thermal stability. As an important complement to experiments, molecular modeling tools have been applied in the screening and selection of suitable porous materials for the applications of molecular adsorption and separation. Previously, we developed first principles-based transferable force fields for CO_2 adsorption in siliceous and Na-exchanged zeolites.[1] In this work, we extend our approach to CO_2 adsorption in other alkali cation (K, H, Li and mixed K/Na-exchanged) zeolites. The force fields are validated by comparing simulated adsorption isotherms and heats of adsorption with experimental data for both pure and mixed cation zeolites. Then we use the derived force fields in molecular simulations to screen 10-member ring zeolites for CO_2 capture. The effects of zeolite topology, Si/Al ratio, and cation concentration on adsorption properties are investigated, and the relationship between adsorption performance and geometrical properties such as pore volume, surface area, and pore sizes of zeolite materials are developed. These results enable us to rationally design optimal zeolite materials for CO_2 capture under specified process conditions and elucidate structure-property relationships.

Reference:

1 H. J. Fang, H. Demir, P. Kamakoti and D. S. Sholl, *J. Mater. Chem. A*, 2014, **2**, 274.

[*] Corresponding authors. Email: preeti.kamakoti@exxonmobil.com, david.sholl@chbe.gatech.edu

[§] These authors contribute equally to this work

391751 Understanding Complex Structured Adsorbent Contactors Via 3-D CFD Modeling

Abstract:

The primary objective of this work is to carry out a comprehensive modeling, theoretical study and partly experimental evaluation of the performance of a structured adsorbent contactor for gas separation applications. A rigorous two dimensional model including mass, momentum and energy balances and appropriate boundary conditions will be developed and used to investigate the adsorption and desorption dynamics of CO_2 capture from flue gas in a parallel passage adsorbent contactor covered with zeolite 13X under different conditions. The model will be expanded to a very rigorous 3-D model in order to deal with real applicable systems of triangular and square channels. Results will be compared to the parallel channel. Behavior of the system in terms of break through curves, loading, pressure, velocity, concentration and temperature profiles needed to understand the fundamentals of the structured bed adsorbent contactor will be investigated. Then, experimental studies will be used to develop a one dimensional pressure drop correlation which can be substituted with Ergun equation for packed bed systems used in the Dynamic Adsorption Process Simulator (DAPS), developed at USC, to be able to include structured adsorbent contactors in the developed in house one dimensional code which requires much less calculation and time cost. The simulation results will be compared to the full 2-D and 3-D models. Lastly, the most important concern about the use of structured adsorbent contactors is to ensure plug flow conditions will be sustained to avoid premature breakthrough. The design and operational requirements necessary to satisfy this condition will be developed. This presentation will discuss the latest results obtained by Ritter and his group.

Abstract

Pressure reactors are used extensively in petroleum, polymer, biofuels, pharmaceutical, and chemical processing industries. But very little information on reactor design and specifications has been published. In this study, we describe 10 and 210L batch reactors designed to scale-up heterogeneous catalysis of insoluble solids in an aqueous phase. These reactors were designed, developed, and tested at Advanced Biofuels Process Demonstration Unit (ABPDU), Lawrence Berkeley National Lab (LBNL) to rapidly reach high temperatures (10 to 350^0 C) and high pressures (0 to 1800 psig) and efficiently handle high concentrations of insoluble solids (up to 10,000 cP) in the aqueous phase substrate. Reactor vessels were built with Hastelloy C-22 to provide resistance to acidic and other corrosive chemical catalysts. To ensure improved mass transfer, the insolubles and solid catalysts should remain suspended in the aqueous phase throughout the reaction. Also, flow patterns in the reactors were expected to alter continuously due changes in material behavior during conversion of insoluble reactants to soluble products. Three impellers were selected to provide effective mixing and suspension of the solids in high to low viscosity aqueous phase substrates. The scale-up reactors were custom fitted with both direct (catalyst) and indirect (oil in an external dimpled jacket) heating. Direct heating is particularly useful for the scale-up of reaction chemistries that require rapid heating of reaction contents to target temperature. Other features of 210L reactor include a special discharge valve that opens under pressure to simulate explosion-like studies, head space for solvent expansion, mass flow meters to measure the loss of solvent during venting process, etc. We were successfully able to commission the reactors with a fast heating rate of 3.5^0 C/ minute and achieved a 95% overall mass balance closure. In this presentation, we will discuss the learnings from the reactor design, development, and commissioning of the reactors and also share the results from scale-up of heterogeneous catalysis of lignocellulosic biomass to produce cellulosic sugars and levulinic acid. The findings from this study will benefit researchers in industry and academia that are intending to scale-up reaction chemistries that require special considerations for improved mass and heat transfer and also intend to achieve 95+% mass balance closure at a scale-up level operation.

Keywords: pressure reactor, reactor design, insolubles, cellulosic sugars, levulinic acid

387407 Scale-up and Commercialization of a Novel Gas Fermentation Process for the Production of Fuels and Chemicals

Abstract:

The ever increasing demand for renewable feedstock-based fuels and chemicals is driving the development of new technologies and processes. An entire industry is emerging to meet the demands for renewable energy with a significant portion of these new technologies based on microbial conversion processes. The challenge these new processes face is that they have to compete with the scale and/or economics of conventional petroleum derived fuel and chemical processes. To compete on a plant footprint and capital basis, many of these bio-technologies are developing continuous operations for processes that are conventionally operated in batch-mode, such as fermentations. The rising field of renewable biotechnology not only faces the challenges of developing processes for new reactions and products, but also designing and scaling-up novel processes to compete with incumbent fossil-based industries.

One such emerging biotechnology offering is gas fermentation which is an innovative hybrid thermochemical/biochemical process that converts gaseous carbon to fuels and chemicals through a microbial conversion process. The field of gas fermentation represents a transformative technology offering that could reduce emissions while producing fuels and chemicals from a renewable microbial source. The gas feedstock can come from a variety of sources including gaseous carbon residue from industrial processes or from thermochemical processes such as gasification of biomass/MSW or reformed natural gas. The main feedstocks are CO, H2, and/or CO2 which face the challenge of poor water solubility thus requiring high mass transfer multi-phase bioreactors. Along with the design and scale-up challenges of a novel biotechnology, innovative bioreactor concepts are also required for the successful deployment of high productivity continuous processes.

The LanzaTech's integrated thermochemical/biochemical gas fermentation process is described including the scale-up path from lab to pilot to pre-commercial demonstration. Various pilot and demonstration activities are described highlighting the design and scale-up challenges faced to date. The LanzaTech technology can integrate with a number of various industrial processes which presents additional challenges designing for the integration of a gas fermentation process with commercially established industrial processes. Successful scale-up activities and operations are presented to demonstrate the design and scale-up of this innovative biotechnology.

354791 Standardization of Cocoa Butter Extraction in the Cocoa Pressing Process in a Chocolate Factory

Abstract:

At chocolate factories, dry cocoa beans are roasted, milled, sieved to remove seed shells, and crushed to produce a viscous mass (cocoa paste) which by heating melts into the cocoa liquor. This liquor contains roughly equal proportions of fat (cocoa butter) and solids. In order to reduce the fat content in cocoa liquor a hydraulic pressing process is commonly used. By compressing cocoa liquor under high pressure against a filter media, cocoa butter is extracted, and the remaining solid cake is pulverized to obtain cocoa powder. Fat content is a major specification in cocoa powder, being 12% and 16% w/w commercial standards in the chocolate market. In order to control this specification, a trial an error method is generally applied at industrial scale. Depending on the required fat content and the liquor characteristics, extraction time in the press is adjusted. This operation is not desirable because it reduces productivity and affects quality of the product. In order to standardize operating conditions during pressing process, in this work it was studied the effect of cocoa liquor fat content and pressing time in the final fat content of powdered cocoa. By mean of a statistical approach based upon preliminary experiments, and using the statistical software R, it was established a reliable method for cake sampling at the industrial scale press. Using this technique, and by taking the appropriate average over different solid cakes from the press, a standardized method for the measurement of fat content in powdered cocoa was established. By using this characterization technique, a series of extractions experiments were carried out (using a box-behnken design) varying liquor fat content and pressing time, and observing the effect over solid cake fat content. Results were correlated with a linear expression and a good agreement between model and experiments was observed. By implementing obtained model during extraction process in the chocolate factory increased cocoa powder and cocoa butter daily production, and significantly reduced the amount of non-conforming product.

Use of Deep Eutectic Solvents for Liquid-Liquid Aromatic/Aliphatic Hydrocarbon Pilot Plant Separation

Bessadok A. J., Hadj-Kali M. K., Sourwano, M. d AlNashef I. M.
Chemical Engineering Dept, College of Engineering, King Saud University, KSA.

The separation of aromatic/aliphatic hydrocarbons is a difficult task in unit operation because the range of boiling points is very similar. In addition, azeotropes could be formed. Indeed, no adequate separation technology is available for process streams having an aromatic content under 20 wt/wt %. Recently, we demonstrated that low cost ionic liquid analogues, called Deep Eutectic Solvents (DESs), can be utilized for the separation of aromatics using liquid-liquid extraction method. Both the distribution ratio and selectivity were higher than that for sulfolane which is the most used solvent in industry. Laboratory scale results demonstrated the benefits of DESs in this separation operation and a patent application has been filed. In this work we propose to use a pilot plant scale centrifugal extraction multi-stage unit for the separation of aromatics from a mixture of aliphatics and aromatics. The pilot plant is being used for the validation of previous laboratory scale results and for the optimization of the liquid-liquid separation operation. This could constitute a very important step towards the implementation of this method in industrial scale.

390485 Systematic Evaluation of Triethyl Citrate Purification By Using a Falling Film Evaporator System

Abstract:

The triethyl citrate is a bio-based plasticizer for different polymers, and also a fixative agent widely used in cosmetic and food industry. As fragrance and flavor ingredient, triethyl citrate must fulfill certain organoleptic properties such as transparency, clarity, brightness and neutral odor. Because esterification of citric acid requires large excess of ethanol to overcome solubility and chemical equilibrium limitations, purification of the ester requires ethanol stripping. In general citric acid esterification is carried out under batch operation, where the jacketed reactor is also used as reboiler for the stripping stage. During ethanol removal, the ester gets undesirable color and odor mainly because large residence time is required to reduce volatiles concentration up to commercial specifications, and also because of the low heat transfer area to volume ratio.

In order to improve organoleptic characteristics of triethyl citrate during ethanol removal, a falling film evaporation system was evaluated for product purification. In this system the heat transfer area to volume ratio is much larger than in traditional heat transfer equipments, and also reduced residence times are obtained. A simple lab scale methodology was developed to evaluate the preferable operating conditions for a continuous falling film evaporation system. Time dependent evaporation profiles were obtained gravimetrically at different temperatures (50, 80, 100 and 120 °C) in a vacuum oven. Based upon organoleptic panels, this methodology allowed establishing the range of operating conditions where the product remained stable and no degradation was observed. Besides, It was possible to estimate the residence time required in a continuous evaporation system.

In subsequent experiments, lab scale (40 cm high and 5 cm diameter, electrically heated), and pilot scale (1m high and 12 mm diameter, steam heated) falling evaporators were ran under different operating conditions (residence time and temperature). It was possible to demonstrate that the developed methodology can be used to predict best operating conditions for falling film evaporation systems.

A Novel Carbonized Polydopamine Adsorbent (C-PDA) with High CO_2 Capacity and Low Water Vapor Adsorption

Sikai Xian, Junjie Peng, Zhenxia Zhao, Qibin Xia, Haihui Wang and Zhong Li*
School of Chemistry and Chemical Engineering, South China University of Technology, Guangzhou, 510640, China

*Corresponding author e-mail address: cezhli@scut.edu.cn

Abstract

This work is mainly involves the study of adsorption performance of a novel carbonized-PDA polydopamine adsorbent (C-PDA) for CO_2/CH_4 under condition of atmosphere pressure and high pressure at atmosphere temperature. The carbonized-PDAs were prepared, and then characterized by ASAP2020, XRD, SEM, DRIFTS, and Boehm titration. Their adsorption performance for CO_2/CH_4 was tested by static adsorption method and fixed bed adsorption. Results showed that the BET surface area and pore volumes of the prepared C-PDAs were up to 3514 m^2/g and 1.78 cm^3/g, respectively. Its CO_2 adsorption was up to 31 mmol/g, being higher than that of MIL-101 At 25 atm and 298 K. In addition, its CO_2 adsorption was up to 6.2 mmol/g at 1 atm and 298 K , being nearly equal to that of HKUST. More interestingly, the C-PDA is more stable at humid atmosphere than HKUST-1 and MIL-101. Isotherms of water vapor on the C-PDAs were much lower than those on the HKUST-1 and MIL-101. Fixed-bed breakthrough experiments showed that CO_2 working capacities of the C-PDAs were much higher than those of MIL-101 or HKUST-1 under the humid atmosphere of relative humidity 30-80%. They are promising adsorbents for CO_2 capture or separation from CO_2/CH_4 gas mixture.

Experimental and modelling studies of CO_2 capture via Pressure Swing adsorption for zeolites and amine modified mesoporous silicas

E. Tsalaporta, J.M.D. MacElroy

UCD School of Chemical and Bioprocess Engineering, University College Dublin, Belfield, Dublin 4, Ireland

Corresponding Email Address: eleni.tsalaporta@ucdconnect.ie don.macelroy@ucd.ie

The capture of carbon dioxide via Pressure Swing Adsorption (PSA) has been examined experimentally and mathematically. The adopted method was a two bed/four step process, known as the Skarstrom Cycle.

Experimentally, the performance of zeolites and mesoporous silicas has been investigated, for different experimental conditions (cycle time, pressure ratio, feed/purge ratio). Recently, amine modified mesoporous silicas have received attention as an alternative option to zeolites for carbon dioxide capture. Several studies have been conducted in this area over the last few years. In the majority of these studies, the adsorption capacity of modified SBA-15 was measured by the injection of CO_2 into modified SBA-15 powder, under pressure, which is a much simpler approach than that normally encountered in more complex PSA processes. The relationship between the diffusion rates within and uptake characteristics of powder beds and the stationary columns of pelletised materials employed in PSA is not trivial and it was felt that a more direct investigation of PSA (and its variants) employing pellets of amine modified SBA-15 was warranted.

In this study, pelletised and calcined SBA-15 powder, modified with APTES ((3-Aminopropyl)triethoxysilane) has been tested in a PSA configuration. In order to investigate the diffusion mechanism (macropore/micropore) of carbon dioxide into the calcined SBA-15 pellets (unmodified), three diameters were selected for the production of the cylindrical pellets (cylinder diameters 1.2 mm, 1.7mm and 2.5mm each of length XX mm). A comparison of the breakthrough curves (Figure 1) indicated that macropore diffusion could be neglected and that the micropore diffusion mechanism plays a more important role in the control of the overall adsorption kinetics. The adsorption performance (Figure 2) of the APTES modified material was compared to the unmodified material, only to demonstrate that the poor performance of the calcined SBA-15 can be more than doubled with the presence of one amino group (APTES modified SBA-15).

For comparative purposes, PSA tests with APTES modified SBA-15 and zeolite 13X, which is a commercial adsorbent, were employed. In the case of zeolite 13X, capture of up to 99.9% of the CO_2 in a gas feed containing 10% carbon dioxide was achieved with a three-fold increase in the average composition of the CO_2, for a cycle time of 800 seconds, 6.25 bars pressure and 20°C temperature. APTES SBA-15 was less efficient than zeolite 13X in PSA, while RPSA (Rapid Pressure Swing Adsorption) was found to be more appropriate.

Capture of up to 45% of the CO_2 in a gas feed containing 10% carbon dioxide was achieved with a three-fold increase in the average composition of the CO_2, for a cycle time of 30 seconds, 3 bar pressure and 20°C temperature. This difference could be easily explained due to the different adsorption characteristics. Physisorption is the mechanism for CO_2 adsorption onto zeolites, while chemisorption is considered to be the primary mechanism for adsorption onto amine modified mesoporous silicas. Additional studies will be conducted in order to investigate the improvements in the limits of APTES SBA-15. A small amount of moisture will be introduced in the feed stream (to encourage the formation of carbonates in contrast to carbamates during adsorption) and higher operating temperatures will be tested. Moreover, the modification of SBA-15 with polyamines (more amino groups) is expected to extend the adsorption capacity of the material, so polyamine modified materials will be produced and tested.

In order to examine the temperature profile of PSA, three thermocouples were placed along the two columns of the PSA apparatus. In the case of zeolite 13X the profile demonstrated the existence of significant non-isothermal behaviour (Figure 3). The temperature profile for the APTES modified SBA-15 exhibited no clear non-isothermal trend, in view of its significantly different adsorption behaviour. The thermal characteristics of latter are under investigation both experimentally and theoretically.

For comparative purposes, two mathematical models have been employed in the numerical simulations of the two bed/four step process: the first model was originally suggested by Doong and Yang (1987)* and is based on uncoupled linear Fickian diffusion within bidisperse porous media subject to the linear driving force (LDF) approximation; the second approach is a more advanced model of linear coupled diffusion with a non-linear driving force approximation.

The experimental and simulated PSA results for both the zeolite and for the modified SBA-15 systems will be compared and discussed.

*Doong, S.J., Yang, R.T., *AIChE J.*, **33**, 1045-1049 (1987).

Keywords: Carbon dioxide capture, Pressure/Vacuum Swing Adsorption, Skarstrom Cycle, Mesoporous Silica, Zeolites, Modelling-Simulations

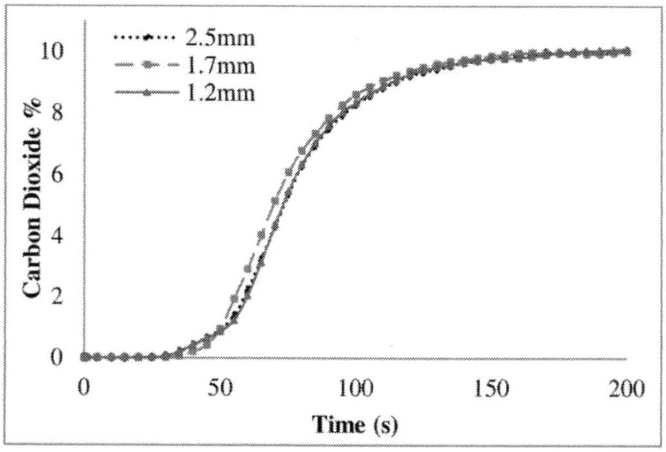

Figure 1: Breakthrough curves of SBA-15 for the experimental examination of micropore and macropore diffusion in PSA (feed flow: 1000cm^3/min, feed stream composition: 10% CO_2)

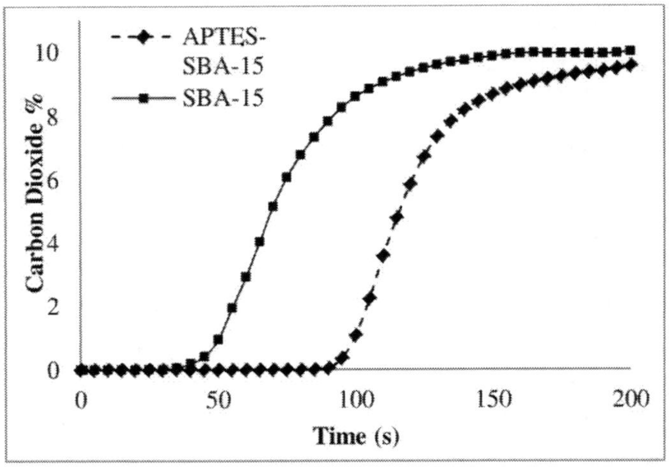

Figure 2: Breakthrough curves of unmodified and APTES modified SBA-15 pellets
(feed flow: 1000cm^3/min, feed stream composition:10% CO_2)

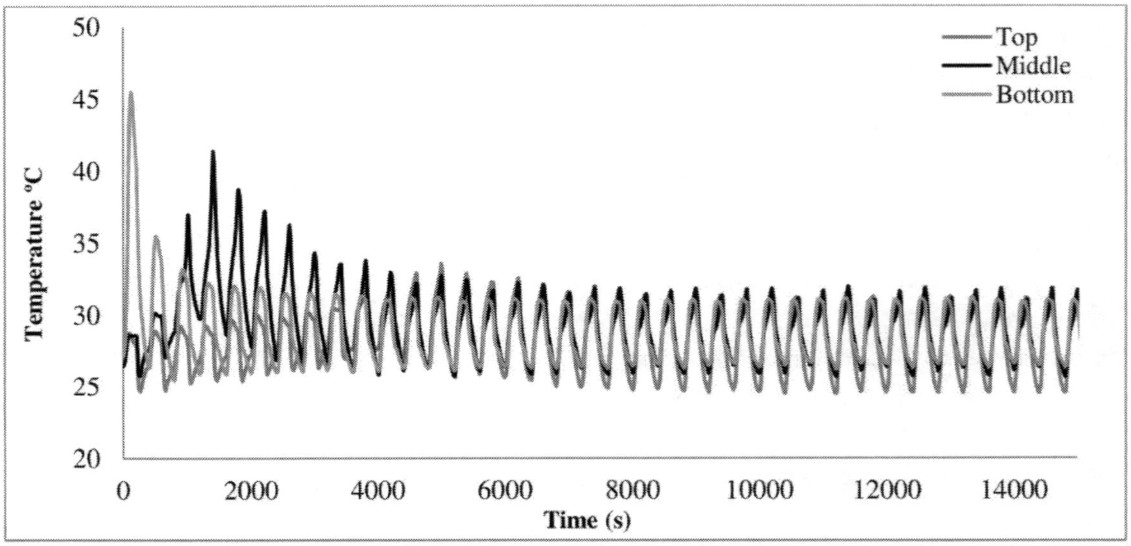

Figure 3: Temperature profile of carbon dioxide capture in PSA for zeolite 13X - Non isothermal behaviour. The feed enters at the bottom of the column and the purge at the top.

Kinetic Studies of Carbon Dioxide Capture by Potassium Carbonate Supported on Activated Carbon Using a Fluidized Bed Reactor

Aditya Prajapati, T. Renganathan and K. Krishnaiah

Department of Chemical Engineering, Indian Institute of Technology Madras, Chennai, India

Abstract

CO_2 is a major anthropogenic gas contributing to global warming. The growing concerns for climate change have encouraged research activities towards developing more-efficient processes for CO_2 capture. Some of the excellent adsorbents like zeolites, carbon nanotube-based solid sorbents and carbon molecular sieves have been suggested for adsorption of CO_2 in a dry gas stream. However, presence of moisture content creates a deteriorating effect on these sorbents [1].

CO_2 can be chemically adsorbed on dry regenerable alkali-metal carbonate-based sorbents (M_2CO_3, where M = K, Na, Li) by the reaction: $M_2CO_{3(s)} + CO_{2(g)} + H_2O_{(g)} \leftrightarrow 2MHCO_{3(s)}$. This shows that this class of sorbents have an inherent advantage as moisture is a necessity for this reaction. It is seen that K_2CO_3 gives the best performance and has a wide range of carbonation temperature where the sorbent efficiency is almost 100% [2].

In the present work, Activated Carbon (AC) is chosen as a support for potassium carbonate, due to its highly porous structure and high surface area. A 30 wt % of K_2CO_3 is coated on AC by wet impregnation method [3]. This sorbent is then characterized by XRD, EDX, SEM and BET Surface area.

Kinetic studies are done in a fluidized bed column heated in a furnace, using a mixture of air or N_2 and CO_2, humidified to desired moisture content as the fluidizing medium. The reaction is followed by measuring the concentration of CO_2 in the exit gas as a function of time. The effect of various parameters such as flow rate of gas, composition of gas, height of the bed and temperature on the conversion of K_2CO_3 is studied. The reaction is carried out in the temperature range of 60-90°C. The composition of both CO_2 and water vapor is varied between 5-20% with gas velocities ranging from 1-3 times of minimum fluidization velocity.

References:

1. Samanta, A., A. Zhao, G.K.H. Shimizu, P. Sarkar and R. Gupta, "Post-Combustion CO_2 Capture Using Solid Sorbents: A Review", *Ind. Eng. Chem. Res.*, **51**, 1438-1463 (2012)

2. Hayashi, H., J. Taniuchi, N. Furuyashiki, S. Sugiyama, S. Hirano, N. Shigemoto and T. Nonaka, "Efficient Recovery of Carbon Dioxide from Flue Gases of Coal-Fired Power Plants by Cyclic Fixed-Bed Operations over K_2CO_3-on-Carbon", *Ind. Eng. Chem. Res.*, **37**, 185-191 (1998)

3. Lee, S.C., B.Y. Choi, T.J. Lee, C.K. Ryu, Y.S. Ahn, and J.C. Kim, "CO_2 Absorption and Regeneration of Alkali Metal-Based Solid Sorbents", *Catal. Today,* **111**, 385-390 (2006)

378671 A Post-Combustion CO2 Capture Rapid Pressure/Vacuum Swing Adsorption Process and Its Application for Decarbonizing a Coal-Fired Power Plant

Abstract:

Up to now, relatively less attention has been paid to adsorption-based carbon capture research than conventional amine processes and other promising alternatives, such as solid- or chemical- looping and oxy-combustion, when it is aimed to achieve 90+% CO_2 recovery and 95+% CO_2 purity for decarbonising fossil fuel-based power plants. However, both numerical simulation studies and lab- and pilot- scale experimental campaigns has been demonstrating that a well-designed CO_2 capture adsorption process can spend less energy than a conventional amine process for post-combustion carbon capture. There is still a concern with practicality of a commercial-scale CO_2 capture adsorption process lingering on due to its requiring a huge volume of adsorbents relating to the performance of currently available best CO_2-selective adsorbents. But this issue can be overcome to some extent by process design given the fact that the size of the adsorption column can be drastically reduced by operating the adsorption process in a fast cyclic mode.

In this study, a simple one-column Rapid Pressure/Vacuum Swing Adsorption (RPSA) process was designed and its performance was evaluated by a high-fidelity adsorption process simulation using gPROMS. The mathematical model contains a rigorous energy balance incorporating temporal and spatial changes of the kinetic energy of a gas flow on a fixed-bed adsorption column. The effects of the kinetic energy terms are in most cases deemed negligible in simulating conventional cyclic adsorption processes for gas separation. However, the terms relating to the kinetic energy change become influential in case of a cyclic adsorption process experiencing an abrupt velocity change during the pressure-varying steps, i.e. RPSA. The simulation of the CO_2-enriching RPSA unit demonstrated that the use of the rigorous energy balance in combination with a full momentum balance could generate the bed temperature and concentration profiles that are different from what commonly used simplified energy balances predict, leading to different CO_2 purity and recovery.

In addition, a novel strategy to configure the CO_2 capture RPSA unit will be presented in order to achieve the aggressive CO_2 capture target, i.e. 95+% CO_2 purity and 90+% CO_2 recovery at the same time.

Poly(ethyleneimine) Impregnated γ-Alumina Monoliths for CO_2 Capture from Air

Miles Sakwa-Novak, William P. Addiego and Christopher W. Jones
School of Chemical & Biomolecular Engineering, Georgia Institute of Technology, 311 Ferst Drive, Atlanta, GA 30332, USA

Abstract

Amine functionalized porous oxide materials have shown great potential for next generation post combustion CO_2 capture applications. In addition, these materials have been shown to be highly effective for removing CO_2 from ultra-dilute streams, such as air. The removal of CO_2 from air, or air capture, has been increasingly discussed as a complementary emissions mitigation strategy to post combustion capture and alternative vehicle power sources. For air capture to be practiced on a large scale, however, enormous volumes of gas must be processed at low cost. Practically, for amine oxide adsorbents, this means that the sorbent material must be incorporated into a contactor to facilitate such processing constrains, but maintain the high capacities and rapid kinetics that make them attractive. Here, we report on the incorporation of poly(ethyleneimine) into extruded mesoporous γ-alumina honeycomb monoliths. We will discuss the procedure used for incorporation of the polyamine functionalities, as well as the performance of the monolith in a practical processing cycle where steam stripping is used for sorbent regeneration. Differences in equilibrium and working capacities will be also be discussed. Finally, areas of critical unresolved topics will be identified in this important effort to scale these promising materials from the lab to an intermediate scale.

389023 A Dual-Adsorbent, Two-Bed VSA Process for CO2 Capture from Wet Flue Gas

Abstract:

Various options for the capture and concentration of CO_2 from a wet flue gas at 25°C containing 15% CO_2 in 82% N_2 and 3% moisture have been analysed through detailed simulation and optimization based on equilibrium data of CO2 and N2 that were measured in our laboratories. Dynamic column breakthrough experiments were performed to validate the binary equilibrium between CO2 and N2. The data for water adsorption was obtained from the literature [1, 2].

First, the best cycle for dry flue gas, consisting of 4 steps including light product pressurization in a column packed with zeolite 13X established in an earlier communication [3] and demonstrated at a pilot scale [4], was applied to wet flue gas. The minimum energy consumption in this process, in order to achieve 95% purity (dry basis) and 90% recovery was 195.1 kWh/tonne CO_2 captured with a productivity of 1.32 tonne CO_2/m^3 adsorbent/day. This energy consumption was considerably higher and productivity considerably lower than those reported for dry flue gas. Next, in order to improve the energy performance, a new dual adsorbent, 4-step VSA process with silica gel and zeolite 13X. The process was then simulated and optimized. With the proposed new cycle, it is shown that 95% purity and 90% recovery target is achievable at a lower energy penalty (177 kWh/tonne CO_2 captured) while also improving the productivity (1.82 tonne CO_2/m^3 adsorbent/day).

References

1. Wang, Y.; LeVan, M. D., Adsorption equilibrium of carbon dioxide and water vapor on zeolites 5A and 13X and silica gel: Pure components. *J Chem EngData* **2009,** 54, (10), 2839-2844.
2. Wang, Y.; LeVan, M. D., Adsorption equilibrium of binary mixtures of carbon dioxide and water vapor on zeolites 5A and 13X. *J Chem EngData* **2010,** 55, 3189-3195.
3. Haghpanah, R.; Nilam, R.; Rajendran, A.; Farooq, S.; Karimi, I. A., Cycle synthesis and optimization of a VSA process for post combustion CO_2 capture. *AIChE J* **2013,** 59, 4735-4757.
4. Krishnamurthy, S.; Ramarao, V.; Guntuka,S.; Haghpanah,R.; Sharratt, P.; Rajendran, A.; Karimi, I.A.; Ullah, M.A.; Farooq, S., Post combustion CO_2 capture by vacuum swing adsorption: A pilot plant study. *AIChE J* **2013,** 60, 1830-1842.

391709 Novel Structured Adsorbent and Flowsheet for CO2 Capture from Flue Gas By PSA

Abstract:

A cost effective pressure swing adsorption (PSA) process is under development for post-combustion CO_2 capture from coal fired power plants. Several research groups from around the world are showing that PSA is still a viable option, as no separation process has emerged as a clear winner for CO_2 capture. PSA is also attractive because it requires only electricity; it does not require any steam from the power plant. This project is being carried out with funding from DOE NETL and four collaborative partners:

The key challenge to the success of this project is two-fold: 1) an attrition resistant and low pressure drop structured adsorbent must be developed based on commercial zeolite that is compatible with the high velocities associated with shorter PSA cycle times, and 2) a PSA cycle configuration must be developed in concert with the structured adsorbent so that the resulting PSA process delivers an exceptional performance at reduced capital and operating costs. Since the rapid PSA concept is conceptually simple but difficult to implement, a select team has been assembled to carry out this ambitious project and demonstrate the feasibility of the PSA concept for CO_2 capture at the bench scale. This team includes researchers from USC, W. R. Grace, Catacel, and Battelle.

To date, a key development includes the development of a low energy, high feed throughput PSA cycle for CO_2 capture via PSA process simulation. This was an incredible achievement when considering that the bulk density was reduced from 710 kg/m^3 (typical for a packed bed of zeolite beads) to 400 kg/m^3 (entirely possible with a Catacel structured adsorbent core). It boasts around 3,000 L(STP)/hr/kg feed throughput and a separations energy of less than 18 kJ/mol CO_2 captured. This presentation will discuss the latest results obtained from Ritter and his group.

Structure & Synthesis of Cage-Based Small-Pore Zeolites

Tracy Davis, *Dan Xie, Christopher Lew, Stacey Zones, Robert Saxton*
Chevron, Richmond, CA, USA.

Cage-based small pore zeolites (e.g., CHA, AEI, AFX) serve as catalysts and adsorbents in a number of commercial applications. Specifically, CHA and AEI have been shown to perform very well as deNOx catalysts [1, 2], and CHA is used commercially in the conversion of methanol-to-olefins [3]. The high silica forms of CHA, AEI and AFX were first discovered at Chevron [4,5,6], and research surrounding their synthesis continues. Additionally, work is ongoing to determine the potential of SSZ-52, the structure of which was recently shown to bear important similarities to CHA [7].

In this paper, the structures of CHA, AEI, AFX and SSZ-52 will be considered and compared in detail. While all of these zeolites are characterized by multi-dimensional small-pore structures and belong to the ABC-6 family, differences in the stacking sequence of the double 6-rings result in cages of varying dimensions. Such differences are a direct result of both the inorganic and organic species present during synthesis. This relationship between synthesis conditions and structure formation will be highlighted and specific attention will be paid to SDA-zeolite interaction energies. Following on a careful analysis of these zeolite structures, their performance as deNOx catalysts will be presented.

1. M. Jacoby, *Chem. Eng. News* 90 (2012) 10-16.
2. M. Moliner, C. Franch, E. Palomares, M. Grill, A. Corma, *Chem. Commun*. 48 (2012) 8264-8266.
3. P. Barger, B. Vora, U.S. Patent 6531692, 2003.
4. S. Zones, U.S. Patent 4544538, 1985.
5. S. Zones, Y. Nakagawa, S. Evans, G. Lee, U.S. Patent 5958370, 1999.
6. S. Zones, U.S. Patent 5194235, 1993.
7. D. Xie, L. McCusker, C. Baerlocher, S. Zones, W. Wan, X. Zou, *JACS* 135 (2013) 10519-10524.

378025 One-Step Dual Template Synthesis of Hierarchical Layered Zeolite with Tailored Textural and Catalytic Properties

Abstract:

A one-step dual template strategy was employed to achieve facile synthesis of meso-/microporous layered zeolites with tailored morphology, mesoporosity and catalytic activity. Tetrapropyl ammonium hydroxide (TPAOH) and diquaternary ammonium surfactant ($[C_{22}H_{45}\text{-}N^+(CH_3)_2\text{-}C_6H_{12}\text{-}N^+(CH_3)_2\text{-}C_6H_{13}]Br_2$, $C_{22\text{-}6\text{-}6}$) with a composition of $100SiO_2/xC_{22\text{-}6\text{-}6}/y$TPAOH/$4000H_2O$ were used as the templates in meso-/microporous layered zeolite synthesis. The influences of several parameters on the zeolite formation were investigated, including concentration of TPAOH and $C_{22\text{-}6\text{-}6}$, hydrothermal synthesis time, and ageing temperature and time prior to hydrothermal crystallization. The sizes of mesoporous and microporous domains of the zeolite and thus its meso- and microporosity can be modulated by simply tuning these parameters. The catalytic studies showed that the catalytic activity and selectivity can be systematically modulated by varying these parameters in the synthesis. The ability to form hierarchical zeolite particles with tunable meso-/microporosity is important to keep their potential applications as broad as possible for a specific zeolite structure. The dual template synthesis approach presented is potentially applicable to form other types of meso-/microporous zeolite materials.

386832 From Self-Assembled Mesoporous Thin Films to Kesterite Solar Cells

Abstract:

Given the terawatt-scale of future energy needs, the most promising future photovoltaic materials should be Earth abundant with their primary mineral resources distributed across several geographic regions and their supply chains robust to reduce concerns of price volatility. In addition, the process of forming the solar cell should be scalable, low-cost, and not utilize dangerous or toxic materials. One of the strongest candidates appears to solution processed kesterite structures of $Cu_2ZnSn(S,Se)_4$ or similar alloys. Our most recent result is the development of benign-solvent molecular-inks that yield ~11% efficient solar cells from these kesterites. This presentation will focus on the materials science and physics that led from a starting point focused on the development of nanostructured thin film photovoltaics based on self-assembled mesoporous ceramic films to the current state-of-the-art. We will show the results of composition spread experiments coupled with spatially resolved photoluminescence and a new theoretical framework to analyze photoluminescence that reveal the effects of native point defects and doping on material quality and device performance.

368256 Modeling the Formation of Ordered Nanoporous Silica Materials

Abstract:

Porous materials are ubiquitous in modern technology. Traditional applications include catalysis and adsorption separations, while more recent applications include membrane separations, biosensors, drug delivery and low-k dielectrics for microelectronics. The development of new porous materials with controlled pore structure that can be tailored for specific applications is a very active area of materials research. Understanding how such materials form and how to control the pore geometries will help facilitate advanced applications of these materials. Molecular modeling of syntheses of porous materials is challenging because large system sizes are needed to have representative sampling of the processes, and long time scales are required to observe the formation of ordered structures. In this presentation we describe the development and application of models of porous material assembly, focusing on silica materials. We show how quite simple models based on the assembly of corner-sharing silica tetrahedra can describe the self-assembly of a wide range of ordered nanoporous materials, including ordered mesoporous silicas and all-silica zeolite frameworks.

363545 Site-Specific Placement of Phases in Porous Nanocomposites

Abstract:

Templating methods developed over the past two decades have enabled syntheses of nanoporous materials with well-controlled architecture, suitable for a variety of applications that include energy storage and conversion, catalysis, opto-electronics and photonics, sensing, sorption and separation, controlled storage and release, and more. Control over dimensions and morphology of both the pore space and the solid wall components is achieved by hard templating and/or soft templating. Hard templating involves confinement of precursors in a solid mold, such as colloidal crystals, anodic alumina membranes, track-etch membranes, lithographic patterns, and other porous solids. Soft templating relies on dynamic structures formed from emulsion droplets, surfactants, block-copolymers, ionic liquids and other micelle-forming substances to direct the organization of precursor components. Multiple templating methods can be combined to achieve hierarchical pore structures with defined pore size distributions and pore interconnectedness.

Next-generation nanostructured materials will require multiple components, including multiple active phases (catalytic, electroactive, conductive, magnetic, etc.) and phases that provide a structural backbone, stabilization, and perhaps transport pathways. It will be critical to have reliable control over component placement, ideally through self-organization and in a single step. Site-specific placement of component phases extends the level of achievable complexity in nanocomposite materials and governs the performance of porous nanocomposites, because it determines component spacing (dispersion, interactions), accessibility, and grain growth/aggregation. Positioning of phases also affects relative dimensions of components and relative amounts that are in contact with interactive species. For example, in catalyst materials, the location of active components on a supporting matrix impacts their accessibility and long-term stability. In electrode materials for electrical energy storage devices, such as rechargeable batteries and supercapacitors, it is often necessary to disperse active material throughout a conductive phase or coat it with a protective phase. The relative location of such multiple phases affects conductivity, accessibility to electrolyte, and the ability of active material to maintain nanosized dimensions during insertion or deinsertion of charged species.

This presentation will provide on overview of some methods that allow placing of individual components in porous nanocomposites at specific sites, such as the pore surfaces, within the pore walls, at specific nodal points of a porous skeleton, or uniformly distributed throughout the structure. These methods include capillary condensation of volatile precursors between closely spaced surfaces of the hard template, or modulating the interactions between precursors and a hard template. The interactions are tuned through precursor selection, choice of ligands that complex with the precursor and may influence polymerization of sol-gel precursors, and through surface modification of the hard template. Interactions between multiple components confined in a template can also be exploited to control the morphology of the porous microparticles and hierarchically structured extended porous solids.

369556 Structure-Reaction Property Relationships for Lewis Acid Molecular Sieves

Abstract:

Lewis acidic tetravalent heteroatoms (Sn^{4+}, Ti^{4+}, Zr^{4+}) isolated within silica-based molecular sieves catalyze sugar isomerization and epimerization reactions as well as Diels Alder reactions of ethylene with substituted furans. Here, I will discuss the atomic arrangements and local environments of the Lewis acid centers and their reaction properties.

369125 Multiscale Ab Initio Modeling of Catalysts

Abstract:

In this talk, multiscale simulation will briefly be introduced as an enabling technology that bridges the gap between scales. Understanding the emergent properties of heteroepitaxial catalysts and the in silico prediction of suitable bimetallic catalysts will be elucidated as an application domain. We will demonstrate this methodology for the specific example of ammonia decomposition on single metals and bimetallics. It will be shown that incomplete monolayer bimetallics form an important class of catalytic materials.

Modeling CO_2 adsorption in metal organic frameworks

Laura Gagliardi

Department of Chemistry, Supercomputing Institute, and Chemical Theory Center, University of Minnesota, 207 Pleasant Street SE, Minneapolis, Minnesota 55455-0431, USA

During the formation of some metal organic frameworks (MOFs), the metal centers coordinate with both linker and solvent molecules. Subsequent activation removes the solvent molecules and creates open metal sites, which have strong affinity for CO_2. At present the interactions of CO_2 with these open metal sites are so poorly understood that commonly used force fields underestimate the adsorption of CO_2 in a recently synthesized MOF, Mg-MOF-74 at carbon capture conditions by as much as two orders of magnitude. Here we present a systematic procedure to develop a force field for Mg-MOF-74 and related materials using high-level quantum chemical calculations.[1] This transferable force field accurately takes into account the chemistry of the open metal sites. Recently we have employed a similar procedure to study also Fe-MOF-74.[2]

In the second part, results of a recent quantum chemical study[3] on the mechanism of carbon dioxide adsorption in an alkylamine functionalized metal-organic framework will be presented. Our calculations demonstrate that in these materials an ammonium carbamate species is formed via the insertion of CO_2 into the Mg-N_{amine} bonds.[4]

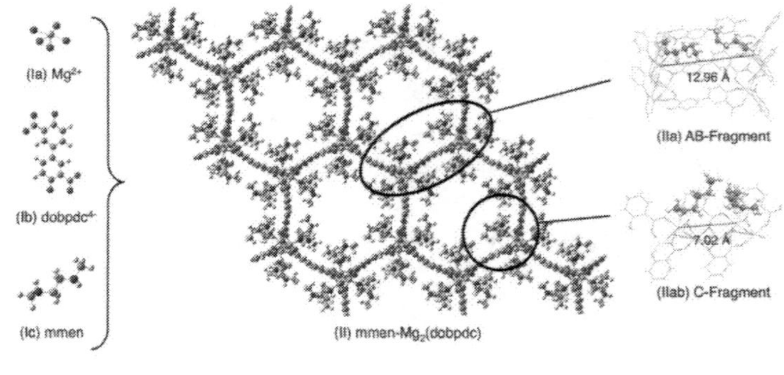

[1] A. L. Dzubak, L.-C. Lin, J. Kim, J. A. Swisher, R. Poloni, S. N. Maximoff, B. Smit, and L. Gagliardi Ab initio Carbon Capture in Open-Site Metal Organic Frameworks Nature Chemistry 4, 810-816 (2012)

[2] J. Borycz, L.-C. Lin, E. Bloch, J. Kim, A. L. Dzubak, R. Maurice, D. Semrouni, B. Smit, and L. Gagliardi; CO_2 Adsorption in Fe_2(dobdc): A Classical Force Field Parameterized from Quantum Mechanical Calculations J. Phys. Chem. C. **DOI:** 10.1021/jp500313j (2014)

[3] N. Planas, A. L. Dzubak, R. Poloni, L.-C. Lin, A. McManus, T. McDonald, J. Neaton, J. Long, B. Smit, L Gagliardi The mechanism of carbon dioxide adsorption in an alkylamine functionalized metal-organic framework *J. Am. Chem. Soc.*, *135* (20), 7402–7405 (2013)

[4] S. O. Odoh, A. L. Dzubak, B. Vlaisavljevich, S. Schnell, N. Planas, K. Lee, J. Neaton, B. Smit, and L. Gagliardi, submitted (2014)

AICHE, Atlanta, GA, November 16-21, 2014
Division: Separations, **Program:** MEMBRANE BASED SEPARATIONS
Session: Membrane Research and Innovation Activities Around the World (2D24)
Chair: Zhongyi Jiang <zhyjiang@tju.edu.cn>

Concentration Polarization and Fouling: The Achilles Heels of Membrane Filtration

Georges Belfort, PhD

Department of Chemical and Biological Engineering and Center for Biotechnology and Interdisciplinary Studies, Rensselaer Polytechnic Institute, Troy, NY 12180-3590

Both concentration polarization (CP) and fouling (F) limit the performance of pressure-driven membrane processes such as microfiltration, ultrafiltration, nanofiltration and reverse osmosis. They generally decrease optimal performance by limiting permeation flux, effect selectivity and increase energy requirements. In this presentation, we will review the contributions of the Belfort research group over the past 40 years toward understanding the dynamic behavior of CP of ions in reverse osmosis (1, 2) and F of suspended colloids in all of the processes listed above (3-5). In addition, mitigation strategies for both of these applications from published papers and patents from the Belfort Group will be discussed. This includes a non-moving membrane module design that produces self-cleaning vortices that reduces F when filtering colloid suspensions (6, 7). Unfortunately, rational understanding of membrane fouling such that caused by proteins dissolved in solution is wanting. So we have designed and will present a powerful evolutionary non-rational method for identifying the optimal membrane chemistry for a particular application with specific feed characteristics in one afternoon (8, 9). It allows high throughput synthesis, screening and selection of optimal membrane chemistries that exhibit low protein F using industrially available monomers or new unavailable monomers using combinatorial chemistry. Most other methods for choosing monomers are based on past experience from the literature or just guesses.

References

1. Mahlab D, Yosef NB, Belfort G. Concentration polarization profile for dissolved species in unstirred batch hyperfiltration (reverse osmosis). 2. Transient case. Desalination. 1978;24(1-3):297-303.
2. Mahlab D, Yosef NB, Belfort G. Interferometric measurement of concentration polarization profile for dissolved species in unstirred batch hyperfiltration (reverse osmosis). . Chemical Engineering Communications. 1980;6(4-5):225-43.
3. Altena F, Weigand, R. J., and Belfort, G. . Lateral migration of spherical particles in laminar porous tube flows: Application to membrane filtration. Physico Chem Hydrodyn. 1985;6(4):20.
4. Otis JR, Altena, F. W., Mahar, J. J., and Belfort, G. . Measurement of single spherical particle trajectories with lateral migration in a slit with one porous wall under laminar flow conditions. Experiments in Fluids. 1986;4:10.
5. Belfort G, Davis, R. and Zydney, A. . The behavior of suspensions and macromolecular solutions in crossflow microfiltration,. J Membrane Sci. 1994;96:59.
6. Brewster ME, Chung, K.-Y. and Belfort, G. . Dean vortices with wall flux in a curved channel membrane system: 1. A new approach to membrane module design. J Membrane Sci. 1992;81:11.
7. Gehlert G, Luque, S. and Belfort, G. . Flux enhancement due to secondary flow in a helical tubular module: 14 ultra and microfiltration of polysaccharides, proteins and yeast suspensions. Biotechnology Progr. 1999;14:12.
8. Zhou M, Liu, H., Venkiteshwaran, A., Kilduff, J. C., Anderson, D. G., Langer, R. and Belfort G. . High throughput discovery of new fouling-resistant surfaces,. J Mater Chem. 2011;21:12.
9. Gu M, Vegas AJ, Anderson DG, Langer RS, Kilduff JE, Belfort G. Combinatorial synthesis with high throughput discovery of protein-resistant membrane surfaces. Biomaterials. 2013;34(26):6133-8.

390809 Thermally Rearranged Polymer Membranes for Energy-Related Applications

Abstract:

Microporous polymers are a class of polymeric materials with high free volumes and large surface areas. Microporous polymers have received much attention for various energy-related applications in gas separation, gas storage, and for clean energy resources due to their easy processability as well as microporosity for high performance. As gas separation membrane materials, microporous materials exhibit high flux based on the fast gas transport through membrane matrix. Moreover, the easy processability of polymer materials has been a great advantage for expansion into industrial applications.

Reported rigid polymer membranes, so-called thermally rearranged (TR) polymer membranes have been developed from ortho-functionalized polyimides by a post thermal conversion process. They demonstrate bimodal cavity size of 0.3-0.4 nm and 0.7-0.9 nm and narrow cavity size distributions characterized by PALS measurement which is an efficient transport path for small molecules, especially for CO_2 molecules. As a result, TR polymer membranes exhibit both high permeability and high selectivity based on high diffusion of gas molecules. TR polymer membranes are also suitable for water treatment applications due to their microporous characteristics. In this presentation, we will report on the recent progress of thermally rearranged microporous polymer membranes for applications such as in energy and water.

The development of multi-bore membranes for water reuse, desalination and biofuel separation

Lin Luo[a,b] , Dan Hua [a,b], Peng Wang [b] , Tai-Shung Chung [a,b]*

[a] NUS Graduate School for Integrative Science and Engineering, National University of Singapore, Singapore 117456, Singapore

[b] Department of Chemical & Biomolecular Engineering, National University of Singapore, Singapore 117585, Singapore

* Corresponding author. Tel.: (+65) 6516 6645; fax: (+65) 6779 1936.

E-mail address: chencts@nus.edu.sg (T. S. Chung).

Abstract

Advances in membrane technology are the most direct, effective and feasible approaches to solve various issues in water reuse, desalination and biofuel separation. Among all the advances, hollow fiber membranes with the multi-bore configuration have great potential and advantages because of their mechanical strength, easy module fabrication, and excellent stability.

In this presentation, we will present a new generation multi-bore hollow fiber (MBF) membrane with excellent mechanical durability for vacuum membrane distillation (VMD). Credit to the multi-bore configuration, a 65% improvement was obtained on the anti-wetting property. Using a synthetic seawater feed, the new membrane with the optimal fabrication condition exhibits a high flux and the salt rejection is consistently greater than 99.99%.

Novel tri-bore hollow fiber (TBF) membranes with regular triangle-shape outer geometry were fabricated for ultrafiltration (UF). The fabricated Matrimid® and Polyethersulfone (PES) triangle TBFs were advantageous in terms of uniform mechanical strength and enhanced permeation properties due to the evenly distributed membrane wall thickness. Experimental results suggest that the combined effects of rapid phase inversion, die swell, outer surface shrinkage and stress balance during the membrane formation are responsible for the unique membrane geometry.

In addition, thin-film composite (TFC) TBF membranes were also developed for forward osmosis (FO). The new developed Matrimid® TFC TBF membranes exhibit high water fluxes of 50.5 LMH with salt leakages as low as 3.5 gMH in pressure retarded osmosis (PRO) modes, when using 2 M NaCl as the draw solution and pure water as the feed. In addition, theoretical calculations suggest that much more fibers of triangle shape can be packed into a FO module, water output of the module consisting of triangle membranes would double the one comprising round membranes at the same packing density.

The TBF membranes were also successfully extended to pervaporation applications via the polyethyleneimine coating followed by interfacial polymerization. These fabricated outer-selective TFC TBF membranes were then used for pervaporative dehydration of isopropanol (IPA). The optimal TFC TBF membrane shows a flux of 2.65 kg m^{-2} $h^{-1}$and a separation factor of 246 for the dehydration of IPA/water (85/15 wt%) mixture at 50 °C. Besides, the designed TBF membrane not only owns better separation performance and mechanical strength than the conventional TFC TBF, but also exhibits long-term stability, which suggesting the TFC TBF has a great potential in pervaporation.

Acknowledgements

These research were funded by the Singapore National Research Foundation under its Competitive Research Program for the projects entitled, "Development of Hybrid Desalination Processes using Cold Energy from LNG Regasification" (Grant Number: R-279-000-291-305), "Advanced FO Membranes and Membrane Systems for "Wastewater Treatment, Water Reuse and Seawater Desalination" (Grant nos.: R-279-000-336-281 and R-279-000-339-281) and "New Biotechnology for Processing Metropolitan Organic Wastes into Value-Added Products" (CRP Award No. NRF-CRP 5-2009-5 (NUS grant number R-279-000-311-281)).

385457 Advanced Membranes for Future Separations Requirements

Abstract:

Today membrane based separation processes find numerous applications in a range of industries e.g. bioseparations, biomedical devices, water treatment, biomedical devices. Membranes are attractive as often they are cheaper to operate compared to existing technologies; membrane devices are modular and therefore easy to scale up; they often have a much smaller foot print than competing technologies. In addition for some niche applications membrane based separations are the only viable technology.

In this presentation, the unique advantages of membranes for liquid separations will highlighted.

Often, in membrane based separation processes, the properties of the membrane surface that contact the feed stream can have a significant effect on membrane performance. For example, membrane surface properties have been shown to affect the rate of fouling during processes such as microfiltration, ultrafiltration, nanofiltration and reverse osmosis. Modifying just the membrane surface has the potential to retain the properties of the bulk membrane material but modify the interactions between components of the feed and the membrane surface. Research being conducted in our group that focuses on advanced membranes for future separations requirements will be discussed. In addition, the University of Arkansas recently became a site of the Membrane Science, Engineering and technology center, a National Science Foundation Industry/University Cooperative Research Center. A major focus at the University of Arkansas is the development of advanced membranes for applications in water treatment. The importance of obtaining industrial input when developing novel membranes for future separations needs will be highlighted.

375664 Graphene-Based Membranes for Separation

Abstract:

This talk will try to reveal permeation properties of graphene-based material for filtration and purification systems and try to provide insights on if graphene-based material could be used to make the perfect filters/membranes. The talk will focus on the following aspects:

- What are the properties of graphene-based materials that make them so suitable for filtration?
- Evaluating graphene/graphene oxide's role as a water filter
- How to process graphene-based flakes into large scale permeation membranes/filters?
- What are the real world applications of these properties, including the effective filtration of water from aviation engines and preventing engine failure? Desalination?
- Scrutinizing the potential for use in oil and gas applications – oil/water separation? natural gas purification?

391713 Exploring Virus Retention during Virus Filtration Using Confocal Microscopy with Fluorescently-Labeled Bacteriophage

Abstract:

Although virus filtration is a well-established method for combating the inherent risk of viral contamination in the production of therapeutic proteins, there are still significant uncertainties regarding the underlying mechanisms controlling virus retention. Previous studies have reported a significant decline in virus retention during constant pressure virus filtration with certain filters (but not others). Several studies showed that a temporary release in the transmembrane pressure can cause a significant transient increase in virus transmission. However, there is currently no detailed understanding of the factors controlling these phenomena or the underlying physical basis for the changes in virus retention. The objective of this work was to employ confocal microscopy with fluorescently labeled bacteriophage to directly probe virus capture in several different virus filters (Viresolve Pro, Viresolve NFP, and DV20) having different pore morphology. A model bacteriophage (phix-174) was labeled with either SYBR Gold (green fluorescence) or Cy5 (red). The use of dyes with different excitation and emission wavelengths makes it possible to distinguish between phage that were captured during the initial challenge and after the pressure release. Bacteriophage retention during the normal flow filtration was quantified using a plaque forming assay. Images obtained with the DV20 filter showed two distinct bands of phage at the end of a pressure release experiment, with the band that penetrated deeper into the filter consisting primarily of phage that were captured before the pressure release. The behavior of the Viresolve membranes was very different, with the phage captured in a more diffuse region deep within the filter upstream of the retentive skin layer of the membrane. The confocal images and retention data were analyzed using an internal polarization model that accounts for the accumulation of phage within a reservoir zone within the filter. The results provide important insights into the mechanisms controlling virus retention as well as the effects of membrane morphology on virus capture during virus filtration.

376645 A New Multimodal Membrane for Monoclonal Antibody Purification

Abstract:

This presentation will describe our latest research efforts to develop a new class of multimodal membranes (MMM) with high binding capacities, salt-tolerance, and high-productivity for the chromatographic purification of therapeutic proteins. The membranes incorporate ligands that can bind proteins through a combination of Coulombic interactions, hydrophobic interactions, and/or hydrogen bonding, thus widening the operation ranges for separation.

Our design approach is to graft functional polymer tentacles from the pore surfaces of macroporous membranes using a controlled surface-initiated polymerization technique. One advantage of this membrane preparation technique is that it greatly enhances the membrane binding capacity without severely decreasing its permeability. Here, we present work on the preparation of multimodal cation-exchange membranes for bind-and-elute purification of immunoglobulin G (IgG). Compared to commercial multimodal adsorbents, our membranes have 4 times higher ligand density, which contributes to the high static binding capacities (180 mg IgG/mL, 310 mg lysozyme/mL).

Protein binding experiments with bovine IgG showed binding capacities of the multimodal membranes were dependent on pH but had very good salt tolerance. The multimodal membranes maintain significant binding capacities in excess of 90 mg IgG/mL at ionic strength values that are typical for elution buffers used in multi-stage bioseparation processes. For sodium citrate, a conventional salt used in elution buffers of Protein A columns, increasing ionic strength had only a minor effect on the IgG binding capacity.

To further understand protein adsorption on the multimodal membrane, a thermodynamic adsorption isotherm model was employed that provides a unique set of physically meaningful parameters. This model can be used to explain the effect of salt type (kosmotropes, neutral, and chaotropes) on protein adsorption.

Finally, protein dynamic binding capacities were determined from breakthrough curve analysis. Dynamic binding capacities of >50 mg IgG/mL were found near the isoelectric point of the protein. A range of flow rates was used to study the effect of volumetric throughput on dynamic binding capacities. Protein elution using pH adjustment, chaotropic salts, and organic displacers will be described.

Taken together, the results to be presented indicate that the newly developed multimodal membranes have great potential to compete with more traditional cation-exchange materials following the Protein A purification step in the downstream processing of antibody products.

Discovery of Anti-Fouling Membranes via Combinatorial Chemistry and a Novel High Throughput System

Joseph Imbrogno and Georges Belfort

Howard P. Isermann Department of Chemical and Biological Engineering and Center for Biotechnology and Interdisciplinary Studies, Rensselaer Polytechnic Institute, Troy, NY 12180

Ultrafiltration membranes are widely used in the biotechnology, food, and water treatment industries. The major drawbacks of membranes are fouling and concentration polarization, which ultimately lead to membrane degradation that negatively affects performance. A major source of fouling is non-specific protein binding to the surface of a membrane as well as physical adsorption on the membrane surface. As upstream feed titers increase, especially in the biotechnology industry, the majority of the operating costs of the entire process are shifted heavily to the downstream processes. This suggests that new synthetic membranes are needed to replace current technology. High-throughput surface chemistry modification of membranes is an effective method to develop new membranes, especially if the produced surfaces can be screened efficiently.

In this work, we utilize a high-throughput (HTP) platform with an Atmospheric Pressure Plasma (APP) polymerization system to modify irradiation-sensitive commercial poly(ether sulfone) (PES) membranes. APP is attractive because it can easily be incorporated into membrane production lines, is flexible (monomer and material variability), and is economical (no vacuum required for the substrate). A vast number of model surfaces can be produced using combinatorial chemistry to expand our library of monomers. Eleven new monomers, including acids, zwitterionic, mixed matrix, chiral, azo, and heterocyclic classes were synthesized. The reaction is conducted at room temperature and neutral pH. Then, a 96 well membrane plate is bombarded with high energy oxygen and helium plasma, which induces radical sites in the membrane. The library of 11 novel monomers, at a concentration of 0.2 M, is then added to separate filter wells of a 96 well membrane plate to react via free radical polymerization. The buffer flux through the modified and as-received membranes (or inverse resistance to flow) before and after exposure to a fouling challenge, consisting of 1 mg/ml BSA solution for 24 hours is then measured using spectrophotometry. Additionally, we measure selectivity of the membranes using a dynamic fouling test consisting of 1 mg/ml BSA in phosphate buffered saline (PBS).

Many different chemistries are screened in a single afternoon with reproducibility and efficiency. The selected winners are determined by measuring fouling index, selectivity, resistance, and calculating Hansen solubility parameters. Previous work has indicated that water interactions at surfaces play a critical role in membrane protein fouling[1]. Special focus will be given to the role of water interactions in order to gain a fundamental understanding of membrane fouling and concentration polarization. The Hansen solubility parameter calculations (δ_d, δ_p, δ_h) indicate that hydrophilic monomers, that have solubility parameters close to those of water, will perform the best. From our new monomers, these are the carboxylic acid, and ester terminated monomers. Additionally, we have shown previously that fouling index is exponentially correlated with the distance between the water and the monomer solubility parameters[1]. That is,

[1]Gu, M. *et. al.* Combinatorial synthesis with high throughput discovery of protein-resistant membrane surfaces, Biomaterials 34 (2013) 6133-6138

the farther away from water the Hansen solubility parameters are, the higher the fouling index is. Our results were consistent with the Hansen solubility parameter predictions. The carboxylic acid, zwitterionic, ester, and amide terminated monomers performed the best with the lowest fouling index, lowest resistance, and highest permeability.

385309 Overcoming Diffusional Limitations for Protein Binding in Polymer Brush Grafted Nonwovens to Reduce Equilibrium Binding Times

Abstract:

Polymer brushes grafted to nonwoven membranes create a 3-dimensional matrix capable of achieving very high equilibrium protein binding capacities. Polybutylene terephthalate (PBT) nonwovens can readily be grafted with glycidyl methacrylate (GMA) via UV induced radical polymerization creating polymer brushes that can further be functionalized to become ion exchangers. The capacities achieved by these materials when used as an ion exchange platform are many times that of monolayer coverage according to the specific surface area of the material. However, very long residence times are required to reach equilibrium due to the diffusion of protein through a dense protein-polyGMA matrix. Increasing the specific surface area of the material can effectively reduce the polyGMA graft thickness while still observing a high equilibrium binding capacity, resulting in shorter times to reach equilibrium. A novel islands-in-the-sea (I/S) PBT nonwoven having a specific surface area of 2.5 m^2/g was compared to a commercially available PBT nonwoven having a specific surface area of 0.9 m^2/g, as a base material for polyGMA grafting and ion exchange capture of target proteins. Islands-in-the-sea nonwoven PBT was provided by Dr. Behnam Pourdeyhimi at the Nonwovens Cooperative Research Center. I/S PBT nonwovens are made of bicomponent fibers containing a removable polymer ("the sea") that encapsulates discreet fiber entities of a permanent polymer ("the islands") that are substantially smaller in diameter. Both materials were successfully grafted to achieve conformal uniform polyGMA graft layers of varying degrees of grafting. Following grafting, the membranes were either derivatized with diethyl amine or phosphoric acid to become weak anion exchangers or strong cation exchangers respectively. It was observed for both PBT nonwovens in both ion exchange formats that equilibrium binding capacity increased linearly with the degree of polyGMA grafting, and it was shown that the initial surface area of the PBT material did not have an impact on the final equilibrium binding capacity. Capacities of 800-900 mg/g of protein were observed for all systems when grafted to 18-20% polyGMA weight gain. At these weight gains, equilibrium binding was reached on an order of minutes for the I/S PBT nonwoven compared to hours required for the commercially available PBT nonwoven. The enhanced binding performance of the I/S PBT nonwoven is due to the lower graft thickness of the polyGMA layer at specific degrees of grafting. The diffusion of protein through a dense matrix of protein bound to functionalized polyGMA was modeled using a shrinking core model. Using this model a relative diffusion coefficient can be estimated for the transport of protein through the polymer/protein layer. the diffusion of protein through this system was found to be several orders of magnitude slower than bulk diffusion of protein in an aqueous solution.

Title:

A case study in Polysorbate-80 adsorption to 500kDa polysulfone filters used in a tangential flow filtration process for protein purification

Authors:

Matthew Woodling*, Adam Kristopeit*

* Merck Sharp and Dohme Corp., Merck Research Laboratories, Vaccine Process Development, West Point, PA, United States

Abstract:

Surfactants are commonly used to prevent aggregation and nonspecific binding during membrane steps of biopurification processes. Such biopurification processes often require control of surfactant concentration during processing and at final product. However, control of surfactant concentration throughout membrane processing can prove difficult due to adsorption of the surfactant to the membrane. To better understand these interactions for a specific vaccine process using Polysorbate-80 (PS-80), experiments were performed to characterize the binding of PS-80 to 500 kDa polyethersulfone hollow fiber membranes. Subsequently, a model utilizing elementary adsorption kinetics was proposed to predict the kinetic relationship between PS-80 mass and filter lumen area. Through fitting the proposed model to experimental data, the model parameters were fully defined, and it was found the model was a good predictor of PS-80 levels during lab and commercial scale processes utilizing such membranes.

380774 The Effect of Salt Ions on Responsive Membrane Based Hydrophobic Interaction Chromatography

Abstract:

Recent efforts have been devoted to develop responsive membrane based hydrophobic interaction chromatography (HIC) to achieve high recovery and resolution for protein purification at preparative and analytical scale respectively. Our group has designed novel responsive membrane based HIC by grafting the thermo-responsive poly(N-vinylcaprolactam) (PVCL) from the surfaces of micro-porous regenerated cellulose membranes (0.45μm) via atom transfer radical polymerization. It is known that the lower critical solution temperature (LCST) of PVCL is sensitive to both salt type and salt concentration in solution. In order to optimize the binding capacity and recovery using responsive membrane based HIC, systematic investigations have been conducted to understand the effects of salt ions on the HIC behavior based on this responsive membrane system. Dynamic binding capacities and recoveries for immunoglobulin G_4 (IgG_4) and bovine serum albumin (BSA) were determined varying solution pH, salt ion type and salt concentration at room temperature. In addition, the cationic effects were studied through static binding tests using various monovalent (Na^+, NH_4^+), divalent (Mg^{2+}, Zn^{2+}) and trivalent (Al^{3+}) sulfate salt solutions. BSA adsorption isotherm as well as adsorption kinetics were obtained. Moreover, the adsorption mechanism was elucidated. Finally, the effect of salt on the PVCL ligand was investigated and correlated with the performance of this responsive membrane based HIC system.

382593 A Comparison Between Membrane Adsorbers and Packed Bed Columns

Abstract:

Despite the development of new membranes with improved binding capacity, the use of membrane adsorbers is still limited to niche applications, and packed bed chromatography remains the workhorse of industrial downstream processing. While membrane chromatography, in which mass transfer is dominated by convection, is a fast process, packed bed chromatography exhibits a higher binding capacity. In this work a systematic comparison between the two processes has been developed using an integrated approach that combines both theory and experiments.

The detailed performance of a membrane adsorber and a packed bed column has been studied experimentally in detail using the same pilot scale chromatography system. The results obtained using BSA as a model protein have been used to validate a mathematical model of the two processes and a critical comparison has been performed including the role of binding capacity, throughput, selectivity, yield and cost. From the results obtained, it is apparent that the fluid dynamics of the membrane module can play a major role in the overall performance of membrane adsorbers when compared to chromatographic columns.

Purification and concentration of wood extracts by membrane filtration

By Yipeng Xie
Advisor: Dr.Shijie Liu
Department of Paper and Bioprocess Engineering
SUNY college of ESF

Outline

1.Background
2.Equipment and Procedures
3.Mathematical model
4.Results analysis and discussion
5.Conclusion
6.Reference

1. Background

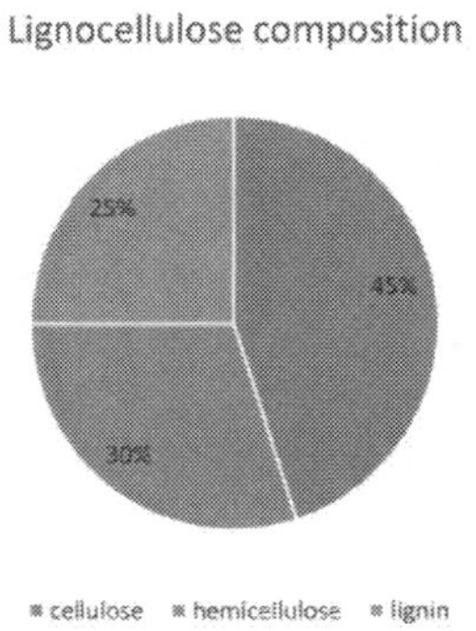

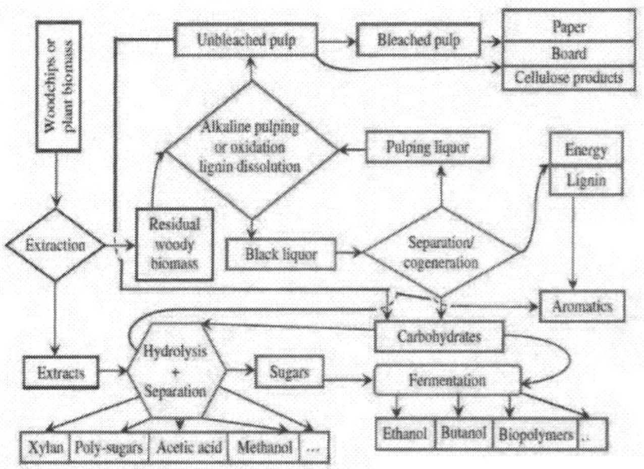

Composition of hot water extracts

Hot water extraction of *Paulownia* woodchips was performed at 160 °C for 2 h, with a solid to liquor ratio of 1:8.
Composition: lignin, polysaccharides, furfural, acetic acid, etc.

Fermentation-Inhibitors	Methanol	Hydroxyl methyl-furfural(HMF)	Formic acid	furfural	free acetic acid, mol/L	bound acetic acid, mol/L
Concentration, g/L	0.8339	0.4613	0.3092	0.5186	0.03189	0.004113
sugars	galactose	glucose	xylose	arabinose	rhamnose	mannose
Concentration, g/L	1.603882	1.381168	1.513271	0.796788	trace	trace

Separation methods

Separation Method	Equipment cost	Operation cost	Recover cost for valuable products	Efficiency	Environmental harm
Distillation	High	High	Low	High	Low
Chemical extraction	Low	Medium	High	Medium	High
Filtration	Medium	Medium	Low	High	Low
Centrifugation	High	High	Low	Low	Low
Vacuum evaporation	High	High	Low	Medium	Low
Chromatography or liquid adsorption	Medium	Medium	Low	Depends	Low

Comparison of different membrane filtration means

Class	Pore size, Å	Applications
Reverse osmosis (RO)	1-10	Desalination, sugar concentration
Nanofiltration (NF)	8-50	Sugar concentration, water softening, pesticide removal, sulphate removal
Ultrafiltration (UF)	10-200	Electro-coat paint concentration, food processes, cold sterilization, waste recovery: latex &milk production
Microfiltration (MF)	200-1000	Sterilization, particle removal
Electro Dialysis (ED)	NA: Charge-oriented	Salt recovery/desalination, cheese whey purification, acid recovery

Comparison of different membrane modules

Module type	Channel spacing(cm)	Packing density (m^2/m^3)	Energy costs	Particulate plugging	Ease of cleaning
Hollow fiber	0.02-0.25	1200	Low	High	Fair
Tubular	1.0-2.5	60	High	Low	Excellent
Flat plate	0.03-0.25	300	Moderate	Moderate	Good
Spiral wound	0.03-0.1	600	Low	Very high	Poor-fair
Rotating	0.05-0.1	10	Very high	Moderate	Fair

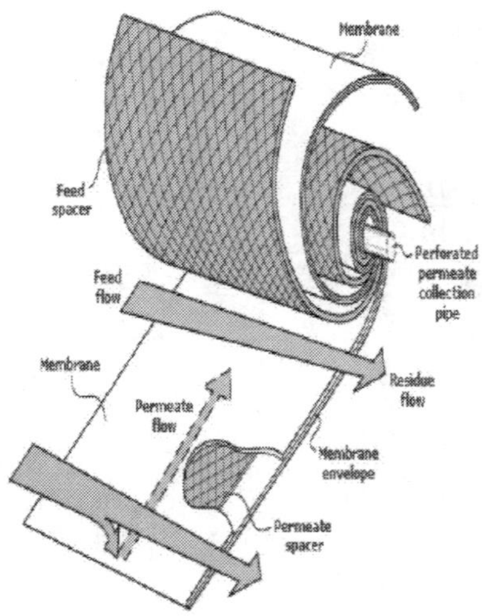

2. Equipment and procedures

Ultrafiltration: ceramic hollow fiber membrane with 200nm pore size

Process: keep 80℃ while pressure drop within 20psi. Permeate streams go for nanofiltration

Purpose: remove macromolecules (big lignocellulose fragments) and solids

Membranes and device used for nanofiltration

Parameter	DK4040f1020	AG90
manufacturer	GE-High rejection NF	GE-High rejection RO
configuration	Spiral wound	Spiral wound
Filtration area ft²(m²)	98(9.1)	90(8.4)
Average permeate flow gpd(m³/day)	2,000(7.6)	2,200(8.3)
minimum salt rejection	98%	99.3%
Working pH range	3-9	4-11

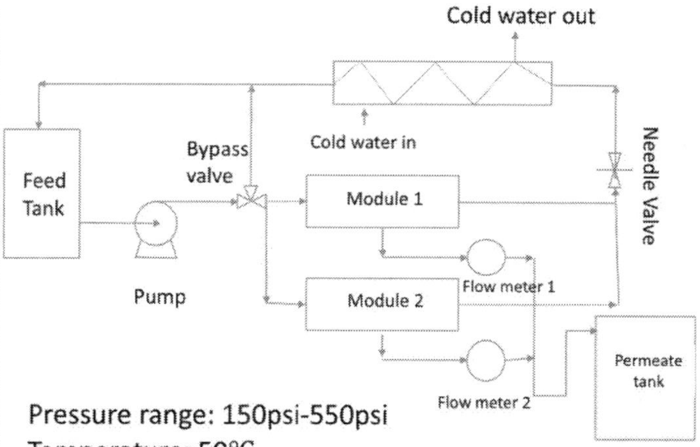

Pressure range: 150psi-550psi
Temperature: 50°C

Composition analysis

1. For feed samples, use both ^1H and ^1H-^{13}C HSQC NMR images
2. For permeate samples, use ^1H NMR image to measure inhibiters first. Then use vacuum evaporator to concentrate so that we can use ^1H-^{13}C HSQC NMR imagesto test sugar concentrations

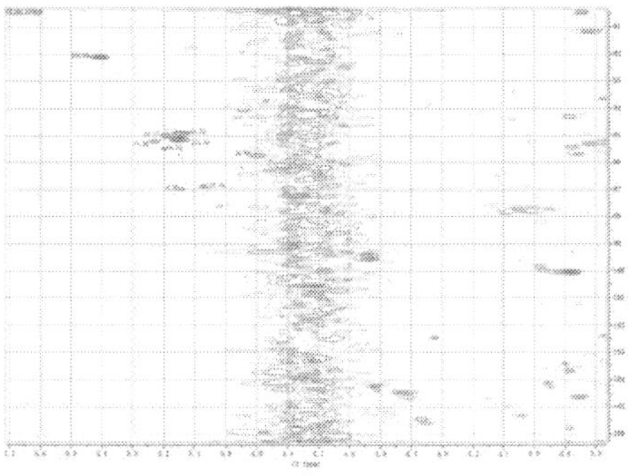

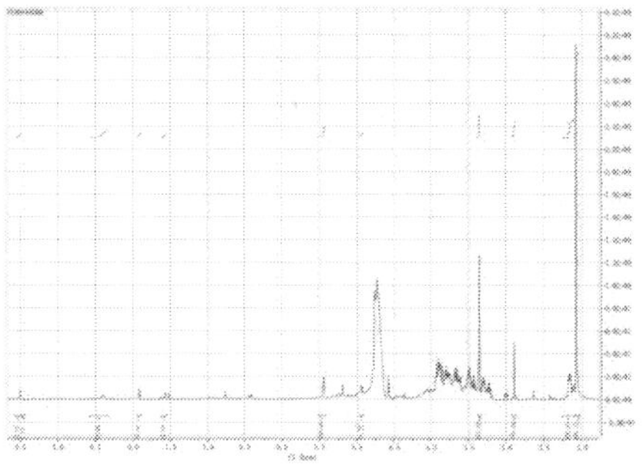

3. Mathematical model

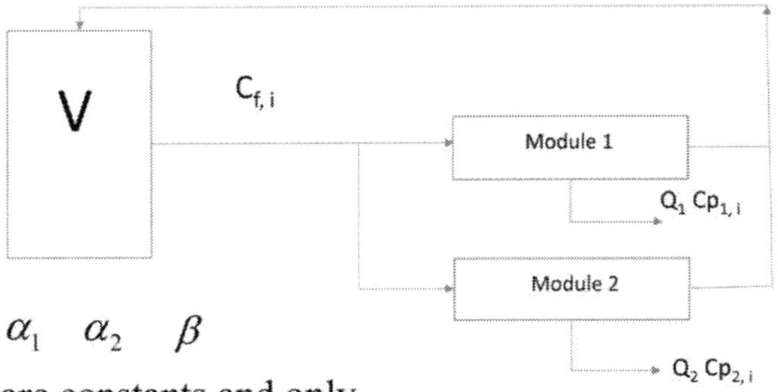

$\alpha_1 \quad \alpha_2 \quad \beta$

are constants and only determined by membrane and solutes

$$\frac{dV}{dt} = -(Q_1 + Q_2) \qquad (1)$$

$$C_{P1,i} = \alpha_{1,i} C_F \qquad (2)$$

$$C_{P2,i} = \alpha_{2,i} C_F \qquad (3)$$

$$\frac{d(VC_{F,i})}{dt} = -(\alpha_{1,i} Q_1 + \alpha_{2,i} Q_2) C_{F,i} \qquad (4)$$

$$Q_1 = \beta Q_2 \qquad (5)$$

$$\frac{dV}{dt} = -(\beta + 1) Q_2 \qquad (6)$$

$$\frac{d(VC_{F,i})}{dt} = -(\alpha_{1,i}\beta + \alpha_{2,i}) Q_2 C_{F,i} \qquad (7)$$

$$Q_2 = \frac{1}{-(\beta + 1)} \frac{dV}{dt} \qquad (8)$$

$$V \frac{dC_{F,i}}{dt} = -\Theta_i C_{F,i} \frac{dV}{dt} \qquad (9)$$

Where $\Theta_i = 1 - \frac{\alpha_{1,i}\beta + \alpha_{2,i}}{\beta + 1}$ (10)

Solution: $C_{F,i} = C_{F0,i} (\frac{V_0}{V})^{\Theta_i}$ (11)

Retention: $R_i = 1 - \frac{C_{p,i}}{C_{f,i}}$ (12)

4. Results analysis and discussion

	Methanol	HMF	Formic acid	furfural	free acetic acid, mol/L	bound acetyl, mol/L
Final concentration g/L	0.76	1.685	1.168	0.84	0.0729	0.425
concentrated ratio	0.912	3.652	3.778	1.62	2.287	103.235
α1 (dk)	0.999	0.821	0.561	0.92	0.79	0.0166
α2 (ag)	0.971	0.163	0.397	0.586	0.419	0.0539
	Galactose	**Glucose**	**Xylose**	**Arabinose**	**Rhamnose**	**Mannose**
Final concentraion g/L	22.7	27.52	26	9.43	10.38	-
concentrated ratio	14.1	19.93	17.2	11.84	9.25	-
α1 (dk)	0.0146	0.0204	0.0276	0.0127	0.015	0.015
α2 (ag)	0.0137	0.0191	0.0165	0.0125	0.0144	0.01

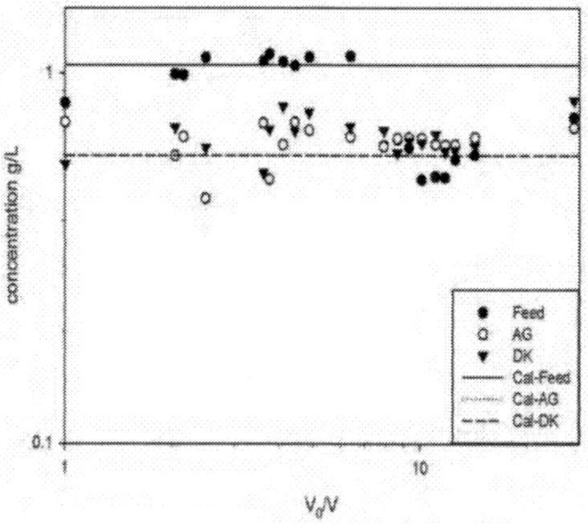

Variation of methanol concentration during NF process

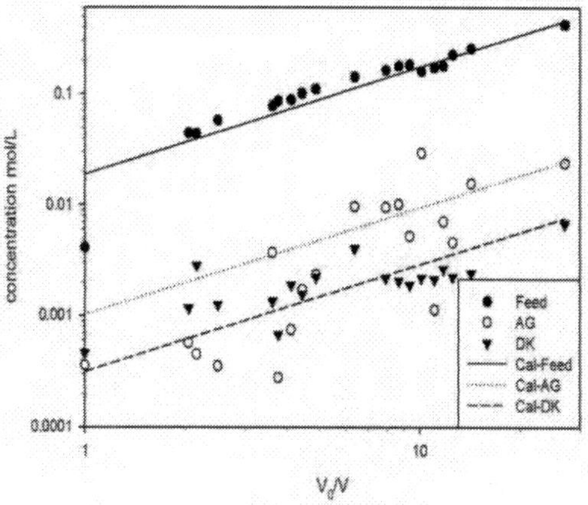

Variation of bound acetyl concentration during NF process

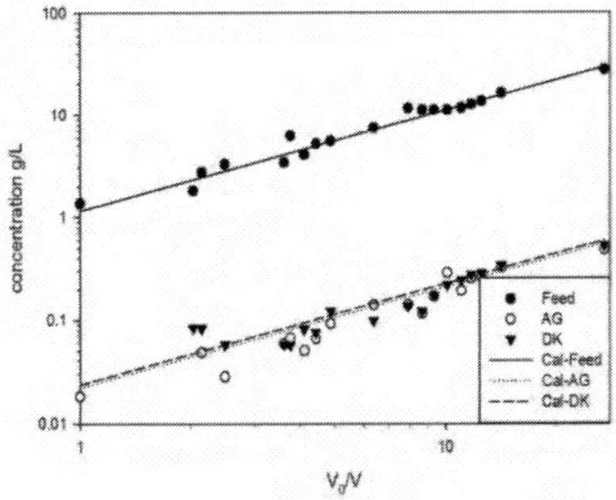

Variation of glucose concentration during NF process

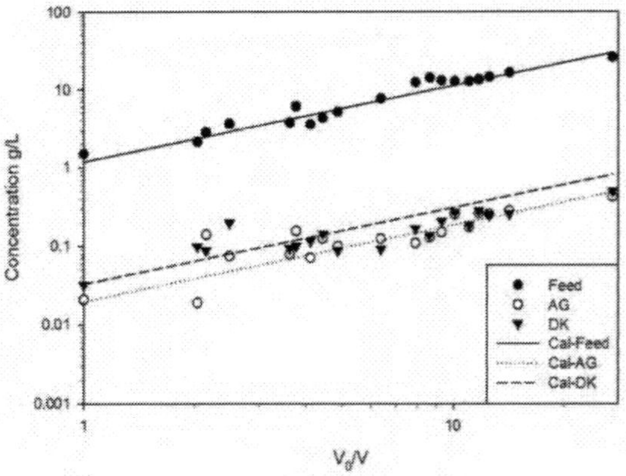

Variation of xylose concentration during NF process

5. Conclusion

1. Nanofiltration is an effective mean to remove fermentation inhibitors while preserving sugars.
2. Retentions of each species are subject to the difference of molecular weight, hydrophobicity and membrane salt rejection.
3. Nanofiltration is not efficient in separating hexoses and pentoses.

6. References

1. Shijie Liu, Thomas E. Amidon, and Christopher David Wood. Membrane filtration: concentration and purification of hydrolyzates from biomass. Journal of biobased materials and bioenergy. Vol. 2, 121-134, 2008.
2. Huajiang Huang, Shri Ramaswamy, U.W. Tschirner, B.V. Ramarao. A review of separation technologies in current and future bio-refineries. Separation and purification technology 62(2008), 1-21.
3. J. Yan, N. Joshee and S. Liu 2013 "Kinetics of the Hot-Water Extraction of Paulownia Elongata Woodchips" J. Bioproc. Eng. Biorefinery. 2(1):1-10.
4. R.D. Noble and S.A. Stern. Membrane separations technology, principles and applications. Elsevier Science B.V., New York (1995).
5. Marcel Mulder (1996). Basic Principles of Membrane technology. Kluwer Academic Publisher, Norwell, MA.
6. Fanglei Zhou, Cunwen Wang, Jiang Wei. Separation of acetic acid from monosaccharides by NF and RO membranes: Performance comparison. Journal of Membrane Science, 429(2013): 243-251.

Thank you for your attention!
Question?

373981 Design of an Organic-Inorganic Amine Monomer to Prepare Thin Film Composite Membranes for Bioethanol Purification

Abstract:

A novel hybrid organic-inorganic amine monomer has been designed and synthesized to prepare thin film composite (TFC) membranes for bioethanol dehydration via pervaporation. Different from the conventional amine monomer, m-phenylenediamine (MPD), used for interfacial polymerization, the new monomer was molecularly designed to sandwich an inorganic component nonafluorohexylmethyldichloro silane (ClSi) between two MPD molecules. Nuclear magnetic resonance (NMR) and Fourier transform infrared (FTIR) spectrometry were used to characterize and verify the chemical structures of the new monomer and the polyamide film. As a result, the advantages of polymeric and inorganic materials are effectively integrated in the newly developed TFC membranes, and these membranes exhibit a superior pervaporation separation performance as compared to most reported membranes in the literature for ethanol dehydration. Their best pervaporation performance shows a flux of 1.9 kg/m^2-h and a separation factor of 108 for the dehydration of the 85/15 wt% ethanol/water mixture at 50 °C. These novel TFC membranes show a great potential to compete with commercially available membranes.

379632 Distillation and Isolation of Commodity Chemicals from Bio-Oil Made By Tail-Gas Reactive Pyrolysis

Abstract:

Owing to instabilities, very little has been accomplished with regards to simple cost-effective separations of fast-pyrolysis bio-oil. When heated, the resulting depletion of valuable compounds (already at low concentrations) renders bio-oil distillation as a futile effort. A recently developed pyrolysis process based upon tail-gas reactive pyrolysis (TGRP) significantly improves bio-oil quality through reduction of oxygen to < 10 wt%, reduction of viscosity, increased hydrocarbon concentrations, and increased thermal stability.

We used fractional distillation to isolate compounds from bio-oil produced by TGRP. All bio-oils produced from TGRP contained significantly less acid (0.2 – 4% total), trace amounts of aldehydes, and significantly higher concentrations of hydrocarbons and phenolics (5 – 20% each). Organic yields from distillation of TGRP bio-oil ranged from 55 – 65%. On a dry basis, organic yields were more than 3 times that from traditional bio-oil; discounting acetic acid and furfural, TGRP distillation yields jumped to 10-fold. We categorized distillates according to boiling temperature and chemical components: benzene-toluene-xylenes (BTX), styrene, indene, pyridine, phenols, naphthalenes, acetic acid, and fluorene/anthracene. The relatively narrow product distributions allowed for the isolation of pure naphthalene by recrystallization from a naphthalene-rich fraction. Using GC-MS compositions, a thermodynamic computation of the TGRP distillation profiles correctly predicted the experimental results when acid content was < 1%.

Effect of operating conditions on the pervaporation of butanol from aqueous binary solutions for bio-butanol production

Hoda Azimi, Peter Gaudet, Jules Thibault, F.Handan Tezel

Handan.Tezel@uottawa.ca

Bio-butanol can be used as a renewable fuel, replacing gasoline to be used in existing car engines without modification, since it can be produced from agricultural waste. In this study, recovery of bio-butanol from acetone- butanol-ethanol (ABE) fermentation broth has been considered by using pervaporation.

A commercial PDMS (poly dimethyl siloxane) membrane was used to investigate the effect of operating conditions on performance of pervaporation for the separation of butanol from aqueous binary solutions. The effect of the feed composition and temperature on the permeation flux, separation factor and pervaporation separation index (PSI) was studied using binary solutions of butanol and water. Over the butanol concentration range of 0.5-5 wt % in the feed, the permeation flux increased with feed butanol concentration, whereas the separation factor and pervaporation separation index decreased. An increase in the operating temperature increased the permeation flux, and the increase in butanol flux was more than the associated increase in water flux, resulting in an increase in the separation factor and pervaporation separation index. Future work will include the pervaporation of ABE fermentation broths for more realistic separation applications for the production of bio-butanol.

387654 Strategies for Efficient and Cost-Effective Ionic Liquid Recycle and Product Recovery

Abstract:

Pretreatment with certain ionic liquids (IL) is highly effective at reducing the recalcitrance of a wide variety of lignocellulosic biomass to subsequent hydrolysis to release fermentable sugars. The key to an economically viable and scalable IL pretreatment technology is to develop efficient and cost competitive processes to recovery sugars and lignin and recycle IL for reuse. We have developed and tested several IL recycle and recovery strategies at the Joint BioEnergy Institute (JBEI). We have developed an improved liquid-liquid membrane extraction method that recovers >90% glucose and >95% xylose at ambient temperature and the recycled IL retains >95% of its original pretreatment efficiency. Additionally, we evaluated membrane (micro, ultra, and nano-) filtration and electro-dialysis separation technologies in conjunction with our newly developed one-pot IL pretreatment and saccharification process to enable IL recycle and recovery of lignin and fermentable sugars. Finally, we demonstrated a new regeneration protocol that use solvents precipitating cellulose but are immiscible with IL, a process that eliminates the use of water and facilitates lignin and IL recovery. Comparing with the costly distillation/chromatography, the membrane or phase separation technologies allow easier operation and less energy consumption and show great potentials for developing cost effective IL-based biomass pretreatment technologies.

386330 Separation and Purification of Biobutanol during Bioconversion of Biomass

Abstract:

Biofuels from biomass are becoming increasingly more important, due to the need for reduction in GHG emissions, energy independence, and limited global availability and increasing demand and costs of fossil fuels. Butanol has several advantages over ethanol as a drop-in bio-fuel such as higher energy content, potential for higher blending percentage with gasoline, lower vapor pressure, and lower hygroscopy. It can be used in existing transportation fuel distribution infrastructure. Butanol can be produced by fermentation of carbohydrates derived from biomass using *Clostridium acetobutylicum* or *C. beijerinckii* under anaerobic conditions. There are still many unsolved challenges for making biobutanol technically, and economically viable. The unsolved challenges lie in severe product (especially butanol) inhibition during bioprocessing, which leads to low butanol yield and productivity, and very low final product concentration (<3 wt%), causing expensive downstream processing (product separation) costs. There are two ways for solving these problems. One is the modification of microorganisms for ABE (Acetone, Butanol and Ethanol) fermentation by genetic engineering, which could keep the microorganisms alive and active under higher concentration of products in the broth. This could significantly increase the product yield, productivity, and concentration and hence reduce the production costs. However, this is still an unrealized long term goal. Another approach is the development of efficient separation and purification processes for product recovery. And, even if the modification of microorganisms becomes a reality, product separation and purification will still remain a major critical challenge. In this study, an extensive review of separation and purification of butanol from fermentation broth is provided, including gas stripping, vacuum flash, liquid-liquid extraction, membrane solvent extraction or perstraction, membrane pervaporation, membrane distillation, thermopervaporation, reverse osmosis, adsorption, and integrated bioprocessing with various separation methods. It is concluded that membrane pervaporation, solvent extraction, and adsorption are the most energy-efficient approaches for removal of butanol from the ABE fermentation broths. Integration of bioreactors with these separation methods could significantly increase the product yield, productivity and concentration and hence lower the production cost. Butanol dehydration is also discussed. This review could be helpful in the research and development and commercialization of renewable drop-in butanol biofuels and biochemicals.

361474 An Integrated Systematic Framework to Assist the Development of Pharmaceutical Processes

Abstract:

An integrated systematic framework to assist the development of pharmaceutical processes.

Emmanouil Papadakis, Gürkan Sin, Krist V.Gernaey, Rafiqul Gani*

CAPEC-PROCESS - Department of Chemical and Biochemical Engineering, Technical University of Denmark, Søltofts Plads, Building 229, DK-2800 Kgs. Lyngby, Denmark, tel. +45 45252882, e-mail: rag@kt.dtu.dk

The pharmaceutical industries are dominated by batch wise processes, which have served well the pharmaceutical industries as well as the regulatory bodies. Batch vessels provide the flexibility of handling different unit operations in a multipurpose plant. However, batch processes may not be very efficient for product quality assurance and may have drawbacks. Over the last decade the regulatory bodies required the pharmaceutical companies to demonstrate more process understanding. Continuous pharmaceutical manufacturing (CPM) has become an attractive option, which naturally eliminates some of the drawbacks of the batch processing while maintaining the flexibility [1].

Pharmaceutical industries are therefore looking for opportunities to evaluate the feasibility to apply continuous manufacturing. Methods and tools, which have been applied to other industries (such as chemical and petrochemical) may not be directly applicable for pharmaceutical processes. Therefore Process Systems Engineering (PSE) tools and methods tailor-made for pharmaceutical applications can have an important role in the transition from batch to continuous manufacturing in the pharmaceutical industries [2].

An integrated computer-aided framework for the development of pharmaceutical processes has been developed. The objective of the framework is to assist the pharmaceutical companies to facilitate the move from batch to continuous manufacturing and the development of batch or continuous pharmaceutical processes. The integrated framework consists of four parts: (Part A) identification of the reaction mechanism, (Part B) reaction analysis, (Part C) separation analysis and (Part D) evaluation of the developed process. Identification of the reaction mechanism is the first part and it consists of two steps. Step A.1 is the selection of the active pharmaceutical ingredient (API) or the intermediate to be produced and step A.2 the identification of the reaction pathway. The reaction analysis (part B) consists of six steps and it aims to identify all the relevant information for the reactor design. Step B.1 is the general analysis of the reaction, in this step the basic information for the reaction (such as reactants, solvent, kind of reaction, catalyst, properties and by-product) are collected (if available). In step B.2 the kinetic analysis is performed and the reaction limitations are identified. In the next step (Step B.3) the reaction variables are evaluated with respect the reaction performance. Based on the information which have been collected (or predicted through reliable models), a decision (Step B.4) whether continuous operation is beneficial or not for the system is taken. In the step B.5 decision on the reactor design is taken considering the collected information and finally, if separation of the reaction stream is needed (step B.6), the separation analysis (part C) is performed. The separation analysis consists of seven steps, in the beginning the objective of the separation is defined. In the Step C.1 the mixture to be separated is analyzed and the separation tasks are defined. In the step C.2 the alternatives of the separation are generated and in step C.3 they are evaluated in order to minimize the number of alternatives. Finally, in the steps C.4-C.7 the separation process is designed (unit operation design parameters and operation mode). The final part of the framework (part D) is the evaluation of the developed process and consists of two steps, simulation/evaluation of the process (Step D.1) and optimization, control and monitoring of the developed process (Step D.2).

Development of the framework requires the development and the use of model-data based computer aided methods and tools such as a knowledge database, a model library, modeling tools, solvent selection tools, process analytical technology (PAT) and quality by design (QbD) approaches. The knowledge library should provide all the relevant information for pharmaceutical product/process development such as reaction mechanisms, reactants (substrates, catalysts, and solvents), reaction conditions, and process efficiency. The model library should contain information to describe many unit operations involved in pharmaceutical processes, reaction kinetic models and properties prediction models. The modeling tools must assist in the development of new models and model-based simulations. Solvent selection tools must assist for the selection of an appropriate solvent when needed and to optimize the use of solvents in pharmaceutical development. PAT and QbD approaches are needed to improve and ensure high the quality of the product.

The application of the proposed integrated framework is highlighted through the BHC (Boots and Hoechst-Celanese) patented synthesis pathway for ibuprofen. The improved BHC synthesis of ibuprofen consists of three reaction steps [3] and it is used to verify the proposed integrated framework. The objective here is to demonstrate that the framework plus the methods and tools can be applied for new products (and their processing paths) as well as validate or retrofit existing ones.

References:

[1] K. Plumb, 2005. Chem. Eng. Res. Des., 83, 730.

[2] K.V. Gernaey, A. E. Cervera- Pardell, J. M. Woodley, 2012. Comput. Chem. Eng., 42, 30

[3] Elango, V. Murphy, M., Smith, B.L., Davenport, K.G., Mott, G.N., Zey, E.G., Moss, G.L. (Hoechst Celanese Corp.), 1990. European Patent 4.981.995

364734 Modeling of Disintegration and Dissolution of Formulated API

Abstract:

The absorption rate and extent of absorption of an active pharmaceutical ingredient (API) in patients often depends strongly on the mechanisms of the API release from the formulation into absorbable form. These mechanisms include coating dissolution, tablet mechanical disintegration into granules or aggregates, granule disintegration into component powder particles, and API dissolution from available surfaces. These phenomena can occur simultaneously and can affect each other. Additionally, each of these phenomena is affected by a number of chemical and physical parameters stemming from the molecular properties of the API and excipients, the nature of the dissolution medium, and the processing parameters for tablet formation such as compaction force and blending shear. In this work, a mathematical model is derived to relate API tablet dissolution profile to API and medium properties, tablet preparation parameters, and measurable tablet properties. Case studies are presented illustrating the application and utility of this model for formulations.

A practical computational tool to predict formulation and process variables during the development of spray-dried amorphous solid dispersions

Íris Duarte, Hovione Farmaciência SA & University of Lisbon, Faculty of Pharmacy, Portugal

José Luís Santos, Hovione Farmaciência SA, Loures, Portugal

João F. Pinto, University of Lisbon, Faculty of Pharmacy, Lisbon, Portugal

Márcio Temtem, Hovione Farmaciência SA, Loures, Portugal

Abstract

The aim of this work is to demonstrate the implementation of a computational tool to guide the early identification of potential physically stable amorphous dispersion formulations. The practical utility of this tool will be demonstrated as part of a new screening protocol that has been developed to accelerate the development of pharmaceutical amorphous solid dispersions produced by spray-drying. This methodology enables the narrowing of the scope of formulations, identifying more rapidly suitable systems to be scaled- up.

Introduction

Today, between 70 to 90% of the new chemical entities (NCEs) in the pharmaceutical pipeline belong to BCS Class II or IV, thus presenting at least solubility constraints [1]. Promising therapeutic molecules with limited aqueous solubility are a major concern, especially when considering the development of oral-dosage forms. To overcome this issue enabling platforms such as, particle-size reduction, use of surfactants, salt formation, complexation with cyclodextrines, production of self-emulsifying drug delivery systems or amorphous solid dispersions have been used.

The development of new amorphous solid dispersions (ASDs) has increased exponentially since its appearance in the late 90's [2]. ASDs have become popular among the pharmaceutical industry and their efficacy is supported by the increased number of marketed products (*e.g.* Incivek® (2011), Zelboraf® (2011), Kalydeco® (2012)). The requirements for the successful development of an ASD are related with performance and chemical/physical stability aspects. In what regards the performance requirements, an amorphous dispersion formulation should present an improved dissolution profile compared with the crystalline reference and should be capable of sustaining drug supersaturation in solution for a longer time. Both parameters will contribute to an increased amount of drug available for absorption. In what accounts physical stability, this concept is related with the inherent perceived risk of drug recrystallization from the amorphous state, followed by consequent loss of the solubility advantage.

The development of ASDs with the desirable long-term stability and performance is a challenging process, due to the wide number of formulation and process variables that influence both physical and chemical properties of the product (*e.g.* several existing polymeric stabilizers, surfactants, different drug-polymer ratios, solvents, preparation methods, temperatures, etc). For a long time, the selection of the best formulations was simply based on trial and error experiments, combined with the experience of researchers themselves, which demonstrated to be costly, time-consuming and API demanding. At a time, in which the pharmaceutical industry demands for fast turnaround times, lower costs and to reduce the risks associated with the development of new drugs, it became critical the development of screening programs for narrowing the scope of formulations and to rapidly identify suitable systems for subsequent clinical evaluation.

In the scope of this work the applicability of a computational tool to guide the early identification of potential physically stable systems, based on molecular miscibility predictions between the drug and the polymer, was evaluated. The final goal is to include this tool in a screening program that has been developed for the production of stable ASDs.

Materials and Methods

Materials

Drug A (Tm~200ºC, poorly water soluble) was evaluated with differently selected polymers, ranging from PVP K29/32, PVPVA 64, Eudragit® L100, HPMCAS and HPMC [drug load up to 65% (w/w)], according to the screening protocol developed. By opposition, Itraconazole (ITZ)-polymers systems were also evaluated, but the results only served to validate the computational tool. ITZ was combined with structurally different polymers, namely PVPVA 64, Eudragit® EPO and HPMCAS up to 85% (w/w) drug load.

Methods

A) Screening Protocol Overview

Figure 1 represents a workflow demonstrating the implementation of the screening protocol under development.

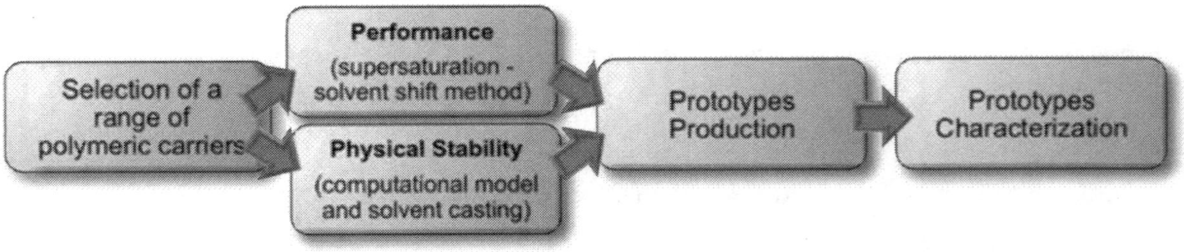

Figure 1: Workflow for the early development of a new spray dried amorphous dispersion.

The implementation of the screening program for a given API first consists in the analysis of drug physicochemical properties, target drug release profile and target dose. According to these characteristics a suitable number of polymeric carriers were selected, based on chemical structure diversity, their physicochemical properties and potential to promote drug-polymer interactions (*e.g.* H-bonding). In order to evaluate the precipitation inhibition effect of the selected polymers on the recrystallization of the API from solution, supersaturation screening studies were performed in a small-scale dissolution apparatus using a solvent-shift method, under fasted biorelevant conditions and at a pH value where the API presents lower solubility [3]. Concomitantly, drug-polymer miscibility estimates were obtained using a computational tool that will be explained in detail in the next section. Still related to the latter, a small-scale solvent casting protocol was implemented. Solvent casting experiments proved to be useful to get early information on experimental stability trends, using small quantities of API [4]. Moreover, it can also be useful to fine-tune formulation variables such as the drug loading, total solids fraction, solvents, among others.

At this point of the development work, polymers selections should be narrowed based on their ability to inhibit drug precipitation in solution, but also based on the drug-polymer miscibility capacity and physical stability. However, in this work and in order to evaluate the applicability range of the screening methods herein presented, different polymers presenting different performances and compatibilities with Drug A were selected to proceed with the solvent casting experiments, lab-scale production of spray-drying prototypes and respective analytical characterization, namely physical stability (DSC, XRPD) and powder dissolution (performance). The analytical results obtained for the spray-dried dispersions were compared with those obtained during the screening phase.

A.1) In silico analysis of drug-polymer miscibility

The computational tool herein presented is based on diffuse interface theories and presents advantages over commonly applied strategies to guide rationale polymer selection and drug load. Examples of current approaches are the analysis of the Hildebrand and Hansen Solubility Parameters (SPs) to infer about drug-polymer compatibility [5] or, more recently, the implementation of lattice-based solution theories, such as the Flory-Huggins (F-H) theory to study the thermodynamics of drug-polymer mixing [6]. The most significant differences of this model over other approaches is (1) the potential to evaluate a ternary system made of drug, polymer and solvent, by comparison with the traditional two-component system and (2) the consideration of time-dependent phenomena, such as components mass diffusion and solvent evaporation (Figure 2).

As can be seen in Figure 2, the model used in this work takes into account different thermodynamic, kinetic and process variables. Those were estimated according to the methods described in Table 1.

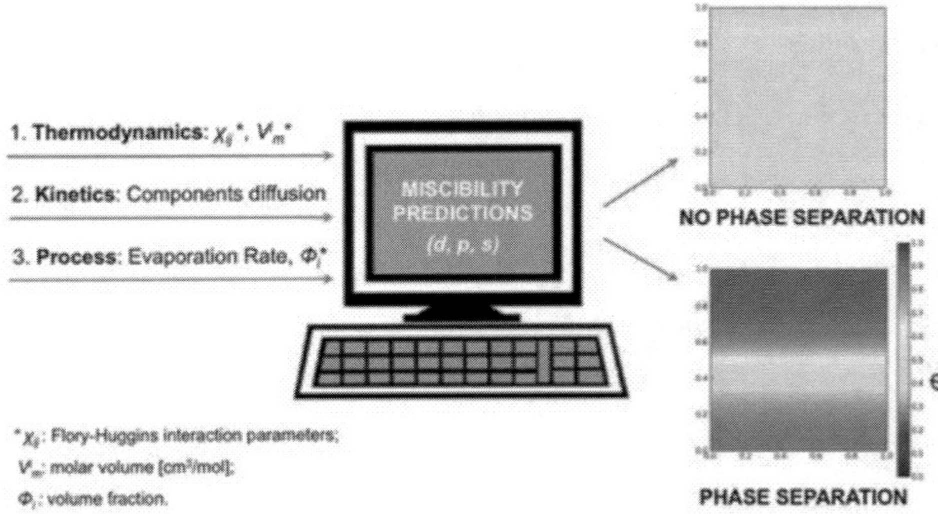

Figure 2: Input parameters and two limiting situations given by the model (outputs).

Table 1: Input parameters and respective methods of calculation.

Input Variable	Type of calculation
Flory-Huggins interaction parameters (χ_{ij})	Inverse gas chromatography (SEA, from SMS) [7]
Molar volume (cm^3/mol)	Molecular weight (g/mol) divided by the true density (g/cm^3)
Solid's diffusivity	Wilke-Chang equation
Evaporation Rate	Correlation for the evaporation of a single droplet in still air

For more details on the derivation and implementation of the computational tool readers are referred to the work of Duarte *et al.* [4].

Results and Discussion

The methodology herein reported is divided in two main parts: performance and stability considerations. First supersaturation profiles were used to assess the potential of the polymers to inhibit Drug A precipitation in fasted biorelevant conditions and at a pH value where the API presents lower solubility. The following polymer ranking by ascending order of precipitation inhibition was obtained: PVP K29/32 < PVP/VA 64 < HPMC < HPMCAS < Eudragit L100. Second, miscibility computational simulations were conducted complementarily to predict the influence of thermodynamic and kinetic material properties and process conditions on the phase behavior limits of Drug A-polymer systems. The following polymer ranking by ascending order of miscibility capacity was obtained: Eudragit L100 < PVP/VA 64 < HPMCAS < HPMC. Both results were compared, and four polymers - PVP/VA 64, Eudragit L100, HPMCAS and HPMC - were selected to proceed with the solvent casting experiments, lab-scale production of spray-drying prototypes and respective analytical characterization.

Validation of the screening methodologies implemented

Figure 3 compiles the predicted and experimental miscibility results given by the computational model, solvent casting and spray drying experiments. In the same figure, the results previously obtained for the ITZ-based systems are also presented [4]. Both results served to validate the screening methodology.

In general, cast films and spray-dried dispersions generated were amorphous, and true solid solutions with drug concentrations up to 65 and 85% w/w of Drug A and ITZ were obtained, respectively. Comparing the experimental and the computational data, the results showed an acceptable correlation with the ability to maintain the rankings initially defined. Indeed, when the simulations indicated the formation of homogenous amorphous systems for higher drug loads the experimental references indicated high miscibility capacity; when the simulations indicated phase separation for lower drug loads the real references presented their maximum of experimental miscibility at lower drug loads.

However, among the systems evaluated two false-negatives were observed (Drug A:PVPVA 64 and Drug A:Eudragit® L100). It should not be neglected the fact that the theoretical basis of the computational model may not fully capture the complexity of drug-polymer particle formation [4].

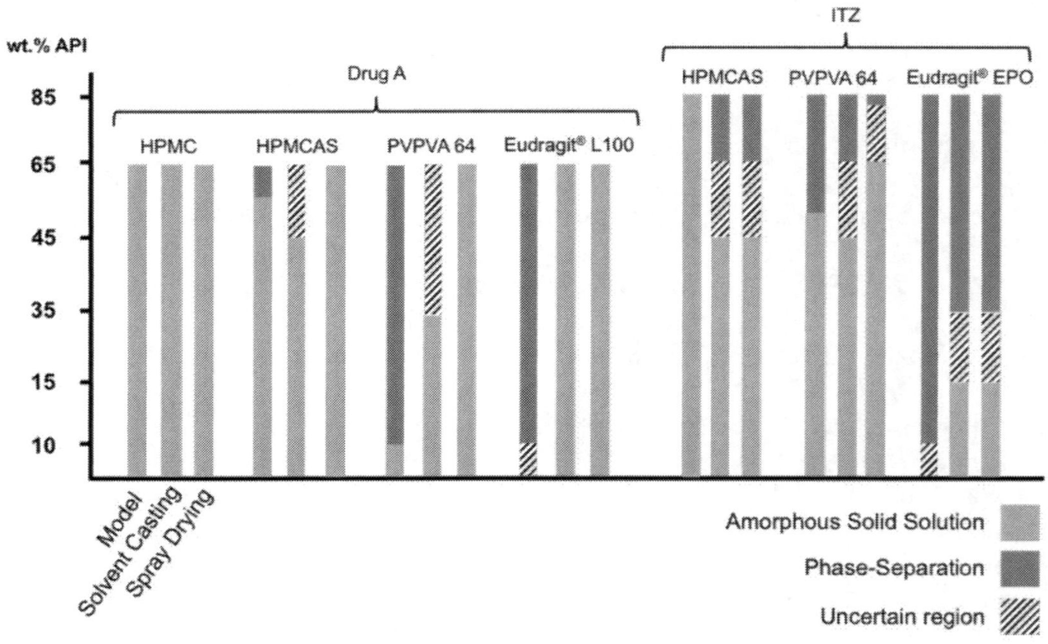

Figure 3: Summary of the results obtained: miscibility and phase behavior predictions (Model) and experimental miscibility assessment by mDSC (solvent casting+spray drying).

The potential occurrence of false-negative results corroborates the importance of implementing a small-scale solvent casting methodology; otherwise one may be neglecting the real solubilization capacity offered by the polymer. Despite these shortcomings, the results show that one can still use the information given by the computational model not only to rank the most stable amorphous formulations with the highest API/polymer ratio, but also to reduce the number of experiments, reducing time and resources in the following phases of product development.

In what regards the evaluation of the performance of the Drug A-based spray dried systems amorphous systems, powder dissolution profiles were obtained for all samples with 65% (w/w) drug load under constant pH (the same pH as in the early supersaturation experiments). The high API load was selected based on the fact that no signs of amorphous-

amorphous phase separation or drug recrystallization were observed. Figure 4 compares the results obtained during the screening phase and the powder dissolution characterization of the spray-dried dispersions. The results are presented in terms of the accumulated drug concentration (AUC) obtained during 240 minutes of testing.

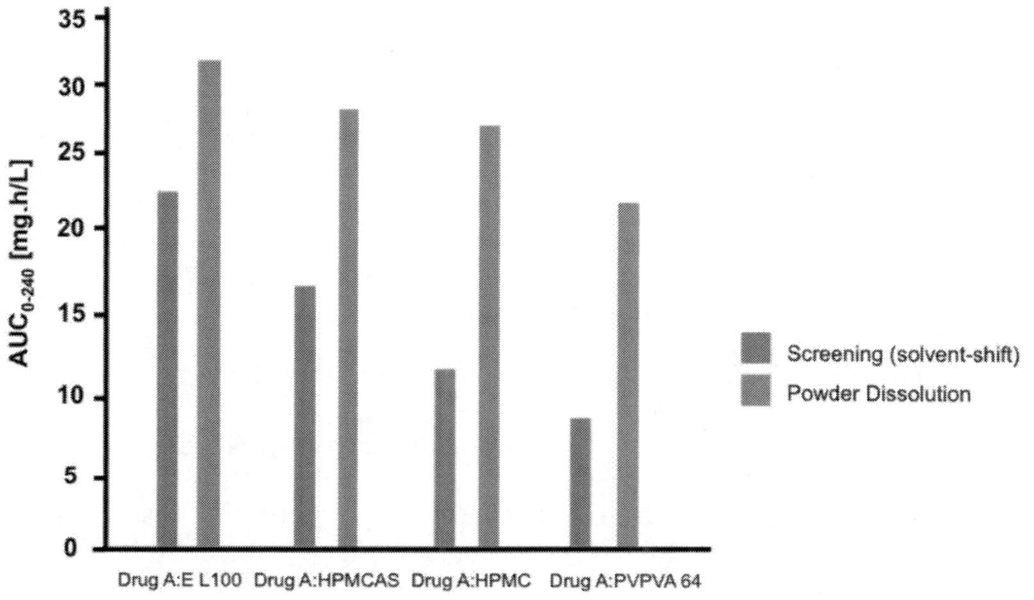

Figure 4: Accumulated drug concentration (AUC) for Drug A systems (up to 240min) in the screening phase and powder dissolution experiments. The AUC was computed using the trapezoidal rule.

As can be seen, both results from the solvent-shift *versus* powder dissolution experiments are aligned. The polymer ranking by ascending order of potential to inhibit Drug A precipitation in solution (PVP/VA 64 < HPMC < HPMCAS < Eudragit L100) agreed with the overall performance and powder dissolution profiles of ASDs of Drug A and the respective polymers.

Conclusions

The implementation of screening programs to support and to expedite the rationale development of new amorphous formulations is of utmost importance. In the present work a screening program was described, which considers performance and physical stability aspects.

The results obtained after implementation of the screening methodologies tested (*i.e.*, solvent-shift method, computational model, solvent casting) provided valuable insights on the performance in solution and miscibility assessment of Drug A amorphous dispersion formulations. Results should be analyzed concurrently to guarantee proper selection of the most promising systems along the different phases of formulation development. The computational model implemented not only allowed the ranking of the most promising amorphous formulations in terms of physical stability, but also to narrow down the range of drug-polymer ratios to be tested in the following phases.

Acknowledgements

The authors would like to thank Anett Kondor, Majid Naderi and Daniel J. Burnett from Surface Measurement Systems Ltd for analyzing the raw materials used in this work using the iGC surface energy analyzer. Íris Duarte would like to thank the financial support from Hovione Farmaciência SA and from Fundação para a Ciência e Tecnologia through the doctoral grant BDE/51422/2011.

References

[1] Löbenberg R. and Amidon GL., *Eur J Pharm Biopharm* (2000), 50(1), pp. 3-12;

[2] Sekiguchi K. and Obi N., *Chem Pharm Bull* (1961), 9, pp. 866-872;

[3] Vandecruys R. *et al.*, *Int J Pharm* (2007), 342, pp. 168-175;

[4] Duarte I. *et al.*, *Pharm Res* (2014), online;

[5] Greenhalgh DJ. *et al.*, *J Pharm Sci* (1999), 88 (11), pp. 1182-1190;

[6] Flory PJ., *Cornell University Press:Ithaca* (1953);

[7] Thielmann F. and Naderi M., Surface Measurements Systems, Application Notes.

375130 Method for Prediction of Pharmaceutical Solubility for Solvent Selection

Abstract:

A method is described for modeling the solubility of organic, non-electrolyte solutes in common solvents and mixtures. This has applications for solvent selection for reaction, crystallization and extraction operations as well as in formulation design. A small number of solubility measurements are required to fit model parameters; thereafter, solubility predictions can be made in a wide range of single, binary and ternary solvent systems (e.g. for identification of conditions for a solubility maximum in mixtures).

These calculations assist scientists and engineers in selecting solvent systems and ensuring that promising candidates that may not be 'on the shelf' are given due consideration.

The method comprises adding an extra group representing the solute molecule to the UNIFAC (ref 1) tables, including its UNIFAC R and Q parameters, then fitting the interaction parameters between the solute and the functional groups of the solvents used when measuring solubility. An alternative tuning of standard UNIFAC has been described elsewhere (ref 2). Heat of fusion and melting temperature are inputs (preferably) and can also be fitted. Predictions can then be made of solubility in any solvent system where these functional groups are present.

The method overcomes some traditional weaknesses of UNIFAC while maintaining the standard table of parameters intact. These include:

- Application for solutes that are pharmaceutically relevant and cannot be fragmented into UNIFAC groups
- Much improved accuracy compared to normal UNIFAC, which precludes regression of parameters.

Some recent applications of the method will be reviewed to indicate its performance relative to other techniques for solubility prediction (refs 3, 4 and 5).

Utilizing the functional group contribution approach (rather than conceptual segments) maintains advantages that may be leveraged in future development of the method to predict solubility of structurally similar impurities as well as primary solutes.

The current implementation considers 163 solvents, enabling potential prediction in thousands of binary solvent mixtures and several million ternary solvent mixtures. Users can limit the scope to solvents that are desirable for e.g. regulatory, cost or environmental considerations.

References:

1. Fredenslund, A., Jones, R. L., Prausnitz, J. M., Group-Contribution Estimation of Activity Coefficients in Nonideal Liquid Mixtures. *AIChE J.*, 1975. **21**(6): p. 1086-1099
2. Modarresi, H., Conte, E., Abildskov, J., Gani, R., & Crafts, P. A. (2008). Model-Based Calculation of Solid Solubility for Solvent Selection A Review. Industrial & Engineering Chemistry Research, 47(15), 5234–5242. Retrieved from http://pubs.acs.org/doi/abs/10.1021/ie0716363
3. E. Sirota and T. Lin, Modelling Small Molecule Synergistic Solubility Behavior, Presented at DynoChem User Meeting, Chicago, May 2011: available online at https://dcresources.scale-up.com/Default.aspx?id=388
4. N. Chennamsetty, O Lyngberg & J. Qiu, Determining Solubility of APIs and Intermediates From Automated Parallel Experiments and Modeling, AIChE Annual Meeting 2011, Paper 56c: http://www3.aiche.org/proceedings/Abstract.aspx?PaperID=236416
5. New solubility prediction utility for solvent selection in DynoChem, Scale-up Systems blog, March 2010: http://blog.scale-up.com/2010/03/new-solubility-prediction-utility-for.html

376951 Impact of Continuous Drug Product Processing on Formulation and Process Development Workflow

Abstract:

Continuous drug product processing can significantly reduce the cycle time from powder dispensing to product formation. However, the output of process development is not product but rather knowledge. The manufacture of product is a relatively small percentage of the knowledge generation cycle time. Non-production related activities include: planning experiments, defining safety measures, collecting samples, analyzing samples, integrating multiple data streams, and analyzing data. This must then be combined with experience and science and compared with experimental objectives. Typically new insights are gained and the next cycle of experiments is planned and executed.

It was observed that the production cycle time sets the tempo for the other aspects of the work. When experiments are measured in weeks (as in batch), the preceding and following steps are also measured in weeks. When it is days or hours (as in continuous), the preceding and following steps are also measured in days or hours. Continuous processing also greatly improves the reliability of the system at the critical interface between sample generation and sample analysis. Improved reliability in delivering the samples to the lab at the promised time improves efficiency in the lab. Taken to its extreme, this becomes real time analytical. This creates the ability to make decisions during the run rather than after the run. This causes decision makers to spend more time in and around the processing suite / lab. Additionally, because of the short time to perform an experiment the impact of a failed experiment is reduced. This means higher risk / higher return experiments can be performed. These experiments can lead to unique insights. Finally, rapid or real-time data enables adaptive design of experiments that can shrink or grow as the runs proceed. If managed well that can be a powerful technique. However, poorly managed adaptations can destroy a well designed experiment.

377737 CFD-Predicted and PIV-Measured Velocity Profiles in a USP Dissolution Testing Apparatus 2 Equipped with an Arch-Shaped Fiber Optic Probe and Their Impact on Tablet Dissolution

Abstract:

Dissolution testing is a critical tool in drug formulation and product quality assessment. This test is a routinely used in pharmaceutical development to evaluate the *in vitro* performance of solid dosage forms, determine the dissolution behavior of formulations and provide an optimization of drug release dosage forms, and as a quality control tool to insure that solid dosage forms have consistent dissolution properties. The USP Dissolution Testing Apparatus 2 is the device most commonly used for this purpose. In a typical test, a tablet is added to the vessel and samples are manually withdrawn from the vessels over time and analyzed for their drug content. An approach to overcome the limitations of manual sampling requires using sampling probes, such as fiber optic probes, permanently inserted in the dissolution medium and continually monitoring drug concentration as the solid dosage form dissolves. Despite their advantages, permanently inserted fiber optic probes can alter the normal fluid flow within the vessel and produce different dissolution testing results.

In this work, the hydrodynamic changes introduced by an arch-shaped fiber optic probe in a USP Dissolution Testing Apparatus 2 filled with 900 mL of dissolution medium were studied by (1) experimentally determining the velocity profiles in the vessel, with and without the probe, using Particle Image Velocimetry (PIV) and quantifying changes in the flow velocities on selected horizontal iso-surfaces; and (2) predicting the velocity distribution using Computational Fluid Dynamics (CFD). These results so obtained were used to analyze the dissolution data from of actual dissolution tests conducted with and without the probe using Prednisone tablets fixed at nine different locations at the bottom of the vessel.

In general, substantial agreement was found between the PIV data and the CFD predictions. The velocities in the upper region of the vessel where the probe is located were clearly affected by the presence of the probe. More importantly, the some velocity components near the vessel bottom where the tablet is usually located, were found to be appreciably affected by the presence of the probe. Specifically, both the simulation results and PIV data show that the axial velocities near the bottom are larger in the probe system than in the standard system. In addition, in the region just under the probe, the differences in axial velocities between the two systems are the largest in both of the simulation and PIV measurement, but they become progressively smaller at locations further away from the probe.

The results of this work indicate that that the arch shaped fiber optic probe introduces as small but measurable baffling effect on the hydrodynamics in the dissolution vessel. This effect results in changes in the flow velocities which parallel the results of dissolution tests, where the dissolution profiles were typically found to be higher in the presence of the probe, but not sufficiently high to fail the tests, according to the FDA criteria (f_1 and f_2 values).

In summary, the hydrodynamic effects generated by the arch-shaped fiber optic probe are small but clearly measurable. The changes in velocity profiles in the dissolution vessel result in in detectable differences in the dissolution profiles, although not high enough to cause test failures per se. However, these differences could contribute to amplify the difference in dissolution profiles in those cases in which tablet has an intrinsically higher release rate.

394222 Slow Growth Increases Myo-Inositol Availability for Glucaric Acid Production in S. Cerevisiae

Abstract:

Glucaric acid has been classified as a top value-added chemical from biomass for potential uses ranging from cancer therapies to sustainable nylons by the Department of Energy. A biosynthetic pathway was constructed in *E. coli* to convert glucose to glucaric acid but activity of the *Mus musculus* myo-inositol oxygenase (MIOX) enzyme was rate-limiting. New expression constructs enabled a one-step knock-in of the four pathway genes to *S. cerevisiae.* Protein sequence similarity mapping paired with simple structure-function hypotheses enabled efficient sampling of over 300 MIOX enzymes. Substituting a plant MIOX in the pathway increased glucaric acid titers detected in the YPD culture broth at 72hrs. Supplementation with 60mM myo-inositol tripled glucaric acid titers for yeast expressing either the plant MIOX (1.5 g/L) or the mouse MIOX (0.94g/L). However, myo-inositol supplementation did not affect glucaric acid titers in yeast expressing the mouse MIOX during slower growth on non-fermentable carbon sources such as glycerol or ethanol. Myo-inositol is a component of yeast membrane phospholipids and carbon-limited chemostat experiments demonstrated that slow growth rates lead to increased myo-inositol availability and glucaric acid titers. We pursued a fed-batch strategy in YPD with yeast expressing the plant MIOX and achieved final titers of 1.2g/L.

Development of Tissue Phantoms for a New Breast Cancer Detection Technique

Matthew A. Conrad
Caitlin B. Douglas
Adam J. Nolte, PhD

Rose-Hulman Institute of Technology | Department of Chemical Engineering

Background Information

- Healthy breast tissue has a Young's modulus of ~3 kPa and tumorous tissue has a Young's modulus of ~20 kPa

⟹ Gelatin hydrogels seem to be acceptably soft

- 250 Bloom B Gelatin was utilized

Rose-Hulman Institute of Technology | Department of Chemical Engineering

Research Goals

- To prepare gelatin tissue phantoms of a Young's modulus of healthy tissue with tumors of various sizes and geometries
- To characterize the mechanical properties such as the effects of aging, multiple tests, strain rate, and sodium alginate as an additive

Rose-Hulman Institute of Technology | Department of Chemical Engineering

Tissue Phantom Preparation

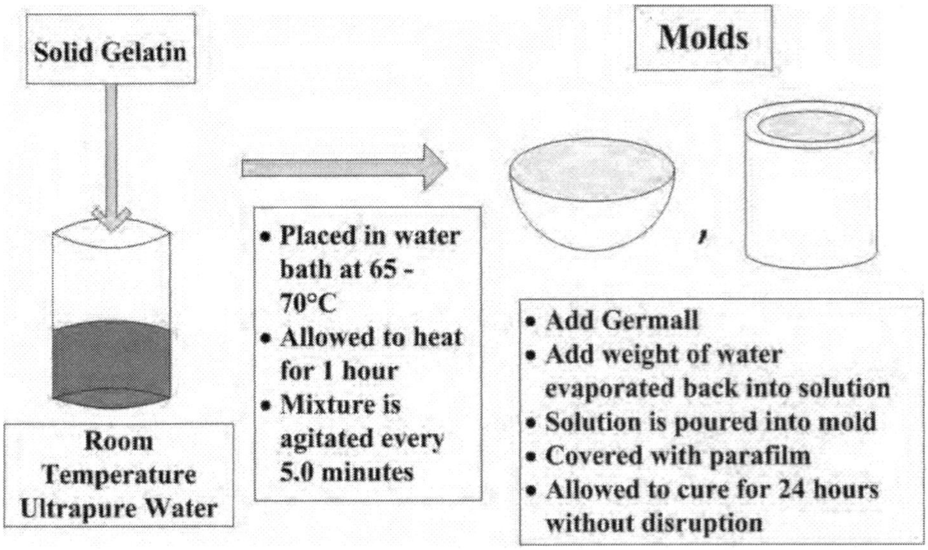

Rose-Hulman Institute of Technology | Department of Chemical Engineering

Tissue Phantom Preparation

Rose-Hulman Institute of Technology | Department of Chemical Engineering

Tumor Inclusion

Two proposed methods:

Liquid Injection via syringe

- 12.0% w/w gelatin liquid solution syringe-injected into congealing 4.0% w/w gelatin solution

⟹ • Not effective

Pre-made inclusions

- 12.0% w/w solid gelatin inclusion placed in congealing 4.0% w/w gelatin solution (T ~ 23.5°C)

⟹ • More effective approach

Rose-Hulman Institute of Technology | Department of Chemical Engineering

Mechanical Testing

Mechanical properties measured:

Young's Modulus
Fracture Strength
Toughness

- Samples tested in compression with an Instron with a 10 kN LC
  - Young's Modulus from strains less than 10%
  - Some samples tested to failure

Rose-Hulman Institute of Technology | Department of Chemical Engineering

Concentration Dependence

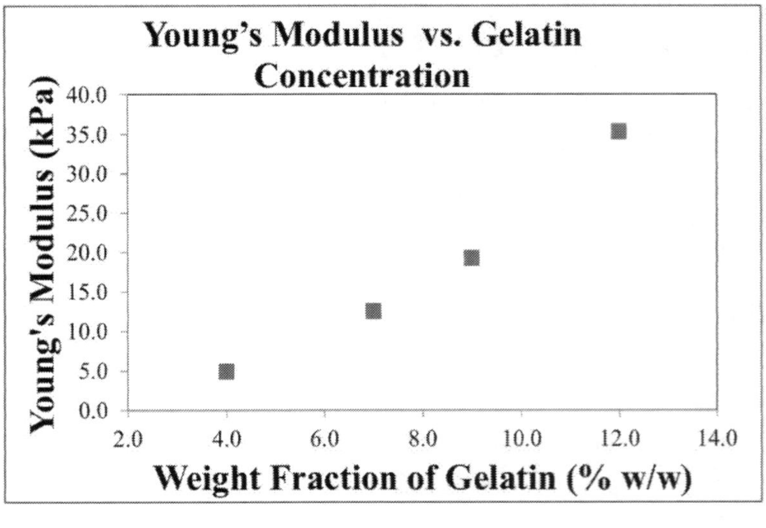

Rose-Hulman Institute of Technology | Department of Chemical Engineering

Aging Effects

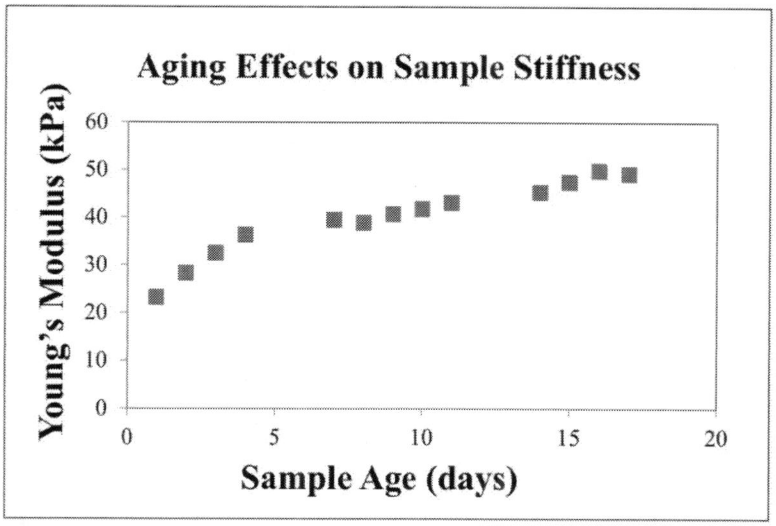

Rose-Hulman Institute of Technology | Department of Chemical Engineering

Testing Speed

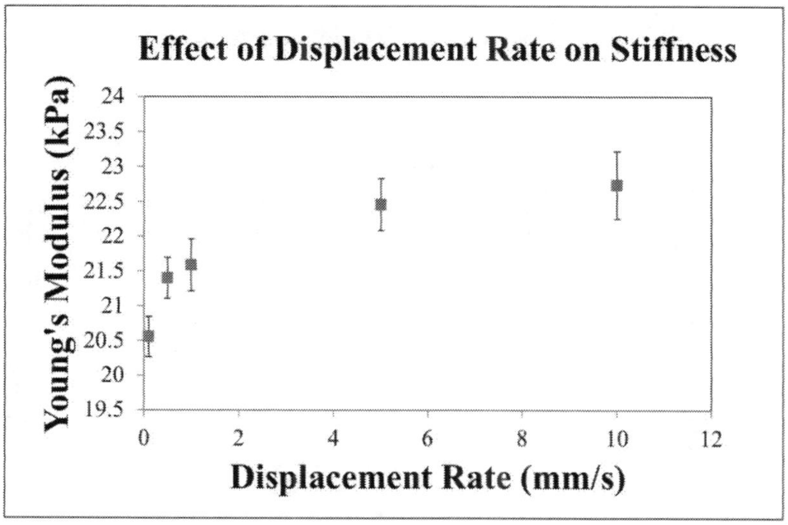

Rose-Hulman Institute of Technology | Department of Chemical Engineering

Multiple Test Effect on Stiffness

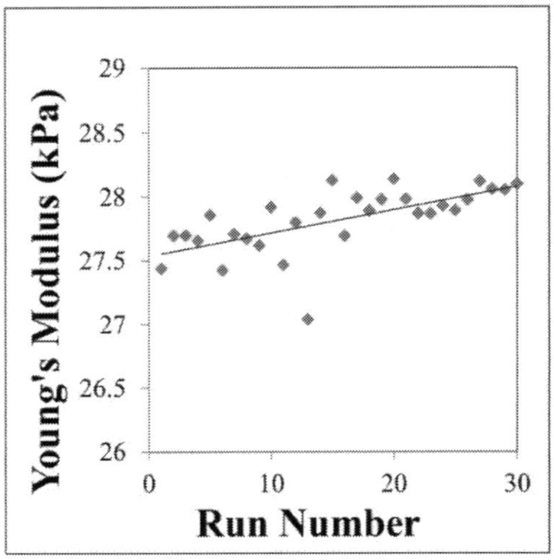

Uncovered during testing

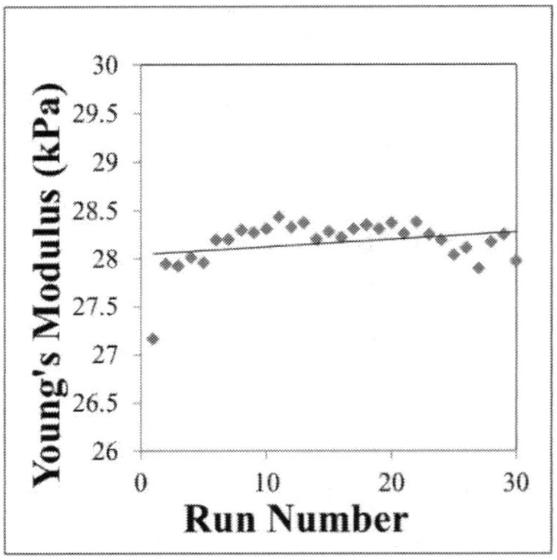

Covered during testing

Rose-Hulman Institute of Technology | Department of Chemical Engineering

Additive Effects: On Stiffness

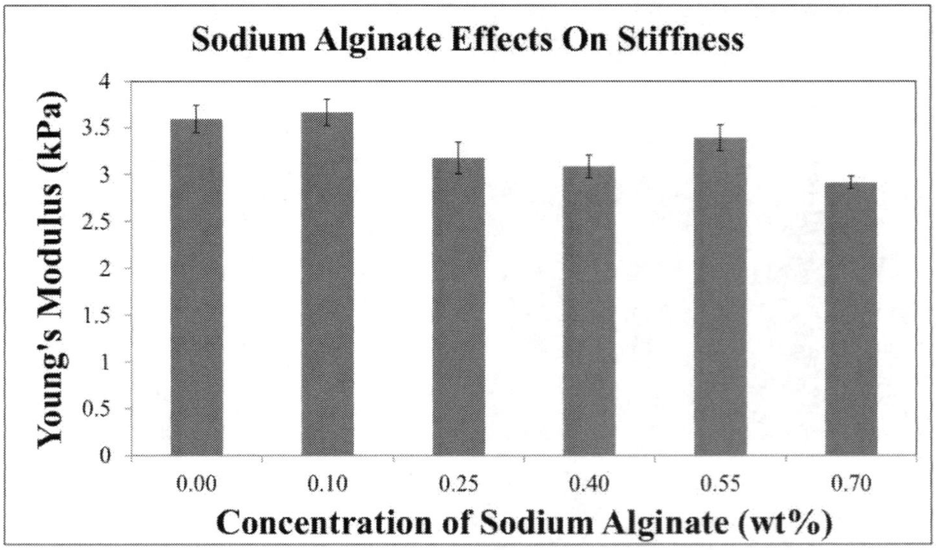

Rose-Hulman Institute of Technology | Department of Chemical Engineering

Additive Effects: On Fracture Strength

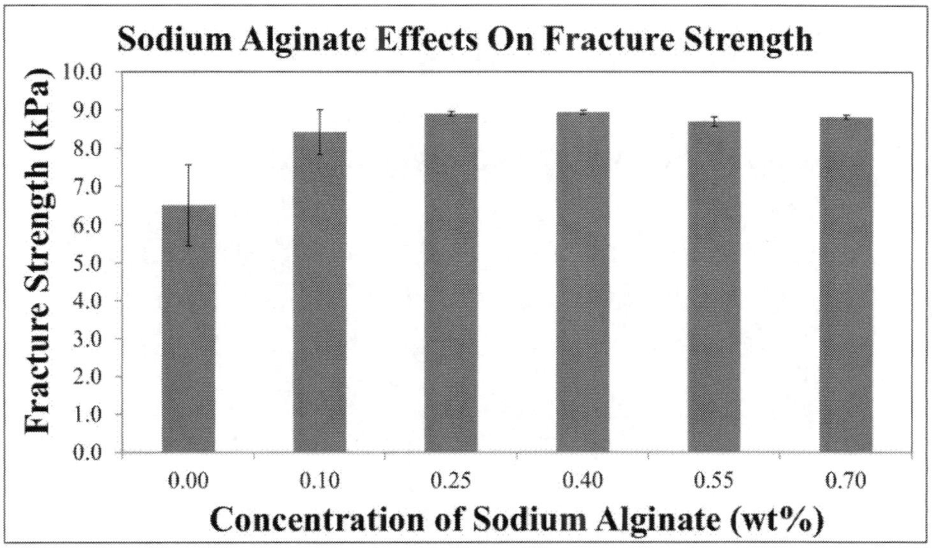

Rose-Hulman Institute of Technology | Department of Chemical Engineering

Additive Effects: On Toughness

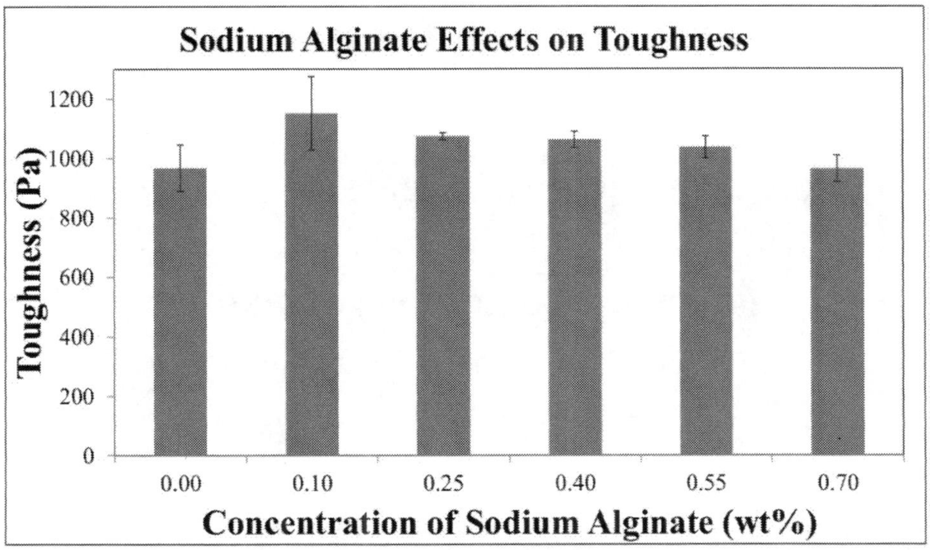

Rose-Hulman Institute of Technology | Department of Chemical Engineering

Conclusion

- Gelatin hemispheres can be made to mimic healthy breast tissue with tumor inclusions by inserting a pre-made tumor into a congealing gelatin solution
- Increasing the sodium alginate concentration appears to improve the fracture strength of gelatin hydrogels while decreasing the stiffness

Rose-Hulman Institute of Technology | Department of Chemical Engineering

Future

- Confirmation of results with texture analyzer
- More accurate relaxation tests
- Optimizing concentrations of additives

Rose-Hulman Institute of Technology | Department of Chemical Engineering

Acknowledgements

- *This research was funded by the National Science Foundation under Grant No. ECSS-1306808*
- *Professor Michael Insana and research group members (UIUC)*
- *Professors Olson and Throne (RHIT)*

Rose-Hulman Institute of Technology | Department of Chemical Engineering

Relaxation Effects

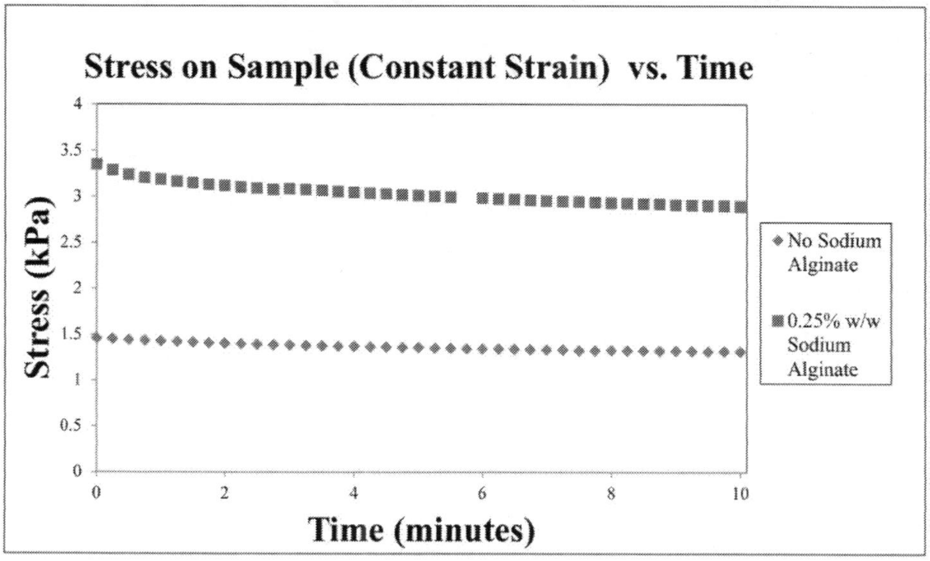

Rose-Hulman Institute of Technology | Department of Chemical Engineering

394192 Effective Antisense Design Using Ensemble of Energetically Sub-Optimal Secondary mRNA Structures

Abstract:

There is a deficit of effective therapies in the current research against bacterial infection. Many therapeutic strategies seek to use small molecules to target the infectious pathogen. One approach to disease inhibition involves direct manipulation of the pathogen at the RNA level. Part of the central dogma, messenger RNA (mRNA) is a transcript of genetic information that encodes the fundamental instruction for protein production. Inhibiting translation of mRNA effectively prevents synthesis of proteins.

In order to effectively block the RNA's message from being read, it must be physically accessed by the therapeutic agent. A field of study has arisen in which a complementary nucleic acid binding strand, called antisense[1], is generated to impede protein synthesis. One of the issues that arises in creating effective antisense RNA is the ability to find appropriate target sites on the mRNA to be inhibited. This problem is in part due to fluctuating secondary structure sequestering target sites. In fact, the kinetics of physical accession of the antisense to the target strand are suggested to be the rate-limiting factor[2] and thus the cause of inefficiency of this method. However, the fluctuations among suboptimal conformations are energetically inherent to the nucleic acid sequence and can be predicted. A software package, GenAVERT, has been developed to thermodynamically determine effective target sites by identifying the ensemble of the most probable energetically suboptimal states and determine which regions of mRNA are most accessible[3]. Once properly identified, a successful and effective antisense can, in theory, be synthesized.

In this work, a proof of principle will be demonstrated in *E. coli* with the downregulation of a short-lived variant of the green fluorescent protein through construction of an inducible antisense molecule. Statistical reduction in the fluorescence upon antisense induction will affirm the efficacy of this model.
Supported by fundamental energetic principles, this method of producing an antisense sequence complement for the most accessible mRNA target region has the potential to greatly reduce the strife of pathogenic infection.

---

[1] Lee, L. and Roth, C. (2003). Antisense technology in molecular and cellular bioengineering. *Current Opinion in Biotechnology*, [online] 14(5), pp.505-511.

[2] Patzel, V. and Sczakiel, G. (2000). In vitro selection supports the view of a kinetic control of antisense RNA-mediated inhibition of gene expression in mammalian cells. *Nucleic Acids Research*, [online] 28(13), pp.2462-2466.

[3] Johnson, E. and Srivastava, R. (2013). Volatility in mRNA secondary structure as a design principle for antisense. *Nucleic Acids Research*, [online] 41(3), pp.e43-e43.

Nanotechnology REU
Summer 2014

Raghav Malik
AIChE National Conference: Undergraduate Research Forum II: Bioengineering, November 17th
Supervising Graduate Student: Giulia Suarato
Principal Investigator: Dr. Yizhi Meng
Department of Materials Science and Engineering

NSF

Stony Brook University

Proanthrocyanidin (PAC)

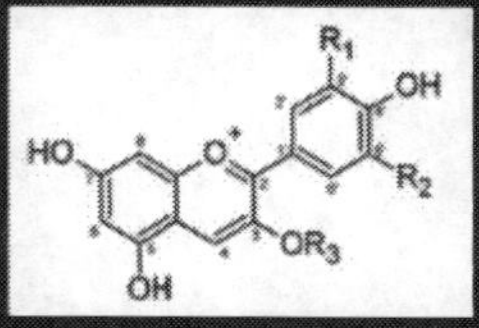

- Organically derived (from fruits and wines)
- Digests in saliva
- Promising chemotherapeutic effects, novel cancer cytotoxicity
- Encapsulation to improve bioavailability

NSF

Kom Kamonpatana, et. al. Susceptibility of anthocyanins to ex vivo degradation in human saliva, Food Chemistry. 2012: Vol. 135, 738-747

Stony Brook University

3

Chemotherapeutic Effects of PAC

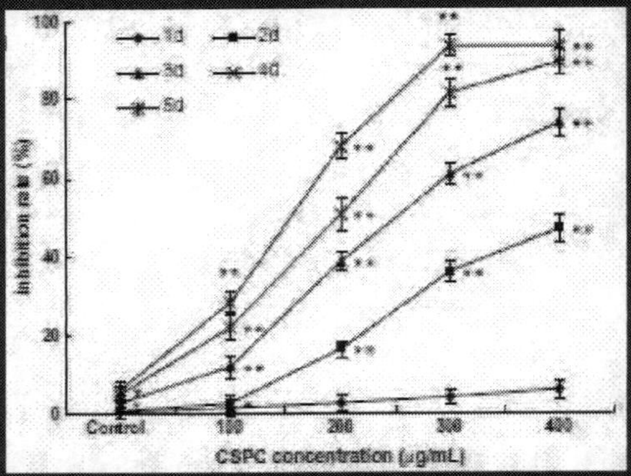

- Human Hepatoma G2 Cells
- Inhibition Rate: (Control – Treated / Control)*100
- Inhibition Rates Dose-Dependent
- Inhibition Time Dependent

NSF

Stony Brook University

H. Zhang, J. Ke, T. Shao, J. Li, Y. Duan, Y. He, C. Zhang, G. Chen, G. Sun, X. Sun. Food and Chemical Toxicology. 2014: Vol. 64, 166-176

4

Objective

Encapsulate PAC using chitosan and characterize the resulting nanoparticle as well as study the effect on MG-63 Osteosarcoma cells *in vitro*.

NSF

Stony Brook University

Why Chitosan?

- High Bioavailability
- Organic, derived from Chitin
- Biodegradable
- Hydrophilic

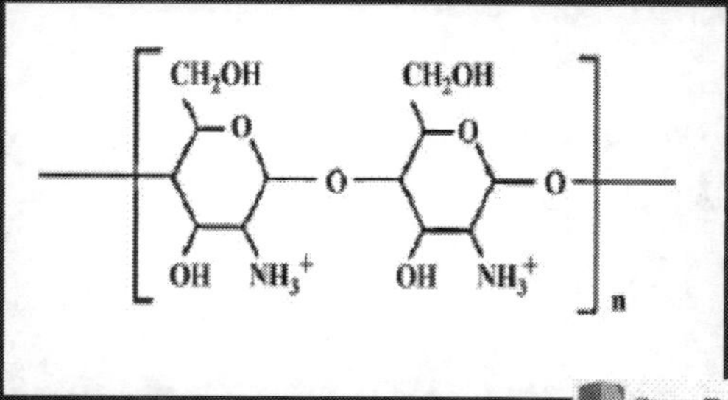

Chitosan

Encapsulation Potential

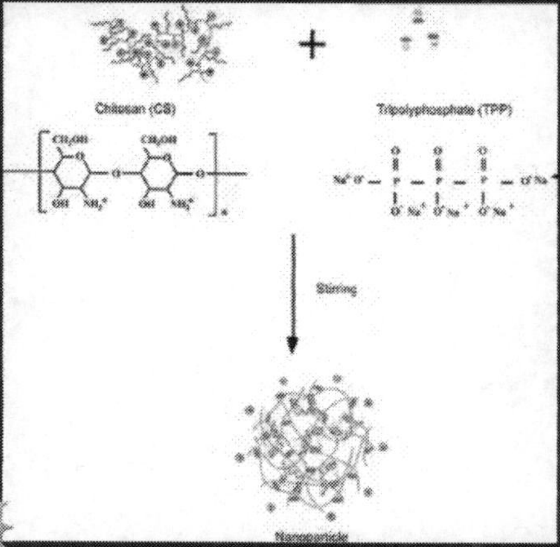

- Ionic Gelation Method
  - Pluronic F68 (detergent)
  - GCS:TPP 8.7:1 and 2.9:1
- Sizes are typically 200-300 nm

Adriana Trapani, Johannes Sitterberg, Udo Bakowsky, Thomas Kissel. International Journal of Pharmaceutics, 2009: Vol. 375, 97-106

7

Structure of Glycol Chitosan

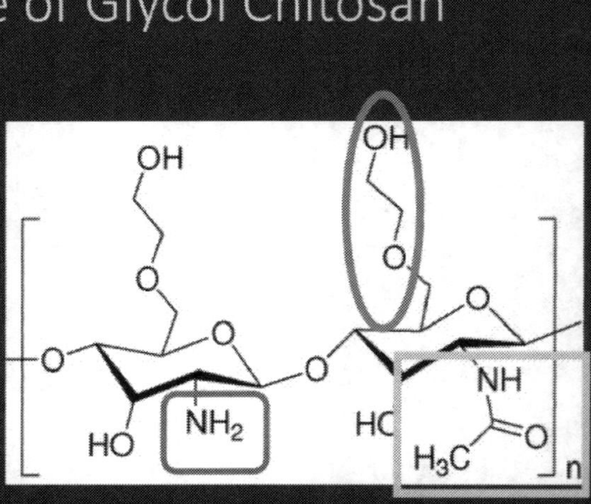

8

Approach

Synthesis

Characterization

Delivery *in vitro*

9

Fabrication and Characterization of NPs

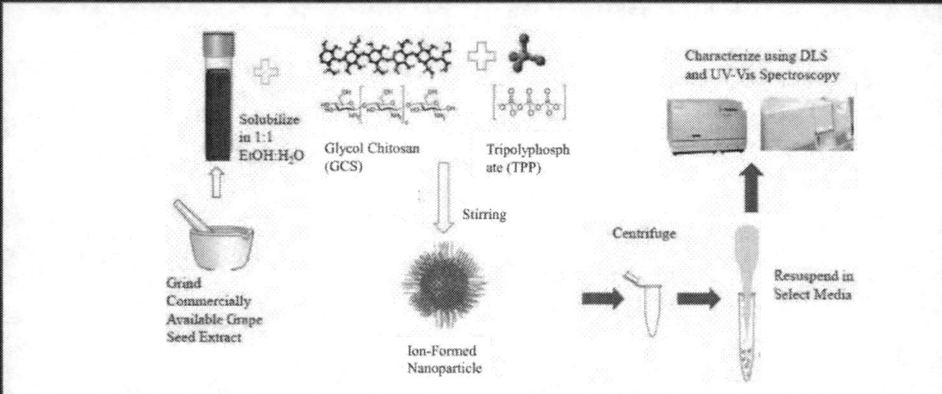

Stony Brook University

10

Nanoparticle Characterization

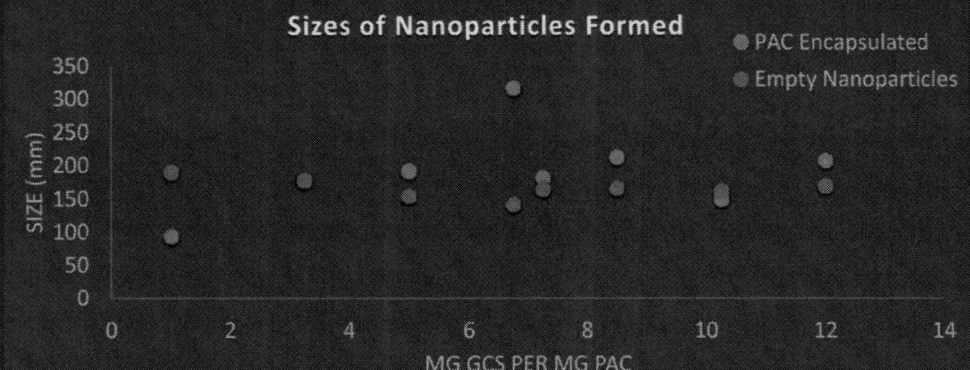

Stony Brook University

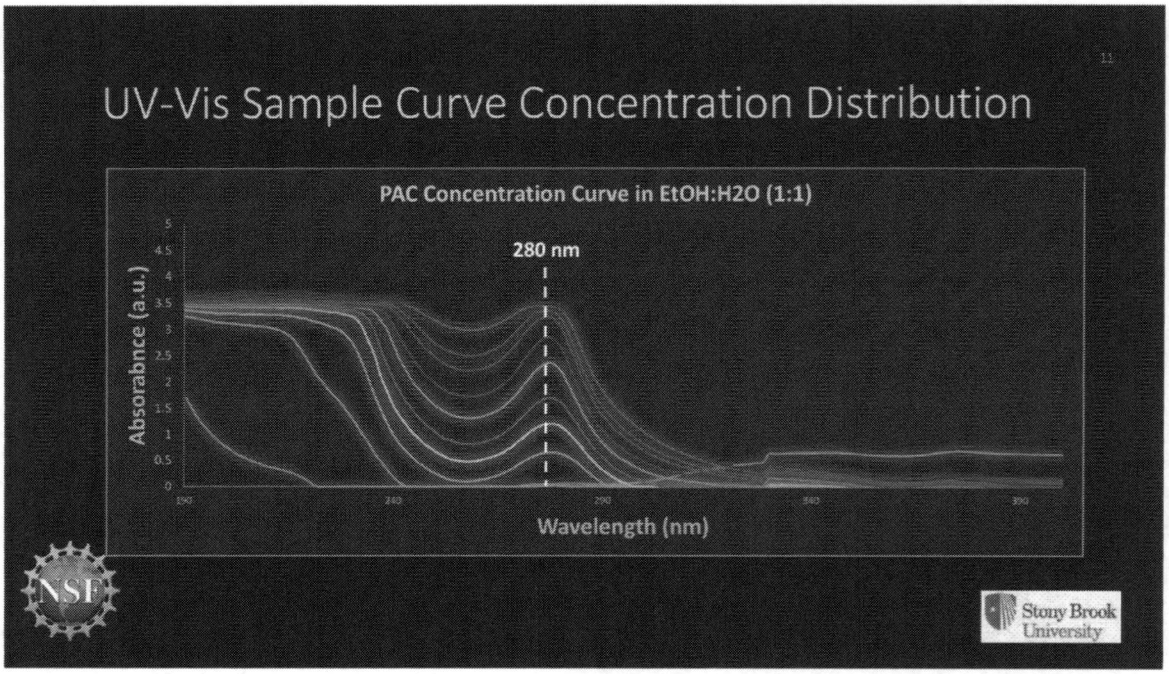
UV-Vis Sample Curve Concentration Distribution
PAC Concentration Curve in EtOH:H2O (1:1)
280 nm
Absorabnce (a.u.)
Wavelength (nm)
NSF
Stony Brook University

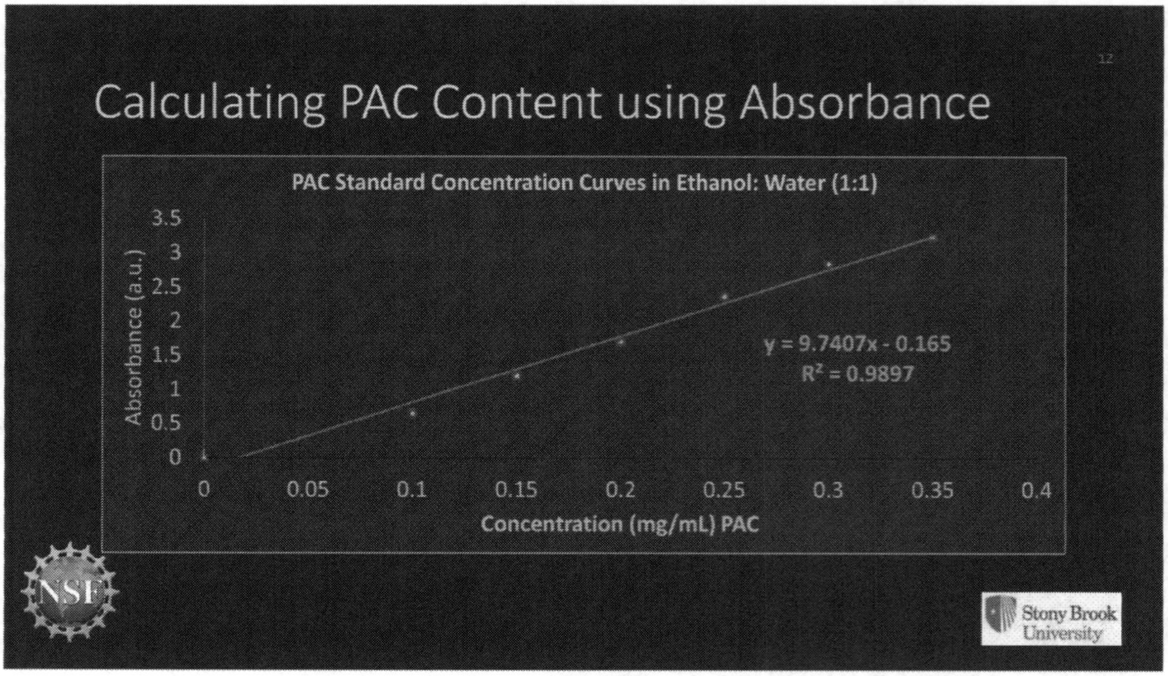
Calculating PAC Content using Absorbance
PAC Standard Concentration Curves in Ethanol: Water (1:1)
Absorbance (a.u.)
3.5
3
2.5
2
1.5
1
0.5
0
0
0.05
0.1
0.15
0.2
0.25
0.3
0.35
0.4
y = 9.7407x - 0.165
R² = 0.9897
Concentration (mg/mL) PAC
NSF
Stony Brook University

13

Encapsulation Efficiencies of GCS NPs

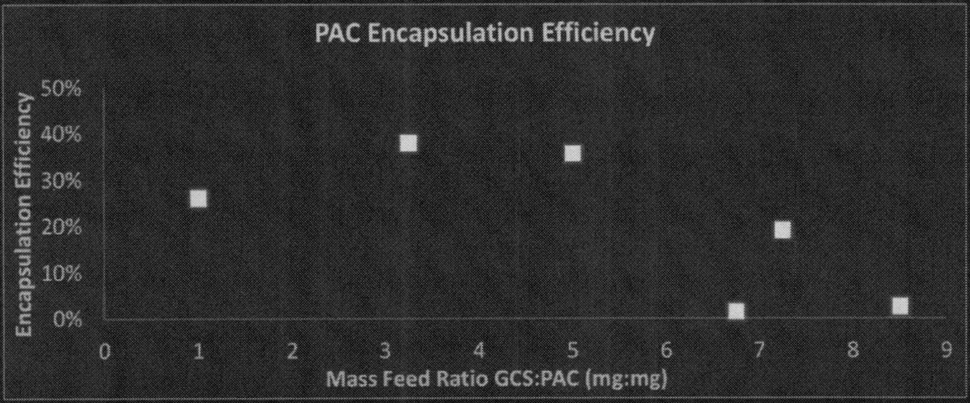

NSF

Stony Brook University

14

Delivery

- Cells used: MG-63 Human Osteosarcoma Cells
- Morphological study of cells after application
  - Comparing free and encapsulated PAC
- Cell Viability Assay done using MTS Assay
  - Comparing free and encapsulated PAC

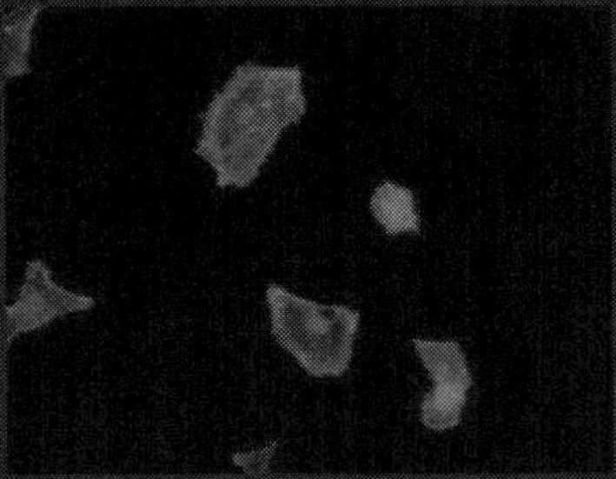

NSF

Stony Brook University

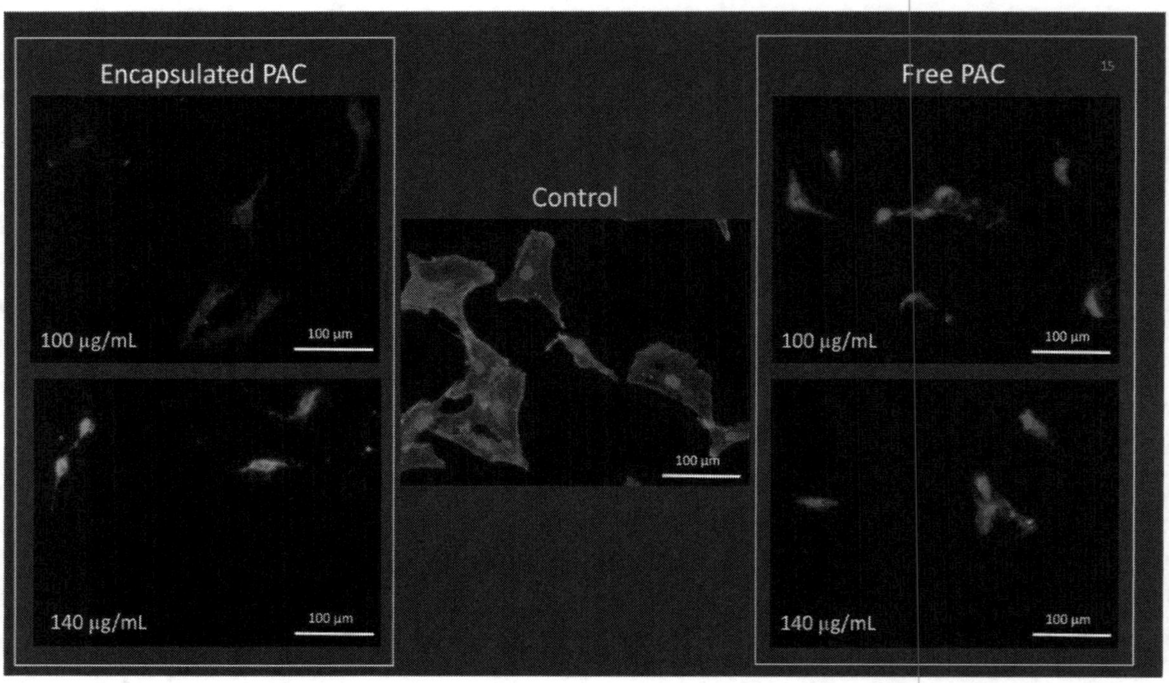

16

Cell Viability after 72 Hours

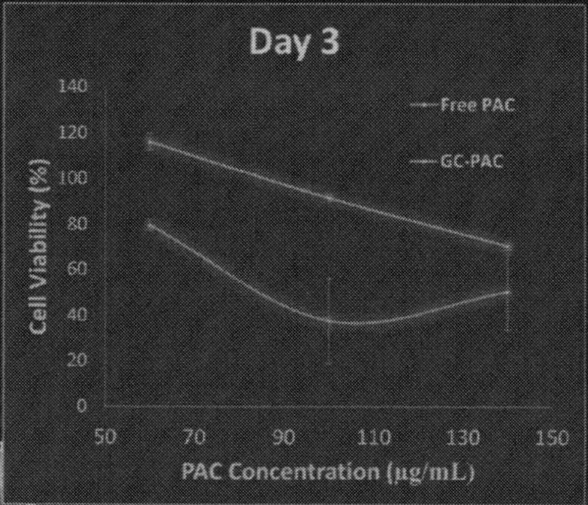

- MG-63 Osteosarcoma cells doubling time <72 hours
- Cell Viability is dose dependent
  - Can be adjusted for desired effect

Stony Brook University

Percent Cell Viability

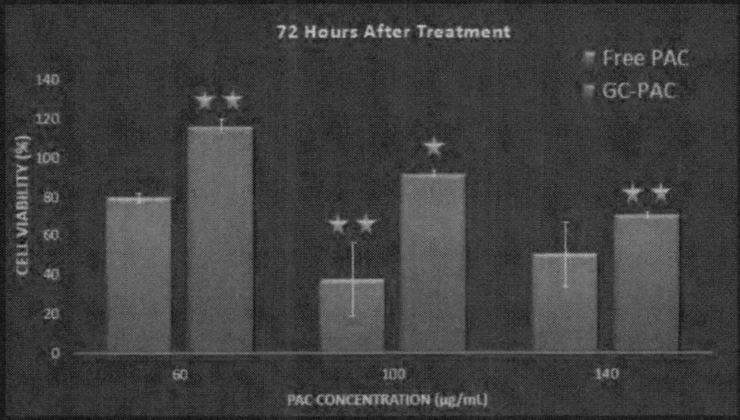

★ = P < 0.05
★★ = P < 0.01

Conclusions

- Most Efficient Mass Feed Ratios: 5 mg : 1 mg GCS:PAC
- Morphological analysis shows similar effects by both free and encapsulated PAC
- MTS Analysis indicates encapsulated PAC is advantageous over free PAC in its dose-dependency as well as effectiveness with greater amounts of PAC

Future Studies

- Release Kinetics of GCS-PAC NPs
- Amount of GCS in formed NPs

Acknowledgements

I would like to thank the National Science Foundation, Giulia Suarato, Dr. Yizhi Meng, the Center for Inclusive Education, and all other students and mentors involved in the REU for making this research possible by providing the resources necessary

NSF

Stony Brook University

394191 Single Cell Isolation Via Microfluidic-Based Droplet Delivery

Abstract:

Commercial technologies such as fluorescent-activated cell sorting (FACS) for isolating and delivering single cell samples are often inappropriate for smaller research facilities due to their size and cost. Researcher labs conducting experiments on single cells require an inexpensive and accessible method for creating samples. We demonstrate the practicability of using a gravity-driven microfluidic device under a bench-top optical trap to generate droplets containing isolated single cells for delivery onto a 96 well plate. The microfluidic device dispenses droplets of constant volume at fixed intervals. We establish functionality by delivering droplets containing glutaraldehyde fixed red blood cells (RBCs) onto glass slides. A fluorescent confocal microscope capable of imaging the glutaraldehyde-induced fluorescence of the RBCs is employed to count the number of cells contained per droplet. We attained 64% fidelity of single-cell samples, indicating that delivery from our device is not stochastic. Biomedical researchers often expect droplet delivery methods to be compatible with a 96 well plate standard; to this end we achieved automated delivery into each well of a 96 well plate using a Teensy 3.0 encoded motorized stage. Additionally, we implemented a side and forward scatter detection scheme around the optical trap to show the feasibility of fully automating the cell isolation and droplet delivery process. The entire apparatus is compact and portable enough to fit within a sterile environment to prevent contamination of samples. We propose that our setup is an inexpensive and compact alternative to flow cytometers that is accessible to most labs and can reliably sort single cells into droplets for sterile delivery.

394190 Self-Assembling Zwitterionic Nanogels As Immune Isolating Coatings for Stem Cell Derived Pancreatic Islet Transplantation

Abstract:

The concept of pancreatic islet transplantation to patients with diabetes to restore normoglycemia is a viable clinical strategy. However, widespread clinical application of islet transplantation remains limited because of the harmful side effects of the necessary immunosuppressive therapy to prevent host rejection of transplanted cells. Here we report the fabrication of a superbiocompatible nanogel and a novel coating strategy for protection of stem cell derived pancreatic islets against host rejection in vivo. Zwitterionic nanogels (100-300nm in size) were fabricated using an inverse emulsion polymerization synthesis scheme and further surface modified with CLICK reactive functional groups. Stem cell derived pancreatic islet cells were engineered to express azide functionalities on their cell membranes. Confocal fluorescence microscopy was used to verify covalent reaction and assembly of dual-layered nanogel coatings on engineered islet cell clusters. Cytocompatibility of coating and islet cell functionality were examined using a combination of in vitro glucose tolerance tests, live/dead staining, and in vivo studies on mice treated with streptozocin, a small molecule drug that destroys the beta cells. Our results indicate that the developed nanogel formulation and encapsulation strategy provides a viable strategy for prolonged protection and isolation of transplanted islets against host rejection

394187 Modeling of Selenium Nanoparticle Deposition for Optimized Production of Antibacterial Surfaces

Abstract:

The Webster Nanomedicine lab designs and explores applications of nanomaterials to inhibit bacterial growth and cancer cell responses. Due to its beneficial effects on the body, and known chemopreventive and chemotherapeutic properties[1], selenium has been used in its nano-form to impart nano-roughness and chemical effects in these experiments. To ensure precise and consistent selenium nanoparticle (SeNP) coverage, we aim to tabulate several relevant aspects of the particle. By adjusting the parameters of the reaction producing SeNP, a variety of data can be collected and analyzed. Characteristics that are pertinent include the nanoparticle size, coverage, surface energy, and roughness. Eventually, a cubic response surface model will be developed using a central composite design (CCD), which will be able to predict the interaction of the particles with bacteria and cells.

SeNP are produced through a reaction of glutathione (GSH) with sodium selenite (Na_2SeO_3). The reactants are introduced in a 4:1 molar ratio, respectively. Thin discs of poly(l-lactic acid) (PLLA) are added to the reaction, providing a surface for the cells to grow. In order to precipitate out the SeNP, 2M NaOH is added to the mixture. The reaction is then halted by the addition of DiH_2O. The experimental design used four factors: time in GSH/ Na_2SeO_3, time in NaOH, reaction volume, and the number of samples in the volume. Three results are compared: nanoparticle size, nanoparticle coverage, and surface energy of each sample. Both the size and the coverage of the produced nanoparticles can be measured using scanning electron microscopy. The surface roughness is portrayed through atomic force microscopy. Surface energy can be calculated utilizing the contact angle, which is defined using a Goniometer.

Previous studies show longer time increments (in both cases) increase the deposition of SeNP on the substrate. However, as the increments reach larger values of time (30-60 seconds), there is less of a change in coverage. This suggests the existence of a saturation point. On the PLLA, there is a finite number of binding sites for the nanoparticles. The size of the nanoparticle, with time being the only deviation, remains consistent. It is predicted that development time and reaction volume directly correlate. Recent observations of previous studies have shown that increased coverage correlates with development time in both steps, and therefore, roughness.

Through characterization of the SeNP product, a cubic response surface will be calculated relating the input parameters to expected results, optimizing the coverage of nanoparticles on substrates. These efforts conserve both time and resources because reactions can be designed with focused specifications.

Quality-by-Control (QbC) Technologies for Batch and Continuous Crystallization Systems

Zoltan K Nagy

Purdue University

The advent of process analytical technologies (PAT) more than a decade ago has brought the applications of advanced control of pharmaceutical manufacturing processes in the realm of possibility. The presentation will review key challenges and recent developments in the measurement, modeling and control of crystallization processes. The role of advanced feedback control concepts in improving the performance of batch processes, as well as enabling technologies in the paradigm shift from batch to continuous manufacturing will be corroborated. The benefits of modern model-free and model-based feedback control approaches will be exemplified in the case of batch and continuous pharmaceutical crystallization systems. Recent advances in the monitoring, modeling and control of crystalline product properties, such as purity, size and shape distribution as well as polymorphic form will be corroborated with industrially relevant examples. Novel control approaches including aspect ratio control (ARC), competitive purity control (CPC) and active polymorphic feedback control (APFC) will be described. These approaches use tailored additives, optimally designed seed recipe and dynamic control of crystallization trajectories in the phase diagram to produce crystalline materials with tailored maid properties with excellent consistency. The role of rigorous mathematical modelling in the systematic optimal design, start-up and control of continuous crystallization processes to achieve and maintain the desired controlled stage of operation will be illustrated through examples for both cascade of mixed suspensions mixed product removal (MSMPR) as well as plug flow crystallization systems. Anti-fouling control (AFC) will be presented as an illustrative example demonstrating how active feedback control can be the key enabling technology for the implementation and smooth operation of continuous crystallization processes. A novel view for implementing continuous drug manufacturing processes that are based on spatially distributed control of various particle formation processes within a single continuous plug flow system, rather than the integration of individual unit operations, will be presented. The system uses spatially distributed antisolvent addition, heating/cooling cycles and/or binder/excipients addition to control the individual mechanisms of nucleation, growth and spherical agglomeration to produce particles with controlled internal and external size and micromeritic properties. The talk intends to provide motivating examples of the next stage of innovation in pharmaceutical manufacturing, illustrating the potential benefits of the new quality-by-control (QbC) framework in improving product quality and process efficiency while reducing costs and time-to-market.

The material properties and performance of crystalline inorganic solids are often determined by the planes that are exposed on the crystal surface, e.g., catalysts, PV materials, nonlinear optical materials, etc. Currently, the protocols for synthesizing (and hence manufacturing) these materials are invented by chemical intuition coupled with trial-and-error experiments. It would be helpful to have a computer-assisted method whereby "experiments" could be carried out rapidly in a computer so that a short list of potentially successful experiments could be identified and tested in the laboratory. As a step towards creating such an approach, we have developed a mechanistic model for predicting the growth of inorganic crystals grown from solution. The model is based on the classical spiral growth mechanism of Burton, Cabrera and Frank [1].

The concept of a building unit of a periodic bond chain (PBC) is applied to simplify the identification of the PBC directions in the bulk crystal using the Hartman-Perdok theory [2]. The long-range nature of the electrostatic interactions necessitates a Madelung-type approach for the calculation of the interaction energies along the PBC directions. The variation in the electronic structure of the atoms in the surface layer is captured using the bond valence method [3], that can predict the partial charges of surface atoms, provided the local surface and solvent structure around the atoms is known. A space partitioning method is applied to sum up the 3D electrostatic interactions that involve non-bulk partial charges of the growth units situated in the kink sites on a spiral edge.

The step velocity of a spiral edge depends on the density of kink sites present along the edge and the net rate of incorporation in those kink sites [4]. A general expression is presented here to calculate the kink incorporation rate along a spiral edge on an AB-type crystal surface containing cations and anions in various orientations. The attachment and detachment fluxes from a kink site depend on experimental/engineering deign parameters (e.g., supersaturation, ionic activity ratio) and potential energy barriers, that are calculated from kink site potential energies.

The growth model successfully predicts the dependence of the step velocity on supersaturation and ionic activity ratio for the spiral edges on the (10-14) surface of calcite crystals, as compared to the *in situ* AFM measurements reported in the literature [5,6]. This high-fidelity crystal growth model can be applied to a wide class of inorganic crystals grown from solution, such as barium sulfate, titanium dioxide, potassium dihydrogen phosphate (KDP), etc. Ultimately, the model will provide a tool for predicting crystal morphology so that suitable growth media can be designed (e.g., solvent(s), additives, supersaturation) to deliver inorganic crystals of desired shapes depending on their applications.

References

1. Burton, W. K., Cabrera, N., and Frank, F. C. *Phil. Trans. Roy. Soc. A* **243**(866), 299-358 (1951).

2. Hartman, P. and Perdok, W. G. *Acta Crystallogr.* **8**(1), 49-52 (1955).

3. Brown, I. D. *The Chemical Bond in Inorganic Chemistry: The Bond Valence Model.* Oxford University Press, (2002).

4. Kuvadia, Z. B. and Doherty, M. F. *Cryst. Growth Des.* **11**, 2780-2802 (2011).

5. Teng, H., Dove, P. M., and De Yoreo, J. J. *Geochim. Cosmochim. Acta* **64**(13), 2255-2266 (2000).

6. Stack, A. G. and Grantham, M. C. *Cryst. Growth Des.* **10**(3), 1409-1413 (2010).

391766 Structure and Properties of Aqueous Methylcellulose Solutions and Gels

Abstract:

Cellulose represents the most abundant polymer across the planet. This polysaccharide can be chemically modified to impart solubility in water resulting in a plethora of commercially important products ranging from construction materials to delivery vehicles for pharmaceuticals. Substitution of a fraction of the hydroxyl groups on the native glucose rings with alkoxy moieties leads to the versatile class of polymeric materials known as cellulose ethers. Methylcellulose is especially important owing to a lower critical solution temperature (LCST) when mixed with water leading to gelation at elevated temperatures at relatively low concentrations. This nontoxic, biodegradable material, generally recognized as safe by the US Food and Drug Administration, finds widespread uses in food applications and other commodity products. Only recently has the mechanism of gelation been elucidated, based on cryogenic transmission electron microscopy (cryo-TEM), small-angle neutron scattering (SANS), and dynamic mechanical spectroscopy (DMS) measurements. This lecture will describe the discovery of a remarkable fibrillar structure that accompanies gelation and the associated temperature dependent viscoelastic properties that make methylcellulose so useful.

371668 Energy Efficient Gas Separations: Engineering Problems and Solutions

Abstract:

In principle, advanced membrane approaches that minimize energy intensive phase change-driven separations offer an order of magnitude reduction in energy intensity per unit of product purified. Implementing such approaches, however, requires coupling advanced materials and advanced processing to control costs of the final devices. Polymer-derived membranes based on crosslinked glassy polymers, polymer-inorganic nanoparticle hybrids and carbon molecular sieves are particularly attractive materials that cover different areas in the broad separation landscape. Examples of these new materials, devices based on them, as well as their manufacturing and the savings they enable will be discussed for high impact separations cases.

391911 Confinement, Electron Transfer and Lewis Acidity: Growing Facets of Zeolite Catalysis

Abstract:

Most of the zeolite research in the area of catalysis centers on the acid/base properties of zeolites and on the catalytic properties of extra-framework cations for redox and hydrogenation/dehydrogenation reactions on zeolites with, for example, Fe/Cu or Ga/Zn in extra-framework positions. Yet there are other zeolite properties that can be exploited to carry out catalytic reactions. This talk will describe recent investigations in our laboratory that highlight specific instances of catalysis by confinement and electron transfer in reactions such as NO oxidation and alkane activation. We will also discuss the catalytic properties of beta zeolites containing titanium, tin, zirconium and hafnium coordinated to the zeolite framework. This class of materials, having Lewis acidic properties but no Bronsted acids, has been investigated for the *Meerwein–Pondorf–Verley* reduction of carbonyl compounds and for the Oppenauer oxidation of alcohols. We will show that these materials are also effective catalysts for other 'classical' Lewis-catalyzed reactions such as the Diels-Alder reaction.

370231 Double Membrane Redox Flow Batteries for Economical and Efficient Renewable Electricity Storage

Abstract:

Redox flow batteries (RFBs) are considered as the most promising technology for economical and scalable electricity storage for wind and solar energy. By decoupling the energy storage and the power delivery from each other, RFBs hold a level of design flexibility and system scalability unavailable to traditional rechargeable batteries. Since their invention, RFBs have stayed with single ion-exchange membrane (single-IEM) configuration that cannot effectively use redox pairs with mixed charges and/or supporting electrolytes with different pHs. In this work, we show a universal design concept based on a double-IEM configuration that brings unprecedented freedom in choosing both redox pairs and supporting electrolytes. Two aqueous RFB examples are featured: (1) ultra-high voltage zinc-cerium RFB with a standard cell voltage of 3.08 V, the highest among all known aqueous RFBs; and (2) ultra-low cost sulfur-iron RFB with a standard cell voltage of 1.22 V, based on two highly available electroactive elements (iron and sulfur being the 1st and 5th most produced elements worldwide, respectively). The double-IEM concept is general and can be applied to non-aqueous RFBs.

Multiscale Systems Engineering for the Discovery of Cost-effective Zeolites for Natural Gas and Biogas Purification

Christodoulos A. Floudas, Eric L. First, and M. M. Faruque Hasan

Department of Chemical and Biological Engineering, Princeton University, Princeton, NJ 08544, USA

Significant methane sources in the U.S. are undeveloped due to the high cost of separating CO_2. These sources include natural gas, with 10% of U.S. reserves [1], or 30 trillion cubic feet [2], requiring purification to reduce CO_2 content to meet pipeline specifications. Shale gas and coalbed methane are other major fixed sources of methane that often require CO_2 removal. Biogas, which is released by landfills and animal waste, is primarily a mixture of methane and CO_2. In addition to being a major source of greenhouse gas emissions, biogas is a continuously-generated, untapped energy resource.

To take advantage of these opportunities for enhanced domestic energy production, cost-effective technologies are needed for methane/CO_2 separation. Pressure swing adsorption (PSA) is a promising option that utilizes a microporous sorbent, such as a zeolite or metal-organic framework, to selectively capture CO_2 from methane. We have developed a novel computational framework to identify the best zeolites, optimal PSA process topology, and operating conditions to minimize the total separation cost, including investment, operating, and material costs [3]. The importance of a combined material screening and process optimization approach was demonstrated previously for CO_2/N_2 separation [4].

Our multiscale methodology utilizes zeolite pore characterizations from ZEOMICS [5] to calculate metrics including shape, size, pore, and adsorption selectivities [6–9]. A detailed nonlinear algebraic and partial differential equation (NAPDE) model [10] is used to simulate the temperature, pressure, and composition profiles throughout the three-step PSA cycle. We employ an efficient Kriging-based grey-box constrained optimization approach [3,4,10,11] to find the minimum-cost solution to the complex NAPDE model.

We have considered a variety of feed compositions representative of methane sources with CO_2 content ranging from 5–50%. Constraints for 97% purity and 95% recovery are imposed to meet pipeline specifications and minimize losses. Eight novel zeolites are identified for economically separating CO_2 from methane, including the top zeolite WEI, which has a total separation cost ranging from $0.15/MMBTU for 5% CO_2 sources to $1.44/MMBTU for 50% CO_2 sources. Given the latest natural gas Henry Hub price of $4.90 (March 2014) [12], the proposed technology may be highly desirable for developing previously uneconomical energy sources.

References:

[1] Baker RW, Lokhandwala K. Natural gas processing with membranes: an overview. Ind Eng Chem Res. 2008;47(7):2109–2121.

[2] U.S. Energy Information Administration. Proved Reserves, Reserves Changes, and Production. 2013. http://www.eia.gov/naturalgas/data.cfm. Accessed on December 20, 2013.

[3] First EL, Hasan MMF, Floudas CA. Discovery of novel zeolites for natural gas purification through combined material screening and process optimization. AIChE J. 2014;60(5):1767–1785.

[4] Hasan MMF, First EL, Floudas CA. Cost-effective CO_2 capture based on in silico screening of zeolites and process optimization. Phys Chem Chem Phys. 2013;15(40):17601–17618.

[5] First EL, Gounaris CE, Wei J, Floudas CA. Computational characterization of zeolite porous networks: an automated approach. Phys. Chem. Chem. Phys. 2011;13(38):17339–17358.

[6] Gounaris CE, Floudas CA, Wei J. Rational design of shape selective separation and catalysis–I: Concepts and analysis. Chem. Eng. Sci. 2006;61(24):7933–7948.

[7] Gounaris CE, Wei, J, Floudas CA. Rational design of shape selective separation and catalysis–II: Mathematical model and computational studies. Chem. Eng. Sci. 2006;61(24):7949–7962.

[8] Gounaris CE, Wei J, Floudas CA, Ranjan R, Tsapatsis M. Rational design of shape selective separations and catalysis: Lattice relaxation and effective aperture size. AIChE J. 2009;56(3):611–632.

[9] First EL, Gounaris CE, Floudas CA. Predictive framework for shape-selective separations in three-dimensional zeolites and metal-organic frameworks. Langmuir. 2013;29(18):5599–5608.

[10] Hasan MMF, Baliban RC, Elia JA, Floudas CA. Modeling, simulation, and optimization of postcombustion CO_2 capture for variable feed concentration and flow rate. 2. Pressure swing adsorption and vacuum swing adsorption processes. Ind. Eng. Chem. Res. 2012;51(48):15665–15682.

[11] Hasan MMF, Boukouvala F, First EL, Floudas, CA. Nationwide, Regional, and Statewide CO_2 Capture, Utilization, and Sequestration Supply Chain Network Optimization. Ind. Eng. Chem. Res. 2014;53(18):7489–7506.

[12] U.S. Energy Information Administration. Natural Gas Spot and Futures Prices. 2014. http://www.eia.gov/dnav/ng/ng_pri_fut_s1_m.htm. Accessed on May 16, 2014.

In 2003, we introduced the idea of using exfoliated zeolite layers for thin film and nanocomposite applications.[1] For such applications, a high-aspect ratio, preservation of crystal (and pore) structure, thickness and aspect ratio uniformity, and elimination of amorphous and non-exfoliated contaminants are of paramount importance. These properties could not be fulfilled by the previously reported, solution-based, exfoliation methods and require an exfoliation methodology based on melt blending with a polymer, which ensured structure preservation of the exfoliated layers.[2] Removal of the polymer and purification of the 2-D zeolites from non-exfoliated particles turned out to be a challenging problem, which was solved by the development of gradient centrifugation methods combined with selection of appropriate solvents. With this approach, single-unit-cell thick zeolite nanosheets with intact crystal and micropore structure were shown to be possible.[3, 4, 5, 6] The structural integrity and unprecedented purity and uniformity of these microporous nanosheets, open exciting possibilities for technological breakthroughs in molecular sieve membrane fabrication, synthesis of hierarchical catalysts and polymer-zeolite nanocomposites. Moreover, zeolite nanosheets enable for the first time zeolite pore mouth adsorption and catalysis to be studied by traditional uptake methods as well as surface science techniques. However, along with the exciting possibilities, challenges abound. For example, the in-plane dimensions of the existing nanosheets are in the sub-micrometer range limiting potential applications and processability as thin films. Moreover, the two exfoliated zeolites currently available are only a small fraction of zeolite topologies one would like to have available for a representative set. Earlier attempts to exfoliate other layered zeolites, including certain layered silicates and aluminophosphates with microporous layers, did not preserve the crystallographic order of the layers. Synthesis of high aspect ratio zeolite and other crystalline nanoporous nanosheets, methods to characterize their structure and properties, along with their processing and assembly to create functional devices will be the focus of this talk.

---

[1] Jeong H, Nair S, Vogt T, Dickinson L, Tsapatsis M. A highly crystalline layered silicate with three-dimensionally microporous layers. *Nature Materials*. 2003;2(1):53-58.
http://dx.doi.org/10.1038/nmat795

[2] Maheshwari S, Jordan E, Kumar S, Bates FS, Penn RL, Shantz DF, Tsapatsis M. Layer structure preservation during swelling, pillaring, and exfoliation of a zeolite precursor. *J Am Chem Soc*. 2008;130(4):1507-1516.
http://dx.doi.org/10.1021/ja077711i

[3] Xu D, Swindlehurst GR, Wu H, Olson DH, Zhang X, Tsapatsis M. On the synthesis and adsorption properties of single-unit-cell hierarchical zeolites made by rotational intergrowths. *Advanced Functional Materials*. 2014;24(2):201-208.
http://dx.doi.org/10.1002/adfm.201301975

[4] Zhang X, Liu D, Xu D, Asahina S, Cychosz KA, Agrawal KV, Al Wahedi Y, Bhan A, Al Hashimi S, Terasaki O, Thommes M, Tsapatsis M. Synthesis of self-pillared zeolite nanosheets by repetitive branching. *Science*. 2012;336(6089):1684-1687.
http://dx.doi.org/10.1126/science.1221111

[5] Agrawal KV, Topuz B, Jiang Z, Nguenkam K, Elyassi B, Francis LF, Tsapatsis M, Navarro M. Solution-processable exfoliated zeolite nanosheets purified by density gradient centrifugation. *AIChE J*. 2013;59(9):3459-3467.
http://dx.doi.org/10.1002/aic.14099

[6] Varoon K, Zhang X, Elyassi B, Brewer DD, Gettel M, Kumar S, Lee JA, Maheshwari S, Mittal A, Sung C, Cococcioni M, Francis LF, McCormick AV, Mkhoyan KA, Tsapatsis M. Dispersible exfoliated zeolite nanosheets and their application as a selective membrane. *Science*. 2011;334(6052):72-75.
http://dx.doi.org/10.1126/science.1208891

369131 New Membranes for CO2 Separation/Capture and Water Purification

Abstract:

This presentation covers two types of new membranes: (1) carbon dioxide-selective membranes for hydrogen purification for fuel cells and carbon capture from flue gas and (2) interfacially polymerized reverse osmosis membranes for water purification. We have synthesized CO_2-selective membranes by incorporating amino groups into polymer networks. The membranes have shown high CO_2 permeability and selectivity vs. hydrogen, carbon monoxide and nitrogen up to 180°C. We have elucidated the effect of amine steric hindrance in the solid membrane, showing significant enhancement for CO_2 transport. Hydrogen sulfide permeates through the membrane much faster than CO_2, allowing H_2S removal in the treated synthesis gas before water-gas-shift (WGS) reaction. Our initial experiments have shown a nearly complete removal of H_2S from 50 ppm in synthesis gas to about 10 ppb in the hydrogen product. Using the membrane, we have obtained <10 ppm carbon monoxide in the hydrogen product in WGS membrane reactor experiments via CO_2 removal. The data have been in good agreement with our modeling predictions. The membrane is being scaled up for commercial development. We will also discuss a new concept of inorganic/polymer composite membranes for CO_2 capture from flue gas in coal-fired power plants for sequestration. On interfacially polymerized reverse osmosis membranes, the state-of-the-art membranes in the thin-film-composite (TFC) structure prepared by interfacial polymerization are reviewed and discussed. Recent advances have been in high flux TFC interfacially polymerized membranes for increasing water productivity and decreasing energy consumption. We have synthesized high flux membranes by incorporating a hydrophilic additive in the aqueous amine solution during interfacial polymerization, resulting in increasing water flux significantly through an additional pathway for water transport while maintaining high salt rejection via charge repulsion. The membranes synthesized for seawater desalination showed a very high flux of 1.81 m^3/m^2/day (44.4 gallons/ft^2/day (gfd)) and a salt rejection of 99.41% using 3.28% NaCl solution. The high-flux membrane was further tested using seawater from Port Hueneme, CA and exhibited a very good and stable desalination performance for 30 days. FTIR confirmed the good membrane stability. The membrane synthesized with hydrophilic additive showed significantly improved fouling resistance. Analysis using atomic force microscopy (AFM) showed a smoother membrane surface for the membrane incorporated with the hydrophilic additive.

Recent Advances in Evapoporometry, Hydrogel Membranes and Energy-Efficient RO

William B. Krantz

Department of Chemical and Biological Engineering, University of Colorado
Singapore Membrane Technology Center, Nanyang Technological University

Annual Meeting of the AIChE
Atlanta, Georgia
November 17, 2014

1

Development of Evapoporometry for Determining the Pore-Size Distribution of Membranes

1. A.R. Greenberg, W.B. Krantz, S.S. Hosseini, E. Kujundzic, A. Yeo, "Determination of Pore Size in Porous Materials by Evaporative Mass Loss," U.S. Patent Application No. US 2013/0042670 A1, Pub. Date: February 21, 2013.
2. E. Akhondi, A.G. Fane, W.B. Krantz, F. Wicaksana, "Determination of the Porosity and Pore-Size Distribution of Hollow Fibers and Tubular Shape Materials by Evaporative Mass Loss," U.S. Provisional Patent Application No. 61/329,593, filed on June 4, 2013.
3. W.B. Krantz, A.R. Greenberg, E. Kujundzic, A. Yeo, S.S. Hosseini, "Evapoporometry: A novel technique for determining the pore-size distribution of membranes," *J. Membrane Sci.* 438 (2013) 153-166.
4. S. Goh, Q. Zhang, J. Zhang, D. McDougald, W.B. Krantz , Y. Liu, A.G. Fane, "Impact of a biofouling layer on the vapor pressure driving force and performance of a membrane distillation process," *J. Membrane Sci.* 438 (2013) 140-152.
5. J.-W. Chew, W.B. Krantz, A.G. Fane, "Effect of a macromolecular- or bio-fouling layer on membrane distillation," *J. Membrane Sci.* 456 (2014) 66-76.
6. E. Akondi, F. Wicaksana, W.B. Krantz, A.G. Fane, "Evapoporometry determination of pore-size distribution and pore fouling of hollow fiber membranes," *J. Membrane Sci.* 470 (2014) 334-345.

2

Focus of this Study

- Develop simple inexpensive technique for characterizing full range of membrane pore size from 4-200 nm
- Test this technique using model flat sheet membranes with a well-defined pore size
- Apply this technique to commercial flat sheet and hollow fiber membranes

3

Operating Principle

- Kelvin effect – vapor pressure lowering owing to negative curvature of interface between wetting liquid and gas
- If saturated gas phase is maintained above membrane, the pores will drain sequentially from largest to smallest

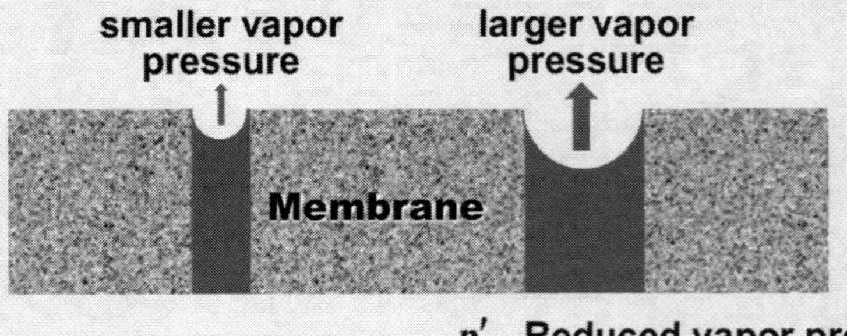

$$\frac{p'_A}{p^\circ_A} = \frac{x_A}{x^\circ_A} = e^{-\frac{4\gamma_A V_A}{dRT}}$$

Kelvin Equation

- p'_A Reduced vapor pressure
- p°_A Normal vapor pressure
- x_A Mole fraction
- γ_A Surface tension
- V_A Molar volume of liquid
- R Gas constant
- T Temperature
- d Pore diameter

4

Evapoporometry Apparatus

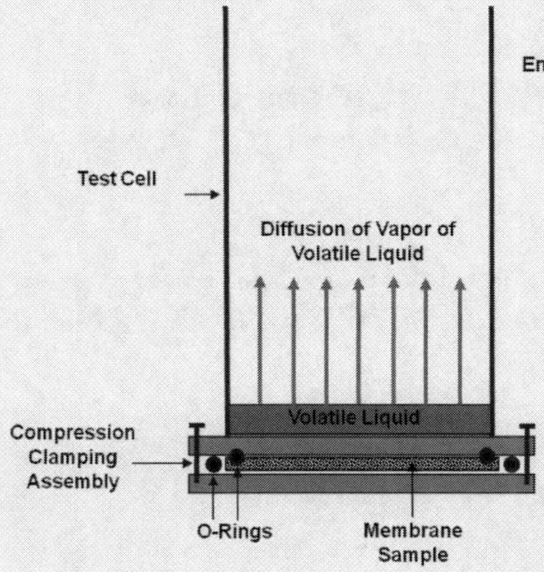

Test Cell

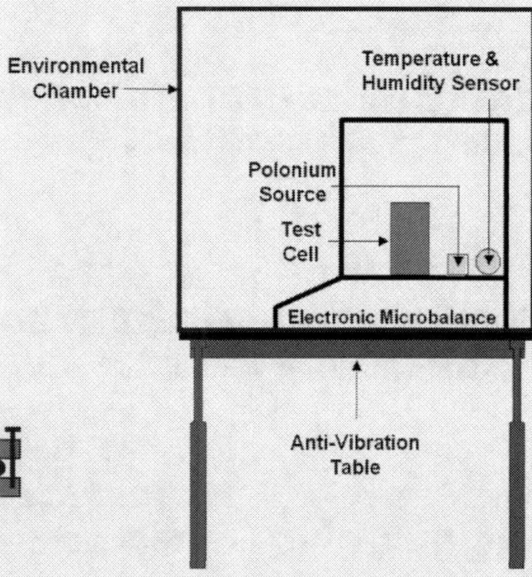

Microbalance Setup

5

Representative Evaporation Rate vs. Time Data for UF Membrane using IPA as the Volatile Wetting Liquid

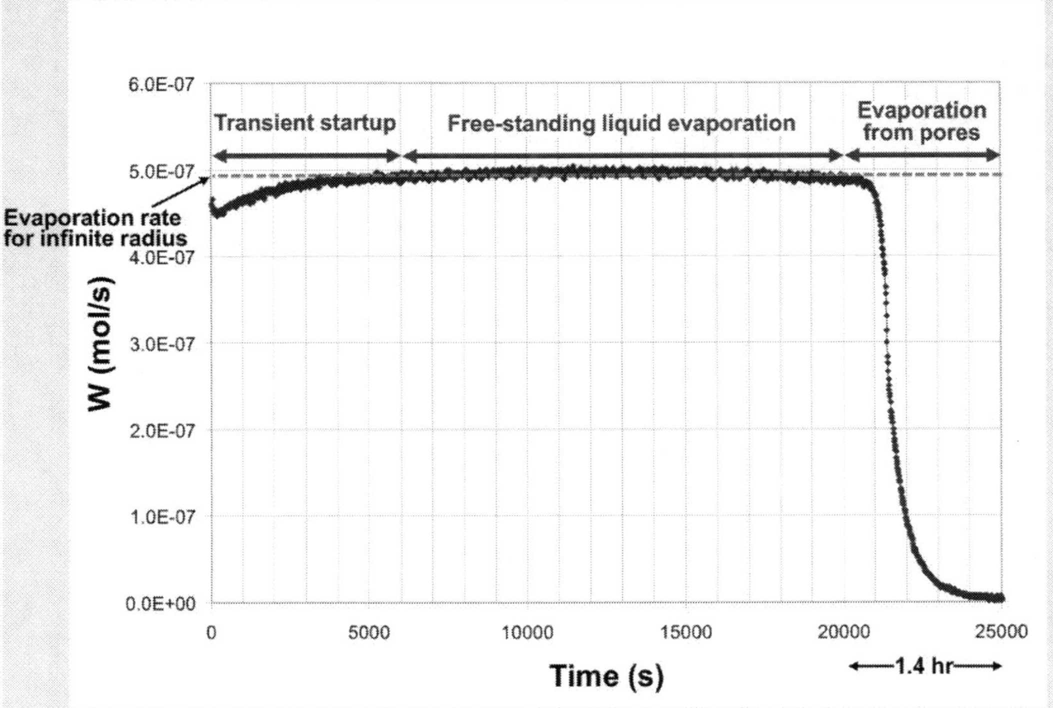

6

Data Analysis

- ➢ **Determine mass-transfer coefficient for test cell from evaporation rate of free standing liquid layer**
- ➢ **Determine vapor pressure at membrane surface from evaporation rate from porous membrane and mass-transfer coefficient**
- ➢ **Determine pore diameter from vapor-pressure depression using Kelvin equation**

7

EP Characterization of a 'Model' 100 nm Anopore® Membrane

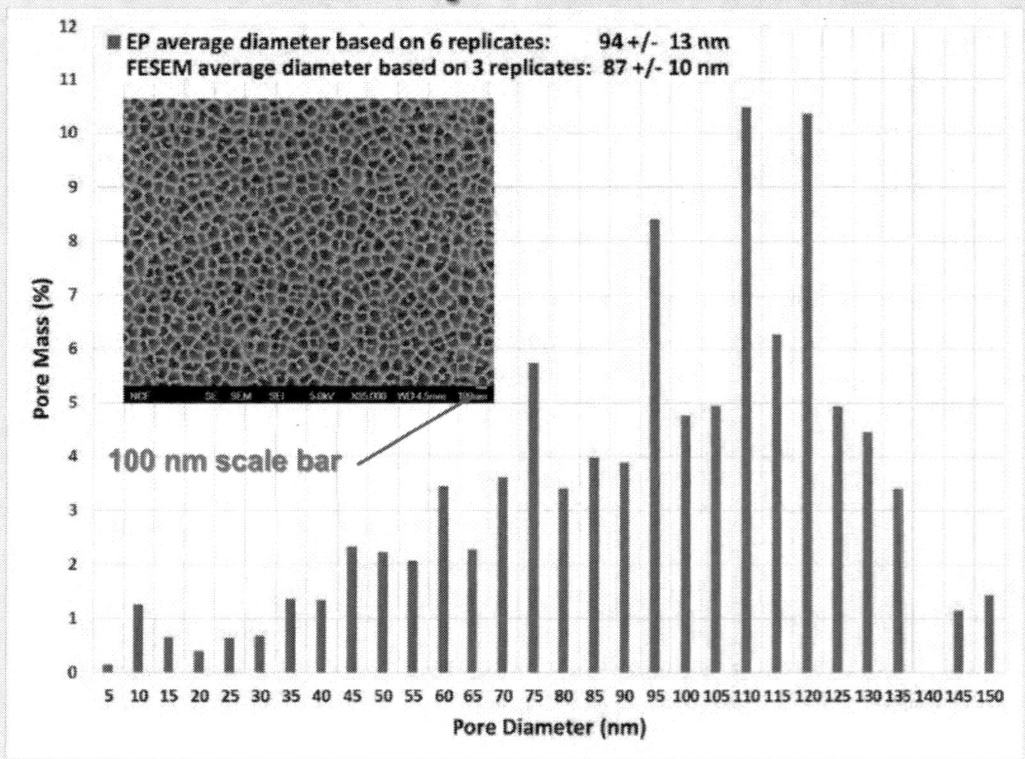

8

EP Determination of PSD for 35 nm PVDF HF Membrane – Replicate Runs Using IPA

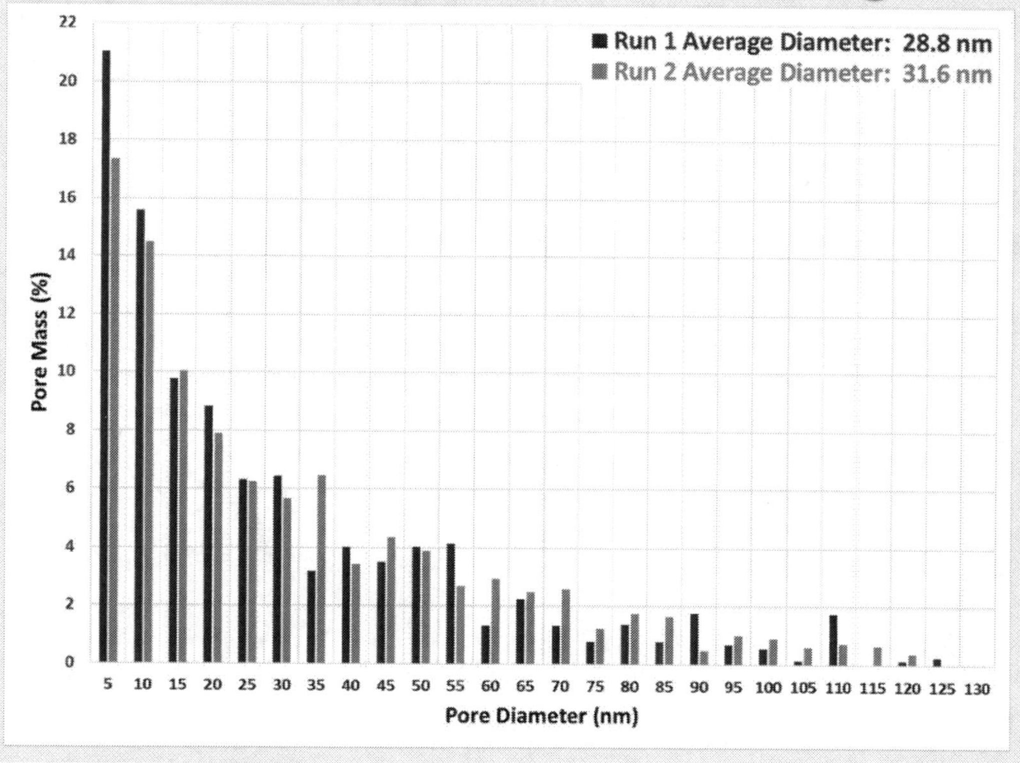

9

Average Pore Diameters Determined by EP & LDP for PVDF & PAN Membranes

PSD Method	Nominal 35 nm PVDF	Commercial PAN
Evapoporometry using IPA	30.2 ± 1.4 nm	11.8 ± 1.0 nm
Evapoporometry using H_2O	28.9 ± 0.3 nm	12.4 ± 1.9 nm
Liquid displacement porometry	24.4 ± 1.4 nm	27.2 ± 1.5 nm

- Isopropyl alcohol (IPA) and water give the same average pore diameter within the error bounds
- LDP gives smaller pore diameter for PVDF due to membrane compaction at high pressures (~30 bar)
- LDP gives larger pore diameter for PAN because it cannot measure pores smaller than 14 nm

10

Summary of Evapoporometry

- ➢ Sensitive to pore sizes in the range of 4-200 nm
- ➢ Good replicability for isopropyl alcohol and water
- ➢ Reasonably fast characterization ~ 2–2.5 hr
- ➢ Does not require specialized dedicated apparatus
- ➢ Inexpensive (< $1000) and small footprint (< 1 m^2)
- ➢ Does not damage the membrane samples
- ➢ Can characterize flat sheet & hollow fiber membranes
- ➢ Can detect internal pore fouling
- ➢ Can determine pore-size distribution of biofilms

11

Development of Novel Cryogels for Potable Water Production in Disaster-Relief Applications

1. A.G. Fane, X. Hu, W.B. Krantz, T.T. Lim, S.L. Loo, "Method of Disinfection of Drinking Water using Silver Nanoparticles Modified Cryogels," U.S. Provisional Patent Application No. 61/834,052, filed on June 12, 2013.
2. A.G. Fane, X. Hu, W.B. Krantz, T.T. Lim, S.L. Loo, "Antibacterial Cryogels and Porous Hydrogels: their Preparation Method and their Use for Disinfecting Water," U.S. Provisional Patent Application No. PCT/SG2014/000273, filed on June 12, 2014.
3. S.-L. Loo, A.G. Fane, W.B. Krantz, T.-T. Lim, X. Hu, "Emergency water supply: A review of potential technologies and selection criteria," *Water Research* 46 (2012) 3125-3146.
4. S.-L. Loo, W.B. Krantz, T.-T. Lim, A.G. Fane, X. Hu, "Design and synthesis of ice-templated PSA cryogels for water purification: Towards tailored morphology and properties," *Soft Matter* 9 (2013) 224-234.
5. Y.F. Cai, W.M. Shen, S.L. Loo, W.B. Krantz, R. Wang, A.G. Fane, X. Hu, "Towards temperature driven forward osmosis desalination using Semi-IPN hydrogels as reversible draw agents," *Water Research* 47(11) (2013) 3773-3781.
6. S.-L. Loo, A.G. Fane, T.-T. Lim, W.B. Krantz, Y.-N. Liang, X. Liu, X. Hu, "Superabsorbent cryogels decorated with silver nanoparticles as a novel water technology for point-of-use disinfection," *Environmental Science & Technology* 47 (2013) 9363-9371.
7. Loo, S.-L., T.-T. Lim, W.B. Krantz, A.G. Fane, X. Hu, "Potential evaluation and perspectives on using sponge-like superabsorbent cryogels for onsite water treatment in emergencies," accepted for publication in *Desalination and Water Treatment* (2014).
8. Loo, S.-L., T.T. Lim, W.B. Krantz, Y. Gao, A.G. Fane, X. Hu, "Ultrafast water disinfection by silver nanoparticles on porous cryogels: Bactericidal mechanisms revealed," submitted to *Environmental Science and Technology* (2014).

12

Providing Potable Water for Disaster Relief

SMTC
Member of the NEWRI Ecosystem

- Providing potable water in sufficient quantity is one of the first priorities in the aftermath of disasters
- Challenging due to lack of infrastructure and resources

13

Focus of this Study

SMTC
Member of the NEWRI Ecosystem

To prepare PSA/Ag hydrogels (cryogels) having:

- Fast and substantial water sorption
- Good water recovery via simple compression
- Effective antibacterial properties
- Ability to sustain repeated sorption-compression cycles

14

Synthesis of Microporous PSA Cryogels

SMTC
Member of the NEWRI Ecosystem

Synthesis of poly(sodium acrylate) (PSA) Cryogels

$[\text{-CH}_2\text{-CH(COONa)-}]_n$

Monomer (SA)	*Crosslinker (MBA)*	*Initiator (APS)*	*Activator (TEMED)*
sodium alginate	*methylenebis(acrylamide)*	*ammonium persulfate*	*tetramethylethylenediamine*

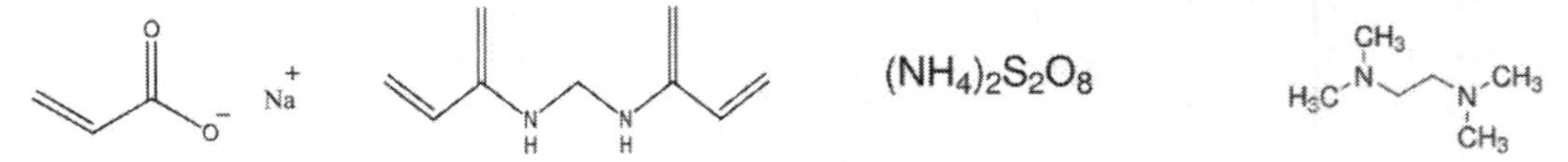

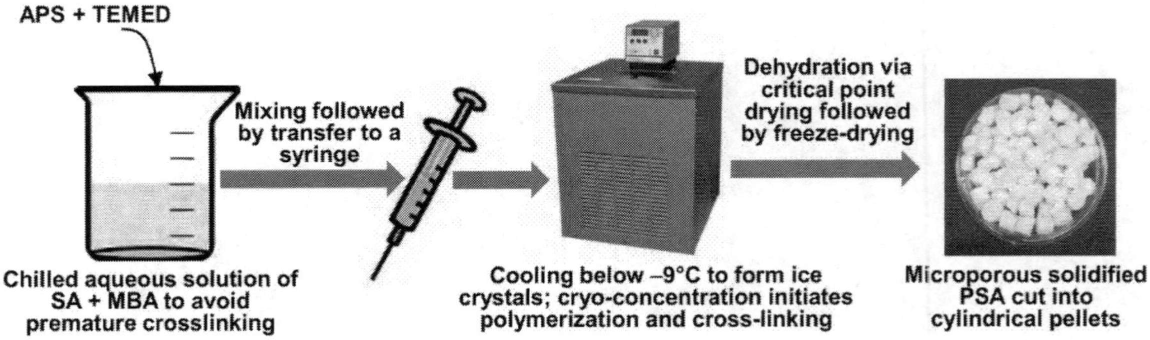

15

Impregnation of Cryogels with Silver NPs

SMTC
Member of the NEWRI Ecosystem

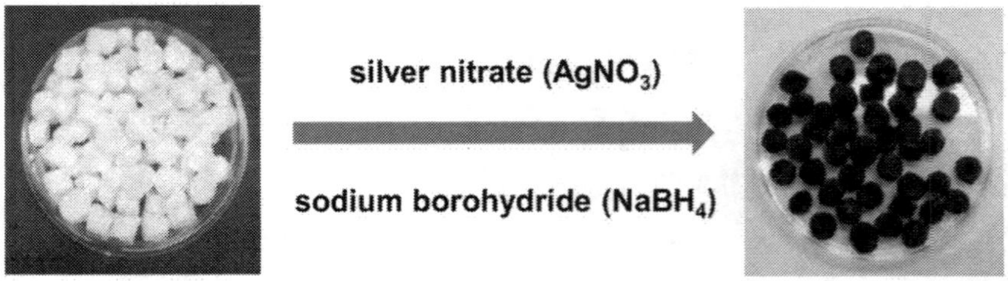

poly(sodium acrylate) cryogels | silver nanoparticle impregnated cryogels

Load PSA cryogels with aqueous solution of silver nitrate

$$\text{R–COO}^-\text{Na}^+ + \text{Ag}^+ \rightarrow \text{R–COO}^-\text{Ag}^+ + \text{Na}^+$$

Reduce silver ions sorbed within the cryogels with sodium borohydride

$$\text{R–COO}^-\text{Ag}^+ + \text{NaBH}_4 + 3\text{H}_2\text{O} \rightarrow \text{R–COO– Na}^+ + \text{Ag}^0 + \frac{7}{2}\text{H}_2 + \text{B(OH)}_3$$

16

Silver NP Impregnated PSA Cryogel Morphology

Member of the NEWRI Ecosystem

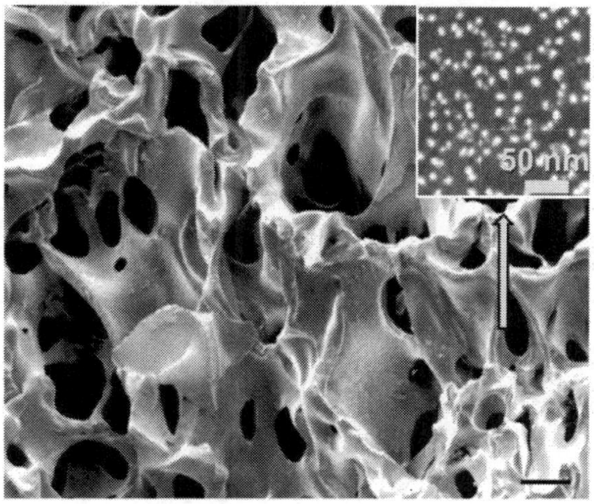

Good dispersion of the silver nanoparticles that are embedded on the pore walls of the cryogel

17

Water Sorption and Release

Member of the NEWRI Ecosystem

AgNC-X: X denotes the mg of silver per g of cryogel

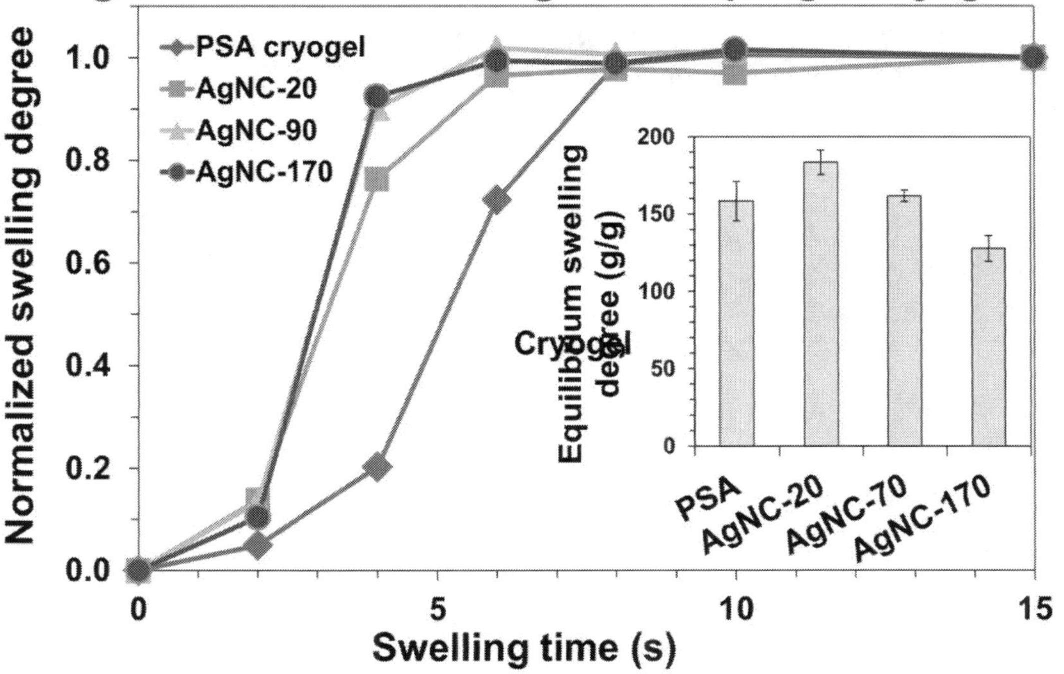

18

Bacteriological Pathogen Reduction

SMTC
Member of the NEWRI Ecosystem

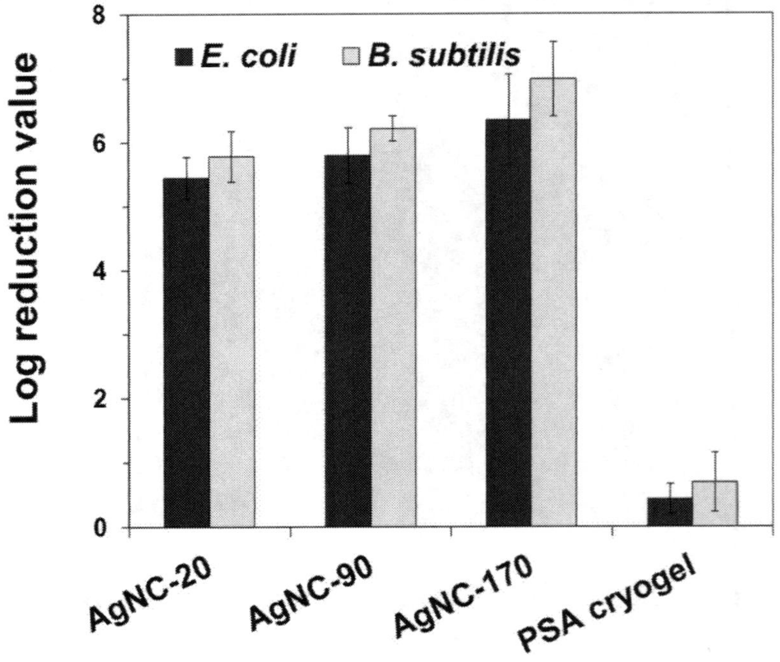

19

Summary

SMTC
Member of the NEWRI Ecosystem

- Silver nanoparticle impregnated cryogels can absorb water up to nearly 200 times their dry weight
- Reach equilibrium absorption within 7 seconds
- Release 85% of sorbed water via compression by a hand-operated device
- Six-log reduction in bacteriological pathogens
- No degradation over 100 cycles of sorption and release

20

A Novel Energy-Efficient Reverse Osmosis Desalination Process

1. T.H. Chong, S.-L. Loo, W.B. Krantz, "Energy-Efficient Reverse Osmosis (EERO) Process," U.S. Provisional Patent Application No. 61/972,718, filed on March 31, 2014.

2. Chong, T.H., S.-L. Loo, W.B. Krantz, "Energy-efficient reverse osmosis desalination process," *J. Membrane Sci.*, 473, 177-188 (2015).

3. Chong, T.H., W.B. Krantz, A.G. Fane "Novel reverse osmosis cascade design for enhanced water recovery and reduced energy demand," submitted to *Desalination* (2014).

21

3-Stage EERO Process

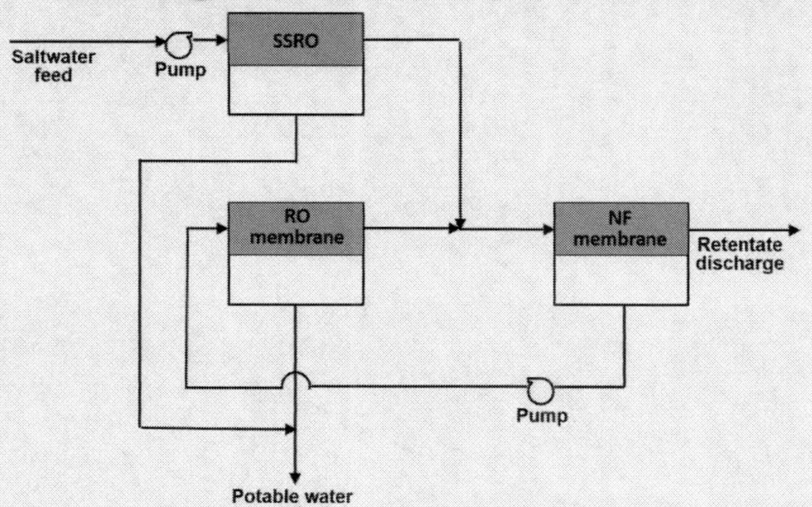

- Capitalizes on performance of single stage RO
- Employs CMCR to increase recovery and reduce pressure
- Employs NF membrane to facilitate retentate self-recycling

22

4-Stage EERO Process

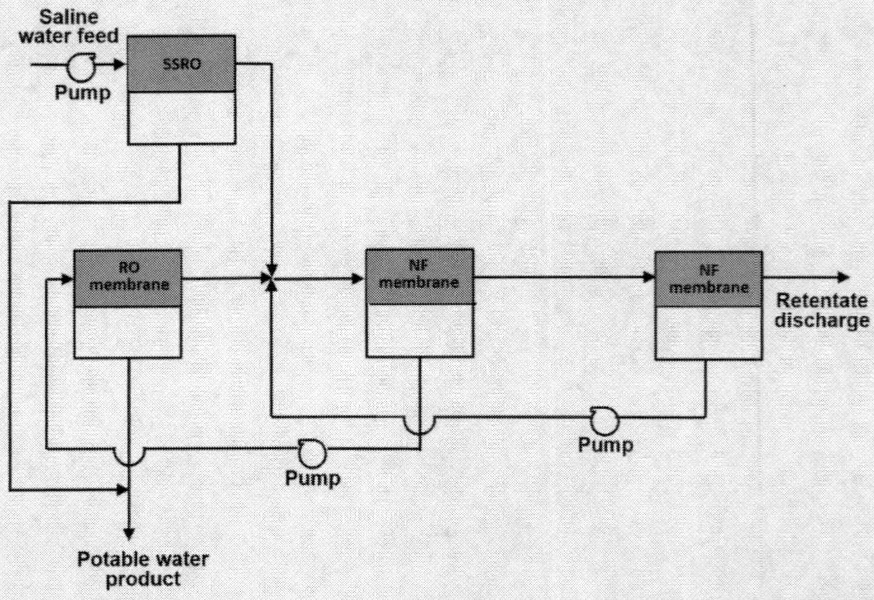

➢ **Additional stage increases recovery and reduces pressure but increases specific energy consumption (SEC)**

23

OPD for SSRO and EERO Processes

Pump Efficiency = 85%; ERD Efficiency = 90%

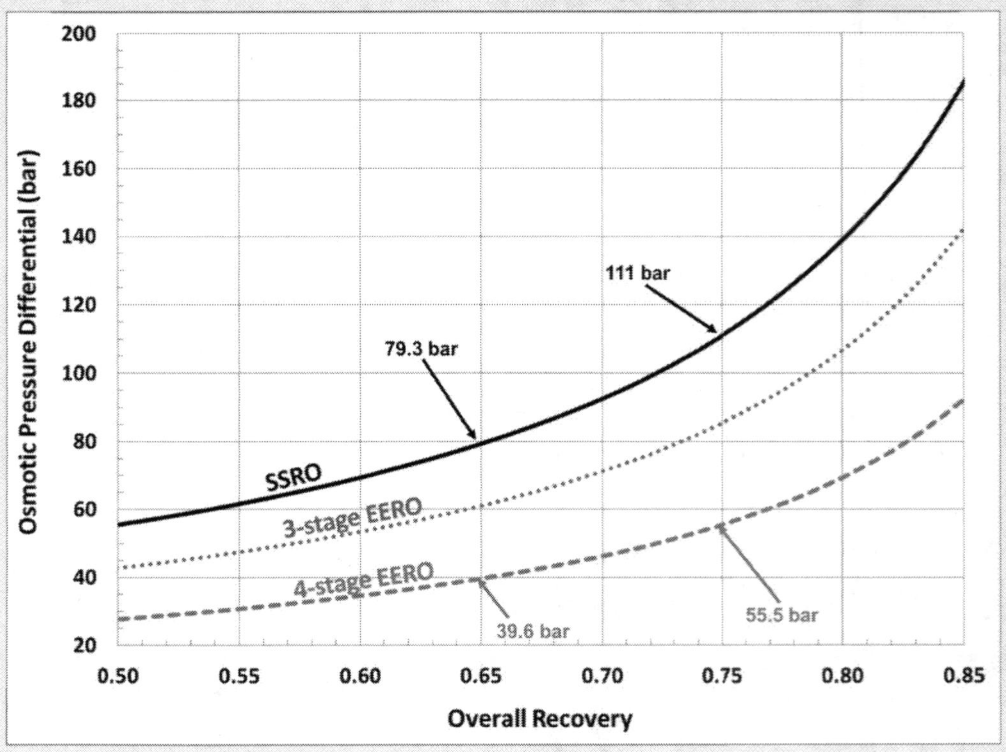

24

SEC_{net} for SSRO and EERO Processes

Pump Efficiency = 85%; ERD Efficiency = 90%

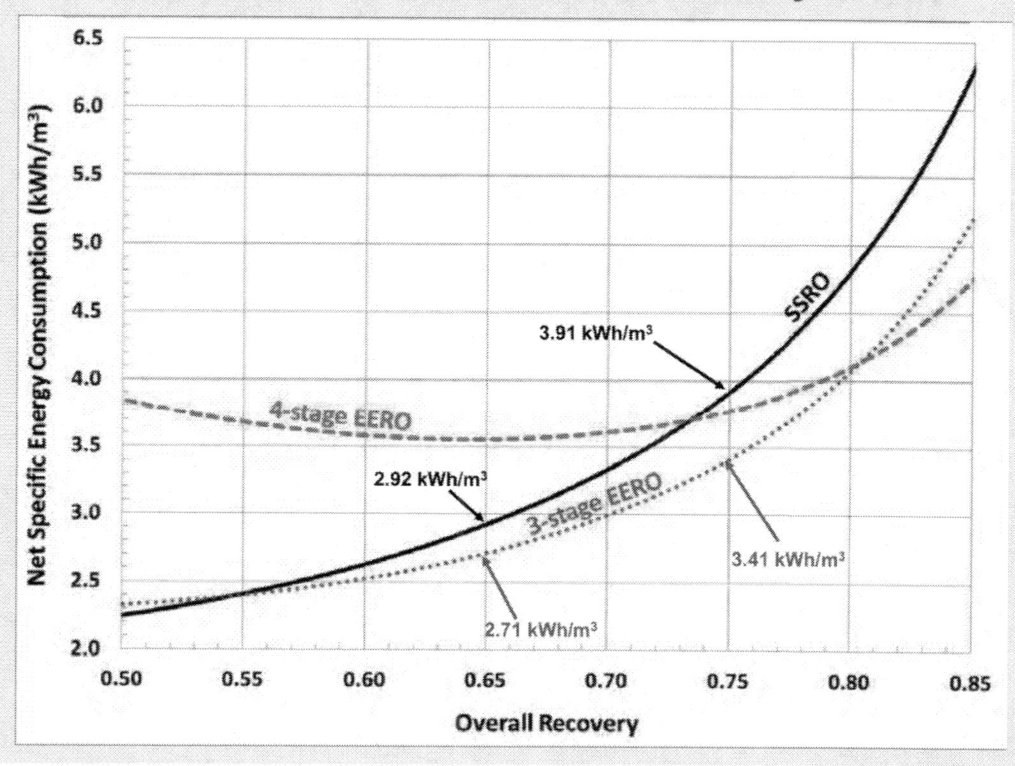

25

SUMMARY OF PERFORMANCE METRICS

35 g/L feed and 0.35 g/L product

Pump efficiency: 85%; ERD efficiency: 90%

Process	65% Recovery		75% Recovery	
	Osmotic Pressure Differential (bar)	Net Specific Energy Consumption (kWh/m³)	Osmotic Pressure Differential (bar)	Net Specific Energy Consumption (kWh/m³)
Single Stage RO	79.3	2.92	111	3.91
3-Stage EERO	61.0	2.71	85.4	3.41
4-Stage EERO	39.6	3.56	55.5	3.77

26

Summary of EERO

- 3-stage EERO: 75% recovery at reduction of 33% in pressure and 13% in SEC_{net} relative to single stage RO
- 4-stage EERO: 75% recovery at reduction of 50% in pressure and 3.6% in SEC_{net} relative to single stage RO
- Can be done without high pressure booster pumps on the retentate side
- Terminal RO stage in EERO can be operated at higher recovery owing to foulant removal by the NF stage(s)
- Reduced operating pressure in EERO will decrease fixed and maintenance costs
- Concentrated brine from EERO process can be used as draw solution for osmotic energy recovery via the pressure-retarded osmosis (PRO) process

27

Acknowledgments

- The Singapore Membrane Technology Center (SMTC) is supported by the Environment and Water Technologies Strategic Research Program administered by the Environment and Water Industry Program Office of the Singapore Public Utilities Board
- Professor Wang Rong, Director of SMTC, for her encouragement in this research and Professor Tony Fane, Director Mentor of SMTC, for helping the speaker become more knowledgeable in membrane process technology
- Professor Ng Wun Jern, Executive Director of Nanyang Environment and Water Research Institute, for his support of SMTC

SMTC

Member of the NEWRI Ecosystem

28

Thank You!
29

RESAERCH INTO MOLECULAR SEPARATIONS USING MEMBRANES AT IMPERIAL COLLEGE

Andrew Livingston*
Professor of Chemical Engineering
Department of Chemical Engineering
Imperial College, London SW7 2AZ UK
a.livingston@imperial.ac.uk

Organic liquids are ubiquitous in chemical science based industries, which range in scale from refining to pharmaceutical production. It is generally accepted that 40-70% of capital and operating costs in these industries are dedicated to separations; and a substantial fraction of this cost is related to processing of organic liquids, both as product streams and solvents. Membrane technology has the potential to provide game changing alternatives to conventional concentration and purification technologies such as distillation, liquid extraction, adsorption and chromatography. The obvious benefits of membrane processing have been apparent for many years, and have attracted research, development and commercialisation efforts from academic groups, end users, and membrane suppliers. In the last few years these efforts have resulted in Organic Solvent Nanofiltration (OSN) emerging as a new membrane technology.

This presentation will describe some key recent research activities in the Livingston group at Imperial College London into OSN. These include: (i) formation and function of new OSN membranes based on Polyetheretherketone (PEEK); Tthese have outstanding chemical resistance and can be applied to acidic and basic organic solutions at elevated temperatures; (ii) Formation of thin film nanocomposite (TFN) membranes embedding 50-150 nm MOF nanoparticles [ZIF-8, MIL-53(Al), NH_2-MIL-53(Al) and MIL-101(Cr) via interfacial polymerisation on top of a cross-linked P84 support. MOF nanoparticles were homogeneously distributed in the polyamide layer, and their presence was inferred by a combination of measurements. TFN membranes showed dramatically increased MeOH and THF permeances when compared to the same membranes with no MOFs added, without sacrificing rejection. A direct trend was observed between the MeOH permeance and the porosity of the MOFs, suggesting preferential flow paths provided by the MOF pores

Simulation and process design allied to validation using spiral wound module data allows the prediction of process performance. There are a growing number of applications of OSN to industrial separation problems, and two main areas will be highlighted, the purification of pharmaceutical API, and the formation of peptides and oligonucleotides using OSN technology.

Peeva LG , Da Silva Burgal J, and Livingston AG, "Experimental strategies for increasing the catalyst turn-over number in a continuous Heck coupling reaction" Journal of Catalysis (2013) pp.190-201

Sorribas S, Gorgojo P, Tellez C, Coronas J and Livingston AG "High Flux Thin Film Nanocomposite Membranes Based on Metal-Organic Frameworks for Organic Solvent Nanofiltration" J.Am.Chem.Soc. 135 (2013) pp.15201-15208

Peshev D and Livingston AG, "OSN Designer, a tool for predicting organic solvent nanofiltration technology performance using Aspen One, MATLAB and CAPE OPEN" Chem. Eng. Sci. (2013) 104C pp.975-987

389887 Membrane Research and Technology in Australia

Abstract:

Membrane research and implementation has expanded enormously in the Australian and Asia-Pacific region in the last 15 years. Major desalination plants have been built in major Australia cities in parallel with widening interest in membrane technology across the board in water and wastewater treatment. Recently the development of unconventional gas resources within Australia has further spur applications in produced water treatment and concentrate management using membranes. At the Centre for Membrane Science and Technology at the University of New South Wales, research has focussed both on challenges of emerging technologies and issues in maturing membrane areas in both gas and water arenas. Functionalization for antifouling, superhydrophobicity/superhydrophilicity, and biocatalytic properties have focussed on nanocomposite and additivation approaches. Ageing and cleaning of membranes along with reuse and disposal of reverse osmosis membrane are other major endeavours in the Centre relevant to the maturing membrane industry. Membrane distillation, assisted forward osmosis, CFD modelling, validation of membrane processes, hybrid processes, and resource recovery are other major themes in the water space. The prevalence of coal fired power plants and large natural gas reserves within Australia and abroad has also driven ongoing research in CO_2 capture and utilization with both pressure driven membranes and new hybrid enzymatic membrane systems.

379255 Membrane Science and Engineering at the University of Toledo

Abstract:

The University of Toledo, located in Toledo, Ohio, is a comprehensive university with colleges of engineering, medicine, nursing, pharmacy, and law as well as colleges for the arts and sciences. The University is among the largest in Ohio in terms of student population and budget. The University also possesses one the largest clusters of faculty working in membrane science and engineering in the United States. Faculty are involved in myriad application areas from reverse osmosis to gas separation to energy production. They work on fundamentals of membrane science and engineering ranging from material synthesis to membrane formation and modification to module design. A summary of past research accomplishments and current research activity is provided.

390454 Triaxial Electrospun Fibers: Fabrication and Characterization

Abstract:

Triaxial Electrospun Fibers: Fabrication and Characterization

Abdurizzzgh Khalf, Sundar Madihally Chemical Eng, Oklahoma State

University, Stillwater, OK, United States.

Triaxial electrospinning provides an effective method for fabrication of desirable biocomposite for particular applications. Modified process (based on core-sheath fluid flow where the enforced sheath polymer encapsulates the core liquid) can enable different features such as hydrophilicity, hydrophobicity, and mechanical strength. Selection of solvents with appropriate boiling points for each layer is critical as rapid solvent evaporation from the jet surfaces could cause instabilities in fiber formation. However, the effect of solvent volatility and relative polymer molecular weights on uniform encapsulation of the core polymer in triaxial electrospinning process is not well understood. In this study, we investigated the effect of solvent volatilities and MW on uniform fiber formation. Formed fibers were characterized for distribution of component via hydrophilicity, mechanical, rheology and biology properties. We explored combinations of polycaprolactone (PCL), cellulose acetate (CA), and polyvinyl alcohol (PVA) and mineral oil (as the inner core) of various molecular weights. Hollow fibers were developed by selectively removing the mineral oil. Different solvent mixtures were tested based on their boiling points, determined using a process simulator ChemCAD. Solution viscosities were evaluated at various shear rates. Obtained fibers were analyzed by scanning electron microscopy and fiber sizes were also characterized using digital micrographs. Differential scanning calorimetry (DSC) and FTIR were performed to characterize various components in the triaxial fibers. Tensile tests (both wet and dry) were also measured to assess the uniform distribution of polymers. 24-h viability of human umbilical vein endothelial cells was also evaluated. The hydrophilicity of electrospun fiber was also measured by contact angle. Micrographs indicated the formation of triaxial structured fiber of outer hydrophobic PCL/CA/hollow, PCL/PVA/hollow and outer hydrophilic CA/PCL/hollow fibers. Fiber sizes were uniform and ranged in micrometers size in all the formed configurations. While switching the configuration of polymers, rapid solvent evaporation (volatility) of the inner core caused phase separation, leading to failure in obtaining fibers. Solvents mixture (chloroform and methanol) of outer sheath solution PCL with low boiling point of 54 C^o caused the polymer molecules to evaporate faster from the charged jet surface and diffuse easily from the inside of the charged jets to the outer surface, which significantly enhanced the outer sheath solution wrapping ability and gave the jet the capability to maintain the core-sheath jet fabrication. However, fibers did not form using chloroform and methanol as solvent mixture for intermediate polymer (PCL). The instability region (stretching and elongation) of the core-sheath liquid jets occurred during electrospinning process that prohibited the formation of the fibers was due to the interactions between the viscoelastic forces and surface tension that governor the solution evaporation rate. The high volatility of the intermediate solution (PCL) weakens the outer sheath wrapping ability and results wet fibers formation. It was required to regulate its solvent mixture boiling point by adding acetic acid to moderate the spinning rate of PCL by lowering the solvent evaporation rate. This significantly strengthened the sheath jet stability allowing full control of CA polymer to encapsulate the intermediate polymer (PCL) and raising the wrapping ability of CA spinning solution. 80 kDa and 45 kDa PCL solutions molecular weight were tested while keeping all other parameters constant. Solution of 45 kDa MW showed better electrospinning performance compared to 80 kDa solution. At high molecular weight (80 kDa), the strength of sheath solution increased eventually making it tough for electrical force to drug and enforce the sheath polymer to encapsulate the core liquid. The jet appeared to be broken along the length of spinning distance indicating an inability of sheath polymer to wrap the core polymer. Moreover, better electrospinning performance was obtained when the inner core MW was smaller or equal to that of the outer sheath. Stripping outer CA in CA/PCL/hollow fibers confirmed the distribution of inner PCL. Endothelial Cells attached and spread on these fibers suggesting no toxicity from the solvents. Tensile tests (both wet and dry) indicated that CA hollow fibers with the inner PCL showed significant improvement in load carrying capacity and stiffness in hydrated conditions. DSC thermographs showed presence of both PCL and CA components. The rheology results suggest that the fibers could be formed when the viscosity of the core is less than that of sheath. Fibers produced from CA/PCL/hollow and PCL/CA/hollow had an average diameter of 11.6 (±3.9) and 9.5 (±1.5 μm) respectively. In summary, the triaxial structured fiber fabricated in this work offers new features, such as hydrophobicity, hydrophobicity and mechanical strength. Enhance the structural morphology by increasing the fiber thickness (breaking the limitation in fiber size produced by traditional electrospinning process). Upon formation of the fiber, outer sheath can be extracted, exposing the desired inner sheath the feature serve polymers that cannot electrospun in their own. Solvents properties study showed that outer sheath should have a boiling point less than that in the inner cores. Better control in the structure and wall thickness of formed fibers could be obtained when the inner core MW is smaller or comparable to that of the outer sheath.

ELECTROSPINNING SILK WITH SELENIUM NANOPARTICLES FOR ANTIBACTERIAL SKIN APPLICATIONS

Stanley Chung[1], Phong Tran[2], Batur Ercan[1], Thomas J. Webster[1,3,4].
1. Chemical Engineering Department, Northeastern University, Boston, MA
2. Chemical and Biomolecular Engineering Department, University of Melbourne, Melbourne, Australia
3. Bioengineering Department, Northeastern University, Boston, MA
4. Center of Excellence for Advanced Materials Research, King Abdulaziz University, Saudi Arabia

Abstract

Healthy skin with robust skin regeneration provides the most effective defense against microbes, viruses, and other environmental agents. Silk has been studied for skin regeneration and has been widely used as an additive for cosmetics. Silk improves collagen synthesis, re-epithelialization, wound healing, atopic dermatitis alleviation, and scar reduction. However, purified silk promotes microbial growth. Here, we propose for the first time, an electrospun silk scaffold doped with selenium nanoparticles to address this issue. Electrospun silk scaffolds have smaller interstices and higher surface areas, allowing for efficient nutrient transfer to skin. In addition, selenium nanoparticles have been shown to have excellent antibacterial properties. By incorporating selenium nanoparticles into silk, we expect to retain silk's beneficial skin healing properties while increasing its antibacterial ability.

Introduction: 7-10% of all hospitalizations in the United States are caused by skin infection, many due to bacterial infections.[1] The rise of antibiotic resistant bacteria is of particular concern: the CDC estimates that there are more than two million cases of antibiotic-resistant bacterial infections, causing 23,000 deaths.[2] Current practices of addressing bacterial infection by increasing antibiotic dosage have accelerated the growth of antibiotic resistant bacteria. Thus, there is a need for an alternative solution to this growing epidemic. The best preventative measure against infection is to maintain healthy, intact skin tissue, which wards off not only bacterial infection but also other environmental agents. Silk is well characterized and has been demonstrated to possess beneficial properties for skin applications: silk promotes collagen synthesis, epithelialization, wound healing, atopic dermatitis alleviation, and scar reduction[3]. However, silk has poor antibacterial efficacy[4] while selenium nanoparticles demonstrate good antibacterial properties[5]. By incorporating the selenium nanoparticles into electrospun silk scaffolds, we hope to address a major drawback to an otherwise ideal skin graft material.

Materials and Methods: Silk was extracted from *Bombyx mori* by Rockwood's protocol[6] while selenium nanoparticles were synthesized by Tran's protocol[7]. Extracted silk was dissolved in formic acid at 8-14%, w/v, and spun at flow rate of 2-3 mL/hr, distance to a collector at 10cm, and at a voltage at 20kV. *Staphylococcus aureus* (ATCC -10832D-5) was cultured on the substrates with 0, 7.8, 15.5, and 31 μg/mL selenium nanoparticles in 0.3% tryptic soy broth (Sigma-Aldrich). After 3-5 hours after inoculation, the plates were read at 562 nm to determine cell density. All experiments were conducted in triplicate and repeated at least three times. The electrospun scaffold was characterized by SEM and goniometry to determine the physical make-up of the scaffolds, with and without selenium nanoparticles. Selenium nanoparticles were characterized by DLS to determine particle size and dispersity.

Results and Discussion: Electrospun scaffolds possessed fiber diameters of 200-300 nm and pore sizes of ~5μm (Figure 1). Surface contact angles were 50°, showing a hydrophilic surface. Results of this study showed that *S. aureus* growth was inhibited in the presence of selenium nanoparticles (Figure 2).

Figure 1. SEM image of electrospun silk. 3500 magnification. Scale bar is 10 μm

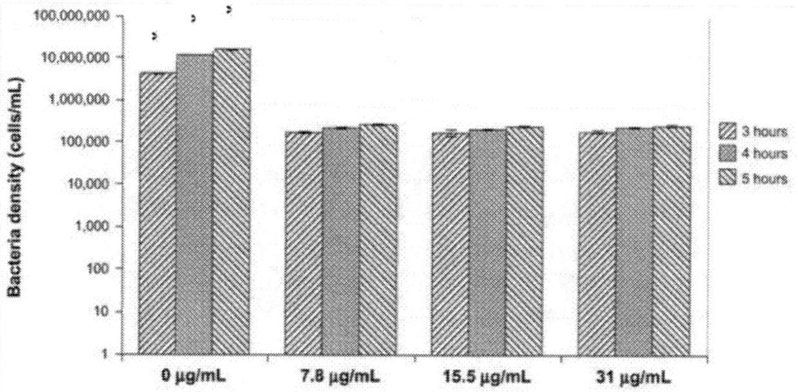

Figure 2. Bacterial density after contact with 7.8, 15.5 and 31 µg/mL concentration of selenium nanoparticles. All selenium treated samples displayed significant differences from control. Data = mean±standard deviation, n=3. (*P<0.01)

Conclusions: Selenium nanoparticles were effective at decreasing *S. aureus* growth at even low time periods of exposure. Electrospun silk generated fiber diameters ~200 nm and micron sized pore sizes, ideal for mammalian cellular adhesion.

Acknowledgements: The authors thank William Fowle, Scott McNamara, and the Northeastern University Department of Chemical Engineering for facilities and funding.

References: 1) Vinh DC et al. The Lancet Infectious Diseases. 2005; 5(8), 501-513. 2) CDC. Antimicrobial resistance threats in the United States. 2013. 3) Roh DH. J Mat Sci Mat Med. 2006; 17 (6): 547-552. 4) Kaur J et al. Biopolymers. 2013; 101(3): 237-245. 5) Tran PA et al. I J Nanomed. 2011; 6: 1553-1558. 6. Rockwood DN, et al. Nat. Protocols. 2011; 6(10): 1612-1631.

390686 PCL-Gelatin Electrospun Fibers: Fabrication and Characterization

Abstract:

Popular biodegradable electrospinable polymers are polycaprolactone (PCL), gelatin (Gel). PCL is non-toxicity, biocompatible and biodegradable. PCL synthetic polyester has gained much attention due to its low melting point (60 C^{o}) and elastomeric properties. PCL has been studied to form many medical devices, or scaffolds for tissue regeneration of in vivo and in vitro cell culture using serum added media. To alter the hydrophobic characteristics of the surface, PCL has been blended with many natural polymers and also has been extensively evaluated in electrospinning process. Gelatin natural polymer has been widely studied and exposed to various biomedical applications due to its excellent biocompatibility and biodegradability. Blending natural and synthetic polymers provides a new biomaterial with proper biocompatibility and improved mechanical, physical and chemical properties which is beneficial for cell adhesion and degradation rate. In this study, we addressed the fabrication of biocompatible and biodegradable electrospun fibers by utilizing two different solvent systems to dilute the spinning solution for a possible increase in process productivity and fibers size. We also varied the PCL-Gelatin ratio in the spinning solution to evaluate the influence of PCL-Gelatin continent on the biological, mechanical properties and stability of the fabricated scaffolds and cellular behavior. We tested the ratio of PCL-Gelatin component under two different combinations 70:30 and 50:50 PCL-Gelatin respectively. Two different solvent system were tested Trifluoroethanol (TFE) and Hexafluoro-2-propanol (HFP) based on solution electrospinability and fiber uniformity. Distribution and stability of gelatin was studied in two-week incubation under physiological condition study using carboxyfluorescein diacetate-succinimidyl ester (CFDA-SE) staining and incubation. Fibers fabricated from PCL-Gelatin in HFP and TFE were compared and characterized for their size, structural and morphology using scanning electron microscopy and digital micrographs. In addition to in situ analysis via fluorescence microscopy, differential scanning calorimetry (DSC) and FTIR were performed to characterize various components in the fabricated fibers. Tensile tests (both wet and dry) were also measured to assess the uniform distribution of polymers. 24-h viability of human umbilical vein endothelial cells was also evaluated. The hydrophilicity of electrospun fiber was also measured by contact angle. SEM images of electrospun fiber of PCL-Gelatin in TFE solvent system revealed that the fibers are slightly different in size, while fibers fabricated from PCL-Gelatin in HPE were more uniform in size, production process is significantly slow. With the use of TFE, production rate significantly increased. Combination of PCL-gelatin was stable for two-weeks at physiological conditions. The results also showed that gelatin distribution in PC-Gelatin fiber was uniform. This provided improved cellular response due to distributed gelatin not only provides uniform cell binding domains to cells but also increases the fiber stability while cell culture study in aqueous media.

Single Cell Force Measurement Using Suspended Fiber Nanonets

J. Wang, K. Sheets, B. Behkam and A. Nain
Virginia Tech

Introduction: Mechanical communication between a cell and its substrate occurs bi-directionally. Cells generate forces which act on their surroundings to interrogate substrate stiffness and topography (inside-out) but also respond to forces transmitted through the ECM which elicit reactionary cellular response (outside-in)[1–4]. Force measurement platforms that are able to capture both modes can allow one to distinguish the forces that cells apply from those that they can withstand, with implications to disease states where progression from normal function to failure either occurs rapidly or with minimal warning or detection (bone fracture, muscle or ligament tears, blood vessel aneurysms). In recent years, numerous force measurement (*inside-out* and *outside-in*) techniques have been developed to investigate single cell behavior. While these methods have yielded fundamental insights, they lack ECM mimicking fibrous environment for capturing cell-fiber interactions. Fiber based biophysical interactions, involving parameters of curvature, structural stiffness (N/m), alignment and hierarchy, have been shown in studies by others and us to play key roles in disease and developmental biology. Here, using our non-electrospinning fiber manufacturing platform (STEP: Spinneret based Tunable Engineered Parameters[5–7]), we demonstrate development and application of fused-fiber nanonets as force probe sensors for measuring single and population level cell forces. Using these nanonets, we demonstrate the ability to measure contractile forces exerted by single cells attached to and migrating along the fiber axis (*inside-out*). This is followed by an investigation of force response of single cells attached to two parallel fibers using a probe capable of deflecting the leading fiber at different strain rates (*outside-in*). In both techniques, forces are calculated using beam deflection within the elastic limit.

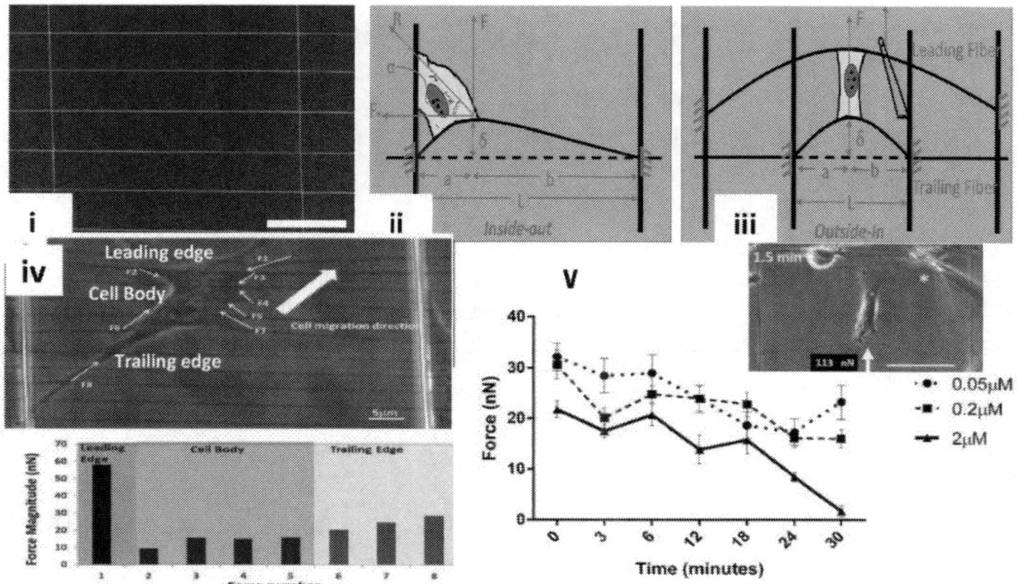

Figure 1. Scanning electron microscopy (SEM) image of suspended STEP nanonets showing with scale bars of 100 µm. STEP based force measurement probes: (iv) inside-out method showing single cell force measurement during migration, and (v) outside-in. Effect of cytochalasin D drug dosage on cell force (iv) inside-out, and (v) outside-in with inset respective optical images of single cell force response.

Materials and Methods: STEP nanonets of varying diameters were deposited in parallel and cross hatch patterns and fused at the intersections using an in-house fusing technique (Fig. 1). Cells of different lineages were investigated for force response with and without the presence of cytoskeletal altering drugs using a microscope with incubating capability (Zeiss AxioObserver Z1, Thornwood, NY).

Results and Discussion: One of the important goals in this cell force study is to achieve a force map for a single cell by increasing the spatial force resolution. Microdimensional force sensor arrays (µFSA) with large pillar densities and polymer gels with densely embedded beads have been used to achieve detailed single cell force maps. Currently, the spatial force resolution ranges from 2-9 µm depending on center to center distance between pillars and beads. In our study, densely spaced nanofiber arrays (fiber spacing ~5 µm) have been prepared for single MSC force mapping as shown in **Figure 1 (iv)** (A)-(D). It was found that the largest contractile forces always localized at the edge of the cell, indicating that much of the mechanical work underlying cell locomotion is generated near the cell boundary. This observation of traction force decline with distance away from the edge is in agreement with previous cell force studies based on micropillar arrays and polymer gels.[8,9] In addition, we find that the advancing edge of the cell is always exerting larger forces than the mid body and posterior. Similar results have been reported by Dembo *et al*, who argues that the mid-body and posterior are more like passive structures being pulled forward by the contractions generated at the leading edge.[9] Note that our current spatial resolution (5 μm) is comparable to most micro-pillar arrays (2-9 μm) and polymer gels (5μm). For the highly aggressive glioma model (DBTRG-05MG), we investigated single cell force modulation with and without the presence of a cytoskeleton altering drug (cytochalasin D) using outside-in perturbation. Cytochalasin D compromises the spreading ability (area) of single glioma cells, eventually decreasing their 'inside-out' contractile forces, and 'outside-in' force response to external strain (Figure 1). Additionally, we have recently demonstrated measurement of cell-cell junctional forces and forces involved in cellular blebbing using *outside-in* perturbation.

Conclusions: Through time lapse microscopy and immunofluorescence imaging, we present our findings on the role of suspended fiber geometry (curvature and structural stiffness) on single and population level cell force modulation. The suspended and aligned fiber networks act as ultra-sensitive force probes (~ nanoNewtons) for investigating temporal drug response and efficacy in disease models at the single cell resolution. Furthermore, measurement of stem cell forces involved in life-cycle of a single and population of cells provides new insights in the role of forces in cell differentiation and development biology. Combined, these two platforms provide new pathways to investigate cell behavior on synthetic ECM fiber mimics which present cells with simultaneous 1, 2 and 3D environments. Through engineering and study of these simple systems of exceedingly complex interactions, we hope to contribute to development of mechanistic in vitro assays combined with relevant biochemical growth factors, thus approaching in vivo physiological relevance for studying development biology and disease.

Acknowledgements: The authors would like to acknowledge support received from the Institute for Critical Technology and Applied Sciences (ICTAS) at Virginia Tech. This research was funded in part by the Bill and Andrea Wade Research Fund.

[1] Goldfinger, L. E., 2013, Integrin Signaling, Elsevier Inc.

[2] Holle, A. W., and Engler, A. J., 2011, "More than a feeling: discovering, understanding, and influencing mechanosensing pathways.," Curr. Opin. Biotechnol., **22**(5), pp. 648–654.

[3] Eyckmans, J., Boudou, T., Yu, X., and Chen, C. S., 2011, "A hitchhiker's guide to mechanobiology.," Dev. Cell, **21**(1), pp. 35–47.

[4] Chen, C. S., 2008, "Mechanotransduction - a field pulling together?," J. Cell Sci., **121**(20), pp. 3285–3292.

[5] Nain, A. S., Sitti, M., Jacobson, A., Kowalewski, T., and Amon, C., 2009, "Dry Spinning Based Spinneret Based Tunable Engineered Parameters (STEP) Technique for Controlled and Aligned Deposition of Polymeric Nanofibers.," Macromol. Rapid Commun., **30**(16), pp. 1406–12.

[6] Nain, A. S., and Wang, J., 2013, "Polymeric nanofibers: isodiametric design space and methodology for depositing aligned nanofiber arrays in single and multiple layers," Polym. J., **45**(7), pp. 695–700.

[7] Sheets, K., Wunsch, S., Ng, C., and Nain, A. S., 2013, "Shape-dependent cell migration and focal adhesion organization on suspended and aligned nanofiber scaffolds.," Acta Biomater., **9**(7), pp. 7169–77.

[8] Du Roure, O., Saez, A., Buguin, A., Austin, R. H., Chavrier, P., Silberzan, P., Siberzan, P., and Ladoux, B., 2005, "Force mapping in epithelial cell migration.," Proc. Natl. Acad. Sci. U. S. A., **102**(7), pp. 2390–5.

[9] Dembo, M., and Wang, Y. L., 1999, "Stresses at the cell-to-substrate interface during locomotion of fibroblasts.," Biophys. J., **76**(4), pp. 2307–16.

An electric field responsive nanopore with ion selectivity: Design based on controlling transport resistance

Yudan Zhu[a,*], Linghong Lu[b], Ximing Wu[c], Xiaohua Lu[d,*]

College of Chemistry and Chemical Engineering, State Key Laboratory of Materials-Oriented Chemical Engineering, Nanjing Tech University, Nanjing 210009, P. R. China

[a] ydzhu@njtech.edu.cn; [b] linghonglu@njtech.edu.cn; [c] toby10222224@sina.com; [d]xhlu@njtech.edu.cn

Using external stimuli to control ionic transport behavior confined in the nanopore has many widely potential applications ranging from water desalination to fuel cells and becomes an important focused topic for membrane material designers. Conventionally, the idea that the intensity of electric field could change the ion selectivity with same charge seems impossible. In this work, we proposed a novel strategy that controlling the resistance of ionic transport utilizing the different structural responses of modified pore wall to the electric field intensity to regulate the ionic selectivity.

Molecular simulation has been an integral part in the field of nanofluidics and has an important role in discovering novel view of nanomaterial design[1]. In order to verify the above notion, molecular simulation studies were performed to investigate the transport behaviors of K^+ and Na^+ in the inner wall modified nanopore under electric field. The results showed that the flux and K^+/Na^+ selectivity increased in the modified nanopore. The observation reinforces the notion that both the driving force and the resistance force can be simultaneously affected to external stimuli under nanoconfinement, which is useful for smart nanodevice designing.

References

1. Y.D. Zhu, J. Zhou, X.H. Lu, X.J. Guo, L.H. Lu, *Microfluid Nanofluid,* **15**, 191 (2013).

Preferred presentation type: oral

Topic: Membrane-Based Separations(02D)

02D20 Carbon Based Nano-Structured Membranes

388737 Novel in Situ Methods for Floc Size Determination in Cell Culture Seperations

Abstract:

Recently flocculation has emerged as a viable option for the improvement of cell culture separations. The controlled flocculation of cells and cell debris can increase filtration and centrifugation speed, enhance throughput and improve clarification performance. Flocculation is particularly important for high density cell cultures where large amounts of biomass, especially in the sub-micron range, can prove challenging to effectively separate.

The particle size of flocculated cell debris is critical to consider when developing and implementing a flocculation step since it is often directly related to separation performance and will define requirements for the separation technology (pore size of depth filters for example). Offline particle size analysis tools have failed to gain broad acceptance for this task since flocs are inhernetly delicate and can break apart easily during the sampling, preparation and analysis phases of an offline analytical method. Also, the time delay between sampling and receipt of results can mean that expensive flocculants are added to a process for longer than needed

In this paper examples of how *in situ* measurement of floc size and count have being used to improve cell culture flocculation and separations will be presented. *In situ* monitoring using ParticleTrack technology will be shown to eliminate the most common errors associated with sampling and provide a more reliable method for the determination of flocculation onset, kinetics and endpoint. Examples relating flocculent dosage and type with floc size and stability will be presented and correlations between floc size and separation performance will be shown.

392128 Progress on a Mixed-Mode Hydrogel Membrane – a New Tool for Mab Purification

Abstract:

In today's high-titer biopharmaceutical mAb processes, there is a need for innovative separation approaches for capture or polish of biomolecules. Challenges exist, for example when the conductivity of the feed stream is high or when the aggregates level needs to be reduced immediately. Natrix is developing a mixed-mode chromatography membrane, which combines the properties of strong cation exchange and hydrophobic interaction. This mixed-mode chromatography membrane offers fast separation in a disposable format without sacrificing process performance. It opens up new opportunities for solving challenging purification problems as an alternative to conventional chromatography columns. This paper discusses the characteristics of the mixed-mode membrane together with performance across the design space to-date. The preliminary data shows high mAb binding capacity (up to 85 mg/mL), good clearance of HCP and aggregates at high conductivity (up to 15 mS/cm) and all at 6 seconds residence time. In addition, the low salt elution capacity (6 mS/cm) facilitates direct load onto a quaternary amine adsorber for mAb polishing.

388551 Rapid Characterization of HCP Removal on Adsorptive Membranes Using Protein Surrogate Biomarkers

Abstract:

Anion exchange membrane adsorbers are used for clearance of impurities such as viruses, host cell protein (HCP), and DNA in flow-through polishing steps for manufacture of monoclonal antibody therapeutics. HCP can be very complex in nature. CHO host cell protein composition shows that the majority of HCP in general, and in Protein A eluates in particular, are acidic proteins with pI ~ 4-6 and molecular weight between 10-75 kDa. In this study, surrogate biomarkers or model proteins that are easily available, easy to quantify and have properties comparable to typical host cell proteins, are utilized for rapid characterization of anionic membranes. Choice of these model proteins allowed for a large number of experiments to be carried out at moderate costs, thus enabling study of a broad range of conditions for determining binding constants and capacity for HCP adsorption for Q membranes. The results of these studies will be presented. Use of surrogate biomarkers for rapid and reproducible scale-up of membrane adsorbers will also be discussed

385271 Optimization of Tangential Flow Ultrafiltration Processing Step with High Concentration Self-Buffering Proteins

Abstract:

High protein concentration biopharmaceuticals are becoming more common due to the increased viability of these formulations for subcutaneous administration. Formulation of a biopharmaceutical product is most often performed using tangential flow ultrafiltration (TFUF) to concentrate the product to the desired level and, then exchange buffer components into a more stable formulation. As protein concentrations increase, self-buffering becomes a contributing factor in achieving the targeted formulation conditions. This case study will examine methods to formulate a self-buffering protein at commercial scale. The developed process demonstrates both robustness to manage expected parameter variation, as well as flexibility to allow potential future changes that may be deliberately implemented upstream of the step.

378644 Purification of Adenovirus Serotype 5 By Two-Column, Open-Loop, Size-Exclusion, Simulated Countercurrent Chromatography

Abstract:

The purification of complex biopharmaceuticals, such as recombinant proteins, antibodies or vaccines, is gaining increasing importance to fulfill the requirements of a reliable downstream purification process with high purity and yield, and, most importantly, cost efficiency. Viruses, in particular, play an important role in the vaccine and gene therapy fields.

The downstream biopurification train of viral vectors has been extensively developed in the past years by combining different chromatography steps, namely ion-exchange (IEX)) and size-exclusion chromatography (SEC), and, less frequently, affinity chromatography, intermingled with concentration and ultra/diafiltration steps.

To be more specific, the classical approach for adenovirus purification consists of three major steps—clarification, concentration/purification, and polishing—applied sequentially: (i) clarification of the harvested bioreaction bulk to remove cell and cell debris; (ii) ultra/diafiltration; (iii) anion exchange; (iv) ultra/diafiltration to concentrate the product; and (v) SEC as a final polishing step. The SEC step is usually carried out last, mainly because of its low productivity and low product titer. Usually in this final step the amount processed is reduced by 50–100 fold.

The current paradigm in the downstream processing of viral vectors is to operate each chromatographic step using classical single-column batch chromatography. However, the cost of goods and overall processing time can be reduced by switching to multicolumn, (semi-)continuous chromatography, which gives higher throughput, has lower buffer consumption, provides for higher capacity of utilization of the stationary phase (or smaller column volume), and hence gives increased productivity. Unfortunately, unlike other industrial fields, biotechnology is late to embrace continuous processing or multicolumn preparative chromatography.

The present work reports on the design and experimental validation of a simple quasi-continuous, open-loop, two-column countercurrent chromatographic process for size-exclusion purification of adenovirus serotype 5 (Ad5). As mentioned above, SEC has been used in the past for both polishing and intermediate purification of adenoviruses. However, it is often claimed that the main drawbacks of SEC, namely low productivity and high product dilution, make it a costly purification step. Here, it is shown that these drawbacks can be eliminated to a large extent by switching from single-column batch operation to two-column SMB-type operation.

Because of the anticipated boost in performance obtained by switching to multi-column (quasi-)continuous operation, the Ad5 purification train employed for assessing the effectiveness of the enhanced SEC step was streamlined: the SMB-SEC step was applied after the first ultra/diafiltration step and replaced the last three steps of the standard purification train. The main reason for working with the streamlined purification train was to challenge the SMB-SEC step with a less purified bulk in order to prove that, once the classical limitations of SEC are alleviated, the standard purification train can be changed to meet specific needs in terms of cost reduction or purity requirements. Moreover, running the process with a less clean bulk gave more confidence on the performance assessment of the SMB-SEC step.

In a two-column SEC process the cycle is divided into two equal-length time intervals, which we shall henceforth refer to as switching intervals. To design the cycle it suffices to define the sequence of flow-line configurations for one of the switching intervals; the sequence for the other twin switching interval is similar, but the positions of the two columns are exchanged. In practice, it is easier to keep the columns fixed in space and to move the positions of the inlet and outlet lines by means of an automated arrangement of two- or three-way valves. Over a full cycle, each column undergoes the same sequence of steps but phased out in time by one switching interval.

In the developed process, each half-cycle is divided into the following sequence of steps:

(1) Elution of the upstream column while the effluent of the downstream column is diverted to the waste line.

(2) Elution of the upstream column but its effluent is diverted to the waste line instead of being directed to the downstream column; the latter column is fed with the clarified bioreaction bulk while is effluent is collected as purified product;

(3) Operation of the system as in step 1, but the effluent of the downstream column is collected as product.

(4) Elution of the upstream column but its effluent is again diverted to the waste line instead of being directed to the downstream column; the latter column is kept frozen, that is, the flow through it is temporarily halted.

At the end of every half-cycle the positions of the two columns are exchanged to implement the simulated moving-bed contact between the solid and the fluid.

Clearance of impurities, namely DNA and host cell protein (HCP), were experimentally assessed. The obtained performance compares very favorably against single-column batch chromatography for the same volume of size-exclusion resin. In particular, our two-column SEC process achieves about the same DNA and HCP clearances as the standard batch process. However, and most importantly, the virus yield is increased from 57% for the batch process to 86% for our two-column because of internal recycling of the mixed fractions of contaminated Ad5 even though the process is operated strictly in an open-loop configuration. And last, but not least, the productivity is boosted by 6-fold with the two-column process. This is a clear demonstration that with this increased productivity, size exclusion can do much more than just final polishing and buffer exchange in the downstream train of a biopurification.

In conclusion, the main drawbacks of size-exclusion chromatography, namely low productivity and low product titer, were easily overcome by an innovative two-column configuration that keeps the mixed fractions inside the system at all times.

377231 Centrifugal Separation of Disk Centrifuge in Downstream Processing

Abstract:

Disk centrifuges[1] have been widely used for downstream harvesting of yeast, mammalian and microbial cells. Heretofore, the operation of the centrifuges is largely based on ad-hoc trial-and-error experiences.

A model of separating cells using a disk centrifuge is presented. The model is validated against experiments on a pilot-scale centrifuge. An interesting example is used to illustrate how the model can be used to predict performance of inclusion-body separation.

Reference

[1] Wallace Woon-Fong Leung, Centrifugal Separations in Biotechnology, Academic Press, UK, 2007.

Effect of CO_2 on Cryogenic Moisture Requirements and Application for the Controlled Freeze Zone™ Technology

AIChE Annual Meeting – November 2014

Authors: David W. Maher, Jaime A. Valencia, P. Scott Northrop, Ananda Krishna Nagavarapu
ExxonMobil Upstream Research Company, Houston, TX

Abstract

Prevention of ice and hydrates in processing facilities is of paramount importance in the upstream oil and gas industry. Ice or hydrates can form when insufficiently dehydrated gas is chilled below ambient temperatures, resulting in precipitation of a solid-phase which can cause significant blockage to flow paths and disruption of operation. Thus, expensive dehydration units are often installed upstream of processing plants to reduce the feed gas moisture content such that ice and hydrate formation is avoided in the remaining process.

Prior studies have indicated that liquid CO_2 at relatively warm temperatures is capable of dissolving large amounts of water, and that its presence in natural gas can lower hydrate formation temperatures. While the same effect is expected at lower temperatures, few tests have been performed to confirm and quantify this. ExxonMobil presents laboratory and novel field test data that demonstrate this moisture carrying capacity, and lower hydrate formation temperatures for hydrocarbon mixtures containing CO_2 at cryogenic temperatures. The results indicate that when CO_2-containing natural gas is chilled and partially condensed, the resulting CO_2-rich liquid can absorb sufficient water vapor to meaningfully contribute to the dehydration of the vapor phase. The combination of these effects can significantly reduce upstream dehydration requirements.

The dehydration requirements of high-CO_2-containing natural gas mixtures are of particular interest to ExxonMobil for their Controlled Freeze Zone™ (CFZ™) technology. CFZ™ is an efficient, single-step, cryogenic distillation process for removal of high concentrations (>10%) of CO_2, H_2S, and other impurities from natural gas. CO_2 present in natural gas freezes at cryogenic temperatures, and therefore, most separation processes have tried to circumvent this phenomenon using solvents, solidification inhibitors, or other methods. CFZ™ allows for solidification of CO_2, in a controlled fashion. The resulting purified natural gas can be sent to the pipeline or used as feed for LNG liquefaction, while the impurities are removed as a high pressure liquid stream that is ready for underground injection for enhanced oil recovery applications or acid gas injection (AGI) disposal.

Sufficient water must be removed in the dehydration system upstream of CFZ™ to prevent hydrate formation anywhere in the cryogenic process. The ability to use glycol dehydration, as opposed to molecular sieves, upstream of a commercial CFZ™ facility furthers the economic advantages of the technology.

Laboratory test results and operational experience from a 14 MMSCFD CFZ™ Commercial Demonstration Plant are discussed, with a focus on the correlation between water- and CO_2-contents in the dehydrated feed and the formation temperatures and pressures of hydrates at cryogenic conditions.

391609 Technical Challenges and Solutions for Processing Distressed Crude Oils

Abstract:

The growing global demand for energy and shrinking supply of "easy oil" lead to ever-increasing production of distressed crudes. Most of those distressed crude oils contain impurities such as unstable sulfur species, high TAN, and high metals, etc, which affects properties of crude oil and creates many operational and quality issues. If those challenging properties are not properly addressed, those distressed crudes may not be safely produced or transported to the market, and typically suffer a value penalty. As a consequence, there is an increased need of field processing of the crude produced to ensure safe operation, avoid loss of production and improve crude value. This presentation will first examine the key challenges, such as complex feed composition, large throughput, low concentration of impurities to be removed, etc, and then highlight technical gaps for processing various distressed crudes. Potential technical solutions to improve the quality and value of distressed crude oils will be reviewed by using examples from both recent literatures and authors' experiences. The need for better understanding of fundamental phenomena and steps involved in various processes to develop innovative separation technologies will also be discussed.

387709 Analysis of Commercial and Synthesized H2S Scavengers for High Performance, Safe and More Economical Asphalt Roads

Abstract:

In recent years, the product quality and related processing conditions have been challenged by the increasing sulfur levels of crude oil resources. In addition to the recent strict legal regulations on allowable sulfur amounts in the products, those sulfur compounds have tendency to accumulate in the bottom products such as bitumen and fuel oil which makes it harder than ever to reach the required sulfur limits for the heavy bottom products. In addition to the naturally existing sulfur, it can also be artificially added into the bitumen at quite low amounts in order to cope with the bitumen-polymer incompatibilities caused by the properties of crude oil blends being processed in the refinery. Sulfur is known to trigger the cross-linking reactions and thereby improve the compatibility and storage stability of polymer modified bitumen that is used for long lasting and high performance asphalt roads. Since the storage and pavement temperatures of bitumen reaches up to 180 ¢ªC, this may results in the release of poisonous, flammable and colourless gas of hydrogen sulphide (H_2S). It can cause the road workers to die during pavement if above 300 ppm or make them suffer from several illnesses when exposed to concentrations as low as 10 ppm for longer than 8 hours. This study focuses on the H_2S scavenger performances of different commercial and synthesized chemicals by on-line monitoring of the concentrations of H_2S and other sulfur based chemicals. A home-made apparatus was made with a stainless steel column (1 m in long and ¼" in diameter) packed with ceramic beads. One end of the column is connected with Gas Chromatography - Sulfur Chemiluminescence Detector (GC-SCD). The other end is connected with 316 L stainless steel, 0.5 liter bomb reactor having a pressure gauge at the top and containing refinery gas sample with known amount of H_2S concentration and taken from an absorber of a fluid catalytic cracking, FFC, unit. Calibration curves are obtained by using FCC unit gas of high H_2S concentration and commercially available standard gases of low H_2S concentration, details of which are given elsewhere (*M. Yasar et al., 2013 " New Technique for Quantitative Determination of Sulfur Compounds in Refinery Gases by using GC-SCD" American Chemical Society (ACS) 2013 Proceedings, Hydrocarbon Resources, Pub# 500*). During the experiments, different H_2S scavengers are loaded into the column in equal amounts at each measurement. Then refinery gas with known amount of H_2S is passed through the column at room temperatures and the scavenging performances are observed by GC-SCD simultaneously. For the next step, scavenging performances of these chemicals in the bitumen is studied. Firstly, thermal stability of these chemicals at high temperatures are analysed by Thermal Gravimetric Analysis (TGA). Later on, 0.5-25 %(wt.) sulfur and 1-5 % (wt) scavenger containing bitumen samples are prepared and analysed at 150 ¢ªC. The results indicated that the third scavenger in a form of supported and pre-determined mixture of metal oxides has the best performance among synthesized ones and has a comparable performance with the commercial ones both in the refinery gas and in the bitumen phase. Scavenging performances up to 99 % decrease in H_2S concentration at the outlet are observed. As a further study, sulfur and scavenger added bitumen samples are further modified with varying amounts of Styrene-Butadiene-Styrene (SBS) diblock co-polymers addition. Rheological properties of those samples are studied by Dynamic Shear Reometer (DSR) to understand the improved performance and compatibility of prepared modified bitumen samples. Comparable performances together with decreased amount of required polymer have been found indicating that the prepared chemicals are potential bitumen additives for high performance, safe and more economical asphalt roads.

378621 Optimization of Gas-Oil Three Stage Separator's Pressure to Increase Stock Tank Oil Using Aspen Plus

Abstract:

The separation processes are one of the most significant components of any chemical industry and specially when we consider the crude industry. There are several types of separator involved in the industry, which can deviate from a single point to multi stage separators. The idea of multi stage separation is to produce the maximum amount of hydrocarbon liquid from the well fluid, and offer the high stability of the two currents of liquid and gas. To make the more efficiency of the process in terms of separation quality, several separation processes at decreasing pressures are used. The choice of operating pressures in surface separators can have a notable impact on the amount and quality of the oil produced in the store tank.

The present work is focusing on the optimization of pressures of the separator installed in the crude oil production unit using ASPEN Plus software. The proposed method determines the optimum pressures of separators in different levels of separation and, therefore, optimize the operating conditions. In this study, the optimum pressures of separators in a 3896 Kg mole/hr oil production unit were determined. As a consequence, the oil recovery was improved about 467 Kg mole/hr (130 bbl/day), and the worth of total produced oil was enhanced by 0.3°API.

393304 Numerical Simulation on the Feasibility of Bulk CO€2 Separation in Natural Gas via Supersonic Nozzle Expansion System

Abstract:

The application of supersonic nozzle separator has been proven for dehydration and hydrocarbon dew pointing. A study was conducted on the potential of the nozzle supersonic expansion technology for CO_2 separation from natural gas. A proof of concept testing of the nozzle prototype is being validated via a multiphase flow loop. The setup consists of a multiphase pump, gas-liquid separators and heat exchangers. Pressure transducers were installed along the nozzle prototype and the pressure reading is recorded by the data acquisition system. Besides pressure, the temperature inlet, primary and secondary outlets of the tube were also monitored. Two vibration sensors were installed to monitor the effect of the high vibration to the prototype. An on-line gas chromatography was installed at the inlet and the outlet of the nozzle tube to monitor the inlet composition and the effect of the phase separation occurred in the tube to the outlet composition. The pressure profile plotted along the prototype was compared with the CFD simulation result and significant flow of liquid CO_2 was produced during the test. The result of temperature reduction at the nozzle outlet shows that the thermodynamic prediction on gas expansion using de-Laval nozzle was proven as per simulation. The outcome from the experiment provides the improvements to the CFD simulation numerical modelling as well as the in-house coding for CO_2 separation efficiency.

394223 Exploiting Polymer-Nanoparticle Interactions to Create a Hydrogel with Biomedical Applications

Abstract:

Injectable hydrogels have recently become a relevant material platform in applications such as in vivo drug delivery or as cell carriers because they can be implanted in a minimally invasive manner and closely mimic biological tissue. Attractive properties for such injectable hydrogels are that they easily flow under applied stress (*shear-thinning*) and rapidly heal upon relaxation of the applied stress (*self-healing*). Previously, we developed a shear-thinning and self-healing injectable hydrogel driven by non-covalent interactions between hydrophobically modified cellulose and PEG-PLA nanoparticles. [1] The nature of this gel enables facile loading with a variety of therapeutics as well as tunable release profiles of multiple therapeutics. Specifically, the PEG-PLA nanoparticles allow for loading with hydrophobic therapeutics while a hydrophilic therapeutic can be entrapped within the aqueous phase of the gel.[2] In this work, we have quantified the factors that affect the mechanical properties and release profiles of the gels, demonstrating the range of their tunability. Additionally, we explored other nanoparticles and biopolymers for the fabrication of similar materials. Lastly, we studied the use of this injectable hydrogel as a cell carrier with osteogenic properties. Human mesenchymal stem cells (hMSCs), which are capable of differentiating into bone, cartilage, and fat cells, were encapsulated in the gel and survived injection through a 28 gauge needle into three-dimensional collagen hydrogels. Additionally, PEG-PLA nanoparticles were loaded with dexamethasone to stimulate osteogenic differentiation of the encapsulated hMSCs. We are currently exploring the use of this drug-loaded cell carrier as an injectable system for bone healing. Overall, we present an injectable hydrogel with tunable properties for the controlled delivery of therapeutics with facile synthesis and minimally invasive application *in vivo* that is capable of serving as a stimulatory cell carrier.

---

[1] Appel, E.A., Tibbitt, M.W., Langer, R., Nature Communications., **2014**, *in revision*.

[2] Greer, J.M., Appel, E.A., Tibbitt, M.W., Langer, R., *manuscript in preparation*.

Validation of High Throughput Electrochemical Gas Sensing Screening System

Zixuan Wang

November 17, 2014

UNIVERSITY OF SOUTH CAROLINA

Background

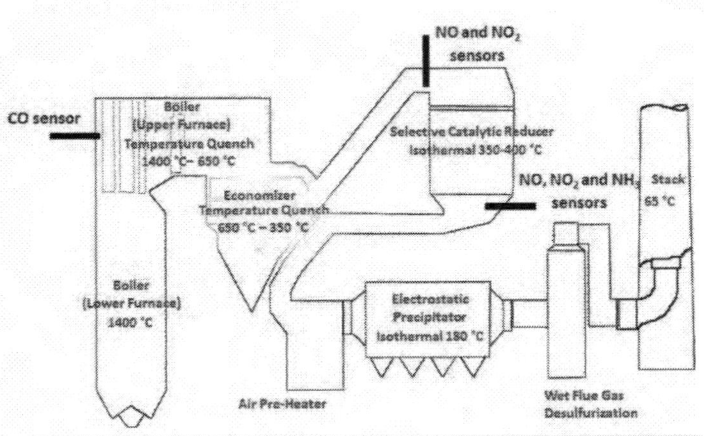

Harsh combustion environment - 800 ° C
Accurate and selective to target gas

UNIVERSITY OF SOUTH CAROLINA

High Throughput Methodology

Subatomic difference affect large scale

↓

Not feasible to individually test material

↓

High throughput synthesis and characterization methods

UNIVERSITY OF SOUTH CAROLINA

High Throughput Methodology

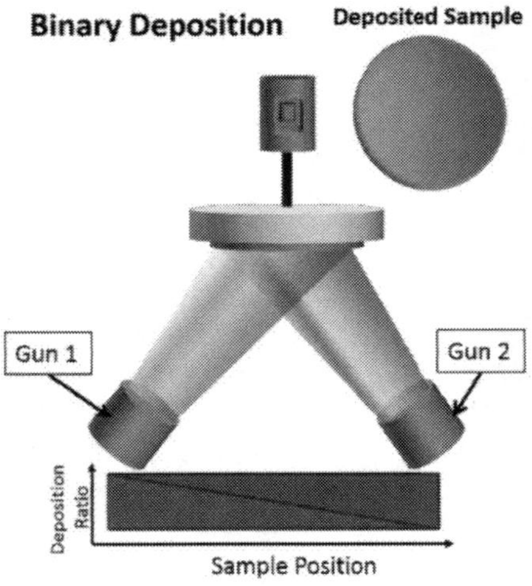

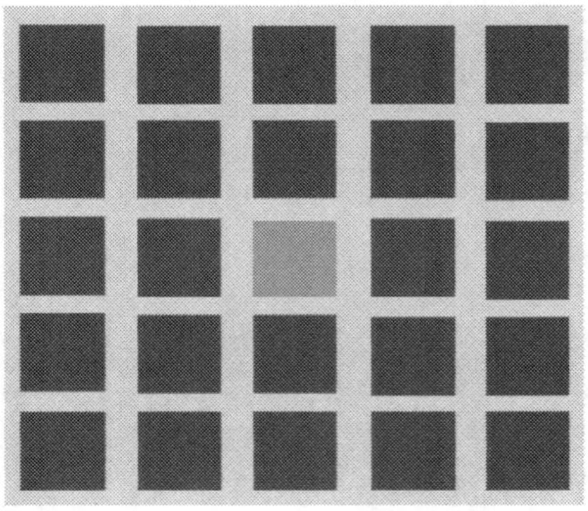

- Normalize affect of distance on the potential

UNIVERSITY OF SOUTH CAROLINA

Development of Uniform Standard Sample

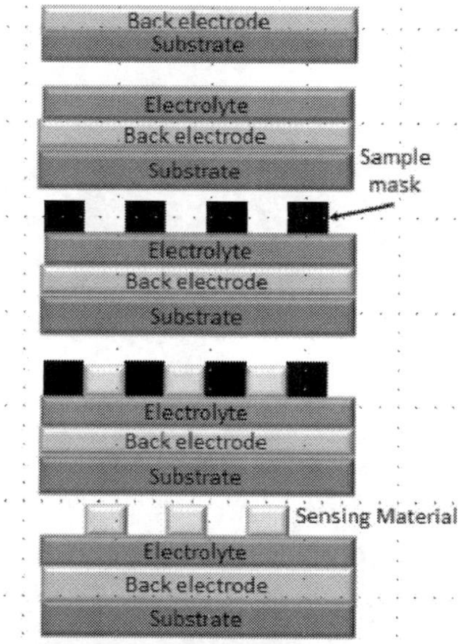

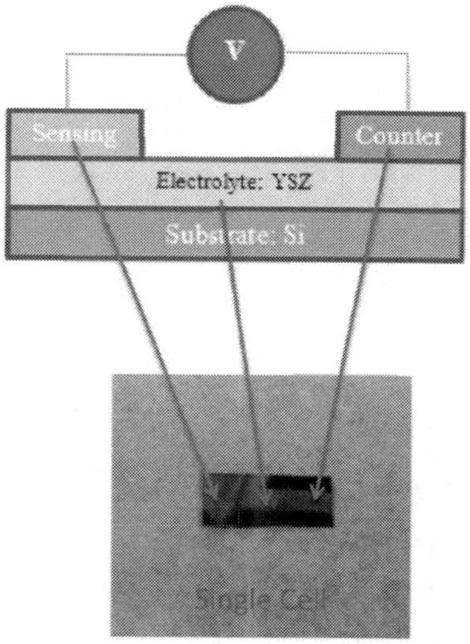

UNIVERSITY OF SOUTH CAROLINA

Surface Mechanisms

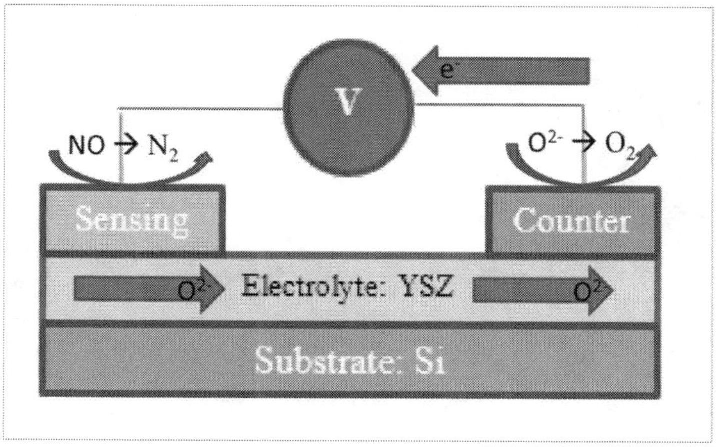

UNIVERSITY OF SOUTH CAROLINA

Sensors and Actuators B 121 (2007) 652-663

Reflectometry Results - YSZ

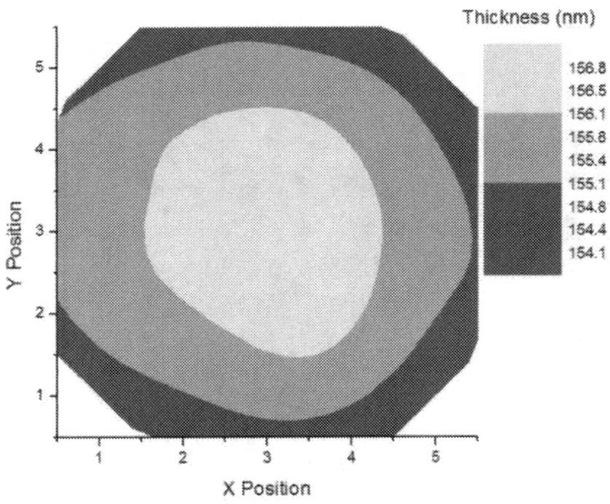

- Results confirm relative uniformity of sample (1.8% difference)

UNIVERSITY OF SOUTH CAROLINA

XRD Results - YSZ

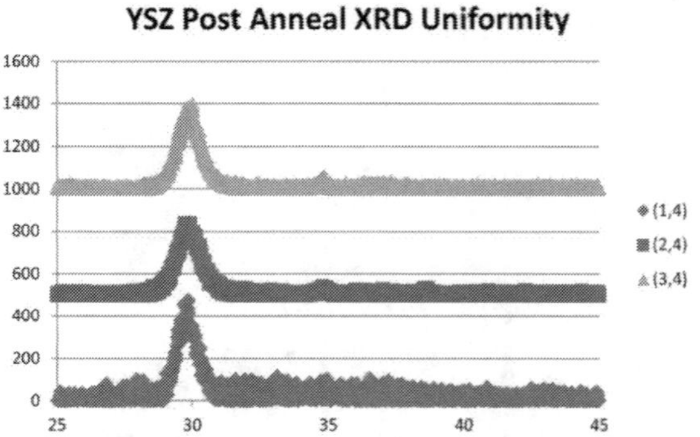

(1,4)	29.745	
(2,4)	29.836	34.848
(3,4)	29.876	34.859

- XRD analysis using Jade
- YSZ Cubic:
  - (111): 29.675°
  - (200): 34.398°
- Confirm identical phases

UNIVERSITY OF SOUTH CAROLINA

Reflectometry Results – WO_3

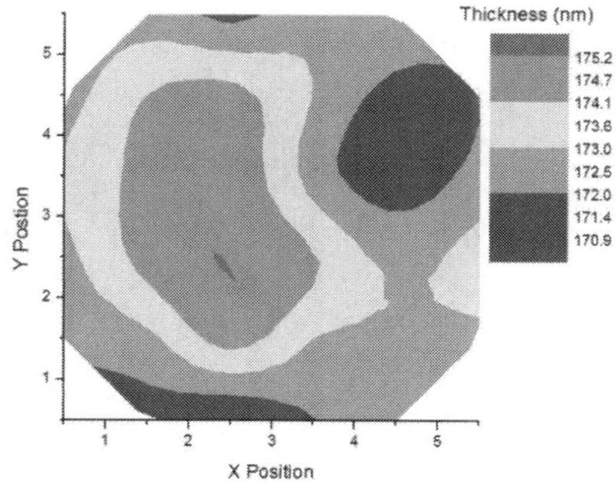

- Confirms relative uniformity (2.5% difference)
- Evidence some parts of the plate that are slightly thicker or thinner

XRD Results – Post Annealed WO_3

(1,4)	23.143	23.653	24.362
(2,4)	23.140	23.671	24.397
(3,4)	23.127	23.641	24.377

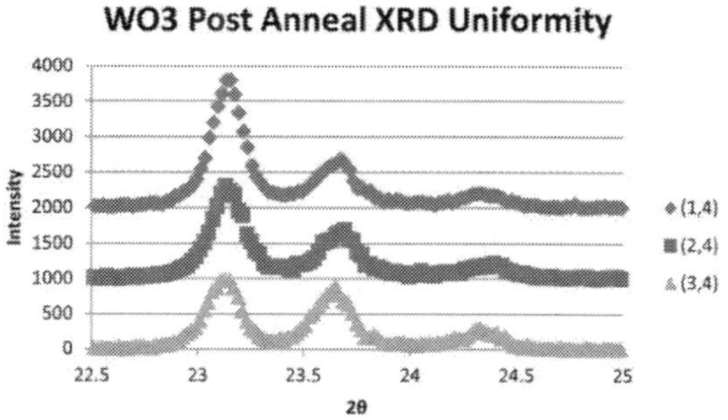

- XRD analysis using Jade
- WO_3 monoclinic:
  - (002): 23.118°
  - (020): 23.582°
  - (200): 24.365°
- Confirm identical phases with phase shifts present

Reflectometry Results - WO_3-YSZ

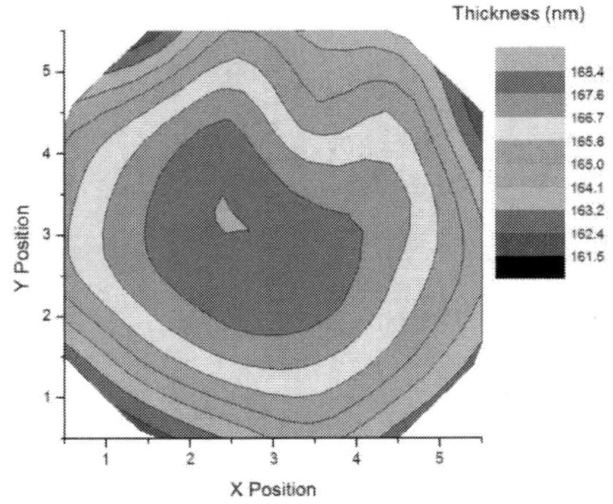

- Sputtering conditions: same
- YSZ held constant
- Confirms relative uniformity (4.3% difference)

UNIVERSITY OF SOUTH CAROLINA

XRD Results – WO_3-YSZ Post-annealed

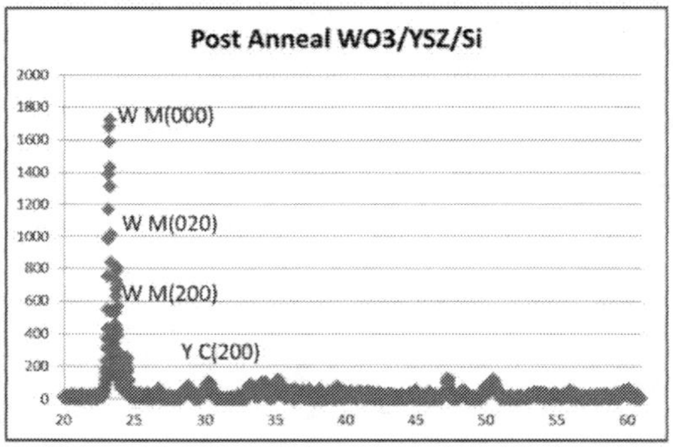

WO_3 monoclinic:
- (002): 23.118°
- (020): 23.582°
- (200): 24.365°

YSZ Cubic:
- (111): 29.675°
- (200): 34.398°

UNIVERSITY OF SOUTH CAROLINA

Electrochemical Cell Setup

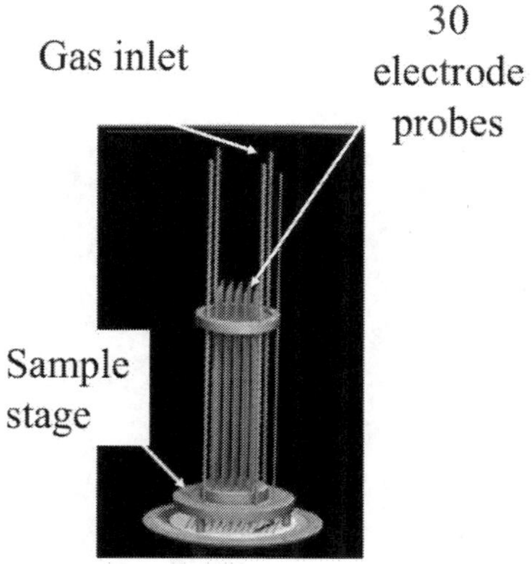

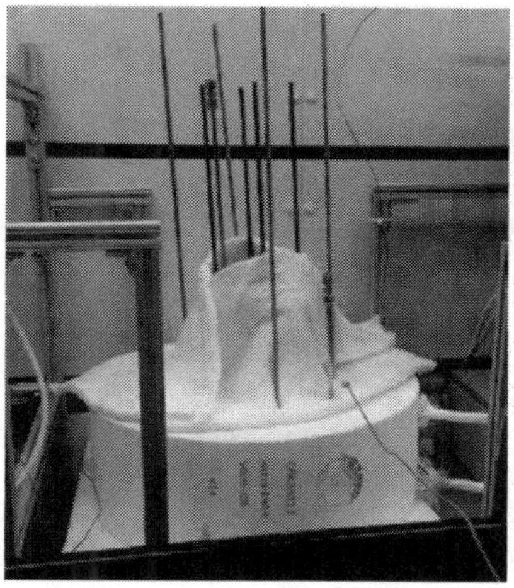

Single Electrochemical Cell Tests

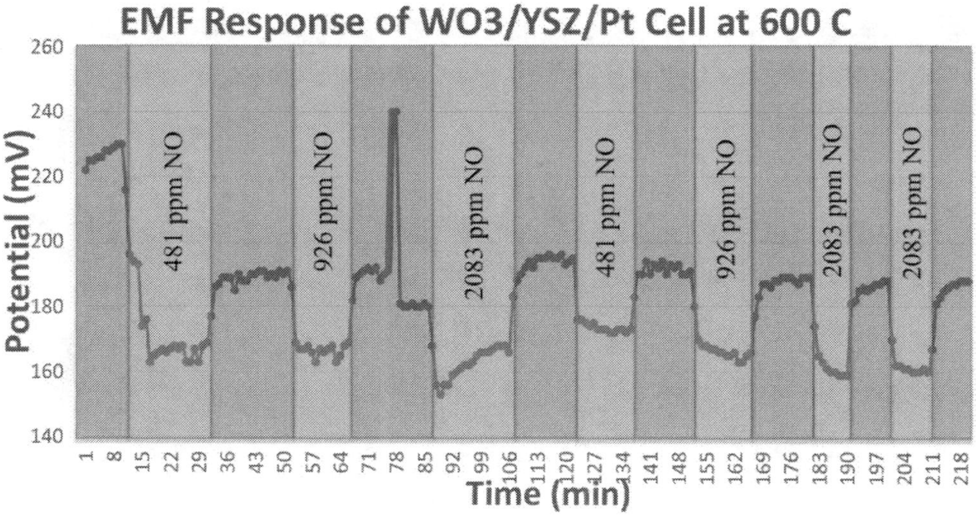

Electrochemical Cell Tests

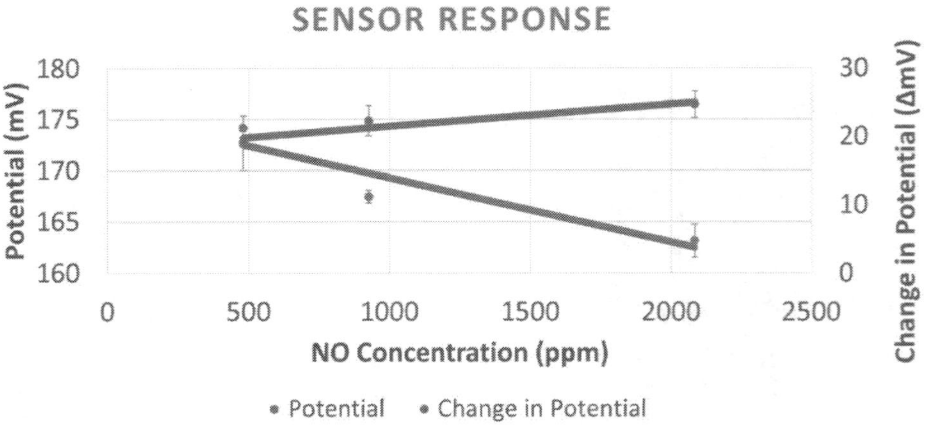

Multielectrode Cell Tests

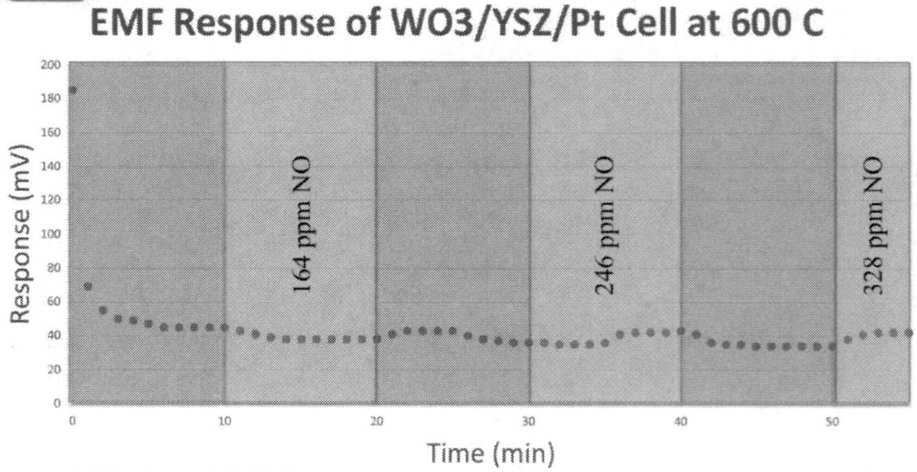

Conclusions

- Designed planar sensor array
  - Use for combinatorial screening of electrochemical gas sensors
- Confirm deposition uniformity
  - Reflectometry: thickness
  - XRD: phases
- Cell potential
  - Changes valid use of planar sensor design

UNIVERSITY OF SOUTH CAROLINA

Future Work

- Create multielectrode with increased surface area/perimeter of the electrodes
- Normalize potential to distance of electrodes
- Create multielectrode with different composition gradient

UNIVERSITY OF SOUTH CAROLINA

Acknowledgements

- National Science Foundation (NSF)
- University of South Carolina – College of Engineering and Computing
- Strategic Approaches to the Generation of Electricity (SAGE)
- Principal investigators: Dr. Jochen Lauterbach and Dr. Jason Hattrick-Simpers
- Graduate mentor: Ben Ruiz-Yi

UNIVERSITY OF SOUTH CAROLINA

394216 Long-Range Correlations in Liquid Water

Abstract:

The purpose of this research project was to observe water molecules' environments to determine whether or not the system forms identifiable clusters of similar environments. The different clusters aid in proving the hypothesis that different water patches at differing high and low densities values form in a size of 1nm in a system at room and lower temperatures. Molecular dynamics computer simulations were done using the dl_poly classic program with the water model potential TIP4P/2005. The simulations were run at various configurations for density, temperature, and pressure. The results were read by a program, which took the file from the simulations and determined the coordinate positions of the water molecules. Different clusters were found based on the results. These results are indicative of distinctive structures formed in supercooled water that may lead to the water patches of varying densities.

394155 Frictional Study of Polyethylene Glycol Monolayers on Silica Substrate

Abstract:

Lubricating micro- and nano-electromechanical systems (MEMS and NEMS) has become increasingly challenging as devices are continuously miniaturized. As the dimensions of these systems decrease, their surface area to volume ratios increase significantly and frictional forces become dominant where moving parts exist. Solid-state lubrication via self-assembled monolayers presents a promising alternative to traditional macro-scale lubricants which are too bulky to use in such small systems.[1,2] Previous studies have shown that self-assembled monolayers containing alkylsilanes and mixed alkylsilane and perfluoroalkylsilane chains reduce friction coefficients compared to equal length systems. Although, hydrocarbon-based monolayers have been widely studied and shown to improve frictional properties and device life, mixed monolayers impart the beneficial properties of perfluoroalkylsilanes (low critical surface tension and high thermal and mechanical stability) to the film. Furthermore, a mixed length configuration with longer alkylsilane chains relative to the perfluoroalkylsilane chains leads to a liquid-like layer formed by the taller chains that fold over and align with the shearing direction.[3]

In this study, we investigate polyethylene glycol (PEG) monolayers using molecular dynamics simulations as a potential solid-state lubricant for NEMS. The presence of oxygen atoms within the chains render these chains more flexible than their alkyl- and perfluoroalkyl-counterparts. Thus, even sparsely functionalizing a surface leads to a uniform film. We have investigated the tribological properties of such films across a range of surface coverages and chain lengths. Preliminary results show the effects of surface coverage on friction coefficients is diminished as chain length is increased. This suggests that equivalent friction coefficients could be achieved while reducing separation distances by using longer chains on sparser surfaces.

394001 A Machine-Learning Model to Predict Activation Energies of Hydrogenation Reactions

Abstract:

The goal of this research is to develop a machine-learning model to predict activation energies associated to the hydrogenation of organic compounds in the presence of specific catalytic substrates. The current method of predicting activation energies involves a linear relationship with the change in Gibbs free energy of the reaction, which can be very inaccurate. Other methods include experimentation and simulation, both of which have significant cost and/or time drawbacks. This research seeks to improve on these methods to develop a means to predict activation energies quickly, accurately, and with minimal cost.

393764 Microscopic Modeling of the Self Assembly of Poly(ethylene oxide)-Poly(propylene oxide)- Poly(ethylene oxide) (PEO-PPO-PEO) Block Copolymers: Critical Micelle Concentrations

Abstract:

Single chain mean field theory (SCMFT) is applied as a coarse-grain simulation model for Pluronics, a trademarked branch of triblock copolymers. The microscopic model is used to quantify the micellization process of the nonionic class of surfactants at 37 degrees Celsius for drug delivery applications. The simple model with only three interaction parameters is shown to be able to reproduce with high accuracy experimental data for the critical micelle concentrations of different chain lengths of the polymers. These parameters are physically significant and are close to the experimental values from Flory models. In addition, aggregation number of the surfactants and volume fraction profiles are explored. The effects of the chain lengths of hydrophobic tails and hydrophilic heads on CMC values are discussed. SCMFT is transferable in surfactants of different ABA block conformations and other more complex polymer structures where previous simulations have not excelled.

Recommendations for Legislative Actions to Reduce Carbon Emissions in the Electricity Production Sector

Sam White

American Institute of Chemical Engineers
University of Michigan
Washington Internships for Students in Engineering

August 2014

Executive Summary

There is currently a movement in America pushing for environmental effects to be taken into consideration when assessing new and existing business ventures. As emphasized by President Obama"s 2014 State of the Union Address, a major area for this movement to make progress is within the energy sector [1]. Electricity production currently produces the most greenhouse gas emissions of any energy sector in the US, making it the energy sector where the largest environmental impact can be made with proper policy action [2].

Luckily, currently-available, low-emission renewable energy technologies are most well suited for electricity production over any other energy sector application. As a result of recent government regulations combined with high coal prices, coal plants, which currently dominate the electricity production landscape, are being retired at an accelerated rate [8]. This, combined with increased electricity demand, has opened up the door for new electricity production technology to be implemented. However, due to the low price of natural gas resulting mainly from the onset of fracking, efforts to implement new renewable-sourced electricity production facilities have been stalled in favor of natural gas power plants. While natural gas plants produce only half the carbon emissions of coal, and do not produce harmful byproducts such as sulfur dioxide or mercury, they are not as environmentally friendly as the other available alternatives, such as nuclear, solar, or wind [11] [12] [13] [14].

The fact that natural gas dominates the shift away from coal is not surprising, however, given the current economic landscape. The figure below shows the levelized cost of electricity (LCOE) for select available electricity production technologies, giving a $/Mwh cost of production by factoring in all applicable costs (construction, fuel, maintenance, operation, etc.) [23].

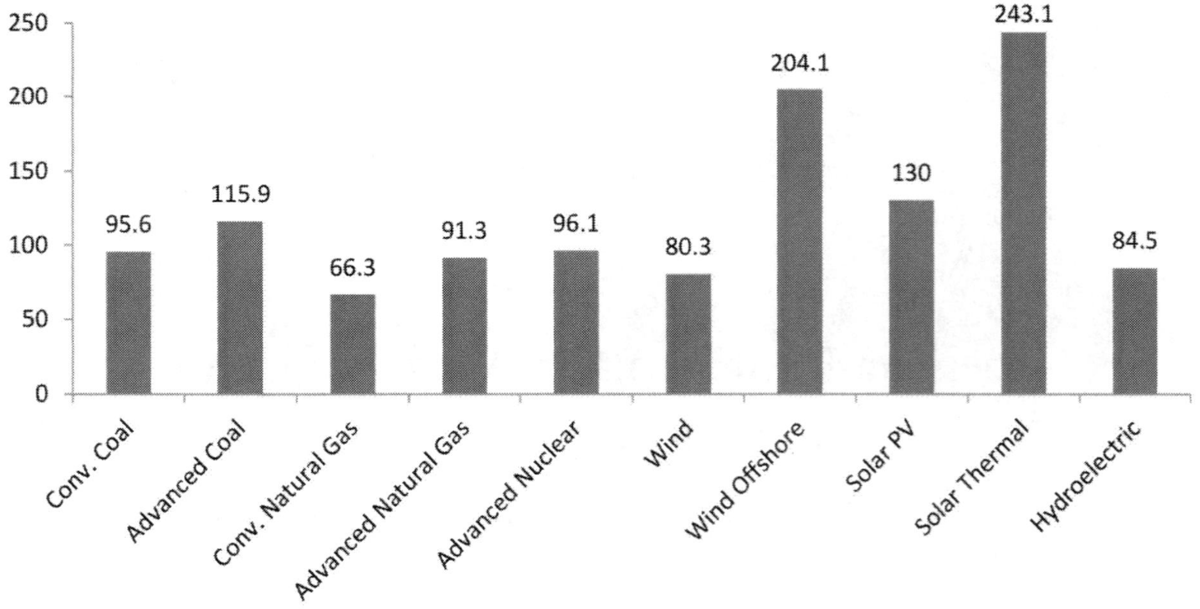

US Average LCOE for Plants Entering Service in 2019, in 2012 $/Mwh [23 EIA LCOE]

As seen in figure above, conventional natural gas currently offers the cheapest form of electricity production. It is important to note, though, that some alternatives like nuclear or wind are already competitive with other conventional electricity production methods like coal. Other alternatives, like solar or offshore wind, are still too expensive to be competitive at the moment, mainly due to the fact that electricity production from renewable sources is still too variable. With the advent of advanced electricity storage technology this issue would be eliminated, but such technology is currently unavailable or too expensive to be implemented economically [19].

Because of their positive environmental impacts, alternative electricity production methods are given some assistance by the federal government in order to make them more economically competitive. These economic aids mainly come in two forms: tax credits and loan programs. On the tax side, an installment tax credit allows solar production facilities a credit equal to 30% of the cost of implementation, while a production tax credit allows wind, hydro, and geothermal plants to receive a .23 cents/kwh credit. However, the installment tax credit is only available until the end of 2016, and the production tax credit already expired at the beginning of 2014 [26]. On the loan program side, the government has extended loans to help both nuclear and renewable power plants, though the renewable plant loan program expired in 2011 [27] [28].

On top of aid being given to alternative electricity production methods, the federal government can also level the playing field by putting a burden on conventional electricity sources. This mainly comes in one of two forms, a carbon tax or a cap-and-trade program, both of which seek to charge electricity producers for the amount of carbon emissions they produce. On top of producing revenue for the federal government, these programs help incentivize electricity producers to innovate new ways to reduce their carbon emissions. Though no such program has ever been implemented, a cap-and-trade program was able to at least pass the House in 2009 [24] [25].

Even small actions by the federal government can have noticeable effects in shifting the production methods used in the electricity production sector. Just a relatively small tax such as $10/metric ton of carbon emissions would cause significant carbon emission reductions compared a scenario in which no government actions are taken [33]. Additional analysis specific to natural gas and renewables shows that any carbon tax would actually increase the amount of natural gas-based electricity production for the next couple of decades, and increase the amount of renewable-based electricity production indefinitely. Additionally, if current favorable tax policies for renewables are renewed long-term, the use of renewables is significantly higher than those projected given current tax policy deadlines, especially after 2030 [34] [35].

As a result, it is recommended that a two-pronged approach be taken to limit the amount of economic disturbance while ensuring the electricity production sector makes a greater shift towards renewables given the reduction of coal use. First, a carbon tax equivalent of $10/metric ton of carbon emissions should be enacted to reflect the social cost of emissions. Each year this tax would increase 5% per year after inflation in order to push companies to innovate new carbon reduction technologies. Second, the production tax credit should be extended for all renewable electricity production methods using the current credit of .23 cents/kwh. This credit would decrease 5% per year after inflation to mirror the carbon tax in order to account for the increased rate of improvement in renewable technologies and the relative advantage given by the increasing carbon tax.

It is important that such a program be long-term in order to allow industry to accurately model the economic landscape their projects will enter. Given the current gridlock in Washington and the long-term nature of implementing new electricity production facilities, a long-term tax plan is the only way to ensure economic stability. Factoring in only revenue from taxes on natural gas use and costs from credits extended to renewables use, such a program would pull in an average surplus of $5.5 billion per year between 2015 and 2040; factoring in coal would increase this figure significantly, particularly in the near future. Feeding at least part of this surplus into loan programs for nuclear plants would ensure that an increased amount of low carbon electricity production facilities are able to come to market.

Table of Contents

Table of Figures

Table of Tables

1.0 Introduction

In his 2014 State of the Union address, President Obama stated that "[O]ne of the biggest factors in bringing more jobs back is our commitment to American energy"[1]. He stressed the importance of "a cleaner energy economy" and acknowledged that "[it] won't happen overnight, and it will require tough choices along the way." Though he touted the success of his "all-of-the-above energy strategy", and highlighted the advantages of natural gas and solar, the President laid out no definitive plans for how the US would approach energy production in the future. However, with the release of recent carbon emission regulations, the executive branch seems to be taking at least some action to direct the country towards a cleaner energy economy. However, if a cleaner energy economy is going to happen in America, the legislative branch needs to get more involved by developing and implementing a long-term energy policy plan.

Of all the areas of energy production in the America, the most promising area for change is the electricity production sector, since a vast majority of alternative energy sources are most viable for electricity production instead of applications like transportation or residential heating. Luckily, this is also the energy sector in which the most impactful changes can be made: in 2012, electricity production was the largest single contributor to greenhouse gas emissions in the US, accounting for 32% of greenhouse gas emissions as seen in Figure 1 below [2].

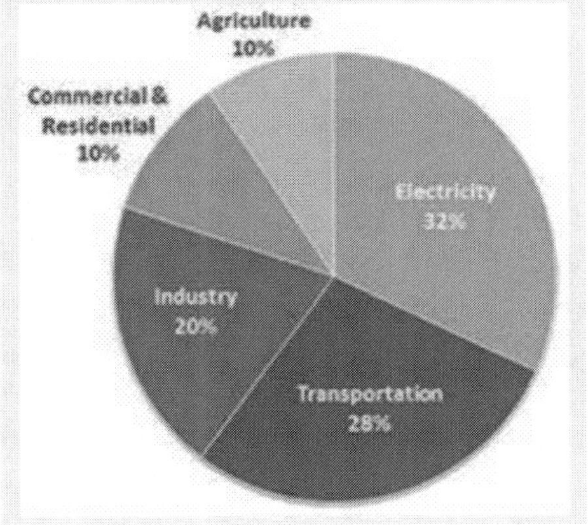

Figure 1. Total Greenhouse Gas Emissions by Economic Sector in 2012 [2]

The implementation of a cleaner energy economy has shown to be a popular policy among voters from both parties. According to a Washington Post – ABC News poll from June 2014, 70 percent of voters say that the federal government should require limits to greenhouse gases from existing power plants. 57 percent of Republicans, 76 percent of Independents, and 79 percent of Democrats support state-level limits on greenhouse gas emissions. Most significantly, voters are willing to pay extra for energy if it means a reduction in greenhouse gas emissions. "Asked whether Washington should still go forward with limits [on energy-sector greenhouse gas emissions] if „they significantly lowered greenhouse gases but raised your monthly energy bill by 20 dollars a month," 63 percent of respondents say yes, including 51 percent of

Republicans, 64 percent of Independents and 71 percent of Democrats" [3]. These results suggest that Americans on both sides of the aisle are in support of implementing a cleaner energy economy.

In addition to the popularity of a cleaner energy economy, it also makes sense from an environmental and health stand point. On top of the fact that the reduction of greenhouse gas emissions reduces the rate of climate change, it brings along health benefits as well. With the release of its Clean Power Plan in June 2014, which seeks to cut carbon pollution from the power sector by 30% by 2030 by imposing stricter emission limits on existing plants, the EPA projects that "annual public health benefits will total somewhere between $55 billion and $93 billion by avoiding up to 100,000 asthma attacks and 2,100 heart attacks each year." By contrast, they predict compliance cost of $7.3 to $8.8 billion per year, significantly lower than the expected public health benefits [4]

However, despite the ongoing executive branch actions regarding energy policy, the legislative branch has taken relatively little action regarding energy policy lately, mainly due to party-politics-driven gridlock. At the beginning of 2014, the Production Tax Credit for renewables was allowed to expire for the first time since 2004, and only the fourth time since its enactment in 1992 [5]. In May of 2014, the Senate allowed a bipartisan energy efficiency bill to die due to disagreement over other energy related issues, such as the fate of the Keystone XL pipeline and possible legislative response to President Obama"s climate change regulations. In fact, it has been 7 years since Congress passed a major energy bill, which means there has been no major legislative action in response to the changing energy landscape as a result of the shale fracking boom [6].

A plan for implementing a cleaner energy economy becomes even more important as energy demands increase, particularly when it comes to electricity production. As electricity demand increases, more electricity production facilities are built, and how these facilities produce electricity greatly affects their environmental impact. According to the EIA"s Annual Energy Outlook 2014 report, US electricity demand will increase from anywhere between 14% and 41% between 2012 and 2040, though their best estimates suggest that electricity demand will increase by 29% over this time period. Additionally, the growth of electricity demand is projected to increase sometime around 2025, marking the beginning of an upwards slope in electricity demand growth for the first time in 75 years [7].

In order to meet this demand and fill the gap left behind by plant retirements, new electricity production facilities need to be built. As a result of recent emission regulations like the Clean Power Plan, coal power plants are being retired at an accelerated rate. However, due to the low price of natural gas in recent years thanks to fracking, much of this gap has been filled by natural gas fired plants instead of alternatives, like nuclear or renewables. This is mainly due to economics: even the cheapest alternative energy sources, which are on the verge of being if not already cost competitive to coal, simply cannot compete with the low price of natural gas. While natural gas is a lower-emissions alternative to coal, it has emissions nowhere near as low as alternative electricity sources.

The onset of natural gas is not a bad thing, however: using natural gas as a bridge fuel to reduce emissions in the immediate future while hoping that renewables become more cost competitive in the future is not unreasonable given the rate of development and investment in renewable technologies. Natural gas also offers the unique advantage of "spinning reserve", a source that can quickly be powered up to meet rapid fluctuations in demand. This means natural gas can offer a flexibility that base load electricity sources like coal or nuclear cannot, while acting as a buffer against the variability of renewables like solar or wind.

Embracing natural gas in the here and now is also economically responsible, as it keeps electricity prices low by offering a cheap alternative to coal and helps support the growing American natural gas production market. However, without government action, this shift towards natural gas will dwarf any efforts to make renewables a larger share of our electricity production sector. If the government is serious about using natural gas as a bridge fuel to a cleaner energy economy in the electricity sector, a long-term policy needs to be enacted to slowly guide the market in the right direction.

2.0 Overview of Electricity Production Methods and Their Environmental Impacts

In order to discuss energy policy, it is first important to establish the electricity needs of the country, as well as the current capabilities of various electricity production methods and their effects of the environment. As seen in Figure 2 below, coal and natural gas currently dominate the electricity production sector, with nuclear and renewables making up the bulk of the remaining production.

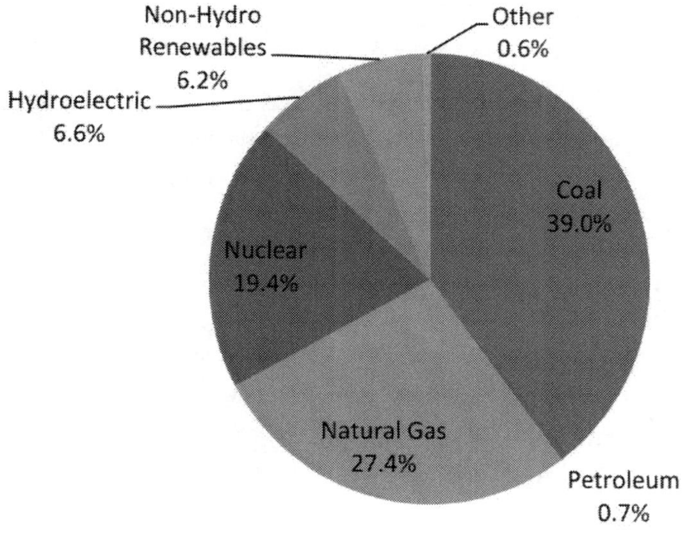

Figure 2. Share of Total US Electricity Production by Source [Adapted from 8]

2.1 Coal and Natural Gas

The two main fossil fuel based / conventional energy producing sources, coal and natural gas power plants are currently the largest source of electricity in America, accounting for nearly two-thirds of the country"s total electricity production [8]. Coal plants are relatively easy to keep in operation over a long time scale, but it is difficult for them to make relatively quick changes to production capacity in response to the changing needs of the grid, particularly during peak-usage hours. As a result, coal plants are very useful in providing a baseline power supply somewhere between peak and off-peak requirements.

Natural gas, on the other hand, allows for a very flexible energy production method. Gas flow rates into a furnace can be controlled with relative precision, unlike solid input streams found in coal plants. As a result, natural gas plants can be very responsive to the changing needs of the energy grid. In fact, natural gas plants are typically used to augment other energy production facilities for this very reason.

From an environmental perspective, the EIA considers conventional electricity sources like coal and natural gas as being the only contributors to CO_2 emissions in the electricity production sector. In 2010, the EIA reported that 1969 million metric tons of CO_2 were released by coal plants nation-wide, while 1285 million metric tons of CO_2 were release by natural gas plants [9].

However, coal tends to have other unique environmental impacts. First, burning coal tends to release nitrogen oxides, which tend to have larger warming potentials than CO_2. For example, N_2O has a SAR 100 year warming potential of 310, meaning that one ton of N_2O has 310 times the impact on global warming compared to one ton of CO_2 over the course of 100 years [10]. From a life cycle assessment standpoint, coal plants produce 974 metric tons of CO_2 equivalent per gigawatt-hour, which accounts for this increased warming potential as well as CO_2 emissions due to non-production factors, such as plant construction or fuel transportation [11]. Coal also tends to produce SO_2, which, along with nitrogen oxides, is the major man-made cause of acid rain. According to the EPA, "In the United States, roughly two-thirds of all SO_2 and one-quarter of all nitrogen oxides come from electric power generation that relies on burning fossil fuels, like coal" [12]. Additionally, coal plants are a major source of mercury, a highly toxic element. According to the EPA, "Coal-burning power plants are the largest human-caused source of mercury emissions to the air in the United States, accounting for over 50 percent of all domestic human-caused mercury emissions" [13].

In the case of natural gas, environmental impacts are lower than that of coal. According to the EPA, "… the burning of natural gas produces nitrogen oxides and carbon dioxides, but in lower quantities than burning coal". In fact, from a life cycle assessment standpoint, natural gas plants produce 469 metric tons of CO_2 equivalent per gigawatt-hour, less than half that of coal [11]. Additionally, the EPA states that "[e]missions of SO_2 and mercury compounds from burning natural gas are negligible". [14]. While the EPA also mentions that unburned natural gas being released either from a power plant or during transportation is a concern, though in 2009 the

EIA stated that "[i]ncreases in energy-related methane emissions [were] largely from underground coal mining", suggesting that methane releases are more prevalent in the procurement of coal than in the use and transport of natural gas [14] [15].

2.2 Nuclear

In the US, nuclear energy makes up nearly one-fifth of all electricity production [8]. According to the World Nuclear Association, "it [nuclear power] is especially suitable for large-scale, continuous electricity demands, which requires reliability (i.e. base load)" [16]. In this way, nuclear power plants are analogous to coal plants in the sense that they are very useful in meeting baseline power needs, but have trouble making quick changes in response to fluctuating peak-hour demand.

From an environmental standpoint, "nuclear power plants do not emit CO_2, SO_2, or nitrogen oxides as part of the power generation process" according to the EPA [17]. However, as with all power production facilities, the construction of nuclear power plants leads to non-negligible greenhouse gas emissions. Based off a compilation of life-cycle emissions studies gathered by the Nuclear Energy Institute (NEI), nuclear energy"s life-cycle emissions are significantly lower than those of coal or natural gas plants, and are in fact comparable to those of renewable like solar, wind, or hydro [18]. From a life cycle assessment standpoint, nuclear produces 15 metric tons of CO_2 equivalent per gigawatt-hour, which is the second-lowest behind wind, and only about 1.6% that of coal plants [11]. Much of these emissions come from the construction of plants, which are significantly larger than other electricity production facilities, and the transportation of fuel.

2.3 Solar and Wind

As shown in Figure 2, only about 6% of electricity production in the US comes from non-hydro renewable sources. Of this 6%, a majority comes from solar (in the form of both photovoltaic cells and solar thermal) and wind [8]. Of all of the available technologies, solar and wind are the most challenging to integrate into the grid due to the fact that they are considered variable sources: if the sun doesn"t shine or the wind doesn"t blow, these technologies are not actively putting energy into the grid, meaning demand has to be met by some other means [19] [20]. In order to offset this, large-scale production facilities are looking into using some form of energy storage system. This would allow excess energy collected during times of sunshine or wind gusts to be used when the sun doesn"t shine or the wind doesn"t blow [19]. However, the storage capacity of current electricity storage technology is limited. As a result, any useful amount of electricity storage is currently too expensive, which limits the viability of renewables.

From an environmental standpoint, neither solar nor wind produce carbon emissions during their operating lifetime, since their energy is derived from the sun and not from burning fossil fuels [19] [20]. However, like nuclear, solar and wind do have some emission impacts due to the processes used to manufacture materials and construct plants: wind is responsible for 14 metrics tons of CO_2-equivalent per gigawatt-hour, while solar photovoltaics are responsible for

39 metric tons of CO_2 equivalent per gigawatt-hour [11]. As with nuclear plants, many of the emissions related to renewables come from construction of production facilities. In the case of solar photovoltaics, higher life-cycle emissions are seen as a result of the emissions that result from the manufacturing process of the solar cells themselves.

2.4 Other Renewables

Solar and wind are not the only renewable sources of energy available to the US. As seen in Figure 2, nearly 6.6% of electricity production comes from hydro plants. The main issue with hydro is that most of the available sites for dams in the US have already been dammed, meaning there is little room for expansion. As a result, the Obama administration has identified a collection of dams that currently do not produce electricity, but outfitting these dams with turbines would only increase the current US hydro capacity by roughly 15% [21]. Another alternative energy source is geothermal, which utilizes naturally-occurring underground sources of heat to power turbines. While this energy is clean, it is only available on a regional basis: Hawaii and the western US have a large geothermal production potential, while areas like New England do not [22]. Since hydropower does not have a lot of potential for growth, and geothermal is regionally specific, they will not be explored as in-depth in this paper.

2.5 Environmental Impact Summary Table

Table 1 below summarizes the life-cycle environmental impact per gigawatt-hour of the main electricity production methods highlighted in this paper.

Technology	Metric Tons of CO_2 released / GWh
Coal	974
Natural Gas	469
Nuclear	15
Solar	39
Wind	14

Table 1. Life-Cycle Environmental Impact Summary of Electricity Production Technologies [Adapted from 11]

3.0 Economics in the Electricity Production Sector

On top of technological limitations and environmental concerns, economics plays a major role in determining which technologies get used for energy production. On the production side, economic factors tend to play a major role in determining which electricity production technologies are utilized by producers. On the consumer side, the electricity production technology used by electricity providers plays a role in the price of electricity sold to consumers.

3.1 Comparing Costs of Electricity Production Technologies

Due the wide range of variables present in energy production, it can be difficult to compare the cost of one electricity production technology to another. However, a useful metric, levelized cost of electricity (LCOE), lays out the cost of given technologies on a per-MWh basis, factoring in such costs as construction, fuel, operating costs, maintenance, and financing costs [23]. Figure 3 below shows the US average LCOE for plants entering service in 2019, in 2012 dollars/MWh, and does not factor in any public financial support (tax credits, subsidies, etc.).

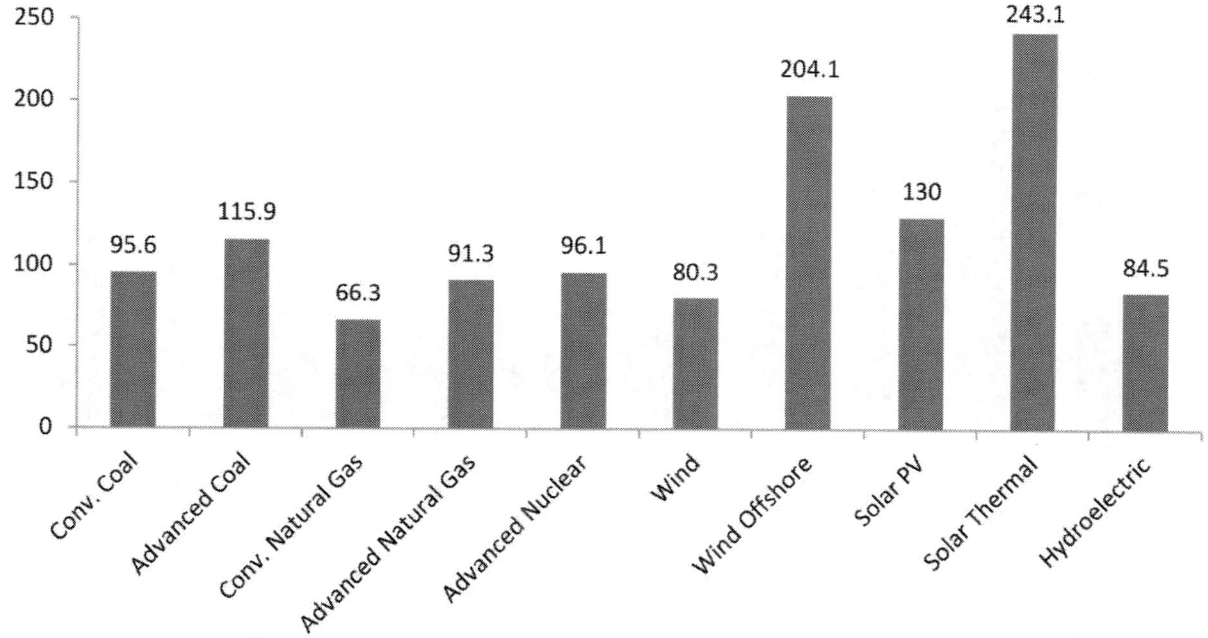

Figure 3. US Average LCOE for Plants Enetering Service in 2019, in 2012$/MWh [23]

As Figure 3 shows, the conventional energy production sources, coal and natural gas, are currently the cheapest available electricity production technologies, which explains why they currently account for such a large portion of US electricity production. Much of this can be contributed to the low price of fuel. Natural gas in particular is benefitting greatly from low fuel prices as a result of the recent American shale fracking boom. However, advanced coal and natural gas facilities, which incorporate carbon capture and sequestration technologies, are significantly more expensive than their conventional counterparts.

While coal and natural gas remain the cheapest electricity production technologies, it is important to note that there are alternatives that are already competitive. For example, the LCOE of nuclear energy is nearly the same as conventional coal and advanced natural gas, and is in fact less expensive than advanced coal. This is particularly noteworthy since nuclear fills the same base load electricity production role as coal, and produces no greenhouse gas emissions during its operational lifetime. It is also important to note that, while many of the renewable energy sources are significantly more expensive than conventional sources or nuclear, wind built on land has the second lowest LCOE, higher than only conventional natural gas.

3.2 Capacity Factor

One factor contributing to the relative expense of renewable sources, such as offshore wind or solar, is their variable nature. Since they are not able to produce electricity at all times of the day, renewable sources suffer from prolonged periods of not operating at peak capacity. The easiest way of quantifying this is through capacity factors, the ratio of actual output to the theoretical output if the plant could operate at full capacity indefinitely. A comparison of capacity factors for each electricity production technology can be seen in Figure 4 below.

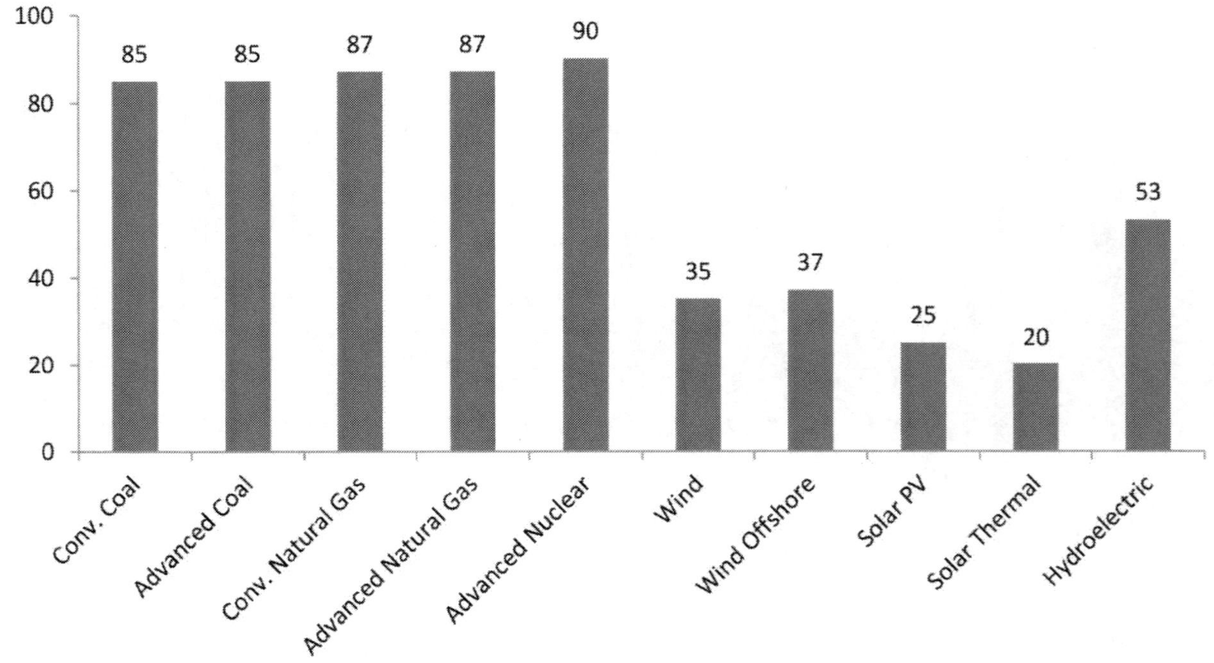

Figure 4. Capacity Factor (%) of Various Electricity Production Technologies [23]

It is important to note that there is no electricity production technology with a capacity factor of 100%: plants do not always run at maximum capacity, and require downtime for things like routine maintenance or system upgrades. However, it is easy to see that renewable sources have significantly lower capacity factors, mainly due to the fact that they cannot produce electricity at all times of the day. Because of this, capital and operating costs of renewable energy plants factor significantly more into their LCOE.

However, with advancements in technology, particularly in the energy storage field, the capacity of renewable sources could significantly increase, and possibly even become comparable with conventional sources. This would cut LCOE by anywhere from 40-70% (depending on current capacity factor). If this electricity storage is cheap enough, even the most currently expensive renewable energy technologies would be competitive with conventional sources.

4.0 Recent Policy Strategies

In recent years, legislators have used a wide variety of strategies to help shape the electricity production sector, most notably by offering some form of economic incentive to desirable technologies. These economic incentives typically come in the form of tax credits, though they may also come in the form of subsidies or guaranteed loan programs. Congress has also made efforts in the past to enact a carbon tax program, but such a program has never been able to make it out of Congress [24] [25].

4.1 Tax Credits for the Electricity Production Sector

The major economic incentive congress has implemented is by extending tax credits to electricity producers, which reduces the amount of taxes they have to pay to the federal government and thus improving their bottom line[1]. Recently, these tax credits have been targeted at renewable electricity resources, most notably solar and wind, though to a lesser extent hydroelectric and geothermal as well [26]. These tax credits have taken on two different forms: an investment tax credit and a production tax credit.

The investment tax credit allows qualifying electricity producers to receive a tax credit equal to a certain percentage of their initial cost of implementation. Solar energy installments are eligible for a 30% credit, while geothermal plants are eligible for a 10% credit. These credits have no limit. As of right now, these tax credits extend to any qualifying facility put in service before January 1, 2017. Wind power facilities put in service between 2009 and 2012 were also eligible for a 30% investment credit, but that has not been renewed [26].

The production tax credit, on the other hand, allows qualifying electricity producers to receive a tax credit based off the amount of electricity produced. As of the end of 2013, this tax credit was 2.3 cents / kWh, and extended to wind, geothermal, and hydropower facilities [5] [26]. If a facility qualifies for both the investment and production tax credit, they are only permitted to claim one of them, at their choosing. The production tax credit extended to any qualifying facility under construction by January 1, 2014, and is available for qualifying facilities for the first 10 years of operation. However, since January 1, 2014, the production tax credit for renewable electricity production facilities has not been extended [26].

4.2 Loan Programs

The federal government also offers loan programs for electricity producers. These loan programs differ from tax cuts mainly in the fact that producers are required to pay back their loan; while there is a cost to the federal government, it is only in the risk of borrowers defaulting,

---

[1] It is important to note that non-profits, such as municipal utilities and utility cooperatives, do not pay taxes, meaning that federal tax cuts do not have an effect on their bottom line. While alternative programs to incentivize non-profits do exist, they are not covered in the scope of this paper.

which has historically been low. Under Title XVII of the Energy Policy Act of 2005, two loan programs were outlined that are still used today: section 1703 and section 1705 [27] [28].

Section 1703 provides loans for "a new or significantly improved technology that is NOT a commercial technology" [27]. Qualifying facilities must also be located in the US. In the current state of the electricity production sector, the only facilities that really meet these requirements are related to nuclear energy; the most recent loans are an $8.3 billion loan for two new nuclear power plants in Georgia and a $2 billion loan for a uranium enrichment plant in Idaho [27]. This program is essential to nuclear power plants, as the most prohibitive cost to entering the market for nuclear plants is upfront capital costs. Due to relatively low operating costs once the plant is built, loan repayment is usually not an issue.

Section 1705, on the other hand, focuses on loan programs for renewable energy production facilities. Unlike section 1703 rules, section 1705 has a time limit, and was only available to projects that had begun construction before September 30, 2011. Since then, this project has not been renewed. Section 1705 loans tend to be lower than their section 1703 counterparts: one of the more expensive section 1705 loans was for $1.3 billion to an Oregon-based wind farm [28].

The major significant difference between section 1703 and 1705 loans is how the Credit Subsidy Cost (CSC) is handled. By law, the CSC, which is "the expected long-term liability to the Federal Government by issuing the loan guarantee", must be paid either by budget appropriations or by the borrower directly. For section 1703, which tend to have more expensive loans and thus larger CSCs, the CSC is paid by the borrower. For section 1705, the CSC was paid for by appropriations [27] [28]. In the American Recovery and Reinvestment Act of 2009, $6 billion was appropriated towards section 1705 CSCs. [29]

4.3 Carbon Tax and Cap-and-Trade Program

On top of offering economic benefits to desirable electricity production methods, Congress can also place economic burdens on undesirable electricity production methods. In the electricity production sector, this can be practically done in one of two ways: either a carbon tax or a cap-trade-program.

A carbon tax works very simply: electricity producers pay an additional tax for each unit of carbon produced during electricity production. This means that carbon-heavy electricity production facilities like coal (and to a lesser extent, natural gas) power plants would pay an additional amount for their electricity production, thus increasing their LCOE. In the same way that tax credits and subsidies seek to level the economic playing field by making renewables cheaper, a carbon tax would seek to level the playing field by increasing the cost of conventional electricity production methods. However, there has never been a carbon tax approved by either chamber of Congress.

An alternative method to a carbon tax is a cap-and-trade program, which seeks to place an artificial limit on the amount of carbon that can be emitted. Once this limit has been determined, electricity producers can buy permits to emit certain amounts of carbon dioxide. The government only sells as many permits as it takes to reach the cap, and then monitors the emissions of each producer to make sure they do not exceed their permitted limit. Producers are able to buy or sell permits to or from other producers as needed.

Cap-and-trade programs create economic incentive to reduce carbon emissions in two ways: first, reducing carbon emissions means that electricity producers need to buy fewer permits, saving them money. Second, if producers reach their permitted cap, their competitors determine the cost of additional permits (since each company now control its own excess permits, not the government). This means exceeding the permitted carbon emission limit can be disproportionately costly, either through the purchase of competitor permits or paying government fines.

In order to ensure long-term carbon emission reduction, cap-and-trade programs also incorporate a permit allowance reduction system. This means that, each year, the amount of permits allowed is reduced, effectively lowering the acceptable amount of carbon dioxide that can be emitted by electricity producers. While a cap-and-trade program did make it out of the House in 2009, the bill was never heard by the Senate, and died when that session of Congress ended [24] [25].

5.0 Projected Electricity Sector Growth: Electricity Sector Capacity Additions for 2013

In order to establish a reference case, it is useful to look at how the electricity production sector has changed recently. Figure 5 below shows the EIA projections for electricity sector capacity additions from 2008.

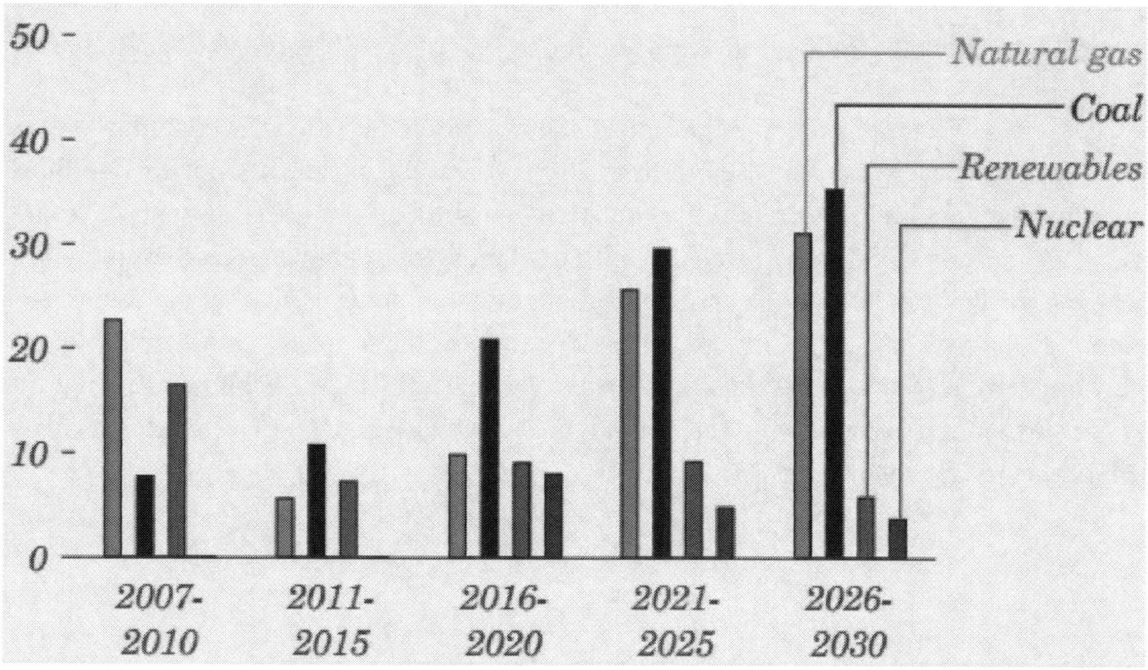

Figure 5. Electricity Generation Capacity Addition Projections by Fuel Type, 2008 (Gw) [30]

As Figure 5 shows, in 2008, it was projected that coal would begin to dominate electricity production additions starting in 2011. Much of this can be attributed to the increasing cost of natural gas, as shale fracking was still in its infancy in 2008. Additionally, renewables, which see a large spike in 2007-2010, were projected to fall by the wayside after 2010 as they begin to be dwarfed by both natural gas and coal.

However, as a result of the shale fracking boom, the reduced cost of renewables, and increased government pressure to reduce carbon emissions, these 2008 projections do not reflect what has actually happened in recent years. Figure 6 below shows the electricity generation capacity additions for 2013.

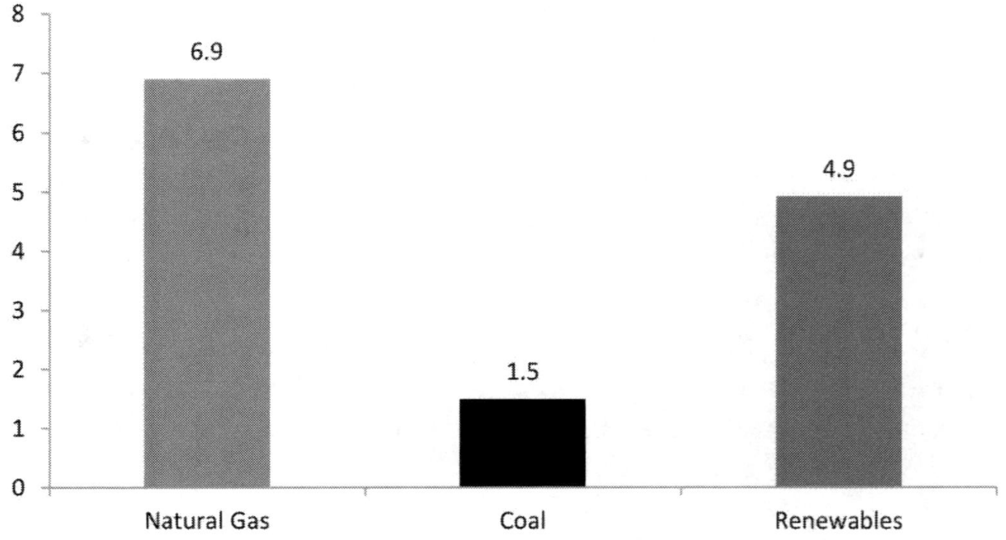

Figure 6. Electricity Generation Capacity Addition by Fuel Type 2013 (GW) [Adapted from 31]

As seen in Figure 6, natural gas dominated new capacity additions, accounting for more than half of the electricity capacity additions in 2013. Renewables made up roughly one-third of new additions, which is attributed to the completion of a number of large scale solar thermal plants in Arizona and California as well as "falling [photovoltaic] costs, aggressive state renewable portfolio standards, and continued federal investment tax credits" [31]. Coal makes up a meager 11% of new capacity additions, in stark contrast to the projections shown in Figure 5.

Long term projections have also changed to reflect the changing economic and political landscape surrounding electricity production. Figure 7 below shows the EIA projections for electricity sector capacity additions from 2014.

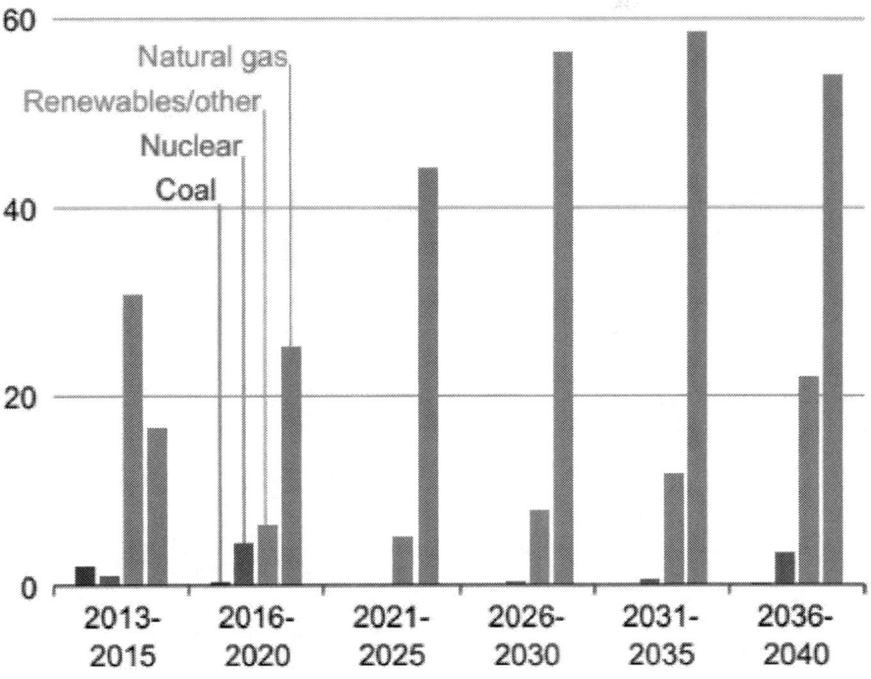

Figure 7. Electricity Generation Addition Projections by Fuel Type 2014 (GW) [32]

As Figure 7 shows, natural gas dominates new additions through 2040, mainly as a result of low gas prices. Coal additions fall virtually to 0 as a result of increased emissions regulations. Most alarmingly, the addition of renewables is actually lower than those projected in 2008 for the period between 2016 and 2035. This is mainly due to the extremely low cost of natural gas relative to any other fuel source, making it take up a larger share of the new additions in the 2014 projections than conventional sources did in the 2008 projections. It is important to note, however, that these projections only show a reference case where no additional government actions are taken.

5.1 Policy Action Effects: Overall Carbon Emissions

Despite the recent transition from coal to natural gas, as well as the increased onset of recently-cheap renewables, carbon emissions from the electricity production sector are projected to increase over the next several decades if current laws and regulations remain in effect. This is mainly attributed to the shift towards natural gas-based electricity production and the shift away from renewables and nuclear as a result of cheap natural gas for the foreseeable future [33]. However, if some form of carbon tax is implemented, carbon emissions could decrease instead increase, as seen in Figure 8 below.

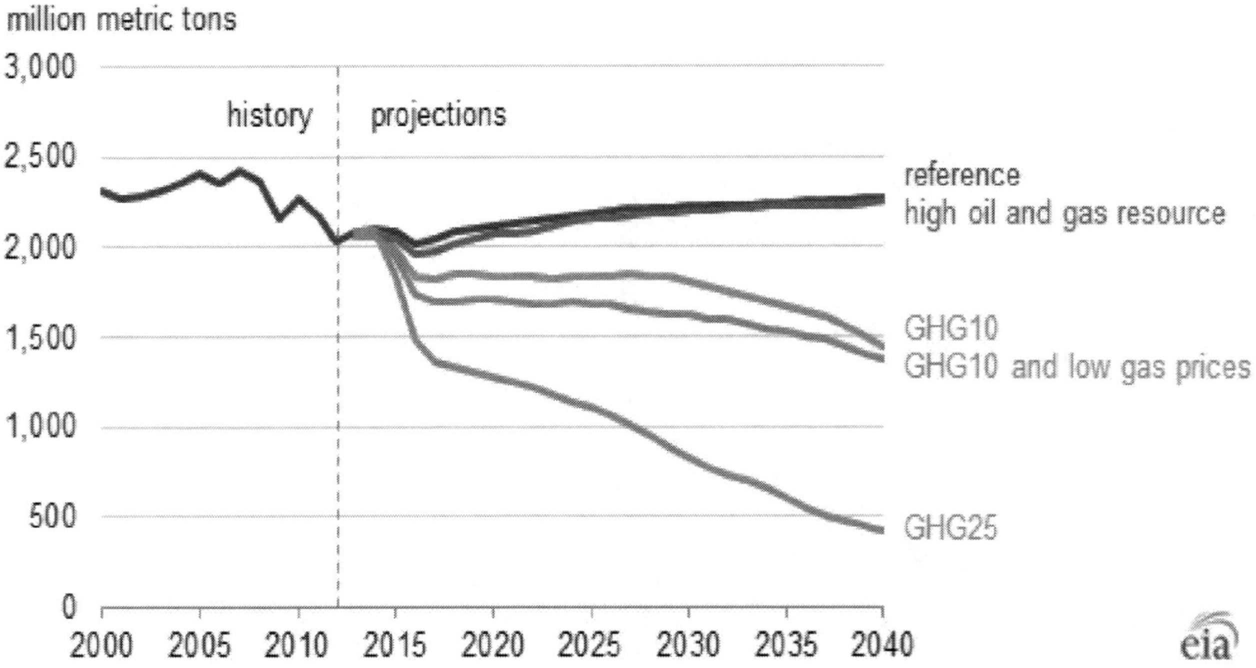

Figure 8. Electricity-Related CO_2 Emissions in Five Cases, 2000-40 [33]

Figure 8 takes into account five different potential cases. The reference case reflects no change in laws or regulations surrounding electricity production and assumes relatively constant economic factors relative to those in 2014. As seen in Figure 8, the reference case shows a 12% increase in emissions up through 2040 [33].

The GHG 10 and GHG 25 cases both refer to the implementation of a carbon tax (or equivalent cap-and-trade permit cost) in 2015 of $10 / metric ton of CO_2 or $25 / metric ton of CO_2, respectively. The GHG 10 case shows a 16% lower CO_2 emission rate in 2025 and a 36% lower CO_2 emission rate in 2040 relative to the reference case. The GHG 25 case shows an even greater decrease in CO_2 emission rates, with a 49% reduction in 2025 and an 82% reduction in 2040. In both cases, there is a large dip in emission rates before 2020, partially due to an accelerated shift away from coal and partially due to an increased implementation of renewables [33].

The High Oil and Gas Resource case assumes the same conditions as the reference case, except that oil and gas resources become more abundant than currently projected. In this case, CO_2 emissions decrease ever so slightly more than in the reference case as a result of increased shift towards natural gas, but by 2040 emission rates are roughly the same as the reference case as a result of producers expressing and increased favor towards natural gas relative to renewables [33].

Finally, the GHG 10 and Low Gas Prices case effectively combines the GHG 10 and High Oil and Gas Resources cases. In this case, reduction of CO_2 emissions before 2020 is greater than those seen in the GHG 10 case, mainly due to an even greater shift from coal to natural gas. However, because a greater share of this shift away from coal is given to natural gas and not renewables, CO_2 emissions do not decrease as drastically after 2030 compared to the

GHG 10 case; it can be extrapolated that CO_2 emissions would actually begin to be higher than the GHG 10 case sometime between 2040 and 2050 [33].

Overall, these cases suggest that some form of action needs to be taken in order to produce a marked reduction in carbon emissions from the electricity production sector. If nothing changes, CO_2 emissions will increase slightly over the next 40 years, in contrast to the significant emission reductions we"ve seen since 2005. Emissions get even worse with an increased gas resource and lower gas prices, which may result with the advancement of fracking and other extraction technologies. Based off of this projection, the only way to continue a trend of CO_2 emission reduction is to accelerate the shift away from coal and ensure that an increased share of that shift is made up of renewables or nuclear.

5.2 Policy Action Effects: Conventional Fuels

When assessing the viability of the cases listed in the previous section, it is important to understand the effects of each action on each specific electricity production method. As discussed previously, there is already a shift towards natural gas and away from coal in response to lower gas prices and increased carbon emission regulations. Figure 9 below shows how possible policy actions may augment that shift by showing the projected natural gas electricity generation in the five previous cases.

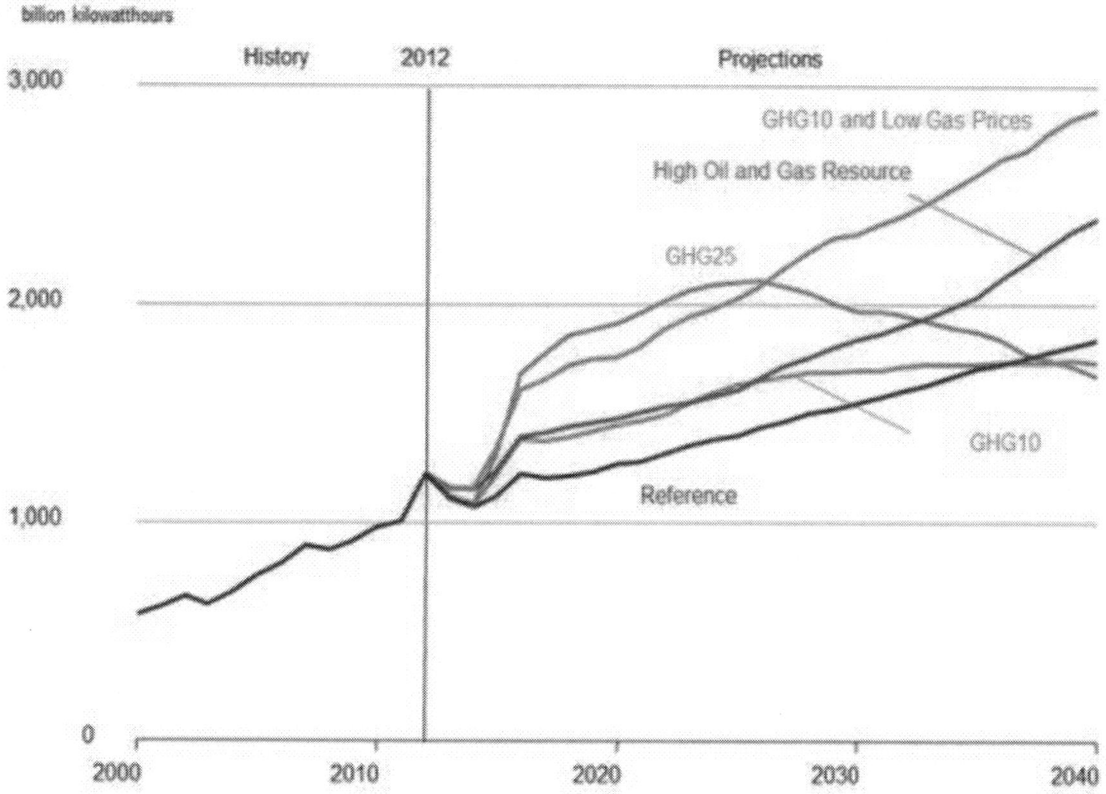

Figure 9. Natural Gas Electricity Generation in Five Cases, 2000-40 [34]

As seen in Figure 9 above, electricity production via natural gas is projected to increase over the next several decades in all cases. In the reference case, we see a relatively stable and constant increase of natural gas fired plants as they replace retiring coal plants; in 2012, 37% of electricity generation comes from coal, and in 2040 that share is 32%, marking a relatively slow shift [34]. With a high gas resource, this shift is more dramatic in the near future, and the long term rate of increase in natural gas plants is also slightly higher than the reference case.

The most interesting part of this projection, however, is the effects of a carbon tax. In all GHG cases, the shift towards natural gas in the near future is increased relative to the reference case. In the GHG 10 case, the share of electricity from coal is 28% in 2025 and 19% in 2040. For the GHG 25 case, the shift is even more drastic, with the coal share being 10% in 2025 and 1% in 2040 [34]. In both cases, the increase in natural gas production is curbed by the increased implementation of nuclear of renewables: in the GHG 10 case, natural gas electricity production levels off around 2030, while in the GHG 25 case, it actually decreases starting in 2025. However, if gas prices remain low, natural gas electricity production continues to increase through 2040, reaching the highest level of any of the five test cases.

5.3 Policy Action Effects: Renewables

Despite the fact that natural gas dominates new electricity production additions, renewables are also seeing an increase in implementation as the electricity production shifts away from coal. However, without some form of government action, the rate of renewable additions will be dwarfed by those of natural gas, leading to the increased carbon emissions seen in Figure 8. Figure 10 below shows the effects of some possible government actions by plotting the projected renewable electricity generation in four cases.

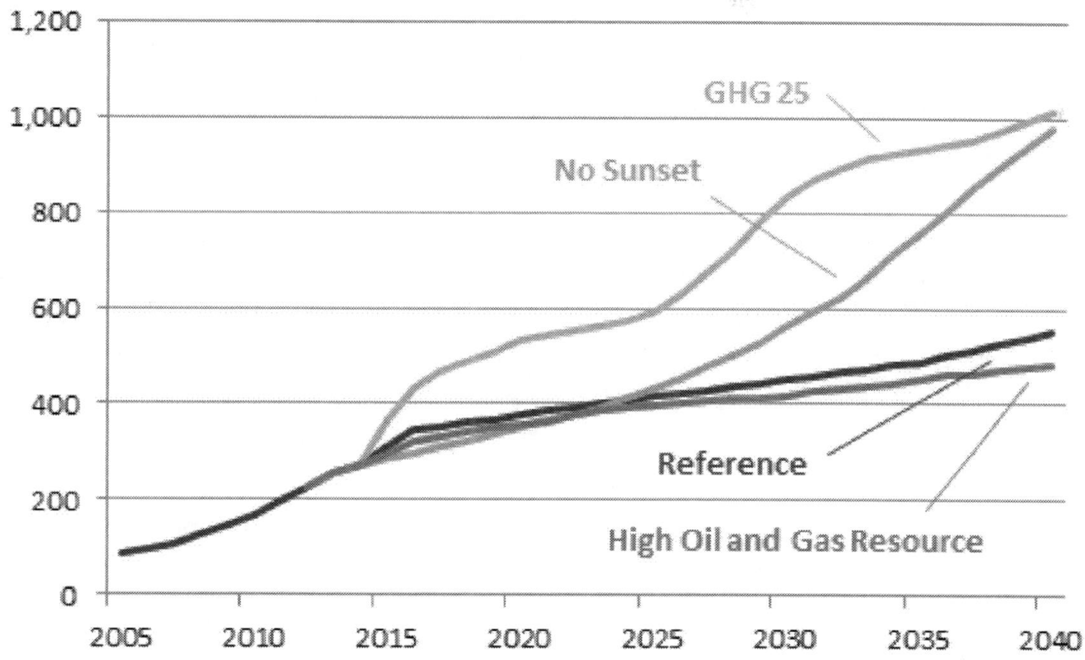

Figure 10. Renewable Electricity Generation in Four Cases, 2000-40 (billion kWh) [Adapted from 35]

In all cases, electricity generation from renewables increases over the next few decades. However, for the reference and high gas resource cases, the rate of increase in renewables is less than that seen between 2000 and 2014, mainly due to the increased implementation of natural gas power plants. The no sunset case shows the effects of extending renewable tax credits (i.e. the installment tax credit and the production tax credit) indefinitely in their current forms. In this case, the growth seen between 2000 and 2014 remains relatively constant through 2025, at which point the rate of implementation of renewables begins to increase steadily. In the GHG 25 case, a large increase is seen just after 2015 in response to the implementation of the carbon tax, after which the rate of implementation of renewables begins to level off. However, around 2025, electricity generation from renewables begins to increase dramatically once again, as more projects come into completion in response to the increased cost of natural gas power plants, whose production begins to decline as seen in Figure 9.

Compared to the reference case, both the no sunset and GHG 25 cases see disproportionate increases in solar and wind relative to other renewables like hydro or geothermal. When it comes to solar, electricity production in both cases is triple that of the reference case in 2040; for wind, it"s double. In both cases, geothermal actually sees slightly lower electricity production in 2040 compared to the reference case, and hydro sees relatively little growth between 2014 and 2040 in all cases [35].

6.0 Policy Recommendations

The electricity production sector in the US is rapidly changing as a result of the shale fracking boom. While the large shift from coal to natural gas will improve carbon emissions on a per kwh basis, the low price of natural gas is also reducing the amount of renewable sources entering the market. If no government action is taken, the growth of electricity demand will still mean an increase in carbon emissions from the electricity production sector over the next 30 years due to the low prices of natural gas. If the government truly wishes to use natural gas as a bridge fuel to a low-carbon electricity sector, a carbon tax needs to be implemented and tax credits for renewables need to be extended.

A carbon tax is necessary to appropriately reflect the societal cost of carbon-emissions-heavy technology. However, a large carbon tax is not necessary, and would cause unnecessary instability as the market acted swiftly to adjust. Instead, a low carbon tax that progressively increases is the most responsible way to implement such a change. Based off the projections in Figures 8 and 9, a carbon tax of $10 / metric ton of CO_2 would lower carbon emissions in the electricity production sector without drastically changing the economic viability of natural gas facilities. This tax would increase at a rate of 5% per year in order to incentivize investment in further carbon emission reduction technology. For a cap-and-trade program equivalent, this means the acceptable level of carbon emissions each year would be reduced, effectively increasing the cost of emissions a proportional amount.

On the other side of things, renewable technology still needs some assistance to be competitive with conventional fuels, as seen in Figure 3. However, the current split between the investment tax credit and the production tax credit makes the development of new and improved renewable technologies difficult, as it is more difficult to predict the economic environment these technologies will emerge into. A single, long-term tax strategy would offer economic stability and predictability that promotes private investment that has been yet unseen. In this way, a single, long-term tax strategy should be used to incentivize the use of all types of renewables in the electricity sector.

Of the two programs currently available, fully embracing the production tax credit makes the most sense, as it incentivizes the implementation of energy efficient and energy dense technologies[2]. It also incentivizes the development of improvements that reduce the upfront cost of these technologies, which help open them up to small scale residential and commercial consumers. The current rate of 2.3 cents per kwh is an appropriate start-point for such a tax program, but with the implementation of a carbon tax, indirect incentive towards renewables increases proportionately. The increasing nature of the carbon tax combined with the increased rate of improvement for renewable technologies actually means that the production tax credit can be slowly reduced. As such, a 5% reduction per year to mirror the carbon tax rate increase would be appropriate.

In both the carbon tax and the production tax credit, both programs need to be comprehensive and long term in order to allow the electricity producers to accurately model the economic environment that they will be entering with new projects. Due to the long-nature of construction of electricity production facilities, as well as the long operating life-time of these

[2] Again, this credit would not incentivize non-profit utility providers. Alternative programs, not covered in the scope of this paper, could be used to incentivize these utility providers.

facilities, stability in the tax structure from the government is paramount. The unexpected expiration of the production tax credit in 2013 mainly to political gridlock is unacceptable, and should be avoided in the future if at all possible.

All in all, if such a two pronged approach of a $10 / metric ton carbon tax combined with an extension of the production tax credit is used, the government will turn a net profit as a result. Based on the projections in Figures 9 and 10, there will be an average profit of $5.5 billion per year between 2015 and 2040 under such a program, with yearly profits ranging between $4.5 billion and $6.1 billion [adapted from 40 and 41]. Keep in mind that these projections only factor in the revenue from natural gas facilities; factoring in the revenue of coal should increase revenue further, despite the increase in cost from the increased use of renewables.

It is further recommended that the money made from this program be reinvested in the loan programs for electricity producers. While the reintroduction of the section 1705 program may be beneficial, it seems more beneficial to use this money to offset or eliminate the CSC cost burden on borrowers of section 1703 loans. With the accelerated retirement of coal, reduced or eliminated CSC costs would incentivize increased use of nuclear power, which acts as a more practical and carbon neutral base load power supply alternative compared to natural gas.

7.0 References

[1] Obama, B. (2014, January 28). *President Barack Obama's State of the Union Address.* Speech presented at the House Chamber, Capital Building, Washington, DC. Retrieved from http://www.whitehouse.gov

[2] EPA. (2014, June 2). *Learn About Carbon Pollution From Power Plants*. Retrieved from http:// www2.epa.gov

[3] Clement, S. (2014, June 2). A huge majority of American support regulating carbon from power plants. And they're even willing to pay for it. *The Washington Post*. Retrieved from http://www.washingtonpost.com

[4] Koch, W. (2014, June 3). EPA carbon emission limits may shift energy mix. *USA Today*. Retrieved from http://www.usatoday.com

[5] Union of Concerned Scientists. (2014, January 31). Production Tax Credit for Renewable Energy. Retrieved from http://www.ucsusa.org

[6] Davenport, Coral. (2014, May 12). Amid Pipeline and Climate Debate, Energy-Efficiency Bill Is Derailed. *New York Times*. Retrieved from http://www.nytimes.com

[7] EIA. (2014, May 7). *Annual Energy Outlook 2014*. "Market Trends: Electricity demand". Retrieved from http://www.eia.gov

[8] EIA. (2014, May 21). *Electric Power Monthly*. "Table 1.1. Net Generation by Energy Source: Total (All Sectors), 2004 – March 2014". Retrieved from http://www.eia.gov

[9] EIA. (2013, May). *State-Level Energy-Related Carbon Dioxide Emissions, 2000-2010.* "Table 2. 2010 state energy-related carbon dioxide emissions by fuel". Retrieved from http://www.eia.gov

[10] EIA. (2011, February). *Emissions of Greenhouse Gases in the United States 2009*. "Table 5. Greenhouse gases and 100-year net global warming potentials". Retrieved from http://www.eia.gov

[11] Meier, P.J. (2002, August). *Life-Cycle Assessment of Electricity Generation Systems and Applications for Climate Change Policy Analysis*. pg 70. Madison, Wisconsin: University of Wisconsin, Fusion Technology Institute. Retrieved from fti.neep.wisc.edu/pdf/fdm1181.pdf

[12] EPA. (2012, December 4). *What is Acid Rain?*. Retrieved from http://www.epa.gov

[13] EPA. (2014, March 10). *Basic Information | Mercury*. Retrieved from http://www.epa.gov

[14] EPA. (2013, September 25). *Natural Gas | Clean Energy*. Retrieved from http://www.epa.gov

[15] EIA. (2011, March 31). *Emissions of Greenhouse Gases in the US*. Retrieved from http://www.eia.gov

[16] World Nuclear Association. (2014, May). *World Energy Needs and Nuclear Power*. Retrieved from http://www.world-nuclear.org

[17] EPA. (2013, October 22). *Nuclear Energy | Clean Energy*. Retrieved from http://www.epa.gov

[18] Nuclear Energy Institute. (2014). *Life Cycle Emissions Analyses*. Retrieved from http://www.nei.org

[19] EPA. (2012, February 28). *Solar Energy | Energy and Global Climate Change in New England*. Retrieved from http://www.epa.gov

[20] EPA. (2014, May 1). *Wind Energy | Energy and Global Climate Change in New England*. Retrieved from http://www.epa.gov

[21] DOE. (2012, April 17). *Energy Department Report Finds Major Potential to Increase Clean Hydroelectric Power*. Retrieved from http://energy.gov

[22] EPA. (2014, May 1). *Geothermal Energy | Energy and Global Climate Change in New England*. Retrieved from http://www.epa.gov

[23] EIA. (2014, April). *Levelized Cost and Levelized Avoided Cost of New Generation Resources in the Annual Energy Outlook 2014*. Retrieved from http://www.eia.gov

[24] American Clean Energy and Security Act of 2009. H.R. 2454 of the 11th Congress. Retrieved from http://www.opencongress.org/bill/111-h2454/actions_votes

[25] Hulse, Carl. (2010, July 22). Democrats Call Off Climate Effort Bill. *New York Times*. Retrieved from http://www.nytimes.com

[26] 26 U.S.C. §§46, 48

[27] DOE. *Section 1703 Loan Program*. Retrieved from http://energy.gov

[28] DOE. *Section 1705 Loan Program*. Retrieved from http://energy.gov

[29] Wilson Sonsini Goodrich & Rosati. *Status Update – Department of Energy Title XVII Loan Guarantee Program*. Retrieved from http://www.wsgr.com

[30] EIA. (2008, June). *Annual Energy Outlook 2008*. "Early Capacity Additions Use Natural Gas, Coal plants Are Added Later". Retrieved from http://www.eia.gov

[31] EIA. (2014, April 8). *Half of power plant capacity additions in 2013 came from natural gas*. Retrieved from http://www.eia.gov

[32] EIA. (2014, May 7). *Annual Energy Outlook 2014*. "Most new capacity uses natural gas and renewables". Retrieved from http://www.eia.gov

[33] EIA. (2014, June 9). *Power sector CO_2 emissions sensitive to policy changes and natural gas supply*. Retrieved from http://www.eia.gov

[34] EIA. (2014, May 7). *Market Trends: Emissions*. "Energy-related carbon dioxide emissions are sensitive to potential policy changes". Retrieved from http://www.eia.gov

[35] EIA. (2014, April 29). *Renewable electricity projections show growth under alternative assumptions in AEO2014*. Retrieved from http:// www.eia.gov

381133 Membrane-on-Membrane -Development of Liposome-Entrapped Hydrogel

Abstract:

The biological systems have complex environmentally-responsive separation and conversion process. It is therefore necessary to construct a bio-inspired system. If the artificial systems and biological systems are accreted, it is possible to develop new conceptual response system, inspired by biological systems. Liposome membrane is amphipathic molecule and self-assembly organization and metastability condition, and was constructed non-polar and polar membrane interface bilayer, and was changed phase condition with temperature and lipid mixture ratio. Liposome can design the lipid interfacial surface. It was reported that liposome was utilized for Drug Delivery System (DDS) as capsule, in addition, it was reported that misfolded protein was refolded on liposome membrane, liposome was able to control gene expression, and chiral recognition of amino acid and peptide. Liposome was reported close link with vital function expression. However, there are some disadvantages, such as lower immobilized amount of liposome and multi-step for operation immobilization. In this study, development of novel method utilizing liposome membrane entrapped matrix in order to utilize biological membrane function. In order to subject characteristic analysis and functional evaluation, optical resolution of amino acid was investigated. In this study, liposome immobilized method was focused on direct entrapping method of physical containment method.

INTRODUCTION OF PMMA-TAG TO VHH FOR IMPROVING RECOVERY AND IMMOBILIZATION RATE OF VHHS

Dr. Bongmun Kang, Kyoto Institute of Technology, Kyoto-si, Kyoto-hu

Introduction

Camelid antibodies are composed of heavy-chain dimers and are devoid of light chains while nevertheless has an extensive antigen-binding repertoire. Variable domain of heavy chain (VHH) represents the smallest antigen-binding unit with a molecular size of~ 15 kDa in comparison with scFv fragments (30 kDa), Fab fragments (60 kDa), and whole antibodies (150 kDa). They have been expressed in microorganisms and have a high stability and solubility due to simpler structure of beta strand protein with only one disulfide bond. For these reasons, VHH have raised the possibility of various industrial applications such as neutralizing drugs, as well as diagnostic agents for specific detection of biomarkers. Especially, for use of VHH in clinical diagnosis, there is considerable interest in denser immobilization of VHHs on the surfaces of solid materials, such as plastic plates. Michiel et al reported passive adsorption of VHH antibodies onto the surface of polystyrene (PS) plate for use in enzyme-linked immunosorbent assays (ELISA). The highly immobilized VHHs on the plate could offer the potential to increase the sensitivity and specificity of VHHs to various type of antigen. In general, VHHs have been obtained by purification of soluble protein located in periplasmic fraction in a E. coli cell. However, the production yields of soluble VHH are considerably low and it is significantly changed with aminoacid sequences of VHHs that solubility, stability of VHH and protease resistance were highly related. In contrast, larger amounts of insoluble and inactive VHH were obtained from inclusion body. Thus, the refolding of denatured VHHs is necessary to prepare soluble VHH with original antigen-binding affinity and specificity. We recently reported that a newly found affinity peptide retained the ability to interact with the surface of poly(methyl methacrylate) (PMMA-tag) [1]. Consequently, an efficient refolding method for PMMA-tag-fused anti-RNase scFv (scFv-PM) were developed in our lab [2]. However, there is no study about the PMMA-tag-fused VHH for improving refolding and immobilization of denatured VHHs. Thus, this study examined whether PMMA-tag was available for efficient refolding and immobilization of VHHs.

Title

Introduction of PMMA-Tag to VHH for improving recovery and immobilization rate of VHHs

Author/Byline Information

Bongmun Kang, Kyoto Institute of Technology, Kyoto, Kyoto-Fu, Japan
Kagnari Yamakawa, Kyoto Institute of Technology, Kyoto, Kyoto-Fu, Japan
Yoshihisa Hagihara, AIST, Ikeda, Osaka, Japan
Michimasa Kishimoto, Kyoto Institute of Technology, Kyoto, Kyoto-Fu, Japan
Yoichi Kumada, Kyoto Institute of Technology, Kyoto, Kyoto-Fu, Japan

Material and Methods

hCG (Human chorionic gonadotropin) purchased from SIGMAwas biotinylated with biotinamidocaproate N-hydroxysuccinimide ester (Nacalai Tesque Co. Ltd). Biotinylated hCG (biotin-hCG) was used as a model antigen for sandwich ELISA. HRP-labeled streptavidin was purchased from VECTOR Laboratories. As a model, Anti-hCG VHH antibody that the structure was determined by Renisio et al was used in this study. And the gene of anti-hCG VHH was chemically synthesized with optimization of codons to these used in *E. coli*. Other chemicals were of reagent grade unless specified. The VHH gene was amplified by polymerase chain reaction (PCR) with the synthesized DNA as a template and the sense and antisense primers that recognition sitea for restriction enzymes, *Nde* I (catatg) and *Not I* (gcggccgc) were additionally introduced at 5' terminal ends, respectively. . The amplification of VHH gene was confirmed by agarose gel electrophoresis and purified using illustra™ GFX™ PCR DNA and Gel Band0 Purification Kit (GE Healthcare Co. Ltd). The vector pET-PMOMP25 that the nucleotide sequence encoding PMMA-tags (DVEGIGDVDLVNYFEVGATYTFNK) was introduced into the commercially-available expression vector, pET22 (Merck) was constructed in the previous study. The pET22 and pET-PMOMP25 vectors were used for cloning of amplified VHH gene for construction of expression vectors for VHH and PMMA-tag-fused VHH (VHH-PM), respectively. The vectors and DNA fragment of VHH gene were digested with the restriction enzymes *Not* I, and *Nde* I at 37 °C overnight.The digested vectors and DNA fragment of VHH gene were purified, and then they were ligated by addition of Ligation High (TOYOBO Co. Ltd). *E. coli* (HST08 Premium Competent Cell) was transformed with the ligated mixture, and transformant cells were selected on LB-agar plate containing 50 μg/ml ampicilin. The sequences of VHH genes introduced into the vectors were confirmed by DNA sequencing with the primer of T7 promoter. The constructed vectors for VHH and VHH-PM were introduced to *E. coli* Rosetta (DE3) (Merck), and transformants were selected on LB-agar plate containing 50□g/ml ampicillin (Amp) and 34 μg/ml chloramphenicol(Cam). On the basis of the method reported previously, the production and purification of VHH and VHH-PM were performed with some modification. Firstly, single colonies of transformed *E. coli* cells were innoculated in the 200ml Erlenmeyer flask containing 10 ml of LB medium (Difco™ LB Broth, Miller;0.25g per 10ml) supplemented with Amp and Cm, and the flask was shaken at 200 rpm and 37℃ overnight. A portion of the pre-cultured *E. coli* cells was transferred to 500ml shake flask containing 50 ml of Overnight Express™ Instant TB medium (Merck Co. Ltd)containing Amp and Cam, and the final cell density was adjusted to be $OD_{600} = 0.1$. The flasks were shaken at 200 rpm and 37℃ for 24 h. Cells were harvested by centrifugation for 20 min at 10,000 g and 4℃, and then suspended with 10 ml of 1×PBS containing 1 mg/ml Triton X-100 (Nacalai Tesque Co. Ltd) and 1xProtease inhibitor (Nacalai Tesque Co. Ltd.). The sonication of suspension was repeated 4 times by a probe-type sonicator (ULTRASONIC DISRUPTOR UD-211; TOMY Co. Ltd) at 30 W (Duty level: 30%) for 4 min with cooling by iced water. Insoluble aggregates were collected by centrifugation for 15 min at 10,000 g and 4℃ and were washed three times with distilled water followed by lyophilization (FREEZE DRYER FD-1000; EYELA Co. Ltd). They were then dissolved in 5 ml of denaturation buffer (pH 7.5) containing 6 M guanidine-HCl, 5 mM DTT, 300 mM NaCl and 40 mM sodium phosphate. After the removal of cell debris by centrifugation and filtration, the supernatant was directly applied to 5 cm^3 of a HisTrap HP affinity column (GE HealthCare Co. Ltd) equilibrated with buffer A (pH 7.5) containing 8 M urea, 20 mM imidazole and 2xPBS. After washing the column with buffer A, the bound proteins were eluted with buffer B (pH 7.5) containing 8 M urea, 400 mM imidazole and 2xPBS. The flow rate of both buffer A and B was 1 ml/min. The eluate was dialyzed against 1xPBS containing 8 M urea (8 M urea-PBS). Protein concentration in the eluate was determined by DC-protein assay (Bio-rad Co. Ltd), using BSA dissolved in 8 M urea as a standard. Both

VHH and VHH-PM obtained were stored at 4℃ until use. The aggregation properties of VHH, as well as those of VHH-PM, in the presence of 0.5 M urea were evaluated. VHH and VHH-PM were diluted to 24 kinds of refolding buffers that pH and NaCl concentrations were comprehensively different. For adjustment of pH from 6.0 to 8.5, 6 types of good's buffers were adopted. (MES; pH 6.0, ADA; pH 6.5, MOPS; pH 7.0, HEPES; pH 7.5, EPPS; pH 8.0, and TAPS; pH 8.5) Final concentrations of VHH, good's buffer and urea after dilution were 400 □g/ml, 50mM and 0.5M, respectively. Hundred □l portions of the mixture prepared were added to microtiter plate made of PS, the plate was incubated for 6 h at 25°C. Turbidity of mixture was evaluated by measurement of absorbance at 450 nm using microplate reader. A stock solution of VHH and VHH-PM dissolved in 8M urea-PBS was rapidly diluted to 1mg/ml, while urea and TAPS (pH 8.5) were presented at 0.5 M and 50 mM, respectively after dilution. The solutions were dialyzed twice against 1L of 50 mM TAPS (pH 8.5) to remove the residual urea from the solution. The aggregates formed were sedimented by centrifugation for 5 min at 10,000 g and 4℃. The refolding yield (*Y*) was determined by the following Eq (1).

$$Y = (VHH_{conc.\ after} / VHH_{conc.\ before}) \times 100 \qquad (1)$$

Where, *Y* is the refolding yield (%); $VHH_{conc.\ after}$ and $VHH_{conc.\ before}$ are the concentrations of VHHs after and before centrifugation, respectively.

The immobilization conditions of VHHs refolded were evaluated by sandwich ELISA. The microtiter plates made of PMMA was originally-manufactured and used for the immobilization of VHH and VHH-PM. The VHH solutions containing 50mM 6 different good's buffers (MES; pH 6.0, ADA; pH 6.5, MOPS; pH 7.0, HEPES; pH 7.5, EPPS; pH 8.0, and TAPS; pH 8.5) and 4 different concentrations of NaCl (0 ~ 300) were prepared. Final concentrations of VHHs were adjusted to 100 □g/ml. The 100μl solution of each well were transferred to PMMA plate, and incubated at 4□C during overnight. The plates were washed 5 times with PBS containing 0.1% Tween 20 (PBST), and then 300 μl of PBST solution containing 2% BSA (2%-BSA-PBST) were added to each well. The plate was incubated at 25°C for 1 h, followed by washing 5 times with PBST. Hundred μl portions of 0 or100 ng/ml of biotinylated hCG were added, and then the plate was incubated for 1 h at 25℃. After washing 5 times with PBST, 100μl portions of HRP-labeled streptavidin 5,000-fold diluted with 2% BSA-PBST were added to each well. Then the plate was incubated for 1 h at 25℃ followed by washing 5 times with PBST. Hundred μl portions of TMB solution were added to each well followed by incubation for 15 min at 25℃ for color development. The reaction was stopped by the addition of 100 μl of 0.5 M H_2SO_4 and then absorbance at 450 nm (reference at 650 nm) was measured via a microplate reader. The immobilization density of VHH and VHH-PM adsorbed in PMMA plate was measured by Micro BCA™ Protein Assay Kit (Thermo scientific Co. Ltd). Hundred μl portions of 0 ~ 400 □g/ml of VHH and VHH-PM in the optimized buffer for immobilization (VHH: 50mM MES, pH 6.0, VHH-PM: 50mM MES, pH 6.0 containing 50mM NaCl) that were determined by the experiments described in XX, were added onto the PMMA plate and the plate was incubated overnight at 4°C. The plates were washed 5 times with PBST, and 75 □l of PBST was added to each well and, then colored signals for VHHs were developed by addition of 75 □l reagent mixture of microBCA kit (Thermofisher Scientific). The 75 μl of 0 ~ 30 mg/ml BSA in PBST was used as a standard. The plates were incubated at 37°C for 2h and then, absorbance at 562 nm was measured by microplate reader. Hundred μl portions of 100 μg/ml VHHs diluted in the optimized buffer for immobilization (VHH: 50mM MES, pH 6.0, VHH-PM: 50mM MES, pH 6.0 containing 50mM NaCl) were added to each well of a microtiter plate followed by incubation overnight at 4℃. After washing the plate, different concentrations of biotinylated hCG (0 – 1000 ng/ml) were detected by the method described in section.

Results

1. Production and purification of anti-hcG VHH and VHH-PM

In this study, single-domain antibody (VHH) which specifically recognizes -chain of human chorionic gonadotropin (hCG), was used as a model. When both VHH and PMMA-tag-fused VHH (VHH-PM) were produced using *E. coli* Rosetta (DE3), both of them were mainly localized in an insoluble aggregates, inclusion body. Amounts of VHH and VHH-PM recovered from 1L cultures were range of 0.3 g ~ 0.5 g, and more than 10 mg/ml of VHHs were finally obtained after one-step IMAC purification in the presence of 8M urea. They were analyzed by SDS-PAGE under the reduced and non-reduced conditions. As shown in Fig. 1 (a), main protein bands of VHH (lane B) and VHH-PM (lane D) were detected at approximately 20 kDa. Therefore, both the VHHs were highly purified from inclusion bodies by one-step IMAC purification in the presence of 8M urea. Interestingly, positions of protein bands for the VHHs shifted to downwards in the absence of reducing agent (lane A and C), in comparison with those of reduced VHHs (lane B and D), indicating that apparent molecular size of VHHs became larger to some extent under the reduced condition. These results suggested that a disulfide bridge inside a VHH molecule was already formed during purification and dialysis processes in the presence of 8M urea. Oligomer of VHH molecules through intermolecular disulfide bridge was also detected (lane A and C) at the position more than 31kDa. In the purification process, both VHHs was first solubilized with 6M guanidine-HCl, and reduced with 5mM DTT. Therefore, the denatured and reduced forms of VHHs were eluted from the IMAC resin, and their concentration just after purification were significantly more than 10mg/ml. These results suggested that intermolecular disulfide bridges of VHHs were formed after IMAC purification because high concentrations of VHHs as reduced forms would be intramolecularly or intermolecularly oxidized in the absence of reducing agent during dialysis against 8M urea-PBS. It was found that formation of intramolecular disulfide bridge in VHHs was completed within 24h by incubation at 25°C after removal of reducing agent. Furthermore, the oxidization was zero-order reaction that the reaction rate was not changed with protein concentration, while reaction rates of intermolecular polymerization through the disulfide bridge might be a function of VHH concentrations. Consequently, as shown in Figure 1 (b), denatured and monomeric VHHs with an intramolecular disulfide bridge were successfully prepared by repeating reduction and re-oxidation of the VHHs at the concentrations less than 1mg/ml.

2. Effects of refolding conditions on aggregation during the refolding of VHH and VHH-PM

Turbidity induced by aggregation of VHH and VHH-PM in their refolding processes was evaluated using 24 types of refolding buffers with different pH and NaCl concentrations. As shown in Fig.2, aggregation of VHH-PM was considerably suppressed with increase in pH and decrease in NaCl concentration. Especially, the degree of aggregation was hardly detectable when scFv-PM was diluted to 50mM TAPS (pH 8.5) without NaCl. A high degree of aggregation was, however, observed in VHH regardless of the range of pH and NaCl concentrations. These tendencies were consistent with the results obtained in the previous study that scFvs genetically-fused with PMMA-tag (scFv-PM) were refolded. [3] Based on these results, the refolding of VHH-PM was conducted with 50 mM TAPS (pH 8.5) solution without NaCl. As a results, high recovery yield of soluble VHH-PM was shown and significantly higher than that of VHH in this refolding conditions (Fig.3). The genetic fusion of PMMA-tag to C-terminal region of VHHs significantly affect the recovery of their soluble protein during refolding by 50 mM TAPS at pH 8.5. PMMA-tag used in this study consist of 5 acidic amino acids, including Asp (D) and Glu (E), and one basic amino acid (Lys (K)). The apparent differences were observed in isoelectric point (pI)s between VHH and VHH-PM tested in present study (Fig.4). The pI of VHH-PM is 6.03, which is lower than that of VHH (7.83). The pI of VHH-PM was decreased by approximately 25%, compared to that of VHH. Thus, the genetic fusion of the PMMA-tag at the C-terminal region of VHHs lost their positive charge at alkaline pH (approximately 8.5) during refolding. Consequently, there are negative charged at the surface of VHH-PM. As a results, it prefers to interact with water rather than with other protein molecules. Finally, there are significant differences between pI values of VHH-PM compared to that of VHH.

3. Determination of the optimal immobilization condition of VHH-PM on the surfaces of PMMA plate

Firstly, the degree which both of VHH and VHH-PM were immobilized on the surface of PMMA microtiter plate was evaluated at 24 different immobilization conditions of pH (6.0, 6.5, 7.0, 7.5, 8.0 and 8.5) and NaCl concentration (0, 50, 150 and 300 mM), and their antigen-binding abilities in adsorption states were compared by a sandwich ELISA.When the refolded VHHs were immobilized on PMMA plates, significantly higher antigen-binding signals were obtained from the wells on which VHH-PM was immobilized, as shown by a sandwich ELISA (Fig.5C). Furthermore, their S/N ratio, which represent the level of a desired signal to the level of background noise, was larger at pH 6 and low NaCl concentrations (Fig.5D). However, the significant differences of antigen-binding signals and their S/N ratio was not observed with the wild-type VHH without PMMA-tag, regardless of various range of pH and NaCl concentration (Fig.5A and B). Based on these results, we decided to immobilize refolded VHHs on the surface of PMMA plate at pH 6 and 50mM NaCl to minimize the level of background noise.

4. Evaluation of the immobilization density of VHH-PM on the surfaces of PMMA plate.

According to the measurements of immobilization density, scFvs genetically fused with the PMMA-tag highly immobilized on the surface of PMMA plate. Thus, we investigated the VHH-PM could be immobilized to the same extent as scFv-PM, even though different type of antibody. The immobilization density of VHHs on the surface of PMMA plate was measured in Micro BCA™ protein assay (Thermo Scientific Co. Ltd) at pH 6 and 50mM NaCl. As a results, the immobilization density was larger in VHH-PM than those in VHH. Particularly, the VHH-PM showed approximately 0-0.45μg/cm^2, which were about 2-fold greater than those detected with VHH (Fig.6). These data suggest that the immobilization density could be highly improved in VHHs by genetic fusion of PMMA-tag to C-terminal region of VHHs at the same rate as scFvs.

5. Evaluation of the antigen-binding activity of VHH-PM immobilized on the surfaces of plastics

Results presented in 3.4 demonstrated that the VHHs genetically fused with the PMMA-tag lead to increase immobilization density on the surface of PMMA plate. Consequently, we investigated whether the antigen-binding activities in the adsorption state could be affected by the genetic fusion of PMMA-tag in VHH (Fig.7). Obviously, the antigen-binding activities in the adsorption state was significantly high in VHH-PM compared to wild type of VHH. There are approximately 8-fold differences in the antigen-binding activities in the adsorption state between VHH-PM and VHH. These results suggested that the introduction of PMMA-tag on the C-terminal region of VHH lead to high-density immobilization on the surface of a PMMA plate, thereby retaining the antigen-binding activities in the adsorption state. In this study, we practically screened an efficient refolding conditions for PMMA-tag-fused anti-hCG VHH (VHH-PM). It could be refolded with a recovery of more than 95% by dialysis against 50 mM TAPS buffer (pH 8.5). By using this method, the solubility of VHH-PM was significantly increased due to a repulsion of negatively charged VHH-PM molecules. Based on these results, further study is directed toward the genetic fusion of PMMA-tag to another type of single domain antibody, for example, Fab, for efficient recovery of denatured single-domain antibodies during refolding and high immobilization rate of refolded single-domain antibodies.

Figure 1. SDS-PAGE of Anti-hCG VHH (A) and Anti-hCG VHH-PM (C) purified from inclusion bodies. (B) and (D) indicate Anti-hCG VHH and Anti-hCG VHH-PM with reduction under 5mM DTT, respectively.

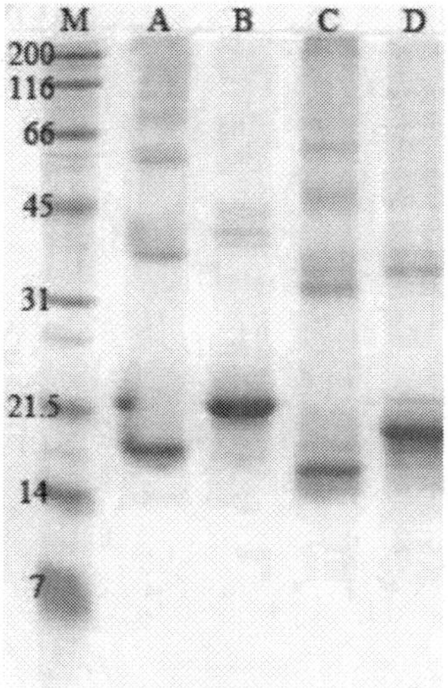

Figure 2. Aggregation test of VHH (A) and VHH-PM (B) diluted in 24 different refolding buffers.

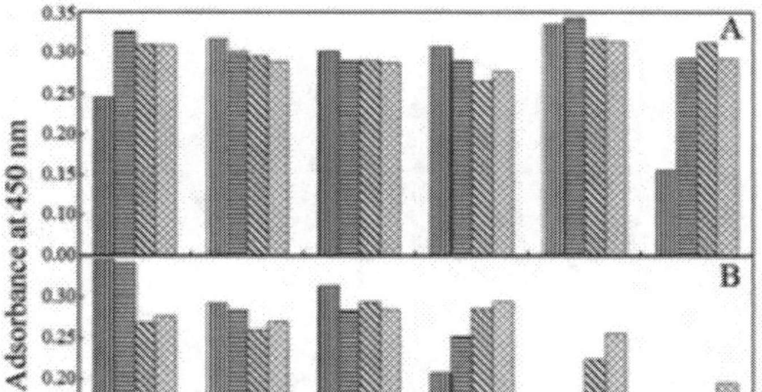

Figure 3. Recoveries of VHH and VHH-PM after dialysis refolding using 50mM TAPS

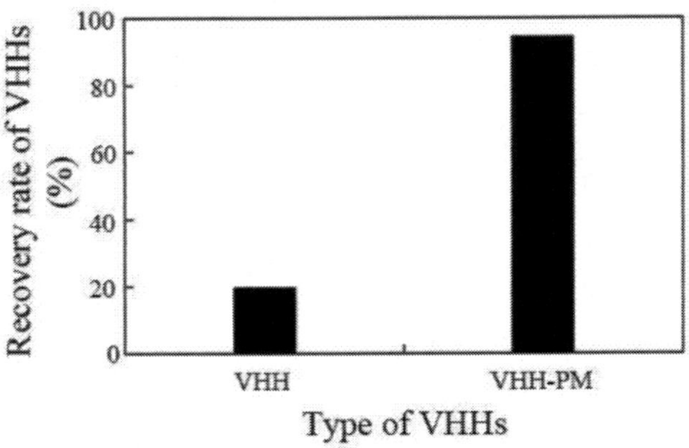

Figure 4. Isoelectric point of VHH and VHH-PM tested in this study

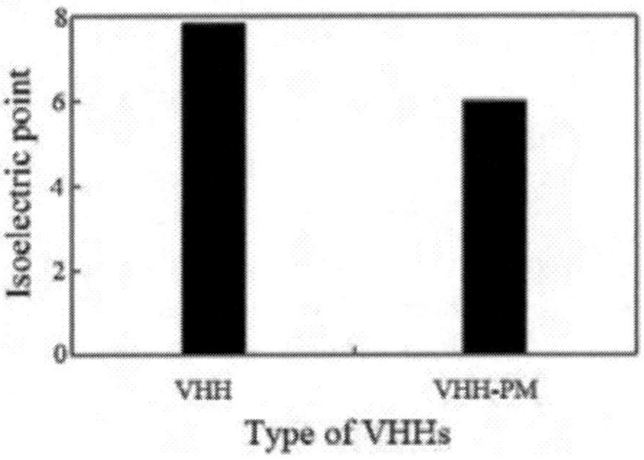

Figure 5. The signal and noise obtained from different immobilization conditions of Anti-hcG VHH after refolding using 50mM TAPS. (A) The signal (S) and noise (N) of Anti-hCG VHH, (B) The S/N ratio of Anti-hCG VHH, (C) The signal (S) and noise (N) of Anti-hCG VHH-PM, (D) The S/N ratio of Anti-hCG VHH-PM.

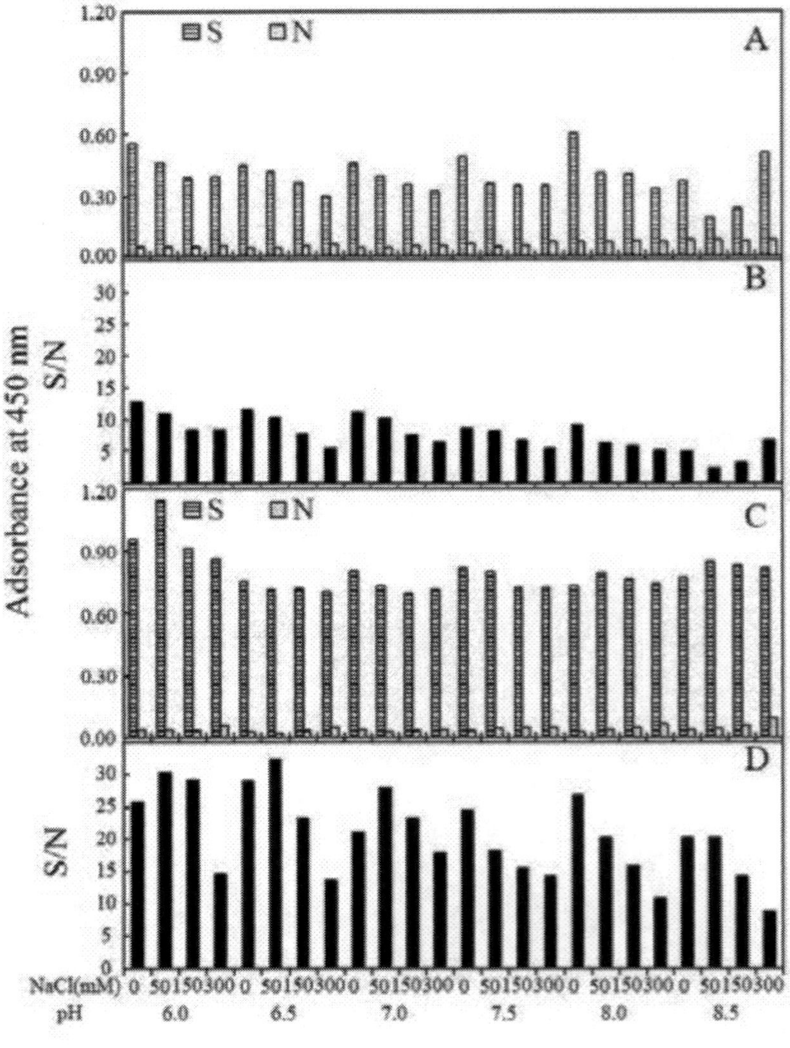

Figure 6. The immobilization density of Anti-hCG VHH and Anti-hCG VHH-PM after dialysis refolding using 50mM TAPS

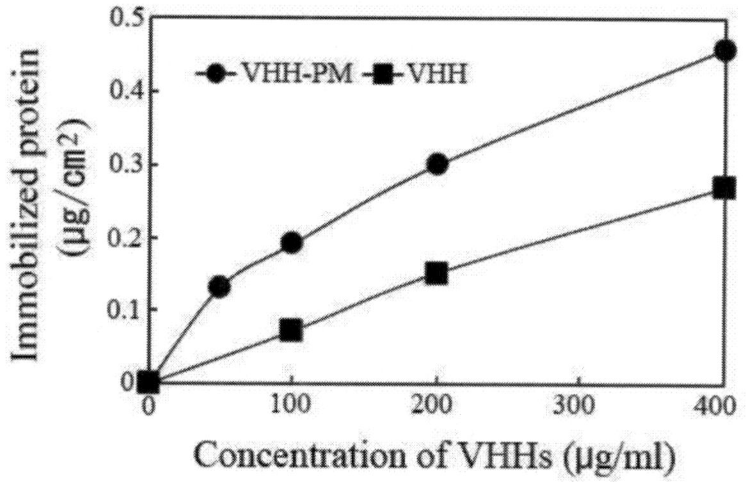

Figure 7. The antigen-binding activity of VHH and VHH-PM after dialysis refolding

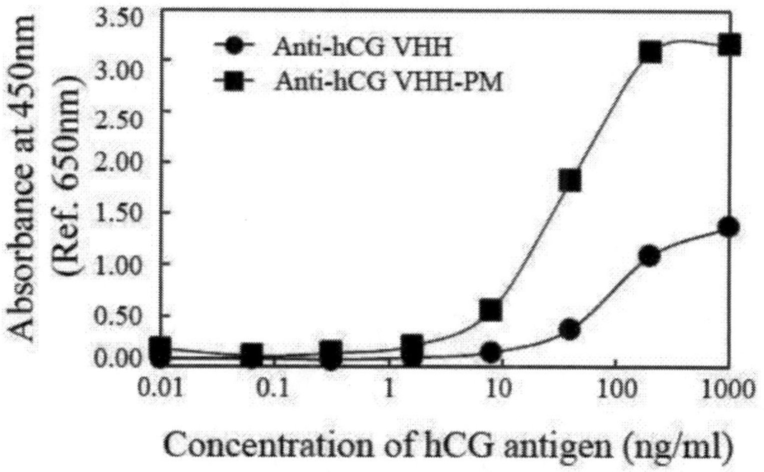

References

1. Kumada, Y, Murata, S., Ishikawa, Y., Nakatsuka, Y. and Kishimoto, M. (2012a) "Screening of PC and PMMA-binding peptidesfor site-specific immobilization of proteins". *Journal of Biotechnology*, 160 , pp. 222-228.
2. Kumada, Y., Shiritani, Y., Hamasaki, K., Nakagawa, A., Sasaki, E. and Kishimoto, M. (2010b) "Novel solid-phase refolding method for preparation of scFv-immobilized polystyrene plates with high-antigen-binding activity". *Analysis and Bioanalysis Chemistry*, 398, pp. 295-303.
3. Ewert, S., Cambillau, C., Conrath, K. and Plückthun, A. (2002) "Biophysical Properties of Camelid VHH Domains Compared to Those of Human VH3 Domains". *Biochemistry*, 41 pp. 3628-3636.

Design and Simulation of an Automated Rare Blood Cell Detector

Zhixi Qian[a], Eugene D. Boland[2], Paul W. Todd[2], and Thomas R. Hanley[1]
[1]Department of Chemical Engineering, Auburn University, AL 36849
[2]Techshot, Inc., 7200 Highway 150, Greenville, IN 47124

Abstract

Today the research of circulating tumor cells (CTC) plays an essential role in cancer's detection and treatment. To detect the cancer at an early stage, a point-of-care diagnostic system for detecting and counting cancer cells in whole blood is required. Our work focus on the simulation of flow pattern in the device, and during the research, computational fluid dynamics (CFD) is used as an effective method for simulating flow in the automated rare blood cell detector. The result showed a well-developed laminar flow inside the magnetic trap, with an experimental work it confirms a flow rate of 1ml/min is best for the magnetic separation. And the simulation of the static mixer showed the mixing pattern and properties of blood and reagents.

Key words: circulating tumor cells; CFD simulation; magnetic separation; point-of-care system; static mixer

Introduction

In recent years there's an increasing demand of using circulating tumor cell (CTC) counts in cancer management. The primary objective of this research is to design, simulate and test an accurate, cost-effective, user-friendly point-of-care CTC detection device. The preliminary design includes three syringes for blood sample, buffer and reagents, a static mixer, a magnetic filter, a microfluidic "chip" and a waste receptor. The device is smaller and cheaper than traditional detection devices. Also, device operation is easier with only one operator required. The device can be used in clinics, nursing homes or residences thus allowing earlier detection of most fatal cancers, including pancreatic cancer, ovarian cancer and melanoma. Our recent research focused on the analysis and simulation of static mixer and the flow cytometer and a further analysis of factors affecting the fluid mixing.

The fluid dynamic analysis of the staged mixer was accomplished using the FLUENT™ analysis package, which performs modeling of fluid flow and tracking particles' movement in complex geometries. The simulation showed that inside the magnetic trap, the fluid performs a well-developed laminar flow. And it's estimated most magnetic particles will be captured at first pair of magnets.

An experiment was set up to test the magnetic trap's separation ability. Various conditions were tested, including flow rate, magnetic size and type. In addition, we tested the blood cell's sedimentation inside the magnetic trap. Then it's possible to optimize the device's flow rate.

Devices and methods

During the research task CFD software simulation and mathematic methods were applied. Computational fluid dynamic analysis of the channel design was accomplished using the FLUENTTM analysis package, which performs modeling of fluid flow and tracking particles' movement in complex geometries. With the help of ICEM CFDTM, a mesh file of the channel geometry was developed, and a laminar flow model was employed for calculation. For the simulation of magnetic particles, the channel was simplified to a 2-D rectangle shape. A discrete model was applied to simulate the particle movement. Tracking of particles movements was performed to show the particle trapping efficiency.

The trajectories of magnetic particles were simulated by choosing a set of coordinates within the channel and adding the flow and magnetophoretic velocity vectors as a function of position in the mesh. At each step in time the velocity is recalculated to account for simultaneous changes in streamline velocity, dy/dt, and magnetophoretic velocity, dx/dt.

To test the magnetic trap's capture efficiency, a hemacytometer is used in counting and calculation of magnetic particle captured. The red blood cell (RBC) and magnetic beads are put in sample tubes. After counting the origin concentration, the sample is diluted and changes its concentration to around 10^7 particles/ml. Then take 5ml of sample, and pump through the magnet trap then collect the outlet. After that, use hemacytometer to count particles' concentration of outlet sample and compare it with original value.

Two types of magnet particle is used, the diameter of magnetic particle include: 2.7μm, 3.13μm and 5.01μm. And the flow rate is changed from 0.25ml/min to 1ml/min.

Results and discussion

Ansys Fluent is used in the simulation of magnetic trap's flow, the result showed the flow's pattern inside the channel, the magnet trap is 15cm long and divided by 2 channels, the size of the channel is 4mm *3mm. The fluid is flowing through the trap at a rate of 1ml/min.

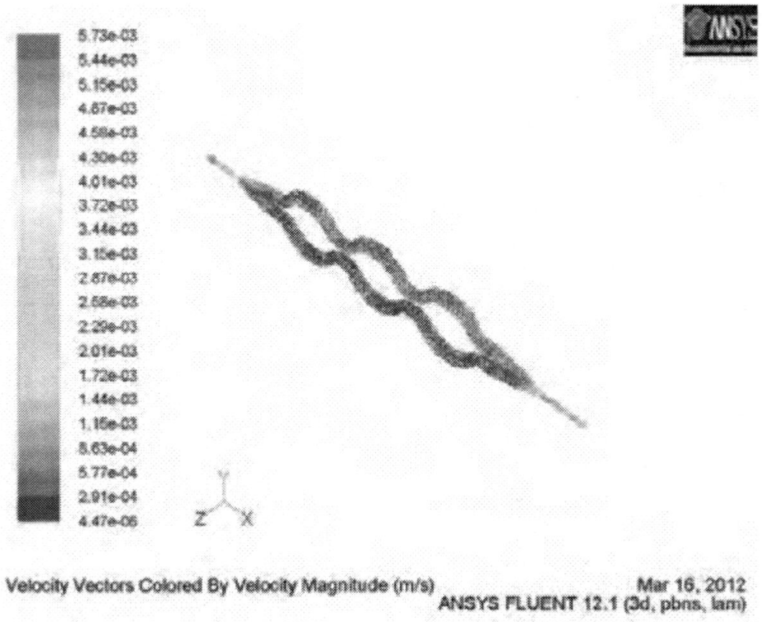

The velocity magnitude of magnetic trap shows that at the middle of static mixer, the velocity is less than 1mm/s, compare with the size of static mixer, it's a rather slow velocity.

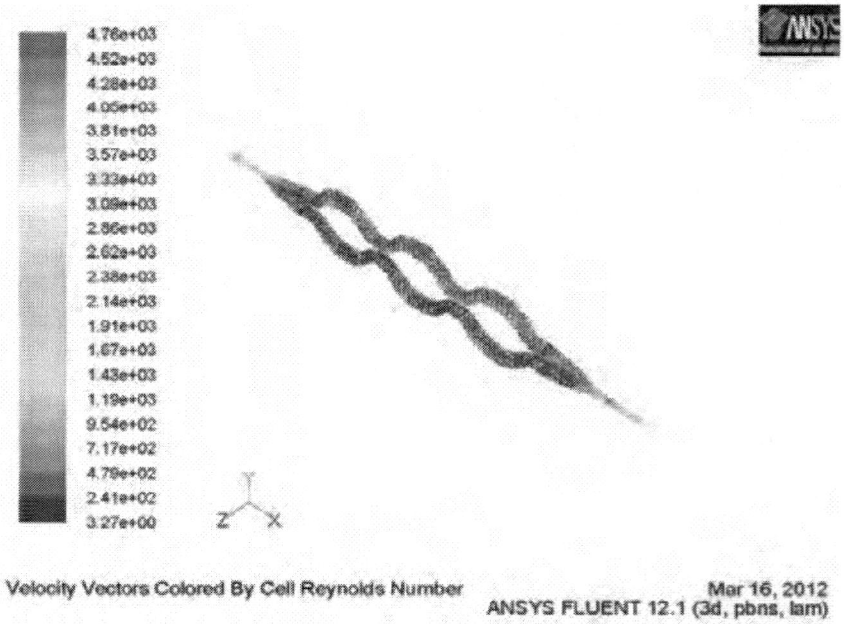

The Reynolds number inside the magnetic trap is generally less than 100, thus it's clear that the fluid inside the magnetic trap is a laminar flow, so the following model set up and calculation can be simplified.

The result of simulation showed the flow pattern of the magnetic trap, figure below showed the velocity vector of partial channel at position of a magnet for magnet particle analysis, to simplify the calculation, the channel is considered as a 2-D model.

It shows the velocity profile of fluid, and inlet average velocity is around 5.32 x 10^{-3}m/s.

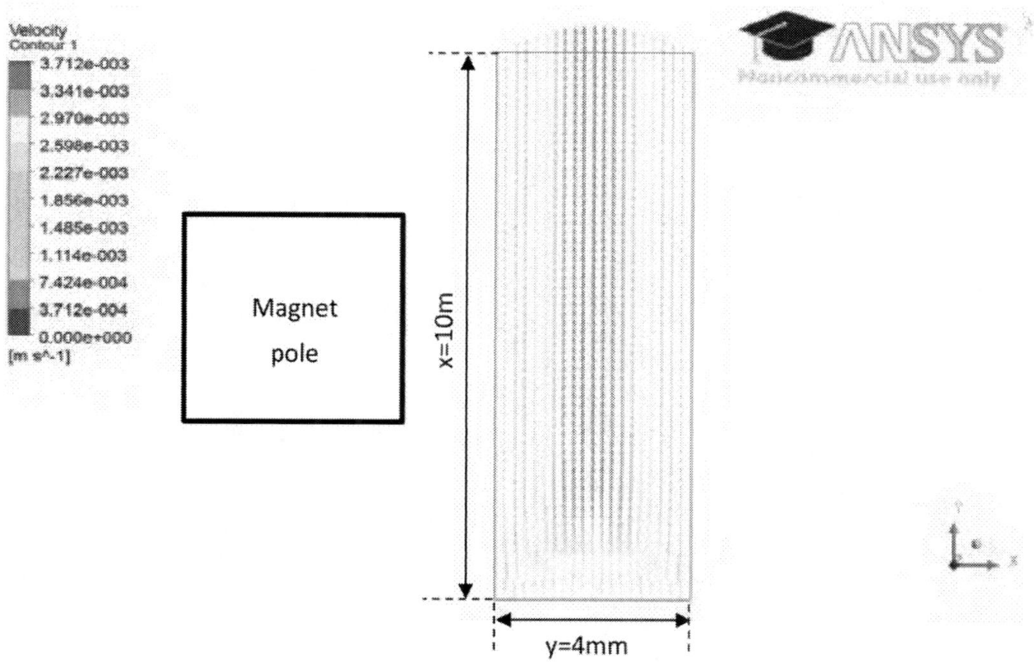

The result of particle track showed the first pair of magnet's able to capture all magnet beads at a flow rate of 1ml/min. However, with a higher flow rate, the flow velocity will dominate the particle movement and most magnetic particles will escape the magnet field.

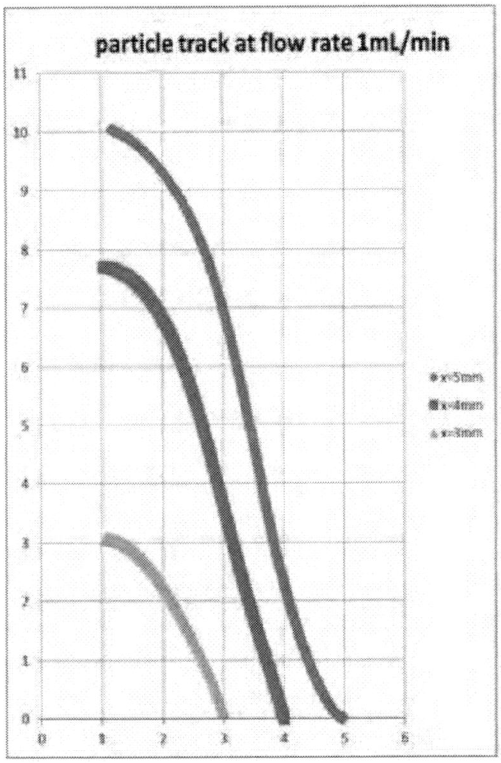

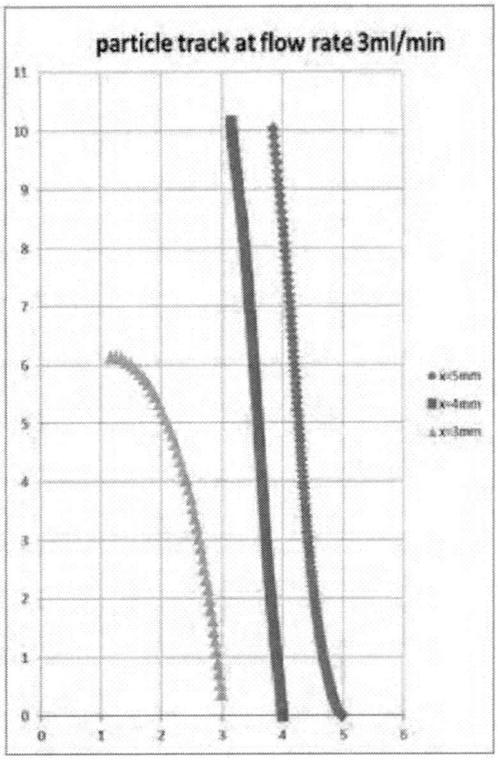

Experiments have been performed to test the magnetic trap's separation ability. The table below showed the result of magnet particle capture.

Magnetic particle size	**Flow rate**	**Initial concentration**	**Concentration after separation**	**Capture Ratio**
2.6μm(V1)	0.25ml/min	$2.66*10^7$	$8.00*10^5$	96.99%
2.6μm(V1)	0.5ml/min	$2.04*10^7$	$2.80*10^6$	86.28%
2.6μm(V1)	1.0ml/min	$2.04*10^7$	$6.20*10^6$	69.61%
2.7μm(V2)	0.25ml/min	$1.03*10^7$	$1.22*10^4$	99.98%
2.7μm(V2)	0.5ml/min	$1.40*10^7$	$1.00*10^4$	99.93%
2.7μm(V2)	1.0ml/min	$1.67*10^7$	$4.44*10^3$	99.97%
3.13μm (V2)	1.0ml/min	$2.32*10^7$	$5.56*10^3$	99.98%

5.01μm(V2)	1.0ml/min	$1.04*10^7$	$1.11*10^3$	99.99%

As the result of table indicates, the capture ratio is higher with larger magnetic particles and a slower flow speed, while in our magnetic trap, with the designed flow rate and choice of magnetic beads, the capture ratio is high enough to capture more than 99.9% of magnet particles.

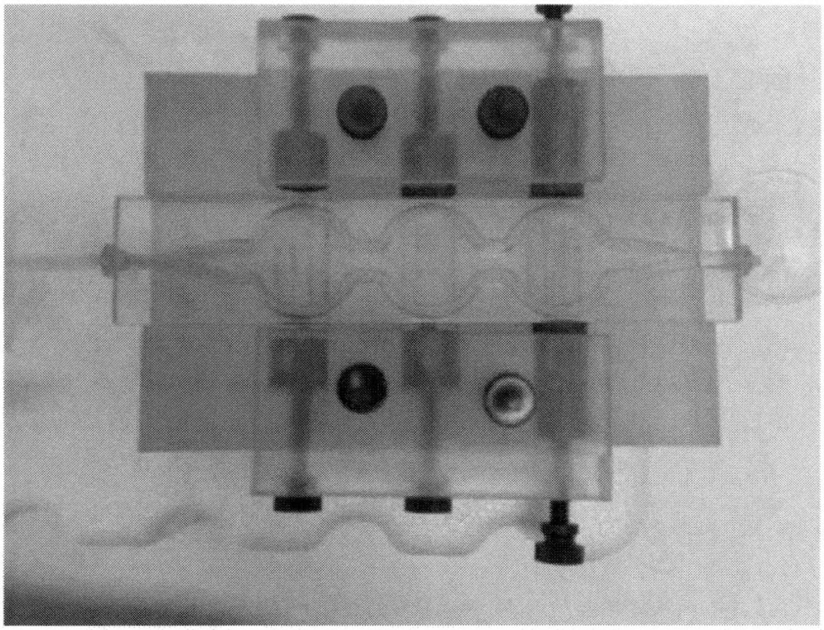

As the photo above shows, most of the inlet magnet particle can be trapped at the first pair of magnet, while the few escaped can be captured successfully in following two pairs of magnet.

The sedimentation is another factor to be considered. For the red blood cells (RBC) left inside the magnet trap will affect following detection, so it's necessary to reduce the sedimentation of unlabeled blood cell inside the mixer.

Type	**Flow rate**	**Initial concentration**	**Concentration after separation**	**Ratio for mass loss**
Magnet particle (5.01μm)	1.0ml/min	$1.39*10^7$	$1.30*10^7$	6.83%
RBC	0.25ml/min	$2.95*10^7$	$1.28*10^7$	56.78%
RBC	0.5ml/min	$2.26*10^7$	$1.89*10^7$	16.37%

RBC	1.0ml/min	2.14*10^7	2.11*10^7	1.40%

The table showed particle's sedimentation, at a low flow rate (0.25ml/min), almost half of RBCs will stay inside the mixer, which is an unacceptable result, increasing the flow rate will solve the problem. At a flow rate of 1ml/min, the sedimentation rate drops to 1.4%, thus it will not affect the detection of CTCs.

Conclusion

The simulations completed confirm that the proposed magnetic trap design successfully traps positively labeled cells. The magnetophoretic velocity of all beads and cells analyzed to date is much greater than the fluid flow velocity in the magnetic trap and is sufficient to cause cells to be captured by the lowest-strength magnet in the current design. However, a high velocity will reduce the capture ratio considerably thus the flow rate should be no higher than 1ml/min.

The experiment work showed that the magnetic trap has a good performance in particle separation, the first pair of magnets is able to capture most magnetic beads and with a proper choice, the capture rate can be more than 99.9%. The flow rate is also an important factor, when flow rate is less than 1ml/min, the sedimentation will affect the separation process greatly as RBCs will stay inside the trap. So the flow rate should be high enough to avoid particle sedimentation.

Consider both particle sedimentation and residence time, the choice of flow rate should keep a balance between efficiency and sedimentation, so a flow rate of around 1ml/min is a proper choice.

References

Afshar, R., Moser,Y, Lehnert, T and Gijs, M. A. 2011. Three-dimensional magnetic focusing of superparamagnetic beads for on-chip agglutination assays, Annul. Chem., 83(3) : 1022-1029.

Allard, W. J., Matera, J., Miller,M.C., Repollet, M., Connelly, M. C., Rao, C., Tibbe, A. G. J., Uhr J. W and Terstappen, L. W. M., 2004. Tumor cells circulate in the peripheral blood of all major carcinomas but not in healthy subjects or patients with nonmalignant diseases, Clin. Cancer Res.10: 6897-6904.

Cancer Data and Statistics, http://www.cdc.gov/cancer/dcpc/data/

Etzioni, R., Urban N. and Ramsey S., 2003. The case for early detection, Nature Reviews (Cancer), 3: 1-10.

Fan, T., Q. Zhao, J. J. Chen, W-T. Chen and M. L. Pearl, 2009. Clinical significance of circulating tumor cells detected by an invasion assay in peripheral blood of patients with ovarian cancer, *Gynecologic Oncology*, **112**, 185-191

Haun, J. B., Yoon, T-J., Lee H. and Weissleder, R. 2010. Magnetic nanoparticle biosensors, Nanomedicine and Nanobiotechnology, 2: 291-304.

Hsieh, H. B., Marrinucci, D., Bethel, K., Curry, D. N., Humphrey, M., Krivacic, R. T., Kroener, J., Kroener, L., Ladanyi, A., Lazarus, N, Kuhn, P., Bruce R. H. and Nieva, J. 2006. High speed detection of circulating tumor cells, Biosensors and Bioelectronics, 21(10): 1893–1899.

A Split Intein-based Self-cleaving Tag for Recombinant Protein Purification

In a Mammalian Cell Expression system

Tzu-Chiang Han

Dr. David W. Wood

Inteins are genetic elements that can self-excise from various host proteins in a post-translational protein splicing process. With a simple mutation of the intein, one can yield isolated, controllable intein cleaving activities at either the N- or C-terminus. By combining self-cleaving inteins with conventional affinity tags, self-cleaving affinity tags have been developed. The self-cleaving tags benefit the tag removal process by a simple, inducible cleaving reaction without any addition of protease. The phenomenal tag removal strategy works successfully in prokaryotic cell systems in our previous studies; however, the *in vivo* premature cleaving of the precursor protein remains the bottleneck of its application in eukaryotic cell systems. Here we present our most recent work in developing a split intein-based recombinant protein purification strategy in a mammalian system. Split intein consists of two complementary fragments of intein that can undergo reconstruction to form a fully-active intein by *in vitro* assembling, while no splicing activity shown individually. The affinity tag-fused intein (N-fragment) was first expressed in *E.coli* and further conjugated with the beads in a typical affinity column. The target protein that fused with the C-fragment of intein was expressed in a mammalian system and then mixed thoroughly with the tagged N-fragments of the intein. After an affinity-based purification, the target protein can be eluted by inducing the self-cleaving reaction of the re-assembled inteins. With the split-intein system, we provide a feasible scheme to overcome the unwanted *in vivo* premature cleaving and generate a potential platform technology for the manufacture of non-antibody glycoprotein therapeutics at large scale.

378777 On-Column (solid-phase) Pegylation and Separation of Pegylated Proteins

Abstract:

PEGylation is known to be an effective method for making a smart protein drug. It prolongs the residence time in the body and makes proteins more stable and soluble. However, producing a PEGylated protein is not an easy process since the reaction yield is not very high and consequently, the reaction mixture contains non-reacted protein and PEG as well as PEGylated proteins. In addition, due to over PEGylation reaction unfavorable, mutli-PEGylated proteins are produced. Separation of a target mono-PEGylated protein from other PEGylated protein forms and native protein is also a difficult process.

Solid phase or on-column PEGylation is claimed to be a promising method for improving the reaction selectivity. We carried out random PEGylation of lysozyme on various cation-exchange chromatography columns. After the PEGylation reaction, linear gradient elution experiments were performed to separate PEGylated protein forms.

Production of multi-PEGylated forms was reduced significantly. However, the yield was relatively low. In order to understand the reason for the low yield, various different chromatography media as well as mobile phase conditions were tested. Porous beads of different particle size (SP-Sepharose FF and HP), hydroxyapatite, Phosphate-Cellufine, Sulfate-Cellufine, SP-Toyopearl650M, Resource S and SO3-monolithic disk were used. Large beads (Sepharose FF) resulted in much lower yield values most likely due to the mass transfer resistance. However, convention-aided chromatography media, monolithic disk did not improve the yield significantly. By choosing a suitable salt concentration of the mobile phase, lysozyme was rather weakly adsorbed onto the disk entirely before PEGylation. This condition increased the yield.

378808 Diffusion Coefficients of DNAs and Pegylated Proteins for Estimating the Performance of Chromatography

Abstract:

Pore diffusion governs the separation or adsorption performance of chromatography. Especially, for large biomolecules such as DNA and PEGylated proteins low pore diffusion coefficients result in low performance of chromatographic separations. Monolithic chromatography, a type of convection-aided chromatography, is best suited for such large biomolecule separations. In order to understand the advantage of monolithic chromatography against porous-bead chromatography quantitatively the diffusion coefficients of large biomolecules are needed. Unfortunately, diffusion data (pore diffusion coefficient D_s and molecular (free) diffusion coefficient D_m) for large biomolecules are not readily available.

In this study, the molecular diffusion coefficient D_m (free diffusion) values were determined by using the Taylor dispersion method for various DNAs and PEGylated proteins as well as standard globular proteins. Necessary conditions for measuring D_m precisely such as tube length, tube diameter, coil diameter and flow velocity were first checked. Single-strand (ss) DNAs showed much lower molecular diffusion coefficients compared with those for globular proteins.

The pore diffusion coefficient D_s values were measured by using isocratic elution at non-binding conditions. D_s of ss-DNA decreased with increasing DNA molecular weight much more sharply compared with D_m. Similar trends were observed for PEGylated proteins. Based on these data we calculated the separation performance of typical porous bead ion-exchange chromatography (IEC) in comparison with that of monolithic IEC.

383947 Simulated Pressure Drop in Structured Packing: Experimental Validation and Impact of Packing Geometry

Abstract:

Computational fluid dynamics (CFD) simulations of single-phase flow through structured packing were performed using STAR-CCM+. Pressure drop data using a detailed geometry generated via X-ray CT compared favorably to experimental results. The steady-state data showed the Realizable k-ε turbulence model yielded better predictions than the SST k-ω model.

With a validated model, multiple small-scale, periodic geometries were utilized to evaluate the impact of fundamental packing geometry on overall pressure drop. Corrugation angle, packing specific area, and channel dimensions (e.g. channel height and channel opening angle) were varied with F-factors ranging from 0.5-3.0 $Pa^{0.5}$. The impact of density and viscosity was examined.

Pressure drop contributions were broken into two terms: a pipe flow (wall friction) term and a mixing/shear layer term. The pipe flow term was accounted for by running simulations of vertical channels. The pipe flow term was non-dimensionalized using the hydraulic diameter and diagonal (effective) velocity through the triangular channels. The shear pressure drop term was non-dimensionalized using the channel side length and horizontal velocity. The regressed model predicted the CFD data with an average error of 5% and a maximum error of 20%.

The mixing layer term was found to account for over 80% of the pressure drop in packings with a corrugation angle of 45° and over 60% of the pressure drop in packings with a corrugation angle of 60°. The shear pressure drop was also found to be a strong function of the packing channel opening angle. For a constant specific surface area and corrugation angle, pressure drop was maximized with a channel opening angle of 90°.

386492 Two-Phase Flow Pattern Measurements on a 1m Distillation Tray

Abstract:

Even after a century of tray technology research, the flow on distillation trays is still not fully understood. A method to investigate the flow distribution on 1m diameter chordal downcomer trays during operation is presented. The method is based on high frequency sampling of a temperature tracer. The technology, the evaluation method and several results are shown. The measured results are compared to a full-scale two-phase CFD simulation run on a high performance cluster

391198 The Political Exploitation of the Scientific Illiteracy of the American Public

Abstract:

Despite the attempt to develop a national emphasis on technology too much of the American public is scientifically illiterate. This means that not only are they ignorant of too much of the technical world around them but they are psychologically unable to properly process technical information and productively learn about technical issues because of political, religious, and/or other biases, and for other reasons not well understood, but possibly related to a low level contempt for people who have chosen STEMs careers and an unwillingness to accept that such people could know more than they do about anything related to ordinary life. It almost seems that some people think and/or can be easily convinced that the laws of physics and the laws of nature can be changed or overridden by charismatic rhetoric, signatures on a petition, good public relations skills, or the actions and attitudes of a noisy mob.

392225 Enhance Tray Anti-Fouling with Larger Tray Deck Openings

Abstract:

Severe fouling services have been a problem in the Refining, Ethanol and Olefins Industry for decades. It's not a matter of eliminating the fouling, but rather how long can Operations prolong the interval between cleanings. There are specialized packings (grids) that help with those applications that need low pressure drop, but for trayed applications, people are still looking for a device that can extend the operational life even further than before. Grosserichter and Stichlmair pointed out in 2003 that larger openings are better.

This paper will talk about Sulzer's endeavor to take this philosophy to heart. We will discuss how Sulzer is providing, to the above mentioned industries, a new type valve unit that has already exhibited outstanding results.

374726 Bibliography of Mr. Mark Pilling

Abstract:

Mr. Mark Pilling is a globally-known distillation expert. His background includes trays, packings, process simulations, programming and troubleshooting. He is a Professional Engineer (PE) in the state of Oklahoma.

After receiving a B.S. in chemical engineering from the University of Oklahoma, Mark joined Total Petroleum in Ardmore, Oklahoma. For six years, he was responsible for the day-to-day operations of the following refinery units: Crude, vacuum, saturated gas separations, vacuum gas oil hydrotreater, distillate hydrotreater, naptha hydrotreater, catalytic reformer and C_5/C_6 isomerization.

After a brief stint with Fluor-Daniel in Tulsa, Oklahoma. Mark joined Nutter Engineering in Tulsa, OK in 1995. That company was acquired by Sulzer Chemtech in 1996. Mark continues to contribute on that Tulsa staff. At present, Mark's title is Manager of Technology. His responsibilities include R&D, Technology Transfer and Process Engineering.

Mr. Mark Pilling has contributed appreciably to Fractionation Research, Inc. (FRI) where he has served on the Technical Committee (TC) ever since 2001. He has also contributed appreciably to the AIChE where he is an Elected Fellow. He was a Director of the Separations Division for 5 years; its Chair, in 2012.

This presentation reviews Mr. Pilling's bibliography and summarizes several of his most important papers.

391786 My work with Winston

Abstract:

The first project that Winston and I worked together was liquid membrane at Exxon's Corporate Lab. Specifically, on two applications, one was a novel process of extracting uranium from phosphoric acid and the other was a new fracturing fluid for drilling oil wells. The second project we worked together was the formation of polymeric membranes for water treatment. Specifically, we were trying to develop high flux reverse osmosis membranes. Our past work as well as the current development in this area will be discussed.

Winston is an exceptional researcher with solid fundamental knowledge and insights about his research. It is a real pleasure to have him as my colleague over the past 40 years.

378151 Ultrathin High Flux Membranes: Importance of Microporous Support

Abstract:

Membrane separation has enjoyed significant growth in the last three decades achieved by continual improvement in membrane separation performance and cost reduction. Industrial membranes are often thin film composites comprised of a thin, dense skin layer (0.1 μm) performing molecular separation and a porous bulk of the membrane (150-200 μm) providing mechanical strength and no resistance to mass transport. As the selective layer is produced thinner to enhance the permeance and thus reduce the capital cost, the geometric structure of the microporous support becomes more important, because the porosity and pore size may restrict the concentration profile of the permeant in the selective layer. There is a crucial need to understand the effect of the support layer on the thin film composite membranes to further increase the competitiveness of the membrane technology over conventional separation technologies such as distillation and absorption.

This presentation will discuss the effect of the support geometry on the membrane permeance using an integral approach combining computational fluid dynamics simulations and experiments. The performance of the membrane structure is simulated using a three-dimensional (3D) computational model. The steady-state flux distribution through the separating membrane is studied as a function of various membrane parameters including the thickness of the thin, selective layer and the pore size and porosity of the microporous support. Thin film composite membranes comprised of support membranes with well controlled pore size and porosity are prepared and characterized. Approaches to mitigate the adverse effect of support membranes will be discussed.

387821 Responsive Membranes for Advanced Separations

Abstract:

Stimuli responsive membranes have been developed for many applications some of the original being controlled release of drugs. Stimuli responsive membranes change their physical properties in response to changes in their environmental conditions such as pH, ionic strength and temperature or due to changes in external field due to photo-irradiation or changing electric and magnetic fields. Changes in physical properties of the membrane in response to changed environmental conditions can lead to changes in the mass transfer and interfacial properties of the membrane. Today stimuli responsive membranes find many potential applications.

Often, in membrane based separation processes the properties of the membrane surface that contact the feed stream can have a significant effect on membrane performance. For example, membrane surface properties have been shown to affect the rate of fouling during processes such as microfiltration, ultrafiltration, nanofiltration and reverse osmosis. Modifying just the membrane surface has the potential to retain the properties of the bulk membrane material but modify the interactions between components of the feed and the membrane surface. Here we have developed responsive membranes by surface modification. Two examples will be discussed in this presentation

We have developed novel, magnetically responsive membranes. Magnetically responsive nanobrushes have been grafted to the barrier surface of commercially available nanofiltation membranes. In an oscillating magnetic field, movement of the magnetically responsive nanobrushes leads to suppression of concentration polarization resulting in higher permeate fluxes and better rejection during filtration. In addition, we have grafted magnetically responsive nanorbrushes to the pore surface of track etched microfiltration membranes. Movement of the nanobrushes can lead to changes in pore size and hence change the rejection properties of the membrane. These novel magnetically responsive membranes could lead to a new class of fouling resistant membranes for separation applications.

In the second example we have developed responsive membrane for hydrophobic interaction chromatography. Poly N-vinylcaprolactam (PVCL) chains were grown from the surface of regenerated cellulose membranes using atom transfer radical polymerization (ATRP). PVCL is a thermo-responsive polymer with a lower critical solution temperature (LCST) that depends on the concentration of salt ions present in solution. The LCST decreases below room temperature in buffer containing 1.8 M $(NH_3)_2SO_4$ where the polymer adopts a more hydrophobic/collapsed conformation. At low ionic strength, the LCST remains above room temperature. Adsorption and desorption of bovine serum albumin (BSA) and immunoglobulin G (IgG) have been investigated. Loading was conducted in high ionic strength buffer. Elution was conducted in low ionic strength buffer. By using a responsive ligand that changes its conformation during loading and elution, high protein recoveries were obtained. Ligands that respond to solution ionic strength show promise for high performance hydrophobic interaction chromatography.

Polymeric membranes play a critical role during the generation of electricity in hydrogen and direct methanol proton-exchange membrane (PEM) fuel cells. The membrane in such devices performs three roles: It physically separates the positive and negative electrodes, prevents mixing of the fuel and oxidant, and provides a conduit for ion (proton) transport between the electrodes. For hydrogen proton-exchange membrane fuel cells, the membrane must exhibit low gas permeability and high proton conductivity. For a direct liquid methanol PEM fuel cell, the ion-exchange membrane must conduct protons and be a good methanol barrier. DuPont's Nafion® (a perfluorosulfonic acid polymer) is the membrane of choice for H_2/air PEM fuel cells that operate at a temperature ≤ 80°C. For a reformate fuel, it is desirable to operate the fuel cell at a temperature greater than 100°C (to reduce the effects of CO poisoning), but Nafion looses water and the conductivity degrades at such high temperatures unless the water activity in the fuel gas feed is near unity. Nafion has also been used in direct methanol fuel cells, but high methanol crossover leads to a low power output due to cathode depolarization by methanol and unwanted fuel losses. There is considerable research and development around the world to develop new membrane materials for PEM fuel cells with tailored physical and transport properties. Alternatively, researchers are studying modified forms of Nafion (e.g., Nafion blends and composites) in order to selectively improve certain membrane properties. In general, current membrane fuel cell work is focused on: (i) proton conductors that operate at a temperature of 100-120°C under low humidity conditions (25% relative humidity) for hydrogen/air fuel cells, (ii) polymeric membranes that conduct protons in the dry state at high temperatures (160-200°C), and (iii) proton conducting ion-exchange membranes with low methanol permeability for operating temperatures in the range of 60-140°C.

Recent fuel cell membrane work by the present authors has focused on: (i) stretched recast Nafion and Aquivion® membranes and (ii) proton conducting membranes based on interconnecting nano-fibers. The over-riding objective of these studies is to control and manipulate the membrane morphology in such a way as to improve membrane properties for fuel cell applications, such as increasing proton conductivity at high temperature and low relative humidity or lowering methanol permeability. In this talk, an overview of the various membrane systems under investigation in the authors' lab will be presented. Membrane fabrication methods and the results of membrane characterization experiments will be described. Physical property data relevant to PEM fuel cell applications will be related to the membrane morphology. Available fuel cell performance data will be also discussed.

Recovery and separation of rare-earth metal ions by using supported liquid membranes with strip dispersion

I-Wei Chang[1], Jia-Yun Han[1], Yu-Wei Chen[1], Tzu-Yang Hsien[2], Hsyue-Jen Hsieh[1], Juin-Yih Lai[3], Da-Ming Wang[1*]

[1] Department of Chemical Engineering, National Taiwan University, Taipei, Taiwan, 10617

[2] Center for General Education, China University of Technology, Taipei, Taiwan, 11695

[3] R&D Center for Membrane Technology, Chung Yuan University, Chungli, Taiwan, 32023

*daming@ntu.edu.tw

Because of the fast demand growth of rare-earth metals, recycling of them for reuse has attracted much research attention. The present research exploited the technique of supported liquid membranes with strip dispersion to recover and separate rare-earth metal ions from acidic solution. Two systems were investigated: the solution of neodymium and dysprosium and that of europium and yttrium. Neodymium and dysprosium are used to manufacture high-performance permanent magnet motors for hybrid and electric cars and wind turbines. And europium and yttrium are widely used to prepare fluorescent lamps and phosphors for LEDs. The extractant used for both systems was di-(2-ethylhexyl)phosphoric acid, D2EHPA. We first performed batch extraction experiments to determine the equilibrium constants for the formation of the metal complexes (metal ions with D2EHPA). On the basis of the obtained extraction equilibrium constants, models were developed to describe the permeation of the ions through the supported liquid membrane. The validity of the models was examined by comparing the model results with experimental data. With the models, we were able to determine the operation conditions that gave excellent selectivity for the metal ions.

Efficient separation of ionic liquids using by pervaporation

Jian Sun [a,b], Jian Shi[a,b], Seema Singh[*a,b], Blake Simmons[a,b]

[a] Deconstruction Division, Joint BioEnergy Institute, Emeryville, CA, USA. E-mail: seesing@sandia.gov; Fax: +1 510-486-4252; Tel: +1 925 294 4551

[b] Biological and Materials Science Center, Sandia National Laboratories, Livermore, CA, USA

Abstract:

Recently, the efficient separations of ILs have become an increasingly topic accompany their wide applications since many of ILs are expensive and have potential environmental influences.Generally, ILs could be separated by distillation, extraction and electrodialysis. Among those methods, distillation might be a widely choice in practical applications due to its high efficiency and large capability.However, the drawback of distillation is its high energy consumption. To improve this item, pervaporation has been taken into consideration, which is a highly selective and versatile membrane separation process for quantitatively recovering volatile solutes directly from non-volatile compound (e.g. ILs). It could prove more versatile than solvent extraction processes as well as more efficient and energetically more advantageous than evaporative techniques. Compared to traditional distillation, it could be more effectively and needs milder operating conditions. Thus, this kind of method has attracted increasing attentions recently.

Herein, separations of typical ILs used for biomass pretreatment were investigated by using pervaporation from different IL/H_2O, and IL/H_2O/glucose systems, respectively. Effects of various parameters such as temperature, sweeping gas flow rate, feed mass and IL concentration on the separation as well as the conductivity changes on the IL systemswere studied in detail. It was found that: 1) a little separationdifferencebetween different type of ILs; 2)temperature played akey role on the separation efficiency; 3) within 4 hours, IL could be separated with a >99% purity from almost all of the samples; 4) with the presence of glucose, the separation efficiency of ILdecreased. Based on the experimental results, an energy consumption comparisonbetween traditional vacuum distillation and pervaporation was discussed. The present work showed that pervaporation could be used as an efficientandsimple method for the separation of ILs in bio-processes.

Keywords: Pervaporation, Ionic liquids, Biomass, Separation, Sugar

384268 Optimization of Direct Contact Membrane Distillation Based on Regressive Rotation Design

Abstract:
Membrane distillation (MD) is a promising separation technology for desalination. It is of significance for the practical application of MD to improve process efficiency and to reduce energy consumption through process design and parameter optimization. From literature, the study on the modelling and optimization of MD process considering both permeate flux and thermal efficiency integrating both operating conditions and module dimension parameters is rare. The purpose of this study is to try to search for both optimum operating conditions and module dimension parameters as well as to know their interaction effect on MD performance to achieve high permeate flux, large water production and low energy cost. The statistical experimental design and response surface methodology were applied for the modeling and optimization of direct contact membrane distillation (DCMD) process. The method is useful to understand the interaction effects between factors and to reduce the total number of experimental runs.
In DCMD process, polyvinylidene fluoride (PVDF) hollow fiber membranes were employed to desalting 3.5wt% sodium chloride aqueous solution and all of salt rejection in this study is above 99.9%. The operating conditions including feed inlet temperature, permeate inlet temperature and velocity of feed and the module configuration parameters including module packing density and length-diameter ratio were taken into account. The average permeate flux , water productivity per unit volume of module, water production per unit exergy loss and an comprehensive index (CI) of them were taken as optimization objectives and their regression models were developed.
It is concluded that the optimal set of operating conditions and membrane module parameters has been determined and the significant increase of MD performance was achieved. The coupling and interaction effect between the operating conditions and the module parameters were significant on the performance of DCMD. The models for the prediction of DCMD performance were established. Good agreement was observed between the simulated and the experimental results.

387632 Experimental and Analytical Investigation of the Effect of Membrane Thickness Together with a Critical Commentary on MD Analysis

Abstract:

An analytical model was created to estimate the optical membrane thickness for DCMD operations. It is found that the value of optimal thickness increases with reduced heat transfer coefficients; decreased feed inlet temperature; increased membrane permeability; and increased salinity. PVDF electro-spun membranes, consolidated with a heat press process, as thin as 27 microns were fabricated and tested in a DCMD cell. A transmembrane flux as high as 60 L/hr m^2 was observed. Except for the thinner membranes with high inlet feed temperatures, the fit of the model was excellent provided due allowance was made for the measured variation in porosity with thickness. The utility of describing deviations from ideal performance in terms of a Temperature Polarisation Coefficient is questioned and an alternative approach will be proposed.

386651 Free Surface Electrospun Polyvinylidene Fluoride Membranes for Direct Contact Membrane Distillation

Abstract:

Direct Contact Membrane Distillation (DCMD) can be utilized as a desalination process similar to traditional distillation and reverse osmosis. In DCMD a porous, hydrophobic membrane is used to desalinate water by separating nonvolatile molecules such as salt, and other impurities from seawater/brackish water. However, the DCMD process employs a temperature gradient to create a vapor pressure gradient; driving the separations of chemical species. Here, free surface electrospinning was considered to produce membranes for the DCMD process in large quantities while maintaining high porosity and controllability over pore size. A solution of Polyvinylidene Fluoride (PVdF) in Dimethylacetamide (DMac) was electrospun over a range of high voltages and times to produces varying fiber morphologies and membrane thicknesses. The membrane's morphology was examined by scanning electron microscopy. In order to investigate the mechanisms responsible for efficient productivity, a DCMD setup was assembled. Various temperature gradients and flow rates were investigated with the DCMD setup. It was determined that increased temperature gradients and flow rates led to an increase in productivity. These observations suggest that higher temperature gradients provide a higher driving force for transport. This study aims to find a membrane's maximum productivity by varying operational parameters.

390216 Fouling and Crystallization with Hydrophobic and Superhydrophobic Hollow Fibers Membranes Using Direct Contact and Vacuum Membrane Distillation

Abstract:

Superhydrophobic nano-composite polypropylene hollow fibers were fabricated for membrane distillation and crystallization by using a low temperature hydrothermal approach for TiO_2 coating and fluoro-silane compounds. The membrane distillation performance was investigated during direct contact membrane distillation and vacuum membrane distillation, and the crystallization behaviors of these membranes were observed in real-time. Crystallization on hydrophobic and superhydrophobic membranes differed in terms of structure and uniformity. With the introduction of humic acid to the feed, the crystal deposition for both membranes increased in terms of. The rate of deposition was found to be lower for superhydrophobic membranes, and deposited solutes were more easily removed. Crystal nucleation and growth behaviour was consistent with those observed flat sheet membranes, indicating similar crystallization mechanisms for both membrane modules.

390560 Novel Aerogel Membranes for Membrane Distillation

Abstract:

Membrane Distillation (MD) technology, which substitutes low-temperature waste heat for electricity as the energy input to the desalination process, fills a unique niche in a water treatment portfolio. MD is able to treat brines over a much higher salinity window than other membrane based processes, thereby improving recovery rates from standard reverse osmosis processes and enhancing the viability of inland desalination. Second, MD is capable of using low-temperature waste heat as a feed source. This will enable step-changes in efficiency of water treatment if proximally located to a waste-heat source. The proposed aerogel membrane is expected to simultaneously probe the upper bounds of flux and the lower bounds of exergy loss for MD systems. Here we present highly successful preliminary results, and share the design implications for future MD membrane systems.

391762 On the Development of a Novel Pressure Swing Reactor for the Production of Ammonia

Abstract:

It has always been desirable to carry out reaction and separation in the same vessel. The concept of a pressure swing reactor that can do reaction and separation in the same vessel has been around for several decades. However, this concept has yet to be commercialized, although several patents have been issued and some novel approaches have been tested at various scales.

A pressure swing reactor combines a pressure swing adsorption (PSA) process and catalytic gas phase reaction in such a way that a higher conversion can be achieved usually via Le Chatelier's principle. The difficulty is to find a suitable adsorbent that operates at or near the conditions of the catalytic reactor, i.e., at high temperature and pressure. High pressure is conducive to adsorption, but high temperature is not. In fact, most adsorbents are regenerated at temperatures far below the conditions of a catalytic reactor. Nevertheless, there are some catalytic systems that present an excellent opportunity for the design of a pressure swing reactor. The synthesis of ammonia from hydrogen and nitrogen is one such system.

Process simulations of a pressure swing reactor have shown that very high conversions can be obtained by using an adsorbent that reversibly adsorbs ammonia at high temperature by carrying out a relatively simple pressure swing cycle. The goal of this work is to experimentally validate the simulation results. For this purpose a bench-scale pressure swing reactor has been set-up and is currently being utilized to study the conversion of hydrogen and nitrogen into ammonia using a commercial catalyst and in the absence and presence of an adsorbent. This presentation will present the latest results obtained from this effort.

365546 Hot-Gas Pressure Swing Adsorption for Drying Reactive Monomers

Abstract:

A wide variety of unit operations may be used to break an organic + water azeotrope in order to produce a dry organic product. These include temperature-swing and pressure-swing adsorption; membrane-based pervaporation and vapor permeation; and azeotropic, extractive, and pressure-swing distillation. However, in applications involving a polymerizable or otherwise reactive monomer, the viability and operability of these processes can come into question. This presentation will describe the use of Hot-Gas Pressure Swing Adsorption (HGPSA) to dry a reactive monomer from 3 wt% water (a typical azeotropic amount) down to trace ppm levels. HGPSA is well-known as a commercial process for breaking alcohol + water azeotropes such as ethanol-water and isopropanol-water. It differs from classic PSA processes in that the feed is not a gas but a liquid that must be completely vaporized and superheated prior to processing. Our HGPSA pilot work shows how very dry monomer can be produced with no detectable polymerization – by avoiding any condensation during processing and by taking measures to ensure inhibitor is present during the initial vaporization step and on condensation of the dry product. Results of process simulation using Aspen ASDIM, process design, and pilot-scale operation will be presented, and the applicability of the technology for processing reactive monomers in general will be discussed. Key process design considerations include selection of adsorbent, the amount of superheat in centrigrade degrees above the dew point, adsorption and desorption pressures, the desorption factor (a kind of reflux ratio), and PSA cycle time.

362608 Pilot Plant Evaluation Featuring High Temperature PSA-Oxygen Using Perovskite-Type Oxygen Adsorbent

Abstract:

As described in the previous report on pressure swing adsorption using perovskite-type oxygen adsorbent, oxygen/nitrogen separation from air at high temperature (HTPSA-O_2) shows substantial reduction in the electric power consumption rate.

In this study, we built a $5m^3$N-O2 / h pilot plant as a step towards adoption for practical use. At high temperatures in the range of 500-800 degrees Celsius, 1) the heating method for the adsorption tower, 2) temperature control, and 3) the optimum temperature for HTPSA-O2 were considered during investigation of power consumption reduction by heat recovery.

In particular, in order to achieve a higher heat recovery ratio, the front and rear parts of the adsorption tower are filled with the heat storage material, and in the adsorption step (Step 11) of the first cycle, the feed air is supplied from the front part to contact with the heat storage material for heat exchange between the feed air and the desorbed oxygen. The lack of heat content is compensated for by means of an electric heater. Given that effluent nitrogen comes into contact with the heat storage material at the rear part of the adsorption tower, the heat in the nitrogen can be recovered. In the desorption step (Step 12), since the recovered oxygen comes into contact with the heat storage material in the front part, the heat in the desorbed oxygen can be recovered.

In the next (second) cycle, by supplying the feed air from the rear part of the tower in the adsorption step (Step 21), the feed air contacts the heat storage material in the rear part and heat exchange between the feed and the heat storage material proceeds. Heat insufficiency is again supplemented by an electric heater. The heat in the effluent can be recovered with the heat storage material in the front part. Heat recovery from the desorbed oxygen can be undertaken with the heat storage material in the rear part in the desorption step (Step 22).

For each cycle, the directions of the feed air and the desorbed oxygen are changed to confirm the increase in the heat recovery ratio between the feed air and the effluent nitrogen, and between the feed air and the desorbed oxygen.

During heat exchange with the heat storage material, the relationship is assessed between a) the desorption pressure and b) cycle time on the one hand, and 1) the electric power consumption rate of the rotatory equipment and heater and 2) the oxygen production rate on the other.

Based on these results, HTPSA-O_2 in this study was compared with conventional processes such as PSA-O_2 and VPSA-O_2.

The overall results of this study are summarized as follows:

1. As the adsorption amount of oxygen on LSCF1991 shows a strong Langmuir-type adsorption isotherm, the lowest electric power consumption rate of the oxygen production and the greatest oxygen production rate versus the unit adsorbent weight are shown at a relatively low level of vacuum pressure (1-5kPa)

2. To maintain the practical oxygen production rate versus the unit adsorbent weight, the adsorption temperature needs to be higher than 500 degrees Celsius.

3. The heat recovery ratio was improved dramatically with the two cycles in different directions.

   While the heat exchange ratio of the shell & tube heat exchanger mechanism is about 80%, and that of the plate fin type heat exchanger mechanism is about 92%, the heat exchanger in the two-cycle operation mode 2 showed 95%.

   In conclusion, the pilot unit in this study showed achievement of a heat recovery ratio for HTPSA of greater than 95% at ($5m^3N$ / h). At a rapid cycle of 100 seconds (40 seconds of adsorption, 40 seconds of desorption and 20 seconds of other operations), the electric power consumption rate was 1.46kWh / m^3N, and the product amount of oxygen/adsorbent weight ratio was measured at $30m^3N$-O_2/h/ton (per 500kg/tower, in a 2-tower system).

   As the performance of larger scale units such as $1{,}000m^3N$/h-O_2 can be predicted from the results of this study, the electric power consumption rate is assumed to be 0.2kWh/m^3N, while the product amount of oxygen/adsorbent weight ratio is assumed $30m^3N$-O_2/h/ton.

   Compared with cryogenic separation, which exhibits the smallest value of 0.32kWh/m^3N at the present time, HTPSA-O_2(as in this study) indicates a much lower electric power consumption rate in the future.

   It is assumed that the initial cost of HTPSA-O_2 is as expensive as VPSA-O_2 with nitrogen adsorbent.

388099 Syngas Purification Using MIL-125(Ti)_NH2

Abstract:

Parallel to the removal of a specific component (H_2S or CO_2) from a gas mixture; the purification of H_2 or tuning of (H_2+CO) composition by adsorbing all its impurities except the target gas(es) is the main objectives in syngas purification. Depending on their origin, the feed gases can contain a variety of impurities: N_2, CH_4, C_2H_6, C_2H_4, CO, CO_2, H_2S, NH_3, H_2O. In commercial applications for hydrogen purification, PSA cycles use layered beds with activated carbons and 5A zeolites. Gas purity can reach 99.999% at 70-80% recovery. Syngas itself can also be processed into high value products, such as methanol or synthetic fuel via Fischer-Tropsch (FT). However, H_2/CO ratio in syngas is about ~0.7 while the ideal H_2/CO ratio of the incoming syngas for Fischer-Tropsch (FT) synthesis is 2 and the required (H_2–CO_2)/(CO+CO_2) ratio for methanol production is 2.1. Therefore, syngas must be conditioned and PSA processes are alternatives to amine absorption processes. Another aim in syngas treatment can be to purify CO_2 to purities above 95% required for Carbon Capture and Storage (CCS). This is a relatively new research field without much experimental data and would support the new EU directive on CO_2 quality required for geological storage. The EU project "Dynamis , taking into account health and safety considerations as well as technical and storage issues, recommended the following target values: H_2S < 200ppm, CO < 200ppm, SOx < 100ppm, NOx < 100ppm, H_2O < 500ppm, O_2+CH_4+N_2+Ar+H_2 < 4%vol (all non-condensable gases). Thus, the challenge is to find an adsorbent that simultaneously removes all impurities, eliminating the need for a multi-layered separation column.

In this work, MIL-125(Ti)_NH2 supplied by KRICT - Korea Research Institute of Chemical Technology as granulates is proposed as potential adsorbent for this task, aiming an adsorption based process. The titanium(IV) 1,4-benzenedicarboxylate (or terephthalate) MIL-125(Ti)_NH2, with the chemical formula $Ti_8O_8(OH)_4$-(O_2C-$C_6H_3NH_2$-CO_2)$_6$ has a three-dimensional pseudo-cubic structure with highly interconnected porous. This possesses two types of cages corresponding to the octahedral (12.5 Å) and tetrahedral (6 Å) vacancies of a cc packing accessible through triangular narrow windows of ca. 6 Å. The carbon monoxide, carbon dioxide, methane, nitrogen, and hydrogen adsorption equilibrium data and adsorption dynamics (single, binary, and ternary adsorption and desorption breakthrough curves) on this adsorbent were assessed at 4.5 bar and 323 K. MIL-125(Ti)_NH2 granulates present an high selectivity for CO_2 and exhibited easy regeneration. The single component, binary, and ternary fixed-bed adsorption experiments were simulated in order to validate the mathematical model. The same column, employed for the fixed-bed breakthrough curves described above, was used for PSA experiments, previously designed by simulation using the validated model. A PSA cycle of four elementary steps was performed. The PSA cycle started with the co-current pressurization with feed, followed by feed, blowdown and purge steps. In the purge step, a stream of pure hydrogen was used countercurrently at low pressure to regenerate the column.

389596 Rapid Swing Adsorption/Desorption Using Polymer Supported Amine/Silica Hollow Fiber Sorbents for CO2 Capture

Abstract:

Rapid temperature swing adsorption/desorption (RTSA) using novel polymeric supported amine/silica hollow fiber sorbents represents a new approach to post combustion CO_2 capture [1]. Hollow fiber sorbent contactors function as a "nanoscopic heat exchangers," where flue gas is fed through the shell-side of fiber module and heating/cooling media is fed through the bore. The unique structure of the porous hollow fiber sorbent allows coupling of efficient heat transfer with effective gas contacting, achieving essentially isothermal sorption while providing a lower parasitic load compared to a traditional fixed bed process. In this work, we report the synthesis and application of cellulose acetate/silica supported amine hollow fiber sorbents [2-4] that demonstrate high dynamic CO_2 capacities in the presence of water along with outstanding cyclic stability over 60 cycles. Additionally, robust dual layer polyamide-imide (PAI, Torlon®, Solvay Advanced Polymers©) silica supported amine hollow fiber sorbents (Torlon-S-PEI) were successfully developed as well, with the successful addition of impermeable lumen layer in the fiber spinning process [5]. A CO_2 breakthrough capacity of up to 1.33 mmol/g-fiber (2.66 mmol/g-silica) was observed when the first generation fibers were actively cooled by flowing cooling water in the fiber bores (with efficient heat removal), whereby the sorption enthalpy recovery can reach ~ 72%. Deactivated fibers were regenerated via post-spinning amine infusion, allowing recovery of the full sorption capacity of fiber sorbents after deactivation due to oxidation or poisoning by SO_2 or NO_2, The potential for significant savings in terms of energy, capital and operational costs exists with this scalable CO_2capture technique based on polymer/silica supported amine fiber sorbents.

1. Lively, R.P., et al., *Hollow fiber adsorbents for CO_2 removal from flue gas.* Industrial & Engineering Chemistry Research, 2009. **48**(15): p. 7314-7324.

2. Labreche, Y., et al., *Post-spinning infusion of poly(ethyleneimine) into polymer/silica hollow fiber sorbents for carbon dioxide capture.* Chemical Engineering Journal, 2013. **221**(0): p. 166-175.

3. Rezaei, F., et al., *Aminosilane-grafted polymer/silica hollow fiber adsorbents for CO_2 capture from flue gas.* ACS Applied Materials & Interfaces, 2013. **5**(9): p. 3921-3931.

4. Fan, Y., et al., *Evaluation of CO_2 adsorption dynamics of polymer/silica supported poly(ethylenimine) hollow fiber sorbents in rapid temperature swing adsorption.* International Journal of Greenhouse Gas Control, 2014. **21**(0): p. 61-71.

5. Labreche, Y., et al., *Dual layer spinning with lumen layer containing PAI/Silica/PEI hollow fiber sorbent for CO_2 separation by rapid temperature swing adsorption.* In preparation.

390390 Do Performance Indicators Predict Adsorption Process Performance? the Case of Post-Combustion CO2 Capture

Abstract:

Finding the right adsorbent for adsorptive post-combustion carbon capture has been a research topic that has been actively pursued around the globe. Most studies in the literature have focused on materials development. Several new metal organic framework (MOF), zeolites and carbon-based materials have been reported. Based on equilibrium and kinetic parameters, researchers have argued the superiority of one material over the other. Various performance metrics have been proposed in the literature, e.g., various forms of selectivities, loading ratios, figures of merit etc. Firstly, many of the parameters were developed for light-product purification while separations such as CO2 capture deal with the purification of the heavy component. Secondly, it is not known which performance parameter(s), e.g., purity, recovery, productivity, energy consumption etc. do they correlate to. Very often when trying to rank materials, the various metrics identify a different material to be superior.

We consider the case of CO2 capture from dry post-combustion flue gas. Through detailed process design and optimization, that takes into account key phenomena such as mass and heat transfer, pressure drop in the bed, we show that that most properties are poor indicators of large-scale process performance. We propose a framework for rational material selection through detailed process optimization, which can be an effective tool for material selection. The framework is illustrated by comparing the performance of three adsorbents, viz. zeolite 13X, Mg-MOF-74 and UTSA-16.

360998 Development and Optimization of Temperature Swing Adsorption Cycles for CO2 Capture and Utilization

Abstract:

Cyclic adsorption processes for gas separations have been extensively studied for the purification of the light components, since purities as high as 99.999% can be achieved with limited energy consumption. Examples of these processes are pressure swing adsorption (PSA) for H_2-purification, or temperature swing adsorption (TSA) for the drying of gases [1]. Contrary to that of the light product, the purity of the heavy product is thermodynamically limited in traditional (stripping) cycles which are characterized by the production of the light product at high pressures/low temperatures during the feed step, and by the desorption and production of the heavy product at low pressures/high temperatures [2]. Therefore, the design of cycles aimed at enriching the heavy product still remains a challenge.

A niche of applications for TSA cycles for CO_2 recovery lies in the chemical industry, where CO_2 is often emitted together with N_2 in the flue gas of reformers or boilers, but where CO_2 also represents a valuable reactant for the synthesis of chemicals. Examples of which are the synthesis of urea and that of methanol, where CO_2 is limiting with respect to the ammonia or hydrogen, respectively. In such applications the choice of a thermal regeneration allows a partial recovery of the sensible heat of the flue gas, which will lower the effective energy consumption of the process.

In this work we present a comprehensive investigation of TSA cycles for the recovery of CO_2 in post-combustion applications with downstream CO_2 utilization based on a detailed 1D adsorption column model [3]. Appropriate performance indicators are established, which define a theoretical performance limit: the minimal total volume of adsorbent required, and the minimal thermal energy requirement. The influence of the operating step duration and operating conditions on the region of attainable specifications and on the process performance are elucidated, and the effect of constraints such as the temperature of the flue gas is reported. This extensive analysis leads to the elaboration of heuristics for the development and optimization of TSA cycles for CO_2 recovery. Since the total volume of adsorbent required depends on the idle times of the cycle, hence on its scheduling, we further report the effect of different scheduling constraints on the limits of the attainable performances for a novel TSA process.

References

1. Ruthven, D., *Ind. Eng. Chem. Res.* (2000), **39**, 2127—2131
2. Ebner, A. D. and Ritter, J. A., *AIChE Journal* (2002), **48**(9), 1679—1691
3. Casas N., Schell J., Pini, R. and Mazzotti M., *Adsorption* (2012), **18**, 143—161

391926 Reactive Separation of CO2 with Liquid Absorbents: Novel Approaches

Abstract:

Growth of the global economy, industry and population lead to increasing demand for energy. Until emission-free renewable energy resources account for a significant portion of the market, traditional fossil fuels (e.g., oil, gas and coal) will remain the main sources of energy on the planet; likely for the next several decades. Before transport to the consumer or direct utilization in power generation or by the chemical industry, fossil fuels, especially natural gas, requires removal of undesirable acid gases including CO_2. Additionally, CO_2 is an unavoidable product of fossil fuel combustion. To meet the specifications and potential upcoming regulations on emissions of CO_2 into the atmosphere, development and worldwide utilization of efficient CO_2 capture and sequestration technologies are required. Certain microporous molecular sieves are proven to selectively remove CO_2, but reactive separation of acid gases can offer the lowest losses and highest product purity.

This presentation will specifically discuss the CO_2 reaction mechanisms with various amine blends in aqueous, non-aqueous and ionic liquids solutions targeting highly selective and more economically efficient CO_2 scrubbing from, e.g., flue gas. We developed the experimental capabilities to revisit the classic acid-base (CO_2-amine) chemistry and gain a more in-depth fundamental understanding of CO_2 reaction mechanisms with various reactive solutions. *In-situ* ^{13}C and ^{1}H Nuclear Magnetic Resonance (NMR) spectroscopy using a built-in micro reactor was used to provide real time insights on reaction mechanisms and product speciation under conditions for CO_2 absorption and regeneration.

A new approach to non-aqueous CO_2-amine carbon capture is based on the utilization of a combination of a nucleophilic amine CO_2 sorbent (Lewis base) with a second, non-nucleophilic Brønsted base; a "Mixed Base" system. The effects of amine type, amine basicity, and the use of non-aqueous solvents on the reaction pathway and product stability will be discussed. It will become apparent that both Lewis and Brønsted basicity are key components in the chemistry. Proper choice of these bases and non-aqueous solvent allows capture of CO_2 molecules to tailor reaction conditions (T, P) to specific applications. The implications of these findings for further amine studies will be outlined.

380397 De-Entrainment Characterization of a Mist Elimination Device Using Phase Doppler Interferometry

Abstract:

The use of liquid distribution devices in the presence of a counter flowing gas can result in droplet entrainment. In industrial processes, a loss of equipment operating efficiency, damage to downstream gas handling equipment, and corrosion/fouling can result from liquid entrainment. Mist elimination devices serve as an effective way to remove entrained liquid and negate these effects. These devices are commonly composed of a wire mesh pad placed just above the liquid distribution device and use droplet momentum to divert entrained liquid to the column wall. The design and effectiveness of de-entrainment devices to remove liquid droplets is dependent on the entrained droplet size, velocity, and physical properties. In this study, Phase Doppler Interferometry (PDI) is utilized to characterize the effectiveness of a de-entrainment device to remove droplets in an air-water system. PDI measurements are reported from both above and below the de-entrainment device. A spray nozzle is used for liquid distribution in an 8 inch I.D. absorption column, where gas is introduced in a counter current flow relative to the liquid. A variety of nozzle sizes are used to produce small (10 μm) and large (100 μm) droplets under various liquid and gas flow rates.

Sorption Modeling and Verification for Off-Gas Treatment

JORGE F. GABITTO[1] AND COSTAS TSOURIS[2]

[1]Prairie View A&M University , TX

[2]Oak Ridge National Laboratory, Oak Ridge, TN

OAK RIDGE National Laboratory

Off-gas Modeling and Verification Team

- Participants:
  - I_2, CH_3I, and HTO equilibrium and kinetic data acquisition
    - Ronghong Lin, Yue Nan, and Lawrence L. Tavlarides (PI)
  - HTO and I_2 column experiments and data acquisition
    - David W. DePaoli (Co-PI)
  - Modeling of adsorption of HTO, I_2, CH_3I, Kr, and Xe
    - Austin P. Ladshaw, Sotira Yiacoumi (Co-PI) and Costas Tsouris (Co-PI)
  - Transport modeling of gases
    - Veronica Rutledge and Jack Law, INL
  - CO_2 absorption experiments and modeling
    - Jorge Gabitto (Co-PI)

Relevance

- A model to simulate dynamic absorption of carbon dioxide by a liquid alkaline solution is important for the design of processes to capture radioactive gases (^{14}C) generated in nuclear fuel reprocessing.
- It is also important in normal industrial applications to decrease greenhouse gases emissions.

PRAIRIE VIEW A&M UNIVERSITY 1876

OAK RIDGE National Laboratory

Project Goals

- The task is to develop a dynamic absorption model for CO_2 based on current literature for both steady state and transient state operations.
- In addition, the experimental part aims to determine the procedure to calculate gas-liquid mass transfer coefficients.

PRAIRIE VIEW A&M UNIVERSITY 1876

OAK RIDGE National Laboratory

Reactive Absorption

- Absorption is a process in which molecules move from one phase and enters a bulk phase of either gas, liquid, or solid.
- In this project, we will be modeling the reactive absorption of gaseous CO_2 by a liquid alkaline (NaOH) solution.

Theoretical Model-System

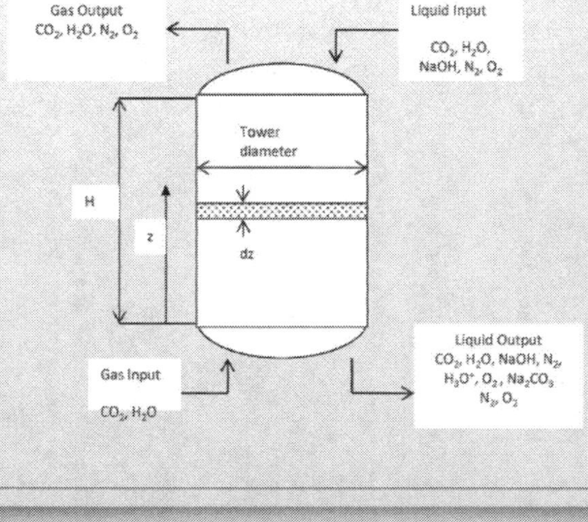

Theoretical Model-Mass Transfer

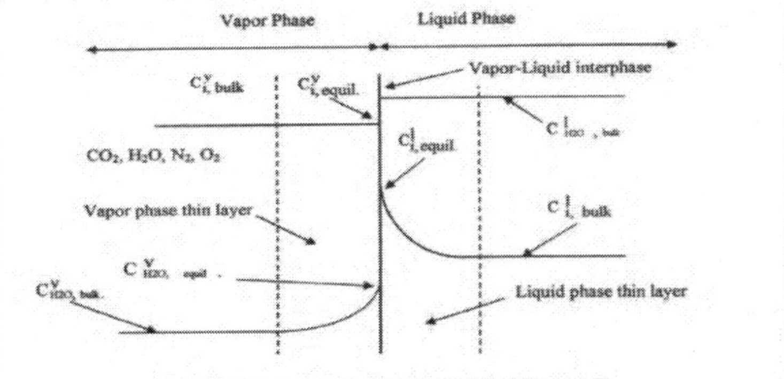

Figure 2. Concentration gradients at vapor-liquid interphase.

OAK RIDGE National Laboratory

Theoretical Model-Equations

- $\frac{\partial C_i^l}{\partial t} = u_L \frac{\partial C_i^l}{\partial z} - \dot{N}_{diff}$, liquid phase mass balance
- $\frac{\partial C_i^v}{\partial t} = -u_v \frac{\partial C_i^v}{\partial z} + \dot{N}_{diff}$, gas phase mass balance
- $\dot{N}_{diff} = k_l a_w \left(C_i^b - C_i^*\right)$, interphase mass fluxes
- $y_i = H x_i = \frac{P_i^s \varphi_i^l}{P_T \varphi_i^v} x_i$, interphase equilibrium model

OAK RIDGE National Laboratory

Theoretical Model-Equations

- Fast reaction confined to liquid layer
- $r_{CO2} = k_1 [CO_2] [OH^-]$. , reaction rate
- $\dot{N}_{CO2} = - k_l E a_w H^{cc} C^g_{CO2}$, CO_2 mass flux
- $HA = \frac{\sqrt{k_{ef} D_{CO2} C_{OH-}}}{k_L} = E$, enhancement factor

Theoretical Model-Equations

Liquid phase energy balance

$$\frac{\partial T^l}{\partial t} = - u_l \frac{\partial T^l}{\partial z} - \dot{N}_{diff} \Delta H^{vl} / C_p^l - \dot{N}_{d,CO2} \Delta H_R / C_p^l - U_T A_T \left(T^l - T^v\right) / C_p^l$$

Gas phase energy balance

$$\frac{\partial T^v}{\partial t} = - u_v \frac{\partial T^v}{\partial z} + U_T A_T \left(T^l - T^v\right) / C_p^v$$

Solution Procedure

- Model implemented in a computer program
- Physical properties subroutine developed
- Numerical algorithm to solve mass differential equations developed
- Numerical algorithm to solve temperature differential equations developed
- Combination of subroutines into complete code

Numerical Procedure

- All equations are hyperbolic partial differential equations given by,

$$\frac{\partial u}{\partial t} + \alpha \frac{\partial u}{\partial z} + N_i = 0, \alpha = \text{const.} > 0,$$

- An explicit finite differences numerical scheme was used for solving hyperbolic partial differential equations,

$$\frac{u_j^{n+1} - u_j^n}{\Delta t} + \alpha \frac{u_j^n - u_{j-1}^n}{\Delta z} + N_i^n = 0$$

- This scheme is consistent with the original differential equation and conditionally stable. Thus, is stable if the Courant's condition is met:

$$\left| \alpha \frac{\Delta t}{\Delta z} \right| \le 1$$

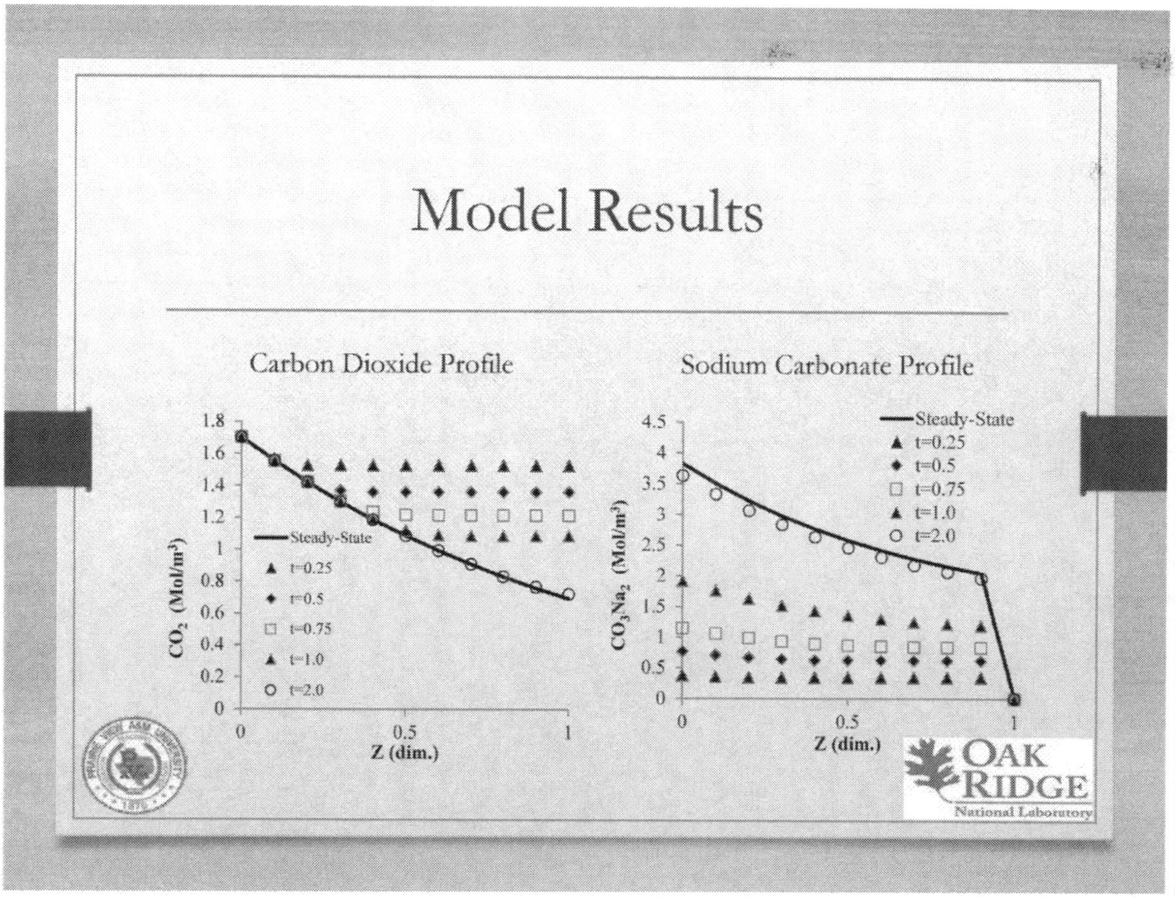
Model Results
Carbon Dioxide Profile
Sodium Carbonate Profile
CO2 (Mol/m3)
Z (dim.)
Steady-State
t=0.25
t=0.5
t=0.75
t=1.0
t=2.0
CO3Na2 (Mol/m3)
OAK RIDGE
National Laboratory

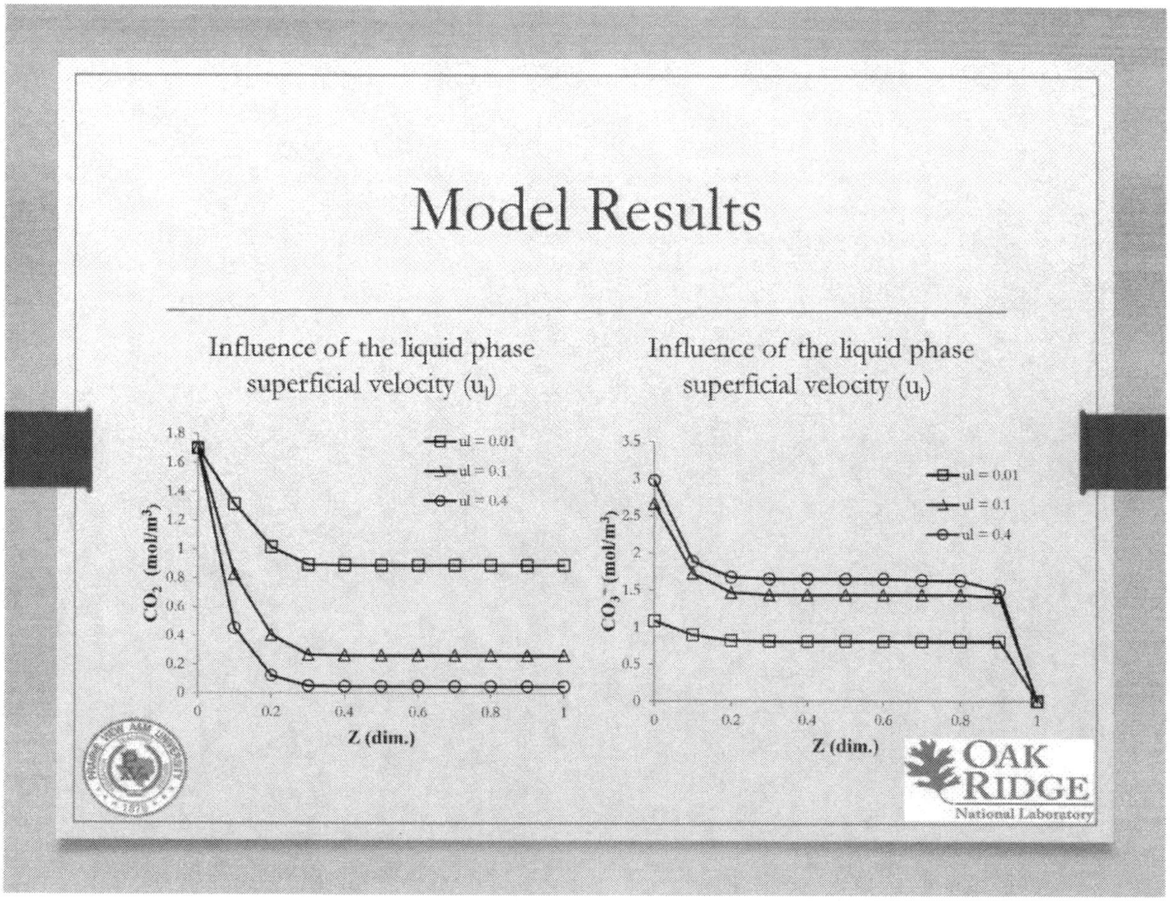
Model Results
Influence of the liquid phase superficial velocity (ul)
Influence of the liquid phase superficial velocity (ul)
CO2 (mol/m3)
ul = 0.01
ul = 0.1
ul = 0.4
CO3= (mol/m3)
Z (dim.)
OAK RIDGE
National Laboratory

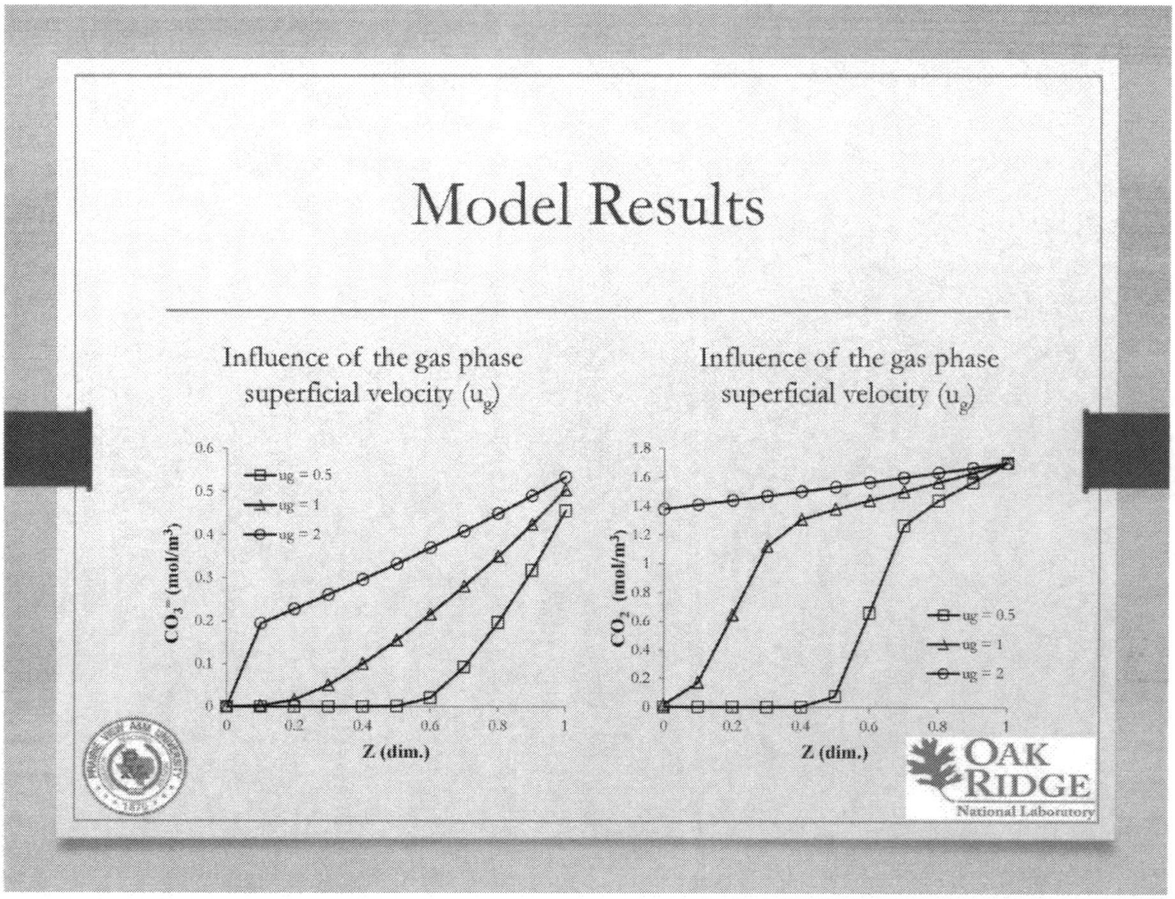
Model Results
Influence of the gas phase superficial velocity (u_g)
Influence of the gas phase superficial velocity (u_g)
ug = 0.5
ug = 1
ug = 2
CO3= (mol/m3)
CO2 (mol/m3)
Z (dim.)
Z (dim.)
OAK RIDGE
National Laboratory

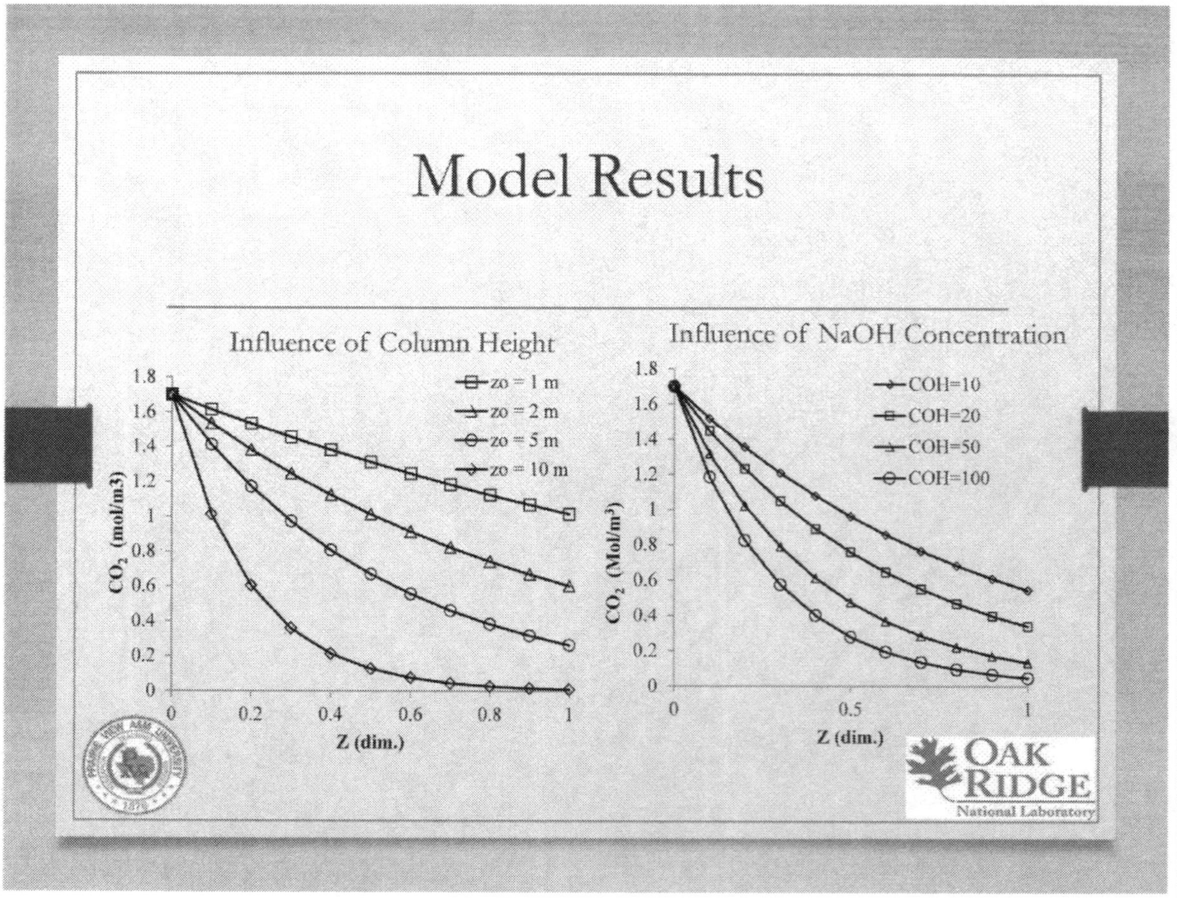
Model Results
Influence of Column Height
Influence of NaOH Concentration
zo = 1 m
zo = 2 m
zo = 5 m
zo = 10 m
COH=10
COH=20
COH=50
COH=100
CO2 (mol/m3)
CO2 (Mol/m3)
Z (dim.)
Z (dim.)
OAK RIDGE
National Laboratory

Model Results

Influence of NaOH amount on vapor phaseTemperature

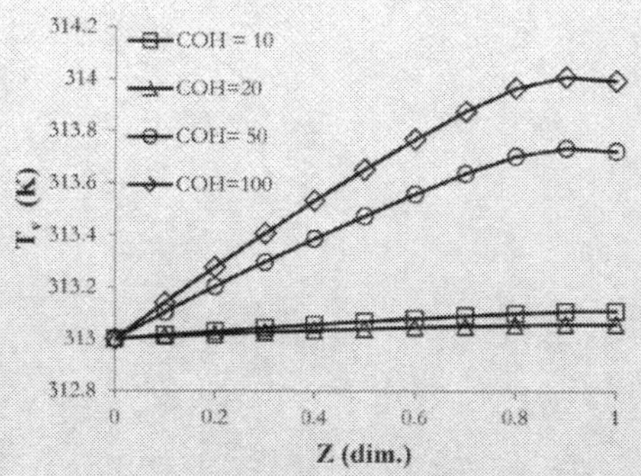

Influence of NaOH amount on liquid phase Temperature

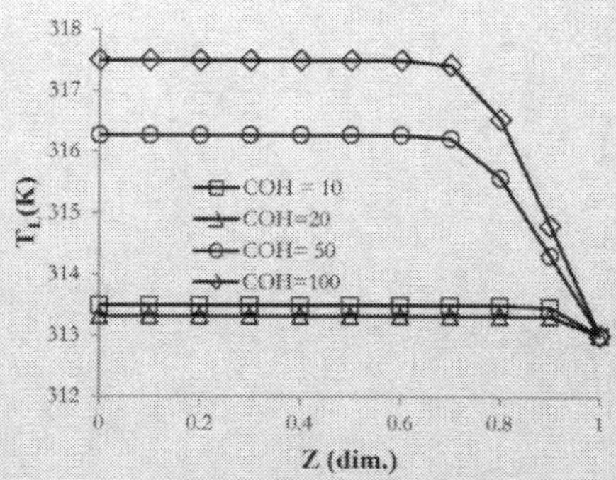

Model Results

Step Input

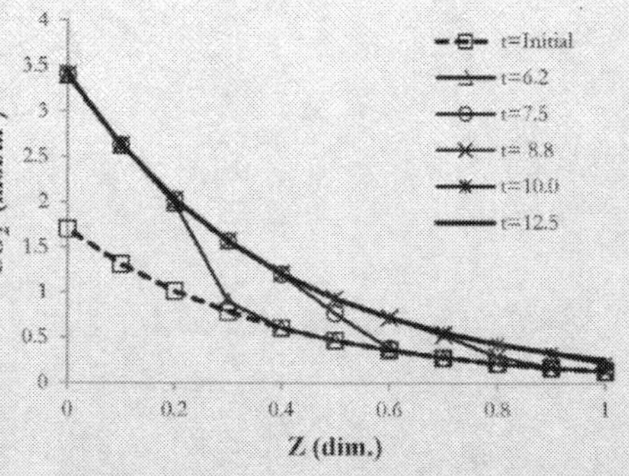

Step input

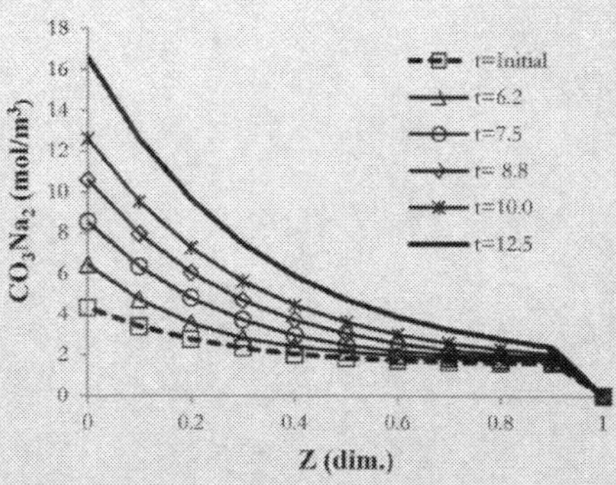

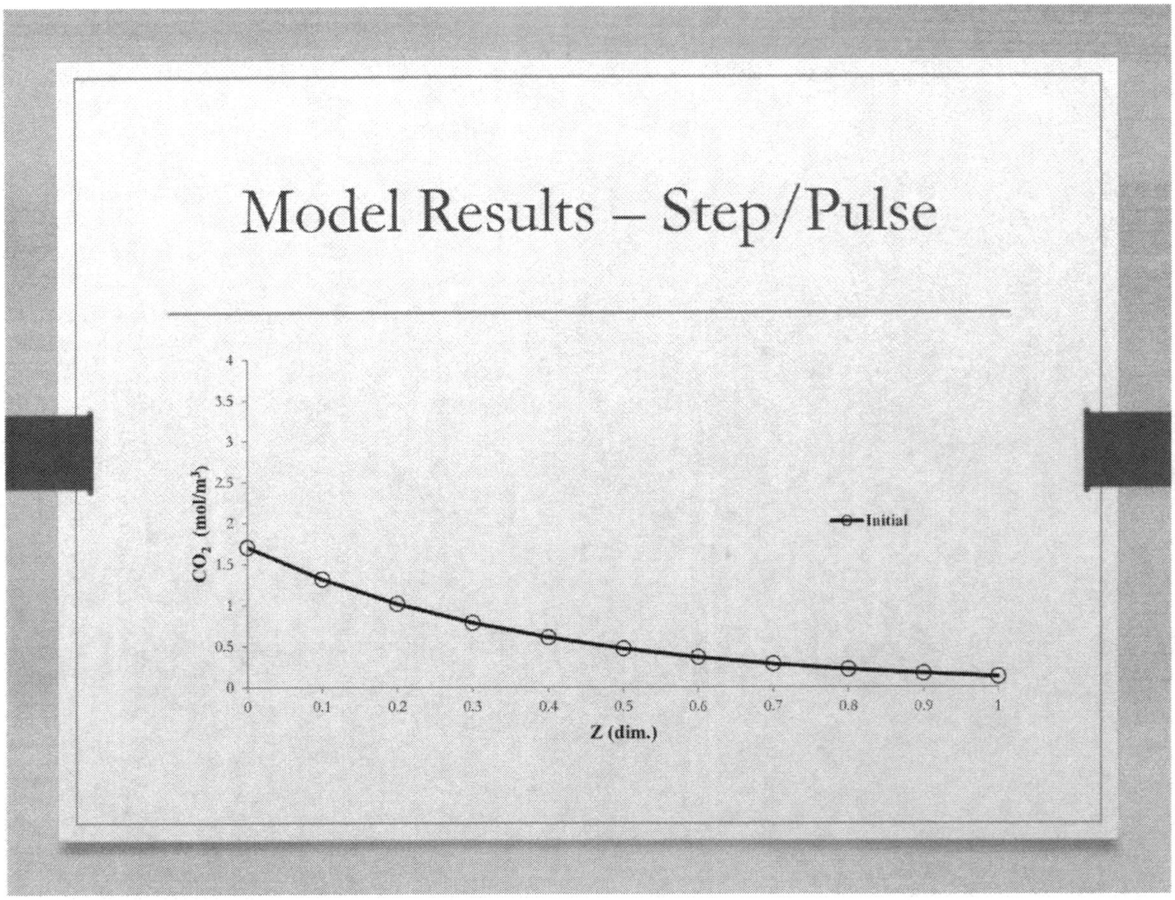
Model Results – Step/Pulse
Initial
CO2 (mol/m3)
Z (dim.)
0
0.5
1
1.5
2
2.5
3
3.5
4
0
0.1
0.2
0.3
0.4
0.5
0.6
0.7
0.8
0.9
1

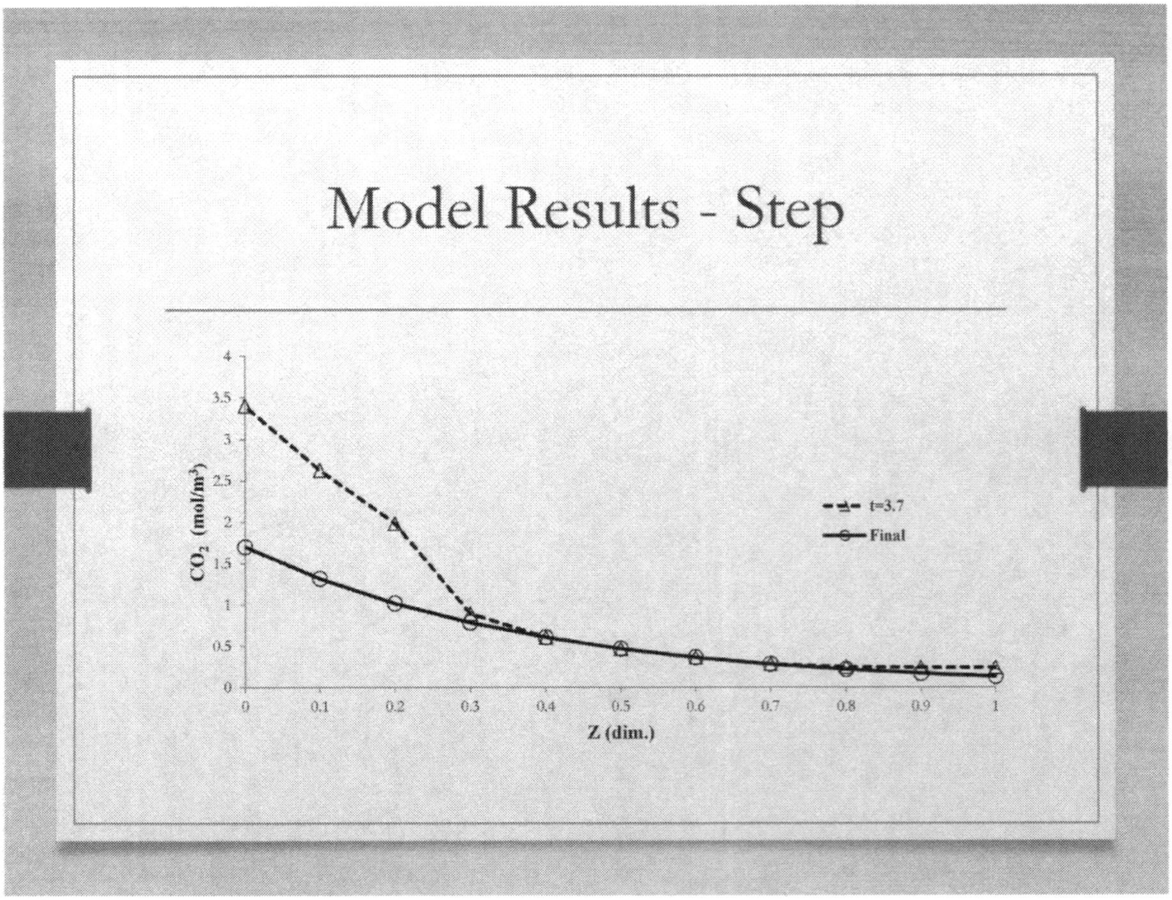
Model Results - Step
t=3.7
Final
CO2 (mol/m3)
Z (dim.)
0
0.5
1
1.5
2
2.5
3
3.5
4
0
0.1
0.2
0.3
0.4
0.5
0.6
0.7
0.8
0.9
1

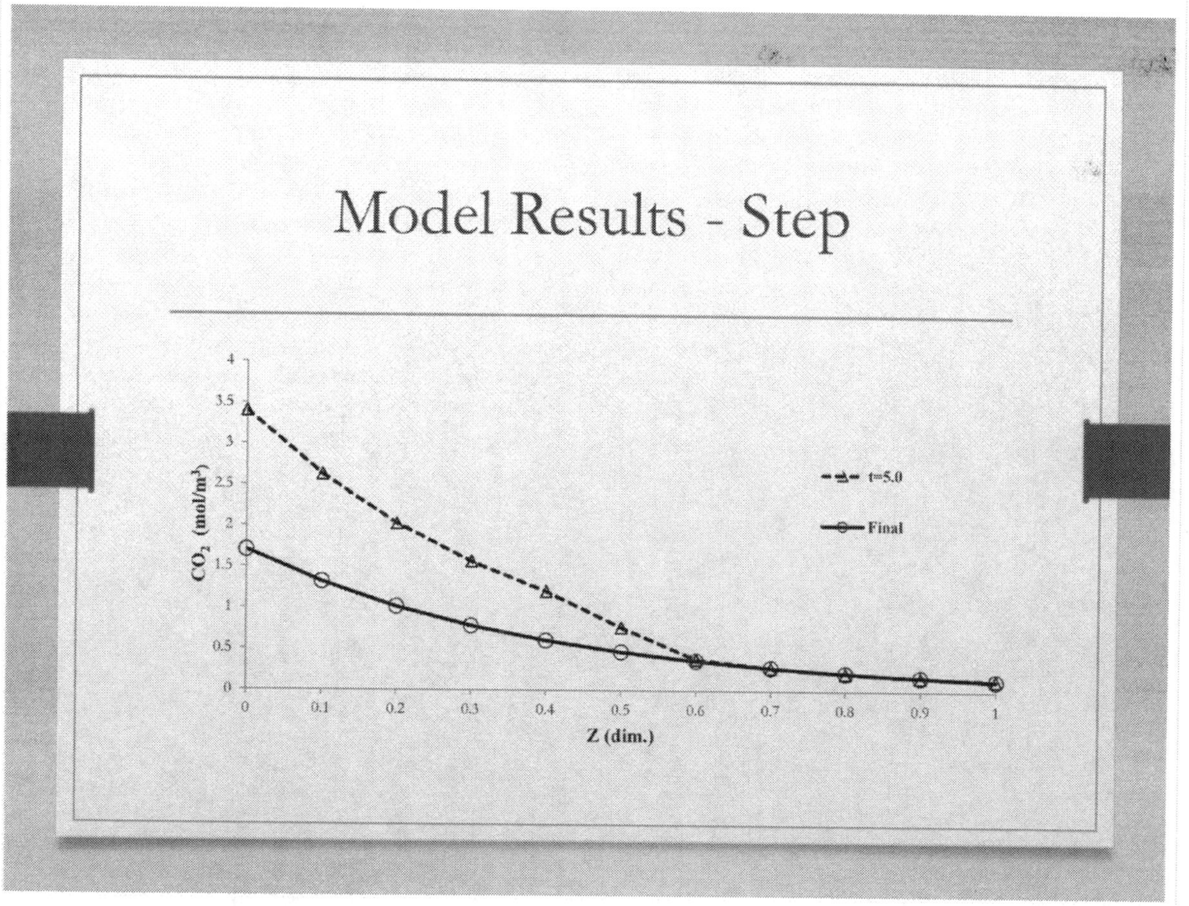
Model Results - Step
CO2 (mol/m3)
0
0.5
1
1.5
2
2.5
3
3.5
4
0 0.1 0.2 0.3 0.4 0.5 0.6 0.7 0.8 0.9 1
Z (dim.)
t=5.0
Final

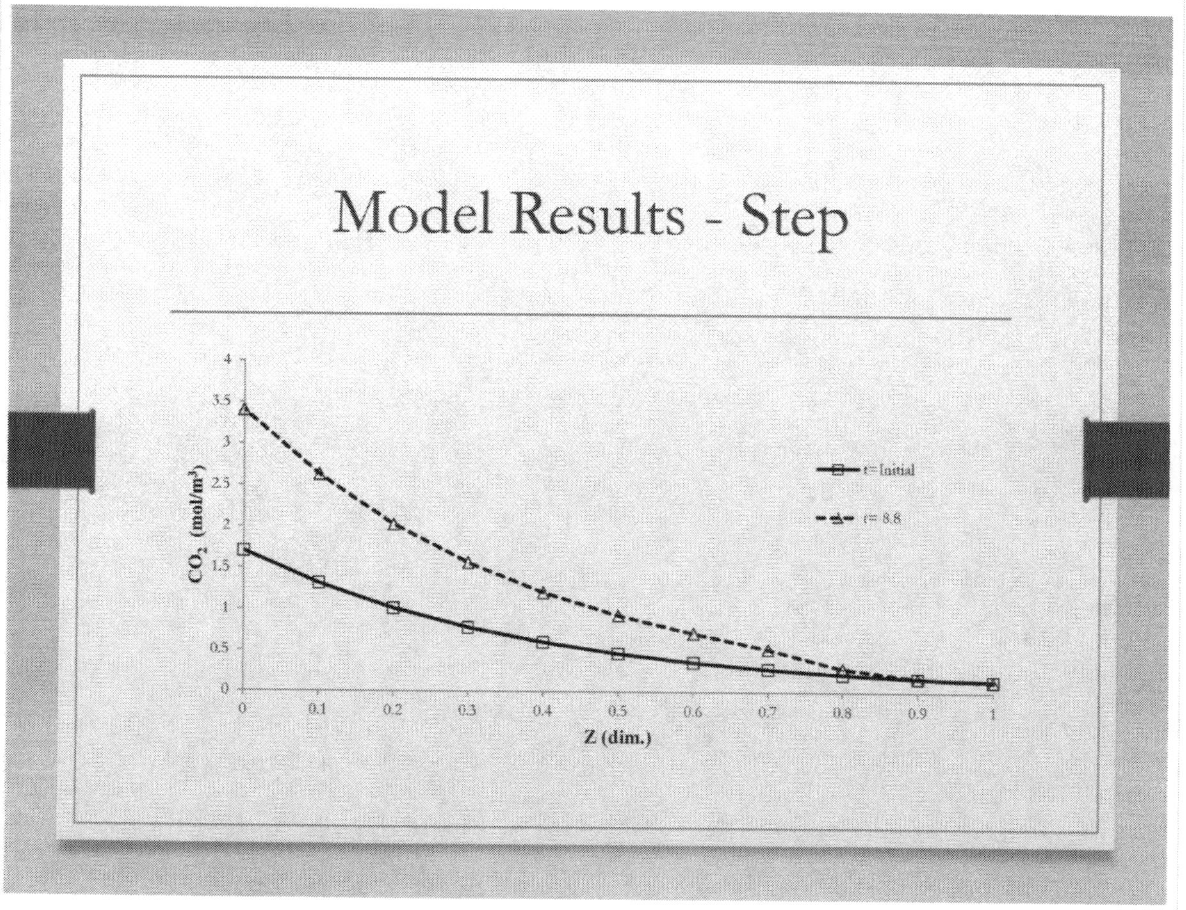
Model Results - Step
CO2 (mol/m3)
0
0.5
1
1.5
2
2.5
3
3.5
4
0 0.1 0.2 0.3 0.4 0.5 0.6 0.7 0.8 0.9 1
Z (dim.)
t=Initial
t= 8.8

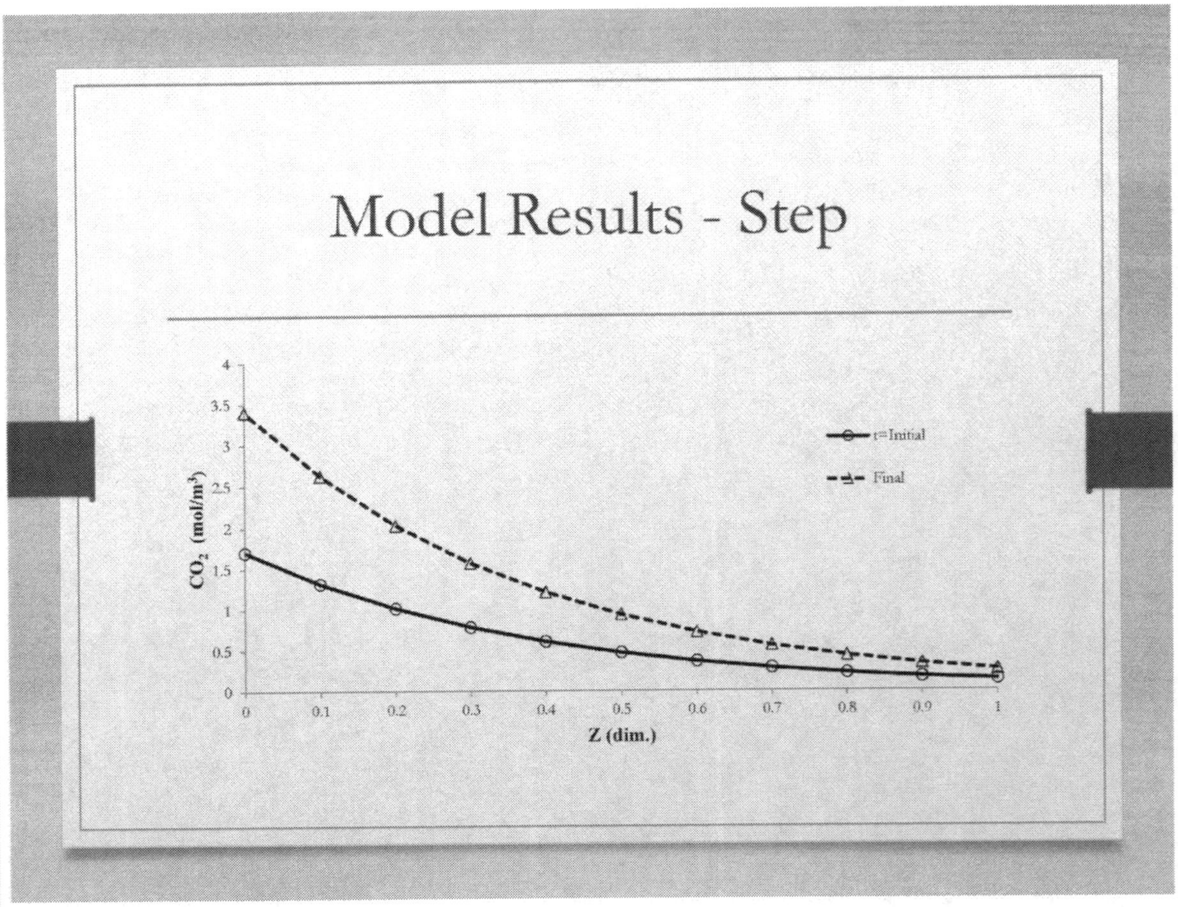
Model Results - Step
4
3.5
3
2.5
2
1.5
1
0.5
0
0
0.1
0.2
0.3
0.4
0.5
0.6
0.7
0.8
0.9
1
CO2 (mol/m3)
Z (dim.)
t=Initial
Final

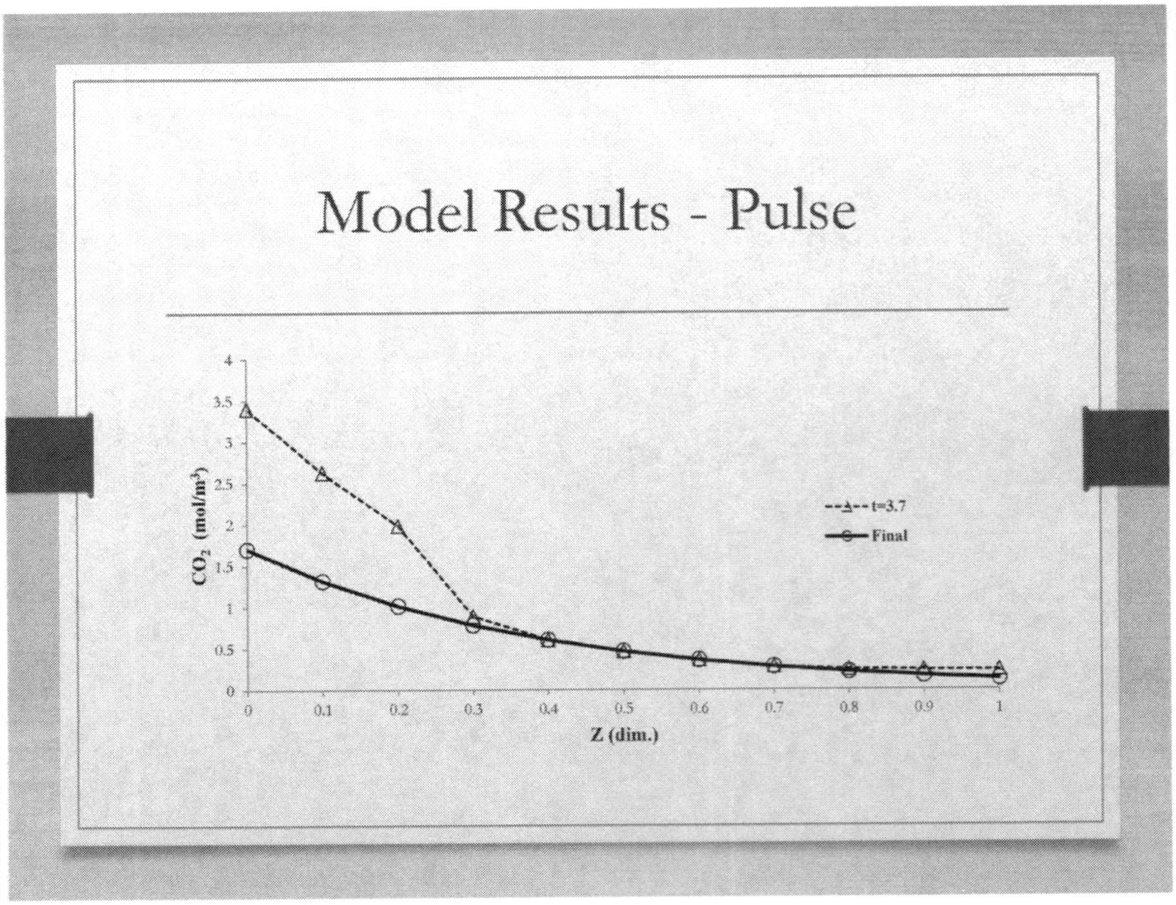
Model Results - Pulse
4
3.5
3
2.5
2
1.5
1
0.5
0
0
0.1
0.2
0.3
0.4
0.5
0.6
0.7
0.8
0.9
1
CO2 (mol/m3)
Z (dim.)
t=3.7
Final

Model Results - Pulse

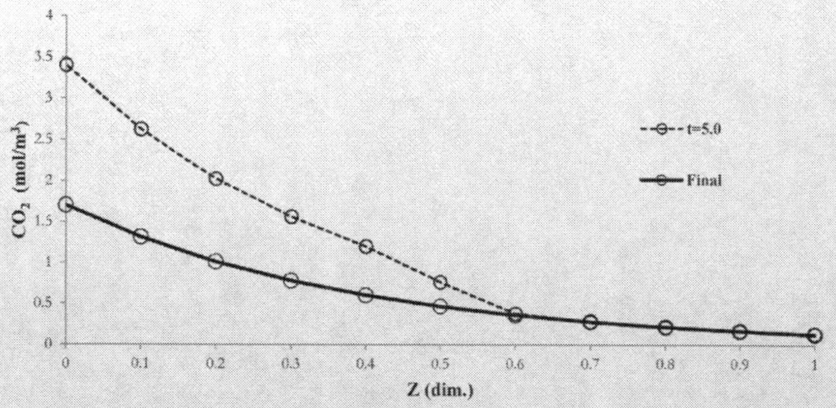

Model Results - Pulse

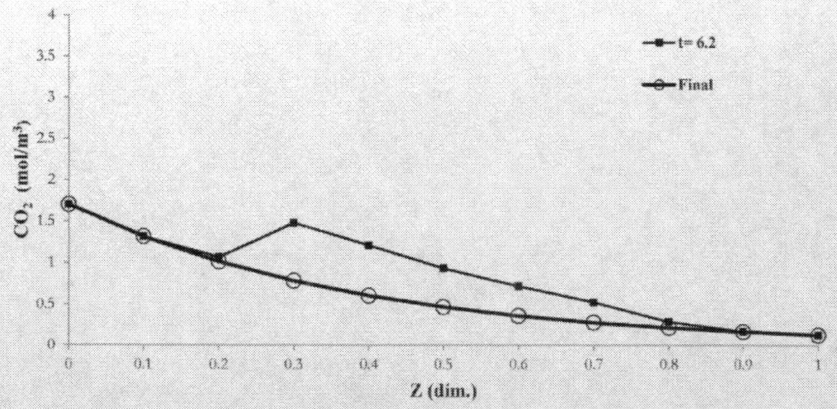

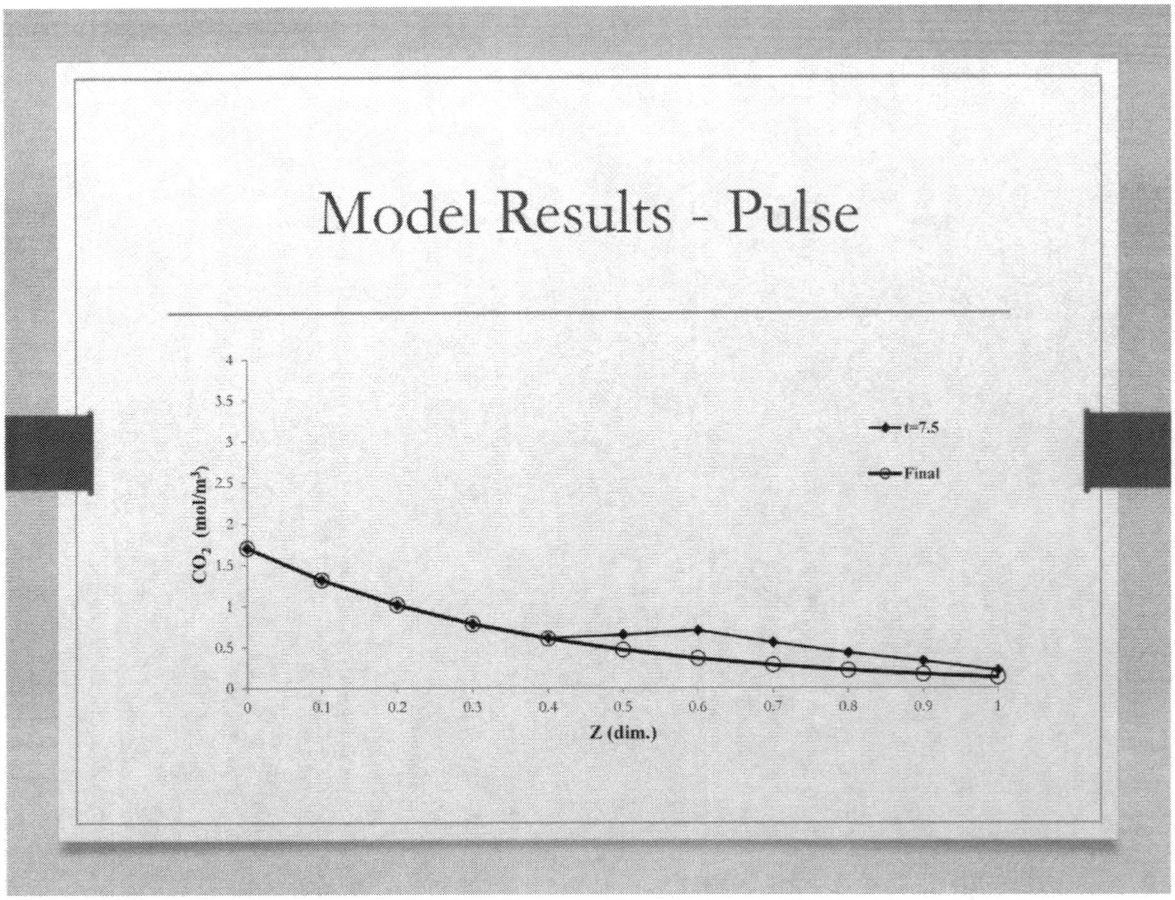
Model Results - Pulse
CO_2 (mol/m^3)
Z (dim.)
t=7.5
Final
0
0.5
1
1.5
2
2.5
3
3.5
4
0
0.1
0.2
0.3
0.4
0.5
0.6
0.7
0.8
0.9
1

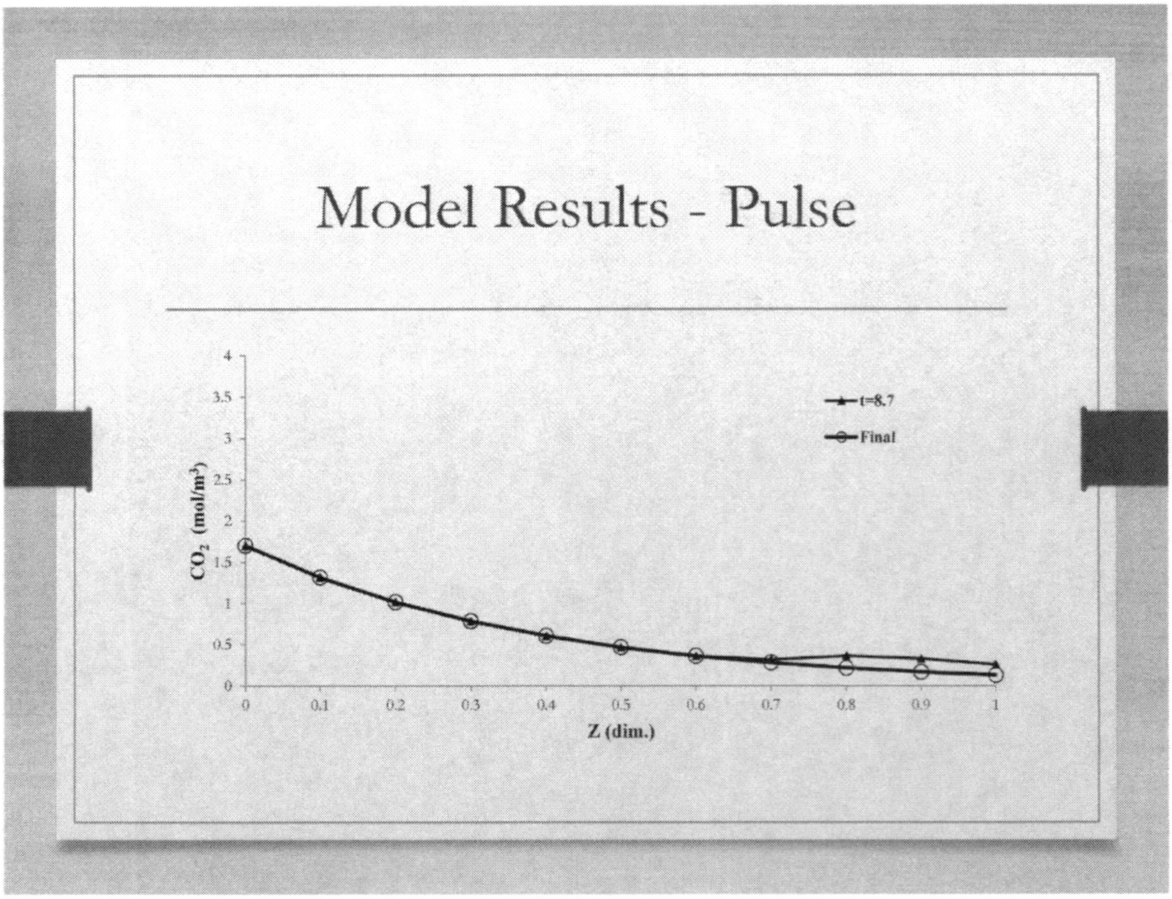
Model Results - Pulse
CO_2 (mol/m^3)
Z (dim.)
t=8.7
Final
0
0.5
1
1.5
2
2.5
3
3.5
4
0
0.1
0.2
0.3
0.4
0.5
0.6
0.7
0.8
0.9
1

Model Results - Pulse

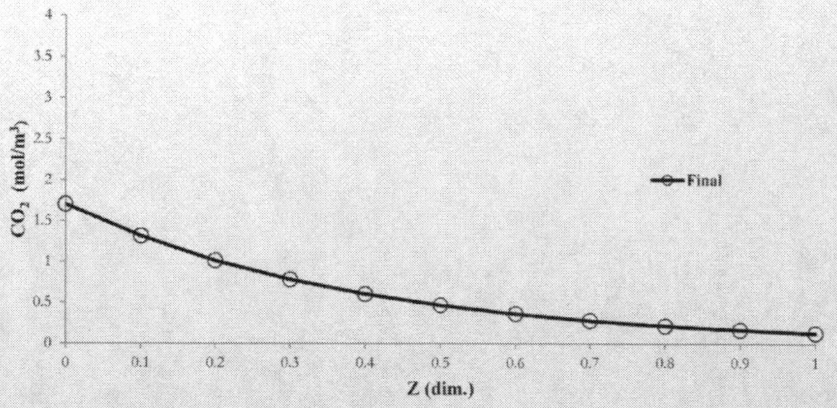

Conclusions

- A model for the dynamic behavior of reactive CO_2 absorption developed.
- All the figures show that the dynamic results converge to the steady state results after a transient period.
- The model has been used to study the influence of operating parameters.

Oak Ridge National Laboratory

Conclusions

- Results demonstrate that the absorbed concentrations of carbon dioxide and carbonate ion both increase as the gas and liquid velocities increase.
- These results are logical and comply with physical expectations. Further validation is needed to ensure required accuracy on the developed models.

Acknowledgments

- This work is supported by the Office of Nuclear Energy, US Department of Energy, under Nuclear Energy University Program, Standard Research Contract 00120924
- We are thankful to graduate and undergraduate students at PVAMU who helped on the project
- We are also thankful to the Off-Gas Treatment Modeling and Verification team

Carbon monoxide separation by reactive absorption in chlorocuprate(I)-based ionic liquid

Gabriel Zarca, Inmaculada Ortiz, Ane Urtiaga

Dep. Chemical and Biomolecular Engineering, Universidad de Cantabria, Santander 39006, Spain

zarcag@unican.es

1. Introduction

High amounts of CO are often release in the waste gas streams of industrial processes when a partial combustion of hydrocarbons is performed, e.g. carbon black manufacturing, steel-making and silicon carbide processes. The selective recovery and valorisation of this CO could be sought as a means to obtain a valuable by-product such as hydrogen or chemicals, e.g. methanol, formaldehyde, etc. Available technologies at commercial scale to address the CO separation are based on cryogenic distillation, liquid absorption, and adsorption processes. Nevertheless, if nitrogen is present in the gas mixture, the use of an absorption process may be considered as the most attractive technology at medium scale given that cryogenic distillation becomes less efficient with increasing N_2 content and pressure swing adsorption process has a lower yield [1]. Nowadays, the CO absorption process is based on the complexation/decomplexation reaction between CO and copper(I) aluminum chloride salt ($CuAlCl_4$) dissolved in an aromatic hydrocarbon, usually toluene, and has several concerns regarding the solubility of the complexing salt and its long-term activity. In this work, a novel approach to the separation and purification of CO is explored taking advantage of the outstanding properties offered by ionic liquids, namely, their negligible vapour pressure, high solvent capacity, wide liquid range and thermal and chemical stability.

2. Thermodynamic study

Taking that into account, we proposed employing the ionic liquid 1-hexyl-3-methylimidazolium chlorocuprate(I) as an efficient medium to provide an improved CO solubility over N_2. In the Lewis basic regime, that is when there is a chloride excess, chlorocuprate(I) anionic species are formed, which react with CO in a 1:1 ratio leading to a copper(I)-CO adduct [2]. In this way, the CO solubility in this medium was greatly enhanced, whereas the N_2 physical solubility was very low; therefore, this chemical system provides high CO/N_2 solubility-selectivity, up to 90 at 303 K with a 2.7 mol copper(I) L^{-1}, and contrary to the COSORB process, it employs a non-volatile and non-toxic solvent in which copper(I) can be solubilised in high amounts without the need of adding another salt such us $AlCl_3$ [3].

3. Kinetic study

Afterwards, the reaction kinetics between CO and the chlorocuprate(I) species have been studied in a batch reactor with flat gas-liquid interface. First, the mass transfer coefficients were assessed in the absence of chemical reaction employing carbon dioxide as an inert model gas with almost identical diffusion properties to CO. The influence of several process variables on the mass transfer coefficients were analysed, namely, temperature and stirring speed. In addition, the effect of copper(I) concentration was studied since it significantly affects the physical properties of the ionic liquid and thus, variable mass transfer coefficients have to be considered. Eventually, a potential proportionality between the mass transfer coefficients and the ionic liquid viscosity to the power of -0.5 was found and a dimensionless correlation to estimate the mass transfer coefficients in the studied range of operating conditions was proposed [4].

Then, in order to characterize the reaction kinetics, reactive absorption experiments were performed with CO in the pseudo-first order kinetic regime in which the absorption rate is not diffusion-controlled. The experimental enhancement factors that stand for the way in which the mass transfer is enhanced by the chemical reaction were determined as function of temperature and copper(I) concentration and the Hatta number was obtained with two different approaches: first, assuming that the reaction took place in the fast pseudo-first order regime; and secondly, employing the relationship between the enhancement factors and Hatta number proposed by DeCoursey. The results showed that the reaction apparently was placed in the pseudo-first order regime and therefore, the overall kinetic constants were directly obtained from the absorption rates [5].

In summary, the information on the thermodynamics and reaction kinetics derived in this study, together with the characterization of the mass transport, allow the design and optimization of a novel ionic liquid-based separation process that represents a promising alternative to selectively recover CO from industrial flue gases.

References

1. J.A. Hogendoorn, W.P.M. van Swaaij, G.F. Versteeg, The absorption of carbon monoxide in COSORB solutions: absorption rate and capacity, Chem. Eng. J. 59 (1995) 243-252.

2. D.R. Smith, J.A. Quinn, Facilitated transport of carbon monoxide through cuprous chloride solutions, AIChE J. 26 (1980) 112-120.

3. O.C. David, G. Zarca, D. Gorri, A. Urtiaga, I. Ortiz, On the improved absorption of carbon monoxide in the ionic liquid 1-hexyl-3-methylimidazolium chlorocuprate, Sep. Purif. Technol. 97 (2012) 65-72.

4. G. Zarca, I. Ortiz, A. Urtiaga, Carbon monoxide recovery from flue gases by reactive absorption in imidazolium chlorocuprate(I) ionic liquid: Mass transfer coefficients, Chinese J. Chem. Eng. Accepted manuscript (2014).

5. G. Zarca, I. Ortiz, A. Urtiaga, Kinetics of the carbon monoxide reactive uptake by an imidazolium chlorocuprate(I) ionic liquid, Chem. Eng. J. Accepted manuscript (2014).

377222 Multilayer Nanofiber Filter Technology for Providing High Filtration Efficiency and Low Pressure Drop

Abstract:

Nanoaerosols in form of viruses or pollutants produce serious health problems. Filters made of nanofibers (100-300 nm in diameter) can effectively remove these nanoaerosols by diffusion for particles less than 100 nm (referred as nanoparticles by ISO) and interception for particles greater than 100 nm. Unfortunately, the pressure drop across the filter can be enormous as the pressure drop varies inversely as the fiber diameter is raised to the second power.

We have developed a multilayer nanofiber technology whereby the pressure drop can be kept under control while the filtration efficiency is not sacrificed. A number of examples are used for the demo including nanofibers made of various materials.

361491 Dissolution Kinetics of Witwatersrand Far West Rand Gold Ore in Sodium Cyanide Leaching

Abstract:

A study on the dissolution kinetics of gold ore in a solution of NaCN has been undertaken. The dissolution kinetics of the gold was found to depend on the stirring rate, reaction temperature, particle diameter and the leachant concentration. The dissolution rate increase as the stirring rate, reaction temperature and leachant concentration increases and decreases with an increase in particle size. The activation energy of ~3 kcal/mol was obtained for the dissolution, and it suggests diffusion model for leaching reaction. Further, the linear relationship between the rate constant and there reciprocal of square of particle size supports the fact that gold dissolution of the process is diffusion controlled.

Temperature-Responsive Polymers for Flocculation and Dewatering

Qing Xu, James Swan and T. Alan Hatton
Department of Chemical Engineering
Massachusetts Institute of Technology
Cambridge MA 02139

Dewatering, an important process in mineral processing, is often facilitated by the use of flocculants such as long chained polymeric reagents. Conventionally-used polymer flocculants usually form large and loose, water-laden flocs. It is difficult to reduce the water content in the sediment beds by mechanical compaction due to the highly-attractive interparticle forces induced by the flocculants.

Temperature-responsive polymers that undergo physical changes in reaction to easily-controlled external thermal stimuli fall into the category of smart materials. One such polymer is poly (N-isopropylacrylamide) (PNIPAM), which is known to switch from being hydrophilic to hydrophobic when the temperature is changed from below to above the lower critical temperature (LCST).

In this work, we report the use PNIPAM as a temperature-responsive flocculant for kaolinite clay and silica suspensions. The LCST was determined by differential scanning calorimetry (DSC) and UV-Vis light absorbance. Settling tests were conducted at both room temperature and 60 °C, with significantly higher settling rates and smaller sediment volumes observed at the higher temperature. The particle size distribution of the kaolinite suspension was inferred by focused beam reflectance measurement (FBRM). The mean particle size was found to increase when the temperatures was above the LCST. The flocculation process could be reversed by reducing the temperature. The dewatering of a silica suspension in a bench scale pressure filter with a temperature control was also studied. The smart behavior of PNIPAM has possible applications in temperature controlled particle flocculation and dewatering.

Additionally, Brownian dynamics simulations were used to study particle aggregation via polymer bridging in shear flow. The attractive force between particles was modeled after atomic force microscopy experiments. Temporal evolution of the size distribution and fractal dimension of flocs were measured. Simulation results are consistent with experiment data.

Effects and mechanism research of the filtering separation for semi-crystalline silicate

Tan Wei, Zhu Guorui, Liu Ying, Liu Liyan[*]

School of Chemical Engineering &Technology, Tianjin University, Tianjin 300072, People's Republic China

Semi-crystalline silicate materials have special crystalline state and physical attributes because crystal nucleus generate slowly and protective film generate on the surface in the course of crystallization, which results in the traditional filtration separation theory which cannot clarify the filtration mechanism of this material. Based on delay formation theory of the crystal nucleus and the protective film theory, different crystalline state silicate was synthesized by hydrothermal synthesis method. According to the crystalline states, they can be named as three categories: active calcium silicate, hydrate of calcium silicate, kind of xonotlite. Then the suspension was filtered and the filtration rate, filter cake moisture content and the resistance were measured. The property parameters of the filtration separation were tested and analyzed by XRD, SEM, BET and the theoretical equation of the filter cake resistance can be modified. Finally, filtration mechanism model of semi-crystalline silicate materials is built by researching the crystalline state at the molecular level and the coupling relationship between the microstructure of the filter cake and macro filtering operation in the progress of filtration. The results of the study can provide theory and guidance to change the structure of semi-crystalline silicate and make some improvements for filtering separation.

390705 Exergy and Energy Analysis for Solar Vs Membrane Water Desalination

Abstract:

Separation of salt from water is necessary for many applications in several industry fields. Because of the high energy demand of the operation, many approaches have been intended.

The use of solar energy to accomplish this operation has been successfully intended and is now applied in a commercial scale. The technology for solar panels design and construction has evolved enormously in the last few decades. Nevertheless, not many researchers have considered and energy-exergy analysis of the operation which has the potential to bring many insights that may become useful in the field of solar panels technology.

This paper is intended to present such analysis and shows a comparison between the technology that uses membranes for separation (more specifically, reverse osmosis) and solar panels.

CO_2-Selective Membranes for High-Pressure CO_2/H_2 Separation

Jian Zou[1] and W.S. Winston Ho[2]
[1]Momentive Performance Materials Inc.
Friendly, WV 26146-7511
[2]William G. Lowrie Department of Chemical and Biomolecular Engineering
The Ohio State University
Columbus, OH 43210-1180

The present paper is based on the author's research during his doctoral study at The Ohio State University between 2002 and 2007 under the advisory of Dr. W.S. Winston Ho.

High purity hydrogen has seen increasing demands as an important raw material in many industrial processes and the fuel for fuel cells. In the United States, most hydrogen is produced by steam reforming of natural gas followed by the water gas shift reaction. The resulting syngas contains hydrogen, carbon monoxide and carbon dioxide. The separation of CO_2 from syngas is a critical step in obtaining high purity hydrogen, and the current industrial technologies for CO_2 removal from syngas include amine scrubbing and pressure swing adsorption. The typical industrial process conditions for CO_2removal from syngas require feed pressures higher than 200 psia and temperatures ranging from 100 to 200°C, which have been challenging for polymeric CO_2-removal membranes.

In the present study, polymeric CO_2-selective membranes were synthesized and studied specifically for the high-pressure syngas purification. CO_2/H_2 separation at feed pressures higher than 200 psia and temperatures ranging from 100 to 150°C was investigated. The effects of feed pressure, type of porous support, temperature, and permeate pressure were investigated using a simulated syngas containing 20% carbon dioxide and 80% hydrogen. The membranes synthesized showed the best CO_2 permeability and CO_2/H_2 selectivity at 110°C. At a feed pressure of 220 psia, the CO_2 permeability and CO_2/H_2 selectivity reached 756 Barrers and 42, respectively, whereas at a feed pressure of 440 psia, the CO_2 permeability was 391 Barrers and CO_2/H_2 selectivity was about 25.

387897 Helium and Hydrogen Sorption, Diffusion, and Transport in Hydrocarbon, Silicon, and Fluorocarbon-Based Polymers

Abstract:

The permeability-selectivity upper bounds show that perfluoropolymers have uniquely different separation characteristics than hydrocarbon-based polymers. For separating helium from hydrogen, these differences are particularly large. At a given helium permeability, the upper bound defined by perfluoropolymers has helium/hydrogen selectivities that are 2.5 times higher than that of the upper bound defined by hydrocarbon-based polymers. Robeson hypothesized that these differences in transport properties resulted from the unusual sorption relationships of gases in perfluoropolymers compared to hydrocarbon-based polymers, and this paper seeks to test this hypothesis experimentally. To do so, the gas permeability, sorption, and diffusion coefficients were determined at 35°C for hydrogen and helium in a series of hydrocarbon-, silicon-, and fluorocarbon-based polymers. Highly or completely fluorinated polymers have separation characteristics above the upper bound for helium/hydrogen separation because they maintain good diffusivity selectivities for helium over hydrogen and they have helium/hydrogen sorption selectivities much closer to unity than those of hydrocarbon-based samples. The silicon-based polymer had intermediate sorption selectivities between those of hydrocarbon-based polymers and perfluoropolymers. Comparisons of hydrogen and helium sorption data in the literature more broadly extend the conclusion that helium/hydrogen sorption selectivity is rather different in hydrocarbon and fluorocarbon-based media.

377641 Novel Membrane Absorption-Based Processes for CO2 Capture

Abstract:

This presentation will deal with novel membrane absorption and stripping based processes and devices for CO_2 absorption and stripping for carbon capture and sequestration. Two feed streams are considered: simulated flue gas at around 40-50^{0}C; lower temperature (L-T) post-shift reactor synthesis gas at high pressures and temperature up to 100^{0}C where He has been used as a surrogate for H_2.

Conventional methods for CO_2 capture generally employ an absorption and stripping process. The CO_2 is absorbed for example in an aqueous solution of amine e.g., monoethanolamine and then stripped out in a separate tower with steam at 120^{0}C. Such a process is highly energy intensive. Further the amine used is volatile. The low partial pressure of CO_2 in the flue gas inhibits the application of CO_2-selective membranes unless methods are employed to increase the CO_2 partial pressure in the flue gas to be treated. Adsorption processes using novel sorbents having high CO_2 adsorption capacity as well as other separation methods are being studied. We will describe novel membrane-based techniques to potentially bypass most of the shortcomings of many existing approaches. We have employed a simulated humidified flue gas containing around 14% CO_2 and demonstrated successful removal of bulk of the CO_2 and its recovery in a CO_2-concentrated stream.

To separate hydrogen from the low temperature (L-T) post-shift reactor synthesis gas and simultaneously obtain a purified CO_2 stream, an advanced pressure swing membrane absorption (PSMAB) technology utilizing a hollow fiber membrane contactor device and a cyclic absorption-desorption process is being developed. Studies have been carried out with around 40%CO_2-rest He feed gas mixture coming in at temperatures up to 100^{0}C and pressures varying from 100 to 250 psig (689-1723kPag). The nonvolatile absorbent liquid consists of an ionic liquid 1-butyl-3-methylimidazolium dicyanamide ([bmim][DCA]) containing a suitable amine. Studies were carried out with porous hydrophobized polyether ether ketone (PEEK) hollow fiber membrane modules of small sizes having different characteristics. It was determined that reducing the dead volume on the module tube side and decreasing the fiber packing density greatly enhanced gas absorption and improved product qualities while increasing temperature had a negative impact on product qualities with pure [bmim][DCA] as absorbent. Adding nonvolatile amines to the ionic liquid showed an improvement for CO_2 absorption capability especially at higher temperatures. The measured CO_2 concentration in the CO_2-rich product stream from the less-than-optimum membrane modules has varied between 90.7-92.9% depending on the temperature. A mathematical model has been developed and numerically solved to predict the cyclic process performance of pure ionic liquid as the absorbent. Solubility and diffusion coefficient data for the gases needed in this model were determined for the pure ionic liquid absorbent using a pressure-decay-dual transducer apparatus at temperatures of 323, 353, 363, and 373 K and at pressures up to 1.38 MPa (~200 psig). The solubility selectivity of CO_2 over He was also determined for various amine concentrations in the ionic liquid with or without water.

Gas and vapor permeability of gases and plasticization effects in glassy membranes

Matteo Minelli, Giulio C. Sarti

Interdipartimental Center for Industrial Research - Advance Mechanics and Materials (CIRI-MAM)
Alma Mater Studiorum - Università di Bologna

Department of Civil, Chemical, Environmental and Materials Engineering (DICAM)
Alma Mater Studiorum - Università di Bologna

The transport properties of gaseous or vapor penetrants in glassy polymer systems are analyzed following a fundamental and simple approach based on general thermodynamic theory for glassy phases. This study aims at the description of the permeability behaviors of various probes with respect to penetrant upstream pressure and or temperature.

The model utilizes a the fundamental expression for the diffusive flux given as the product of a purely kinetic factor, the mobility, and a thermodynamic factor, which in turn is related to the dependence of the penetrant chemical potential on its concentration within the glassy polymer system. The penetrant solubility as well as the thermodynamic factor are provided by the nonequilibrium thermodynamics for glassy polymers, NET-GP (and the nonequilibrium lattice fluid - NELF - in particular), which represents accurately the thermodynamic behavior of polymer/penetrant systems at the different T and p conditions, in a predictive fashion. Conversely, a simple exponential function of penetrant concentration is used for the mobility coefficient, thus introducing the only two adjustable parameters entering this model, i.e.: the penetrant mobility at infinite dilution and the plasticization factor, namely the coefficient that characterizes the dependence of mobility on the concentration.

The model provides a simple and effective expression for the description of penetrant permeability in glassy polymer membranes, using only two adjustable parameters. The model describes accurately all the types of permeability dependence on penetrant pressure, which may be either constant, decreasing or increasing with pressure itself. Furthermore, the model is able also to represents accurately the non-monotonous behaviors observed in the presence of plasticization effects.
Remarkably, the same transport model can also provide the diffusion time-lag in transient permeation experiments, either with or without the occurrence of plasticization.

The model has been applied successfully to describe gas transport in many different glassy polymers, considering as penetrants CO_2, light gases (CH_4, N_2, Ar or He), as well as heavier hydrocarbons or solvents, which can show very different behaviors for the permeability – pressure relationship.

Interestingly, the model provides, for the case of highly swelling probes (e.g. CO_2), a simple and straightforward procedure for the determination of both plasticization pressure and the corresponding permeability value, based only on low pressure transport data. The mechanism of plasticization of the polymer matrix is thus effectively described based on simple and fundamental concepts (thermodynamic factor and mobility), pointing out the factors which originate such phenomenon.

361389 Separation of V(V) and Cr(VI) By Primary Amine in Case of Resource Utilization of Industrial Residue Containing Vanadium and Chromium

Abstract:

Solvent extraction of Vanadium(V) and Chromium(VI) by primary amine was firstly applied to high selectively separate V(V) and Cr(VI) from residue containing Vanadium and Chromium to make high value product respectively. Fundamental research work was systematically carried out in lab, then pilot scale and commercial scale experiments were operated in factory, and it is the first commercial application of V-Cr separation by primary amine. The experiments in lab indicated that, kerosene solution of LK-N21 could separate Cr(VI) and V(V) from nearly neutral aqueous solutions with high efficiency: LK-N21 without acidify as extractant could extract >85% V(V) and >90% Cr(VI) left in aqueous phase with operation conditions as material liquors pH=5, phase ratio=1:1, mixing time >5min. The loaded organic phase could be easily stripped in 15-30min under 45 , nearly 100% V(V) transferred into aqueous phase. The analysis results showed the mechanism of Cr(VI) and V(V) solvent extraction by primary amine from nearly neutral aqueous solutions was solvation with molecular association between primary amine and V(V)/Cr(VI) through hydrogen bonding. The pilot scale and semi-commercial scale experiments also showed V(V) and Cr(VI) could be separated very well in the conditions above: after solvent extraction separation process, liquor containing 28~35g/L V(V) and 0.3~2g/L Cr(VI) was produced with separation factor over 70, more than 98% V(V) was recovered, the loss of extractant less than 0.2%. The purity of Vanadium pentoxide produced by semi-commercial scale process is over 99.5%.

373497 Membrane Assisted Vapor Stripping: Reduced Cost and Energy Usage for Alcohol-Water Separation

Abstract:

Alcohols, including ethanol and butanols, are receiving increased attention as renewable liquid biofuels. Fermentative production of alcohols, particularly butanols like 1-butanol and isobutanol, is limited to low concentrations due to product inhibition and limited lignocellulosic-based substrate concentrations. The result is high separation energy demand by conventional distillation approaches, despite favorable vapor-liquid equilibrium and, for butanol, partial miscibility with water. In previous work from our group, a process integrating steam stripping, vapor compression, and vapor permeation separation was proposed and evaluated for separating ethanol, 1-butanol, and acetone/butanol/ethanol (ABE) mixtures from water. Such a Membrane Assisted Vapor Stripping (MAVS) process was found to be 60-70% more energy efficient than conventional distillation approaches. In such hybrid distillation/membrane MAVS systems, the stripping column provides high alcohol recovery and low effluent concentration while the vapor compression and membrane steps enable the efficient recovery of latent and sensible heat from both the retentate and permeate streams from the membrane system. In this work, capital costs and energy usage for simulated 2nd generation MAVS designs will be discussed.

376060 Techno-Economic Analysis of Hybrid Membrane/Distillation Column Systems for Ethanol-Water Separation

Abstract:

We have carried out a comprehensive analysis of the technical and economic challenges associated with the implementation of ethanol selective membrane systems for the separation of ethanol from water.

Membranes have the potential to displace current technologies such as distillation. Such a displacement is only attractive if energy and costs savings are possible. Our goal was to investigate the minimum membrane performance characteristics needed to accomplish such a savings. The case study chosen was the ethanol-water distillation in the dry-grind corn ethanol process.

The distillation column in the base case was modeled using the McCabe-Thiele method. The ethanol product purity and recovery of the base case provides the performance targets for the novel configurations suggested here.

We have modeled the membrane unit as a hollow fiber counter-current membrane system with nitrogen sweep gas. An initial analysis of a stand-alone membrane system shows that this is incapable of performing the required separation.

Next, we considered a hybrid system composed of a membrane and a distillation column. Our analysis has shown that significant economic savings are possible when the membrane follows a small distillation column. This hybrid system has the capability to better capitalize on the VLE characteristics of the ethanol-water system than systems composed uniquely of distillation columns or membranes.

We have analyzed the effect of different values of membrane ethanol flux and selectivity on the performance of the hybrid system. It was found that changes in selectivity have a greater effect on the product purity than changes in flux and that changes in flux have a greater effect on the required membrane area (and therefore cost) than changes in selectivity. We have found that a minimum selectivity of 30 is required for the target product purity of 95 wt% ethanol.

Finally, we have formulated a net present value analysis of the resulting hybrid system. This has allowed us to generate a set of performance targets in terms of the membrane's flux, selectivity and costs.

Separation of glucosinolates from Camelina seed meal via membrane and acidic aluminum oxide column

Yuhe Cao, Zhengrong Gu*, Kasiviswanathan Muthukumarappan

Agricultural and Biosystems Engineering Department, South Dakota State University, Brookings, SD, USA

***Correspondence:**

Zhengrong Gu

North Campus Drive 1400, Agricultural and Biosystems Engineering, South Dakota State University, Brookings, SD 57007, USA

Email: Zhengrong.gu@sdstate.edu

Abstract

Heavy metal salt is usually used to precipitate protein during extraction of glucosinolates from Camelina seed meal. That precipitant is expensive and environmental harmful . Molecular weight cutoff 5000 Dalton ultrafiltration (UF) membrane was employed to remove protein after ethanol extraction of glucosinolates. Palladium chloride was used to react with glucosinolates and colored reactant was produced. Absorbance at 540 nm was measured to determine the glucosinolate content according to the absorbance of model glucosinolate Sinigrin. Preparative ion exchange chromatography, based on acidic aluminum oxide, was used to further purify glucosinolates. 0.2 mol/l, 0.5mol/l and 1.0 mol/l NaCl buffer were used to recover the glucosinolates absorbed by acidic aluminum oxide. The results show that 1.0 mol/l salt solution can get 97.1% glucosinolates recovered for permeate from UF. The glucosinolate content of the seed meal is 125.5 mg/g. High performance liquid chromatography was employed to analyze the efficiency of FPLC. It indicates that the major glucosinolates were appeared at retention time 22.2 and 24.6 min. Most impurities of UF permeate were removed by using FPLC.

Keywords: ultrafiltration; glucosinolate; HPLC; FPLC; Spectrophotometric Method

Introduction

Glucosinolates are anionic, hydrophilic and sulfur-containing secondary metabolites which are present in the meal residue of Camelina seed after oil extraction. This meal has been utilized as animal feed, fuel for oilseeds processing and fertilizer. In those applications, glucosinolates can produce ionic thiocyanate (SCN-), which can act as a bioherbicide against redroot pigweed and wild oat (Borek, V et al., 2005). Moreover, Camelina seed meal was used to control soil-borne plant pests (Morra, M.J., 2004). Meanwhile, this meal can be fertilizer for organic vegetables (Snyder, A. et al., 2009). Normally Camelina seed meal can be sold at hundred dollars per ton as animal feed, while only $20/ton as fertilizer or fuel (Datu B. Agusdinata. et al., 2011). Although Camelina meal, is a good substitute of soybean meal, to supply high quality protein and fatty acid in animal feed (Kakani R et al., 2012; Junfeng Qian et al., 2013), glucosinolates in Camelina seed meal, as anti-nutrients (M.K. Tripathi et al., 2007), significantly limits efficient application of those bio-jet fuel co-products in animal feed.

In addition, to obtain insights of the nutritional and other bioactive effects of glucosinolates on animals, such as poultry (Khajali, F. et al., 2012), fish (George Francis et al., 2001), pig (F. Schöne et al., 1994; T.G. Rowan et al., 1991) and bovine (A.M.H. Subuh et al., 1995), purified glucosinolates is needed to quantify their effects on different animals. On the other hand, glucosinolates and their metabolics have been recognized as beneficial nutrient in human diet for their anti-cancers activities (Zhiyong Xie et al., 2012). Glucosinolates are of particular interest due to their role in the prevention of cancer (Ze-You Li et al., 2012) and other chronic and degenerative diseases (Antone R. Opekun et al., 2003). The organic compounds found in glucosinolates reportedly detoxify the human body of potential carcinogens. Therefore, separation of glucosinolates from Camelina seed meal will enable production of those bioactive compounds as value added co-products, as well as improve nutrition efficient of seed meal in animal feed (Adarsh Pal Vig and Amandeep Walia, 2001).

Aqueous extraction (Quinsac, A et al., 1994; Charpentier N et al., 1998), high performance liquid chromatography (Helboe, P. et al., 1980), flash chromatography (Peterka, S. et al., 1988), and slow rotary countercurrent chromatography (Qizhen Du et al., 2008) were used to separate glucosinolates. Most of them used heavy metal salt such as $Pd(Ac)_2$ and $Ba(Ac)_2$ to precipitate the protein from the extraction. And these protein cannot be used anymore. Moreover, the heavy metal is difficult to handle. It is dangerous that the residual heavy metal ion is not completely removed after this step.

Ultrafiltration (UF) has been widely used in pharmaceutical processes for many years primarily for concentration and buffer exchange. More and more studies have been focused on the use of UF for protein purification recently (R. van Reis and A. Zydney, 2007).

In order to find an eeconomic viable and environment friendly way to get relatively pure glucosinolate from Camelina sativa seed meal, membrane separation and acidic aluminum oxide column were used to separate the glucosinolate in this report. Moreover, reverse phase chromatography was used to analyze the elution fractions from fast protein liquid chromatography (FPLC).

Materials and methods

2.1 Materials

Camelina sativa seed meal was obtained through cold pressing. Sinigrin as a model compound for different glucosinolates, was purchased from Sigma-Aldrich Chemical Company. 95% ethanol (medicinal grade), acidic aluminum oxide (100-200 meshes, analytical reagent), palladium chloride (analytical grade), sodium carboxymethyl cellulose (CMC,analytical grade) and tetrabutylammonium hydrogen sulfate (analytical grade) were purchased from Thermo Fisher Scientific Inc. Methanol and water were HPLC grade. Polyethersulfone (PES) membrane (molecular weight cut-off, 5000 Dalton) was purchased from Sartorius Stedim North America Inc.

2.2 Extraction of glucosinolates

Camelina sativa seed meal (5 g) was ground to a fine powder (< 1 mm) and heated at 105 ℃ for one hour. Then the meal was defatted by with hexane (50 ml) at 92 ℃ using soxhlet extraction for one hour. The residual hexane was removed from the solid residue at 25 ℃ under vacuum. The solid residue was extracted with 50 ml ethanol at 60 ℃ for two hours. After filtration by 0.22 μm membrane, the ethanol was recycled by rotary vacuum evaporator, then the residue was dissolved in 50 ml DI water, which would be filtered by ultrafiltration membrane.

2.3 Membrane filtration

The extraction solution contained protein and high molecular weight compounds. Usually glucosinolates were smaller molecules (any evidence through analysis?). In order to remove the impurities, ultrafiltration was used. The schematic diagram of dead-end filtration process is shown in Fig.1. Membrane filtration module was purchased from Millipore, with 50 cm^3 volume cell and 13.4 cm^2 membrane area. The pressure in the cell, adjusted by manually operated valve, was measured using a pressure gauge. Dead-end experiments of dissolved solution was carried at room temperature (20-23℃) and for 40 psi transmembrane pressure, with moderate agitation speed (350 rpm). Permeate was collected with beak and the weight was measured by balance.

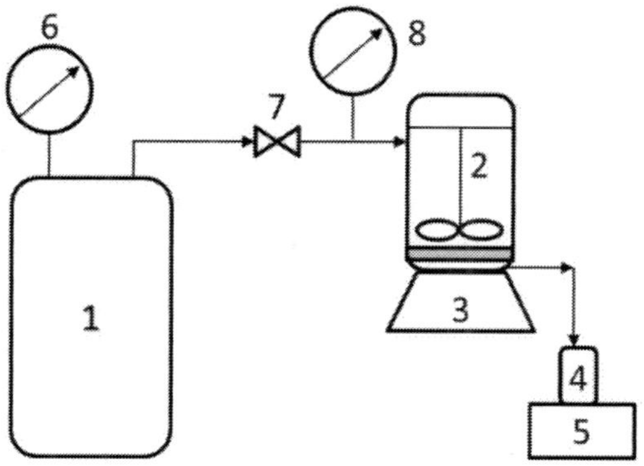

Fig. 1 – Schematic diagram of dead-end filtration apparatus:

1 – compressed air tank; 2- membrane cell; 3 - magnetic stirrer; 4 – beaker; 5 – digital balance; 6 – pressure gauge; 7 – valve; 8 – pressure gauge.

2.4 Column chromatography

Permeate from membrane was further purified with acidic aluminum oxide column connected in a Biologic DuoFlow system (Bio-Rad Laboratories, Inc.). Clarified permeate (1.0 ml) was loaded onto 1.3 ml glass column packed with acidic aluminum oxide column that had been equilibrated with 15 ml Buffer A (water). After sample loading, the column was continuously washed with 25 ml of Buffer A. Glucosinolate was then eluted by using Buffer B (NaCl solution) in 20 ml. The flow rate was 0.5 ml/min. BioFrac fraction collector collected the fractions according to the UV absorbance at 280 nm.

2.5 HPLC analysis

A liquid chromatograph model HPLC Agilent 1200 series was equipped with a vacuum degasser, a binary pump, an autosampler, a thermostated column compartment, and a multiple wavelength detector. Spherisorb ODS1 with guard column was purchased from Waters, which was silica-based, reversed-phase C18 column. This column was used to analyze the glucosinolates at 228 nm, 30 ℃. The initial mobile phase conditions were 40% methanol and 60% aqueous 0.005 M tetrabutylammonium hydrogen sulfate at a flow rate of 0.6 ml/min. after injection of 20 μl of sample, the initial conditions were held for 10 minutes, and then up to 100% methanol over an additional 25 minutes and held at 100% methanol for 15 minutes. All the solutions were filtered through a 0.45 mm filter before use.

2.6 Quantification of Total Glucosinolates ()

The spectrophotometric method (**Wang Ninghui, 2009**), will be used to quantify all glucosinolates. Absorbance was measured using a Shimadzu UV–visible spectrophotometer. A reference solution (blank) is obtained from a mixture of 1.0ml palladium chloride (8μmol/l), 2.0ml CMC (0.15% wt.) solution and 1.0ml water or related salt solution. Measure the absorption at 540 nm the color produced as a result of the Pd-glucosinolate complex formation after 2 hours at 25 ℃ and against the blank. Standard curve can be used to calculate the glucosinolate amount in the solution, and thereby in the seed meal sample.

3 Results and discussion

3.1 Membrane filtration based extraction and purification

Ultrafiltration (UF) membrane was employed to separate the protein from the extraction of glucosinolates by using ethanol. As can be seen from Fig. 2, the extractions with ethanol is transparent. The residual was dissolved by water after ethanol was removed by vacuum evaporation. Then the color was not transparent and just like colloidal solution. After the membrane filtration the magnitude of color decreased and permeate was transparent. Finally, 46.5 ml permeate was obtained. And the membrane flux vs time is shown in Fig. 3. It indicates that the membrane flux drop off slowly. After one round filtration, we clean the membrane with 50% ethanol solution for 30 minutes, which was assisted with ultrasonic. Then we compared the recovered flux with the non-fouling membrane flux. It indicates that the flux has been recovered nearly 94%, shown in Fig. 4.

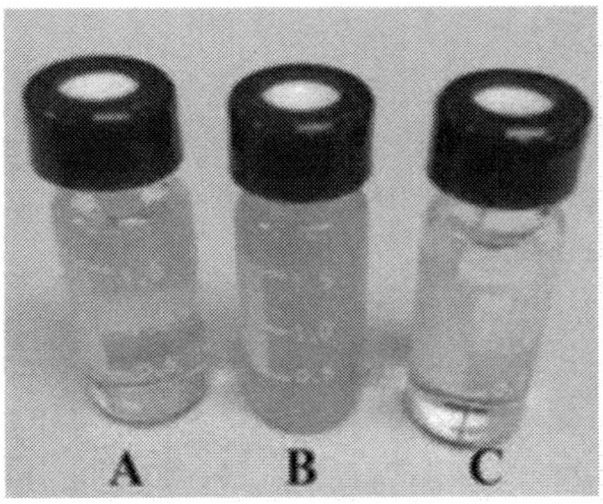

Fig. 2 Pictures of different extractions

A: extraction from ethanol (filtrated by 0.22μm membrane) B: removed ethanol from A, then dissolved by water C: permeate of membrane (5000 Dalton) for B

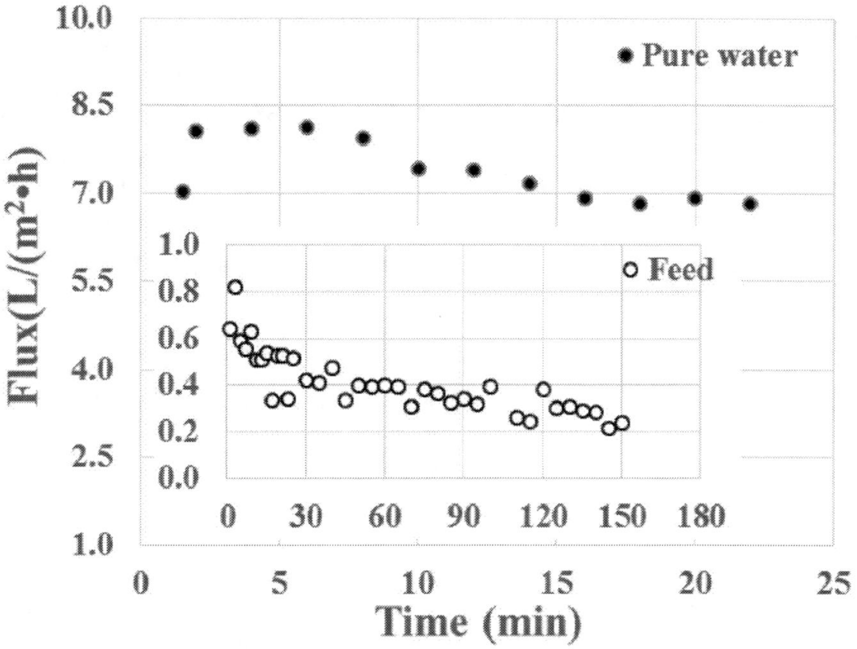

Fig.3 Membrane flux vs operation time.

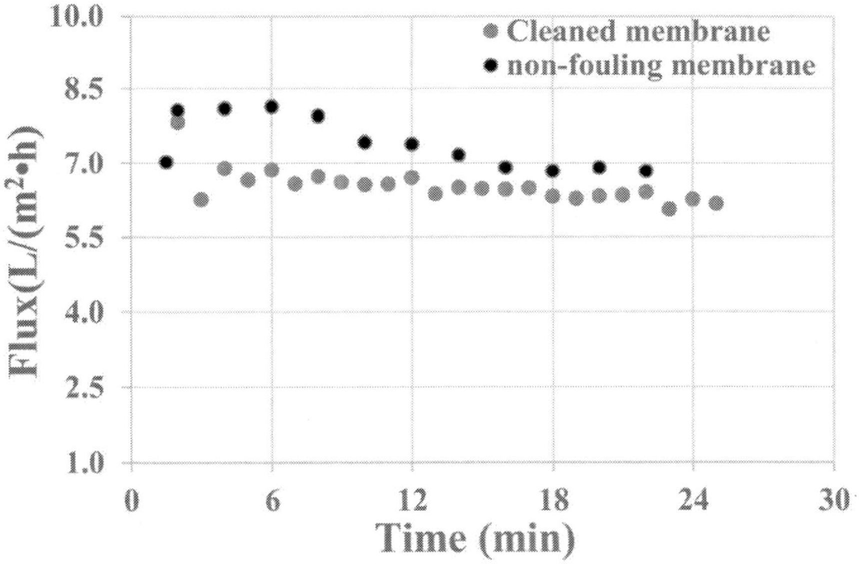

Fig. 4 Membrane flux recovery after cleaning.

3.2 FPLC analysis results

In order to remove impurities, FPLC was employed to further purify permeate from membrane. Fig. 5 shows the formation of acidic aluminum oxide (Elio Santacesaria and Davino Gelosa, 1977). Since acidic aluminum oxide has positive charge from the aluminum atom it can bond with the anion. Glucosinolate is easy to ionize into a cation and an anion, and Fig. 6 shows the structure of glucosinolate. Glucosinolate is also easy to hydrolyze in the strong acid condition. Therefore, glucosinolates can be separated by acidic aluminum oxide, whose acidity is weak.
After the protein was removed by membrane, acidic aluminum oxide column was used to separate glucosinolates from permeate according to their acidic groups. As described for the

column chromatography, 1ml permeate from membrane filtration was injected into FPLC system each time. The first step in FPLC separation is the binding of glucosinolates onto the acidic aluminum oxide. After that, unbound compounds are washed out with water. Then the separation of glucosinolates is carried out by elution with NaCl solutions. As shown in Fig. 7, compounds pass through the column without interaction by using 25ml water. After that, the buffer changed into different salt concentrations 0.2mol/l, 0.5mol/l and 1.0mol/l respectively. The intensities of the UV absorbance peaks of fraction (elution from buffer B) is increasing with the buffer concentration increasing. It indicates that the salt concentration is higher, glucosinolates are easier to exchange and eluted from acidic aluminum oxide.

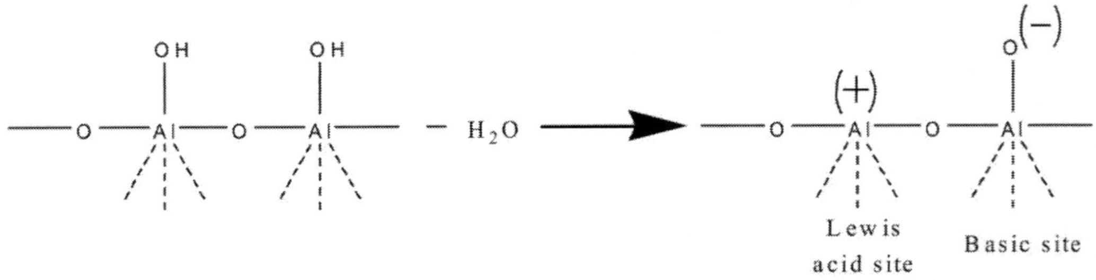

Fig. 5. Formation of acidic aluminum oxide.

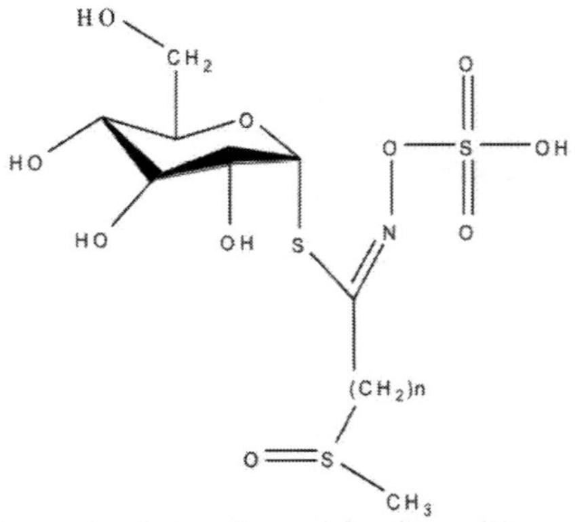

Fig. 6. Basic structure of the glucosinolates found in Camelina sativa seed (Mark A. Berhow et al., 2013).

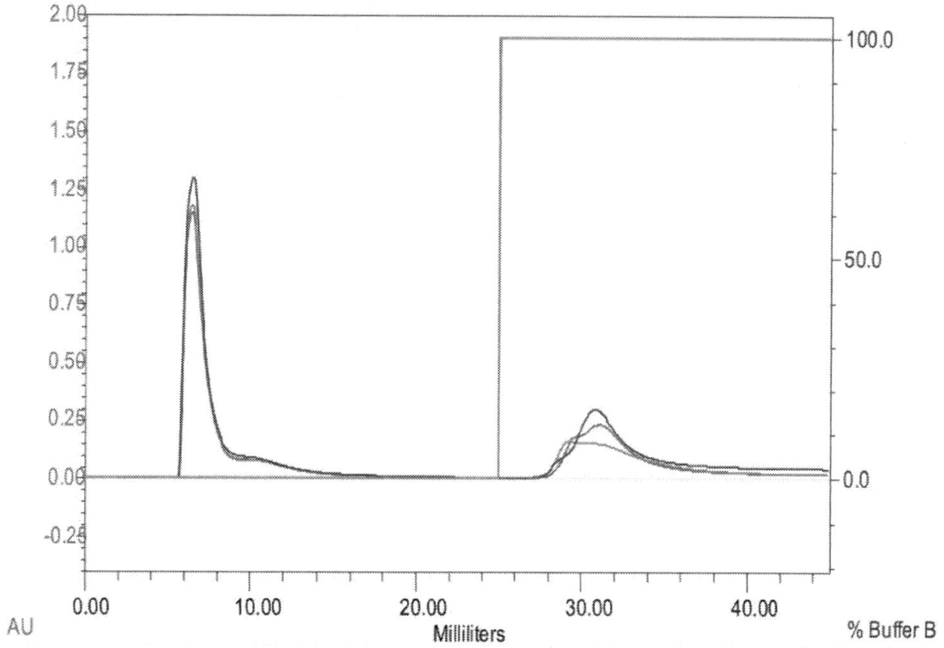

(——UV 0.2mol/LNaCl ——UV 0.5mol/LNaCl —— UV 1.0mol/LNaCl ——% Buffer B)

Fig. 7. FPLC graph of permeate by using different sodium chloride concentration.

3.3 Total glucosinolates analysis results

3.3.1 Model compound analysis

Glucosinolates can react with palladium (Kumar S et al., 2004) and form Pd-glucosinolate, which is colored can be detected at 540 nm. In order to determine the weight of glucosinolate obtained from Camelina seed meal, model glucosinolate sinigrin solutions with different concentration (0.4mg/ml, 0.6mg/ml, 1mg/ml, 2mg/ml, 4mg/ml, 8mg/ml and 16mg/ml) were used to make the standard curve. Firstly, add 1.0ml solution into a 10.0 ml tube, then added 2.0 ml CMC and shake it well, after that add 1.0 ml palladium chloride and shake the mixture well. Keep the mixture reacting at 25 ℃ for 2 hours and measure the reactant's absorbance at 540 nm. Then standard curve was plotted according to the concentration and absorbance, which is shown in Fig. 8. Regression of the standard absorbance with concentration typically resulted in an $R^2 \geqslant 0.99$. Sample glucosinolate concentrations were determined from the standard curve and expressed as mg of glucosinolates.

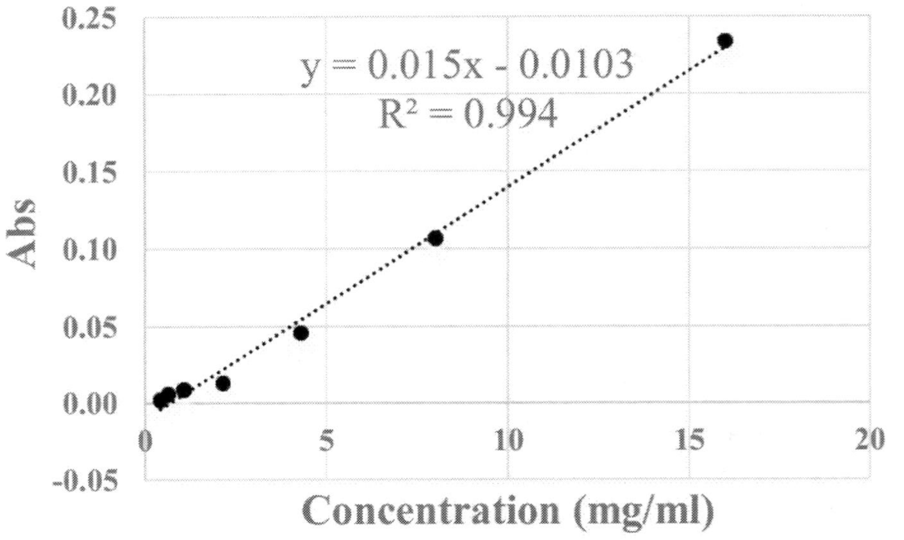

Fig. 8 Standard curve for palladium-glucosinolate complexes by using model glucosinolate (sinigrin)

3.3.2 Quantification of total glucosinolates from FPLC fractions and UF permeate

Different fractions from FPLC and permeate from UF were reacted with palladium chloride under the conditions described in standard curve experiments. The only thing changed is the reference solution is the related salt solution. For example, when measuring the fraction from 0.2 mol/l NaCl the reference solution should be 1.0 ml 0.2 mol/l NaCl, 2.0 ml CMC and 1.0 ml palladium chloride mixed solution. After reaction with palladium chloride, the absorbance for the reactant from different factions and permeate were obtained, which is shown in Table 1. The absorbance of permeate is 0.192. With different salt concentration, absorbance of fractions are 0.011, 0.025, and 0.029 for 0.2, 0.5 and 1.0 mol/l NaCl respectively. According to the equation (y=0.015x-0.0103) obtained from the standard curve, the glucosinolate concentration can be calculated, which are 1.42 mg/l, 2.35 mg/l and 2.62 mg/l for the fractions respectively. The glucosinolate concentration of permeate is 13.49 mg/l. all the fractions were come from 1.0 ml permeate, which produced 5.0 ml fractions for each column separation. Therefore, total glucosinolate for different elution methods were obtained. 13.49 mg glucosinolates were injected into the FPLC system, and 13.10 mg glucosinolates were recovered by elution with 1.0 NaCl solution. Moreover, 7.10 mg and 11.77 mg glucosinolates were recovered by using 0.2 and 0.5 mol/l NaCl solution. It is concluded that high salt concentration (1.0 mol/l) will get higher recovery rate of glucosinolates, which can be 97.1% shown in Fig. 9. And lower concentration of salt 0.2 and 0.5 mol/l can get lower recovery rate 52.6% and 87.2% respectively. Moreover, 5 g meal produced 46.5 ml permeate. Hence, we can conclude that the glucosinolate content of the meal is 125.5 mg/g.

Table 1.Total glucosinolates in different fractions from HPLC and permeate from UF.

Fractions	0.2mol/l NaCl elution	0.5mol/l NaCl elution	1.0mol/l NaCl elution	Permeate

Items				
Abs	0.011	0.025	0.029	0.192
Concentration(mg/l)	1.42	2.35	2.62	13.49
Volume(ml)	5.00	5.00	5.00	1.00
Total glucosinolates(mg)	7.10	11.77	13.10	13.49

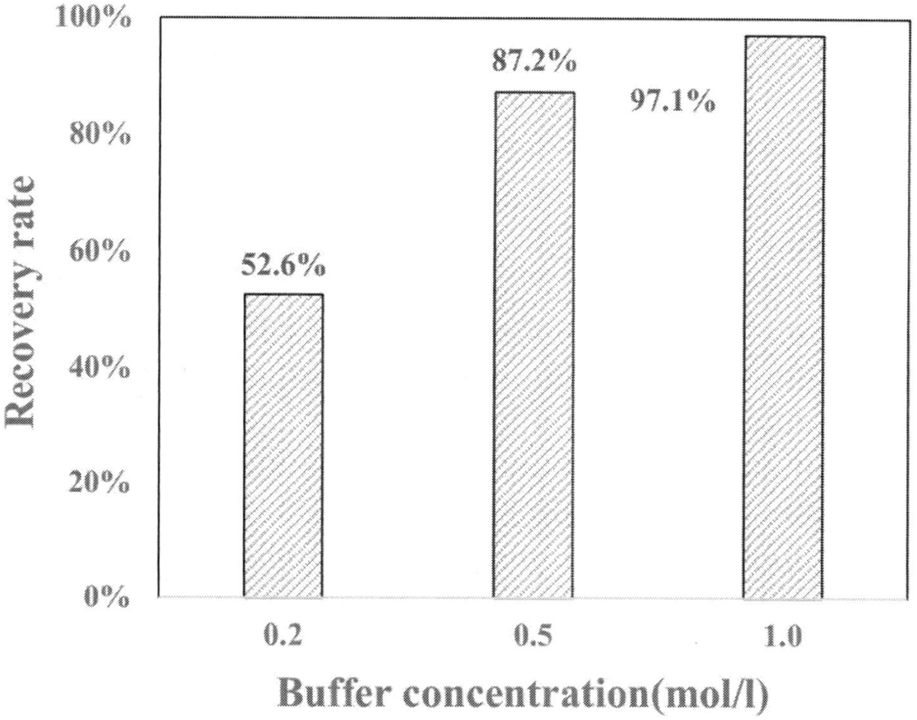

Fig. 9. Recovery rate for the FPLC separation.

3.3.2 FPLC efficiency analysis by HPLC

Since 1.0 mol/l NaCl can get highest recovery rate of glucosinolates from membrane permeate, we used HPLC to analyze the separation efficiency of the acidic aluminum oxide column using 1.0 mol/l NaCl buffer. Fig. 10 shows the UV absorption peaks with different retention time.

The major peaks (a, b, c, d, and e) for fraction eluted by 1.0 NaCl are at retention time of 4.3, 5.7, 7.5, 22.2 and 24.6 min. And there are 12 major peaks for permeate, which are much more than these of fraction from elution. Their peak areas are shown in Table 2. As we have known, the glucosinolates have been recovered for 97.1% from permeate. Peak d and e from the fraction have the same retention time as the peak 11 and 12 from permeate. Moreover, the sum peak area of peak d and e is 958.2 mAU*s, which is almost 20% of that of peak 11 and 12 (4609.4 mAU*s). However, the volume of the fraction is 5 ml, which is 5 times than that of permeate. It indicates that glocosinolates are almost appeared at retention time 22.2 and 24.6 min. Fig. 10

shows the HPLC peaks from the fraction and permeate. We can conclude that most of the impurities appeared at the early stage of the HPLC procedure. And these impurities (retention time at 4.6, 5.1, 5.5, 6.3 and 7.3 min) were removed a lot by FPLC.

Table 2. HPLC results for fraction from 1.0 mol/l NaCl elution and permeate from UF.

Solutions \ Items	Peak #	Retention time (min)	Area (mAU*s)
Permeate from UF	1	4.1	100.4
	2	4.3	623.7
	3	4.6	705.5
	4	5.1	926.2
	5	5.5	5475.7
	6	6.3	5078.8
	7	7.3	6063.4
	8	15.3	1954.2
	9	17.3	454.4
	10	18.2	388.5
	11	22.2	3399.7
	12	24.6	1209.7
Fraction from 1.0 mol/l NaCl elution	a	4.3	714.3
	b	5.7	124.0
	c	7.5	97.6
	d	22.2	637.0
	e	24.6	321.2

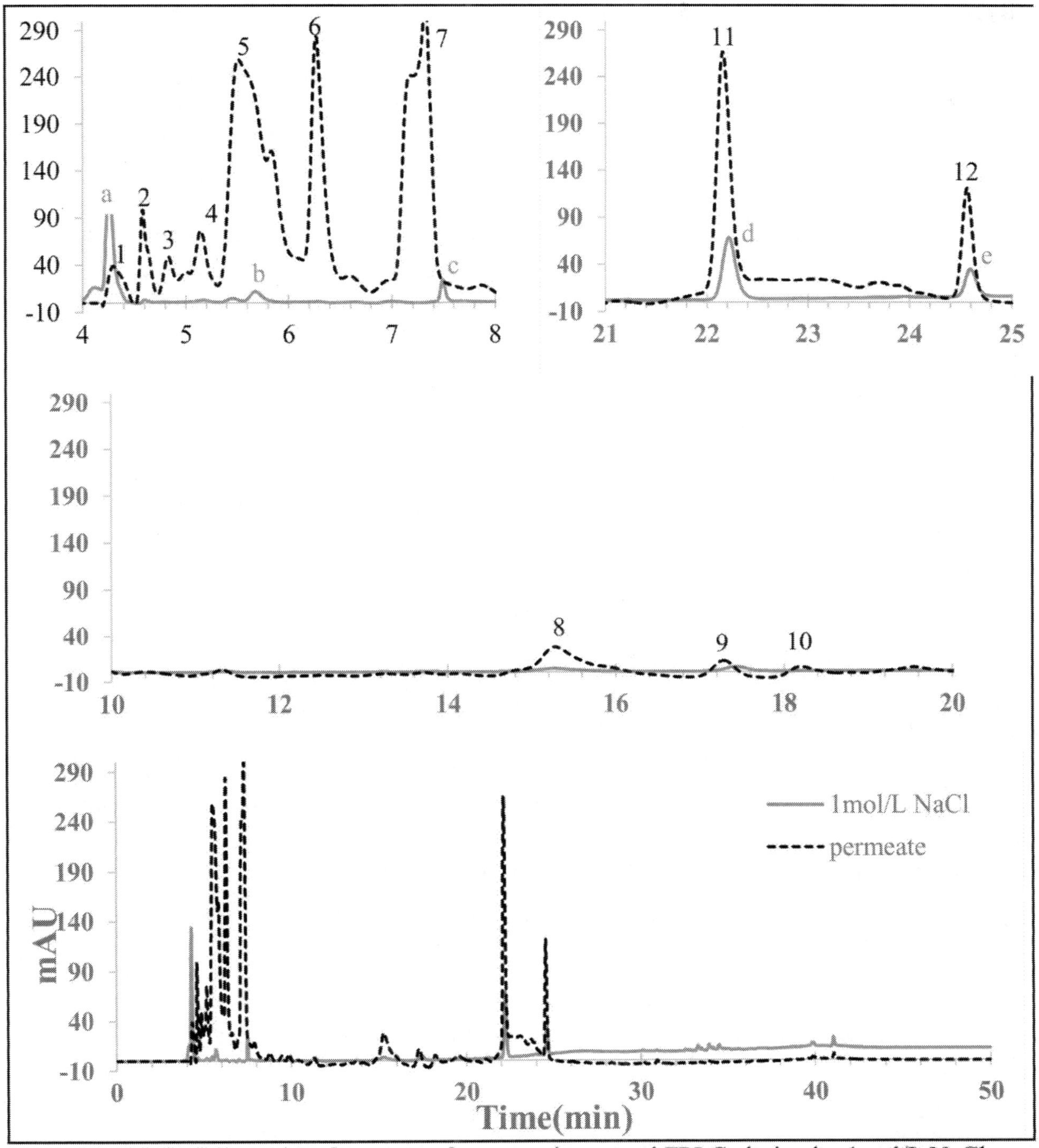

Fig. 10. HPLC results of permeate from membrane and FPLC elution by 1mol/L NaCl.

Conclusion

Membrane separation is a promising method for extraction of glucosinolates without using heavy metal salt. The glucosinolates were separated chromatographically by utilizing FPLC technique which is based on ion exchange. 97.1% recovery is obtained by using 1.0 mol/l NaCl solution as buffer. And the glucosinolate content of the seed meal is 125.5 mg/g. HPLC results indicates that the FPLC has removed most of the impurities from permeate from UF. Moreover, glucosinolates appeared at retention time 22.2 and 24.6 min under the experimental mobile phase condition.

Reference:

[1] Borek, V., M.J. Morra., 2005. Ionic thiocyanate (SCN-) production from 4-hydroxybenzyl glucosinolate contained in Sinapis alba seed meal. J. Agric. Food Chem. 47, 3837-3842.

[2] Morra, M.J., 2004. Controlling soil-borne plant pests using glucosinolate-containing tissues. Agroind. 3, 251-255.

[3] Snyder, A., M.J. Morra, J. Johnson-Maynard, D.C. Thill., 2009. Seed meals from Brassicaceae oilseed crops as soil amendments: Influence on carrot growth, microbial biomass N, and N mineralization. HortScience. 44, 354-361.

[4] Datu B. Agusdinata, Fu Zhao, Klein Ileleji, Dan DeLaurentis, 2011. Life Cycle Assessment of Potential Biojet Fuel Production in the United States. Environ. Sci. Technol. 45 (21), 9133-9143.

[5] Kakani R, Fowler J, Haq AU, Murphy EJ, Rosenberger TA, Berhow M, Bailey CA, 2012. Camelina meal increases egg n-3 fatty acid content without altering quality or production in laying hens. Lipids. 47 (5), 519-26.

[6] Junfeng Qian, Qiuhui Yang, Fuan Sun, Mingyang He, Qun Chen, Zhi Yun, Lizhen Qin, 2013. Cogeneration of biodiesel and nontoxic rapeseed meal from rapeseed through in-situ alkaline transesterification. Bioresource Technol. 128, 8-13.

[7] M.K. Tripathi, A.S. Mishra, 2007. Glucosinolates in animal nutrition: A review. Anim. Feed Sci. Tech. 132, 1-27.

[8] Khajali, F.; Slominskif, B. A., 2012. Factors that affect the nutritive value of canola meal for poultry. Poultry Sci. 91 (10), 2564-2575.

[9] George Francis, Harinder P.S Makkar, Klaus Becker, 2001. Antinutritional factors present in plant-derived alternate fish feed ingredients and their effects in fish. Aquaculture. 199 (3-4), 197-227.

[10]F. Schöne, U. Kirchheim, W. Schumann, 1994. Glucosinolate degradation by rapeseed myrosinase and effect on rapeseed acceptability by growing pigs. Anim. Feed Sci. Tech. 48 (3-4), 229-235.

[11]T.G. Rowan, T.L.J. Lawrence, S.J. Kershaw, 1991. Effects of dietary copper and a probiotic on glucosinolate concentrations in ileal digesta and in faeces of growing pigs given diets based on rapeseed meals. Anim. Feed Sci. Tech. 35 (3-4), 247-258.

[12]A.M.H. Subuh, T.G. Rowan, T.L.J. Lawrence, 1995. Toxic moieties in ruminal and duodenal digesta and in milk, and hepatic integrity in cattle given diets based on rapeseed meals of different glucosinolate contents either untreated or treated with heat or formaldehyde. Anim. Feed Sci. Tech. 52 (1-2), 51-61.

[13] Zhiyong Xie, Rui Wang, Yingchun Wu, Li Yang, Zhengtao Wang, Yiming Li, 2012. An efficient method for separation and purification of glucosinolate stereoisomers from radix isatidis. J. Liq. Chromatography R. T. 35 (1), 153-161.
[14] Ze-You Li, Yong Wang, Wen-Tao Shen, Peng Zhou, 2012. Content determination of benzyl glucosinolate and anti–cancer activity of its hydrolysis product in Carica papaya L. Asian Pacific J. Tropical Medicine. 5 (3), 231-233.
[15]Antone R. Opekun, Julia L. Opekun, David Y. Graham, 2003. Broccoli (Sulforaphane glucosinolate) for the treatment of H. pylori infection in humans. Gastroenterology. 124 (4, Supplement 1), A362.
[16]Adarsh Pal Vig, Amandeep Walia, 2001. Beneficial effects of Rhizopus oligosporus fermentation on reduction of glucosinolates, fibre and phytic acid in rapeseed (Brassica napus) meal. Bioresource Technol. 78 (3), 309-312.
[17]Quinsac, A., Lessire, M., Krouti, M., Ribaillier, D., Coïc, J.-P., Fauduet, H., Rollin, P., 1994. Improvement in the nutritive value of high and low glucosinolate rapeseed meal by aqueous extraction. Anim. Feed Sci. Technol. 48, 265-272.
[18]Charpentier N, Bostyn S, Coic J P., 1998. Isolation of a rich glucosinolate fraction by liquid chromatography from an aqueous extract obtained by leaching dehulled rapeseed meal (Brassica napus L.). Ind. Crop. Prod. 8, 151-158.
[19]Helboe, P., Olsen, O., Sorensen, H., 1980. Separation of glucosinalites by high performance liquid chromatography. J. Chromatogr. 197, 199-205.
[20]Peterka, S., Fenwick, G.R., 1988. The use of flash chromatography for the isolation and purification of glucosinolates (mustard oil glycosides). Fat Sci. Technol. 90, 61-64.
[21]Qizhen Du, Jie Fang, Shijun Gao, Qiaoyun Zeng, Cehui Mo, 2008. A gram-scale separation of glucosinolates from an oil-pressed residue of rapeseeds using slow rotary countercurrent chromatography. Sep. Purif. Technol. 59 (3), 294-298.
[22]R. van Reis, A. Zydney, 2007. Bioprocess membrane technology, J. Membrane Sci. 297, 16-50.
[23]Wang Ninghui, 2009. A quantitative analysis method of glucosinolates in rapeseed meal. Sci. Technol. Qinghai Agr. Forestry. 3, 58-59.
[24]Elio Santacesaria, Davino Gelosa, 1977. Basic Behavior of Alumina in the Presence of Strong Acids. Ind. Eng. Chem. Prod. Res. Dev. 16, 45-46.
[25]Mark A. Berhow, Umit Polat, Jan A. Glinski, Michal Glensk, Steven F. Vaughn, Terry Isbell, Ivan Ayala-Diaz, Laura Marek, Candice Gardner, 2013. Optimized analysis and quantification of glucosinolates from Camelina sativa seeds by reverse-phase liquid chromatography. Ind. Crop. Prod. 43, 119-125.
[26]Kumar S, Yadav SK, Chauhan JS, Singh AK, Khan NA, Kumar PR, 2004. Total glucosinolate estimation by complex formation between glucosinolates and tetrachloropalladate (II) using ELISA reader. J. Food Sci. Technol. 41, 63-65.

381282 Industrial Investigation of Milling and Leaching Process: An Attainable Region Approach

Abstract:

- Size reduction is an integral and expensive part of almost all mining processes especially for the downstream leaching process. Therefore, significant energy and cost savings exist even with slightest increase in milling efficiency to a certain degree of liberation. In this work, the attainable region (AR) is used to explore the effect of grinding media fill level, grinding media size and grinding time on the resulting size distribution. The resulting particle sizes are then leached to determine the optimal size for maximum gold recovery as a function of operation time (grinding and leaching), energy consumption and other operational variables. Preliminary leaching results show that the majority of gold is found in the smaller size classes (-150 microns) and the difference in grade between -150+75microns and -75+25microns is small, suggesting that a size of -150 microns is an adequate cut-off for milling, rather than the -75microns currently employed. Therefore an integrated grinding and milling process is investigated.

Hydrogenation of Hydroxymethylfurfural using a Ruthenium Coated Polymeric Membrane Reactor

John Stanford, Peter Pfromm, and Mary Rezac*

Department of Chemical Engineering, Kansas State University
Manhattan, Kansas, United States

Catalytic membrane reactors afford an alternative and potentially more efficient method for performing three-phase heterogeneous chemical reactions. Traditional three-phase reactors often present mass transfer limitations, namely gas solubility in the liquid phase. Membrane reactors can alleviate the inherent mass transfer limitations by directly and abundantly supplying gas to the catalytic sites on the membrane surface, which acts as a gas/liquid contactor. The focus of this study is the application of a polymeric catalytic membrane reactor to the hydrogenation of hydroxymethylfurfural (HMF), which is a biomass-based platform chemical derived from fructose. The various partially hydrogenated HMF intermediates have a variety of manufacturing uses, and a notable product of hydrogenated HMF, 2,5-dimethylfuran (DMF), has many attractive characteristics as a liquid fuel.

Polyimide polymers are used for membrane synthesis because of their good chemical and high temperature resistance. The polymer solutions are cast as asymmetric integrally-skinned flat sheet membranes and are then coated with a ruthenium catalyst on the non-porous surface. Post-fabrication modifications such as thermal annealing and chemical crosslinking are being explored to enhance the chemical and temperature resistances of the membranes. The completed membrane is positioned in a flow over configuration maintaining liquid contact on the metal coated surface while allowing hydrogen gas to permeate from the porous support side to the catalytic sites on the non-porous surface. This study explores reaction kinetics and the influence of reaction temperature and gas phase pressure on system performance. Reaction products are identified and quantified using HPLC and GC-MS. Scanning electron microscopy is employed to examine the membrane and catalyst morphology before and after the reaction. An important parameter of this study is the reaction solvent choice, which is crucial for maintaining a functional membrane reactor as the necessary elevated temperatures for the chemical reaction also increase the solubility of the solvent in the membrane, leading to swelling and possible membrane failure. Solvent solubility in the polyimides of this study at different solvent vapor activities and temperatures is measured using a quartz crystal microbalance. Solvents of interest include water and short-chain alcohols. Sorption isotherms are described using a diffusion-relaxation model and concentration-dependent diffusion coefficients and relaxation parameters are calculated. This research on the development and characterization of a functional polymeric catalytic membrane reactor for HMF hydrogenation may serve as an example of the methodology needed for developing membrane reactors for other heterogeneous catalysis applications.

Material Properties and Operating Configurations of Membrane Reactors for Propane Dehydrogenation

Seung-Won Choi, Christopher W. Jones, David S. Sholl, and Sankar Nair

School of Chemical & Biomolecular Engineering, Georgia Institute of Technology, Atlanta, GA 30332-0100

Jason S. Moore, Yujun Liu, Ravindra S. Dixit, and John G. Pendergast

Engineering & Process Sciences, The Dow Chemical Company, Freeport, TX 77541

Abstract

We consider the problem of defining physically realistic and technologically interesting ranges for catalytic and transport properties, and operating configurations, of packed-bed membrane reactors (PBMRs) for propane dehydrogenation (PDH). This work employs an isothermal PBMR simulation with simplified kinetics and transport terms to investigate the influence of a broad range of reaction and transport parameters on the PBMR performance, and thereby determine operating windows for such reactors when constructed with industrial PDH catalysts and nanoporous (*e.g.*, zeolitic) membranes. First, dimensionless analysis of the simulation results is carried out using Damköhler (Da) and Péclet (Pe) numbers to represent the material catalytic properties, transport properties, and the reactant throughput. We also consider the effects of operating parameters such as different sweep flow rates and countercurrent versus cocurrent modes. An increased Da and a decreased Pe are useful for enhanced conversion in the PBMR. However, the conversion reaches a maximum and then decreases with further decrease in Pe if the membrane allows permeation of the propane reactant in addition to hydrogen, due to reactant depletion along the length of the PBMR. This is significant because of the consideration of nanoporous zeolite membranes as a leading candidate for PDH PBMRs. A highly hydrogen-selective, small-pore zeolite membrane (*e.g.*, SAPO-34 or DDR) will show better performance over a wide range of weight-hourly space velocity (WHSV) conditions than a higher-flux, lower-selectivity medium-pore zeolite membrane (*e.g.*, MFI) or a Knudsen-selective high-flux mesoporous membrane. H_2-selective membranes show a plateau region of conversion at low Pe. This limit can be overcome by using a larger sweep flow rate (which enhances the driving force for membrane permeation) and countercurrent operation. However, due to a complex trade-off between the reaction kinetics and forward/back-permeation across the membrane, the countercurrent mode can be recommended only in a limited (but nevertheless realistic) window of material properties. The preference for small-pore H_2-selective zeolitic membranes requires the fabrication of thin (~1 μm or thinner) membranes with low defect densities in order to achieve an adequately high transport flux. Low-selectivity, high-flux membranes can only provide moderate enhancement of PBMR performance at low WHSV (< 1 hr^{-1}). The above analysis is useful in quickly identifying desired material properties For example, in order to enhance the PDH conversion from 48 % (PBR) to 52 % (PBMR) at a WHSV of 1 hr^{-1} using Pt-Sn catalyst, the required dimensionless numbers in the operating window are Pe ~ 1.4 and Da ~ 28, which can be realized with a membrane surface area in the range of 4.5×10^{-3}-4.5×10^{-4} m^2/g catalyst and a small-pore zeolite membrane with H_2 permeance in the range of 10^{-8}-10^{-7} $mol.m^{-2}.s^{-1}.\ Pa^{-1}$.

392000 An economic performance evaluation study under uncertainty for Pd-based catalytic membrane reactor modules used for H2 production

Abstract:

A comprehensive economic performance evaluation framework for palladium-based catalytic membrane reactor (CMR) modules potentially integrated into hydrogen production (HP-CMR) through methane steam reforming process systems has been developed. The HP-CMR technology option offers a promising technical pathway towards hydrogen production with enhanced environmental performance in a carbon-constrained world, and since no significant operating experience has been accumulated at the industrial/commercial scale any preliminary attempt to assess its economic viability is certainly justified. Within the above context, the development of comprehensive baseline models for fixed capital investment (FCI), total capital investment (TCI), as well as total product cost (TPC) in order to evaluate the economic performance of the above industrial scale membrane reactor module represents a natural starting point in our study. Various sources of uncertainty (raw material market prices, utilities costs, labor costs, membrane lifetime and maintenance costs, financing costs, etc.) are explicitly recognized and the effect of these uncertainty drivers on FCI, TCI and TPC are explicitly taken into account using Monte-Carlo techniques. As a result, insightful distribution profiles of FCI, TCI and TPC are derived rather than single-point value estimates, and a more realistic distribution of membrane reactor module economic performance outcomes is generated through which "risks and opportunities" can be identified in the presence of uncertainty. Furthermore, preliminary results have shown that while expected values of the FCI and PC obtained from the above distribution profiles remain higher than the ones associated with the traditional/conventional methane steam reforming-based hydrogen production option (within a meaningful range of permeabilities, membrane length as well as input flow rates), FCI and TPC profiles for HP-CMR become appealing by reducing palladium thickness (in terms of their respective expected values when comparatively assessed against the above conventional option in the presence of uncertainty). Finally, within the proposed economic performance evaluation framework Tornado diagrams have been also developed revealing and establishing the predominant role and effect of financing interest cost, palladium unit price and palladium membrane lifetime on the TCI and PC profiles.

378205 Membrane-Based Thermocatalytic Destruction of Chemical Warfare Agents

Abstract:

The possibility of use of chemical weapons has increased in recent years, both as a result of potential terrorist attacks as well as of ongoing international conflicts. The successful application of a flow-through catalytic membrane reactor (FTCMR) as an individual protection (IP) system for the destruction of dimethyl methylphosphonate (DMMP), which is known as a chemical precursor (and used to simulate its characteristics) for Sarin (GB), a toxic chemical warfare agent (CWA) was previously investigated and reported by this group[1]. As part of the study, the effect of the DMMP concentration in the feed on its complete combustion in the FTCMR was investigated. The studies indicate that the protection (complete conversion) time is a function of the DMMP concentration, with longer times observed for the lower concentrations, and shorter protection times associated with the higher concentrations.

In this current study, the focus is on collective protection (CP) applications. As part of this effort multi-tubular membrane bundles are developed and rendered reactive for the proposed application. The current efforts involve the development and testing of a 1 cfm multi-tubular FTCMR unit, with the eventual goal the development and testing of a 200 cfm full-scale CP system.

In addition, a novel hybrid system is being developed that combines the FTCMR with a surface-flow membrane system (SFMS)[2]. The main advantage of this hybrid system, combining the SFMS (capable of the physical removal of a large fraction of the CWA from contaminated air streams) and the FTCMR (which completely oxidizes the remaining amount) is the continuous CWA destruction for extended time periods which are appropriate for both IP and collective protection (CP) applications. As part of this effort, we also investigate the performance of the SFMS. Experimental and modeling investigations indicate that a high removal rate can be achieved with a proper choice of (feed rate/membrane surface area) and the sweep ratio. The performance of the combined system is currently under investigation, and experimental results and the modeling of the system performance will be presented at the meeting.

References:

1. Motamedhashemi, M.M.Y., Egolfopoulos, F.N., and Tsotsis, T.T., "Application of a Flow-Through Catalytic Membrane Reactor (FTCMR) for the Destruction of a Chemical Warfare Simulant," *J. Membrane Sci.*, 376, 119, 2011.
2. Lee, H. C., Monji, M., Parsley, D., Sahimi, M., Liu, P., Egolfopoulos, F., and Tsotsis, T., "Use of Steam Activation as a Post-Treatment Technique in the Preparation of Carbon Molecular-Sieve Membranes," *Ind. Eng. Chem., Res.,* dx.doi.org/10.1021/ie300261r, 2013.

Performance of reduction-tolerant $Sr_{0.7}Ba_{0.3}Fe_{0.9}Mo_{0.1}O_{3-\delta}$ multichannel hollow fiber membrane reactor for partial oxidation of methane

Jiawei Zhu, Zhengkun Liu, Wanqin Jin *
State Key Laboratory of Materials-Oriented Chemical Engineering
Membrane Science & Technology Research Center,
Nanjing Tech University, Nanjing 210009, P. R. China
* wqjin@njtech.edu.cn

Due to the potential impacts on energy and environment, mixed-conducting dense membranes have attracted considerable interests in partial oxidation of methane (POM) reaction process. Such membrane reactors coupled the POM reaction and oxygen separation in one unit, which makes this reaction more commercially feasible. However, the mixed-conducting membranes applied in POM reaction should meet the following requirements: high oxygen permeation flux, good stability under reducing atmosphere and attractive membrane configuration for industrial application.

So far, large research efforts focused on using single-channel hollow fiber membranes for membrane reaction process because they possess a thin wall and can obtain high oxygen permeation flux. Nevertheless, the conventional single-channel hollow fiber membrane may not meet the requirements of industrialization, suffering from low mechanical strength. In our previous work[1], a mixed-conducting multichannel hollow fiber (MCMHF) membrane with high oxygen permeation flux and superb mechanical strength was proposed to overcome some drawbacks of single-channel hollow fiber membranes.

In this work, to accelerate the industrial application of membrane reactor for POM reaction, a MCMHF membrane was used to construct this membrane reactor successfully. By using phase inversion and sintering technique, the MCMHF membrane has being prepared, and $Sr_{0.7}Ba_{0.3}Fe_{0.9}Mo_{0.1}O_{3-\delta}$ (SBFM)[2], the reduction-tolerant oxide, was selected as the membrane material. The morphology, mechanical strength and oxygen permeability of the MCMHF membrane, the effects of temperature, methane concentration and methane flow rate on the performance of MCMHF membrane reactor and the long-term stability of the membrane reactor were investigated in detail. This work demonstrated that the multichannel hollow-fiber can be an attractive membrane configuration for membrane reactor in the viewpoint of industrial application.

Acknowledgments

This work was supported by the Project of Priority Academic Program Development of Jiangsu Higher Education Institutions (PAPD) and the Innovative Research Team Program by the Ministry of Education of China (No. IRT13070).

References

1. Zhu J, Dong Z, Liu Z, Zhang K, Zhang G, Jin W. Multichannel mixed-conducting hollow fiber membranes for oxygen separation. *AIChE J.* 2014;60(6):1969-1976.
2. Dong X, Jin W, Xu N. Reduction-Tolerant Oxygen-Permeable Perovskite-Type Oxide $Sr_{0.7}Ba_{0.3}Fe_{0.9}Mo_{0.1}O_{3-delta}$. *Chemistry of Materials.* 2010;22(12):3610-3618.

374796 Distillation Column Safety

Abstract:

Distillation columns are dangerous places. It is dangerous to be near them, on them and in them. Production engineers are very aware of the dangers and safeguards. Desk engineers, who only visit production sites occasionally, are less aware.

Whether near, on or in a distillation column, workers should "expect the unexpected" and "unexpect the expected."

390090 Difficulties in Separation: A Unique Thermodynamic Approach

Abstract:

A mixture of hydrocarbons is modeled by the general cubic equation of state equipped with a set of mixing rules. Thermodynamic concepts within the model allow us to understand the separation process at the phase equilibrium level, which provide a major aid from the process design point of view. This thermodynamic model is employed in the design of a multi-column distillation process for the separation of a hydrocarbon mixture. Within the process integration of heat streams and the utilization of flash separators facilitate the utilization columns products in order to achieve optimum cost conditions. Computer aided simulation such as IDEAS & UNISIM are used to provide a sufficiently accurate application of the mixing rules that would enable the model to provide precise mixing behavior. Alternative designs are still being tested.

388140 Avoid Instability in Distillation Using Control Based on the Second Law of Thermodynamics

Abstract:

The distillation column is one of the primary unit operations for separations in the chemical and petroleum industries. Because of their significance, the literature for theory and practice of distillation columns is vast and mature. Most of the literature focuses on the instabilities that may arise when using certain controllers. Despite the identification of the many causes for instability, however, a general stability analysis of the multi-component distillation column has yet to be made.

In this talk we show, for the first time, a rigorous stability analysis that reveals the asymptotically stable operation of the multi-component distillation column. The stability analysis utilizes the thermodynamics availability function, first employed by Alonso and Ydstie [1], to show asymptotic stability of the distillation column using pressure, temperature, and level controllers on the reboiler and condenser. For this analysis, we model the multi-component distillation column as a stack of independent mass exchange units which isolates each phase in each tray of the column. The mass exchange unit is a modeling tool used to isolate the differential equations that describe the dynamics of the vapor and liquid in each tray. A Lyapunov function is then developed for each mass exchange unit. We then stack these mass exchange units to model a distillation column, eventually allowing us to construct the overall Lyapunov function for the distributed system. Ultimately, the overall Lyapunov function is used to show that if the hydrodynamics of the column are well controlled, the thermodynamics will drive the system to asymptotic stability.

References

[1] A. Alonso, B. E. Ydstie, "Stabilization of distributed systems using irreversible thermodynamics" Automatica, 2001, 37, 1739 - 1755

APPLICATION OF INDEPENDENT COMPONENTS ANALYSIS IN DISTILLATION COLUMN HICH PURITY FOR THE IMPROVEMENT OF VARIABLES DECOUPLING

M. Sc. Shirlene K. S. Carmo[1], M. Sc. Mariângela da S. A. Emerenciano[2], Dr. Luis G. S. Vasconcelos[3], [(1,2,3)]Federal University of Campina Grande, Departamento of Chemical Engineering, Campina Grande, Brazil; [(1)]Federal University Rural of Semi-Arid, Pau dos Ferros, Brazil

Abstract

Distillation columns constitute a significant fraction of the investment and production cost of chemical plants. Therefore need to be designed and controlled to keep yourself in optimal operating conditions. The work is in the application of ICA (Independent Component Analysis) in a multicomponent distillation column, high purity, where the mixture presents an unconventional azeotropic behavior. Two components in the plant require a control for the best performance and safe operation of the plant, they are, 1.2 dichloroethane (1.2 EDC) in the top column and carbon tetrachloride (CCl_4) in the base column. The column presents difficulties in stabilizing these components, which can cause problems in subsequent process units. The use of conventional decoupled projects in process control, most often do not have perfect results, especially in this type of column, due to the strong coupling between the variables. The use of artificial intelligence techniques such as ICA technique showed a significant improvement in decoupling of control loops that column, facilitating control of variables in their respective setpoints.

Introduction

The continuous search for improvements in industrial plants is associated with large market competition and the need to increase profit margins and operational safety. In the chemical industries about 80% of the energy operating cost is due to the use of distillation columns (1). Besides the disadvantage of high energy consumption, if the mixture does not presents ideal behavior in relation to the phase equilibrium, there may be a azeotrope formation. The occurrence of azeotropes adds some difficulties for separation, because the liquid and vapor phases present the same composition. The separation of these mixtures is not possible through conventional distillation and they are usually separated by extractive distillation or by azeotropic distillation.

Mainly in high purity distillation columns, the control technique for conventional decoupling cannot eliminate the strong variables interaction totally (2). Thus, the incorporation of artificial intelligence devices in production processes has been showing an increasing trend in recent years. This paper proposes the use of Independent Component Analysis (ICA) in plants of high-purity distillation, thus providing the elimination of the strong interaction between the variables in the process. Furthermore, the work establishes a control strategy for top and bottom compositions of a high purity distillation column of 1.2 EDC purification phase (ethylene dichloride) of the vinyl chloride commercial plant (MVC) production, seen that simultaneous control of top and bottom compositions in a distillation column is a complex problem due to the inherent coupling process.

Review of the State Art

The azeotropic distillation is highly sensitive to disturbances, which can result in deterioration of the products or very difficult to return to normal operation. The most used solution to minimize transient operation is the implementation of advanced control techniques. The ICA technique has been applied in various types of industrial processes in order to reduce or to eliminate this strong interaction between the variables.

Independent Components Analysis

The technique contributes to the reduction of the existing coupling between a set of variables present in a particular system. The aim is to obtain, from mixed captured signals by s particular sensor, the individual signal emitted by each other. It means to separate each source present in the mixture so that they can be separately checked. The separation process from the ICA technique is shown in Figure 1.

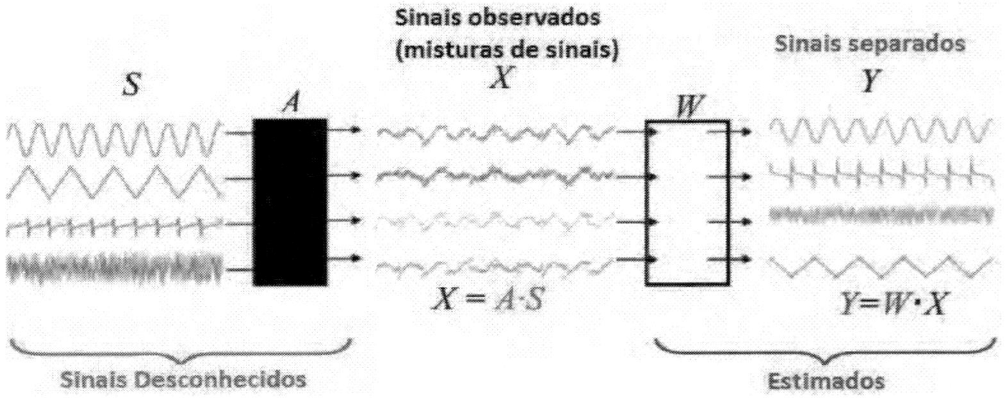

Figure 1. ICA Mixture and Separation Process

X and S are composed by columns vectors which correspond to the original and mixture signals, respectively, and 'A', in this case, is a square matrix, which weights the signals mixture. It is assumed that such matrix coefficients are unknown, because it is not possible to determine their values without knowing the exact physical structure of the environment. Therefore, the purpose of the ICA is, from the captured signals x, to obtain individual signals 's'. The separation matrix W used for the generation of independent signals, is an approximation of the A inverse matrix. To estimate W, it is necessary that the original sources that generated the signals are independent and non-Gaussian. Finally, 'y' vector Independent Component approaching 's'.

Description of Problem

The distillation column studied in this work is located in the area of purification of 1.2 - EDC in an existing commercial plant. The feed stream of the column in question is composed of 98 99% 1.2-EDC, and other organic substances. The purpose of this column is to dry 1.2-EDC and to remove most of the lighter components of the basic product, especially CCl_4 and $CHCl_3$. The appropriate control composition CCl_4 feeding of the furnaces reduces the potential for coke formation which leads to premature stopping of the plant. Keeping the composition of CCl_4 next proper operating range, 3000 ppm, and the composition of 1.2 EDC in 99.5%.

Methodology

The flowchart in this work was developed by Aspen Plus™ (3), and exported to the Aspen Plus environment DynamicsTM. The simulator itself presents the flowchart level controllers to the base of the column and a reflux vessel. In addition, the flow to the feed stream and withdrawal of light controllers have been added to the process flow diagram.

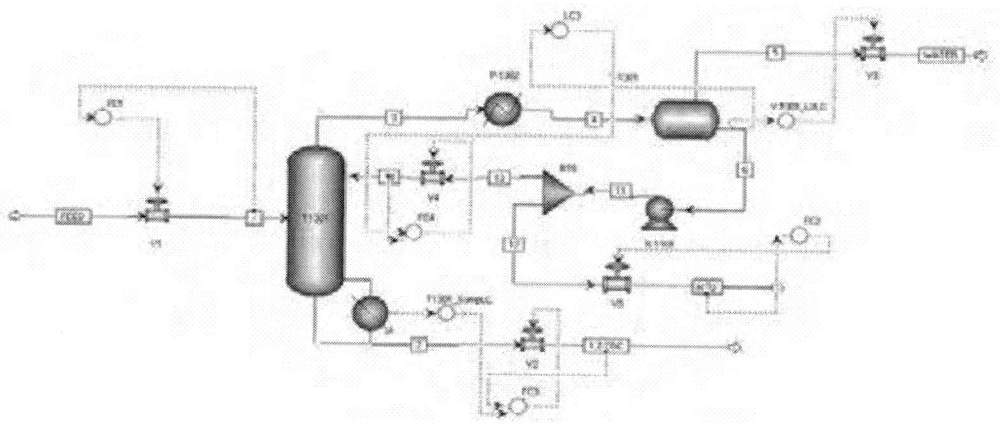

Figure 2 – Flowchart 1.2 EDC column with flow controllers and level

In Aspen Plus DynamicsTM environment tests were performed through disturbances provoked by PRBS components to the input variables of the plant, in order to check the behavior suffered by each output variable to be controlled in the studied process. The input and output variables analyzed in the process were: Input variables: Power Flow - (U1); Flow Withdraw of Light (Distilled) - (U2); Flow EDC 1.2 (Base) in cascade with SumpLC.SP - (U3); Reflux flow cascaded LC3.SP - (U4); Heat Load - (U5). The output variables were: 1,2 EDC composition of the distillate - (xD); CC14 composition of the bottom of the column - (xB); The reflux vessel organic phase level - (nivelorg); Sump level of the column - (nivelsump); The column top pressure - (P_{TOPO}). The PRBS component was initially applied on U1, this component generated a disturbance variable amplitude 5% (random), and 30 minutes at the opening of the valve. Then we developed a block diagram in communication with the Aspen Plus Dynamics™ through the AMSimulation block, contained in sub-library where the information was received directly from the 1.2 EDC plant. In this block signals of the input and output model Aspen Figure 3 are specified.

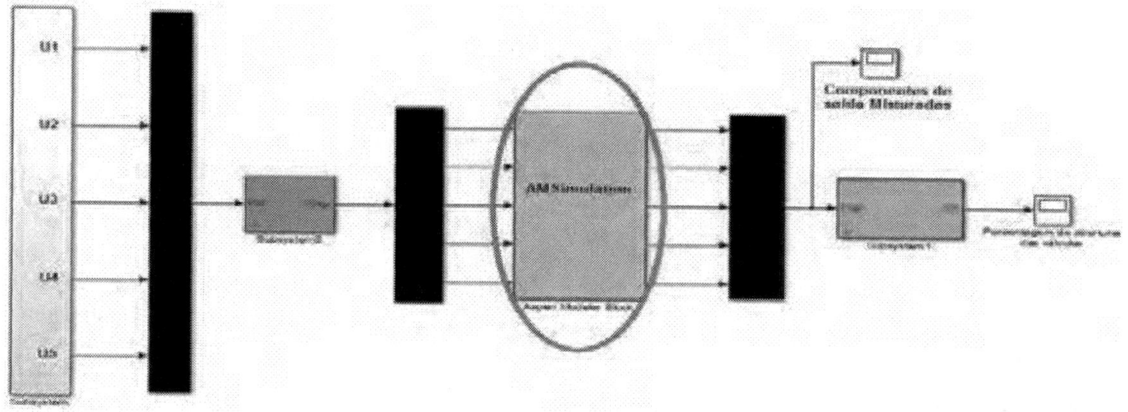

Figure 3 - Block diagram generated in Simulink (Communication Aspen Plus Dynamics x Simulink/Matlab)

After simultaneous and random disturbances applied to the process, it was then possible to obtain the data, to be studied in the output of the process by 'scope' - (Mixed Components Output) Figure 3. The data were saved in two different files. One of the files was used directly for modeling, according to the behavior obtained directly by mixing signals in the plant. The other file was imported into the application containing the ICA, to extract the independent components. Then, the data were modeled in Ident toolbox / matlab for obtaining mathematical models governing the process, and finally their application in a control system considered in this work. Signs of disturbance applied the input variables of the process were: Sine, Sine, Ramp, Sine and Step, respectively, at different frequencies and amplitude of signals. The data obtained in the 'scope' of Mixed Components output, Figure 3, were imported into the application developed in Matlab code and available in (4). This application has in its structure the FastICA algorithm, which is responsible for making the components blended in the independent as possible, thus making it easier and more robust control of the variables in the process. The original signals applied, as well as the mixing signal obtained from the 'scope' of the output Mixed Components are shown in Figures 5 and 6, respectively.

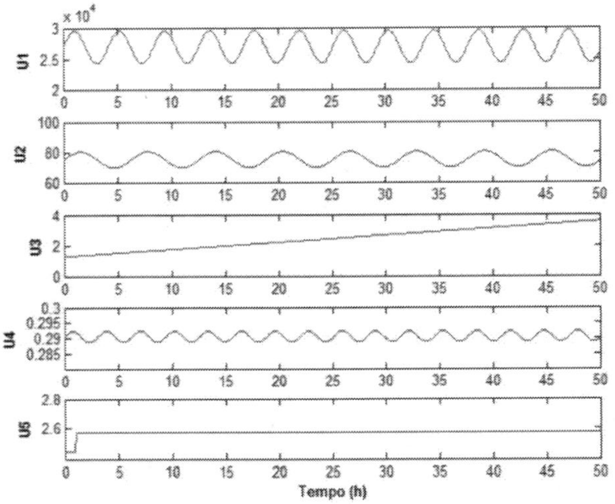

Figure 5 –Originals Signals

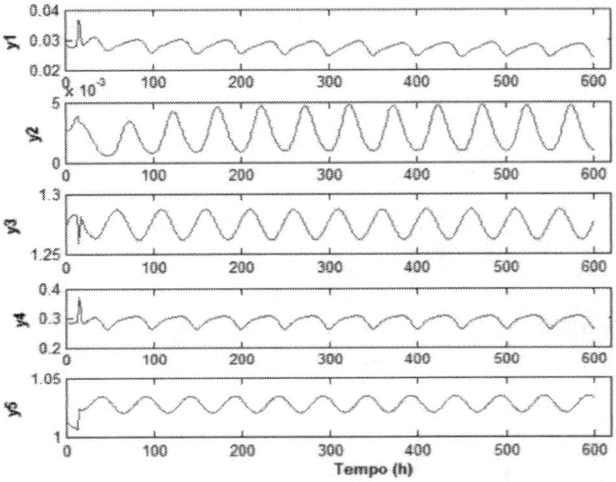

Figure 6 – Mixture Sinals

During the application of the FastICA algorithm, symmetric and deflationary orthogonalization methods were tested to calculate the matrix W. In addition, the nonlinearity function varied. The tested functions were: hyperbolic, Gaussian and cubic tangent. In possession of the data of the output components obtained by communicating directly with Matlab (CASE 1), as well as the data of the independent components after passing through the ICA (CASE 2), the modeling process was initiated. The tool used to identify the models was the System Identification Toolbox / Matlab. Several models were evaluated in order to find those with the best fit to the data corresponding to each 'CV'. The models tested were: Transfer Functions, State Space, Nonlinear Models, Polynomial Models, Models for Correlation. Then, in this paper, two control loops in Simulink / Matlab were proposed:

- Malha de Controle 1 – Baseada no CASO 1.
- Malha de Controle 2 – Baseada no CASO 2.

Both proposed control loops possess the same format, as in Figure 8. The aim of proposing these two fabrics was to be able to make a comparison between the control performed by the conventional method using the PID, and a disengagement control using ICA as well as the PID controller, and thus to verify if indeed the decoupling of variables through the ICA technique favored the control in the process.

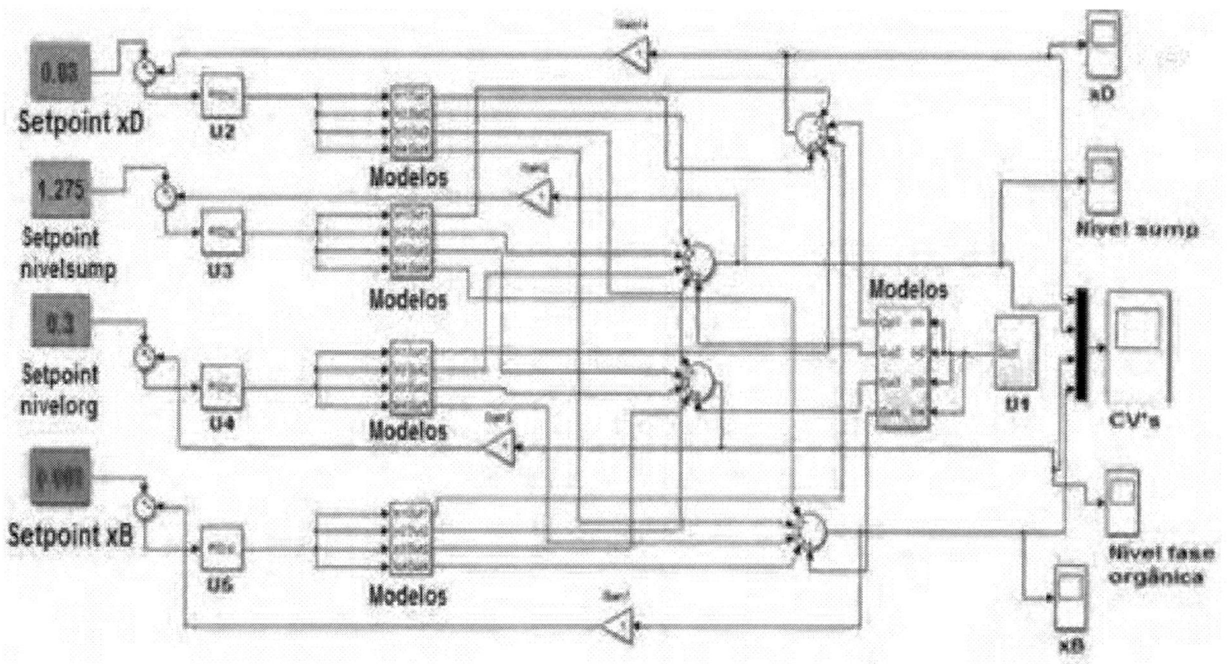

Figure 7. Control Proposal for the 1.2 EDC plant

Results and Discussion

The "U1" flow was chosen as the process disturbance, since this variable may give rise to oscillations at the entrance of the distillation column during the drying process in the purification of 1.2 EDC. After applying the PRBS component on it, it was noticed that the composition of CCl_4 exceeded the value (3000 ppm), a value required for good conversion of 1.2 in the EDC cracking step. Keeping CCl_4 around this reference value, the $CHCl_3$ is automatically maintained below their limits of operation.

Regarding the composition of the distillate 1.2 EDC it is observed that it also suffers significant variations under the influence of the disturbance on U1, and this can reduce its purity in the basic product, which is not interesting to step the purification procedure under study. Hence it was necessary to apply control over xB and xD. Regarding the VM's analyzed, it was possible to conclude, based on the sensitivity presented by each output variable that the best pairing for control is presented in Table 1. The pressure was discarded from the control loop because it does not show any significant change ahead disturbances.

Table 1 – Pairing between VM's and CV's in the control loop

VM	CV
Vazão de Retirada de Leves	1.2EDC composition on topo (xD)
* SumpLC (Setpoint)	Sump Level (sumplevel)
** LC3 (Setpoint)	Vessel organic pahse Level (orglevel)
Carga Térmica	CCl_4 composition at the bottom (xB)

*In cascade with 1.2 EDC output flow
** In cascade with Reflux Flow

Separation of Independent Components

Through the ICA it was possible to obtain the extraction (recovery) of the applied signals on the sources (input variables), through the process output variables (y1, y2, y3, y4 and y5), as shown in Figure 8 to reduce the strong coupling exists between the variables.

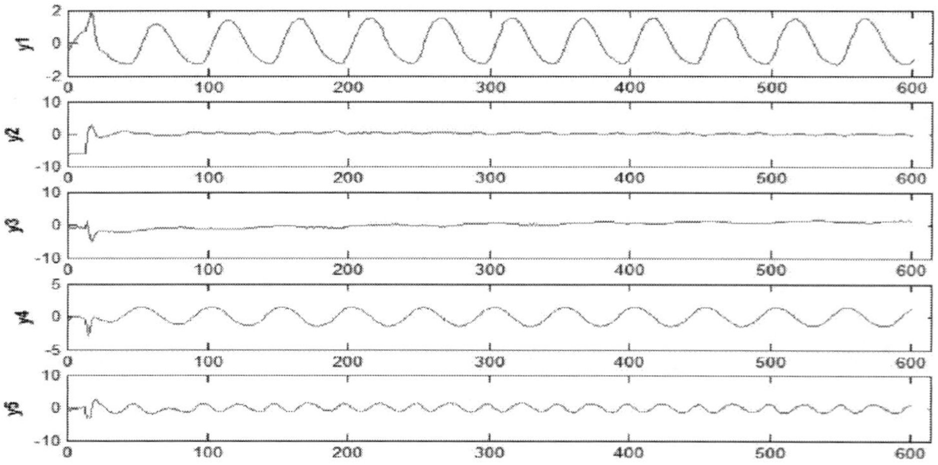

Figure 8 – Components separated by ICA

For maximization of non-gaussianity to negentropy it was used as a way to identify the mutual information between the variables. Alternating the various conditions to try to get the best result in the separation, the method of orthogonalization with best performance was deflation, and by by this method, each component was independently estimated, one by one. As well as non-quadratic function with improved performance in maximizing non-gaussianity is the Gaussian function, Equation 1:

$$G(y) = ye^{\frac{-y^2}{2}} \qquad (1)$$

Models identification

The models found to have the best adjustment were of type ARX (auto-regressive model with exogenous inputs). To these models, a change was made to leave them in the form of transfer function, and then be applied in the control loops, Simulink / Matlab environment. Two criteria were used in selecting the model: the coefficient of the fitted model and the transient response. For Transfer Function models, the analysis of the transient response showed strong instability analyzed on each CV, a factor that directly influences the ability to achieve control in the process.

Control Proposals

The control structure proposed in the work, as observed in Figure 7, was composed of models representing each pair of interaction VM / CV, both for Control Loop 1 as the 2 models corresponding to each mesh are found in (5). In Figures 9 and 10 the xD and the xB are presented when subjected to control by both proposed control loops. Through tests in the 1.2 EDC plant in Dynamics Aspen the importance of controlling the levels of the organic phase and the sump level was verified. In Figures 11 and 12 the behaviors generated for the sump levels and for the organic phase, respectively, are presented, subject to the control meshes 1 and 2.

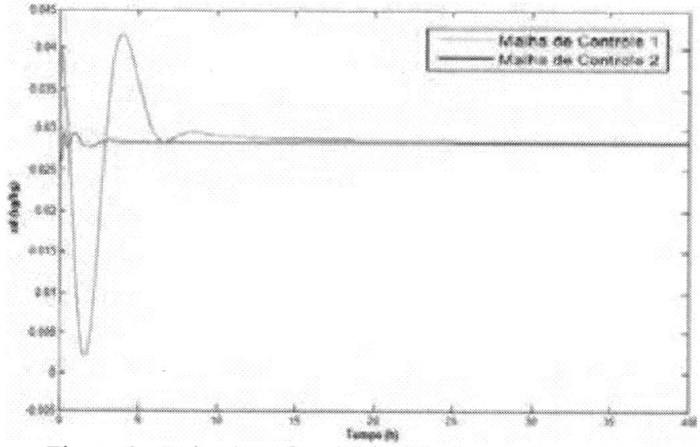

Figure 9 - Behavior of variable xD after tuning controllers in Control Loops 1 and 2

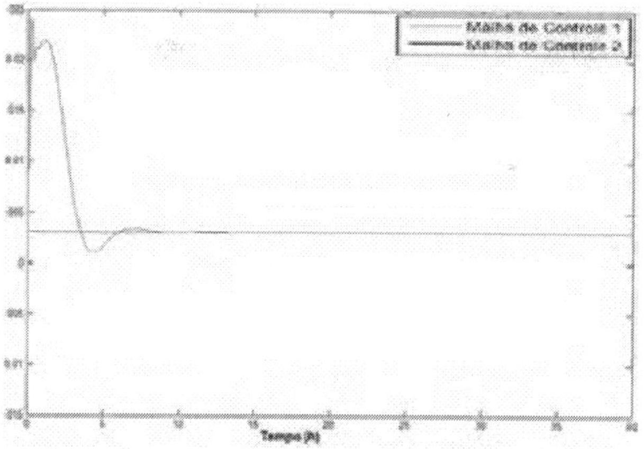

Figure 10 - Behavior of variable xB after tuning controllers in Control Loops 1 and 2

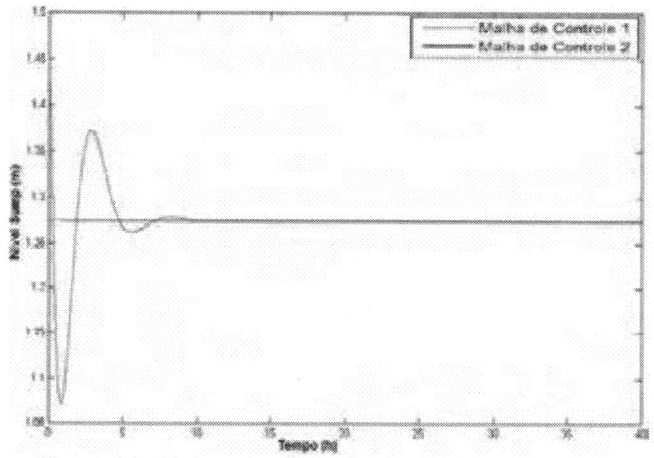

Figure 11 - Behavior of sumplevel variable after tuning controllers in Control Loop 1 and 2

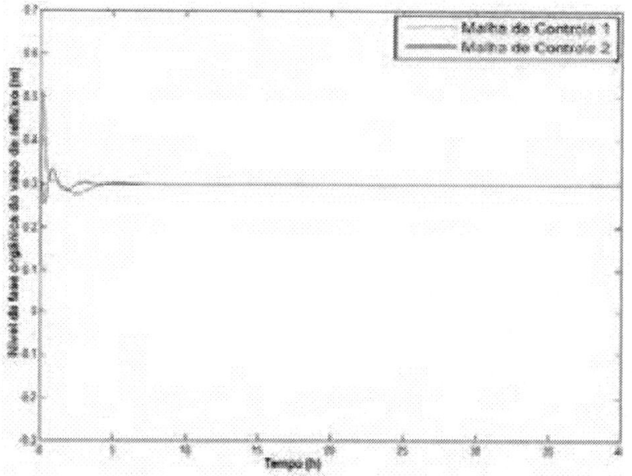

Figure 12 - Behavior of orglevel variable after tuning controllers in Control Loop 1 and 2

In Figure 9, it is seen that xD presents greater instability and requires a longer time to reach the setpoint when subjected to Control Loop 1. As can be seen, applying disturbance in the control loop composition reaches 0.04kg / kg in distillate stream. Thus means a decrease in the purity of 1.2 EDC in the base current, thus reducing process yield. Thus, it is apparent improved results when using the Control Loop 2, wherein the variable stabilization occurs in less than 5 hours of simulation, while the Control Loop 1 takes a time of approximately 20 hours. Analyzing composition xB, Figure 10, is observed in the control loop 1, that xB variable reaches the value 0.024 kg / kg, which corresponds to 24000 ppm, a value well exceeding 3000 ppm, which is the required limit the base current to dry and to purify 1.2 EDC step. Excessive fluctuations in xB, in the product of the base cause variations in the content of this component in the diet of cracking furnaces 1.2 EDC-causing variation in conversion rate and affecting the entire downstream process. The appropriate control composition CCl_4 in feeding of the furnaces reduces the potential for coke formation, which leads to premature stopping of the plant. Regarding the Control Loop 2, it is apparent that it reaches the set point in a short time, and scarcely show instability during the control. In Figure 11, we observe that the level of the sump, under the influence of the Control Loop 1, showed greater fluctuation, as it took a longer time, around 10 hours to reach the setpoint. While in Control Loop 2, the level of the sump hardly showed instability, and took nearly 1 hour time to reach the benchmark. Therefore, a special attention to the sump level should be given, because if there is overflow that level, there will be flooding into the column, and, if it occurs emptying the sump, the pump that allows the movement of the flow through the column burn. Consequently, cause the sudden stop of the plant. In Figure 12, as can be seen, the level of the organic phase under control through the Control Loop 1 presents oscillation reaching 0.7 m, much higher than the value indicated for the setpoint, which is 0.3 m. It is important to give the attention to this variable because, with exceeding the limit specified for the setpoint level, it can cause loss of organochlorines by the output current of the water on top. In the control loop 2, the instability and the time to reach the setpoint were obtained was lower than the control loop 1, avoiding any risk of passage of organochlorine to output the aqueous phase, in addition to achieving the control variable at a time.

Conclusion

This work presented the use of a new azeotropic distillation columns of high purity analysis technique. The initial aim was to separate the different types of random signals acting on the sole of 1.2 EDC, using this ICA methodology. The choice of technique, especially for cases considered complex, was due to the need to reduce the strong interaction between the variables that make up the process and reduce the convergence time in control projects. The application of ICA in 1.2 EDC column allowed efficient extraction of the original signals applied to the input variables of the process, obtained through the output variables. In the control design shown, it was found that the models generated by the data obtained by the extraction of ICA showed more satisfactory results in the control loop than the models used for the direct identification of mixing signals captured at the output of the plant. The technique allowed to keep the top and bottom compositions (xD and xB), levels of organic phase reflux vessel and sump in their setpoints in a shorter simulation time and reduces the instability facing the same process disturbances. Thus it has been found so great benefit that the use of ICA provides to the distillation columns with such specification. Since it is possible to obtain the best results for the control of the compositions of 1.2 EDC and CCl_4 top and bottom of the column, allowing them to remain within the specifications required by the process, 99.5% purity and 3000 ppm (desirable), respectively.

References

1. Delgado, R. D.; Hernández, S.; Muñoz, F. O.; Hernández, J. G.; Montoya, J. C. "From Simulation Studies to Experimental Tests in a Reactive Dividing Wall Distillation Column". *Chemical Engineering Research and Design*, v. 90, p. 855-862, 2012.
2. MAZOCO, B. M., "Identificação e desacoplamento de malhas e sintonia de controladores de uma planta de queima de Usina de Pelotização", Vitória – ES, 2012.
3. Vasconcelos, A. L. S de, Controle de uma coluna de destilação Não Convencional – Um estudo de caso industrial – Dissertação de Mestrado – Universidade Federal de Campina Grande, Campina Grande, 2006.
4. Hyvarinen, A. FastICA for Matlab. Disponível em: http://research.ics.aalto.fi/ica/fastica/code/dlcode.shtml, 2005. Acesso em 04/01/2014.
5. Carmo, S. K. S., Aplicação da Análise de Componentes Independentes (ICA) para melhoria do Controle de uma Coluna de Destilação de Alta Pureza. Relatório interno de pesquisa, UFCG, Departamento de Engenharia Química, 2014.

APPLICATION OF NONLINEAR ANALYSIS TECHNIQUES (CHAOS) ON SIEVE TRAY MULTIPHASE AND TURBULENT FLOWS AS A MODEL VALIDATION INSTRUMENT

Marcelo Muniz, Vinicyus R. Wiggers, Rodrigo K. Decker, Dirceu Noriler, Henry F. Meier, Regional University of Blumenau, Blumenau, SC, Brazil.

Abstract

An analysis of the turbulent and multiphase flow in a sieve trays column was made comparing results obtained by physical experimentation and numerical simulations, using deterministic chaos properties as tools for CFD models validation. A common variable from this kind of process was evaluated, such as total tray pressure drop. A laboratory bench for the physical experimentation, containing an acrylic column with three sieve trays was used. The experimental design is composed by nine operational conditions, varying the gas superficial velocity (0.36, 0.42 and 0.50 m/s) and the liquid flow rate (20.0, 25.0 and 30.0 L/min). A commercial code was used for the numerical simulations, using a single tray stage domain operating in the same conditions from the physical experiments. The chaos analysis was made from an open source code, extracting estimates for the maximum Kolmogorov-Sinai entropy and the correlation dimension. Results obtained by the numerical simulation are in concordance with physical experimentation data, showing to be a new tool for mathematical models validation. It was also noticed the Kolmogorov-Sinai entropy has lower values on the central operating condition chosen, and increases while changing to the corners operating conditions.

Keywords: Kolmogorov-Sinai, Sieve tray, Chaos, Computational Fluid Dynamics.

Introduction

Application of Computational Fluid Dynamics on distillation columns have been commonly performed by using an Eulerian-Eulerian approach, once the phases interaction is very intense and has no clear interface. Many authors have been studying the flow in sieve trays columns, and proposing mathematical models and correlations for the phases interaction applied to them, for example, Krishna *et al.* [1], Krishna *et al.* [2], Krishna, *et al.* [3], Noriler *et al.* [4], Noriler *et al.* [5]. These studies have shown the flow in the sieve trays having a certain behavior, with a pseudo-steady state and nonlinear characteristics.

Deterministic chaos analysis can find patterns in nonlinear signals when sometimes does not seems to have one [1]. A phase diagram is used for a graphical point of view by plotting the system variables trajectories relating themselves along the time in the diagram. Strange attractors were first found by Edward Lorenz [2] in a three variables (3-D) phase diagram. Takens [3] proposed a method for a reconstruction of the phase diagram using only one variable. Grassberger & Procaccia [4] proposed a method for measuring the strangeness of those attractors, later called by correlation dimension (D_2). This method is based on the analysis of correlation sum (C) along with a hypothetical bounding box (ε) and a dimension (m). Lyapunov exponents measures a trajectory escape from a predicted trajectory and are used for estimating the maximum Kolmogorov-Sinai entropy by the Pesyn's identity [5].

The main objective of this paper is applying the chaos analysis for mathematical models validations in the multiphase and turbulent flow in sieve trays by the Kolmogorov-Sinai entropy and the correlation dimension.

Physical Experimentation

The physical experimentation was developed in a laboratory bench which physically simulates the multiphase flow in sieve trays distillation columns. An air-water system represents respectively the gas and the liquid phases. The experimental facility consists in a circular three stages acrylic column (1). A pump (2) feeds water on the top tray, a fan (3) feeds air in the bottom tray, and the middle tray (1) is designated for data acquisition. The experimental bench is illustrated in Figure 1.

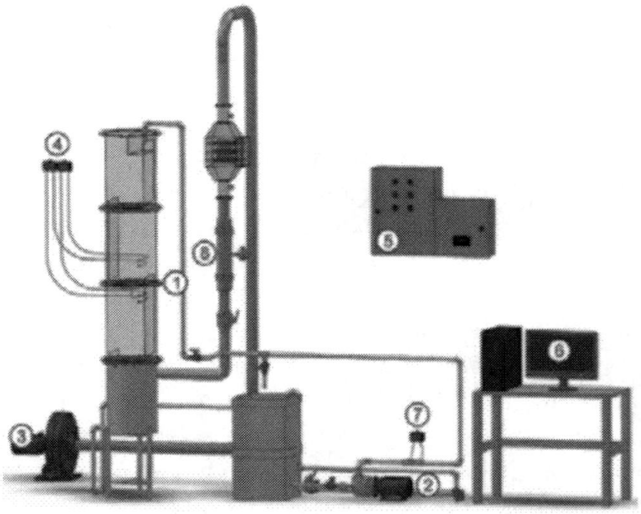

Figure 1. Illustration of the experimental apparatus.

Differential pressure sensors (4) are connected in the middle tray, measuring total tray pressure drop and locals clear liquid heights. Data acquisition and process control are made by a PLC (Programmable Logical Controller) (5) in a frequency of 2 Hz. The communication with the computer (6) is made by a SCADA (Software Control And Data Acquisition) software, in-house made on Indusoft Web Studio 7.1 Educational version. Liquid flow rates are measured in a Venturi tube and a differential manometer connected to it (7), close to the pump. Gas flow rates are measured by a turbine flowmeter (8).

Geometry Characterization

The column diameter is 0.350 m and the tray spacing is 0.480 m. The weir length is 0.234 m and the height is 0.060 m. The liquid entrance section length is 0.234 m, but 0,040 m high. The sieve trays have 68 rectangular holes with a 0.0055 m edge and triangular arrangements between them, as seen in Figure 2.

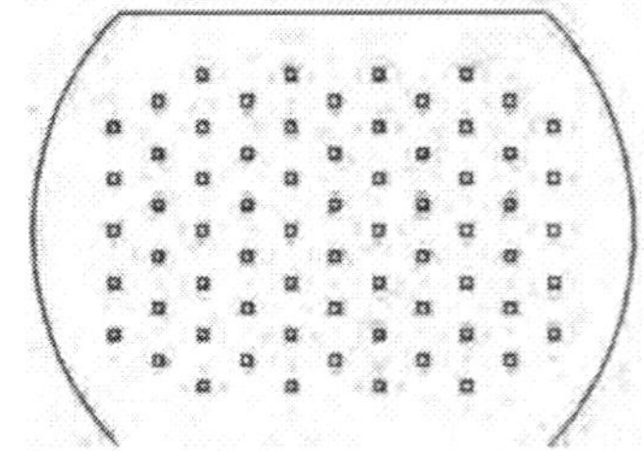

Figure 2. Sieve tray design illustration.

Mathematical Modeling

The Eulerian-Eulerian approach was used for liquid and gas phases, considering them both as interpenetrating fluids and separated momentum conservation equations. Energy and mass transfers were not evaluated, once the experimental system works at isothermal state. The model's conservation equations are as follows:

Continuity Equations

$$\frac{\partial}{\partial t}(f_\alpha \rho_\alpha) + \nabla \cdot (f_\alpha \rho_\alpha \mathbf{v}_\alpha) = 0$$

$$\frac{\partial}{\partial t}(f_\beta \rho_\beta) + \nabla \cdot (f_\beta \rho_\beta \mathbf{v}_\beta) = 0$$

Momentum Equations

$$\frac{\partial}{\partial t}(f_\alpha \rho_\alpha \mathbf{v}_\alpha) + \nabla \cdot (f_\alpha \rho_\alpha \mathbf{v}_\alpha) = -f_\alpha \nabla P_\alpha - \frac{2}{3}\nabla \rho_\alpha k_\alpha + \mathbf{F}_{\alpha\beta}$$
$$+\nabla \cdot \left(f_\alpha \mu_\alpha^{eff}[\nabla \mathbf{v}_\alpha + (\nabla \mathbf{v}_\alpha)^T]\right) - \nabla\left[\nabla \cdot \mathbf{v}_\beta \left(\frac{2}{3}\mu_\beta^{eff}\right)\right] + f_\alpha \rho_\alpha \mathbf{g}$$

$$\frac{\partial}{\partial t}(f_\beta \rho_\beta \mathbf{v}_\beta) + \nabla \cdot (f_\beta \rho_\beta \mathbf{v}_\beta) = -f_\beta \nabla P_\beta - \frac{2}{3}\nabla \rho_\beta k_\beta - \mathbf{F}_{\alpha\beta}$$
$$+\nabla \cdot \left(f_\beta \mu_\beta^{eff}\left[\nabla \mathbf{v}_\beta + (\nabla \mathbf{v}_\beta)^T\right]\right) - \nabla\left[\nabla \cdot \mathbf{v}_\beta \left(\frac{2}{3}\mu_\beta^{eff}\right)\right] + f_\beta \rho_\beta \mathbf{g}$$

$\mathbf{F}_{\alpha\beta}$ represents the phases interaction forces for the momentum.

Constraint Equations

$$f_\alpha + f_\beta = 1$$

$$P_\alpha = P_\beta = P$$

Model Closure

The turbulent viscosity was modeled by k-ε standard model, with the "per-phase" approach, which means two additional conservation equations for each phase.

Phases interaction forces were modeled only by the drag force, based on the coefficient proposed by Krishna *et al.* [2], which applies the Bennett *et al.* [3] correlation for the average gas hold-up in the froth region to estimates the slip velocity, as a strategy for a free bubble diameter model:

$$\mathbf{F}_{\alpha\beta}^{D} = -\frac{3}{4}\frac{f_\beta \rho_\alpha}{d_b} C_D |\mathbf{v}_\beta - \mathbf{v}_\alpha|(\mathbf{v}_\beta - \mathbf{v}_\alpha)$$

$$C_D = \frac{4}{3}\frac{\rho_\alpha - \rho_\beta}{\rho_\alpha} g d_b \frac{1}{V_{slip}{}^2}$$

$$V_{slip} = \frac{V_S}{\overline{f_\beta}}$$

$$\overline{f_\beta} = 1 - \exp\left[-12{,}55\left(V_s\sqrt{\frac{\rho_\beta}{\rho_\alpha - \rho_\beta}}\right)^{0{,}91}\right]$$

Numerical Simulations

The numerical simulations were performed in a commercial software package (FLUENT 14.0 – Ansys Inc. ®). One stage 3-D geometry from the experimental column was chosen (the middle one), which have more established boundary flows than other stages. The geometry and the boundary conditions are presented in Figure 3.

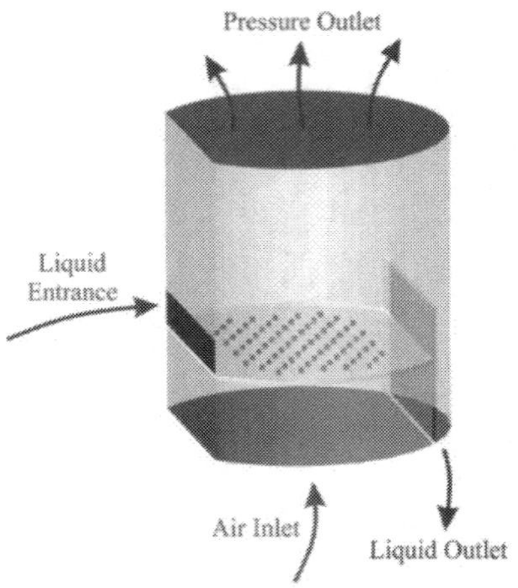

Figure 3. The geometry and the boundary conditions from the chosen stage.

The equations solutions were evaluated in transient regime, once bubbling flow in sieve trays only reaches a pseudo-steady behavior, with non-linear characteristics. A step size of 10^{-3} seconds was chosen to have better convergence solution, around 12 iterations per time step. About the chaos analysis, the literature indicates a minimum sample of 10000 data points to guarantee better confidence on the results obtained. In order to do so, the pressure on several locations was followed during 10 seconds with 1 kHz of acquisition frequency after being at the pseudo-steady state, as in Figure 4. A 10 seconds simulation time costs one week of computing time, distributing the case in 16 processing cores, once the mesh has about 750.000 elements.

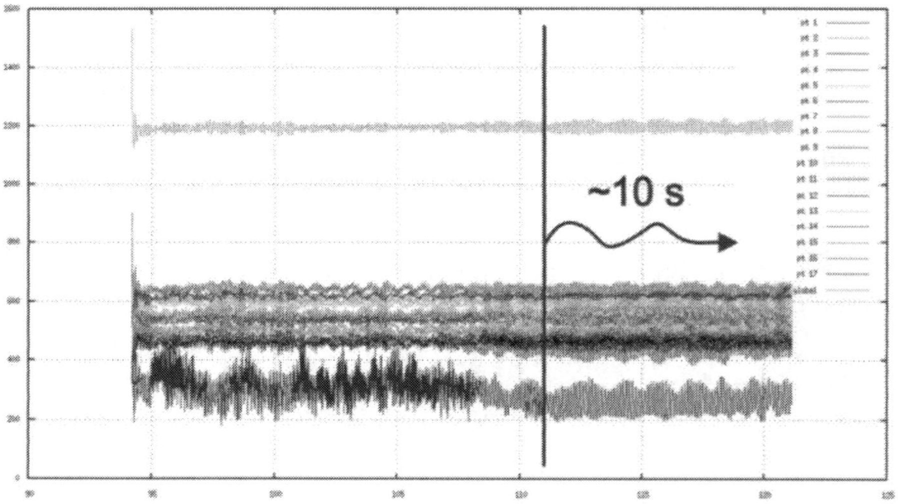

Figure 4. Non-linear characteristics of the pressure time series.

The solution methods chosen in the software were Phase Coupled SIMPLE for the pressure-velocity coupling, and First Order Upwind for all the equations spatial discretization[1]. None of the out standard relaxation factors were used.

Experimental Design

A set of nine operating conditions were proposed varying the gas superficial velocity and the liquid flow rate in three levels each. The experiments were randomly developed and a table with the design's contents is presented as follows:

[1] A Second Order discretization scheme had not a successful solution.

Table 1. Experimental design table.

Operating condition	Vs [m/s]	Ql [L/min]
1	0.36	20.0
2	0.36	25.0
3	0.36	30.0
4	0.43	20.0
5	0.43	25.0
6	0.43	30.0
7	0.50	20.0
8	0.50	25.0
9	0.50	30.0

Results and Discussion

Chaos analysis was developed in an open source code [4] using the Grassberger-Procaccia [5] method for estimating the Correlation Dimension (D_2). The maximum value for KS-entropy was estimated by adding all positive exponents encountered in Lyapunov spectrum [6]. A typical result for the numerical data is presented in Figure 5, and for the experimental ones is presented in Figure 6.

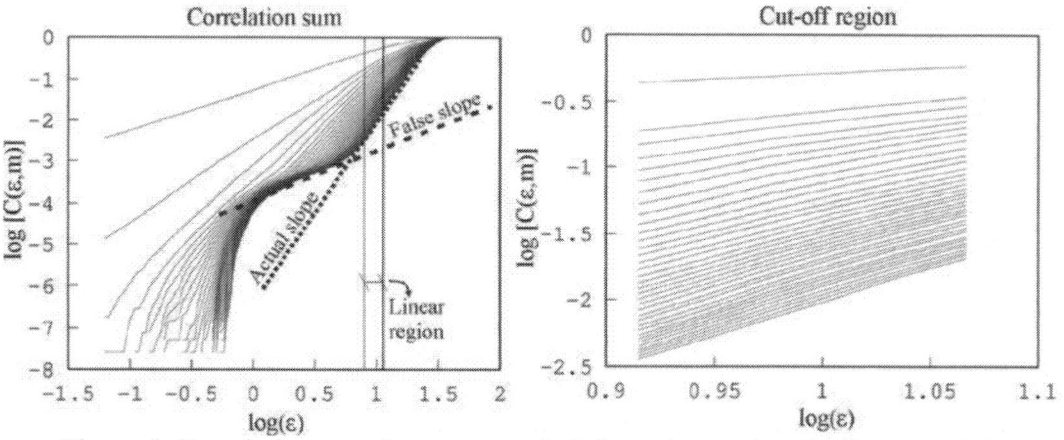

Figure 5. Correlation sum from a numerical data: a) complete; b) linear region.

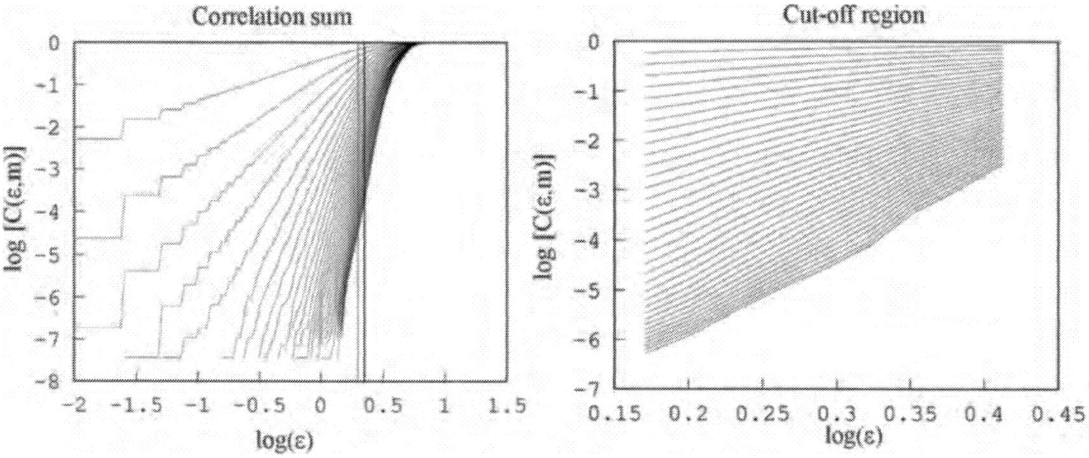

Figure 6. Correlation sum from an experimental data: a) complete; b) linear region.

Continuing to the Grassberger-Procaccia [5] method, a "D_2" relation with the embedding dimension (m) in the linear region is presented in Figure 7.

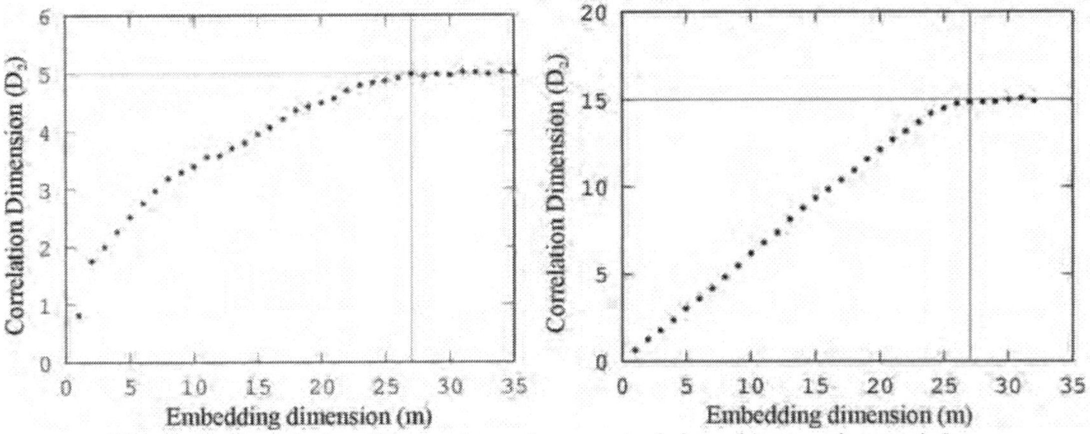

Figure 7. Correlation Dimension: a) numerical data; b) experimental data.

After a certain embedding dimension's value, the correlation dimension's value became almost constant, as is possible to see in Figure 7. So, a mean D_2 value in these points was obtained. Once the Lyapunov spectrum needs the correct embedding dimension, the first value in this same region had been chosen. The overall results of chaos analysis are presented in the Table 2.

Table 2. Results from the chaos analysis.

Operational condition	Vs [m/s]	Ql [L/min]	Num. D_2	Exp. D_2	Num. h_0 [bits/s]	Exp. h_0 [bits/s]
1	0.36	20.0	6.868	16,200 ±1,725	0.417	1,540 ±0,59
2	0.36	25.0	5.901	15,673 ±1,257	0.146	1,790 ±1,04
3	0.36	30.0	6.627	15,098 ±1,740	0.428	1,791 ±0,57
4	0.43	20.0	5.989	14,744 ±0,915	0.413	1,458 ±0,73
5	0.43	25.0	4.426	15,597 ±2,852	0.220	1,276 ±0,59
6	0.43	30.0	4.999	16,364 ±0,428	0.520	0,912 ±0,29
7	0.50	20.0	6.109	15,764 ±0,728	1.397	2,167 ±0,10
8	0.50	25.0	4.620	16,233 ±0,824	0.537	0,848 ±0,32
9	0.50	30.0	4.196	16,184 ±0,392	1.321	2,056 ±0,19

The results obtained in the numerical data clearly seem to be lower than the experimental ones. However, the behavior of maximum Kolmogorov-Sinai entropy along the operational conditions is similar on both cases. This similarity is more intense on two situations: fixing the liquid flow rate to the minor value, and fixing the gas superficial velocity to the major value. These situations are presented in Figure 8.

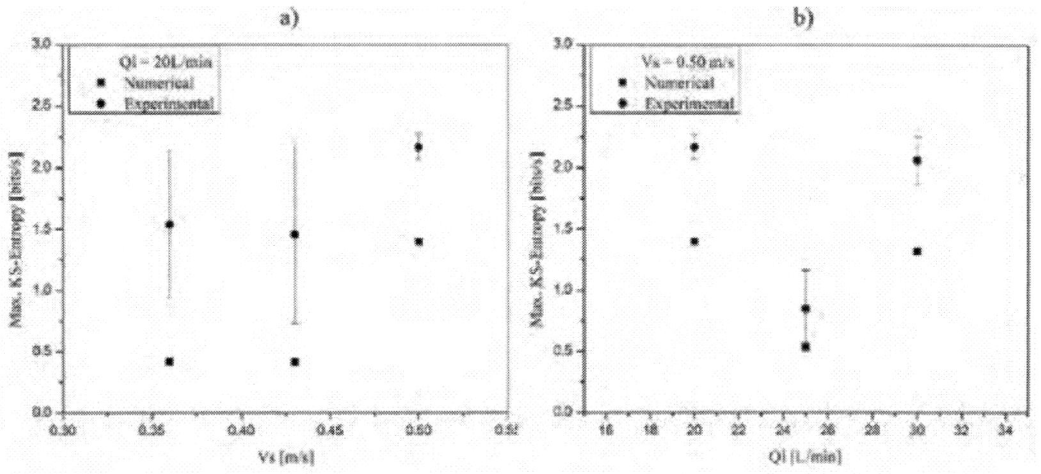

Figure 8. Maximum Kolmogorov-Sinai entropies:
a) minor liquid flow rate; b) major gas superficial velocity.

Even having the great difference between the acquisition frequencies of numerical and experimental data, the maximum entropy obtained has similar behavior. Although the method is not directly influenced by the acquisition frequency, the phenomenon time scales could not be the same, once the difference between them is about three magnitude orders. This could be one explanation for the higher entropy values for the experimental data. Another cause could be the noise during the experimental measuring, which tends to add more entropy on the signal.

Conclusions

The maximum Kolmogorov-Sinai entropy behavior obtained on both cases, numerical and experimental, was very similar. This partially validates the mathematical model and the chaos analysis applied to verification and validation studies in distillation columns. Further experiments must be made with higher frequency acquisition using, for example, an oscilloscope, in order to improve the data quality.

Nomenclature

C	Correlation sum;
C_D	Interface momentum exchange coefficient;
D_2	Correlation dimension;
d_b	Bubble mean diameter;
$\mathbf{F}_{\alpha\beta}^{D}$	Drag force;
f	Volume fraction;
$\overline{f_\beta}$	Mean gas volume fraction;
$\mathbf{g}$	Gravitational acceleration;
h_0	Zero order generalized entropy, maximum Kolmogorov-Sinai entropy;
K	Kolmogorov-Sinai entropy;
k	Turbulent kinetic energy;
m	Embedding dimension;
P	Pressure;
Ql	Liquid flow rate;
t	Time;
$\mathbf{v}$	Velocity vector;
V_S	Gas superficial velocity;
V_{Slip}	Estimated slip velocity between gas and liquid phases;

α Continuous phase (liquid);
β Dispersed phase (gas);
$\beta_{\alpha\beta}$ Fluid-particle interface coefficient;
ε Turbulent kinetic energy dissipation rate, hypothetical bounding box edge;
μ Fluid viscosity;
μ^{eff} Effective fluid viscosity;
ρ Fluid density.

Acknowledgments

The authors acknowledge the Petróleo Brasileiro S/A (PETROBRAS) and the Regional University of Blumenau (FURB) for funding this research. The authors also acknowledge the developers of TISEAN package (Hegger, Kantz and Schreiber).

References

[1] R. Krishna, J. M. v. Baten and M. I. Urseanu, "Three-phase Eulerian simulations of bubble column reactors operating in the churn-turbulent regime: a scale up strategy," *Chemical Engineering Science,* vol. 55, pp. 3275-3286, 2000.

[2] R. Krishna e J. M. v. Baten, "Modelling Sieve Tray Hydraulics using Computational Fluid Dynamics," *Chemical Engineering Science,* vol. 77, pp. 143-151, 2000.

[3] R. e. a. Krishna, "CFD Simulations of Sieve Tray Hydrodynamics," *Chem Eng. Res. Design. Trans IChemE,* vol. 77, pp. 639-646, 1999.

[4] D. Noriler, H. F. Meier, A. A. C. Barros e M. R. Wolf Maciel, "Thermal fluid dynamics analysis of gas-liquid flow on a distillation sieve tray," *Chemical Engineering Journal,* vol. 136, pp. 133-143, 2008.

[5] D. Noriler, H. F. Meier, A. A. C. Barros e M. R. Wolf Maciel, "Prediction of Efficiencies Through Simultaneous Momentum, Mass and Energy Transfer Analyses in a Distillation Sieve Tray by CFD Techniques," *Computer Aided Chemical Engineering,* vol. 27, pp. 1167-1172, 2010.

[6] H. Kantz e T. Schreiber, Nonlinear Time Series Analysis, 2 ed., Cambridge: Cambridge University Press, 2004, p. 369.

[7] E. N. Lorenz, "Deterministic Nonperiodic Flow," *Journal of Atmospheric Sciences,* vol. 20, pp. 130-141, 1963.

[8] F. Takens, "Detecting strange attractors in turbulence," *Springer Lecture Notes in Mathematics,* 1981.

[9] P. Grassberger e I. Procaccia, "Measuring the strangeness of attractors," *Physica D,* vol. 9, pp. 189-205, 1983.

[10] Y. B. Pesin, "Characteristic Lyapunov expoents and smooth ergodic theory," *Russ. Math. Surveys,* vol. 32, p. 55, 1977.

[11] D. L. Bennett, R. Agrawal e P. J. Cook, "New pressure drop correlation for sieve tray distillation columns," *AIChE,* vol. 26, pp. 434-442, 1983.

[12] R. Hegger, H. Kantz e T. Schreiber, *Practical implementation of nonlinear time series methods: The TISEAN package, CHAOS 9, 413,* 1999.

377345 Design and Control of Pressure-Swing Distillation with Partial Heat-Integration

Abstract:

Design and control of pressure-swing distillation with partial heat-integration

Yingzhe Yu, Yamei Sun, Guiming Li, Xiuqin Dong, Minhua Zhang*

Key Laboratory for Green Chemical Technology of Ministry of Education, R&D Center for Petrochemical Technology, Tianjin University, Tianjin 300072, P. R. China

*Corresponding author Tel. & fax: +86-22-27406119, E-mail address: mhzhang@tju.edu.cn

Ethanol and ethyl acetate are separated using partially heat-integrated pressure-swing distillation in this paper. The heat removed from high-pressure column condenser is transferred into the low-pressure column reboiler. An auxiliary reboiler using steam as a heating agent provides the remaining energy. The steady-state simulation and dynamic control is implemented based on Aspen Plus and Aspen Dynamics. Aspen Sensitivity analysis is performed to obtain the vital operating parameters and geometric parameters according to the steady-state simulation results and to choose the suitable temperature control point of distillation column. After the process is designed, three different kinds of control structures are presented and explored in this paper: constant reflux ratio control structure, constant R/F ratio control scheme and improved control structure with QR/F fixed.

For dynamic simulation, feed flow rate disturbance and feed composition disturbance are used to evaluate dynamic performance of each control scheme. Control structure with reflux ratio fixed is quite common industrially and the constant R/F control scheme is proposed by Grassi. However, the dynamic simulation results suggest that, the proposed constant R/F control structure with sensitive plate temperature fixed is unable to maintain the two bottom product at the respect quality. The bottom product concentration deviate their quality specification in the new steady state and more ethyl acetate escape from the low-pressure column. In contrast, the control structure with fixed reflux ratio is more effective. It can handle the feed composition disturbance well. The system spends less time to reach a new steady state, and the bottom product concentration is close to the expected quality. But, the constant reflux ratio control scheme cannot maintain the quality specification when feed flow rate disturbance is added. The problem can be greatly improved by using a feed forward control structure in which the distillation column reboiler heat duty and the feed flow rate are proportional. The QR/F ratio depends on the temperature controller. It can provide good dynamic control for both large feed flow rate and feed composition disturbance for this partial heat-integration pressure-swing distillation. This paper provides guidance to choose appropriate control scheme industrially.

386341 Synthesis of New Semicrystalline Block Co-Polymer Membranes for Artificial Photosynthesis

Abstract:

The dual requirements of product gas separation and high ion conductivity in a solar fuel generator present unique design opportunities for polymer membranes. For this application, the role of the membrane is centered on controlling transport of products and reactants of the reactions responsible for solar fuel generation. New synthetic efforts have focused on improving mechanical properties and decreasing product crossover while retaining the processability of the parent polymeric material. In order to incorporate all of the necessary properties for a solar fuel membrane, our efforts have focused on synthesizing block co-polymer membranes that satisfy various requirements with chemically distinct blocks. We used a combination of ring opening metathesis polymerization (ROMP) and anionic ring opening polymerization (AROP) to synthesize triblock copolymers consisting of a semi-crystalline poly(ethylene) (PE) midblock for low gas crossover and good mechanical properties. We have also explored amine and imidazolium based ion-conducting polyether end-blocks due to their versatile chemical functionality and flexibility. The use of AROP allowed us to precisely control the length of our conductive end blocks. The conductivity and gas permeability of these membranes will be discussed as a function of the mole fraction of the functional end blocks.

381094 The Role of Ion Exchange Membranes and Operating Parameters in Selective Removal of Monovalent or Divalent Ions in Electrodialysis Process

Abstract:

Electrodialysis and electrodialysis reversal (ED/EDR) are two closely-related water purification processes which use an electrical field to propel ions through ion exchange membranes. The most effective use for these technologies has proven to be the production of potable water from brackish water resources; however, the ED/EDR process could also have several benefits in industrial applications, where the process can remove various charged elements. In brackish water resources and industrial water streams, water compositions differ, and the removal of particular monovalent or divalent ions may be more important than the removal of others. For this reason, the ability to selectively remove monovalent or divalent ions would be valuable. This experimental study, which investigates the nature of ion removal and the effects of type ion exchange membrane and operating parameters on the process, was conducted to explore the dominant empirical model on the selective removal of monovalent ions and divalent ions in the laboratory-scale ED process. The ED cell unit used in this set-up was the PCCell ED 64 0 02, a lab-scale electrodialyzer, which has the nominal capacity of 4-8 l/h for dilute and concentrate streams. A complete data acquisition system was used in order to continuously and accurately monitor and record the operating conditions in the experiment. Complete lab-scale experiments, which tested the effects of ion exchange membrane types and different parameters on ion removal, were designed and conducted. The obtained results were used to develop an empirical model for the removal of monovalent and divalent ions in electrodialysis desalination processes using different ion exchange membranes. The developed model can be widely used for full-scale electrodialysis operations as soon as it is verified through pilot-scale experiments, which will be conducted in the next phase of the study at the Brackish Groundwater National Desalination Research Facility (BGNDRF) in Alamogordo, NM, on a pilot-scale EDR system with an influent flow rate capacity of 12 gallons per minute. Brackish feedwaters with different ion compositions and a wide range of initial conductivities, 1700-6500 μS/cm, can be acquired for the experiments from the nearby aquifer in Tularosa Basin.

386407 Structure/Property Relationships in Polymer Membranes for Water Purification and Energy Applications

Abstract:

Providing sustainable supplies of purified water and energy is a critical global challenge for the future, and polymer membranes will play a key role in addressing these clear and pressing global needs for water and energy. Polymer membrane-based processes dominate the desalination market because they are more energy efficient than thermal desalination processes, and polymer membranes are crucial components in several rapidly developing power generation and storage applications that rely on membranes to control rates of water and/or ion transport. Much remains unknown about the influence of polymer structure on even basic intrinsic water and ion transport properties, and these relationships must be developed to design next generation polymer membrane materials. For desalination applications, polymers with simultaneously high water permeability and low salt permeability are desirable in order to prepare selective membranes that can efficiently desalinate water, and a tradeoff relationship between water/salt selectivity and water permeability suggests that attempts to prepare such materials should rely on approaches that do more than simply vary polymer free volume. One strategy is to functionalize hydrocarbon polymers with fixed charge groups, e.g., sulfonate groups, that can ionize upon exposure to water, and this seminar discusses the influence of polymer charge on transport properties. Additionally, in electric potential field-driven applications, the performance of anion exchange membranes (AEMs), containing positively charged functional groups, such as quaternary ammonium moieties, lags behind that of cation exchange membranes, containing tethered negative charges such as sulfonate groups. Improved AEMs are needed to optimize and extend technologies such as electrodialysis and reverse electrodialysis, and processes that rely on thermolytic salts, such as ammonium bicarbonate, will require polymer membranes optimized for those ions. This seminar presents an overview of research aimed at further understanding fundamental structure/property relationships that govern water and ion transport in ion containing polymer films considered for desalination and electric potential field-driven applications that can help address global needs for clean water and energy.

376423 Development of Nanostructured Polyelectrolyte-Based Membranes with Enhanced Permeability and Ion Rejection

Abstract:

Perchlorate ion is a recently found groundwater contaminant which has affected almost 15 million people in United States. Perchlorate-containing salts are commonly used as components of rocket propellants and explosives, which explains the predominance of this ion in defense and military areas. The presence of perchlorate in drinking water causes interference in the normal hormonal activities of the thyroid gland. Removal of this ion can be effectively brought about by commercial reverse osmosis (RO) membranes. However the high pressure requirement of a typical RO process owing to the inherently low water permeabilities of RO membranes leads to high operating costs. Nanofiltration (NF) membranes on the other hand have higher water permeabilities but lower rejections than RO membranes. Our research focusses on the surface modification of existing NF membranes using Layer-by-layer (LbL) assembly in order to fabricate highly perm-selective membranes. LbL is an aqueous based thin film deposition technique which involves the layering of alternately charged polyelectrolytes on a substrate to form nanothin films. The method allows enormous versatility in terms of its fabrication parameters; the tuning of which allows optimization of the membrane performance. The types of polyelectrolytes used, the pH used during their deposition and the number of bilayers deposited are a few of the several tunable parameters available for designing these nanostructured films. These polyelectrolyte-based films when deposited on a NF membrane can serve as nanothin charged barriers to the passage of ions. The most optimized system is expected to have higher permeability than a commercial RO membrane but with equivalent rejection capabilities. All membranes are tested in a cross-flow membrane setup and the perchlorate concentrations are quantified using LC-MS/MS technique.

386464 Investigating Antifouling Properties of Zwitterionic Carboxybetaines from Molecular Dynamics Simulations

Abstract:

Membrane fouling is very complex and is affected by the hydrophobicity, charge and polarity of the foulants and the properties of the membrane surface. A major cause of membrane biofouling is protein (or protein aggregate) adsorption on the membrane and pore surface. Protein deposition on the membrane surface and membrane pores can compromise membrane performance. Designing polymer surfaces that are non-fouling has been a central issue in membrane research. Grafting is frequently used for membrane surface modification to reduce fouling. Hydrophilic polymers are often used for this purpose because of their wettability and biocompatibility. Surface modifications with zwitterionic materials have been found to be effective in resisting protein attachment. The carbon spacer length (CSL) between the cationic and anionic groups influences the hydration of zwitterionic polymers and affects their anti-fouling property. To investigate the effects of CSL on the hydration of zwitterionic polymers, a series of Molecular Dynamics (MD) simulations were conducted for zwitterionic carboxybetaines trimmers with 1-8 and 16 CSL. The atomic charges of the trimmers were derived from quantum chemical calculations with Gaussian09 based on the RESP protocol and the general Amber force field was used in the MD simulations. The atomic charges of the trimmers with CSL varying from 1-8 were directly calculated using Gaussian09. The charges for the trimmer with CSL 16 were based on CSL 8. The atomic charges for the cationic and anionic groups were observed to vary with their separation length. The hydration properties of these zwitterionic polymers were evaluated by determining the average number of hydrogen bonds for each water molecule in their first hydration shell as well as the number of water molecules in the hydration shell and average number of hydrogen bonds between the polymer and water. Our results indicate that as the polymer chain becomes longer, it becomes overall less hydrated. Moreover, besides the separation distance between the positive and negative charges in one polymer chain, the degree of hydration depends also on the overall charge distribution for the other two branches of the trimmers. This has significant implications on the design and fabrication of zwitterionic antifouling polymers.

369212 Adsorbent Screening for Post-Combustion Carbon Capture Based on Optimum VSA Process Performance: A Simple Method Based on Equilibrium Isotherm Properties

Abstract:

New adsorbents are being synthesized in large numbers claiming suitability for post-combustion carbon capture and concentration (CCC). However, no effective method has yet been developed to evaluate their performance in an actual CCC process. Proper screening of an adsorbent for CCC must either be based on extensive process experimentation or process simulation and optimization using realistic inputs for adsorption equilibrium and rate kinetics of fluid transport into the adsorbent pores. Experimental evaluation of adsorbents synthesized is too resource and time intensive to be practically feasible for routine screening. On the other hand, process simulation and optimization is computationally demanding and generally not available to all material researchers.

Since most of the adsorbents for post-combustion capture are equilibrium controlled, their equilibrium isotherm properties strongly influence the process performance. A proper relationship between easily quantifiable isotherm properties on process performance has not been established yet. Performance indicators such as breakthrough times, high loading capacity, etc., for CO_2 have not been proven to be directly correlated to process level performance of the adsorbent.

In the current work, a predictive correlation has been developed for estimating minimum energy penalty for post-combustion CCC yielding 95% purity and 90% recovery for an adsorbent as a function of some easily quantifiable CO_2 and N_2 equilibrium isotherm properties using a 4-step VSA process with light product pressurization (LPP). The 4-step VSA process with LPP has been reported in the literature to be the best cycle for post-combustion capture from dry gas using 13X zeolite. A closer physical insight of this process gives further assurance that its effectiveness should in general apply to adsorbents with favorable CO_2 isotherms. The process simulation is based on a non-isothermal, non-isobaric model, which uses extended dual site Langmuir isotherm to represent binary equilibrium and assumes macropore molecular diffusion controlled transport kinetics. The equilibrium isotherm model has been shown in the literature to reliably capture experimental results for a wide variety of systems, while the transport model is justified for an equilibrium controlled process. Thus, the chosen cycle and features of the simulation model are most appropriate for developing a versatile adsorbent screening tool for CCC application.

A parametric study has been conducted using design of experiments for estimating the effects of the following five CO_2 equilibrium isotherm properties: binary mixture selectivity over N_2, Henry's constant, CO_2 loading, non-linearity and local slope at CO_2 partial pressure in the feed gas on the minimum energy penalty of the VSA process. Each combination of the aforementioned CO_2 isotherm properties, representing a possible adsorbent, has been optimized for the operating process variables using genetic algorithm based optimization. The significance of the independent variables and their interactions were tested using the analysis of variance (ANOVA) with 95% confidence limits (α= 0.05). The correlation has been validated by comparing with the minimum energy predictions from full process optimization for several known and promising hypothetical MOF adsorbents reported in the literature.

It has been found that the optimized process performance of any adsorbent is most sensitive to binary mixture selectivity for CO_2 over N_2. An important non-intuitive observation is that for the same binary selectivity, higher CO_2 loading means increased N_2 loading, which increases energy penalty for the system. The meta-model provides the much needed easy-to-use tool for material researchers for evaluating new adsorbents.

372971 In-Situ Formation of Mesoporous Silica Nanoparticles in Polymer Hollow Fiber Sorbents for CO2/N2 Gas Separation

Abstract:

The concentration of CO_2 in the atmosphere continues to increase at an accelerating rate because of anthropogenic emissions. CO_2 is the largest anthropogenic contributor to global warming [1]. Post-combustion carbon capture and storage (CCS) has been proposed as a method for reducing CO_2 emissions [1,2]. Hollow fiber sorbents are pseudo monolithic materials with the polymer as the 'binder', impregnated with high loadings of selective solid adsorbent particles as 'fillers'. It is desired to optimize the fiber sorbent geometry and operation to achieve a higher surface area for a given bed volume, minimize pressure drop and maximize the mass and heat transfer. We had developed a novel technique for preparing polymeric hollow fiber adsorbents containing class 1 amine (poly(ethyleneimine) (PEI)) adsorbents [3,4]. In this new approach to extend the prior work, polymeric hollow fiber sorbents were formed by using commercial polymer and silica particles were created by an in-situ process. This approach resulted in silica/polymer composite fibers with good amine distribution and accessibility after infusion with PEI, as well as adequate porosity. Moreover, the in-situ formed particles appear to be firmly retained within the fibers and enable rapid mass transfer and adsorption kinetics. Therefore, through this process a low-cost method of fabricating and engineering the amine/silica/polymer hollow fiber sorbents will be described using the new in-situ silica formation/infusion technique and shown to provide a high CO_2 equilibrium capacity from simulated flue gas.

References:

1. B. Smit, J.A. Reimer, C.m. Oldenburg, I.C. Bourg, "Introduction to carbon capture and sequestration", Vol. 1, ICP 2014.
2. C. Azar, K. Lindgren, E. Larson, K. Mollersten, Clim. Change 74 (2006) 47-79.
3. F. Rezaei, R. P. Lively, Y. Labreche, G. Chen, Y. Fan, W. J. Koros, and C. W. Jones, ACS Appl. Mater. Interfaces 5 (2013) 3921-3931.
4. Y. Labreche, R.P. Lively, F. Rezaei, G. Chen, C.W. Jones, W.J. Koros, Chem. Eng. J., 221 (2013)166-175.

374326 Separation of Carbon Dioxide and Methane over Rb- and Cs-ZK-5 Zeolites

Abstract:

Eight-membered ring (8 MR) zeolites hold large potential for industrial CO2 separation such as biogas separations. They offer large selectivity due to the constrained environment for adsorption, especially when large cations are present in the interconnecting windows [1-5]. The relatively small cages and windows of these zeolites increase the interaction strength between the adsorbent and CO_2. At the same time, the diffusion of the slightly larger adsorbates CH4 and N2 through the 8 MR windows can be hindered. As a result, simulations predicted the highest CO2/CH4 selectivities in 8 MR structures among the different types of zeolites. [6] Experimental studies have mainly focused on the RHO, LTA, CHA and KFI structures up to now [7-12].

In the present work, the Rb- and Cs-exchanged ZK-5 zeolites (8 MR KFI type zeolites) were studied for kinetic CO2/CH4 separation. ZK-5 zeolites have the KFI structure, with a slightly higher Si/Al ratio (Si/Al = 3.6). It consists of a three-dimensional network of larger α- cages (11.6 Å in diameter) and smaller γ-cages (6.6 Å x 10.8 Å). The α- and γ-cages are connected through flat eight- membered rings with a diameter of 3.9 Å. A puckered eight-membered ring with a smallest diameter of 3.0 Å connects the γ-cages with each other.

Rb-ZK-5 and Cs-ZK-5 were thoroughly characterized via chemical analysis, argon porosimetry, X-ray diffraction and Rietveld refinements. These refinements showed that a majority of the 8 MR sites were filled with cations. Afterwards, the CO2/CH4 separation potential of both adsorbents was assessed via the measurement of kinetic and equilibrium data (T = 261.15 - 323 K), adsorption and desorption breakthrough measurements at 303 K (P = 1 - 8 bar), and simulations of their performance at lab-scale and in a model pressure swing adsorption (PSA) process. The high occupation of the central 8 MR sites with large cations causes strong diffusional limitations for CH4 on Rb-ZK-5 and Cs-ZK-5. As a result, both zeolites effectively separate CH4 from CO2 with very high selectivities (α = 17). A disadvantage for Cs-

ZK-5 is the occurrence of mass transfer limitations for CO2, yielding lower mass transfer coefficients on Cs-ZK-5 compared to Rb-ZK-5 and a large part of the bed being unused for separation. The global performance of both adsorbents will be compared to a benchmark 13X zeolite.

Acknowledgements

Tom Rémy and Leen Van Tendeloo acknowledge FWO- Vlaanderen for financial support. Joeri F.M. Denayer acknowledges FWO-Vlaanderen for the 1.5.280.11N research grant. Elena Gobechiya and Christine E.A. Kirschhock acknowledge the Belgian Prodex Office and ESA for financial support. Johan A. Martens and Christine E.A. Kirschhock acknowledge the Flemish Government for long-term structural funding, Methusalem.

References

1. J. Shang, G. Li, R. Singh, Q. Gu, K. M. Nairn, T. J. Bastow, N. Medhekar, C. M. Doherty, A. J. Hill, J. Z. Liu and P. A. Webley, J. Am. Chem. Soc., 2012, 134, 19246-19253.

2. T. Remy, S. A. Peter, L. Van Tendeloo, S. Van der Perre, Y. Lorgouilloux, C. E. Kirschhock, G. V. Baron and J. F. Denayer, Langmuir, 2013, 29, 4998-5012.

3. M. M. Lozinska, E. Mangano, J. P. S. Mowat, A. M. Shepherd, R. F. Howe, S. P. Thompson, J. E. Parker, S. Brandani and P. A. Wright, J. Am. Chem. Soc., 2012, 134, 17628-17642.

4. T.-H. Bae, M. R. Hudson, J. A. Mason, W. L. Queen, J. J. Dutton, K. Sumida, K. J. Micklash, S. S. Kaye, C. M. Brown and J. R. Long, Energy Environ. Sci., 2013, 6, 128-138.

5. Q. Liu, T. Pham, M. D. Porosoff and R. F. Lobo, ChemSusChem, 2012, 5, 2237-2242.

6. R. Krishna and J. M. van Baten, J. Membr. Sci., 2010, 360, 323-333.

7. M. Palomino, A. Corma, J. L. Jorda, F. Rey and S. Valencia, Chem.

Commun., 2012, 48, 215-217.

8. J. Zhang, R. Singh and P. A. Webley, Microporous Mesoporous

Mat., 2008, 111, 478-487.

9. F. N. Ridha and P. A. Webley, Sep. Purif. Technol., 2009, 67, 336-

343.

10. J. Yang, Q. Zhao, H. Xu, L. Li, J. Dong and J. Li, Journal of

Chemical & Engineering Data, 2012, 57, 3701-3709.

11. J. Yang, R. Krishna, J. Li and J. Li, Microporous Mesoporous Mat.,

2014, 184, 21-27.

12. M. M. Lozinska, J. P. Mowat, P. A. Wright, S. P. Thompson, J. L.

Jorda, M. Palomino, S. Valencia and F. Rey, Chem. Mater.,

2014.

376979 Evaluation of Polyethylenimine Grafted and Polyallylamine Dendrimer Grafted Mesocellular Silica Foam Promoted By Surfactants for Post-Combustion CO2 Capture

Abstract:

Amino polymer-based adsorbents can significantly reduce the energy penalty for CO_2 capture as compared to amine solutions and offer high CO_2 capture capacities (>2 mmole per g adsorbent). However many amino polymers degrade at low temperatures of 90-110°C, which makes it difficult for desorption.

In this presentation, the technical issues associated with previously reported sorbents such as 1) thermal stability, 2) mechanical stability and 3) resistance to CO_2 diffusion will be addressed. In this study, mesocellular silica foam (MCF) with a pore size range of 20-40 nm and a pore volume 1-2 cm^3/g was synthesized. The control of silica substrate pore sizes is critical to keeping the CO_2 diffusion resistance as small as possible. Adsorbents were synthesized using a double grafting method where anchor agents such as γ-chloropropyltrimethoxysilane help in covalent bond formation between poly-allylamine dendrimer, polyethylenimine and MCF. The thermal stability and CO_2 capture capacities of the sorbent were found to increase by the adding a surfactant. The CO_2 adsorption capacity obtained at 75°C for a surfactant promoted sorbent is 2.3 mmol of CO_2/g adsorbent. The as-synthesized adsorbents with amino polymer (30-50% by weight) and surfactant (5-10% by weight) were evaluated for CO_2 adsorption and desorption. The characterization of the fresh and spent sorbents using BET, SEM, TEM, FTIR, and TGA- will be presented.

An adsorption-desorption study was performed by temperature swing adsorption between 75-85°C using 99.9% CO_2 gas and desorption at 90 °C in 100% N_2 atmosphere using TGA-MS. For desorption, the pressure was reduced from 1 atm to 0.1 – 0.5 atm at 90°C to desorb 99% of the adsorbed CO_2 gas. The equilibrium studies and cyclic performance results of the sorbents for 10 adsorption – desorption cycles will be presented.

381331 MgO-Based Sorbent Promoted with KNO3 for CO2 Capture at Medium Temperatures

Abstract:

The emission control and capture of CO_2 are always being a most challenging environmental issue facing to each country in over the world. The CO2 capture at medium temperature (200-400°C) is one of important area in CCS technology because it can be applied to IGCC effluent gas without cooling step, sorption enhanced reaction or pre-combustion capture. In this study, MgO was prepared by various synthesis methods of MgO (hydrothermal, polyol, aerogel methods) and the sorption capacity was compared by using sulfur compounds in gas mixture. The optimized MgO nanoparticle was used as a base material for CO_2 capture. Then, the developed MgO-based sorbent was promoted by KNO_3 to improve CO_2 capture capacity at medium temperature range. MgO·KNO_3 composites were synthesized using aerogel route and the CO_2 sorption capacity of as-prepared composites was evaluated at the intermediate temperature range of 250-400°C by using a TGA. Effects of MgO/KNO_3 molar ratio and calcination method on CO_2 sorption capacity were compared and their structural and physical properties were investigated by SEM, XRD and N_2 adsorption/desorption isotherm. In addition, the sorption capacity of MgO/KNO_3 composite was compared to other MgO composites promoted by alkali metal salts (K_2CO_3, KOH, $NaNO_3$, Na_2CO_3, Na_2HPO_4 and $LiNO_3$) at the same condition. The sorption rate of CO_2 on MgO composite during initial 30 min was compared at different sorption temperature conditions, using a Fickian model. The cyclic working capacity and stability of MgO composite was evaluated by cyclic test.

385548 CaO Based Sorbents for CO2 Capture – Influence of Metal Ions As Dopants

Abstract:

We are in an era, where environmental concerns have become increasingly prevalent. Increase in the CO_2 concentration widely regarded as a key factor for the global climate pattern shift. Literature reports reveal that installation of currently used amine sorbents in coal gasification plants decreases the plant efficiency by 30%. We have been developing CaO based sorbents for high temperature CO_2 capture for last decade. Our flame synthesized doped Calcium-based novel refractory sorbents have shown excellent performance in CO_2 adsorption. In our previous study we have synthesized Ca/Zr (10:5) sorbents by different synthesis methods. Among the various sorbents, the Ca/Zr sorbent synthesized by sol-gel method exhibits excellent adsorption capacity and remarkable multi-cycle stability until 1200 cycles. The present study mainly aimed at synthesizing various Ca/Zr based sorbents by sol-gel method by varying the ratio between Ca and Zr. Also, in our previous studies the Ca/Al (10:3) sorbent exhibits better adsorption capacity compared to Ca/Zr (10:3) sample. The present also aimed at synthesizing various Ca/Al sorbents by flame spray pyrolysis method by varying the molar ratio between Ca and Al. The sorbents were characterized by BET, XRD, SEM and TEM techniques. CO_2 capture experiments were performed in a TGA at 700 °C. These interesting findings will be discussed in the presentation.

387300 Extending the Concept of Parasitic Energy By Introducing Water Competition

Abstract:

A promising way to significantly reduce the CO_2 emission of power plants is Carbon Capture and Sequestration (CCS). Separating and compressing CO_2, however, impose a large additional energy load on power plants, which in turn is not available for electricity production. The concept of parasitic energy predicts and minimizes the additional energy load on a material-by-material basis and hence, is considered a potential means to evaluate materials for CCS. Besides CO_2 and N_2, water plays a crucial part in CCS, due to its high affinity to bind to the adsorption sites designated for CO_2, which poses a special challenge to the separation process.

On the basis of Grand Canonical Monte Carlo (GCMC) simulations we calculate and implement water isotherms of all-silica zeolites to ensure a more realistic prediction of parasitic energy. Several flue gas compositions are analyzed, including direct air capture, to illustrate the difference in parasitic energy requirement induced by the change in CO_2 concentration.

388986 Assessing the Stability of Amine-Based Hollow Fiber Sorbents to Flue Gas Impurities in Post-Combustion CO2 Capture

Abstract:

For amine-based hollow fiber sorbents to be useful in practical CO_2 capture applications, it is crucial to exhibit a stable performance and a long lifetime in the presence of flue gas impurities. In our earlier investigation,[1] we studied the stability of supported amine sorbents to SO_x and NO_x gas impurities. Here, as a next step, the stability of amine-containing polymer/silica hollow fiber sorbents to SO_2, NO and NO_2 impurities during CO_2 capture is evaluated by conducting cyclic adsorption – desorption experiments in a rapid temperature swing adsorption process. Fiber sorbents containing primary, secondary, and tertiary amine moieties are exposed to different gas mixtures, comprising SO_2/CO_2, NO/CO_2 and NO_2/CO_2 using different SO_x and NO_x concentrations. Furthermore, to assess any synergies in co-adsorption of SO_x or NO_x and CO_2 gases, multi-component adsorption measurements are carried out using a mixture of $SO_2/NO/CO_2$, in addition to the binary-component adsorption measurements listed above. The dynamic cyclic CO_2 capacities of the hollow fiber sorbents are measured accordingly to evaluate any capacity loss over120 cycles due to irreversible binding of SO_2, NO or NO_2 to the amine sites.

Reference

(1) Rezaei, F.; Jones, C. W. *Ind. Eng. Chem. Res.* **2013**, *52*, 12192.

360617 Ion Sorption and Transport in Ion Exchange Membranes

Abstract:

One avenue for improvement of such processes as electrodialysis or reverse-electrodialysis is to design more effective membranes, but this requires knowledge of how membrane structure affects the processes of ion sorption and transport within anion and cation exchange membranes. Despite the long history of literature on such studies and the commercial use of electrodialysis, the current state of understanding of ion sorption and transport in charged polymers is not sufficient to make adequate connections between these fundamental issues and the structure (chemical and physical) of polymer membranes. We have embarked on a long-term program to improve this understanding and have established techniques to characterize both cation and anion exchange membranes by measuring coion and counterion sorption, salt permeation, membrane conductivity, etc. versus external salt solution concentration. This work will eventually involve studies of how the polymer structure, charge density, water content, etc. affect the ion sorption and transport behavior. However, for this presentation, we will focus on a complete characterization of a pair of commercial cation and anion exchange membranes that have similar structures (crosslinked styrenic) except for the nature of the fixed ions attached to the polymer. These membranes have been used to develop our techniques and will serve as benchmarks for future structure-property studies. The presentation will show how the fixed ions affect the sorption and the diffusion of the coions and the counterions of the salt, sodium chloride. By combining ion sorption data with salt permeability and membrane electrical resistance, we deduce the individual diffusion coefficients of sodium ions and of chloride ions within the membrane as a function of the external salt concentration over the range from 0.01 to 1.0 M.

370935 Water Treatment Applications with Integrated Charged NF and Iron Nanoparticle Functionalized Membranes

Abstract:

Nanofiltration (NF) membranes play an important role where selective separations (in contrast to RO) or partial desalinations are needed. NF membranes are used in energy industries and in various water treatment (including virus separations) applications. In various energy area applications it is important to have membranes where one can obtain selective separations of divalent cations (such as, Ca and Mg) from high salt solutions. This allows reduction of osmotic pressure effects and also for reuse of permeate water. On the other hand, functionalization of microfiltration membranes with LbL assembly or metal nanoparticles allows applications where reactive medium may be needed. This presentation will focus on energy related, and groundwater treatment applications. Depending on the types of applications (such as in coal-fired power plant scrubber water reuse) selective partial desalination of mixed salts by NF membranes and trace toxic metal capture (such as, Se) by functionalized MF membranes could be a potential treatment scheme. The special features for membrane processes that make them attractive for industrial applications are their compactness, ease of fabrication, operation, and modular design. We have made highly significant advancements in establishing the effectiveness of nanofiltration membrane systems in terms of salt separation selectivity, and the role of various levels of TDS on water flux and rejection behavior with module systems for low hign water recovery areas. Two types (positively-charged, and negatively-charged) of modules were evaluated. Even at 10 bar applied pressure, water flux as high as 58 LMH (34 gal/ft^2 day) at 25 oC can be obtained. The increase of temperature of feed water to 42 oC showed about 40% increase in water flux (as expected, due to viscosity effect) and insignificant loss of rejection of TDS. Since the water often contains trace toxic metals (such as selenate), pretreatment with functionalized membranes containing iron nanoparticles provides a new approach. Preliminary results starting with 2 ppm Se showed concentration < 5 ppb Se can be achieved with Fe NPs by reduction. The presentation will include: (1) membrane selection approaches, (2) NF results on selectivity and partial desalination, and high water recovery, (3) full-scale PAA functionalization of PVDF membranes and responsive flux behavior, (4) toxic metal and organic removal results. This research has been supported by Southern Company, and by NIH-NIEHS-SRP. Special thanks go to ULTURA Company, Oceanside, CA, for joint collaborative work in developing advanced NF membranes and full-scale functionalized membranes.

Energy-Efficient Reverse Osmosis Desalination Process

William B. Krantz[a,b], Tzyy Haur Chong[b,c], Siew-Leng Loo[b,c]

[a] Department of Chemical and Biological Engineering, University of Colorado
[b] Singapore Membrane Technology Center (SMTC), Nanyang Environment and Water Research Institute (NEWRI), Nanyang Technological University
[c] School of Civil and Environmental Engineering, NTU

Annual Meeting of the AIChE
Atlanta, Georgia
November 18, 2014

1

Professor Winston Ho

Distinguished Professor
Chemical & Biomolecular Engineering
Ohio State University

- National Academy of Engineering
- Gerhold Award of the AIChE Separations Division
- AIChE Institute Award for Excellence in Industrial Gases Technology
- Industrial Research 100 Award
- Chairman, AIChE Separations Division
- AIChE Meeting Program Chair in 2000 and 2009
- Inventor/Co-inventor on 98 Patents & Patent Applications!

2

Overview

- Advance the Energy-Efficient Reverse Osmosis (EERO) process that couples single-stage reverse osmosis (SSRO) with a countercurrent membrane cascade with recycle (CMCR)
- Compare recovery, osmotic pressure differential (OPD) and net specific energy consumption (SEC_{net}) with SSRO
- Consider novel use of highly concentrated retentate product stream from EERO

3

Single-Stage Reverse Osmosis (SSRO)

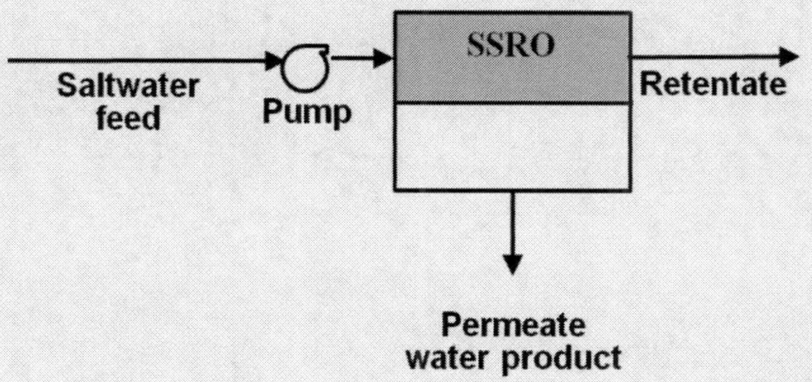

35 g/L feed and 0.35 g/L water product
Pump efficiency: 85%
ERD efficiency: 90%

For 65% recovery, OPD = 79.3 bar, SEC_{net} = 2.92 kWh/m^3
For 75% recovery, OPD = 111 bar, SEC_{net} = 3.91 kWh/m^3

4

Two SSRO Stages in Series

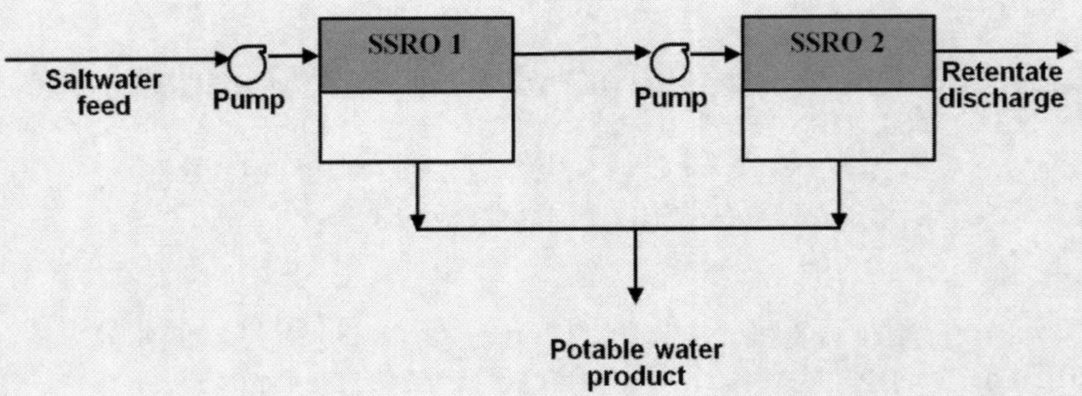

35 g/L feed and 0.35 g/L water product
Pump efficiency: 85%
ERD efficiency: 90%

For 65% recovery, OPD = 79.3 bar, SEC_{net} = 2.32 kWh/m^3
For 75% recovery, OPD = 111 bar, SEC_{net} = 2.70 kWh/m^3

5

Countercurrent Membrane Cascade with Recycle

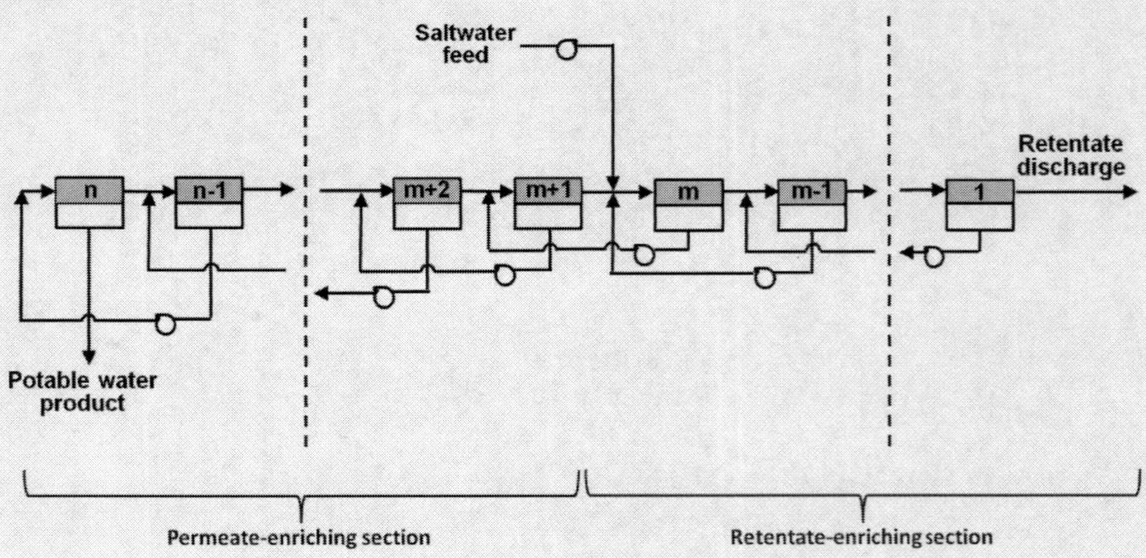

A countercurrent membrane cascade with permeate recycle (CMCR) can increase the recovery and reduce the OPD; however, this increases the SEC owing to re-pressurizing the permeate recycle.

6

1-2 or 2-2 EERO Process

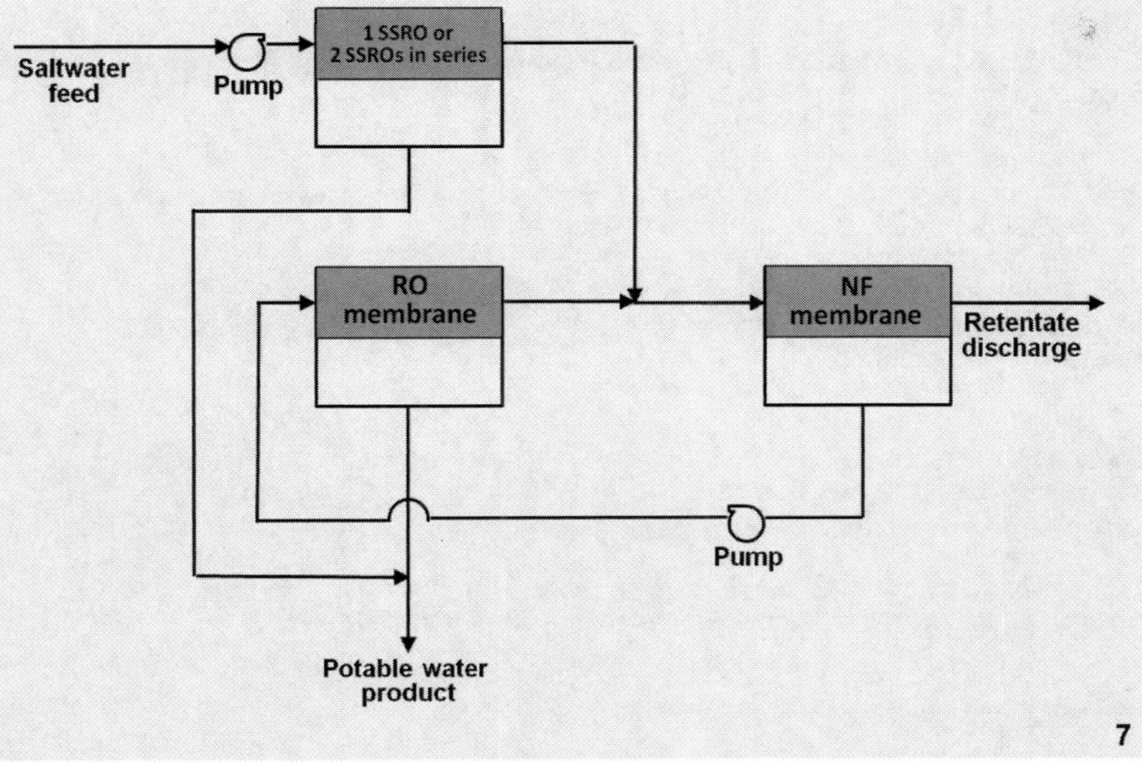

1-3 or 2-3 EERO Process

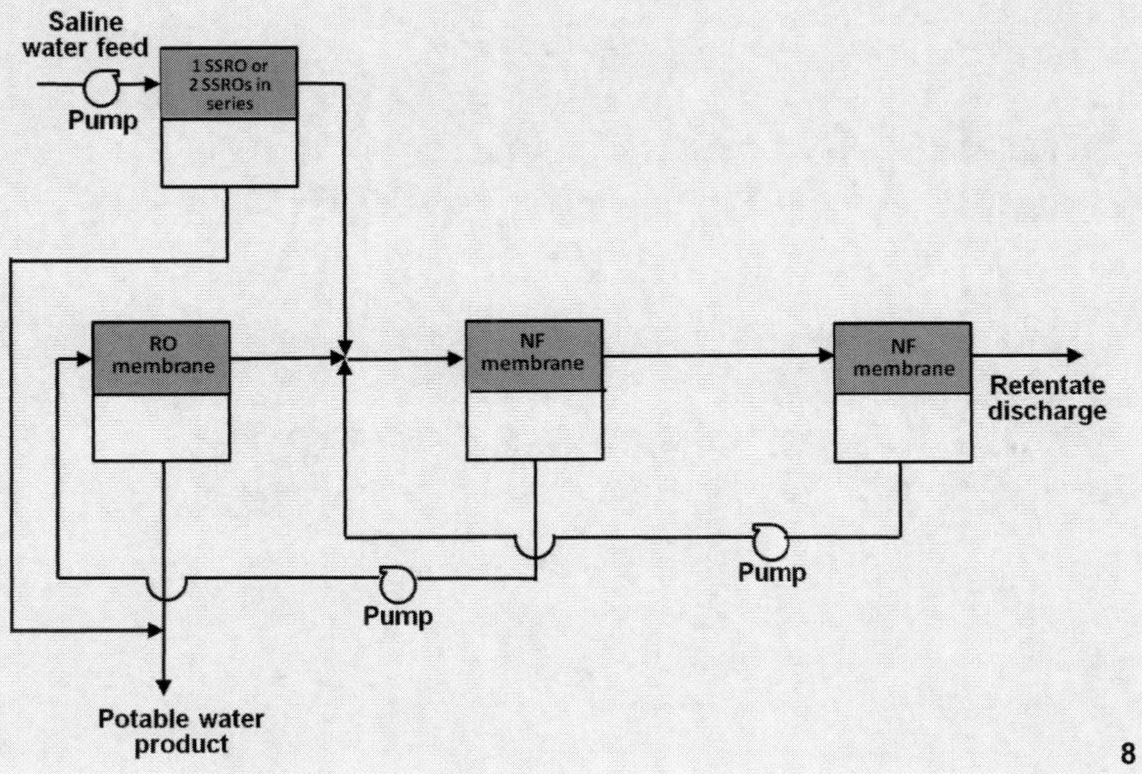

Features of EERO Process

- Capitalizes on performance of single stage RO by using its brine as the feed to a countercurrent membrane cascade
- Reduces osmotic pressure by employing countercurrent retentate and permeate flow in the membrane cascade
- Reduces osmotic pressure by employing permeate recycling in the membrane cascade
- Reduces osmotic pressure by employing one or more nanofiltration stages in membrane cascade that provide retentate self-recycling
- Operates all stages at same osmotic pressure that avoids interstage pumping on retentate side and entropy-of-mixing effects

9

Modeling of EERO Process

- Modeling involves component and overall mass balances on SSRO stage, CMCR stages, and at mixing point
- Resulting set of coupled algebraic equations can be solved analytically

10

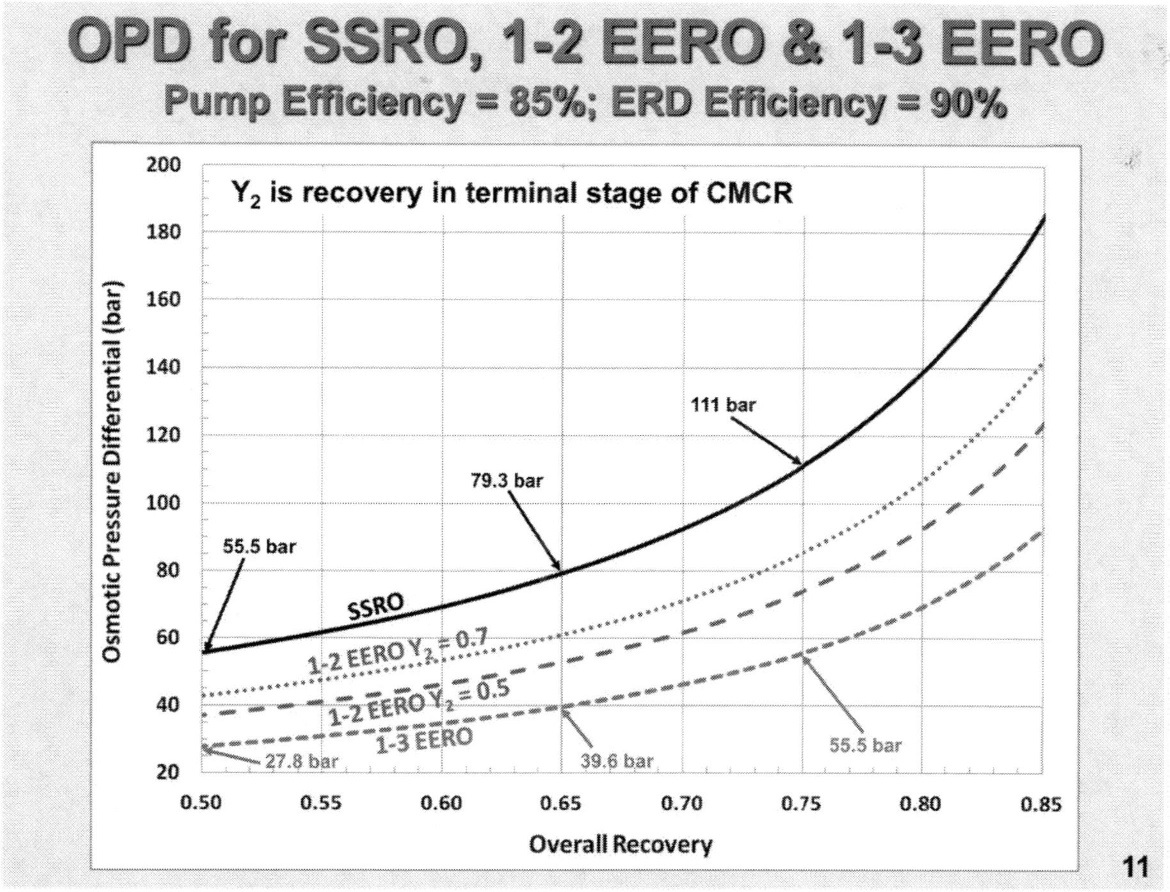
OPD for SSRO, 1-2 EERO & 1-3 EERO
Pump Efficiency = 85%; ERD Efficiency = 90%
Y₂ is recovery in terminal stage of CMCR
Osmotic Pressure Differential (bar)
200
180
160
140
120
100
80
60
40
20
0.50
0.55
0.60
0.65
0.70
0.75
0.80
0.85
Overall Recovery
111 bar
79.3 bar
55.5 bar
SSRO
1-2 EERO Y₂ = 0.7
1-2 EERO Y₂ = 0.5
1-3 EERO
27.8 bar
39.6 bar
55.5 bar
11

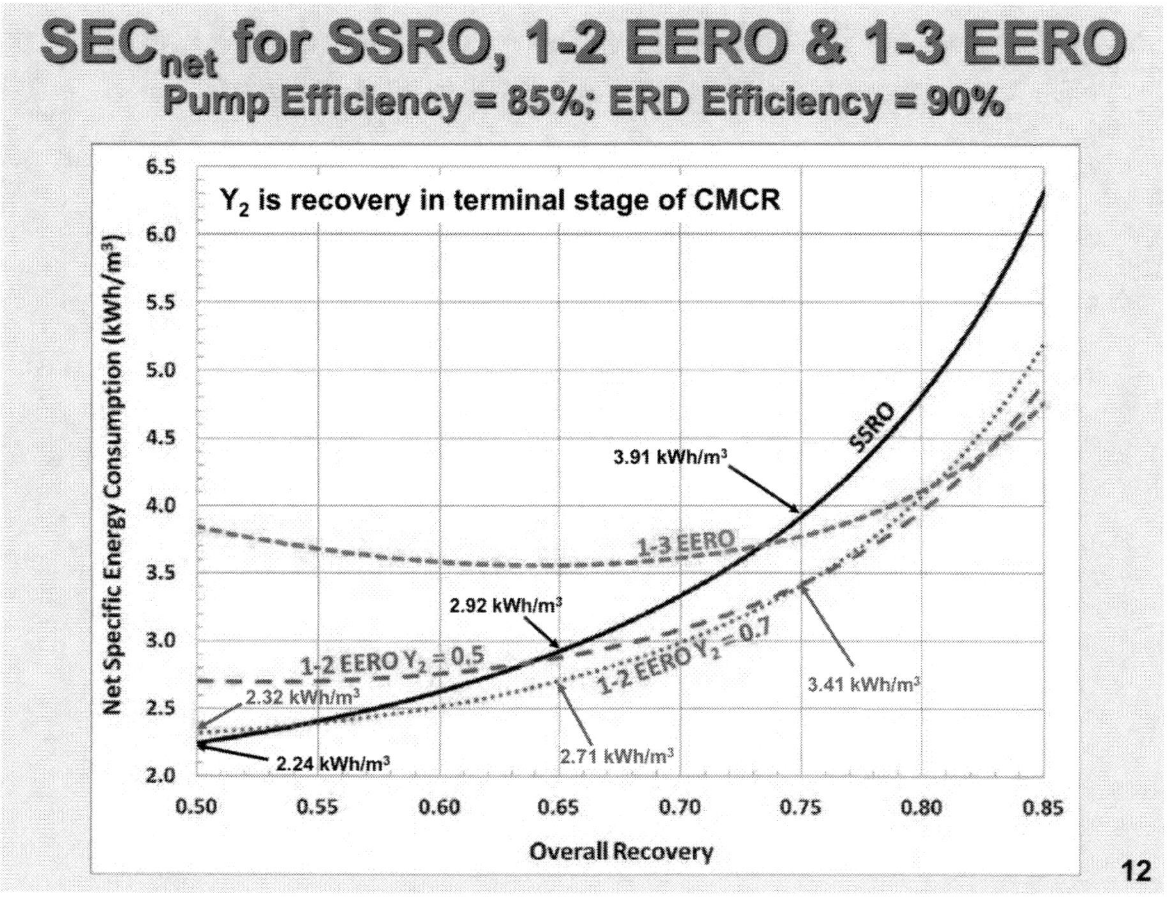
SECnet for SSRO, 1-2 EERO & 1-3 EERO
Pump Efficiency = 85%; ERD Efficiency = 90%
Y₂ is recovery in terminal stage of CMCR
Net Specific Energy Consumption (kWh/m³)
6.5
6.0
5.5
5.0
4.5
4.0
3.5
3.0
2.5
2.0
0.50
0.55
0.60
0.65
0.70
0.75
0.80
0.85
Overall Recovery
SSRO
3.91 kWh/m³
1-3 EERO
2.92 kWh/m³
1-2 EERO Y₂ = 0.5
1-2 EERO Y₂ = 0.7
2.32 kWh/m³
2.24 kWh/m³
2.71 kWh/m³
3.41 kWh/m³
12

OPD for 1-SSRO, 2-SSROs, 2-2 EERO & 2-3 EERO
Pump Efficiency = 85%; ERD Efficiency = 90%

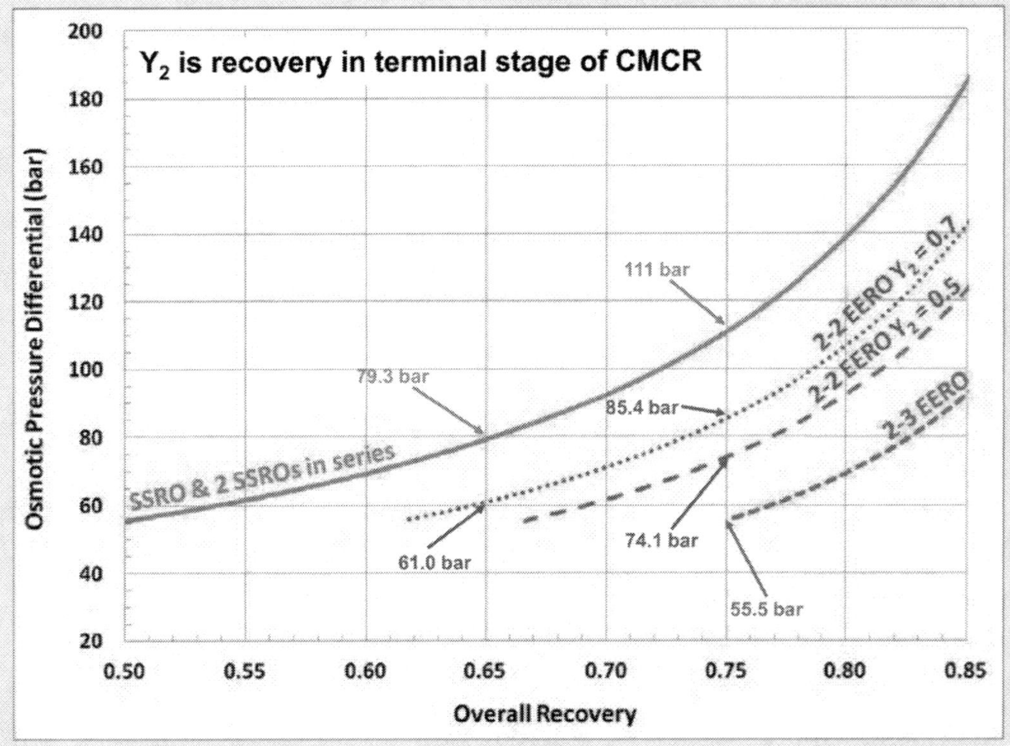

13

SEC_{net} for 1-SSRO, 2-SSROs, 2-2 EERO & 2-3 EERO
Pump Efficiency = 85%; ERD Efficiency = 90%

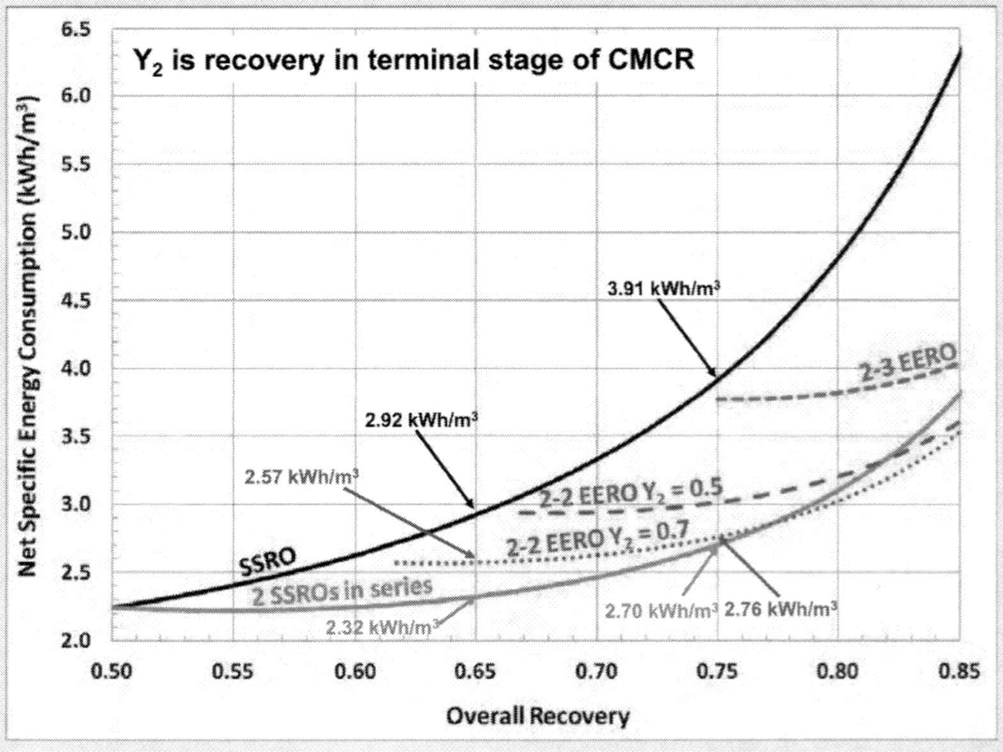

14

SUMMARY OF PERFORMANCE METRICS

35 g/L feed and 0.35 g/L product
Pump efficiency: 85%; ERD efficiency: 90%

Process	65% Recovery		75% Recovery	
	OPD (bar)	SEC_{net}(kWh/m^3)	OPD (bar)	SEC_{net}(kWh/m^3)
1-SSRO	79.3	2.92	111	3.91
2-SSRO	79.3	2.32	111	2.70
1-2 EERO $Y_2 = 0.5$	52.8	2.88	74.1	3.42
1-2 EERO $Y_2 = 0.7$	61.0	2.71	85.4	3.41
2-2 EERO $Y_2 = 0.5$			74.1	3.02
2-2 EERO $Y_2 = 0.7$	61.0	2.57	85.4	2.76
1-3 EERO	39.6	3.56	55.5	3.77
2-3 EERO			55.5	3.77

15

EERO-PRO Hybrid Process

- Pressure-Retarded Osmosis (PRO) process can harvest the osmotic potential energy in the brine from desalination
- PRO must an energy density of at least 10 W/m^2 to be economically competitive
- Typical seawater having a concentration of 35 g/L when used as the draw solution in PRO can achieve 4 W/m^2
- The maximum theoretical energy density is proportional to the square of the OPD
- The EERO process when operated at 75% recovery produces a brine with a concentration of 140 g/L
- A coupled EERO-PRO hybrid process operated at 75% recovery has a theoretical energy density of 64 W/m^2

16

Summary

- The EERO process can achieve higher overall water recoveries at a significantly reduced operating pressure and at a competitive specific energy consumption

- The RO stage in the CMCR can be run at a higher water recovery owing to the foulants removed in the NF stage

- The lower operating pressure in the EERO process will reduce the capital construction and maintenance costs

- A hybrid EERO-PRO process will have an energy density well above the economic threshold for harvesting the osmotic potential energy of the concentrated brine

17

Acknowledgments

- This research is supported by Project 1301-IRIS-41 of the Environment and Water Technologies Strategic Research Program of the Singapore National Research Foundation administered by the Environment and Water Industry Program Office of the Public Utilities Board
- Professor Wang Rong, Director of SMTC, for her encouragement in this research and Professor Tony Fane, Director Mentor of SMTC, for helping the speaker become more knowledgeable in membrane process technology
- Professor Wun Jern Ng, Executive Director of Nanyang Environment and Water Research Institute, for his support of SMTC
- Professor Winston Ho for providing the occasion to have these technical sessions in his honor!

SMTC

Member of the NEWRI Ecosystem

18

Thank You!
19

377986 Permethyl Cobaltocenium (Cp*2Co+) As an Ultra-Stable Cation for Polymer Hydroxide Exchange Membranes

Abstract:

Hydroxide (OH^-) exchange membranes (HEMs) are important polymer electrolytes enabling the use of affordable and earth-abundant catalysts and membranes for electrochemical energy conversion devices such as HEM fuel cells, HEM electrolyzers, and HEM solar hydrogen generators. Many HEM cations exist featuring desirable properties but new cations are still needed to increase chemical stability at elevated temperatures. Here we introduce the permethyl cobaltocenium [$(C_5Me_5)_2Co(III)^+$ or $Cp^*{}_2Co^+$] as an ultra-stable cation for HEMs. Compared with the parent cobaltocenium [$(C_5H_5)_2Co(III)^+$ or Cp_2Co^+], $Cp^*{}_2Co^+$ has substantially higher stability and basicity. We also show that $Cp^*{}_2Co^+$-based HEMs possess very high thermal and alkaline stability, which are useful in designing more durable HEM electrochemical devices.

382248 Continuous Fixed-Bed Hydrosilylation Process Using Heterogeneous Platinum Catalysts

Abstract:

My doctorial research under the supervision of Dr. Winston Ho at The Ohio State University largely contributed to the areas of polymeric gas-separation membranes for fuel processing and proton-exchange membranes for fuel cells [1]. Joining Momentive Performance Materials in 2008, I have been very pleased to be able to contribute to the area of precious metal recovery and recycling using the *in-situ* fixed-bed adsorption technology [2] or the heterogeneous precious metal catalysis technology [3]. My work in separations over the past 10 years has greatly benefited the academic research community, as well as the chemical and organosilicon industries. As part of the separations division for AIChE, I will present the recent process development work at Momentive Performance Materials, focusing on the manufacturing of platinum- and color-free organosilicon products using heterogeneous platinum catalysts [3], particularly in a fixed-bed continuous process, and I will present the potential significant industrial benefits achievable from this innovation.

Currently, Speier's chloroplatinic acid and Karstedt's platinum-vinylsiloxane complex are commonly used for preparation and commercial manufacturing of organosilicon compounds. These platinum compounds are soluble in the organosilicon streams (homogeneous platinum catalysts), which have resulted in tremendous precious metal loss and an undesirable high color of un-distilled organosilicon products. Faced with resource limitations, rising costs and increased demand of precious metals, a significant business effort has been ongoing to recover platinum from hydrosilylation streams within the organosilicon industry.

In this session, Siloxane-A will be used as a model organosilicon product to demonstrate the feasibility of manufacturing this material using a heterogeneous catalysis process, particularly in a continuous fixed-bed process. Data has shown that in a fixed-bed heterogeneous catalysis process, higher Siloxane-A product conversion was observed with (a) increased temperature of the reactor column; (b) increased excess amount of one reactant; and (c) reduced feed rate of reactant mixture (increased residence time of hydrosilylation). These observations are consistent with reaction kinetics of hydrosilylation. With a reactor temperature of 140 °C at 10 mol-% excess of one reactant and a space time (ratio of the empty column space to the reactant mixture flow rate) of 10 minutes, a hydrosilylation conversion of 98 percent was achieved, which can be considered a reaction completion. This is equivalent to a high space time yield (mass of product produced during a certain period of time with a certain amount of catalyst packing) of over 10 grams of product / (hour · gram of solid catalyst).

In terms of the manufacturing production rate, this process would potentially allow very high Siloxane-A product commercial output with a much intensified commercial-scale fixed-bed reactor, in comparison to the traditional batch-wise reactor. In addition, the Siloxane-A from this new process was a Pt- and color-free premier organosilicon product with much improved quality when compared to traditional Siloxane-A produced by homogeneous catalysts. At the time of the catalyst column changeover, such packed heterogeneous platinum catalysts could be easily recovered as they were retained stationary in the fixed-bed reactor, and these used solid catalysts could be processed in a manner (e.g., incineration) in which platinum could be recovered as an elemental metal.

In this honorary session, I would like to thank Dr. Winston Ho for his support, enthusiasm, encouragement and guidance during my doctorial research at The Ohio State University. I would also like to thank Winston Ho for his influence and inspiration throughout the past 11 years, which has led to the success of my careers in both academia as well as the chemical and organosilicon industries.

[1] H. Bai and W.S.W. Ho, "Recent developments in fuel-processing and proton-exchange membranes for fuel cells", *Polym. Int.*, 60, 26-41, **2011** (Review Article).

[2] H. Bai, "In situ platinum recovery and color removal from organosilicon streams", *Ind. Eng. Chem. Res.*, 51, 16457-16466, **2012**.

[3] H. Bai, "Manufacturing of platinum- and color-free organosilicon products using heterogeneous platinum catalysts", *Ind. Eng. Chem. Res.*, 53, 1588-1597, **2014**.

AIChE 2014

EXPERIMENTAL SIMULATION OF STABLE VS UNSTABLE BEHAVIOR OF BAG FILTER UNDER THREE COMMON REGENERATION SCHEMES

MAHMOOD SALEEM[1,], MUHAMMAD SULEMAN TAHIR[2], RAFI ULLAH KHAN[1], GERNOT KRAMMER[3]

1 -UNIVERSITY OF THE PUNJAB, LAHORE, PAKISTAN
2 -UNIVERSITY OF GUJRAT, GUJRAT, PAKISTAN
3 -GRAZ UNIVERSITY OF TECHNOLOGY, GRAZ, AUSTRIA

CORRESPONDING AUTHOR'S E-MAIL:
MSALEEM.ICET@PU.EDU.PK

LAYOUT

- Motivation for the work
- Background of the study
- Experimental
- Results and discussion
- Conclusions

MOTIVATION

- **Fine solids separation from exiting gases in process industries is usually carried out using bag filters.**
- **The jet-pulsed bag filters can be operated for fixed filtration time and fixed number of jet-pulses without regulating upper or lower pressure drop limits (scheme-I), or**
- **at upper pressure drop limit with fixed number of jet-pulses without regulating the lower pressure drop limit (scheme-II), or**
- **between upper and lower pressure drop limits with no control on number of jet-pulses (scheme-III).**
- **The stability of the bag filter is measured by :**
  - (a) steady upper and lower pressure drop, or
  - (b) by the steady filtration cycle time and a steady lower pressure drop, or
  - (c) stable and steady filtration cycle time and number of cleaning pulses per filtration cycle depending on the bag cleaning schemes (I), (II) or (III), respectively.
- While a stable operation is desirable, it is observed that the filter de-rails from stable operation under many operating situations.

BACKGROUND 1

- **assuming jet pulse cleaning at fixed intervals reveals the cake detachment being proportional to pulse over pressure**
- **Changes of dust properties may also influence the filter cake resistance and hence may lead to instability**
- **Cake detachment is accomplished if the detachment force overcomes the adhesive forces between cake and the filter medium or cohesive forces of the cake**
- **detachment forces generated in the cake or at the interface of the cake and the filter medium depends upon many factors**

BACKGROUND 2

- The governing mechanism of cake detachment with jet pulsed fabric filters is not only acceleration and deceleration of the cake and the medium but also the reverse flow of the gas due to the jet pulse
- The thicker cakes are removed easier than the thinner ones.
- Thus the cake area load (w= cake height x cake density) plays important role in the cake detachment.
- The cake detachment function, the fraction of the cake detached, is related to cake height.
- However the relationship is based on measurements with uniform cake formed in a coupon test rig and not on jet pulse filter bag

BACKGROUND 3

- The pulses per cycle and cycle time remain steady for a stable operation.
- However, an increase in pulses per cycle is an indication of incomplete removal of filter cake during regeneration which may lead a situation of continuous jet pulsing, a highly unstable operation
- High pulse pressure, high pressure drop limit or high filtration velocity may hinder regeneration.
- The later may also accelerate re-deposition of dust.
- It is also reported that the cake is thrown away from the bag at higher jet pulse pressures

BACKGROUND 4

- This paper presents the experimental study of jet pulsed bag filter over a range of operating parameters which may or may not lead to unstable filter operation.
- The filter behavior is observed by varying the operating parameters.
- The filter is operated under three regeneration schemes as defined earlier

EXPERIMENTAL

A mixture of dust and air is prepared and allowed to pass through a bag filter. The dust is retained on the bags while the clean air pervades through the filter media to the clean gas side. The flow is induced by a discharge fan and gas flow is regulated by a gas flow controller. Reverse air pulses are used for cake detachment on-line. Separated cake is collected in dust pan at the bottom of the bag.

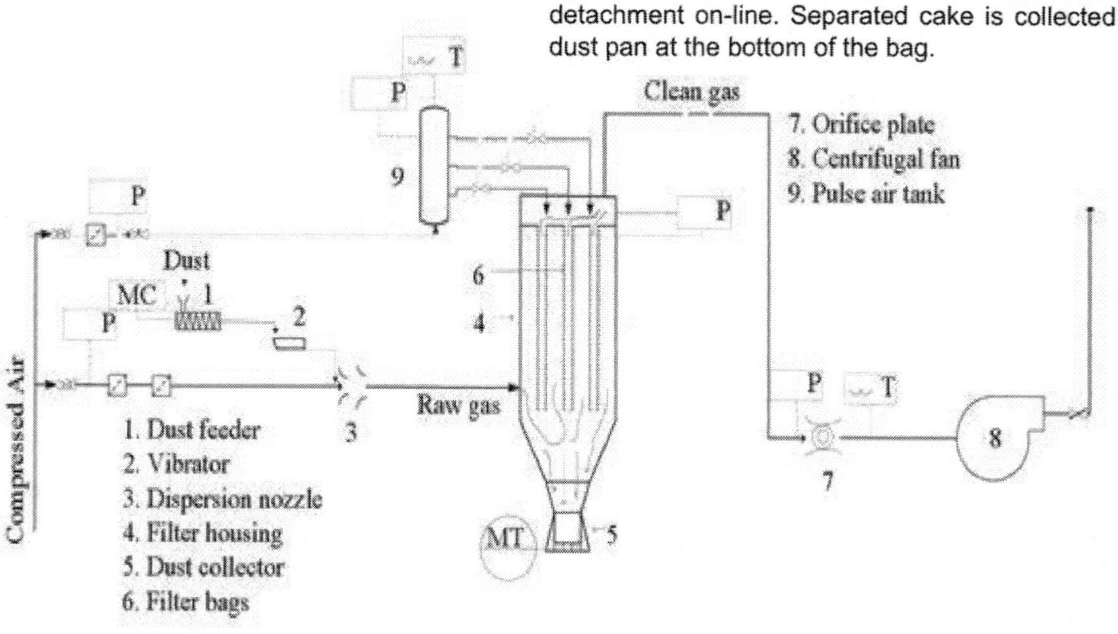

MATERIALS AND CONDITIONS

- Ambient conditions
- Needle felt media made from PI and PPS fibers
- Non-precipitated commercial Limestone dust
- Filtered and dried air

RESULTS & DISCUSSIONS 1

The time per jet pulse can be used for comparing the filter behavior. The upper pressure drop limit and the jet pulse pressure can be combined into the detachment over pressure which is the difference of static pressure developed by the jet pulse in the bag minus the maximum filter pressure drop as a simple approach.

The static pressure developed by the jet pulse is calculated by an empirical equation:

$$P_{static} = 165\ P_{pulse}^{0.6}$$

Detachment Over Pressure= P_{static} - ΔP

RESULTS & DISCUSSIONS 2 SCHEME-I

- The filter bags are regenerated in a simple approach by cleaning a fixed area at fixed time intervals (scheme-I).
- This operation mode is studied at different velocities, jet pulse pressure, and time intervals between jet pulses. Two dust concentration levels are used as well.
- The upper and lower pressure drop levels are decided by the dynamic balance between cake detachment and dust reaching the bags during filtration period.
- Starting with fully regenerated bags the cake area load and pressure drop levels increase. As the cake thickness grows the detachment increases and eventually equilibrium is reached from thereon a steady operation prevails.
- The upper and lower pressure drop levels remain steady under stable conditions.
- Situations arise when such an equilibrium is not established where the limits keep on increasing accompanied by increasing cake area load.

RESULTS & DISCUSSIONS 3 SCHEME-I

- Based on test results from test facility the time per jet pulse, cake detachments over pressure and filtration velocity are presented in Figure 2 .

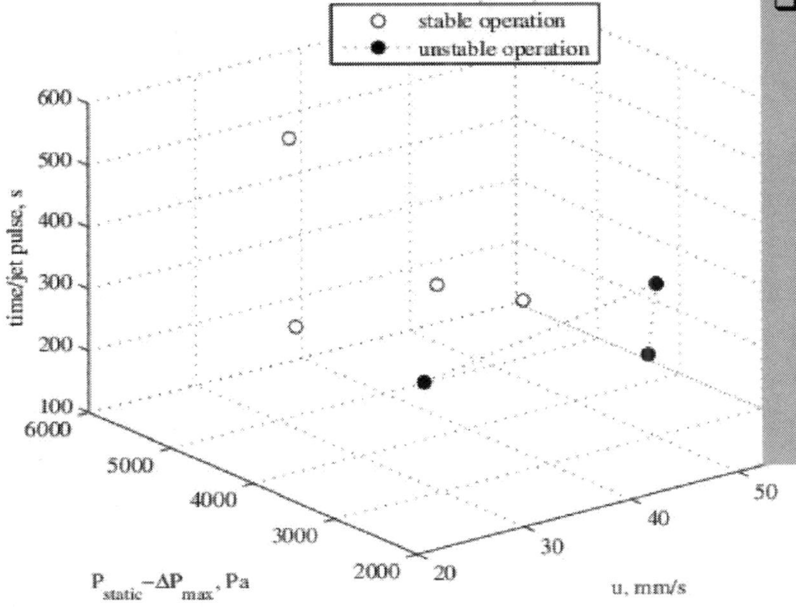

- It observed that the operation is insensitive to time between jet pulses at higher jet pulse pressure and very sensitive to it at lower jet pulse pressure at relatively lower velocity

RESULTS & DISCUSSIONS 4

SCHEME-I

- **Based on test results from test facility the time per jet pulse, cake detachments over pressure and filtration velocity are presented in Figure 2 .**

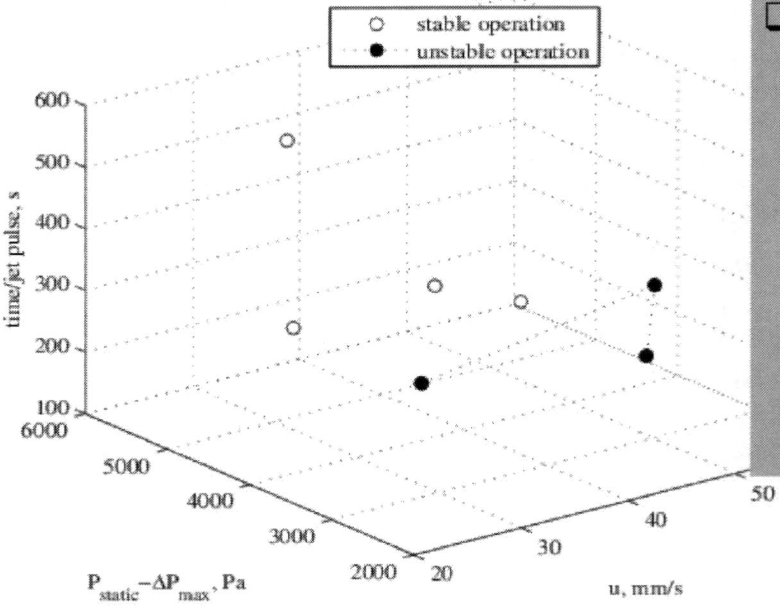

- A higher jet pulse pressure is required at higher velocity (54mm/s) and time between jet pulses is important to decide a stable operation.

RESULTS & DISCUSSIONS 5

SCHEME-I

- **Based on test results from test facility the time per jet pulse, cake detachments over pressure and filtration velocity are presented in Figure 2 .**

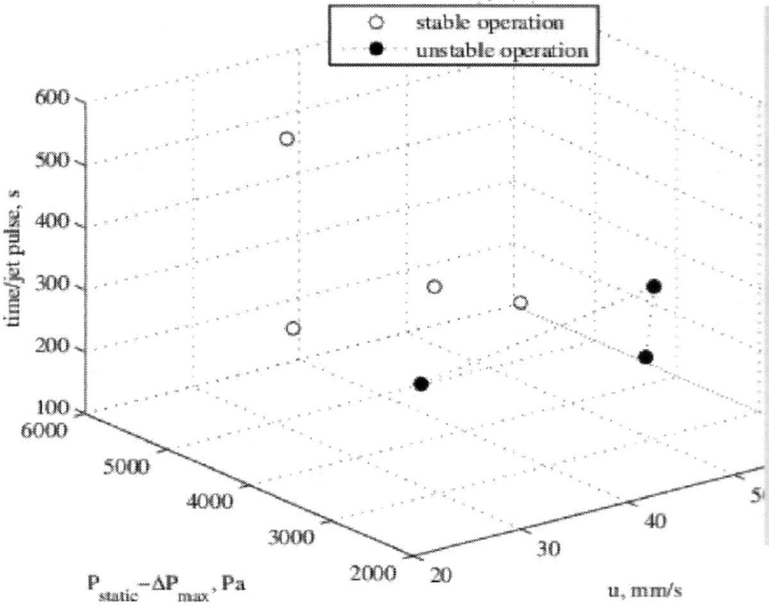

- According to observations, this scheme of regeneration is less flexible to changes in process variables; velocity and dust concentration.

RESULTS & DISCUSSIONS 6 SCHEME-II

- The filter is operated with bag cleaning starting at the upper pressure drop limit regenerating one row of the bags (one 3rd of filter area) with a single jet pulse.
- The filtration velocity is varied over 27- 68mm/s, the jet pulse pressure from 1.5-4 bars, and pressure drop over 700-2300Pa.
- Stable operation is indicated by stable upper and lower pressure drop levels reflected by steady time per jet pulse over time.
- The detachment over pressure as described earlier, time per jet pulse, and velocity are plotted.

RESULTS & DISCUSSIONS 7 SCHEME-II

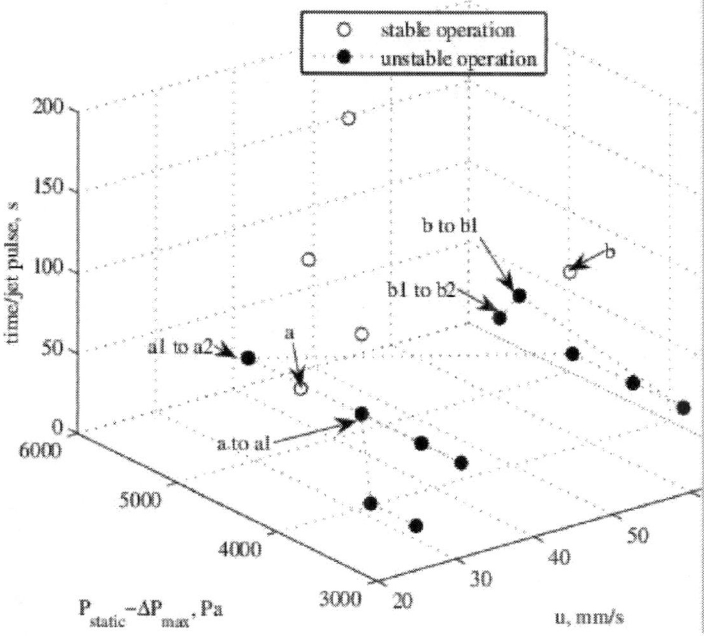

- The operation is observed unstable when the detachment over pressure is below a threshold
- At higher detachment over pressure, the operation is stable
- however, two interesting situations are observed. One At 41mm/s, a stable operation (a) goes unstable (a to a1) because of reducing jet pulse pressure from 4bar to 3bar at 2200Pa.

RESULTS & DISCUSSIONS 8 SCHEME-II

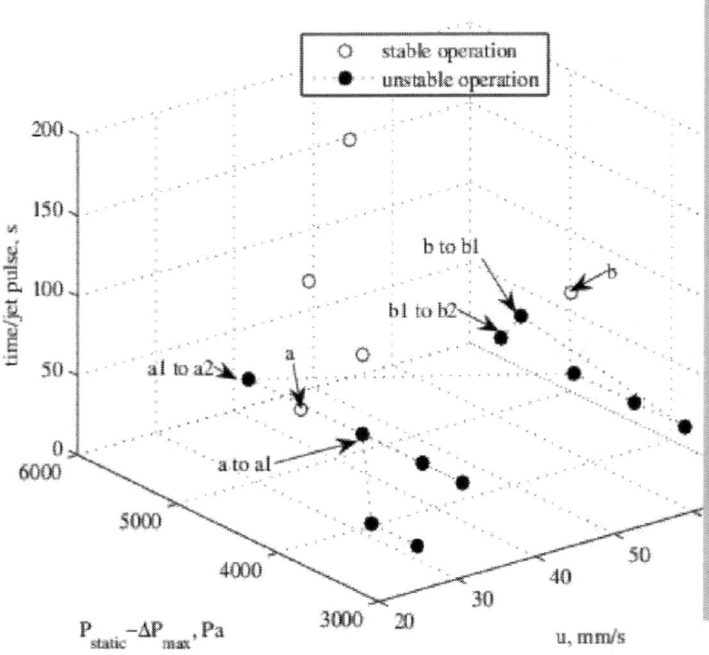

- Reducing the pressure drop could not stabilize the operation (a1 to a2).
- Only increasing the jet pulse pressure from 3 to 4bar stabilizes the operation (a) at the same velocity.
- Further at 68mm/s and 2300Pa, a stable operation (b) turns unstable (b to b1)and (b1 to b2) when pressure drop is reduced.
- The cycle time per jet pulse is already small such that there is very little filter cake formed during a filtration cycle

RESULTS & DISCUSSIONS 9 SCHEME-II

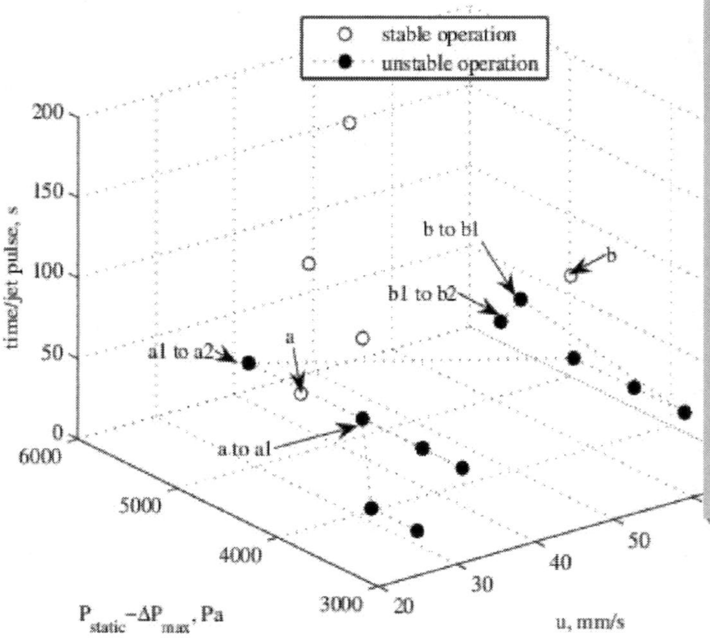

- Decreasing the upper pressure drop limit increases the detachment over pressure but minimized the cake area load further.
- When there is no significant cake on the filter bag, the over pressure cannot remove the cake and thus the operation turns to continuous jet pulsing.

RESULTS & DISCUSSIONS 10 SCHEME-II

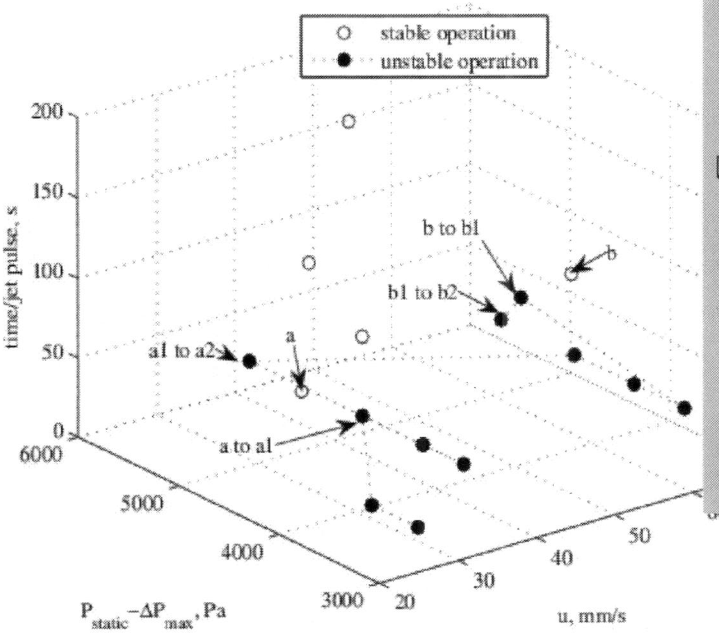

- This mode of regeneration control has less tolerance because the area jet pulsed per cycle is fixed.
- In case the cake detachment becomes ineffective due to some reasons, the lower pressure drop will keep on increasing and may lead to shorter cycles and/or continuous jet pulsing.

RESULTS & DISCUSSIONS 11 SCHEME-III

- **Considering the filter operation according to scheme-III, generally a few pulses may bring the pressure drop to the lower limit.**
- **The pulses per cycle and cycle time are steady under such conditions.**
- **However, an increase in pulses per cycle indicates inefficient regeneration.**
- **The filter cake being detached is lesser compared to the dust reaching the bags which may lead to nonstop jet pulsing**
- **A stable or stable like operation is revealed by higher time per jet pulse.**
- **For unstable or unstable like operation, the time per jet pulse is low and/or declining.**
- **There is a threshold detachment over pressure determined operating conditions at respective filtration velocities.**
- **The threshold increases with filtration velocity.**
- **The filled circles mark a boundary between velocity and the detachment over pressure.**
- **The time per jet pulse reaches values below 50s in most of the unstable cases**

RESULTS & DISCUSSIONS 12 SCHEME-III

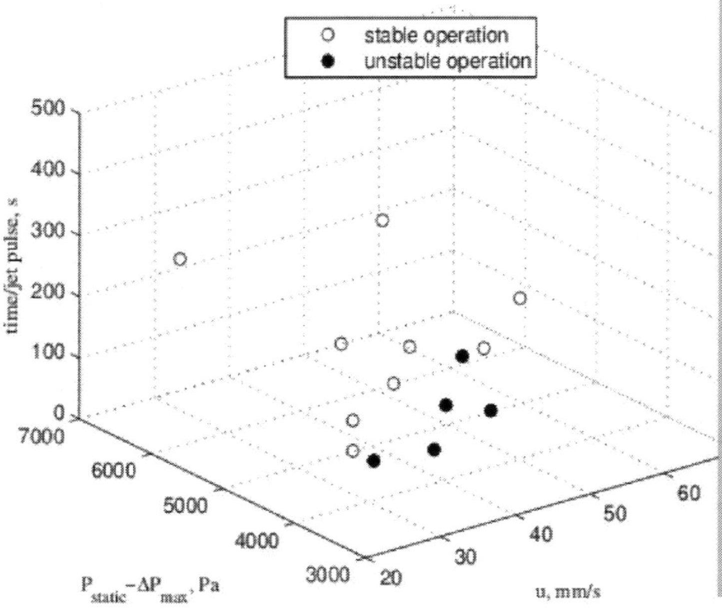

- It is observed that at higher velocity (61mm/s) the operation is unstable at lower pressure drop (1200Pa) and stable at higher pressure drop (1500 and 2000Pa) at the same jet pulse pressure.
- The stable filter operation is not restorable by only increasing detachment over pressure under some situations particularly at high velocities.

RESULTS & DISCUSSIONS 13 SCHEME-III

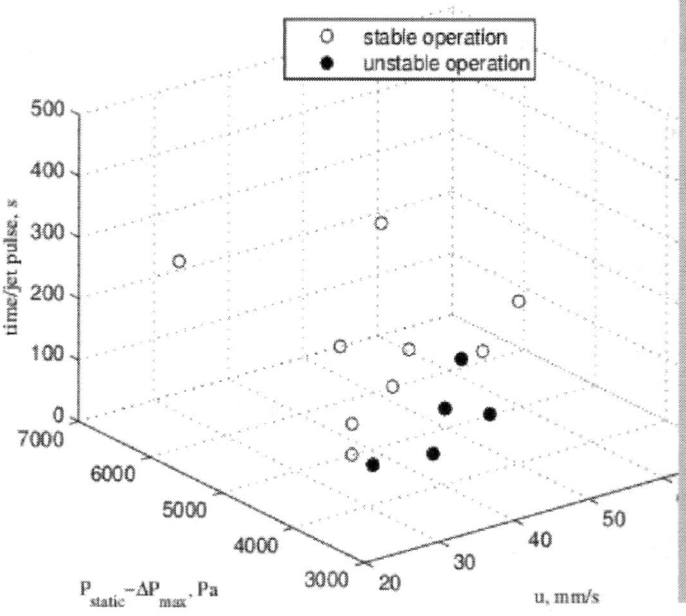

- The cake per unit area at the end of filtration cycle is important especially when the formed cake is too thin at higher filtration velocity and relatively lower value of upper pressure drop limit.
- There allowing formation of a thicker cake before regeneration stabilizes the filter operation.
- It is observed that the filtration time per jet pulse increases from 20s to 50 at 1500Pa and to 160 at 200Pa all other conditions being the same.

RESULTS & DISCUSSIONS 14
SCHEME-III

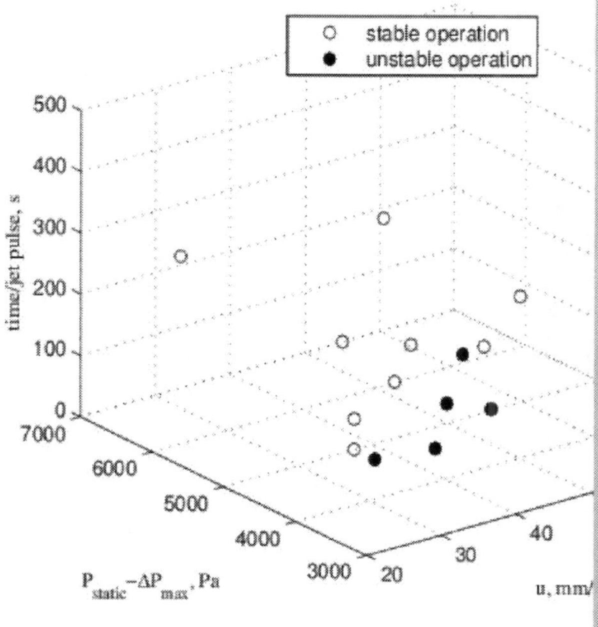

- The experiments reveal that this mode of operation may accommodate wide variations in the operating parameters without running unstable.
- It is also observed that even a feeble jet pulse is capable of restoring the lower pressure drop limit.
- Thus operations still run steady at short duration cycles which are a border line situation.
- At this operating point a minute increase in velocity may drift the plant to an unstable operation.
- Unless the jet pulse pressure is high enough (> 4 bar) there is hardly any risk of instability with scheme-III

CONCLUSIONS

- Scheme-I of regeneration is less flexible to changes in process variables; velocity and dust concentration.
- Scheme-II of regeneration has less tolerance because the area jet pulsed per cycle is fixed.
- Considering scheme-III, at higher velocity, while filtration cycles are already short, reducing upper pressure drop limit may take the system to unstable operation. The detachment of a thicker cake is more efficient as compared to a thinner. There are set of conditions when operation becomes unstable irrespective of the path to reach those conditions. One can say that stable/unstable operation depends on prior set of conditions, therefore.
- Comparisons of the results reveal that the operation is the least flexible with scheme-I and the most flexible with scheme-III, scheme-II being in between.

ACKNOWLEDGEMENTS

The authors acknowledge the financial support of :

- ❑ Austrian Science Foundation,
- ❑ Higher Education Commission of Pakistan, and
- ❑ University of the Punjab, Lahore, Pakistan.

ABSTRACT

Optimization of Neem OIL Extraction from Its SEED Using Response Surface Methodology

Sodeinde O.A & Okanlawon Samuel

A Central Composite Design (CCD) consisting three factors (solvent composition, temperature and lime of extraction) at five levels was used to study the solvent extraction of Neem oil from its seed using n-hexane and ethanol. Neem oil yield, pH, refractive index, acid value, iodine value and saponification value were evaluated as the responses. Forty-two experimental runs resulted from the CCD with a minimum oil yield of 12% and maximum of 41%. Responses surface methodology was used to analyze the results of the CCD of the extraction processes. The optimum values for yield, pH, refractive index, acid value, iodine value and saponification value from the surface plot were 43.48%, 4.99,156, 1.411glg, 89.35glg and 176.64mglg predicted percentage yield was 43.48% at solvent composition of 80.77% n-hexane, 34.93^{0}c and 6 hours duration of extraction. The five validation experiments had optimum oil yield ranged between 32.85% and 37.20% while their accompanying quality characteristics were not significantly different from the simulated values at $p < 0.05$.

Keywords: Neem oil, response surface methodology, central composite design, oil yield

Introduction

Neem (Azadiracha Indica) tree is a native to tropical South East Asia and it is a member of the mahogany family Meliaceae (ICIPE, 1995, Okonkwo, 2004). This tree is popularly known as Dongoyaro in Nigeria. All parts of this tree are very useful in variety of biological activity. The most famous part of this tree is the oil obtained from kernel of its seed. Neem oils have found its use widely in different region of the globe for medicinal purpose (Ranajit, 2002, Awad and Shimaila 2003, Munoz – Valenzenla, 2007, Kumar, 2010). It is used in soap production, as a raw material for producing commercial pesticides and cosmetics, plant protection, stock and textile protection, refining to edible oil, lubricating oil for engines, lamp oil, candle production (Peter 2000). It is found to be of great health use and it has and effective anti-germ property which include the use as an insect repellent and it has shown positive result as pesticide (Koro, 2006). In India, it has been demonstrated that neem oil is a potential new contraceptive for woman (NRC, 1992).

Among various methods available for obtaining neem oil from seeds are mechanical press, super critical fluid extraction and solvent extraction method. Method of extraction is the most widely used to extract oil from neem seed. However, the oil produced with this method is usually turbid, contains significant amount of water and metal contents. Extraction using supercritical fluid produced very high purity oil but operating and investment cost are very high.

Extraction using solvent has several advantages like high oil yield and less turbid oil than mechanical extraction and relative low operating cost compared with supercritical fluid extraction. (Liauw 2008, Soetaredjo, 2008). It is clear from various earlier researches on solvent extraction (Koro, 2006, Liauw 2008, Soetaredjo 2008 and Zahedi, 2010) that the final neem oil quantity and properties are all functions and the process variable of the extraction process. There exist a complex relationship between the oil yield and its properties. The purity and the final properties of the neem oil extract determines its end use. The quest by researches to increase the oil yield has resulted on a great compromise on its properties. The crux of the problem has remained how to combine oil yield with desirable qualities. Hence, the need to apply process optimization as an essential tool that has demonstrated high efficiency in selecting optimal process variables for optimum turnover in many chemical processes.

Among process optimization tools that have been effectively applied to chemical processes is response surface methodology (RSM) (Singh, 2003). RSM is a collection of statistical and mathematical techniques useful for developing,improving, optimizing process and new products as well as in the improvement of existing products design.

Materials and Methods

Neem Seed used in the solvent extraction of Neem Oil was obtained from (Agboju Secondary School) Lagos State in Nigeria. The fruits were de-pulped to obtain the seeds that were washed thoroughly to remove the dirt and impurity.

These seeds were decorticated by winnowing to remove the hull from the seeds. The weight of the seeds (W_1) was measured and then placed in the oven at 50^0c until constant weight (W_2) was obtained using Liauw (2008) method for drying. The dried seed were crushed in a mortar and the sample was expressed on a standard sieve screen of the required mesh size (0.425 – 0.71mm) to obtain the required particle sizes.

Neem Oil extraction and properties analysis

Solvent extraction using various combination of n-hexane and ethanol were used to study the effect of extraction parameter on the yield and properties of neem oil. Other important factors of the extraction processes were the particle size, particle of solvent ratio, temperature of the extraction and time of extraction (Liauw 2008, Kovo 2006).

For this study, the factors considered were temperature of extraction, time of extraction and type of solvent combinations used for extraction because the two other factor has been reported by Liauw (2008) to have optimum performance at the fine particle size (0.425 – 0.71mm) and ratio 1.5 respectively, hence were adopted.

Each extraction run was setup by measuring 10g of the seed powder into 50ml of the solvent mixture in 250ml corked conical flask. This mixture was placed in a thermostatic water bath operating at a preset required temperature. At the set time interval. The samples were taken and centrifuged to separate the solid fraction from the solution. The extracts were heated and evaporated using rotary evaporator apparatus to obtain solvent free oils.

The solvent free oil extracted was analyzed for the PH, refractive index, acid value, saponification value and iodine value using AOAC (2000). The percentage of oil extracted was calculated from the equation.

$$\%\,oil\ extracted = \frac{Mass\ of\ extract}{Mass\ of\ sample}\ x\ 100 - (1)$$

Experimental Design

Central Composite Design (CCD) of RSM was used for the experimental design to optimize the extraction perimeter.

The CCD consisted of three factor solvent composition from O to 100% based on n-hexane, Temperature (30^0c – 50^0c) and time of extraction (2 – 6 hours) and five level. Actual process variable (xi) is related to the coded process variable as shown in table 1 according to equation 1

$$X_1 = \frac{(X_1 - X_0)}{\blacktriangle X}............2$$

Where XI is the dimensionless coded value of the independent variables, X_i is the actual value of the independent variables at the design center point and ▲X is the variation increments about the center point.

The center point chosen for the design were 50% solvent composition base on n-hexane, 40^0c temperature and four hours time of extraction. The coded, actual values of the variable at various levels and responses are given in the matrix. Three replications were carried out for all experimental design conditions and the average.

Recorded Forty-Two experimental runs were carried out and the order of the experiment was fully randomized to reduce the effect of unexplainable variability in the observed responses due to extraneous factor as recommended by Singh (2003).

Analysis of Data and response equation

Regression models were developed for neem oil yield and each of the five properties of the oil as a function of the three process factors. The design expert 6.0 software was used to analyze the extraction data for developing response equation, for analysis variance (ANOVA) generate surface plots and determine optimum extraction conditions using its optimization tool box. In multiple regression as in the present case. R_2, which is the square adjusted coefficient of determination and standard error are the indices. F statistics shows the significance of the overall model while the t statistics tests the significance of each of the variables of the model. The function was assumed to be approximated by a second degree polynomial equation.

$$Y_h = b_{ho} + \Sigma_{i=1}\, bhiXi + \Sigma_{i=1}\, bhiiX^2i + \Sigma_{i\neq}\, j = 1bijX_iX_j.................3$$

Where b_{ho} the value of fitted response at the centre point (O,O,O) and bhi, bhii and bhij are linear, quadratic and cross product regression term respectively. m is the number of factors considered in the study which is equal to 3, Y_1 (oil yield %) Y_2 (PH), Y_3

(refractive index), Y_4 (acid value), Y_5 (iodine value) and Y_6 (saponification value).

Optimization and validation of neem oil extraction

A non linear programming problem of the form of equations 4 to 10 was formed from the vector of equation 3 as follows

$$\text{MAXIMIZE: } Y1\,(Xi) \quad (4)$$

Subject to:

$$Y_2\,(Xi) \leq b_2 \quad (5)$$
$$Y_3\,(Xi) \leq b_3 \quad (6)$$
$$Y_4\,(Xi) \leq b_4 \quad (7)$$
$$Y_5\,(Xi) \leq b_5 \quad (8)$$
$$Y_6\,(Xi) \leq b_6 \quad (9)$$
$$Y - 2 \leq Xi \leq + 2 \quad (10)$$

Where bi are the neem oil property requirement. This constrained maximization problem was solved using the Design Expert 6.0 software. This problem statement was validated with five (5) randomly generated oil quality requirements for optimum extraction conditions. The results obtained theoretically were experimented through extraction and subsequent oil analysis. These theoretical and the experimental results were compared using t-test at p<0.05>.

Results & Discussion

The effect of the CCD on the yield (Y_1), the PH (Y_2), refractive index (Y_3) acid (Y_4), iodine value (Y_5) and Saponification value (Y_6) is as shown on table 3 that was subsequently used to fit the response equations for oil yield and the five oil properties.

Multiple regression analysis was used as tools of assessment of the effects of two or more independent factors on the dependent variables. The co-efficient of determination R^2 is a measure of the total variation of the observed values of the extracted oil about the mean explained by fitted model. The factors of the models, their parameters estimates, and the statistics of the estimates for the best functions adopted taking into considerations all main effects, linear, quadratic and interaction for each model are shown on table 4. The co-efficient of determination (R^2) for the responses, yield, PH, refractive index, acid value were 0.990, 0.890, 0.995, 0.971, 0.992 and 0.987, respectively. The co-efficient of determination were high response surfaces and indicated that the fitted quadratic models accounted for more than 80% of the variance in the experimental data, base on P-values, the regression co-efficient that were significant at P<95% were selected for the models that resulted in Equations (11) to (16).

Analysis of variance (ANOVA) was conducted to evaluate the adequacy and consistency of the model using F-Statistic (Shridhar 2010). The analysis of variance of models is presented in Table 5. As shown in the Table 5, The F-value for the oil yield (225.13), PH (13.72), refractive index (372.38), acid value (74.84), iodine value (120.67) and Saponification value (228.73) were significant at P<0.05 implying good model fit.

Solvent ratio (X_1) had quite high linear positive effects on oil yield than it had on saponification value. Time of extraction (X_3) had positive linear effects on yield, refractive index, acid value and saponification values. While it had negative quadratic effect on iodine value. The interaction of the solvent ratio and temperature had positive effect on iodine value and PH but no negative effect on iodine value and PH but no significant effects on other responses. The interaction of solvent ratio and time had negative effect on the refractive index but no significant effect on all other responses. The interaction of temperature and time also had positive effect on PH but no statistical significant effect on other responses.

Yield (Y_1) $= 33.30 + 27.23 X_1 + 4 X_2 + 1.22 X_3 - 30.28 X_1^2 + 5.72 X_3^2 \quad R^2=0.9903$ (11)

pH (Y_2) $= 3.98 - 0.49 X_1 + 0.5 X_2 + 0.5 X_2 + 0.48 X_2^2 - 0.4 X_1 X_2 + 0.25 X_2 X_3 \quad R^2=0.8900$ (12)

Refractive Index (Y_3) $= 1.53 + 0.049 X_1^2 + 6.756E - 3 X_3^2 - 3.3E-3 X_1 X_3 \quad R^2=0.9951$ (13)

Acid Value (Y_4) $= 1.82 + 0.47 X_1 + 0.075 X_2 + 0.023 X_3 - 61 X_1^2 + 0.2 X_3^2 \quad R^2=0.9711$(14)

Iodine Value (Y_5) $= 89.62 - 2 X_2 - 4.2 X_2^2 - 4.2 X_3^2 - 4.25 X_1 X_2 \quad R^2=0.9920$ (15)

Sapronification Valu e Saponification Value (Y_6) $= 1.762 + 3.59 X_1 + 1.78 X_3 + 5.43 X_2^2 + 5.43 X_3^2 + 7.25 X_1 X_2 \quad R^2=0.987$ (16)

Optimization of the Equation Process

The model (Y_1, Y_2, Y_3, Y_4, Y_5, Y_6) were useful for indicating the direction in which to change the variables in order to maximize yield, PH, refractive index, acid value and saponification value. The multiple regression equations were solved using Design Expert 6.0. The maximum values obtained were as follows. 43.48, 4.99, 1.56, 1.411, 89.345 and 176.643 for yield, PH, refractive index, acid value, iodine value and saponification value respectively. The optimum response is presented in Table 3.0.

As shown in Table 6, the coded levels are within the experimented range and this indicated that the selected variables are valid of the selected variables. The regression equation was optimized for maximum value to obtain the optimum conditions using Design Expert 6.0. The actual value calculated for optimum response as shown in Table 6, were 80.77%, n-hexane solvent, 34.93^{0}c temperature and 6.0 (hours) time for yield, 35.84% n-hexane solvent, 49.06^{0}c temperature and 3.5hours, time of extraction for the maximum PH, 83.22% n-hexane 49.61^{0}c temperature and 5.4 hours, time of extraction for maximum refractive index 20.97% n-hexane solvent, 30.86% temperature and 2.23 hours time of extraction for minimum acid value, 51.31% n-hexane solvent, 38.34^{0}c temperature and 3.23 hours time of extraction for maximum iodine value and 62.58% solvent ratio, 33.49^{0}c temperature and 4.09 hours time of extraction for minimum saponification value.

The response surface plots and the accompany contour plots shown in figures 1 to 6 for the chosen model equations shows the relationship between the independent and the dependent variables. From figures 1, the response surface indicates that the percentage yield of oil increases temperature and solvent composition increase to optimum condition while further increases led to decrease of percentage yield of oil. In addition, there was mutual interaction between the temperature and solvent composition from figure 2. The PH of oil increased with increase solvent composition and temperature for a short period, at a longer period, increasing the temperature of extraction will have negative effects on the PH. From figure 3 and 4, increase in temperature and solvent composition led to increase in refractive index and acid values respectively. It can be seen that the solvent composition was more influential factor that affect the refractive index and the acid value of the extracted oil.

From figure 5, it can be seen that increase in temperature and solvent composition increases the iodine value. Figure 6 shows that saponification value decreased with increase in temperature and solvent composition. The highest saponification value was obtained when all the variables were at the minimum point within the range of study.

Validation of the Optimization of the extraction process.

Table 7 shows result of the optimum process variables for each extraction run, its corresponding oil yield and the Neem oil properties obtained from the experiment of each validation process. The oil yield range is between 32.85% and 37.20%. The corresponding result of the analysis of each extraction were compared with the simulated properties using paired t-test that shows that there are no significant difference between the simulated properties and their corresponding experimental values at $P<0.05$ as presented on Table 8. This result attests to the effectiveness of this frame work for optimum and effective oil extraction.

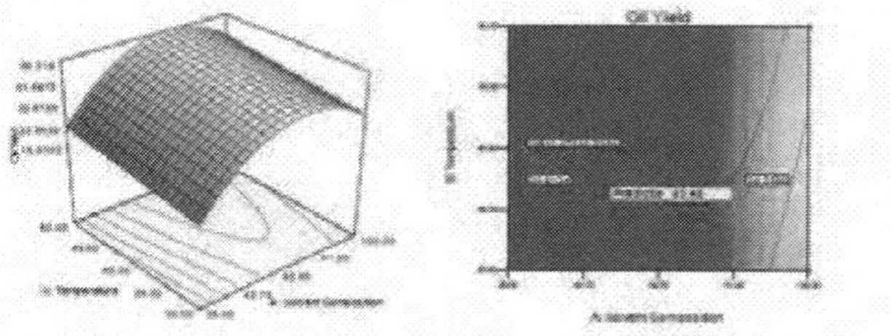

Figure 1: (a) Response surface plot showing the 3D effect of temperature and solvent and their interaction effect on the yield of Neem oil. (b) Contour Plot of Figure 1a

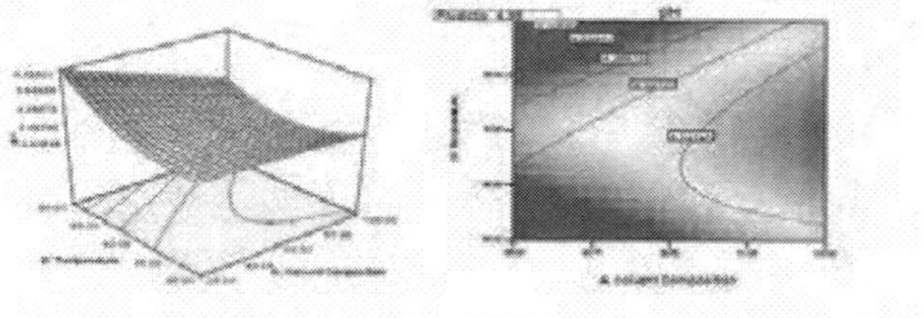

Figure 2: (a) Response surface plot showing the 3D effect of temperature and solvent composition their interactive effect on the pH of Neem oil. (b) Contour Plot of Figure 2a

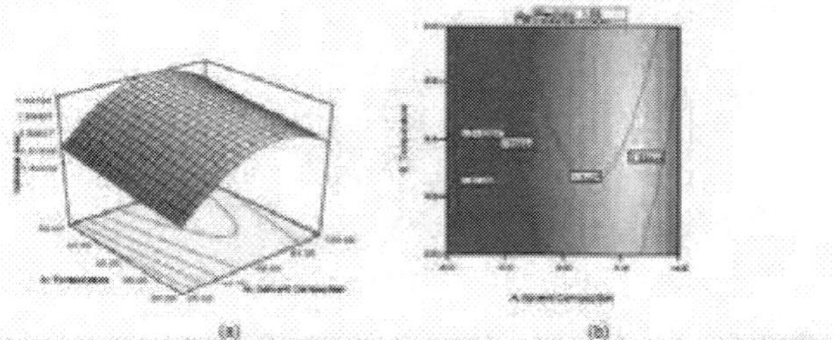

Figure 3: (a) Response surface plot showing the 3D effects of temperature and solvent composition and their interactive effect on the refractive index of the Neem oil. (b) Contour Plot of Figure 3a.

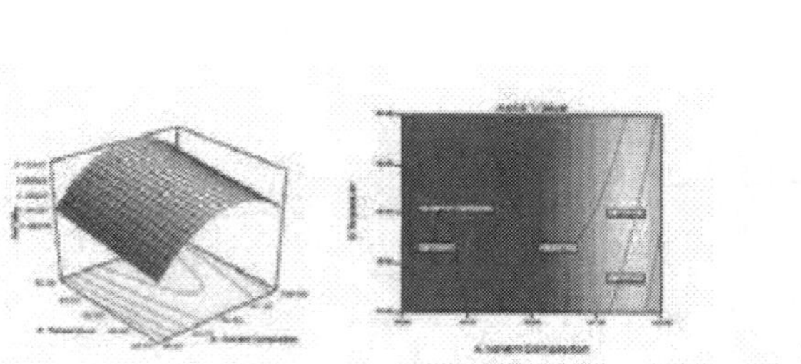

Figure 4: (a) Response surface plot showing the 3D effects of temperature and solvent and their interactive effects on the acid value of Neem oil. (b) Contour Plot of Figure 4a

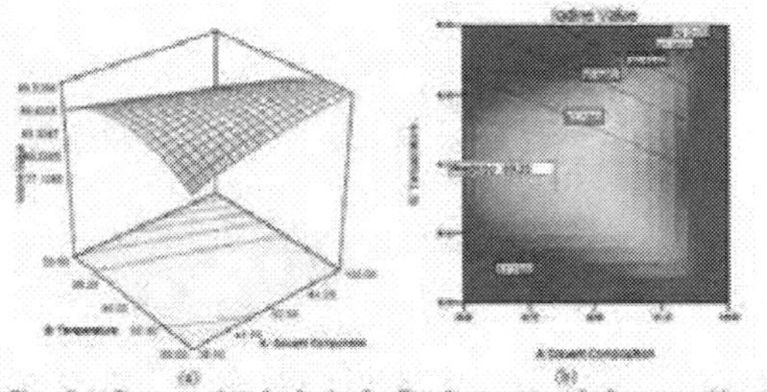

Figure 5: (a) Response surface plot showing the effect of temperature and solvent composition and their interactive effect on the iodine value of the Neem oil. (b) Contour Plot of Figure 5a.

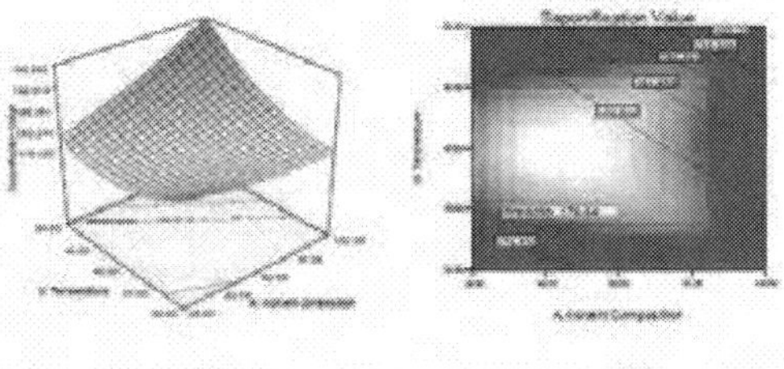

Figure 6: (a) Response surface plot showing the effects of temperature and solvent composition and their interactive effect on the Saponification value of Neem oil. (b) Contour Plot of Figure 6a

REFERENCES

Kumar, P. S., Ghosh, D. G. and Pander C. S. (2010). Biological action and medicinal properties of various constituent of Azadirachta indical (meliaceae a overview). *Annals of Biological research* **1**(3):23-34.

Liauw, M. Y., Natan F. A., Widiyanti P., Ikasiri D. Indraswati N. & Soetaredjo F.E.(2008). Extraction of neem oil (*azadirachta indica a. juss*) using N-hexane and ethanol: studies of oil quality, kinetic and thermodynamic.
ARPN Journal of Engineering and Applied Sciences **3**(3) :49-54

Muñoz-Valenzuela, S., Ibarra-López A. A., Rubio-Silva L. M., Valdez-Dávila H.& Borboa-Flores J. (2007).

Neem tree morphology and oil content. Reprinted from Issues in new crops and new uses. 2007. J. Janick and A. Whipkey (eds.).ASHS Press, Alexandria, VA.

National Research Council (NRC)(1992). *Neem: a tree for solving global problems.* National Academy Press, Washington, DC

Okonkwo, E. M. (2004). Employment creation and opportunities in manufacturing sub-sectors: Acase for neem tree in Nigeria. Bullion, publication of Central Bank of Nigeria. **28**(3):30-35.

Peter, F. (2000). *Global Neem usage.* Deutsche Gesellschaft fdür Technische Zusammenarbeit (GTZ) GmbH Eschborn, Germany.

Ranajit, K. B., Kausik B. IshitaC. & Uday B.(2002). Biological activities and medicinal properties of neem (Azadirachta Indical). *Current Science* **82**(11):1036 -1045.

Soetaredjo F .E., Budijanto G. M., Prasetyo R. I., & Indraswati N.(2008). Effects of pre-treatment condition on the yield and quality of neem oil obtained by mechanical pressing. *ARPN Journal of Engineering and Applied Sciences* **3**(5) :45-49.

Shridhar,B.S. Beena, K.V. Anita,M.V. & Paramjeet K.B.(2010). Optimization and charaterization of castor seed oil. *Leonardo Journal of Science* 17:59-70.

Singh, S., Raina, C. S., Bawa, A. S. & Sexena D. C. (2003). Sweet potato-based pasta product: optimization of ingredient levels using response surface methodology. *International Journal of Food Scienceand Technology* **38**:1-10.

University of Lagos, Department of Chemistry, Chemistry Laboratory.

Zahedi, G., Azizi S., Hatami, T., & Sheikhattar, L. (2010). Gray box modelingof supercritical Nimbin Extraction from Neem Seeds using Methanol as Co-Solvent. *The Open Chemical Engineering Journal* **4**: 21-30.

North Carolina Agricultural and Technical State University

Development of High Selectivity Mixed Matrix Oxygen Enrichment Membrane

Emmanuel Ogbole, Jianzhong Lou, and Shamsuddin Ilias

ABSTRACT

Mixed matrix membranes (MMMs) were prepared from Polydimethylsiloxane (PDMS) filled with surface-treated fumed silica (SiO_2). Techniques (TGA, SEM, and FTIR) were employed to characterize the neat membrane (PDMS) and the mixed matrix SiO_2-PDMS membrane (SiO_2-PDMS). The results confirmed that SiO_2-PDMS has improved thermal property over neat PDMS. Uniform dispersion of SiO_2 within the membrane was observed. Good material interaction between PDMS and fumed SiO_2 was observed. The effects of SiO_2 on the transport pattern and permeability of O_2, N_2, CO_2, and CH_4 were studied. Results showed that the presence of SiO_2 in PDMS matrix had a significant impact on N_2 transport pattern. The incorporation of SiO_2 into polymer matrix gave rise to an increase in solubility dominant flux and a corresponding decrease in the diffusivity of the gases through the MMMs as evident in the permeability trend of: $P_{CO2} > P_{CH4} > P_{O2} > P_{N2}$. 10% SiO_2-PDMS exhibited both improved O_2 permeability of about 640 Barrer and O_2/N_2 selectivity of 3.42 against neat PDMS with O_2 permeability of about 520 Barrer and O_2/N_2 selectivity for O_2 over N_2 of 2.59. However, neat PDMS had a fair performance for the separation of CO_2/CH_4 gas pair with CO_2 permeability of about 3239 Barrer and selectivity of 4.16 against 10% SiO_2-PDMS with 2967 Barrer and selectivity 4.29. This confirmed that SiO_2 as nano filler in PDMS is not a suitable material for separation of CO_2/CH_4 gas pair but could be suitable for the separation of CO_2/N_2 gas pair.

INTRODUCTION

The separation of gases by membranes is a dynamic and rapidly growing field. In membrane- based gas separation process, components are separated from their mixtures by differential permeation through membranes. A number of advantages, including low capital and operating costs, lower energy requirements and, generally, ease of operation are offered by membrane separation. Current applications of membrane-based gas separation include oxygen and nitrogen enrichment, hydrogen recovery, natural gas separation and the removal of volatile organic compounds from effluent streams. The permeation of molecules through membranes is controlled by two major parameters: diffusivity coefficient (D) and solubility coefficient (S). The permeability (P) representing the ability of molecules to pass through a membrane is defined as:

$$P = DS \qquad 1$$

The ability of a membrane to separate two molecules, for example, A and B, is the ratio of their permeabilities, called the membrane selectivity,

$$\alpha_{A/B} = \frac{P_A}{P_B} = \left(\frac{S_A}{S_B}\right)\left(\frac{D_A}{D_B}\right) \qquad 2$$

In this study, nano fumed silica was incorporated into PDMS matrix to develop mixed matrix membrane for oxygen enrichment. In addition to other applications, this membrane would be used in generating oxygen enriched air for fuel cell application.

OBJECTIVES

- Research and select polymer candidate as well as functional nano-filler and suitable solvent;
- Develop defect-free, stable, and high performing mixed matrix membranes (MMM);
- Characterize and document the fundamental microstructure of the membranes using Fourier transform infrared spectroscopy (FTIR), thermo-gravimetric and differential calorimetric analysis (TGA-DSC) and scanning electronic microscope (SEM);
- Perform gas permeation experiments with MMM of different nano-filler compositions;
- Study the influence of nano-filler on the performance of MMM and establish optimal MMM composition.

METHODOLOGY AND EXPERIMENTATION

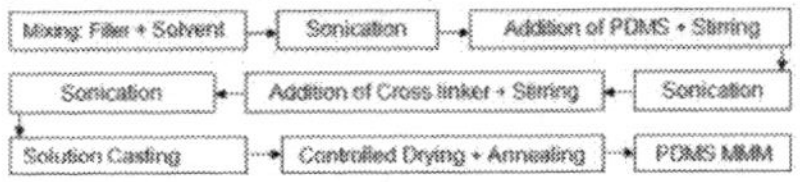

Figure 1. Laboratory Procedure for Preparing PDMS MMM

TGA, SEM, FTIR and pure gas permeation tests were employed in characterization of the neat PDMS and SiO_2 -PDMS samples. Permeability (P) (in Barrers, 10^{-10} cm³-cm (STP)/cm² s cm Hg) and ideal selectivity ($\alpha_{A/B}$) were determined according to Eqs 2 and 3 where J is the gas flux through membrane (cm³/cm² s), l is the thickness (cm) while p_1 and p_2 are the pressures (cm Hg) at the feed and permeate streams, respectively. Data generated from O_2, N_2, CO_2, and CH_4 permeation tests were used to plot graphs of gas flow rates (ml/min) against pressure (Psi), selectivity for O_2 over N_2 and CO_2 over CH_4 were calculated from the ratio of the slopes of the graphs.

$$P = \frac{Jl}{p_1 - p_2} \qquad 3$$

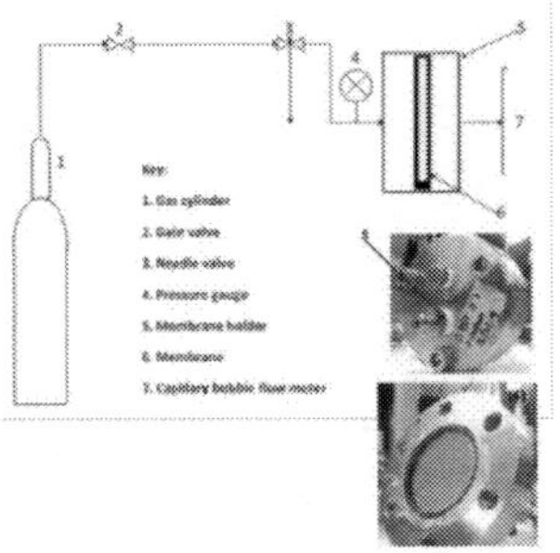

Figure 2. Schematic Laboratory Scale Set-up for Single Gas Permeation Test

RESULTS AND ANALYSIS

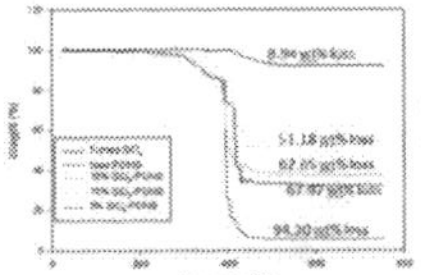

Figure 3. TGA Analysis

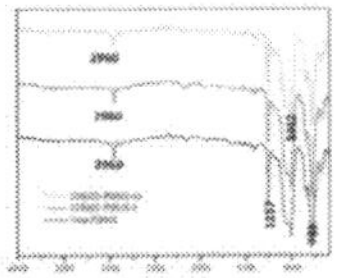

Figure 4. FTIR Spectra

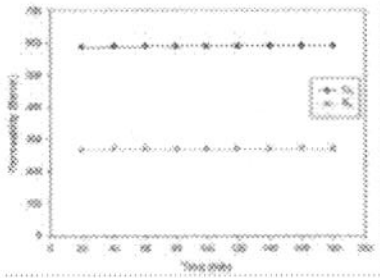

Figure 6. Transport Pattern of O_2 and N_2 through Neat PDMS at 30 psig

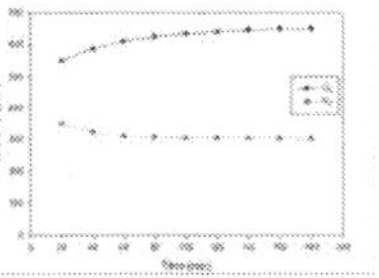

Figure 7. Transport Pattern of O_2 and N_2 through 10% SiO_2-PDMS at 30 psig

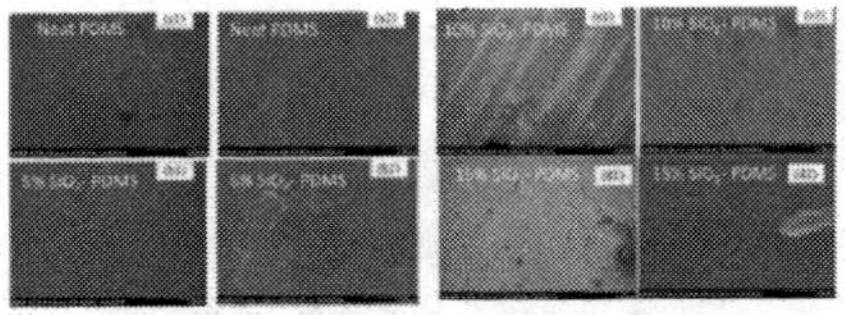

Figure 5. SEM images (1- surface & 2- cross section magnifications)

[illegible]	[illegible]	[illegible]	[illegible]
[illegible]	[illegible]	[illegible]	[illegible]
[illegible]	[illegible]	[illegible]	[illegible]
[illegible]	[illegible]	[illegible]	[illegible]
[illegible]	[illegible]	[illegible]	[illegible]
[illegible]	[illegible]	[illegible]	[illegible]
[illegible]	[illegible]	[illegible]	[illegible]
[illegible]		[illegible]	[illegible]
[illegible]	[illegible]	-	[illegible]

Table 1. Membrane Performance

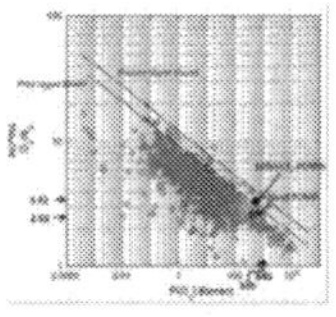

Figure 8. Upper Corr. for O_2/N_2 Sep.

RESULTS AND ANALYSIS

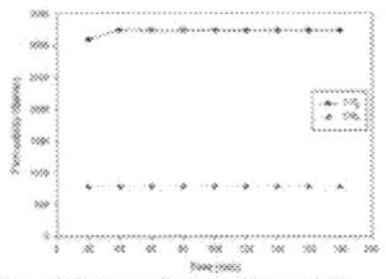

Figure 9. Transport Pattern of CO_2 and CH_4 through Neat PDMS at 15 psig

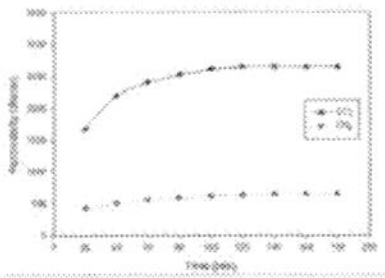

Figure 10. Transport Pattern of CO_2 and CH_4 through 10% SiO_2- PDMS at 15 psig

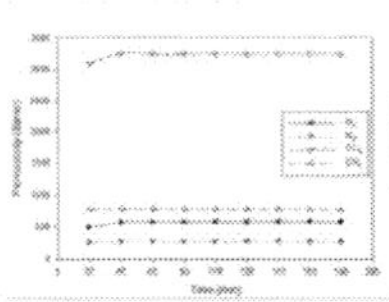

Figure 11. Pure Gas Permeabilities through Neat PDMS at 15 psig

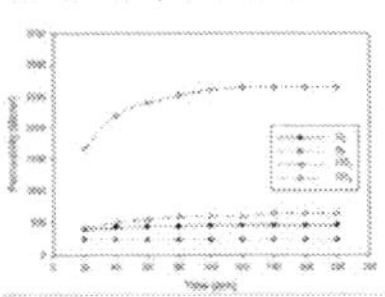

Figure 12. Pure Gas Permeabilities through 10% SiO_2- PDMS at 15 psig

CONCLUSIONS

- Techniques employed to characterize the neat PDMS and SiO_2-PDMS membrane confirmed good material compatibility between PDMS and fumed silica and uniform dispersion of silica in PDMS matrix.
- SiO_2-PDMS exhibited an improved thermal property compare to neat PDMS and the presence of silica in the polymer enhanced the ease of material fabrication.
- Incorporation of fumed silica into PDMS matrix was responsible for the reduction of N_2 gas permeability with time before attaining steady state condition as observed in SiO_2-PDMS
- 10% SiO_2-PDMS exhibited improved performance over neat PDMS for O_2/N_2 gas pair
- Neat PDMS had fair performance than 10% SiO_2-PDMS for the separation of CO_2/CH_4 gas pair.
- The incorporation of SiO_2 into polymer matrix gave rise to an increase in solubility dominant flux and a corresponding decrease in the diffusivity of the gases through the MMMs as evident in the permeability trend of: $P_{CO2} > P_{CH4} > P_{O2} > P_{N2}$.

RECOMMENDATIONS

- With the observed influence the presence of silica in PDMS had on the transport pattern of N_2, further studies should focus on synthesizing SiO_2-PDMS or other membranes with the ability to continuously and progressively slowdown the flux of N_2, while increasing the flux of O_2 for a long runtime before attaining steady state.
- To further confirm the performance of SiO_2-PDMS, the membrane should be tested with mixed gas permeation tests

Acknowledgements

Partial funding from National Science Foundation, NSF grant No. HRD-1242152 is gratefully acknowledged. Analytical support from Labs 132, 232, IRC; Chemistry Lab 210, NSB; Clean lab, Nanoengineering is appreciated with special thanks to Mr. Matthew Sharpe, Dr. Vishwanath Deshmane and Mr. Richard Abrokwah

2014 AIChE Annual Conference, Atlanta, Georgia, November 18, 2014

Continuous Olefin/Paraffin Separation with Imtex Facilitated Transport Membranes

Ali Hamza, Jamie Hughes, and Dale Kline

imtex™ innovative membrane technologies

REVOLUTIONARY MEMBRANE TECHNOLOGY

- Breakthrough Olefin/Paraffin Separation Technology
- Alternative to capital and energy intensive distillation
- High performance separation technology
  - High olefin purities achievable (typ. 99.5%+
  - Commercially viable permeation rates
- Stable Facilitated transport mechanism of olefins
- Suitable for C_2, C_3 and C_4 gas phase separations
- Integrates with existing petrochemical technologies
- Can replace or combine with conventional distillation to optimize feedstock utilization

HOW IT WORKS

- Patented facilitated transport membrane
- Made from chitosan, a natural source polymer
- Silver ions in the chitosan hydrogel matrix complex with double bond in C_2, C_3 and C_4 olefins
- Olefins permeate while paraffins are rejected
- High purity (> 99.5%) achieved

BENCH SCALE SYSTEM

APPLICATIONS

- Olefins production from cracking
- Olefins recovery from purge streams
- On purpose propylene and butenes production
- Oil refining and upgrading
- Speciality chemicals

BENEFITS

- Lower Capital Cost (CAPEX) - relative to distillation
- Lower Energy Cost – relative to distillation
- Reduced Environmental Impact from Emissions
- Recovery of Valuable Olefins from Waste Streams
- Upgrading Value of Olefin Containing Streams
- New Product Opportunities

THE MEMBRANE 'ELEMENT' AND STREAM FLOWS

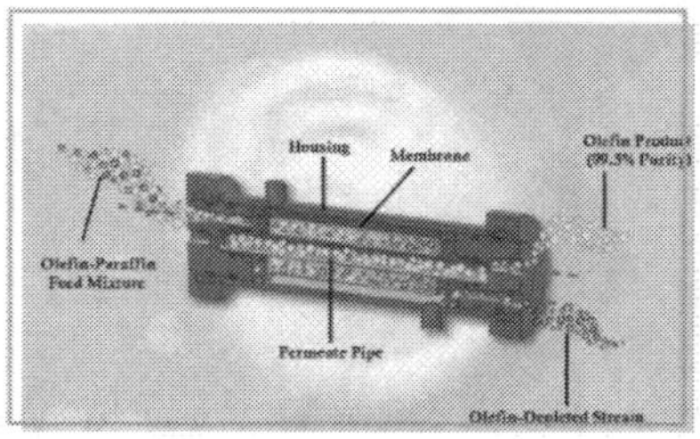

PILOT DEMONSTRATION SYSTEM

COMMERCIAL SCALE SYSTEM

COMMERCIAL MEMBRANE ELEMENT

- Throughput in the range of 50 kg/h to 100 kg/h per element
- Fundamental building blocks for commercial scale systems

SEPARATION OF C_2's, C_3's AND C_4's

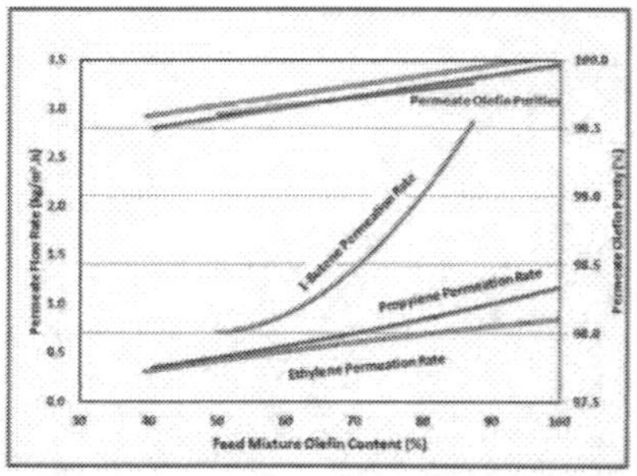

EXTENDED STABLE MEMBRANE OPERATION

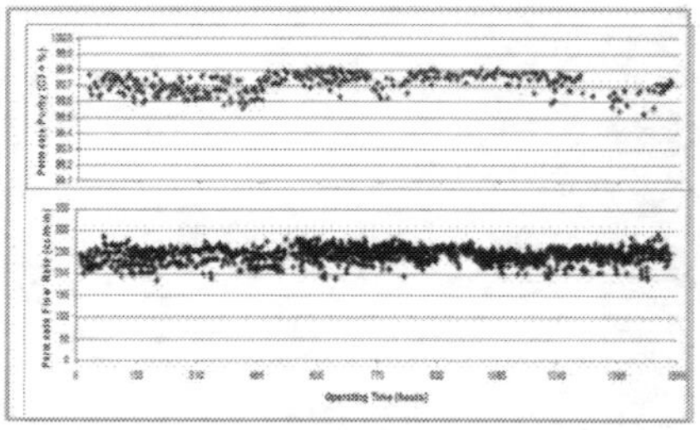

STABILITY OVER 1500 HOURS OF OPERATION

EXAMPLE CASE STUDY

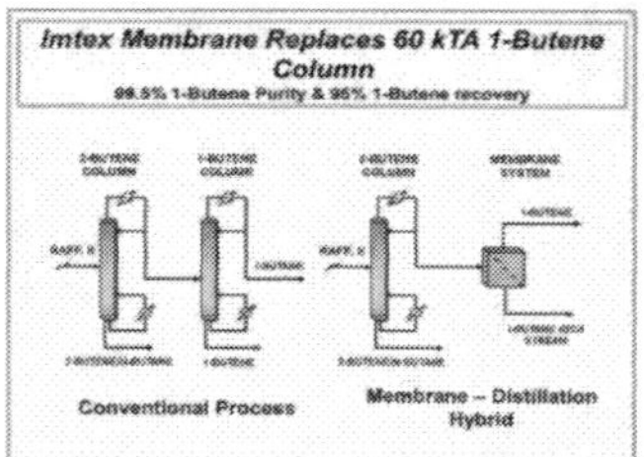

Process Scheme	CAPEX (MM$)	Heating Duty (MMKcal/hr)	Cooling Duty (MMKcal/hr)	Equipment Weight Dry (t)
Distillation	16	8.1	9.2	230
Membrane	2.2	0.7	1.3	40
Benefit	>85%	>90%	>85%	>80%

392324 Energy profile and life cycle assessment of an ethanol dehydration unit using an ionic liquid

Abstract:

To date use of inorganic salts for ethanol dehydration (EDH) process has been studied. Although several benefits for EDH process using an inorganic salt have been reported, the limited solubility of salts is a big drawback which makes the process hardly feasible. Therefore, any improvement in this chemical process must be based on using new solvents having high selectivity. Ionic liquids (ILs) are a relatively new class of chemicals which can be developed to replace molecular compounds (e.g. solvents). They have a wide range of industrial applications from CO2 capture to aliphatic/aromatic separations. Based on their properties, ionic liquids can be a promising alternative for the EDH process. Ionic liquids are powerful solvents and they possess the benefits of inorganic salts and liquid solvents at the same time. Therefore, utilization of ILs for ethanol-water separation is widely being considered. In this work we used COSMO-RS method to design a functionalized ionic liquid which is suitable for EDT process. The designed ionic liquid is in liquid phase, with relatively low viscosity and high solvation power.A process simulation software is used to find the energy profile of the separation process using the selected ionic liquid. Based on raw materials and energy/electricity used for EDH process and all upstream processes prior to ethanol-water separation, a cradle-to-gate life cycle assessment for production of 1 kg ethanol, functional unit, is performed.

375186 Advances in Micro Gas Chromatography (GC) - Fast Analysis of C1 to C8 Hydrocarbons for Mud Logging Applications within 2 Minutes Using a Temperature Programmable Micro GC Fusion

Abstract:

Continuous measurement of hydrocarbon gas concentration is crucial to mud loggers at the drilling site. Hydrocarbon measurement information provides operators with valuable geological formation insight that reveals where gas or oil pockets are. As a result of technological advancement in gas detection, sample composition can now be measured within minutes leading to nearly real-time analysis. Gas chromatography (GC) is a well established technique used to analyze extracted gas to provide accurate, on-site gas composition information. The INFICON Micro GC Fusion is the latest innovation that utilizes temperature programmable columns and micro-machined components to analyze C1-C8 compounds in less than two minutes or C1-C5 compounds in less than 30 seconds on a 24x7 basis. This expanded compound range and speed allow for more data to be acquired compared to traditional GC equipment. The instrument's transportable and compact architecture makes it ideal at drilling sites where space is limited.

376537 Charge-Tunable Polyampholytes for the Enhanced Flocculation of Cellular Biomass

Abstract:

The efficient concentration and separation of microorganisms from dilute culture suspensions is crucial to the success and productivity of biotechnological processes. Previously we have developed a novel approach for the dewatering and harvesting of microalgae using polyampholyte flocculants that can be recovered and recycled in response to their pH-dependent properties, thereby addressing issues of cost and contamination in downstream processing associated with conventional flocculants. Here we characterize these polyamphoteric flocculants with a tunable, mixed charge for the separation of microalgae species *Chlorella vulgaris* (UTEX 395), *Nannochloropsis gaditana* (CCMP526), *Chlamydomonas reinhardtii* (*Chlamydomonas* Genetics Center CC124), a yeast species *Saccharomyces cerevisiae* (BY4741), and wastewater sludge. The broad applicability of the polyamphoteric flocculants is thus demonstrated for biocolloidal suspensions important as feedstocks for renewable fuels, biopharmaceuticals, and ferments. The model ampholytic polyacrylamide-based flocculants have overall molecular charges dependent upon the system pH and provide strong electrostatic attractions to the diverse but primarily negatively-charged cellular surfaces of biological suspensions (*e.g.*, yeast and microalgae). The polyamphoteric flocculants provide enhanced flocculation efficiencies, up to 85% greater than comparable cationic homopolymer flocculants, with their ability to 1) adsorb to the diverse range of charge character in cellular suspensions and 2) operate in cultures over an extended range of pH.

Solvent Extraction of Nickel from Ni Sulfate solution using Mixer-Settler

Chul-Joo Kim*, Ho-Sung Yoon, Hyun-Ho Kim, Chul-Woo Nam

Mineral Resources Research Division,
Korea Institute of Geoscience and Mineral Resources, Daejeon, Republic of Korea
*cjkim@kigam.re.kr

Abstract: The objective of the present study is aimed to investigate for the extraction of nickel from Ni sulfate solution in a continuous scale mixer settler at the predetermined experimental conditions obtained from batch scale test study. The counter current continuous mixer settler composed of four stages of extraction and three stages of stripping run experiment was carried out at following optimized conditions: organic phase concentration is 0.8M D2EHPA(60% saponified with NaOH), phase ratio O : A =1.5: 1.0 for extraction process, phase ratio O : A = 1 : 1.5 in stripping stage, and operating time 72 hours. The retention time for total solution (aqueous + organic) in mixer were kept 13 min. In these conditions, the equilibrium state of extraction reached after 10 hours. The nickel extraction efficiency was determined to be above 99.9%. The final raffinate with equilibrium pH 5.19, obtained at equilibrium conditions reads to contain 0.005g/L Ni, confirming the nickel extraction was at about 99.9% with 0.05% loss of nickel. The nickel loaded in organic phase was recovered in aqueous phase by stripping of 3 stages, and the stripping efficiency was determined to be above 99.97%.

Keywords□solvent extraction, D2EPHA, nickel, leaching, manganese nodule

381416 A Method for Determining the Practical Limits of Polymer Pre-Concentration in Flash Evaporators: Modeling and Simulation

Abstract:

The polymerization of monomers is a process that produces commercially valuable materials that in their concentrated state typically have very high viscosities. These polymerization reactions are often exothermic and require large amounts of heat to be removed during the polymerization process. Due to the high viscosity of the concentrated polymer and the exothermic nature of the reactions, these polymerization reactions are often conducted using a large volume of solvents or a large excess of monomers. The solvents or excess monomers act as a heat sink to absorb the heat of reaction while also reducing the viscosity of the reaction mixture to a level that can be processed. The monomers and solvents are relatively volatile materials so they are referred to as volatiles. These volatiles must be reduced to very low levels to meet the final requirements of the polymer. The preferred method for removing these volatile materials from the polymer mixture is through several direct devolatilization steps. The direct devolatilization steps can be classified by the relative amount of volatile material removed during the step. Pre-Concentration refers to the first step where the largest amount of volatile material is removed, Final Devolatilization refers to the final step where the volatile level is reduced to the meet the final requirements, and Main Evaporation refers to the step(s) between the Pre-concentration step and the Final Evaporation step.

There are several commercially available systems used for the direct devolatilization steps, and in the very broadest terms these systems can be divided into two categories: Mechanically Agitated systems and Non-Mechanically Agitated systems. The most common Mechanically Agitated systems include High Volume Kneaders, Vented Extruders, and Thin-Film Evaporators. The Non-Mechanically Agitated systems include Flash Evaporators, Forced Circulation Evaporators and Falling Strand Evaporators.

Flash Evaporators have been used for over 80 years for the devolatilization of polymers. This technology is very robust and has been used for Pre-Concentration, Main Evaporation, and even for Final Devolatilization. However, this technology does not mechanically agitate the mixture so it is not suitable for all polymers / volatile mixtures. Using model transport, equilibrium and rheology equations, as well as available data for the main components of Flash Evaporators, a study was performed that shows the practical limits of the Flash Evaporator technology for use in Pre-Concentration systems.

The Acid Pressure Leaching of a Complex(Cu-Ni-Co-Fe) Matte

Ho-Sung Yoon*, Chul-Joo Kim, Hyun-Ho Kim, Chul-Woo Nam

Mineral Resources Research Division,
Korea Institute of Geoscience and Mineral Resources, Daejeon, Republic of Korea
**hsyoon@kigam.re.kr*

Abstract: Depletion of high grade sulphide/oxide ores containing base metals (Cu, Co and Ni) leads to the processing of low grade ores for growing demand in industry for several applications. Polymetallic sea nodules containing the above metals could be an alternate resource to meet the demand. Author's research groups of KIGAM, are being associated with the recovery of valuable metals from Pacific ocean Sea nodules since last one and half decades. Substantially, they have developed a successful flow diagram by following to Pyro-cum-Hydrometallrgical route which includes smelting of sea nodules to produce the matte and acid pressure leaching of matte followed by SX-EW of Cu, Co and Ni from the leach liquor. This route produces low volume of matte having high metal content, which is suitable to generate leach liquor with high concentration of Cu, Ni and Co. Present study describe about the pressure oxidative leaching of Cu, Co, Ni and Fe from complex artificial matte (Cu-Ni-Co-Fe) in sulphuric acid medium. The leaching efficiencies of the above metals with respect to acid concentration, pulp density and time were described. The dissolution behavior of the complex (Cu-Ni-Co-Fe) matte in sulfuric medium was studied at P_{O2} 15 atm and temperature 150°C. It was shown that, under comparable conditions, the dissolution of the complex matte at 5% (w/v) pulp density is faster than that at 10% (w/v) pulp density and finishes in 2 h in contrast to the dissolution at 10% (w/v) pulp density, which takes 4 h in both 1.0 L and as well as 15.0 L scale. The leaching yield of Cu, Co, Ni and Fe was 91.4%, 98.9%, 96.8% and 1.9%, respectively in the case of 15 L scale.

Keywords□acid pressure leaching, matte, base metal, sea nodule

AUBURN UNIVERSITY
SAMUEL GINN COLLEGE OF ENGINEERING

Comparison of Performance for Sea Salt Particles Filtration Between Nonwoven and Woven Activated Carbon Fiber Media

Pengfei Zhao*, Bruce J.Tatarchuk
Center of Microfibrous Material Manufacturing (CM³)
Department of Chemical Engineering, Auburn University, AL, 36849
pzz0005@tigermail.auburn.edu, Tel: 334-750-5816

Motivation

- ❑ Sea salt removal is critical need for a number of current and future air breathing ocean platforms, especially for SOFCs, which are currently being considered to be applied as power sources
  - ✓ *Naturally from wind stresses over whitecaps*
  - ✓ *Wake or the bow spray by ship itself*

- ❑ Harmful effects of sea salt particles on fuel cell systems
  - ✓ *Clog the finely divided and highly tortuous flow channels*
  - ✓ *Significantly degrade the performance of fuel cells (SOFCs)*

Objective

- ❑ Significantly enhance filtration performance for sea salt particles
  - ✓ *Higher filtration efficiency*
  - ✓ *Lower pressure drop across the filter media*
- ❑ Significantly improve sea salt particles loading capacity
  - ✓ *Longer filter lifetime*
  - ✓ *Lower cost for changing filters*
- ❑ Compare the performance between nonwoven and woven ACF filter media and decide what kind is better and could be applied in real applications

Experimental Setup

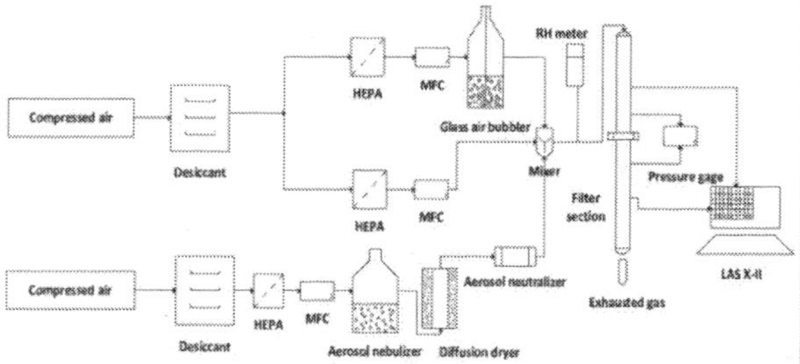

Nonwoven ACF Filter Media

Wet-laid Method

- ❑ Pre-form Preparation
  - ✓ *Disperse fibers in water using a blender*
  - ✓ *Preforms created by draining water away using TAPPI hand-sheet former*
- ❑ Sintering
  - ✓ *Further increase the bonding between fibers to make media strong and robust*

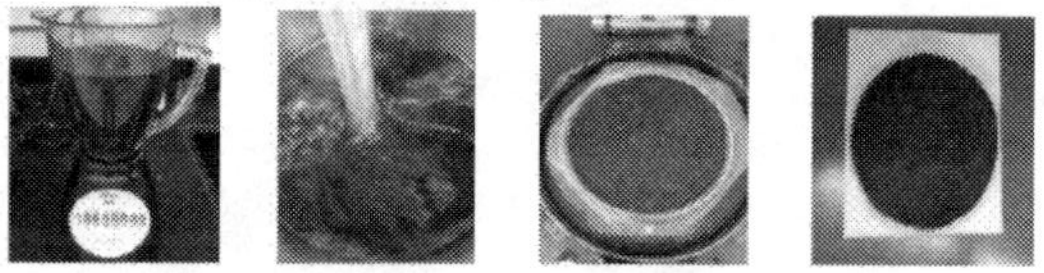

Advantages of ACF Filter Media

- ❑ Both particulates and gas impurities removal
  - ✓ *Cost saving*
- ❑ Higher particles filtration efficiency
  - ✓ *Smaller average fiber diameter and voidage*
- ❑ Both surface and depth filtration
  - ✓ *Lower pressure drop when particles loaded*

Nonwoven Knit Single weave Felt

Filtration Efficiency and Pressure Drop Results

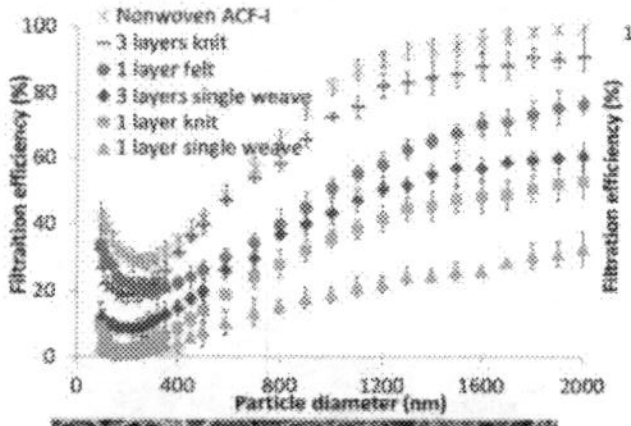

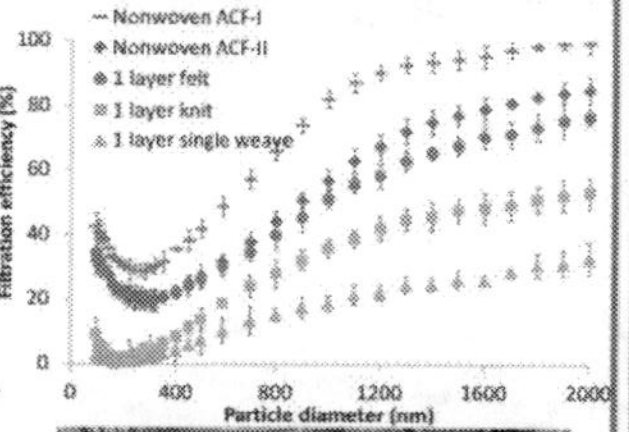

	ACF portion (%)	Number of layers	Thickness (mm)	Overall Filtration efficiency (%)	Pressure drop at 50 cm/s (Pa)	Pressure drop per thickness (Pa/mm)
Nonwoven ACF-I	66.7	1	3.8	60.76	274.02	72.11
Nonwoven ACF-II	66.7	1	2.2	46.18	130.50	59.32
Knit	100	1	0.50	21.92	78.12	156.24
		2	1.0	36.59	163.84	163.84
		3	1.45	50.49	247.20	170.48
Single weave	100	1	0.5	12.02	42.80	85.60
		3	1.45	23.91	140.40	96.83
Felt	100	1	2.0	40.21	114.00	57.00

Quality Factor (QF) Results

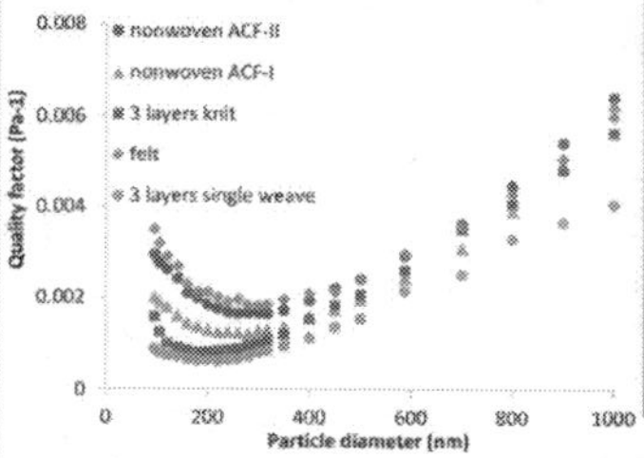

	Overall quality factor (Pa-1)
Nonwoven ACF-I	0.00216
Nonwoven ACF-II	0.00264
Felt	0.00277
3 layers knit	0.00179
2 layers knit	0.00156
1 layer knit	0.00150
3 layers single weave	0.00137
1 layer single weave	0.00132

Comparison of Particles Loading Performance

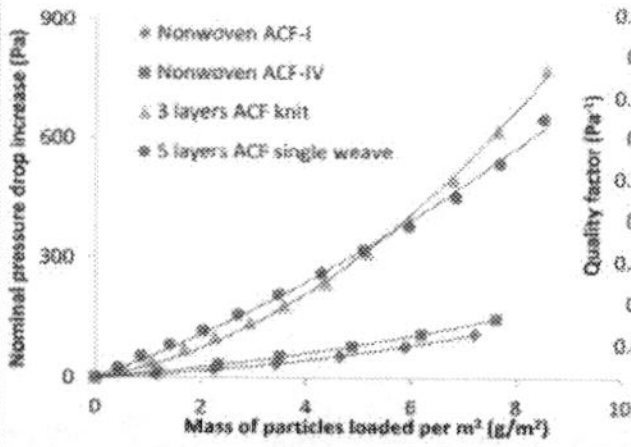

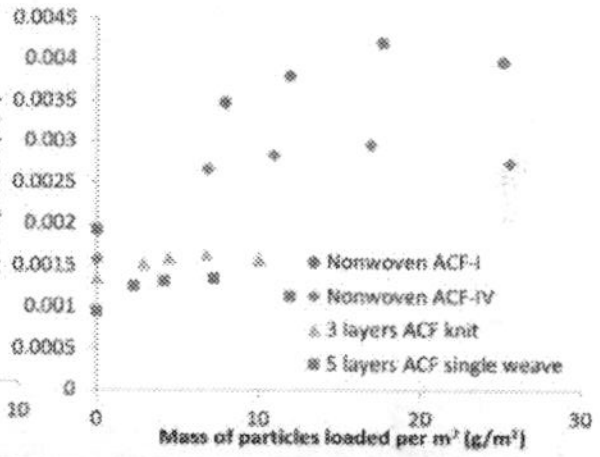

	Initial pressure drop (Pa)	Final pressure drop (Pa)	Particles loading capacity (g/m²)	Filter media life time (Days)
Nonwoven ACF-I	280.0	996	21.52	40
Nonwoven ACF-IV	471.7	996	17.47	31
3 layers ACF knit	247	996	8.52	14
5 layers ACF single weave	293	996	9.30	15

Conclusions

- ❑ Nonwoven filter media have shown much higher filtration efficiency and lower pressure drop across filter media compared to woven filter media, which is due to the homogeneous pore structure
- ❑ Meanwhile, nonwoven filter media could have greater particles loading capacity and longer filter media lifetime
- ❑ In real applications, nonwoven ACF filter media could be applied instead of woven filter media

387555 A Rapid Analytical Method for Quantification of Volatile Organic Acids in Fermentation Broth

Abstract:

Short chain fatty acids or volatile fatty acids (VFAs) (i.e. acetic, butyric, propionic, lactic, succinic, itaconic) are common products of biochemical conversion of biomass into useful chemicals. This group of compounds is considered to be one of the top chemical platforms for the production of high value products from renewable sources. Liquid chromatography is the commonly used method for analyzing VFAs in aqueous medium. However, the presence of microbial cells in the medium adds to the sample work-up prior to analysis. The sample work-up if not done correctly could cause clogging which could cause process downtime as well as reduce column lifespan. Thus, an alternative procedure was proposed in this study.

A simple and rapid analytical method for quantification of VFAs in fermentation broth was developed using solid phase microextraction (SPME) coupled to gas chromatography (GC). The rapid determination is useful in monitoring fermentative production of VFAs and/or their utilization to produce chemicals such as microbial lipids. Using the proposed analytical method, acids including acetic, propionic, butyric, valeric, citric, lactic, itaconic, and succininc were separated and quantitated. Detection limits and recovery for these organic acids in various fermentation broths using different waste substrates were determined. This method allows fast determination with little to no sample preparation requirement, therefore allowing fast turnaround times to conduct kinetic rates for fermentative VFA production and utilization..

389318 Water Filtration Using Moringa Oleifera Functionalized Sand

Abstract:

Popular methods currently used for cleaning water in developing communities include slow sand filtration and chlorination. However, these methods can be complicated and are not always practical to implement in many of the communities that need them. Seed proteins, such as the *Moringa oleifera* Cationic Protein (MOCP), can be used for water sanitation by flocculating sediment and bacteria from contaminated water. MOCP has a large number of positively charged residues, which makes it effective in binding to negatively charged bacteria and sediment creating large floccs that can settle or be filtered. In addition, the protein has a hydrophobic, proline rich region that gives the protein antibacterial properties. The cationic MOCP can be irreversibly bound to negatively charged sand particles and used in a continuous flow sand column. This approach is similar to the iron, aluminum, and quaternary amine functionalized sands that have been previously used to create a positive charged sand interfaces for removal of bacteria and viruses. This presentation will discuss the adsorption of MOCP to sand, characterization of the structure and stability of the protein, and characterization of the ability of the functionalized sand to adsorb bacteria and sediment. The use of the functionalized sand in columns for removal of bacteria and turbidity will also be discussed.

Reactive distillation for ETBE production: Setting the focus on the thermodynamic model

Gisela Durruty, María Soledad Díaz and Patricia Hoch. Chemical Engineering, Planta Piloto de Ingeniería Química (PLAPIQUI), Universidad Nacional del Sur - CONICET, Bahía Blanca, Argentina.

ETBE is a gasoline additive traditionally obtained from exothermic reaction of isobutene and ethanol. The conventional process includes the pretreatment of the butene and isobutene feed flows, reaction, purification, and recovery of non-reacted products, which renders an expensive process with high capital and operating costs (Weber de Menezes et al., 2008). A significant reduction in capital and operating costs can be achieved with reactive distillation (RD). This process combines reaction and distillation in one vessel, where products are continuously withdrawn, this resulting in a great benefit of the overall reaction, due to conversion increase. Reactive distillation also allows to reduce downstream processing and has a higher energy efficiency than conventional processes (Sharifzadeh, 2013).

In this work we propose a model for the optimization of the RD unit for ETBE production, with both a rigorous thermodynamic model and hydraulic constraints together with MESH equations (mass, equilibrium, summation and enthalpy). The column consists of a reactive section with two feedflows (one of ethanol and the other of the n-butene and isobutene mix) and a rectifying and a stripping section at the top and bottom, respectively. The reactive section is modeled using the correlations for reaction kinetics from Al-Arfaj & Luyben, 2002. The thermodynamic model proposed here includes models for calculation of activity for the liquid phase (UNIFAC, Fredenslund, Gmehling, & Rasmussen, 1977) and fugacity for the vapor phase (SRK, Soave, 1972) with quadratic mixing rules. To the best of our knowledge, there is no reliable full set of parameters for these thermodynamic models. A parameter estimation problem is then posed, to adjust the parameters from experimental data. Thermodynamic predictions show good agreement with the data. The reactive distillation column model is then posed using the obtained set of parameters. The parameter estimation and the optimization are formulated in an equation oriented environment within GAMS (Brooke et al., 2011). In the parameter estimation problem the objective function is to minimize the sum of squares of the difference between experimental and predicted pressure and in the optimization the objective function is to minimize the total annual cost. Data for column economics are taken from Sharifzadeh, 2013. Results show that reactive distillation is a good alternative to reduce capital and operating costs.

Keywords: reactive distillation, ETBE, UNIFAC, SRK.

References

Weber de Menezes, E, Cataluña, R. (2008). Optimization of the ETBE (ethyl tert-butyl ether) production process. Fuel Processing Tehnology, 89, 1148-1152. doi:10.1016/j.fuproc.2008.05.006

Sharifzadeh, M. (2013). Implementation of a steady-state inversely controlled process model for integrated design and control of an ETBE reactive distillation. *Chemical Engineering Science*, *92*, 21–39. doi:10.1016/j.ces.2013.01.026

Al-Arfaj, M.A., Luyben, W.L. (2002). Control study of ethyl tert-butyl ether reactive distillation. Ind. Eng. Chem. Res. 41 (16), 3784–3796.

Fredenslund, Gmehling, & Rasmussen. (1977). Vapor-liquid equilibria using UNIFAC - a group-contribution method. *Elsevier scientific publishing company*, 2800. doi:10.1016/B978-0-444-41621-6.50001-6

Soave G. (1972). Equilibrium constants for a modified Redlich-Kwong equation of state. *Chem. Eng. Sci.27*, 1197-1203.

Brooke, A., Kendrick, D., Meeraus, A., Raman, R. (2011), *GAMS, A User Guide*. Washington, DC, USA.

389666 An Evaluation of Processes for the Capture of Lipids for Later Use As Feedstocks for Fuels and Other Chemicals Production

Abstract:

Lipids are essential compounds in the production of bio-based diesels and other valuable chemicals. Lipids are often disposed into waters resulting in the production of vast quantities of wastewaters containing appreciable levels of lipids. Examples include both municipal and industrial wastewaters. The oil and grease (lipid) content of these wastewaters depends on the source of the wastewater (food processing vs. slaughter houses vs. municipal wastes); however, they typically range from 50 mg/L to 8,000 mg/L as fats, oils, and grease (FOG). Additionally, the chemical makeup of the lipids in these wastewaters can vary in terms of percent C-C saturation, free fatty acid content, phosphorous content, and carbon backbone branching – all of which greatly impact the end-use options and overall value of the lipids. Generally, these wastewaters are treated biologically or chemically and the lipids are simply degraded within these processes. The end result is that highly valuable lipids are literally wasted each day, and in fact, money is being expended to treat these wasted lipids as opposed to generating a profit from them. Another interesting source of lipids within aqueous media is when lipids are generated by cultivated oleaginous organisms that extra-cellularly release the lipids into fermentation broths. Examples organisms that have been proven to release lipids include *Saccharomyces cerevisiae* and *Escherichia coli* which can be cultivated for extracellular lipid secretion into the fermentation broth. However, there are currently very limited proven, cost-effective options for recovering these free lipids (whether it is in wastewaters or fermentation broths). Hence, a study was initiated by our team to evaluate economically and technically viable methods to capture free lipids and fatty acids that can be found in wastewaters and/or fermentation broths. This presentation will focus on processes indicating promise with preliminary results being discussed on the evaluation of some of the selected options along with profiles of the types of free lipids identified.

390775 Efficiency of Separation of Macromolecules Using Electrochromatography Technique

Abstract:

Recent years have seen an increasing interest in use the electrochromatography technique for separation of macromolecules. Electrochromatography is a chemical separation technique in analytical chemistry, biochemistry and molecular biology used to resolve and separate mostly large biomolecules such as proteins. This technique is a combination of size exclusion chromatography (gel filtration chromatography) and gel electrophoresis. These separation mechanisms operate essentially in superposition along the length of a gel filtration column to which an axial electric field gradient has been added. The molecules are separated by size due to the gel filtration mechanism and by electrophorectic mobility due to the gel electrophoresis mechanism. Additionally there are secondary chromatographic solute retention mechanisms.

In this work we have developed a modified electrochromatography system in order to able to provide useful estimates of protein retention time and velocity in a column packed with Sephadex gel and subjected to an electric field. Uniformly sized polymer particles and with different shapes were utilized as the shape template to compare their suitability as a uniformly sized packing material for small-scale high-performance liquid chromatography (HPLC) or capillary electrochromatography (CEC).

A correlation of electrophoretic mobility of peptide and proteins with respect to their charge, molecular mass, and asymmetry enables the selection of solute target molecules for electrochromatographic separations. The results regarding the efficiency of separation of protein mixtures, and the characterization of the different columns packed are compared and discussed in order to understand the influence of the particles (used as template) in the separation process. Directions for future research are also included.

Improving Propylene/Propane Separation Performance of Zeolitic-Imidasolate Framework (ZIF-8) Membranes Synthesized by a Counter Diffusion-Based *In Situ* Method

Hyuk Taek Kwon[1], and Hae-Kwon Jeong*[1,2]

[1]Artie McFerrin Department of Chemical Engineering and [2]Materials Science and Engineering Program

Texas A&M University, College Station, TX 77843-3122

* Corresponding author: hjeong7@mail.che.tamu.edu

Zeolitic-imidazolate frameworks (ZIFs) are a sub-set of metal organic frameworks (MOFs) with zeolite topologies, comprising of divalent metal cations (Co^{2+} or Zn^{2+}) linked by imidazole-based ligands.[1] Their exceptional thermal and chemical stabilities coupled with ultra-microporosity make them desirable candidates especially for gas separation membranes.[2]

Particularly, zeolitic-imidazolate framework ZIF-8 has shown great potential for high resolution separation of propylene/propane mixtures based on the size exclusion principle. Recently, we have developed a simple and advantageous protocol for ZIF-8 membrane synthesis based on a counter diffusion concept, leading to highly propylene-permselective ZIF-8 membranes. Considering its potential for industrial-scale applications, it is material to further improve the performance of ZIF-8 membrane for process economic.

To this end, we report that the separation performances of ZIF-8 membranes can be further improved by controlling membrane microstructures via manipulating important synthesis variables as well as activation processes. Altering important synthesis variables such as the catalyst to ligand ratios, nature of zinc salts and water contents yielded the membranes with discernible microstructures, which led to notable enhancement in propylene/propane separation factors. Furthermore, solvothermal activation processes were found to be effective in significantly increasing the propylene permeance while maintaining decent propylene/propane separation factors.

References

1. K. S. Park, Z. Ni, A. P. Côté, J. Y. Choi, R. Huang, F. J. Uribe-Romo, H. K. Chae, M. O'Keeffe and O. M. Yaghi, Proc. Nat. Acad. Sci. U.S.A. **103** (27), 10186-10191 (2006).
2. M. Shah, M. C. McCarthy, S. Sachdeva, A. K. Lee and H.-K. Jeong, Ind. Eng. Chem. Res. **51** (5), 2179-2199 (2011).

High performance hollow fibre membranes for hydrogen separation based on lanthanum tungstate

Chen Yan, Liao Qing, Cheng Shunfan, Wang Yanjie, Wang Haihui*

School of Chemistry & Chemical Engineering, South China University of Technology, No. 381 Wushan Road, Guangzhou 510640, P. R. China

* hhwang@scut.edu.cn

Nowadays, lanthanum tungstate type high temperature mixed electron and proton conductors are proposed as a promising oxide for hydrogen separation from hydrogen containing gases. Because of there are not alkali earth metals in the lanthanum tungstate oxides, we can speculate that they are stable under CO_2 atmosphere. However, the hydrogen permeation flux of the lanthanum tungstate membranes is very low. In order to improve the hydrogen permeation flux, in this study, the Mo-substituted lanthanum tungstate hollow fibre membranes were prepared and studied. The U-shaped hollow fibre membranes based on Mo-substituted lanthanum tungstate $La_{5.5}WO_{11.25-\delta}$ have been prepared using wet-spinning / sintering technology. The dense hollow fibres were obtained, and the wall thickness was 225 μm. Furthermore, the effective thickness was less than 80 μm due to the presence of the fingerlike structure. The hydrogen permeation and stability of the hollow fibres were investigated systematically. The effects of the sweep gas flow rate, the hydrogen partial pressures in feed side, the operation temperatures and the presence of water vapour in both sides were experimentally studied. A relative high hydrogen permeation flux of 0.35 ml/min.cm^2 was obtained at 950 °C when the water vapour was presented in sweep side (feeding with 80% H_2 in He). The hydrogen permeation flux was almost steady during 700 min operation. XRD and SEM analysis of the spent hollow fibre membrane showed the stability of Mo-substituted lanthanum tungstate $La_{5.5}WO_{11.25-\delta}$ hollow fibre membranes. The results indicated that the Mo-substituted lanthanum tungstate hollow fibre membranes have a promising application in the separation of hydrogen.

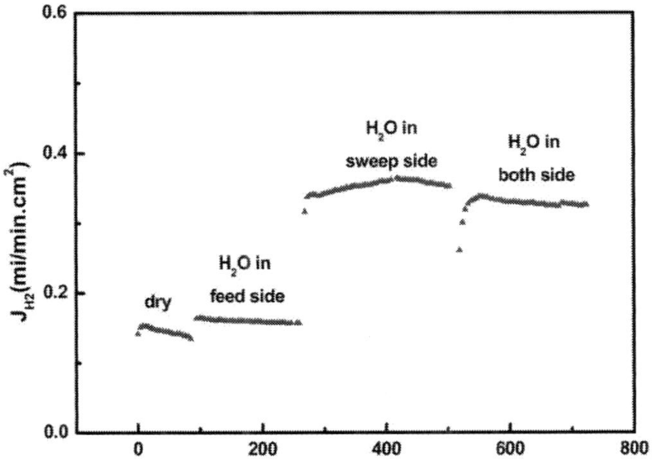

Figure 1. Hydrogen permeation flux as a function of time under different conditions at 950 °C(P_{H2O} = 0.025 atm). Conditions: 80% H_2 in He as feed gas, F_{Ar} = 80 ml/min, F_{H2+He} = 60ml/min.

References

1. S. Escolastico, J. Seeger, S. Roitsch, M. Ivanova, W.A. Meulenberg, and J.M. Serra, *ChemSusChem*, 6, 1523(2013).
2. M. Amsif, A. Magraso, D. Marrero-Lopez, J.C. Ruiz-Morales, J. Canales-Vazquez, and P. Nunez, *Chem. Mater.*, 24, 3868(2012).
3. S. Escolastico, S. Somacescu, and J.M. Serra, *Chem. Mater.*, 26, 982(2014).
4. S. Escolastico, C. Solis, T. Scherb, G. Schumacher, J.M. Serra, *J. Membr. Sci.*, 444, 276(2013).
5. S. Escolastico, C. Solis, J.M. Serra, *Int. J. Hydrogen Energy*, 36, 11946(2011).
6. V. Gil, J. Gurauskis, C. KjΦlseth, K. Wiik, M. A. Einarsrud, *Int. J. Hydrogen Energy*, 38, 3087(2013).

381606 Membranes Containing Cation-Exchanged Zeolites for CO2 Separation

Abstract:

As concerns over the impact of rising concentrations of atmospheric CO_2 on climate change continue to mount, there is an urgent need for the development and implementation of technologies that capture CO_2 in anthropogenic sources. The CO_2 separation technology can also be implemented in many energy-production processes such as natural gas treatment, hydrogen purification in steam reforming process, and the recovery of CH_4 from biogas produced by anaerobic digestion of wastes. Meanwhile, membrane-based molecular separations, energy-efficient and environment-friendly processes that can replace conventional energy-intensive separation methodologies, have attracted a great amount of interests. However, current polymeric membranes which can be produced in a large scale due to their excellent processibility often showed the limited performance. To design high-performance membranes while utilizing the advantages of polymeric materials, various nanoporous materials have been incorporated into polymer membranes.

Most recently, we have found that cation-exchanged zeolites exhibit outstanding CO_2 separation properties. For example, the Ca-A zeolite outperforms Mg_2(dobdc) metal-organic framework, one of the best performing CO_2-capture materials reported so far, in terms of the CO_2 adsorption capacity and CO_2/N_2 selectivity at the conditions relevant to dry flue gas. Furthermore, compared to metal-organic frameworks, zeolites are inexpensive nanoporous materials, which are currently produced in large scales for many industrial applications. Motivated by these facts, membranes containing cation-exchanged zeolites were fabricated for applications in CO_2/N_2 and CO_2/CH_4 separations. Not only gas permeation properties of the membranes but also the methodology to control the zeolite-polymer interfacial contact in mixed-matrix membranes, which has been a key prerequisite in designing high-quality zeolite/glassy polymer composite membranes, will also be discussed.

384547 Layer-By-Layer Assemblies As Size-Selective Polymeric Membranes

Abstract:

The production of inexpensive, high purity hydrogen remains a critical challenge to improving the sustainability of fossil fuels and for realizing renewable and clean energy sources capable of displacing fossil fuels. Polymeric gas separation membranes combining low production costs with robust mechanical properties have found broad industrial application for O_2 and N_2 enrichment of air, upgrading of natural gas and hydrogen recovery from ammonia. In the absence of structure (i.e. in amorphous glassy or rubbery films), increasing the free volume of size-selective polymer membranes improves the permeability at the cost of reducing the permselectivity. This translates to an upper bound in separation performance achievable by dense, homogeneous and amorphous polymeric films. [1,2] The authors have recently demonstrated that layer-by-layer (LbL) deposition techniques provide a robust, low-cost means to introduce size-selective functionality via manipulation of polymer processing conditions.[4]

Extension of the LbL assembly technique to gas purification membranes promises the ability to tune overall gas permeabilities and permselectivities through manipulation of film nanostructure by varying deposition conditions. In this presentation, the authors report the use of a branched polyethylenimine (PEI) / poly (acrylic acid) (PAA) system is used for achieving breakthroughs in light gas separations. This PEI/PAA assembly appears to have a "scrambled salt" structure, resulting in a highly interpenetrating network of high density which demonstrates size-selective gas separations in excess of Robeson's 'upper boundary,' [1,2] which we attribute to the structure imparted by a high degree of 'ionic-crosslinking.' Details of materials analysis and separation measurements will be presented. The potential of the LbL approach to gas separation coatings to impact to several other significant applications of social and industrial importance will also be discussed.

[1] L. M. Robeson, *J. Membr. Sci.* **1991**, 62, 165.

[2] L. M. Robeson, *J. Membr. Sci.* **2008**, 320, 390.

[3] D. Kim, P. Tzeng, K. Barnett, Y-H. Yang, B.A. Wilhite, J. Grunlan, Adv. Mater. 2014, 26, 746-751.

Engineering Advanced Membranes For Non-CO_2 Greenhouse Gas Removal

Seung hwan Park[1,2], Wooram Kang[3], Sang Hyup Lee[3], Jong Suk Lee[1,*]

[1]Center for environment, health and welfare research, Korea Institute of Science and Technology, Hwarangno 14-gil 5, Seongbuk-gu, Seoul, 136-701, Republic of Korea
[2]Department of chemical and biological engineering, Korea University, Seoul, 136-731, Republic of Korea
[3]Center for water resource cycler, Korea Institute of Science and Technology, Hwarangno 14-gil 5, Seongbuk-gu, Seoul, 136-701, Republic of Korea

Perfluorinated gases including SF_6, CF_4 and NF_3 have been used in the various applications such as insulating gas, inert gas, and etchant gas, etc. The global warming potential is defined as a relative extent of global warming induced by the gas of interest compared to the same mass of carbon dioxide over a given time span. Even though the absolute amount of perfluorinated gas emissions to atmosphere is much smaller than CO_2, their global warming potential is substantially higher (i.e. 22,200, 16,600, 6,500 in the order of SF_6, NF_3 and CF_4). Conventionally, their emissions were mitigated by adsorption, distillation or metal-ion exchange; however, they are quite energy-intensive. Membrane-based gas separation technique can be a potential alternative due to its better energy efficiency and environment-friendly characteristics. Polymer-based membrane especially has been actively investigated due to its good cost efficiency and easy processability. However, the inherent limit of their separation performance, *so-called* the upper bound limit has given us restrictions and challanges. This presentation will introduce advanced membranes with their separation-associated challenges.

Gas separation of PEBAX blended membranes with poly (ethylene glycol) and glutaraldehyde

K. Y. Kim[1], S. Y. Kang[1], H. S. Shin[2], C. S. Kim[2], J. W. Rhim[1]

[1]*Department of Chemical Engineering, Hannam University, Daejeon, Korea*
[2]*Techwin Co. Ltd., Cheongju City, Choongcheongbuk-do, Korea*
(jwrhim@hnu.kr)

Combustion of sulfur containing fuels, such as coal, in power plant is the major emission source of sulfur dioxide(SO_2). SO_2 emitted into atmosphere is the main cause of acid rain which is one of the most pervasive air pollutants. In addition, the emission and concentration of SO_2 in the emitted gas tend to be increased due to the low-quality coal use. In case where the emitted SO_2 is incoming to CO_2 capture process, such as absorption process, the deterioration in process performance by the furious flames of absorbents and the contamination in the captured CO_2 would take place. Therefore, the inflow of SO_2 to the absorption process is strictly restricted and, in general, must not exceed 10 ppm.

In this study, PEBA 1657, which is known to have high bonding strength to acid gases, was used blended with polyethylene glycol (PEG) to enhance permeability and again with glutaraldehyde (GA) to improve permeation selectivity and durability. Poly(ether-block-amide) 1657 (PEBAX 1657) blended membranes with molecular weight 400 polyethylene glycol (PEG 400) were prepared and their permeability tested for the gases N_2, O_2, CH_4, CO_2, and SO_2 by the time-lag method. The permeation characteristics were investigated in terms of diffusivity and solubility, which are dominant factors for gas transport. With the addition of PEG 400, the permeability of all the gases increased and also the ideal selectivity for several pair gases was enhanced. In particular, selectivity for CO_2/N_2 ranged from 53.2 (pristine PEBAX 1657 membrane) to 84.1 (50% PEG 400 added), for SO_2/CO_2 from 38.9 to 50.7, and for CO_2/CH_4 from 17.7 to 31.4. The increase of both permeability and selectivity is mainly because of the increase of solubility of the gases, especially CO_2 and SO_2. To obtain durability against water vapor, glutaraldehyde (GA) was added to the PEBAX 1657/PEG 400 blended membranes. As a result, permeability decreased owing to a reduction of the free volume and ether oxide units, which are the main factors in elevating the permeability for the blended membranes, and also selectivity decreased; however, we believe that the durability of the resulting membranes would increase. The details will be discussed at the presentation.

COMPARATIVE PREPARATION OF ZIF-7 AND ZIF-8 MEMBRANES SUPPORTED ON TEZONTLE AND α-ALUMINA.

Fernando Romero-Romero[1*], Yadira Frasco-Castro[1], Carlos Barrera-Díaz[2], Victor Varela-Guerrero[2] and Hae-Kwon Jeong[3].

(1) Universidad Autónoma del Estado de México, Estado de México, México, (2)Ingeniería Química, Universidad Autónoma del Estado de México, México, (3)Chemical Engineering, Texas A & M University

*mail: fe_rom2@yahoo.com.mx

Abstract

The Zeolitic Imidazolate Framework (ZIFs), represent an important alternative because they have thermal and chemical stability over 700 °C. According to research conducted mainly using α-alumina ceramic having features such as mechanical and chemical resistance that makes it a suitable material for the manufacture of supports; although it is important to note that the materials can be found in nature such as volcanic rock (Tezontle) which can perform this function as it is characterized by high porosity, surface area represents plus it is a low cost and high availability. The proposal for the construction of a solid absorbent material such as zeolitic imidazolate framework (ZIF-7, ZIF-8), that can be supported on solid materials such as α-alumina and tezontle based on support surface modification and in situ solvothermal growth to in this paper; in order to compare to get free of fractures that may be applicable in membrane separation processes. The results by the scanning electron microscope (SEM) was done to check the morphology of the membrane and the crystals and DRX for phase structure.

Keywords: Zeolitic Imidazolate Framework, Tezontle, α-alumina.

Introduction

Metal Organic Frameworks (MOFs) are crystalline materials consist of metal ions or metal clusters with organic ligands coordinated[1-3]. It´s synthesis have been aroused great interest because your remarkable properties as the large surface area, accessible pore volume and functionalities allowing their utilization in various applications as in gas separation, storage, catalysis and sensor[4-7].

According at the information of the last years for MOFs, there are strong interests to fabricate uniform, defect-free films thin for continuous membrane of based separations, membrane reactors, and other advanced applications[8]. There are various techniques for the fabrication of MOF membranes and films in the actuality, some of these are the direct synthesis and secondary growth methods; and coordination polymer films with Langmuir–Blodgett and layer-by-layer thin-film preparation techniques on which the literature is abundant. The importance of membranes technology has increased during the last two decades, MOF membranes and films have been studied intensively, and several excellent reviews have been published[9, 10]. However for membranes synthesis not only the problem

with growth of the crystal of form homogeneous on support but also the thermal and chemical stability[11, 12].
The Zeolitic Imidazolate Frameworks (ZIFs), a new subclass of MOFs, have emerged as a novel type of crystalline porous, are three-dimensional structures constructed from tetrahedral metal ions (Zn, Co) bridged by imidazolate (Im). Due to the M-Im-M angle is similar to the Si-O-Si angle (145°) in zeolites (Fig. 1), combines highly desirable properties of both zeolites and MOFs, such as crystallinity, microporosity, high surface , and exceptional thermal and chemical stability[13, 14].

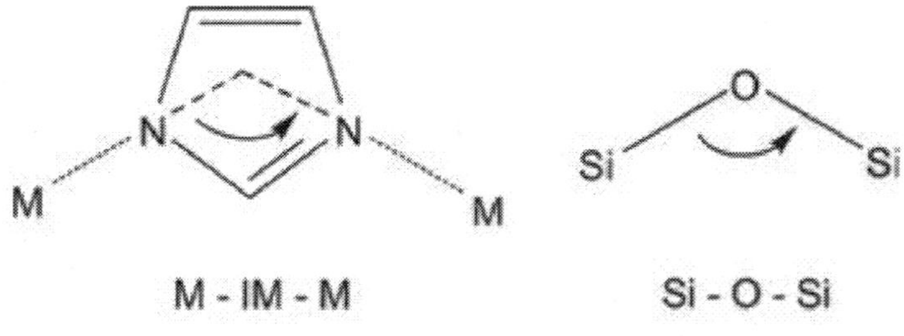

Fig. 1. The bridging angles in metal IMs (**1**) and zeolites (**2**)[13].

Most of the recently published studies related on the ZIF Film growth are focusing on porous supports (α-alumina[15-17], titania[18], zinc oxide[10, 19]), this inorganic materials generally possess the desired mechanical strength, defined pore sized, high thermal and chemical stability, but for many technical applications they are too expensive[10]. Therefore the tezontle has been considered here because naturally available for combines the economical processing with the promising properties of the zeolitic imidazolate frameworks (ZIFs).
Here, we report the proposal for the construction of a solid absorbent material such as zeolitic imidazolate framework (ZIF-7, ZIF-8), that can be supported on solid materials such as α-alumina and tezontle based on support surface modification and in situ solvothermal growth to in this paper; in order to compare to get free of fractures that may be applicable in membrane separation processes.

2. Experimental

2.1 Chemicals

Zn nitrate hexahydrate ($Zn(NO3)_2$-6H2O, Sigma-Aldrich, >98%) as metal source, 2-methylimidazole (HmIm, $C_4H_6N_2$, Sigma-Aldrich, >99%) and benzimidazole ($C_7H_6N_2$, >99%) were used as ligands, sodium formate (NaCOOH, Sigma-Aldrich, >95%) and methanol (MeOH, >95%) were used as received without further purification for ZIF-8 and ZIF-7 synthesis.

2.1 *α* –alumina and tezontle support

10g of Al_2O_3 99.7 % Sigma–Aldrich were storage in a flask and after was hydrated with 1g of distilled water and mix. Subsequently, one sample of 2g of this mixture was taken to be pressed inside of a holder of stainless steel using 6kg/cm^2 to form a disk substrate with 20mm of diameter and 0.5mm of thickness. This support was entered into the furnace at 1100 °C for 24h, and after cooled down until room temperature slowly.

To support tezontle used 10g and after was hydrated with 1g of distilled water and mix. Subsequently, one sample of 2g of this mixture was taken to be pressed inside of a holder of stainless steel using 6kg/cm^2 to form a disk substrate with 20mm of diameter and 0.5mm of thickness. This support was entered into the furnace at 1000 °C for 24h, and after cooled down until room temperature slowly.

2.2 Synthesis of the membrane of ZIF-7

The support was entered into the oven to 120°C/2h, and the linker solution with 3.62g of benzimidazole ($C_7H_6N_2$ 99 % Sigma-Aldrich, dissolved in 50ml of ethanol) was dropped on the surface of the support and evaporated during 20 min, this process was based on the method reported by McCarthy et al. 2010. After the support was sonicated in ethanol for 30s and got back into the oven, the process was repeated about six times.

For the membrane growth, one solution was prepared with 0.81g of the linker benzimidazole 99% Sigma-Aldrich, 0.70 g of sodium formate and 1.53g of $ZnNO_3 * 6H_2O$ 98% Sigma-Aldrich, dissolved in 40ml of ethanol, this solution was placed in the teflon autoclave of 45ml at 120 °C during 3 hours in contact with the support. Then this was cooled down at room temperature and finally the support was rinsed with ethanol and dried.

2.3 Synthesis of the membrane of ZIF-8

The support was entered into the oven to 120°C/2h, and the linker solution with 3.62 g of 2-methylimidazole (99 % Sigma-Aldrich, dissolved in 50ml of methanol) was dropped on the surface of the support and evaporated during 20 min at room temperature. After the support was sonicated in ethanol for 30s and got back into the oven, the process was repeated about six times.

For the membrane growth, one solution was prepared with 1.61g of the linker 2-methylimidazole 99% Sigma-Aldrich, 2.87 g of sodium formate and 1.08g of $ZnNO_3 * 6H_2O$ 98% Sigma-Aldrich, dissolved in 40ml of methanol, this solution was placed in the teflon autoclave of 45ml at 120 °C during 4 hours in contact with the support. Then this was cooled down at room temperature and finally the support was rinsed with ethanol and dried.

2.4 Characterization

2.4.1 Scanning electron microscopy (SEM)

The films and membranes were simply broken and were pre-coated with a thin gold film to improve conductivity. The size and morphology of the ZIF samples were investigated on a JEOL JSM- 6510LV SEM operated at 15 kV.

2.4.2 X-ray diffraction (XRD)

X-ray diffraction (XRD) measurements of all ZIF films were taken with a Rigaku Miniflex 600 diffractometer with Cu KR radiation (λ=1.54 A °) with 40kV and 15 mA.

3. Results and discussions

Fig. 2 shows the scanning electron microscopy (SEM) image of the ZIF-7 sample prepared in this work. It can be seen that the crystals of ZIF-7 are in cubic or rectangular shape, with length 5–20 μm (Fig. 2a), whereas in Fig. 2b,c the phase structure of the ZIF-7 sample on α-alumina or tezontle respectively depicts a continuous film with larger, well-intergrown grains (approximately 5-10 μm) and Fig. 2d confirmed the interaction between the support and the crystal of ZIF-7 (approximately 10 μm thick).

Fig. 2. a) crystals of ZIF-7, b) the top of view of the membrane of ZIF-7 on α –alumina, c) the top of view of the membrane of ZIF-7 on tezontle and d) cross-section of the membrane d) crystals of ZIF-8, e) the top of view of the membrane of ZIF-8 on α –alumina, f) the top of view of the membrane of ZIF-8 on tezontle.

In Fig. 2e, the SEM image of ZIF-8 of crystals synthesized, confirm the expected morphology and the Fig. 2f,g SEM images of continuous ZIF-8 films produced using the surface modification and subsequent in situ solvothermal growth technique are shown. Although both films consist entirely of ZIF-8 crystals, they exhibit drastically different microstructures. The film in Fig. 2g is continuous but composed of smaller, poorly intergrown crystals (approximately 1-5 μm).

The phase structure of the ZIF-8 sample was confirmed by the powder X-ray diffraction patterns shown in Fig. 3. The close match of the locations of the main peaks (2θ) from simulated and the experimental XRD pattern suggests that ZIF-8 sample synthesized in this work has the correct phase structure with a good crystallinity.

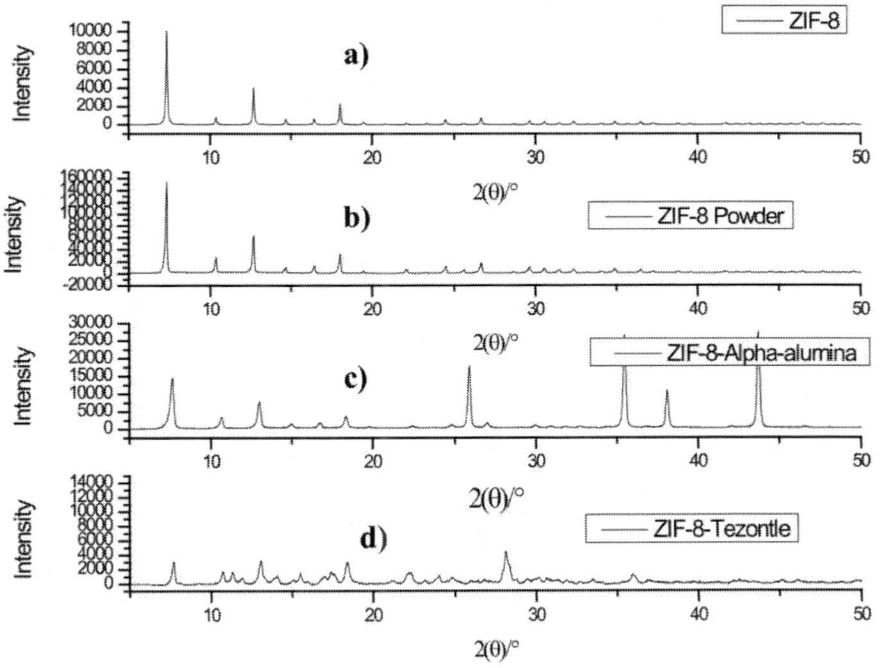

Fig. 3 a) XRD pattern of ZIF-8 b) XRD pattern of ZIF-8 synthetized c) XRD pattern of ZIF-8 membrane on α-alumina, well intergrown crystals on tezontle and (b) XRD pattern of ZIF-8 membrane on α-alumina.

4. Conclusion

In conclusion, we have compared the form of growth of crystal ZIF-7 and ZIF-8 with different supports for ZIF membrane, , according to the result set presented SEM and XRD through synthesis that consists of modifying porous supports with the organic ligand of the ZIF of interest followed by conventional solvothermal synthesis, but it is necessary to evaluate the interaction of the crystal with XPS Tezontle specifically for comparison with alumina which has already been reported, in addition to the permeation tests.

Acknowledgment
We are thankful for Victor Varela Guerrero, Rafael Vilchis Nestor,for their helpful comments that improved the quality of the paper.

References

1. Yaghi, O. M., Metal-organic Frameworks: A tale of two entanglements. *Nat Mater* **2007,** *6,* 92-93.
2. Shekhah, O.; Wang, H.; Kowarik, S.; Schreiber, F.; Paulus, M.; Tolan, M.; Sternemann, C.; Evers, F.; Zacher, D.; Fischer, R. A.; Wöll, C., Step-by-Step Route for the Synthesis of Metal−Organic Frameworks. *Journal of the American Chemical Society* **2007,** *129,* 15118-15119.
3. Rowsell, J. L. C.; Yaghi, O. M., Metal–organic frameworks: a new class of porous materials. *Microporous and Mesoporous Materials* **2004,** *73,* 3-14.
4. DeCoste, J. B.; Peterson, G. W., Metal–Organic Frameworks for Air Purification of Toxic Chemicals. *Chemical Reviews* **2014,** *114,* 5695-5727.
5. Dhakshinamoorthy, A.; Opanasenko, M.; Cejka, J.; Garcia, H., Metal organic frameworks as heterogeneous catalysts for the production of fine chemicals. *Catalysis Science & Technology* **2013,** *3,* 2509-2540.
6. Dincă, M.; Long, J. R., Hydrogen Storage in Microporous Metal–Organic Frameworks with Exposed Metal Sites. *Angewandte Chemie International Edition* **2008,** *47,* 6766-6779.
7. Gücüyener, C.; van den Bergh, J.; Gascon, J.; Kapteijn, F., Ethane/Ethene Separation Turned on Its Head: Selective Ethane Adsorption on the Metal−Organic Framework ZIF-7 through a Gate-Opening Mechanism. *Journal of the American Chemical Society* **2010,** *132,* 17704-17706.
8. Yoo, Y.; Jeong, H.-K., Rapid fabrication of metal organic framework thin films using microwave-induced thermal deposition. *Chemical Communications* **2008**, 2441-2443.
9. Bétard, A.; Fischer, R. A., Metal–Organic Framework Thin Films: From Fundamentals to Applications. *Chemical Reviews* **2011,** *112,* 1055-1083.
10. Barankova, E.; Pradeep, N.; Peinemann, K.-V., Zeolite-imidazolate framework (ZIF-8) membrane synthesis on a mixed-matrix substrate. *Chemical Communications* **2013,** *49,* 9419-9421.
11. Bux, H.; Chmelik, C.; van Baten, J. M.; Krishna, R.; Caro, J., Novel MOF-Membrane for Molecular Sieving Predicted by IR-Diffusion Studies and Molecular Modeling. *Advanced Materials* **2010,** *22,* 4741-4743.
12. Centrone, A.; Yang, Y.; Speakman, S.; Bromberg, L.; Rutledge, G. C.; Hatton, T. A., Growth of Metal−Organic Frameworks on Polymer Surfaces. *Journal of the American Chemical Society* **2010,** *132,* 15687-15691.
13. Park, K. S.; Ni, Z.; Côté, A. P.; Choi, J. Y.; Huang, R.; Uribe-Romo, F. J.; Chae, H. K.; O'Keeffe, M.; Yaghi, O. M., Exceptional chemical and thermal stability of zeolitic

imidazolate frameworks. *Proceedings of the National Academy of Sciences* **2006,** *103*, 10186-10191.
14. Tan, J. C.; Bennett, T. D.; Cheetham, A. K., Chemical structure, network topology, and porosity effects on the mechanical properties of Zeolitic Imidazolate Frameworks. *Proceedings of the National Academy of Sciences* **2010,** *107*, 9938-9943.
15. McCarthy, M. C.; Varela-Guerrero, V.; Barnett, G. V.; Jeong, H.-K., Synthesis of Zeolitic Imidazolate Framework Films and Membranes with Controlled Microstructures. *Langmuir* **2010,** *26*, 14636-14641.
16. Kwon, H. T.; Jeong, H.-K., In Situ Synthesis of Thin Zeolitic–Imidazolate Framework ZIF-8 Membranes Exhibiting Exceptionally High Propylene/Propane Separation. *Journal of the American Chemical Society* **2013,** *135*, 10763-10768.
17. Li, Y.; Liang, F.; Bux, H.; Yang, W.; Caro, J., Zeolitic imidazolate framework ZIF-7 based molecular sieve membrane for hydrogen separation. *Journal of Membrane Science* **2010,** *354*, 48-54.
18. Bux, H.; Chmelik, C.; Krishna, R.; Caro, J., Ethene/ethane separation by the MOF membrane ZIF-8: Molecular correlation of permeation, adsorption, diffusion. *Journal of Membrane Science* **2011,** *369*, 284-289.
19. Dong, X.; Huang, K.; Liu, S.; Ren, R.; Jin, W.; Lin, Y. S., Synthesis of zeolitic imidazolate framework-78 molecular-sieve membrane: defect formation and elimination. *Journal of Materials Chemistry* **2012,** *22*, 19222-19227.

377131 Computational Screening of Metal-Organic Frameworks for Energy Applications

Abstract:

The demand for developing energy-efficient methods to separate CO_2 from flue gas and natural gas has been rapidly increasing. Capturing CO_2 from coal-fired power plants is crucial for reducing greenhouse gas emissions. Separating CO_2 from natural gas is also required prior to pipeline delivery because CO_2 reduces the energy content of the natural gas.[1] The most widely used conventional method for CO_2 separation is absorption of CO_2 in aqueous solutions of alkanolamines. However, this method is expensive due to the solvent exchange process for regeneration.[2] Adsorption-based and membrane-based separations methods have been recently shown to be promising for capture of CO_2 from flue gas and natural gas streams. Developing new materials that can efficiently and economically separate CO_2 has significant importance in industrial applications. In this study, we aim to assess the potential of new nanoporous materials, bio-metal organic frameworks (bio-MOFs) for adsorbent-based and membrane-based CO_2 separation. Bio-MOFs have been recently synthesized using biomolecules such as amino acids, nucleobases, sugars as linker molecules and biocompatible metal cations.[4] Bio-MOF-1, composed of adenine as a biomolecular ligand and zinc salts as metal centers, was synthesized as the first member of bio-MOF family reported to show high CO_2 adsorption capacity due to the its linker molecule's high affinity for CO_2.[5]

The storage capacities of bio-MOFs have been reported to exceed those of many traditional nanoporous materials due to the large pore sizes and high pore volumes of bio-MOFs. For example, the CO_2 adsorption capacity of bio-MOF-11 at 1 atm and 298 K (4.06 mmol/g MOF) was reported to be higher than that of many other widely studied MOFs including IRMOF-1, HKUST, ZIF-69, CUK-1 due to the high interaction energy between CO_2 molecules and bio-

MOF-11.[6] Large surface areas, high pore volumes, good thermal and mechanical stabilities of bio-MOFs have attracted the interest of reserachers and studies on bio-MOFs have been recently started. There are currently nine bio-MOFs (1, 11, 12, 13, 14, 100, 101, 102, and 103) that have been synthesized in the literature. In a recent study, bio-MOFs-12, -13 and -14 were reported as promising materials for CO_2/N_2 separation due to strong adenine-CO_2 interactions.[7] Bio-MOF-101, 102 and 103 were demonstrated to be the most porous MOFs that have been reported up to date.[9] The CO_2 selectivity of these bio-MOFs was evaluated based on the single-component adsorption isotherm data rather than the mixture adsorption data, which is much more significant in industrial applications. There is currently no adsorption and transport data for gas mixtures in the pores of bio-MOFs and this lack of information limits the assessment of gas adsorption and separation performance of bio-MOFs.

In this study, we investigated the potential of bio-MOFs for CO_2 separation from CO_2/CH_4, CO_2/N_2 and CO_2/H_2 mixtures using Grand-canonical Monte Carlo and Equilibrium Molecular Dynamics simulations. We first validated the accuracy of our molecular simulations by comparing simulated gas uptake results with the experimental ones and found a good agreement between the predictions of our simulations and experimental measurements. Motivated from this agreement, adsorption-based and membrane-based separation performances of bio-MOFs were calculated based on the results of molecular simulations and compared with the performances of other nanoporous materials. Our results showed that bio-MOF-11, bio-MOF-12 and IZUMUM show higher CO_2 selectivities and working capacities compared to other bio-MOFs and traditional zeolites such as MFI, ERI, DDR and TON. Bio-MOFs that were studied in this work can be used as promising membranes especially in CO_2/H_2 separations due to their high CO_2 permeabilities and high CO_2/H_2 permeation selectivities. Results of this study will be helpful to gain molecular level understanding of bio-MOFs' performance in CO_2separation applications and direct the experimental researchers to focus on more promising bio-MOFs for adsorption-based and membrane-based gas separations.

References:

[1] Baker, R. W. Future directions of membrane gas separation technology. *Ind. Eng. Chem. Res.*, **2002**, *41*(6), 1393.

[2] Arstad, B.; Fjellvåg, H.; Kongshaug, K.; Swang, O.; Blom, R. Amine functionalised metal organic frameworks (MOFs) as adsorbents for carbon dioxide. *Adsorption*, **2008**, *14*(6), 755..

[3] Robeson, L. M. Polymer membranes for gas separation. *Curr. Opin. Solid State Mater. Sci.*, **1999**, *4*, 549.

[4] Imaz, I.; Rubio-Martinez, M.; An, J.; Sole-Font, I.; Rosi, N. L.; Maspoch, D. Metal-biomolecule frameworks (MBIOFS). *Chem.Commun.*, **2011**, *47*(26), 7287.

[5] An, J.; Geib, S. J.; Rosi, N. L. Cation-triggered drug release from a porous zinc-adeninate metal-organic framework. *J. Am. Chem. Soc.,* **2009**, *131*(24), 8376.

[6] Keskin, S.; Heest, T. M. V.; Sholl, D. S. Can metal–organic framework materials play a useful role in large-scale carbon dioxide separations? *ChemSusChem*, **2010**, *3*, 879.

[7] Li, T.; Chen, D.-L.; Sullivan, J. E.; Kozlowski, M. T.; Johnson, J. K.; Rosi, N. L. Systematic modulation and enhancement of CO_2 : N_2 selectivity and water stability in an isoreticular series of bio-MOF-11 analogues. *Chem. Sci.*, **2013**, *4*(4), 1746.

[8] An, J.; Farha, O. K.; Hupp, J. T.; Pohl, E.; Yeh, J. I.; Rosi, N. L. Metal-adeninate vertices for the construction of an exceptionally porous metal-organic framework. *Nat. Commun.*, **2012**, *3*, 604.

[9] Li, T.; Kozlowski, M. T.; Doud, E. A.; Blakely, M. N.; Rosi, N. L. Stepwise ligand exchange for the preparation of a family of mesoporous MOFs. *J. Am. Chem. Soc.*, **2013**, *135*(32), 11688.

[10] Bohrman, J. A.; Carreon, M. A. Synthesis and CO_2/CH_4 separation performance of Bio-MOF-1 membranes. *Chem.Commun.*, **2012**, *48*(42), 5130.

[11] Xie, Z.; Li, T.; Rosi, N. L.; Carreon, M. A. Alumina-supported cobalt–adeninate MOF membranes for CO_2/CH_4 separation. *J. Mater. Chem. A*, **2014**, *2*(5), 1239.

378762 Constructing CO2-Transport Passageways in Matrimid Membranes By Incorporating Hydrogel Nanoparticles

Abstract:

Constructing CO_2-transport passageways in Matrimid membranes by incorporating hydrogel nanoparticles

Authors: Xueqin Li, Hong Wu*

Reporter: Xueqin Li

Key Laboratory for Green Chemical Technology of Ministry of Education School of Chemical Engineering and Technology, Tianjin University, 300072, P. R. China

Collaborative Innovation Center of Chemical Science and Engineering (Tianjin), Tianjin 300072, China

[*] Corresponding author: wuhong@tju.edu.cn

Abstract: Composite membranes were fabricated by blending of Matrimid 5218 polymer with poly(N-isopropylacrylamide) hydrogel nanoparticles (PHNs, ~250 nm). The PHNs were prepared by precipitation polymerization of poly(N-isopropylacrylamide) (NIPAM) as the monomer and N, N-methylenebisacrylamide (MBA) as the crosslinker. Since PHNs can absorb large quantities of water and have high water-retention capability, the incorporation of PHNs into Matrimid matrix increased water uptake, bound water, and water retention capacity in the composite membranes. Matrimid/PHNs composite membranes have both high CO_2 permeability, CO_2/CH_4 and CO_2/N_2 selectivities due to the higher water content in composite membranes and additional CO_2-transport passageways constructed by incorporating PHNs. It was found that the gas permeability increased with increasing PHNs loading, and the CO_2 permeability of the composite membranes doped with 20 wt% PHNs increased by the fourfold compared with pure Matrimid membrane.

Keywords: Matrimid; Hydrogel nanoparticles; Composite membranes; Water content; CO_2 separation

379956 Hexane Isomer Permeation and Separation Properties of Silicalite-1 Membrane Under Pressurized Conditions

Abstract:

Membrane separation has been studied extensively for gas separation and water purification as energy-saving processes in the recent decades. Some of them have already been put to practical use, for example, desalination of seawater and dehydration of ethanol. Petroleum refinery and petroleum chemistry can be the next target of membrane separation because separation and purification occupy the largest part of energy consumption in petroleum industry.

Zeolites draw attention as membrane material for hydrocarbon separation because of their high thermal and chemical resistances. Flux improvement is one of the issues of hydrocarbon separation with zeolite membrane. In general, the difference of partial pressures between feed and permeate sides strongly affects permeation flux. In this study, we investigated hydrocarbon separation property of silicalite-1 membrane which has high separation ability based on molecular sieving effect. In addition, we discuss the dependency of flux on the difference of partial pressures in the feed and permeation sides.

Silicalite-1 membrane was prepared on a tubular a-alumina support (i.d. = 7.0 mm, o.d. = 10 mm, average pore size = 150 nm) by secondary-growth method. Silicalite-1 seed crystals were supported on the outer surface of support by dip coating method. The seeded support was immersed in a synthesis gel and hydrothermal treatment was carried out. Vapor permeation property of silicalite-1 membrane was investigated under pressurized conditions. Hexane isomers (*n*-hexane, 2-methylpentane and 2,2-dimethylbutane) were used for model substances of hydrocarbons. Hydrocarbon vapor was fed to the outer surface of tubular membrane and the permeate side was swept with inert gas. The membrane temperature and pressure of feed side were adjusted from 373 - 673 K and 1 - 4 atm, respectively.

n-Hexane permeation flux was investigated under pressurized conditions in the unary system. The *n*-hexane flux increased with increasing difference of partial pressures between feed and permeate sides over the whole temperature range tested. Besides, the flux increment was larger at higher temperature.

Permeation flux is determined by the adsorption amount and diffusion rate of permeation molecule. Consequently, diffusion rate at lower temperature and adsorption amount at higher temperature mainly dominate the permeation flux. Therefore, it is deduced that the *n*-hexane flux increment under pressurized conditions is elucidated in increasing of adsorption amount. In particular, the effect was greater in higher temperature range where the adsorption value was sufficiently low. In this experiment, *n*-hexane flux reached 8.2 kg m^{-2} h^{-1} at 673 K at 4 atm of partial pressure difference.

In a similar way, *n*-hexane permselectivity from hexane isomers was investigated. The *n*-hexane flux under pressurized conditions increased with pressure and reached a plateau above 2 atm at 573 K. On the other hand, *n*-hexane flux increased linearly with increasing partial pressure up to 4 atm at 673 K. This trend was able to be explained with adsorption inhibition by other components. Since other components also adsorbed easily at lower temperature, the adsorption amount of *n*-hexane was less increased even under pressurized conditions. Accordingly, the *n*-hexane flux had a maximum at relatively low pressures at lower temperature. The flux and the separation factor of *n*-hexane and 2,2-dimethylbutane at 673 K and 4 atm were 2.9 kg m^{-2} h^{-1} and 279, respectively.

Thermodynamic model for gas solubility in glassy polymers: analysis of mixed gas sorption behavior

Matteo Minelli, Ferruccio Doghieri

Interdipartimental Center for Industrial Research - Advance Mechanics and Materials (CIRI-MAM)
Alma Mater Studiorum - Università di Bologna

Department of Civil, Chemical, Environmental and Materials Engineering (DICAM)
Alma Mater Studiorum - Università di Bologna

The solubility of low molecular weight species in polymeric systems has been widely analyzed by many authors, both from the experimental and modeling point of view, as this topic is crucial for several purposes in industrial applications, from gas separation to barrier materials for packaging. Hence, the modeling analysis of gas solubility has received a great attention in the last decades, leading to the development and tuning of various approaches able to provide the description of the thermodynamic behavior of polymer-penetrants systems.

Equation of state models, developed following different routes, provide reliable means for the calculation of gas solubility in rubbers or polymer melts [1-4]. On the other hand, the general approach called nonequilibrium thermodynamics of glassy polymers (NET-GP) [5-7] gives the required extension to EoS models for the representation of the out-of-equilibrium behavior of polymer glasses. Although most of the modeling analysis is carried out in combination with Sanchez-Lacombe EoS [1] (giving rise to the so-called nonequilibrium lattice fluid - NELF), other equation of state approaches have also been employed to the same aim with profitable results. NET-GP relies on the use of polymer mass density and compressibility in glassy state as additional material parameters for the representation of the nonequilibrium behavior of the polymer-penetrants systems, which ultimately allows the model to predict gas solubility at different pressure and solute fugacity. The model is also able to describe the solubility of gaseous mixture in the glassy polymeric system with no introduction of further parameters and this specific feature is in focus in this work. Therefore, the NELF model is a reliable tool for the analysis of the effect of competitive sorption as well as plasticization related phenomena, commonly involved during real separation process due to the presence of multiple penetrants in the gaseous streams.

Therefore, the role of the parameters characterizing the polymers in the glassy state, such as the excess of free volume and reduced compressibility, in determining the solubility of different penetrants are thoroughly investigated. These parameters, indeed, are the key quantities utilized by the NELF model representation to describe the thermodynamic behavior of the glassy states, as characteristics of the pure polymer only.

Test and model cases will be considered and studied by the NELF model, for the determination of the gas solubility and the selectivity of specific gaseous mixtures,

chosen as reference. The influence of operative parameters such as the concentration of the penetrant mixtures, as well as the total pressure, on the solubility and selectivity behaviors will also be investigated.
The results of the present modeling study will be analyzed in order to determine the required characteristics for the selection of the most effective and efficient material properties for gas separation using membranes. The solid thermodynamic model considered will thus be used to provide general trend lines for the profitable design of membrane materials with improved separative performances.

References

[1] R.H. Lacombe, I. Sanchez. J. Phys. Chem. 80 (1976) 2568.
[2] W.G. Chapman, et al., Fluid Phase Equilib. 52 (1989) 31.
[3] Y. Song, et al. Macromolecules 27 (1994) 441.
[4] J. Gross, G. Sadowski. Ind. Eng. Chem. Res 40 (2001) 1244.
[5] F. Doghieri, G.C. Sarti. Macromolecules 29 (1996) 7885.
[6] M.G. De Angelis, G.C. Sarti, Annu. Rev. Chem. Biomol. Eng. 2 (2011) 97.
[7] M. Minelli, F. Doghieri, Ind. Eng. Chem. Res. 51 (2012) 16505.

366826 Fixed-Bed Study of MG Adsorption on Chitosan Hydrogels

Abstract:

Dyes are popularly used in textile and printing industries. Most dyes are toxic and even carcinogenic to aquatic life and are harmful to aquatic systems. Industrial waste water containing dyes is difficult to treat as dyes are very stable under such conditions of aerobic digestion, light, heat and oxidizing agents.

Adsorption technology has been popularly used in the treatment of dye containing waste water. Low cost adsorbents like chitosan are particularly attractive in this application. However, most of the literature studies used the batch experiments only. In this research, chitosan hydrogels were prepared and examined for the MGO (Malachite Green oxalate or MGO, a basic dye) removal in the batch (equilibrium) adsorption as well as the fixed bed breakthrough (kinetic) experiments. The effect of pH value, dye concentration, and flow rate were investigated. It was found that the hydrogels present a good adsorption capacity while the fixed bed breakthrough curves are s-shaped in general. Analysis showed that the mass transfer mechanisms are complicated for the kinetic experiments, although it can be roughly fitted to some simple equation like Bohart and Adam model.

376734 A New Multimodal Polymeric Membrane for Monoclonal Antibody Purification

Abstract:

This presentation will describe our latest research efforts to develop a new class of multimodal membranes (MMM) with high binding capacities, salt-tolerance, and high-productivity for the chromatographic purification of therapeutic proteins. The membranes incorporate ligands that can bind proteins through a combination of Coulombic interactions, hydrophobic interactions, and/or hydrogen bonding, thus widening the operation ranges for separation.

Our design approach is to graft functional polymer tentacles from the pore surfaces of macroporous membranes using a controlled surface-initiated polymerization technique. One advantage of this membrane preparation technique is that it greatly enhances the membrane binding capacity without severely decreasing its permeability. Here, we present work on the preparation of multimodal cation-exchange membranes for bind-and-elute purification of immunoglobulin G (IgG). Compared to commercial multimodal adsorbents, our membranes have 4 times higher ligand density, which contributes to the high static binding capacities (180 mg IgG/mL, 310 mg lysozyme/mL).

Protein binding experiments with bovine IgG showed binding capacities of the multimodal membranes were dependent on pH but had very good salt tolerance. The multimodal membranes maintain significant binding capacities in excess of 90 mg IgG/mL at ionic strength values that are typical for elution buffers used in multi-stage bioseparation processes. For sodium citrate, a conventional salt used in elution buffers of Protein A columns, increasing ionic strength had only a minor effect on the IgG binding capacity.

To further understand protein adsorption on the multimodal membrane, a thermodynamic adsorption isotherm model was employed that provides a unique set of physically meaningful parameters. This model can be used to explain the effect of salt type (kosmotropes, neutral, and chaotropes) on protein adsorption.

Finally, protein dynamic binding capacities were determined from breakthrough curve analysis. Dynamic binding capacities of >50 mg IgG/mL were found near the isoelectric point of the protein. A range of flow rates was used to study the effect of volumetric throughput on dynamic binding capacities. Protein elution using pH adjustment, chaotropic salts, and organic displacers will be described.

Taken together, the results to be presented indicate that the newly developed multimodal membranes have great potential to compete with more traditional cation-exchange materials following the Protein A purification step in the downstream processing of antibody products.

377862 Cell Separation Using Microfluidic Devices

Abstract:

Separation of white blood cells from red blood cells is the first step in the analysis of blood. Separation of cells in microfluidic devices is performed by different approaches such as hydrodynamic, microfiltering, acoustic, electrokinetic, dielectrophoretic, and magnetic approaches. Microfiltration is the simplest of these approaches. Different microfiltering approaches are: weir, pillars, membrane, cross-flow, tortuous channels designs, between others. This project presents a review on the different designs that have been previously used and the efficiencies obtained. These devices are fabricated using photolithography techniques with silicon and glass. It has been shown that cross-flow filters avoid the clogging and jamming problem encountered with other designs. Moreover, higher efficiencies were obtained using the cross-flow filters. Separation efficiencies of around 70-95% for trapping WBCs have been achieved.

380947 Roles of Physicochemical Properties of "Nano-Platform of Membrane" in Asymmetric Recognition Function of Liposome Membrane

Abstract:

The techniques of separation and purification of chiral molecules are very important for synthesis of optically-active drugs because of a therapeutic effect and an adverse effect. Traditional methods of chiral separation requires "tailor-made" fitting of molecular binding state between host and guest molecules, but it is relatively expensive. In order to solve these problems, many researchers have been interested in the use of self-assembly systems, which form a flexible membrane with high molecular order. Liposome is a self-assembly of the phospholipid molecules, prepared in an aqueous solution, and can be regarded as a model biomembrane. Recent research has showed the possibility of the chiral recognition function of the liposome membrane through the binding of the dipeptide on the liposome membrane. It was revealed that chiral recognition was achieved by liposome membrane, which showed very drastic difference about the adsorption of amino acids at the difference of its chirality. In this research, the dynamic behavior of polarity of liposome membrane surface induced by adsorption of L-tryptophan (L-Trp) was investigated by some molecular probes. At the initial step of adsorption (around 24 hours incubation), surface polarity of liposome membrane changed by the adsorption of L-Trp, but did not by D-Trp. Finally, the adsorption of L-Trp showed Langmuir adsorption model around 48 hours incubation, so it was proposed that there were adsorption sites at the surface of liposome membrane with specific conformation formed by multiple phospholipids. From these results, it is expected to contribute to understanding of the mechanism of chiral recognition function induced by liposome membrane.

381122 Combined Effect of Saline and Organic Modifiers on HIC and Rpc Separation of Insulin Variants

Abstract:

Introduction

With the introduction of the first biopharmaceuticals in the early 1980s, the pharmaceutical industry was transformed. Since then, the market for medicines based on recombinant proteins has grown rapidly, and today approximately a fourth of all new pharmaceuticals belong to this group. [1] However, the complexity of the fermentation broth from the production of recombinant proteins calls for sophisticated purification methods. One of the most widely used and most successful purification methods in this area is preparative chromatography. [2] Since proteins are prone to denaturize at increasing levels of organic modifier, the relatively mild conditions in hydrophobic interaction chromatography (HIC) are favorable, and this chromatographic mode is very commonly applied for protein purification. Reversed-phase chromatography (RPC) is the chromatographic mode-of-choice for analytical purposes, but it can also be a powerful tool for separation of proteins which can tolerate the harsher conditions, e.g. insulin and peptides. [3, 4]

Both HIC and RPC are based on hydrophobic interactions between the ligands and the proteins, but the two modes differ in the degree of hydrophobicity, and in the type of mobile phase modifier used. In HIC, the hydrophobicity of the resin is low, and thus, addition of a salt is sufficient to modify the retention. In RPC, on the other hand, the resin is highly hydrophobic and elution is achieved by adding an organic modifier, such as ethanol, to the mobile phase. Nevertheless, HIC and RPC are based on the same fundamental physical phenomena, and in principle, both saline and organic modifiers can be used to modify both types of chromatography. These similarities suggest that a common mechanistic model for HIC and RPC can be developed.

As chromatographic experiments are expensive as well as time-consuming, model-aided process development and optimization is becoming increasingly attractive to the pharmaceutical industry. An undisputable prerequisite for model-aided process development and optimization is the existence of a model structure which describes the process well. In contrast to ion-exchange chromatography (IEC), for which the steric mass action (SMA) [5] and the Langmuir mobile phase modulator (MPM) [6] models are well-established, the debate on how to describe the hydrophobic phenomena involved in HIC and RPC is still on-going. The effects have, for example, been assigned to changes in the mobile phase surface tension [7], in the salt-solute interactions [8], and in the solute activity coefficient [9].

Assuming that the latter is correct, a change of resin would only affect the adsorption equilibrium constant, and for the same combination of solute and solvent system, the effect of the modifier concentrations would be the same. One interesting route towards a more general and predictive model is to attempt to develop a combined model for HIC and RPC, in which the modifier effect is attributed to changes in the activity coefficient of the solute. This is the motivation for this work; we want to investigate, understand and describe the common factors and dividers for the two modes of chromatography. The aim of our present study is to investigate whether or not similarities in modifier effects can be observed, and thus if the suggested route is feasible.

Chromatographic Experiments

The proteins chosen for this study were three insulin variants, with very similar physicochemical properties. A mock feed was prepared by adding the insulins to a buffer without any salt and organic modifiers. All experiments were linear range isocratic pulse experiments, performed at constant temperature and pH. The HIC resins used were two methacrylate-based ones, one with butyl ligands and one with phenyl ligands. A C4 and a C18 resin, on silica backbone, were used for the RPC experiments. The mobile phase modifiers chosen were ethanol and KCl.

For the HIC resins, three different experiment series with varying KCl concentrations were run; at 0, 5, and 10 wt% ethanol. For some of the series, the solubility of KCl in the ethanol-water system was limiting, and in these cases, high retention volumes could not be reached. For the RPC resins, five different experiment series were run within the range 0.1-0.7 mol KCl/kg solvent, and with varying ethanol content. Each series ranges from almost non-retained conditions, up to retention volumes above 30 CV. The total column porosity and the system dead volume were determined from salt pulse experiments.

Data Analysis

From the retention volumes, the natural logarithm of the thermodynamic retention factor (A) was calculated for each solute in each experiment. For the HIC data, lnA was plotted against κ^2, the squared inverse of the Debye length, which is directly proportional to the ionic strength. All series display a positive linear dependence on κ^2, but with different slopes for different ethanol content. This is consistent with Kirkwood's theory for the effect of salt on proteins [10]. Many salts have a salting-out effect at high concentrations and a salting-in effect at low concentrations, resulting in a minimum in retention. However, for the combination of solutes and solvent system studied, no salting-in effect was observed on the HIC resins.

For the RPC data, lnA was plotted against the volume fraction of ethanol in the mobile phase. The result was parallel, slightly convex curves with a negative dependence on the ethanol content. In accordance with the assumption that the modifier concentrations only affect the solute activity coefficient, quadratic polynomials with different intercepts but the same slope and curvature could be fitted to all RPC series, independent of resin. Additionally, the intercepts for each resin decrease linearly with increasing salt concentration. Thus, the solutes display a salting-in behavior on the RPC resins, within the interval 0.1-0.7 mol KCl/kg solvent.

The next step is to calibrate a linear isotherm, in which the modifier effects are attributed to changes in the solute activity coefficient, to the data collected. As the results agree well with Kirkwood's salting-in and salting-out theories, the ability of these theories to describe the changes in the solute activity coefficient will be investigated. When the linear isotherm has been established, we will study the non-linear range, with the aim to develop a full model which can be used as a tool in process development and optimization.

Acknowledgements

The Swedish Foundation for Strategic Research (SSF) and the Process Industry Centre at Lund University (PIC-LU) are gratefully acknowledged for financial support. Nanna Mikkelsen at Novo Nordisk in Bagsværd, Denmark has provided invaluable help with the experimental equipment.

References

1. Walsh, G., *Second-generation biopharmaceuticals.* Eur. J. Pharm. Biopharm., 2004. **68**: p. 185-196.

2. Jungbauer, A., *Chromatographic media for bioseparation.* J. Chromatogr. A, 2005. **1065**: p. 3-12.

3. Freitag, R., *Chromatographic Techniques in the Downstream Processing of Proteins in Biotechnology*, in *Animal Cell Biotechnology - Methods in Molecular Biology*, R. Pörtner, Editor 2014, Humana Press: New York, NY, USA. p. 419-458.

4. Graumann, K. and A.-A. Ebenbichler, *Development and Scale up of Preparative HIC for the Purification of a Recombinant Therapeutic Protein.* Chem. Eng. Technol., 2005. **28**(11): p. 1398-1407.

5. Brooks, C.A. and S.M. Cramer, *Steric Mass-Action Ion Exchange: Displacement Profiles and Induced Salt Gradients.* AIChE J., 1992. **38**(12): p. 1969-1978.

6. Melander, W.R., S. El Rassi, and C. Horváth, *Interplay of hydrophobic and electrostatic interactions in biopolymer chromatography: Effects of salts on the retention of proteins.* J. Chromatogr., 1989. **469**: p. 3-27.

7. Melander, W. and C. Horváth, *Salt Effects on Hydrophobic Interactions in Precipitation and Chromatography of Proteins: An Interpretation of the Lyotropic Series.* Arch. Biochem. Biophys., 1977. **183**: p. 200-215.

8. Arakawa, T. and S.N. Timasheff, *Preferential Interaction of Proteins with Salts in Concentrated Solutions.* Biochemistry, 1982. **21**: p. 6545-6552.

9. Mollerup, J.M., *A Review of the Thermodynamics of Protein Association to Ligands, Protein Adsorption, and Adsorption Isotherms.* Chem. Eng. Technol., 2008. **31**(6): p. 864-874.

10. Kirkwood, J., *Protein, Amino Acids and Peptides as Ions and Dipolar Ions*, in *American Chemical Society Monograph Series*, E.J. Cohn and J.T. Edsall, Editors. 1943, Reinhold Publishing Corporation: New York, USA.

387745 The Development of Cellulose Monolith with Anion Exchange/Pseudo-Affinity Functionalization for the Separation of Influenza Virus

Abstract:

The purification of virus or virus-like particles (VLP) from host cells such as mammalian cells and *E. coli* is a time consuming procedure and has been undergoing extensive study. A monolith is a single-piece, interconnected, and porous stationary phase material which is widely applicable for separation technologies such as gas and liquid chromatography, high performance liquid chromatography, and capillary electro-chromatography [1, 2]. To make the virus separation technology faster and more efficiently, a new cellulose monolith with a tunable anion exchange/pseudo-affinity surface was developed. We propose a method to prepare cellulose fiber enhanced monolith for the separation of the influenza virus directly from mammalian cell culture media. The interconnected structure of the monolith was a result of the porogen generating pores and fiber interactions. To address the variance in expression of surface glycoproteins among different strains, as well as to maintain a specific separation, chromatography utilizing pseudo-affinity and anion-exchange immobilization was prepared for use with influenza virus. The cellulose monolith structure was characterized by Scanning Electron Microscopy. The two different modifications of anion-exchange and pseudo-affinity were quantified by element analysis. An H1N1 influenza virus strain (A/WSN/33) was replicated and harvested in MDCK cells and used directly for separation efficiency evaluation, after a freeze-thaw operation [3]. The result shows for the H1N1 strain from MDCK cell culture, the pseudo-affinity modification has higher virus separation efficiency.

Reference

[1] Potter OG, Hilder EF. Porous polymer monoliths for extraction: Diverse applications and platforms. Journal of separation science. 2008;31:1881-906.

[2] Namera A, Nakamoto A, Saito T, Miyazaki S. Monolith as a new sample preparation material: Recent devices and applications. Journal of separation science. 2011;34:901-24.

[3] Wolff MW, Reichl U. Downstream Processing: From Egg to Cell Culture Derived Influenza Virus Particles. Chemical engineering & technology. 2008;31:846-57.

388279 Adsorption and Ion Exchange Processes for Cephamycin C Purification

Abstract:

Cephamycin C (cepC) is a beta-lactam antibiotic produced by the actinomycetes *Streptomyces clavuligerus* and *Nocardia lactamdurans*. It is an important compound for the pharmaceutical industry because it is a raw material for some commercial antibiotics, named as cefoxitin and cefotetan. There are few studies in the literature about purification processes of this antibiotic, and most of the information available is described in patents. It has already been demonstrated that the use of the anionic resin Q Sepharose XL (QXL) in a fixed-bed column was effective in extracting cepC from the broth during column loading. However, the study did not show the separation of any contaminants during elution step, as separated peaks were not observed. Therefore, the method of column operation should be optimized and a deeper study should be carried out.

In this study, three operations in sequence for cepC purification were evaluated, including a different operational method of the fixed-bed column using the resin QXL. The operations consisted of adsorption onto the neutral resin Amberlite XAD4 in a stirred reactor; fixed-bed column process using the resin QXL; and adsorption onto a C18 SPE cartridge. Ultrafiltered broth obtained after fermentation with *S. clavuligerus* was used. Contaminants were monitored during all steps, by different techniques: absorbance measurements between 310 to 400 nm, which are wavelengths that cepC does not absorb; biological assays using *Escherichia coli* ESS, that is a bacteria sensitive to cepC; mass spectrometry analyses.

During adsorption in the stirred reactor with XAD4 resin, 27% of cepC and 44% of contaminants were adsorbed onto the resin. The purification factor obtained was of 1.5.The resulting broth of this step (clarified broth) was used in the fixed-bed column process. Breakthrough curves of the column process showed that cepC and contaminants competed for the biding sites on the resin. Two separated peaks with antibacterial activity were obtained during bed elution, by using 0.1 and 0.5% NaCl solution. CepC was present only in the first fraction. The compounds present in the second peak were not known. Fractions from the center of the peak containing cepC were collected and used in the next step, adsorption in the SPE cartridge. CepC completely adsorbed onto the SPE cartridge, and was eluted using 50% methanol solution. Mass spectrometry analyses of a sample collected in the elution fraction and of the clarified broth showed a reduction in the peaks corresponding to the m/z ratios of contaminants that are difficult to separate, because of the similarity between their molecular structures and that of cepC. The contaminants monitored were penicillin N, deacetylcephalosporin C, deacetoxycephalosporin C and lysine. At this final operation, a partially purified fraction of cepC was obtained. This result is of great relevance since the establishment of a suitable process for cepC purification from fermentation broth has been strongly desired.

388586 Understanding How Membrane Materials Influence Protein Adsorption and Membrane Fouling at the Single-Molecule Level

Abstract:

Membrane fouling, due to protein adsorption and accumulation on membrane surfaces, leads to decreased efficiency and selectivity of filtration processes. Certain membrane materials, especially hydrophilic materials, have shown reduced protein accumulation and decreased membrane fouling. However, a molecular-level mechanistic understanding of protein layer formation and macroscopic fouling behavior is still lacking and has been heavily debated in the literature. Using single molecule (SM) imaging techniques, based on total internal reflection fluorescence microscopy (TIRFM), we are able to directly observe adsorption, desorption, and interfacial diffusion for millions of individual molecules. This SM technique uniquely allows us to identify heterogeneous surface behavior and distinct dynamic rates (e.g. adsorption vs. desorption rates rather than *net* adsorption). We investigated the interfacial dynamics of bovine serum albumin (BSA) and a human monoclonal antibody (IgG) on several commonly used membrane materials on both bare and protein crowded surfaces using SM TIRFM. In particular, we were interested in the role of protein-surface interactions vs. interfacial protein-protein interactions and the independent effects of adsorption and desorption rates. In the case of BSA, surface chemistry had little effect on adsorption and desorption rates on different bare surfaces. Once a BSA layer formed, further protein adsorption was reduced and desorption was insensitive to the underlying surface chemistry, suggesting that protein-protein interactions play a large role in BSA monolayer and multi-layer formation. In contrast, IgG interfacial dynamics showed some dependence on surface chemistry (adsorption and desorption varied) and desorption rates in most cases were insensitive to protein concentration (e.g the same for bare and protein crowded surfaces). This suggests that for IgG surface chemistry directly influences protein-surface influences and can regulate protein-protein interactions. Ultimately, membrane material effects on protein adsorption dynamics and membrane fouling appear to be highly protein dependent.

BIASED BROWNIAN MOTION IN ANNULAR CAPILLARIES

Abbas Motamedilamouki, Parvin Golbayani, J. Robby Sanders and Pedro E. Arce[1]

Department of Chemical Engineering, Tennessee Technological University, Cookeville, TN 38505.

Abstract

Biased Brownian motion is of crucial interest in many biotechnological applications including clinical diagnostics, tissue scaffolding, and drug delivery. In this contribution, an area averaging approach is used to derive the hydrodynamic-biased dispersion (usually known as Taylor-Aris dispersion) in a capillary. This study focuses on an annular system typically found in porous or fibrous media within the biotechnology domain with a Newtonian, incompressible, isothermal fluid and under steady state conditions. The system used here has been described by the convective-diffusive transport differential model for which the spatial area-averaging approach has been applied to identify key parameters in the system without the actual solution of the species-continuity equation. An explicit analytical expression is derived for the effective parameter, i.e., diffusivity. This parameter plays a significant role in controlling the optimal separation time in technologies that could lead to, for example, the improvement of present clinical diagnostics applications.

[1] Author for correspondence (PArce@tntech.edu)

368972 A Competitive Kinetic Model for Polymorph Nucleation

Abstract:

A competitive kinetic model for polymorph nucleation is developed in this work to describe the time evolution of two different polymorphic cluster size distributions in a supersaturated solution controlled by the agglomeration and disruption of the solutes. Although there is only one type of the solute molecules, it is assumed that the agglomerated solute clusters due to the Brownian motion are divided into two types – A form and B form, which resemble the structures and morphologies of the different mature polymorphs and eventually lead to the formation of two polymorphic crystals. Recently, Shiau and Lu (2009) developed a discrete binary random disruption kernel to account for dissolution of the solute clusters. In this work, a disruption kernel is incorporated into the proposed model to account for gradual dissolution of the solute clusters smaller than a critical nucleus size due to the thermodynamic instability. By fitting the experimental induction period data and the final measured weight fractions of eflucimibe polymorphs with the proposed model, the agglomeration and disruption rate constants for two polymorphs are recovered. It is found that the developed model is satisfactory to describe the competitive mechanism of nucleation for eflucimibe polymorphs in a supersaturated solution.

Ref.

[1] Shiau, L.D., Lu, Y.F., 2009. Modeling solute clustering in the diffusion layer around a growing crystal. J. Chem. Phys. 130, 094105.

Isosteric Heats of Adsorption on Mesoporous Amine Adsorbents at Low Pressures Relevant to Post-Combustion CO_2 Capture and Direct Air Capture

Mustafa A. Alkhabbaz, Praveen Bollini, Christopher W. Jones
School of Chemical & Biomolecular Engineering, Georgia Institute of Technology, Atlanta, GA, 30332 USA

Supported amine adsorbents have emerged as promising materials for post-combustion CO_2 capture from flue gas [1-2] as well as from ambient air.[3-4] However, the molecular basis for CO_2 adsorption is not yet well developed, with most proposals for the adsorption mechanism based on only in-situ IR analyses. Understanding and quantifying the strength of the interactions between CO_2 and amine sites is critical to the development of efficient amine adsorbents.

Most of the reported heats of adsorptions in the literature are calculated via the Clausius-Clapeyron equation using isotherm measurements at multiple temperatures, which often deviate substantially from directly measured calorimetric heats.[5-8] A few studies have reported heats of adsorption using calorimetry, whereby the heats of adsorption were measured at pressures up to few bars. [9-11] However, the CO_2 pressure region that is of interest to post-combustion CO_2 capture and direct air capture (DAC) is below 0.1 bar, where few data exist.

In this work, we report experimentally measured calorimetric heats of adsorption of supported amine materials at low CO_2 pressures (0-0.1 bar). The measurement of the isosteric heats of adsorption was completed using a customized Setaram Sensys Evo calorimeter designed to dose very low pressures of CO_2 to the sorbent. Aminosilica adsorbents with different amine loadings were investigated to study the effect of amine density on heats of adsorption. Furthermore, a series of aminosilane-grafted silica materials were studied to explore the role of amine structure on the measured adsorption heats. The role of enthalpic vs. entropic factors in CO_2 adsorption on supported amine sites will be discussed.

References

[1] Choi, S.; Drese, J. H.; Jones, C. W.; *ChemSusChem* **2009**, *2*, 796–854.
[2] Bollini, P.; Didas, S. a.; Jones, C. W.; *J. Mater. Chem.* **2011**, *21*, 15100–15120.
[3] Goeppert, A.; Czaun, M.; May, R. B.; Prakash, G. K. S.; Olah, G. A.; Narayanan, S. R.; *J. Am. Chem. Soc.* **2011**, *133*, 20164–20167.
[4] Jones, C. W.; *Annu. Rev. Chem. Biomol. Eng.* **2011**, *2*, 31–52.
[5] Leal, O.; Bolívar, C.; Ovalles, C.; García, J. J.; Espidel, Y.; *Inorganica Chim. Acta* **1995**, *240*, 183–189.
[6] Wang, L.; Yang, R. T.; *J. Phys. Chem. C* **2011**, *115*, 21264–21272.
[7] Zukal, A.; Jagiello, J.; Mayerová, J.; Čejka; J. *Phys. Chem. Chem. Phys.* **2011**, *13*, 15468–15475.
[8] Watabe, T.; Yogo, K.; *Sep. Purif. Technol.* **2013**, *120*, 20–23.
[9] Knöfel, C.; Descarpentries, J.; Benzaouia, A.; Zeleňák, V.; Mornet, S.; Llewellyn, P. L.; Hornebecq, V.; *Microporous Mesoporous Mater.* **2007**, *99*, 79–85.

[10] Knöfel, C.; Martin, C.; Hornebecq, V.; Llewellyn P. L. *J;. Phys. Chem. C* **2009**, *113*, 21726–21734.
[11] Mello, M. R.; Phanon, D.; Silveira, G. Q.; Llewellyn, P. L.; Ronconi, C. M.; *Microporous Mesoporous Mater.* **2011**, *143*, 174–179.

380579 Improvement of Sorption Enhanced Water Gas Reaction Performance By Using Multi-Section Packing Concept

Abstract:

Fossil fuels generate a variety of pollutant causing serious environmental problems during combustion. To overcome the drawbacks of using fossil fuels, hydrogen are considered as an alternative energy carrier which has an advantage of generating only water as a product. Hydrogen can be produced in many different ways. Among various techniques, water gas shift (WGS) reaction ($CO + H_2O \leftrightarrow H_2 + CO_2$) has been highlighted for an efficient method of mass hydrogen production. However, WGS reaction generates CO_2 as a by-product and additional separation processes are required, which increase the cost and degrade the efficiency of the entire process. Recently, sorption enhanced water gas shift (SE-WGS) reaction has been developed to solve this problem and to improve performance. In SE-WGS reaction, CO_2 generated from the WGS reaction is removed by using adsorbent simultaneously with reaction in a single reactor. However, the reactor temperature increases significantly due to the exothermic WGS reaction. The increased temperature may degrade catalyst and adsorbent, and eventually decrease the efficiency of sorption enhanced reaction. In this study, the temperature profile in the SE-WGS reactor is numerically simulated and a concept of multi-section packing is applied to alleviate temperature increase and to optimize the SE-WGS reaction performance.

Formatting Issue - Best Available Paper/Graphic

Research on removal and recovery of the chlorinated organic compound by a pressure swing adsorption process

Kazuyuki Chihara, teruhisa kato, Takashi chiba, Meiji University, Kawasaki, Japan

Introduction

The PSA method is a method of collecting the specific ingredients in mixed gas, by making the specific ingredient in the mixed gas pressurized using the high adsorbent of adsorptive power to a certain gas in mixed gas adsorb preferentially, and subsequently reducing pressure. If vapor exists in the gas used as these materials, it will be known from before that adsorptive capacity will fall and, in such a case, advanced dehumidification of source gas will have been made indispensable.It investigated by having used two or more adsorbent in order to investigate what kind of influence water vapor has on an absorption-and-desorption action, and this research examined whether simplification of a pre-dehumidification process would be possible.

Moreover, action prediction of the real process was tried by creating the simulation using the Stop & Go method about an action in case water vapor is included, and comparing with an experimental result.

Experimental

Experimental apparatus

An experimental device figure, as shown in Fig.1.It is a 2 tower type PSA process, and it is each tower and the adsorption tower capacity of adsorbent fill volume is 90g in 27.0 mm × 300 mm. Source gas is mixed gas which consists of dry air, dichloromethane, and water vapor, and it controlled the flow of dichloromethane and water vapor by the massflow controller, respectively.

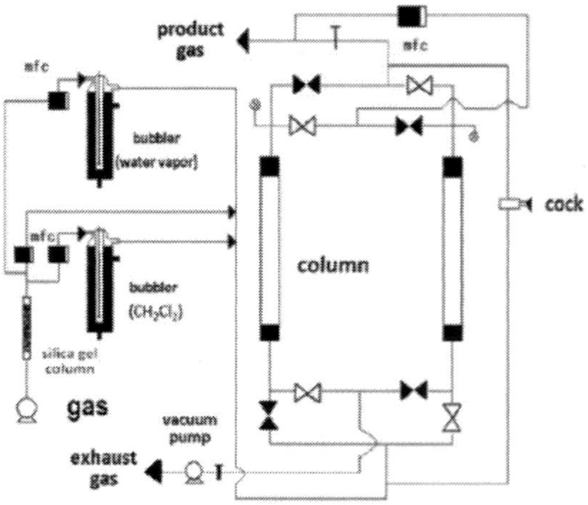

Figure 1. Basic concept of Stop & Go method

The PSA cycle used for this research consists of 4 of pressurization, adsorption and decompression / purge-desorption process processes. The cycle time of each process was set up and the pressure control screw performed the pressure change. The organic-solvent concentration of refining gas and enrichment gas discharged from an adsorption tower for every cycle fixed as 1 cycle in these four processes (360 [sec]) was measured using the gas chromatography equipment. Moreover, about measurement of humidity, the dew point recorder was used.

Adsorbents

Silica gel which is hydrophile-property adsorbent at adsorbent: CARiACT Q-6 (FUJI SILYSIA CHEMICAL) was used.

Moreover, high silica zeolite:HSZ (TOSOH) which is hydrophobic adsorbent was used as a candidate for comparison. It investigated to the difference in the absorption-and-desorption action by the hydrophile property under water vapor coexistence, and hydrophobicity.

Experimental condition

Each process time of PSA operation, as shown in Table 1, and an experimental condition, as shown in Table 2. About adjustment of source gas, dichloromethane concentration was unified at 8000 ppm, an arbitrary quantity of vapor was added, and the flow was adjusted to 1000 ml/min in the amount of dry air. In addition, the candidate for consideration of this research is separation performance for which it asks from the adsorption exit after repeating a PSA cycle 400 cycle and reaching a stationary state, and desorption reproduction outlet gas concentration.

Table 1 Each process time of PSA operation

left column	right column	process time [sec]
pressurization	decompression	20
adsorption	purge-desorption	160
decompression	pressurization	20
purge-desorption	adsorption	160

Table 2 Experimental condition

absorbent	CARiACT Q-6
	HSZ
volume	27.0mmΦ × 300mm
porsity	0.54
solvent	CH_2Cl_2
organic-solvent consentration	8000 [ppm]
supply gas mass flow	1000 [ml/min]
pressure of adsorption	2.0 [atm]
pressure of desorption	0.4 [atm]
ratio of purge	1
cycle time	400

Formatting Issue - Best Available Paper/Graphic

Simulation theory

By this research, the simulation using the Stop & Go method which considers the material balance by a cell model to prediction of the action of the amount of adsorption within the packed bed in a PSA process and gaseous phase concentration was created, and the calculation result was compared with the experimental result. The Stop & Go method is one of the simulation methods based on calculation. This method is explained below. The key map of the Stop & Go method, as shown in Fig.2.

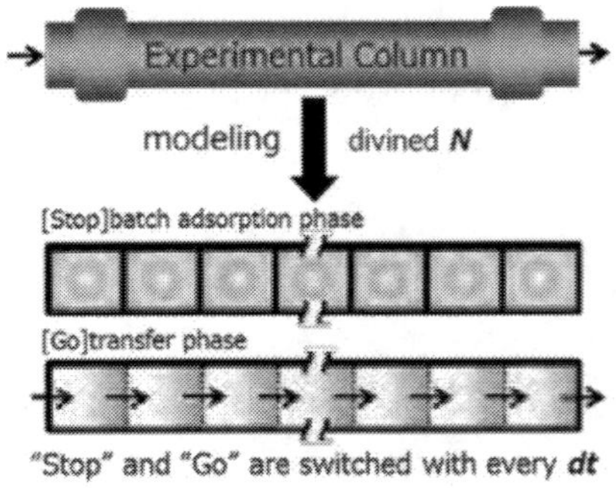

Figure 2. Basic concept of Stop & Go method

N division of is first done towards the shaft orientations of a packed bed, and N cells are considered. A flow is stop(ed) in a certain time t, and only the minute time dt makes an adsorption ingredient adsorb by a time within the cell of each cell. It asks for the pressure of the cell after time part adsorption (time dt+t), and the flow between cells required for each cell to reach is considered, and it is made to move immediately after the dt time by the income and outgo of the gas volume into which the column within dt time which was stop flowed in or flows, and the amount of gaseous phase gas increase and decrease by adsorption and desorption in each cell. And it is the calculation method which makes stop it and repeats a flow again in next dt time. However, the quantity into which gas flows in or flows is kept from exceeding the capacity of the adjacent cell. The material balance of movement in time part adsorption in each above cell and the following cell is considered from the following three formulas.

$$\frac{m}{N}\frac{dq_{n,j}}{dt} + \frac{\varepsilon V}{N}\frac{dC_{n,j}}{dt} = Q_n C_{n-1,j} - Q_{n+1} C_{n,i} \qquad -(1)$$

$$q_i^* = \frac{K_i P_i}{1 + \sum K_i P_i} q_{i,\infty} \qquad -(2)$$

$$\gamma \frac{dq_{n,i}}{dt} = K_{s,i} a_v \left(q_{n,i}^* - q_{n,i}\right) \qquad -(3)$$

Formatting Issue - Best Available Paper/Graphic

Result and Discussion

Examination of the adsorption performance under vapor existence

Comparison of the organic-solvent concentration in refining gas shows the case where hydrophobic adsorbent (HSZ) is used to Fig.4, as a result of using hydrophile property adsorbent (CARiACT) for Fig.3.By containing water vapor also a small quantity shows that separation performance falls. However, when hydrophobic adsorbent is used, at 20% or less, relative humidity is not subject to the influence of water vapor, and is maintaining separation performance.

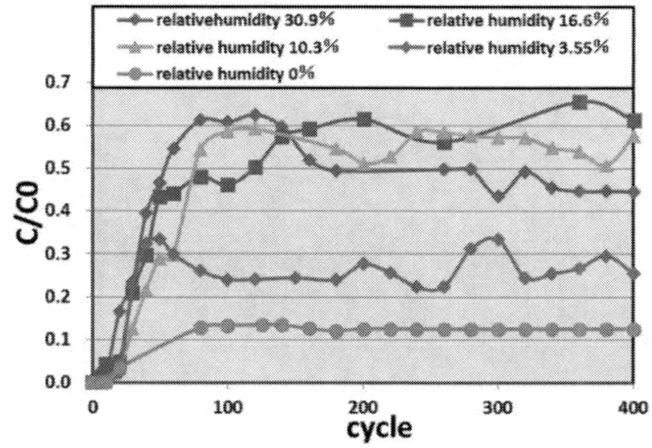

Fig.3 Refining gas concentration (CARiACT)

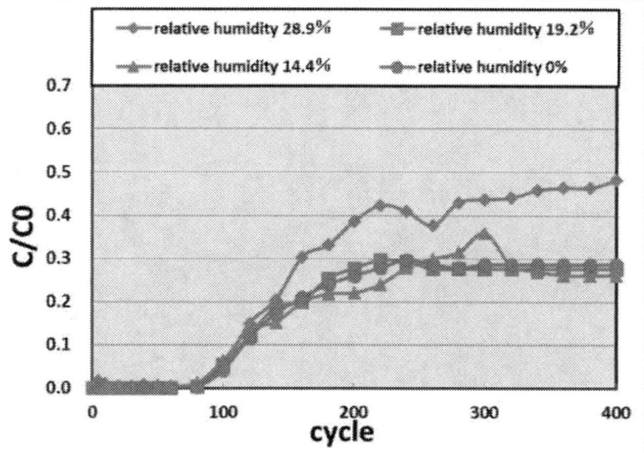

Fig.4 Refining gas concentration (HSZ)

Comparison with hydrophile property and hydrophobic adsorbent was performed about the separation performance in refining gas and enrichment gas. About refining gas, enrichment gas, as shown in Fig.5 Fig.6. The separation performance in refining gas excels [relative humidity / 0% of] in CARiACT. However, the direction of the separation performance of HSZ is excellent in the case where it is contained, in water vapor. About the separation performance of enrichment gas, even if vapor was contained, it turned out that separation performance becomes high from the case where the direction using hydrophile property adsorbent uses hydrophobic adsorbent.

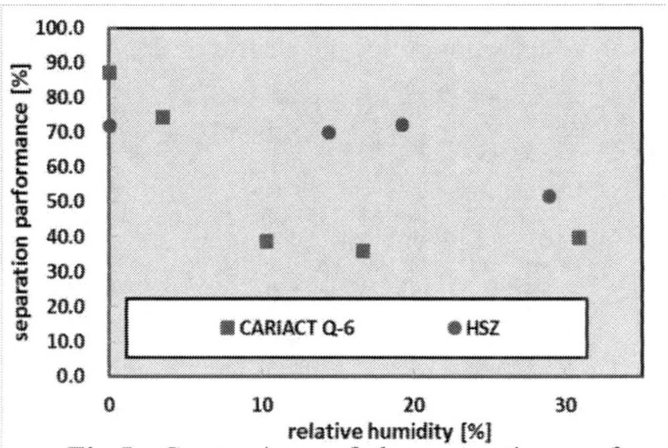

Fig.5 Comparison of the separation performance performance in refining gas

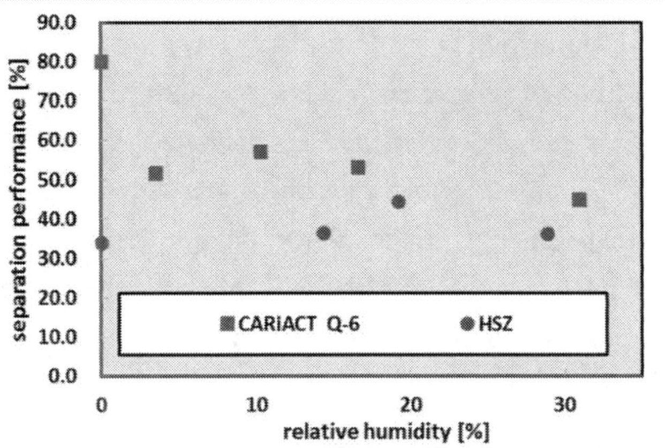

Fig.6 Comparison of the separation in enrichment gas

Moreover, consideration by absolute dry ratio was performed about each adsorbent. About refining gas, Cgas, as shown in Fig.7, Fig.8. Fig.7 and Fig.8 show that hydrophobic adsorption material is not subject to most influences of water vapor to 20% of relative humidity. Moreover, it turned out about hydrophile property adsorbent that water vapor affects strongly in refining gas, i.e., the process of adsorption.

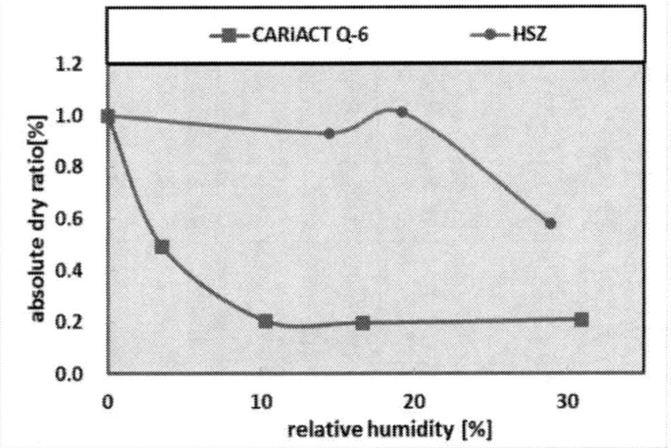

Fig.7 Refining gas

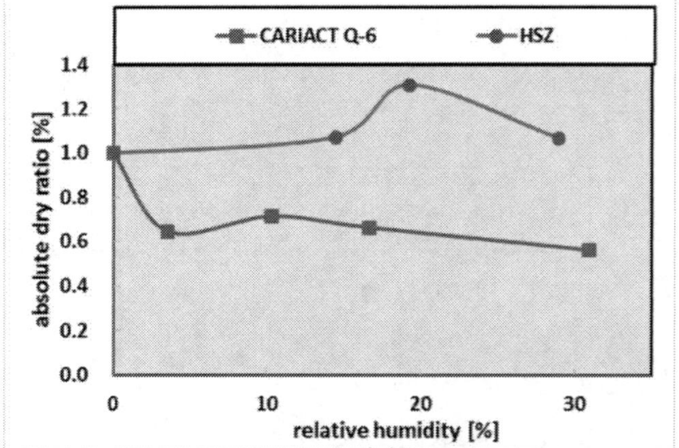

Fig.8 Enrichment gas

Although it is desirable to use hydrophile property adsorbent (CARiACT) from the above result on the conditions with which 0% of relative humidity, i.e., water vapor, does not coexist, when using the source gas containing water vapor, it turns out that a pre-dehumidification process is indispensable. However, by using the hydrophobic adsorbent which is not subject to the influence of vapor, if it is a certain amount of relative humidity, this pre-dehumidification process may be able to be simplified and energy saving can be expected.

Prediction by a simulation

Comparison of a simulation and an experimental result is shown in Fig.9. In addition, the refining gas concentration at the time of using hydrophile property adsorbent (CARiACT) for adsorbent as an example this time is indicated. In the concentration prediction in the stationary state, the experimental value and the simulation value were in agreement in general, and it has checked that a simulation was appropriate. However, the error arose in prediction of the concentration action in front of a stationary state about the prediction in water vapor high concentration. Moreover, separation performance was predicted. It is shown in Fig.10. It succeeded in predicting the fall of the separation performance by water vapor. However, while becoming a steady-state value in the experiment with the relative humidity more than fixed, a result which continues falling was brought in the simulation. Although a subject remains about the prediction in water vapor high concentration from the above thing, since the concentration change under the influence of water vapor can be predicted, it can be said that there is usefulness of the simulation using the Stop & Go method.

Formatting Issue - Best Available Paper/Graphic

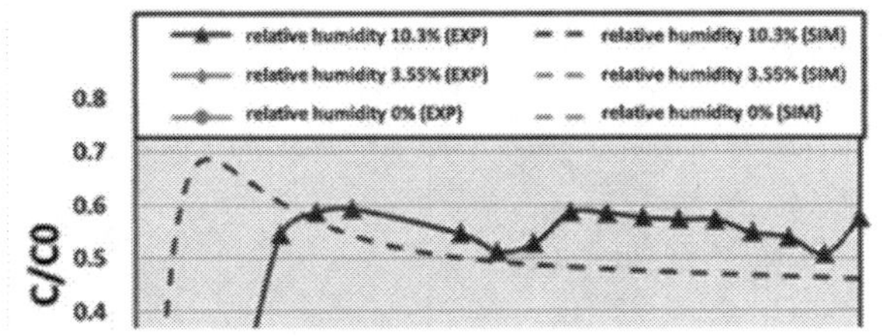

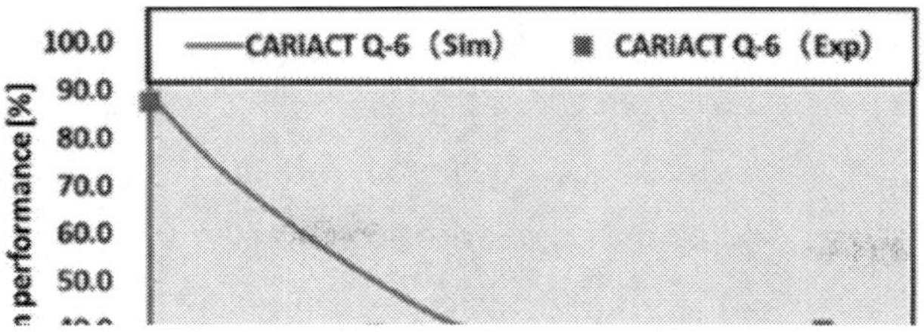

Fig.9 Comparison of a simulation and an experimental result

Fig.10 Separation performance

Conclusion

Although water vapor will strong-stick to hydrophile adsorbent if water vapor is contained, adsorption of the organic solvent will be checked and separation performance will fall, when using hydrophobic adsorbent, it turned out to about 20% of relative humidity that water vapor does not affect absorption and desorption. When water vapor was included in source gas and hydrophile adsorbent was used from this, there was the necessity for advanced dehumidification, but since source gas did not need to be dehumidified to the low dew point by using hydrophobic adsorbent, the possibility of simplification of a pre-dehumidification process was suggested. It checked that it was possible to express the action of a real process with comparing with an experimental result quantitatively about simulation prediction.

Nomenclature

Formatting Issue - Best Available Paper/Graphic

C	= concentration	[g/ml]
C_0	= frist concentration	[g/ml]
i	= component (benzene, toluene, CO_2)	[-]
K	= adsorption equilibrium constants	[1/atm]
K_s	= overall adsorption rate constant	[cm/s]
m	= mass of MSC packed in the bed	[g]
N	= division number of the column	[-]
n	= cell number in the divided column	[-]
P	= column pressure	[atm]
Q	= flow rate	[mol/s]
q^*	= amount adsorbed	[mol/kg]
q_∞	= amount of saturated sorption	[mol/kg]
t	= time	[s]
γ	= packing density	[g/cm^3]

References

1. A.Kodama, M.Seo,Y.Miyashita and Y.Osaka (2013), "Separation of a Simulated Biogas Mixture (Methane-Carbon Dioxide-Water Vapor) by Pressure Swing Adsorption with SAPO-34" *Kagaku Kogaku Ronbunshu* p.504-506.
2. C.L.Mantel (1969) "*kyuuchaku oyobi Kyuuchakuzai*" p8,9,161.
3. K. Chihara, et al "*Saishinn kyuuchaku binnrann*" p.793-799.
4. M.Yosida, P.Koompai, Y.Yamashita and S.Matsumoto (2005), "Analysis and Optimization of a Pressure-Swing Adsorption Process for Air Separation" *Kagaku Kogaku Ronbunshu* p.441-448

Kinetic Separation of light Isotope Mixtures on Microporous Materials

Daiki Minami, Katsumi Kaneko*

CREST (JST)
Center for Energy and Environmental Science, Shinshu University, 4-17-1, Wakasato, Nagano-City, Nagano, Japan

*Corresponding author email: kkaneko@shinshu-u.ac.jp

Light isotope separation with quantum molecular sieving using microporous materials has been actively studied by many researchers [1-3]. Beenakker et al. reported that the quantum molecular sieving can be applied on the separation of deuterium (D_2) and hydrogen (H_2) mixture on a cylindrical nanotube [1]. Until now, the quantum molecular sieving effect has been investigated by adsorption equilibrium experiments using a pure isotope gas at lower temperature and by quantum simulations [4-8]. Bhatia et al. showed importance of kinetic separations on D_2 and H_2 with diffusing effect [3, 9]. We have studied kinetic quantum molecular sieving effect using H_2-D_2 mixed gas [10].The separation with the quantum molecular sieving can be also applied to the separation of methane isotope mixtures such as $^{12}CD_4$-$^{12}CH_4$ or $^{13}CH_4$-$^{12}CH_4$. The separation of methane isotopes by the quantum molecular sieving should be a prospective method from the stand point of effective separation of carbon isotopes.

We have investigated separation behaviors of light isotopes by experimental and molecular simulation techniques. Methane isotope mixed gases of $^{12}CH_4/^{12}CD_4$ and $^{12}CH_4/^{13}CH_4$ were used as a model to evaluate the separation behavior of methane isotopes. Kinetic adsorption behaviors of the methane isotope mixed gases on activated carbon fiber (ACF) and zeolite were directly measured at 112 K. We evaluated the separation behavior using selectivity which calculated by the ratio of isotope molecules on the gas and adsorption phases. Even $^{12}CH_4/^{13}CH_4$ mixed gas which has extremely small difference of mass number can be separated by the kinetic adsorption measurement. In addition, we found a time dependence of the selectivity of methane isotopes. The time dependence of adsorption occurs by kinetic effect of the isotope molecules in the micropore. Also we have studied the kinetic behavior on isotope separation caused by quantum sieving effect with molecular dynamics (MD). Kinetic adsorption behaviors of $^{14}CH_4/^{12}CH_4$ mixed gas in carbon slit pores which have physical pore widths (w) of 0.5 – 1.5 nm were simulated by MD simulation techniques. The interaction between adsorbed molecules and pore wall was evaluated by quadratic Feynman-Hibbs (FH) effective potential. We also found the time dependence of adsorption selectivity on the MD results. At the initial step of adsorption (around 20 pico second), the selectivity of $^{14}CH_4/^{12}CH_4$ in the slit pore of w = 0.7 nm is higher than those of slit pores of 1.0 and 1.5 nm widths. However, the selectivities decreased with time. Thus, the kinetic effect plays an important role in separation dynamics of methane isotope molecules. We also simulated the time dependence of selectivity using conventional Lenard-Jones (LJ) potential model. The selectivity simulated by FH potential model including quantum effect was higher than that obtained by classical LJ potential model. Therefore, the kinetic effect in methane isotope separation should come from quantum effect.

[1] J. J. M. Beenakker, V. D. Borman, S. Y. Krylov, Chem. Phys. Lett., 232, 379 (1995).
[2] Q. Wang, S. R. Challa, D. S. Sholl, J. K. Johnson, Phys. Rev. Lett., 82, 956 (1999).
[3] A. V. A. Kumar, H. Jobic, S. K. Bhatia, J. Phys. Chem. B, 110, 16666 (2006).
[4] S. R. Challa, D. S. Sholl, J. K. Johnson, Phys. Rev. B, 63, 245419 (2001).
[5] H. Tanaka, H. Kanoh, M. Yudasaka, S. Iijima, K. Kaneko, J. Am. Chem. Soc., 127, 7511 (2005).
[6] A. V. A. Kumar, S. K. Bhatia, Phys. Rev. Lett., 95, 245901 (2005).
[7] R. B. Rankin, J. Liu, A. D. Kulkarni, J. K. Johnson, J. Phys. Chem. C, 113, 16906 (2009).
[8] D. Noguchi, H. Tanaka, T. Fujimori, H. Kagita, Y. Hattori, H. Honda, K. Urita, S. Utsumi, Z. Wang, T. Ohba, H. Kanoh, K. Hata, K. Kaneko, J. Phys.: Condens. Matter, 22, 334207 (2010).
[9] T. X. Nguyen, H. Jobic, S. K. Bhatia, Phys. Rev. Lett., 105, 085901 (2010).
[10] S. Niimura, T. Fujimori, Y. Hattori, L. Abrams, D. Corbin, K. Hata, K. Kaneko, J. Am. Chem. Soc., 134, 18483 (2012).

Trinary Component PSA Experiments and Influences of Water Vapor

Kazuyuki Chihara,[a] Katsuhira Kawahara,[a] Jun Kobayashi,[a]
Satoshi Nakagawa,[b] Kuniyasu Ogawa[c]

[a]Applied Chemistry, Meiji Univ., Kawasaki, Kanagawa 214-8571, Japan
[b]IKIKEN CO.,LTD., Sayama, Saitama 350-1331, Japan
[c]Mechanical Engineering, Keio Univ., Yokohama, Kanagawa 223-8522, Japan
*Corresponding author email: chihara@isc.meiji.ac.jp

When an organic solvent drains it from a factory, the organic solvent adversely affects the human body and the environment. Therefore , we study the adsorption method called the PSA method to collect the disposal gas. Water vapor obstructs the adsorption of an adsorbate and gradually results in deterioration of an adsorbent. we examined how water vapor affects the organic solvent recovery. As a method of removing the organic solvent, we determined to use adsorption in the chemical field. It is PSA method. Adsorption of PSA depend on pressure , each operation pressure is alternately changed with the high pressure and the low pressure. Accoding to two kind of operation pressure , unit device of PSA is conducted adsorption and desorption at the same time.

As is Figure 1, adsorption and desorption alternatively turn in flow courses column which filled an adsorbent separate the gas well. Trinary components were chosen as an adsorbate. Silica gel (CaRiACT Q-6) was chosen as an adsorbent. we tried to optimize the experiment method about PSA method for solvent recovery by process simulation , named Stop & Go method. Process simulation using this method was able to compare the experimental data. In the same way as experiment, simulation operation sequence of PSA can carry out tentative experiment to process. And, to check how the column is adsorbed, we conducted a study to know the adsorption distribution. Compared with simulation, we considered whether there is plastically.

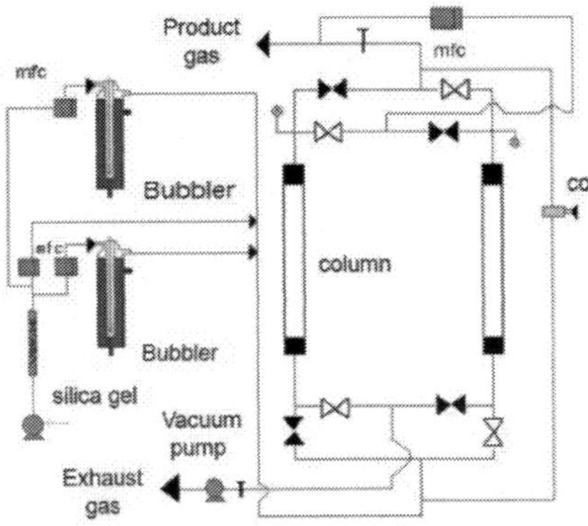

Fig.1

Stripping Re-flux PSA unit

Evaluating an adsorbent for adsorption separation process: A Case Study.

ABSTRACT:

New adsorbents are being synthesized in large numbers claiming suitability for various separation processes based of simple criteria like selectivity or uptake capacity. The performance of an adsorbent in a separation process is not a simple function of these criteria, but also depends on many operating process parameters. Hence, adsorbent screening based on such simple criteria may often be misleading. Proper screening of an adsorbent for a given separation process must be based on at least simulation and optimization using realistic inputs for adsorption equilibrium and rate kinetics of fluid transport into the adsorbent pores.

The current study focuses on the methodology for conducting a meaningful investigation of any adsorbent for a separation based on limited available information. Post-Combustion carbon capture from power plant flue gas by Vacuum Swing Adsorption (VSA) has been chosen to illustrate the procedure. Flue gas is composed of 12-15% CO_2, 70-85% N_2 and 3-15% H_2O. The performances of the known adsorbents that are suitable for CO_2/N_2 separation are significantly affected by the presence of moisture. For such adsorbents, it is desirable to remove moisture prior to CO_2/N_2 separation. These adsorbents can also handle wet flue gas when the moisture content is low. Either way, the consequence is increased capital cost and energy penalty. The desirable alternative is to develop hydrophobic adsorbents without compromising CO_2 capacity and CO_2/N_2 selectivity so that water can practically pass through the bed as an inert without affecting CO_2/N_2 separation.

A recently reported hydrophobic MOF adsorbent, SIFSIX Cu_i[1] will be evaluated for its performance for CO_2 capture and concentration (CCC) from wet flue gas. It will be compared with the performance of 13X, which is the current benchmark adsorbent for CCC from dry flue gas, but can also handle low moisture levels with higher energy penalty.

How to exploit available limited experimental information on the hydrophobic adsorbent to extract useful equilibrium and kinetic information will be discussed. These constitute essential inputs for modeling, simulation and optimization of the VSA cycle for the hydrophobic adsorbent. In this way, a realistic indication of the energy required for the separation process using the newly discovered hydrophobic adsorbent is obtained, thus shedding light on its potential for CCC from wet flue gas.

Predicting Adsorbent performance for Post-combustion carbon capture using VSA process based on Equilibrium isotherm properties

ABSTRACT:

Large number of adsorbents claiming suitability for post-combustion carbon capture are being synthesized. Evaluation of the adsorbents, however, still remains irregular due to two main reasons. Firstly experimental evaluation of the numerous adsorbents synthesized consumes lot of resource and time and at times is not feasible at all. Secondly, process simulations/optimization is computationally intensive and generally not available to all material researchers.

Since most of the adsorbents for post-combustion capture are equilibrium based, adsorbent equilibrium properties are the most critical component affecting the adsorbent performance. The translation of these properties, however, on process performance still remains to be captured. The perceived performance indicators such as breakthrough times, high loading capacity have not been proven to be directly correlated to process level performance of the adsorbent.

Adsorption separation process, apart from the adsorbent, also depends on various other process configurations and operating conditions. While there have been a few predictive models developed on the basis of process simulations, they lack the use of proper cycle configuration and process conditions suitable for adsorbents. Using either a simple adsorption cycle or non-ideal process conditions for evaluating an adsorbent might be misleading and should be avoided.

In the current work, a predictive model has been developed for estimating minimum energy penalty of post-combustion capture for an adsorbent using 4-step VSA with light product pressurization (LPP) process, which has been reported in the literature to be best for post-combustion capture from dry gas using 13X zeolite. A parametric study has been conducted using design of experiments for estimating the effects of adsorbent equilibrium properties: Binary mixture selectivity, Henry's constant, CO_2 loading, Isotherm non-linearity (λ) and local slope of CO_2 isotherm for a given adsorbent on the minimum energy penalty of VSA process. Each point in the study has been optimized for process variables using genetic algorithm based optimization. The significance of independent variables and their interactions were tested by means of the analysis of variance (ANOVA) with 95% confidence limits ($\alpha = 0.05$).

It has been found that binary mixture selectivity plays the most important role in performance of any adsorbent. Also, it is observed, that higher CO_2 loading results into increased N_2 loadings and subsequently into higher energy penalty for the system which is counter-intuitive and deviant to existing notions. The meta-model acts as an easy-to-use tool for material researchers for evaluating new adsorbents. The meta-model can also be used as screening tool for hypothetical adsorbents and selecting the promising adsorbents for further inspection.

---

[1] P. Nugent, Y. Belmabkhout, S.D. Burd, A.J. Cairns, R. Luebke, K. Forrest, T. Pham, S. Ma, B. Space, L. Wojtas, M. Eddaoudi, M.J. Zaworotko, Nature 495 (2013) 80-84.

Study about Adsorption Equilibrium of Azeotropic Mixture on FAU Type High Silica Zeolite - Interpretation By Molecular Simulation and Do's Method –

Kazuyuki Chihara, Shintaro Kume, Natsuki Yamada, Tatsuharu Taira, and D. D. Do*

Department of Applied Chemistry, Meiji University, Kawasaki, Kanagawa, 214-8571, Japan ,
The University of Oueenslamd, St Lucia QLD 4072, Australia
Fax: +81 -44 -934 -7917   E-mail:chihara@isc.meiji.ac.jp

INTRODUCTION

Many of the cases in chemical engineering industry, we must use separation processes. But, there are in multi component mixture called “azeotropic mixture”, thus it is too difficult to separate objective substance perfectly. So, we tried to analyze behavior of adsorption in one of the azeotropic mixture system”2-propanol-trichloloethylene mixture” (following, we describe them “IPA” and “TCE) performed by GCMC simulation.

THEORY

Do's Method [1]

We utilized Do’s Method in analyzing behavior of adsorption. According to the theory, binary components mixture system is classified dominant adsorbate and inferiority adsobate. Inferiority adsobate is adsorbed the portion called “supercage” in adsorbent model selectively. The model of supercage is shown in Fig.1 and Fig.2. We defined them “Phase 2”,and the other area determined “Phase 1”. Then, in these simulations, Phase 2 is only occupied by IPA which is inferiority adsobate. and all TCE is adsorbed into Phase 1.

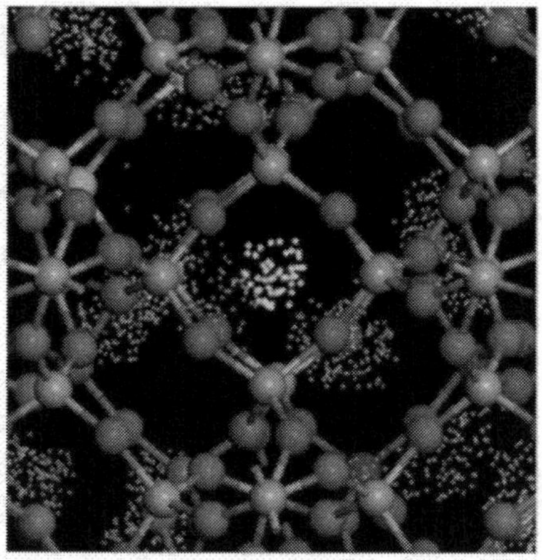

Fig.1 The shape of a spercage (The enlarged view of an unit cell)

Fig.2 The view of an unit cell (It has 8 supercages which were colored)

Physical properties of adsorbent model and adsorbates are shown Table1 and Table2.

Table1. Physical propaty of an unit cell of adsorbent

	Number of atoms	Amount of substance [mol]	Mass [g]
O	384	6.38×10^{-22}	1.02×10^{-20}
Al	0	0	0
Si	192	3.19×10^{-22}	8.96×10^{-21}
H	0	0	0
Total	576	9.57×10^{-22}	1.92×10^{-20}

The shape of an unit cell	Length of an unit cell [Å]	Volume of an unit cell [$Å^3$]	Voxel volume of an unit cell [$Å^3$]
	25.028	15678	0.062698
Cube	Number of partitions	Total number of partitions	Avogadro's number [mol^{-1}]
	63	250047	6.0221×10^{23}

Table.2 Molecular weight of adsorbate

Dominant adsorbate	Inferiority adsorbate
TCE [g/mol]	IPA [g/mol]
131.39	60.10

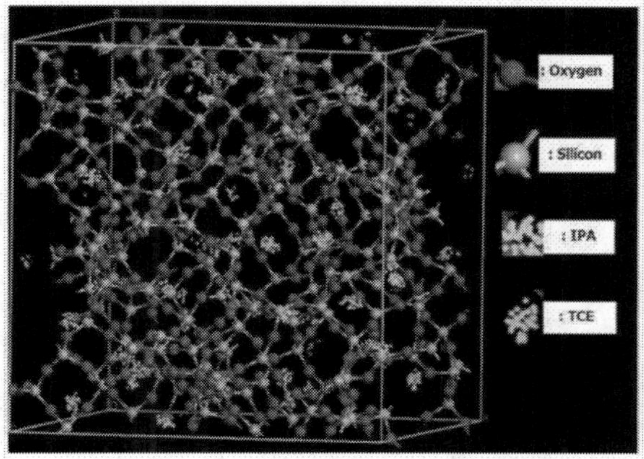

Fig.3 The distribution diagram after GCMC simulation

Experimental condition

Pressure : 7.000 kPa (constant)

Temperature : 298.15 K (constant)

Composition of adsorbate : IPA-TCE

= 0.1 : 0.9, 0.2 : 0.8, ~, 0.8 : 0.2, 0.9 : 0.1

EXPERIMENT

Model of Adsorbent

We simulated adsorption equilibrium of IPA-TCE binary components by using Material studio 7.0. Fig. 3 is an example of the distribution diagram after GCMC simulation. Then, we used a function of density distribution in order to calculate the coordinates of IPA and TCE. Finally adsorbed molecular was distributed to Phase1 to Phase 2. Condition of calculation is as described above. We performed simulation only the ratio of injection adsorbate changed step to 0.1.

Single-component Isotherm

We simulated and created single-component isotherm of IPA and TCE respectively. But, we assumed single-component isotherm follows Langmuir Formula. Then, we calculated Langmuir constant K and q_{∞}. Fig.4 and Fig.5 show the result.

Langmuir Formula : $$q = \frac{q_\infty KP}{1+KP} \qquad \frac{P}{q} = \frac{1}{q_\infty}P + \frac{1}{Kq_\infty}$$

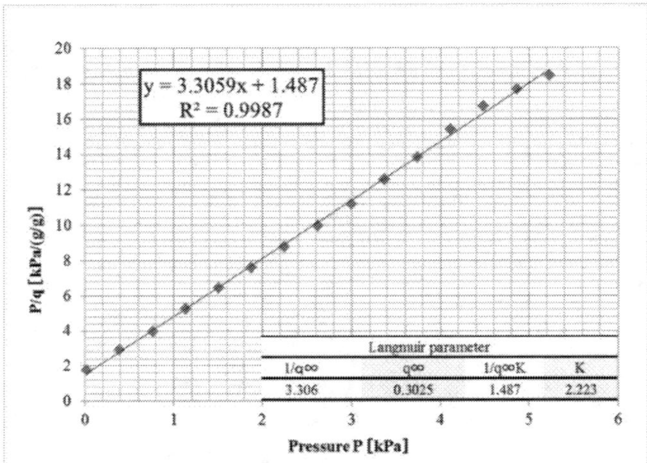

Fig.4 Langmuir plot (Temperture = 298K , FAU zeolite-IPA system)

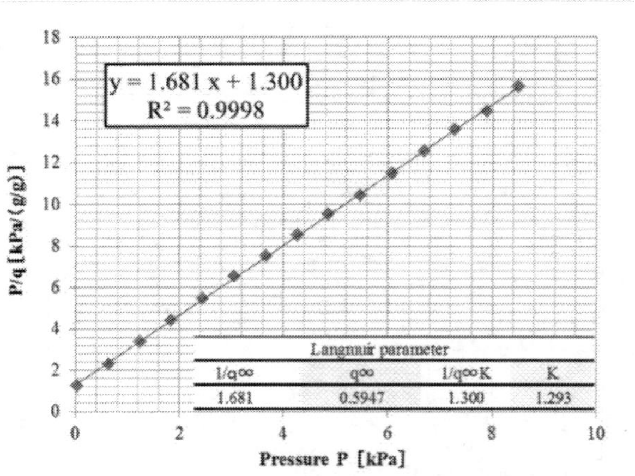

Fig.5 Langmuir plot (Temperture = 298K , FAU zeolite-TCE system)

Adsorption Isotherm of Two-component System

According to Do's Method, the amount of adsorbed quantity follows below formula.

Amount of Adsorbed TCE q_{TCE}[g/g]

$$\frac{q_{TCE}}{q_{\infty TCE}} = \frac{K_{TCE}P_{TCE}}{1 + K_{TCE}P_{TCE} + \varphi_{IPAphase\,1}K_{IPA}P_{IPA}}$$

Amount of Adsorbed IPA q_{IPA}[g/g]

phase 1

$$\frac{q_{IPAphase\,1}}{\varphi_{IPAphase\,1}q_{\infty IPA}} = \frac{\varphi_{IPAphase\,1}K_{IPA}P_{IPA}}{1 + K_{TCE}P_{TCE} + \varphi_{IPAphase\,1}K_{IPA}P_{IPA}}$$

phase 2

$$\frac{q_{IPAphase\,2}}{\varphi_{IPAphase\,2}q_{\infty IPA}} = \frac{\varphi_{IPAphase\,2}K_{IPA}P_{IPA}}{1 + \varphi_{IPAphase\,2}K_{IPA}P_{IPA}}$$

$$q_{IPA} = q_{IPAphase\,1} + q_{IPApase\,2}$$

We calculated the amount of adsorbed quantity by utilizing above formula used Langmuir constant determined by single-component isotherm and the ratio of adsorbate calculated by function of density distribution respectively.

RESULT AND DISCUSSION

Gas-liquid equilibrium of IPA-TCE binary ideal solution's system [2]

We assumed the system of the simulation follows The Low of Raoult, calculated saturated pressure of respective adsorbates in conditions of simulations by utilizing Antoine Formula. Then, we created the line of gas-liquid equilibrium respectively. Finally we plotted the data of the ratio of adsorption calculated by simulating above the graph. Table 3 and Fig.6 shows them.

Equation of Binary Components of Ratio of Molecule

$$x_i + x_j = 1$$

$$y_i + y_j = 1$$

The Low of Raoult

$$\mathrm{P}_{all}\, y_i = P_i^s x_i$$

$$x_i = \frac{\mathrm{P}_{all} - P_j^s}{P_i^s - P_j^s}$$

Antoine Formula

$$\log P_i^s/\mathrm{kPa} = A_i - \frac{B_i}{T/\mathrm{K} + C_i}$$

Relative Volatility

$$\alpha = \frac{K_i}{K_j} = \frac{(y_i/x_i)}{(y_j/x_j)}$$

Table3. Gas-liquid equilibrium of IPA-TCE binary ideal solution's system (P_{all} = 7.000 kPa, T = 298.15 K , α = 1.669)

Gas-liquid equilibrium of TCE	x_{TCE}	0	**0.1**	**0.2**	0.3	0.4	0.5	0.6	0.7	0.8	0.9	1
	y_{TCE}	0	**0.1564**	**0.2944**	0.4170	0.5267	0.6253	0.7146	0.7957	0.8697	0.9376	1
Gas-liquid equilibrium of IPA	x_{IPA}	1	**0.9**	**0.8**	0.7	0.6	0.5	0.4	0.3	0.2	0.1	0
	y_{IPA}	1	**0.84357**	**0.70559**	0.58299	0.47334	0.37467	0.28543	0.20432	0.13028	0.06242	0
Molecular ratio of injection gas	IPA	0	**0.1**	**0.2**	0.3	0.4	0.5	0.6	0.7	0.8	0.9	1
	TCE	1	**0.9**	**0.8**	0.7	0.6	0.5	0.4	0.3	0.2	0.1	0
Molecular ratio of adsorbate	IPA	0	**0.1569**	**0.2600**	0.1633	0.1301	0.2339	0.1923	0.1346	0.2315	0.2232	1
	TCE	1	**0.8431**	**0.7400**	0.8367	0.8699	0.7661	0.8077	0.8654	0.7685	0.7768	0

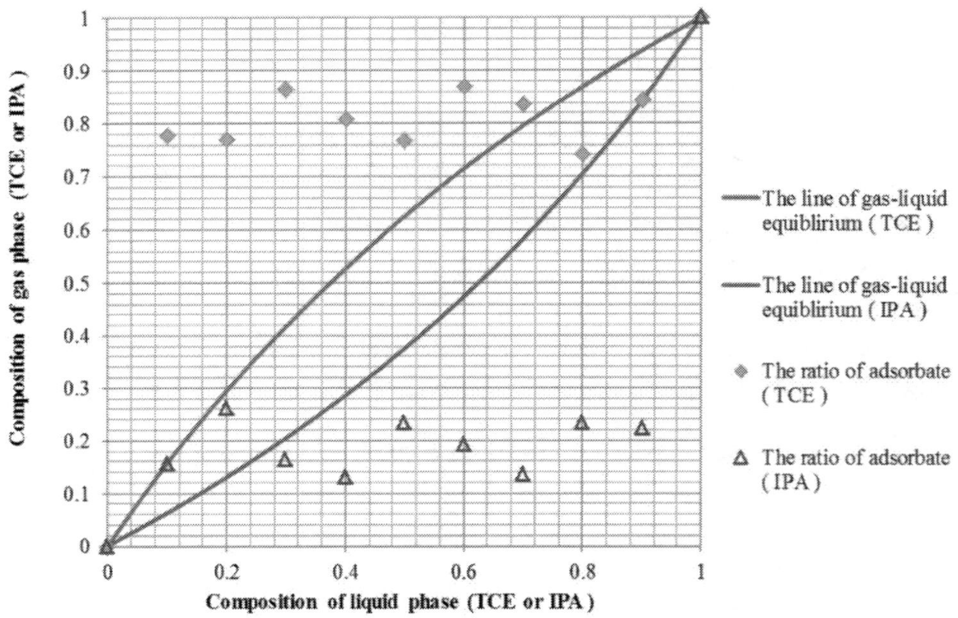

Fig.6 The relationsip of gas-liquid equilibrium and its ratio of adsorbate (P_{all} = 7.000 kPa, T = 298.15 K , α = 1.669)

Please pay attention to the result on Fig.6 of the ratio of molecular IPA vs TCE = 0.1 : 0.9 , 0.2 : 0.8 The ratio of adsorption of IPA is on the line of gas-liquid equilibrium of TCE. On the other hand, the ratio of adsorption of TCE is on the line of gas-liquid equilibrium of IPA. By the results, we assumed in the condition of this pressure and temperature, when injection of ratio of molecular of IPA is greater than 0.8, TCE behaves like condition of IPA gas phase, IPA behaves like condition of TCE gas phase. And, without injection of the molecular ratio of IPA vs TCE, It is roughly constant that the ratio of adsorbate respectively. If it were not priority of adsorption between IPA and TCE, in the situation of simulation, The ratio of adsorbate of IPA vs TCE was equal about 38 : 62. (See Table 4) But, in actually, that was about 20 : 80 . (See Table 5) It is seemed that the extent TCE behaves like condition of IPA gas phase, IPA behaves like condition of TCE gas phase concern with the determination the ratio of the ratio of adsorbate of IPA vs TCE.

Table4. The relationsip of gas-liquid equilibrium of TCE-IPA binary ideal solution's system

Ratio of composition of liquid phase x (IPA : TCE)	
0.2607	0.7393
Ratio of composition of gas phase y (IPA : TCE)	
0.3805	0.6195

(P_{all} = 7.000 kPa, T = 298.15 K , α = 1.669)

Table5. Molecular ratio of adsorbate (average)

Ratio of composition of adsorbate (IPA : TCE)	
0.1918	0.8082

Respective result of simulations

Table 6 and Fig.7 show the result of respective molecular ratio of adsorbate.

Table6. Molecular ratio of adsorbate (respective)

IPA : TCE	IPA	TCE
0.1:0.9	0.1569	0.8431
0.2:0.8	0.2600	0.7400
0.3:0.7	0.1633	0.8367
0.4:0.6	0.1301	0.8699
0.5:0.5	0.2339	0.7661
0.6:0.4	0.1923	0.8077
0.7:0.3	0.1346	0.8654
0.8:0.2	0.2315	0.7685
0.9:0.1	0.2232	0.7768
Average	0.1918	0.8082

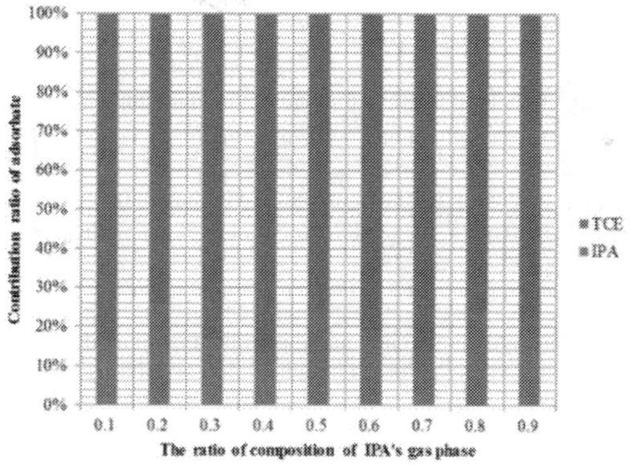

Fig.7 The relationsip of the ratio of composition of IPA's gas phase and contribution ratio of adsorbate (P_{all} = 7.000 kPa, T = 298.15 K , α = 1.669)

When the ratio of molecular IPA vs TCE = 0.3 : 0.7, 0.4 : 0.6, 0.6 : 0.4 , 0.7 : 0.3 ,the ratio of adsorbed IPA is less than 20% .(But, the result of IPA vs TCE = 0.1 : 0.9 is the reason TCE behaves like condition of IPA gas phase, IPA behaves like condition of TCE gas phase, we don't consider the situation of them)

Adsorption's distribution of IPA and TCE

Table 7 ~8, Fig.8 ~ 9 show that the ratio of adsorption's distribution calculated by using function of density distribution respectively.

Table7. The ratio of adsorption's distribution (IPA)

IPA : TCE	Phase1	Phase2
0.1:0.9	0.5000	0.5000
0.2:0.8	0.4665	0.5335
0.3:0.7	0.1250	0.8750
0.4:0.6	0.2185	0.7815
0.5:0.5	0.3311	0.6689
0.6:0.4	0.0076	0.9924
0.7:0.3	0.1429	0.8571
0.8:0.2	0.3248	0.6752
0.9:0.1	0.3511	0.6489

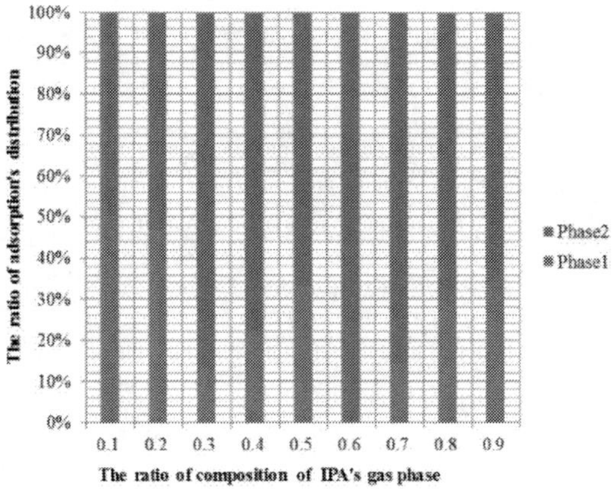

Fig.8 The relationsip of the ratio of composition of IPA's gas phase and its adsorption's distribution (Adsorbate = IPA, P_{all} = 7.000 kPa, T = 298.15 K , α = 1.669)

When the ratio of molecular IPA vs TCE = 0.3 : 0.7, 0.4 : 0.6, 0.6 : 0.4 , 0.7 : 0.3 ,It is larger that of the ratio of distribution to Phase2 than the others situations'。 These extents are equal to the extents that of the ratio of adsorbate are less than the others. So, it seems that adsorbed to Phase2 selectively is related to reduction of amount of adsorbed quantity of inferiority adsorbate totally.

Table8. The ratio of adsorption's distribution (TCE)

IPA : TCE	Phase1	Phase2
0.1 : 0.9	1.000	0.000
0.2 : 0.8	1.000	0.000
0.3 : 0.7	0.9948	0.0052
0.4 : 0.6	0.8785	0.1215
0.5 : 0.5	1.000	0.000
0.6 : 0.4	0.9580	0.0420
0.7 : 0.3	1.000	0.000
0.8 : 0.2	1.000	0.000
0.9 : 0.1	1.000	0.000

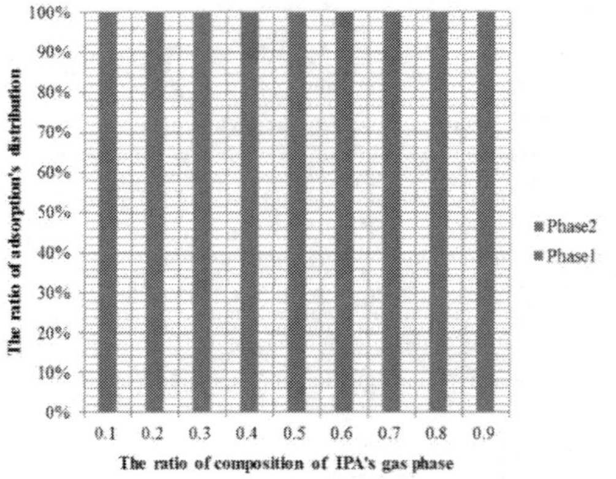

Fig.9 The relationsip of the ratio of composition of IPA's gas phase and its adsorption's distribution (Adsorbate = TCE, P_{all} = 7.000 kPa, T = 298.15 K, α = 1.669)

When the ratio of molecular IPA vs TCE = 0.3 : 0.7, 0.4 : 0.6, 0.6 : 0.4, TCE was adsorbed to Phase2. According to Do's Method, All TCE is adsorbed into only Phase1.So, It is seemed that TCE was adsorbed into Phase2, TCE had property of inferiority adsorbate, and it has been azeotropic mixture in those situations.

CONCLUSIONS

In this condition (Total pressure =7.000 kPa, Temperature = 298.15 K constant), It could not confirm azeotropic phenomena of IPA - TCE binary components by Do's Method. But ,calculated in the situations of the ratio of molecular IPA vs TCE = 0.3 : 0.7, 0.4 : 0.6, 0.6 : 0.4, TCE was adsorbed into Phase2 ,so it is seemed that IPA-TCE binary mixture system becomes azeotropic mixture in those situations. And, we confirmed tendency that the larger the ratio of adsorption's distribution of Phase2, the lesser the amount of adsorbed quantity of inferiority adsorbate. Interestingly, When the injection of the ratio of molecular IPA vs TCE = 0.1 : 0.9, 0.2 : 0.8, The ratio of adsorption of IPA is on the line of gas-liquid equilibrium of TCE. On the other hand, the ratio of adsorption of TCE is on the line of gas-liquid equilibrium of IPA. However, the other conditions of injection of the ratio of molecular IPA vs TCE , The ratio of adsorbate of IPA vs TCE was equal about 20 : 80 constantly. But, through the simulations, we assumed that the extent TCE behaves like condition of IPA gas phase, IPA behaves like condition of TCE gas phase concern with the determination the ratio of the ratio of adsorbate of IPA vs TCE. Hence, we want to study about the relationship between extents the one component behave like the other component and the ratio of adsorbate.

Notation

q	= amount of adsorbed	[g/g]
q_∞	= amount of saturated sorption	[g/g]
K	= Langmuir constant	[kPa^{-1}]
P	= pressure	[kPa]

x_i, x_j	= molecular composition of liquid phase	[−]
y_i, y_j	= molecular composition of gas phase	[−]
P_{all}	= total pressure	[kPa]
P_i^S, P_j^S	= saturated pressure	[kPa]
A_i, B_i, C_i	= Antoine constant	[−]
K_i, K_j	= equilibrium constant	[−]
α	= relative volatility	[−]

Greek letters

$\varphi_{IPA,phase1}$, $\varphi_{IPA,phase2}$ = distribution coefficient [−]

References

1. Jeong-Gil Choi, D. D. Do, and H. D. Do (2001), "Surface Diffusion of Adsorbed Molecules in Porous Media: Monolayer, Multilayer, and Capillary Regimes" *Ind. Eng. Chem. Res*, 40, 4005−4031
2. Yasushi Takeuchi, Masakuni Matsuoka, Kenji Ochi, Kazuyuki Chihara (2007), "Kaisetsu kagaku kougaku " Bfukan, pp116-123

Acknowledgment : Special research fund of 2013 fiscal year by Meiji univ. is appreciated.

Adsorption experiment and simulation to MSC3A of the multicomponent system gas in the volumetric method

1.Introduction

We counducted an experiment of adsoption equilibrium by using the capacity method.
MSC3A is used as an adsoption agent.This experiment of “single component (N2,CH4,CO2) and two-component(CO2-CH4,N2-CH4)” is conducted under conditions of 303,313,323K. We performed two-component adsorption equilibrium,examination about the adsorption speed from data provided in single component.

2.Experiment□simulation

In the single component,gas, we used three kinds of gas(CO2,CH4,N2). In two-component gas,we used mixed gas(CO2-CH4,N2-CH4). The column temperature reached at 303,313,323K. We introduced gas to save gas holder and recorded a pressure descent in a data logger from a point in time when We opended every direction cook.
After reach the equilibrium, We gradually repeated the operation that a vacuum deaerated gas collection. to reach 5 atm. In two-component gas experiment, changing a ratio of introduction gas to 3:7,5:5,7:3,I experiment. We use a circulation pump to keep mixed gases uniformly. In addition, we used a gas chromatograph to find the vapor phase ingredient ratio. From reading introduction pressure, initial pressure, equilibrium pressure, the quantity of adsorption of the result that we measured was calculated using an ideal gas equation. In two-component gas, we calculated quantityof adsorption equally by considering the gas ingredient ratio of the vapor phase.
About the analysis of the adsorption speed, we demanded micro aperture diffusion, a generalization mass transfer coefficient by demanding t[s] for the number of the Bi and real time at the age of $\tau F=1$ by letting a change have theory curve and curve fitting in the time for pressure descent.

374944 A New Commercial Instrument for Gas Sorption Measurements Under Extreme Conditions

Abstract:

A new commercial microbalance has been developed in order to characterize the gas sorption properties of adsorbents under a range of challenging conditions. The instrument, called the XEMIS, features a new type of microbalance in which the sensitive detection and control components are located outside the main chamber of a symmetrical beam balance. The design allows the gravimetric measurement of the adsorption and desorption of a range of gas species, including corrosives, at pressures of up to 200 bar and temperatures of 500°C. To demonstrate its operation, we have performed measurements on a series of samples under different conditions of temperature and pressure. We have selected a Metal-Organic Framework (MOF), a shale and a zeolite. We present hydrogen adsorption and desorption isotherms on a 50 mg sample of Na-X zeolite up to 150 bar at 77 K, high resolution SO_2 adsorption and desorption isotherms on a novel MOF (NOTT-300) at sub-ambient pressure and a range of near-ambient temperatures, and CH_4 adsorption isotherms up to 100 bar on a shale. These data demonstrate the operation of the XEMIS for the characterization of different types of adsorbent for a range of practical gas adsorption applications.

Study on Sludge Volume Reduction and Activated Carbon Regeneration Using Liquefied Dimethyl Ether Gas

Tadashi SANO, Hitachi Research Laboratory, Hitachi Ltd.
Mitsuhiro MATSUZAWA, Hitachi Research Laboratory, Hitachi Ltd.
Yoko KOKUGAN, Hitachi Research Laboratory, Hitachi Ltd.

At oil wells, muddy water that include oil and suspended solids is produced as a byproduct along with crude oil, and is called produced water. The produced water was discarded to tailing pond or old wells without treatment previously, however, the introduction of the water treatment equipment has been demanded in the crude oil production industry due to the improvement of the recent environmental awareness. Therefore, Hitachi Ltd. developed the Flocculation Magnetic Separation (FMS) system that can remove suspended solids as sludge from produced water and investigated application method of activated carbon for removal of dissolved organic matter.

In this study, we investigated of method for reduction the volume of sludge generated in FMS system and for regeneration of the spent activated carbon by common technology. In order to achieve the two objectives□dimethyl ether (DME) which is gas at ordinary temperatures and pressures was used. As liquefied DME by being pressurized can dissolve water and oil, it is easy to remove water from brown coal [1]. Therefore we study on new separation process by utilizing these characteristics. As water and oil was extracted from the sludge by DME, weight of the sludge was reduced by 76%. Also, organic matter which adsorbed to activated carbon was desorbed with water by DME, so, adsorption ability of the regeneration activated carbon recovered to 80%. On the other hand, as liquefied DME which was used for extraction was separated from the oil and the water because of evaporation by a slight heat, we consider the amount used of DME in this method is very small when it runs in circle.

[1] H. Kanda, H. Makino, and M. Miyahara, Energy-saving drying technology for porous media using liquefied DME gas, Adsorption, 14, 467-473 (2008.10)

380222 Development of Highly Porous Carbon from Polyvinylidene Fluoride for CO_2 Adsorption

Abstract:

In the increasing awareness of abnormal climate changes, carbon dioxide (CO_2) emissions from the use of fossil fuels have been considered as the major contributing greenhouse gas. Therefore, the reduction of CO_2 has become an important environmental concern, and various approaches such as absorption, adsorption, and membrane separation have been employed in CO_2 capture. In recent times, technology designed to capture $CO_2$by adsorption has received considerable attention due to its easy regeneration and low energy consumption without producing unfavorable by-products or any polluted sorbent.

Commercial adsorbents such as activated carbon, zeolite, and metal-organic frameworks (MOFs) have been widely used in CO_2 adsorption. Although zeolites and MOFs have high CO_2adsorption capacities, they substantially lose their adsorption abilities in humidified conditions due to their hydrophilic characteristics [1-2]. On the other hand, porous carbonaceous materials are hydrophobic and have additional advantages of high surface area, thermal and chemical stability, and relatively low cost.

In this study, polyvinylidene fluoride (PVDF) was used to synthesize porous carbon for CO_2 adsorption. PVDF has been known to process porous structure with high surface area and pore volume during carbonization at high temperature [3], indicating great potential in CO_2 adsorption. Therefore, highly porous carbons were prepared for CO_2 capture from the PVDF carbonized at high temperature in the range 400–800 °C, and the effects of carbonization temperature on the characteristics and CO_2 adsorption behavior of prepared carbon materials were investigated. The carbonized PVDFs showed high surface area and pore volume more than 1000 m^2/g and 0.4 cm^3/g, respectively, and the significant increase of these textural properties was predominantly attributed to the evolution of narrow micropores (< 0.7 nm).

The CO_2 adsorption uptake of the carbonized PVDF was measured by CO_2 adsorption isotherm at different temperatures. The maximum CO_2 adsorption capacities at the temperature of 0 and 25 °C were 5.27 and 3.59 mol/kg, respectively under atmospheric pressure. The carbonized PVDFs also showed excellent recyclability, easy regenerability, and rapid adsorption/desorption kinetics, providing potential possibility in continuous CO_2capture processes. The characteristics of the carbonized PVDFs were further analyzed by transmission electron microscopy (TEM), thermogravimetric analysis (TGA), Raman spectroscopy, and element analysis.

[1] Sunho Choi, Jeffrey H. Drese, and Cristopher W. Jones, Adsorbent materials for carbon dioxide capture from large anthropogenic point sources. ChemSusChem **2009**, 2, 796-854.

[2] Abdelhamid Sayari, Youssef Belmabkhout, and Rodrigo Serna-Guerrero, Flue gas treatment via CO_2 adsorption. Chemical Engineering Journal **2011**, 171, 760-774.

[3] Bin Xu, Shanshan Hou, Mo Chu, Gaoping Cao, and Yusheng Yang, An activation-free method for preparing microporous carbon by the pyrolysis of poly(vinylidene fluoride). Carbon **2010**, 48, 2812-2814.

374130 The Effect of Desorption Pressure on Breakthrough Behavior of CO2 and N2 Mixture on Zeolites 3A, 4A, 5A, and 13X

Abstract:

The effect of desorption pressure on CO2/N2 breakthrough behaviors for 4 different adsorbents were studied at a fixed bed. Zeolite 3A, 4A, 5A, 13X pellets were used as adsorbents. Cyclic operations were executed with varying desorption pressure from vacuum (0 bar) to 3 bar without changing other conditions such as adsorption step pressure (3 bar), temperature (293K), composition (CO_2:N_2=10:90vol%) and flow rate (400ccm). Adsorption step time was 80 min and desorption step time was 80 min for 5 Cycles. After the experiment, breakthrough time, saturation time, and adsorption amount were measured and compared in order to find an optimum adsorbent and a proper operating condition for a post combustion CO_2 capture process.

Steam Stability of Alumina Supported Amine Materials for CO_2 Capture from Ultra Dilute Gas Streams

Miles Sakwa-Novak and Christopher W. Jones
School of Chemical & Biomolecular Engineering, Georgia Institute of Technology, 311 Ferst Drive, Atlanta, GA 30332, USA

Abstract

Silica supported amine materials are a promising class of solid adsorbents for applications in CO_2 removal from power plant flue gasses and also from ultra-dilute streams such as air. While these materials have been demonstrated to have high CO_2 uptakes and kinetics, their degradation under practical operating conditions (amine oxidation, steam/water degradation) remains a drawback. Alumina supported amine materials are hypothesized to provide enhanced steam stability. To better understand their long term stability, a detailed understanding of the degradation mechanisms of candidate sorbents under relevant environments is a critical research need. Here we present a study of an alumina supported amine adsorbent upon cyclic exposure to low temperature steam, as might be encountered in a steam-stripping regeneration cycle. A mesoporous γ-alumina impregnated with poly(ethyleneimine) was subjected to steam treatments for times ranging from 5 minutes to 24 hours, and its stability was evaluated based on CO_2 capacities relevant to CO_2 capture from air. It was found that in the absence of significant amine leaching, the material retained its CO_2 capacity after steam treatment. XRD and ^{27}Al NMR characterizations showed that the γ -alumina partially rehydrated to the mineral phase boehmite upon treatment with steam for times of 90 minutes and greater, the extent of which may be nonlinearly dependent on steam exposure time. It will be explicitly shown that this partial rehydration of the alumina support does not directly affect the CO_2 capacity of the sorbent.

Further subtle changes in the textural properties of the sorbent upon treatment with steam for extended periods were observed as well.

AUTHOR INDEX

AUTHOR INDEX

AUTHOR INDEX

AUTHOR INDEX

AUTHOR INDEX

AUTHOR INDEX

AUTHOR INDEX

AUTHOR INDEX

AUTHOR INDEX

AUTHOR INDEX

AUTHOR INDEX

AUTHOR INDEX

AUTHOR INDEX

AIChE
120 Wall Street, FL 23
New York, NY 10005-4020

ISBN 978-1-5108-1272-7

CURRAN ASSOCIATES INC.
proceedings.com

Separations Division 2014

Core Programming Area at the 2014 AIChE Annual Meeting

Atlanta, Georgia, USA
16-21 November 2014

Volume 2 of 2

AIChE
The Global Home of Chemical Engineers

Separations Division 2014

Core Programming Area at the 2014 AIChE Annual Meeting

Atlanta, Georgia, USA
16-21 November 2014

Volume 2 of 2

ISBN: 978-1-5108-1272-7

Printed from e-media with permission by:

Curran Associates, Inc.
57 Morehouse Lane
Red Hook, NY 12571

Printed by Curran Associates, Inc. (2015)

For permission requests, please contact AIChE
at the address below.

AIChE
120 Wall Street, FL 23
New York, NY 10005-4020

Phone: (800) 242-4363
Fax: (203) 775-5177

www.aiche.org

Additional copies of this publication are available from:

Curran Associates, Inc.
57 Morehouse Lane
Red Hook, NY 12571 USA
Phone: 845-758-0400
Fax: 845-758-2634
Email: curran@proceedings.com
Web: www.proceedings.com

TABLE OF CONTENTS

VOLUME 1

VOLUME 2

381115 Adsorption Equilibria and Kinetics of CO2, CH4, and N2 on Zeolite 13X up to 1 Mpa

Abstract:

Adsorption equilibria and kinetics of carbon dioxide, methane, and nitrogen on zeolite 13X were studied at (293, 308, and 323) K and pressure up to 1 MPa. The experimental data were measured by a high pressure volumetric system. The physical properties of zeolite 13X were analyzed by N_2 adsorption/desorption isotherms at 77K with automatic volumetric sorption analyzer. The preferential adsorption capacity of carbon dioxide on zeolite 13X was much higher than that of the other gases. Each experimental isotherm was correlated by the Langmuir and Sips and temperature dependent Sips models, and the deviation of each model was evaluated. The Sips model showed smaller deviation than the Langmuir model. Isosteric enthalpy of each component were evaluated by using the temperature dependent Sips model and presented along with surface loading. And it was analyzed and compared at 0.1 MPa range as well as 1.0 MPa range. Diffusion rate of each adsorptive on zeolite 13X were estimated by the isothermal and non-isothermal diffusion models. The non-isothermal model provided a theoretical model which can be used to determine the significance of thermal effects in a transient adsorption rate measurement and to analyze the uptake curves obtained when diffusion is too fast for the isothermal approximation to be valid. The adsorption kinetics of each component was compared at various conditions.

378678 Carbon Dioxide Capture Using Amine-Functionalized Multi-Walled Carbon Nanotubes: Adsorption/Desorption Kinetics Studies and Effects of Flue Gas Contaminants

Abstract:

Industrial grade multi-walled carbon nanotubes (IG-MWCNTs) impregnated with tetraethylenepentamine (TEPA) are promising materials for solid sorbent-based CO_2 capture from large point sources. In this work, the adsorption/desorption kinetics of amine-impregnated IG-MWCNTs were examined at various temperatures, CO_2 partial pressure, and amine loadings. After evaluating four kinetic models for CO_2 adsorption, it was found that Avrami's fractional order kinetic model provided the best fitting for the adsorption behavior of CO_2. In order to find the optimal regeneration method, three desorption methods were evaluated for the regeneration of solid sorbents. Moreover, the tolerance of amine-impregnated IG-MWCNTs to trace flue gas contaminants was also investigated. It was found that H_2O, NO, and SO_2 had a minimal impact on CO_2 adsorption capacity at 323K, while CO_2 adsorption capacity decreased with increasing SO_2 concentration at high temperature. Further analysis showed that irreversible sulphate/bisulphite deposited onto the solid sorbent might contribute to the decrease of CO_2 adsorption capacity. Compared with silica-based adsorbents, the loss of CO_2 capacity of IG-MWCNTs-50 (9.530%) was much lower than that of MCM-41-50 (40.430%) under SO_2 concentration of 200 ppm in the presence of water vapor after 5 cycles.

Modeling of Diffusion Resistance of CO_2 Adsorption on Poly(ethylenimine) Impregnated Hollow Fiber Adsorbents

Jayashree Kalyanaraman, Yanfang Fan, Ryan P. Lively, William J. Koros, Christopher W. Jones, Yoshiaki Kawajiri and Matthew J. Realff

School of Chemical & Biomolecular Engineering, Georgia Institute of Technology - Atlanta, GA 30332, United States

Introduction

- ❑ CO_2 capture by post combustion has high impact on coal fired power plant
- ❑ Adsorption - Energy efficient compared to liquid amine scrubbing
- ❑ Hollow fiber- Novel sorbent contactor with potential for high heat recovery compared to packed bed sorbent
- ❑ Amine based sorbents efficient under humid conditions

  - ➢ Two classes of amine sorbents can be loaded in hollow fiber
  - ➢ Class1 : molecular basket sorbents (MBS) with amine impregnated in silica support – PEI (Polyethylenimine)
  - ➢ Class2:silica is functionalized by grafting amine (APS)-3-aminopropyltrimethoxysilane

Mass Transfer Kinetics of Amine Sorbents

- ❑ Mass transfer kinetics limit the adsorption process
- ❑ *Xu et al.* experimentally found capacity increases with temperature in MBS- diffusion limitation
  - ➢ 'concave trend' of breakthrough capacity
- ❑ CO_2 breakthrough curve has long tail- asymmetric shape

Objective: Modeling of CO_2 diffusion in amine sorbents (MBS)-Polyethylenime impregnated silica sorbent

Experimental setup

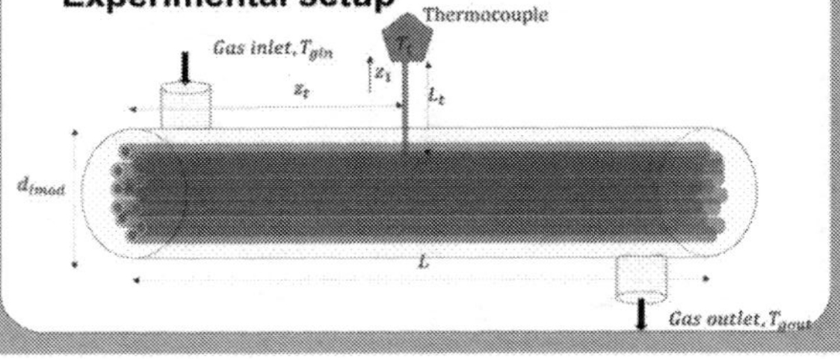

Hollow Fiber Mass Transfer Model

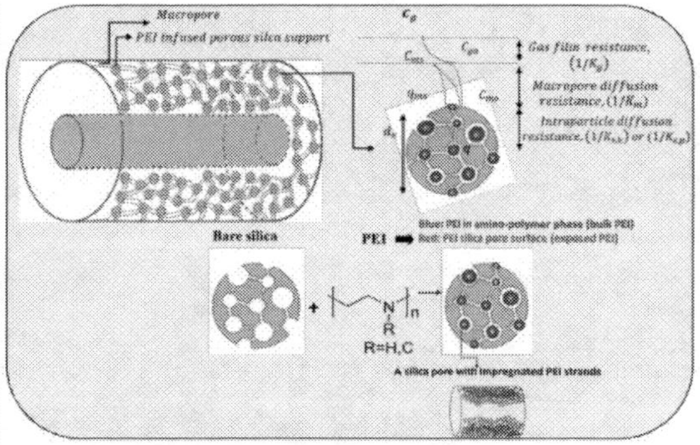

Process Governing Equations

Gas mass balance

$$\frac{\partial C_g}{\partial t} + u_g \frac{\partial C_g}{\partial z} + \frac{A_o \rho_f v_s}{V_{fs}} \frac{\partial q}{\partial t} = 0$$

Sorbent mass balance

$$\frac{\partial q}{\partial t} = K_s(\psi q_{eq} - q_1) + K_p((1-\psi) q_{eq} - q_2)$$

$$K_p = \frac{1}{\frac{1}{K_g} + \frac{1}{K_m} + \frac{1}{K_{bulk}}}; \; K_{bulk} = \frac{15 D_p \epsilon_s}{r_s^2 \frac{\partial q_{eq}}{\partial t} (1-\psi) \rho_s}$$

$$D_p = D_{p0} \exp(-\alpha q_2) \exp(-E/RT_g)$$

Adsorption isotherm equilibrium

$$q_{eq} = q_{max} \frac{bP}{(1 + (bP)^n)^{1/n}}$$

Gas heat balance

$$\rho_g C p_g \frac{\partial T_g}{\partial t} + u_g \frac{\partial T_g}{\partial z} + \frac{h_g A_o}{V_{fs}} (T_g - T_f) + q_{thm} = 0$$

Fiber heat balance

$$\rho_f C_{pf} \frac{\partial T_f}{\partial t} = \frac{\lambda_f}{r} \frac{\partial^2 T_f}{\partial z^2} + (-\Delta H_{ads}) \frac{\partial q}{\partial t} v_s \rho_{sorb} + \frac{h_g A_o}{V_f} (T_g - T_f)$$

Experimental Validation

❑ Different Flue gas flow rates

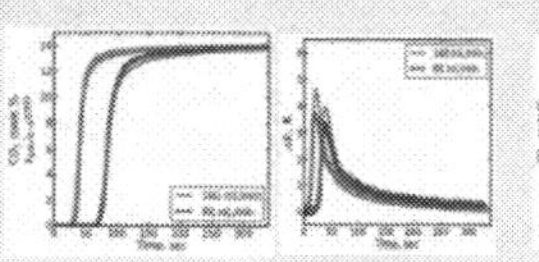

❑ Different Fiber packing

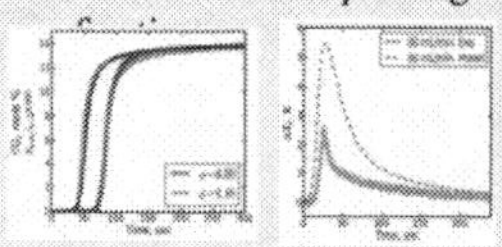

❑ Different silica particle size

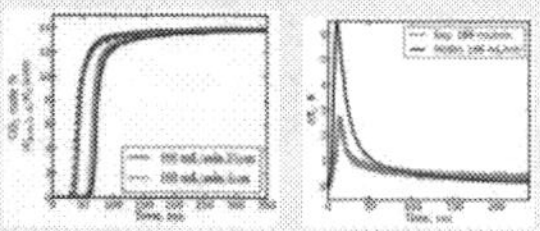

Model based analysis of process behavior

❑ 'Concave Trend' of breakthrough capacity: **equilibrium vs. mass transfer resistance**

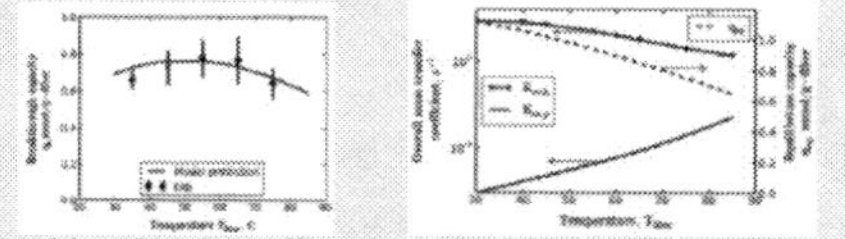

❑ Packing fraction effect on mass transfer resistance components

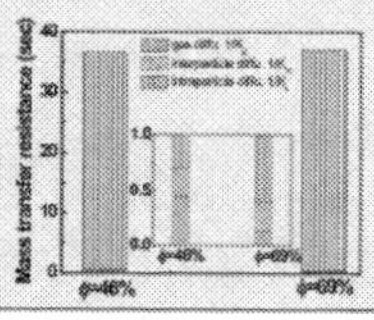

Acknowledgment: US Department of Energy, Grant no: DE-FE007804

References

❑ Kalyanaraman et al. Modeling and Experimental validation of carbon sorption on hollow fibers loaded with silicasupportd poly(ethyl Chem.Engg. J. 2015, 259, 737-751.
❑ Rezeai et al. Modeling of rapid temperature swing adsorption using holk sorbents, Chemm.Engg.Sci 2014, 113, 62-76.

Contact Information
Dr. Matthew Realff
School of Chemical & Biomolecular Engineering
Georgia Institute of Technology
Email: Matthew.Realff@chbe.gatech.edu

Georgia Institute of Technology
School of Chemical & Biomolecular Engineering

ADSORPTION BEHAVIOR OF BTX IN SUPERCRITICAL CARBON DIOXIDE ON MOLECULAR SIEVING CARBON USING CHROMATOGRAPHY AND NUMERICAL SIMULATION METHOD

Kazuyuki Chihara ,Takaki Yasuda, Masashi Shinsako*
Meiji University, Kawasaki, Japan

Introduction

Separation method using chromatograph measurements were made for the adsorption of volatile organic compounds (VOCs) on binary porous structure possessing molecular sieving carbon (MSC) in supercritical fluid CO_2 (SC-CO_2)mixed with organics. And adsorption of organics on adsorbents, such as activated carbon, in supercritical CO_2 could be mild [1] [2] [3] compared with adsorption in gas phase [4]. This is because of co-adsorption of CO_2 [5].

In this investigation, adsorption behaviors for BTX which is benzene, toluene, *o*-xylene, *m*-xylene or *p*-xylene on adsorption column packed molecular sieving carbon 5Å (MSC 5A) in SC-CO_2 were studied. Supercritical fluids chromatograph was used to detect impulse response peaks of BTX in SC-CO_2. Apparent adsorption equilibria and micropore diffusivity for BTX in SC-CO_2 were obtained by moment analysis of the peak. Moment analysis considering function of micropore, macropore, axial dispersion and external mass transfer was applied to the response peak with experimental technique, and examined dependences of adsorption equilibrium constants, K*, and micropore diffusivity, D, on the amount adsorbed. These experimental equilibrium data were correlated by Markham-Benton equation. It was showed specific behaviors of BTX existed in SC-CO_2 phase by the parameters of Markham-Benton equation. Amount adsorbed of organics can be changed by changing temperature and pressure of supercritical CO_2.So, adsorption and desorption of organics could be easy and mild in supercritical CO_2 condition.

The experimental peak could be almost equated by numeriacal simulation using Stop&Go method. Adsorption behaviors for each component in CO_2 – BTX system were predicted by this numeriacal simulation .New process of adsorptive separation using supercritical fluid might be possible [6]. However, equilibria and dynamics for co-adsorption of organics and CO_2 have not been studied so much [7].

Experimental

Chromatographic measurement

The schematic of experimental apparatus was shown in Figure 1. The carrier fluid of the chromatograph was supercritical CO_2 (critical temperature 304 K, critical pressure 73 atm) and it's mixture with the above-stated organics (benzene or toluene) respectively. The adsorbates used in the form of impulse were the same of organics mixed with supercritical CO_2.

In the case of CO_2 mixed with benzene, the organic used in the form of pulse was benzene. The volumes of the impulse were fixed to be 4.0 x 10^{-9} m^3 as liquid. MSC 5A (HGK882, Takeda chemicals Co.) was crushed and screened to obtain particle size between 1.49 x 10^{-4} - 1.77 x 10^{-4} m (an average particle radius of 8.12 x 10^{-5} m). Properties of MSC 5A was shown in Table 1 [8]. 1.45 x 10^{-3} kg of these particles were packed into the chromatographic column of 15 x 10^{-2} m long and 4.6 x 10^{-3} m in diameter. The void fraction, ε, of the bed was determined to be 0.3548. Flow rate of supercritical CO_2 was 1.33 x 10^{-7} m^3/s at 268 K and at 150, 200 and 250 atm respectively and flow rate of adsorbate (benzene or toluene) was 0.00 m^3/s, 5.00 x 10^{-11} m^3/s, 8.33 x 10^{-11} m^3/s, 1.67 x 10^{-10} m^3/s, 5.00 x 10^{-10} m^3/s and 1.00 x 10^{-9} m^3/s as liquid at room temperature (298 K).

The column pressure was kept at 150, 200 and 250 atm respectively using back pressure regulator (BP-2080 Plus, Japan Spectroscopic Co., LTD.). The pressure drop across the adsorbent bed was estimated to be about 1.0 atm and was assumed to be negligible. The experimental column temperature was kept at 313, 333 and 353 K respectively using intelligent column oven (co-2065 Plus, Japan Spectroscopic Co., LTD.). Before experimental runs started, the adsorbent particles were regenerated and stabilized by feeding pure CO_2 for 2 hours at the experimental pressure and temperature. Impulse was introduced into the carrier. Introduction of impulses was performed by injection. The impulse size was 4.0 x 10^{-9} l. Response data were processed by a personal computer. Impulse responses were detected using a multi-wavelength UV detector (MD-2010 Plus, Japan Spectroscopic Co., LTD.) (195 - 650 nm), as shown in Figure 2. In addition, the definition equation obtained by the impulse response method is as this.

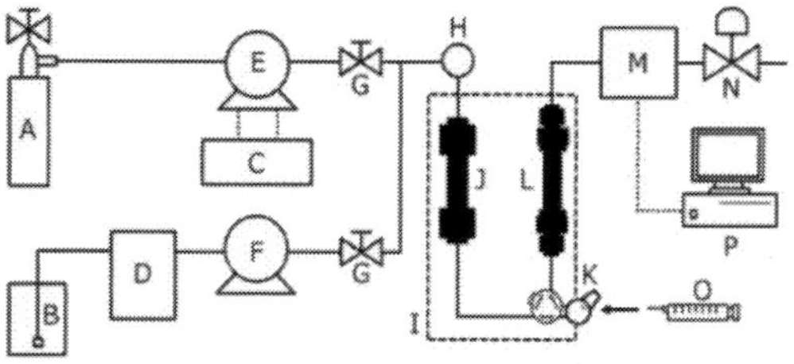

Figure 1. Experimental apparatus (schematic): A; CO_2 cylinder, B; organics, C; cooler, D; degasser, E; CO_2 pump, F; solvent delivery pump, G; valve, H; safety valve, I; column oven, J; mixed column, K; injector, L; packed column, M; multi-wavelength detector, N; back pressure regulator, O; organics, P; PC

Table 1. Properties of MSC 5A

Property	Value
True density [g/cm^3]	1.8
Particle density [g/cm^3]	0.90
Macropore volume [cm^3/g-carbon]	0.38
Micropore volume [cm^3/g-carbon]	0.18
Porosity of macropore [-]	0.342
Macropore radius [μ]	2.0
Micropore radius [Å]	5.0

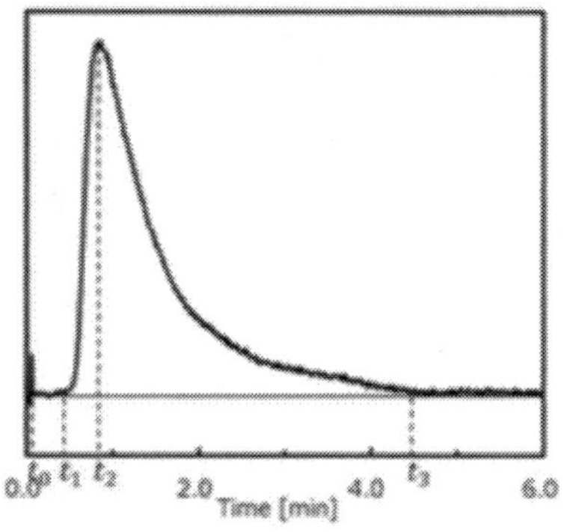

Figure 2. Typical impulse response peak: t_2–t_0; average retention time, t_3–t_1; intraparticle diffusion

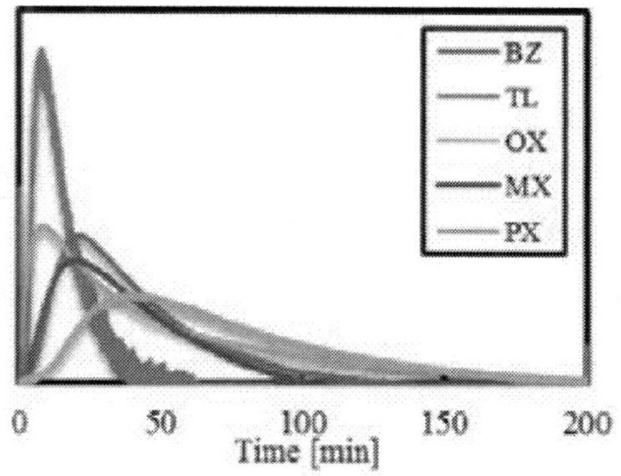

Figure 3. Typical impulse response peak of adsorbate

Moment Analysis

The first absolute moment of the chromatographic peak [s], μ_1 ,which is related to the adsorption equilibrium ,and second central moment [s^2], μ_2' which is related to the diffusion were obtained by the moment analysis of supercritical fluid chromatogram was tried. The analysis equations for μ_1 and μ_2' are as follows:

$$\mu_1 \equiv \frac{\int_0^\infty C_e(t)t\,dt}{\int_0^\infty C_e(t)dt} \quad -(1)$$

$$\mu_2' \equiv \frac{\int_0^\infty C_e(t)(t-\mu_1)^2 dt}{\int_0^\infty C_e(t)dt} \quad -(2)$$

The apparent adsorption equilibrium constant, K* , and time constant of micropore diffusivity,

D, obtained from first and second moment of response peak, as in references [8]. The resultant moment equations of the impulse response based on the basic equations are in ref. (Kumar et al. 1982) [9] are as follows:

$$\mu_1 = \frac{Z}{u}(1+\delta_0) \quad -(3) \qquad \mu_2' = \frac{2Z}{u}(\delta_d+\delta_f+\delta_a+\delta_i) \quad -(4)$$

which were defined as

$$\delta_0 = \frac{1-\varepsilon_0}{\varepsilon}\left\{1+\frac{\rho_P}{\varepsilon_a}(1-x)K^*\right\} \quad -(5) \qquad \delta_a = \frac{1-\varepsilon}{\varepsilon}\frac{R^2}{15D_a}(1-x)\rho_P^2K^{*2} \quad -(6)$$

$$\delta_d = \frac{E_z}{u^2}\left\{1+\frac{1-\varepsilon}{\varepsilon}(1-x)\rho_P K^*\right\}^2 \quad -(7) \qquad \delta_f = \frac{1-\varepsilon}{\varepsilon}\frac{R}{3k_f}(1-x)\rho_P^2K^{*2} \quad -(8)$$

$$\delta_i = \frac{1-\varepsilon}{\varepsilon}\frac{a^2}{15D_i}(1-x)\rho_P K^* exp(\sigma^2) \quad -(9)$$

Equations (3) and (5) ware used along with the experimental μ_1 (Eq. 1) to obtain the adsorption constant K^*. Equations (4), (6) to (9) were used along with the experimental μ_2' (Eq. 2) to obtain the diffusivity in micropores, D,. Equations (10) was used along with above K^* and D to obtain the overall mass transfer coefficient, K_s,.

Stop&Go method

In the numerical simulation, attempt to predict the adsorption behavior in the column by Stop&Go method, and were compared to reproduce the peak response. Stop&Go method was used as simulation method of perturbation chromatography for investigating adsorption equilibrium and rate. Figure 4 showed Stop&Go method basic concept. We thought as an aggregate of cells that are continuous in the N column axis direction. Here, by using considered to be divided into batch adsorption in the cell in the flow of fluid between cells each, the behavior of the column and to the calculation of each component continuous adsorption and desorption of the cells in very short time for each , we have to calculate the adsorption behavior. As a formula, we're using the adsorption rate equation mass balance equation (Eq. 10) , Markham-Benton equation (Eq. 11) and linear driving force (LDF) approximation (Eq. 12)

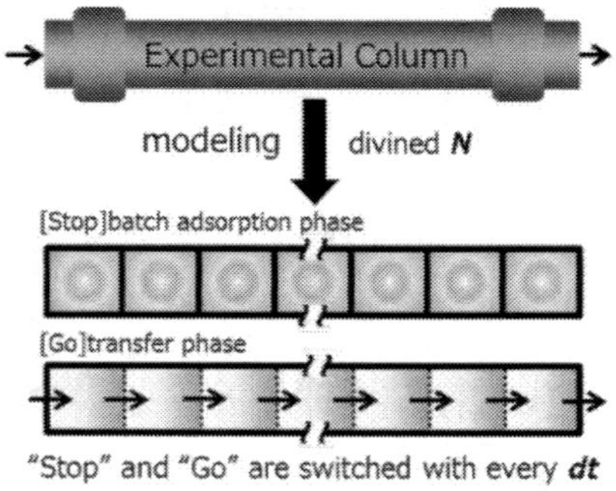

$$\frac{m}{N}\frac{dq_{n,j}}{dt} + \frac{\varepsilon V}{N} = Q_n C_{n-1,j} - Q_{n+1}C_{n,i} \quad -(10)$$

$$q_i^* = \frac{K_i P_i}{1+\sum K_i P_i} q_{i,\infty} \quad -(11)$$

$$\gamma \frac{dq_{n,i}}{dt} = K_{s,i} a_v (q_{n,i}^* - q_{n,i}) \quad -(12)$$

Figure 4. Basic concept of Stop&Go method

Result and Discussion

Experimental Data

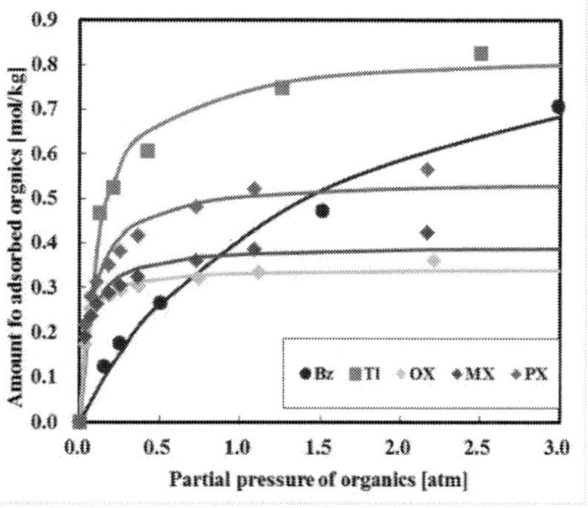

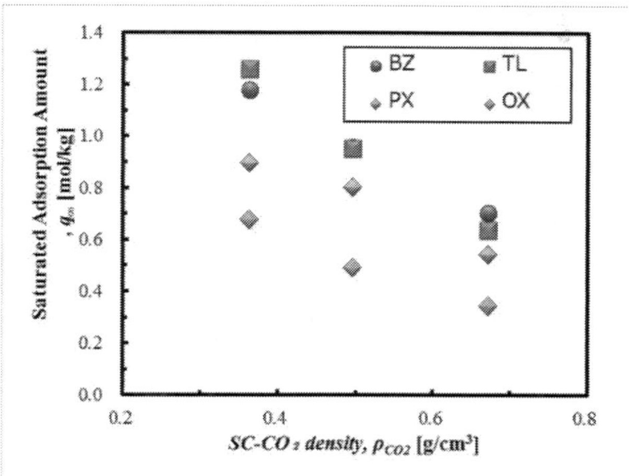

Figure 4. Adsorption isotherm of VOCs

Figure 5. Dependency of q_∞ for VOCs on the SC-CO_2 density

Figure 4 show adsorption isotherm of VOCs at 150atm 313K. Then, to use Markham-Benton equation for obtaining the saturation adsorption amount and equilibrium constant (BTX and CO2), the adsorption isotherms was carried out the curve fitting method of least squares by the solver. Basically, the amounts adsorbed became larger with decreases column pressure and increases temperature which decrease supercritical CO_2 density. This is because of co-adsorption of CO_2

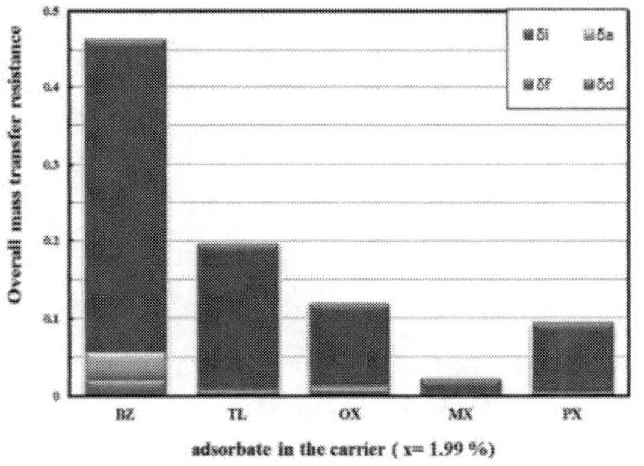

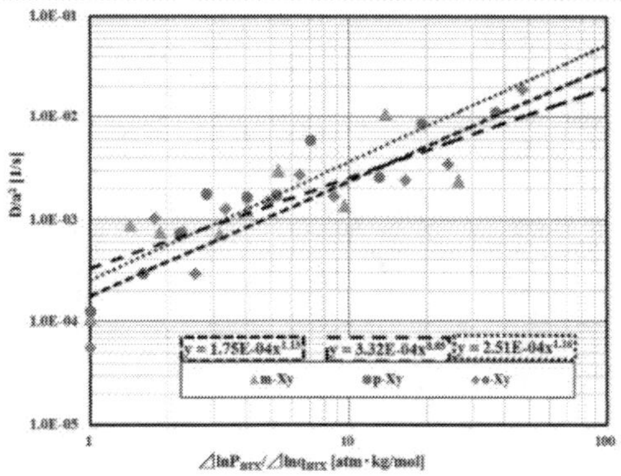

Figure 6. Contribution of mass transfer for VOCs

Figure 7. Correlation of micropore diffusivity for VOCs

Figure 6 show contribution of axial dispersion, δ_d, external mass transfer, δ_f, macropore diffusion, δ_a and micropore diffusion, δ_i in impulse response ($\mu_2{}'$) at 150 atm, 313 K.It can be said that the contribution rate of diffusion on the activated carbon in SC-CO_2 is dominated by micropores and the rate-limiting step.

Figure 7 shows the correlation of micropore diffusivity of organics with chemical potential driving force at 150 atm, 333K. It was applied chemical potential model (Eq.13) with micro pores and means that micropore diffusion is based on chemical potential driving force.

$$\frac{D^2}{a\,i} = \frac{D^2}{a\,0}\frac{\Delta lnP}{\Delta lnq} \qquad -(13)$$

Stop & Go Simulation

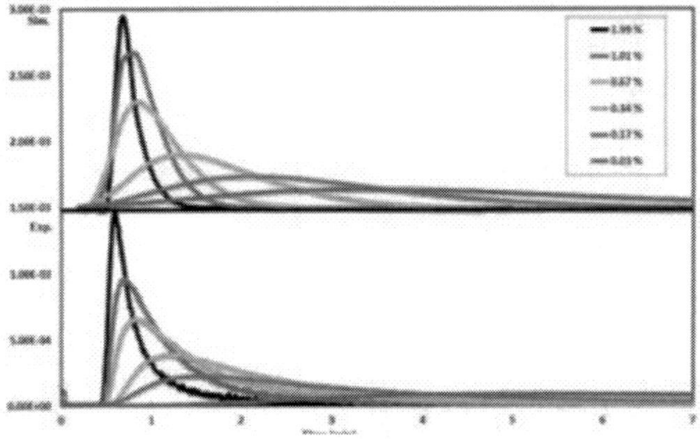

Figure 8. Corresponding Stop & Go Simulation and chromatographic measurements

A simulated chromatogram can be compared with the experimental chromatogram to determine the adsorption equilibrium and rate parameters.To run the simulation considering experimental conditions similar to the impulse response method, the experimental method was tried to reproduce the response peak.

Figure 8 show comparison between experimental and simulated pulse response at constant column temperature and total pressure. Organics (benzene) and CO_2 were used for adsorbates. Conditions (Fig. 8) were 313 K, carrier concentration of benzene 0.03~1.99% and benzene + CO_2 mixed carrier with benzene pulse.

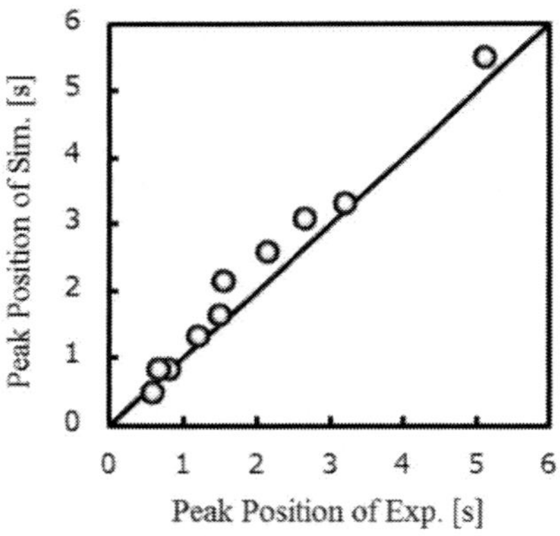

Figure 9. In comparison with peak position of experiment and simulation

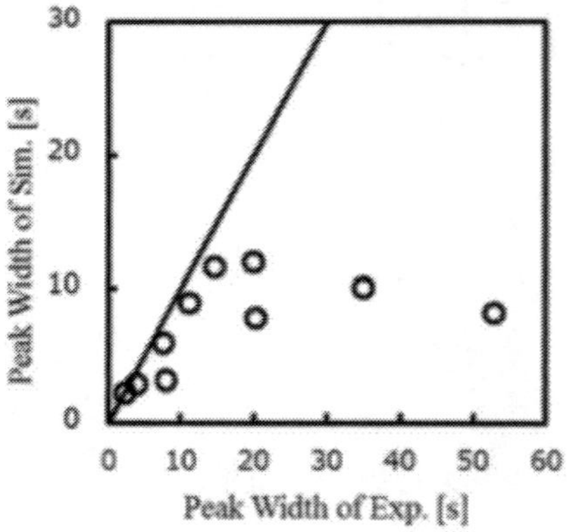

Figure 10. In comparison with peak width of experiment and simulation

Figure 9 and 10 show correlation of peaks position(average retention time) and width (intraparticle diffusion) between experiment and simulation. It is able to reproduce the position of the peak indicating the adsorption equilibrium between experiments and simulations. However, as a result of width of a peak indicating the diffusion, the smaller the concentration of the carrier flow rate shows increased result error. In order to reduce the error, we have to consider diffusion of CO_2 – BTX system.

Conclusion

Chromatographic measurements were made for the adsorption of benzene, toluene and m-xylene on molecular sieving carbon in supercritical fluid CO_2 mixed with organics. Adsorption equilibrium and adsorption dynamics on MSC were evaluated for each organics in supercritical CO_2 fluid mixed with adsorbate by chromatographic measurement. Firstly, the dependencies of amount of saturated sorption of benzene and toluene on supercritical CO_2 density were obtained. The amounts adsorbed became larger with decreases column pressure and increases temperature which decrease supercritical CO_2 density.Secondly, the contribution rate of diffusion on the activated carbon in SC-CO_2 is dominated by micropores and the rate-limiting step. Thirdly, micropore diffusion is based on chemical potential driving force. Finally, good coincidence of peaks between experiment and Stop&Go simulation were observed as for peak retention. If it will be able to reproduce the simulation of SC-CO_2-BTX system and then find the optimum solution, new process of adsorptive separation using supercritical fluid might be possible.

Nomenclature

a	= radius of microparticle	[cm]
a_v	= specific surface area of adsorbent	[cm^2/cm^3]
C, C_e	= concentration in the fluid phase	[g/ml]
D	= micropore diffusivity	[cm^2/s]
D_a	= macropore diffusivity	[cm^2/s]
E_z	= axial dispersion coefficient	[cm^2/s]
i	= component (benzene, toluene, CO_2)	[-]
K	= adsorption equilibrium constants	[1/atm]
K^*	= apparent adsorption equilibrium constants	[ml/g]
k_0	= van't Hoff constant	[1/atm]
k_f	= external mass coefficient	[cm/s]
K_s	= overall adsorption rate constant	[cm/s]
m	= mass of MSC packed in the bed	[g]
N	= division number of the column	[-]
n	= cell number in the divided column	[-]
P	= column pressure	[atm]
Q	= flow rate	[mol/s]
q^*	= amount adsorbed	[mol/kg]
q_∞	= amount of saturated sorption	[mol/kg]
R	= gas constant	[l·atm/mol/K]
r	= radius of particle	[cm]
T	= absolute temperature	[K]
t	= time	[s]
u	= interstitial velocity of fluid	[cm/s]
V	= column volume	[l]
x	= fraction of Benzene ,Toluene or Xylene in carrier	[-]
Z	= longitudinal position in the bed	[cm]
γ	= packing density	[g/cm^3]
δ_0	= function defined by Eq. (5)	[s]
δ_a	= function of macropore diffusion defined by Eq. (6)	[s]
δ_d	= function of axial dispersion defined by Eq. (7)	[s]
δ_f	= function of external mass transfer defined by Eq. (8)	[s]

δ_i	= function of micropore diffusion defined by Eq. (9)	[s]
ε	= void fraction in the bed	[-]
ε_a	= macropore fraction in the particle	[-]
μ_1	= first absolute moment of the chromatographic peak	[s]
μ_2	= second central moment of the chromatographic peak	[s^2]
ρ_P	= particle density	[g/cm^3]
σ	= standard deviation	[-]

References

1. H. Shojibara, Y. Sato, S. Takishima and H. Masuoka (1995), "Adsorption Equilibria of Benzene on Activated Carbon in the Presence of Supercritical Carbon Dioxide," *Journal of Chemical Engineering of Japan*, 28, 3, pp. 245-249.
2. K. Chihara, K. Oomori, T. Oono and Y. Mochizuki (1997), "Supercritical CO_2 Regeneration of Activated Carbon Loaded with Organic Adsorbates," *Water Science and Technology*, 35, 7, pp. 261–268.
3. K. Chihara, S. Ito, H. Nagashima, M. Hikita, and R. Suzuki (2011), "Adsorption of Organics on MSC5A in Supercritical CO_2: Chromatographic Measurements & Stop&Go Simulation," *diffusion-fundamentals. org*, 16, 83, pp. 1-6.
4. J. Benkhedda, J. Jaubert, and D. Barth (2000), "Experimental and Modeled Results Describing the Adsorption of Toluene onto Activated Carbon, *Journal of Chemical and Engineering Data*, 45, 4, pp. 650-653.
5. K. Tanida, Y. Sato, H. Masuoka (1996),"Measurement and Correlation of Supercritical Carbon Dioxide Adsorption on Activated Carbon," *Kagaku Kogaku Ronbunshu*, 22, 2, pp. 385-391.
6. M. Sato, M. Goto, A. Kodama and T. Hirose (1998), "New Fractionation Process for Citrus Oil by Pressure Swing Adsorption in Supercritical Carbon Dioxide," *Chemical Engineering Science*," 53, 24, pp. 4095-4104.
7. K. Chihara, S. Ito, H. Nagashima, M. Hikita, and R. Suzuki (2012), "Adsorption of Organics on MSC5A in Supercritical CO_2, Chromatographic Measurements and Stop and Go Simulation," *Adsorption* 18 pp. 417-422.
8. K. Chihara, M. Suzuki, K. Kawazoe (1978), "Adsorption Rate on Molecular Sieving Carbon by Chromatography," *AIChE Journal*, 24, 2, pp. 237-246.
9. R Kumar, R. C. Duncan and D. M. Ruthven (1982), "A Chromatographic Study of Diffusion of Single Components and Binary Mixtures of Gases in 4A and 5A Zeolites," *The Canadian Journal of Chemical Engineering*, 60, 4, pp. 493-499.
10. R.Ash and Barrer,R.M.,(1967),Surf. Sci. 8, 461

379365 Approaches to Synthesize New Zeolitic Imidazolate Framework Type Metal Organic Frameworks

Abstract:

Zeolitic imidazolate frameworks (ZIFs) are subclass of metal organic frameworks (MOFs) which are structurally comprised of ditopic imidazolate linkers and tetrahedrally coordinated transition metal ions as nodes. As metal-linker-metal bonds of ZIFs make the same angle as Si-O-Si of zeolites, 145°, the ZIFs can attain the same topologies of zeolites. In addition of high surface area which is typical property of MOFs, ZIFs exhibit prominent thermal and chemical stability. Therefore, they are proper candidates for industrial applications like gas separation and storage. Alike to other MOFs, solvothermal method is common *de novo* approach to produce ZIFs by which large number of ZIFs have been synthesized by variation of metal salt, imidazole, and solvent. This approach has inherent challenges when:

- The linker solubility in the solvent is limited,
- The linker is not chemically or thermally stable at reaction condition,
- Side reaction between functional group of the linker and metal ion leads to exclusion of MOF formation (especially when multi-functional linker is used),
- The reaction lead to formation of kinetic product (i.e. less stable polymorph) rather than thermodynamic product.

Post synthesis methods are capable of mitigate these shortcomings and enable one to achieve an MOF with desired linker, metal, and/or topology. In post synthesis method linker or metallic node can be modified (post synthesis modification) or exchanged by solvent assisted linker exchange and transmetalation methods, respectively. The main future of post synthesis methods is preserving the topology of the parent MOF in the modified one. The goal of this study is to synthesize new ZIF type MOFs by applying *de novo* and post synthesis approaches.

383427 Hybrid ZIF-7-90 Zeolitic Imidazolate Frameworks: Controlling Framework Porosity By Mixed-Linker Synthesis

Abstract:

Hybrid zeolitic imidazolate framework (ZIFs), a subclass of metal-organic frameworks (MOFs) with zeolitic topologies, can be synthetically fine-tuned in terms of their microporosity and functionality by using a mixed-linker approach.[1] However, the reliable synthesis and characterization of mixed-linker ZIFs is a relatively new area of research.[2] In this work, the synthesis and characterization of hybrid ZIF-7-90 frameworks with mixed-linkers of ZIF-7 (benzimidazole, Bz-IM) and ZIF-90 (carboxaldehyde-2-imidazole, OHC-IM) is demonstrated. Different synthesis methods based on variations of the solvent system, temperature, and crystallization technique at different linker ratios with either the ZIF-7 or ZIF-90 parent framework are investigated to better understand the structural and chemical properties of this new hybrid ZIF-7-90 framework. The talk will discuss in detail the effects of the solvent system and temperature on the inclusion of the ZIF-7 and ZIF-90 linkers in the ZIF-7-90 framework. ^{1}H-NMR spectroscopy, X-ray diffraction, SEM, carbon dioxide physisorption, and nitrogen physisorption allow for detailed insights into the tunable crystallinity, morphology, and porosity of these materials.

(1) Thompson, J. A.; Blad, C. R.; Brunelli, N. A.; Lydon, M. E.; Lively, R. P.; Jones, C. W.; Nair, S. *Chemistry of Materials* **2012**, *24*, 1930.

(2) Thompson, J. A.; Brunelli, N. A.; Lively, R. P.; Johnson, J. R.; Jones, C. W.; Nair, S. *Journal of Physical Chemistry C* **2013** *117*, 8198.

Synthesis and Characterization of Microporous Metal–Organic Frameworks /Graphene Oxide (HKUST-1/GO) by Mechanochemical Grinding

Jinpeng Miao, Yujie Li, Xue jiao Sun, Qibin Xia*, Hongxia Xi, Zhong Li
School of Materials Science and Engineering, South China University of Technology, Guangzhou, 510640, P. R. China

New composites of a copper-based metal–organic framework and graphite oxide were synthesized via a fast room-temperature solvent-free mechanochemical grinding. MOF/graphite materials were characterized by X-ray diffraction (XRD), sorption of nitrogen, water stability test, FT-IR spectroscopy, surface area and porosimetry analyzer, thermal analyses and scanning electron microscopy(SEM). Through the characterization analyses and gas adsorption experiments, we found that mechanochemical synthesis is a very fast method for the quantitative formation of MOF/graphite with complete synthesis achieved in as little as 10 min of grinding. HKUST-1/GO composites show more remarkable stability than HKUST-1 upon exposure to the high humidity environment. While GO existed in HKUST-1/GO materials possessed of epoxy and hydroxyl groups exist within the interlayer space, the water molecules were attracted there via hydrogen bonding. That could prevent the interaction between functional group in HKUST-1 and water molecules. Epoxy groups on the GO layers also act as seed sites for Cu-BTC crystallization, the facile method leads to HKUST-1/GO with BET surface areas of 1574.5m^2/g. In addition, the maximum uptake of methanol in HKUST-1/GO synthesized by mechanochemical method was up to 9.96 mmol/g at 298 K、168mbar, was much higher than the conventional adsorption materials, such as HKUST-1、activated carbon and NaA zeolite. The results also indicated that mechanochemical synthesis is environmentally friendly, rapid and highly scalable, allowing cost-effective, large-scale deployment of MOFs /or composite materials.

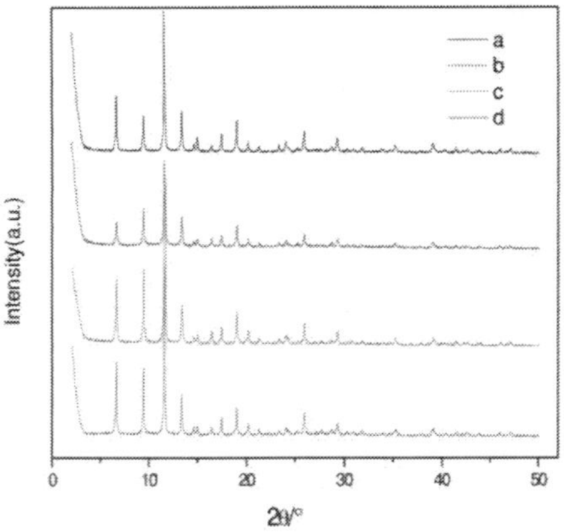

Figure1. XRD patterns of mechanochemically prepared materials: (a)HKUST-1 as received (dehydrated form) and (b) after prolonged exposure to ambient moisture (hydrated form) ; (c)HKUST-1/GO as received (dehydrated form) and (d) after prolonged exposure to ambient moisture (hydrated form).

Figure2. SEM images of HKUT-1/GO powder samples prepared via mechanochemical grinding method.

*Corresponding author email: qbxia@scut.edu.cn.

388305 Encapsulation of Gold Nanoparticles in Uio-66

Abstract:

Metal-organic frameworks (MOFs) are porous, crystalline materials consisting of a metal or metal oxide center connected by organic linkers. Their high surface areas and pore volumes, uniform pore size distributions, and chemical tunability give them potential in applications such as gas storage and separation, drug delivery, biomedical imaging, air purification, and catalysis. Recently, incorporating metal and metal oxide nanoparticles within the MOF has been explored. MOF composites will utilize the interesting chemical and physical properties of the nanomaterials, coupled with the selectivity of MOFs. These materials have potential for applications in sensing, improved adsorption and storage, and catalysis. The controlled preparation of MOF composites can be accomplished using an encapsulation procedure which grows the MOF around the pre-formed nanomaterials.

UiO-66 is a zirconium based material that is chemically resistant in humid, acidic, and basic environments. It can also be tuned to target specific molecules by a simple linker substitution. This chemical resistivity and tunability make it an attractive nanomaterial support for a range of potential applications. Unfortunately, the production of HCl during the conventional UiO-66 synthesis negatively interacts with many nanomaterials including gold nanoparticles. Gold nanoparticles are particularly attractive materials because of their low temperature catalysis capabilities of toxic gases such as CO. This presentation will introduce work on encapsulating gold nanoparticles in UiO-66 using a novel synthesis procedure that eliminates HCl formation and addition.

379007 Hydrous Cerium Oxide Nanoparticles Modified Activated Carbon(HCO-AC) As a Multifunctional Adsorbent for As(III), As(V) & Pb(II) Removal

Abstract:

In the current study, a multifunctional adsorbent based upon hydrous cerium oxide nanoparticles modified activated carbon (HCO-AC) was successfully synthesized, and systematically investigated for As (III), As (V) and Pb (II) removal from aqueous solutions. The adsorption kinetics, adsorption isotherm, pH and co-existing ions effects were studied with repect to As (III), As (V) and Pb (II) removal. Characterization techniques such as SEM, XPS and FT-IR were as well performed to study the adsorbent structure and adsorption mechanisms. It was found that adsorption equilibrium could all be reached within the first 12 hours, and the adsorption capacity could reach up to 34.25 mg-As (III)/g, 44.64 mg-As (V)/g, and 45.45 mg-Pb (II)/g, respectively. Generally the adsorption process would be retarded owing to the existence of common co-existing anions and humic acid. However, it is noteworthy that the co-existence of humic acid could considerably enhance the adsorption capacity of HCO-AC on Pb (II).

The magnetic iron oxide nanoparticles (MNP) are excellent adsorbents for both As(V) and Cr(VI) oxyanions in water because of their proven advantages in terms of huge adsorption capacities and magnetic recovery feature. The understanding of competitive adsorption between As(V) and Cr(VI), which is important for determining adsorption mechanism, is not far from adequate. In this study, the competitive adsorption between As(V) and Cr(VI) oxyanions on MNPs has been studied by using both adsorption experiments and molecular simulation. It is shown that for system only containing As(V) or Cr(VI), there is little effects of pH (3.0-11.0) and ionic strength (0-50 mmol/L NaCl) on the adsorption for both As(V) and Cr(VI), revealing that the inner-sphere complexing between MNPs and oxyanions is the dominating driving force. For the system containing both As(V) and Cr(VI), it is shown that when the increase of amount of MNPs, Cr(VI) oxyanions are not absorbed until As(V) oxyanions are completely removed, indicating that As(V) oxyanions is priority adsorbed on the surface of MNPs than Cr(VI) oxyanions. When MNPs absorbing the equal amount of As(V) and Cr(VI) oxyanions is regenerated by treatment of 0.5 M NaOH, Cr(VI) oxyanions are firstly washed out, indicating that the interaction of Cr(VI) oxyanions with MNPs is weaker than that of As(V) oxyanions. Furthermore, the above conclusion is also proved by the experiment that MNPs pre-saturated with As(V) oxyanions could not nearly remove Cr(VI) oxyanion while MNPs pre-saturated with Cr(VI) oxyanion could adsorb As(V) oxyanions leaving Cr(VI) oxyanions in water. In conclusion□the research about competitive adsorption between As(V) and Cr(VI) will be contributed to explore thoroughly the removal process of heavy metal pollutions by MNPs and the treatments of wastewater with multiple contaminations in the future.

Continuous Cesium Separation Using a Silica Microhoneycomb Supporting Ammonium Molybdophosphate

Seiichiro Yoshida, Yoshinao Kimura, Isao Ogino, Shin R. Mukai
Graduate School of Chemical Sciences and Engineering, Hokkaido University
e-mail: smukai@eng.hokudai.ac.jp

Introduction

○ Morphologies of adsorbents in flow systems

Adsorbents with "microhoneycomb" structure can process a large amount of fluid while enabling fast mass transfer without causing a serious hydraulic resistance

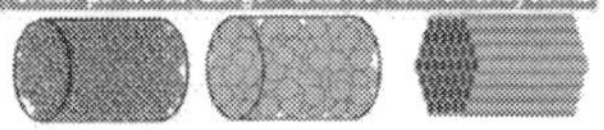

	Small particles	Large Particles	Microhoneycomb
Hydraulic resistance	High	Low	Low
Mass transfer	Fast	Slow	Fast

○ Uptake of radioactive cesium using adsorbents

Fukushima daiichi nuclear power plant

- A large volume of ^{137}Cs contaminated water (> 230,000 m^3) was generated at Fukushima daiichi nuclear disaster in Japan

➡ It is urgent to decontaminate Cs-containing wastewater

○ Example of Cs^+ adsorbent

Ammonium molybdophosphate (AMP)

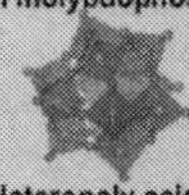

- Heteropoly acid salt
- High selectivity for Cs^+
- Fine particles (~ 10 μm)

Aim of this work

To investigate the potential of the adsorbent with a microhoneycomb structure to separate Cs^+ in continuous systems

- Synthesis of silica microhoneycomb-supported AMP (AMP-SMH)
- Separation of Cs^+ using AMP-SMH in batch and flow systems

Experiments

○ Synthesis of AMP-SMH using ice crystals as the template

Preparation of a silica sol → AMP addition → Gelation → Unidirectional freezing → Immersion in *tert*-butyl alcohol → Freeze-drying at -10 °C

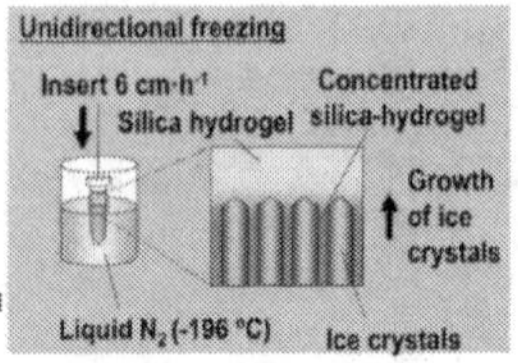

○ Evaluation of breakthrough curves

Length of unused bed (LUB) :
The length of region where an adsorption takes place in a packed bed adsorbent

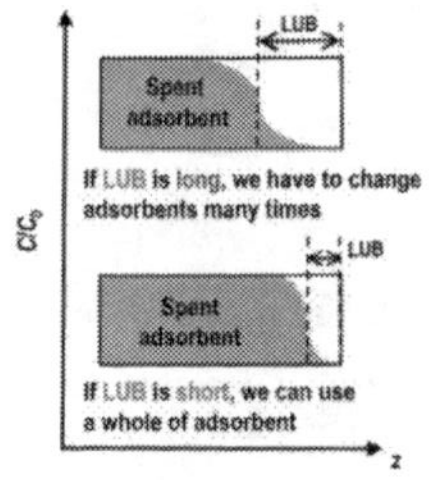

○ Breakthrough Experiments of AMP-SMH in flow systems

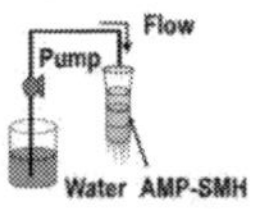

- Conditions
  - Feed concentration C_f : 10 ppm
  - Monolith length L : 0.9 ~ 2.7 cm
  - Monolith diameter : 1.0 cm
  - Superficial velocity u_s : 3.1 ~ 11.2 cm·min^{-1}

Results and discussion

○ Morphology of AMP-SMH

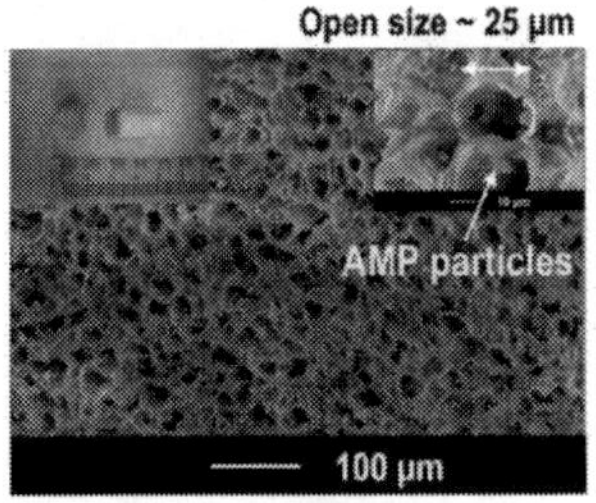

Cross section SEM image

- Channel size was 20-25 μm
- AMPs are immobilized on the channel walls

AMP-SMH is a microhoneycomb

○ Measurement of pressure drop of water

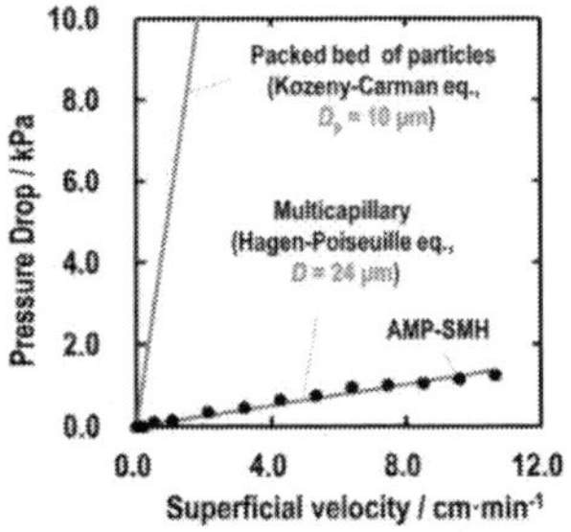

The pressure drop caused by AMP-SMH is 46-fold less than that the packed bed causes

AMP-SMH can process a large amount of fluid without causing severe hydraulic resistance

○ Cs^+ adsorption isotherm

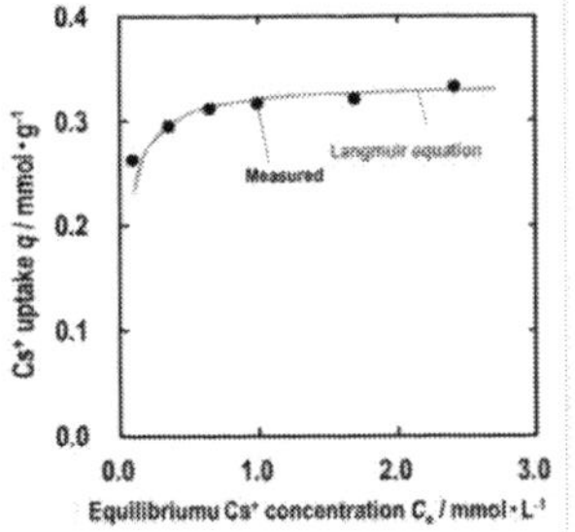

The isotherm can be modeled using the Langmuir model

Cs^+ uptake by AMP-SMH is a typical ion exchange reaction

○ Breakthrough curves for Cs^+

- Effect of linear velocity

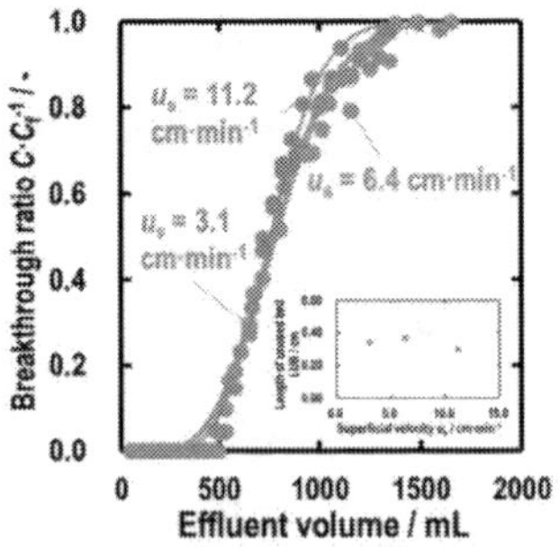

- Short length of unused bed (LUB) : < 0.4 cm
- LUBs are not affected by superficial velocity

AMP-SMH can process Cs^+ at high superficial velocities

- Effect of monolith length

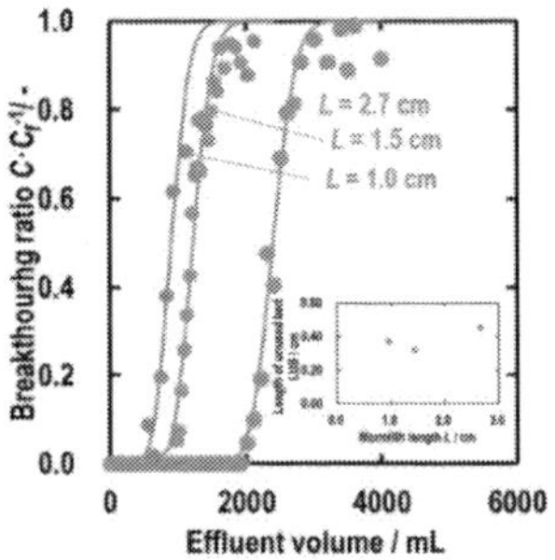

LUBs are short (< 0.5 cm) at a long monolith

Utilization of the monolith is very high

- Comparison with a packed bed particles

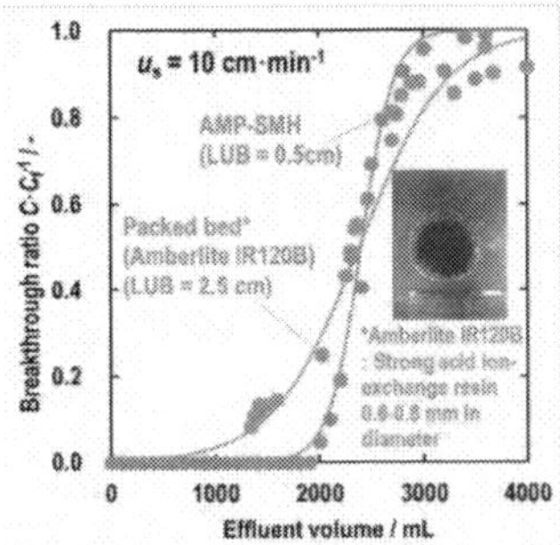

LUB is shorter than that of the ion-exchange resin

AMP-SMH can effectively separate Cs^+ compared to packed bed particles

Conclusion

- AMP-SMH was successfully synthesized by using the Ice-Templating method
- AMP-SMH shows high performance for Cs^+ separation in flow systems without causing a severe hydraulic resistance
- AMP-SMH shows a low LUB during Cs^+ separation compared to packed bed particles

AMP-SMH has a high potential to separate Cs^+ in continuous systems

378681 Preparation and Characterization of Yttrium-Based Magnetic Adsorbent for Arsenic Removal

Abstract:

In this study, the adsorption behavior of a Y-magnetic adsorbent (developed by a co-precipitation technology) is systematically investigated. Studies have indicated that better adsorption can be obtained at lower pH and the optimum pH is around 4. The maximum adsorption of As (III) and As (V) were found to be as high as 84.22 mg-As/g and 170.48 mg-As/g respectively at pH 7, which is much higher than many reported adsorbents. Adsorption equilibrium for As (III) and As (V) could be reached within 12 hours and 24 hours respectively. Compared with other co-existing substances such as carbonate, sulfate and fluoride, phosphate ions were found to have significant impact on the adsorption of arsenic.

H_2S Chemisorption on Copper-ETS-2: Experiment and Modeling

Sabereh Rezaei, Adolfo M. Avila, Steven M. Kuznicki*

Department of Chemical and Materials Engineering, University of Alberta, Edmonton, Canada

OBJECTIVE

To protect inorganic membranes used in syngas processing we need to remove the last traces of H_2S at room temperature.

Loading active metal ions on a high external surface area substrate enhances the utilization of metal species.

Cu oxide shows promising results for deep desulphurization at room temperature[1].

Engelhard Titanosilicate-2[1]

- Non-porous, high density titanate
- sodium titanate on the surface of the particles and TiO_2 in the bulk[2];
- High external surface area (>200 m²/g)
- Strong affinity for metal cations

Figure 1- SEM image of ETS-2.

Cu-ETS-2

Table 1- Elemental composition during the Cu exchange procedure.

PH	1 ml solution	Cu (mmol)		Na (mmol)	
2.9	$Cu(NO_3)_2$	0.764	0.186	0.000	0.368
3.7	$Cu(NO_3)_2$ + ETS-2	0.578		0.368	

➢ Na ions on the surface of ETS-2 have been well replaced by Cu ions.

EXPERIMENTAL

➢ 10 ppm H_2S in He, 100 sccm gas flow
➢ Sulfinert tubes

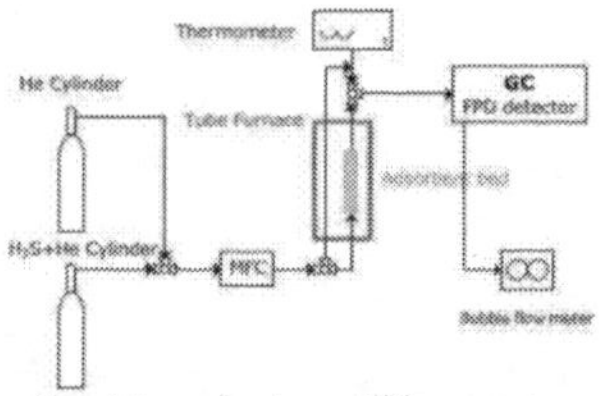

Figure 2- Process flow diagram of H2S removal set up.

Cu-ETS-2 shows the best breakthrough capacity in comparison with different ion exchange forms of ETS-2 and a fully developed commercial sample from BASF[3].

MASS TRANSPORT PHENOMENA

Model assumptions

➢ Ideal gas law.
➢ 1D problem.
➢ No heat effect, negligible pressure drop along the bed .
➢ No concentration gradient in the gas phase.
➢ Steady state condition on the solid surface.

Continuity equation

$$\varepsilon_b \frac{\partial c_i}{\partial t} + \varepsilon_b \nabla(c_i v) = \rho_b r_i \qquad (1)$$

Non-catalytic system of H_2S removal by Cu-ETS-2 chemisorbent

High external surface area and low micro pore volume of Cu-ETS-2 as well as its amorphous structure[4] and high affinity of the Cu sites for the reaction with H_2S:

- H_2S molecules get adsorbed on the surface easily
- Adsorbed H_2S is able to react with copper sites

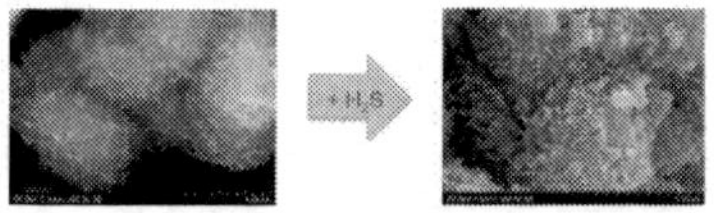

Figure 3- SEM images of Cu-ETS-2 before and after H_2S exposure.

Sequential steps:

1. Dissociative adsorption[2] of H_2S molecules on the copper site.
2. Chemical reaction between the adsorbed H_2S molecule and the copper site

At low H_2S concentration:

$$\theta = k_H c_i \qquad (2)$$

$$r_i = -k_r \theta a \qquad (3)$$

Deactivation term[5,6]:

$$\frac{da}{dt} = -k_d c_i a \qquad (4)$$

Nomenclatures

- c_i H_2S concentration in gas phase
- a Deactivation rate term because of sulfidation
- θ Occupied fraction by the chemisorbed specie at time t
- k_H Henry constant
- K Kinetic parameter ($\rho_b k_r k_H / \varepsilon_b$)
- ρ_b Bulk density
- k_d Deactivation constant
- ε_b Bed void fraction
- k_r Chemical reaction rate constant
- τ Deactivation time constant

Proposed model:

$$\varepsilon_b \frac{\partial c_i}{\partial t} + \varepsilon_b v \frac{\partial c_i}{\partial z} = -\rho_b K c_i a \qquad (5)$$

$$@ \begin{cases} t = 0 & c_i = 0 \quad for \quad all\ z \\ z = 0 & c_i = c_{in} \quad for \quad t > 0 \end{cases}$$

➢ Model has been converted to a system of numerical ordinary differential equations and been integrated via an optimization process using MATLAB_R2013a.

RESULTS

Chart 1 – Experimental data vs proposed model

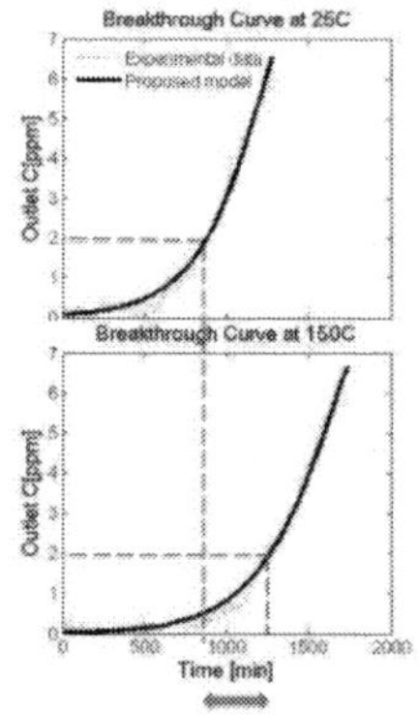

Table 2- Kinetic parameters at 25 and 150C.

Temperature (°C)	K^*10^{-4}[min⁻¹]	$K_d^*10^4$[min⁻¹]	τ^*10^{-3}[min]
25	2.66	5.27	1.89
150	3.30	4.80	2.08

DISCUSSION

➢ Higher temperature, higher breakthrough capacity: chemisorption
➢ Higher kinetic parameter, higher interaction of H_2S with the solid
➢ Lower breakthrough capacity, faster deactivation

TEMPERATURE DESORPTION PROGRAM

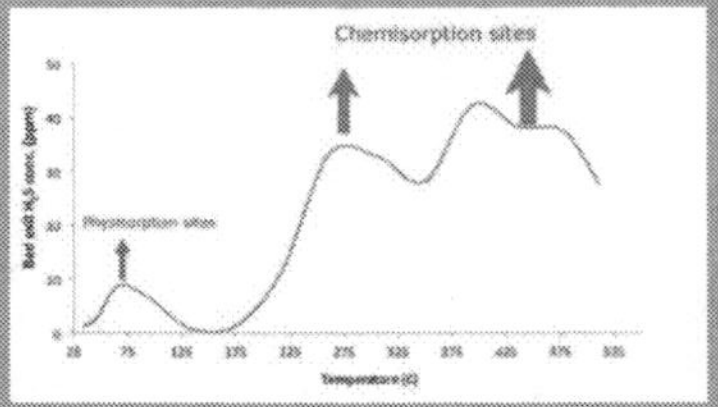

Chart 2 - Temperature program desorption test under inert He at 10C/min.

➢ Cu-ETS-2 surface includes multi adsorption sites[7].
➢ As expected, adsorption sites with stronger binding energies need more energy input during desorption period.

CONCLUSIONS

ETS- 2 has a high surface area and a high cation exchange capacity making it a suitable support to allow Cu ions a high degree of accessibility toward H_2S.

Cu-ETS-2 has displayed the most promising adsorbent in comparison to a fully developed commercial H_2S adsorbent and other tested metal-exchanged ETS-2 for deep H_2S removal (to sub-ppm levels) for gas purification applications at room temperature.

H_2S column breakthrough experiments on Cu-ETS-2 are analyzed by a new dynamic model based on the rigorous mass balance equations for the fluid and solid phases.

The proposed model includes the chemical reaction term, which is affected by the deactivation of the solid surface.

REFERENCES

[1] Rezaei et al. Ind. Eng. Chem. Res. 51 (2012) 12430.
[2] Yazdanbakhsh et al. Ind. Eng. Chem. Res. 53 (2014) 11734.
[3] Rezaei et al. Chem. Eng. Sci submitted Aug. 2014.
[4] Choi et al. Chemical Physics Letters 421 (2006) 179.
[5] Yasyerli et al. Chem. Eng. and Processing 41 (2002) 785.
[6] Garsec et al. Micropor. Mesopor. Mat. 127 (2010) 190.
[7] Leary et al. AIChE 34(2) (1988) 263.

Acknowledgements

Helmholtz-Alberta Initiative
The research team members

385502 UV-Assisted Adsorptive Desulfurization (ADS) of Liquid Fuels Using TiO2 Based Adsorbents

Abstract:

Organic sulfur compounds in liquid hydrocarbon fuels are converted into SO_x during combustion, which may cause severe environmental pollution such as acid rain and smog. Sulfur compounds can also damage the electrodes of fuel cells. As a result, desulfurization has attracted much attention nowadays. The conventional hydro-desulfurization (HDS) process cannot meet the stringent environmental regulation. Thus several alternative technologies have been investigated in order to replace HDS. Among these technologies, adsorption desulfurization (ADS) is one of the most promising methods to achieve an ultra-low sulfur level under ambient operating conditions. Ag/TiO_2 and Ag/TiO_2—Al_2O_3 adsorbents developed by our research group showed high selectivity and capacity and can be regenerated. The desulfurization mechanism was investigated by previous work based on the interaction between sulfur compounds and surface hydroxyl groups. In the present work, UV assisted adsorptive desulfurization technique was developed for the first time by applying TiO_2/UV system into dynamic desulfurization (breakthrough) process. According to our acknowledgement, sulfur removal assisted by UV using TiO_2 and Ag-TiO_2 is unique in adsorptive desulfurization area. Experimental results showed that sulfur capacities of TiO_2 based adsorbents were improved during UV-assisted breakthrough experiments using model fuels and jet fuel. The breakthrough capacity of blank TiO_2 increased from 2.45 mgS/g to 4.05 mgS/g under UV irradiation. H_2O impurities can decrease the activity of TiO_2 based adsorbents. The sulfur capacity of Ag/TiO_2 decreased from 5.84 mgS/g to 4.63 mgS/g in the presence of H_2O. However, the capacity was increased to 6.40 mgS/g upon UV illumination using H_2O as fuel additive. No photo-oxidative reactions were observed during sulfur removal process. Relative concentrations of active OH radicals on TiO_2 under UV were measured by fluorescence technique using terephthalic acid. Photo-induced charge carries on TiO_2 surface were characterized by *in situ* infrared (IR) spectroscopy. The effect of UV irradiation on surface hydroxyl groups and possible desulfurization pathway were discussed based on research results.

374125 The Adsorption Characteristics of Hydrogen Chloride Gas Mixture on Zeolite 13X Pellet at High Pressure and Low Temperature Condition

Abstract:

The adsorption breakthrough test of hydrogen chloride was performed by loading zeolite 13X pellet (molecular sieve 13X, Aldrich Co., USA), a commercial absorbent, onto a fixed bed reactor (height 15 cm, inside diameter 0.5 cm). The breakthrough test was carried out by changing various experimental conditions such as inlet flow rate (0.5, 1, 1.5, 2 l/min), HCl concentration (200, 350, 500, 650 ppm), reaction temperature (30, 50, 70, 90°C), and reaction pressure (1, 5, 10, 15, 20 bar) with considering the retention time and space velocity. After evaluating the adsorption and the desorption characteristics of HCl on the MS-13X, the consecutive cyclic test was carried out with optimum condition. The optical, physical, and chemical characteristics of MS-13X were analyzed using BET, XRD, SEM, EDX, and TGA before and after breakthrough tests.

Factors Affecting H_2S Desulfurization Performance of ZnO based Adsorbents for Wide Temperature Applications (20-400°C)

Effect of COS Formation and Moisture Adsorption on H_2S Removal Efficiency

AUBURN UNIVERSITY
SAMUEL GINN COLLEGE OF ENGINEERING

Achintya Sujan and Bruce J Tatarchuk*
*Department of Chemical Engineering, Auburn University, Auburn, AL 36849 (USA)
*tatarbj@auburn.edu, Tel: +1-334-844-2023

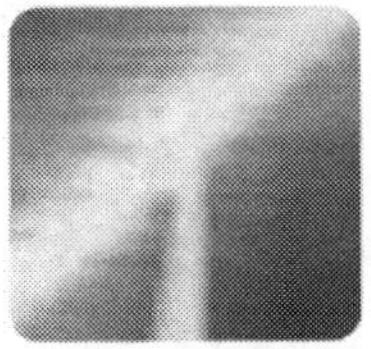

Objective

The aim of this work is to understand the effect of COS formation on the H_2S removal efficiency of ZnO based adsorbents during wide temperature desulfurization of a wet and dry simulated reformate stream for applications in logistic fuel cell systems.

Introduction

- ZnO based adsorbents have been widely applied for H_2S removal in the chemical industry via the reaction: **$ZnO + H_2S \rightarrow ZnS + H_2O$**
- Presence of CO_2 or CO along with H_2S in a variety of hydrocarbon streams lead to the undesired formation of carbonyl sulfide during desulfurization[1][2][3].

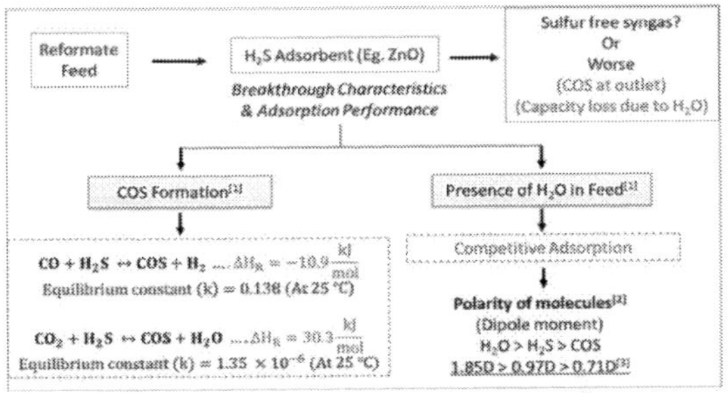

- Presence of moisture prevents COS formation due to a shift in the equilibrium reaction between COS and H_2O to form H_2S and CO_2
- Even worse, reduced metal sulfides (For instance, ZnS) formed during desulfurization act as catalysts for COS formation[4].

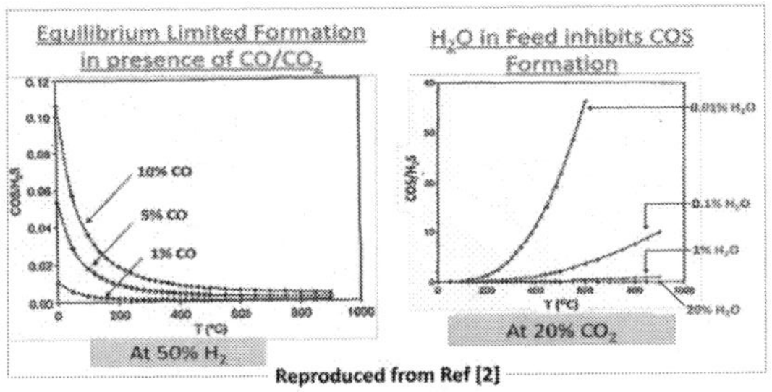

Reproduced from Ref [2]

- COS is less acidic than H_2S and ZnO sorbents suffer from negligible capacity for COS adsorption at mild conditions.
- COS formation during the desulfurization process is a highly undesirable side reaction and causes degradation of downstream process catalyst, equipment and fluid pipelines.

Lab-Scale Experimental Methods

Simulated Reformate	30 vol% CO + 30 vol% CO_2 + 0.6 vol% H_2S Balance H_2
Adsorbent examined	21 wt% ZnO loaded on porous (inert) SiO_2 support
Support Characteristics	BET Surface Area = 300 m^2/g ; Pore Volume = 1 ml/g ; Pore Size = 5 nm
Desulfurization temperature	22°C (Ambient), 200°C (Low), 400°C (High)
Sulfur Analysis	Thermal Conductivity Detector on GC (VARIAN CP3800)

Simulated Feed is typical outlet from methane reformer, water gas shift reactor & composition is based on average values from literature[1-8]

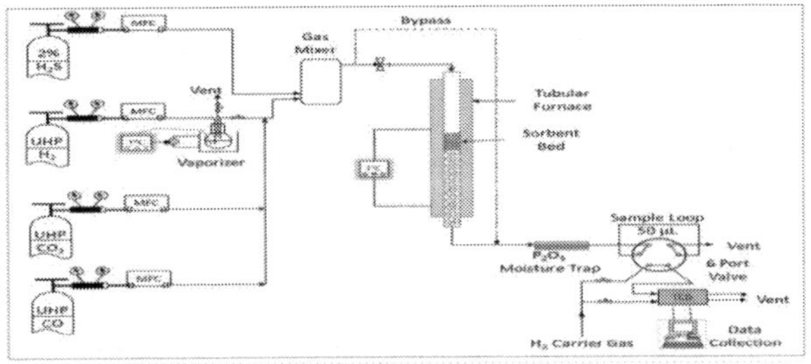

COS Formation as a function of Feed Operating Temp

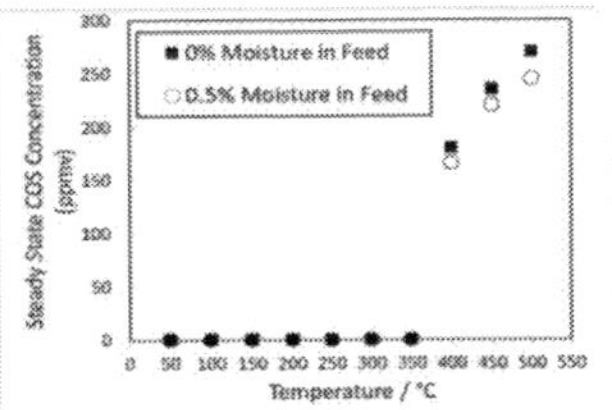

Fig. 1 COS Formation vs Feed Temperature
Inlet = Simulated Reformate (See Experimental)
Feed Flow = 100 ml/min through empty reactor

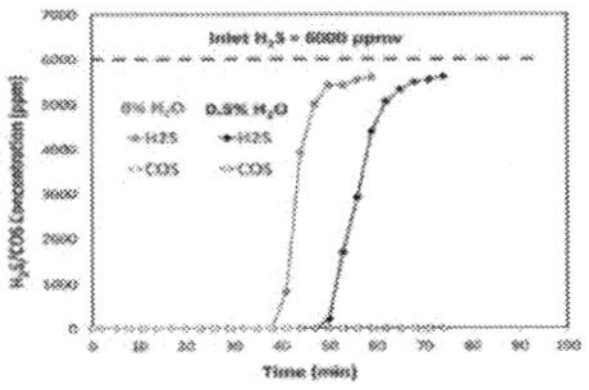

Fig. 2 Breakthrough Curves on 21% ZnO/SiO_2
T = 22°C ; GHSV = 1700 h^{-1}
Inlet = Simulated Reformate (See Experimental)

COS Breakthrough Characteristics on ZnO/SiO_2

200°C Desulfurization

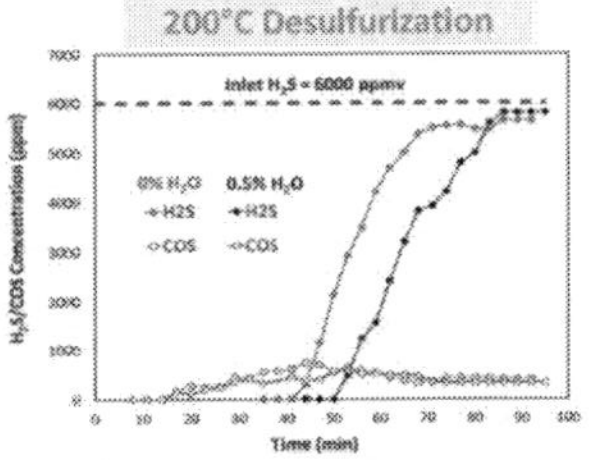

Fig. 3 Breakthrough Curves on 21% ZnO/SiO_2
T = 200°C ; GHSV = 1700 h^{-1}
Inlet = Simulated Reformate (See Experimental)

400°C Desulfurization

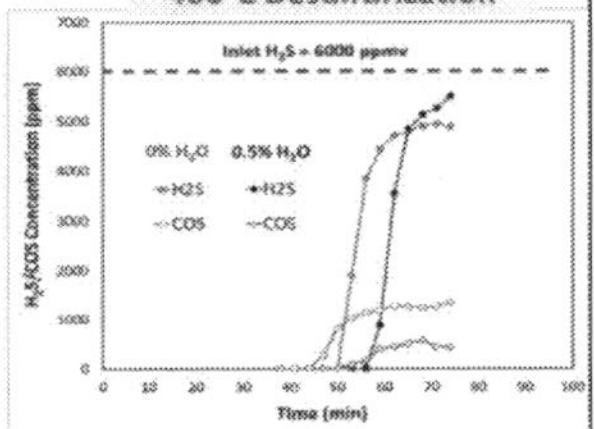

Fig. 4 Breakthrough Curves on 21% ZnO/SiO_2
T = 400°C ; GHSV = 1700 h^{-1}
Inlet = Simulated Reformate (See Experimental)

COS Formation through Metal Sulfides (ZnS)

Step 1: Pre-sulfidation

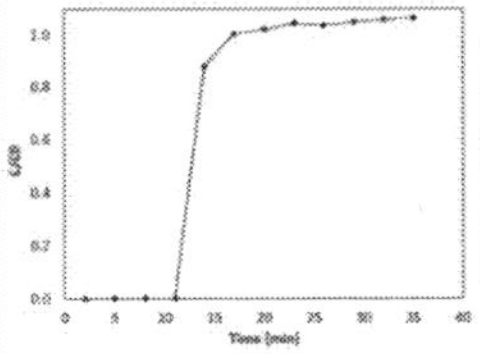

Fig. 5 Pre-sulfidation of 21% ZnO/SiO_2
With 2% H_2S/H_2 ; T = 400°C
GHSV = 6800 h^{-1}

Step 2: Pass UHP CO/CO_2

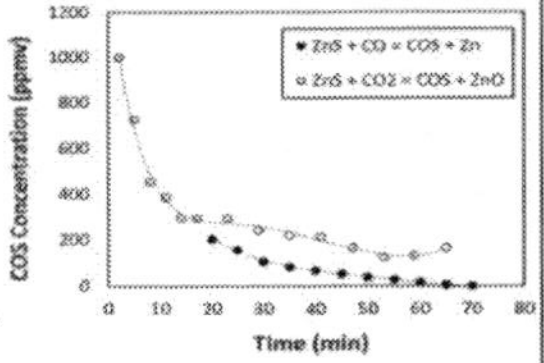

Fig. 6 Pass UHP CO/CO_2
T = 400°C
Flow = 50 ml/min

Summary & Conclusions

- Carbonyl Sulfide (COS) formation during desulfurization process contradicts the aim of desulfurization operation and the formed COS is a severe poison for fuel cell anodes.
- Restricting the feed stream to low operating temperature means avoiding the reaction between CO_2 and H_2S
- At temperature >200°C, COS breakthrough always occurs before H_2S breakthrough suggesting the low capacity for COS adsorption on ZnO based sorbents
- Moisture had a COS inhibition effect most likely due to hydrolysis of COS (in this case, ZnS is known to be a catalyst)
  $$COS + H_2O = H_2S + CO_2$$
- At 400°C, COS formation through CO route is negligible as this temperature is insufficient for reduction of ZnS to elemental Zn
- At 400°C, COS formation through CO_2 route occurs most likely due to the following reaction:
  $$ZnO + H_2S = ZnS + H_2O$$
  $$ZnS + CO_2 = ZnO + COS$$

References

1. *Yang et al. I&ECR. 2008,47 (24)*
2. *Song et al. Hydrogen and Syngas Production Technologies, 2010, Wiley & Sons*
3. *D. Shekhawat et al. Fuel Cells: Technologies for Fuel Processing, Elsevier B.V., UK, 2011.*
4. *E.C. Rupp et al. Fuel, 92 (2012) 211-21*
5. *Novochinskii et al. Energy & Fuels, 18 (2004) 576-583*
6. *E. Sasaoka et al. I&ECR, 35 (1996) 2389-239*
7. *J.P. Wakker et al. I&ECR, 32 (1993) 139-149*
8. *S. Ju et al., Fuel, 108 (2013) 80-84.*

390881 Diffusion Studies of Linear Paraffins in Activated Carbon

Abstract:

As the search for greener processes are growing in recent years, Fischer-Tropsch (FT) reaction has become attractive due to its possibility to produce cleaner fuels. The FT reaction uses syngas, a mixture of CO, CO_2 and H_2, provided by coal or biomass gasification, with metallic catalysts, to produce a wide range of hydrocarbons. The efficiency of this reaction can be improved by modifying the catalyst properties and therefore deeper investigations to relate the syngas conversion and selectivity with the catalyst properties, which include adsorption kinetics, are needed. Activated carbon has been pointed as an attractive metal support for the FT reaction due to its higher specific surface area, low cost, wide pore size distribution and the chemistry complexity on its surface. This study presents diffusion measurements of linear paraffins (C_7-C_{16}) on activated carbon, as a support, impregnated with iron or cobalt, using the Zero Length Column (ZLC) technique.

Development of a High Throughput Ion Exchange Resin Characterization Workflow

Authors: Chun Liu,[*] Dan Dermody,[*] Keith Harris,[*] Jeff Sweeney,[#] Daryl Gisch,[$] Bob Goltz[$]

*** Formulation Science, Core Research and Development, The Dow Chemical Company, Midland, Michigan 48667**

 # Engineering Sciences, Core Research and Development, The Dow Chemical Company, Midland, Michigan 48667

$ Dow Water & Process Solutions, The Dow Chemical Company, Midland, Michigan 48674

Abstract (limited to 1500 words) submitted to "*Separation Division*"

High throughput (HT) tools and methods have been explored and developed by the Dow Chemical Company since 2000. A typical HT research workflow consists of a collection of specialized automatic laboratory tools that perform liquid and solid handling, formulation development, and parallel batch reactions and polymerizations. These HT workflows can be integrated with various HT analytical methods such as UV/Vis, conductivity, pH, high-performance liquid chromatography (HPLC) etc. Overall, the HT tools/workflows have been successfully applied to a variety of application areas in the Dow Chemical Company, including cleaning formulations, coatings, thermosets, and others.

In this talk, we focus on the development of a HT ion-exchange (IEX) resin characterization workflow that can be applied to provide extensive isotherm tests data against a range of different applications where the water environment differs from site to site. The developed HT IEX resin characterization workflow consists of resin pre-processing, resin-in-solution formulations, resin-in-solution post-processing and resin-ion exchange analysis using ion chromatography (IC) for anions and inductively coupled plasma atomic emission spectroscopy (ICP-AES) for cations. Owing to the much higher throughput than traditional research, Design Of Experiment (DOE) methodology can be easily applied for studying the effects of multiple factors on the resin performance. Two case studies will be presented here to illustrate the applications of the above HT IEX resin workflow and DOE method. In case study one, a series of anion exchange resins have been screened for the selective removal of nitrates in simple water environment only consisting of nitrate (NO_3^-) and sulfate (SO_4^{2-}) anions. Then, the selected anion exchange resins are further tested against complex water environment consisting of multiple competing anion species (SO_4^{2-}, Cl^-, HCO_3^-/HCO_3^{2-}), varying pH and ionic strength. The DOE method is applied in the second part of the study to produce a statistical model from which the best resin candidate can be predicted based upon the specific water environment. In case study two for high TDS (Total Dissolved Salt) water softening, the DOE method is applied throughout the study to screen the performance of difference cation exchange resins towards the removal of magnesium (Mg^{2+}), calcium (Ca^{2+}) and barium (Ba^{2+}) divalent cations in the present of high sodium levels. Individual DOE models were built for each resin, from which the dependence of resin performance on pH level, Na^{+} background and initial level of divalent cations were revealed. Then, a master DOE model involving all the cation exchange resins studied was created to better understand the influence of the resin structures on IEX performance and identify the most promising IEX candidate for specific conditions.

The adoption of high throughput workflows and methods for IEX resin evaluation reduces the development and response time to address industry and application needs. The ability to take existing resins (or to generate libraries of new functionalized resins) and quickly test them for customer specific needs could significantly reduce upfront investment of resources and time. The reduced R&D investment coupled with the businesses existing market development and commercial scale manufacturing capabilities could allow Dow to produce differentiated products on a customer or application basis.

Standing Wave Design of a four-zone non-isothermal SMB fractionators and concentrator for non-linear systems

Abstract

It has been well established that the productivity and column efficiency limitations of chromatography can be overcome using a conventional, isothermal SMB system. However, despite its significant improvements, the challenge of overcoming product dilutions and desorbent usage still persists with isothermal SMB. One alternative to overcome these challenges is to apply temperature gradient in an SMB system[1, 2]. The current common method to determine the operating parameters of such systems is the triangle theory, which is based on the assumption of local equilibrium. The drawback of the triangle theory is that it does not take into account mass transfer effects. To meet desired purity and yield requirements, time-consuming searches are required to determine the operating parameters. The standing wave design (SWD) was developed with the purpose to overcome the drawbacks of triangle theory method[3, 4]. By implementing the SWD method to the non-isothermal SMB system, the optimum operating parameters can be easily determined through a direct calculation, which leads to the discovery of thermal SMB fractionators and concentrator (4-zone TSMB-FC) for linear systems in the previous work[5]. In this study, we extended the SWD method to the 4-zone TSMB-FC for non-linear system. We will use fructose-glucose separation as an example. The results show that a 4-zone TSMB-FC can produce two products with high purity (>99%). The extract product fructose is enriched by 1.5 fold, and the raffinate product glucose has no dilution. Moreover, a 60% reduction in desorbent was achieved compared to the isothermal SMB system.

References

1. Migliorini C, Wendlinger M, Mazzotti M. Temperature gradient operation of a simulated moving bed unit. *Ind Eng Chem Res*. 2001;40:2606.
2. Jin W, Wankat PC. Thermal Operation of Four-Zone Simulated Moving Beds. *Ind Eng Chem Res*. 2007;46:7208.
3. Ma Z, Wang NHL. Standing wave analysis of SMB chromatography: linear systems. *AIChE J*. 1997;3:2488.
4. Mallman T, Burris BD, Ma Z, Wang NHL. Standing wave design of nonlinear SMB systems for fructose purification. *AIChE J*. 1998;44:2628.
5. Soepriatna N, Wang NHL, Wankat PC. Standing wave design of a four-zone thermal SMB fractionators and concentrator (4-zone TSMB-FC) for linear systems. *Adsorption*. 2014;20:37.

380870 Experimental Study of Feedcol Strategy on Simulated Moving Bed Chromatography with Mal-Functional Column Configuration

Abstract:

Simulated Moving Bed (SMB) Chromatography has been considered as important separation process due to its effective separation process. Therefore, it has been widely used in various industries including petrochemicals, fine chemicals, pharmaceuticals and biotechnology etc. In the previous study, the FeedCol strategy was introduced for improving performance parameters by applying one short chromatographic column to the SMB process. The experimental study of the FeedCol strategy was performed in the SMB with partially mal-functional column configuration (two of eight columns are malfunctioning) and two sets of the FeedCol experiments were performed with various injection times at the fixed injection length of 50%. The experimental performance of FeedCol strategy was compared with the experimental results of conventional SMB process (2-2-2-2 configuration) with the same columns. In addition, the experimental results of FeedCol strategy and conventional SMB were compared with the simulation results of conventional SMB with normal columns. It was proved experimentally that the FeedCol strategy showed better performance than the conventional SMB at the same condition and even the conventional SMB with normal column. As a result, the FeedCol operation was proved as one of strong operational strategies to improve the SMB performance.

Preparation and its performance of CuCl@MIL-100(Fe) for CO/N_2 adsorption

Junjie Peng, Jing Xiao, Hongxia XI, Yingwei LI and Zhong Li*
School of Chemistry and Chemical Engineering, South China University of Technology
Guangzhou, 510640, China

*Corresponding author e-mail address: cezhli@scut.edu.cn

Abstract

In this work, we synthesized MIL-100(Fe) and then prepared composites CuCl@MIL-100(Fe) with different loading of CuCl. The prepared composites CuCl@MIL-100(Fe) were characterized by ASAP2020, XRD, SEM and DRIFTS. The CO, N_2 and water vapor isotherms on the composites were measured by a gravimetric method. XRD analysis showed that a high dispersion of CuCl was obtained on the surfaces of the composites. The BET surface area and pore volumes of the composites CuCl@MIL-100(Fe) decreased slightly with loading of CuCl. Isotherms of CO and N_2 on the composites can be well fitted by Freundlich equation. More importantly, CO adsorption capacity of CuCl@MIL-100(Fe) increased with the loading of CuCl. Maximun CO adsorption capacity of CuCl@MIL-100(Fe) was up to 1.39 mmol/g at 1 atm and 298 K, having an increase of 267% in comparison MIL-100(Fe) (0.378 mmol/g), and at the same time its N_2 adsorption capacity became somewhat lower. It made its adsorption selectivity of CO/N_2 be further improved. In addition, Isotherms of water vapor on the CuCl@MIL-100(Fe) were lower than those on the MIL-100(Fe), suggesting that its hydrophobic property had been greatly improved. The CuCl@MIL-100(Fe) will be a promising adsorbent for CO capture or separation from CO/N_2 gas mixture.

Formation of Freestanding ZIF-8 membrane in Microfluidic Channel

Tsz Nok Ng[1,2] Yakub FAM[1],King Lun Yeung[1,3]

[1]Department of Chemical and Biomolecular Engineering, [2]Energy Technology Program, [3]Division of Environment, The Hong Kong University of Science and Technology, Clear Water Bay, Kowloon, Hong Kong, P.R. China

Email: kekyeung@ust.hk

Session: 02 | Separations Division

Topic: 02D07 Poster Session: Membranes

Metal-organic frameworks (MOFs) consist of metal centers connected by organic ligands that form networks of pores, giving the material a rich topological and chemical diversity. MOFs have potential applications in chemical conversion, separation and storage as well as in sensing, electronics and medicine. Zeolitic imidazolate frameworks (ZIFs), are considered as superb candidate materials for membranes. However, the fabrication of high-quality MOF membranes remains a challenge due to poor nucleation, adhesion and intergrowth of the MOF crystals. Different strategies including in-situ and seeding-regrowth methods have been used to prepare supported MOF membranes. Few works have been done to prepare freestanding MOF membrane despite the benefits of lower transport resistance and larger surface area.

Freestanding ZIF-8 micoromembrane with average thickness of 40 microns was successfully grown along the gap between the microchannels. This is achieved by establishing two parallel flows of reactants within the channel and inducing nucleation, crystallization and growth at fluid interface. The permeation result of dye molecules has proved the obtained ZIF-8 membrane defect-free Based on the characterization results of confocal fluorescence microscope and scanning electron microscope, the formation mechanism of ZIF-8 membrane is also proposed.

Enhanced Proton Conductivity of Nafion Hybrid Membrane over the Entire Relative Humidity Range By Incorporating Metal-Organic Frameworks with High Content Phytic Acid Loading

Authors:Zhen Li[1,2], Guangwei He[1,2], Zhongyi Jiang[1,2]*, Hong Wu[1,2],

[1] *Key Laboratory for Green Chemical Technology, Ministry of Education of China, School of Chemical Engineering and Technology, Tianjin University, Tianjin 300072, China*
[2] *Collaborative Innovation Center of Chemical Science and Engineering (Tianjin), Tianjin 300072, China*

[*] Corresponding author: zhyjiang@tju.edu.cn

Abstract□Proton exchange membranes (PEMs) often suffer from sharply declined proton conductivity at elevated temperature because of undesirable dehydration. The introduction of phosphate groups into membrane is deemed a quite promising method to improve the proton conductivity of membranes in low relative humidity. For stable and long-term utilization of phosphate groups, phosphate group donation molecule should be in immobilized form rather than in free form, which required to find a suitable math of supporter and phosphate group donation molecules (guest molecules). Phytic acid (myo-inositol hexakisphosphate) and MIL101 were employed as guest molecules and supporter in this study, respectively. Phytic acid could be immobilized in MIL101 well because of the chelation between CUSs and phytic acid, and the similar diameter of pore in MIL101 and phytic acid molecule. In addition, MIL101 is endowed with potential of high load capacity for phytic acid since its characters like gigantic pore volume of 1.5~1.9 cm3 g^{-1} and huge Langmuir surface area of ~5900±300 m^2 g^{-1}. Herein, We impregnated phytic acid into MIL101 by vacuum-assisted method and the loading content of phytic acid in MIL101 was up to 11 wt. %. The corresponding Nafion/MIL101 hybrid membranes were prepared. Because of unique properties of phytic acid and the good match between phytic acid and MIL101, the Nafion/phytic@MIL hybrid membranes displayed excellent proton conductivity under different RHs, especially under low RHs. For instance, the proton conductivity was up to 0.228, 0.0608 and 7.63×10-4 S cm-1 at 100%, 57.4 % RH and 10.5 % RH (0.46, 2.8 and 11.0 times higher than that of pristine membrane), respectively.

Keywords□Metal-organic framework; MIL101; Nafion; Phytic acid, Proton exchange membrane; Proton conductivity.

378679 Enhanced Gas Separation Properties Performed By Ag+-Containing Mixed Matrix Membrane Utilizing Organic-Inorganic Janus Composite Nanoparticle

Abstract:

Enhanced gas separation properties performed by Ag^+-containing mixed matrix membrane utilizing organic-inorganic Janus composite nanoparticle

Tiantian Zhou, Liu Luo, Shen Hu, Shaofei Wang, Hong Wu†, Zhongyi Jiang

[a]*Key Laboratory for Green Chemical Technology, School of Chemical Engineering and Technology, Tianjin University, Tianjin 300072, China*

[b]*Collaborative Innovation Center of Chemical Science and Engineering (Tianjin), Tianjin 300072, China*

wuhong@tju.edu.cn

Abstract: An Ag^+ containing Janus composite nanoparticle were synthesized by adhering Ag onto as-prepared dopamine nanosphere which was embedded in Pebax® MH 1657 polymer matrix. The fabricated mixed matrix membrane (MMM) was utilized in CO_2 capture process. Dopamine, acted as bioadhesive, metal chelator as well as reductant, carried Ag^+ and elemental Ag nanosphere. The high Ag^+ loading (38.36 wt% of total Ag element) endowed the Janus nanoparticle with CO_2 facilitated transport privilege; while the structural improvement created by Ag-dopamine neighboring bestowed benign interface gap on the membrane nanostructure, ensuring the high gas permeation flux. The membrane separation properties were investigated by two different protocols: dopamine nanofillers with and without silver modification in order to comprehend the role which silver played in MMM. The maximum value of the MMM separation performance could be obtained at Janus nanoparticle loading of 10 wt%, which was 149.89 barrer for CO_2 permeability (82.9% higher), 25.91 for CO_2/CH_4 selectivity (40.04% higher), and 72.48 for CO_2/N_2 selectivity (39.30% higher) compared with that of pure Pebax® 1657. Effect of both feed pressure and operating temperature on separation properties were investigated through gas separation process with CO_2/CH_4binary system. And results suggested that the Janus particles with composite structure had a great potential merit in MMM utilization.

Keyword: Gas separation; mixed matrix membrane; Janus nanoparticle; facilitated transport; structural improvement

380619 Analytical Model of Shell-and-Tube Dehydrator Using Tubular Zeolite Membrane

Abstract:

Isopropanol (IPA) is an important chemical used as a solvent for dye paint and cleaning agent for precision machinery and semiconductor electronics. Dehydration of IPA has been usually carried out with multi-step distillation process because IPA-water mixture forms azeotrope at IPA concentration of 87.4 wt%, thereby requiring a large amount of energy. Energy saving will be expected by the IPA dehydration with vapor permeation (VP) when direct dehydration of IPA-water azeotrope vapor discharging from top of a distillation tower is carried out. Recently, a hybrid process of distillation and zeolite membrane separation has been attracting a lot of attention in terms of lower energy consumption compared with the conventional process and higher separation performance. Concentration polarization, well known as a phenomenon reducing separation performance significantly, should be eliminated by introducing a membrane separation process into the conventional process. In this study, an effect of concentration polarization on permeation flux in laminar flow condition is analyzed by comparing numeric calculation and experimental results.

A porous α-alumina tube with outer diameter of 2.5 mm and average pore size of 0.15μm was selected as support, on which polycrystalline zeolite layer was synthesized by the secondary growth method. The zeolite membrane was observed with a scanning electron microscope and identified with an X-ray diffraction. VP test was carried out using the tubular membrane at IPA concentration of 80 wt% in laminar flow condition to determine the separation performance of membrane such as permeation flux and separation factor. The composition of permeate was analyzed with a gas chromatograph.

In order to evaluate the dehydration performance, mathematical modeling and its numeric calculation were carried out for the shell-and-tube dehydrator. In addition, plug flow model as ideal flow was developed and compared with experimental results.

As the result of VP tests with the synthesized membrane, it was found that separation factor and permeation flux of water was reduced as feed rate of mixture decreased, while the permeation flux of IPA increased. It made clear that permeation flux of water decreased because water concentration was significantly reduced toward the membrane surface by selective water permeation through the membrane.

381651 Energy Consumption Optimization of Brackish Water Reverse Osmosis Membrane Desalination, Based on Operating Factors

Abstract:

Desalination is a key to provide water for various sectors in today's world water crisis. There are different types of desalination methods, including thermal and membrane desalination technologies. Both methods are energy intensive, they consume thermal energy and electricity, respectively. In inland region due to low salinity of water it is common to use membrane desalination technologies. Brackish Water Reverse Osmosis desalination (BWRO) system is technically applicable worldwide, but has technical challenges such as disposal and management of the concentrate as well as economic barriers due to the final cost of treated water.

Energy cost can be a significant portion of the total cost of treated water and the high pressure pumps utilized in the treatment of the BWRO consume a major portion of the total energy used during the processing. In order to address the affordability of BWRO, energy optimization is required. While some work has continued in energy minimization, our investigation is considering a number of effective parameters such as pressure, temperature, different feed concentrations as well as their interactions on the optimized performance of the system.

381850 The Preparation of Sulfonated Poly(ether ether ketone) Membranes and Application for CO2/CH4 Separation

Abstract:

Sulfonated poly(ether ether ketone)(SPEEK) with sulfonic acid group(–SO_3H) and ether bond(–O–) functional groups have special effect toward CO_2 molecular. In this paper, SPEEK membranes with different degrees of sulfonation (DS) were prepared by post-sulfonation and used for CO_2/CH_4 gas mixture separation. The prepared membranes were characterized by FTIR, TGA, PALS, water content. The results indicate that the weight loss of SPEEK membranes and the content of free water in membrane increased with increasing DS of membranes. Gas permeation experiments indicate that the permeation flux of CO_2 decreased with increasing DS of membranes, separation factor increased with the DS of membrane from 44% to 57%, then decreased with the DS of membrane from 57% to 80%. The membrane with DS of 57% exhibited the highest CO_2 permeability (101 GPU) and separation factor 904. There are two mainly reasons for this: The free water in membrane combined with CO_2 to form HCO_3^- , then CO_2 transport in this way through the membrane. In addition, the hydrophilic –SO_3H on the benzene ring combined with the water surround and form a water channel, the CO_2 molecular transporting through the channel, result in the improving of separation performance.

Thermodynamic model for gas solubility in glassy polymers: prediction of dual mode sorption parameters

Matteo Minelli, Ferruccio Doghieri

Interdipartimental Center for Industrial Research - Advanced Mechanics and Materials (CIRI-MAM)
Alma Mater Studiorum - Università di Bologna

Department of Civil, Chemical, Environmental and Materials Engineering (DICAM)
Alma Mater Studiorum - Università di Bologna

A great attention has been devoted in the last decades on the description of the solubility of low molecular weight species in glassy polymeric systems, and different models have been developed following a variety of approaches. Among the others, the thermodynamics of glassy polymers (NET-GP) [1] is a general approach that proved to be effective in the representation of gas and vapor solubility in different glassy polymers. The NET-GP approach provides an extension to the usual equation of state models for the description of the nonequilibrium characters of polymer systems below T_g, based on the assumption that the actual polymer mass density is a further internal state variable, and in turn it measures the departure from the equilibrium. Hence, the lattice fluid theory after Sanchez and Lacombe [2] is considered, giving rise to the so-called nonequilibrium lattice fluid (NELF) model.
The recent modification of this model allows the prediction of the dilation induced by the penetrant to the polymer matrix, with the punctual estimation of the volumetric properties of the polymer/penetrant mixtures [3]. This approach is based on a simple rheological scheme that considers a bimodal distribution of relaxational times, short and long characteristic time, and on the assumption that only the first ones are involved within the usual experimental times of sorption experiments. This model has been applied to a variety of different cases, showing its accuracy in the predication of the solubility of many gaseous or vapor penetrants in polymeric glasses.
However, in the common practice, the description of gas solubility in glassy polymers is still performed following empirical and semi-empirical approaches, and among the others, the most used is the dual mode sorption model (DMS) [4]. The DMS model counts on adjustable parameters depending on the penetrant/polymer couple, but also on operative conditions, temperature, pressure, as well as on the previous history of the glassy polymer. This model provides a simple and explicit expression to correlate penetrant solubility with its partial pressure, coupling two different contribution to sorption given by physical dissolution (Henry's law) and adsorption (Langmuir type):

$$c = c_D + c_H = k_D p + \frac{c'_H \, b \, p}{1 + b \, p} \qquad \text{(Eq. 1)}$$

in which k_D represents the Henry's law dissolution constant, C'_H is the adsorption site capacity and b the surface affinity parameter.

In view of its simplicity, DMS is still one of the most employed models for the representation of the experimental data, but its empirical character prevents any use in predictive manner, and it has to be considered as a mere correlative tool.

In this work, the NELF model is employed to derive simple and effective correlations for the a priori determination of the dual mode parameters, k_D, C'_H and b, on the basis of the pure polymer and penetrant characteristics.

The Henry's law dissolution constant measures the ability of the gas molecules to be solved in the polymer matrix, and it is related to the penetrant solubility in the polymer at the corresponding theoretical equilibrium state. On the other hand, the Langmuir-type adsorption capacity accounts for the adsorption in the excess of free volume in the polymer, and it is correlated to its departure from the equilibrium state, estimated as the difference in the polymer specific volume. The surface affinity parameter is then deduced from the comparison of the solubility behaviors of the penetrant analyzed in the glassy polymer in the nonequilibrium state and in the hypothetical corresponding equilibrium.

The dual mode parameters returned from the NELF model analysis are able to provide the effective description of the solubility isotherms of the different penetrants, mainly CO_2, N_2, Ar and CH_4, in various glassy polymers inspected. This approach is also able to distinguish for the different nonequilibrium initial states of the polymer, resulting in different sorptive capacities.

In conclusion, the determination of gas solubility of in glassy polymers can be carried out using two separate approaches. The first, more accurate and precise, requires the resolution of the complete phase equilibrium problem following the usual scheme provided by the NELF model. The second, simpler and handier although with more limited capabilities, uses general correlations derived in the framework of the NELF model to estimate the DMS parameters on the basis of a few of pure penetrant and polymer properties, which will be then used to evaluate the gas solubility.

References

[1] F. Doghieri, G.C. Sarti. *Macromolecules* **29** 7885 (1996).

[2] R.H. Lacombe, I. Sanchez. *J. Phys. Chem.* **80** 2568 (1976).

[3] M. Minelli, F. Doghieri, *Ind. Eng. Chem. Res.* **51** 16505 (2012).

[4] A.S. Michaels, et al. *J. Appl. Phys.* **34** 1 (1963).

384117 Pore-Size Tuning of Plasma-Deposited Organosilica Membranes for Gas Separation

Abstract:

Organosilica membranes for gas separation were prepared by plasma-enhanced chemical vapor deposition (PECVD) using hexamethyldisiloxane (HMDSO), trimethylmethoxysilane (TMMS), and methyltrimethoxysilane (MTMS) as the silicon precursor. The gas permeation measurement revealed that MTMS-derived membrane had the highest He/N_2 selectivity, followed by the TMMS-derived and HMDSO-derived membranes. FT-IR characterization indicated that the HMDSO-derived membrane has the highest content of methyl group, while the methyl group content for the MTMS-derived membranes was the lowest. These results suggested that the pore size of organosilica membranes can be tuned by changing the chemical structure of the silicon precursor.

384259 Gas Permeability in a Polymer of Intrinsic Microporosity (PIM-1): Effect of Time and Heat Treatments

Abstract:

In this work we studied the permeability of different gases (N_2, CH_4, CO_2, O_2) in films of a polymer of intrinsic microporosity (PIM-1) at various temperatures (from 30 to 60°C) and upstream pressures (from 0.5 to 5 bar). The gas permeability decreases with pressure, in particular for the case of CO_2, and increases with temperature.

The experiments were carried out on samples subject to different prior history, as follows :

a) Vacuum heating at 100°C and fast cooling

b) Vacuum heating at 80°C and slow cooling

c) Ageing at room temperature over 135 days

The experiments allowed to detect that the gas permeability decreases rather markedly with both age and thermal treatment. The two phenomena seem to produce a similar effect on the gas transport rate of all penetrants investigated, so that the ideal CO_2/gas selectivities are much less affected than the single gas permeabilities. On the other hand, the physical ageing at room temperature alters the gas permeability mostly in the first 20 days of observation: subsequent measurements show that the gas permeability decreases much more slowly with time.

The determination of the rates of transport at different temperature allows the estimation of the activation energy, E_p, whose values depend on the history of the sample, likely due to the fact that in the course of the experimental investigation the polymer free volume decreases as time and/or temperature increase. As the different phenomena occurring in the polymer matrix are difficult to decouple, such results indicate that the E_p value must be calculated using values measured on samples of the same age and at a maximum temperature lower than the one experienced by the sample in its prior history, in order to avoid physically meaningless situations like negative E_p values.

385405 Water Vapor Permeation through Amine-Containing CO2-Selective Facilitated Transport Membranes

Abstract:

Abstract

We have synthesized carbon dioxide-selective membranes by incorporating amino groups into polymer networks. The membranes have shown high carbon dioxide permeability and selectivity vs. hydrogen, carbon monoxide and nitrogen up to 180°C. Hydrogen sulfide permeates through the membrane much faster than carbon dioxide due to much higher reactivity as a result of small-size proton transfer, allowing nearly complete removal of hydrogen sulfide to about 10 ppb in the treated synthesis gas before water-gas-shift (WGS) reaction. In some applications, it is also required to selectively separate CO_2 by restricting the amount of water permeated. The amine-containing membranes are highly hydrophilic, and they exhibit not only high CO_2 permeance but also high water vapor permeance. This work demonstrates various water and CO_2 transport results exhibited by these thin-film composite amine membranes. Water vapor permeance higher than 3500 GPU has been experimentally demonstrated with >3000 Barrers of CO_2 permeability and >150 CO_2/H_2 ideal selectivity. Gas permeation experiments showed water vapor permeance to be not sensitive to the amine-containing selective layer thickness. The water transport resistance offered by the polymeric substrate appeared to be the controlling mass transfer resistance, independent of the resistance offered by the selective amine layer. Water vapor permeance was experimentally demonstrated to reduce on increasing the polymeric substrate resistance via increasing the number of substrate layers used to synthesize the membrane. The hydrophilic amine-containing selective layer was expected to retain moisture and served as the governing layer for CO_2 and H_2 permeation. Effect of the extent of water retention by the polymer matrix on the overall transport result was also investigated by use of different fixed-site amine carriers. The use of commercial polyvinylamine (containing >60% salts) as a fixed carrier was shown to retain more moisture and aid the CO_2 permeance as compared to the laboratory synthesized pure polyvinylamine (without containing salts). Water vapor permeance was shown to be related to the moisture retention capacity of the matrix.

389436 Polydopamine Modified PVDF Nanofibers Supported Thin Film Composite Membranes for Osmotically Driven Membrane Processes

Abstract:

Polydopamine modified PVDF nanofibers supported thin film composite membranes for osmotically driven membrane processes

Malgorzata Chwatko, Jason T. Arena, Liwei Huang, Jeffrey R. McCutcheon

One key area of research in the development of osmotically driven membrane processes (ODMPs) is the development of new membranes. Though viable commercial membranes have been featured in industrial processes [1], few membranes such as the cellulose triacetate and thin film composite forward osmosis (FO) membranes manufactured by Hydration Technology Innovations are available for study. All of the membranes used industrially are built from wet-dry cast cellulose acetate or thin film composite structures.

Thin film composite (TFC) membranes, comprising the newest generation of membranes for ODMPs, are composed of a three-tiered structure, with a very thin selective layer on top of a support layer and lastly backing layer. This design offers high flexibility of optimizing the selective and support independently. A new membrane platform which has gained interest in recent years expands upon the classical thin film composite structure, but replaces the normal phase-inverted polymer support with one prepared via the electrospinning technique. Electrospun supports characteristically have a highly porous and interconnected pore structure. Some of the most recent work on nanofiber support membranes was done by Huang and McCutcheon using hydrophilic Nylon 6,6 nanofiber support and identified a unique swelling behavior which hindered performance [2]. Hydrophobic materials have reduced interaction with water and as a result have a tendency for less swelling. One such polymer recently reported on by Tian, et al. used a polyvinylidene fluoride (PVDF) support in their preparation new membrane for FO [3].

The membrane prepared in this work expands on this approach by also using electrospun PVDF nanofibers as a support of an interfacial polymerized TFC membrane. Here seeking to address a limitation of hydrophobic supports identified in work by McCutcheon, et al. that observed the importance of support layer hydrophobicity and the resulting lack of support layer wetting which dramatically impairs water flux due to severe internal concentration polarization [4]. Work later done by Arena, et al. hydrophilized the hydrophobic supports by applying polydopamine (PDA) to the surface of commercial TFC membranes. Modified membranes illustrated significant improvements in water flux following PDA modification [5, 6]. The work presented here will detail the preparation and performance of the PVDF supported TFC membrane, which have had a hydrophilic character imparted to them through the addition of PDA to the surfaces of the PVDF support layer combining the mass transport advantages of a nanofiber support, the low swelling propensity of a hydrophobic support and thanks to the PDA modification a hydrophilic character essential for good performance in osmotically driven membrane processes.

Figure 1: Scanning electron micrograph a cross-section of the PVDF membrane support

Works Cited:

[1] McGinnis RL, Hancock NT, Nowosielski-Slepowron MS, McGurgan GD. Pilot demonstration of the NH_3/CO_2 forward osmosis desalination process on high salinity brines. *Desalination* (2013) **312**: pp. 67-74

[2] Huang L, McCutcheon JR. Hydrophilic nylon 6,6 nanofibers supported thin film composite membranes for engineered osmosis. *Journal of Membrane Science* (2014) **457**: pp. 162-169

[3] Tian M, Qiu C, Liao Y, Chou S, Wang R. Preparation of polyamide thin film composite forward osmosis membranes using electrospun polyvinylidene fluoride (PVDF) nanofibers as substrates. *Separation and Purification Technology* (2013) **118**: pp. 727-736

[4] McCutcheon JR & Elimelech M. Influence of membrane support layer hydrophobicity on water flux in osmotically driven membrane processes. *Journal of Membrane Science* (2008) **318**: pp. 458-466

[5] Arena JT, McCloskey BD, Freeman BD & McCutcheon JR. Surface modification of thin film composite membrane support layers with polydopamine: enabling use of reverse osmosis membranes in pressure retarded osmosis. *Journal of Membrane Science* (2011) **375**: pp. 55-62.

[6] Arena JT, Manickam SS, Reimund KK, Freeman BD & McCutcheon JR. Solute and water transport in forward osmosis using polydopamine modified thin film composite membranes. *Desalination* (2014) **343**: pp. 8-16.

389945 Preparation and Characterization of Sulphonated Poly(styrene-ethylene/butylene-styrene) Block Copolymer Membranes for Desalination By Pervaporation

Abstract:

Pervaporation is a potential low energy membrane technology. It has been extensively used for separation or concentration of mixtures of aqueousorganic
or organic liquids. Using dense hydrophilic pervaporation membranes for desalination has been proved to be an efficient way to obtain potable water as it has the advantage of near 100% of salt rejection and energy consumption being independent of the feed salt concentration. The pervaporation of an aqueous salt solution can be regarded as separation of a pseudoliquid
mixture containing free water molecules and bulkier hydrated ions formed in solution upon dissociation of the salt in water. Not too many open literatures involved desalination by pervaporation. In this work, hydrophilic sulphonated poly(styrene-ethylene/butylene-styrene) (S-SEBS) triblock copolymer membranes based on poly(styrene-ethylene/butylenestyrene)
(SEBS) thermoplastic elastomer were synthesized via chemically modified of grafting a sulfonic group into the polystyrene chain. A range of techniques such as FTIR, SEM, TGA were used to characterize the structure and properties of S-SEBS membranes. Results revealed that the sulfonate groups were grafted into the para-position of PS endblocks
successfully. Desulfonation was happened at around 200–300 proved by a weight loss in TGA curve, which indicated a good thermal stability of S-SEBS.
Taking advantages of the ion channel generated by phase separation due to thermodynamic incompatible of S-SEBS and its highly hydrophilic properties, S-SEBS membranes were ted for pervaporation of aqueous salt solution with NaCl concentrations of 1wt%-3wt% at temperatures 50-
75 , feed flow rate 16L/h and vacuum 0.095MPa. The salt rejection remained high (up to 99%) under all operating conditions because of the intimacy between water and sulfonate groups in S-SEBS membranes and the non volatilization of salt. The water flux of SSEBS membranes increased with the increase of the sulfonated degree and feed temperature while the flux decreased with the increase of feed salt concentration. A high water flux of 22.87 kg/m2·h could be achieved when the sulfonated degree was 54.1% at the feed temperature of 65
and salt concentration of 1 wt%. The activation energy for water permeation was found to be 58.9 kJ/mol. In a word, the results indicated that the pervaporation process with S-SEBS membrane is a promising candidate for desalination.

390711 Optimization of a Community-Scale Reverse Osmosis Desalination Plant

Abstract:

From chemical engineering point of view, performance of reverse osmosis process for water purification applications could be improved in two completely different ways: improving efficiency of each part of the process (e.g. pump, membranes, and energy recovery devices) or developing efficient layout for the process and changing operating conditions. For given project specifications the latter would become more important to design the most efficient system with current technology level. Rules of thumb and commercial preferences which may not be accurate are currently dominant in decision making procedure for reverse osmosis design.

Community scale photovoltaic powered desalination system has recently gained attention of government sector and researchers to provide clean and potable water in remote areas (Out of grid). Taking into account specifications of such projects, a comprehensive model needs to be developed to provide accurate information. Taking advantage of a reliable model will be helpful to decide whether it is feasible to use PVRO for a specific condition or not. The aim of this work is to consider numerous variables of reverse osmosis process and produce an optimization algorithm for final design based on the specific conditions of community scale water production system.

390834 Highly Durable Composite Palladium Membrane Prepared By Introducing a Metallic Interlayer and a Surface Alumina Coating Layer below 573 K

Abstract:

For the last two decades, several ways such as electroless-plating, electroplating, sputtering and chemical vapor deposition have been proposed to prepare a composite type of thin palladium membrane, whereby improving the hydrogen permeability significantly. However, it has been pointed out that the durability of palladium composite membrane especially at low temperatures is not enough for practical use. Thin palladium film is well known to expand and contract by hydrogen absorption and desorption, leading to peel-off from its support and crack. These defects, in general, are apt to result from an increase in the hydrogen solubility with lowering temperature. Furthermore, the Kirkendall void, which is known as small voids produced around a heterogeneous metal grain attached on palladium film, results in a leakage of gases other than hydrogen.

In this study, an approach is explored to avoid such occurrences of defects. We propose here a method to depress a deformation of thin palladium film by introducing a metallic interlayer. The hydrogen-absorbed palladium film produces an inner stress accompanied by expansion, which induces cracks and peel-off. We infer that such defects should be depressed if the palladium film adheres tenaciously to the porous support. For that, metallic layer like platinum is thinly deposited on the support, then followed by palladium deposition, showing an important role of preventing the deformation completely. As the countermeasure to the Kirkendall void, surface coating with alumina sol onto the palladium film is employed, thereby enabling to prevent direct contact with other metal particle from catalyst etc.

An alumina coated Pd/Pt/porous anodic alumina disk membrane was prepared. Hydrogen permeation tests in the range of 373 k – 573 K proved that a new type of composite membrane could maintain its high hydrogen permeability and selectivity without an increase of leakage.

390931 Fouling Behavior of Mycobacterium

Abstract:

Microbial fouling is an ongoing issue in membrane science. Typically, hydrophilic components are added to the membrane surface in order to reduce the degree of fouling. To determine if this method is effective, we are actively testing the effect of surface hydrophobicity on membrane fouling using a quartz crystal microbalance (QCM) . QCM is highly sensitive to variations in the mass of a substrate, making it an ideal tool for the quantitative study of fouling behavior. We consider four surfaces with a water contact angle range of zero to ninety degrees, enabling us to study the full spectrum of surface hydrophobicity. We then flow cultures of Escherichia coli or Mycobacterium mucogenicum over the surface of the crystal, and monitor the bacterial adhesion in real time using our QCM apparatus. We have determined that there are potentially key differences in the adhesion character of these two strains, suggesting that microbial fouling may not be readily controlled through a simple variation of surface hydrophobicity.

390993 Gas Permeability in PTMSP/Graphene, Pmtsp/Graphene Oxide Mixed Matrix Membranes

Abstract:

Poly(trimethyl silyl propine) (PTMSP) is a glassy polymer with high free volume and high gas permeability suitable for gas and vapor separation, that is also subject to physical aging which can drastically decrease the permeability with time.

We have studied the effect of addition of small amounts (below 1wt%) of graphene (G) and graphene oxide (GO) on He, N_2, CH_4 and CO_2 permeability in PTMSP. A specific analysis was carried out to compare the ageing of composite membranes with that of pure polymeric ones, within a time interval of 30 d.

The addition of GO has an opposite effect with respect to the addition of graphene in PTMSP membranes on the permeability, and in particular it enhances the gas permeability while the presence of graphene lowers it. The effect of the gas kinetic diameter on the variation of gas permeability after filler addition was also quantitatively analyzed.

Both fillers have a significant impact on the ageing of PTMSP, i.e. they affect the variation of gas permeability with time. A high temperature treatment was performed on the composite membranes at 200°C, in order to study the possible reduction of graphene oxide and its effect on gas permeability.

Preparation and pervaporation of HTPB based polyurethane membranes

Wang Yu[1] , Ye Hong[1, 2*], Li Rong[1] , Ding Yu Ling[1]

（Beijing Engineering and Technology Research Center of Food Additives, Beijing Technology & Business University, Beijing 100048,2. Beijing Higher Institution Engineering Research Center of Food Additives and Ingredients, Beijing Technology & Business University, Beijing 100048）

ABSTRACT:

In recent years, pervaporation attracts the worldwide attention in the separation of aromatic compounds form aqueous solution. It has many advantages such as cleanness, energy-saving, and high efficiency. Phenol is a toxic contamination to water. The pervaporation membranes for separating phenol from water have been researched and should be further developed. In this study, the HTPB based polyurethane (PU) membrane and modified PU membrane were prepared. Their structures, thermal properties were characterized by FTIR, DSC and TGA. The pervaporation performances of these two kinds of PU membranes were investigated and discussed. Total flux and selectivity were increased with temperature. The HTPB based PU showed the high selectivity of 65 at 80 ^{0}C, while modified PU membrane obtained a higher total flux of 111.23kg· μ m·m^{-2}·h^{-1}.

KEY WORDS: Pervaporation; Polyurethane; β-cyclodextrins; carbon nanotube

1. INTRODUCTION

Phenols play an significant role in chemical industry. They are important raw materials in the petrochemical, pharmaceutical, plastic and pesticide industries. But phenols are common pollutants in wastewater of paper pulp industries , petrochemical, coal processing, and so on; They pollute environment severely. They are also toxic

* Corresponding to: Hong Ye, E-mail: yehcn@163.com

to people. So it is very necessary to separate and recover phenol from wastewater.

Pervaporation is a kind of energy saving and environment-kind technology. In recent years , it has attracted a lot attention as technique for the removal of phenol from water. Membranes play a vital role in the process of pervaporation. By now, pervaporation with siloxane (especially polydimethylsiloxane[1] (PDMS)) membranes, polyurethane[2](PU) membranes, and poly (ether block amide) [3](PEBA) membranes have been investigated widely. PU membranes have unique polymer chain structure and morphology which is easy for modification. So pervaporation with PUmembranes catch wide attation.

In this study, HTPB based PU membranes and modified HTPB based PU membranes were used for the separation of phenol /water mixtures. The structures and thermal stability of membranes were characterized by means of FTIR , DSC and TGA. Then the effects of the feed temperature on pervaporation performance were discussed.

2. EXPERIMENTAL

2.1 Materials

HTPB (hydroxyl value = 0.82mmolKOHg^{-1}) with average molecular weight of 2500 (Mn, g/mol) was obtained from qilonghuagong Co. Ltd. (Shandong, China), and dehydrated under vacuum at 110℃ with constant stirring for 1h before used; H_{12}MDI(80%) was obtained from TCI Reagent Factory(Shanghai, China), and used as received; Dibutyltindilaurate(DBTDL, catalyst) was used as received. BDO, β-CDs and N, N- dimethyl acetylamide (DMAc)were purchased from huihaikeyi Co. Ltd.(Beijing, China) ; BDO was stored with molecular sieves (4Å);β-CDs was dried at 110℃ in vacuum for 12 h before used; DMAc was purified by distillation under reduced pressure and stored with molecular sieves (4Å); carboxylic multi-walled carbon nanotubes (MWCNT-COOH) were obtained from nanjingxianfeng (Nanjing,China).

2.2The preparation of non-covalent carbon nanotube

The dry carbon nanotubes (MWCNT-COOH) and β-CD were mixed in a ratio of 1:30(weight) in the mortar, continuous grinding 2h with the dropwise addition of ethanol (1ml) over the fist 10min . Then the carbon nanotubes was rinsed with DMF and drying in a vacuum oven for 24h at 110℃.

2.2 Preparation of PU membranes

The Polyurethane (PU) membranes were synthesized by a two-step method including prepolymerization and chain extension reaction. At first, HTPB reacted with H12MDI in DMAc at 75℃ for 1 h with 0.1 wt% DBTDL as catalyst. Secondly, The modified carbon nanotubes which was dispersed in DMAc by ultrasonic for 1h and the chain-extender BDO was added to the prepolyurethane under mechanical stirring for 15min, and the membranes with 30% solid content were prepared by solution casting in a Teflon mould. The pure PU membrane was obtained with the addition of BDO only at the second part. And then two kinds of PU membranes were left in oven at 80℃ for 10h..

2.3 Characterization of PU membrane

The structure of PU membranes was characterized by FT-IR spectra (Nicolet IR560).

The Differential Scanning Calorimetry (DSC) curves of PU membranes were obtained by DSC(TA company) under nitrogen (N_2) atmosphere in the -90–250℃ range. The temperature increase with a rate of 10℃/min .

A thermogravimetric analyzer (TGA, model 2050, TA Instruments) was used to determine the thermal stability of the membranes in the 20–400℃ range. The temperature increase with a rate of 20℃/min.

2.4Pervaporation experiment

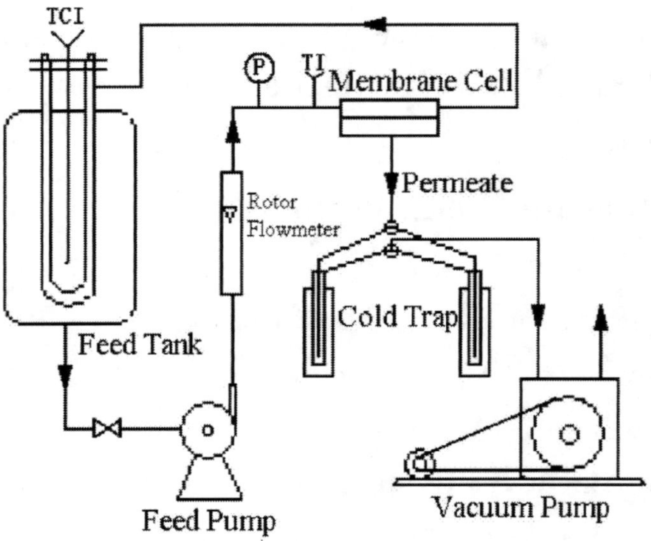

Fig.1. Pervaporation membrane separation equipment

$$J = \frac{Q}{A \cdot T} \cdot l \qquad \alpha = \frac{Y_A \cdot X_B}{X_A \cdot Y_B}$$

Where Q was the permeation weight collected in cold traps during the operation time T; A was the membrane area (m^2); l was the thickness of the membranes (μm); Y_A and Y_B(X_A and X_B) were the weight fractions of ethanol and water in the permeate(feed), respectively.

3. Results and discussion

3.1 Structure characterization of PU membranes

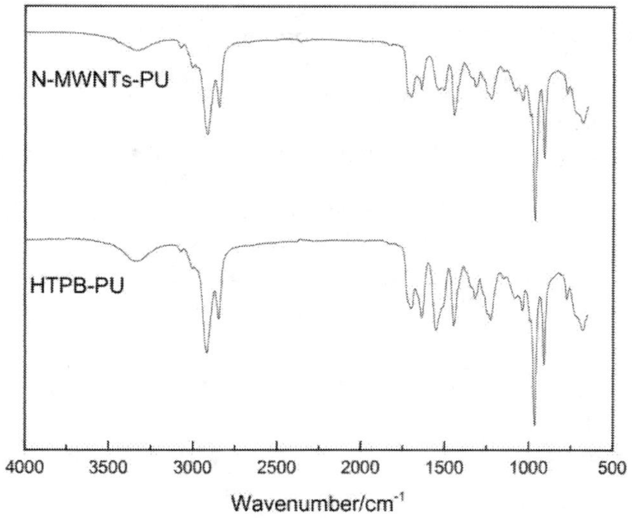

Fig.2. FTIR spectra of PU membranes

Fig.2. showed the FTIR spectra of two kinds of PU membranes. The absorption peak were at 3300 cm^{-1} (stretching vibration of N-H), 1720 cm^{-1} (stretching vibration of C=O), 1540 cm^{-1} (flexural vibration of N-H) ,which was the characteristic peaks of the urethane groups[4]. There is no significant difference between the FTIR spectra of these two kinds PU membranes, which may due to the lower content of carbon nanotubes in this study (0.1% wt).

3.2 Thermal properties of the PU membranes

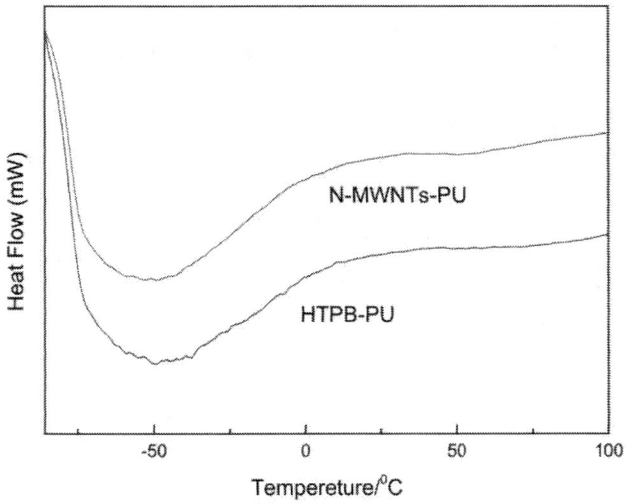

Fig.3. DSC curves of two samples of PU membranes

Fig.3. showed the DSC curves of two samples of PU membranes. Only one glass transition temperature (Tg) was observed, which were -77.2℃ of modified PU membrane and -77.6℃ of pure PU membrane. The Tg of hard segment can not be observed in the DSC curves , which may due to the larger molecular weight of soft segment [5] .

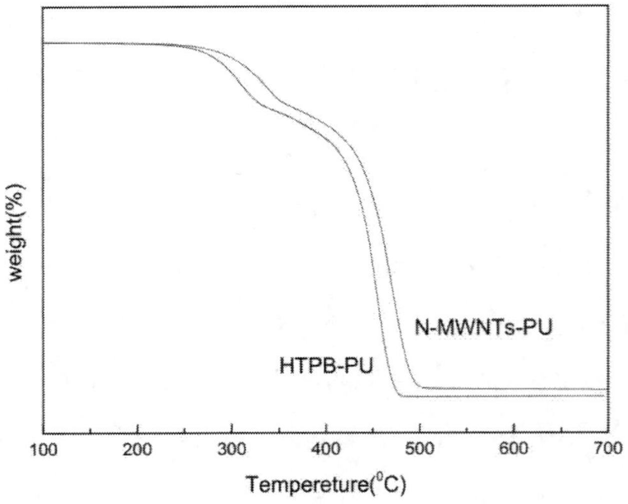

Fig.4. TGA curves of two samples of PU membranes

Fig.4. showed the TGA curves of two samples of PU membranes which were divided into two stages. The first decomposition occurred between 280-350 ℃ due to -NHCOO- decomposition of hard segment; the second decomposition occurred between 350-500 ℃ due to the decomposition of the soft segment.

3.3 Pervaperation performances

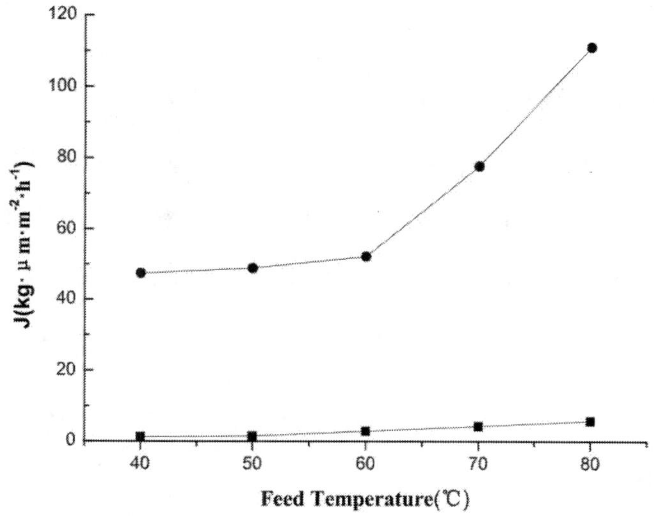

Fig.5. Effects of feed temperature(℃) on permeation flux

Fig.5. showed that the total flux increased with the increase of feed temperature. The chain mobility and the solubility of molecules on the surface was increased with feed temperature, which induced a higher permeation flux. On the other hand, the increase of feed temperature increased the free volume and softness of polymer chain.

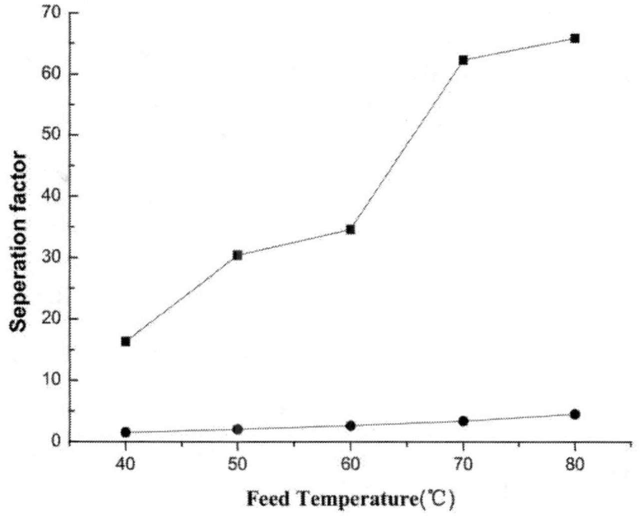

Fig.6. Effects of feed temperature(℃) on separation factor

Fig.6. showed that the separation factor increased with the increase of feed temperature. Gupta[6] used HTPB-PU membrane to separate phenol from water, also found a similar conclusion . These may due to the hydrophobic molecules of soft segment had stronger affinity to phenol.

4. CONCLUSION

Two kinds HTPB-based PU membranes were successfully synthesized by a two-step method. The total fluxes of both membranes increased with the increase of feed temperature from 30 to 80℃, meanwhile the separation factors of PU membrane also increased with feed temperature. The modified PU membranes obtained a higher total flux, while the pure PU membrane had a higher separation factor.

Acknowledgements

The authors greatly appreciate the financial supports of National Natural Science Foundation of China (No. 20906001), Beijing talents training subsidy（2009D005003000008）, the Research Foundation for Youth Scholars of Beijing Technology and Business University.

REFERENCES

[1] M Bennett, B.J Bridson, R. England, et al. Performance of PDMS and organofunctionalised PDMS membranes for the pervaporation recovery of organics from aqueous streams[J].Journal of Membrane Science, 1997, 137(2):63-88.

[2] W. Kujawski, A.Warszawski,W. Ratajczak, etal. Application of pervaporation and adsorption to the phenol removal from wastewater[J]. Separation and Purification Technology,2004,40(2): 123–132.

[3] M. Hoshi, M. Kogure, T. Saitoh, T. Nakagawa. Separation of aqueous phenol through polyurethane membranes by pervaporation[J]. Journal of Applied Polymer Science1997,65 (3) :469–479.

[4] Y L Luo, B X Wang, F Xu. Effect of soft–hard segment structure on vapor responsiveness of polyurethane conducting composite thin films loaded with multi-walled carbon nanotubes[J].Sensors and Actuators B, 2011, 156 (1):12–22.

[5] R W Seymour, S L Cooper. Thermal analysis of polyurethane block polymers[J]. Macromolecules, 1973, 6(1): 48-53.

[6] T. Gupta, N.C. Pradhan, B. Adhikari. Synthesis and performance of a novel polyurethaneurea

as pervaporationmembrane for the selective removal of phenol from industrial waste water[J].Bulletion of Materials Science,2002,25(6):533–536.

Performance of multi-tube palladium membrane reactor and its CFD analysis

Naotsugu Itoh, Utsunomiya University
7-1-2 Yoto, Utsunomiya 321-8585, Japan

Introduction

Hydrogen can be employed as a fuel for fuel cells, vehicle engines and power plant etc. However, there is no well spouting hydrogen on the earth differently from natural gas. In this sense, it is the fact that hydrogen is a secondary energy and therefore is necessary to be produced via water electrolysis. Nevertheless, a great expectation has been concentrating on hydrogen because of its cleanness, that is, the product by combustion is water only. Further, if water could be decomposed to hydrogen by any help of natural energy like solar, wind, hydro and geothermal power, a closed recycle system with energy would be completed, thereby constructing a hydrogen-based energy society.

Cyclohexane and methylcyclohexane promise to be chemical hydrides [1-4] because of having high hydrogen content (7.19wt% and 6.16wt%) and being a liquid state at room temperature. The hydrogen recovery by the dehydrogenation of the hydrides, however, is essential to be limited by chemical equilibrium. To release from the equilibrium limitation, an application of membrane reactor has been concerned. However, one must pay attention to a large endothermic heat of dehydrogenation, without which the reaction cannot proceed at all.

A multi-tube membrane reactor using palladium, which could produce about 1L/min of hydrogen, was designed and used for evaluating the performance for cyclohexane dehydrogenation. Physical and chemical phenomena occurring inside the membrane rector was analyzed employing a CFD (Computational Fluid Dynamics) code, whereby visualizing the cross-sectional concentration and temperature profiles in the catalyst-packed bed. As a comparison between experimental and calculated results, it was found that effects of heat transfer rate on the conversion of cyclohexane were very large. By changing the heating media for the reactor, it is shown that the reactor performance can be much improved.

Three Dimensional CFD Model for the Multi-tube Palladium Membrane Reactor

Figure 1 shows the cross-sectional view of a proto type of membrane reactor used in this study. Based on this configuration, 3-D CFD model was constructed as shown in **Fig.2**, with a total of 140000 meshes (10700 for the reaction side and 33000 for the permeation side) [5].

Simulation Results

1) Conversion of Cyclohexane

Figure 3 shows a comparison between the experimental (plots) and simulated (lines) results for cases 1–3 in Table 1, where both conversion and hydrogen permeation rate can be found to be in good agreement. This indicates that the CFD model developed here is valid for analyzing the multi-tubular membrane reactor designed in this study.

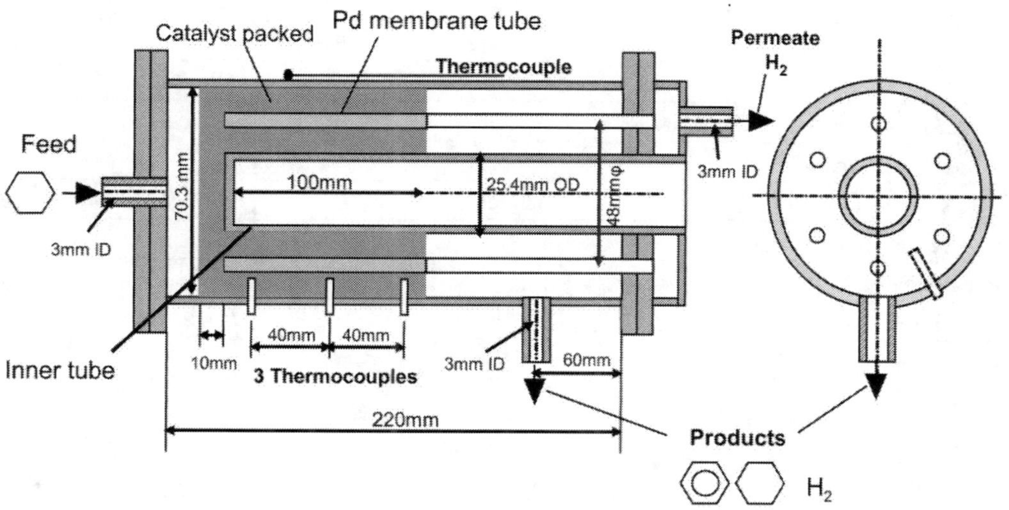

Fig.1 A multi-tubular palladium membrane reactor for dehydrogenation of cyclohexane using six thin palladium composite membrane tubes

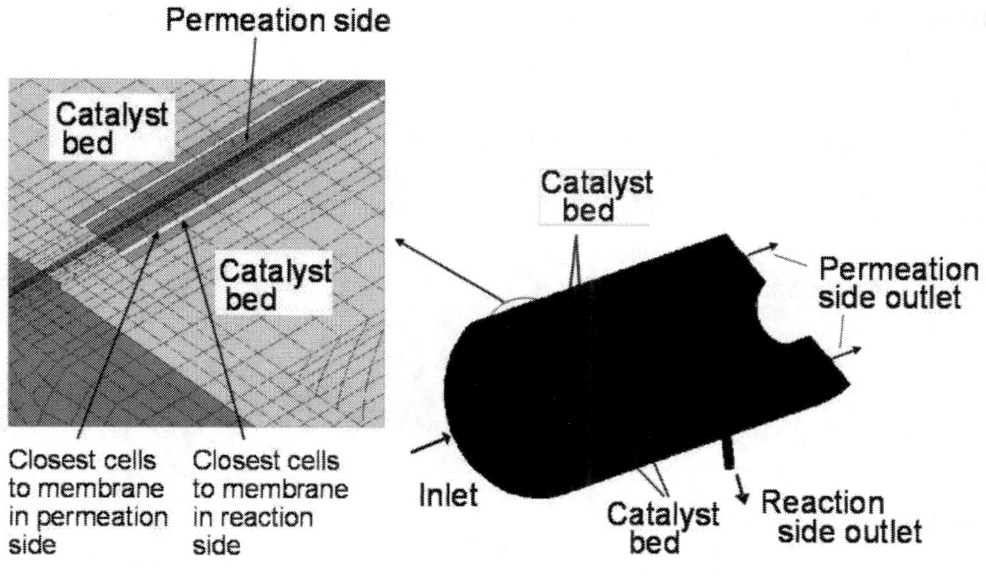

Fig.2 CFD model for the membrane reactor shown in Fig.13

Table 1 Experimental conditions and results for cyclohexane dehydrogenation using the multi-tubular palladium membrane reactor.

Case		Case 1	Case 2	Case 3	Case 4
1) Reaction Side					
Inlet Temperature	°C	300	300	300	300
Inlet Pressure	bar	2	2	2	2
Liquid Feed Rate	ml/min	2.20	1.80	1.55	2.20
Feed Composition					
C_6H_{12}	vol%	100	100	100	100
2) Permeation Side Pressure	bar	0.1	0.1	0.1	0.1
3) Heating condition					
Shell wall temperature	°C	340	340	340	340
Inner pipe wall	°C	300	300	300	300
4) C_6H_{12} conversion					
Experiment	%	80	83	89	-
Simulation	%	79.7	85.2	89.9	94.9
5) Permeated H_2					
Experiment	cm^3(STP)/min	800	733	702	-
Simulation	cm^3(STP)/min	801	755	718	1154

Simulation Results

1) Conversion of Cyclohexane

Figure 3 shows a comparison between the experimental (plots) and simulated (lines) results for cases 1–3 in **Table 1**, where both conversion and hydrogen permeation rate can be found to be in good agreement. This indicates that the CFD model developed here is valid for analyzing the multi-tubular membrane reactor designed in this study.

2) Changes of Temperature and Composition along the Reactor Length

The contour maps of temperature and molar fractions of three components, C_6H_{12}, C_6H_6 and H_2, in the axial

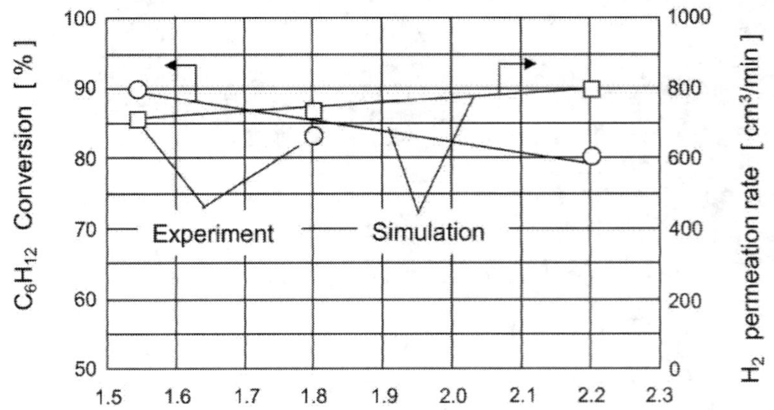

Fig.3 Cyclohexane conversion and hydrogen recovery with changing feed rate

cross-section of the membrane reactor (Case 1) are presented in **Fig. 4**. It is found that the temperature drops quickly near the inlet and then gradually rises toward the outlet. This is because the heat transfer rate is not large enough to supply the endothermic reaction heat. Concerning the composition inside the palladium membrane tube, it is clearly seen that the hydrogen mole fraction inside the membrane tube becomes 1 and those for cyclohexane and benzene are zero. Interestingly, one can see that the hydrogen fraction around the membrane tube is small due to its permeation, and instead the fractions of cyclohexane and benzene become relatively high. Therefore, one can expect that the dehydrogenation will be enhanced more near the membrane if the heat transfer manner is improved.

3) Radial Profile of Temperature and Composition

The radial contour maps with the catalyst-bed temperature are shown in **Fig. 5**. In all maps, as the radial position approaches to the center, the temperature drops because the shell wall is heated but the inner tube wall is not. The temperature near the membrane tube is found to be somewhat lower compared with the same radial position apart from the membrane. This is, as mentioned above, because the reaction with an endothermic heat can be accelerated in a lower hydrogen atmosphere around the membrane tube. Actually, it is recognized from **Fig. 6** that the hydrogen molar fraction around the tube is considerably lower. Further, when one sees the contour lines of hydrogen around the tube carefully, it becomes clear that those are not formed like concentric circles.

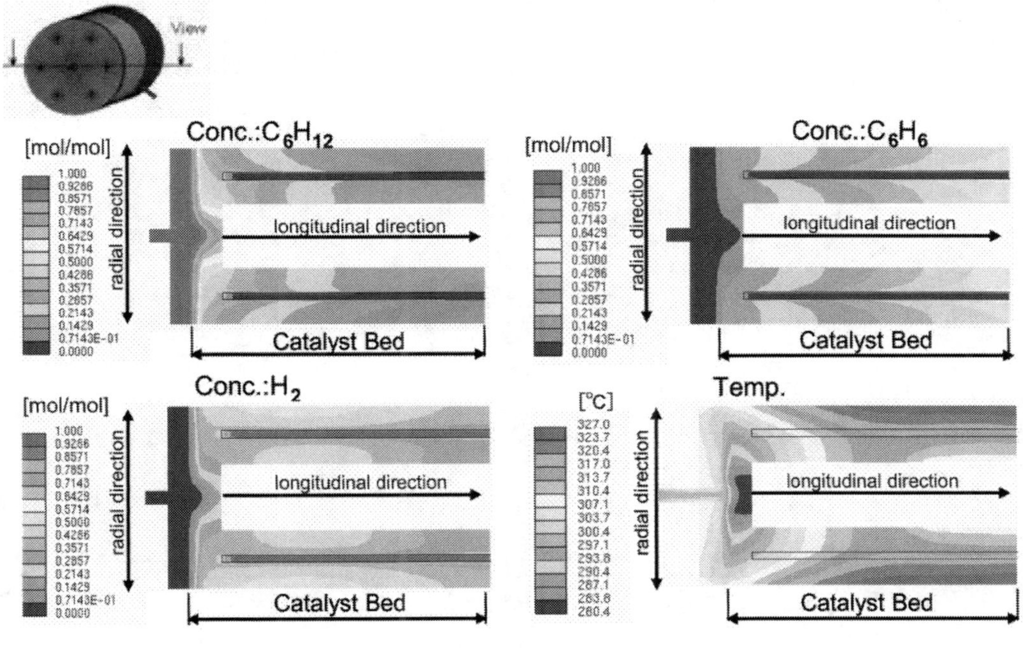

Fig.4 Temperature and concentration contour map of the longitudinal plane (Case 1)

4) Improvement of Membrane Reactor Performance

It has been shown so far that the CFD model describing the multi-tube membrane reactor very well can be constructed successfully. Similarly, it has been shown that the 3-D CFD models developed for decomposition of ammonia [6] and water gas shift reaction [7] in palladium membrane reactors are powerful tool for explaining the experimental results. The developed CFD model, however, as more essential and valuable way to use, should be applied to a new simulation, in which the reactor size and the reaction conditions are varied. In this section, as one of example, it is examined that how much the reactor performance can be improved when the membrane area is enlarged.

The hydrogen permeation rate is directly proportional to the tubular membrane area given by $\pi d_m L$, where d_m is the diameter of tubular membrane and L is the tube length. Accordingly, to use a membrane tube with larger d_m should lead to higher conversion by separating more hydrogen evolved. When the present diameter of 3mm is enlarged to 6 mm

(Case 4 in Table 1), the conversion increases by 15.2% and the hydrogen recovery by 354cm^3(STP)/min. In this case, although the catalyst volume was decreased by 4% by increasing the tube diameter, there was no particular influence on the reaction progress. This suggests that to increase the membrane area is the most effective for improving the performance.

Literature Cited

1) Cacciola, G., Giordano, N., Restuccia, G., Int. J. Hydrogen Energy, 9, 411 (1984)
2) Gretz, J., Baselt, J.P., Ulimann, O., Wendt, H., Int. J. Hydrogen Energy, 15, 419 (1990)
3) Itoh, N., Tamura, E., Hara, S., Takahashi, T., Shono, A., Satoh, K., Namba, T., *Catal. Today*, **82**, 119-125 (2003)
4) Itoh, N., Watanabe, S., Kawasoe, K., Sato, T., Tsuji, T., *Desalination*, **234**, 261 (2008)
5) Itoh, N., *J. Jpn. Petrol. Inst.*, **55**, 160 (2012)
6) Di Carlo, A., Dell'Era, A., Del Prete, Z., *Int. J. Hydrogen Energy*, **36**, 11815 (2011)
7) Smith, R.J.B., Muruganandam, L., Shekhar, S.M., *Chem. Eng. Res. Des.*, **89**, 2448 (2011)

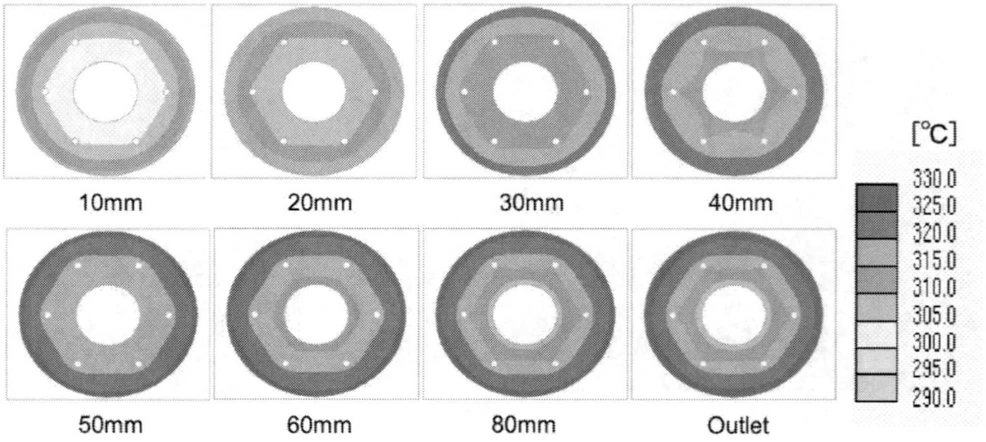

Fig.5 Temperature contour map of the several transverse planes (Case 1)

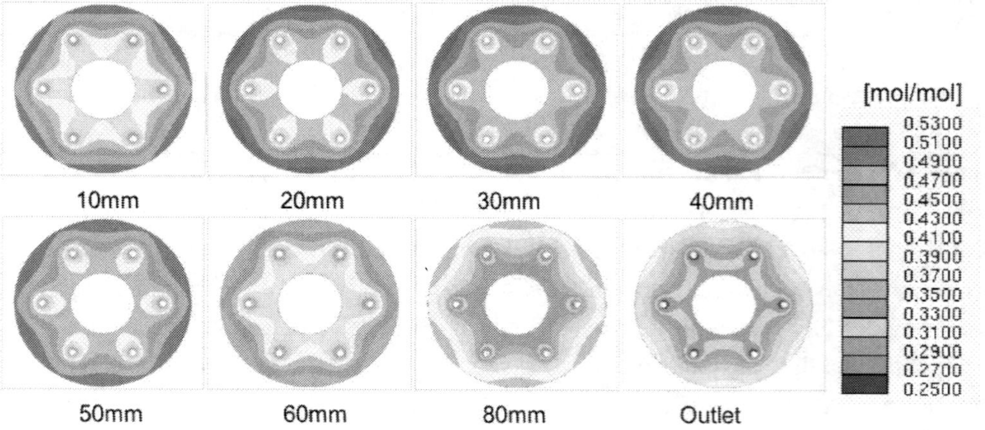

Fig.6 Hydrogen concentration contour map of the several transverse planes (Case 1)

Thermoregulation of Gas Permeation in Electrospun Fiber Membranes

Sangil Han[1] and Gregory C. Rutledge[2]

[1]Department of Chemical Engineering, Changwon National University, Changwon 641-773, South Korea

[2]Department of Chemical Engineering, Massachusetts Institute of Technology, Cambridge, MA 02139 USA

The thermoresponsive gas transport of electrospun fiber mats are of great importance in a large number of applications including combustion control, sensors, and safety control. Most previous research on the thermoresponsive gas transport has been focused on liquid phase transport for drug delivery. Even the studies of electrospun fiber mats are limited to gas transport properties without thermoresponsive gas transport properties. In this work, we report the first experimental demonstration of the thermoregulated gas transport using electrospun fiber membranes. Three-layered membranes with thermoresponsive properties were developed with polystyrene electrospun mat as a phase changing layer and cellulose acetate electrospun mat as a support using a hot pressing technique. The membrane exhibited oxygen flux reduction with temperature increase when polystyrene fibers were melted.

An integrated model of a spiral-wound membrane module for reverse osmosis considering the effects of winding and spacers

Boram Gu, Claire Adjiman, Yun Xu
Department of Chemical Engineering, Imperial College London

Keywords: Reverse osmosis, spiral-wound module, spacer, computational fluid dynamics

Reverse osmosis (RO) membrane technology has been considered to be more effective in terms of separation capability, its simple installation and less energy requirement than any other water purification technologies [1]. For this reason, RO processes have been successfully applied to water purification and seawater desalination for several decades, particularly in large-scale plants. Of various commercially available RO membrane modules, the spiral-wound module (SWM) has been accepted as most suitable for large-scale use due to its high packing density, low fouling propensity and easy cleaning [2]. From the viewpoint of system design, it is crucial to predict the performance of SWM because it is directly related to the specific energy consumption and required footprint of plants, which are also correlated with cost.

An SWM for RO is made by assembling folded flat sheet membranes and spacers inserted in between and rolling them around a perforated central pipe. By gluing permeate spacers to the membrane, the feed stream flows in the longitudinal direction, while the permeate flows in a spiral direction without mixing with the feed stream. Since the set of folded membranes and spacers is rolled around the central pipe, the geometry of an SWM becomes far more complex than a plate-and-frame module; the latter is simply a stack of membranes and spacers, with alternating feed and permeate channels. Therefore many existing computational models of SWM have been developed under the assumption that feed and permeate channels are rectangular and two sides of the folded membrane are symmetrical in order to simplify the geometry. However the assumptions are inadequate for describing a unique feature, which differentiates an SWM from a plate-and-frame module.

In the present work, a detailed spiral model is developed in order to describe the winding geometry and corresponding geometric parameters more accurately, which may affect the prediction of the module performance. By introducing a third parameter, the winding angle ϑ, the cross-sectional spiral geometry of an SWM can be reconstructed as shown in Figure 1. Based on the reconstructed spiral geometry, more accurate cross-sectional area and flow path length can be obtained. Interestingly, it is found that a part of the outer membrane (coloured in grey in Figure 1) is not used to generate permeate flux. This is due to the absence of a concentration difference between the two sides where the feed stream flows. Therefore the effective area is less than the total membrane area used to assemble a module.

Considering variations in the channel height, the flow path length and the cross-sectional area along the membrane, the permeate fluxes through an inner and outer membrane are calculated separately by employing a solution-diffusion model, which is the most widely used theory to describe water and solute transport through a semi-permeable membrane. Differential equations describing the overall and component material balances and pressure drop are solved simultaneously by means of MATLAB, in order to obtain flow velocity, concentration and pressure distribution for the feed and permeate channel.

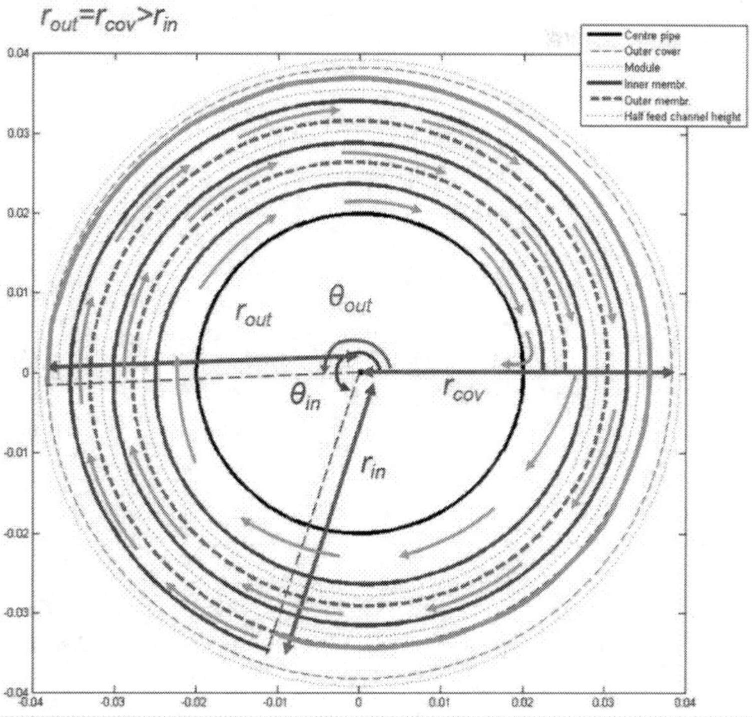

Figure 1. Reconstructed spiral geometry. Blue solid line: inner membrane, green dashed line: outer membrane, blue dotted line: the half feed channel, red dashed line: outer cover, magenta dotted line: arbitrarily chosen actual outer cover added due to numerical instability.

Apart from the spiral geometry, the performance of an SWM is strongly influenced by spacers, especially, feed spacers. Not only do the spacers prevent the two sides of a folded membrane from being in contact with each other but also mitigate concentration polarisation, which is caused by the accumulation of rejected solutes on a membrane surface and consequently results in reduced driving force for water permeation [3]. However, a drawback with the spacers is that they cause elevated pressure drop along the feed channel compared to an empty channel. Since concentration and pressure are the most important variables used to calculate the permeate flux and pumping energy, the concentration polarisation and pressure drop inside the module should be accurately predicted by taking into account the presence of the spacers. There are many publications which deal with spacer-filled channels in two- and three-dimensions in order to study the effects of spacers on mass transfer and pressure drop using computational fluid dynamics (CFD) packages. However, most of the CFD simulations focus on qualitative comparisons of various spacer-filled channels. There is a lack of detailed quantitative study which mathematically correlates the concentration polarisation and pressure drop with variables involved in the CFD simulations. Although a number of mass transfer correlations have been proposed, they are limited to a specific type of spacers [4].

In the present work, therefore, correlations for a mass transfer coefficient and frictional coefficients are developed with respect to design parameters of spacers such as the distance between parallel filaments, the angle of crossed filaments, the variation in the diameter of a filament and the type of spacers (woven/non-woven). Figure 2 shows a small segment of a spacer-filled channel surrounded by non-woven spacers with filaments of a uniform diameter, which represents the most basic geometry. A CFD package, COMSOL, is used to simulate flow, concentration and pressure patterns inside a spacer-filled channel in three-dimensions. Finally, the developed correlations for concentration polarisation and pressure drop are implemented into the detailed spiral model described earlier.

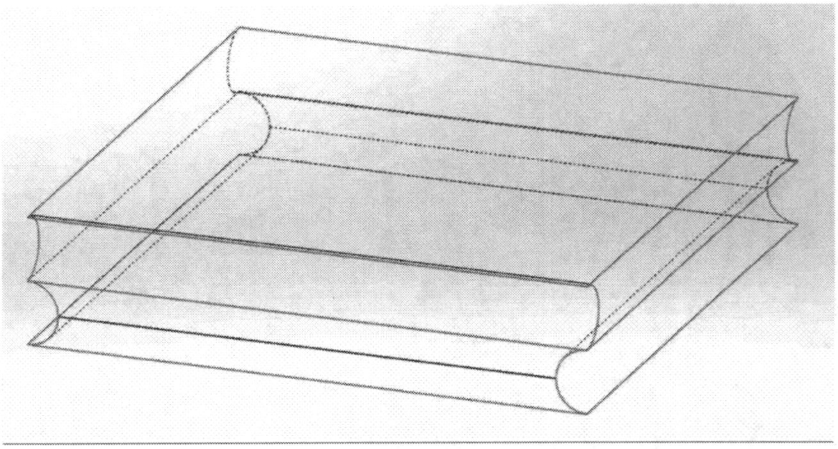

Figure 2. A small segment of the spacer-filled channel with non-woven and uniform-diameter filaments.

By integrating the MATLAB-based two-dimensional model for SWM and the detailed three-dimensional CFD model for an elemental unit of the space-filled channel, it is possible to take into account both the entire dimensions of the module, which cannot be easily handled in a CFD model due to the high computational demand, and the presence of spacers, which is impossible to be considered in a lumped or two-dimensional model due to the complex geometry. Furthermore, the integrated model can predict the performance more accurately and provide more information and insights into the module design because of the approach developed here allow the spiral geometry to be reconstructed in a more realistic manner.

References

[1] M. Elimelech, W.A. Phillip, The future of seawater desalination: energy, technology, and the environment, Science 333 (2011) 712–717

[2] J. Schwinge, P.R. Neal, D.E. Wiley, D.F. Fletcher, A.G. Fane, Spiral wound modules and spacers: Review and analysis, Journal of Membrane Science 242 (2004) 129–153

[3] E. Matthiasson, B. Sivik, Concentration polarization and fouling, Desalination 35 (1980) 59–103

[4] C.P. Koutsou, S.G. Yiantsios, A.J. Karabelas, Direct numerical simulation of flow in spacer-filled channels: Effect of spacer geometrical characteristics, Journal of Membrane Science 291 (2007) 53–69

376982 Multiple-Effect Membrane Distillation for Deep Concentrating Brine By Using PTFE Hollow Fiber Membrane

Abstract:

For the pollution and uneven distribution of fresh water resource, water shortage has become a serious problem in the world. Desalination of seawater and brackish water is attracting increasing attention and has been developed rapidly. At present, commercially available desalination technologies are multi-stage flash (MSF), multiple-effect distillation (MED) and reverse osmosis (RO). Due to technical limits and production cost, those desalination technologies have low water recovery which usually ranges from 30% to 50%, and the concentrated brine with high salinity and many harmful chemicals is also generated. The brine by-produced is usually discharged directly into the sea and even a bay, which has negative impacts on the ecological environment. The concentrated brine can also been used to produce sea salt. In the other side, the seasalt-related industry in China is facing the problem of insufficient area of saltpan. Therefore, it is important and necessary to deeply concentrate the brine drained from desalination plants.

Nowadays, membrane distillation (MD) for desalination is receiving more and more attention. Because MD process is not seriously affected by concentration polarization and feed salinity, a high-salt concentration can be reached in MD process. However, the thermal efficiency of the traditional membrane distillation is extremely low. When the performance ratio (*PR*) is used to evaluate the thermal efficiency of evaporation-based process, the *PR* of traditional MD process is usually in the range from 0.3-1 which is much lower than that of MSF and MED. In order to solve this problem, multiple-effect membrane distillation (MEMD) with the function of heat recovery is developed. In MEMD process, the heat of condensation can be recovered in AGMD module with function of internal heat recovery. So the *PR* of MEMD can reach up to 7 to 20.With its unique advantages, it is ideal for the desalination of seawater and further concentration of the high salinity solutions.

A problem with MEMD is the wetting of hydrophobic membrane. When the hydrophobic membrane is wetted, the hydrophobic membrane can't prevent the feed solution from entering into its dry micropores, which leads to the leakage and contamination of the permeate product. So the membrane material used in MEMD process must have good hydrophobicity. The common membrane materials for MEMD process include polypropylene (PP), polyvinylidene fluoride (PVDF) and polytetrafluoroethylene (PTFE). Compared to PP and PVDF, the hydrophobicity, chemical resistance, thermal stability and mechanical strength of PTFE is more outstanding, thus PTFE porous membrane is an attractive membrane for MEMD process.

In this study, MEMD process by using AGMD module of PTFE hollow fiber membrane was developed to deeply concentrate the brine drained from the conventional desalination plant. The inner diameter and outer diameter of the PTFE porous hollow fiber were 0.45mm and 1.0mm, and the porosity was 60%. The performance of MEMD was evaluated by the water vapor flux (J), performance ratio (PR) and thermal efficiency (η). Experiments were conducted to investigate the performance of PTFE-MEMD process by using the brine (4.8°Bé) from a MED desalination plant in Tianjin, China. The brine could be directly concentrated to a content of 12°Bé, furthermore, a modified de-calcium technology was used to remove >90% of calcium ions, since the precipitation of insoluble calcium salt on membrane surface leaded to severely decaying of MEMD performance when the salt concentration of brine was over 12°Bé. The decalcified brine could be concentrated from 12°Bé to 24°Bé. In the MEMD process to concentrate a brine of 4.8°Bé to a concentrate of 24°Bé, the value of J decreased from 6.8L/(m^2•h) to 2.6 L/(m^2•h), the value of decreased from 0.91 to 0.53, and the value of PR decreased from 12.8 to 4.5. When the concentration of brine was reached up to 24°Bé, the electrical conductivity of the permeate was still less than 150μs/cm. Then a long-term test on the operational stability of the MEMD process was investigated with the deep concentration of the de-calcium brine of 12°Bé for 80 days, in which the PTFE module remained in good condition. All these results demonstrated that MEMD process with PTFE hollow fiber membrane can be efficiently used to deeply concentrate the brine from desalination plants. Compared with the traditional salt making technologies with the brine from desalination plants, high-quality fresh water can be obtained and most land occupation is saved with the MEMD process. Moreover, the concentrated brine achieved in the MEMD process is highly purified, from which high purity solid sodium chloride, high purity magnesium salts and other sources could be obtained.

References

1. Kun Yao, **Yingjie Qin,** Yingjin Yuan, Liqiang Liu, Fei He, Yin Wu. A Continuous-Effect Membrane Distillation Process Based on Hollow Fiber AGMD Module with Internal Latent-Heat Recovery, **AIChE J.**, 2013, 59(4):1278-1297.

2. Rongling Liu, **Yingjie Qin**, Xiaojun Li, Liqiang Liu, Concentrating Aqueous Hydrochloric Acid by Multiple-Effect Membrane Distillation, **Front. Chem. Sci. Eng.**, 2012, 6(3):311-321.

3. Rui Ban, Miaomiao Liu, **Yingjie Qin**, Huan Wang, Dongsheng Cui. Enrichment of Semi-volatile Organic Acids from Aqueous Solutions by Multiple-Effect Membrane Distillation, **Transactions of Tianjin University**, 2012, 18:320-329

4. Xiaojun Li , **Yingjie Qin**, Rongling Liu, Yanping Zhang, Kun Yao. Study on concentration of aqueous sulfuric acid solution by multiple-effect membrane distillation, **Desalination**, 2012, 307:34-41.

365788 Microscopic Diffusion of Ethylene in 6FDA/DAM – ZIF-8 Mixed Matrix Membranes Using NMR Diffusometry

Abstract:

Mixed-matrix membranes (MMMs) are hybrid porous membranes, which consist of molecular sieve particles dispersed throughout a polymer matrix. In recent years, MMMs have been the subject of intense interest as they represent promising membranes for light gas separations including olefin/paraffin separations. In this work, we present the results of microscopic measurement for self-diffusion of ethylene in a mixed-matrix membrane using NMR-based diffusometry. The mixed-matrix membrane was derived from 6FDA/DAM polymer containing ZIF-8 particles. In addition to the MMM, transport in the constituent components, 6FDA/DAM polymer and ZIF-8, were also investigated which allows direct comparison of transport in the MMM and its precursors. In our diffusion NMR studies we combine the advantages of using large magnetic field gradients and high magnetic field. As a result, observations for diffusion inside the mixed-matrix membrane system for a broad range of microscopic displacements were possible. Three different microscopic diffusion regimes were resolved: ethylene diffusion inside the molecular sieve particles, transport in the polymer matrix between the particles and exchange of ethylene molecules between the molecular sieve particles and the polymer matrix. The results of these measurements will be presented and compared for diffusion of ethylene inside the MMM and the precursor polymer as well as molecular sieve particles.

376987 Highly Hydrophobic Polyvinylidene Fluoride (PVDF) Membrane for Direct Contact Membrane Distillation through Diffusion Induced Phase Inversion

Abstract:

PVDF membrane for direct contact membrane distillation (DCMD) with good mechanical stability was fabricated through diffusion induced phase inversion (DIPS). The scalability and simplicity of the phase inversion and among other hydrophobic polymer the tractability of PVDF in common organic solvents makes PVDF an interesting candidate for the MD membrane fabrication. Nonetheless there exist significant challenges in tuning the important structural parameters that influence the MD membrane performance. To understand the influence of the major parameters contributing to the performance of the flat sheet membranes in direct contact mode operation, the pores size and the surface roughness of the membranes were controlled by using different solvents and different coagulation bath compositions, and the membrane performance were evaluated using a laboratory-scale DCMD unit. Dope solutions with polymer concentration ranging from 8 to 15 weight % in *N,N*-dimethylacetamide (DMAc), *N,N*-dimethylformamide (DMF), *N*-Methyl-2-pyrrolidone (NMP), and triethyl phosphate (TEP) were used for casting of the membranes. A mixture of isopropanol in water was used as the coagulation medium. By adjusting isopropanol to water ratio, PVDF phase inversion was directed toward incipient precipitation.. Minor improvement in the contact angle was observed when NMP, DMAc, and DMF were used as solvents. However, by using TEP as solvent, it was found that the surface roughness can be enhanced without the penalty of losing the porosity and the contact angle found to be tunable and sensitive to the isopropanol content in the coagulation bath. By exceeding a specific volume fraction of isopropanol in the coagulation medium, the incipient precipitation becomes the dominant process and films with poor mechanical integrity and low porosity were obtained. Nevertheless, at the optimal condition, free standing membranes with good mechanical stability and contact angles as high as 143° and 140° on the top and the bottom side, respectively, were fabricated. By reducing the thickness to 70 μm and increasing the porosity to 70%, we were able to achieve high water flux and complete salt rejection. The maximum water flux achieved was 45 LMH at temperature difference of 40 °C and a cross-flow velocity of 12.8 cm/s. Here we demonstrate, that controlling the wetting by increasing the contact angle and keeping the porosity high are the major parameters for designing an efficient membrane for DCMD. Increasing the liquid entry pressure by reducing the surface pore size accompanied by decline in the water flux, so regardless of the bulk porosity of the membrane, effective porosity of the membrane at the surfaces is a critical design parameter for the DCMD application.

A combined blocking model for cross flow membrane filtration

Sourav Mondal and Sirshendu De*

Department of Chemical Engineering, Indian Institute of Technology Kharagpur, Kharagpur 721302, India

*corresponding author email: sde@che.iitkgp.ernet.in

Introduction

Membrane fouling during filtration is one of the major concerns in the field of membrane technology. Fouling is caused by the development of the concentration polarization layer over the membrane surface, formation of a cake layer and/or by blocking of the membrane pores. The pore blocking can further be classified as complete, intermediate and partial (Field et al., 1995). Theoretical description of the blocking mechanisms has been first described by Hermia (1987). Several researchers (Bowen et al., 1995; Ruohomaki and Nystrom, 2000; Ho and Zydney, 2000; Yuan et al., 2002; Taniguchi et al., 2003; Orsello et al., 2006; Bagga et al., 2008) have analyzed the fouling mechanisms in batch filtration mode considering a single mechanism or simultaneous occurrence of the cake and pore blocking phenomenon. The blocking principle for continuous cross flow filtration has been developed by Field et al. (1995), which has been analyzed during ultrafiltration of depectinized pineapple juice (Barros et al., 2003), mosambi juice (Rai et al., 2006) and blood orange juice (Cassano et al., 2007). In all these studies it has been observed that only one mechanism is prevalent at a time. However, it is likely that more than one fouling mechanism can occur during the filtration. In another study by Mondal and De, (2009, 2010), sequential occurrence of complete pore blocking followed by cake and intermediate followed by cake has been presented.

Theoretical development of sequential or simultaneous presence of these mechanisms for a continuous steady state filtration is not available. In the present work, a generalized formulation for steady state continuous filtration is attempted, considering simultaneous occurrences of complete pore blocking and intermediate pore blocking followed by cake filtration. Various parameters are combined in non-dimensional numbers and flux decline analysis is carried out in terms of these non-dimensional numbers. A complete map in parameter space is presented to identify the relative dominance of fouling mechanism. The analysis presents the domain of dominant fouling mechanism that can take occur during continuous filtration. Apart from identification of the dominant fouling mechanism, in cases of enzyme immobilization over membrane surface (Hibino et al., 1989), where cake formation is particularly essential or impregnation of catalyst in hollow fiber membrane (Caro et al., 2007) for gas phase catalytic reaction, where pore blocking is essential, the present analysis provides useful information about selection of parameters to operate the filtration in their respective regimes.

Theoretical Development

In the combined model, it is reasonable to consider that complete pore blocking proceeds from the beginning of the experiment upto a time of operation (t_1) when cake formation starts. Once the cake formation starts, solutes start depositing over the membrane surface, there is hardly any scope of pore blocking to take place. Therefore, the governing equation of flux decline due to pore blocking can be written directly from Hérmia's model (Field et al., 1995).

$$-\frac{dJ}{dt} = k_0 J^n \qquad (1)$$

where, $n = 1$ for complete pore blocking, $n = 2$ for intermediate pore blocking and $n = 3$ for cake formation.

Flux decline for $t < t_1$

Substituting $n = 1$ and 2 we obtain,

$$J = J_0 \exp(-k_1 t) \qquad (2)$$

for complete pore blocking and,

$$\frac{1}{J} = \frac{1}{J_0} + k_2 t \qquad (3)$$

for intermediate pore blocking, where J_0 is the pure water flux $\left(J_0 = \frac{\Delta P}{\mu R_m} \right)$. In terms of resistance, the pore blocking resistances were defined as,

$$R^*_{CPB} = \frac{R_{CPB}}{R_m} = \exp(k_1 t) - 1 \qquad (4)$$

$$R^*_{IPB} = \frac{R_{IPB}}{R_m} = J_0 k_2 t \qquad (5)$$

Flux decline for $t > t_1$

Growth of cake resistance is restricted due to external cross flow of the feed. In this case, the flux decline equation is obtained from the modified equation of Field et al., (1995)

$$\frac{dJ}{dt} = -k_c (J - J_{ss}) J^2 \qquad (6)$$

where, J_{ss} is the steady state flux. In terms of resistance, the expression of the flux becomes at $t > t_1$,

$$J = \frac{\Delta P}{\mu[R_m + R_{CPB}(t_1) + R_{IPB}(t_1) + R_c(t - t_1)]} \qquad (7)$$

The above equation can be expressed in terms of non-dimensional resistances as,

$$J = \frac{J_0}{[1 + R^*_{CPB} + R^*_{IPB} + R^*_c]} \tag{8}$$

Permeate flux at time $t = t_1$ is obtained as

$$J_{t_1} = \frac{J_0}{\left[1 + R^*_{CPB}(t_1) + R^*_{IPB}(t_1)\right]} \tag{9}$$

Combining the above two equations, the following expression is obtained,

$$J = \frac{J_{t_1}}{\left[1 + R^{**}_c\right]} \tag{10}$$

where, R^{**}_c is defined as, $R^{**}_c = \dfrac{R^*_c}{[1 + R^*_{CPB}(t_1) + R^*_{IPB}(t_1)]}$ (11)

Taking the derivative of equation Eq. (11) with respect to t, the rate of flux change is obtained as,

$$\frac{dJ}{dt} = -\frac{J_{t_1}}{(1 + R^{**}_c)^2} \frac{dR^{**}_c}{dt} \tag{12}$$

Using Eqs. (8) and (14), the governing equation of cake resistance is obtained,

$$\frac{dR^{**}_c}{dt} = \frac{J_{t_1} k_c}{(1 + R^{**}_c)} [J_{t_1} - J_{SS}(1 + R^{**}_c)] \tag{13}$$

At the steady state, $\dfrac{dR^{**}_c}{dt} = 0$ and therefore, from Eq. (13), the following criterion at the steady state is obtained, $J_{t_1} = J_{SS}(1 + R^{**}_{CS})$ (14)

Considering Eq. (8-11), we can transform Eq. (14) as,

$$R^*_c = \frac{J_0}{J_{ss}} - (1 + R^*_{CPB} + R^*_{IPB}) \tag{15}$$

Formation of non-dimensional groups

From the continuity equations at $t = t_1$, the flux obtained through both the mechanisms would be equal, which implies,

$$\left.\frac{dJ}{dt}\right|_{t=t_1-\delta} = \left.\frac{dJ}{dt}\right|_{t=t_1+\delta} \tag{16}$$

Using Eqs. (2), (3) and (6) in Eq. (18), we obtain,

$$k_1 \exp(k_1 t_1) + J_0 k_2 = J_0 k_c (J_{t_1} - J_{ss}) \tag{17}$$

Selecting non-dimensional parameters as $k_1 t_1 = \tau$, $\frac{k_c J_0^2}{k_1} = G_1$ and $\frac{J_0 k_2}{k_1} = G_2$, we obtain Eq. (17) as

$$\exp(\tau) + G_2 = G_1 \left(\frac{1}{\exp(\tau) + G_2 \tau} - \frac{J_{ss}}{J_0} \right) \tag{18}$$

Comparison of resistances

Since, from Eq. (4) and (5), complete and intermediate pore blocking resistances are quantified. Therefore,

$$\frac{R_c^*}{R_{IPB}^*} = \frac{1}{J_0 k_2 t_1} \left[\frac{J_0}{J_{ss}} - \exp(k_1 t_1) \right] - 1 \tag{19}$$

In terms of non-dimensional terms (using Eq. 18) Eq. (19) can be transformed to

$$\frac{R_c}{R_{IPB}} = \frac{(e^\tau + G_2)(e^\tau + G_2 \tau)^2}{G_2 \tau [G_1 - (e^\tau + G_2)(e^\tau + G_2 \tau)]} \tag{20}$$

Similarly, $\frac{R_c}{R_{CPB}}$, $\frac{R_c}{R_{CPB} + R_{IPB}}$ and $\frac{R_{CPB}}{R_{IPB}}$ can be obtained as

$$\frac{R_c}{R_{CPB}} = \frac{1}{(e^\tau - 1)} \left[\frac{G_1 G_2 \tau + (e^\tau + G_2)(e^\tau + G_2 \tau)}{G_1 - (e^\tau + G_2)(e^\tau + G_2 \tau)} \right] \tag{21}$$

$$\frac{R_c}{R_{CPB} + R_{IPB}} = \frac{1}{(e^\tau + G_2 \tau - 1)} \left[\frac{(e^\tau + G_2)(e^\tau + G_2 \tau)^2}{G_1 - (e^\tau + G_2)(e^\tau + G_2 \tau)} \right] \tag{22}$$

$$\frac{R_{CPB}}{R_{IPB}} = \frac{e^\tau - 1}{G_2 \tau} \tag{23}$$

Relative dominance of the each of these resistances can be compared by setting the values of the ratios to be greater or less than unity.

Infeasibility criteria

Infeasible solution will result when $\frac{R_c}{R_{CPB}}$, $\frac{R_c}{R_{IPB}}$, $\frac{R_c}{R_{CPB} + R_{IPB}}$ and $\frac{R_{CPB}}{R_{IPB}} < 0$. Considering Eq. (20-23), the necessary and sufficient condition for infeasibility is

$$\frac{G_1}{e^\tau + G_2} < e^\tau + G_2 \tau < 1 \tag{24}$$

for all positive real solutions of τ from Eq. (17).

Fig.1 shows the infeasibility boundary of τ in the parameter space by solving Eq. (18). The region above the curves shows the realistic solution of τ for different combinations of G_1 and G_2. So, for existence of the fouling mechanisms one has to select G_1, G_2 and J_{ss}/J_0 suitably. Solution of τ in the feasible domain has been present in Fig. 2. Using the values of τ and the parameters (G_1, G_2 and J_{ss}/J_0) relative dominance of the blocking mechanism during filtration can be identified. For the instances of only complete pore blocking followed by cake formation (Mondal and De, 2009) and only intermediate pore blocking followed by cake filtration (Mondal and De, 2010), Fig. 3 represents the regimes of dominant fouling mechanisms. An operator can preset these values of the parameters within the feasible boundary of τ to operate the cross flow filtration in a preferential fouling regime.

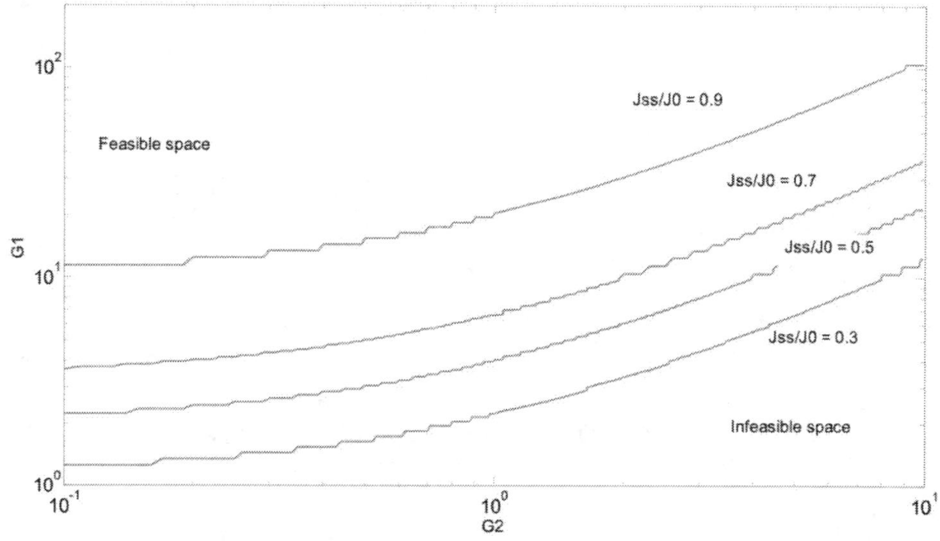

Fig. 1: Infeasibility regimes of solution of τ in the parameter space.

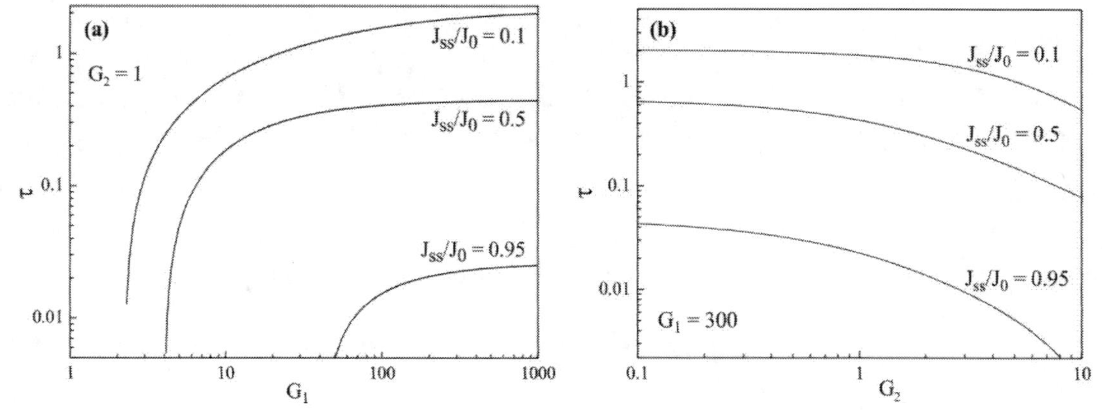

Fig. 2: Real solution of τ in the feasible domain of G_1, G_2 and J_{ss}/J_0.

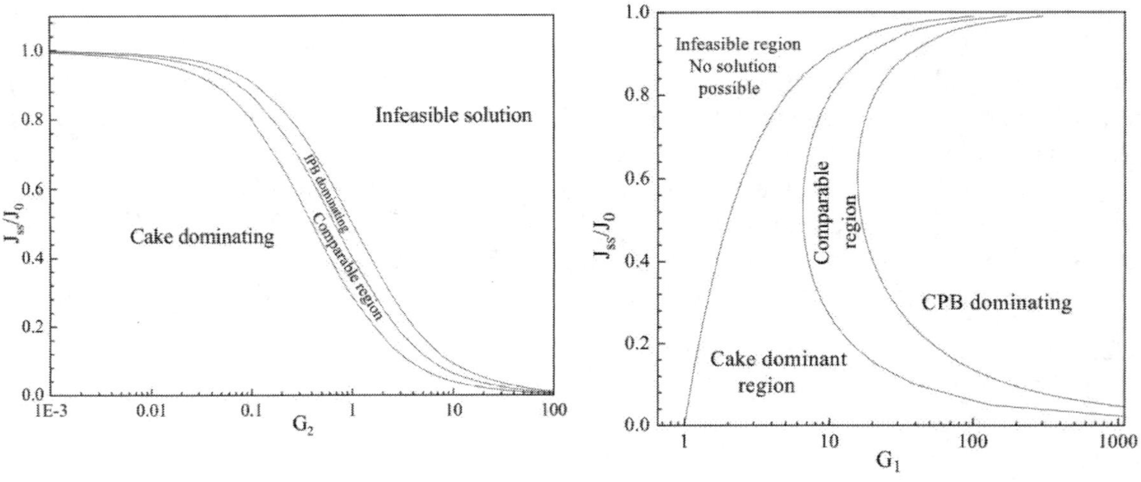

Fig. 3: Dominant filtration regimes in case of intermediate pore blocking followed by cake formation and complete pore blocking followed by cake formation.

References

Bagga, A., Chellam, S., Clifford, D.A. (2008) Evaluation of iron chemical coagulation and electrocoagulation pretreatment for surface water microfiltration, J. Membr. Sci. 309, 82–93.

Barros, S.T.D.d., Andrade, C.M.G., Mendes, E.S., Peres, L. (2003) Study of fouling mechanism in pineapple juice clarification by ultrafiltration, J. Membr. Sci. 215, 213–224.

Bowen, W.R., Calvo, J.I., Hernández, A. (1995) Steps ofmembraneblocking in flux decline during protein microfiltration, J. Membr. Sci. 101, 153–165.

Caro, J., Caspary K.J., Hamel, C., Hoting, B., Kolsch, P., Langanke, B., Nassauer, K., Schiestel, T., Schmidt, A., Schomacker, R., Morgenstern, A.S., Tsotsas, E., Voigt, I., Wang, H., Warsitz, R., Werth, S., Wolf, A. (2007) Catalytic Membrane Reactors for Partial Oxidation Using Perovskite Hollow Fiber Membranes and for Partial Hydrogenation Using a Catalytic Membrane Contactor, *Ind. Eng. Chem. Res.*, 46, 2286–2294.

Cassano, A., Marcia, M., Drioli, E. (2007) Clarification of blood orange juice by ultra filtration: analyses of operating parameters, membrane fouling and juice quality, Desalination 212, 15–27.

Field, R.W., Wu, D., Howell, J.A., Gupta, B.B. (1995) Critical flux concept for microfiltration fouling, J. Membr. Sci. 100, 259–272.

Hérmia, J. (1982) Constant pressure blocking filtration laws: applications to power-law non-Newtonian fluids, Trans. Inst. Chem. Eng. 60, 183–187.

Hibino, K., Okada, T., Kazno, N., Sahashi, Y. (1990) Enzyme immobilization in an anisotropic ultrafiltration membrane, *US Patent* 4,963,494.

Ho, C.C., Zydney, A.L. (2000) A combined pore blockage and cake filtration model for protein fouling during microfiltration, J. Colloid Interface Sci. 232, 389–399.

Mondal, S., De, S. (2009) Generalized criteria for identification of fouling mechanism under steady state membrane filtration. J. Membr. Sci. 344, 6-13.

Mondal, S., De, S. (2010) A fouling model for steady state crossflow membrane filtration considering sequential intermediate pore blocking and cake formation. Sep. Purif. Technol. 75, 222-228.

Orsello, C.D., Li, W., Ho, C.C. (2006) A three mechanism model to describe fouling of microfiltration membranes, J. Membr. Sci. 280, 856–866.

Rai, P., Majumdar, G. C., Sharma, G., DasGupta, S., De, S. (2006) Effect of various cutoff membranes on permeate flux and quality during filtration of mosambi (Citrus sinensis L.) Osbeck juice, Food Bioprod. Process. 84, 213–219.

Ruohomaki, K., Nystrom, M. (2000) Fouling of ceramic capillary filters in vacuum filtration, Filtr. Sep. 37, 51–57.

Taniguchi, M., Kilduff, J.E., Belfort, G. (2003) Modes of natural organic matter fouling during ultra filtration, Environ. Sci. Technol. 37, 1676–1683.

Yuan, W., Kocic, A., Zydney, A.L. (2002) Analysis of humic acid fouling during microfiltration using a pore blockage–cake filtration model, J. Membr. Sci. 198, 51–62.

Separation and Recovery of Platinum, Palladium and Rhodium from Wasted Auto-Catalysts by Three-Liquid-Phase Extraction Approach

Kun HUANG, Huizhou LIU

Key Laboratory of Green Process and Engineering, Chinese Academy of Sciences, Beijing 100190, China

Institute of Process Engineering, Chinese Academy of Sciences, Beijing 100190, China

(E-mail: khuang@ipe.ac.cn; hzliu@ipe.ac.cn, Tel : +86-10-82544910, Fax: +86-10-62554264)

Abstract

The main purpose of present article is to summarize our previous efforts to develop a new three-liquid-phase extraction strategy for recovery and recycling of platinum, palladium and rhodium from the hydrochloric acid leach solutions of various wasted auto-catalysts. It was found that Pt, Pd and Rh can be separated and selectively enriched into the three different liquid phases of the proposed three-liquid-phase system (TLPS), respectively. The partitioning behaviors of Pt(IV), Pd(II), Rh(III) are controllable and have a close correlation with the phase-forming behaviours of TLPS. A microphase mass transfer mechanism was proposed to explain the differences in partitioning behaviours of three platinum metal ions. The self-assembly of amphiphlic extractant molecules in aqueous solutions result in the controllable enrichment of target metal ions by those separated microphase aggregates. The differences in interfacial mass transfer of three metals are due to a controllable microscopic interaction of platinum metal ions with the microphase aggregates. The TLPS provide a unique separation medium to control the interphase mass transfer of different metal ions. The separation selectivity of the three platinum metals was improved remarkably by using the three-liquid-phase extraction technique in comparison with the traditional liquid-liquid two phase extractions. The present researches highlight a potential possibility for future industrial application of the multiphase extraction technique for one-step's group separation of different target metals from complex mixtures.

375677 Study of the Effects of Physicochemical Properties and Surface Crystalline Structure on the Electrosorption of Metal Ions By Titania Nanotubes

Abstract:

The separation and recovery of metal ions and ionic compounds from aqueous solutions is most challenging due to the stability of these charged species in aqueous environments.

The possibility of exploiting physicochemical properties of ions and specific interactions with separation media to trigger selective electrosorption of ions was explored in this work. The model materials used were titanium dioxide nanotubes, grown via anodization with well-defined diameters and lengths. The model ions used were sodium, lithium, cesium, and calcium in order to explore three physicochemical characteristics that could affect ion separation via electrosorption: (1) ion charge, (2) ion size and (3) ion solvation strength. Two different crystalline structures for the titanium nanotubes were explored: amorphous and anatase, achieved via annealing. Choronoamperometry experiments to quantify electrosorption were performed at applied potential windows between 0 mV and 600 mV, as to prevent the occurrence of any Faradaic reactions in the system. A three-electrode set-up was used, with titania nanotubes acting as the working electrode, a platinum wire acting as the counter electrode and a Ag/AgCl reference electrode.

In contrast with classical theory predictions, electrosorption of ions was achieved not solely in response to electrostatic interactions between ions and electrode surface. Selective separation of ionic species bearing the same charge was achieved most likely via desolvation and specific ion-surface interactions triggered by the difference in crystalline structure of the TiO_2-electrodes used in this work. For example, selective electrosorption of lithium ions over sodium and cesium ions (all monovalent) was achieved by changing the crystalline structure of titanium dioxide nanotubes via annealing.

371450 Effect of Length of Ligands on Permeation of Ions and Water and Lipid Flip-Flop in Nanoparticle Transport through a Lipid Membrane

Abstract:

Functionalized nanoparticles are generally considered excellent candidates for targeted drug delivery systems. However, the ion and water leakage induced by permeation of these nanoparticles is a challenge in these drug delivery methods due to cytotoxic effects of some ions. In this study, we have carried out a series of coarse-grained molecular dynamics simulations to investigate the effect of length of ligands on permeation of a nanoparticle across a protein free phospholipid bilayer membrane. Water and ion penetration as well as incidence of lipid flip flop are explored in this study while varying the ion concentration gradient, pressure differential across the membrane, nanoparticle size, length of ligand, and nanoparticle permeation velocity. Some conclusions from our studies include: (1) The number of water molecules in the interior of the membrane during ligand-coated nanoparticle permeation increases with nanoparticle size, ligand size, pressure differential, and permeation velocity but is not sensitive to the ion concentration gradient. (2) Frequency of lipid molecule flip-flop during ligand-coated nanoparticle permeation increases with nanoparticle size, but is not sensitive to ligand size, ion concentration gradient, permeation velocity, or pressure differential. (3) Ion transport during ligand-coated nanoparticle permeation is sensitive to the size of the nanoparticle but does not vary with ligand length, permeation velocity, or pressure differential; no anion/cation selectivity is observed for small nanoparticle permeation (2 nm), while anions are preferentially translocated through the membrane when the size of the nanoparticle increases (3 nm). Finally, we will also report on the effect of ligand coverage on water and ion leakage and the frequency and mechanism of lipid flip-flop.

Selective Separation of Hydrogen from C2H6 and CO2 through Pressed Natural Zeolite Membranes

A.Farjoo, A.M. Avila, S. Amadaref, W. An, S.M. Kuznicki
Department of Chemical & Material Engineering, University of Alberta, AB, Canada

Abstract

Pressed natural clinoptilolite and copper-clinoptilolite composite disk membranes are used for hydrogen separation. Their permselectivity properties were evaluated using hydrogen, carbon dioxide and ethane. Permeances and selectivies were quantified and the two types of membrane disks were characterized based on the relative averaged pore size.

Introduction

In order to minimize the non-zeolite pores in the pressed clinoptilolite material, copper metal was mixed with clinoptilolite particles to make clinoptilolite-copper composite membranes. Copper was added to clinoptilolite powders so that it can work as a binder among the clinoptilolite particles.

Having no chemical reaction with the zeolite being processed, copper provides effective mechanical seal among clinoptilolite particles in the membrane.

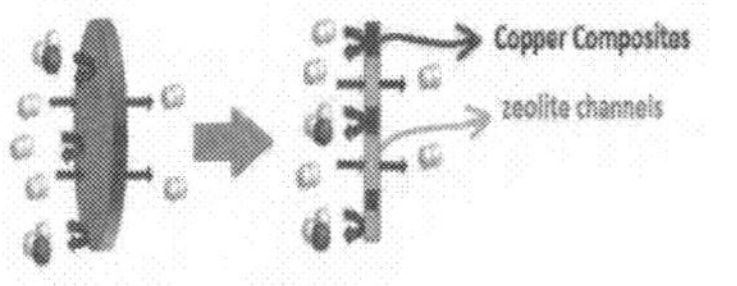

Fig-1:Separation mechanism for composite membrane

Results and Discussion

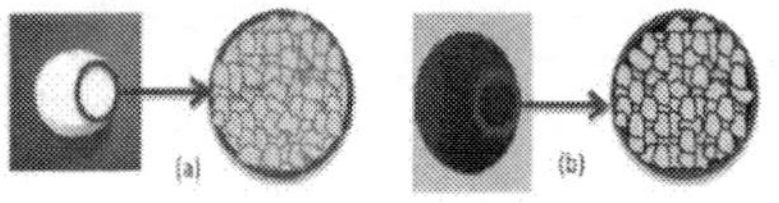

Fig-2:Schematic of Pressed clinoptilolite(a) and composite copper-clinoptilolite (b)

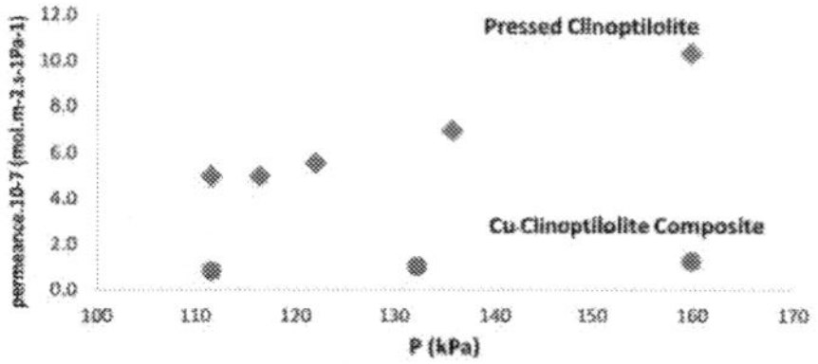

Fig-3: H2 permeance through membranes at different feed pressures (T=25 ° C)

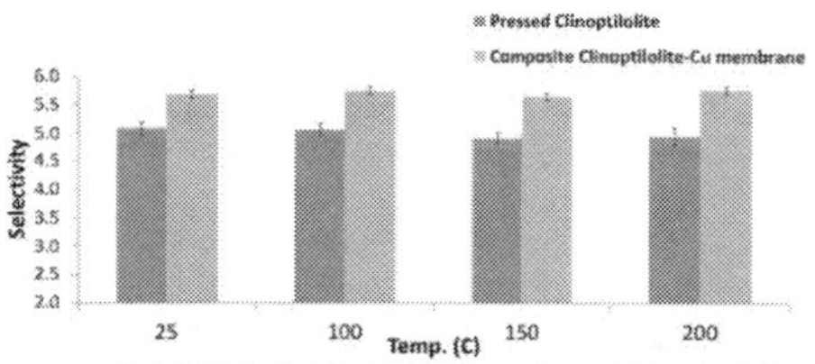

Fig-4: H2/C2H6 selectivity at different temperatures and P_{feed}=111.16 kPa

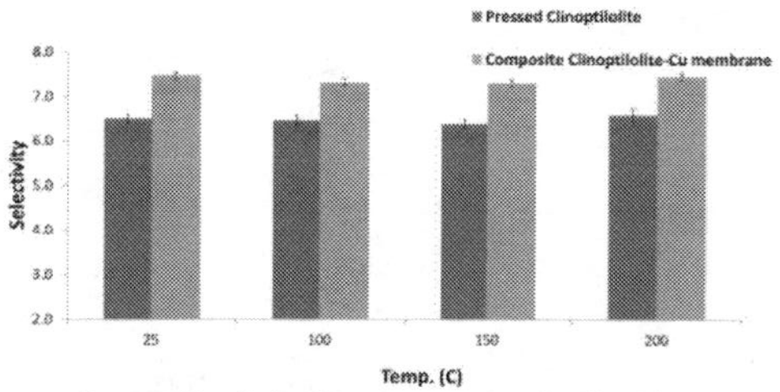

Fig-5: H2/CO2 selectivity at different temperatures and P_{feed}=111.16 kPa

The hydrogen permeance of the pressed and composite natural zeolite membranes with approximately 1.3 mm thickness used for these experiments had permeance values that can be compared to the permeation results reported in the literature

Membrane	Temperature (° C)	H_2 Permeability Mol/(m.s.Pa)	Ref.
MFI	450	1.1×10^{-12}	Gu et al.
Pressed Clino	**200**	$\mathbf{6.32 \times 10^{-10}}$	**This Work**
Molecular Sieve Silica	220	2.275×10^{-12}	Huang et al.
MOF (ZIF-8)	25	1.8×10^{-12}	Bux et al.

Conclusions

H_2 Separation selectivity through pressed clinoptilolite membranes were substantially influenced by applying copper and binder amongst the zeolites. By using a composite membrane more stability of the membrane over different temperature and pressures were observed. The higher selectivity for the composite membranes is attributed to the better connection of zeolite particles and filling up the potential gaps between the pressed particles.

Achnowlegment

- Nova Chemicals Co. ,Calgary, Canada
- University of Alberta, Canada

NOVA Chemicals

365938 Electrospray Assisted One-Step Synthesis of ZIF Membranes

Abstract:

Zeolitic imidazolate frameworks (ZIFs) are promising materials for advanced membrane applications due to their desirable properties, which enable them to recognize specific target molecules based on molecular sieving mechanism and/or adsorption affinity. However, their large scale commercial applications still remain challenge because of the absence of a synthesis method that offers reproducibility, scalability, and cost-effectiveness. To this end, here we report our work of synthesizing supported zeolitic imidazolate framework ZIF-7 films and membranes in a facile and time-efficient manner at ambient pressure using a simple electrospray deposition technique. A governing parameter, which dominantly affected the microstructures and crystal integrity of the membranes, was deposition temperature while precursor flow rate, and applied voltage to the precursor solution were important variables to obtain a stable spray mode which is essential to have a uniform coating. Our method, which is an electrostatic force-assisted coating approach, has the following attractive advantages over conventional synthetic routes such as *in situ* and secondary growth methods: (1) dramatic reduction in synthesis time and precursor consumption, (2) simplification in the activation process, and (3) potential scalability. Furthermore, the membranes obtained under optimized conditions outperformed previously reported ZIF-7 membranes for H_2/CO_2 separation while showing rather noticeable improvement in H_2 permeance which is roughly 4 to 10 times higher than the previous counterparts.

378601 Correlation of Plasticization Pressure Versus Permeability Parameters at Plasticization for Polymeric Membranes Used for CO2 Removal from Natural Gas

Abstract:

Semi-empirical model was developed to provide a relationship between the plasticization pressure and permeabilities of polymeric membranes used for high pressure CO_2 removal from natural gas. A theoretical model was first derived based on the solution – diffusion mechanism. The model derived was then used to perform a multiple regression using permeability and plasticization pressure data from literature. Fair agreement between the model predictions and the experimental values were realized by introducing two model parameters (κ_1 and κ_2). The values of these parameters are 0.20 and 0.60 respectively. The predicted values of permeability compare favorably well with the experimental values with absolute errors ranging from 0.30 to 35.66%. The coefficient of determination was obtained as 0.987.

381887 Sulfuric Acid Concentration By Vacuum Equipped Direct Contact Membrane Distillation

Abstract:

Production of every ton of titanic oxide (TiO_2) will generate around 8 tons of sulfuric acid (~20wt.%) with various inorganic salts (e.g. $FeSO_4$, $TiOSO_4$, $Al_2(SO_4)_3$). Recovery of the above sulfuric acid will lead to both environmental and economic benefits. In current work, vacuum equipped direct membrane distillation (VEDCMD) was used to concentrate the simulated sulfuric acid. The inorganic salt in the solution is $FeSO_4$. The process delivered purified water that can be recycled and iron vitriol crystals were formed during the cooling crystallization step. The experiments were carried out keeping the warm feed temperatures at 323 to 343, while cold solution 290 K, respectively. Effects of feed temperature, feed flow rate and distillate flow rate on the separation performance were investigated. Surface contamination and membrane wetting were studied by SEM, XPS, XRD etc. Chemical cleaning and the recovery of the membrane separation properties were also investigated.

SUSTAINABLE MEMBRANE PROCESSES AND PRODUCTS: ASSESSMENT

A.Irabien

Departamento de Ingenierías Química y Biomolecular Universidad de Cantabria. Avenida de los Castros s/n. 39005 Santander, Spain. E-mail: angel.irabien@unican.es

Membranes have been developed as an interphase for separation processes or as membrane products for different uses. Membrane technologies have been developed for the separation/concentration of components from a fluid phase or in order to integrate separation and reaction in the "so called" membrane reactors, but also membranes have been introduced as final products for example in construction, medical care, fuel-cells, etc.

In the nineties the word "sustainable" has been incorporated to the technical literature as an added value nickname to the technologies and we manage different "renewed unit operations", such as: sustainable distillation, sustainable extraction, sustainable absorption, etc. but the meaning of the sustainability qualification referred to a separation process, reactor or product remains unclear.

The first interpretation of sustainability in the membrane technology refers to the introduction of membrane processes for a sustainable industrial growth [1] as an innovative approach to the separation/concentration requirements in the process industries, the second interpretation for sustainable membrane technology is related to the sustainable use of natural resources: energy, water and raw materials [2].

Membrane products are related to different endusers and they may be certified according to standards as sustainable products following regulated procedures, for example in construction: a wide range of sustainable roofing membranes can be found in the market, they are suitable for flat, pitched or barrelled rooftops. Providing essential protection, these products guard structures against the dangers of water, root and climate damage, ensuring a long lifetime to the substrate.

With life expectancies of over thirty years, many of the products are fully recyclable and may be overlaid when replaced. Combined with a reliance on recycled and sustainable materials and an avoidance of harmful agents such as plasticisers, this market proven membranes offer superior protection with minimal environmental impact, and they are sold as sustainable membranes.

The presentation deals with the qualitative, semiquantitative and quantitative assessment of membrane processes and products taking into account the Supply Chain (SC) and the Life Cycle Assessment (LCA) respectively.

A set of environmental, economic and social indicators will be described in order to qualify membrane processes and products as sustainable, sustainability indicators allow a quantitative assessment of processes and/or products and need to be clarified in order to help decision making .

References

[1] Membrane processes for a sustainable industrial growth M G Buonomenna RSC Adv., 2013,3, 5694-5740.

[2] Ismail, Ahmad Fauzi / Matsuura, Takeshi (eds.) Sustainable Membrane Technology for Energy, Water, and Environment 1. Edition March 2012, 352 Pages, ISBN 978-1-118-02459-1 - John Wiley & Sons

Acknowledgements I thank to the Spanish Ministry of Economy and Competitiveness (MINECO) for funding the project ENE2010-14828.

Insights into Unsteady Shear Strategies for Fouling Mitigation in Highly Permeable New-Generation Membrane Applications – A review

Farhad Zamani[a], Jia Wei Chew[a,b], Ebrahim Akhondi[b], Anthony Gordon Fane[b]
[a]*School of Chemical and Biomedical Engineering, Nanyang Technological University, Singapore*
[b]*Singapore Membrane Technology Center, Nanyang Technological University, Singapore*

New technologies have enabled scientists to develop novel highly permeable membranes. Specifically, the permeability of RO membranes has been improved by orders of magnitude using embedded carbon nanotubes or aquaporins or replacing the selective layer by a graphene layer[1,2]. Such membranes herald the opportunity for significantly higher fluxes, leading to much lower costs and more compact plants with a reduced footprint.

Unfortunately, high fluxes are inevitably accompanied by worsened concentration polarization and fouling. The increased flux through the membrane needs to be coupled with a larger mass transfer coefficient (*k*) in order to counter the problem of more severe concentration polarization and fouling. For conventional module design with steady state hydrodynamics, namely cross-flow modules, *k* is typically enhanced by increasing the tangential flow. However, to keep the same concentration polarization ratio for a membrane offering permeability of merely an order of magnitude higher, we need six orders of magnitude more pump energy in these modules, thereby increasing cost considerably and negating the benefit of the high fluxes.

Fortunately, unsteady state shear stress to the membrane surface may hold the key to better means to enhance *k*, which is the prime determinant for the successful implementation of the new-generation membranes. Hence, a comprehensive review of feasible unsteady state strategies is warranted to provide insight into the flux enhancement that can be achieved in practice. This review considers a spectrum of unsteady state flow techniques that have been, or could be, applied to membranes, with the goal of assessing the physical limits to improvements in mass transfer. Unsteady shear techniques for fouling mitigation include air sparging, vibration of the membrane, fluidization of particles, pulsatile flows, and ultrasound waves. The hydrodynamic effects conferred by each of these techniques are evaluated, and the associated mass transfer enhancements are compared using a unified parameter, namely the enhancement factor (*E*). To address the practicality of the high fluxes conferred by the new-generation membranes, the energy demands of the techniques are assessed against the enhancement possible and recommendations are made regarding the most feasible unsteady state sheer method.

References

1. Cohen-Tanugi D, Grossman JC. Nanoporous graphene as a novel desalination membrane: Insights from molecular dynamics. *Abstr Pap Am Chem S.* Mar 25 2012;243.
2. Cohen-Tanugi D, Grossman JC. Water Desalination across Nanoporous Graphene. *Nano Lett.* Jul 2012;12(7):3602-3608.

ION EXCHANGE MEMBRANE BIOREACTORS FOR TREATMENT OF GROUNDWATER HIGHLY CONTAMINATED WITH PERCHLORATE

Shalom Fox, Alon Zelichower, Yoram Oren, Zeev Ronen, Jack Gilron

Zuckerberg Institute for Water Research, The Jacob Blaustein Institutes for Desert Research, Ben Gurion University of the Negev, Sede Boqer Campus, 84990, Israel.

Perchlorate contamination of ground water is a worldwide concern. The treatment of high levels of perchlorate in groundwater depends mostly on biological processes. In order to improve and increase the acceptance of this treatment there is a need to reduce the contact between the bacteria in the treatment unit and the final effluent. An Ion exchange membrane bioreactor (IEMB), in which treated water is isolated from the bioreactor was suggested for this purpose. In this study, the IEMB performance was studied at a concentration range of 1 to 250 mg L^{-1} which is two to three orders of magnitude greater than in previous studies. Two system designs were studied and compared: a CSTR and a PFR design, both systems being studied for perchlorate Donnan dialysis and for complete operation with bacteria in the bio-compartment. Glycerol was used as a low cost and nontoxic carbon source for the biodegradation of perchlorate in the bio-compartment. We found that high perchlorate concentrations in the feed render the anion exchange membrane significantly less permeable to perchlorate. When comparing both systems the PFR perchlorate specific flux was five times higher than for the CSTR for the same water compartment perchlorate concentrations, furthermore, the PFR designs allows us to attain high removal rates and efficiency while keeping the HRT two orders of magnitude lower than the required HRT for the CSTR. In addition, results suggested in both systems a minimal secondary contamination (< 3mg C L^{-1}) of the treated water with optimum feed of carbon substrate. Nitrates, which are also found in the contaminated ground water were not found to significantly affect the transport of the perchlorate across the ion exchange membrane and this is seen as being related to the relative affinity of each of the anions for the ion exchange membrane material.

Our results show clear advantages in using IEMB to remediate groundwater highly contaminated with perchlorates.

SYNTHESIS OF ION EXCHNAGE MEMBRANES AND THEIR APPLICATION TO MCDI

J. S. Kim, S. H. Kwon, K. Y. Kim, S. Y. Kang, J. W. Rhim

Department of Chemical Engineering, Hannam University, Daejeon, Korea (jwrhim@hnu.kr)

Capacitive deionization (CDI) which concept was first introduced by Caudle et. al.[1] is an electrochemically controlled method for removing salt from aqueous solutions by utilizing the principle of electrosorption consisting of adsorption and desorption of ions at the electric double layer of an electrode's surface. CDI uses carbon electrodes that have large surface areas to increase the adsorption capacity of the electrodes. However, because carbon electrodes have a porous structure, there can be a problematic decline in efficiency during operation of the process. To address this problem, membrane capacitive deionization (MCDI) has been recently introduced, which combines ion exchange membranes with the carbon electrodes.

In this presentation, the followings will be introduced. For ion exchange membranes, the aminated polysulfone membranes as the anion exchange membrane and the sulfonated polyphenylene oxide as the cation exchange membrane were synthesized, respectively, and their membrane characterization were carried out by using FT-IR, swelling degree, contact angles, ion exchange capacity, and ion conductivity. The coated carbon electrodes with the synthesized polymers were applied to MCDI experiments. The influent salt concentration of NaCl was varied from 100 ppm to 500 ppm, and the applied cell voltages was from 0.8 to 1.2, and the adsorption time/desorption time of 3/1, 5/1, 5/2 was used and also the feed flow rate was 15 and 23 ml/min. From this parameters, the salt removal efficiency, conductivity of the treated water as a function of cell potential, changes in current at different cell potentials, and the relative concentration of the effluent, etc., will be discussed in more detail.

References

1. D.D. Caudle, J.H. Tucker, J.L. Cooper, B.B. Arnold and A. Papastamataki, Electrochemical demineralization of water with carbon electrodes, Research Report, Oklahoma University Research Institute, 1966.

Effect of Ionic Structure on Performance of Ionic Liquids in Electrodialytic Product Recovery

Alexander M. Lopez[1], Jamie A. Hestekin[1,*]

[1]Ralph E. Martin Department of Chemical Engineering, University of Arkansas Fayetteville Fayetteville, AR 72701

* jhesteki@uark.edu

Organic acids are ubiquitous in the chemical, pharmaceutical, and plastics industries. Their production has come mainly from petroleum sources; however, production from fermentation pathways has gained great support due to socioeconomic pressures insistent upon sustainable, energy efficient synthesis methods. Fermentation does lead to desired products, yet the optimization of these processes are currently ongoing. Substantial research is needed in the development of efficient and cost effective separation methods capable of producing organic acids in high yields. This project presents an investigation of ionic liquid structure on separation performance of electrodialysis. Electrodialytic separations allows the selective separation of desired organic products while maintaining stable conditions in the fermentation systems; however, large amounts of water are often required and low product concentrations can lead to complicated separations with less than desirable production levels. Recently, the use of ionic liquids in electrodialytic separations demonstrated the possibility of performing organic acid separation and purification out of aqueous solutions. Unfortunately, separation efficiencies were very low (~10-30%), thus further optimization was needed. This research discusses how variations in ionic liquid structure through cation and anion replacement can affect product and solvent affinity, resulting in drastic changes in separation efficiency. Through these developments and advances in membrane science, a simple and cost-effective method for organic acid separation from wastewaters and fermentation broth was developed that allows for solvent recycle and high product recovery.

FACILITATED TRNSPORT SEPARATION OF CO_2 USING AMINATED POLYSULFONE MEMBRANES

S. H. Kwon[1], K. Y. Kim[1], J. S. Kim[2], S. Y. Kang[2], J. W. Rhim[1]

[1]*Department of Chemical Engineering, Hannam University, Daejeon, Korea*
[2]*Techwin Co. Ltd., Cheongju City, Choongcheongbuk-do, Korea*
(jwrhim@hnu.kr)

The first facilitated transport membranes(FTM) for gas separations were introduced by Ward and Robb[1] by filling the pores with a carrier solution resulted in 4,100 times more permeable to carbon dioxide than oxygen. Today, we call these membranes the supported liquid membranes(SLMs) which are known to have the serious problem, loss of carrier solution, usually accompanied by the loss of the chemical complexing agent. To overcome this fatal weakness of SLMs, LeBlanc[2] introduced the ions as mobile, membrane-bound carriers and the counter ions in ion-exchange membranes. In this type membranes, the ion-exchange membranes act as supports while ions in membranes do as complexing agents which are exchanged into the membrane and then held there by electrostatic forces. Also this electrostatic force between carrier and ion exchange membrane contributes to the prevention of the carrier loss to some extent, but this prevention is not perfect especially in the presence of other ions leading to the carrier loss. Therefore, to exclude the possibility of carrier loss, the direct inoculation of carriers into polymer membranes, so-called fixed site carrier(FSC) was introduced accompanying the use of ion exchange membranes as supports.

In this presentation, the polysulfone (PSf) was aminated by varying the ion exchange capacity, and the gas permeation characteristics of this aminated PSf (A-PSf) were measured by using time-lag method to look into which factor is more dominant between the two of the diffusivity and solubility. And then the gas permeation was measured under the various relative humidity conditions. It will be discussed in more detail.

References

1. Ward W. T., Robb W. L., Science, 156:1481-1484, 1967.
2. LeBlanc Jr. O.H., Ward W.J., Matson S.L., Kimura S.G., J. Membrane Sci., 6:339, 1980.

Mathematical description of water vapor influence on gas separation performance in hydrophobic polyimides

Luca Ansaloni[1], *Matteo Minelli*[1,2], *Marco Giacinti Baschetti*[1,2], *Giulio Cesare Sarti*[1,2]

[1]Dipartimento di Ingegneria Civile Chimica, Ambientale e dei Materiali (DICAM)
[2]Centro Interdipartimentale per Ricerca Industriale - Meccanica Avanzata e Materiali (CIRI-MAM)
Alma Mater Studiorum - Università di Bologna, Bologna, Italy

Water vapor is frequently present in membrane-based gas separation applications[1,2] and its influence must be taken into account for a reliable characterization of the separation process. In case of hydrophobic polymers, such as aromatic polyimides, the low water uptake observed in sorption experiments[3,4] does not correspond to a negligible effect on the membrane gas transport properties, which can be severely affected[5,6]. Although a small literature is dedicated to the description of the physical mechanisms of gas permeation in the presence of water, the implementation of a theoretical model in membrane processes simulations represents a great opportunity to improve the separation process design.

In the present study, the influence of the water vapor on the gas transport properties of several hydrophobic polyimides, characterized in our lab (6FDA-6FpDA, Matrimid, 6FDA-mHAB and 6FDA-pHAB), or obtained from literature (CARDO-ODA and NTDA-BAPHF) has been analyzed, in different experimental conditions and for several penetrants (CH_4, N_2, CO_2, H_2 and He).

Interestingly, very similar trends have been observed at increasing the relative humidity (R.H.) of the feed stream: the gas permeability decreased at increasing the water vapor concentration in the membrane, with different extent of variation for the considered polymers.

The monotonic decrease of permeability coefficient is explained in terms of competitive sorption; indeed, due to its higher condensability, H_2O is absorbed preferentially in the polymer matrix, reducing the amount of free volume available to the other gas to permeate across the membrane. The physical phenomenon can be explain from the mathematical point of view, with a simple but effective model based on the free volume theory[7]. Indeed, starting from the solubility of water vapor in the polymer, retrieve from sorption isotherms, the relative humidity of the feed stream is correlated with the water concentration in the membrane. The residual fraction free volume available for the less condensable penetrants to diffuse across the selective layer is then evaluated accounting for the presence of water molecules within the polymer. Finally, the gas permeability calculation is obtained by means of a simple approach based on the free volume theory[7], using only two adjustable parameters.

In addition, for those materials whose pressure-volume-temperature data are available, a more accurate degree of description could be achieved, considering the effect of the presence of water on solubility and diffusivity, separately. Solubility of different gases and water vapor in the polymer matrix can be predicted by means of the non-equilibrium thermodynamic model[8] (NELF, non equilibrium lattice fluid), whereas the free volume theory could be used in its original formulation to describe the variation of the diffusion coefficient at different water vapor concentration in the membrane.

In spite of its extraordinary theoretical simplicity, the model proposed is able to describe consistently and accurately the experimental data, providing a physical sound representation of the mechanism of gas transport in the presence of water, at all the temperatures and activities values inspected.

References:

1. Baker, *Ind. Eng. Chem. Res.* **41,** 1393–1411 (2002).
2. Scholes *et al., Energy Procedia* **4,** 681–687 (2011).
3. Minelli *et al.,Ind. Eng. Chem. Res.* **52,** 8936–8945 (2013).
4. Lokhandwala *et al., J. Polym. Sci. Part B Polym. Phys.* **33,** 965–975 (1995).
5. Piroux *et al., Desalination* **145,** 1–4 (2002).
6. Chen *et al., J. Memb. Sci.* **379,** 479–487 (2011).
7. Vrentas & Vrentas, *Macromolecules* **27,** 5570–5576 (1994).
8. Doghieri & Sarti, *Macromolecules* **29,** 7885–7896 (1996).

Functionalized Polyaniline Based Membranes for Water Purification

Yong Min [1,2*], Youhai Yu[1], Dewei Wang[1], Tianju Fan[2], Katie Kinstedt[3], Harold W. Walker[3], and Yidong Liu[2,4*]

[1] Xi'an Institute of Optics and Precision Mechanics of Chinese Academy of Science, Xi'an, China

[2]Key Laboratory for Organic Electronics &Information Displays(KLOEID), Institute of Advanced Materials(IAM), Nanjing University of Posts and Telecommunications(NUPT), 9 Wenyuan Road, Nanjing, 210046, China, E-mail: yong686@126.com

[3]Department of Environmental Engineering, The Ohio State University, Columbus, OH 43210, USA

[4] State Key Laboratory of Organic Electronics& Information Displays at Yan Cheng, YanCheng 224005, China, TEL:+86-18651590909, E-mail: yidong.liu@fangyuanyc.com

INTRODUCTION

Water scarcity and the health issues arising from poorly treated water necessitate the development of improved methods of water purification. One promising field to meet these pressing needs is membrane technologies, but widespread use is limited due to complications with membrane fouling [1,2]. Hydrophilic membranes have been shown to reduce fouling due to NOM species [2]. The emerging field of nanotechnology offers great promise to specifically tailor membranes to exhibit ideal properties for purification such as hydrophilicity. New super-hydrophilic membranes, with a water contact angle of less than 5°, membranes have been created with polyaniline coated-carbon nanofibers [3]. The super hydrophilic nature of these membranes is attributed to the surface roughness effect of the carbon nanofibers [4].

Polyaniline can self-assemble to form elongated 1-D nanostructures under certain synthesis conditions [6]. Promising applications of the polyaniline nanofibers explored have included nano/microelectronics, metallic corrosion protection, sensors, and actuators [6]. Using self-assembly techniques, membranes are created with polyaniline (PANI), creating a super-hydrophilic surface [3].

This work focuses on three main forms of PANI nanofiber membranes: a thermally-cured thin film, a composite membrane, a PANI nanofiber-coated membrane. For the composite membranes, a thin active layer, about twenty nanometers thick, is formed on top carbon nanofibers supported by another substrate. The hydrophilic properties of these membranes offer great promise in the field of water purification to significantly reduce fouling, and thus reduce the cost of operation [2, 6].

EXPERIMENTAL

The Polyaniline Nanofiber is synthesized from aniline, dopant acid, peroxydisulfate (APS) and ammonium hydroxide.The Composite Membranes is prepared by a dispersion of 0.50 wt% CNF, 0.5 wt% sodium dodecylsulfate (SDS) and the dispersant and the membranes can be coated with polyaniline nanofibers in the process outlined above. This process is seen in **Figure 1**. The formation of Thermally Cured Thin Film Membranes is that the nanofibers were applied on the coated substrate and thermally cured using a hot plate at 260°C for one minute. After the wafer reaches room temperature, the wafer is placed in water so the thin film can be removed from the substrate surface using a razor and a micro-spatula.

The measuring device of Pure water flux is shown in **Figure 2.** The vessel is the 10L vessel used in the fouling trials. The change in mass was then measured over time at a given pressure using an electronic balance. When small flow rates were observed, a graduated cylinder was used to measure the volumetric flow rate. The pure water flux was determined by using MilliQ deionized water (18.2MΩ). Fouling trials were run in the same manner described in the Pure Water Flux section, but with foulant water and a 10L vessel instead of a 4L vessel. The fouling of the membranes was determined by recording both the flow rate and the time of the reading to provide a measurement of the decrease in flux over time.

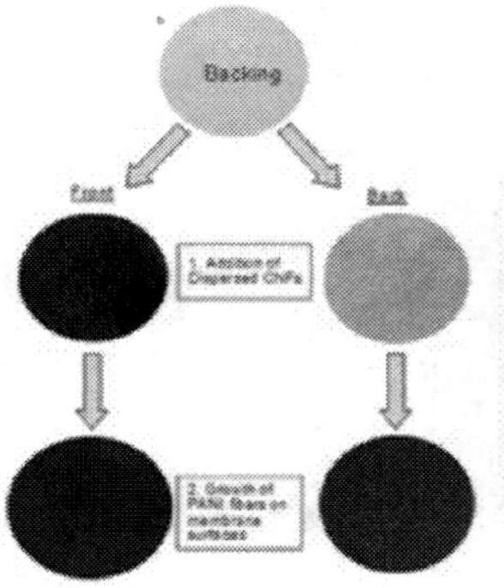

Figure 1. Formation of Composite Membranes

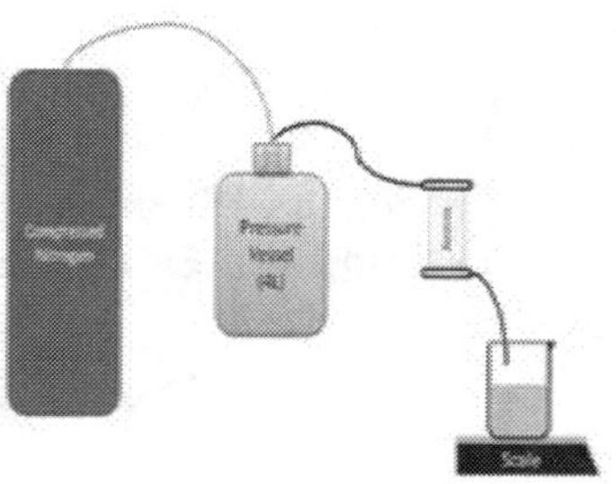

Figure 2. Experiment Set-up Schematic

DISCUSSION

Composite Membrane Results

Pure water flux results from different trials can be seen in **Table 1**.

It was found that one batch of membranes exhibited mechanical failure while testing due to the pressure differential. This failure indicates that the CNF thickness may play a key role in mechanical integrity of the membranes. The fouling was also measured for these membranes. This data is presented below in Figure 3. Due to the difficulties in obtaining consistent CNF deposition between batches, further trials to determine the impact of different parameters affecting the polyaniline nanofiber coating were determined using commercially available membranes.

Table 1. Summary of pure water flux data

Membrane	Pressure (psi)	Flow (g/min)	Std. Dev
Whatman #4	10	513.36	5.93
CNF #1	10	6.36	0.21
CNF #2	10	6.19	0.60
CNF & PANI #1	10	6.71	0.38
CNF & PANI #1	10	6.83	0.28
CNF & PANI #1	15	9.00	0.45
CNF & PANI #1	20	9.71	0.45
CNF & PANI #2	10	4.70	0.28
CNF & PANI #2	20	11.80	0.28

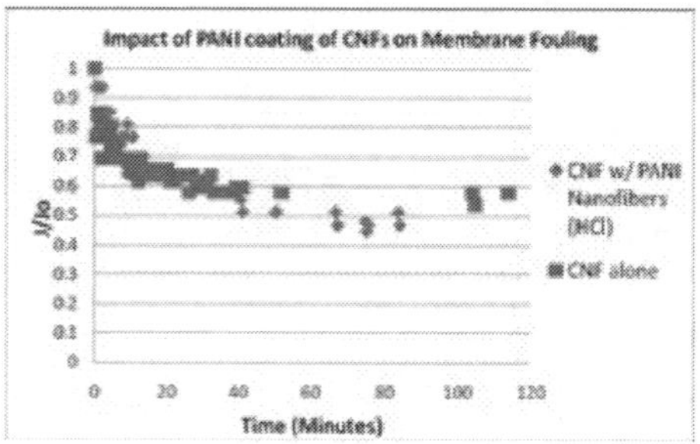

Figure 3. Coated vs. Uncoated Composite Membrane Fouling

SEM Results: SEM images of the three membranes can be seen below in Figures 4. The coating of the Anodisc ceramic membrane resulted in a significant reduction of flux, indicating a significant blockage of pores (seen also in the SEM). Thus, it was decided not to investigate the performance of the ceramic membranes in greater detail. In general, the SEM images show there are only small changes on the morphology of Cellulose Ester membrane after comparing the SEM pictures of before polyaniline coating and after polyaniline coating. However, this is significant morphology changes on the Anodisc Ceramic Membrane.

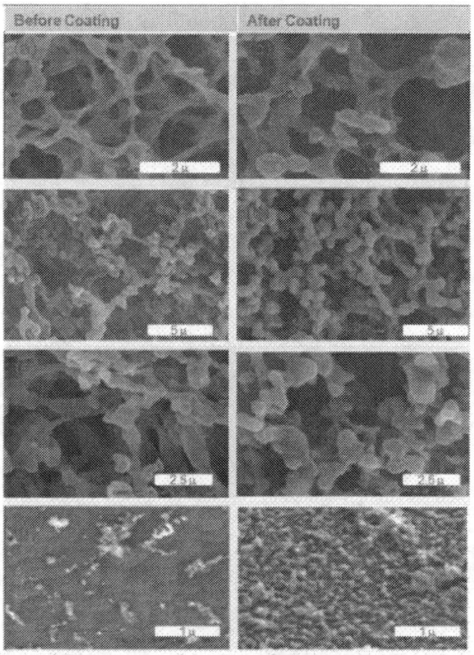

Figure 4. (a-b) SEM Image of 0.45μm Cellulose Ester Membrane (a) before and (b) after polyaniline coating; (c-f) SEM Image of 1.2μm Cellulose Ester membrane (c, e) before and (d, f) after polyaniline coating, (g-h): SEM Image of 0.02μm Anodisc Ceramic Membrane (g) before and (h) after polyaniline coating.

Fouling Trials with Humic Acid. In addition to trials with Olentangy River water, fouling trials were carried out using Suwannee River Humic Acid (SRHA) reference. Three dopants were evaluated in this experimentation, hydrochloric acid (HCl), PSA Poly (4-styrenesulfonic acid) solution (PSA), and 4-Dodecylbenzene sulfonic acid (DBSA) to compare the impact of different PANI nanofiber coatings on membrane performance. The membrane evaluated for these coatings was a 0.45μm Polyethersulfone (PES) membrane. In addition to testing the impact of three dopants, the impact of the doped state, as well as the pH of the foulant solution was investigated. The pure water flux for the 7 membrane types were determined and are listed below in **Table 2**.

Table 2. Impact of Doped State on Pure Water Flux of Membranes

	Flow (mL/min)		
	Dedoped	Doped	Change/%
PES 0.45μ with PANI (HCl)	26.02 ± 0.90	36.95 ± 2.25	-29.6%
PES 0.45μ with PANI (PSA)	11.55 ± 0.18	10.43 ± 0.06	8.7%
PES 0.45μ with PANI (DBSA)	101.46 ± 3.11	26.02 ± 0.90	139.0%
PES 0.45μ Reference	97.13 ± 2.57		

The impact of the pH was determined by the addition of sodium bicarbonate to raise the pH of the fouling solution to approximately 8. The impact of the two pH values on the reference PES 0.45 μm membranes is seen below in Figure 5c. While increasing the pH was expected to decrease the fouling, no significant trend was observed. This is perhaps due to the fact that a lower concentration of SRHA, only 5 milligrams per liter, was used. The impact of the change in pH on the PES 0.45μm membranes coated with PANI (HCl dopant) was also investigated. A comparison of fouling at pH values of 6 and 8.6 can be seen in Figure 5d. Increasing the pH of the PANI coated membranes made no difference in the fouling during the first five minutes and only slightly improved fouling after five minutes. Again, a greater impact of pH on the fouling of the PANI coated PES membranes may be observed with a higher concentration of SRHA.

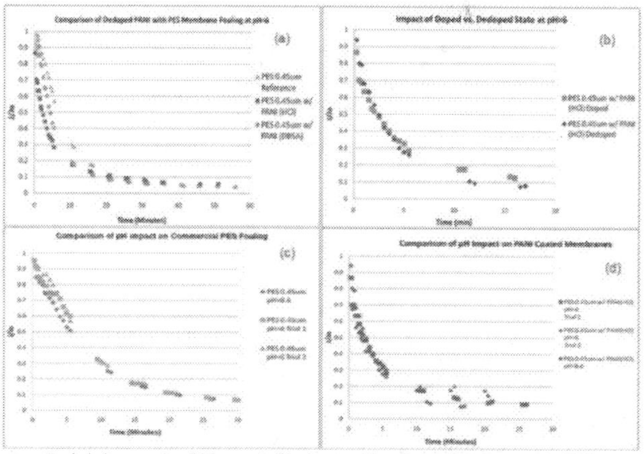

Figure 5. (a) Impact of Doped State on Fouling; (b) Impact of Doped State on Fouling; (c) Impact of pH on Fouling of PES Membranes, (d) Impact of pH on Fouling of PANI Coated Membranes.

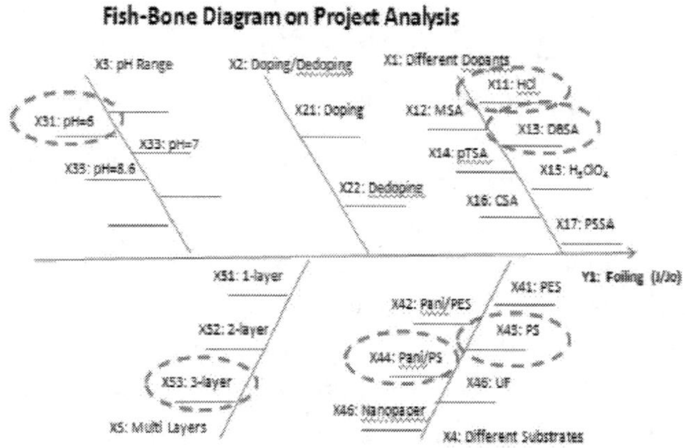

Figure 6: The summary of "Fish-Bone Diagram" on the impacts for foiling.

CONCLUSION

We summarized all the factors, which may impact to the foiling (J/Jo) results as a "Fish-Bone Diagram" as listed in Figure 6, where the "red cycle" areas indicated there is some impact on the foiling results. However, we must realize that the quality of laboratory film is not uniform enough for the conclusive results. Scale up experiment is ongoing to father investigate the results. In addition, the in-situ polymerized polyaniline coating may not have enough adhesion to the membranes substrate, which may wash out by the high pressure water during the test. Polymer solution coated polyaniline could be a better candidate to continue the experiments.

REFERENCES

1. Christian, B.; Hsiao, B.S.; Chu, B. *Annu. Rev. Mater. Res.* **2006,** *36* , 333-368..
2. Hong, L., Elimelech, M. *J. Membr. Sci.* **1997,** *132*, 159-81.
3. Chiou, N.; Lu, C.; Guan, J.; Lee, L.J.; Epstein, A.J. *Nature Nanotechnology* **2007**, *2*, 354-357..
4. Chiou, N. *Thesis*. **2006**.
5. Chiou, N.; Lee, L.J.; Epstein, A.J. Chem. Mat. **2007**, *19*, 3589-591..
6. Fan, Z.; Wang, Z.; Duan, M. *J. Membr. Sci.* **2008**, *310*, 402-08.

Separation Characteristics Using Multi-layer Hollow Fiber Nanofiltration Membranes for Ultra-low Pressure

S. Y. Kang, J. S. Kim, S. H. Kwon, K. Y. Kim, Ji Won Rhim[1]

[1]*Department of Chemical Engineering, Hannam University, Daejeon, Korea*
(jwrhim@hnu.kr)

Abstract

Environmental issues and shortages in energy resources due to abrupt urbanization, population increases, and weather changes have led to increased interest about the environment specifically in economic, social, and ecological areas in the water industry caused by the unbalance between demand and supply. Various researches for the development of separation membranes and related materials, which are key materials in the water industry, suggest that multi-layer membrane technology is the most preferable method.

In this study, polysulfone, which is a supporting material for nanofiltration membranes and possessing excellent physiochemical characteristics, was selected as subject. A composite membrane was prepared through interfacial polymerization in the hollow fiber membrane. External pressure type filtration was evaluated by carrying out tests while changing the concentration of the coating solution and drying temperature. The process was intended to improve flux and rejection using test results. The characteristics of PSf hollow fiber membrane and coating qualification were examined by SEM photos. The porous surface and spongy structure of the PSf hollow membrane were, thus, confirmed. Also, it was found that pores on the surface were all blocked and coating was formed.

In PART 1 of the test, a flux test was performed by preparing a composite membrane through the interfacial polymerization between PVAm and aqueous phase PEI having functional group –NH, and organic phase TMC having functional group –COCl. The test was carried out by mixing PVAm and PEI macromolecules to obtain adequate flux and to improve rejection. The highest flux 38LMH and salt rejection 70% for 100ppm NaCl could be achieved when 10,000ppm PVAm and 4,000ppm PEI were mixed.

In PART 2 of the test, interfacial polymerization was tried by replacing PEI and having functional group –NH used in PART 2 with aqueous phase piperazine. As the concentration of PIP was increased to 1-2wt.%, flux was decreased from 37LMH to 23LMH. Meanwhile, rejection was increased until 55-70%, further increasing to a maximum of 84% when it was coated with 2 wt.% of PIP. This might be due to the enhancement of reaction while the aqueous functional group –NH was increased, enabling interfacial polymerization with TMC.

References

[1] R. J. Peterson, "composite reverse osmosis and nanofiltration membrane", J. Membr. Sci. 83 (1993) 81-150

[2] P. Erlksson, "Nanofiltration extends the range of membrane filtration", Environmental progress 17 (1) (1998) 58-62

Gas separation of PEBAX blended membranes with poly(ethylene glycol) and glutaraldehyde

G. Y. Kim, S. Y. Kang, S. H. Kwon , J. S. Kim, J. W. Rhim
[1]*Department of Chemical Engineering, Hannam University, Daejeon, Korea*
(jwrhim@hnu.kr)

Combustion of sulfur containing fuels, such as coal, in power plant is the major emission source of sulfur dioxide(SO_2). SO_2 emitted into atmosphere is the main cause of acid rain which is one of the most pervasive air pollutants. In addition, the emission and concentration of SO_2 in the emitted gas tend to be increased due to the low-quality coal use. In case where the emitted SO_2 is incoming to CO_2 capture process, such as absorption process, the deterioration in process performance by the furious flames of absorbents and the contamination in the captured CO_2 would take place. Therefore, the inflow of SO_2 to the absorption process is strictly restricted and, in general, must not exceed 10 ppm.

In this study, PEBA 1657, which is known to have high bonding strength to acid gases, was used blended with polyethylene glycol (PEG) to enhance permeability and again with glutaraldehyde (GA) to improve permeation selectivity and durability. Poly(ether-block-amide) 1657 (PEBAX 1657) blended membranes with molecular weight 400 polyethylene glycol (PEG 400) were prepared and their permeability tested for the gases N_2, O_2, CH_4, CO_2, and SO_2 by the time-lag method. The permeation characteristics were investigated in terms of diffusivity and solubility, which are dominant factors for gas transport. With the addition of PEG 400, the permeability of all the gases increased and also the ideal selectivity for several pair gases was enhanced. In particular, selectivity for CO_2/N_2 ranged from 53.2 (pristine PEBAX 1657 membrane) to 84.1 (50% PEG 400 added), for SO_2/CO_2 from 38.9 to 50.7, and for CO_2/CH_4 from 17.7 to 31.4. The increase of both permeability and selectivity is mainly because of the increase of solubility of the gases, especially CO_2 and SO_2. To obtain durability against water vapor, glutaraldehyde (GA) was added to the PEBAX 1657/PEG 400 blended membranes. As a result, permeability decreased owing to a reduction of the free volume and ether oxide units, which are the main factors in elevating the permeability for the blended membranes, and also selectivity decreased; however, we believe that the durability of the resulting membranes would increase. The details will be discussed at the presentation.

374116 Permeation and Separation Behaviors of Simulated Flue Gases on Polymer and Inorganic Membranes at Lab Scale Equipments

Abstract:

In order to apply membranes to the real CCS processes, permeation and separation characteristics of each membrane should be evaluated with appropriate working conditions and rigorous regulations. 0.2 Nm^3/hr and 1.0 Nm^3/hr scale test equipments and 0.2 Nm^3/hr multi stage modules were built at KIER combustion-gas processing complex laboratory building. Those test equipments can control temperature, pressure, flow rate, steam amount, gas composition, stage cut, and also real flue gas from 10MW_{th} scale CFB(circulating fluidized bed) boiler can be supplied. In this study, hollow fiber type and tubular type membranes were used for establishing standardization of membrane processes for CCS and certificating membrane performance such as permeance, selectivity, chemical-physical resistance and long term operability.

374988 CO_2-Selective Membranes Containing Amino or Ethylene Oxide Groups Potentially for Carbon Capture from Flue Gas

Abstract:

This poster presentation covers two types of CO_2-selective membranes for carbon capture from flue gas: (1) amine-based membranes and (2) ethylene oxide (EO)-based membranes. The former is based on the facilitated transport mechanism, while the latter is based on the solution-diffusion transport mechanism. For the membranes containing amino groups, a high-molecular-weight of polyamine was successfully synthesized. Different selected aminoacid salts were incorporated into the polyamine solution for membrane preparation. The polyamine served as a fixed carrier, and the amino acid salts acted as mobile carriers. The amine-containing membranes showed a CO_2 permeance of more than 1100 GPU and a CO_2/N_2 selectivity of at least 200 at 57°C. For the membranes containing EO groups, we have synthesized membranes with polyamide-polyethylene oxide (PA-PEO) copolymer blended with polyethylene glycol dimethyl ether (PEG-DME) or polyethylene glycol (PEG). We have also synthesized a polydimethylsiloxane (PDMS) membrane reproducibly with a very high CO_2 permeance, which can be used as a caulking layer for our thin membranes. The membrane containing EO groups we synthesized was SO_2-insensitive but had a low CO_2/N_2 selectivity. The resulting membranes showed a CO_2 permeance of 940 GPU and a CO_2/N_2 selectivity of 25 at 57°C. The syntheses and transport properties of the membranes will be presented. The effects of different aminoacid salt mobile carriers, EO-containing small molecules and membrane thicknesses on the membrane transport performance will be discussed. We are continuing to improve the transport performance of both types of CO_2-selective membranes.

Supercritical fluid extraction and reaction to modify metabolites from biomass

Universidad Nacional de Colombia sede Manizales

Idárraga, A.M.; Dávila, J.A, Cardona, C.A.+

Biotechnology and Agroindustry Institute, Chemical Engineering Department, Universidad Nacional de Colombia sede Manizales.

+Corresponding Autor. Tel.: +57 6 8879300x55880 E-Mail: ccardonaal@unal.edu.co

14AIChE Annual Meeting, Atlanta, GA

Introduction

Currently, there is growing interest in metabolites produced by the plant kingdom, Thanks to these features are of interest to the cosmetic, food, chemical and pharmaceutical industry. (Miron, Plaza, Bahrim, Ibáñez, & Herrero, 2011) Some features that can provide a benefit to industry are related with the antioxidant capability to generate more stable foods with longer shelf life. The health sector, including the pharmaceutical and cosmetics industries have been increased the interest in the production of molecules with the ability to prevent disease or delaying the onset of these. Additionally, the trend of environmental protection, creates the necessity to use "Green Technologies". It allows to achieve less impacts than conventional technologies.(Cerón, Higuita, & Cardona, 2012)

Figure 1. Avocado

Natural raw materials are composed by a lot of interesting metabolites, among which can be founded the anthocyanins, flavonoids and polyphenols, which can be extracted using various technologies. In this work were performed experimentally the extraction of compounds with antioxidant capacity using supercritical fluid from *Persea americana* (avocado). The inhibition percent was performed through free radical DPPH (1, 1-diphenyl-2-picrylhydrazyl) assay. (Brand-Williams, Cuvelier, & Berset, 1995; Cerón et al., 2012)

Figure 2. Colombian Avocado production

Methodology

Physicochemical composition data was obtained from avocado fruit. The value added products after supercritical extraction processes and enzymatic hydrolysis were performed also experimentally.

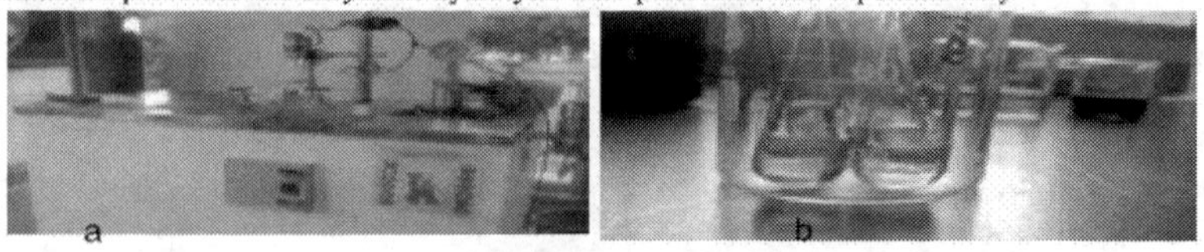

Figure 3. Experimental work

Table 1. Avocado Chemical composition

Compound	Content (wt.%) Pulp	Seed	Peel
Water	70.03	55.87	71.76
Fatty Acids	22.21	2.16	0.43
Glucose	0.49	3.31	1.92
Cellulose	2.67	18.18	10.58
Hemicellulose	1.70	11.57	6.73
Lignin	1.56	5.08	1.14
Protein	0.62	1.94	7.00
Ash	0.72	1.86	0.44
Procyanidins	0.00	0.01	0.00

Using process engineerring strategiesTwo different products were proposed to obtain from Avocado Seed, ethanol and PolyHydroxyButyrate (PHB). The Avocado seed characterization was used to calculated the conditions of each one of the processes and finally could be found the price of production of PHB and Ethanol.

The simulations are performed in the Aspen Plus software v8.0, then the results in the form of material and energy balances are used in the WAR algorithm (EPA), in order to find the environmental impacts.

Enzymatic Hydrolysis Results

Different pretreatments were made with aim to compare the results with respect to fermentable sugars content and compared with those obtained without treatment. The antioxidant potential of the extract obtained by supercritical fluid extraction was 62.55 g DPPH / gAvocado seed

Pretreatment	Ratio of Reduced Sugars (RS)
Without Pretreatment	0.02 (gRS/gsample)
SFE	0.09 (gRS/gsample)
Acid Hydrolysis	0.17(gRS/gsample)

Techno-economic and Environmental Analysis

The production of Ethanol and PHB was simulated and economically evaluated using the commercial software Aspen Plus V8.0, the cost production results are consigned on Table 2, this result evidence that the Ethanol production doesn't compete with market prices, while the PHB production shows a better price than the commercial but the profit margin is lower .

Table 2. Product Cost Results

Product	COST (US/Kg)
Ethanol	13, 21
PHB	2,91

The PHB production revealed a better production price than Ethanol production. On figure is showed the economic distribution on PHB production.

Economic Evaluation

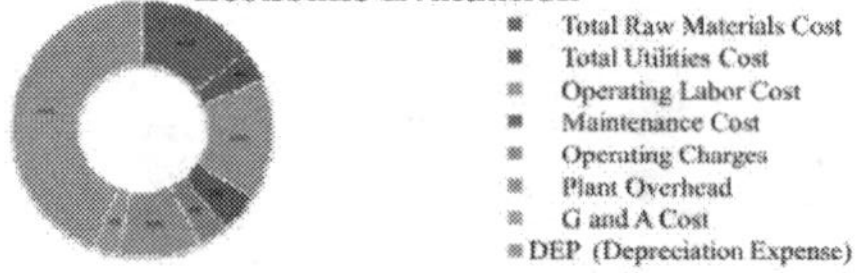

Figure 2. Economical evaluation

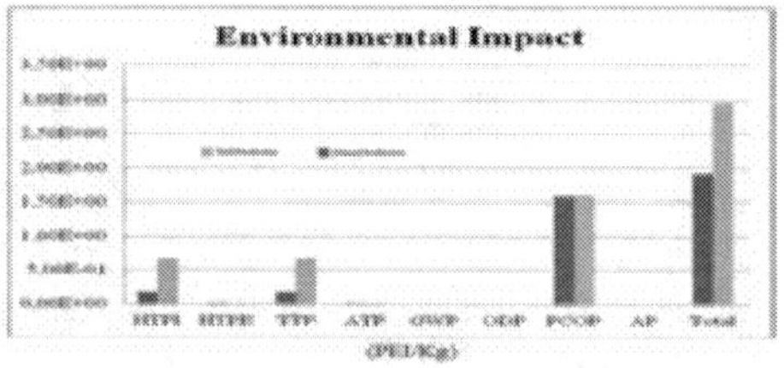

Figure 3. Potential Enviromental Impact

Indicators: Human toxicity by ingestion (HTPI), human toxicity by dermal exposition or inhalation (HTPE), terrestrial toxicity potential (TTP), aquatic toxicity potential (ATP), Global warming (GWP), Ozone depletion potential (ODP), Photochemical oxidation potential (PCOP) and acidification Potential (AP)

The impact factors was higher on PHB production, the reason is linked to this process solid reagents that impact soil toxicity when ingested and used. This is a process where more reagents for the purification thereof involved in contrast ethanol production

Conclusions

Enzymatic hydrolysis after different pretreatments were performed, it was found that the carbon sources are more affordable after supercritical extraction fluids compared with any pretreatment. Due to the acidified co-solvent effect. The fermentation process most profitable was the PHB production.

References:I. Amendola, D., De Faveri, D. M., Egües, I., Serrano, L., Labidi, J., & Spigno, G. (2012). *Bioresource Technology, 107*(2012), **II.** Cerón, I.X; Higuita, J.C, Cardona C. A. (2012) Journal of Supercritical Fluids 62 96-101 **III.** Cissé, M., Bohuon, P., Sambe, F., Kane, C., Sakho, M., Dormier, M. (2011) *Journal of food Engineering* 109, 16-21 **IV.** Garcia-Perez, J. V., García-Alvarado, M. a., Carcel, J. a., & Mulet, a. (2010) *Journal of Food Engineering, 101*(1), 49–58 **V.** Moncada, J; EL-HALWAGI , M; Cardona, C.A. (2013) Bioresource Technology. **V.** Ping, L., Brosse, N., Sannigrahi, P., & Ragauskas, A. (2011). *Industrial Crops and Products, 33*(1), 200–204. **VII.** Prozil, S. O., Evtuguin, D. V., & Lopes, L. P. C. (2012). *Industrial Crops and Products, 35*(1), 178–184 **VIII.** Spigno, G., Maggi, L., Amendola, D., Dragoni, M., & De Faveri, D. M. (2013). *Industrial Crops and Products, 46*, 283–289. **IX.** Zhang,W., Seki, M., Furusaki, S (1998) Journal of Fermentation

395304 Development of Anti-Fouling Membranes By Chemical Patterning

Abstract:

Over 1 billion people lack access to clean drinking water. Membranes are a tool that can help provide clean water to these people. However, treatment of impaired waters for beneficial use exposes the membranes to feed waters containing biological and abiotic species, which leads to fouling and loss of membrane productivity over time. Since reduction in flux due to fouling is one of the largest costs associated with membrane processes in water treatment, new coatings that limit fouling would have significant economic and societal impacts. Developing these advanced coatings is the focus of our work.

Prior studies in this area largely have focused on chemical modifications to the membrane surface that either prevent microorganisms from attaching or that kill the microorganisms once they do attach. Based on considerable literature and experience, we know that chemical modification strategies alone can be effective but not sufficient for controlling biofouling. A more recent area of research is nano-patterning the membrane surface. Nature provides many examples where structured features can prevent accumulation of 'foulant' materials on the surface (i.e., shark skin, lotus leaves, etc.). Our hypothesis is that combining these two methods (chemical coating and patterning) will yield membrane surfaces that are more effective at biofouling control than either method alone. This strategy is unique in its use of both chemical coating and patterning to combat biofouling.

In this contribution, we will introduce the methodology used to pattern membrane surfaces with the anti-fouling coating. We will demonstrate the feasibility of this methodology through AFM, SEM, and attenuated total reflectance (ATR)-FTIR measurements. Subsequently, we will test our hypothesis in bacterial deposition and membrane biofouling experiments using appropriate model systems.

An extraction process for fluorinated compounds from water

Wolfgang Arlt, Detlef Freitag, Liudmila Mokrushina, Martin Drescher
Institute of separation science and technology
Friedrich-Alexander-Universität Erlangen-Nürnberg, Germany

Keywords: PFOA, PFOS, liquid extraction, airport

Fluorinated anionic surfactants have drawn considerable attention because of significant concentrations in surface and ground waters, among them perfluorooctanoic acid (PFOA) and perfluorooctane sulfonate (PFOS). PFOS and PFOA are extremely persistent in the environment, cannot be treated in a water treatment plant and accumulate in fat 100 times more than DDT. Both are used as surface-active agents in various high-temperature applications, among them in fire fighting foams. Thus, more or less all airports fight the problem of cleaning the surface and ground water from these compounds. The present procedure includes the adsorption on suited adsorbents, among them active carbon. Because of the adsorption isotherm, the total amount of contaminated material is not significantly reduced by adsorption. Further, the loaded adsorbent cannot easily be recycled.

The authors developed an extraction scheme that delivers the fluorinated anionic surfactants in a pure form. In this way, the amount of contaminated material is drastically reduced and thus the cost, because there are no recovery expenses. The lecture will highlight the process details and show some results with water from Nuremberg airport.

382972 Investigation of Ionic Liquid Systems for Extraction of Benzoic Acid from Hydrocarbons

Abstract:

Ionic liquids are regarded as green solvents and therefore their application in the field of separation processes such as liquid–liquid extraction is increasing substantially. In petroleum industry there is a need of separating the aromatic compounds from the crude oil. Currently, most of the oil producing countries are beginning to produce heavy crude oil with high contents of carboxylic acids. The presence of benzoic acid in crude oil could cause severe corrosion to refineries processing units especially those operating at temperature above 230ºC. A new class of solvents, namely the ionic liquids, have recently shown promising application for reducing the acid content in crude oil. This study will investigate the ability of ionic liquids to extract aromatic compounds like benzoic acid from hydrocarbons such as dodecane, cyclohexane and hexane. The selectivity and capacity of ionic liquids for extracting benzoic acid can be determined using COSMO*therm* simulation software. For calculating various parameters, this software takes into account the interactions of different molecules surface charges. It can also be used as a tool for the determination of thermo-physical properties of ionic liquids. In this study the effect of solvent, anions and cations of ionic liquids, and the length of carbon chain on the selectivity and capacity are examined. The results indicate that imidazolium based ionic liquids are better than phosphonium based ionic liquids for the extraction of benzoic acid. It was also observed that with the increase in the length of carbon chain the selectivity decreases where as the capacity increases.

384896 Extraction of Furfural from It's Aqueous Solution with a Novel Solvent

Abstract:

Furfural is a platform chemical produced from pentoses. It is utilized in the chemical industry for synthesizing solvents, adhesives, medicines, and plastics. The main sources of pentosans are corn cob, bagasse, wheat and rice straw. Hardwoods can be used as alternative source for furfural production as they are rich in pentosans and prehydrolysis of birch allows the separation of almost 9% of xylan on wood.

Furfural is produced by acid catalyzed dehydration of pentoses which are released during the hydrolysis of pentose-rich lignocellulosic substrates such as birch wood. When the dehydration reaction is carried out in purely aqueous solution, the formed furfural is present in a rather low concentration. In this case, the liquid-liquid extraction (LLE) is an attractive way for the selective concentration of furfural. In the literature many solvents, such as supercritical CO2, methyl-isobutyl ketone (MIBK), tetrahydrofuran (THF) and others have been studied for the extraction of furfural from its aqueous solutions.

In this study, a novel solvent has been applied for the selective extraction of furfural from a highly diluted aqueous solution. The effect of the initial furfural concentration and the presence of organic acids, such as formic and acetic, on the extraction efficiency was investigated. The liquid-liquid equilibrium for ternary system furfural-water-organic solvent has been measured. In addition, vapor-liquid equilibrium and excess enthalpy for the binary system furfural-organic solvent has been measured. The results obtained in this research will lead to a development of novel method for furfural extraction from its diluted aqueous mixture.

Extractive recovery of low molecular weight fatty acids from aqueous effluents

D. Painer, S. Lux, M. Siebenhofer

Outline

- Introduction
- Liquid-liquid-equilibrium
  - Single acid extraction
  - Binary acid extraction
- Vapor-liquid-equilibrium
- Reactive distillation
  - Solvent recovery via reactive distillation
- Conclusion

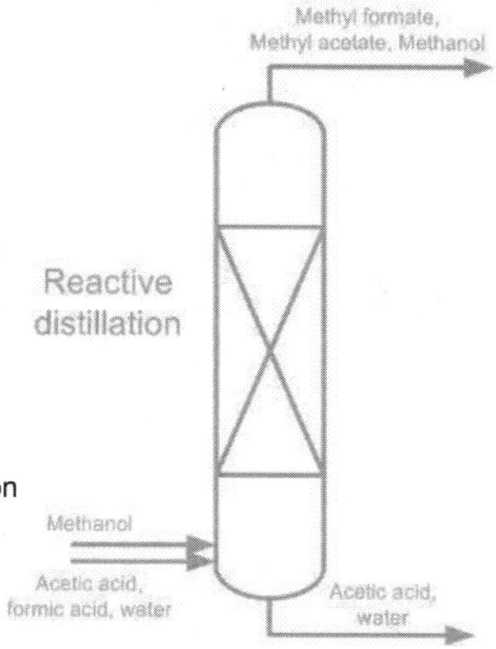

Extractive recovery of low molecular weight fatty acids from aqueous effluents

TU Graz

Motivation

- Biomass as feedstock material
- Target: total utilization of biomass

- Multicomponent mixtures
- Low concentration of valuable products
- Thermodynamic limitations - azeotropes

3 TU Graz | **Institute of Chemical Engineering and Environmental Technology** Atlanta, 19/11/2014 CEET

Extractive recovery of low molecular weight fatty acids from aqueous effluents

TU Graz

Extraction of carboxylic acids

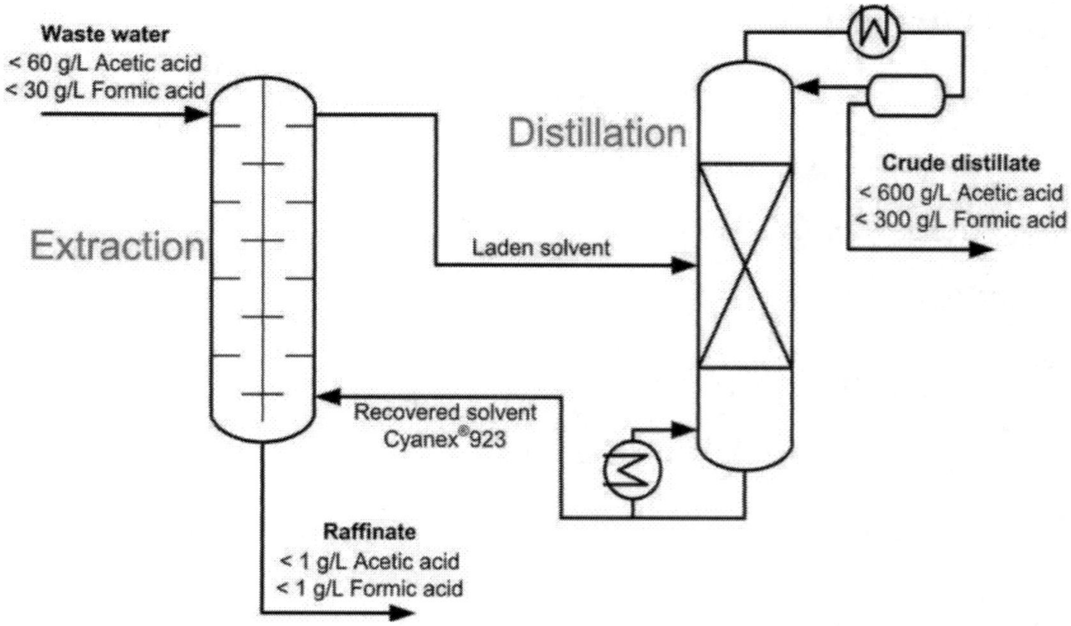

4 TU Graz | **Institute of Chemical Engineering and Environmental Technology** Atlanta, 19/11/2014 CEET

Solvent

- 70 wt.% Cyanex®923
- 30 wt.% Undecane

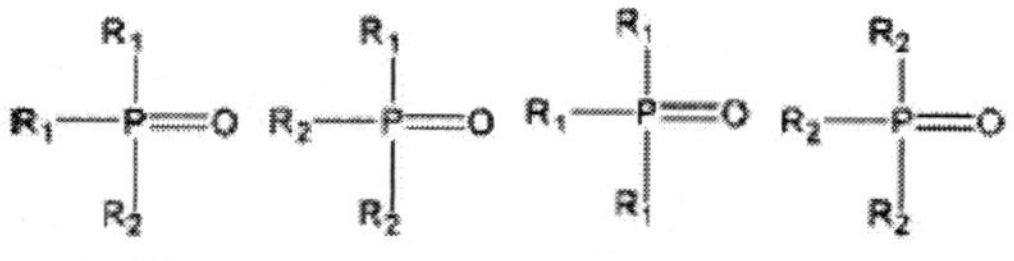

30.4% 37.4% 16.1% 8.5%

where: $R_1 - C_6H_{13}$, $R_2 - C_8H_{17}$

Wisniewski, M.; Pierzchalska, M.; J Chem Technol Biotechnol, 2006

$$m \cdot \mathrm{HAc} + \mathrm{S} \leftrightarrow \mathrm{HAc_m} \cdot \mathrm{S} \qquad K_{\mathrm{HAc}} = \frac{[\mathrm{HAc_m}\cdot\mathrm{S}]_{\mathrm{org}}}{[\mathrm{HAc}]_{\mathrm{aq}}^{\mathrm{m}}\cdot[\mathrm{S}]_{\mathrm{org}}}$$

$$n \cdot \mathrm{HFo} + \mathrm{S} \leftrightarrow \mathrm{HFo_n} \cdot \mathrm{S} \qquad K_{\mathrm{HFo}} = \frac{[\mathrm{HFo_n}\cdot\mathrm{S}]_{\mathrm{org}}}{[\mathrm{HFo}]_{\mathrm{aq}}^{\mathrm{n}}\cdot[\mathrm{S}]_{\mathrm{org}}}$$

Liquid-liquid-equilibrium T = 25 °C

$$a_{\mathrm{A,org}} = \frac{n \cdot K_{\mathrm{A}} \cdot a_{\mathrm{A,aq}}^{\mathrm{n}} \cdot a_{\mathrm{S,org}}}{1 + K_{\mathrm{A}} \cdot a_{\mathrm{A,aq}}^{\mathrm{n}} + K_{\mathrm{B}} \cdot a_{\mathrm{B,aq}}^{\mathrm{m}}}$$

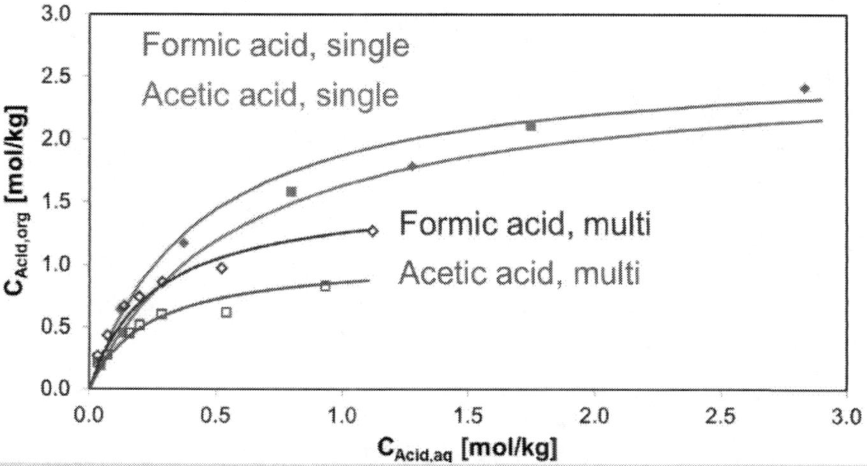

Boiling point surface

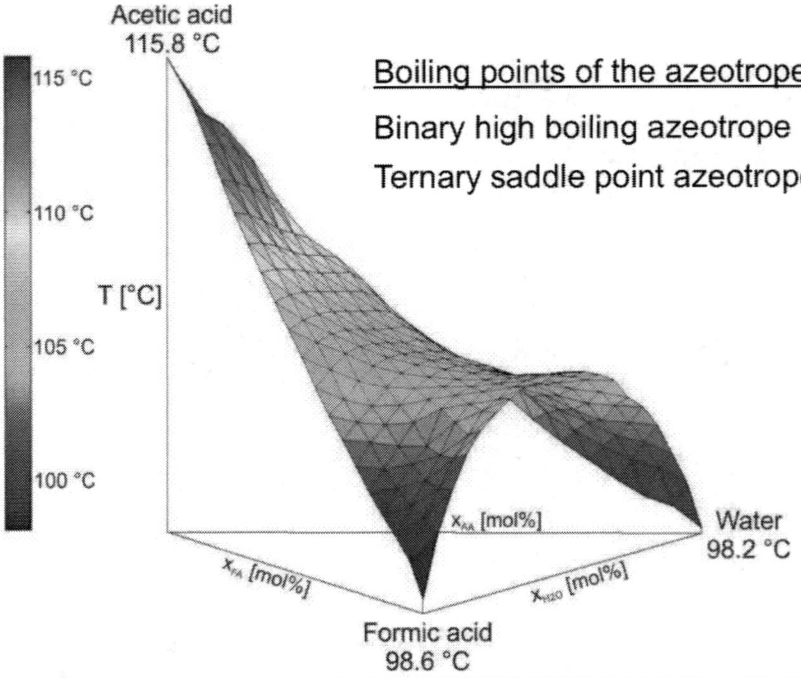

<u>Boiling points of the azeotropes at 950 mbar:</u>

Binary high boiling azeotrope	105.6 °C
Ternary saddle point azeotrope	105.2 °C

Extractive recovery of low molecular weight fatty acids from aqueous effluents

Boiling point surface

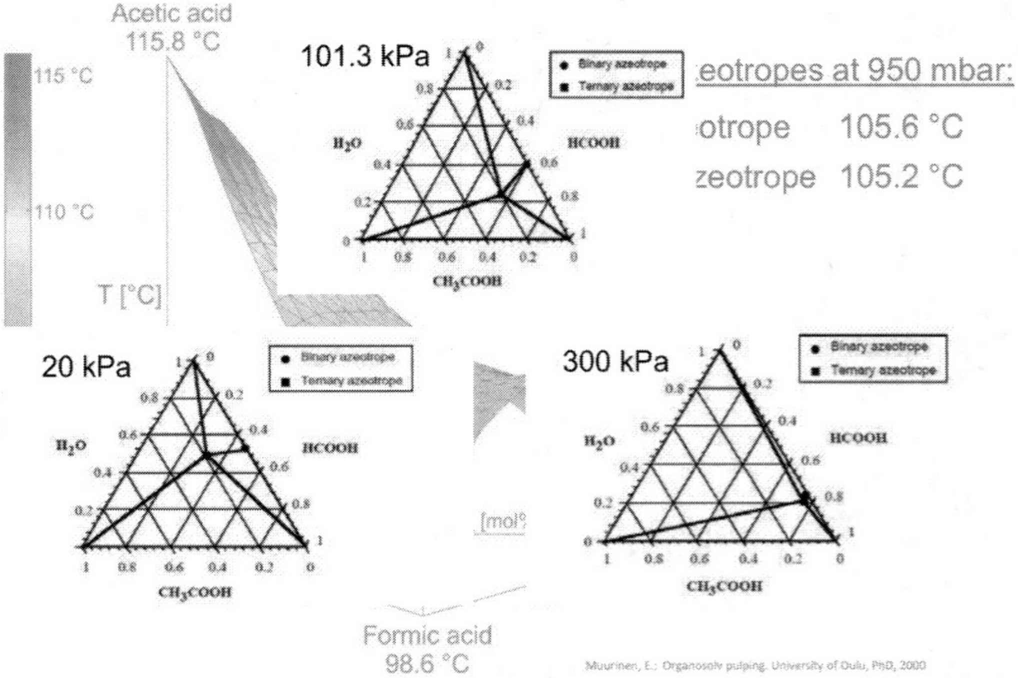

Muurinen, E.: Organosolv pulping. University of Oulu, PhD, 2000

Vapor-Liquid-Equilibrium

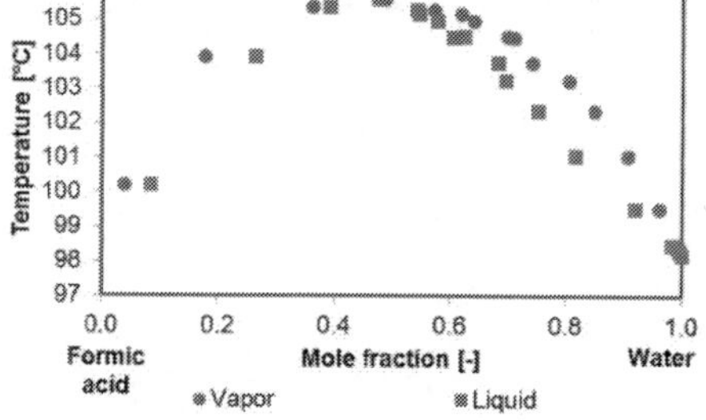

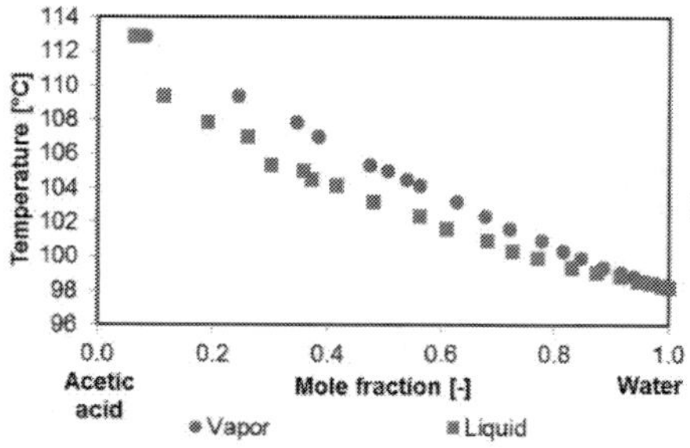

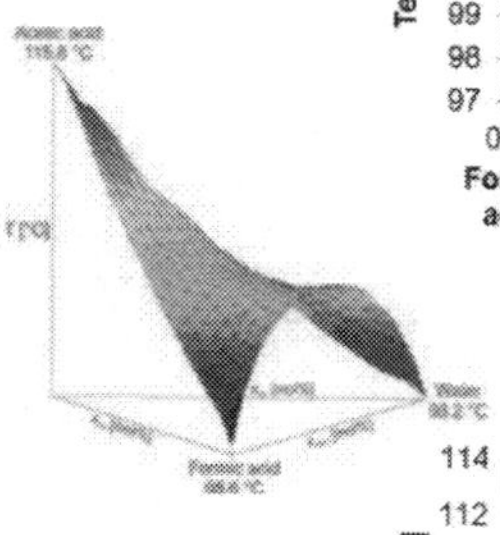

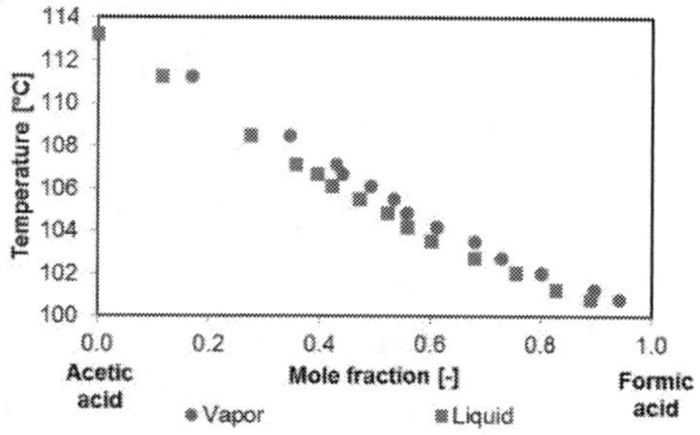

Boiling point surface

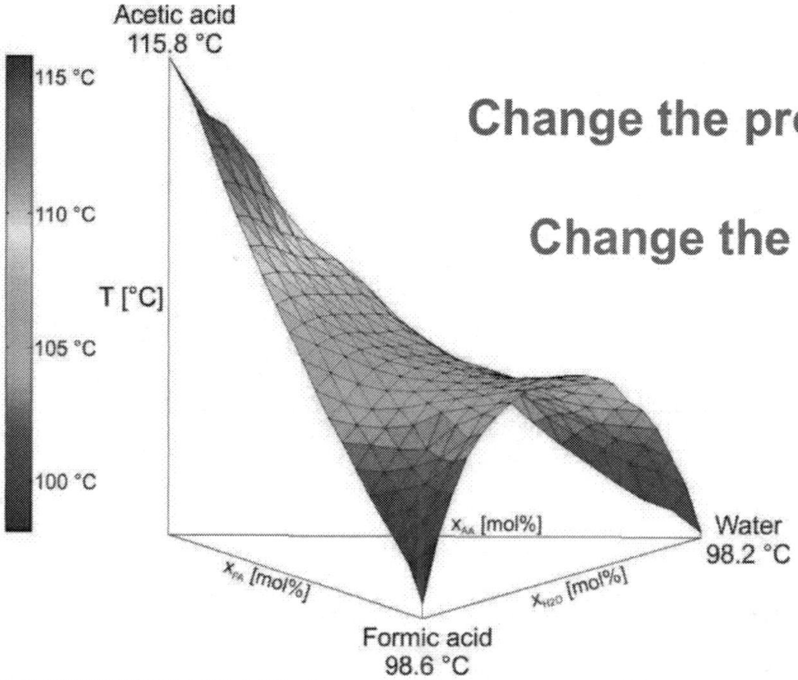

Change the properties!

Change the state!

Change the properties

- Esterification with methanol

$$HCOOH + CH_3OH \overset{H^+}{\longleftrightarrow} HCOOCH_3 + H_2O$$

$$CH_3COOH + CH_3OH \overset{H^+}{\longleftrightarrow} CH_3COOCH_3 + H_2O$$

	Δhv [kJ/kg]	Boiling point [°C]
	Bp, 1.013 bar	1.013 bar
Water	2265	100
Formic acid	479	101
Acetic acid	398	117.9
Methanol	1097	64.6
Methyl formate	469	31.8
Methyl acetate	411	57

Esterification: Formic acid vs. Acetic acid

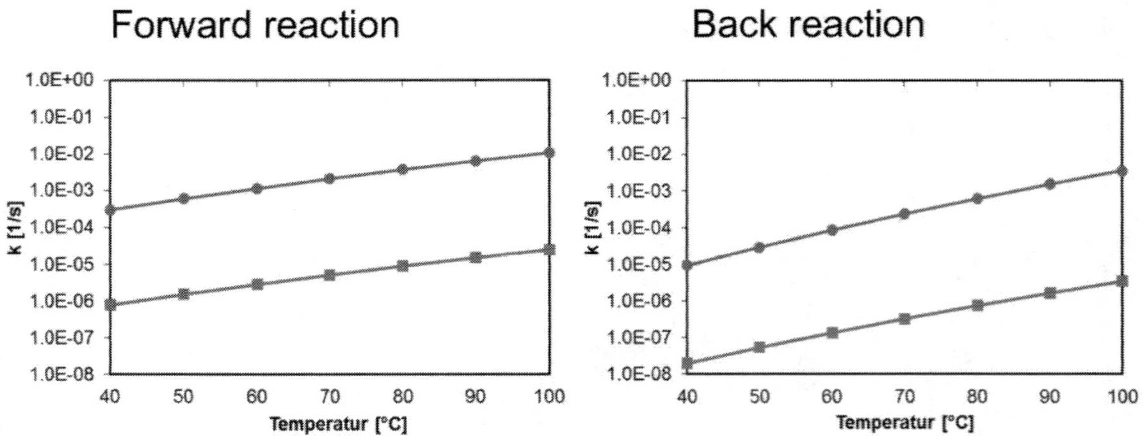

Formic acid esterification

Acetic acid esterification

Reactive Distillation

Continuous reactive distillation

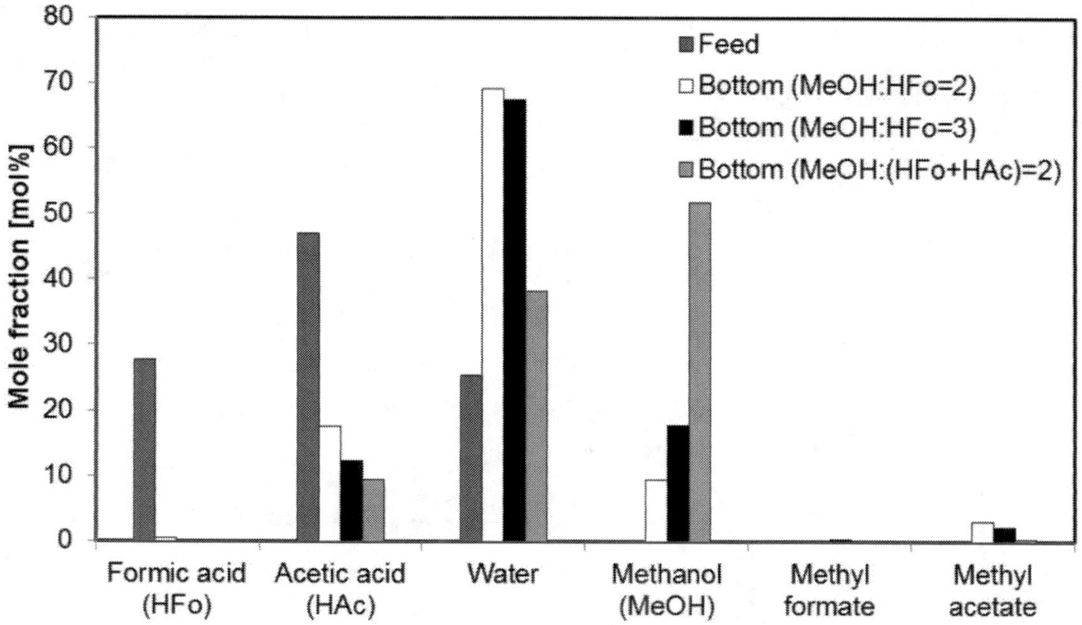

Continuous reactive distillation

	MeOH:HFo=2	MeOH:HFo=3	MeOH:(HFo+HAc)=2
Distillate composition	**mol%**	**mol%**	**mol%**
Methanol	20.4	24.4	46.0
Methyl formate	41.7	33.0	27.6
Methyl acetate	38.0	42.3	26.2
Conversion	**%**	**%**	**%**
Formic acid conversion	98.0	99.3	98.7
Acetic acid conversion	60.8	71.5	64.4

Extraction of carboxylic acids

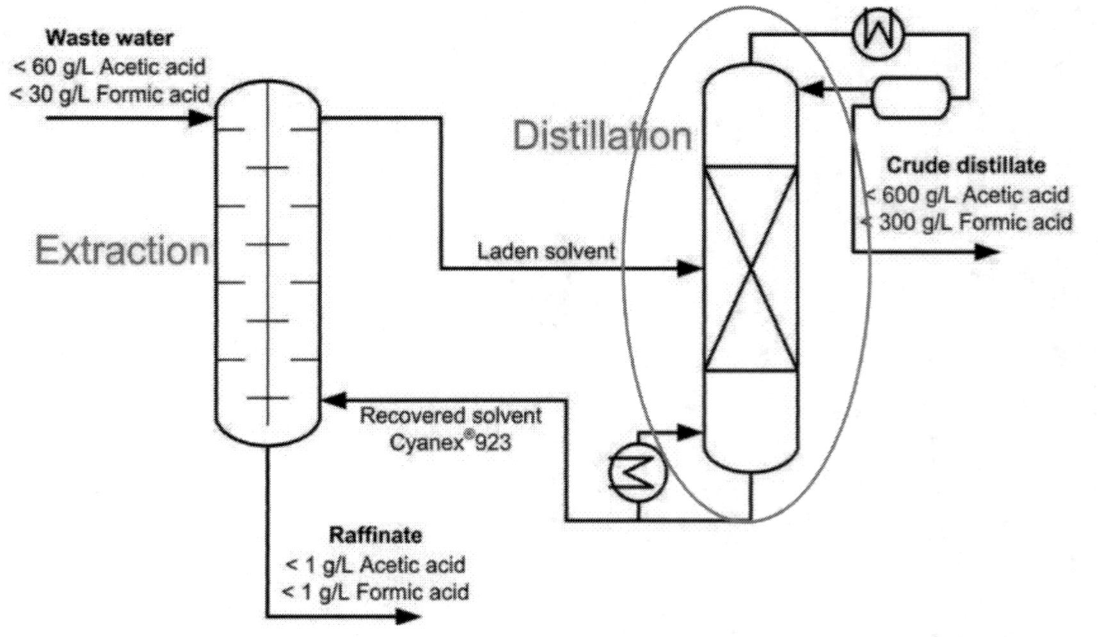

Extraction of carboxylic acids

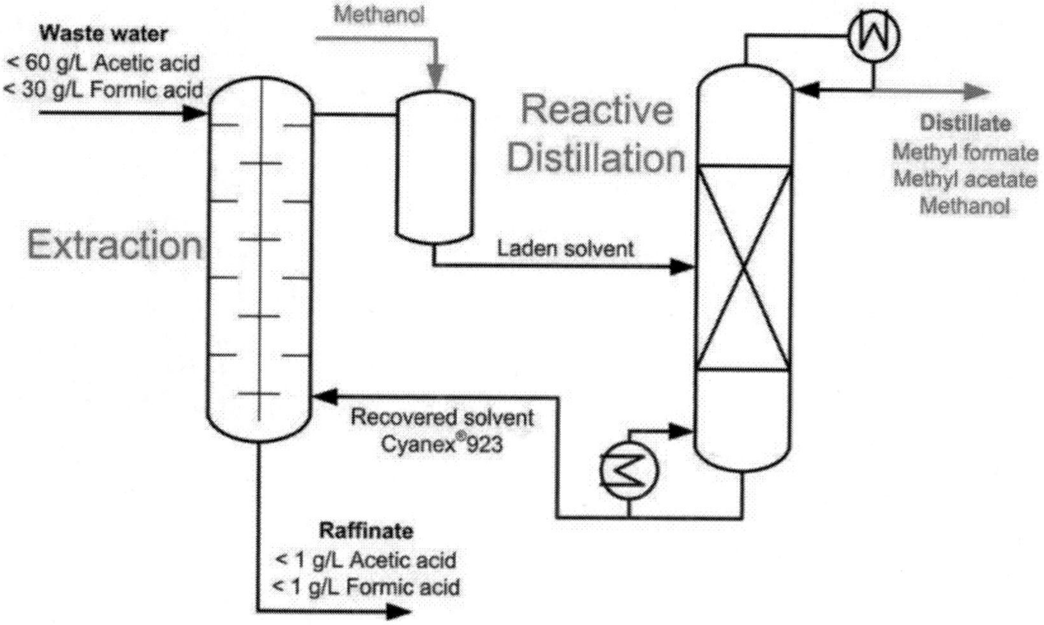

Esterification: Temperature effect

Formic acid esterification | Acetic acid esterification

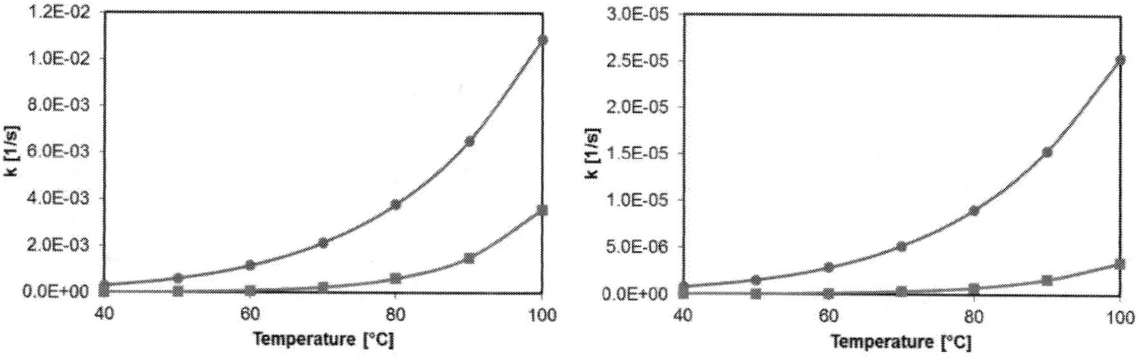

Forward reaction

Back reaction

Solvent recovery via reactive distillation

PROOF OF CONCEPT

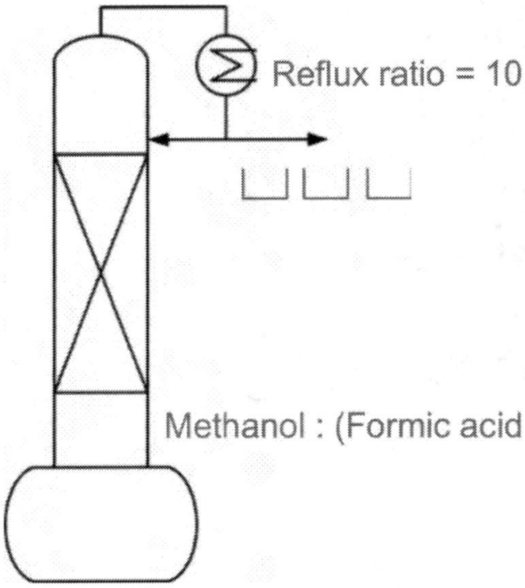

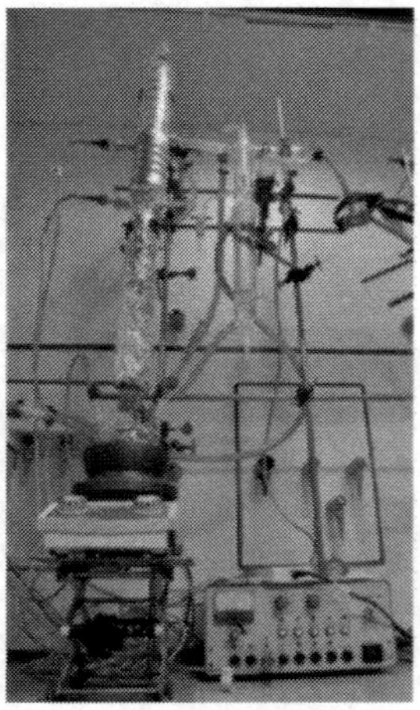

Solvent:

70 wt.% Cyanex®923
30 wt.% Undecane

Solvent Load:
13 wt.% Formic acid
24 wt.% Acetic acid
3 wt.% Water

Solvent recovery via reactive distillation

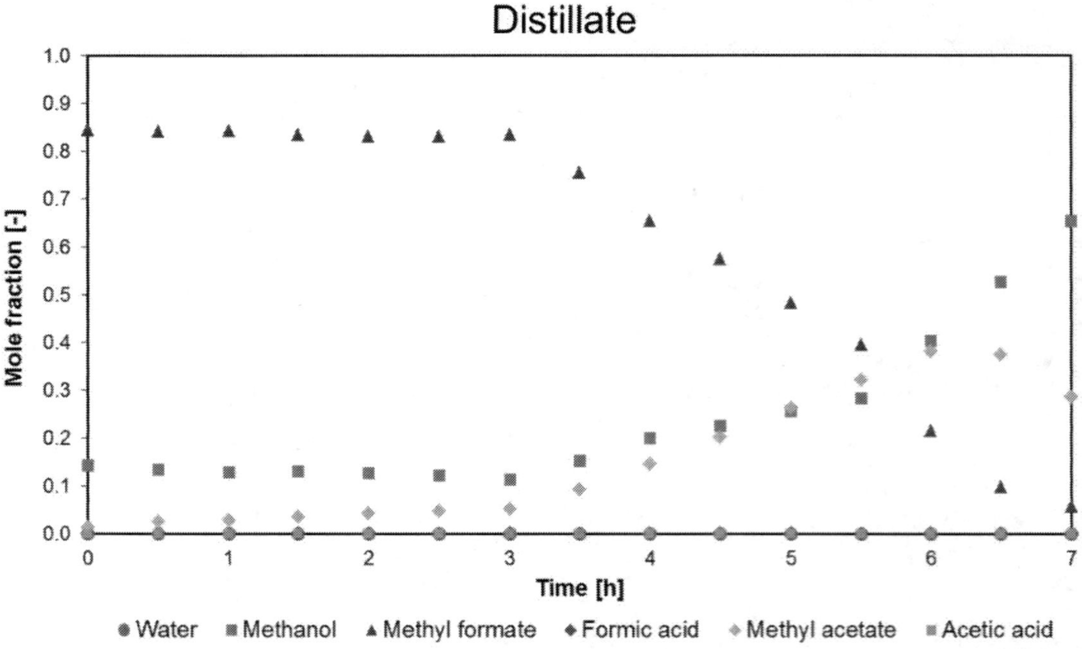

Summary

- Challenge:
  - Binary high boiling azeotrope
  - Ternary saddle point azeotrope
- Process intensification solver strategy:
  - Change the properties
  - Change the state
- Reactive distillation
  - Conversion of formic acid = 99.3 %, Conversion of acetic acid = 71.5 %
  - Solvent recovery via reactive distillation is possible

Extractive recovery of low molecular weight fatty acids from aqueous effluents

D. Painer, S. Lux, M. Siebenhofer

Graz University of Technology
Institute of Chemical Engineering and Environmental Technology, NAWI Graz
Inffeldgasse 25c/I, A-8010 Graz
d.painer@tugraz.at

Selective recovery of formic and acetic acids from their aqueous solutions by reactive extraction using non-polar solvents

Following to the pretreatment of biomass, acetic and formic acids appear in the aqueous media. These carboxylic acids also co-exist at several biochemical productions. Both have broad applications and significant roles in various industries. Thus, their selective recovery from aqueous media would be useful and may reduce the need for their re-manufacture and the cost in industry. However, recovery of carboxylic acids from their dilute solutions keeps its challenge and high cost of separation processes should be decreased. Reactive extraction is one of the most efficient methods for the recovery of value-added products and its relatively low cost could be further reduced without decreasing the extraction efficiency by the appropriate selection of the organic phase constituents (solvents and extractants-concentrations) and their use at the optimized conditions (aqueous pH and compositions).

In the present study, a selective recovery of formic and acetic acids from aqueous solutions by reactive extraction was explored. Separation factor (α_{AF}) values were calculated using the distribution coefficient (K_D) of both acids. The most efficient organic phase condition was determined according to the highest K_D and α_{AF} values obtained. Both acids were extracted from their aqueous solutions with initial concentration of 0.1 or 0.5 M. While using oleyl alcohol (a relatively green and non-polar solvent) as the diluent against high acid concentration levels, a third emulsion phase, which is keeping most of the acid in the matrix, was formed between the organic and aqueous phases. Similar and negligible K_Ds were obtained with the non-polar solvents, xylene and hexane, during the extraction of acetic and formic acids. With the addition of Alamine 336 -a tertiary amine (C_8-C_{10}) mixture- into the organic phase as the extractant, the K_Ds were significantly increased for both acids, however the recovery levels were still low compared to the ones obtained with 1-octanol, a relatively polar organic solvent. The modification of non-polar diluents by a relatively polar alcohol provided a robust effect on the recoveries, especially for xylene. Even they are both non-polar inert diluents; its negligible but relatively higher polarity presented higher extraction efficiencies and provided a synergy for xylene compared to hexane. Therefore, the polarity of the solvent is playing a critical role at reactive extraction of carboxylic acids. Using an organic phase made of 30% modified xylene (with 1-octanol) and having 0.5 M Alamine 336; almost 70% of acetic acid was recovered via single reactive extraction stage. For formic acid, this extraction level was obtained by using only 0.3 M of the same tertiary amines in the same solvent mixture. The loading ratio values were representing the formation of 1:1 acid-amine complex in the modified organic phase.

In conclusion, the results showed that the K_Ds decreased with the increase in the initial pH of aqueous phase, as expected, however the difference in the extent of the decrease varied and was useful for the selective separation. Moreover, having different pK_A values (3.77 for formic acid and 4.75 for acetic acid) facilitated their selective recovery. Besides the initial pH of the aqueous phase; other parameters such as concentration of the extractant, percent of the modifier and type of the solvent also affected the K_D and α values while achieving a selective recovery. However, the former has the highest effect among them. In summary, the modification was shown to enable a relatively cheaper inert diluent to be used for the selective recovery of carboxylic acids from dilute aqueous solutions.

389971 Extracting Ultrapure Lignin from Biomass By-Product Streams with Tunable, Renewable Solvent Systems

Abstract:

Lignin is one of the most common organic compounds on earth, comprising about 30% of all organic carbon. Lignin is also unique among biopolymers in having significant aromatic character, which makes it potentially attractive for a wide range of uses, including as a replacement for petroleum-derived polymers and as a precursor for carbon fibers. Unfortunately, the commercial-grade Kraft lignins available today are severely limited in their applications because of their high metals (primarily sodium) and ash content. Thus, less than 0.2% of the 50 million tons/yr of available lignin is being recovered. Cost-effective separations processes must be devised to dramatically improve the purities of today's commercial lignins, if lignin is to achieve its potential as a biopolymer.

Clemson researchers have discovered a powerful, versatile, and renewable solvent system that can be "tuned" to reduce the sodium content of lignin derived from the Kraft pulping process from several thousand ppm down to less than 50 ppm in two simple steps. Furthermore, this process can also be used to control the molecular weight of the recovered lignin. Properties such as temperature, solvent composition, and solvent-to-lignin ratio have been studied to determine their effect on yields, metals selectivities, and molecular weights.

A softwood lignin derived from a Kraft paper-mill black liquor with a sodium content of 1500 ppm and a polydispersity index of 4.1 was used as the feed in this work. Low-ash lignins with sodium contents ranging from 25 to 200 ppm and PDIs from 1.9 to 3.1 were obtained, with the recovered lignin fractions encompassing a range of molecular weights, from under 1000 to over 10,000 Daltons.

This process has also been shown to work for both batch and semi-batch processes, and continuous processing tests are also planned.

382531 Fundamental Investigation of Divided Wall Distillation through Pilot Column

Abstract:

Fundamental Investigation of Divided Wall Distillation through Pilot Column

Roach, B., University of Texas at Austin; Eldridge, R. B., University of Texas at Austin;

In the United States, distillation accounts for approximately 5% of total energy consumption. Dividing wall columns (DWC) offer an innovating approach to distillation, which can minimize energy usage and capital costs compared to traditional separation methods. Despite the advantages of a DWC, the technology has not been widely implemented due to the lack of fundamental research available in open literature.

A project has been initiated by the University of Texas' Process Science and Technology Center (PSTC) with industry support to study and document the operation and control of DWCs. A pilot scale dividing wall distillation column has been built and is fully operational. The pilot column is investigating a variety of mixtures with a broad range of pressure and temperature ratings. The column expands upon the work previously done by the Fieg group[1]. Their group studied an alcohol system at vacuum conditions with a 2.4" column diameter. Their model's utility in DWC operation and design has not been determined on its applicability to larger columns and other chemical systems. The PSTC's 6" pilot column will allow for a study of DWC scale-up from Fieg's column studies.

Preliminary results show the validation of process models and simulations. The DWC was modeled with initial mass transfer and hydraulic coefficients, using AspenPlus and HYSYS. Initial models show 30-50% energy savings for DWC versus traditional two column configurations.

Both MPC and PID control are being employed in the pilot plant to determine optimal control strategies. Since little is publically known about the DWC process the locations along the column that best reflect changes in composition must be identified. The interactive nature of the DWC process presents challenges in composition control. Stable divided wall column control has been achieved over a wide range of operation parameters. Studies for various disturbances in the feed flow rate and composition as well as for start-up and shutdown have been performed to model industrial applications.

1) Niggemann, Gerit, Christoph Hiller, and Georg Fieg. "Experimental and Theoretical Studies of a Dividing-Wall Column Used for the Recovery of High-Purity Products." Industrial & Engineering Chemistry Research 49.14 (2010): 6566-577.

Plethora of Dividing Wall Columns for Fully Thermally Coupled Distillation

Gautham Madenoor Ramapriya[#], *Mohit Tawarmalani*[$] *and Rakesh Agrawal*[#]

[#]*School of Chemical Engineering, Purdue University, West Lafayette, IN 47907*

[$]*Krannert School of Management, Purdue University, West Lafayette, IN 47907*

Fully thermally coupled (FTC) distillation is known to offer large energy savings in comparison to conventional distillation.[1,2] To further reduce the capital costs associated with FTC distillation, it can be performed inside a single shell with a vertical partition, generally termed a dividing wall column (DWC). In these DWCs, there is a vapor split at the bottom of every vertical partition which is unregulated during operation. Thus, because of this lack of external control on vapor splits during operation, there is a potential threat that the DWCs may operate far away from optimality, and the full benefits of FTC distillation may not be realized. This is one major reason why the DWCs, especially of four components and higher, despite offering very promising energy and cost savings, may find limited application in a green economy.

In this work, we present a multitude of hitherto unknown DWCs for *n*-component FTC distillation by converting a thermal coupling to a liquid-only transfer stream. Based on this strategy of converting a thermal coupling to a liquid-only transfer stream, we realize that only a small fraction of possible DWCs have been so far presented for three and four components.[3] In this work, we aim to synthesize all the useful DWCs that result from the above conversion strategy for any *n*-component FTC distillation. To do so, we introduce an easy-to-use exhaustive enumeration method. Using the method, for example, we have identified 20, 210 and 3024 DWCs for four-, five- and six-component FTC distillation. A distinct feature of all the novel

DWCs that have been presented in this work is that the vapor flow in each parallel zone of any DWC can be controlled by external means during operation, unlike the conventional DWC, and hence, can be operated close to optimality. We believe these novel DWCs will make possible the successful industrial implementation and optimal/close-to-optimal operation of the *n*-component ($n \geq 3$) FTC distillation in a DWC, thereby supporting a greener economy.

1. Petlyuk FB, Platonov VM, Slavinskii DM. Thermodynamically Optimal Method for Separating Multicomponent Mixtures. *Int Chem Eng*. 1965;5(3):555-561.
2. Fidkowski ZT, Krolikowski L. Thermally coupled system of distillation columns: optimization procedure. *AIChE J*. 1986;32:537–546.
3. Madenoor Ramapriya G, Tawarmalani M, Agrawal R. Thermal Coupling Links to Liquid-only Transfer Streams: A Path for New Dividing Wall Columns. *AIChE J*. DOI: 10.1002/aic.14468

THERMODYNAMIC EFFICIENCY AND REMIXING EFFECT ON A THERMALLY COUPLED EXTRACTIVE DISTILLATION SEQUENCE

K. D. Brito, M. F. Figueirêdo, W. B. Ramos E R. P. Brito
Federal University of Campina Grande – Campina Grande, PB, Brazil

Abstract

Thermally coupled distillation sequences are considered an alternative with the potential to reduce high energy consumption and capital costs of the conventional distillation sequences. According to the literature, this reduction occurs because the remixing effect decreases, which increases the thermodynamic efficiency of the system. In this work, a thermally coupled extractive distillation sequence with a side rectifier was evaluated. When compared to a conventional sequence, in fact, the energy consumption decreased. However, the thermodynamic efficiency did not decrease as expected and the remixing effect increased.

Introduction

In order to mitigate the high energy consumption in distillation, the design of new distillation sequences has been the subject of investigation by many researchers. One of the most promising alternatives is the use of thermally coupled sequences, which are obtained by interconnecting process streams, one in the vapor phase and other in the liquid phase, between two columns in series. Studies have proven the effectiveness of these settings in the reduction of energy consumption by 30% and increase in thermodynamic efficiency compared to conventional distillation systems (ERRICO et al., 2013; SHIRSAT et al., 2013). However, some studies still question the increase in thermodynamic efficiency resulting from the use of thermally coupled sequences (FINN, 1987; AGRAWAL and FIDKOWSKI, 1998).

The thermodynamic efficiency is related to an effect called remixing (TRIANTAFYLLOU and SMITH, 1992). In conventional distillation sequences intended for separating mixtures, at least ternary mixtures, the remixing effect, which is the "dilution" of the component with intermediate volatility, occurs. In fact, the composition of this component (in liquid phase) reaches a maximum at some point in the column and then decreases (as it is not removed) until it reaches a certain value. In order to purify again this component, more energy is needed in the column downstream of the main column. For this reason, the remixing effect is identified as the main source of irreversibility, reducing the thermodynamic efficiency of distillation systems. According to the literature, the use of thermally coupled sequences reduce (or eliminate) this effect, which increases the thermodynamic efficiency (DUNNEBIER and PANTELIDES, 1999; HERNÁNDEZ and JIMÉNEZ, 1999). However, studies that relate the remixing effect with thermodynamic efficiency are still incipient. In the literature, only two papers studying further the remixing effect were found (LING et al., 2011; MÉNDEZ et al., 2013), both considering ideal mixtures. The study of the remixing effect in the separation of azeotropic mixtures was not found in the literature.

The fact that azeotropic mixtures present non trivial behavior allied to differences on thermodynamic efficiency of thermally coupled sequences indicates that further studies should be conducted to increase the understanding of thermally coupled extractive distillation sequences. The focus of this paper is to deepen the understanding of the relationship between thermodynamic efficiency and remixing effect. Thus, the thermally coupled extractive distillation sequence with a side rectifier (TCEDS-SR) and conventional extractive distillation sequence (DS, direct sequence) were compared from the viewpoint of thermodynamic efficiency and remixing effect. The azeotropic mixture of ethanol (ETOH) and water (H2O), using ethylene glycol (EG) as solvent, was used as a case study.

Simulation

Both configurations, conventional and thermally coupled, were evaluated at steady state (Figures 1 and 2) using the Aspen Plus® simulator. RadFrac® block was chosen to simulate each configuration. Due to the operating pressure of the distillation columns, the vapor phase was considered ideal. For the liquid phase, the NRTL model was used to represent its non-ideality.

Considering the same purity specifications for both configurations, the methodology of analysis proposed by FIGUEIRÊDO et al. (2014) was used to evaluate various operating scenarios. This procedure takes into account the solvent plateau (or content), a feature common to every concentration profile of the less volatile component in extractive distillation columns. The authors noted that for each plateau, it was possible to obtain different operating conditions for the same purity specifications. In this sense, for each fixed plateau, a different pair of reflux ratio (R) and solvent flowrate (S) - variables that have a high impact in an extractive distillation, for a fixed design - is obtained by providing a different value of the heat duty (QR).

The calculation of efficiency, according to the Second Law of Thermodynamics, was performed as suggested by SEADER et al. (2011). The thermodynamic properties, such as enthalpy and entropy, of the streams were obtained from simulations performed using the Aspen Plus® simulator. The remixing effect is evaluated by the composition profile of the component with intermediate volatility in the liquid phase.

Please use this example paper as your primary reference in completing your paper. Your assistance in helping follow the styles and guidelines outlined for the following parts of your paper will help your association publish a more "professional looking" publication in the most cost-effective and timely manner possible.

Figure 1. Conventional extractive distillation process diagram (extractive and recovery columns).

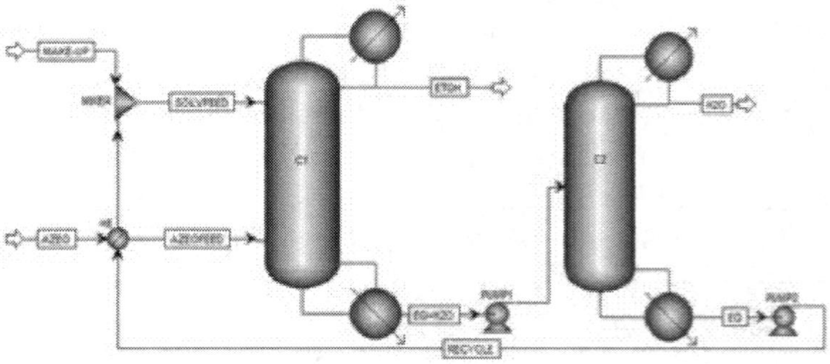

Figure 2. Separation process diagram for ETOH/H2O using a TCEDS-SR.

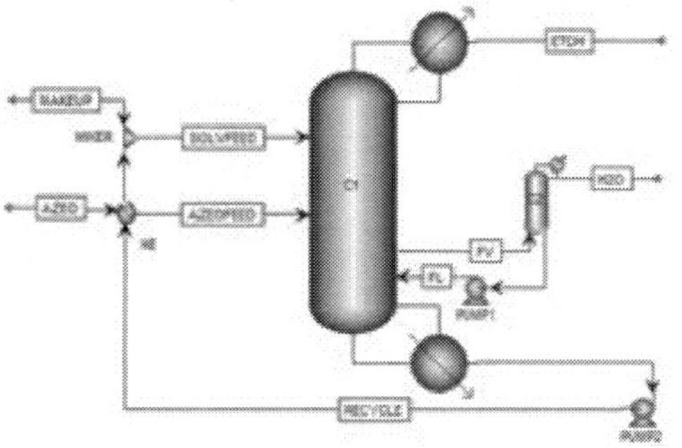

Results

Figure 3 shows the results of the simulations for S, R1 and QR1 in the two sequences evaluated as a function of different solvent content (xEG), where it is possible to observe that the minimum energy consumption of the extractive column C1 (Figure 3c) occurs for the lowest reflux ratio (Figure 3a) and maximum solvent flowrate (Figure 3b).

However, the extractive distillation system is comprised of two columns and therefore the second column should also be evaluated. The column C2 feed is directly proportional to the S flowrate, which justifies the similarities of the curves as showed in Figure 4 (R2, QR2 and S). The dependence of the recovery column to the amount of solvent is already known, as stated by LASTARI et al. (2012) and LUYBEN (2013).

When considering the total energy consumption (QRtotal) it is possible to see that for the DS there is a minimum. It was expected that the increase in the solvent flowrate would decrease the energy consumption; and that is what happens in the extractive column C1. However, the increase in the solvent flowrate supercharges the recovery column (C2), and at one point (from xEG = 0.75, see Figure 5) negatively impacts on QRtotal. This drawback does not occur for the TCEDS-SR sequence (because of the absence of the second column), the minimum QRtotal occurs for the maximum solvent flowrate. Therefore, according to Figure 5, only for high levels of solvent, it is more advantageous to use TCEDS-SR for dehydrating ethanol. Otherwise, the DS is the most appropriate configuration.

Figure 3. Reflux ratio of C1 **(a)**, solvent flowrate **(b)** and heat duty of C1 **(c)** versus EG content.

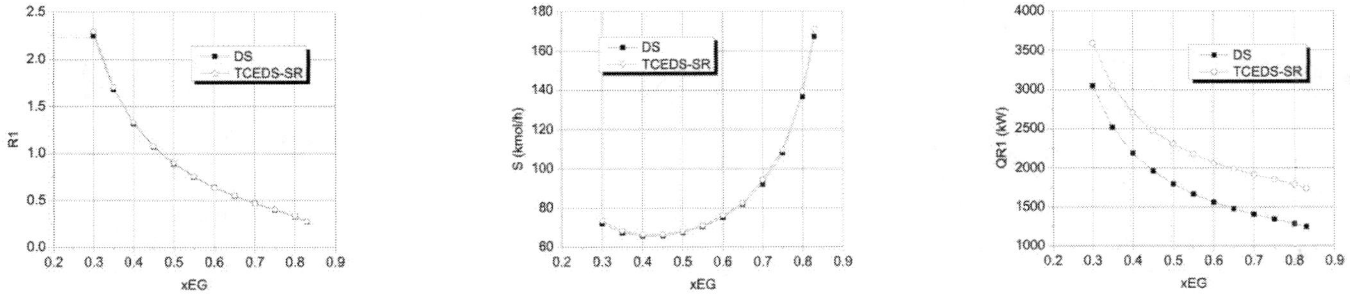

(a) (b) (c)

Figure 4. Reflux ratio R2 **(a)** and heat duty of C2 **(b)** versus EG content: DS.

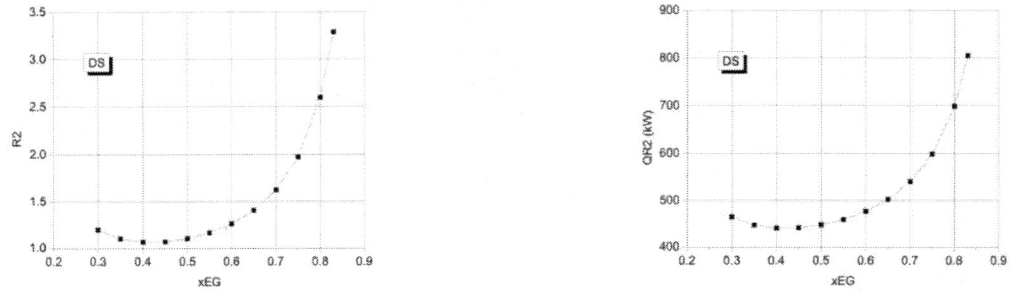

(a) (b)

Figure 5. Total heat duty versus EG content for DS and TCEDS-SR.

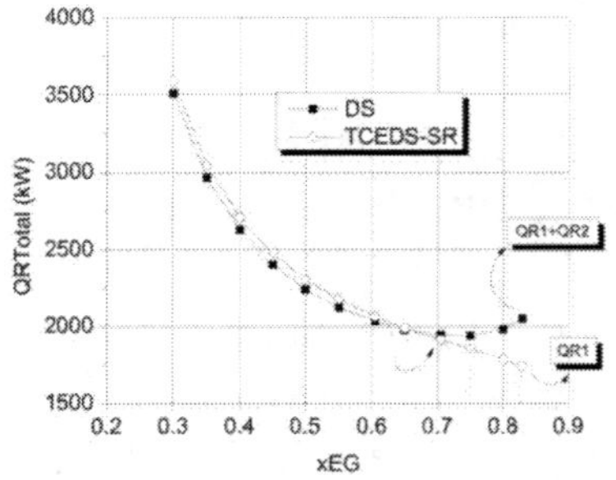

Most of the studies on thermally coupled distillation sequences are for ideal ternary mixtures involving hydrocarbons. GUERRA et al. (2008) were the first to apply thermal couplings in the separation of azeotropic mixtures and although they did not investigate the remixing effect deeply, they also justified that the low efficiency of conventional extractive distillation sequences was caused by it. The authors also stated that the use of a TCEDS-SR eliminates this effect. However, the results of this study (Figure 6) shows that just the opposite occurs: besides from not being eliminated, the remixing effect increases with the use of TCEDS-SR, independent of the solvent content used.

Figure 6. Remixing effect as function of xEG.

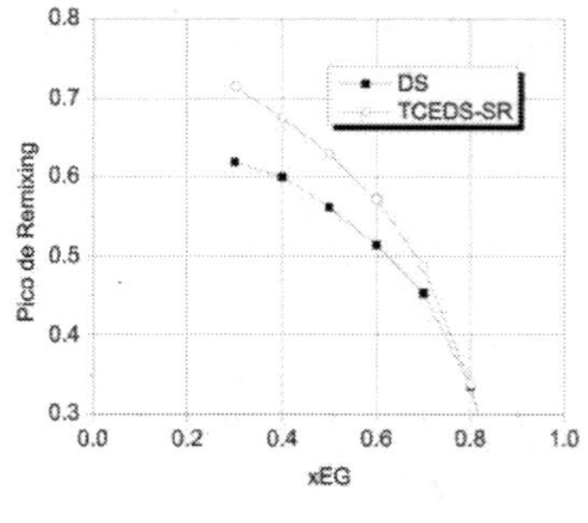

The investigation to understand what has occurred was deepened through thermodynamic efficiency according to the Second Law. Figure 7 shows the results on the thermodynamic efficiency. In both sequences, the results for efficiency are low (maximum 10%), as expected for a distillation process. It is possible to see that the efficiency of the TCEDS-SR reached a maximum value for xEG = 0.83, which corresponds to the lowest energy consumption of this sequence. The efficiency of the DS configuration reached a maximum value for xEG = 0.75, exactly the point where the DS achieves the lowest energy consumption. After this point, the efficiency decreases. This reversal behavior of the efficiency of DS is the result of the overloading that the recovery column C2 takes with the increase of solvent flowrate, S.

Figure 7. Thermodynamic efficiency as a function of solvent content.

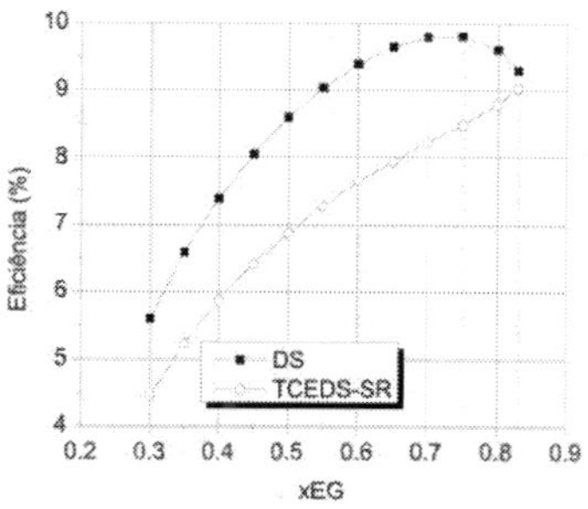

The literature suggests that the use of thermally coupled sequences must reduce the remixing effect and cause an increase in thermodynamic efficiency. It was not what happened: the DS showed lower remixing effects and higher thermodynamic efficiencies. This point is contrary to the assertion of most of the studies consulted, including the study of GUERRA et al. (2008), which presented higher thermodynamic efficiencies for the TCEDS-SR sequence.

FIDKOWSKI and AGRAWAL (1998) compared different distillation sequences for separating ternary mixtures with different relative volatilities and, although they did not include azeotropic mixtures in their studies, the authors studied ternary mixtures that exhibited two of its components with relative volatilities close to unity. The authors noted that in these situations the thermally coupled distillation sequence with a side rectifier showed extremely poor thermodynamic efficiency. The fact that the value of the relative volatility between two substances of the ternary mixture is small requires a great amount of steam to separate them, which decreases the efficiency value.

In the case of DS, it was found that a significant amount of the energy needed to perform separation is provided at 171°C in the reboiler of the column C1 and a smaller share is being supplied at 185°C in the reboiler of the column C2. In the TCEDS-SR configuration, all the necessary energy is supplied at a higher temperature level (203°C). In this regard, the temperature difference between the top (boiling point of pure component) and bottom of the column is greater in the TCEDS-SR sequence which causes the irreversibilities in the separation process. In fact, temperatures at higher levels are more "valuable" in terms of exergy, and the TCEDS-SR consumes more this kind of energy. The optimal thermodynamic minimizes the entropy or, equivalently, the cost of exergy. And this is accomplished by minimizing the energy quality required by the column.

Figure 8 shows the thermodynamic efficiency of each sequence as a function of remixing effect. Considering only the TCEDS-SR, it is clear relating the reduction of the remixing effect with the increasing in thermodynamic efficiency. However, the same reasoning does not apply to the DS.

Figure 8. Thermodynamic efficiency versus remixing effect.

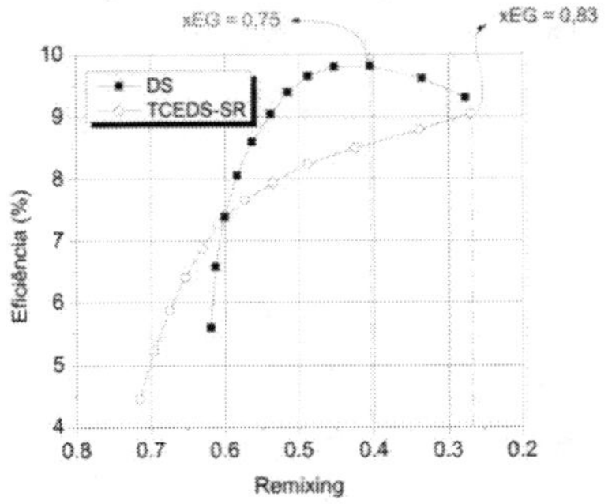

Conclusions

An analysis on the remixing effect was performed by comparing the two configurations, and in both cases, this effect is reduced with the increase of the solvent content. However, the use of thermal coupling did not reduce the remixing effect of the conventional sequence; an outcome contrary to that found in the literature. However, in the literature, the remixing effect was not evaluated for extractive distillation columns.

The result of the study on the remixing effect was corroborated by evaluating the thermodynamic efficiency of each configuration. In the whole space of possible solutions, the efficiency of the conventional configuration is larger than in the thermally coupled sequence, despite of its energy consumption, that is higher. The efficiency of the conventional sequence reaches a maximum, and then begins to decrease, showing that the efficiency is more dependent upon the levels at which the temperatures are provided and rejected in the reboilers and condensers than a possible dilution of the intermediate component (remixing). In contrast, the efficiency of thermally coupled sequence is an increasing function for the increasing of the xEG content, which is not always true in the conventional sequence.

References

1. AGRAWAL, R.; FIDKOWSKI, Z. T. Are Thermally Coupled Distillation Columns Always Thermodynamically More Efficient for Ternary Distillations. Industrial and Engineering Chemistry Research, v. 37, p. 3444-3454, 1998.
2. DUNNEBIER, G.; PANTELIDES, C. C. Optimal Design of Thermally Coupled Distillation Columns. Industrial Engineering Chemical and Research, v. 38, p. 162-176, 1999.
3. ERRICO, M.; RONG, B.; TOLA, G.; SPANO, M. Optimal Synthesis of Distillation Systems for Bioethanol Separation. Part 1: Extractive Distillation with Simple Columns. Industrial Engineering Chemical Research, v. 52, 1612−1619, 2013.
4. FIGUEIRÊDO, M. F.; BRITO; K. D.; BRANDÃO, W.; BRITO, R. P. Effect of Solvent Content on the Separation and the Energy Consumption of Extractive Distillation Columns, Chemical Engineering Communications, 2014.
5. FINN, A. J. A New Approach to Distillation Sequence. Chem. Engineering. Symposium Series, 1987.

6. GUERRA, R. G.; HERNÁNDEZ, J. G. S.; HERNÁNDEZ, S. Reducing Energy Consumption and CO2 Emissions in Extractive Distillation. Chemical Engineering Research and Design, v. 87, p. 145-152, 2008.
7. HERNÁNDEZ, S.; JIMÉNEZ, A. Design of Energy-Efficient Petlyuk Systems. Computers and Chemical Engineering, v. 23, p. 1005-1010, 1999.
8. LASTARI, F.; PAREEK, V.; TREBBLE, M.; TADE, M.; CHINN, D.; TSAI, N.; CHAN K. I. Extractive distillation for CO2–ethane azeotrope separation. Chemical Engineering and Processing: Process Intensification, v. 52, p. 155–161, 2012.
9. LING, H.; CAI, Z.; WU H.; WANG, j.; SHEN, B. Remixing Control for Divided-Wall Columns. Ind. Eng. Chem. Res., v.50, p.12694–12705, 2011.
10. LUYBEN, W. Distillation Design and Control Using Aspen Simulation. 2. Edition Wiley, 2013.
11. MENDEZ, S.; J. SR. Minimum Remixing Effect in Distillation Sequences. 2013 Spring Meeting & 9th Global Congress on Process Safety. Kister Distillation Symposium, 2013.
12. SEADER, J. D.; HENLEY, E. J.; ROPER, D. K. Separation Process Principles: Chemical and Biochemical Operations. 3. ed. [S.l.]: WILEY, 2011.
13. SHIRSAT, S. P.; DAWANDE, S. D.; KAKADE S. S. Simulation and optimization of extractive distillation sequence with pre-separator for the ethanol dehydration using n-butyl propionate. Korean J. Chem. Eng., v. 30, n.12, p.2163-2169, 2013.
14. TRIANTAFYLLOU, C.; SMITH, R. The Design and Optimization of Fully Thermally Coupled Distillation Columns. Transactions of the Institution of Chemical Engineers. v. 70, p. 118, 1992.

Process Modeling of Hybrid Forward Osmosis-Distillation Desalination Systems

Xingshi Zhou, Daniel B. Gingerich, Jeffrey J. Siirola, Meagan S. Mauter
Chemical Engineering Department, Carnegie Mellon University, Pittsburgh, PA 15213, USA

Forward osmosis (FO), also known as engineered osmosis, is two step process for desalinating seawater and other complex aqueous feed streams. The first step, the passive diffusion of water across a polyamide thin-film composite membrane (similar to that used in reverse osmosis) is driven by an osmotic gradient between the feed stream and a high osmotic pressure draw solution. When using a thermolytic salt, the second step of draw solute recovery uses low-temperature heat to recycle the draw solution and produce a permeate stream of fresh water. Compared to alternative techniques for water desalination this process does not require electricity to power high-pressure pumps (as in reverse osmosis) or high grade thermal energy to evaporate the water (as in multi-stage flash distillation). One promising draw solute/recovery pair are the thermolytic salts created by ammonia and carbon dioxide (NH_4HCO_3, $(NH_4)_2CO_3$, and $NH_2CO_2NH_4$) which can be driven out of water at relatively low temperatures (<70°C) in distillation columns.

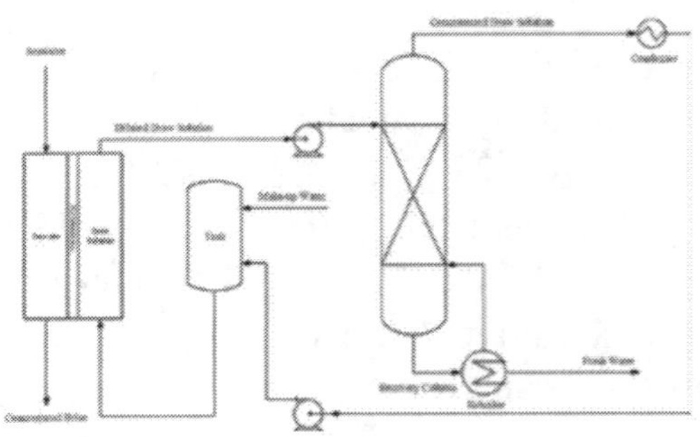

Figure 1. Process Flow Diagram of a Hybrid Forward Osmosis-Distillation Desalination System.

Despite five decades of research on the membrane separation step in FO processes, there are only a handful of papers that have been published on the draw solute recovery step (McGinnis and Elimelech 2007, Kim et al. 2012, Kim et al. 2013). These papers have potential flaws in their modeling and costing approach: neglecting to include the energy consumption of the vacuum in vacuum-assisted distillation, ignoring intermediate equipment, or simply using RO processes for reconcentration. Establishing rigorous models for the temperature and heat duty of the draw solute recovery system is the primary barrier to FO deployment and integration with infrastructure, like power plants, with abundant low-temperature heat.

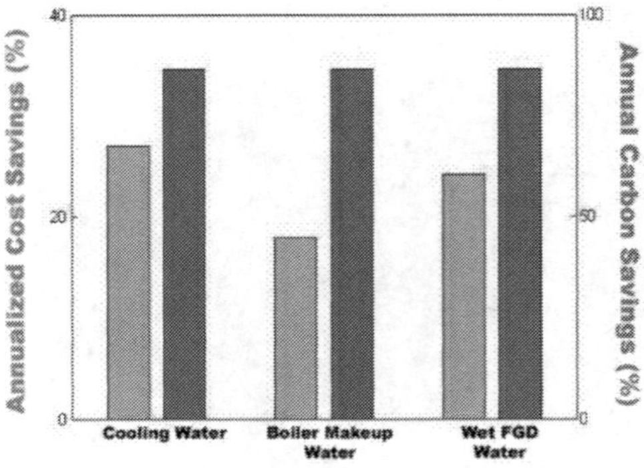

Figure 2. Annual Costs and Carbon Dioxide Emissions Savings for Water Treatment for a 550MW PCC Power Plant (NETL, 2009) using Forward and Reverse Osmosis Processes.

In this study, we design a model to assess the techno-economic feasibility of FO integration with a 550MWH coal fired power plant to treat both on-site boiler make-up water and wastewater effluent from a wet flue gas desulfurization system. The draw solute recovery module for recycling thermolytic ammonia-carbon dioxide salts was modeled using fundamental process models and was compared to a simulation built in commercial software (i.e. ASPEN® Plus). Rigorous calculations considering the existence of dissociations and reactions has made the model more plausible. We then optimized the system to consider heat availability at the power plant, methods for heat capture and utilization, and different column configurations. Finally, we perform an optimization case study for total annual cost and environmental metrics (CO_2 emissions, wastewater effluent composition) by varying membrane area, heat load of distillation column, and necessity of vacuum operation. According to the optimization results, with all possible forms of energy consumption taken into consideration, the hybrid forward osmosis-distillation desalination systems still saved much energy from the ordinary RO or thermal desalination technologies and the integration with waste heat from power plant makes it stand out as a green separation scheme for the future.

Chemical reactions take place in practically any chemical process in order to manufacture a wide range of products. Unfortunately, these reactions produce by-products, which affect the chosen purity and the value of the final product. This means that some types of separation task must be designed to get rid of the undesired by-products and to reach the required purities of the desired components. Nowadays, there have been developed and studied some different technologies to carry out the separation task in a chemical process such as enhanced distillation, membrane separations and solvent extractions, among others [1]. However, of all these separation techniques, distillation is by far the most understood and its economic behavior on an industrial scale is at the moment the most advantageous. Due to the enormous energetic costs associated to distillation, it is necessary that chemical engineers can fully understand the process at work. As evident from a chemical process, the energy consumption of a separation problem has a decisive impact on the viability and reliability of a process.

The energy requirements for a simple binary distillation to achieve desired product compositions can be enormous, not even to reference a multicomponent system where a sequence of distillation columns might be involved to achieve the final desired product compositions. Luckily, currently there are some advanced techniques to overcome these limitations, for example by using thermal coupling of distillation columns as side rectifiers and side strippers. The thermal coupling of distillation columns basically use a linking of streams between units to simplify a utility input, which could be a reboiler or a condenser. The most common examples of fully thermally coupled columns are the Petlyuk distillation columns and the Dividing-wall columns (DWCs). The principal cause for the energetic efficiency of this type of arrangements is the prevention of the remixing of internal streams, which in configurations of two columns in series, shows concentration differences between either above or below the feed stage. Petlyuk and Dividing-wall columns (DWCs) are used for the separation of a broad spectrum of multicomponent systems [2].

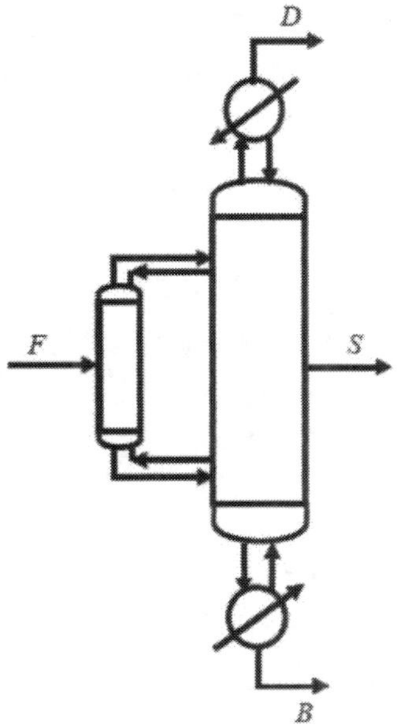

Figure 1. Standard Petlyuk Distillation Column

Due to the possibilities of cost savings and yield improvement, the present work analyzes the modelling and use of Petlyuk columns and DWCs on the separation of a mixture composed by dimethylether, water and methanol, evaluating their thermodynamic behavior as a first tool for the design of both types of columns.

This work presents an integrated approach to the solution of Petlyuk and DWCs distillation synthesis, design and analysis problems. Integration is achieved by combining mathematical models based on thermodynamic insights and structural designing, coupled with simulations and experimental procedures. At first, employing the relationships between physicochemical properties and separation principles, the physical limits for a specified separation problem are analyzed and the corresponding process flowsheets are determined. Successively shortcut approximations such as the Fenske–Underwood–Gilliland-Kirkbride model are used for the preliminary design of the pre-fractionator, which is modeled as a column with a partial condenser and a partial reboiler. Also constant molar overflow approximations of the distillation profiles are used to calculate the minimum number of equilibrium stages, the minimum reflux ratio, the stage requirements at chosen operating reflux ratio and the feed stage for the main column. This allowed a more precise assessment of the separation possibilities and costs for a ternary mixture, without recurring to the constant volatility assumption [3]. To verify the results obtained, an experimental evaluation was followed. The aim of this was to corroborate different possibilities in operation conditions provided for the mathematical models for both types of columns in an experimental way. Nevertheless, this could be considered as an important tool to show that also Petlyuk columns and DWC have the potential to separate complex mixtures allowing the understanding of the phenomena, providing a guide for the design procedures of these processes which could lead to pilot lab scale demonstrations [4].

[1] Pisarenko, Y. A., L. Serafimov, et al., Reactive distillation design: analysis of the process statics. Reviews in Chemical Engineering, 2001, 17(4):253-327.

[2] Petlyuk, F.B., V.M. Platonov, and D.M. Slavinskii, Thermodynamically optimal method for separating multicomponent mixtures. International Chemical Engineering, 1965, 5(3):554–561.

[3] Stupin,W.J. and F.J. Lockhart, Thermally coupled distillation—a case history. Chemical Engineering Progress, 1972, 68(10):71.

[4] Westerberg, A.W., The synthesis of distillation based separation systems. Computers and Chemical Engineering, 1985, 9(5):421.

366942 Microbial Desalination Cells for Simultaneous Wastewater Treatment and Water Desalination: Status, Challenges and Perspectives

Abstract:

Microbial desalination cells (MDCs) are a newly emerged technology that takes advantage of microbial metabolism with organic compounds (e.g., organics in wastewater) to generate electrons, which drive salt migration to achieve desalination. An MDC integrates microbial fuel cells (MFCs) and electrodialysis (ED). Compared with an MFC, an MDC contains both cation and anion exchange membranes and has additional compartment for desalination. Unlike ED, an MDC does not require additional voltage or is applied with a small external voltage (less than 1 V) that is not sufficient for water oxidation. The bioenergy-driven desalination in an MDC makes this technology attractive as a low-energy desalination process. However, desalination efficiency in an MDC is generally low; therefore, the potential application of MDC technology is either as pre-treatment of conventional desalination processes or desalination of brackish water. Since the first MDC was reported in 2009, there has been some progress towards MDC development. Our team at Virginia Tech is one of the most active and productive groups in the MDC research. We have investigated membrane fouling, desalination of both brackish water and seawater, system configuration and operation. Especially, we have scaled up the MDC system from bench scale to a transitional scale (~100 L), which represents the largest MDC system worldwide. We have also synergistically link MDCs to osmotic bioelectrochemical systems or forward osmosis to accomplish water reuse and improve desalination performance. In this presentation, we will introduce the current status of MDC research, discuss the challenges in its development, and provide some perspective on its application in wastewater treatment, water desalination, and water reuse.

Novel mixed-matrix-membrane with activated carbon for steel-industry effluent treatment

Raka Mukherjee, Sirshendu De

Chemical Engineering, Indian Institute of Technology, Kharagpur

1. Introduction

Waste water effluents from various iron and steel industries contain a lot of harmful chemicals. Especially the by-products from coke-oven plant contain phenols, which is very toxic and carcinogenic [1, 2]. This group of compounds is in the 11th position out of 126 chemicals, which are termed as priority pollutants by United States Environmental Protection Agency. Exposure to phenol has many bad effects like high irritation to eyes, skin and mucous and also causes headache and dizziness. Long term exposure can even result in high blood pressure, liver and kidney damage. Present technologies applied for removal of phenolic compounds are solvent extraction, adsorption, chemical oxidation, biological treatment and distillation [3-7]. However, the need for cost-effective, simple and high throughput technology still persists. It has been found that, activated carbon is very good adsorbent of phenolic compounds [8]. But, this process is slow and less economic. On the other hand, membrane separation technology is a physical separation process with high throughput and it is also external chemical free. Mixed matrix membrane combines the goodness of adsorptive property of inorganic additives and simplicity of membrane technology. Mixed matrix membrane of some base polymer and activated carbon can cater to the need of phenol removal technology. In this work, one flat sheet mixed membrane of cellulose acetate phthalate (CAP) and powdered activated carbon (PAC) has been fabricated. Various properties of this membrane are compared with the undoped CAP membrane.

Effluent of a steel industy is filtered using the doped membrane. It has been found that the doped membrane has higher permeability, better hydrophilicity, lesser porosity and better phenol adsorption properties. This special MMM is found to be capable of reducing the phenol concentration to a great extent.

2. Materials

CAP was purchased from M/s, GM Chemicals Company Limited, Mumbai, India. Dimethylformamide (DMF) and PAC were obtained from M/s, Merck (India) Ltd. Mumbai, India. Waste water was collected from M/s, Tata steel, Jamshedpur, India after the secondary treatment stage. For determination of total phenol (TP) by American public health association (APHA) method, analytical grade hydrochloric acid (HCl), ammonium hydroxide (NH_4OH), di-potassium hydrogen phosphate (K_2HPO_4), potassium di-hydrogen phosphate (KH_2PO_4) were procured from M/s, Merck (India) Ltd. Mumbai, India. 4-aminoantipyrine and potassium ferricyanide were purchased from M/s, Loba Chemie, Mumbai, India.

3. Experimental methods and results

Flat sheet mixed matrix membrane of cellulose acetate phthalate and activated carbon was cast by phase inversion method. Casting solution was prepared by mixing PAC (25%) and CAP(15%) with solvent di-methyle formamide (DMF) (60%), and the solution was sonicated for one hour by probe-type sonicator. Then one nonwoven fabric was attached with a glass plate. Using a doctor's knife, the membrane was cast by manual draw down at a speed of 20 mm/min. The membrane was immediately transferred to water bath. Then for complete phase inversion, the membrane was kept in water bath for 12 hours.

The morphology the membrane was analysed using scanning electron microscopy (SEM) and atomic force microscopy (AFM). SEM images of the top surface of the undoped and dopes membranes are given in Fig. 1(a) and 1(b), respectively. Presence of carbon particles on the top can be observed from SEM image of the top surface of the doped membrane, which is absent in the undoped membrane. Also, AFM image analysis confirmed that average surface roughness increases with addition of carbon. The average r.m.s. roughness of undoped membrane is 12 nm, where as that is 247 nm in dopes membrane. These findings corroborate that activated carbon particles are accumulated mainly on the surface of the membrane. PAC particles are hydrophilic in nature. During the phase inversion process, activated carbon particles come toward the surface. Cross-sectional SEM images of the undoped and the doped membranes are presented in Fig. 1(c) and 1(d), respectively. From the cross-sectional image, we can see that pores are more open for the doped membrane. As PAC is hydrophilic in nature, addition of PAC induces thermodynamic instability, delaying instantaneous demixing. Hence the pores are larger.

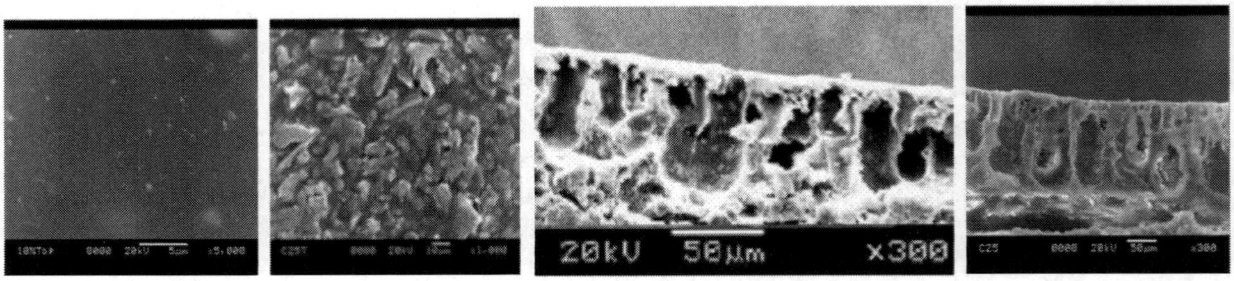

Fig 1: SEM image of (a) undoped membrane top surface, (b) doped membrane top surface; (c) undoped membrane cross-section, (d) doped membrane cross-section.

Porosity of the two membranes were measured by gravimetric analysis. The mixed-matrix membrane has lesser porosity (48%) than the undoped membrane (68%). Carbon particles sit inside the pores and block those. Hence, porosity is reduced. Hydrophilicity of the membrane increased significantly. Contact angle of distilled water with the membrane was 69° for the undoped membrane, which reduced to 56° for doped membrane. Activated carbons are accumulated on the surface of the doped membrane. As, PAC is hydrophilic, overall hydrophilicity of the membranes are increased. Distilled water permeability of the membrane increased because of higher hydrophilicty. The linear permeability increases fron 5.3 to 7.8 $m^3/m^2.Pa.s$ for the composite membrane. Hydrophilic surface draws more water toward the membrane, as a result, permeability of the mbrane is increased. The mechanical strength of the membrane was measured by Universal Testing Machine. The doped membrane has a tensile strength 19.5 MPa, indicating that it can with stand high pressure.

To study the effect of pH on phenol adsorption on activated carbon, phenol adsorption isotherms of different pH values, like 5.5,7,8 and 10 were prepared. It was observed that at lower pH, adsorption of phenol on membrane is higher. As the pH was increased, adsorption capacity goes down. Some basic sites on the activated carbons are developed because of the basic surface oxygen complexes and/or π-electron reach complexes at the basal plane. Among these basic sites and the aromatic rings of the phenols, electron donor-acceptor complexes are formed. At lower pH, this chemistry is favoured, hence higher adsorption is observed. However, at a pH, higher than pK_a value of the phenol, phenol compounds dissociate in phenolate ion, having residual negative charge. Electrostatic repulsion between phenolate ions and negatively charged membrane results in lower adsorption capacity [9].

To test the applicability of the membrane in phenol removal process, real life effluent from TATA Steel India Limited was collected which contains phenol in significant amount (around 17-24 mg/l). The effluent stream was passed over the membrane in cross-flow filtration unit, where the fluid flow is in parallel with the membrane surface. The result of which is presented in Fig. 2.

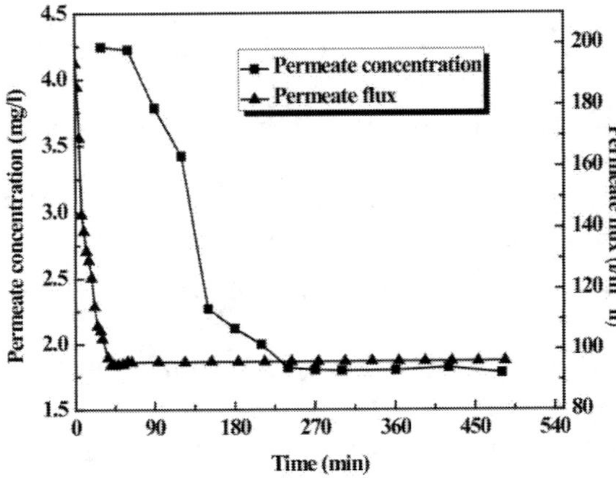

Fig. 2: Permeate flux and concentration of effluent, after cross-flow filtration, by the doped membrane at 70 kPa and 20 lph flow-rate.

It can be observed that, in the initial time period of the filtration process, total phenol concentration in the permeate stream is rapidly reduced. After 240 minutes of filtration total phenol content in permeate reduces down to 1.8 mg/l, and remains almost steady there-after. Phenol compounds are adsorbed by the membrane, hence permeate contains less pheol. Initial stage of rapid decline occurs because of unsteady state adsorption. After the steady state (dynamic equilibrium) is reached, steady concentration of phenol in permeate can be observed.

From Fig. 2, we can see that permeate flux declines rapidly from initial 195 l/m^{2}h to 96 l/m^{2}h, after 60 minutes. In the initial stage, solid particles rapidly form a cake layer on the membrane which offers additional resistance to permeate stream in its flow path. As more solids are filtered, the layer grows in depth. After sometime, (in this case, it is around 60 minutes) the cake layer does not grow further, as equilibrium is achieved. This cake layer, as well as the carbon in the membranes aids to the filtration of phenolic compounds. The phenolic compounds are adsorbed by the membrane and resisted by the cake layer. Once the cake layer stops growing, flux becomes almost steady at a certain value (in this case, around 96 l/m^2 h).

4. Conclusion

Novel mixed matrix of PAC-CAP composition offers an excellent solution for treatment of industrial effluent.. Addition of PAC enhances the property of adsorption of phenol.

Reference:

[1] Canadian Environmental Protection Act, Priority Substances List Assessment Report: Phenol, 1999, www.ec.gc.ca.

[2] R. J. Lifton, The Nazi Doctors, first ed., Basic Books, New York, 1986, pp. 254–268.

[3] Z. Li, M. Wu, Z. Jiao, B. Bao, S. Lu, Extraction of phenol from wastewater by N-octoyl-pyrolidone, J. Hazard. Mater. B114 (2004) 111–114.

[4] L. Damjanović, V. Rakić, V. Rac, D. Stošić, A. Auroux, The investigation of phenol removal from aqueous solutions by zeolites as solid adsorbents, J. Hazard. Mater. 184 (2010) 477-484.

[5] Y. Yavuz, A. S. Koparal, Electrochemical oxidation of phenol in a parallel plate reactor using ruthenium mixed metal oxide electrode, J. Hazard. Mater., 136, (2006) 296-302.

[6] Y. Wang, J. Song, W. Zhao, X. He, J. Chen, M. Xiao , In situ degradation of phenol and promotion of plant growth in contaminated environments by a single Pseudomonas aeruginosa strain, J. Hazard. Mater. 192 (2011) 354-360.

[7] H. G. Franck, J. W. Stadelhofer, Industrial Aromatic Chemistry, Springer Verlag, Berlin, (1989) 148–157.

[8] A. Da'browski , P. Podkos´cielny, Z. Hubicki, M. Barczak, Adsorption of phenolic compounds by activated carbon—a critical review, Chemosphere 58 (2005) 1049–1070.

[9] **C.** Moreno-Castilla, J. Rivera-Utrilla, M. V. Lopez-Ramon, F. Carrasco-Marin, Adsorption of some substituted phenols on activated carbons from a bituminous coal, carbon 33 (1995) 841-851.

369968 Graphene Oxide Membranes with Hierarchical Roughness for High Flux, Anti-Fouling Oil/Water Separation

Abstract:

Underwater superoleophobic surfaces have shown increasing importance in a variety of applications among which oil/water separation is of great interest due to the increasing amount of oily wastewater from industries, a shortage of clean water, and frequent oil spill accidents. Despite recent advances in nano/ultrafiltration membranes in oil/water separation, fully recoverable oil/water separation performance has rarely been demonstrated, especially for stable, difficult-to-separate oil emulsions. A promising strategy of designing superoleophobic surfaces for underwater applications is to incorporate desirable hierarchical surface roughness, as inspired by the natural formation of fish scales. Mimicking the structure of fish scales has been attempted to for the preparation of superoleophobic surfaces and has been studied for oil/water separation. Compared with conventional antifouling coating agents, bio-inspired underwater superoleophobic materials have improved oil/water separation performance and enhanced fouling resistance. Most previous studies, however, focused merely on replicating the structure of fish scales. The function of the mucus layer on the surface of fish scales was usually neglected. In fact, the mucus layer replicates the hierarchical roughness of the fish scales underneath, and it preferentially attracts water molecules in oil/water mixtures, forming a superoleophobic oil/water/solid interface. To mimic the excellent anti-fouling and self-cleaning performance of fish scales, a combination of the hierarchical surface roughness and the conformal, underwater superoleophobic coating layer is apparently necessary. However, it is highly challenging to perfectly combine these two aspects, and little success has been reported to date. Graphene oxide (GO) is a promising two-dimensional (2-D) membrane/coating material, owing to its one carbon atom thickness, ease of conformation to substrates, excellent chemical stability, and mechanical strength. GO membranes have shown great potential for gas and liquid separations through size-sieving by either interlayer spacing or selective structural defects. Here, we report, for the first time, the utilization of mucus-mimicking GO for fabricating ultrafiltration membranes with desirable hierarchical surface roughness and their application for high flux, anti-fouling oil/water separation.

REVERSE OSMOSIS – THE STATE OF THE ART OF WATER TREATMENT

Richard L. Stover, Ph.D.
Executive Vice President, Desalitech, Inc.
rstover@desalitech.com, +1-510-333-2767

ABSTRACT

Water scarcity is one of the most serious global challenges of our time. Desalination and water reuse are effective and reliable means to provide new water resources. Among the many methods available to treat water, reverse osmosis has widely demonstrated superior reliability and cost-effectiveness at removing dissolved species such as salts and trace contaminants. This paper reviews the development and state of the art of reverse osmosis for industrial water treatment, wastewater treatment and drinking water production. Technological advances that have improved process performance, reliability and reduced cost are reviewed, currently-available components and methods are described, and the prospects for future advances are considered.

Aqueous Electrolyte Solutions Near Carbon Surfaces: How Atomistic Simulation Studies Led to the Design of a Continuous Desalination Cell with High Performance Predicted for Brackish Water

Tuan A. Ho and Alberto Striolo*

University College London
Department of Chemical Engineering
Torrington Place, London WC1E 7JE, United Kingdom

* Author to whom all correspondence should be addressed:

Email: a.striolo@ucl.ac.uk
Phone: +44 020 7679 3826
Fax: +44 020 7383 2348

Abstract

Our group, and many others have investigated aqueous electrolyte solutions near graphitic electrodes and/or through graphene membranes under various conditions. The results have been quantified in terms of structure of water, orientation of interfacial water molecules as a function of applied voltages, density distribution of electrolytes within the electric double layer, free energy profiles for water molecules and ions as they cross narrow pores carved out of graphene, etc. Using the same molecular dynamics simulation techniques, we managed to design a new desalination cell, which we identify as 'continuous membrane-electrode desalination cell'. This cell is operated taking advantage of both the operating principles responsible for capacitive and membrane-based desalination methods. Our results suggest that the advantages of either technique are retained, while the disadvantages are abated. Under optimum conditions, the desalination cell can continuously desalinate brackish water yielding high water permeability, rejecting 100% ions, recovering up to 70% of salty water, and removing up to ~0.67 monovalent ions per unit charge (electron) applied on the electrodes (~67% electrical energy efficiency). These results suggest that, if our predictions are correct, when implemented in practical devices, the membrane-electrode desalination cell and the process proposed here could help easing the water shortage facing the world today.

Intensification of breaking of water-in -oil emulsions by membranes, treated in the area of corona discharge or in the plasma flow

Dryahlov V.O, Shaihiev T.I., Shaihiev I.G., Galihanov M.F., Bonev B.S.
The Kazan National Research Technological University

Water-in-oil emulsions (WOE) are formed as a result of petroleum processing and utilizing of oil products in the enterprises of chemical and petrochemical profile all over the world at the rate of more millions t/year. WOE form the aggressive stable multicomponent structure which contains surfactant species (SS) and different emulsifiers such as salts, asphaltens, gums, oil-soluble organic acids and minute impurities such as slime and clay.

In recent times ultrafiltration method is used for the breaking process of WOE. The main advantages of this method are: high efficiency, deficiency of existing reagents, small floor space for equipment, energy efficiency, and so the opportunity of close water cycle organizing. Disadvantage of membrane methods is the pollution of membrane surface by the dispersion phase of pollutant. As a result the slowdown of separating process occurs.

In view of the forgoing, high-powered membranes treatment is perspective for the purpose of ultrafiltration intensification of WOE with engineering and ecological side, particularly, in the field of corona discharge or in the flow of low temperature high frequency plasma of low pressure.

Research results of breaking of model WOE by polyacrylonitrile (PAN) membranes with the molecular weight cut-off (MWCO) 25 kDa are introduced in this work. These leaked particles were treated in the field of corona discharge or in the plasma flow. The treatment by the corona discharge was held under the next conditions: productive time (τ) was 0.5, 1, 1.5 min, electrode voltage (U) was 5, 10, 15 kV; plasma treatment was done in gas environments of argon and nitrogen, argon and air during (τ) – 1.5, 4, 7 min for anode voltage (U) - 1.5, 3.5, 5.5 and 7.5 кV.

Productivity and effectiveness are considered as the main parameters of membrane division of emulsion. The first factor is the ratio of quantity of passed flow though membrane of separating environment to production of the process life and area of the filter element, which is $1{,}73 \cdot 10^{-3}$ м^2 in this case. Effectiveness is determined by change of values of chemical oxygen demand (COD) before and after the process of emulsion division. This process was measured on the autotitrator «T70» of «Mettler Toledo» producer company . Results are introduced on the figures 1-3 and in the Table 1, 2.

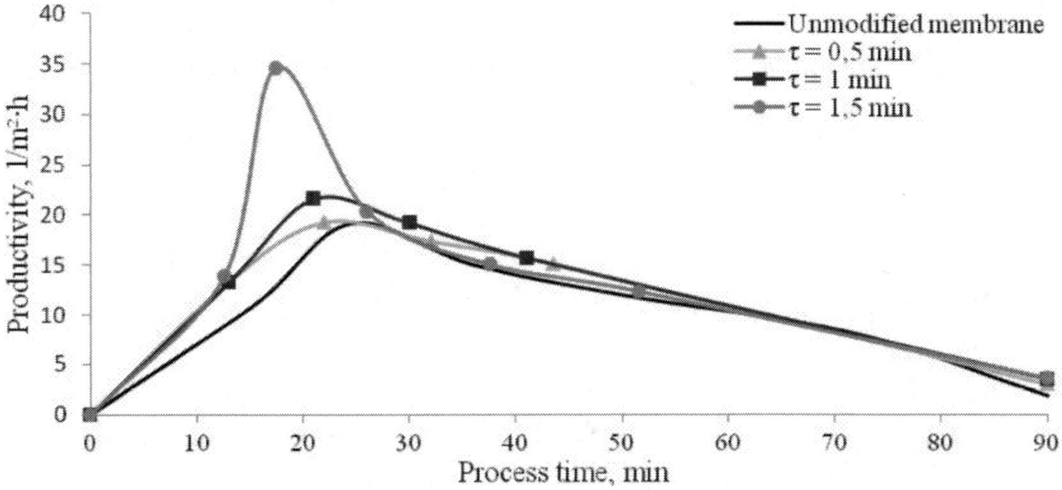

a)

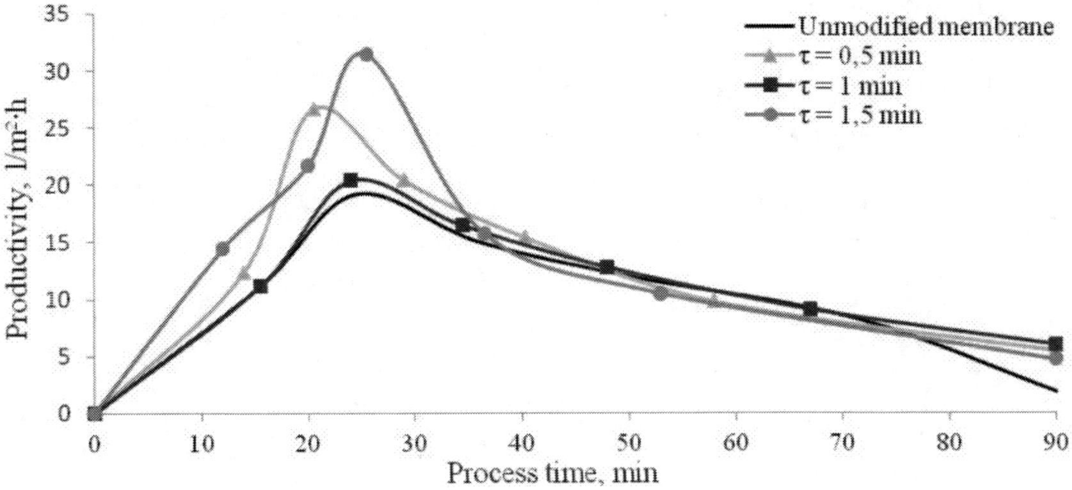

b)

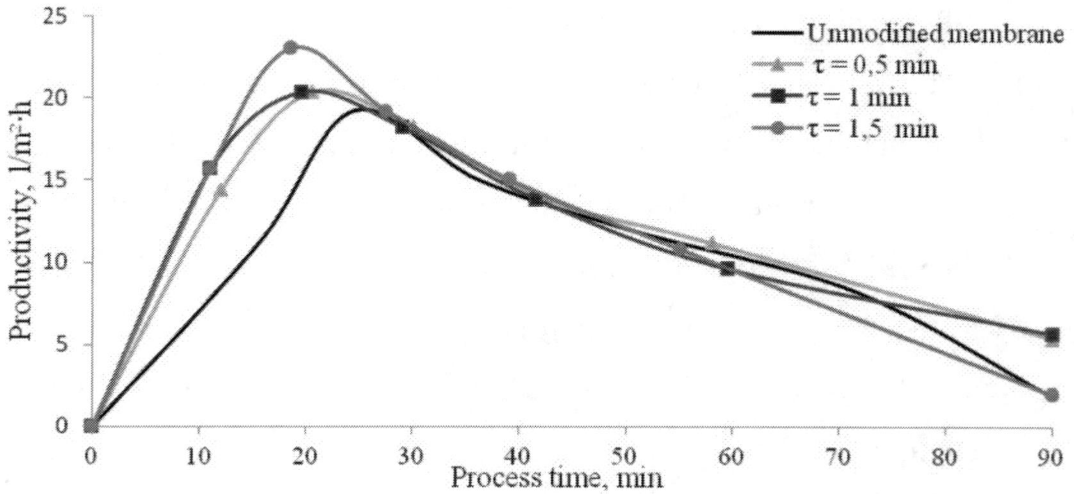

c)

Figure 1 –Productivity of emulsion breaking by PAN membranes with the MWCO 25 kDa, which are electret under the next conditions: a) U = 5 kV; b) U = 10 kV; c) U = 15 kV.

Curves analysis presented on the figure 1 shows the increase in productivity of modified PAN membranes with MWCO 25 kDa till 2 times. These leaked particles were treated in the field of corona discharge in all regimes of electro treatment (Figure 1a-c). In addition of the above the great value of the present parameter for each value U is observed when τ = 1,5 min, and the greatest value was registered when U = 5 кV.

Table 1 – received COD filtrates under division by electret PAN membranes with the MWCO 25 kDa.

U, kV	COD, мg O_2/l		
	Productive time, τ, min		
	0,5	1	1,5
5	10002	7803	7639
10	6578	5642	6810
15	6812	6339	7462
Unmodified membrane	8520		
Emulsion	166550		

Presented values in the table 1 show improving effectiveness of emulsion breaking of modified PAN membranes with the MWCO 25 kDa in 8 cases from 9. The value of COD filtrate received by the use of unmodified membrane is 8520 мg O_2/l, properly, efficiency factor is 95,0 %. The bottom value of COD filtrate is 5642 мg O_2/l when using membrane electreted under U = 10 kV and τ = 1 min, the efficiency factor is 96,5 %.

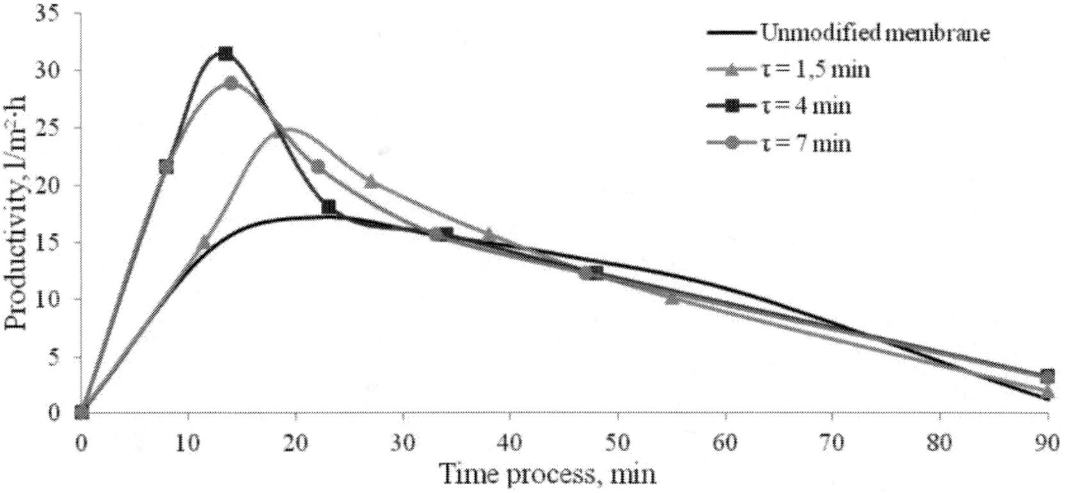

a)

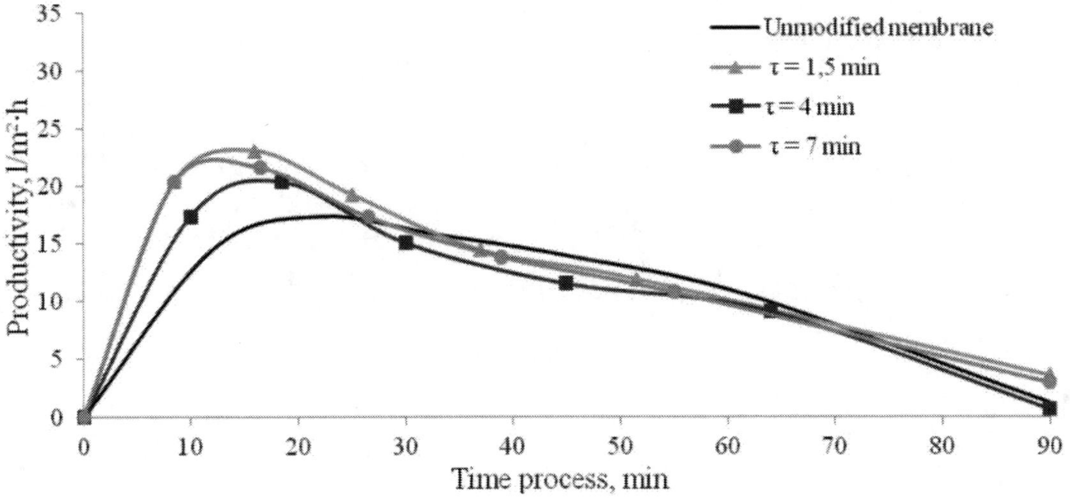

b)

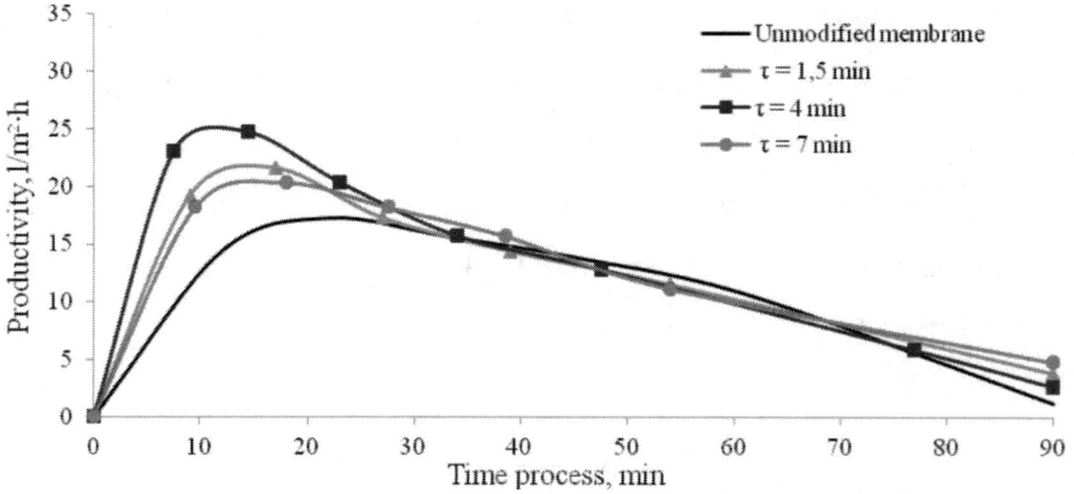

c)

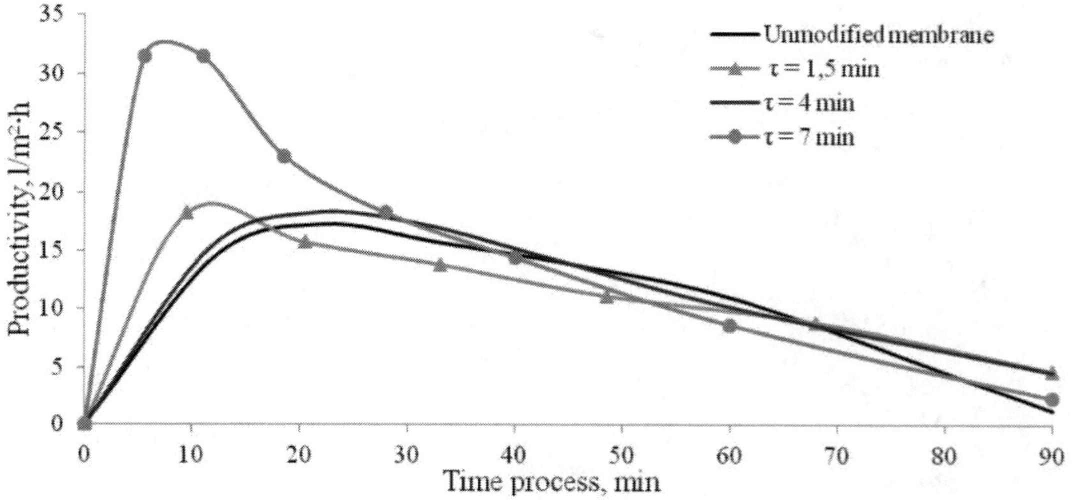

d)

Figure 2 – Productivity of emulsion breaking by PAN membranes with the MWCO of leaked particles 25 kDa, treated in the plasma flow in gaseous medium of argon and nitrogen when the value of anode voltage is: a) U_a = 1,5 kV; b) U_a = 3,5 kV; c) U_a = 5,5 kV; d) U_a = 7,5 kV.

The previous revealed effect of increase in productivity of modified membranes in comparison with unmodified membrane as high as 2 times is registered (Figure 2) as a result of plasma treatment in argon and nitrogen medium of PAN membranes with the MWCO 25 kDa. In addition to above the immense increase of the present parameter was registered when U = 1,5 кV and τ = 4 and 7 min, U_a = 7,5 кV and τ = 7 min, in other cases the increase was feebly marked.

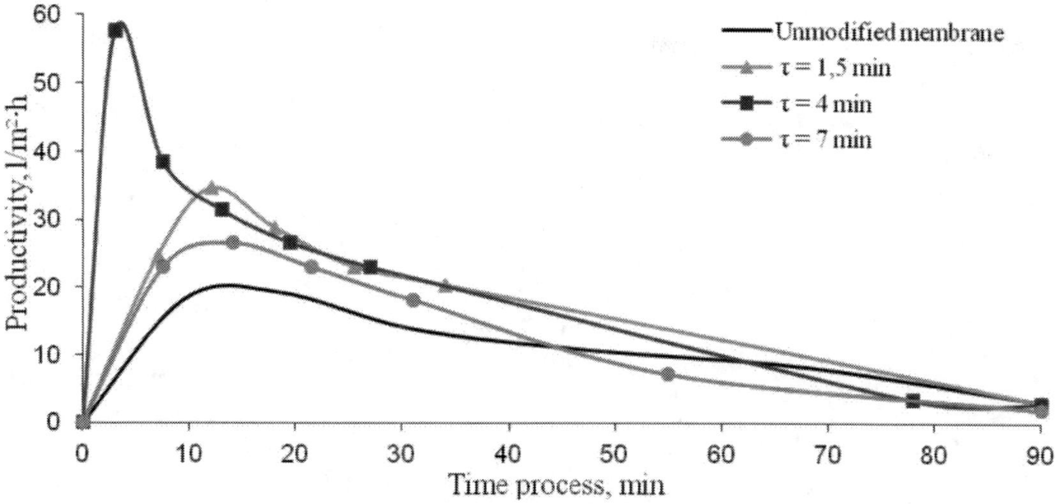
Productivity, l/m²·h
Time process, min
Unmodified membrane
τ = 1,5 min
τ = 4 min
τ = 7 min
a)

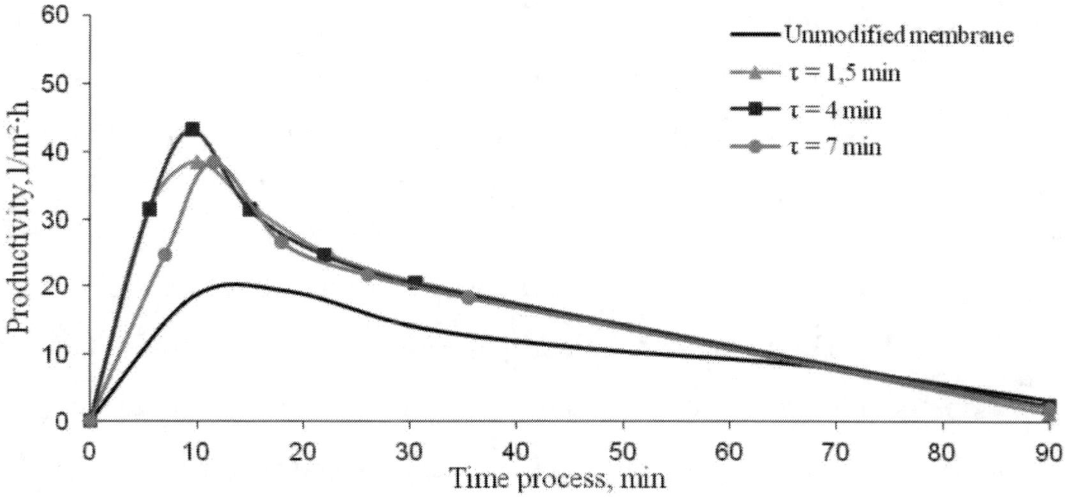
Productivity, l/m²·h
Time process, min
Unmodified membrane
τ = 1,5 min
τ = 4 min
τ = 7 min
b)

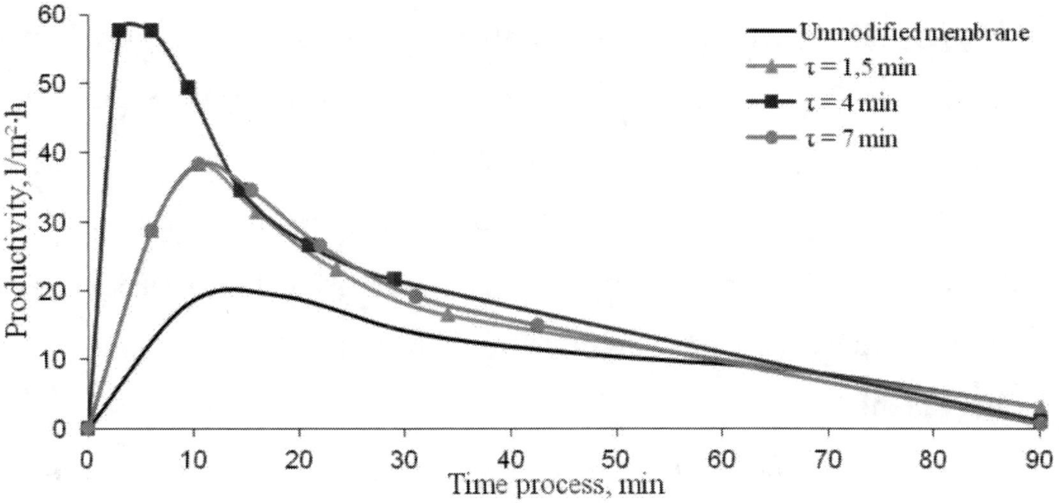
Productivity, l/m²·h
Time process, min
Unmodified membrane
τ = 1,5 min
τ = 4 min
τ = 7 min
c)

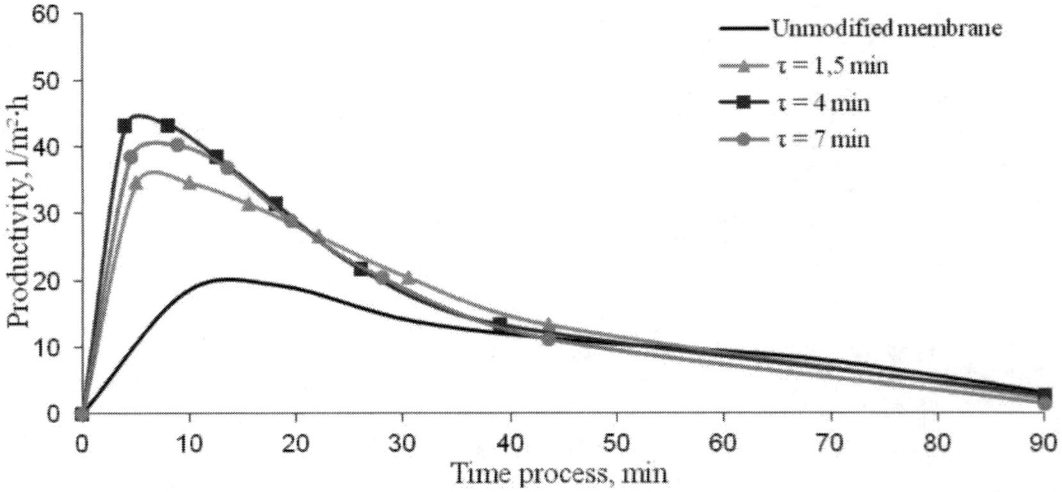

d)

Figure 3 – Productivity of emulsion breaking by PAN membranes with the MWCO 25 kDa, treated in the plasma flow in gaseous medium of argon and air when the value of anode voltage is: a) U_a = 1,5 kV; b) U_a = 3,5 kV; c) U_a = 5,5 кV; d) U_a = 7,5 кV.

Plasma treatment in the argon and air medium of PAN membranes with the MWCO 25 kDa promotes the great increase in productivity by a factor of three in comparison with unmodified membrane. At that on all occassions the largest values of the present parameter are registered when treatment time τ = 4 min, and maximum is registered when U = 5,5 kV.

As follows from the analysis of facts presented in the table 2, it was found that plasma treatment of PAN membranes with the MWCO 25 kDa promoted values decrease of COD filtrates of modified filter elements, in this example in 58,3 % cases. Decrease of present parameter occured with increase of values U and τ in the case of plasma influence in argon with air medium. Analogous dependance was not determined for membranes subjected to plasma treatment in argon and nitrogen medium.

The least value COD = 4610 мg O_2/l was obtained when membrane using, treated in argon and air medium under U_a = 7,5 кV, τ = 7 min; values of COD filtrate, after division of unmodified membrane are 7050 mg O_2/l, properly, the efficiency factor is 97 and 95 %.

Table 2 – received COD filtrates under breaking of emulsion by plasma treated PAN membranes with the MWCO 25 kDa

Gaseous medium	U_a, кV	COD, мg O_2/l		
		Plasma treatment time, min		
		1.5	4	7
Argon+nitrogen 70 : 30	1.5	7717	6130	5530
	3.5	5240	6630	7835
	5.5	5660	6450	7290
	7.5	6680	7330	10030
Argon air 70 : 30	1.5	7340	6500	5730
	3.5	8430	7710	7540
	5.5	8290	6250	5740
	7.5	5820	5130	4610
Unmodified membrane		7050		
Emulsion		166550		

This is the way how positive effect of corona discharge influence and plasma on PAN membranes for the purpose of increase in productivity and selectivity of WOE division was determined based on the results of the research conducted.

385843 Magnetically Responsive Self-Cleaning Micro-Mixing Nanofiltration Membranes

Abstract:

Nanofiltration (NF) is a relatively new membrane separation technology mainly used for removing low molecular weight species from water. NF membranes suffer from concentration polarization thus aggravating membrane fouling and compromising the performance of membrane quickly. Magnetically responsive nanofiltration (NF) membranes with superparamagnetic nanoparticles attached to the chain ends of the grafted polymer nanolayers have been shown to be effective in breaking concentration polarization at the membrane-liquid interface under an appropriate external oscillating magnetic field. NF270 is one of the most commonly used NF membranes for removing low molecular weight species and divalent salt ions from the feed solutions. The effects of polymer grafting degree varying both the chain length and chain density were studied. In addition, the micro-mixing effects of magnetically activated membranes on flux and rejection for various concentrations of NaCl, $CaCl_2$, $MgCl_2$, $MgSO_4$, Na_3PO_4 solutions as well as for feed solutions containing non-charged glucose and sucrose molecules. The magnetic field effect becomes more evident with increasing grafting degree as well as at high feeding concentration or for higher charged species after long period of operation. Moreover, our results also shed significant light on the effects of magnetic field on ion hydration and water structure.

387184 Reduction of Humic Acid Fouling on Polysulfone Ultrafiltration Membrane By Using Pulsed Corona Discharge As the Pretreatment Method

Abstract:

Humic acid (HA) fouling on membrane is a well-known problem in water treatment. The complex chemical structure of humic acids and wide range of molecular weights make HA a persistent foulant leading to irreversible decline of membrane performance. Several modification approaches such as membrane surface chemical modification, incorporation of nanoparticles into membrane and pore morphological modification have been used to increase the antifouling ability of membrane. Pretreatment approaches can be applied to change the physicochemical properties of HA and reduce the fouling potential. In this study the effect of pulse corona discharge at 120 hertz frequency on polysulfone membrane flux ad flux recovery ratio was investigated. The flux decline was reduced for the influent which had been exposed to a pulsed corona streamer discharge batch reactor, with point to point electrodes.

369509 Computational Generation of Low-Energy Configurations of Zeolite NaY

Abstract:

With growing interest in the potential of zeolites to cost-effectively separate gas mixtures of industrial significance [1-2], computational methods are used to model adsorption and diffusion in these microporous materials. Many computational studies focus on pure-silica zeolites due to their readily-available and perfectly-periodic crystal structures [3-4]. Most zeolites, however, have never been synthesized in pure-silica form, and nearly all zeolites used in practice are defined by a Si/Al ratio and the presence of extra-framework cations. Averaged crystal structures of zeolitic materials are available from the International Zeolite Association [5], for example, with atoms described by fractional occupancies. However, to consider such materials in molecular simulations, it is necessary to generate realizations of typical configurations of the aluminum and cations.

We have developed a novel computational framework for modeling cation-modified zeolite structures. The approach consists of dividing a zeolite structure into small clusters centered around each T-atom site, which can take on either Si or Al. The possible states of the neighborhood around each T-atom site are sampled, and for each configuration the energy difference between a Si or Al center is estimated using electronic structure calculations [6]. The cluster energies are combined to construct a larger section of modified zeolite crystal with low energy using mixed-integer linear optimization.

The approach is demonstrated with zeolite NaY, selected for its importance to the petrochemical industry and the simplicity of its faujasite framework in which all T-atoms are symmetrically equivalent. For Si/Al ratios between 2.25 and 2.85, we have generated low-energy configurations of a periodic unit cell of NaY suitable for use in molecular simulations of adsorption and diffusion. This opens the possibility for modified zeolites to be considered in computational screenings of zeolites for applications in separations and catalysis.

References:

1. Hasan, M. M. F., First, E. L., and Floudas, C. A. Cost-effective CO_2 capture based on *in silico* screening of zeolites and process optimization. Physical Chemistry Chemical Physics, 15(40):17601–17618, 2013.

2. First, E. L., Hasan, M. M. F., and Floudas, C. A. Discovery of novel zeolites for natural gas purification through combined material screening and process optimization. AIChE Journal, 60(5):1767–1785, 2014.

3. First, E. L., Gounaris, C. E., Wei, J., and Floudas, C. A. Computational characterization of zeolite porous networks: an automated approach. Physical Chemistry Chemical Physics, 13(38):17339–17358, 2011.

4. First, E. L., Gounaris, C. E., and Floudas, C. A. Predictive framework for shape-selective separations in three-dimensional zeolites and metal–organic frameworks. Langmuir, 29(18):5599–5608, 2013.

5. Baerlocher, C. and McCusker, L. B. Database of Zeolite Structures. http://www.iza-structure.org/databases/, 2014.

6. Frisch, M. J. et al. Gaussian 09, Revision A.02. Gaussian, Inc., Wallingford CT, 2009.

Prediction of Adsorption Properties of Cyclic Hydrocarbons in MOFs Using DFT-Derived Force Fields

Jason A. Gee, Ambarish R. Kulkarni and David S. Sholl*

School of Chemical and Biomolecular Engineering, Georgia Institute of Technology, 311 Ferst Drive NW, Atlanta, Georgia 30332-0100

*Corresponding author: david.sholl@chbe.gatech.edu

We present an extension of previous methods that derive transferable force fields for describing the adsorption of small molecules in zeolites[1] based on Density Functional Theory (DFT) calculations to examine the adsorption of cyclic hydrocarbons in metal-organic frameworks (MOFs). We begin by developing a classical force field based on hundreds of DFT calculations that probe the entire configuration space of the cyclic hydrocarbons in the accessible volume of the MOF. This procedure was repeated for several dispersion-corrected DFT functionals including PBE-D2, PBE-D3, vdW-DF, and vdW-DFT2. These force fields were then used to predict the adsorption properties of these molecules in the MOF MIL-47 using Grand Canonical Monte Carlo (GCMC) simulations. We find that this approach can generate force fields that meet or exceed the performance of generic force fields when compared to literature data for the adsorption of cyclic hydrocarbons in MIL-47. The transferability of our top-performing force field is demonstrated by comparing to adsorption data for cyclic hydrocarbons in other MOFs. While both the generic and DFT-derived force fields can reproduce qualitative trends observed in the experimental literature for these materials, we find that the latter method is able to provide better accuracy predictions for adsorption in these systems.

[1] H. Fang, H. Demir, P. Kamakoti, and D. S. Sholl, J. Mater. Chem. A, 2 (2014) 274

376353 Ionomers of Intrinsic Microporosity (IonomIMs): In silico Development of Ionic-Functionalized Gas Separation Membranes

Abstract:

This work presents the predictive molecular simulations of a polymer of intrinsic microporosity (PIM) functionalized to contain an ionic backbone with extra-framework counter ions for CO_2 gas storage and separation applications. The introduction of the highly CO_2-phillic carboxylate groups maintain high degrees of free volume from the inefficiently packed polymer chains without filling pore volume, with the surface areas increasing with decreased concentration of ionic groups from 100% to 17%. As a result of the ionic groups increasing the CO_2 enthalpy of adsorption, the uptake of the proposed polymers at 293 K and ambient pressure exceeded previous microporous ladder polymer capacities. In addition, the CO_2/CH_4 and CO_2/N_2 mixed-gas separation performances were evaluated at several industrially-relevant conditions, where the IonomIMs are shown to increase both the working capacity and selection performance in certain pressure swing applications, and outperformed many leading metal organic framework materials. These simulations reveal that intrinsically microporous ionomers show great potential as the future of energy efficient gas separation polymeric materials.

384584 Combining *Ab Initio* and Classical Simulation Techniques in a High-Throughput Fashion to Assess Hydrogen Uptake in Metal-Organic Frameworks

Abstract:

New high-throughput techniques, both experimental and computational, are beginning to have a significant impact on materials research and design. One class of materials impacted by this new trend is metal-organic frameworks (MOFs). MOFs are self-assembled from inorganic nodes and organic edges or linkers, resulting in porous, crystalline materials. MOF building blocks (nodes and linkers) can be interchanged and/or altered, resulting in a virtually limitless number of possible structures. Particularly exciting about these materials is the ability to tailor them, through judicious choice of building blocks, for particular applications. Additionally, relevant chemical functionalities such as hydroxyl, amino, and metal alkoxides, may be introduced, therefore adding to the chemical diversity of these materials.

We used novel computational high-throughput techniques to address the hydrogen storage challenge using MOFs. Hydrogen storage remains as, arguably, the biggest challenge towards the development of future fuel cell vehicles. The United States Department of Energy (DOE) has set hydrogen storage targets of 7.5 wt.% and 70 g/L at a minimum temperature of 243 K and a maximum pressure of 100 bar; to date, no material has met these targets. Thanks to their large pore volumes and surface areas, MOFs excel at hydrogen storage near cryogenic temperatures. However, they falter near room temperature due to the low heats of adsorption (~10 kJ/mol) of the hydrogen molecules in MOFs. In order to counteract this, we introduced open metal sites in the form of metal alkoxides. We used quantum mechanical calculations to calculate hydrogen binding energies on a series of metals, and we fit a Morse potential to the interactions to use as a force field in our grand canonical Monte Carlo (GCMC) simulations. We determined Mg alkoxide to be a promising candidate for improving hydrogen storage in MOFs.

Then, we modified our recently published methodology for computationally building MOFs to include the Mg alkoxide groups. Briefly, we took experimentally synthesized structures and disassembled them into their building blocks (inorganic nodes and organic linkers). We then introduced the Mg alkoxide functionalities into the linkers and recombined them with the nodes to form the MOF structures. This resulted in over 18,000 structures with varying Mg alkoxide content, as well as a wide range of physical properties such as void fraction, pore volume, and surface area. We used GCMC simulations to calculate hydrogen uptake in all generated structures at 2 bar and 100 bar at 243 K. The most promising structure in gravimetric storage is predicted to adsorb 9.35 wt.% hydrogen, meeting the DOE target. Volumetrically, the most promising structure is predicted to adsorb 51 g/L of hydrogen, falling just short of the DOE target. We also found trade-offs between gravimetric and volumetric uptake along with new structure-property relationships. Finally, our computational screening techniques allowed us to assess what could be the performance limits of MOFs for hydrogen storage applications near room temperature.

384734 Exploration of the Relationship Between Bare Sorbent Flexibility and Resulting Adsorption Behavior Via Flat-Histogram Monte Carlo

Abstract:

Recent years have seen the discovery and development of adsorbent materials that deform or flex under stress, including gas pressure due to exposure to an adsorbate and external mechanical stress. Such materials are interesting from a fundamental physical point of view, but also for practical engineering applications. In particular, deformable sorbent materials might prove to be superior materials for gas separation applications since their pore space or apertures can expand or contract in a manner that favors the adsorption of one gas in a mixture over the others. Molecular simulation of such materials is advancing, but progress has been limited by the inability of conventional simulation methods to account for framework flexibility in a computationally tractable manner. In this work, we discuss a method for simulating adsorption in flexible porous materials based on flat-histogram Monte Carlo methods. The method, which is applicable in both the "flexible" grand canonical ensemble and the osmotic ensemble, computes a multi-dimensional probability distribution, from which we are able to extract adsorption isotherms, free energies, and limits of stability. We use this method with for a variety of models of adsorbent/adsorbate systems to explore the link between a bare sorbent's flexibility and the range of adsorption isotherms that can result, while keeping the gas-adsorbent interaction fixed. Our results show that a particular mode of flexibility results in qualitatively similar adsorption isotherms for a wide variety of gas-adsorbent interactions. Consequently, one can envision the use of this method for designing new sorbent materials or specifying the material characteristics that an adsorbent must possess to yield desired or specified behavior.

386525 Simulating the Sorption of Small-Molecule Growth Modifiers on Zeolites

Abstract:

Microporous zeolite materials find use in a variety of important industrial applications including fluid-based separations, ion exchange, hydrocarbon cracking, and methanol reformation. Despite their widespread use, fully optimizing zeolites for these applications remains a formidable challenge due to the difficulty in controlling crystal morphology and structure during synthesis. It has been recently demonstrated, however, that such control may be gained through the addition of small-molecule structure directing agents, or zeolite growth modifiers (ZGMs), which bind to specific crystal surfaces and inhibit growth along targeted crystallographic directions [1]. Using small polyols as ZGMs, this strategy has been successfully applied to synthesize zeolite L crystals (LTL type) with controlled pore lengths, suggesting a novel strategy for tuning the transport properties of zeolites [1].

Although ZGMs show tremendous promise in facilitating the development of improved zeolite materials, the underlying molecular-level mechanisms that allow them to regulate crystal growth are not fully understood. To elucidate such mechanisms, we employ molecular simulation techniques to investigate the interactions between small polyol ZGMs and the different crystal faces of zeolite L. While bulk materials have been studied extensively using computational methods, modeling the crystal faces of zeolites poses a unique set of challenges due to the complex surface chemistry involved in terminating the crystal. We discuss these challenges in detail and describe several different approaches for modeling zeolite surfaces. We then demonstrate that state-of-the-art free energy techniques, such a metadynamics and umbrella sampling, can be used to gain significant insight regarding the sorption and binding of ZGMs on zeolites. Our findings support the hypothesis that polyols form hydrogen bonds with the surface of zeolite L, inhibiting growth through a molecular recognition process that is sensitive to both the structure of the polyol and zeolite [1]. Finally, we discuss the boarder implications of our findings and the potential of computational techniques to aid in the judicious selection of ZGMs to target specific crystal faces of zeolites.

[1]. Lupulescu, A.I., Kumar, M., Rimer, J.D., A Facile Strategy to Design Zeolite L Crystals with Tunable Morphology and Surface Architecture, J. Am. Chem. Soc., **135**, 6608-6617

Simulation of the critical conditions of polymer adsorption on porous substrates

R. Cimino[1], C. J. Rasmussen[2], Y. Brun[2], and A. V. Neimark[1,*]

[1]*Department of Chemical and Biochemical Engineering, Rutgers, The State University of New Jersey, 98 Brett Road, Piscataway, NJ 08854, USA.*

[2]*DuPont Central Research & Development, Corporate Center for Analytical Sciences, Macromolecular Characterization, Wilmington, DE 19803, USA*

Polymer adsorption has been simulated extensively over the past sixty years. Until recently, studies have been confined to the cases of adsorption at planar surfaces or within pores with complete confinement (see Figure, states 2 and 3). For adsorption at surfaces, the critical conditions of polymer adsorption (CPA) are well characterized for both ideal and excluded volume (real) chains. For the case of pore confinement, it has been shown that the CPA, while well defined for ideal chains, is ill defined for real chains. At the same time the existence of CPA is implied in practical chromatography of polymers; the regime of liquid chromatography at critical conditions (LCCC) is widely used for separation and characterization of macromolecules and biopolymers. By using the incremental gauge cell MC method and a model representation of a substrate perforated with nanoscale pores, we have extended the thermodynamic definition of critical conditions of polymer adsorption to porous substrates. We found that the overall partitioning is determined by several adsorption mechanisms, including complete adsorption in pores of the stationary phase, adsorption on its external surface, and most importantly by partial adsorption, as illustrated in Figure. As such, we provide an affirmative answer to the long standing problem: whether the critical conditions of

* Corresponding Author: aneimark@rutgers.edu

polymer adsorption may exist in the case of porous substrates for non-ideal polymer chains. Furthermore, we concluded that the critical conditions are determined by partially adsorbed chain conformations, which dominate around the critical point of adsorption.

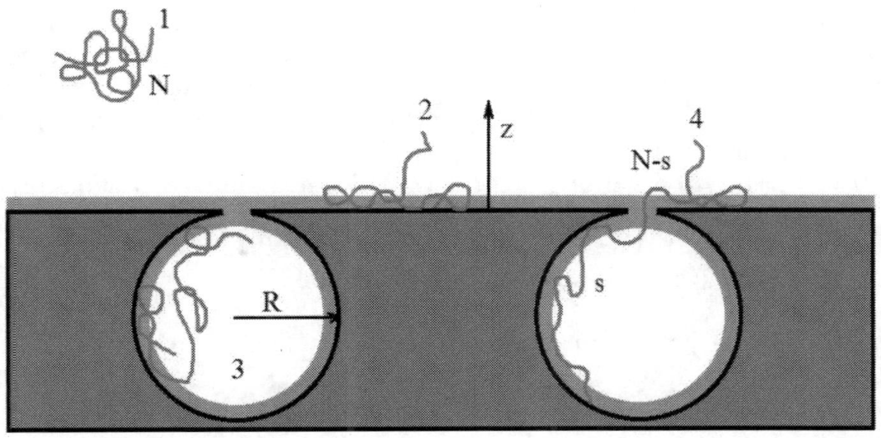

Figure. Different chain conformations; 1 – free chain, 2 – chain adsorbed at the external surface, 3 – chain completely confined within the pore, 4 – partially adsorbed conformation with *s* monomers within the pore and *N-s* monomers outside.

This work was supported by the NSF GOALI Multiscale Modeling of Adsorption Equilibrium and Dynamics in Polymer Chromatography.

391505 Understanding DABCO Nanorotor Dynamics and Water Effects in Isostructural Metal-Organic Frameworks

Abstract:

Flexible framework dynamics present in the subset of metal-organic frameworks known as soft porous crystals gives rise to interesting structural properties that are unique to this class of materials. In this work we use experiments and molecular simulation to provide insight into the highly dynamic nanorotor behavior of the DABCO (1,4-diazabicyclo[2.2.2]
octane) ligand in the pillared Zn-DMOF and Zn-DMOF-TM structures. While DABCO is known to be displaced in the presence of water in the parent Zn-DMOF structure, the Zn-DMOF-TM variation is highly stable even after adsorbing significant amounts of water vapor.[1] The dynamics of DABCO in the presence of water guest molecules is therefore also explored in the Zn-DMOF-TM structure via in-situ experiments along with molecular simulation. This analysis not only provides fundamental insight into the inherent and guest-induced intraframework dynamics in these structures but also has implications in understanding the mechanism of water stability within these materials, an important criterion in the evaluation of metal-organic frameworks for real world applications.[2]

[1] H. Jasuja, N. Burtch, Y.G. Huang, Y. Cai, K. Walton, Langmuir. 29 (2013) 633-642.
[2] N. Burtch, H. Jasuja, K. Walton, Chem. Rev. DOI: 10.1021/cr5002589.

Using PAT to Characterize Scale-up of Mixing Elbows Injectors for Continuous Crystallization of APIs

High shear mixing elbow injectors are used for crystallization processes to improve control of particle size (Polster 2013). Elbow injectors are similar to t-mixers and can be configured in a variety of ways to achieve mixing requirements. Poor mixing can lead to uncontrolled crystallizations with increased secondary nucleation. These mixing elbows can provide rapid controlled micromixing. Previous studies have been performed using the Fourth Bourne Reaction (Bourne 2003) to characterize mixing times (Singh, Spencer et al. 2005) in mixing elbow injectors. This work aims to further characterize mixing times in mixing elbows using the Fourth Bourne Reaction focusing on scale up and the configuration of the mixing elbow. This process was also used to further understand the role of different types of mixing within the elbow as the process is scaled up. A comprehensive lab scale study of a ¼" elbows was undertaken using the Fourth Bourne Reaction as well as Computational Fluid Dynamics. The possibility of using inline PAT, such as ATR-FTIR to monitor acetone concentration resulting from various conditions was investigated in order to increase the efficiency of the study as the scale of the process increases to 3/8" and ½" mixing elbow injectors. Inline PAT such as FBRM and ATR- FTIR is also essential for monitoring the process during case studies of model crystallization systems.

References

Bourne, J. R. (2003). "Mixing and the Selectivity of Chemical Reactions." ChemInform **34**(40): 471 - 508.

Polster, C. (2013). "Design and Scale-Up of the Continuous Reactive Crystallization of an API" AIChE Annual meeting, Pharmaceutical Discovery, Development and Manufacturing Forum (466c)

Singh, U. K., et al. (2005). "Incorporation and Characterization of a Mixing Elbow on the Pilot Plant Scale for a Mixing Sensitive Crystallization of an API." Industrial & Engineering Chemistry Research **44**(11): 4068-4074.

383588 How to Monitor Crystallization Processes in Manufacturing

Abstract:

The *in situ* measurement and imaging of crystals and crystal structures has become ubiquitous in many industrial laboratories during the development of crystallization processes. Standard tools such as ParticleTrack and PVM (formerly Lasentec) provide scientists with deep process understanding allowing them to develop higher quality processes for transfer to manufacturing.

In many cases it is desirable to apply the same tools at the manufacturing scale in order to compare performance from the lab to the plant and to continuously monitor and control the process for variability. Such an approach can also act as an insurance policy should a process fail, allowing a data-based root cause analysis to be performed and corrective measures to be implemented

This presentation will address some of the most common concerns and challenges associated with applying probe-based particle measurement techniques in large scale manufacturing.

- Building a business case for implementation of PAT in manufacturing
- Ensuring representative crystallization monitoring under large scale mixing conditions
- Working with plant managers to ensure effective installation, maintenance and data transfer
- Case studies in industrial crystallization monitoring

384304 Applications of Process Analytical Technologies in the Solid State Development of Pharmaceutical Compounds

Abstract:

In the recent years, process analytical technology (PAT) has been proven to be very useful on the pharmaceutical development and manufacturing for better process understanding and control. For example, the information of the in-situ particle size distribution by Lasentec ® FBRM is invaluable on the crystallization development and scale-up. This paper discussed some less common but very important PAT application cases on the solid-state development of pharmaceutical compounds, for example, 1) the use of on-line mass spectrometer to monitor the form conversion of a de-solvated solid form, 2) the use of the combination of IR and Raman to construct the phase diagram for the crystallization development of pharmaceutical co-crystals, 3) crystallization control using FTIR or Raman in the solvent swap crystallization.

379742 On-LINE Optimal Control of Crystal Properties in NON-Isothermal Antisolvent Crystallization

Abstract:

Controlling crystal properties has been a challenging issue in many crystallization operations due to inevitable disturbances, occuring during the process. This contribution proposes an online strategy to control crystal size distribution in joint cooling antisolvent crystallization systems.

As a first step in the study, digital image texturing analysis in the form of fractal dimension and energy signatures is applied and automated to track crystals' properties at different time of the experiment. The one dimensional Fokker-Planck equation (FPE) represents the dynamic characteristics of the crystal growth. A map of iso-mean and iso-variance curves are reported in an antisolvent flow rate-temperature plane based on evaluating the model at asymptotic condition. This allows identifying the region of input multiplicities in the system so punctual information regarding stability of the process can be obtained. Using controllability tools, regions of singularities needed to be avoided for proper control of the process is determined and it is further shown that the system is ill conditioned in the whole operating map. The problem is investigated experimentally where a multi-loop PID structure is tuned by pairing crystals' mean size with antisolvent feed rate and manipulating temperature to control the size variation. The proposed controller is demonstrated to be incapable of bringing the system into the desired set points and an offset in both crystals' mean size and variance is observed at the end of the batch. In the next step, a feed forward approach is developed in which temperature and flow rate trajectories are obtained from a static optimization procedure. The sensitivity of feed forward control strategy is assessed in the case of disturbances in antisolvent feed rate. Finally, in order to achieve a better performance a two stage controller is set up by combining the feed forward approach with a feedback controller which is designed based on IMC *(Internal Model Control)* algorithm. It is indicated that a perfect control over crystals' mean size can be obtained in this case by renouncing a strict control over size distribution coefficient of variation. Experimental validation of the strategies are carried out for the ternary system of water-ethanol-sodium chloride.

372414 The Evaluation of the Temperature Dependence of Gate-Opening Phenomena in ZIF-8

Abstract:

The onset of gate-opening behavior in ZIF-8 is evaluated through a detailed gas adsorption study. The sharp increase of adsorbent loading at threshold pressures in ZIFs occurs through a structural deformation of the flexible imidazolate linkers at the pore mouth, with potential applications to selective adsorption processes. Increases in temperature show an expected decrease in adsorbent loading along with a transition from Type IV isotherms to Type I isotherms. The adsorbate pressures and loadings at gate-opening are evaluated as a function of adsorbate nature, and the disappearance of the phenomena is related to system variables. Various gas isotherms are presented in a wide temperature range to investigate the thermodynamics of adsorption and structural deformation of ZIF-8.

An experimental and modeling study of the breathing properties of COMOC-2, a vanadium based MOF with very large adsorption capacity

Sarah Couck, Ying-Ya Liu, Karen Leus, Pascal Van Der Voort, Gino V. Baron, Joeri F. M. Denayer

COMOC-2 is a metal-organic framework with V^{4+} centers and biphenyl-4,4-dicarboxylate linkers; due to its larger linkers it can be considered as an extended version of the MIL-47 structure (V^{4+} and benzenedicaboxylate). It exists of lozenge shaped channels with a width of 18.9 Å × 19.4 Å. This results in a structure with a large micropore volume of 0.72 cm^3/g, while MIL-47 for example only has a micropore volume of 0.46 cm^3/g.

We have studied the adsorption properties of this material by measuring isotherms of a range of compounds. Although the material has vanadium as metal center, which normally results in a rigid MOF structure, COMOC-2 shows pronounced flexible behavior and breathing transitions under workable conditions. Molecules like CO_2, ethane, ethene, propane, propene, butane and hexane induce the breathing like behavior, typical associated with the MIL-53 series (M^{3+} and benzenedicaboxylate). For all of these compounds, 2-step isotherms have been observed, with a capacity of about 4 mmol/g in the first plateau and up to 14 mmol/g in the second plateau for CO_2 at high pressure (50 bar). COMOC-2 is initially in a large pore (lp) state and transforms to a narrow pore (np) state upon guest adsorption. At higher pressure, the structure opens again to the lp state (Liu et al; 2013).

Inverse pulse chromatography showed that the material is non shape selective towards *n*- and iso-alkanes, as expected from the large pore size in the lp form. No specific interactions with double bonds or aromatic rings were observed; the adsorption potential is dominated by Van der Waals interactions.

The effect of the pronounced breathing effect (leading to a 3-fold increase in capacity), on separation performance was assessed via breakthrough and gravimetric experiments at different gas mixture compositions and pressures, for CO_2/CH_4 and ethane/propane gas mixtures. A discontinuity in the relation between breakthrough adsorption capacity and total column pressure was observed, which could be attributed to the breathing of the material at a given pressure and gas mixture composition. Ideal Adsorbed Solution Theory predictions of the mixture isotherms reasonably well predicted the adsorption of the individual components in the mixture, although an overestimation of propane adsorption was made in the breathing transition zone. Further experiments and modeling are ongoing to refine these predictions.

Acknowledgements

Sarah Couck acknowledges FWO-Vlaanderen for financial support.

References

New V-IV-Based Metal-Organic Framework Having Framework Flexibility and High CO2 Adsorption Capacity, Y-Y. Liu, S. Couck, M. Vandichel, M. Grzywa, K. Leus, S. Biswas, D. Vollmer, J. Gascon, F. Kapteijn, J.F.M. Denayer, M. Waroquier, V. Van Speybroeck, P. Van der Voort; INORGANIC CHEMISTRY, 52 (1), 113-120, 2013

381374 Mixed-Linker Zeolitic Imidazolate Frameworks: Tuning Molecular Diffusion and Adsorption Properties

Abstract:

Mixed-Linker Zeolitic Imidazolate Frameworks: Tuning Molecular Diffusion and Adsorption Properties

Kiwon Eum, Fereshteh Rashidi, Christopher W. Jones, and Sankar Nair

School of Chemical & Biomolecular Engineering, Georgia Institute of Technology, Atlanta, GA

Metal-organic frameworks (MOFs) are a relatively novel and promising class of nanoporous materials consisting of organic linkers coordinated to metal ions in crystalline structures. They are attractive for use in separations or catalysis, among many other applications.[1,2] A subclass of MOFs, known as zeolitic imidazolate frameworks (ZIFs), form structural topologies equivalent to zeolites.[3] ZIFs have many advantageous properties including good thermal and chemical stability, high microporosity, and high surface area.[1] It is possible to tune the properties of MOFs for specific applications via chemical or structural modifications.[4] A post-synthetic method is one approach for introducing functional groups into MOFs.[5] Structural modification can be also achieved by mixed-linker synthesis, using more than one type of linker. In previous work, the synthesis of ZIF 8-90 mixed-linker frameworks with a combination of ZIF-8 and ZIF-90 linkers was demonstrated.[6]

The present work discusses the synthesis of large (over 50 μm size) ZIF 8-90 mixed-linker crystals varying the ratio of 2-methylimidazole/carboxaldehyde-2-imidazole linkers, and demonstrates that the ZIF-8-90 mixed-linker materials allow tuning of adsorption and diffusion properties of certain hydrocarbons. Volumetric uptake methods are used to study intracrystalline adsorption and diffusion phenomena of different gases in mixed-linker ZIF materials, facilitated by the ability to control the crystal size of the ZIFs. Particle size distributions are measured using dynamic light scattering or optical microscopy, and crystal composition analysis was performed using ^{1}H-NMR spectroscopy and micro-Raman spectroscopy. Specifically, it is shown that: (1) the hydrocarbon adsorption properties of ZIF 8-90 mixed-linker ZIFs change significantly with composition, and (2) the diffusion properties and selectivities show very large changes with mixed-linker composition. This first demonstration clearly shows the potential of this approach for use in a variety of molecular separations.

(1) Shah, M.; McCarthy, M. C.; Sachdeva, S.; Lee, A. K.; Jeong, H.-K. *Industrial & Engineering Chemistry Research* **2011**, *51*, 2179.

(2) Watanabe, T.; Keskin, S.; Nair, S.; Sholl, D. S. *Physical Chemistry Chemical Physics* **2009**, *11*, 11389.

(3) Pan, Y.; Lai, Z. *Chemical Communications* **2011**, *47*, 10275.

(4) Cohen, S. M. *Chemical Reviews* **2011**, *112*, 970.

(5) Demessence, A.; D'Alessandro, D. M.; Foo, M. L.; Long, J. R. *Journal of the American Chemical Society* **2009**, *131*, 8784.

(6) Thompson, J. A.; Blad, C. R.; Brunelli, N. A.; Lydon, M. E.; Lively, R. P.; Jones, C. W.; Nair, S. *Chemistry of Materials* **2012**, *24*, 1930.

382369 Hybrid Zeolitic Imidazolate Frameworks for Separation of Small Molecules

Abstract:

Zeolitic Imidazolate Frameworks (ZIFs) are a relatively new class of materials that fall under the broader classification of Metal Organic Frameworks. They have remarkable chemical and thermal stability and their well-defined crystal structures and pore connectivity make them attractive candidates in many separation processes. Recently, hybrid ZIF materials that incorporate two different organic linkers in the same framework have been reported [1]. By varying the composition of the linkers in these materials, it is possible to tune the porosity and functionality of these hybrid materials. In this work, we will demonstrate the use of gravimetric techniques and PFG NMR diffusiometry to study the adsorption and diffusion behavior of small molecules like water and alcohols in hybrid ZIF 8 – ZIF 90 systems. Results have shown that hydrophobicity and diffusivity of small molecules in hybrid ZIFs can be tuned by controlling the composition of linkers. We will also discuss the use of two-dimensional NMR spectroscopy to gain more insight into the structure of hybrid ZIFs. These experimental results will be correlated with the results from numerical computational techniques such as Monte Carlo simulations to study the impact of ordering of linkers on transport properties. We will discuss the potential of using hybrid ZIFs to economically and efficiently separate water and organic molecules from biomass based feedstock.

1. Thompson, J.A., et al., *Hybrid Zeolitic Imidazolate Frameworks: Controlling Framework Porosity and Functionality by Mixed-Linker Synthesis.* Chemistry of Materials, 2012. **24**: p. 1930-1936.

382830 MOFs for the Storage of Oxygen

Abstract:

Oxygen storage is vital for many applications including medical devices, emergency first responders, space and sea exploration, semiconductor fabrication, and scientific laboratory uses. In many of these areas, there is a desire to increase the amount of oxygen storage without greatly increasing the pressure or volume required. Metal-organic frameworks have been widely studied for the storage of gases such as hydrogen and CO_2 but very little has been done for oxygen storage. Oxygen isotherms from 1 to 30 bar were measured at various temperatures on several MOFs. A Toth isotherm model fit the data well and was used to determine isosteric heats of adsorption. High-pressure oxygen isotherms up to 140 bar at 25 °C showed much higher adsorbed quantities for some MOFs when compared to the zeolite NaX and a high surface area carbon. After undergoing 50 adsorption and desorption cycles, the MOFs showed no loss in oxygen adsorption capacity. Based on the high capacities and oxygen stability, MOFs are promising adsorbents for the storage of oxygen.

Nucleation and Growth of CuBTC MOF on Al_2O_3, ZnO, and TiO_2 ALD Functionalized Fibers

Paul C. Lemaire,[†] Mark D. Losego,[†] Junjie Zhao,[†]
Christopher J. Oldham,[†] Philip S. Williams,[†] Howard J. Walls,[‡]
Sarah D. Shepherd,[‡] and Gregory N. Parsons[*,†‡]

[†]Department of Chemical and Biomolecular Engineering, North Carolina State University, 911 Partners Way, Raleigh, NC 27695
[‡]RTI International, 3040 East Cornwallis Road, Research Triangle Park, NC 27709

Metal organic frameworks (MOFs) are mesoporous, crystalline organic-inorganic materials with chemically functionalized high surface areas. MOFs are attractive for filtration and gas storage applications, but deployment has been limited by the ability to immobilize them onto a useful platform. MOF deposition on bare polymers generally leads to small MOF loading fractions, poor MOF crystal quality, and low surface areas. A key limitation is fundamental knowledge about MOF nucleation and attachment on useful substrates.

Atomic layer deposition (ALD) has been observed to promote hydrothermal growth of CuBTC ($Cu_2(BTC)_3$) on various substrate surfaces. Here, we investigate three ALD oxide coatings (Al_2O_3, ZnO, and TiO_2) for enhancement of MOF growth on polypropylene fiber mats. The metal oxides deposited at 100 C and 1.1 Torr formed conformal coatings on polymer fibers. These coatings act as nucleation-enhancement layers for CuBTC and facilitate MOF growth throughout the fiber mat. ALD coating type is observed to affect MOF growth, crystal morphology, and surface area. Larger CuBTC crystals with a surface area of 509±151 m^2/g grew on TiO_2 coated fibers while ZnO coatings resulted in smaller crystals with a relatively higher surface area of 764±11 m^2/g. CuBTC grown on Al_2O_3 coated fibers had intermediate crystal size and BET surface area of 694±76 m^2/g. We find that CuBTC crystals nucleate more rapidly on ZnO coated substrates than either Al_2O_3 or TiO_3 coated fibers. Uniform CuBTC coverage occurs within 1 hour on ZnO surfaces versus ~8 hours on Al_2O_3 or TiO_2.

Three phenomena have been identified as potential mechanisms for these differences in CuBTC nucleation and growth: electrochemical etching of the oxide layers, relative isoelectric point (IEP) of the ALD coatings, and formation of a MOF reaction layer within the ALD coating. Under the acidic hydrothermal conditions, Al_2O_3 and ZnO ALD coatings are prone to dissolution while TiO_2 coatings are relatively stable. Despite etching, ZnO appears to significantly improve the uniformity and reduce the required time for CuBTC nucleation on fiber mats. ZnO coatings create a basic surface, which promotes copper retention and subsequent CuBTC nucleation. Al_2O_3 and ZnO coatings have also been observed to react with the organic ligand (benzene tricarboxylic acid) used in the hydrothermal synthesis of CuBTC, and form an Al-BTC and Zn-BTC MOFs. In this presentation, we will discuss the implications of these various mechanisms in determining the MOF nucleation and growth on these surfaces.

Investigation of Synthesis Solvent for Large-Pore Stabilized MIL-53 Compounds with Increased CO_2 Adsorption and Decreased Water Adsorption

William P. Mounfield, III[a] and Krista S. Walton[a]

a. School of Chemical & Biomolecular Engineering, Georgia Institute of Technology, Atlanta, GA 30332

In order to understand the effect of solvent on MOF synthesis and CO_2 adsorption properties, a well-studied MOF system must be chosen. MIL-53 is one of the most famous MOFs, with wide investigation into its gas adsorption properties. However, the study of MIL-53 has largely centered on its ability to undergo a large structural transition in the presence of certain guest molecules, leading to promising application in gas adsorption and chemical sensing. While the breathing transition has been probed extensively, the effect of solvent on the synthesis of the material and its breathing behavior is not well studied. Furthermore, MIL-53 has been synthesized with various metal nodes, positioning it as an excellent choice for the analysis in this work as it allows for the elucidation of solvent effects on the structural transition, metal center interactions, and gas adsorption performance.

Large-pore (lp) stabilized MIL-53 metal-organic frameworks prepared via solvothermal synthesis strategies that possess no or only slight breathing behavior will be discussed. PXRD confirms the lp-stabilized materials remain in lp form, while others undergo a gradual breathing transition compared to MIL-53 prepared by traditional methods. SSNMR was employed to elucidate additional structural information and gain insight into the role synthesis solvent plays on breathing behavior. The CO_2 and H_2O adsorption of these lp stabilized materials were studied, and the improvement in CO_2 loadings over traditionally synthesized MIL-53 will be discussed. The different breathing behavior or lack thereof and increased stability of the lp form allows for lower H_2O loadings at less than 50 %RH; an increase in CO_2 adsorption in the 1-5 bar region and overall loading for the lp only materials, and an increase over the entire pressure range for the gradually breathing MOF. The lp stabilized materials' improved hydrophobicity was analyzed and humid CO_2 breakthrough experiments were performed to analyze the water stability of these materials in the presence of large-pore inducing guest molecules.

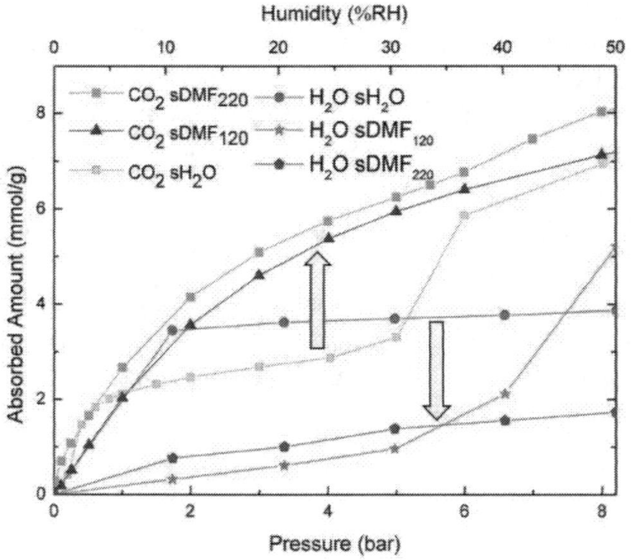

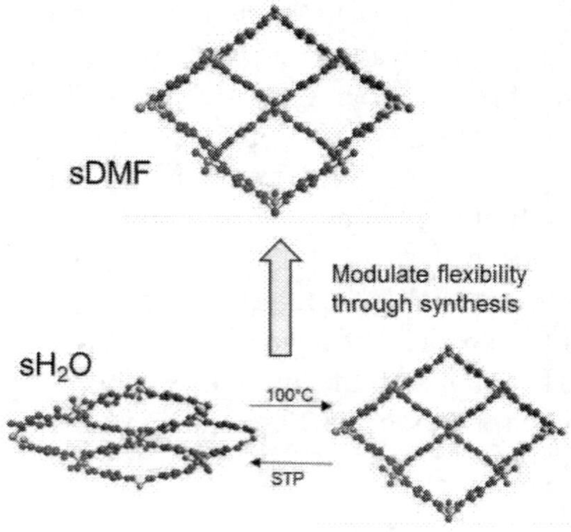

Investigation of fluid dynamic parameters in packed distillation columns using X-ray computed tomography

S. Schug, T. Kohler, W. Arlt

Institute of Separation Science and Technology, University Erlangen-Nuremberg, Erlangen, Germany

The flow behavior of two liquid phases in packed distillation columns is hardly understood. The final aim of this work is to develop a model for the design and process optimization of three-phase distillation, using reliable data for the flow behavior of two liquid phases on a packing.

In this work the X-ray computed tomography (CT) is used as non-invasive measuring technique. The experiments are made with a customized computer tomograph at our institute that enables measurements of flows in vertical objects. The detector and the X-ray source are rotating around the column and the column is not moved, so that the flow inside is not influenced by centrifugal force. As a result of the high spatial (70 µm) and temporal (1000 projections per second) resolution, dynamic processes inside separation columns can be monitored with a high precision. Fluid dynamic parameters in distillation columns (like interfacial area, liquid hold-up and liquid film thickness on the packing or between two liquid phases) can be directly estimated from the recorded CT images of a packing cross section at different axial positions along the column. For the evaluation the images were segmented to represent and identify all occurring phases by a specific grey value.

The experiments were performed in a glass column with an inner diameter of 100 mm and a packing height of 1000 mm. For the study on the influence of the packing structure on the fluid dynamic five different special designs of Mellapak 500.Y packing from Sulzer Chemtech AG were used. The special designs are a W- and L- type corrugation with wholes, a W-type corrugation without wholes, a flat packing with wholes and a flat packing without wholes. Additionally the influence of the rotation of the packing elements to each other by 0° and 90° on the hydrodynamic parameters was determined. The liquids used in the measurements were water and n-heptane. The experiments for the characterization of the packings and the investigation of the influence of the packing structure on the fluid dynamic parameters were performed with one liquid phase by varying the liquid load between 1 and 100 m^3 / m^2h.

For the study on the flow behaviour of the two poorly miscible liquids, water and n-heptane were introduced into the column with a special type of liquid distributor, which ensures that the two phases are equally distributed on the packing. The flow behaviour of the two liquids on the packing was analyzed by varying the total liquid load between 1 and 200 m^3 / m^2h. The feed ratio of water and n-heptane was 1:1, 2:1 and 3:1. Furthermore the flow morphology of the interacting liquids was investigated using a three dimensional reconstruction of five 30 mm long packing sections along the column.

Use of BatchColumn® software for design, analysis and optimization of batch distillation processes

O. Baudouin(1), S. Déchelotte(2), P. Guittard(1), R. Sardeing(1), A. Vacher(1), B. Wincure(2)
1 ProSim SA, Immeuble Stratège A, 51 rue Ampère, F-31670 LABEGE, France
2 ProSim, Inc., Science Center, 3711 Market Street, 8th Floor, Philadelphia, PA 19104, USA

Abstract

Batch distillation is an important unit operation in the batch processing industry and is used in a wide range of applications: fine chemical industries, pharmaceuticals, specialty chemicals, biochemical, essential oils, alcohols... The great flexibility of batch distillation allows for the separation of complex mixtures with a single column. This flexibility, combined with the inherent unsteady state nature of the process, poses challenging design and operation problems [1]. There is no 'rule of thumb' to achieve the optimal solution (reduced operating time, good product separation, minimum energy consumption...) and dedicated tools are required.

BatchColumn® software [2] allows for the detailed representation almost any kind of distillation column, e.g. azeotropic and close boiling separation using heterogeneous or homogeneous entrainers [3, 4, 5, 6], batch distillation with a middle vessel column [7], etc. The software takes into account the associated equipment technology (middle vessels, boiler, condenser...) and column hydrodynamics. It provides the evolution through time of all parameters of interest such as compositions, temperature, pressure, reflux ratio... Through the creation and analysis of operating scenarios, it enables the identification of the best resources management policies, recipes for product quality optimization and cost reduction. BatchColumn® allows modeling of complex systems by taking advantage of the power of Simulis® Thermodynamics (thermodynamic properties server) which offers an extensive set of thermodynamic models with a pure components database of more than 2,000 pure components (AIChE's DIPPR® database [8]). The main features and strengths of BatchColumn® are highlighted on various application examples.

Keywords

Batch distillation; Modeling; Simulation; Thermodynamic server; Complex mixtures.

References

[1] U.Diwekar, Batch Distillation – Simulation, Optimal Design and Control, CRC press, 2nd ed. (2012)
[2] http://www.prosim.net/en/index.php
[3] I. Rodriguez-Donis, V. Gerbaud, X. Joulia, Ind. Eng. Chem. Res., 40, 12, 2729-2741 (2001)
[4] I. Rodriguez-Donis, V. Gerbaud, X. Joulia, Ind. Eng. Chem. Res., 40, 22, 4935-4950 (2001)
[5] S. Pommier, S. Massebeuf, B. Kotai, P. Lang, O. Baudouin, P. Floquet, V. Gerbaud, Chem. Eng. Process., 47, 3, 408-419 (2008)
[6] I. Rodriguez-Donis, V. Gerbaud, X. Joulia, AIChE J., 48, 6, 1168-1178 (2002)
[7] I. Rodriguez-Donis, V. Gerbaud, X. Joulia, Feasibility Analysis and Reflux Policies for the Heterogeneous Batch Distillation in a Middle Vessel Column, SIMO 2002 (2002)
[8] R.L. Rowley, W.V. Wilding, J.L. Oscarson, N.F. Giles, DIPPR® Data Compilation of Pure Chemical Properties, Design Institute for Physical Properties, AIChE, New York, NY (2011)

Advances in Distillation, **481c**
Wednesday, November 19, Hilton Atlanta, 311, 13:30

Advances in Synthesis of Separation Systems for Azeotropic Mixtures

Stanislaw K. Wasylkiewicz, Marco A. Satyro, Yau Kun Li** and **Monika J. Wasylkiewicz

Virtual Materials Group Inc., Computer Modelling Group Ltd., Calgary, Alberta, Canada
Chemical and Materials Engineering, University of Alberta, Edmonton, Alberta, Canada

stan.wasylkiewicz@cmgl.ca

Design Methods

- **Traditional**
  - *Use simulators to design columns*
  - *Trial and error*
  - *Time consuming*

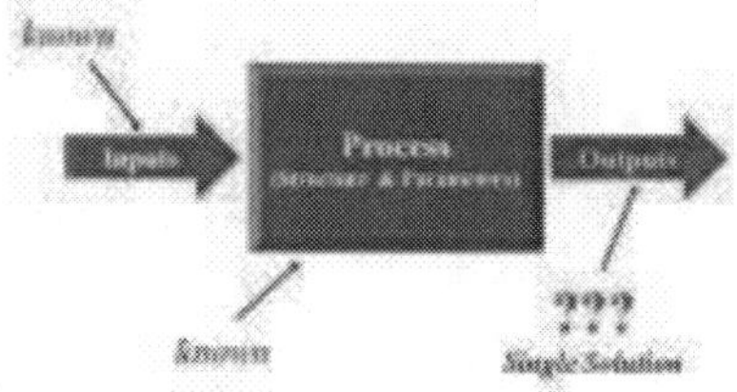

- **Geometric**
  - *Use graphical insights*
  - *McCabe & Thiele*
  - *Boundary Value Design Method*

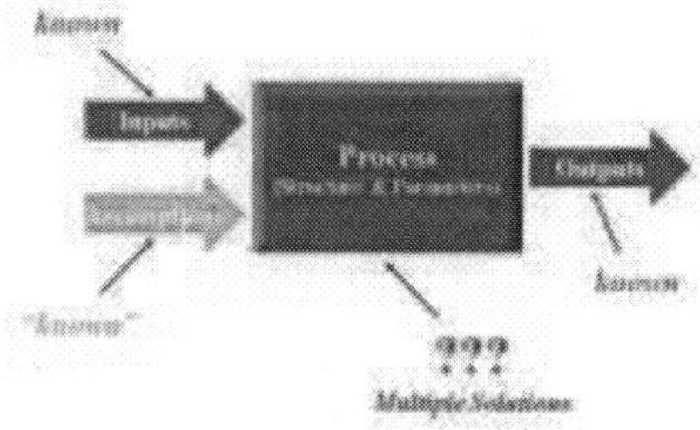

Petlyuk, "Distillation theory and its application to optimal design of separation units", Cambridge University Press, 2004

Design-by-Simulation

- So far the most popular approach
- *[Hassan, Lucia, "An Iteratively Refined Distillation Line Method", AIChE J. 2011]*
  - focus on distillation design feasibility
  - BVDM[1]: rectifying and stripping profiles for four-component mixture do not intersect
  - complete bottom up calculations
  - arbitrary switch from stripping section to rectifying section of the column
  - set the number of stripping and rectifying stages to 300
  - stage-by-stage simulation without SOF[2]

1. *Fidkowski, Doherty, Malone, "Feasibility of separations for distillation of nonideal ternary mixtures", AIChE J. 1993*
2. *Wasylkiewicz, "The Application of the Substitute Objective Function Minimization for Solving Multi-Stage Separation Problems", Mathematical and Computer Modelling, 1992*

3

Design-by-Simulation

- *[Skiborowski, Harwardt, Marquardt, Efficient optimization-based design for the separation of heterogeneous azeotropic mixtures, Comp. Chem. Eng. 2015]*
  - design of heteroazeotropic distillation processes remains a complicated and tedious task
  - severe convergence problems and significant effort to obtain a steady-state design
  - additional DDoF for the rectifying section in the BVDM
  - unidirectional calculation of the column profile, which switches from stripping to rectifying section at the feed tray
  - initialization based on feed flash calculations or shortcut methods
- *[Caballero , "Logic hybrid simulation-optimization algorithm for distillation design", Comp. Chem. Eng. 2015]*
  - design of distillation columns that integrates a process simulator in a generalized disjunctive programming formulation
  - some knowledge about the system is needed
  - it would not be a good idea try to perform a 'blind optimization' from scratch
  - this algorithm can be considered an excellent tool to complement other approaches (conceptual design, shortcut methods)
  - can be used to 'get a rigorous design' from a preliminary design

4

Framework for Synthesis of Separation Systems

- **Thermodynamic tools**
  - rigorous thermodynamic models (activity coefficient and EOS models)
  - comprehensive thermodynamic databases
  - parameter estimation and regression from experimental data
  - rigorous phase equilibrium and stability calculations[1]
  - calculation of all azeotropes[2], pressure sensitivity analysis, topological consistency check
  - generation of RCMs and VLLE diagrams

- **Synthesis tools**
  - calculation of adjacency and reachability matrices[3]
  - identification of basic, continuous and batch distillation regions and boundaries
  - rigorous feasibility test and split generation method
  - identification of recycle options and calculation of recycle mass balances[4]

1. *Wasylkiewicz, Li, Satyro, Wasylkiewicz, "Application of a global optimization algorithm to phase stability and liquid–liquid equilibrium calculations", Fluid Phase Equilib. 2013*
2. *Wasylkiewicz, Doherty, Malone, "Computing all homogeneous and heterogeneous azeotropes in multicomponent mixtures", Ind. Eng. Chem. Res. 1999*
3. *Rooks, Julka, Doherty, Malone; "Structure of Distillation Regions for Multicomponent Azeotropic Mixtures", AIChE J. 1998*
4. *Wasylkiewicz, "Advances in synthesis of azeotropic distillation column sequences: internal secondary recycles", Asia-Pac. J. Chem. Eng. 2008*

5

Structure of Separation Space

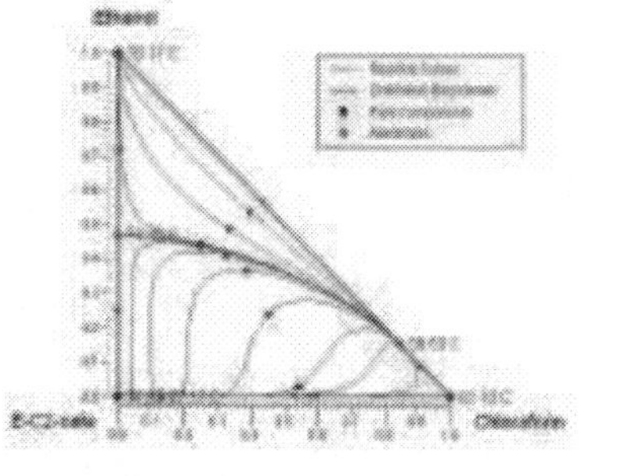

Singular Points (101.325 kPa):
1: Chloroform - Ethanol (59.53 °C)
2: Chloroform (61.10 °C)
3: Ethanol - E-C2-oate (72.19 °C)
4: E-C2-oate (77.05 °C)
5: Chloroform - E-C2-oate (77.17 °C)
6: Ethanol (78.17 °C)

Adjacency matrix:

	1	2	3	4	5	6
1	0	1	1	0	0	1
2	0	0	0	0	1	0
3	0	0	0	1	0	1
4	0	0	0	0	1	0
5	0	0	0	0	0	0
6	0	0	0	0	0	0

Reachability matrix:

	1	2	3	4	5	6
1	1	1	1	1	1	1
2	0	1	0	0	1	0
3	0	0	1	1	1	1
4	0	0	0	1	1	0
5	0	0	0	0	1	0
6	0	0	0	0	0	1

Rooks, Julka, Doherty, Malone; "Structure of Distillation Regions for Multicomponent Azeotropic Mixtures", AIChE J. 1998

6

Distillation Regions

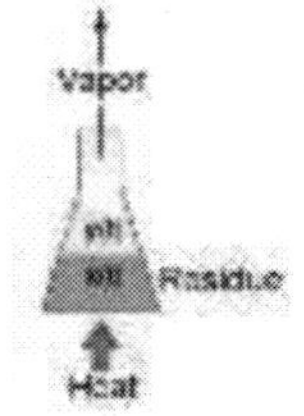

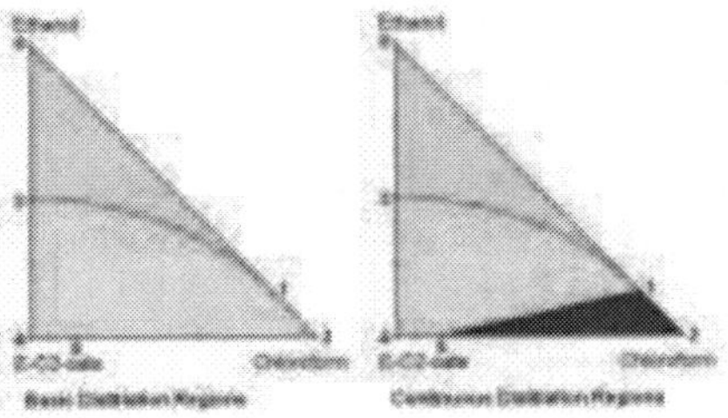

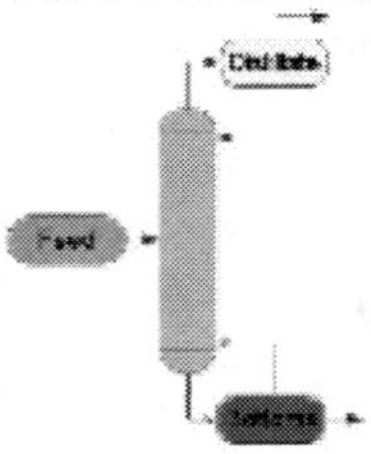

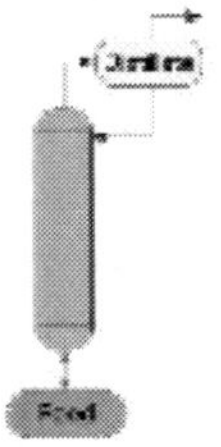

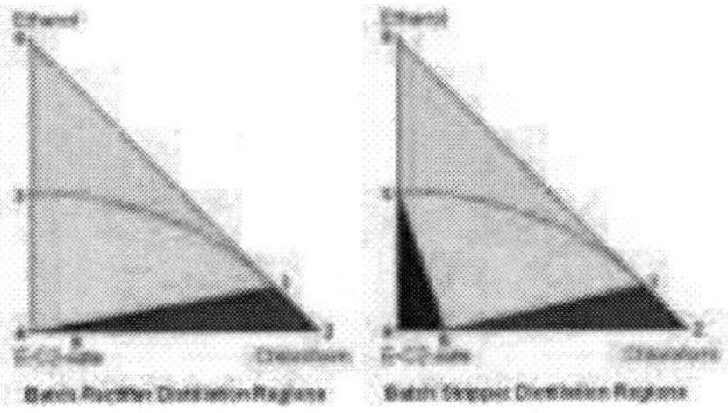

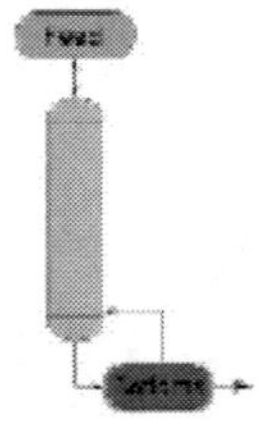

7

Distillation Regions

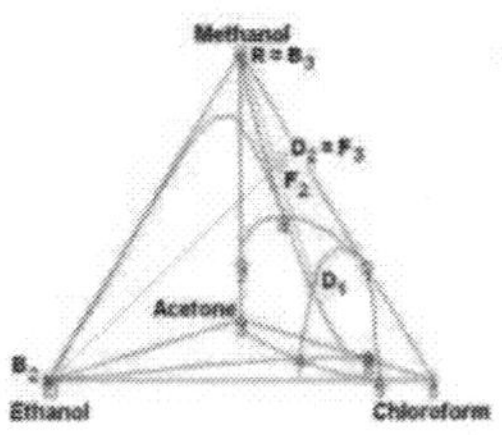

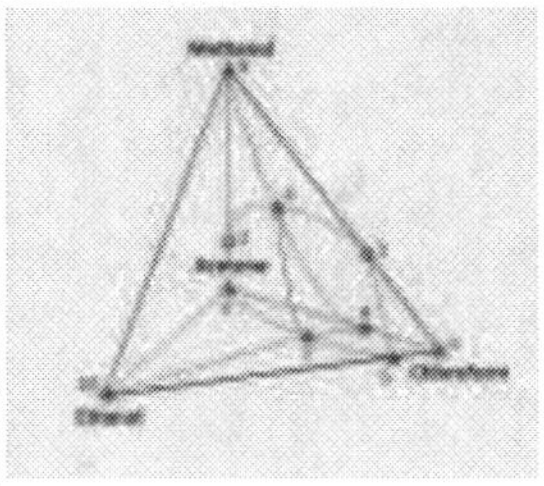

Adjacency Matrix:

0	0	0	1	1	1	0	0	1	0
0	0	1	1	0	0	0	0	1	0
0	0	0	0	0	0	1	1	0	1
0	0	0	0	0	0	1	1	1	0
0	0	0	0	0	1	1	0	0	1
0	0	0	0	0	0	0	1	0	0
0	0	0	0	0	0	0	1	0	1
0	0	0	0	0	0	0	0	0	0
0	0	0	0	0	0	0	0	0	1
0	0	0	0	0	0	0	0	0	0

Reachability Matrix:

1	0	0	1	1	1	1	1	1	1
0	1	1	1	0	0	1	1	1	1
0	0	1	0	0	0	1	1	0	1
0	0	0	1	0	0	1	1	1	1
0	0	0	0	1	1	1	1	0	1
0	0	0	0	0	1	0	1	0	0
0	0	0	0	0	0	1	1	0	1
0	0	0	0	0	0	0	1	0	0
0	0	0	0	0	0	0	0	1	1
0	0	0	0	0	0	0	0	0	1

Basic Distillation Regions

1	1	4	5	6	7	8	**1**	1	4	5	6	7	8
2	1	4	5	7	9	10	**2**	1	4	5	7	9	10
3	2	3	4	7	8		**3**	2	3	4	7	8	
4	2	3	4	7	9	10	**4**	2	3	4	7	9	10

Continuous Distillation Regions

101	1	5	6	8	**101**	1	4	7	8
102	1	5	7	8	**102**	1	5	6	8
					103	1	5	7	8
201	1	4	9	10	**201**	1	4	7	10
202	1	5	7	10	**202**	1	4	9	10
					203	1	5	7	10
301	2	3	7	8	**301**	2	3	7	8
					302	2	4	7	8
401	2	3	7	10	**401**	2	3	7	10
402	2	4	9	10	**402**	2	4	7	10
					403	2	4	9	10

Batch Rectifier Distillation Regions

10101	1	5	6	8	**10101**	1	4	7	8
10201	1	5	7	8	**10201**	1	5	6	8
					10301	1	5	7	8
20101	1	4	9	10	**20101**	1	4	7	10
20201	1	5	7	10	**20201**	1	4	9	10
					20301	1	5	7	10
30101	2	3	7	8	**30101**	2	3	7	8
					30201	2	4	7	8
40101	2	3	7	10	**40101**	2	3	7	10
40201	2	4	9	10	**40201**	2	4	7	10
					40301	2	4	9	10

Batch Stripper Distillation Regions

10101	1	5	6	8	**10101**	1	4	7	8
10201	1	5	7	8	**10201**	1	5	6	8
					10301	1	5	7	8
20101	1	4	9	10	**20101**	1	4	7	10
20201	1	5	7	10	**20201**	1	4	9	10
					20301	1	5	7	10
30101	2	3	7	8	**30101**	2	3	7	8
					30201	2	4	7	8
40101	2	3	7	10	**40101**	2	3	7	10
40201	2	4	9	10	**40201**	2	4	7	10
					40301	2	4	9	10

Wasylkiewicz,, Asia-Pac. J. Chem. Eng. 2008

8

Recovery of Selected Component Preferred Distillation Region

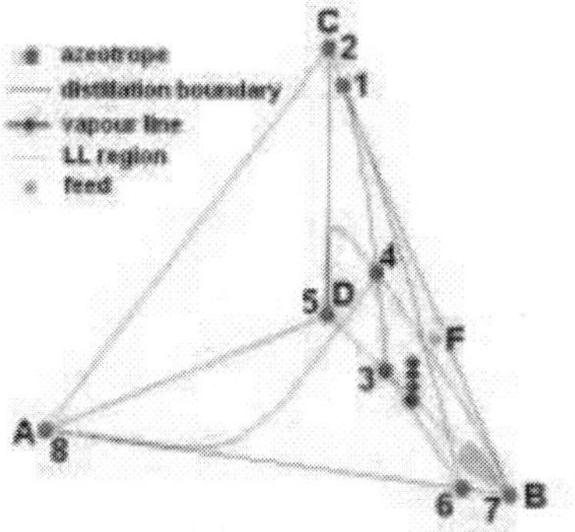

Basic Regions	Continuous Regions	Batch Rectifier Stripper Regions
1: 1 4 6 7	101: 1 4 6 7	10101: 1 4 6 7
2: 1 2 4 5 6 8	201: 1 2 5 8	20101: 1 2 5 8
	202: 1 4 5 8	20201: 1 4 5 8
	203: 1 4 6 8	20301: 1 4 6 8
3: 3 4 6 7	301: 3 4 6 7	30101: 3 4 6 7
4: 3 4 5 6 8	401: 3 4 5 8	40101: 3 4 5 8
	402: 3 4 6 8	40201: 3 4 6 8

Feed: [0.0013, 0.6549, 0.2242, 0.1196]
Goal: completely recover pure **A**
recycle other components

Feed in basic distillation region 1
(batch distillation region 10101)
Preferred batch distillation regions:
20101, 20201 and 40101
where the best recovery of desired product can be accomplished

9

Sequence of Batch Strippers

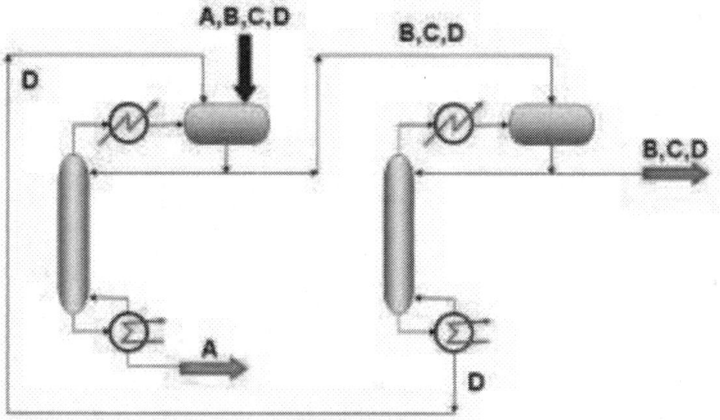

- Add **D** to the original feed (move the overall feed to region **40101**)
- Use batch stripper to recover **A**
- Use batch stripper to produce enough **D** for the recycle

10

Heterogeneous Azeotropic Distillation External Secondary Recycle

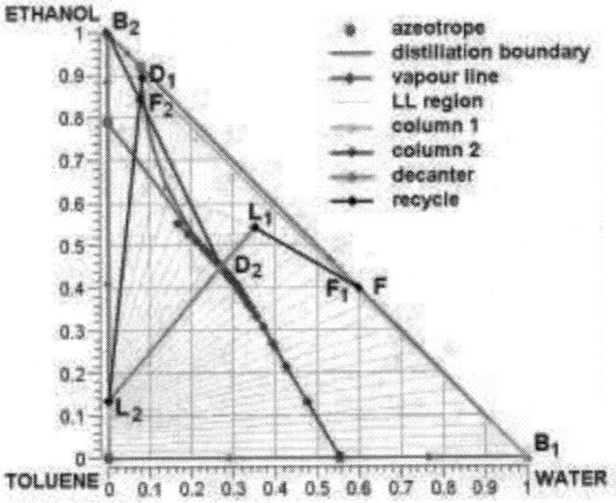

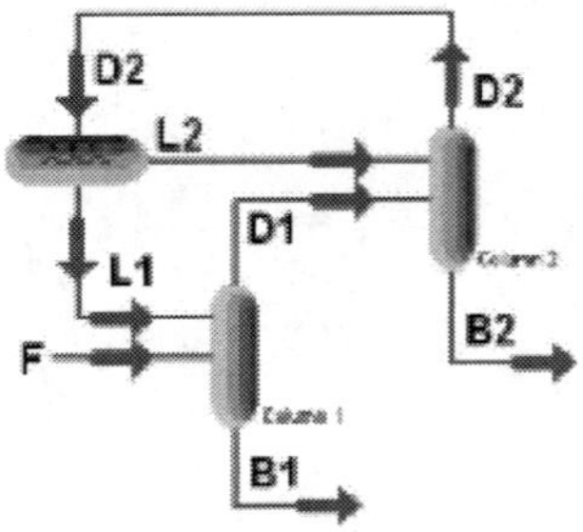

Separation of ethanol – water azeotrope using toluene as an entrainer

11

Generalized Heterogeneous Azeotropic Distillation

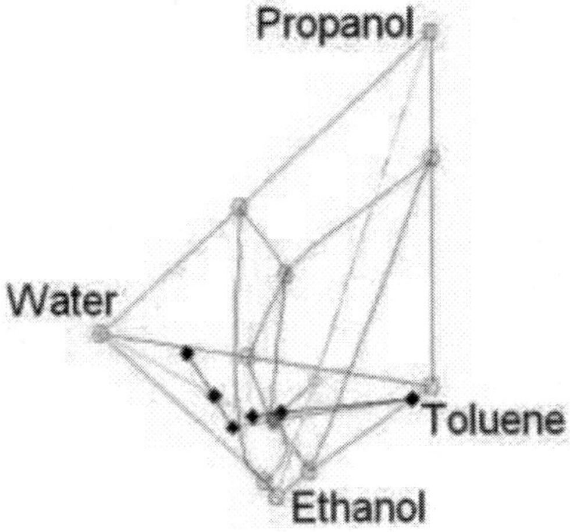

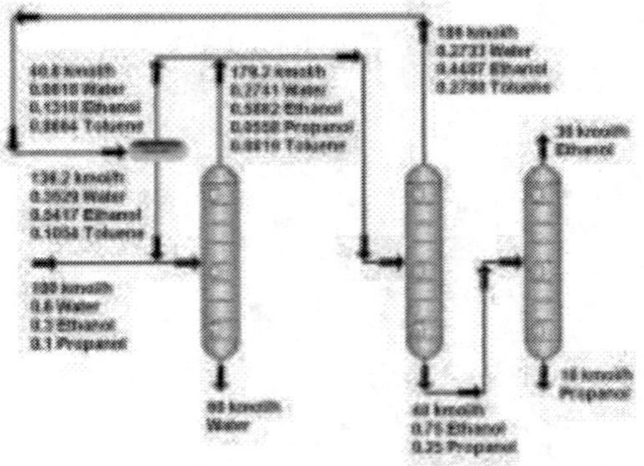

Separation of ethanol – propanol – water mixture using toluene as an entrainer

12

Generalized Heterogeneous Azeotropic Distillation

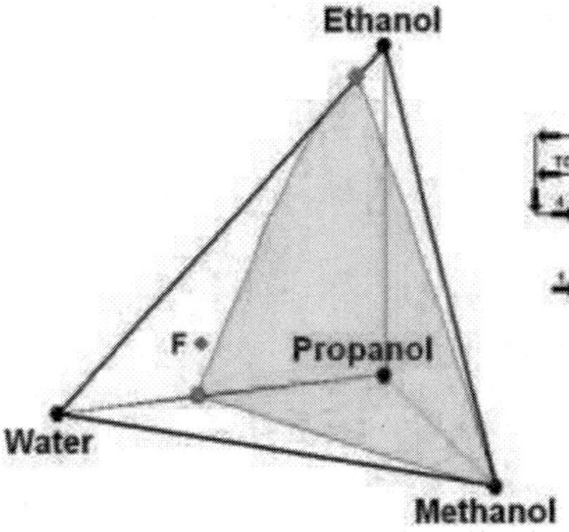

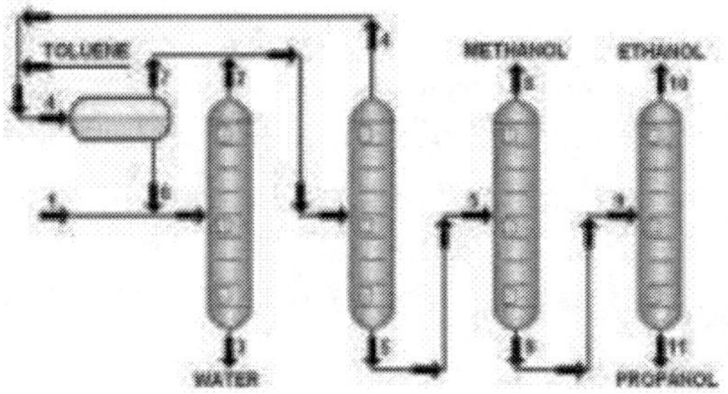

Separation of methanol – ethanol – propanol – water mixture using toluene as an entrainer

13

Internal Secondary Recycle Preferred Distillation Region

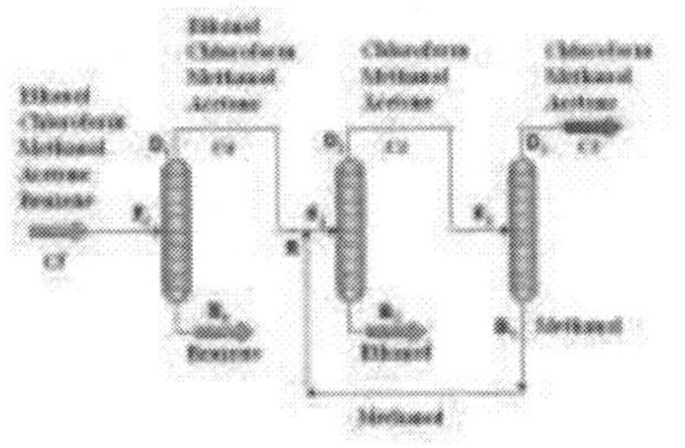

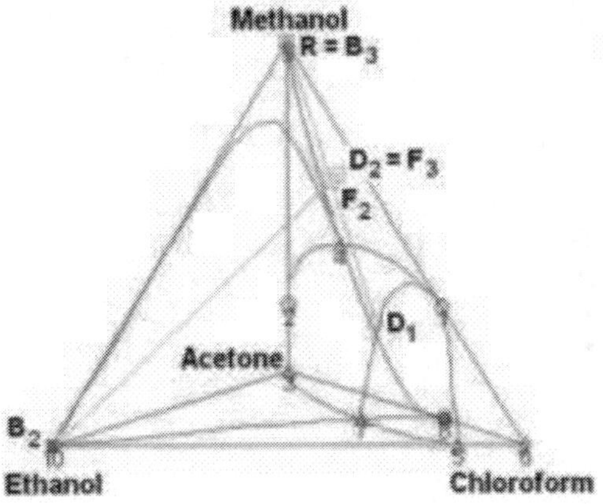

Wasylkiewicz, Asia-Pacific Journal of Chemical Engineering, 2008, 3: 36–45

14

Conclusions

- The **design-by-simulation** is so far the most popular approach to solve distillation design problems
- The **rigorous thermodynamic model** of the mixture, determined from quality experimental data, is of the paramount importance
- Even more important is to use **rigorous VLLE** and **azeotrope** calculation routines that can handle all your model can predict
- Before starting design of a new sequence of azeotropic distillation columns you should know **ALL azeotropes** in the mixture and their **sensitivity to pressure** changes
- Using residue curve technology, **distillation regions** can be calculated for any number of components and **feasibility** of distillation splits can be quickly assessed
- This information allows for reliable screening for the **preferred** distillation regions and rapid generation of feasible distillation separation schemes

15

References

Doherty, Malone, "Conceptual Design of Distillation Systems", McGraw-Hill, 2001

Petlyuk, "Distillation theory and its application to optimal design of separation units", Cambridge University Press, 2004

Kister, "Distillation Troubleshooting", John Wiley & Sons, 2006

Wasylkiewicz, "Troubleshooting Distillation Simulation", Chem. Eng. Progress, June 2006

Wasylkiewicz, "Advances in synthesis of azeotropic distillation column sequences: internal secondary recycles", Asia-Pac. J. Chem. Eng., 2008

Wasylkiewicz, Li, Satyro, Wasylkiewicz, "Application of a global optimization algorithm to phase stability and liquid–liquid equilibrium calculations", *Fluid Phase Equilib., 2013*

Beneke, Peters, Glasser, Hildebrandt, "Understanding Distillation using Column Profile Maps", John Wiley & Sons, 2013

16

364337 Concptual Design of Multi-Side Draw Hybrid Reactive Distillation Columns for Multi-Reactant Multireaction Schemes Involving Multiple Azeotropes

Abstract:

The contribution deals with the development of design algorithm for multi-side draw hybrid reactive distillation (RD) column in order to obtain a desired selectivity in a multireactant multi-reaction system involving multiple azeotropes. This work is the continuation of our earlier work on selectivity engineering with hybrid RD columns [1,2,3,& 4]. Recently Hasan et al. [4] have proposed a conceptual design algorithm for single feed multi-side draw hybrid reactive distillation columns to achieve desired selectivity in single reactant multi-reaction schemes involving zeotropic mixtures. In this work we extend the approch further to show that a single feed multi-side draw hybrid RD configuration is also capable to give a desired selectivity in multi-reactant multireaction schemes involving multiple azeotropes. The developed methodology is based on combined graphical-simulation approach and utilize the concept of visualization of the locus of liquid feed stage compositions of nonreactive distillation column and its intersection with that composition in space at which reaction takes place in a multi-side draw hybrid RD column. Based on the desired product selectivity, we determine the volume slice and hence the surface of reactive stage compositions (SRSCs) in the RD column. If the reaction in the RD column takes place at any of the compositions that belongs to SRSCs then our desired selectivity is ensured. The algorithm developed in this work is applicable to any number of components as well as to the systems with any number of azeotropes but is restricted to single feed configurations and to the cases where both the reactants are saddles. However we believe that the developed approach can be conveniently extended to multi-feed configurations too. The method developed here proves to be very useful for industrially important multi-reactant multireaction systems such as production of diethyl carbonate in which both dimethyl carbonate and ethanol are saddles in the corresponding residue curve map.

References

[1] Hasan, S.U.; Mahajani, S.;Malik, R.K. Selectivity engineering with single-feed hybrid reactive distillation columns. *Chemical Engineering Science*.[Online] **2013,** 99, 324–334.

[2] Hasan, S.U.; Mahajani, S.;Malik, R.K. A conceptual design algorithm for single-feed hybrid reactive distillation column involving azeotropic systems. *Computer Aided Chemical Engineering*. [Online] **2013,** 32, 313–318.

[3] Hasan, S.U.; Mahajani, S.;Malik, R.K. Selectivity Engineering with hybrid reactive distillation column with side draw. *10th International conference on Distillation & Absorption,* [Online] **2014**, September 14-17, Friedrichshafen,Germany. (Abstract Accepted).

[4] Hasan, S.U.; Mahajani, S.;Malik, R.K. Selectivity Engineering with Multi-Side draw Hybrid Reactive Distillation Columns. *23rd International Symposium on Chemical Reaction Engineering (ISCRE- 23) and 7th Asia Pacific Chemical Reaction Engineering Symposium (APCRE-7)*,[Online] **2014**, September 7-10, Centara Grand & Bangkok Convention Centre (BCC) at Central World, Bangkok, Thailand. (Abstract Accepted).

372752 Potential and Limitations of COSMO-RS As a Tool for the Optimization of Extractive Separations

Abstract:

COSMO-RS is a still relatively novel method for the prediction of activity coefficients, solubilities and partial pressures of a very broad range of neutral and charges compounds in pure and mixed solvents. It is based on a sound combination of quantum chemical calculations and thermodynamics, and as such it is able to take into account many molecular details, as intramolecular hydrogen bonding, conformational changes, mutual polarization of functional groups, steric shading, etc., which are out of the scope of most other, usually much more empirical, often group-contribution based approaches.

Since the targets of extraction often are complicated high value molecules, which are out of the scope of the simpler methods, COSMO-RS can be a valuable tool for an improved understanding of the competing intermolecular interactions governing your extraction processes, and for screening hundreds and thousands of possible extraction solvents and mixtures for an optimal extraction yield.

In this talk the COSMO-RS method will be briefly introduced and its potential and limitations for extractive separation will be outline and demonstrated by a few examples.

392325 Solvent selection, modeling, simulation and life cycle assessment of an ionic liquid based liquid-liquid extraction process to produce lubricating base oil

Abstract:

It is necessary to reduce the aromatics content of lube-oil cuts (a petroleum cut) to improve the overall quality of the produced lubricating base oil. A liquid−liquid extraction process carried out in a Rotating Disc Contactor (RDC) is usually used for aromatics removal. Furfural or sulfolane are mainly used as the solvents in this separation process because they show high selectivity towards aromatic compounds. A comparison between several conventional contactor devices shows that RDCs are the most efficient extraction systems in terms of extraction yield. Furfural is a volatile solvent which tends to escape to air when it is recovered from extract phase to be recycled to the main RDC column. Furfural has relatively high human health impacts specially when inhaled by people who work in the industrial site. Therefore, a solvent with lower vapor pressure would be preferable as replacement of such a volatile solvent. Ionic liquids (ILs) are a relatively new class of chemicals which can replace molecular compounds (e.g. solvents). Ionic liquids are made of ion pairs and the strong ionic bonding which accounts for their low vapor pressure, makes them a good alternative to replace volatile solvents. In this work we used COSMO-RS method to screen several ionic liquids which have the potential to replace volatile solvents currently used in lubricating base oil production. A simulation model which enables us to study the performance of RDC column was created. This model was used to find the energy required/raw material for the separation process with ionic liquid as the solvent. A cradle-to-gate life cycle assessment for production of 1kg lubricating base oil, was preformed based on raw materials and energy required. The life cycle assessment was used to calculate the impacts associated with production of lubricating base oil as well as all upstream processes required. In the next step the life cycle environmental impacts of production of 1 kg lubricating base oil with ionic liquid can be compared to those in a process with a volatile solvent.

381538 Enantioselective Extraction in Process Intensified Columns

Abstract:

Living organisms interact with biologically active molecules. Most of the molecules in a metabolism are chiral, such as amino acids, nucleic acids and sugars, as well as composed basic molecules such as peptides, proteins, and polysaccharides. Enantiomers play a key role inside the metabolism of plants and animals. They are important for regulation of processes and for specific interactions inside the biogenesis. Thus, chiral molecules are essential for pharmaceuticals, agrochemicals, foods, aromas, and flavors. The production of pure enantiomers is therefore a necessary requirement.

Due to the same molecular structure, enantiomers have mostly the same physico-chemical properties, like boiling or melting point, solubility, optical spectra, and bonding length. This makes the separation of both molecules complex, and requires delicate separation techniques. Beside common preparative techniques, such as diastereomeric crystallization and chromatography, the enantioselective liquid-liquid extraction (ELLE) is a highly promising alternative to realize a continuous and flexible process that is easily scalable.

The separation is based on a selective complex formation between the enantiomers and a selector molecule, where typically the enantiomers are fed in one liquid phase, and the selector molecule is dissolved in a second liquid phase. In the past decades, a range of selector molecules were identified for various classes of racemates [1], so that nowadays the separation of a broad spectrum of enantiomers is possible. However, the selectivity of the separation is often rather poor, which makes the complete separation more complicated. Therefore, the economic and ecologically sustainable realization of ELLE implies a high selectivity of the selector molecules and high extraction efficiencies for the counter-current extraction process.

Often, the enantioselector molecules are delicate and very expensive, and processing ELLE is therefore preferentially done in compact, highly efficient equipment. We here report the use of miniaturized extraction columns and show that they offer a good solution for the enantioselective extraction in small scale applications. With different test systems, it was demonstrated that the extraction efficiency in these miniaturized columns is more than twice as in conventional miniplant extraction columns [2].

The fractional extraction of racemic mixtures is investigated theoretically and experimentally inside these extraction columns (H_{ac}=1.1 m) with aromatic acids and amino acids. It could be demonstrated that the enantioselective extraction of (R/S)-phenyl succinic acid (PSA) and dinitrobenzoyl-(R/S)-leucine (DNB-Leu) inside this columns is possible. The enantiomeric excess was measured up to 60 % for PSA (the system with a lower intrinsic selectivity), whereas the enantiomers of DNB-Leu could be separated with enantiomeric excesses of over 98%.

[1] Schuur, et al., Organic & Biomolecular Chemistry, 9, 36-51, 2011

[2] Holbach, et al., Chemical Engineering and Processing, 80, 21-28, 2014

Recovery of Polycarbonates from Electronic Wastes Using Mixed-Solvent Extraction

George S. Weeden Jr.*, Nicholas H. Soepriatna*, and N.-H. Linda Wang**

*School of Chemical Engineering, Purdue University, West Lafayette, IN 47907
**Email: wangn@ecn.purdue.edu Phone: (765) 494-4081 Fax: (765) 494-0805

More than 50 million tons of electronic waste is generated each year. Currently, most recycling efforts are focused on the recovery of metals and glass. Polymers compose roughly one-third of the total weight of waste, and valuable polymers have the potential to be recovered with high purity and high yield at low cost. These polymers may be recovered using the differential solubility of the polymers in different solvents and mixtures of solvents. Solvent choices may be aided by theories of polymer solubility based on thermodynamic properties, such as the Hansen solubility parameters. One such valuable polymer is polycarbonate (PC), which was recovered from samples of electronic waste. Sequential, mixed-solvent extraction experiments were performed on the samples at room temperature. The solid waste was mixed with 50/50 (vol.%) acetone/dichloromethane. The solid product was filtered and washed, then dissolved in 100% dichloromethane. The remaining solids were then removed, and acetone was added to precipitate the PC out of solution. The purity of the precipitated PC was > 99% by HPLC and FTIR with a yield > 95% while using < 30 kg of solvent / kg product. Initial cost estimates place the separation cost at 1.50 $/kg (at a scale of 1 tonne PC/day), which is roughly 1.0 $/kg less than the current production cost.

385905 Novel Approach to Study the Supercritical Fluid Extraction Process Continuously Using Evaporative Light Scattering Detection

Abstract:

As a part of the recent trends within analytical chemistry, it's not uncommon to mention the three R's in this context, namely Replace, Reduce and Recycle [1]. That is also why the supercritical fluid extraction (SFE) in theory is the ideal extraction method due to the usage of CO_2 instead of organic solvents. Even though its established use, much work remains in characterizing the fundamentals of the extraction process. The fundamental knowledge gained studying analytical scaled systems can subsequently be used in for example modelling of larger systems.

Modelling in general requires experimental data for calibration which is not easily obtained with good accuracy in large quantities. A popular approach is to sample fractions during the extraction process or terminating the extraction at various time points. Both approaches are time consuming and may cause an inflation of the experimental error. On-line detection for continuously monitoring the process enables the acquisition of accurate data in large quantities with reasonable effort. Previous examples exist using for example UV-Vis and FTIR detectors [2]. These may however limit operating boundaries of the SFE system with regards to pressure due to the flow cell of the detectors. Also the use of UV-Vis detectors requires that the compounds of interest have a chromophore in order to be efficiently detected.

A novel approach has been developed to study the extraction in real-time by coupling an evaporative light scattering detector (ELSD) to a continuous SFE system. The ELSD offers universal detection thus not requiring that compounds of interest readily absorb light in the UV-Vis part of the spectrum. The fact that it is coupled after the depressurization means that it does not limit the maximum pressure of the system.

The SFE-ELSD system was validated by applying it to crushed linseeds with the purpose of quantifying the total amount of oil. A full validation was performed in order to ensure good quality of the measurements. The precision was considered good with a repeatability and reproducibility of 2.6% and the accuracy of the ELSD was in well agreement with a gravimetric determination subsequent the extraction.

The newly developed system was subsequently used to study the extraction process of lipids from crushed linseeds at various extraction conditions. It was also noted that by performing repeated extractions using an analytical scale extraction vessel the yield increased immensely. The accumulated yield was approximately doubled compared to the first single extraction, and the yield was far greater than that of the commonly used reference method using Soxhlet with petroleum ether. Some claims have been made that compression and decompression can rupture the matrix and thus increasing total available content [3]. However by utilizing the proposed SFE-ELSD system we believe that these findings are rather related to channeling effects than rupturing of the sample matrix.

In addition some preliminary evaluation of models using the data acquired from the SFE-ELSD system will be presented.

1. Welch, C.J., et al., *Greening analytical chromatography.* TrAC - Trends in Analytical Chemistry, 2010. **29**(7): p. 667-680.

2. Amador-Hernández, J. and M.D. Luque De Castro, *On-line detection for supercritical-fluid extraction.* Journal of Biochemical and Biophysical Methods, 2000. **43**(1-3): p. 329-343.

3. Barthet, V.J. and J.K. Daun, *An evaluation of supercritical fluid extraction as an analytical tool to determine fat in canola, flax, solin, and mustard.* JAOCS, Journal of the American Oil Chemists' Society, 2002. **79**(3): p. 245-251.

378420 Reactive Extraction of Metals and Organics Using Trioctylphosphine Oxide Impregnated Membranes

Abstract:

Over the past few decades, one of the most significant developments in the realm of solvent extraction is the segregation of the aqueous and the organic phases using semi-permeable membranes. This technique, also known as non-dispersive solvent extraction, facilitates complete recycle and reuse of the non-aqueous partitioning phase, while the downstream phase separation problems associated with conventional extraction are eliminated. The efficacy of membrane-based extraction has also been advanced through the development of other configurations such as supported liquid membranes and solvent encapsulated polymeric microcapsules. Recently, a new technique has also been developed for impregnation of solid organic extractants such as trioctylphosphine oxide (TOPO) into porous membranes. The resulting extractant impregnated membranes (EIM) combine the advantages of extraction, adsorption and membrane separation, which can be harnessed to facilitate solventless extraction of several metals and organic compounds.

So far, the EIMs have only been used in sequential extraction and stripping of phenols. Efforts to carry out simultaneous extraction and stripping could not fructify due to a non-uniform distribution of the extractant within membrane walls. In this research, two approaches were used to obtain a more even distribution of TOPO within the polymer support. In the first approach, the membranes were first soaked in dichloromethane containing high concentrations of TOPO, and dichloromethane was then evaporated in a controlled environment so that TOPO was distributed evenly throughout the membrane walls. In the second approach, membranes were soaked in molten TOPO at 60 °C first and the temperature was brought down quickly to solidify TOPO at the same spot within the membrane walls. Using both the approaches, the EIMs prepared exhibited good TOPO distribution. These EIMs could be used for simultaneous extraction and stripping.

Chromium (VI) and phenol were chosen as the model compounds to demonstrate the extractive performance of the EIMs. In order to understand the factors affecting extraction, and to obtain insight into extraction mechanisms, the effects of ring substitution on phenol extraction was also investigated. It was observed that the most important factor affecting the partition coefficient was hydrophobicity of the molecule. These results suggest that the EIMs can be a promising technique in sustainable and solventless extraction of several metals and organics.

387298 Additive-Free Treatment of High-Strength Agricultural Wastewaters Using Ultrafiltration

Abstract:

The rendering industry protects the environment by processing animal by-products into pet and livestock feed ingredients, materials, fuels, among other products, rather than entering them into a landfill. During this processing, rendering plants generate large volumes of highly impaired wastewater containing high total suspended solids, fats, oils and greases, and proteins. Most facilities use dissolved air flotation (DAF) for primary treatment of their wastewaters, with a very high operating cost of roughly US $3.20/1000 gal of wastewater.

In this contribution, we will discuss our latest work on the use of membrane filtration for the additive-free treatment of high-strength rendering facility wastewaters. The objectives of the present project are to 1) evaluate the performance of ultrafiltration (UF) membranes in the primary treatment of rendering wastewater; 2) understand how cross-flow filtration operating conditions (i.e., cross-flow velocity, transmembrane pressure, and total suspended solids) impact membrane performance; 3) measure the threshold flux for UF membranes; 4) evaluate cleaning protocols for UF membranes.

Our results have demonstrated that UF membranes can efficiently purify the wastewater without the polymeric and pH-adjusting additives required by DAF. Ultrafiltration with a membrane characterized by a pure water permeability of 0.1 L/(m^2·h)/kPa reduced turbidity of the wastewater nearly 100%, chemical oxygen demand by 80%, and total solids by 90%. Operating below threshold flux enabled long-term, steady state measurements. Based on membrane performance, a mathematical model was proposed to predict flux behavior under different operating conditions.

In addition to experimental work, a preliminary cost analysis was done to compare the operating costs (energy and consumables) of the membrane ultrafiltration process to DAF. Assuming a conservative membrane lifetime of 1 year, the results suggest that the application of UF membranes in treating rendering wastewaters can reduce the operating cost to slightly less than $1.00/1000 gal of wastewater. In addition to the significantly lower projected operating costs for the membrane process, there are likely to be other economic benefits associated with the removal of chemical additives from the process.

Threshold flux as an aid to shaping filtration strategy

Robert W. Field

Dept. of Engineering Science,
University of Oxford, Oxford, UK

ABSTRACT

Membrane fouling is the major factor which results in flux decline, or if flux is maintained constant, in increasing Trans-Membrane Pressure. It is widely accepted that the rate of fouling can be controlled if the membrane is operated below the "critical flux" i.e. the flux above which fouling increases remarkably. Whether this should be labelled "threshold flux" to distinguish it from true critical fluxes is still a matter of debate. The link between these concepts of critical/threshold flux and that of sustainable flux has been discussed by Field and Pearce in 2011 [1]. Of course operation below the "threshold flux" does not guarantee *zero fouling* nor is operation above it completely out of the question. For example if fouling below the break-point in the fouling rate *vs* flux curve is very low and above it, the rate is still modest one might want to choose to operate above the threshold flux in order to achieve more productivity[1].

After a review of the three interlinking concepts of critical, threshold and sustainable fluxes, the theoretical element of the presentation will consider the spatial distribution of threshold flux. In particular this will be explored with regard to the direct flow format. Once criticality has seen reached in one region there can be a cascade. This is not a novel observation *per se* but focussing upon the initiation in the sensitive region will probably be more beneficial than analysing an overall response. Secondly the influence of time has been shown to be important. The concept of a *critical time* is not completely new. For example Pollice et al.[2] found that in small-scale experiments, there exists a critical filtration time (t_{crit}) and that limited data is available suggesting that it is a function of flux. Also, more recently Zhang et al. [3] explored the phenomena of a TMP jump mainly based on the feed nature, membrane properties and hydrodynamics. A distinction between coagulation in physical systems and biological changes in biological systems is important and will be discussed.

The application of the concept of threshold flux to some recent pilot-plant ultrafiltration data for produced water treatment using a ceramic membrane will be examined in detail in order to provide some further guidance on filtration strategy.

Lastly some recent work by Miller et al.[4] on membrane fouling at constant flux and constant transmembrane pressure conditions suggests (at least for membrane filtration of emulsified oils) that below a threshold flux the phenomena of fouling is a unique function of permeate volume. However starting above a threshold flux their two behaviours are very different. A mathematical analysis of the change of resistance with volume permeate is presented for both constant flux and constant transmembrane pressure conditions in order to provide some insight into their observations.

---

[1] Robert W. Field, Graeme K. Pearce, Critical, sustainable and threshold fluxes for membrane filtration with water industry applications, Advances in Colloid and Interface Science 164 (2011) 38–44.

[2] Alfieri Pollice, Adam Brookes, Bruce Jefferson, Simon Judd, Sub-critical flux fouling in membrane bioreactors – a review of recent literature, Desalination 174 (2005) 221-230.

[3] J. Zhang, H.C. Chua, J. Zhou, A.G. Fane, Factors affecting the membrane performance in submerged membrane bioreactors, Journal of Membrane Science 284 (2006) 54–66.

[4] Daniel J. Miller, Sirirat Kasemset, Donald R. Paul, Benny D. Freeman, Comparison of membrane fouling at constant flux and constant transmembrane pressure conditions, Journal of Membrane Science 454(2014) 505–515.

388512 pH of Reverse Osmosis Permeate Water and the Chemistry Leading to It's Variability

Abstract:

Every form of life on this Earth is dependent on water. Water covers 70% of our planet, and it is easy to think that it will always be plentiful. However, freshwater—the water we drink, bathe in, irrigate our farm fields with—is incredibly rare. Only 3% of the world's water is fresh water, and two-thirds of that is tucked away in frozen glaciers or otherwise unavailable for our use. According to WHO and Unicef's *Progress on Drinking Water and Sanitation 2014 update*, there are still 748 million people who lack access to proper drinking water source. Around 173 million people rely on untreated surface water for consumption. Though efforts are being made to address this issue, with everyday increase in world population, changes in climate and rising temperature, this problem may soon get out of hand. One way to solve the freshwater scarcity is desalination of brackish groundwater using Photovoltaic Reverse Osmosis. In places where the population has easy access to brackish groundwater, and abundant solar insolation, desalination with the help of Photovoltaic Reverse Osmosis is most suitable. But Reverse Osmosis has many problems associated with it. It needs high pressure pump and it is susceptible to membrane fouling. Another major problem is the lowering of permeate pH. Reverse osmosis is a very simple membrane separation process. Still, for some reason it lowers the pH when water is passed through it. The pH of permeate water can get as low as 5. Given that the pH scale is logarithmic, it means that water with pH 5 is 100 times more acidic than neutral water at pH 7. The rate of decrease of pH is not constant. It varies depending on many factor. Carbon dioxide-bicarbonate equilibrium in the feed water is one of the main factors affecting the pH.

In the present study Photovoltaic Reverse Osmosis system was used for removal of chlorine salts from different types of water and the change of permeate pH were noted. pH of the feed water were changed to notice the change in the permeate pH. Samples were collected from each of these experiments. Studies like Ion Chromatography and Total Inorganic Carbon were performed on these samples to better understand the reactions leading to the change of pH. Relationship between feed water chemistry and permeate pH were noted.

Bacteria mediated Biosynthesis of FeO Nanoparticles for Desalination of Sea Water Using Thin Film Nanocomposite Membrane

Nishant Srivastava[1] *and Mausumi Mukhopadhyay*[1,*]
[1]*Sardar Vallabhbhai National Institute of Technology, Surat, Gujarat, India*

ABSTRACT

The biological method for nanoparticle synthesis emerges as a potential alternative of existing chemical and biological processes. The biosynthesis of nanoparticles is an economical, environment friendly, non toxic and easy to handle process. In the present study, a facile, green, economical synthesis of Iron oxide (FeO) nanoparticles using Gram negative bacteria *Ralstonia eutropha, Zooglea ramigera* and *Erwinia herbicola* reported. All the three bacterial strains found to having potential for the extracellular synthesis of FeO nanoparticles. The nanoparticles were characterized using Transmission Electron Microscope (TEM), Scanning Electron Microscope with energy dispersive X-ray (SEM-EDX), Dynamic Light Scattering (DLS) and Fourier Transform Infra Red spectroscopy (FT-IR) techniques. The TEM analysis revealed that the biosynthesized FeO nanoparticles from *Ralstonia eutropha, Zooglea ramigera* and *Erwinia herbicola* were in the size range of 20-40 nm. Morphologically FeO nanoparticles were spherical in shape. The SEM analysis was in agreement with TEM results. The biosynthesized FeO nanoparticles were utilize to synthesize thin film nanocomposite (TFN) membrane. FeO nanoparticles pre-assembled on a polyethylsulfone (PES) ultrafiltration support that served as seeds for polyamide nanocomposite membrane fabrication. The thin film nanocomposite-FeO (TFN) membrane was further use for desalination of salt water using membrane reactor. The TFN-FeO membrane gives very exciting results for desalination of natural and artificial sea water.

Introduction

The formation of nanoparticles by biological entities are emerges as sustainable, economical and environmental friendly alternative method of chemical and physical processes. The chemical and physical methodology of nanoparticles synthesis widely used is toxic, hazardous, and costly and consumes lots of energy. These procedures are required costly toxic chemicals, surfactants, high temperature, high pressure, costly instruments etc (1-5). Bacteria are the efficient nanofactories capable of reducing toxic metal ions into nontoxic metal nanoparticles. The extracellular enzymes of bacteria act as an electron shuttle for reduction of metal ion into nanoparticles and have excellent redox potential (4). Previously, bacterial species like *Bacillus subtilis* (5), *Shewanella oneidensi* (6), *Actinobacter* sp. (7) are reported for FeO nanoparticles biosynthesis. Gao et al. describes biopolymer assisted biosynthesis of FeO nanoparticles (8).

The rising global water problems motivate worldwide scientific community to find out alternative sustainable majors for water purification. The thin film nanocomposite membranes can be one of the better alternatives for desalination of sea water and full fill global fresh water demands. In the present study, a facile biosynthesis method of Iron oxide (FeO) nanoparticles using Gram negative bacteria *Ralstonia eutropha*, *Zooglea ramigera* and *Erwinia herbicola* is reported. The biosynthesized FeO nanoparticles are further used for fabrication of nanocomposite membrane. The FeO nanocomposite membrane is utilized for the study of desalination of natural and artificial sea water.

Materials and Methods

Biosynthesis of FeO nanoparticles

The freeze dried culture of *Ralstonia eutropha*, *Zooglea ramigera* and *Erwinia herbicola* and bacterial growth medium Nutrient broth (N-broth) was obtained from MTCC, IMTECH, Chandigarh, India and Himedia laboratories, Mumbai, India respectively. The bacterial seed culture was harvested in 100ml N-broth growth medium in 250ml Erlenmeyer flask for 24h at 30°C in incubator (CIS 24 BL, REMI, India) and sub-cultured. The activated *Ralstonia eutropha*, *Zooglea ramigera* and *Erwinia herbicola* cultures (1ml) and 1.5 mM ferrous chloride was added into 100ml sterilized N-broth medium, incubated at 30°C for 48h at 150rpm. These particles were isolated from reaction mixture by centrifugation at 12,000 rpm for 10 min and washed several times with distilled water and acetone. In the control experiment, the bacterial active culture was autoclaved (121°C, 15lbs for 15 min., Obromax, India) and inoculated to 100ml sterilized N-broth medium consisted 1 mM ferrous chloride in same condition as described above with active bacterium. The biosynthesized FeO nanoparticles were characterized by transmission electron microscope (TEM, Philips Holland Tecnai-20) at accelerating voltage of 200kV, dynamic light scattering (DLS), particle size analyzer (Nano ZS 90, Malvern, UK) and scanning electron microscopy (SEM) with an energy dispersive X-ray analysis (EDAX) analyzer (JSM-7600F, JEOL, Japan), for its size and morphology and elemental analysis of product.

Fabrication of Nanocomposite Membrane for Sea Water Desalination

FeO nanoparticles were utilized for fabrication of nanocomposite membrane. The fabrication of nanocomposite thin film membranes was adopted from previously reported methodologies using zeolites by researchers (9). The PES ultrafiltration support was used as the base for fabrication of nanocomposite membrane. In the fabrication process, 1 ml aqueous solution of 2 wt% MPD (m-Phenylenediamine) and 0.15 wt% of SLS (Sodium Lauryl Sulfate) was poured onto the PES (poly-ethyl sulfone) membrane (diameter: 4cm). After 2 min excess solution was removed. 0.2 ml hexane solution of 0.05 wt% TMC, 0.5 and 1 wt% of FeO nanoparticle and 15 wt% ethanol was uniformly coated on PES support. To obtain the well dispersed nanoparticles in hexane solution ethanol was used as a co-solvent. The nanoparticles were well dispersed in ethanol by ultrasonication. The solution was quickly spread on PES support surface because of the positive spreading coefficient of hexane solvent. The pre-seeded support was kept for unforced evaporation of extra solution. Afterwards, the solution containing 0.1 wt% TMC (Trimesoyl Chloride) and hexane was immediately poured onto the pre-seeded PES support and kept for 40-50 s for polymerization reaction. After the completion of polymerization reaction, the membrane was thoroughly washed with de-ionized water. Further the membrane was characterized by using FE-SEM to examine its morphological structure and changes. The nanocomposite membranes were evaluated for natural sea water collected from Arabian Sea, Surat, India and artificial salt water (2000 ppm NaCl, 2000 ppm $MgCl_2$, 500 ppm KCl and 300 ppm $CaCl_2$) desalination with the cross-flow filtration apparatus (Pragati systems, Ahmadabad, India). The membranes were compressed for 1 h at room temperature by DI water at 300 psi. The pressure was controlled by using a pressure knob. The water flux was measured by the weight of the permeate water at a constant trans-membrane pressure. After 1 h compression, natural sea water was added and the conductivity of feed and permeate was measured by a conductivity meter attached in water analysis kit (HPG Systems, New Delhi, India). The rejections were calculated by the following equations:

$$\boldsymbol{Rejection\ (R)(\%)} = \frac{C_f - C_p}{C_f} \times \mathbf{100} \qquad \text{(Eq. 1)}$$

Where, R is the rejection ratio and C_f and C_p are the conductivities of feed and permeate solutions, respectively.

Results and Discussion

A facile, green, economic synthesis of Iron oxide (FeO) nanoparticles using Gram negative bacteria *Ralstonia eutropha*, *Zooglea ramigera* and *Erwinia herbicola* was reported. All three bacterial strains showed potential for the extracellular synthesis of FeO nanoparticles. The TEM analysis (Figure 1) revealed that the biosynthesized FeO nanoparticles from *Ralstonia eutropha*, *Zooglea ramigera* and *Erwinia herbicola* were in the size range of 20-40 nm (Figure 1a, c and e). Morphologically FeO nanoparticles were spherical in shape. The SEM analysis was in agreement with TEM results. The EDX spectrum (Figure 1g, h and i) of biosynthesized FeO nanoparticles was indicated the presence of strong peaks of Fe at 6.5 and 7.2KeV and of oxygen at 0.5 KeV respectively confirms the presence of pure FeO. The weak signals of carbon, phosphorus and nitrogen generated during the EDX analysis were perhaps associated with enzymes or proteins from a bacterium either capped or present near the biologically synthesized FeO nanoparticles. The signal of copper was associated with the glass underlay.

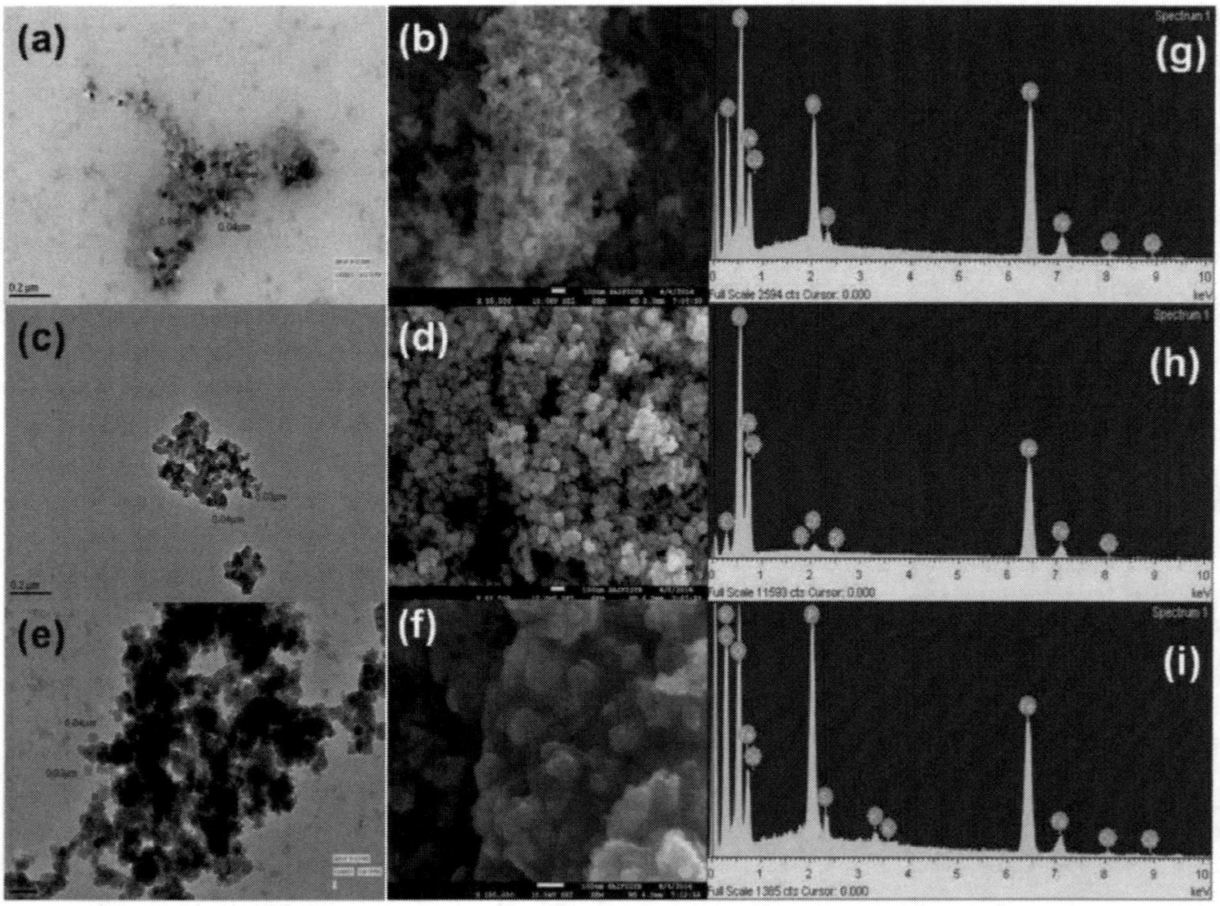

Figure 1: TEM (Scale-200 nm), SEM (100 nm) and EDX images of biosynthesized FeO nanoparticles using bacteria *Ralstonia eutropha* (a), (b), (g); *Zooglea ramigera* (c), (d), (h) and *Erwinia herbicola* (e), (f), (i).

The DLS analysis (Figure 2) of FeO nanoparticles was observed. DLS analysis revealed that the *Ralstonia eutropha* mediated biosynthesized particles were in the size range of 43 nm to 72 nm. The *Zooglea ramigera* mediated nanoparticles were in the range of 43 to 91 nm. Similarly, FeO nanoparticles biologically synthesized using *Erwinia herbicola* was an average size of 32 nm. The size obtained by DLS was bigger than TEM observations, because DLS analysis measured the hydrodynamic diameter of nanoparticles (10).

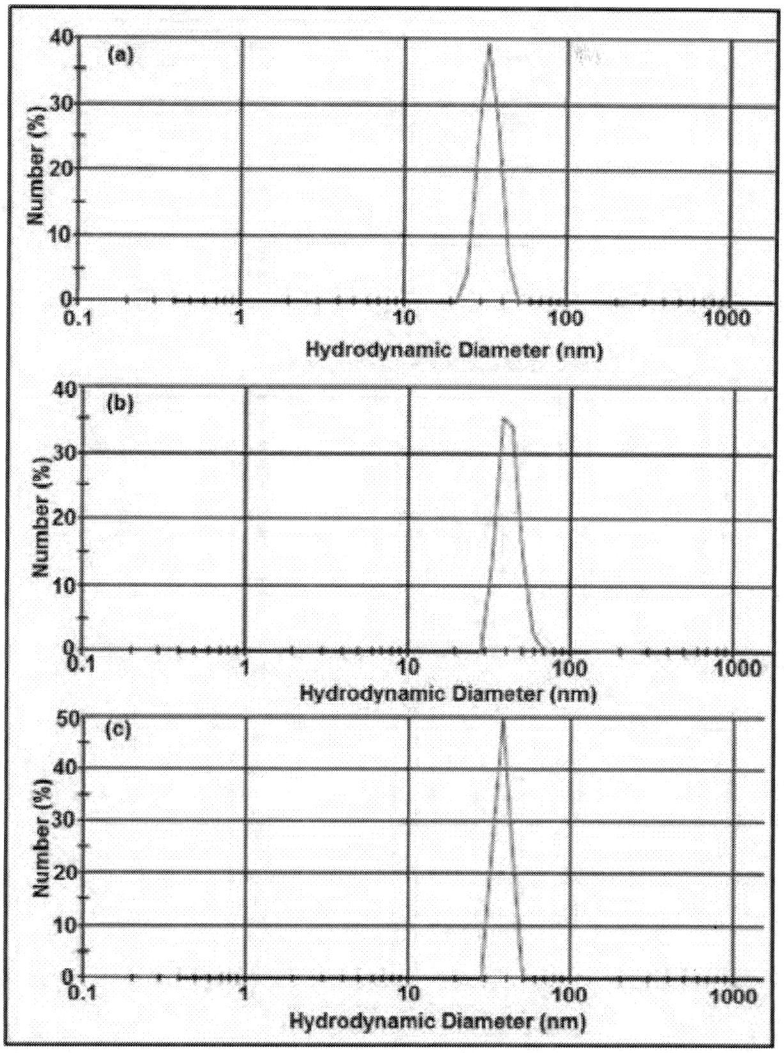

Figure 2: DLS analysis of FeO NPs biosynthesized by using *R.eutropha*; Avg. Size = 34 nm (a), *Z.ramigera*; Avg. Size = 41.72 nm (b) and *E.herbicola*; Avg. Size = 38 nm (c).

The FT-IR spectrum (Figure 3) from *R.eutropha* revealed the functional groups present and responsible for stabilizing and dispersion of FeO nanoparticles. The spectral data revealed the peak at 3269 cm-1 which showed the presence of a primary amine whereas, peak at 16382 cm-1 referred to the carbonyl stretch vibrations in the amide bond of protein. The FT-IR spectrum revealed a peak at 2926 cm-1 stretching vibration of C-H bonds. The peaks at 1273 and 1183 cm-1 related to C-N bonds stretching. C-O stretching vibration peaks at 1047 cm-1 were also observed. The FT-IR spectrum supported the hypothesis of presence of protein on the FeO nanoparticle surface. These protein molecules were responsible for reduction of ferrous ions into the FeO nanoparticles and their stability. The proteins present on the FeO nanoparticles surface acted as capping agent. Similar observation was found for *Z.ramigera* and *E.herbicola.*

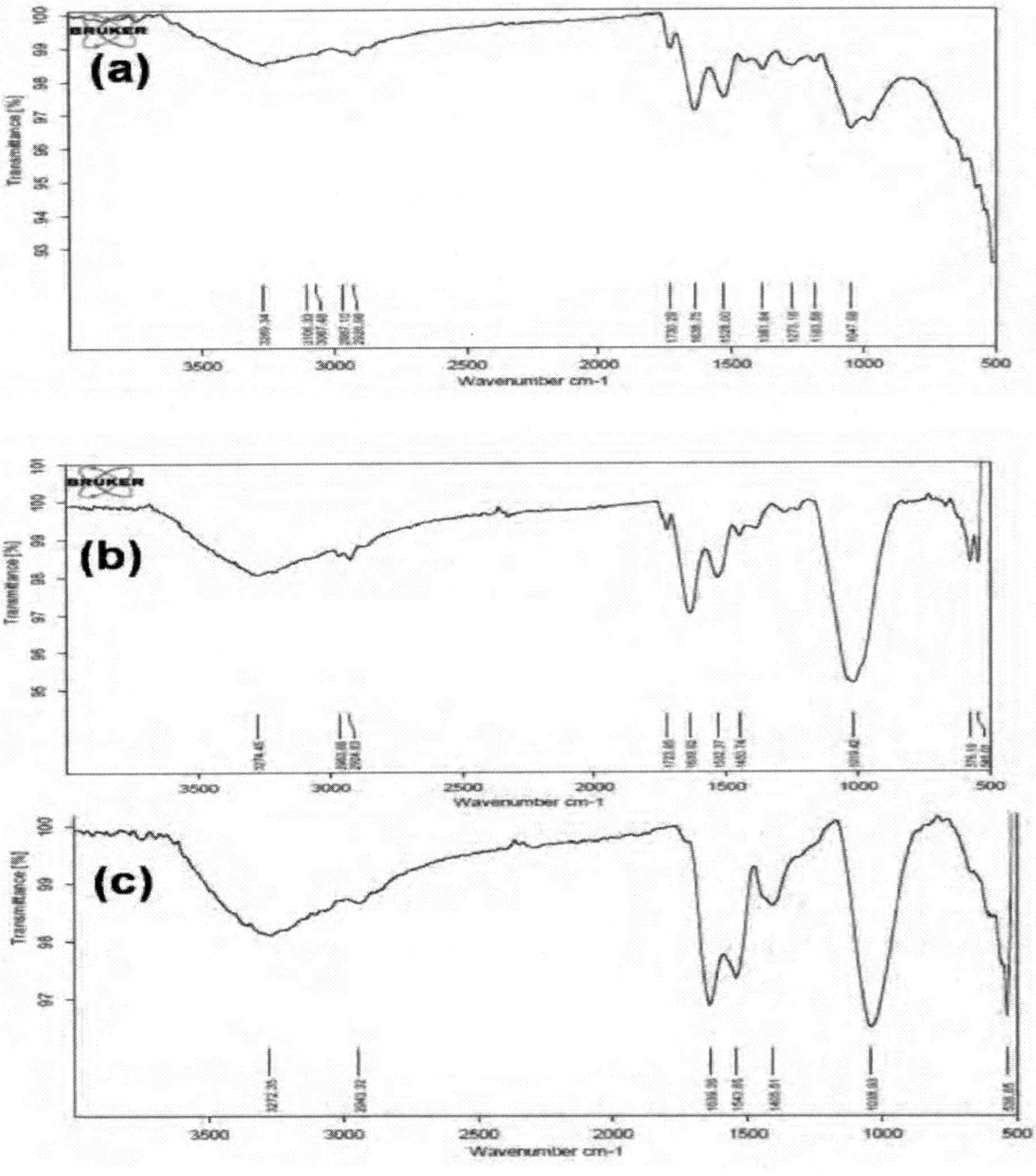

Figure 3: FTIR spectrum of biosynthesized FeO nanoparticles using bacteria *Ralstonia eutropha* (a); *Zooglea ramigera* (b) and *Erwinia herbicola* (c).

The biosynthesized FeO nanoparticles were utilized to synthesize thin film nanocomposite (TFN) membrane. FeO nanoparticles pre-assembled on a polyethylsulfone (PES) ultrafiltration support that served as seeds for polyamide (PA) nanocomposite membrane fabrication. The PA-FeO nanocomposite membrane was further used for desalination of natural and artificial sea water using membrane reactor. In Figure 4 FE-SEM images of PES support and well dispersed FeO nanoparticles assembled on PES support was observed. The clean and clear PES support showed a featureless, smooth morphology of the surface (Figure 4a). Whereas, after interfacial polymerization onto PES support, the thin film containing FeO nanoparticles was formed. This FeO nanoparticle served as seeds for the formation of PA-FeO nanocomposite membrane. The FE-SEM image in Figure 4b clearly showed a compact and flat morphology of PA-FeO nanocomposite membrane in contrast to Figure 4c. The robust morphology indicated that FeO was better dispersed within the nanocomposite membrane. The surface image also revealed some isolated and aggregated FeO nanoparticles. The particle size of FeO nanoparticles in the membrane was larger than original biosynthesized FeO nanoparticles. This was attributed to the covering of interfacial polymerized PA on the surface of FeO nanoparticles. The covering provides stable attachment of FeO in the membrane. Figure 4d were FE-SEM image of FeO nanocomposite membrane after desalination of natural sea water. The image clearly showed change in surface morphology of nanocomposite membrane due to the rejection of salts.

Figure 4: SEM image of (a) PES membrane; (b) and (c) FeO-TFN membrane and (d) FeO- nanocomposite membrane after the desalination process.

The TFN-FeO membrane showed reasonable results for desalination of natural and artificial sea water. The desalination of sea water was monitored by ICP-AES of sodium, magnesium, calcium and potassium and by measuring change in pH, Conductivity, hardness and chloride concentration. About 45 % Na, 46 % K and 40 % Ca removal was obtained with biogenic FeO-TFN membrane process respectively. The conductivity, hardness and chloride concentration was decreased by 34%, 31% and 33% respectively. Approximately 34% salt rejection and desalination was achieved by using FeO-nanocomposite membrane. Whereas, with artificial salt water, the removal of Na was 65%, K was 56% and Ca removal was 58 %. The conductivity was decreased by 63.2%, hardness reduced by 59% and chloride decreased by 61 % respectively. In Table 1 detailed output of result is given for 0.5 and 1 wt % FeO nanocomposite membrane mediated salt rejection. The feed concentration and permeate concentration with final salt rejection in percentage was calculated by using Eq. 1.

Table 1: Concentrations of feed and permeate during permeation of sea water desalination and salt rejection using PA-FeO nanocomposite membrane.

Sea Water	Membrane Concentration and Rejection		Na (ppm)	K (ppm)	Mg (ppm)	Ca (ppm)	Conductivity (μmho/cm)
Natural	PA-FeO 1 (0.5 wt %)	Feed	1565	25	230	467	14440
		Permeate	985	15	142	308	9760
		Rejection (%)	37	40	38	34	32.5
	PA-FeO 2 (1 wt %)	Feed	1565	25	230	467	14440
		Permeate	860	13	126	280	9600
		Rejection (%)	45	46	45	40	34
Artificial water Concentration	PA-FeO 1 (0.5 wt %)	Feed	1610	37	289	471	15954
		Permeate	1094	23	202	315	9521
		Rejection (%)	32	37	30	33	40.4
	PA-FeO 2 (1 wt %)	Feed	1610	37	289	471	15954
		Permeate	565	16	138	197	5871
		Rejection (%)	64.9	56	52	58	63.2

FeO nanoparticles were successfully biosynthesized by the three different bacterial strains. The synthesis method was economical and environment friendly. A new type of PA-FeO nanocomposite membrane was fabricated by a promising pre-seeding method without use of any additional additive and post treatment process. This nanocomposite membrane showed a high-water flux without any loss of salt rejection. The FeO nanocomposite membrane showed encouraging performance for desalination process and utilized for water purification and desalination process.

References

1. Srivastava, N. and M. Mukhopadhyay (2013), *Biosynthesis and structural characterization of selenium nanoparticles mediated by Zooglea ramigera,* Powder Technology, 244: 26-29.
2. Srivastava, N. and M. Mukhopadhyay (2014), *Biosynthesis and Characterization of Gold Nanoparticles Using Zooglea ramigera and Assessment of Its Antibacterial Property,* Journal of Cluster Science, DOI 10.1007/s10876-014-0726-0.
3. Srivastava, N. and M. Mukhopadhyay (2014), *Biosynthesis of SnO2 nanoparticles using bacterium Erwinia herbicola and its photocatalytic activity for degradation of dyes,* Industrial & Engineering Chemistry Research,

dx.doi.org/10.1021/ie5020052.
4. Srivastava, N. and M. Mukhopadhyay (2014), Ralstonia eutropha (Cupriavidus metallidurans) Mediated Biosynthesis of Gold Nanoparticles and Catalytic Treatment of 2, 4 dichlorophenol, Synthesis and Reactivity in Inorganic, Metal-Organic, and Nano-Metal Chemistry, x.doi.org/10.1080/15533174.2011.609230.
5. Sundaram, P.A., R. Augustine, and M. Kannan (2012), Extracellular biosynthesis of iron oxide nanoparticles by Bacillus subtilis strains isolated from rhizosphere soil, Biotechnology and Bioprocess Engineering, 17: 835-840.
6. Roh, Y., et al. (2001), Microbial synthesis and the characterization of metal-substituted magnetites, Solid State Communications, 118: 529-534.
7. Bharde, A.A., et al. (2008), Bacteria-Mediated Precursor-Dependent Biosynthesis of Superparamagnetic Iron Oxide and Iron Sulfide Nanoparticles, Langmuir,. 24:5787-5794.
8. Gao, S., et al. (2008), Biopolymer-Assisted Green Synthesis of Iron Oxide Nanoparticles and Their Magnetic Properties, The Journal of Physical Chemistry C, 112:10398-10401.
9. Kong, C., T. Shintani, and T. Tsuru (2010), "Pre-seeding"-assisted synthesis of a high performance polyamide-zeolite nanocomposite membrane for water purification, New Journal of Chemistry, 34:2101-2104.
10. Jiale, H., et al. (2007), Biosynthesis of silver and gold nanoparticles by novel sundried Cinnamomum camphora leaf, Nanotechnology, 18: 105104.

Metal catalyzed hydrogel synthesis in PVDF membranes for degradation of toxic organic contaminants

Sebastián Hernández, Joseph K. Papp and Dibakar Bhattacharyya
Chemical and Materials Engineering Department
University of Kentucky, Lexington, KY – 40506

Abstract

Water pollution is a serious concern in all the aspects of society. Most of the pollutants have two main origins: organic compounds and heavy metals. One of these contaminants, Trichloroethylene (TCE), has high interest because their persistency in nature and the difficulties related with their treatment and/or elimination. Nano-structured material synthesis or metal capturing using functionalized membranes with polymers that contain ion exchange groups is a promising research field for water pollution, catalytic and mining applications. The polymerization of such materials requires initiation using strong oxidants and heat. However, a redox reaction can take place using metal ions to accelerate it.

This work describes this redox polymerization in order to synthesize a hydrogel of poly (acrylic acid) (PAA) for Polyvinylidene fluoride (PVDF) membrane functionalization using ferrous ions as accelerant and persulfate as initiator. The iron accelerant selected was subsequently used for nanoparticle synthesis and compared with a process of ion exchange and nanoparticle synthesis developed previously by this group. Redox polymerization resulted exothermic with temperatures around 50°C producing ferric ions contributing to the decrease in pH values around 2.0. Iron nanoparticles sizes of 30-60 nm were synthesized. From the use of the ferrous ions in the polymerization of the hydrogel, 9.2 mg Fe /g PAA were produced. This value is lower compared to ion exchange procedure (450 mg Fe/ g PAA) due to the small concentrations used in the polymerization. Since the functionalization of the membranes carry less PAA inside (8.5 to 13.4% wt), the amount of iron in the membranes was lower.

The PAA hydrogel was also characterized by swelling behavior along with the permeability of the PVDF-PAA membranes. The rate of swelling of the cross-linked PAA hydrogel follows a second order swelling kinetics, making the swelling process faster at higher pH values and higher responsive behavior. It was defined an inverse relationship between the hydrogel mesh size (6.30 to 8.34 nm) and the membrane pore size (222 to 110nm). An increase in the mesh size of the hydrogel creates around 93% of decrease in permeability.

Quantitative TCE/chloride generation by Fe/Pd nanoparticles in hydrogel and functionalized membrane were performed in batch process. Iron nanoparticles followed pseudo-first order reaction kinetics with a higher surface-area constant k_{sa} in hydrogel (2.46 L/ ($m^2 \cdot h$)) than functionalized membrane (0.10 to 0.13 L/ ($m^2 \cdot h$)) due to a more open area and larger volume of reaction. The produced chloride using bimetallic Fe/Pd nanoparticles in dechlorination experiments is close to the stoichiometric value, indicating negligible production of intermediates.

The project described was supported by Grant Number P42ES007380 from the National Institute of Environmental Health Sciences. The content is solely the responsibility of the authors and does not necessarily represent the official views of the National Institute of Environmental Health Sciences or the National Institutes of Health.

357157 Application of Non-Dispersive Solvent Extraction in PTA Plant Wastewater Treatment

Abstract:

Purified Terephthalic Acid (PTA) producing process generates about 50 million m^3 wastewater in China every year. PTA plant wastewater (COD 3000-5000 $mg.L^{-1}$) mainly contains intermediate p-Toluic Acid (PT), other small amount of aromatic acids and metal ions. Traditional biodegration and other physico-chemical treatment have been studied since ten yeas ago, however, PTA plant wastewater still cannot reuse at low cost nowadays.

The aim of this work is to develop a complete recycling process for simutenously reusing water and recycling PT and Benzoic Acid (BA) by implementation of Non-dispersive Solvent Extraction (NDSX).

In the first process step appropriate extractant was chosen: p-xylene is the strating material of PTA process and it is an ideal extractant for three reasons: A. The partition coefficient of PT between PX and water is more than 10 which indicates PX has high extraction efficiency. B. After extraction water can direct reuse in hydrogenation process. C. After extraction PX can direct go back to oxidation unit without further separation.

The NDSX process was experimentally studied and simulated. Microporous polypropylene hollow fiber membrane with outer-diameter of 490 μm, thichness of 50 μm, porosity of 0.45 and pore size 20 nm x 200 nm were used. The extraction experiments were conducted in single membrane extraction and a lab scale membrane contactor. PT and BA were separated from simulated wastewater and removal efficiencies were 90% and 40% at beat conditions, respectively.

The NDSX was simulated with a 2D rigrous model which taking into account the axial and radial variation in the fluid properties and an empirical model at the same time. The partition coefficients of PT and BA between PX and water were experimentally obtained. The simulation results fit well with the experiments.

Base on the experimental results the complete recycling route was validated.

362223 Removal of Fluoride from Drinking Water Using Cellulose Acetate Pthalate -Alumina Based Mixed Matrix Membrane

Abstract:

Membrane technology provides attractive alternative to water purification problems. However the cost of membrane production and its operation still has been a challenge. Moreover, removal of heavy metals (like arsenic, cadmium, chromium and lead) by membrane requires the need of operations like reverse osmosis or nanofiltration. This process has their inherent limitations of high power requirements and also the removal is not achieved to a remarkable extent. This demands the rise of a new kind of inorganic additive doped- polymeric membranes which are being termed as mixed matrix membranes (MMM). Here, we exploit the adsorptive capability of the membrane along with the filtration properties. In our work, activated alumina-cellulose acetate phthalate (CAP) mixed matrix membranes was prepared for removal of fluoride from water. Fluoride is an essential micronutrient to our body, but it poses threats of harmful diseases (fluorosis), once its concentration level goes beyond 1.5 mg/l. It was observed that the removal capacity of the membrane increases significantly on alumina addition (average size of 109 μm) to the pristine CAP membrane and the maximum removal percentage was achieved at 35 % (weight) of alumina. Beyond this, the solution becomes too viscous to cast. The permeability of the membranes decreased by 53 %. This was due to the restriction of the continuity of the voids by alumina (as inferred from the SEM images). Moreover, it also decreases the segmental mobility of the polymer chains, thereby decreasing the free volume and hence the cut off. However, hydrophilicity of the membrane was improved due to the presence of particles at the surface (observed from the top view of SEM images). The surface roughness increases with alumina addition, as interpreted from the AFM images. An adsorption capacity of 2.3 mg/g (interpreted from langmuir isotherm at 25°C) was reported. Batch equilibrium was obtained at 27 hours and it follows a second order kinetics. Dynamic run of the selected membrane (area of 0.01 m^2, cut off 24 kDa) was made in continuous cross flow mode. The virgin membrane can provide sustained production of safe drinking water till a time of 11 hours, whereas after five cycles of regeneration, this time decreased only till 8 hours (pH: 7.2). The mechanism of fluoride removal chiefly occurs by a trade-off between adsorption and electric repulsion where the major role is played by the former. Removal efficiency is affected by presence of anions like sulphate, carbonate and bicarbonate whereas chloride and nitrate do not exert any influence. The mechanism of sieving was utilized in the removal of microorganisms like aerogenes and E.coli from tap water medium.

Crucial Role of Blocking Inaccessible Cages in the Simulation of Gas Adsorption in a Paddle-Wheel Metal-Organic Framework

Kang Zhang, Anjaiah Nalaparaju, Yifei Chen, and Jianwen Jiang*

Department of Chemical and Biomolecular Engineering, National University of Singapore, 117576, Singapore

*chejj@nus.edu.sg, +65 65165083

H_2 used in industry is generally produced from steam-methane reformation followed by water-gas shift, and is required to be separated from the effluent gas for production of high-purity H_2 and for pre-combustion CO_2 capture. In the last decade, metal-organic frameworks (MOFs) have emerged as a new family of hybrid porous materials. The large surface area, high porosity, and tunable structure place them at the frontier for CO_2/H_2 mixture separation. Numerous experimental and simulation studies have been reported for gas adsorption in MOFs. In the simulation studies of gas adsorption in MOFs, the predicted capacities are usually compared with available experimental data. The deviations between simulation and experiment are generally attributed to two reasons: impurities or defects present in experimental samples, and the inaccuracy of force fields used to describe gas-MOF interactions. However, a less discussed possible reason is that inaccessible cages existing in MOFs might not be taken into account by simulation. We report a molecular simulation study to investigate the adsorption of CO_2 and H_2 in a recently synthesized paddle-wheel Cu-based metal-organic framework (Cu-MOF). The Cu-MOF consists of three types of cages, and type-III cages are restricted by small windows and inaccessible to gas molecules. Without blocking the inaccessible cages, the simulated adsorption isotherms are much higher than experimental data. However, the agreement is substantially improved after blocking. For CO_2/H_2 mixture, the adsorption uptakes predicted from the ideal-adsorbed solution theory (IAST) are nearly identical to simulation results. The selectivity increases with increasing pressure due to the enhanced cooperative interactions between CO_2 molecules, which is also predicted by the IAST. Furthermore, the breakthrough profiles are evaluated for CO_2/H_2 mixture in a fixed-bed packed with the Cu-MOF. The breakthrough times are estimated to be 2.8 and 85.2 for H_2 and CO_2, respectively, implying the efficient separation of CO_2/H_2 mixture. This simulation study demonstrates that blocking inaccessible cages is crucial to simulate gas adsorption in the Cu-MOF, and the IAST can be applied to the Cu-MOF with inaccessible cages. In new MOFs or other porous materials, one should conduct structural analysis to determine if there exist inaccessible cages before starting the simulation of gas adsorption.

374809 Elasticity of Simple Fluids Confined in Nanopores

Abstract:

Predicting the behavior of fluids in porous media is crucial for hydrocarbons recovery and carbon sequestration in geological formations. Fluid transport in porous medium depends on the fluid elasticity (bulk modulus). Substantial growth of interest in shale gas recovery drew attention to fluids transport in geological media with pores in the nanometer range. Since nanoconfinement is known to significantly affect thermodynamic properties of fluids, one can expect its influence on fluid elasticity as well. Indeed, recently Schappert and Pelster proposed a method to determine the elasticity of the fluid confined in nanopores from ultrasonic experiments [1]. Their findings showed that the bulk modulus of liquid argon confined in Vycor glass is affected by high solvation pressure in the pores, which is also known to cause adsorption-induced deformation [2].

Here we use the Quenched Solid Density Functional Theory (QSDFT) method [3] to calculate the bulk modulus of nanoconfined fluid. QSDFT accurately predicts thermodynamic properties of confined fluids, and is widely used for characterization of porous materials [4]. However, QSDFT has not been employed to calculate the derivative properties, including compressibility which determines the bulk modulus.

Our calculations showed that QSDFT method is capable of predicting the bulk modulus of simple fluids confined in nanopores. Our results showed that the elasticity of confined fluid changes with the solvation pressure in the pore, which is in line with the experimental observations by Schappert and Pelster. Additionally we showed that the bulk modulus of the confined fluid increases with the degree of confinement – fluids adsorbed in smaller pores are noticeably stiffer than those adsorbed in the large ones. These effects need to be taken into account in large scale models for predicting the fluid flow in nanoporous geological formations such as shale.

References

[1] K. Schappert, R. Pelster, Influence of the laplace pressure on the elasticity of argon in nanopores, EPL (Europhysics Letters) 105 (5) (2014) 56001.
[2] G. Y. Gor, A. V. Neimark, Adsorption-induced deformation of mesoporous solids, Langmuir 26 (16) (2010) 13021–13027.
[3] P. I. Ravikovitch, A. V. Neimark, Density functional theory model of adsorption on amorphous and microporous silica materials, Langmuir 22 (26) (2006) 11171–11179.
[4] J. Landers, G. Y. Gor, A. V. Neimark, Density functional theory methods for characterization of porous materials, Colloids and Surfaces A: Physicochemical and Engineering Aspects 437 (2013) 3–32.

Modeling and Visualization of Elastic Layer-Structured Metal-Organic Framework-11 Encapsulating CO_2: Toward a Better Understanding of Gate Adsorption Behavior

Hideki Tanaka, Shotaro Hiraide, Satoshi Watanabe, and Minoru T. Miyahara
Department of Chemical Engineering, Kyoto University, Nishikyo, Kyoto 615-8510, Japan

Elastic layer-structured MOF-11 (ELM-11) shows the gate-opening behavior by CO_2 adsorption, which is promising for CO_2 capture and adsorption heat pump. We have therefore combined the molecular simulations and the Rietveld analysis of a synchrotron X-ray powder diffraction pattern to understand the gate-opening behavior of ELM-11. We obtained the following parameters to determine the open structure of ELM-11 encapsulating CO_2: (1) Tentative atomic charges in the ELM-11 framework by DFT, (2) Modified UFF force field for CO_2-host framework interaction by comparing a theoretical adsorption isotherm from grand canonical Monte Carlo (GCMC) simulations at 273 K and experimental one, (3) Inter-framework interaction of the host by DFT, (4) Position of CO_2 molecules in the framework by GCMC. Then, (1)-(4) were used for canonical Monte Carlo simulation to relax the structure of ELM-11 with CO_2, and the final atomic configuration was used as the initial structure for the Rietveld analysis. The determined structure was used for the free energy analysis with the aid of the GCMC simulations, and a good description for the temperature dependence of the gate-closing pressure was obtained (Fig. 1).

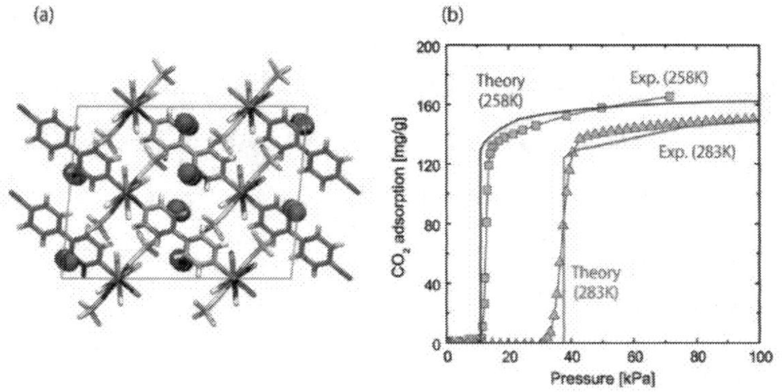

Fig. 1 (a) Rietveld refined structure of ELM-11 with the adsorption of CO_2, (b) A comparison between experimental adsorption isotherms of CO_2 and theoretical adsorption isotherms from the free energy analysis.

Abstract for AIChE 2014 Meeting

Title: Modeling Diffusion of Linear Hydrocarbons in Pure Silica Zeolite LTA Using Transition Path Sampling

Contributors: Salah Eddine Boulfelfel[(a)], Peter I. Ravikovitch[(b)*], and David S. Sholl[(a)**]

Affiliations: [(a)] Georgia Institute of Technology, School of Chemical and Biomolecular Engineering, GA 30332-0100, USA

[(b)] ExxonMobil Research and Engineering, Corporate Strategic Research 1545 Route 22 East, Annandale, NJ 08801, USA

Abstract

The diffusivities of linear hydrocarbons (CH_4, C_2H_6, C_3H_8, and C_4H_{10}) in pure silica zeolite LTA (ITQ-29) are computed at 300 K and infinite dilution. To overcome the time scale problem arising from the slow diffusion process at room temperature we used transition path sampling (TPS). The influence of framework flexibility on diffusion is investigated by combining TPS simulations with fully flexible molecular dynamics performed in the NpT ensemble. The ensemble of the collected reactive trajectories was used to characterize sets of transition states and the corresponding configurations were analyzed to construct window size distributions during the molecular hopping events. The diffusion process is affected by framework flexibility. The influence of framework flexibility on diffusion of propane and butane is much larger than for methane and ethane. We will also demonstrate the extension of TPS to adsorbates at non-zero loadings and adsorbed mixtures.

377242 Computational Screening of Porous Coordination Networks for Adsorption and Membrane-Based Gas Separations

Abstract:

Separation of gas mixtures has social, environmental, industrial and economical importance. Traditional gas separation techniques include distillation, absorption, adsorption and membranes [1]. Among these techniques, adsorption-based and membrane-based gas separation methods have received significant attention from academy and industry since they are energy-efficient and environmentally friendly processes. Various materials such as polymers, zeolites, carbon molecular-sieves, metal organic frameworks have been developed and tested as adsorbents and membranes to achieve high-performance gas separations [2]. Porous coordination networks (PCNs), also known as metal organic frameworks (MOFs), offer high potential for gas storage and separation due to their large surface areas, high porosities, tunable pore sizes, good chemical and thermal stabilities [3]. Various PCNs have been synthesized by different research groups and single-component gas uptake capacities of these PCNs are generally determined by experimental methods or GCMC simulations. However, there has not been any experimental study related with the binary gas adsorption and/or diffusion in PCNs. This lack of information limits the assessment of PCNs' potential in adsorption-based and membrane-based gas separation applications.

We recently used atomic simulations to examine the separation performances of four PCNs, PCN-9-Co, PCN-9-Fe, PCN-9-Mn and PCN-26 for CH_4/H_2, CO_2/H_2, CO_2/CH_4 and CO_2/N_2 mixtures [4]. Our results showed that these PCNs can outperform traditional zeolites and widely studied MOFs in gas separations, especially in CO_2 related separation processes and PCN-26 was identified as a potential membrane material that can exceed the upper bound established for CO_2/CH_4 and CO_2/N_2 separations due to its high CO_2 permeability and selectivity [4]. Motivated from these initial results, we carried out a large scale molecular simulation study in this work for twenty different PCNs, which represent the largest number of PCN materials studied in a molecular simulation work, to examine their potential both for adsorption-based and membrane-based gas separations. Additionally, we examined the structure-performance relationships of materials and developed simple selectivity models based on structural properties. We showed that these simple models can make accurate predictions for adsorption, diffusion and membrane selectivity of PCNs without performing computationally demanding molecular simulations. Therefore, these selectivity models will be extremely useful to screen very large groups of PCNs and other similar nanoporous materials to identify the most promising materials in gas separation applications prior to extensive experiments and calculations. We specifically focused on separation of CH_4/H_2, CO_2/H_2, CO_2/CH_4 and CO_2/N_2 mixtures.

In this work, we examined adsorption-based and membrane-based separation performances of twenty PCNs using molecular simulations to identify the most promising adsorbent and membrane candidates for CH_4/H_2, CO_2/H_2, CO_2/CH_4 and CO_2/N_2 mixtures. Several PCNs such as PCN-9 series and PCN-14 were identified as promising candidates for adsorption-based CH_4/H_2 separations since they exhibit both high adsorption selectivity and working capacity towards CH_4. PCN-9-Co, PCN-9-Mn, PCN-14 and PCN-16 were found to be strong adsorbents for CO_2 capture from CO_2/H_2, CO_2/CH_4 and CO_2/N_2 mixtures because of their high CO_2 working capacities. Several PCNs membranes were shown to exceed Robeson's upper bound established for CO_2/CH_4 and CO_2/N_2 separations. We showed that both adsorption and diffusion favor the same component CO_2 over CH_4 (N_2) in PCN-10, PCN-11, PCN-131' and make these PCNs highly CO_2 selective in membrane-based CO_2/CH_4 (CO_2/N_2) separations.

In addition to computing adsorption selectivity, diffusion selectivity and membrane selectivity of PCNs using molecular simulations, we also developed approximate models that can predict adsorption, diffusion and permeation selectivities of PCNs for CH_4/H_2 and CO_2/H_2 mixtures. These models were mainly based on the structural properties of the materials which can be simply measured or computed such as pore volume, surface area, pore diameter. Model predictions for adsorbent and membrane selectivities were found to be in a good agreement with the direct results of detailed molecular simulations. Therefore, these selectivity models will be extremely useful to screen very large groups of PCNs and other similar nanoporous materials to identify the most promising materials in gas separation applications prior to extensive experiments and calculations.

References

[1] Li, J. R.; Kuppler, R. J.; Zhou, H. C. Selective Gas Adsorption and Separation in Metal-Organic Frameworks. *Chem. Soc. Rev.* **2009**, *38*, 1477.

[2] Li, J. R.; Sculley, J.; Zhou, H. C. Metal-Organic Frameworks for Separations. *Chem. Rev.* **2011**, *112*, 869.

[3] Ma, S.; Simmons, J. M.; Sun, D.; Yuan, D.; Zhou, H. C. Porous Metal-Organic Frameworks

Based on an Anthracene Derivative: Syntheses, Structure Analysis, and Hydrogen Sorption Studies. *Inorg. Chem.* **2009**, *48*, 5263.

[4] Ozturk, T. N.; Keskin, S. Predicting Gas Separation Performances of Porous Coordination Networks Using Atomistic Simulations. *Ind. Eng. Chem. Res.* **2013**, *52*, 17627.

Using Computational Modeling to Identify Functionalization Strategies for Selective Contaminant Removal from Ambient Air with UiO-66

Hakan Demir, Krista S. Walton, and David S. Sholl*

[a]School of Chemical and Biomolecular Engineering, Georgia Institute of Technology, 311 Ferst Drive NW, Atlanta, Georgia 30332-0100

*Corresponding author: david.sholl@chbe.gatech.edu

Metal-organic frameworks (MOFs) provide many opportunities for gas storage and separation applications due to their chemical diversity, high surface area, porosity and linker functionalization. One of the fields that MOFs can be utilized in is selectively capturing contaminant gases such as ammonia from ambient air[1]. In this work, we obtained binding energies of NH_3, H_2S and H_2O with cluster and periodic models of ~40 functionalized UiO-66 materials using quantum chemical methods. For all 3 gases, cluster calculations show that there is a good correlation between lowest binding energies obtained using MP2 and PBE-D2 functionals while M06-2X energies have a less satisfactory correlation with MP2 energies. Thus, PBE-D2 functional is selected for periodic system calculations. Our periodic calculations resulted in a different ranking of gas affinities than that obtained with cluster models, indicating considerable confinement effects. Using Grand Canonical Monte Carlo (GCMC) simulations, we will also discuss the validity of a generic force field, UFF, for predicting gas affinities in these materials.

[1] Kim, K. C.; Yu, D.; Snurr, R. Q. *Langmuir* **2013**, 29, 1446

389762 Understanding the Unusual Adsorption Behavior in Hierarchical Zeolite Nanosheets

Abstract:

Hierarchical zeolites are advanced materials possessing the catalytic and adsorption properties of conventional zeolites while eliminating their transport limitations through the introduction of mesopores. Recent experiments comparing the adsorption in hierarchical self-pillared pentasils (SPP) and silicalite-1 (MFI) revealed an interesting crossover in sorbate loading for branched or long-chain alkanes but not for shorter linear alkanes, but an explanation for this behavior is not readily available through experimental probes due to the complications arising from the presence of multiple adsorption sites. Here we present a molecular simulation study on the adsorption of alkane isomers and show that a multi-step mechanism, found here for all molecules, is responsible for the observed phenomena.

390385 Force Field Development for Gas Adsorption in the M-MOF-74 Series Based on First Principles Calculations

Abstract:

Metal-Organic Frameworks (MOFs) are versatile nanoporous materials that have gained significant interest as low heat capacity, high selectivity sorbents for CO_2 capture applications. Thousands of MOFs have been synthesized and many more have been computationally predicted. Additionally, large-scale atomistic simulations for identifying high-performance MOFs are possible, but are currently limited by the quality of the existing molecular mechanics force field describing the interactions between the guest and framework atoms. This problem is particularly severe for MOFs with coordinatively unsaturated metal centers. It has been previously shown that improved force fields can be derived from first principle calculations.[1] In this work, we have applied this approach to an isostructural series of M-MOF-74, where M = Cr, Mn, Fe, Co, Ni, and Cu. Moreover, we have experimented with various aspects of the force field derivation procedure. Including the choice of electronic structure method and framework model (cluster or periodic) to obtain reference energies, the choice of functional form for the interaction energy and the type of point charges assigned to the framework atoms. Monte Carlo calculations in the Gibbs ensemble were used to calculate CO_2 isotherms in order to assess the quality of the derived force field parameters and to determine an optimal procedure for obtaining a reliable force field for a targeted MOF and adsorbate system. Finally, the computed relative affinities towards CO_2 of the different MOFs in the series agree with experimental observations with Mg-MOF-74 and Ni-MOF-74 interacting with CO_2 the strongest and Cu-MOF-74 the weakest.

1. Dzubak et al., Ab initio carbon capture in open-site metal-organic frameworks. *Nature Chemistry* **4**, 810–6 (2012).

Demonstrating the influence of solvent choice, supersaturation and nucleation temperature on Phenacetin crystallization

Denise M. Croker, B. Kieran Hodnett, Åke C. Rasmuson

Synthesis & Solid State Pharmaceutical Centre, Science & Engineering Faculty, University of Limerick, Ireland.

Keywords: Crystallization, Crystal Shape, Nucleation, Growth, Solvent, Particle Size Distribution

ABSTRACT

Phenacetin was used as a model pharmaceutical compound to demonstrate the impact of solvent choice and crystallization conditions on product crystal shape and size distribution. The BDFH crystal morphology was generated using Mercury ™ software and presented as 3-dimensional hexagonal blocks. Crystallization from solution was used to prepare a series of crystals from different solvents and investigate the impact of the crystallization solvent on the crystal shape of phenacetin. 15 solvents were tested: 1, 4-dioxane, 1-propanol, 1-butanol, toluene, ethyl acetate, ethanol, 2-propanol,butanone, acetonitrile, acetone, methanol, MTBE, dichloromethane, diethyl ether, n-pentane. In 12 of 15 experiments, needle shaped product crystals were recovered. Ethyl acetate and butanone yielded brittle plate shaped crystals and flat blocks were recovered from dichloromethane. Phenacetin solubility was measured in 5 solvents from 10 – 70 °C using a gravimetric method to allow for the design of crystallization experiments. Cooling crystallization experiments at constant nucleation temperature (30°C) were used to investigate the influence of supersaturation on phenacetin crystallization in ethanol. Solution concentration was tracked in-situ using a React IR™ PAT probe. At higher supersaturation, smaller, thinner needles were recovered suggesting that growth in the needle direction is favored. The desupersaturation profiles recorded with the React IR™ illustrated that as supersaturation increases, the onset of nucleation is faster and the subsequent desupersaturation curve is steeper. Cooling crystallizations were also attempted using constant supersaturation at three different nucleation temperatures – 15, 30, and 55 °C. Needles were recovred in all experiments, with thicker needles observed at high nucleation temperature. It is hypothesized that this is due to a combination of greater availability of solute at increased temperature (due to experiment design and inherent dependence of solubility on temperature) and also greater growth in the direction orthogonal to the needle direction. It is surmised that the activation energy for growth in the direction orthogonal to the needle direction is decreased to a greater extent that the activation energy for growth in the needle direction at high temperatures. The desupersaturation profiles show a faster onset of nucleation at higher temperature also, as would be expected. Evaporative crystallization was attempted in toluene, ethanol and acetone. Crystals approximating the BDFH-predicted hexagonal block were recovered from ethanol and acetone, suggesting thermodynamic growth can be realized at very low supersaturation.

377560 A Novel Technique for Continuous Production of Polymer-Coated Drug Crystals

Abstract:

To develop controlled release of a drug as well as to protect the drug from demanding environments, appropriate polymeric coatings are often employed on the drug crystals. Conventional methods for coating or encapsulating micron–sized crystals and nanoparticles utilize dry or wet approaches. Dry methods include physical vapor deposition, plasma treatment, chemical vapor deposition and pyrolysis of polymeric organic materials. Wet methods cover sol-gel processes, emulsification and solvent evaporation techniques. Supercritical fluid (SCF) processes such as Rapid Expansion of Supercritical Solutions (RESS), Supercritical Anti-Solvent (SAS), and Gas Anti-Solvent (GAS) processes employing supercritical CO2 are alternative methods for nanoparticle coating or encapsulation of ultrafine particles/crystals. These SCF-based processes have many shortcomings. Some processes require demanding operating conditions, suffer from low polymer solubility etc. Most of these techniques are also batch processes. Fluidized bed-based processes, which can be continuous, face difficulties with nanoparticles caused by van der Waals and other interparticle forces. In all such processes, scale-up is also quite demanding.

We will illustrate a novel technique allowing a continuous process of producing polymer-coated drug crystals. This process is also easily scalable. It is different from the solid hollow fiber cooling crystallization-based technique we have also recently developed [1]. Polymers used for coating were Eudragit and PLGA.

[1] Chen et al., Ind. Eng. Chem. Res. 2014, 53, 6388−6400.

380659 Pharmaceutical Precipitation on Amorphization and Process Characterization

Abstract:

Most pharmaceutical small molecules are developed as crystalline solids, as a crystalline API confers a number of advantages. These advantages include enhanced chemical and physical stability, reproducibility and better processability which generally leads to crystalline solids being preferred over their poorly crystalline or non-crystalline counterparts. However, the solubility and bioavailability of amorphous drug substances are increasingly compared with their crystalline counterparts during formulation development. When these amorphous materials show enhanced exposure or stability characteristics, they can be of great interest to pharmaceutical development.

In this case study, the material characteristics and processability of the amorphous form of a model compound which intrinsically tends to form crystalline solvates were examined. This was accomplished through solution transformation from crystalline to amorphous solids. Process knowledge leveraged from in-depth crystallization screening and process development provided insights on establishing appropriate conditions and boundaries for producing manageable amorphous material, and avoiding potential process hurdles such as de-mixing or poor filterability. The findings and results that lead to a significant improvement in manufacturability and the selective particle engineering techniques applied will be presented.

389128 Formation of Metastable Polymorphs and Amorphous Structures through Monodisperse Droplet Evaporation

Abstract:

Particle structure and morphology are two important product characteristics of organic molecular compounds with specific applications to the food product and pharmaceutical industries. The structure, whether it is multiple crystalline polymorphs or amorphous forms, directly impacts the performance of the final product because of changes in bioavailability and surface functionality (Bernstein, 2002). The morphology typically affects downstream processes such as filtering and drying. Formation of metastable polymorphs or amorphous structures is often achieved through the addition of polymer excipients or other additives to the system. However, a more fundamental understanding of the choice of the additive and the amount of additive needed to deliver a metastable polymorph or amorphous structure is desired (Eloy and Marchetti, 2013). Additionally, many compounds which are formulated as amorphous dispersions are larger molecules which have a naturally low tendency to form crystals. Previous work has shown that the formation of metastable structures is enhanced by utilizing a monodisperse droplet evaporation technique (Carver and Snyder, 2012).

In this presentation, we highlight our recent work on the formation of metastable polymorphs and amorphous structures of a molecule with a high propensity to crystallize, succinic acid. The structures are generated through monodispersed droplet evaporation using a vibrating orifice aerosol generator, with solution addition of polymer excipient. Succinic acid has two known polymorphs, with a significant tendency to form only the beta structure at room temperature. Furthermore, succinic acid has a very high propensity to crystallize, which is characteristically different from many substances that are formed into an amorphous state. We will show that polyvinylpyrrolidone (PVP) is a compatible excipient for the creation of amorphous dispersions of succinic acid in multiple solvents. In ethanol, the metastable alpha polymorph of succinic acid is formed at low loadings of PVP. At higher loadings, we will show the fraction of PVP required to obtain a purely amorphous sample, as a function of the solvent used and the PVP molecular weight. Additionally, a theory will be presented to explain the PVP fraction required for amorphous dispersion formation with succinic acid. Finally, we will discuss both the morphological impact of PVP addition on succinic acid morphology as well as the potential of this work to extend to more general methods to predict the formation of amorphous structures.

References

Bernstein, J. *Polymorphism in Molecular Crystals*; Oxford, New York, 2002.

Carver, K. M., Snyder, R. C. *Ind. Eng. Chem. Res.* **2012**, *51*, 15720-15728.

Eloy, J. O., Marchetti, J. M. *Powder Technology*. **2013**, 253, 98-106

363225 Uncovering the Growth Mechanism of Hydrothermal Synthesis of Anatase Crystals

Abstract:

Anatase (TiO_2) is a metastable polymorph of titanium dioxide. Anatase crystals with larger surface area of the {001} faces exhibit higher catalytic activity for water dissociation reaction[1]. A better understanding of the crystal growth mechanisms is required to engineer functionally desirable morphology of anatase crystals.

Hydrothermal synthesis of anatase crystals was carried out to obtain large (~1 μm) crystals with the usual bipyramidal morphology dominated by {101} faces. The surface features were characterized using ex situ atomic force microscopy (AFM), and the presence of monomolecular steps suggests that anatase crystals grow by a layered growth mechanism such as spiral growth or birth-and-spread.

Crystal growth under surface integration-limited kinetics proceeds by the attachment of growth units into kink sites along steps present on the surface. These steps are parallel to the lattice directions of strong intermolecular interactions within the solid, also known as Periodic Bond Chains (PBCs) [2]. A systematic method developed earlier[3] was used to identify the PBC directions in anatase crystals. Crystal growth kinetics depends on the long-range electrostatic interaction energies between the kink site growth unit and its solid-state and solvent neighborhood. The kink incorporation rates on the edges of (101) and (001) faces of anatase crystals were calculated from the attachment and detachment kinetics of individual kink sites[4]. The relative growth rates of the crystal faces were calculated using the spiral growth model[5] to accurately predict the steady-state morphology of anatase crystals.

Further refinements to this mechanistic model will allow prediction of the effect of synthesis conditions such as supersaturation, pH, additives, etc. on the morphology of hydrothermally grown materials with important functionalities such as zinc oxide, gallium nitride, etc.

References

[1] H. G. Yang, C. H. Sun, S. Z. Qiao, J. Zou, G. Liu, S. C. Smith, H. M. Cheng, and G. Q. Lu, "Anatase TiO_2 single crystals with a large percentage of reactive facets," *Nature*, 453, 638, 2008.

[2] P. Hartman and W. G. Perdok, "On the relations between structure and morphology of crystals. I," *Acta Cryst.*, 8, 49-52, 1955.

[3] P. Dandekar and M. F. Doherty, "A mechanistic growth model for inorganic crystals: Solid-state interactions," *AIChE J.*, (submitted), 2014.

[4] P. Dandekar and M. F. Doherty, "A mechanistic growth model for inorganic crystals: Growth mechanism," *AIChE J.*, (submitted), 2014.

[5] R. C. Snyder and M. F. Doherty, "Predicting crystal growth by spiral motion," *Proc. R. Soc. A*, 465, 1145-1171, 2009.

Controlling Crystal Habit with Tailored Organic Growth Modifiers

Jihae Chung and Jeffrey D. Rimer

University of Houston, Department of Chemical and Biomolecular Engineering, Houston, TX 77204

Developing novel and reliable methods of controlling crystal habit is relevant to applications in areas such as pharmaceuticals and materials, and plays an integral role in physiologically-relevant processes (e.g., biomineralization). A common method of tailoring the anisotropic rates of crystal growth is the use of growth modifiers, which are additives (i.e. ions, molecules, and/or macromolecules) with specificity for binding to crystal surfaces and reducing the rate of growth in distinct crystallographic directions [1]. Here we will discuss the effects of organic modifiers on the growth of calcium oxalate monohydrate (COM) crystals, which are the primary constituent of human kidney stones. The incidence rate of renal stones is on the rise, yet there has been little improvement in their therapeutic treatments over the past 30 years [2].

A common approach to minimize the risk of stone formation is the use of supplements, such as citrate, which function as inhibitors of COM growth. The design of crystal growth modifiers relies upon a fundamental understanding of the interactions between organic molecules and the crystallographic face(s) of COM – notably, the effect of modifiers on crystal nucleation and growth [3]. Here we will discuss our research efforts focused on elucidating the interfacial interactions between COM crystals and small organic molecules that were selected on the basis of their potential role as inhibitors of crystal growth. In particular, we will discuss the effects of citrate (a moderate growth inhibitor) and various structural derivatives of citrate on COM crystallization.

We have shown that subtle differences in the structure, chemical functionality, and stereochemistry of organic molecules have a significant impact on COM nucleation, growth, and bulk crystal habit. We used *in situ* atomic force microscopy (AFM) to provide microscopic validation of organic molecule binding to specific COM crystal surfaces. AFM has proven to be a useful technique for assessing the dynamics of crystal growth under physiologically relevant conditions. Moreover, we have used AFM to obtain time-resolved images of surface growth in the absence and presence of organic molecules. Collectively, these findings serve as a platform for the rational design of crystal growth modifiers, and a viable method of identifying lead candidates for drug discovery.

References:

[1] Farmanesh, S., Ramamoorthy, S., Chung, J., Asplin, J.R., Karande, P., Rimer, J.D., Specificity of Growth Inhibitors and their Cooperative Effects in Calcium Oxalate Monohydrate Crystallization, *J. Am. Chem. Soc.* 136 (2014) 367-376

[2] Rimer, J.D., An, Z., Zhu, Z., Lee, M.H., Wesson, J.A., Goldfarb, D.S., Ward, M.D., Crystal Growth Inhibitors for the Prevention of L-Cystine Kidney Stones Through Molecular Design, *Science* 330 (2010) 337-341

[3] Farmanesh, S., Chung, J., Chandra, D., Sosa, R.D., Karande, P., Rimer, J.D., High-Throughput Platform for Design and Screening of Peptides as Inhibitors of Calcium Oxalate Monohydrate Crystallization, *J. Cryst. Growth* 373 (2013) 13-19

376990 Protein Adsorption in Functionalized SBA-15 Silica: Influence of pH and Ionic Strength

Abstract:

Ordered mesoporous silica materials have favorable properties as hosts for the immobilization and thermal stabilization of proteins. The ability to control and modify the interaction of protein molecules with the pore walls is of importance in many biotechnical and biomedical application fields, including ultrafiltration, drug delivery and biosensing.

In this work we studied the adsorption of lysozyme, a "hard" globular protein with a high isoelectric point (pI = 11), in the nanochannels of native and surface-functionalized SBA-15. Functionalization with sulfonic acid may help to increase the adsorption of the protein at pH < 7, where the native silica surface is only weakly charged. We also studied the influence of a "protein-repellant" zwitterionic sulfobetaine on the protein uptake. At pH 4, lysozyme adsorption is indeed suppressed in the zwitterionized material and enhanced in the sulfonated material, relative to native SBA-15. At pH 11, on the other hand, a dense protein monolayer is formed in the pores of all materials. We also find that an increase in ionic strength (100 mM) causes an increase in adsorption of lysozyme at low pH but a decrease at high pH. All these findings are consistent with the assumed dominant role of electrostatic interactions in the adsorption of lysozyme in the nanopores of the SBA-15 materials.

Mechanism and Energetics of Bovine Carbonic Anhydrase Adsorption on Ordered Mesoporous Alumina

Shada Salem, Robert Golsby, Stephen W. Thiel

Carbon dioxide is a major greenhouse gas often associated with global climate change [1]. CO_2 emissions can be reduced through the use of alternative energy sources such as wind energy, solar energy, and nuclear energy and through increased efficiency in energy production and use [2, 3]. However, combustion of carbon fuels will likely be the primary source of energy for the foreseeable future, capture and sequestration of CO_2 will be needed [3, 4]. One approach to CO_2 capture from flue gas is wet scrubbing; the rate-limiting step in this process is the hydration of CO_2. Carbonic anhydrase (CA) can be used to catalyze the reversible hydration of CO_2 [5-8]. CA activates hydroxyl ions formed through ligation to a zinc atom incorporated in the enzyme [9, 10]; the enzyme holds CO_2 close to this activated hydroxyl group, while simultaneously keeping water away from the hydroxyl group. There is thus a highly basic microenvironment that fosters rapid formation of carbonate ions.

This work reported here is part of a project to find the immobilization conditions that optimize the highest loading of bovine carbonic anhydrase (BCA) on ordered mesoporous alumina (OMPA). Batch adsorption isotherm measurements of BCA on OMPA show that BCA loadings on OMPA are highly affected by the pH and ionic strength of the adsorption environment. The interactions between BCA and OMPA were studied using flow micro-calorimetry (FMC) to better understand the adsorption mechanism of BCA as a function of pH and ionic strength. FMC allows direct measurements of the heats of adsorption and desorption in dynamic solution conditions, providing vital insights into the fundamental interactions between biomolecules and solid supports [11-18]. These interactions are difficult to assess using only batch adsorption isotherm measurements. FMC measures the enthalpy change of adsorption, often the main contribution to the overall free energy change of adsorption. FMC results indicate that BCA adsorption on OMPA is greatly affected by the ionic strength (0.05, 0.1, and 20 mM) and the pH (7, 8, and 9) of the adsorption environment.

References

[1] Rogner, H.-H., D. Zhou, R. Bradley. P. Crabbé, O. Edenhofer, B.Hare (Australia), L. Kuijpers, M. Yamaguchi, 2007: Introduction. In Climate Change 2007: Mitigation. Contribution of Working Group III to the Fourth Assessment Report of the Intergovernmental Panel on Climate Change [B. Metz, O.R. Davidson, P.R. Bosch, R. Dave, L.A. Meyer (eds)], Cambridge University Press, Cambridge, United Kingdom and New York, NY, USA.

[2] Fisher, B.S., N. Nakicenovic, K. Alfsen, J. Corfee Morlot, F. de la Chesnaye, J.-Ch. Hourcade, K. Jiang, M. Kainuma, E. La Rovere, A. Matysek, A. Rana, K. Riahi, R. Richels, S. Rose, D. van Vuuren, R. Warren, 2007: Issues related to mitigation in the long term context, In Climate Change 2007: Mitigation. Contribution of Working Group III to the Fourth Assessment Report of the Inter-governmental Panel on Climate Change [B. Metz, O.R. Davidson, P.R. Bosch, R. Dave, L.A. Meyer (eds)], Cambridge University Press, Cambridge, United Kingdom and New York, NY, USA.

[3] Reichle, D., Houghton, J., Kane, B., and Ekmann, J. 1999. Carbon sequestration: Research and development. *U.S. DOE report.*

[4] Herzog, H. J. and Drake, E. M. 1996. Carbon dioxide recovery and disposal from large energy systems. *Annual Review of Energy and the Environment.* 21.145-166.

[5] Brown, R.S., in Aresta, M. and J.V. Schloss, editors, *Enzymatic and Model Carboxylation and Reduction Reactions for Carbon Dioxide Utilization, NATO ASI Series, Series C: Mathematical and Physical Sciences, vol. 314,* Kluwer Academic Publishers, Dordrecht (1990), 145-180.

[6] Dodgson, S.J., R.E. Tashian, G. Gros and N.D. Carter, editors, *The Carbonic Anhydrases: Cellular Physiology and Molecular Genetics,* Plenum Press, New York (1991).

[7] Pocker, Y., in Aresta, M. and J.V. Schloss, editors, *Enzymatic and Model Carboxylation and Reduction Reactions for Carbon Dioxide Utilization, NATO ASI Series, Series C: Mathematical and Physical Sciences, vol. 314,* Kluwer Academic Publishers, Dordrecht (1990), 129-143.

[8] Khalifah R.G., The Carbon dioxide hydration activity of carbonic anhydrase I. stop-flow kinetic studies on the native human isoenzymes B and C, The Journal of Biological Chemistry, 1971,246(8), 2561-2573.

[9] Bond G.M., Stringer J., Brandvold D.K., Simsek F.A., Medina .G., and Egeland G.2001. Development of Integrated System for Biomimetic CO2 SequestrationUsing the Enzyme Carbonic Anhydrase. *Energy & Fuels*, 15, 309-316.

[10] Silverman D.N. and S.H. Vincent, "Proton Transfer in the Catalytic Mechanism of CarbonicAnhydrase." *CRC Crit Rev Biochem.* 1983, 14, 207-55.

[11] Dias-Cabral, A. C., J. A. Queiroz, and N. G. Pinto. "Effect of Salts and Temperature on the Adsorption of Bovine Serum Albumin on Polypropylene Glycol-Sepharose under Linear and Overloaded Chromatographic Conditions." *J. Chromatogr. A* 1018 (2003) 137-5

[12] M.E. Thrash Jr, J.M. Phillips, N.G. Pinto, *Adsorption* 10 (2005) 299.

[13] M.E. Thrash Jr, N.G. Pinto, *J. Chromatogr. A* 1126 (2006) 304.

[14] A. Katiyar, N.G. Pinto, *Small* (2006) 644.

[15] J.M. Phillips, N.G. Pinto, *J. Chromatogr. A* 1036 (2004) 79.

[16] M.E. Thrash Jr, N.G. Pinto, *J. Chromatogr. A* 944 (2002) 61.

[17] A.C. Dias-Cabral, N.G. Pinto, J.A. Queiroz, *Sep. Sci. Technol.* 37 (2002) 1505.

[18] M.E. Thrash Jr, N.G. Pinto, *J. Chromatogr. A* 908 (2001) 293.

389188 Monolithic Columns Modified with Ag Nanoparticles for TLC Separations of Fatty Acids

Abstract:

Lipids are a very broad class of biologically relevant compounds, encompassing fatty acids (FA) and their derivatives, as well as substances biosynthetically related to FA. The separation and detection of FA is a key issue for medical diagnostics, as certain FA are associated with pathological conditions, as well as for the food industry, particularly for the hydrogenation of food oils.[1] The separation of FA is traditionally performed by silver-ion chromatography, using stationary phases for column or thin-layer chromatography (TLC), impregnated or modified with Ag ions. The interaction between Ag ions and the unsaturated moieties of FA modulates the separation, on the basis of the number and configuration of the FA double bonds.[2]

Recent computational studies have suggested that Ag nanoparticles (NP) may exhibit a similar type of interaction, which could be utilized for the separation of olefinic molecules.[3] Therefore, stationary phases modified with Ag NP could efficiently perform FA separations, without the disadvantages of Ag ions, such as "bleeding-in" by polar solvents or high sensitivity to light, moisture and mild reducing environments.

To this end, we have prepared polymer monoliths in a thin-layer format[4] and functionalized them uniformly with Ag NP. Such monoliths are indeed able to perform TLC separations of free FA, according to the number of their double bonds (see Figure 1). The effect of the mobile phase composition on the FA retention and selectivity has been investigated. Owing to the flat format of our monoliths, the detection is achieved by Matrix-Assisted Laser Desorption Ionization Mass Spectroscopy (MALDI-MS), by performing automated scans directly on the developed TLC plates. Moreover, Ag NP are able to substitute the function of conventional MALDI matrixes, which are substances that typically need to be additionally applied in order to enable the desorption/ ionization of the analytes towards the MS detector. Thanks to the presence of Ag NP, the detection of FA is made possible from our TLC monoliths immediately after separation.

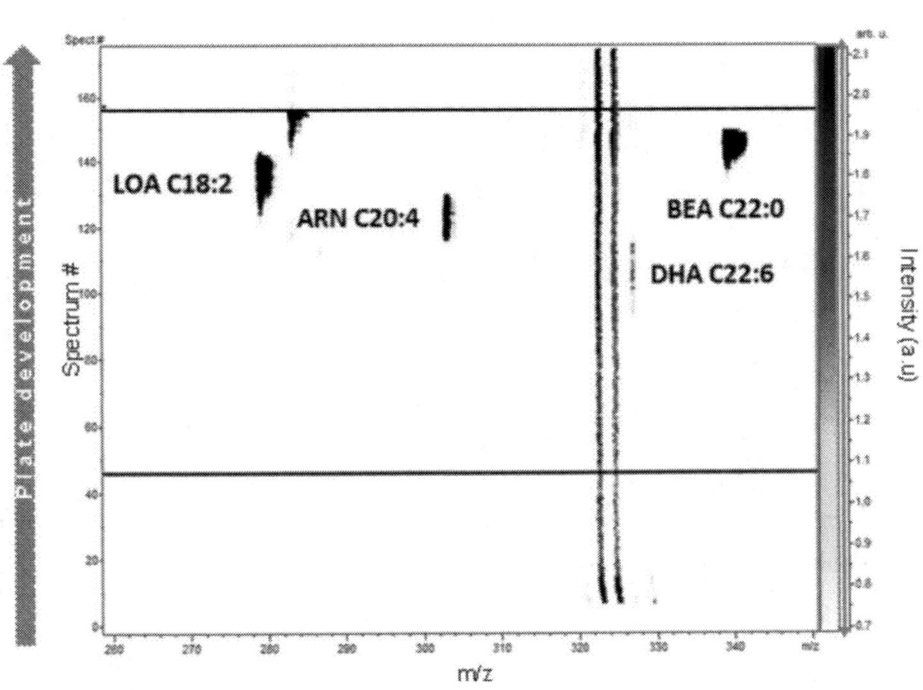

Figure 1. MADLI scan along a single lane on a TLC monolithic plate, functionalized with Ag NP, where FA separation has been achieved; the number of carbon atoms and double bonds is denoted at the positions of the respective FA; development time: 1 min 10 s

1. Fuchs B. et al., Journal of Chromatography A 1218 (2011) 2754
2. Nikolova-Damyanova B. et al., Journal of Liquid Chromatography & Related Technologies 24 (2001) 1447
3. Pozun Z. et al., Journal of Physical Chemistry C 115 (2011) 1811
4. Bakry R. et al., Analytical Chemistry 79 (2007) 486

384564 Protein Elution within Polysaccharidic Stationary Phases for Ion-Exchange

Abstract:

Ion-exchange adsorbents modified with covalently attached or grafted polymer layers have become more widely used in preparative chromatography. The complex networks provided by polymeric carbohydrates such as cellulose and dextran incorporate easily accessible microstructures with substantial binding capacities for biomolecules. However, there exists an abnormal type of elution behavior in stationary phases exhibiting these polymeric particle structures that may result in larger pool volumes or product instability.

The anion and cation exchange moieties of commercially available cellulosic and dextran-grafted agarose materials were characterized by their uptake and elution profiles at differing total ionic strengths and pH conditions. Batch experiments were performed alongside column level experiments to assess the transport and desorption characteristics of several model proteins. Column elution profiles were analyzed to determine the impact of loading factor with regard to elution pool volumes.

Additional mechanistic insight was sought using confocal laser scanning microscopy techniques to gain a physical understanding of protein elution profiles within single chromatographic particles under a variety of loading and elution conditions. Abnormal profiles during elution within certain systems suggest alternative mechanisms of protein expulsion and possibly increased protein-protein interactions within the particle. Comparisons between polymer-modified and non-modified materials were used to gauge the significance of pore structure in dictating functional and mechanical characteristics of these stationary phases during elution.

390231 Aminoglycoside Antibiotic Based Novel Anionic Exchange Resins for Plasmid DNA Binding and Purification

Abstract:

Plasmid DNA is playing a decisive role as a therapeutic option for multiple diseases such as cancer, cystic fibrosis and as DNA vaccines. Future clinical treatments that use plasmid DNA must be able to produce pharmaceutical grade quality of DNA in large amounts. Currently, plasmid DNA is grown in bacterial broths from which is extracted by cell lysis and further downstream processing etc. Although significant techniques have been created for downstream processing and extraction of plasmid DNA in large amounts, there is a significant room for improvement.

We report a novel aminoglycoside based hydrogel microresins that show a very high binding capacity to plasmid DNA. Natural ability of aminoglycosides to bind to nucleic acids was exploited in allowing increased binding to plasmid DNA. Aminoglycoside was converted into hydrogel microbeads using emulsion based polymerization. ~10 um microbeads were found to bind 45 ug of plasmid DNA/mg of beads at a concentration of 300 mg/L of plasmid DNA using a Langmuir adsorption isotherm. To increase the efficiency of plasmid DNA binding, the amines were quaternized using GTMAC (Glycidyl trimethyl ammonium chloride). Quaternized microbeads were found to bind almost 300 ug of plasmid DNA/mg of beads at a concentration of 200 mg/L of plasmid DNA, almost binding 30% of plasmid DNA of their weight.

Confocal microscopy was used to study the location of plasmid DNA binding on to the beads. It was noted that the DNA bound majorly at the surface on the beads. To desorb the pDNA, 1M sodium chloride was enough to desorb almost 90% of bound pDNA from unquaternized beads whereas only 35% from quaternized beads. Recovery of plasmid DNA was greatly improved when 1.25 M sodium chloride solution supplemented with 15% isopropanol at 50C was used. The buffer was sufficient to desorb 75% of adsorbed plasmid DNA from the quaternized beads. Around 200 ug of plasmid DNA was recoverable from 1 mg of the quaternized beads.

Unquaternized beads were also found useful in in-situ cancer cell lysis and DNA extraction and binding. Hence, we show a new microresin that can significantly enhance the binding towards plasmid DNA in a reliable way. Our aminoglycoside microbead platform can also be easily modified with conjugation of multiple ligands that can provide selectivity to the resin. We believe aminoglycoside microbeads provide a multipurpose material that can be used towards different applications related but not limited to plasmid DNA.

381498 Key-Lock Systems for Bioseparations - Molecular Simulation for Kinetic Models

Abstract:

The production of biological material is always connected to separation as the active product substances like virus particles have to be harvested without additional contaminations and byproducts like cell debris and other. We investigate specifically the separation of virus particles from a cell culture based process where mammalian cells are infected by predefined virus [1]. The cells reproduce virus particles inside and release it to the solution. They also desintegrate and many different cell components are part of the final solution including the final product, the virus particles. As virus have special surface proteins like hemagglutinin they can be attached to a surface using a key-lock system.

We investigate the surface adsorption with virus material as well as with artificial nanoparticles. The artificial system resembles the virus particles with an activated key molecule on the surface and an specifically prepared adsorption surface with the corresponding lock molecules. The attachment and detachment of particles to a surface is modeled on a molecular and discretized level.
A BiaCore system is used to obtain dynamic experimental data under variations of process parameters like concentration of nanoparticles, availability of key and lock molecules and other. The molecular simulation of the process is presented which should help to establish an appropriate kinetic model for a follow-up process systems approach.

[1] Wang, W., Wolff, M. W., Reichl, U., & Sundmacher, K. (2014). Avidity of influenza virus: Model-based identification of adsorption kinetics from surface plasmon resonance experiments. Journal of Chromatography A, 1326, 125-129.

383510 Adsorptive Characterization of Porous Solids: Error Analysis Guides the Way

Abstract:

Adsorptive characterization using nitrogen at 77 K is one of the most widely used techniques to assess textural properties of porous adsorbents. Especially the pore volume, specific surface area and pore size distribution are frequently reported. Despite the fact that most popular methods to deduce these quantities, the method posed by Brunauer, Emmet and Teller (BET) [1] for the specific surface area and the method developed by Barrett, Joyner and Halenda (BJH) for pore the pore size distribution, [2] have been used for many decades, there are still inconsistencies in the exact calculation procedure of these methods and still wrong and/or statistically insignificant conclusions therefrom, as will be elucidated in detail in this contribution. Based on a thorough error analysis on the accuracy of volumetric nitrogen adsorption measurements, the influence of experimental uncertainties is analysed. This yields not only insights in the uncertainty in derived properties (pore volume, BET surface area and BJH pore size-distribution) but also generates practical recommendations to optimize accuracy of N_2 adsorption measurements. Consequently, guidelines are developed to obtain representable accurate textural properties. Especially the derivation of specific surface areas requires attention, as erroneous results are easily obtained. A variety of sorbents (MOFs, zeolite, activated carbon and alumina) has been used in this work to ensure that obtained conclusions and posed guidelines hold for a wide range of commonly used porous materials. For (BJH) pore size distributions, statistically irrelevant or highly inaccurate results are frequently obtained. MIL-101, arguably the metal-organic framework (MOF) most reported on, has been characterized by a multitude of researchers and serves thus well to exemplify the current state of characterization using N_2 sorption in literature. In many cases the determination of pore volume and specific surface areas has been executed wrongly and underlines a clear potential for standardized determination conditions for textural properties. In summary, to improve the meaningfulness of derived properties and to minimize statistical uncertainties, practical recommendations and guidelines are proposed for experimental operation variables and data analysis. Some of these are:

- The relative uncertainty in pore volume is lowest when the $V_{manifold}/V_{cell}$ ratio is between 2 and 3.
- A simple two-point BET method is proposed to determine *a priori* the upper relative pressure boundary of the BET window (close to saturation), as alternative to the method reported by Rouquerol *et al.*
- For the lower relative pressure limit determination it is suggested to analyse Studentized residuals. Provided the model isotherm is correct, data points become eligible for possible exclusion when $| res_i^s| > 2\text{-}3$.
- The magnitude of the 95% confidence interval found for BJH-pore size distributions severely impedes drawing quantitative conclusions.

REFERENCES

[1] S. Brunauer, P.H. Emmett, E. Teller, Adsorption of gases in multimolecular layers, Journal of the American Chemical Society, 60 (1938) 309-319.

[2] E.P. Barrett, L.G. Joyner, P.P. Halenda, The Determination of Pore Volume and Area Distributions in Porous Substances. I. Computations from Nitrogen Isotherms, Journal of the American Chemical Society, 73 (1951) 373-380.

Microporous carbons are nowadays used in a wide range of technical applications, such as adsorbents, filters, catalytic components, electrodes, etc. For all these applications knowledge about the carbon pore structure and its respective interaction with fluids is a key factor to develop and optimize the technological potential in terms of adsorption capacity or selectivity. An effect intrinsically connected to the physical process of adsorption is the so-called *adsorption-induced deformation*, which denotes the structural change of the solid during the adsorption of fluid molecules. This expansion-compression behavior of microporous solids usually results in relative volume changes below 1 % and is therefore neglected. However, since the mechanical response of carbons to pore filling is highly sensitive to the micropore size distribution, adsorption-induced deformation can be used as a complementary approach to evaluate the carbon microstructure. Furthermore information about the mechanical modulus of the carbon micropores can be derived.

In this paper we combine experimental data on sorption-induced deformation of microporous carbons measured by in-situ dialtometry and corresponding theoretical modelling by density functional theory (DFT). The dilatometric measurements were performed for argon (77 K), nitrogen (77 K), carbon dioxide (273 K) and water (298 K) on carbon samples of significantly different micropore structures. The DFT calculations were performed for slit-shaped micropores, which are commonly accepted as a reasonable approximation of carbon microstructure. Beyond single pore sorption isotherms the DFT results were used to derive the fluid pressure inside the micropores, i.e. the so-called *adsorption stress*, which causes the pore deformation. By comparing the experimental data sets with calculation results we show how the effect of sorption-induced deformation can be used to evaluate information on pore size distributions and mechanical moduli. Figure 1 shows an example of experimentally measured deformation of a carbon sample and theoretical predictions for the adsorption stress.

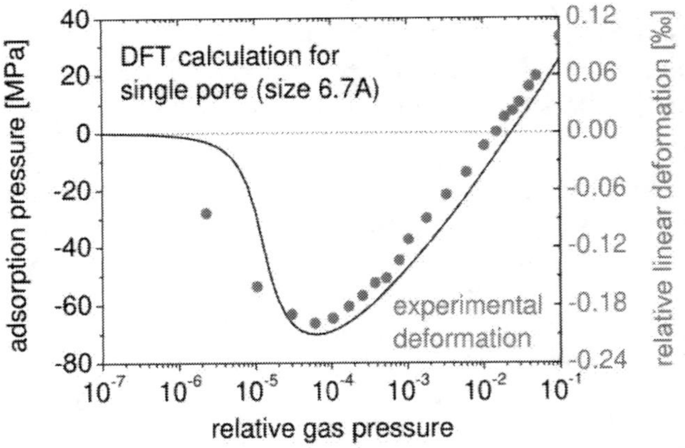

Figure 1: experimental measured linear relative length change of a carbon sample during N_2 adsorption at 77 K compared to the corresponding theoretical prediction of adsorption stress for a slit-shaped pore of width = 6.7 Å.

Acknowledgements:
Partial financial support from the *Deutscher Akademischer Austauschdienst* (DAAD – German Academic Exchange Service) for PhD research studies.

AIChE Annual Meeting - November 19th 2014
Characterization of Adsorbent Materials

Characterization of Lysozyme Adsorption in Cellulosic Chromatographic Particles using Small-Angle Neutron Scattering

Stijn Koshari[1], Norman Wagner[1,2], Abraham Lenhoff[1,3]

[1]Department of Chemical and Biomolecular Engineering, [2]Center for Neutron Science, [3]Center for Molecular and Engineering Thermodynamics
University of Delaware

UNIVERSITY OF DELAWARE

KATHOLIEKE UNIVERSITEIT LEUVEN

Motivation

Characterization of Lysozyme Adsorption in Cellulosic Chromatographic Particles using Small-Angle Neutron Scattering

UNIVERSITY OF DELAWARE

INTRODUCTION 2/39

Motivation

- Chromatography is the main separation technique in downstream processing in the biotechnological industry
- Nature of protein adsorption and mobility is highly dependent on architecture of chromatographic media
- Cellulosic chromatographic particles show good performance, but our mechanistic understanding is limited

UNIVERSITY OF DELAWARE

INTRODUCTION 3/39

Cellulosic particles

- Pall Corporation
  S HyperCel™

- Gel-like, fine, fibrous structure

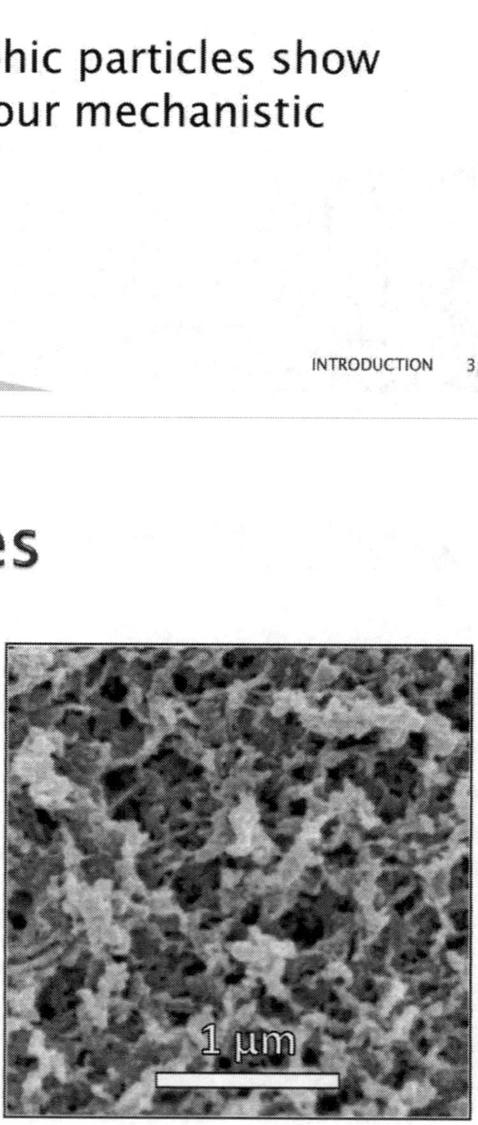

Angelo, J. M., et al. *J. Chromatogr. A* **1319**, 46-56 (2013).

UNIVERSITY OF DELAWARE

INTRODUCTION 4/39

Small-angle neutron scattering

- Protein-scale resolution
- Size, structure, interaction
- Minimally disruptive
- Neutron techniques study static and dynamic aspects

Goal

Study the feasibility of small-angle neutron scattering (SANS) as a characterization technique for protein adsorption on chromatographic particles

Approach

- Scattering spectra from components
- Scattering spectra from protein laden chromatographic particles
- Directly compare scattering spectra of particles with and without adsorbed protein
- Analyze differences to validate method

UNIVERSITY OF DELAWARE

Scattering from components

» · Chromatographic particles
· Lysozyme

UNIVERSITY OF DELAWARE

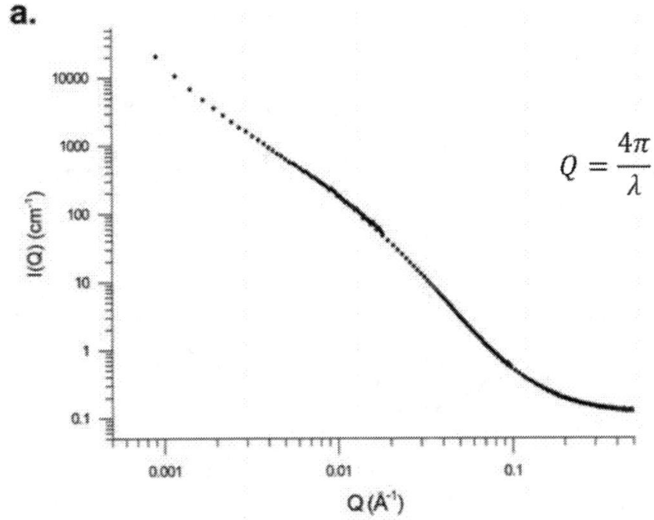

$$Q = \frac{4\pi}{\lambda}\sin\frac{\theta}{2}$$

S HyperCel™

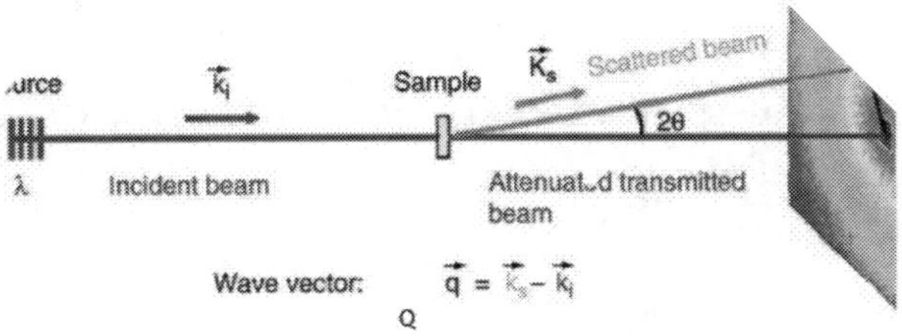

Bragg's law: $L = \frac{2\pi}{Q}$

Grillo, I. in *Soft-Matter Charact.* (Borsali, R. & Pecora, R.) 705-764 (Springer, 2008).

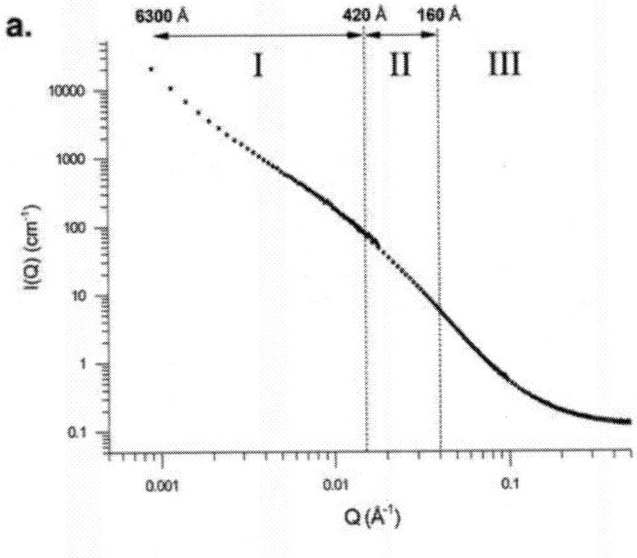

S HyperCel™

Bragg's law: $L = \frac{2\pi}{Q}$

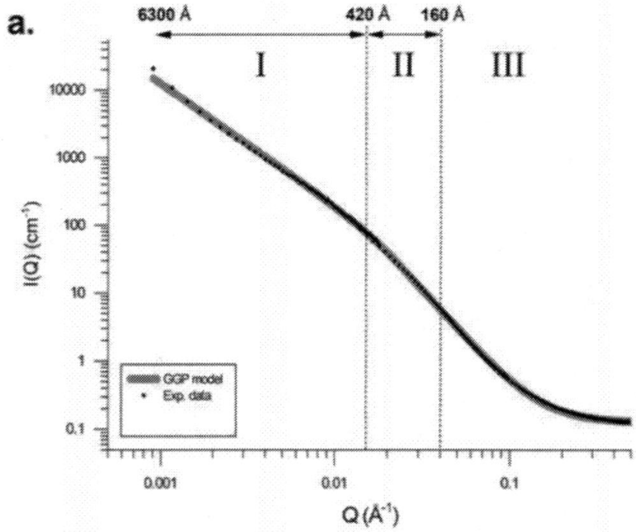

Slopes

s, m

Radius of gyration

R_g

S HyperCel™

Generalized Guinier-Porod model:

$$I(Q) - \mathrm{B} = \frac{G}{Q^s} \exp\left[\frac{-Q^2 R_g^2}{3-s}\right] \quad Q \le Q_1 \qquad \text{where:} \quad Q_1 = \frac{1}{R_G}\sqrt{\frac{(m-s)(3-s)}{2}}$$

$$I(Q) - B = \frac{D}{Q^m} \quad Q > Q_1 \qquad D = G \exp\left[\frac{-Q^2 R_G^2}{3-s}\right] Q_1^{m-s}$$

Hammouda, B. *J. Appl. Crystallogr.* **43**, 716–719 (2010).

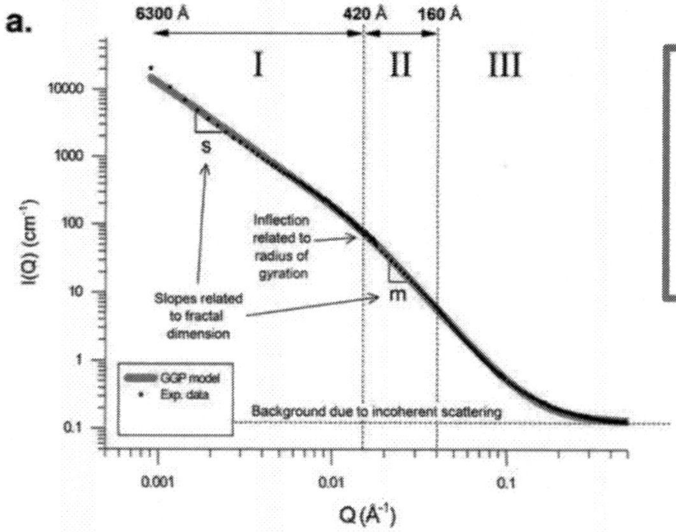

SANS:

$R_g = 35 \pm 1$ Å

ISEC:

$r_{mean} = 44$ Å

S HyperCel™

Generalized Guinier-Porod model:

$$I(Q) - \mathrm{B} = \frac{G}{Q^s} \exp\left[\frac{-Q^2 R_g^2}{3-s}\right] \quad Q \le Q_1 \qquad \text{where:} \quad Q_1 = \frac{1}{R_G}\sqrt{\frac{(m-s)(3-s)}{2}}$$

$$I(Q) - B = \frac{D}{Q^m} \quad Q > Q_1 \qquad D = G \exp\left[\frac{-Q^2 R_G^2}{3-s}\right] Q_1^{m-s}$$

Hammouda, B. *J. Appl. Crystallogr.* **43**, 716–719 (2010).
Angelo, J. M., et al. *J. Chromatogr. A* **1319**, 46–56 (2013).

a.

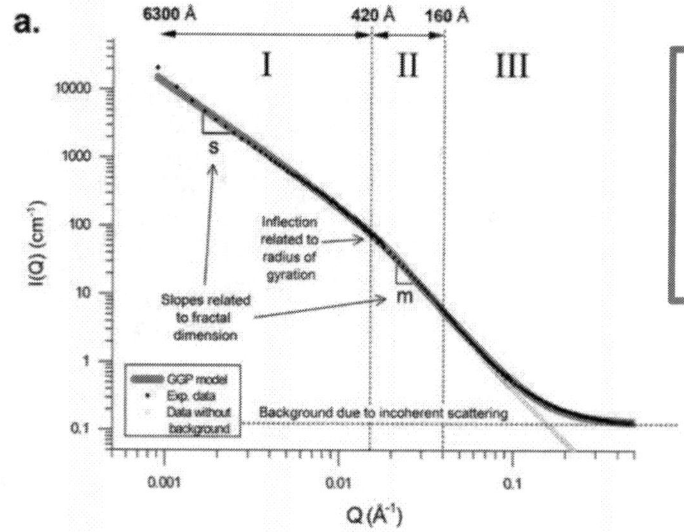

Slopes

$s = 1.8 \pm 0.1$
$m = 2.8 \pm 0.1$

S HyperCel™

Generalized Guinier-Porod model:

$$I(Q) - \mathrm{B} = \frac{G}{Q^s} \exp\left[\frac{-Q^2 R_g^2}{3-s}\right] \quad Q \leq Q_1 \qquad \text{where: } Q_1 = \frac{1}{R_G}\sqrt{\frac{(m-s)(3-s)}{2}}$$

$$I(Q) - B = \frac{D}{Q^m} \quad Q > Q_1 \qquad D = G \exp\left[\frac{-Q^2 R_G^2}{3-s}\right] Q_1^{m-s}$$

UNIVERSITY OF DELAWARE

Hammouda, B. *J. Appl. Crystallogr.* **43**, 716–719 (2010).
Angelo, J. M., et al. *J. Chromatogr. A* **1319**, 46–56 (2013).

SCATTERING FROM COMPONENTS 13/39

a.

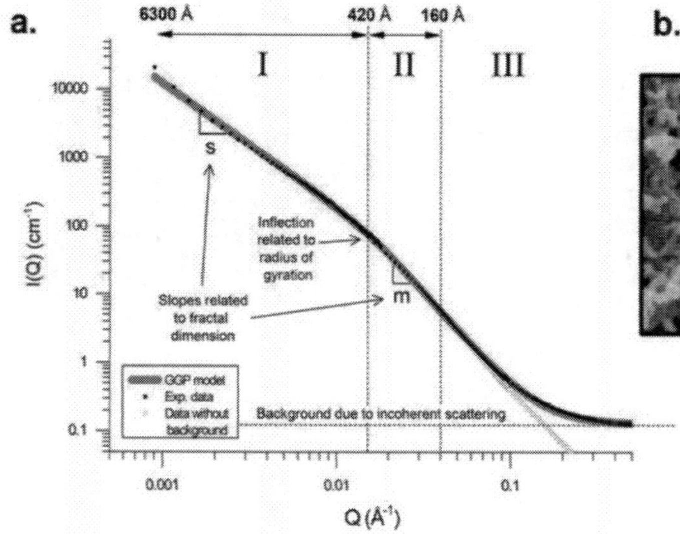

b.

S HyperCel™

UNIVERSITY OF DELAWARE

Angelo, J. M., et al. *J. Chromatogr. A* **1319**, 46–56 (2013).

SCATTERING FROM COMPONENTS 14/39

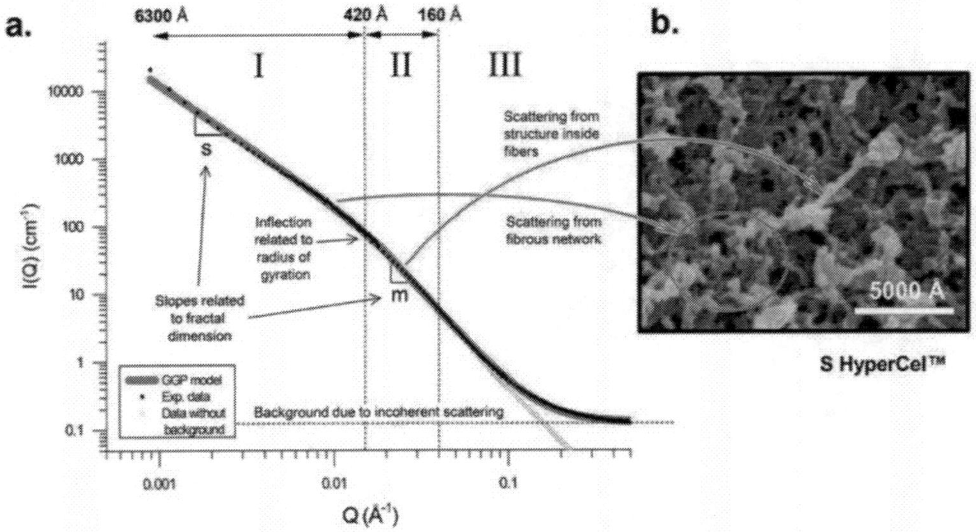

SCATTERING FROM COMPONENTS 15/39

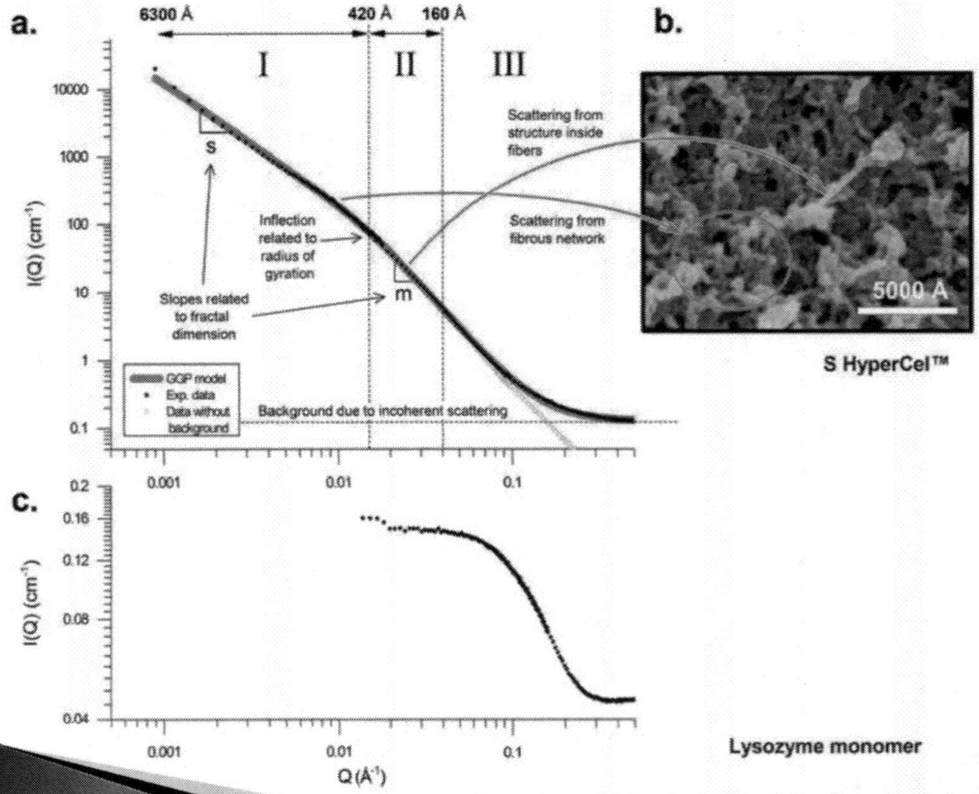

Data courtesy of P. Douglas Godfrin, Wagner Group, University of Delaware

SCATTERING FROM COMPONENTS 16/39

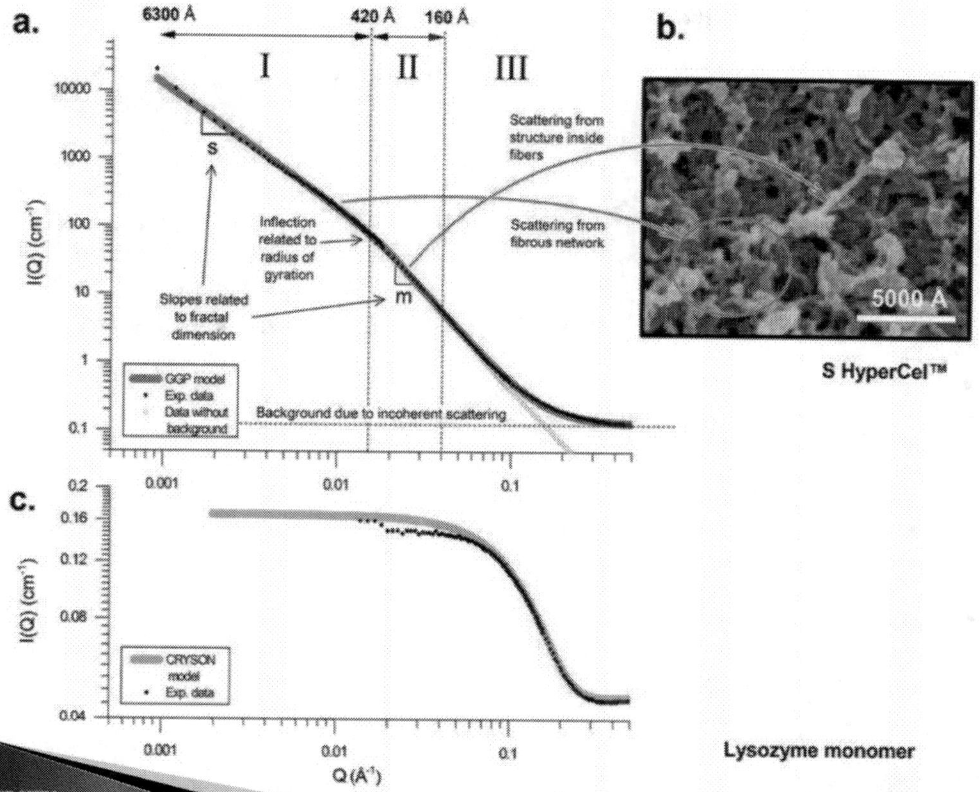

Svergun, D. I. *et al. Proc. Natl. Acad. Sci.* **95**, 2267-2272 (1998).

UNIVERSITY OF DELAWARE

SCATTERING FROM COMPONENTS 17/39

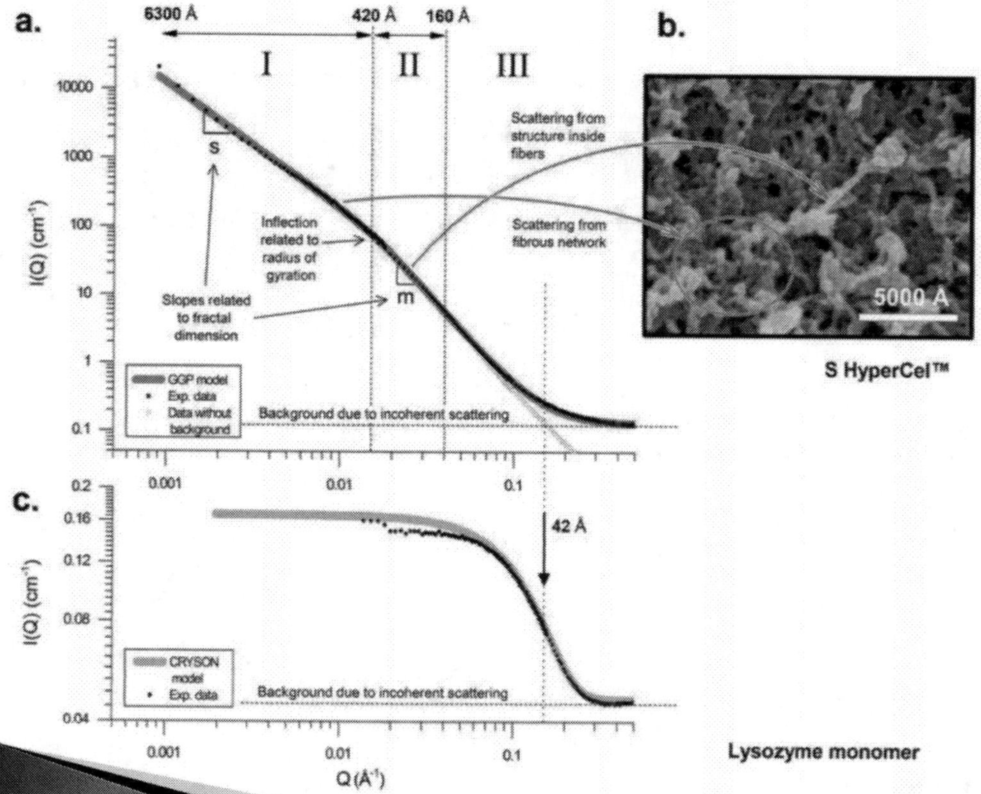

UNIVERSITY OF DELAWARE

SCATTERING FROM COMPONENTS 18/39

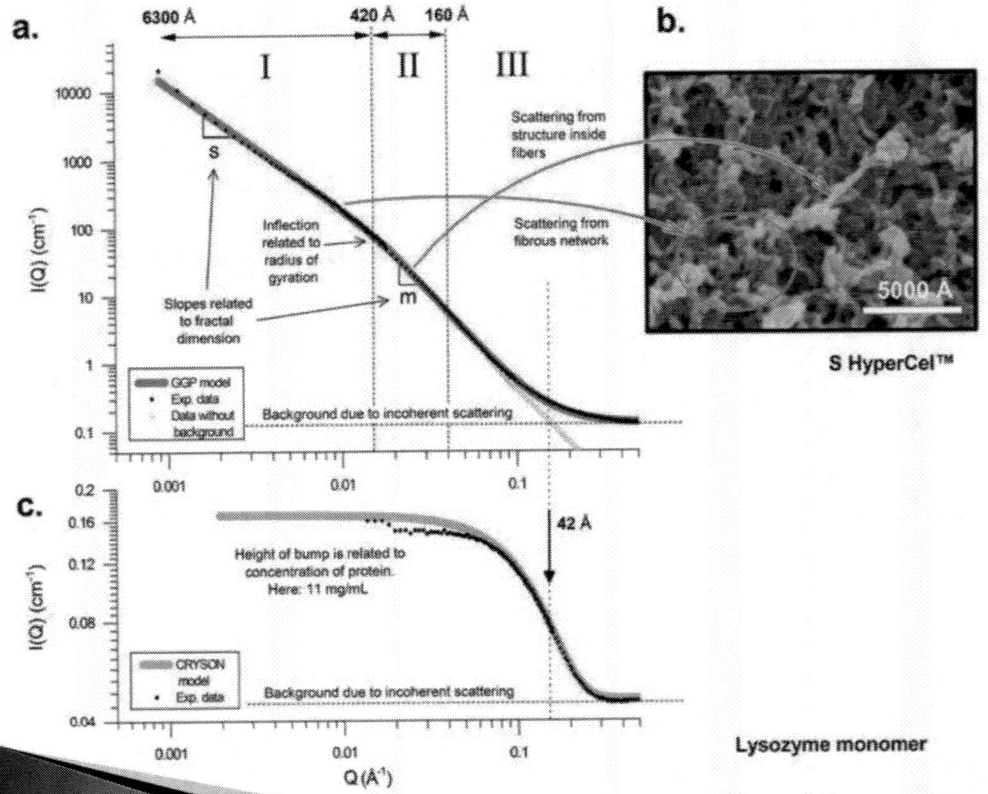

UNIVERSITY OF DELAWARE

SCATTERING FROM COMPONENTS 19/39

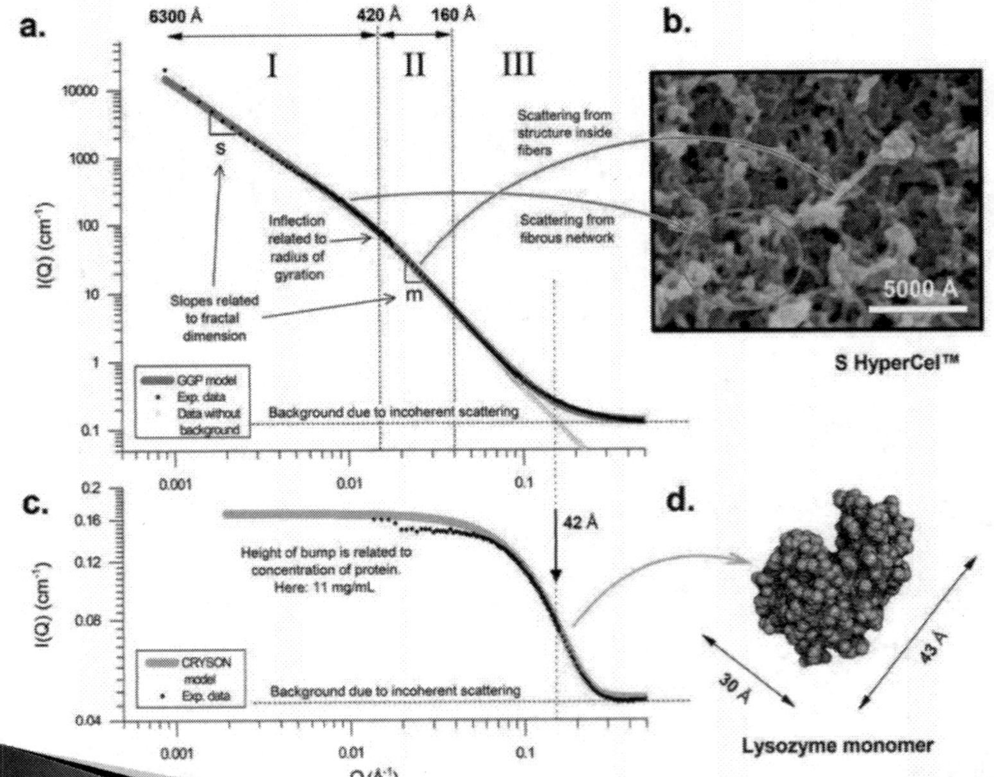

UNIVERSITY OF DELAWARE

DeLano, W. L. (2002). at <http://www.pymol.org>

SCATTERING FROM COMPONENTS 20/39

Scattering from components

- Capable of characterizing particle structure
- Suitable models for chromatographic particles and proteins

UNIVERSITY OF DELAWARE

Scattering from protein laden particles

» • Experimental data
- Reduced data
- Comparison
- Protein contributions

UNIVERSITY OF DELAWARE

Protein adsorption samples

- Preparation of samples with varying amount of adsorbed protein

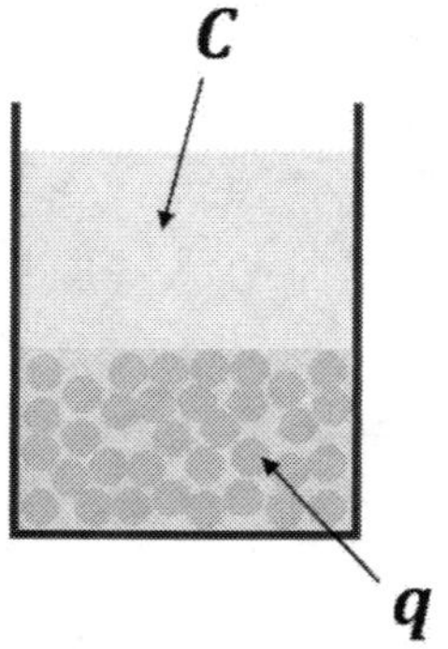

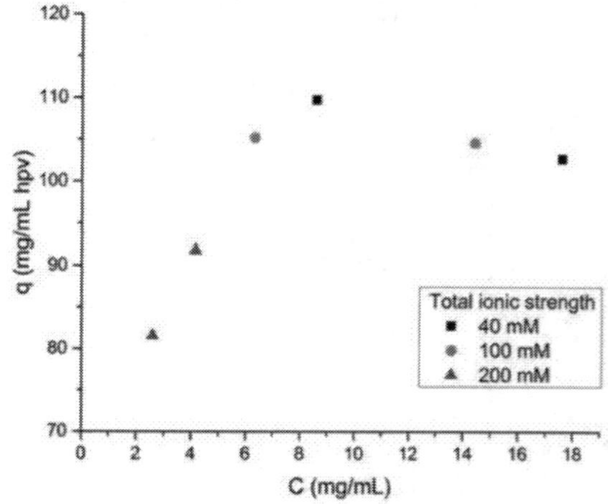

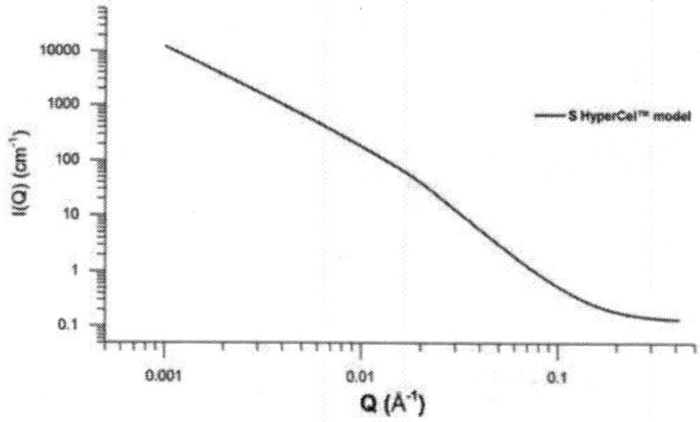

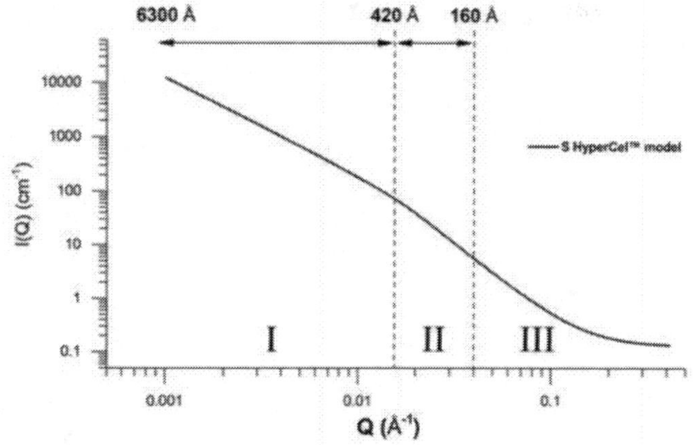

6300 Å
420 Å
160 Å
S HyperCel™ model
I(Q) (cm⁻¹)
Q (Å⁻¹)
I
II
III

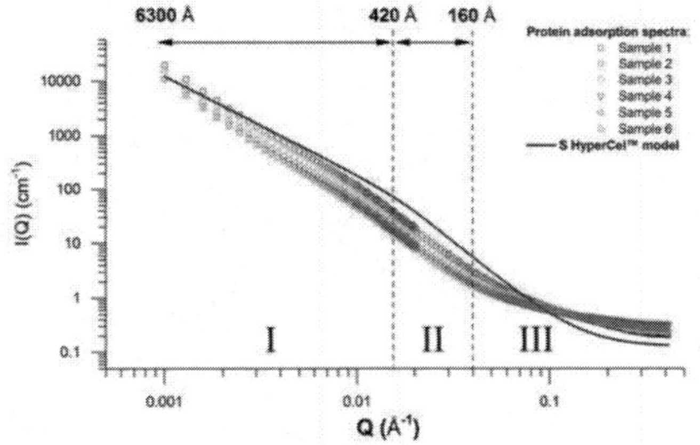

6300 Å
420 Å
160 Å
Protein adsorption spectra:
Sample 1
Sample 2
Sample 3
Sample 4
Sample 5
Sample 6
S HyperCel™ model
I(Q) (cm⁻¹)
Q (Å⁻¹)
I
II
III

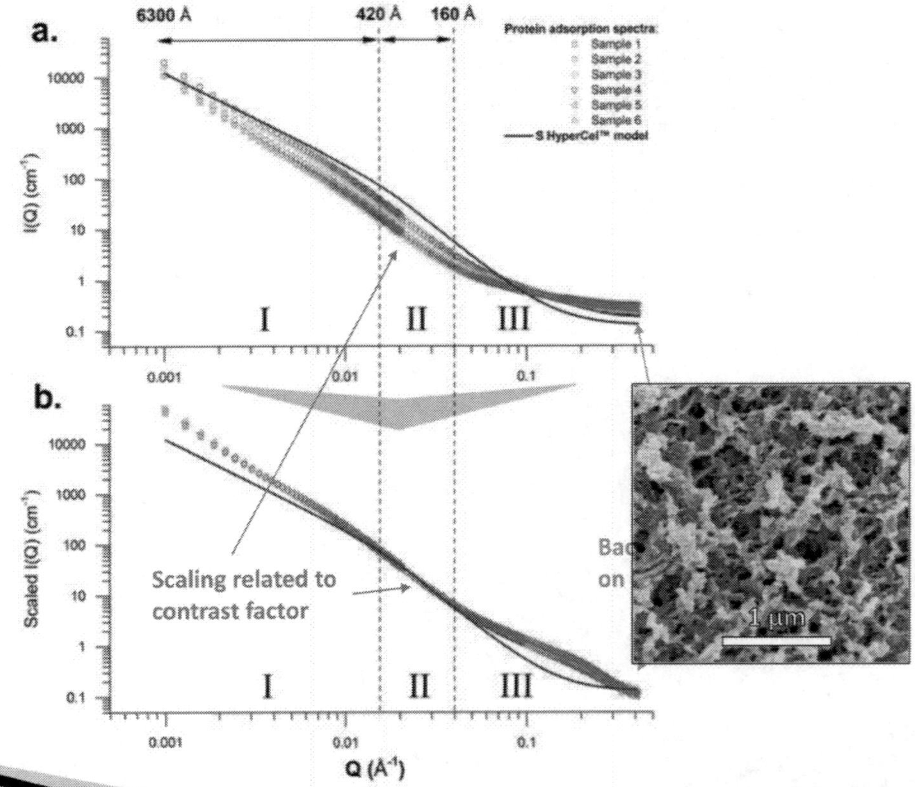

UNIVERSITY OF DELAWARE

PROTEIN LADEN PARTICLES 27/39

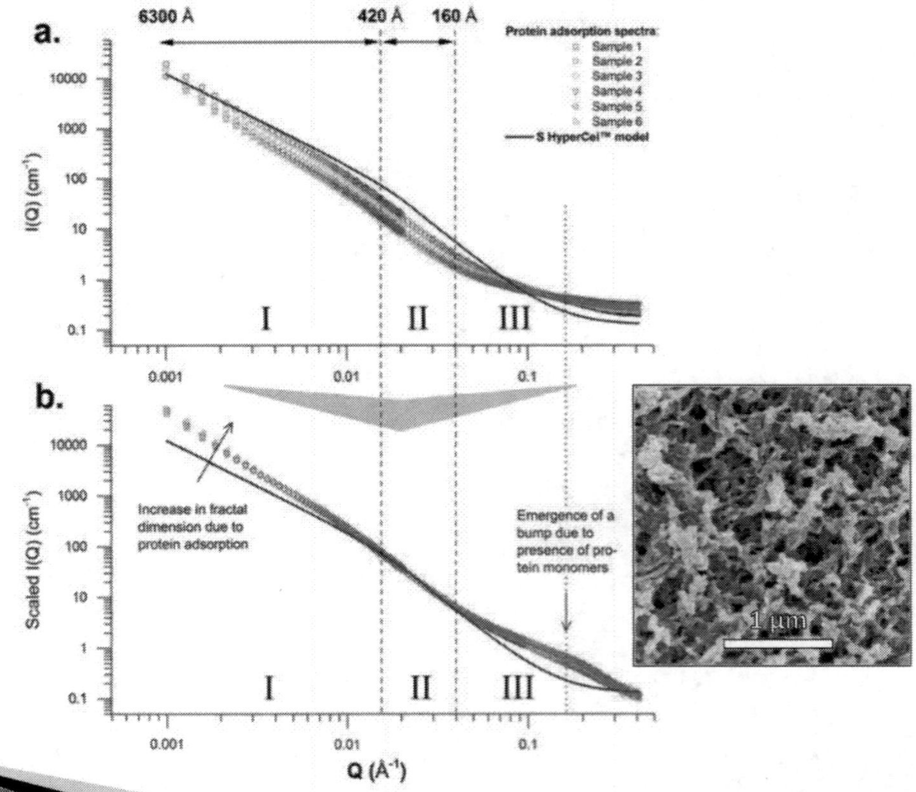

UNIVERSITY OF DELAWARE

PROTEIN LADEN PARTICLES 28/39

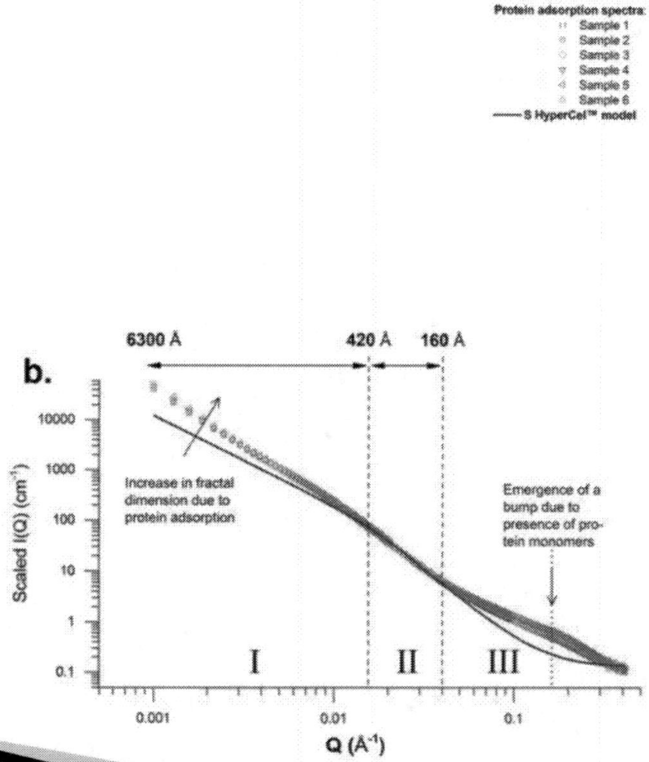
Protein adsorption spectra:
Sample 1
Sample 2
Sample 3
Sample 4
Sample 5
Sample 6
S HyperCel™ model
b.
6300 Å
420 Å
160 Å
Increase in fractal dimension due to protein adsorption
Emergence of a bump due to presence of protein monomers
I
II
III
Scaled I(Q) (cm⁻¹)
Q (Å⁻¹)
10000
1000
100
10
1
0.1
0.001
0.01
0.1

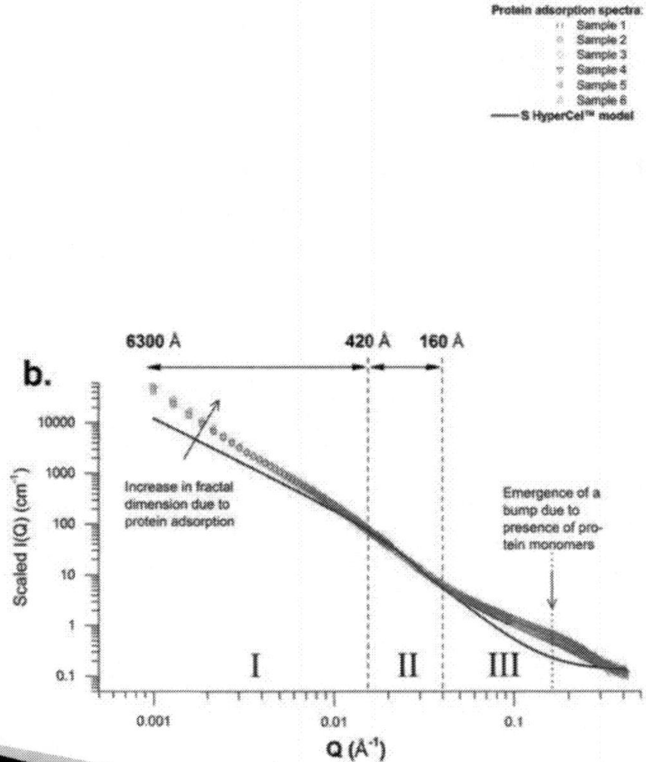
Protein adsorption spectra:
Sample 1
Sample 2
Sample 3
Sample 4
Sample 5
Sample 6
S HyperCel™ model
b.
6300 Å
420 Å
160 Å
Increase in fractal dimension due to protein adsorption
Emergence of a bump due to presence of protein monomers
I
II
III
Scaled I(Q) (cm⁻¹)
Q (Å⁻¹)
10000
1000
100
10
1
0.1
0.001
0.01
0.1

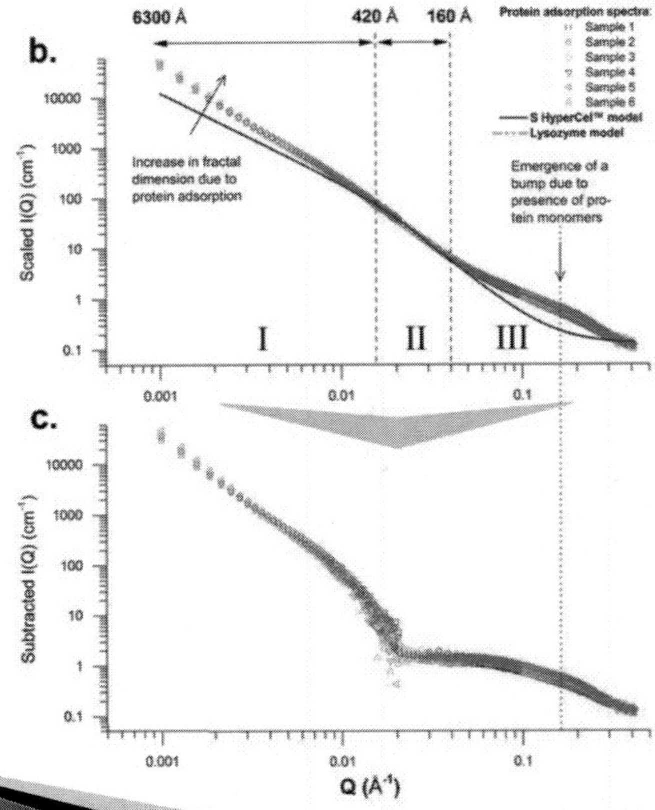

PROTEIN LADEN PARTICLES 31/39

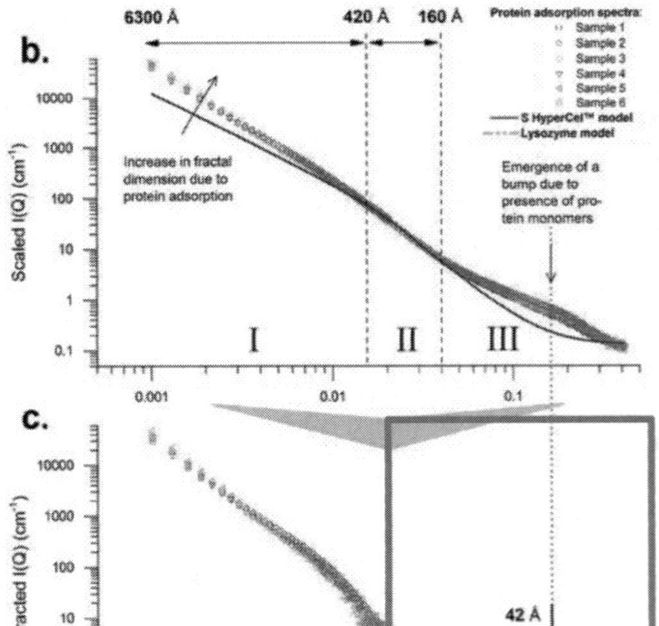

PROTEIN LADEN PARTICLES 32/39

Protein monomers

- Protein monomer contribution can be quantitatively linked to sample composition

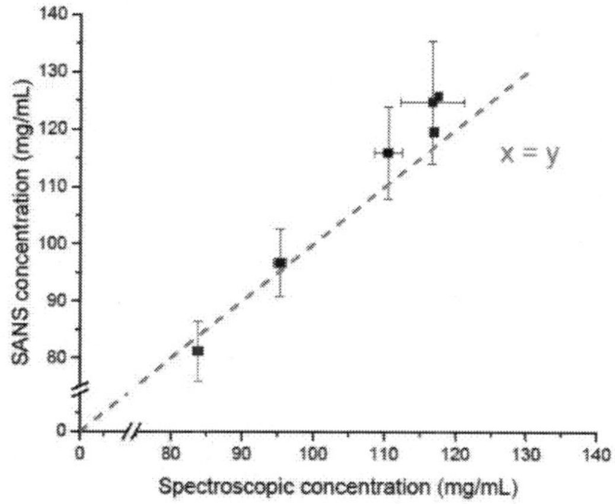

UNIVERSITY OF DELAWARE

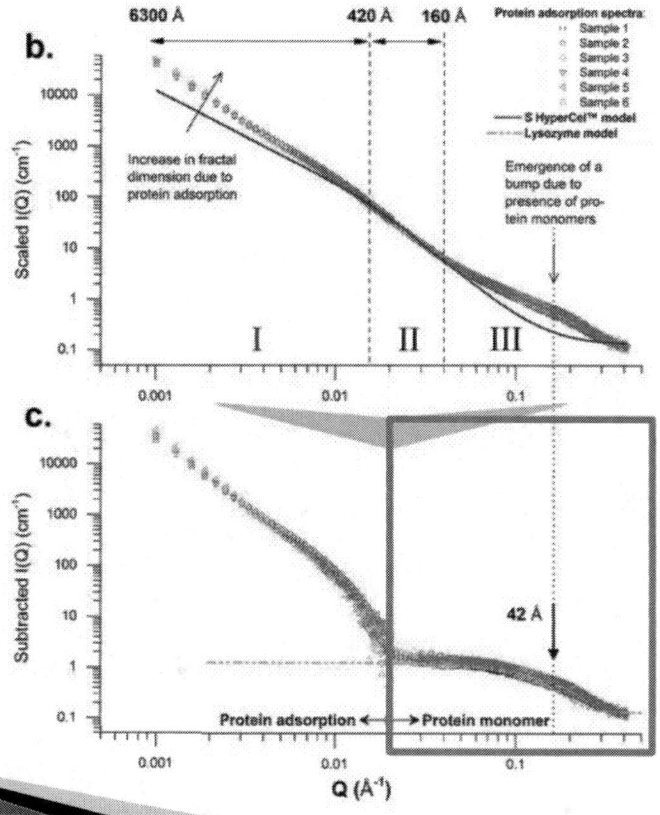

UNIVERSITY OF DELAWARE

Structural change

- Protein adsorption leads to a uniform densification of the structure

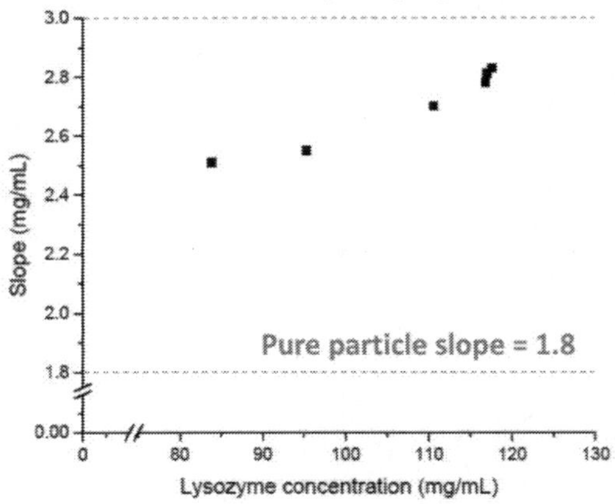

UNIVERSITY OF DELAWARE

Structural change

- Protein adsorption leads to a uniform densification of the structure

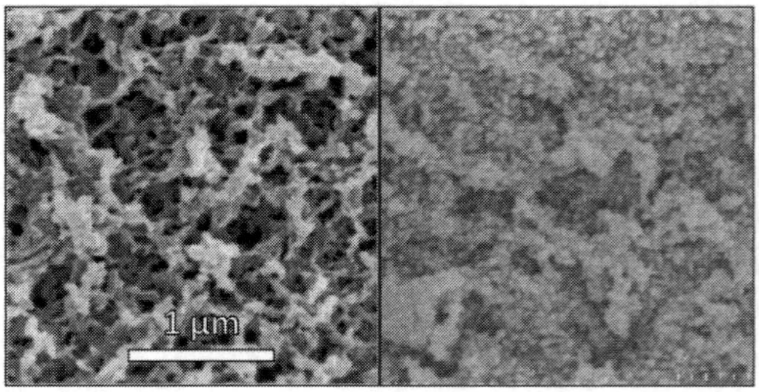

Angelo, J. M., et al. *J. Chromatogr. A* **1319**, 46-56 (2013).
Angelo, J. M., et al. *In preparation* (2014).

UNIVERSITY OF DELAWARE

Summary

- Demonstration of small-angle neutron scattering (SANS) as a novel analytical tool in protein chromatography
- Characterization of lysozyme adsorption in functionalized cellulosic chromatographic particles
- Development of a framework to measure protein adsorption from SANS spectra
- Indication of the uniform adsorption of the protein throughout the chromatographic particles

Koshari, S., et al. *Manuscript in preparation.*

Acknowledgments

- Wagner group
  - Doug Godfrin
  - Simon Rogers
- Lenhoff group
  - Jim Angelo
  - Dan Greene
- NCNR
  - Yun Liu
  - Matt Wasbrough

Materials provided by PALL Pall Corporation

Beam time provided by

NSF NIST Center for Neutron Research CHRNS NIST

Thank you for your applause

Adsorption of Alcohols in Zeolitic Imidazolate Framework-8: Effects of Force Fields, Atomic Charges and Framework Flexibility

Kang Zhang[†], Liling Zhang[‡¶], and Jianwen Jiang[†,*]

[†]Department of Chemical and Biomolecular Engineering, National University of Singapore, 117576, Singapore

[‡] Institute of High Performance Computing, 1 Fusionopolis Way, Connexis, 138632, Singapore

[¶] Jiangsu Key Laboratory for Carbon-Based Functional Materials & Devices, Institute of Functional Nano & Soft Materials, Soochow University, Suzhou, Jiangsu, 215123, China

*chejj@nus.edu.sg, +65 65165083

The intensive combustion of fossil fuels has led to the rise of the atmospheric CO_2 concentration, which is considered as the predominant cause of unequivocal global warming. According to Intergovernmental Panel on Climate Change (IPCC),[1] Europe has warmed more than the global average. The European Commission for Climate Action hopes to reduce greenhouse gas emission for 2020 from 20% below 1990 levels to 30%.[2] Motivated by the misson, European Union (EU) has set out target in 'Renewable Energy Directive' (RED) to increase the utilization of sustainable and renewable energy to 10% of the total energy consumption in transportation by 2020.[3] As alternatives to fossil-fuels, biofuels, which are commonly energy-rich alcohols diluted in water, are environmentally benign that has stimulated great demand in market. To produce water-free fuel-grade alcohol, different processes have been designed, including distillation, pervaporation, extraction, membrane separation and adsorption. Among them, adsorption of the gas-phase alcohols is an energy efficient technique, which does not require large heat input for regeneration process.[4, 5] While quite a few of adsorbent materials have been developed so far, emerging as a new member of nanoporous materials family, metal-organic frameworks (MOFs) have attracted considerable interest in the past two decades. Most experimental and simulation studies for MOFs have been focused on gas storage and separation. While nowadays, an increasing number of studies have examined alcohol adsorption in MOFs towards biofuel purification. However, only few simulation studies have been conducted for alcohol adsorption in MOFs. For the development of novel MOFs for biofuel purification by adsorption technology, first, it is critical to better understand alcohol adsorption in MOFs. We report a molecular simulation study for the adsorption of normal alcohols (methanol, ethanol, propanol and butanol) in zeolitic imidazolate framework-8 (ZIF-8). The effects of force fields, atomic charges and framework flexibility are systematically examined and compared with experimental data. Among three force fields (UFF, AMBER and DREIDING) examined, DREIDING has the best agreement with experiment. The atomic charges and framework flexibility are found to have negligible effects. The isotherms of all the four alcohols are *S*-shaped type V, indicating relatively weak adsorption in a microporous framework. With increasing pressure, three adsorption regimes are observed. At a low pressure, clusters are formed proximal to organic linker (2-methylimidazolate), and the C = C bond of organic linker is identified to be the most favorable site for adsorption. At an intermediate pressure, cage-filling occurs in sodalite cage with sharp increase in adsorption. Finally, saturation is approached at a high pressure. The isosteric heat of adsorption at infinite dilution rises linearly with the chain length of alcohol, as attributed to the enhanced interaction between aliphatic tail and hydrophobic ZIF-8. This simulation study systematically examines the effects of force fields, atomic charges and framework flexibility on alcohol adsorption in ZIF-8. The microscopic insight provided is helpful to better elucidate the adsorption behavior of alcohols in other ZIFs and MOFs, and would facilitate the development of new nanoporous materials for biofuel purification.

Reference

1. Core Writing Team, L.B., Peter Bosch, Osvaldo Canziani, Zhenlin Chen, Renate Christ, Ogunlade Davidson, William Hare, Saleemul, et al., *Climate Change 2007: Synthesis Report*, in *Intergovernmental Panel on Climate Change (IPCC)*. 2007: Geneva, Switzerland. p. 104.
2. *Climate Change: Commission Invites to an Informed Debate on the Impacts of the Move to 30% EU Greenhouse Gas Emissions Cut If and When the Conditions Are Met*. in *European Commission*. 2010. Brussels.
3. *What is the EU Doing About Climate Change?* 2013, European Commission Climate Action.
4. Oudshoorn, A., L.A.M. van der Wielen, and A.J.J. Straathof, *Adsorption equilibria of bio-based butanol solutions using zeolite*. Biochemical Engineering Journal, 2009. **48**(1): p. 99-103.
5. Oudshoorn, A., L.A.M. van der Wielen, and A.J.J. Straathof, *Assessment of Options for Selective 1-Butanol Recovery from Aqueous Solution*. Industrial & Engineering Chemistry Research, 2009. **48**(15): p. 7325-7336.

392294 Effects of Open Metal Site Availability on Adsorption Capacity and Olefin Selectivity in Coordinatively Unsaturated Cu3(BTC)2 Metal-Organic Frameworks

Abstract:

Several metal-organic frameworks (MOFs) have been identified that have metal centers that are not entirely coordinated to framework oxygen atoms. These coordinatively unsaturated MOFs have open metal sites which can be exposed by removing the coordinated solvent molecules during thermal activation. These open metal sites are attractive for electronegative molecules, such as olefins. MOFs such as $Cu_3(BTC)_2$ and MOF-74 are known to separate small olefins from paraffins in the gas phase due to the π-electron interaction with the open metal sites. However, the effectiveness of the open metal sites is strongly dependent on the synthesis and pre-activation of the MOF before its use as an adsorbent. These relationships are not well understood from previous work.

In this work, we study in detail the removal of solvent molecules from $Cu_3(BTC)_2$ MOF as a function of activation temperature, quantify the exposure of open metal sites, and characterize the effects of open metal site availability on the adsorption capacity and olefin selectivity. This study was conducted in a vapor breakthrough adsorption unit with binary mixtures of hexane and 1-hexene In the case of fresh $Cu_3(BTC)_2$ samples separately activated at different temperatures, we observe a trade-off between adsorption capacity and olefin selectivity as the activation temperature is increased. However, in the case of $Cu_3(BTC)_2$ samples cycled repeatedly through adsorption and desorption steps at successively increasing activation temperatures, we observed a different effect. The adsorption capacity and olefin selectivity remained nearly constant up to a limiting activation temperature, beyond which the total adsorption capacity decreased and the olefin selectivity increased dramatically. We discuss these findings, supported by X-ray diffraction, N_2 physisorption, and FTIR spectroscopy, in terms of the accessibility of the pore space and open metal sites.

390700 Nanoporous Materials Explorer: The Database of the Nanoporous Materials Genome Center

Abstract:

Metal-organic frameworks (MOFs), zeolites and other nanoporous materials are potentially useful for many energy-relevant or technologically important processes such as adsorption, catalysis, sensing and drug delivery. Over the past few years a huge number of new nanoporous structures have been synthesized experimentally or predicted computationally[1,2]. These materials, their next-generation computational design and properties prediction offer a high-payoff opportunity. Several large-scale computational studies[2,3] of these materials have been demonstrated to be feasible and capable of identifying candidate materials for various applications generating large datasets of properties. The goal of the DOE funded Nanoporous Materials Genome Center (NMGC) [3] is to discover and study these novel nanoporous materials. Additionally, the NMGC aims to provide a repository of experimental and hypothetical structures and associated properties for the rapidly growing scientific communities that are interested in these materials. In order to accomplish this goal the NMGC has partnered with the Materials Project[4]in order to produce a state-of the art web interface for a comprehensive database of nanoporous materials and their properties, the Nanoporous Materials Explorer. The search functionality of the Nanoporous Materials Explorer is built around a visual representation of the stored properties. One of the available ways to search for materials is by choosing from a list of ever-expanding properties to be plotted in an x-y figure allowing for the selection of a group of materials falling within a range of desired target values. The properties and structures of a group of selected materials can then be extracted for further analysis by the user or by a list of available online tools. This effort aims to make research more effective, consistent and transparent by collecting data from various research groups and eventually expedite the discovery of new high performing nanoporous materials while expanding the knowledge around them. In order to demonstrate the usefulness of such a nanoporous materials database we have identified experimentally synthesized MOFs with coordinately unsaturated metal centers and for a subset of these examined their performance for selective adsorption using first-principles calculations.

1) Deem, Michael W., et al. The Journal of Physical Chemistry C 113.51 (2009): 21353-21360.2)

2) Wilmer, Christopher E., et al., Nature chemistry 4.2 (2012): 83-89.

3) E Haldoupis, S Nair, DS Sholl, Journal of the American Chemical Society 132 (21), 7528-75393)

3) http://www.chem.umn.edu/nmgc/

4) https://materialsproject.org/

Arsenic Removal from Water by Titanium based Adsorbent – Dynamic Adsorption Capacities and Comparison with Other Commercial Adsorbents

Yujun Liu and Bruce S. Holden
The Dow Chemical Company
2301 N. Brazosport Blvd.
Freeport, TX 77581
liuy@dow.com

Abstract

Various processes exist for arsenic removal from drinking water that includes coagulation/precipitation, adsorption, ion exchange and membrane filtrations such as reserves osmosis, electrodialysis reversal, nanofiltration and colloidal floatation, etc. Adsorption is one of the most widely used processes and several types of adsorbents are commercially available for this application: activated alumina, activated carbon, greensand, granular ferric hydroxide, iron oxide coated sand and copper-zinc granules, etc. In an attempt to find cost effective and eco-friendly arsenic removal processes, titanium dioxide and titanium dioxide-based materials have been studied in the past decade as adsorbents due to their physical and chemical stability, negligible toxicity, resistance to corrosion, strong oxidizing power in pores, redox selectivity, high photo-stability, easy preparation, and high affinity to arsenic. Even though much research has been done, dynamic adsorption capacity data are rarely available in the literature on any commercial TiO_2 based adsorbent. This presentation will present dynamic arsenic adsorption capacity data on a commercial TiO_2-based adsorbent, ADSORBSIA^{TM}, that were measured with different feed flow rates, pH values, and different particle sizes. Effects of other impurities such as SiO_2 in the feed water were studied as well as effects of intermittent feed and feed pH swings. Furthermore, comparative studies were conducted between the TiO_2-based adsorbent and other types of adsorbents available in the market place.

From single drop experiments to stirred tanks - Systematic analysis of coalescence in liquid/liquid dispersions

Jörn Villwock[1], *Johannes Kamp*[1], *Felix Gebauer*[2], *Hans-Jörg Bart*[2], *Matthias Kraume*[1]

[1] *Chair of Chemical & Process Engineering, Technische Universität Berlin, Germany*

[2] *Chair of Separation Science and Technology, Technische Universität Kaiserslautern, Germany*

Dispersions of two immiscible liquids are an integral part of several unit operations. The drop size distribution, which is the result of the competing phenomena of drop breakage and coalescence, influences a decisive part of the overall process efficiency. Existing models for the coalescence rate include the physical phase properties, but also hydrodynamic characteristics like energy dissipation rate and relative velocity. However, these influencing parameters are implemented in existent models with different proportionalities and in some cases even contradictorily. As in experimental validation the independent examination of each parameter is necessary, the system is reduced to the fundamental behavior of single droplet coalescence and the impact of the parameters is quantified separately. The gained detailed knowledge can be implemented into the framework of population balance equations which can be applied to describe the transient drop size distribution in technical applications. Following this approach the fundamental behavior of single droplets can be related to the macroscopic behavior of droplet swarms.

In this work the systematic analysis from single drop experiments and batch settling tests to the description of stirred tanks is presented. It focuses on the binary droplet coalescence and two quantities will be discussed in detail: coalescence probability with the addition of ions in varying concentrations and contact/coalescence time depending on the drop size and relative velocity. Additionally the results of standardized batch settling tests for the addition of different ions with varying concentrations are shown.

Using a fully automated test cell (see Fig. 1), serial examinations of binary droplet collisions under different system conditions (varying e.g. drop sizes, mass transfer direction and intensity, pH, ionic strength, disperse and continuous phase) using high speed imaging are possible. With this setup, based on the principle to observe the collision of a pendant drop at a top cannula and a second rising droplet detached from a lower cannula, a significant number of experiments can be conducted to provide a statistically solid data base. Analyzing the recorded images of the coalescence process, important quantities like coalescence probability, contact and coalescence time, relative velocity, momentum and deformation can be achieved. All experimental results were achieved using the EFCE recommended standard test system for liquid/liquid extraction toluene/water.

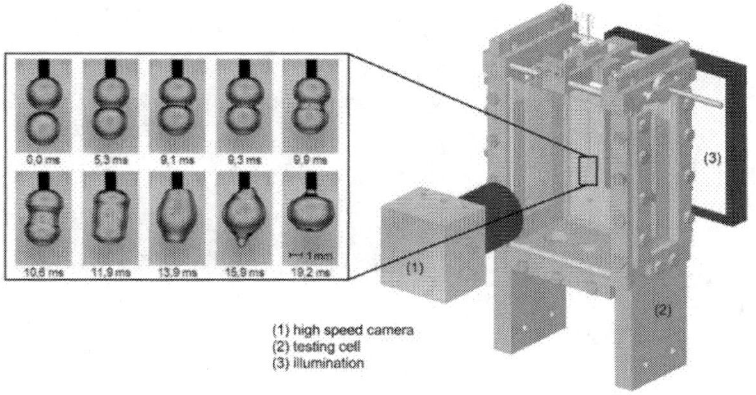

Fig. 1: Sample image sequence for coalescing droplets recorded in the schematically shown test cell

Investigations with NaCl and NaOH in a concentration range of 10^{-5} - 10^{-1} mol/L indicated that the coalescence probability strongly decreases with the addition of the ions. However, for a NaCl concentration of 10^{-1} mol/L, an increase of the coalescence probability with increasing equivalent drop diameter was observed, which could not be found for NaOH at the same concentration.

The contact time was determined in the range of 10 - 40 ms for drop sizes between 1 to 3 mm. With increasing drop size ratio as well as increasing drop rise velocity, contact times became longer. With respect to coalescence time two different types of coalescence can be distinguished: spontaneous coalescence (< 10 ms) and slower coalescence (> 10 ms) where mainly the slower type occurred in the described experiments.

Further investigations will include systematic variations of ion species according to the Hofmeister series with different concentration and the deeper analysis of fluid dynamics. All these results will be used to validate and refine existing models or to develop new models, which describe the drop size distribution in liquid/liquid dispersions.

Financially supported by DFG project KR 1639/19-1 "Coalescence efficiency in binary systems".

Analysis of drop breakage processes in turbulent flow

Matthias Kraume, *Stephanie Nachtigall*
Technische Universität Berlin,
Chair of Chemical and Process Engineering,
Fraunhoferstr. 33-36, 10587 Berlin/Germany

Drop breakages occur in many liquid/liquid mixing applications in chemical, pharmaceutical or food industries. Besides the coalescence process, it directly influences the drop size distribution (DSD) which determines the efficiency of these processes.

It is widely accepted in literature that drop breakage happens when the external forces exerted on particles by the continuous phase exceed the internal forces which keep the particle together. Based on this assumption a multitude of theoretical models describing this process can be found in literature. Nevertheless, a reliable prediction of the DSD when varying for example material and process parameters is not possible even if coalescence is completely hindered. This leads to the conclusion that a deeper understanding of the drop breakage process is still needed.

In this work drop breakage in a turbulent flow comparable to the flow field in a stirred tank is investigated by high-speed imaging (see example pictures in Figure 1). A single stirrer blade, mimicking a part of a Rushton turbine is therefore mounted in a rectangular channel and induces drop breakage.

In order to describe the breakage event properly, the breakage probability, the breakage time as well as the deformations and oscillations leading to the breakage event are determined by automated image analysis. Additionally, the resulting Sauter mean diameter and the number and the size distribution of the daughter drops are evaluated.

To quantify the influence of the stabilizing forces, numerous different droplet substances with and without the addition of surfactants were investigated under the variation of the droplet diameter. Additionally, the fluid flow velocity of the continuous water phase was varied to also determine the influence of the external deforming forces.

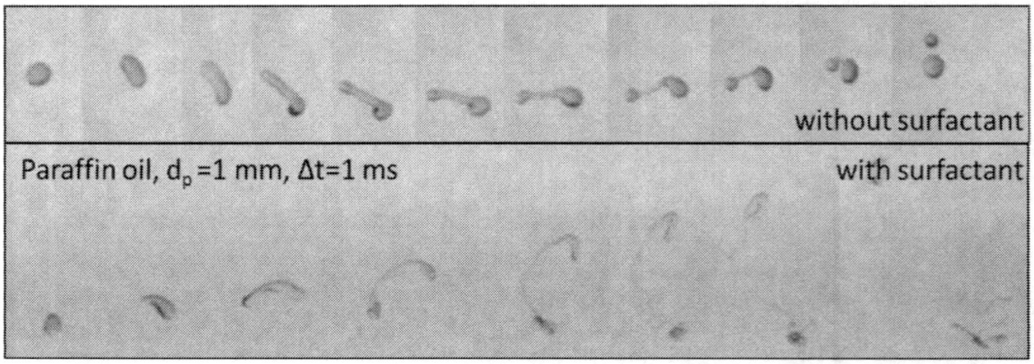

Figure 1: Breakage of a 1 mm paraffin oil droplet in a turbulent flow with different interfacial tensions.

The gained knowledge considering the breakage mechanisms should be scaled up to technical relevant mixing applications. Therefore, experiments in a standard lab scale stirred reactor (6-blade Rushton turbine, d/T=0.33, H/T=1) were conducted. Transient DSD were determined with an endoscope measurement technique and compared to the results of the single drop breakage investigations.

Prediction of Interfacial Properties in Counterflow Extraction Columns

Tommy Grunert, Sabine Enders

Technical University of Berlin, BH 7-1, Ernst – Reuter – Platz 1, 10587 Berlin, Germany.

Email: tommy.grunert@tu-berlin.de; sabine.enders@tu-berlin.de

Key words: Interfacial tension, Interfacial profiles, Density gradient theory, Counterflow extraction columns

Abstract:

For the design and optimization of counterflow extraction columns a detailed knowledge about the mass transfer across the liquid-liquid interface is required. The mass transfer across the interface depends strongly on the interfacial tension itself. The interfacial tension depends on the temperature and the local concentrations of the mixture. Both quantities can change along the extraction column. This knowledge can be provided by division of the column in theoretical stages and assuming liquid-liquid equilibrium (LLE) at every stage. In this contribution, this knowledge is applied and combined with a modified density gradient theory, developed recently [1,2], at every stage. In this way, a theoretical framework will be discussed, which is able to predict the interfacial tension and the concentration profiles at every stage in the counterflow extraction column to the first time. This framework permits the prediction of enrichment of one component in liquid-liquid interface. For the calculation the standard extraction system, composed of toluene + water + acetone, is selected.

The first step along the theoretical framework is the calculation of the LLE using an activity coefficient model. In this contribution, the Koningsveld-Kleintjens model [3] is applied, where the model parameters were fitted to the LLE data measured by Hackl et al. [4]. Using this model permits the calculation of the LLE very close to the experimental data taken from the literature [4]. The interfacial properties are available by combination of the activity coefficient model with the modified density gradient theory [1,2]. The miscibility gap of the ternary system arises from the binary subsystem water + toluene. Therefore, the influence parameter κ, occurring in the theoretical framework, was fitted to one interfacial tension of this binary subsystem at one temperature. The interfacial tensions of the ternary system as function of the feed concentration and of the temperature can be predicted. The predicted interfacial tensions were in excellent agreement with experimental data found in the literature [5]. The interfacial tension depends strongly on the composition; however, only slightly on temperature. The reason for this result is the weak temperature dependency of the miscibility gap. Analyzing the predicted concentration profiles across the interface, it can be found that always acetone enrich at the interface. The amount of acetone in the interface depends strongly from the feed composition, where the relative amount of acetone decreases with increasing amount of acetone in the feed mixture.

The counterflow extraction column was divided in five theoretical stages and operates at constant temperature. The feed, composed of 40 mole per cent acetone and 60 mole per cent water, enters the column on the top and flows to the bottom. The solvent toluene enters the column on the bottom and flows to the top of the column. Assuming liquid-liquid equilibria at every theoretical stage, the concentration profiles can be calculated. The acetone concentration in the feed phase decreases from stage to stage, whereas the water concentration increases simultaneously. The outgoing phase contains 99.75 mole per cent water, if the extraction was performed at 298K. The modified density gradient theory permits the calculation of interfacial properties at each equilibrium stage. On the one hand, the interfacial tension of a steady state working extraction column is experimentally not or almost not accessible. On the other hand, local gradients of the interfacial tension can lead to Marangoni convection [6], which can have an impact on the interfacial mass transfer. The theoretical framework predicts a continuous increasing of interfacial tension from top to bottom of the column due to the larger miscibility gap of the binary subsystem water + toluene in comparison to the miscibility gap in the ternary system. The local gradient reaches its maximum between stage three and five, which increases the possibility of Marangoni convection [6] in this region more than in the rest of the column. Obviously, from stage three to five the interfacial concentration profiles of acetone show a significant enrichment inside the interface. In combination with the local gradients of the interfacial tensions from stage three to five, those effects can have a large impact on the interfacial mass transfer.

By varying the temperature of the incoming feed flow and / or solvent flow, a non-isothermal column simulation is suitable for an optimization with respect to the local gradients of the interfacial tension, as well as the quality of the outgoing water phase. From the theoretical point of view, the excess enthalpy must be taken into account. The calculation of the enthalpy of a mixture at different temperatures was done with constant liquid heat capacities, because the temperature dependency is not crucial in a small range of temperatures. The temperature gradient in the counterflow column is available by the heat balance.

At first, only the temperature of the feed will be varied to show the influence on the interfacial tension inside the column, the solvent temperature is constant at 293.15 K. The temperature of the feed shows the significant influence on the interfacial tension in the counterflow extraction column. While the interfacial tension at the top of the column is almost constant, the interfacial tension at the bottom increases at higher temperatures of the feed due to the temperature dependency of interfacial tension of the binary subsystem water + toluene.

It can be shown that higher temperatures of the feed stream leads to a higher purity of water, while the amount of cleaned water is approximately constant. However, the price is a higher interfacial tension in the lower part of the column, which can have a negative impact on the interfacial mass transfer and thus, leads to a lower mass flow through the interface. Also, the local gradients of the interfacial tension are steeper at higher temperatures of the feed stream, which can increase the possibility of Marangoni convection.

The influence of the temperature of the solvent can also be studied using the theoretical framework. A higher temperature of the solvent stream corresponds to a higher interfacial tension, as well as a higher purity of water, while the amount of water is almost constant. The local gradients of the interfacial tension inside a counterflow extraction column can be reduced by varying both the temperature of the feed and the solvent. The simulation with the lowest local gradient of the interfacial tension required a low feed temperature and high solvent temperature and was accompanied with the highest amount of water and lowest purity of water. In contrast to this, a warm feed and a cold solvent leads to higher local gradients of the interfacial tension. This reduces the amount of water but increases the purity of water. In summary, high water purity can be accomplished with a relative high local gradient of the interfacial tension, which can have impact on the interfacial mass transfer due to Marangoni convection and coalescence behavior.

Literature:

[1] T. Grunert, H. Rudolph, S. Enders, Zeitschrift für Physikalische Chemie 227 (2013) 1–16.

[2] T. Grunert, S. Enders, Fluid Phase Equilibria submitted 2014.

[3] R. Koningsveld, L.A. Kleintjens, Macromolecules 4 (1971) 637-641.

[4] A. Hackl, W. Solar, G. Ziebland, Eur. Fed. Chem. Eng. London (1978).

[5] T. Misek, R. Berger, EFCE Publication Series 46 (1985).

[6] C. Marangoni, Annalen der Physik und Chemie 7 (1871) 337-354.

Design and Operation Characteristics of Liquid-Liquid Extractors with Disc Supported Taylor-Couette-Flow

Enes Aksamija, Matthäus Siebenhofer

Outline

- RDC, *Taylor-Couette* reactor
- Taylor-Couette Disc Contactor (TCDC)
- Comparison: RDC vs. TCDC
  - Onset of operation
  - Hold-Up
  - Flooding limit
  - Drop size distribution
  - Separation efficiency
- Design rules & Summary

Rotating Disc Contactor (RDC)

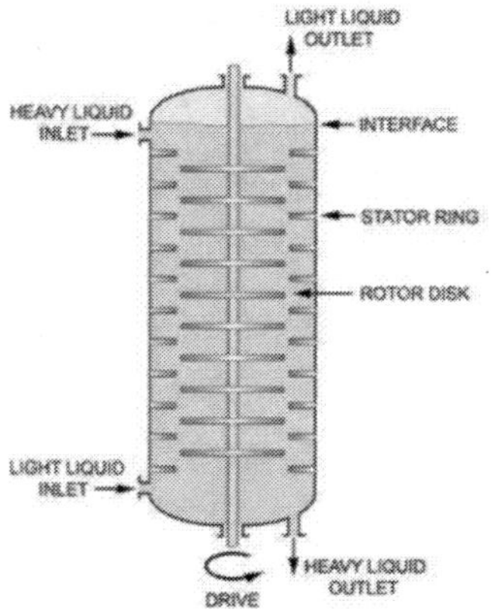

Empiric Design [KOSTERS (1983)]:

$$\frac{D_{Rotor}}{D_{Tower}} = 0.6 \quad \frac{D_{Stator}}{D_{Tower}} = 0.7 \quad \frac{H_{Comp.}}{D_{Tower}} = 0.08\ to\ 0.15 \quad D_{Shaft} = ?$$

- Simple, open design → **high throughput**
- No paddles → **viscous systems**
- Smooth energy input

Problems:

- Empirical design → problematic Scale-Up
- **High axial mixing** → reduced separation efficiency

Taylor-Couette reactors

- **High radial, low axial** mixing
- Homogeneous energy input
- Eddies can be used for **sizing & fluidization** of particles!

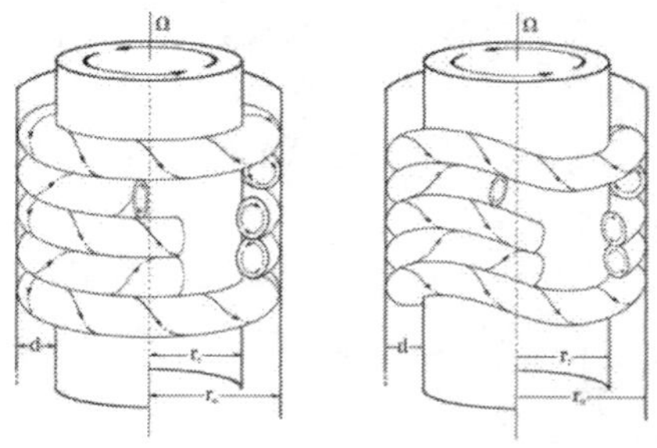

Problems:

- **Instabilities** (~74 flow regimes) → difficult process control
- Gap phenomenon → **small reactor volume**

Source: Richard Lueptow (2009), Scholarpedia, 4(11):6389.

Taylor-Couette Disc Contactor

- No Stator! → D_{Rotor} increased!
- Disc supported Taylor-Couette flow

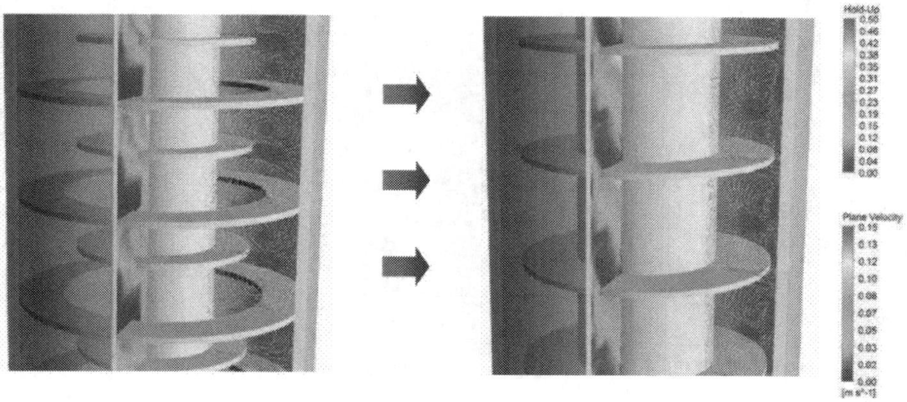

RDC TCDC

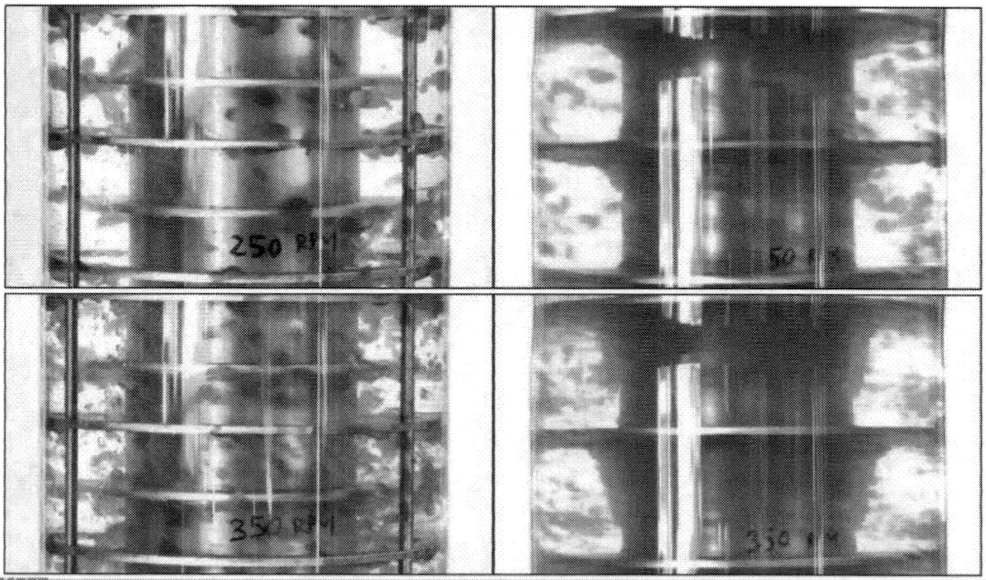

Onset of operation

- Sedimentation experiments with rigid spheres
- Polypropylene, 2mm & 2.5mm
- No axial flow
- Stepwise increase of agitation by 50 [1/min]

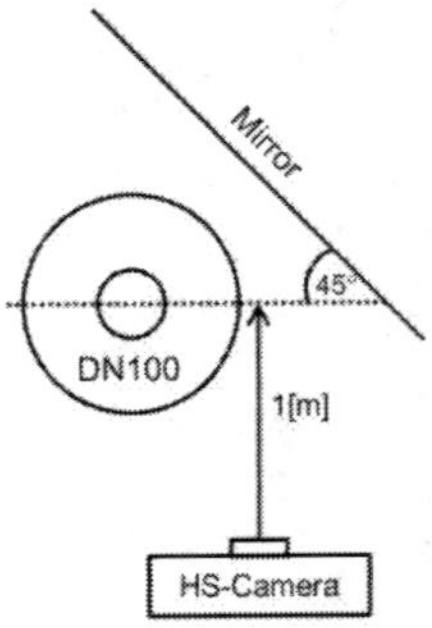

Design and Operation Characteristics of Liquid-Liquid Extractors with Disc Supported Taylor-Couette-Flow TU Graz

Onset of operation

- Residence time distribution, 2mm spheres

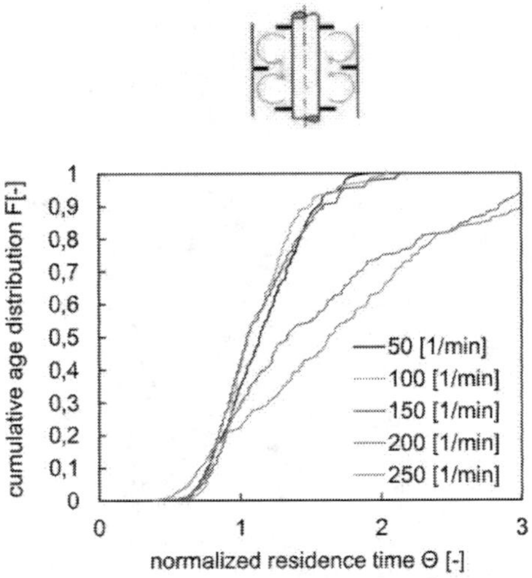

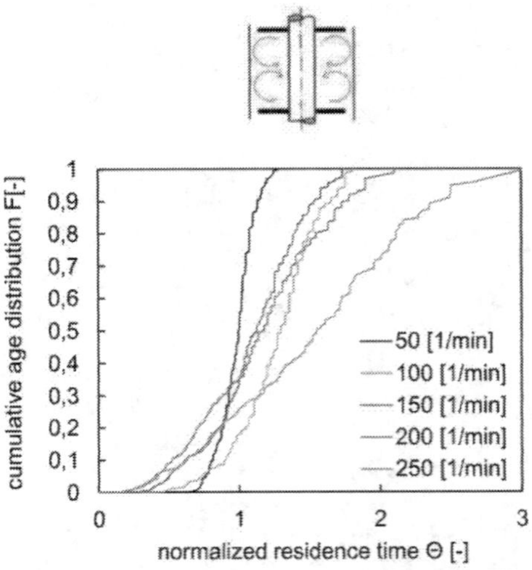

Hold-Up

- High hydraulic load, 30 [m³/(m²h)]
- Position of interface ≙ dynamic Hold-Up

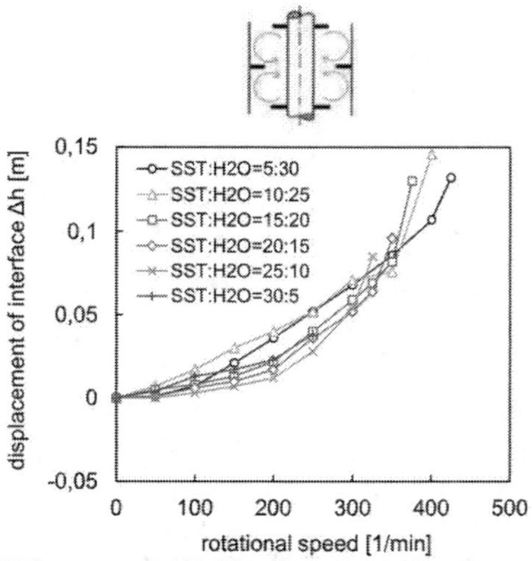

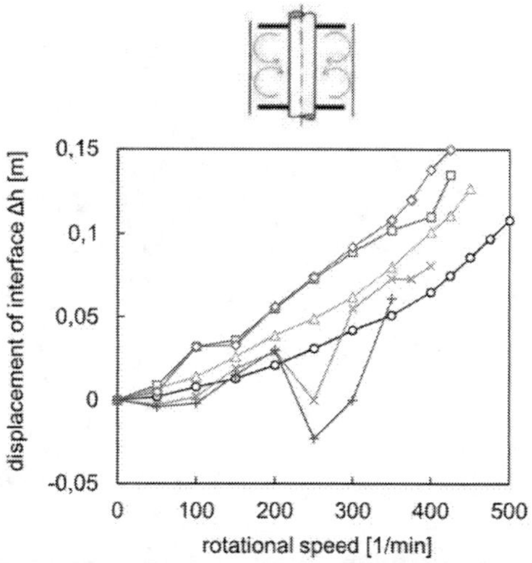

Design and Operation Characteristics of Liquid-Liquid Extractors with Disc Supported Taylor-Couette-Flow

Hold-Up

- Low hydraulic load, 10 [m³/(m²h)]
- c/d = 1

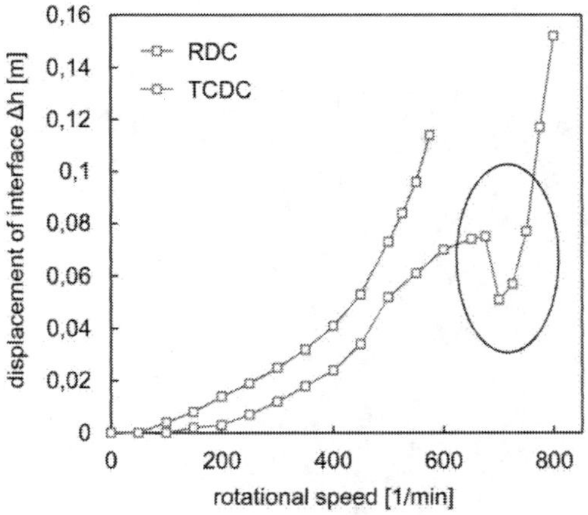

Formation of a second dynamic settler-zone **above** rotor discs!

→ 2 Settlers / Compartment

Flooding limits

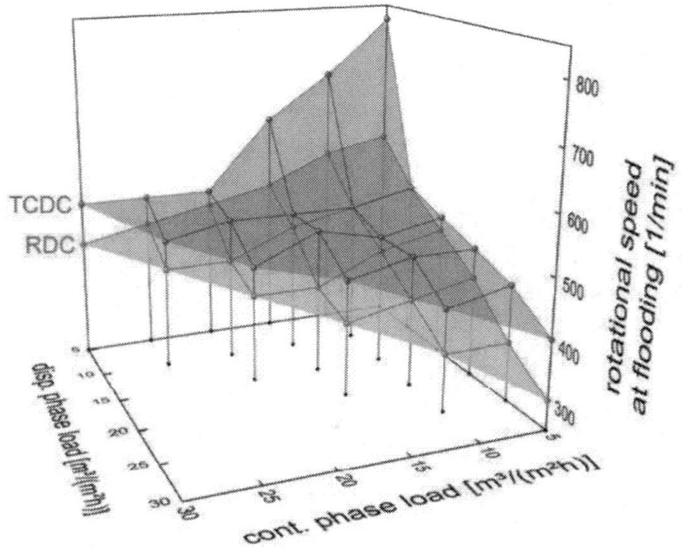

Drop size distribution (DSD)

- Endoscopic probe
- 1[m] active height, Shellsol-T / water
- Variation of B [$m^3/(m^2h)$] and n [1/min]

SOPAT

www.sopat.de

Drop size distribution (DSD)

- B = 20 [m³/(m²h)], c/d = 1

250 [1/min] | 350[1/min] | 450 [1/min]

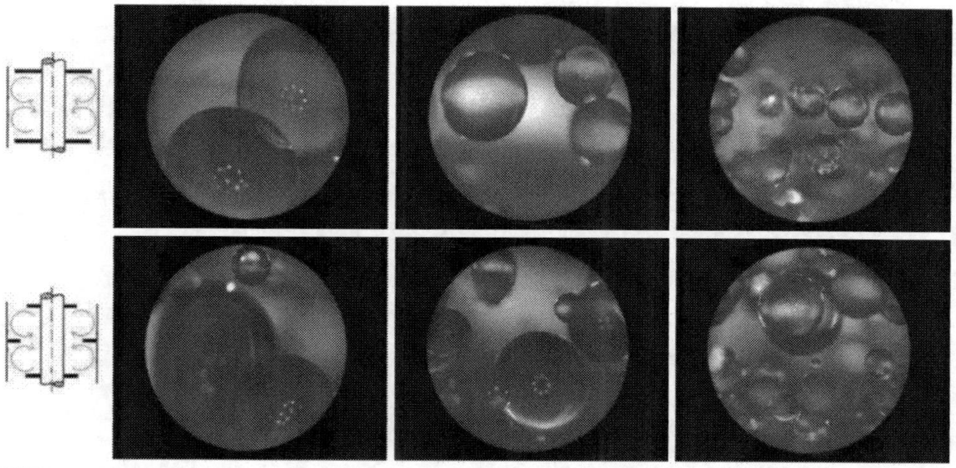

Drop size distribution (DSD)

- 450 [1/min]

B = 15 [m³/(m²h)]

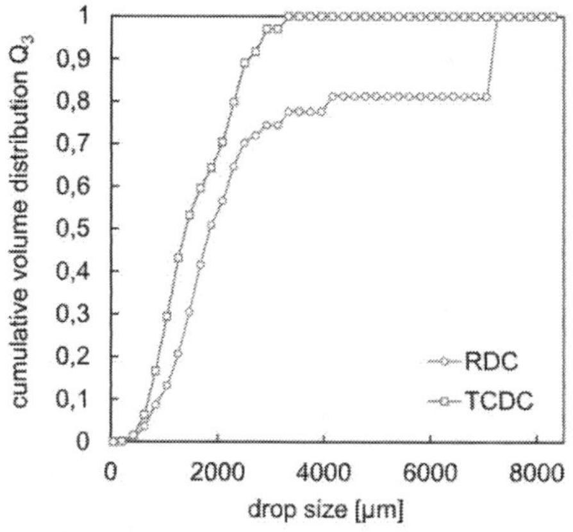

B = 20 [m³/(m²h)]

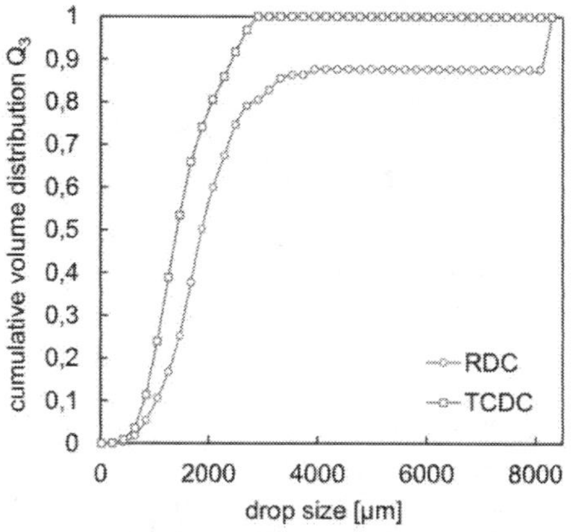

Separation efficiency

- ShellSol-T (disp.) / n-Butanol / Water (cont.)
- $V_{disp.}$ = 40 [l/h], $V_{cont.}$ = 45 [l/h], B = 14 [m³/(m²h)]
- H_{active} = 1 [m], 350 [rpm]

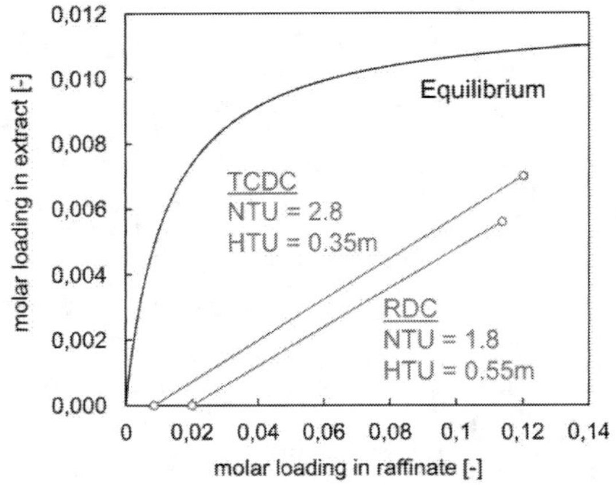

$$NTU = \int_{X_\omega}^{X_\alpha} \frac{dX}{X - X^*}$$

$$HTU = H_{active} / NTU$$

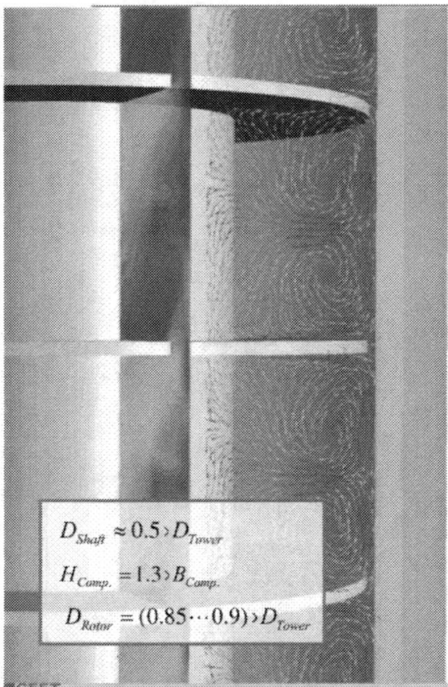

Summary

- Effective compartment separation
  → lower axial mixing of both phases
- Smaller drops & narrower distribution
- Mixer-Settler in each compartment
- Lower investment costs
- Eased cleaning & maintenance
- Higher separation efficiency
- Higher flooding limits

Design and Operation Characteristics of Liquid-Liquid Extractors with Disc Supported Taylor-Couette-Flow

Enes Aksamija, Matthäus Siebenhofer

385484 Rotor-Rotor Spinning Disc for Intensified High Shear High Gravity Extraction

Abstract:

This paper presents a novel centrifugal extractor based on rotor-rotor spinning disc technology. The device consists of an inner and an outer rotor that rotate with individual speeds with a gap distance between the rotors between 1-3 mm. The diameter of the rotors is 17 and 18 cm respectively, the volume of each contacting stage is therefore only 100 mL. The inner rotor has a special design that allows countercurrent contacting of a light and a heavy phase over several contacting stages. The investigated device has three physical contacting stages. Experiments are performed with n-heptane as light phase liquid and water as heavy phase liquid. For inner and outer rotation speeds up to 2500 rpm, a >99.9% physical separation efficiency of light and heavy phase is observed in two distinct operating regions. One region in the area where the outer rotor has a higher speed than the inner rotor and one region where the inner rotor has a higher speed than the outer rotor. Hold-up measurements show that the physical separation efficiency drops when the extractor is almost completely filled with (1) the heavy phase for loss of purity of the light phase and (2) with light phase for loss of purity of the heavy phase. Mass transfer experiments with benzoic acid extracted from the n-heptane phase to the water phase show that three contacting stages are obtained, in accordance with the three physical stages. Liquid-liquid mass transfer coefficients up to 1 m/s were observed at the highest shear conditions. Consequently, extraction can be performed effectively with very short contact times and low solvent volumes. The rotor-rotor spinning disc extractor is especially attractive when quick extraction of unstable (intermediate) reaction products is desired, for pressurized systems, and for corrosive systems. Furthermore, it provides a continuous countercurrent extraction step for flow chemistry, which facilitates continuous production of fine chemicals and pharmaceuticals.

Hydrodynamics of Cocurrent Downward Liquid-Liquid Extraction Column

S. Samdavid, T. Renganathan and K. Krishnaiah

Department of Chemical Engineering, Indian Institute of Technology Madras, Chennai, India

Abstract

Liquid-liquid extraction (LLE), can be carried out using mixer-settlers, spray columns, pulsed columns, rotary disc contactors, fluidized beds etc. Fluidized bed extractors have been proved to have higher efficiency compared to other systems [1]. The operation in conventional fluidized bed extractors, generally, is through cocurrent upward flow of the phases in the presence of a batch of inert solids. In the present work, a three phase liquid-liquid-solid inverse fluidized bed (LLSIFB) extractor, with countercurrent flow of the liquid phases, is proposed based on our experience [2,3] with countercurrent gas-liquid-solid inverse fluidized bed which has high mass transfer characteristics [4]. Similar to the breakage of bubbles due to collision with particles in three phase IFB, particles are expected to break the drops in LLSIFB extractor leading to higher mass transfer rates.

Information on the contacting pattern proposed in the present study viz. liquid-liquid-solid inverse fluidization is not available in the literature to the best of our knowledge. Since this system is first to be studied, it is proposed that experiments can be conducted in cocurrent and countercurrent modes of operation. In the present work, attention is focused on cocurrent downward LLSIFB. As a first step in understanding the hydrodynamics of this system, it is proposed to collect experimental data on the hydrodynamics of two phase cocurrent downward liquid-liquid system (without the presence of solid particles) on flow regimes, pressure drop and phase holdups.

Cocurrent liquid-liquid downflow hydrodynamic experiments were conducted by keeping kerosene flowrate constant and varying water flowrate. Measurements were taken on pressure drop and holdup for various water and kerosene flowrates moving from kerosene continuous to kerosene dispersed regime. Flow regime map is obtained by determining the transition velocities at which phase inversion takes place. The variation of the dispersed phase holdup along the regime boundary is also obtained. Effect of water and kerosene flowrates on pressure drop and kerosene holdup is studied.

References:

1. Roszak, J., and R., Gawronski, "Cocurrent Liquid-liquid Extraction in Fluidized Bed," *Chem. Eng. J.* **17**,101-109 (1979).

2. Krishnaiah, K., S. Guru and V. Sekar, "Hydrodynamic Studies on Inverse Gas-liquid-solid Fluidization", *Chem. Eng. J.* **51**, 109-112 (1993).

3. Renganathan, T., and Krishnaiah, "Prediction of Minimum Fluidization Velocity in 2- and 3-phase Inverse Fluidized Beds", *Can. J. Chem. Eng.* **81**, 853-860 (2003).

4. Nikolov.V., Farag.I., and Nikov.I, "Gas-liquid Mass Transfer in Bioreactor with Three-phase Inverse Fluidized Bed," *Bioprocess Eng.* **23**, 427-429 (2000).

Alkaline Stability Study of Quaternary Phosphonium Cations

Bingzi Zhang, Robert Kaspar, Junhua Wang, Qianrong Fang, Shuang Gu, Yushan Yan*
Department of Chemical and Biomolecular Engineering, University of Delaware, Newark, DE, 19716
* yanys@udel.edu

The cationic group is the functional heart of hydroxide exchange membranes (HEMs), controlling fundamental properties such as stability, conductivity and solubility. Ammonium cations[1,2], exemplified by benzyl trimethyl ammonium (BTMA), have long been used for HEMs, establishing the cation standard for HEMs. Recently, we introduced the phosphonium cation (tris(2,4,6-trimethoxyphenyl) phosphonium (TPQP))[3-6] to polymersulfone backbone and effectively improved the alkaline stability, hydroxide conductivity and solubility of HEMs. We found that in membrane form, TPQP based HEMs show better alkaline stability than BTMA analogs.

In order to find out the reason for the better stability of TPQP based HEMs, we study the alkaline stability of TPQP cation in 1 M alkaline solution at 80 °C and compare it with that of BTMA cation. We find that over the test period (58 days), TPQP degrades slower than BTMA (e.g., 59% degradation of TPQP vs 82% degradation of BTMA after 20 days). We also find that the degradation mechanism of TPQP cation in alkaline environment is different from that of common phosphonium cations[7]. To account for this unexpected behavior, we propose and verify a new multi-step degradation mechanism, including ether hydrolysis, inner salt formation, ketone rearrangement and further hydrolysis. This mechanism demonstrates that due to the strong electron-donating ability of nine methoxyl groups, the essential structure of TPQP becomes much less susceptible to OH^- attack and the methoxyl group on benzene ring becomes the weak point and degrades first so that the overall alkaline stability of TPQP cation is enhanced. This finding indicates that by looking for other benzene substituent which not only has strong electron donating ability but also has sufficient stability in alkaline environment, the alkaline stability of phosphonium cation can be further improved.

Reference

1. Agel, E.; Bouet, J.; Fauvarque, J. F. *Journal of Power Sources* **2001**, *101*, 267.
2. Danks, T. N.; Slade, R. C. T.; Varcoe, J. R. *J Mater Chem* **2002**, *12*, 3371.
3. Gu, S.; Cai, R.; Luo, T.; Chen, Z. W.; Sun, M. W.; Liu, Y.; He, G. H.; Yan, Y. S. *Angew Chem Int Edit* **2009**, *48*, 6499.
4. Gu, S.; Skovgard, J.; Yan, Y. S. *Chemsuschem* **2012**, *5*, 843.
5. Gu, S.; Cai, R.; Yan, Y. S. *Chem Commun* **2011**, *47*, 2856.
6. Gu, S.; Cai, R.; Luo, T.; Jensen, K.; Contreras, C.; Yan, Y. S. *Chemsuschem* **2010**, *3*, 555.
7. Fenton, G. W.; Ingold, C. K. *J Chem Soc* **1929**, 2342.

374494 Polymerized Ionic Liquid Block Copolymers As Anion Exchange Membranes for Alkaline Fuel Cells

Abstract:

Alkaline fuel cells (AFCs) with hydroxide-conducting anion exchange membranes (AEMs) as solid-state electrolytes provide advantages over proton exchange membrane fuel cells, such as the potential to produce high power densities with non-noble metal catalysts. However, long-lasting AFCs require the synthesis of new AEMs that possess high hydroxide conductivity, high chemical and electrochemical stability and mechanical robustness. In this talk, we report on the synthesis and properties of several new polymerized ionic liquid (PIL) block copolymers as hydroxide-conducting AEMs for AFCs. PIL block copolymers combine the advantageous properties of both PILs and block copolymers. Specifically, the unique physiochemical properties of PILs, such as high solid-state ionic conductivity, high chemical, thermal, mechanical and electrochemical stability, and widely tunable physical properties, are incorporated in the block copolymer architecture, which allows for self-assembly into a range of nanostructures, where morphology type and domain size are tunable. The hydroxide conductivity, chemical stability, and morphology of the PIL block copolymers synthesized in this study will be discussed in relation to solid-state AEMs for AFCs.

372017 Electrolyte Membranes for Pemfcs and Solid Alkaline Fuel Cells Using Self-Assembly and Pore-Filling Technologies

Abstract:

Proton exchange membrane fuel cells (PEMFCs) are promising energy devices for residential and motor vehicle applications, largely because of the advantages as high conversion efficiency at low temperature along with no exhaust of carbon dioxide. One of requirements is to simplify the humidification device of the PEMFC electrolyte membranes that provide high proton conductivity under low relative humidity conditions. Solid-state alkaline fuel cells (SAFCs) with the benefits of using non-noble metal catalysts and liquid fuels can directly solve the problems that PEMFCs have. For both of fuel cells, developments of membrane electrolytes are key to achieve the success.

Pore-filling electrolyte membrane is a unique membrane that can suppress the swelling of filled polymer electrolyte because of its rigid porous substrate. Protons can rapidly conduct through this structure without further swelling although fuels cannot transport through the membrane. Also, we found a distinctive phenomenon for proton conduction by generating packed acid structure by using polymer and inorganic electrolytes, and the membrane showed high proton conduction under extremely low humidity condition. The phenomenon is explained by a "packed acid mechanism", proposed after applying *ab initio* molecular simulations. A proton normally shuttles between the proton donors and acceptors, which disrupts proton conduction (pseudo-shuttling). The acid–acid interaction in the packed acids eliminates "pseudo-shuttling" and leads to successive proton conductions. Those results will be discussed.

The anion exchange membranes should be developed to improve the performance of SAFCs and require following properties: high OH^- conductivity, low fuel permeability and thermal stability in alkaline circumstances. To achieve all these goals of the membranes, novel anion exchange membranes have been developed using self assembly technologies, and the nano-sized structures and distance between the ion-exchange groups were designed. Several types of anion exchange membranes will be shown and those performances will be discussed.

390393 The Effects of Confinement and Hydrophobicity on the Transport and State of Protons in PFSA Membranes

Abstract:

Perfluorosulfonic acid (PFSA) based membranes consist of a hydrophobic poly(tetrafluoroethylene) (PTFE) backbone with hydrophilic sulfonic acid-terminated perfluoroether pendant side chains. These materials rely heavily on the absorption of water to conduct protons. Hydration results in the formation of hydrophilic domains through solvation and aggregation of the sulfonic acid groups. Solvation of the acid groups leads to proton dissociation into the aqueous domain which facilitates long-range proton transport through the membrane. High proton conductivity in PFSA membranes is only observed at high levels of hydration and this has led to efforts to develop membranes with high proton conductivity at lower hydration allowing for higher temperature operation. The hydrophilic domains are typically only a few nanometers in diameter and hence the structural and dynamical properties of both the water molecules and the protons differ significantly from their properties when in bulk water. Furthermore, these properties are also influenced by the sulfonic acid group density, the level of hydration, and the hydrophobicity exhibited by the PTFE backbone. Proton transport in these materials is highly dependent on the structure and dynamics of hydrogen bonds, which are considerably impacted by nanoscale confinement. As such, an understanding of how the confined environment affects the nature of confined water is also needed.

Hence, to systematically elucidate all these factors (degree of confinement, acidity, water content, and hydrophobicity) on the transport properties in PFSA membranes we have undertaken *ab initio* molecular dynamics simulations of model systems consisting of single wall carbon nanotubes. This talk will describe our ongoing work to unravel the surprising and yet provocative results in these idealized systems.

Effect of Mechanical Deformation on Structure and Transport in Nafion 117 from Molecular Simulations

Khaldoon Abu-Hakmeh and Seung Soon Jang

Computational Nano Bio Technology Laboratory, School of Materials Science and Engineering, Georgia Institute of Technology, 771 Ferst Drive NW, Atlanta, GA 30332-2045

Nafion is commonly used in polymer electrolyte membrane fuel cells (PEMFC) due to its high proton conductivity and thermal, chemical, and mechanical stability. However, high fuel crossover and cost, along with reduced performance under low humidity and high temperature conditions, have dampened enthusiasm towards Nafion. Studies of Nafion have heavily focused on the relationship between nanophase-segregated structure and transport, revealing that proton mobility is enhanced in more developed nanophase-segregated structures. Recently, hydrated Nafion membrane was uniaxially stretched in experiment, which showed an improvement of transport properties, including higher power densities and reduced methanol crossover. In this study, mechanical properties of Nafion membrane are investigated using molecular dynamics simulation methods. Specifically, the deformation of Nafion is conducted with various water contents to characterize its effect on the structure and transport of hydrated Nafion membrane. First, Nafion 117 membrane is equilibrated at various hydration levels through an annealing procedure. The simulated characteristic correlation length and the diffusion coefficient of water and hydronium are analyzed for comparison with those observed in experiments. Then, the equilibrated Nafion membrane is deformed uniaxially with a constant strain rate up to 300% of strain. The change in nanophase-segregation of hydrated Nafion during the deformation process is characterized using characteristic correlation length as well as the pair correlation function to achieve a fundamental understanding of the relationship of the structural change as a function of strain. We report increased characteristic correlation length and water transport following mechanical deformation.

Multiscale modeling of proton transport in novel PEM fuel cell membranes

Marshall T. McDonnell[1] , Jimmy W. Mays[2,3] and David J. Keffer[4,]*

ABSTRACT

Multiscale modeling techniques are applied to polymeric systems for use as membranes in proton exchange membrane fuel cells (PEMFC). These systems include the block copolymer of cross-linked, sulfonated polycyclohexadiene (xsPCHD) and polyethylene glycol (PEG) and a blend polymer of xsPCHD/PEG. The current industry standard proton exchange membrane, Nafion, and membranes of similar polymer chemistry, have shown a high correlation between charge transport and water transport. The xsPCHD/PEG polymer systems have shown the potential to decouple the charge transport and water transport, promoting increased proton conductivity at high temperatures and low relative humidity. Our modeling tools include classical molecular dynamics simulation, confined random walk theory, percolation theory and reactive molecular dynamics simulation. We use these materials modeling techniques to infer a mechanism for charge transport in order to understand the increase in charge transport and its weak correlation to water transport. Also, we aim to understand and predict what the structure, transport, and conductivity of these systems will be with varying xsPCHD and PEG concentration.

[1]Department of Chemical and Biomolecular Engineering, University of Tennessee, Knoxville, TN 37996-2100

[2]Department of Chemistry, University of Tennessee, Knoxville, TN 37996-2100

[3]Chemical Sciences Division, Oak Ridge National Laboratory, Oak Ridge, TN 37381

[4]Department of Material Science and Engineering, University of Tennessee, Knoxville, TN 37996-2100

373558 Functionalization of Common Non-Selective Polymer to Create Hydrophilic Membrane

Abstract:

Many industrial situations require the transfer of water to or from a process stream. One such example is the dehydration of organic solvent streams to meet dryness specifications. The membrane processes of pervaporation and vapor permeation can be applied in this situation with a suitable hydrophilic membrane material. In this work, a common rubbery polymer was functionalized and cross-linked to form a robust water-selective layer. The performance of the membrane was demonstrated in pervaporation and vapor permeation tests with ethanol-water mixtures. Pervaporation temperatures ranged from 40 to 60 degC while the vapor permeation temperature was 115 degC. Tests of up to 40 days were completed. Concentrations in the pervaporation tests ranging from 99.5/0.5 w/w ethanol/water to 100% water while the vapor permeation tests were carried out at concentrations of 80/20 and 90/10 w/w ethanol/water. Water-ethanol molar permselectivity of the unfunctionalized material was only about 5. Functionalization was able to increase water selectivity by 10-fold and increase water permeability by over 100%. Details regarding the material and functionalization procedures will be discussed.

389886 Control of Microstructure in Polymer Blend Membranes

Abstract:

immiscible polymer blends based on a selective polymer and a highly permeable polymer were prepared as flat membranes and tested for gas separations. Additives such as metal organic frameworks (MOF) function as compatibilizers and were demonstrated to have control over the membane microstructure where the MOF resides at the interface. SEM,TEM amd AFM show a droplect morphology that changes as a function of MOF loading.

Novel hydroxyl functionalized polytriazole-co-polyoxadiazole as substrates for forward osmosis membranes

Phuoc H.H. Duong[1], Stefan Chisca[1], Suzana P. Nunes[1*], Tai-Shung Chung[1,2†]

[1]Water Desalination & Reuse Center (WDRC), King Abdullah University of Science and Technology, 23955-6900 Thuwal, Saudi Arabia

[2]Department of Chemical and Biomolecular Engineering, National University of Singapore, 4 Engineering Drive 4, Singapore 117585, Singapore

Corresponding authors:

[*]Prof. Suzana P. Nunes, e-mail address: suzana.nunes@kaust.edu.sa, tel.: +966 (1) 2808 2771, fax: +966 (1) 2802 0881

[†]Prof. Tai-Shung Chung, e-mail address: chencts@nus.edu.sg, tel.: +65 6516 6645, fax: +65 6779 1936

Abstract

Polyoxadiazoles (PODs) have been demonstrated as highly thermal stable and chemical resistant materials for porous membrane fabrications. However, POD membranes may not be suitable for water treatment applications due to the hydrophobic nature of PODs with water contact angle of >110°. To overcome the hydrophobicity limitation of POD membranes while utilizing their thermal and chemical advantages, hydroxyl functionalized polytriazole-co-polyoxadiazole (PTA-POD) copolymers have been used as substrates for fabricating of thin-film composite (TFC) forward osmosis (FO) membranes. The roles of PTA/POD ratios in the resultant membrane substrate properties and the FO membrane performance have been investigated. This study demonstrates that the copolymer containing 40 % mol PTA is optimal for the forward osmosis substrate fabrication among different PTA degrees of PTA-POD copolymers. Compared to the

POD FO membrane, the 40 % mol PTA FO membrane exhibits remarkable decrease in structural parameter (S) of more than 3.3 times. In addition, the 40 % mol PTA FO membrane shows high water fluxes of 24.9 LMH and 47.2 LMH using 1 M NaCl as the draw solution and DI water as feed under FO and pressure retarded osmosis (PRO) modes respectively. Hence, the use of newly developed hydroxyl functionalized polytriazole-co-polyoxadiazole copolymers for FO processes may open up a new generation of material for membranes in water applications.

378058 Ultrafiltration Membranes Composed of Cellulose Acetate and N-Isopropylacrylamide (NIPAAm)

Abstract:

Ultrafiltration Membranes Composed of Cellulose Acetate and N-Isopropylacrylamide (NIPAAm)

Membrane fouling occurs when there is reversible or irreversible attachment of macrosolutes present in the water to the membrane surface. Membrane replacements due to fouling can add to the operating costs of membrane systems. Reversible fouling can be minimized by crossflow operation, backflushing and/or by chemical pretreatment. On the other hand, irreversible fouling cannot be corrected by backflushing or crossflow filtration, and it results in permanent flux decline. Among irreversible foulants, natural organic matter (NOM) is considered to be a major contributor. NOM is composed of a wide range of hydrophilic and hydrophobic components; hence, any static hydrophobic or hydrophilic membrane can get fouled. Therefore, a dynamic membrane able to alternate between being hydrophobic and hydrophilic is hypothesized to decrease fouling. Stimuli-sensitive polymers can change their conformation from a coiled and hydrophobic state to an uncoiled (or globular) and hydrophilic state in the presence of a stimulus. By using such polymers to make a membrane and then alternating between the phases, the membrane surface can be made dynamic.

The purpose of this study was to develop stimuli responsive membranes to control fouling using membrane casting dopes made of cellulose acetate with N-isopropylacrylamide (NIPAAm). NIPAAm is a stimuli-responsive polymer, which offers the potential to reversibly collapse or expand the membrane as a function of changes in temperature. As a response to a temperature decrease, NIPAAm expands into a hydrophilic state, while a temperature increase causes it collapse into a less hydrophilic state. The phase change arises from the existence of a lower critical solution temperature (LCST) such that the polymer precipitates from aqueous solution as the temperature is increased. By continuously activating the film, it is hypothesized that a dynamic stimuli-responsive surface can prevent foulants from attaching to the membrane surface.

Membranes were cast using phase inversion, and then were characterized chemically, morphologically, and were used in filtration experiments using bovine serum albumin (BSA) and Lipase. Flux studies were conducted at alternating cold and hot temperature cycles, which confirmed the temperature activation of CA-NIPAAm membranes as compared to pure cellulose acetate membranes. Protein rejection was higher in case of CA-NIPAAm membranes. CA-NIPAAm membranes offered an average higher flux during operation and lower protein accumulation on the membrane surface as compared to regular CA membranes. CA-NIPAAm membranes showed higher flux recovery than CA membranes.

Evaluation of gas exchange in hollow-fiber membranes for artificial lungs

Maryam Asgharpour, Neda Mahmoudi, Shannon Servoss, Jamie A. Hestekin

Ralph E. Martin Department of Chemical Engineering, University of Arkansas, Fayetteville, AR 72701, United States

Abstract

There is a substantial need to develop a new design for an artificial lung with enhanced O_2 and CO_2 transfers due to respiratory diseases and also lack of lung donors. The main purpose of this paper is to further improve polymeric hollow-fiber membranes (HFMs) to increase its stability and gas permeability in the artificial lung applications. In this work, polysulfone hollow-fiber membranes have been utilized as the interface between gas pathways and blood. Additionally, a solution of PBS has been used as the liquid phase substituting blood. Oxygen diffuses into PBS through the fibers while CO_2 permeates in a reverse pathway. Hollow-fiber membranes of different concentrations of polysulfone and peptoid, a molecule to reduce bio-fouling, have been used in these applications.

Keywords

Artificial lung, Carbon dioxide, Gas exchange, Hollow-fiber membranes, N-methylpyrrilidone (NMP), Polysulfone.

14AIChE
Annual Meeting, Atlanta, GA

Thin-film Composite Membranes with Modified Polyvinylidene Fluoride Substrate for Ethanol Dehydration via Pervaporation

Yu Zhang[1], Ngoc Lieu Le[2], Yan Wang[3,*]

[1]NUS Graduate School for Integrative Science and Engineering, National University of Singapore, 28 Medical Drive, Singapore, 117456
[2]Department of Chemical and Biomolecular Engineering,National University of Singapore, 4 Engineering Drive 4, Singapore, 117576
[3]School of Chemistry and Chemical Engineering, Huazhong University of Science & Technology, Wuhan, China, 430074

NUS National University of Singapore

Outline

- Introduction
- Thin-film composite membrane preparation
- Results & Discussions
  - Effect of PVDF substrate surface modification
  - Effect of substrate morphology
  - Effect of silicone rubber coating
  - Long-term stability study
- Conclusion
- Acknowledgement

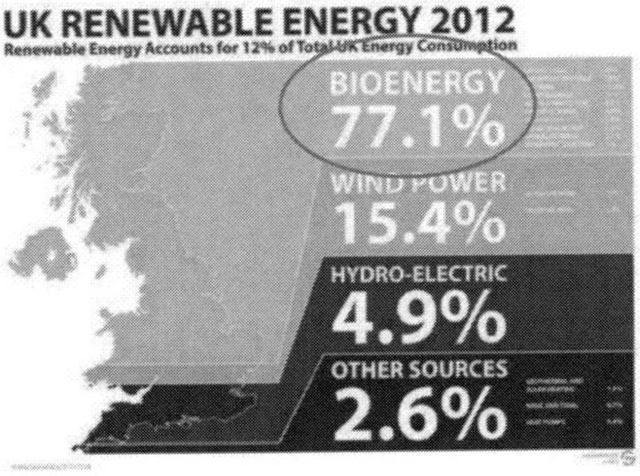

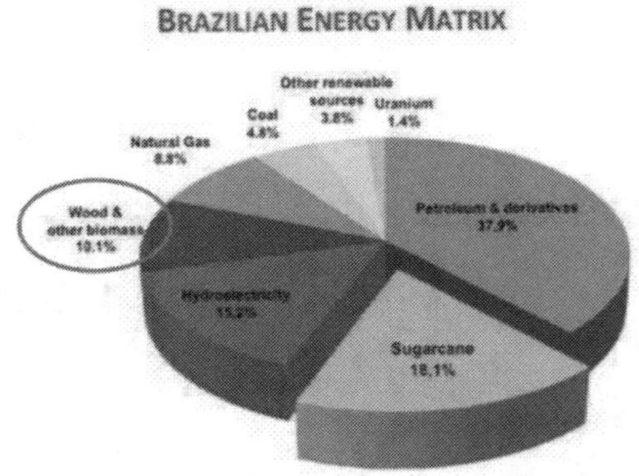

U.S. Energy Consumption by Energy Source, 2011

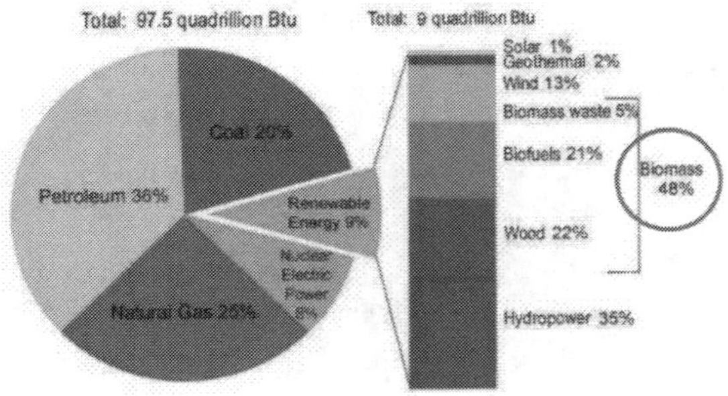

Biofuel is promising renewable energy resources addressing the world energy depletion issues

11/19/2014

Separation process accounts for 40% to 80% processing cost

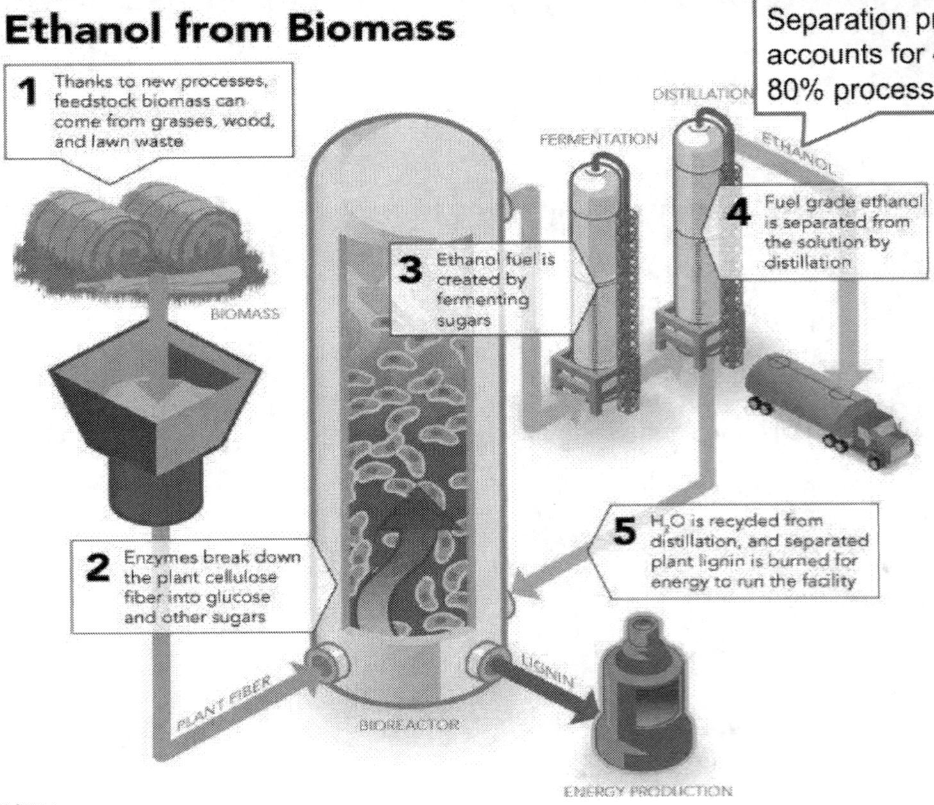

11/19/2014

4

Membrane pervaporation

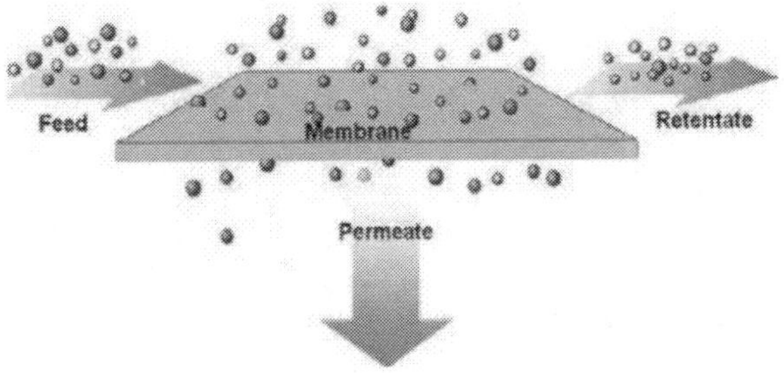

- Pervaporation is the **separation of liquid mixture** by **partial vaporization** via a **membrane**
  - High efficiency
  - Energy saving
  - Easy fabrication and operation

- Biofuel applications
  - Dehydration
  - Recovery
- Membrane type
  - Symmetric membrane
  - Asymmetric membrane
- **T**hin-**f**ilm **c**omposite membrane
  - Cost saving
  - Independently controlled top and bottom layer
- Performance indicator
  - Flux $J = Q/At$
  - Separation factor $\beta_{i/j} = \frac{y_i/y_j}{x_i/x_j}$

TFC membrane

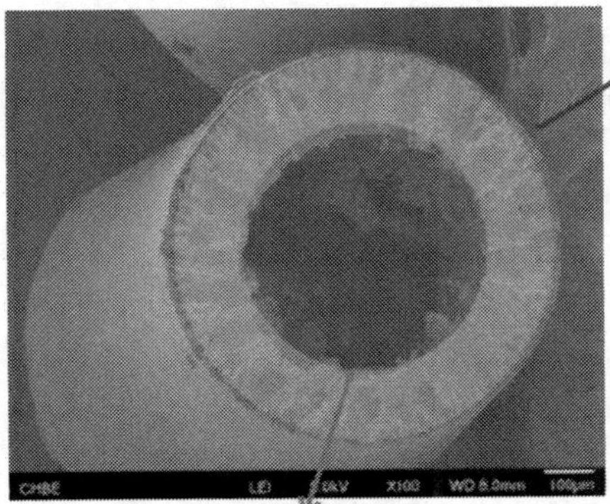

Polyamide selective layer via interfacial polymerization

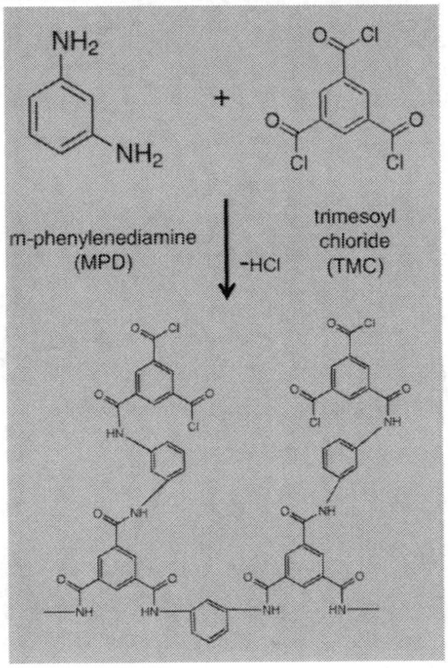

Substrate

PVDF substrate
- Low cost and easy preparation
- High thermal, chemical and mechanical stabilities

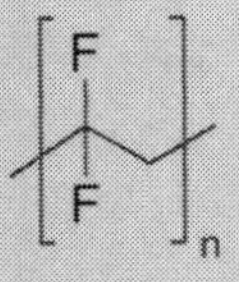

Disadvantages
- Highly hydrophobic, unreactive to aqueous solution in IP process

TFC membrane preparation

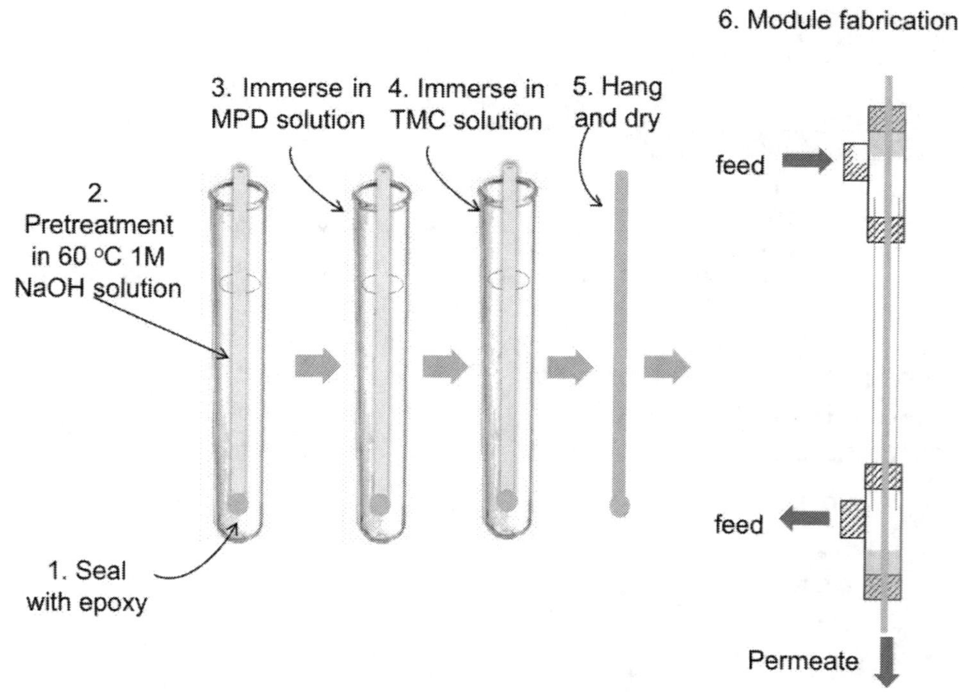

11/19/2014

7

Thin-film Composite Membranes with Modified Polyvinylidene Fluoride Substrate for Ethanol Dehydration via Pervaporation

PVDF surface modification

Mechanism:

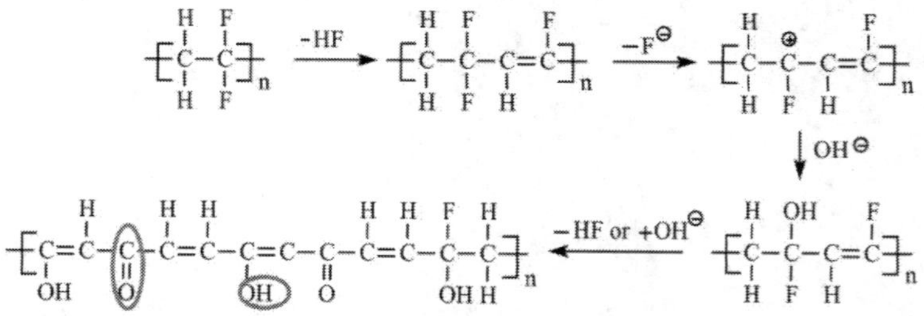

Membrane	PVDF	PVDF+ NaOH
F_{1s}/O_{1s} mass ratio from XPS	5.7	1
Water contact angle (°)	85.3	56.6

Characterizations:

XPS

Water contact angle

FTIR

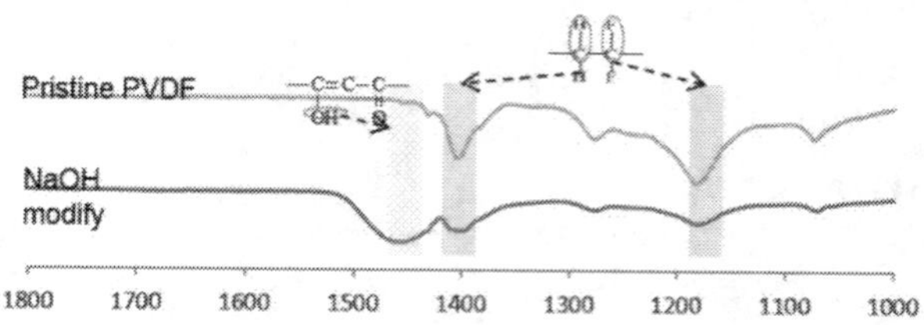

Brewis et al., 1996; Ross et al., 2000

11/19/2014

8

Effect of alkaline surface modification

Time in NaOH solution (h)	MPD conc. (wt%)	TMC conc. (wt%)	Flux (g/m^2h)	Permeate water conc. (wt%)	Separation factor (water/ethanol)
0	N.A.	N.A.	1053	34.0	3
0	2	0.1	742	55.1	7
2	2	0.1	408	76.9	20
4	2	0.1	692	83.4	30
6	2	0.1	Pervaporation tests couldn't be carried due to severe leakage.		
8	2	0.1			

Feed: 85/15 wt% ethanol/water.
Operation temperature: 50 °C.

- Membrane performance ***increases*** with the ***increase*** of modification ***duration*** within 4 hours.
- ***Prolonged treatment*** may degrade the membrane surface morphology.

Thin-film Composite Membranes with Modified Polyvinylidene Fluoride Substrate for Ethanol Dehydration via Pervaporation

Effect of alkaline surface modification

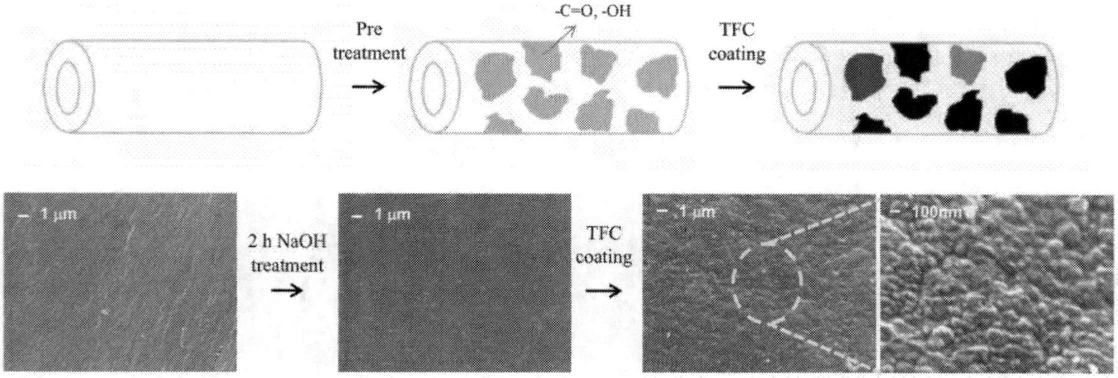

- A short alkaline treatment
  - ***inhomogeneous*** growth of hydrophilic groups on the substrate surface
  - lead to an ***incompletely*** covered polyamide layer

Effect of alkaline surface modification

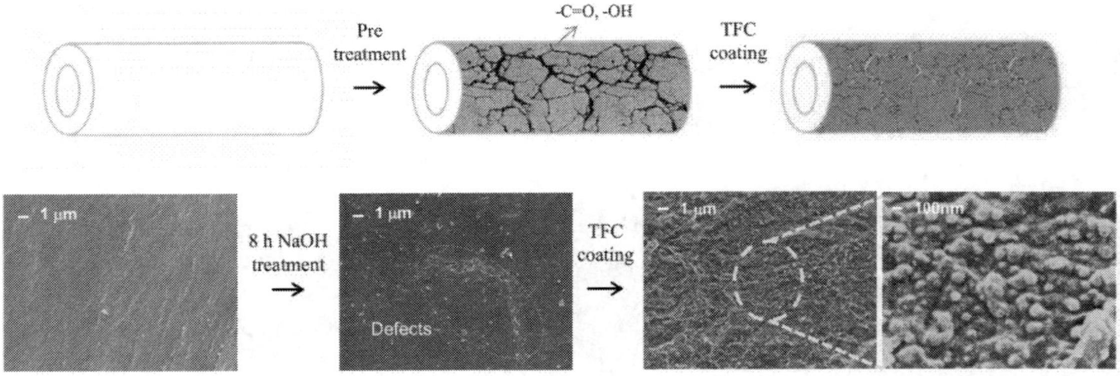

- A prolonged alkaline treatment
  - ***degrade*** the substrate morphology
  - lead to an ***leakage*** during pervaporation test

Thin-film Composite Membranes with Modified Polyvinylidene Fluoride Substrate for Ethanol Dehydration via Pervaporation

Effect of alkaline surface modification

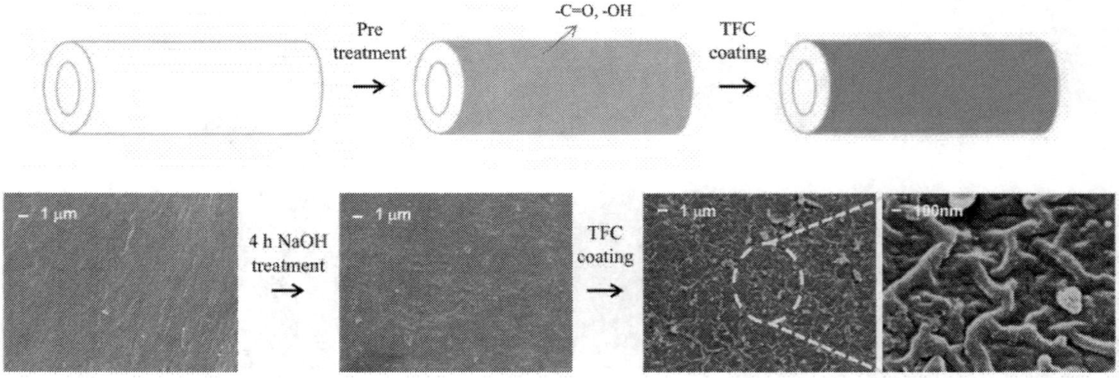

- A ***proper duration*** of alkaline treatment is essential for the formation of a homogeneous polyamide selective layer
  - ***best*** pervaporation performance
  - a typical "ridge-and-valley" morphology

Effect of substrate morphology

NUS National University of Singapore

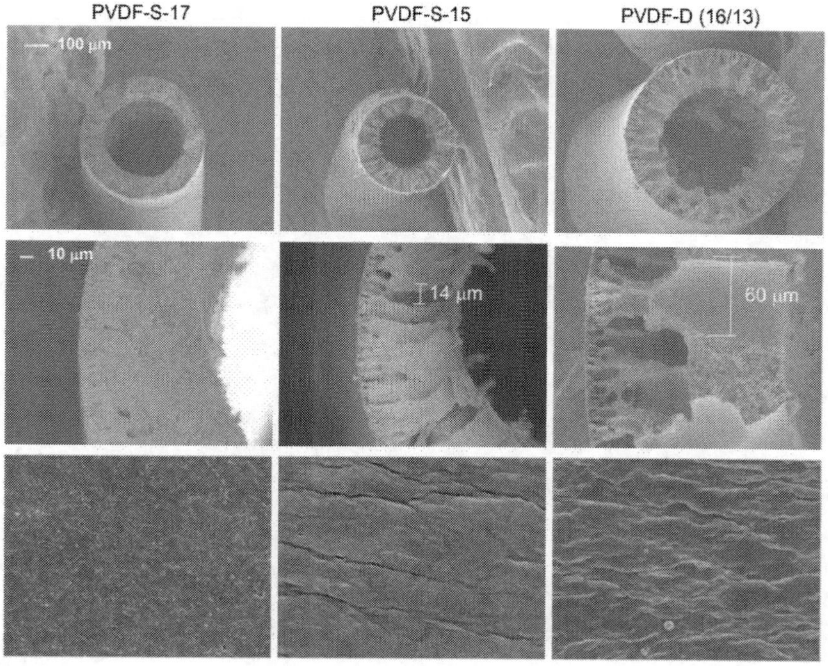

Thin-film Composite Membranes with Modified Polyvinylidene Fluoride Substrate for Ethanol Dehydration via Pervaporation

Effect of substrate morphology

NUS National University of Singapore

Substrate	Time in NaOH solution (h)	Flux (g/m^2-h)	Permeate water conc. (wt%)	Separation factor
PVDF-S-17	4	692	83.4	30
PVDF-S-15	4	2830	39.9	3
PVDF-D	4	3700	54.8	6
PVDF-D	6	3342	62.2	10

Feed: 85/15 wt% ethanol/water.
Operation temperature: 50 °C.

- A higher separation factor but a lower flux for PVDF-S-17 is caused by the ***denser membrane morphology*** mainly resulting from a higher dope concentration.
- The dual-layer PVDF-D hollow fiber may ***mitigate this trade-off*** and exhibit both a ***high flux and a reasonable separation factor***.

NUS National University of Singapore

Effect of silicone rubber coating

Membrane	No. of silicone coating	Time in silicone solution (min)	Flux (g/m^2h)	Permeate water conc. (wt%)	Separation factor (water/ethanol)
PVDF-D	1	3	9520	21.6	1
TFC membrane	1	3	2164	75.7	22
	1	4	1964	77.9	29
	2	3	1288	85.3	40

Feed: 85/15 wt% ethanol/water.
Operation temperature: 50 °C.

- Silicone rubber coating can significantly ***enhance*** the TFC membrane ***selectivity***
  - prevent the direct contact of polyamide layer with liquid feed
  - alleviate extra defect formation

NUS National University of Singapore

Effect of silicone rubber coating

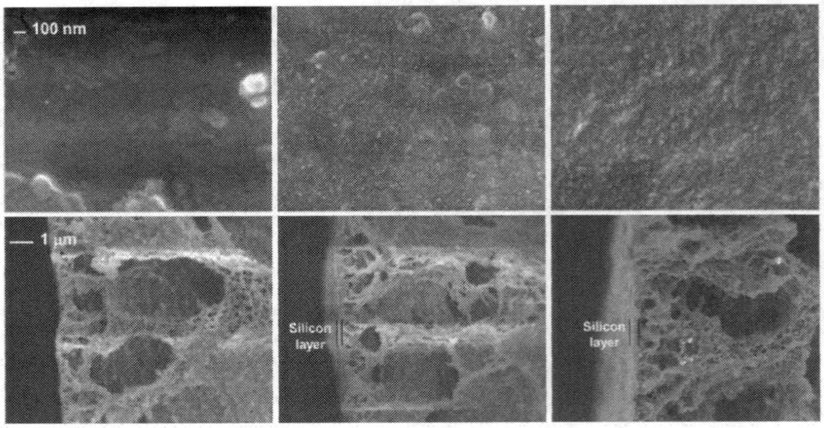

- The silicone layer in the TFC membrane with double silicone rubber coating is obviously ***thicker*** compared with the membrane with single silicone rubber coating

Long-term stabilities

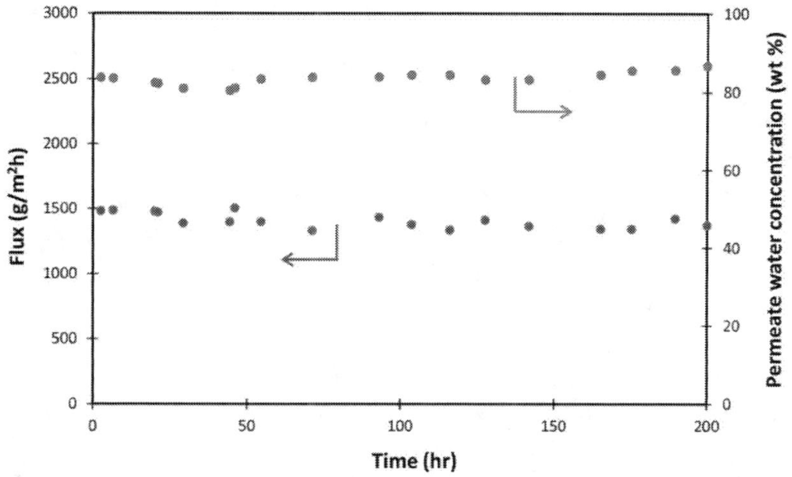

- 200-hr pervaporation test indicates the ***high stability*** of the developed TFC membrane

Thin-film Composite Membranes with Modified Polyvinylidene Fluoride Substrate for Ethanol Dehydration via Pervaporation

Conclusion

- Alkaline surface modification has been investigated to ***hydrophilically modify*** PVDF substrates.
- The dual-layer hollow fiber membrane is the most desirable morphology for the formation of TFC membranes because of its ***dense outer surface*** and ***highly porous inner layer.***
- Silicone rubber coating on the TFC membranes has been proven effectiveness to ***seal membrane defects*** and ***enhance pervaporation performance***.
- This developed membrane has ***high stabilities*** against long term operation.

Acknowledgement

- Singapore National Research Foundation
- Huazhong University of Science and Technology
- National Science Foundation of China
- Prof Chung's membrane group
- NUS Graduate School for Integrative Science and Engineering

NATIONAL RESEARCH FOUNDATION

NUS National University of Singapore

Thin-film Composite Membranes with Modified Polyvinylidene Fluoride Substrate for Ethanol Dehydration via Pervaporation

Thank you!

Q&A

Effect of alkaline treatment on mechanical property

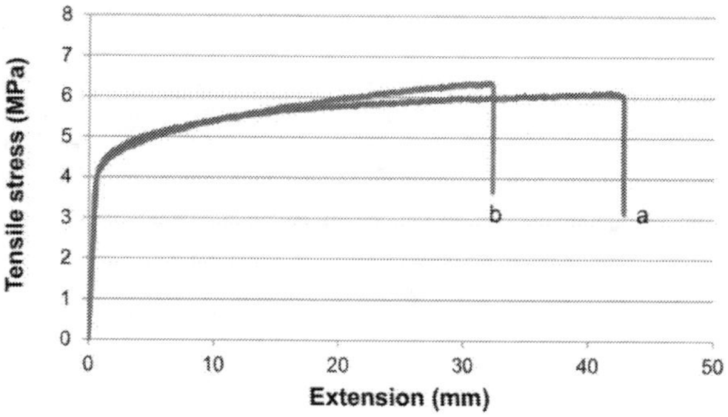

Fiber	Elongation at break (%)
Pristine PVDF	430
PVDF fiber with alkaline treatment	320

11/19/2014

21

Hollow fiber spinning parameter

Spinning parameters	PVDF-S-17 Single-layer	PVDF-S-15 Single-layer	PVDF-D Dual-layer
Outer dope composition (wt%)	--	--	PVDF/NMP 16/84
Inner dope composition (wt%)	PVDF/DMAc 17/83	PVDF/NMP 15/85	PVDF/NMP 13/87
Bore fluid composition (wt%)	DMAc/water 70/30	NMP/water 90/10	NMP/water 95/5
External coagulant	Water	Water	Water
Outer dope flow rate (ml/min)	--	--	1
Inner dope flow rate (ml/min)	2	2	2
Bore fluid flow rate (ml/min)	0.6	0.6	1
Air gap (cm)	5	2	2.5
Take-up speed (m/min)	5	5	5
Spinneret temperature (℃)	25	25	25

11/19/2014

22

NUS National University of Singapore

Interfacial polymerization optimization

Substrate	Time in NaOH solution (h)	MPD conc. (wt%)	TMC conc. (wt%)	Time in silicone solution (min)	Flux (g/m²h)	Permeate water conc. (wt%)	Separation factor (water/ethanol)
PVDF-D	N.A.	N.A.	N.A.	3	9520	21.6	1
PVDF-D	6	2	0.1	3	1951	71.3	19
PVDF-D	6	2	0.15	3	1889	72.5	21
PVDF-D	6	3	0.1	3	2164	75.7	22
PVDF-D	6	3	0.15	3	2204	68.8	16

Feed: 85/15 wt% ethanol/water.
Operation temperature: 50 °C.

NUS National University of Singapore

Bench marker

Membrane	Operation temp. (°C)	Feed water conc. (wt%)	Flux (g/m²h)	Separation factor (water/ethanol)	Refs.
PVA/PAN dense membrane	60	13	200	127	Schehlmann et al. (1995)
PSU hollow fiber	25	10	173	24	Tsai et al. (2002)
Chitosan dense membrane	60	10	250	105	Zhang et al. (2007)
PVA dense membrane	60	10	120	115	Gimenes et al. (2007)
6FDA-ODA-NDA/Ultem® dual-layer hollow fiber	50	15	479	162	Widjojo and Chung (2009)
Torlon®/Ultem® dual-layer hollow fiber	50	15	659	50	Wang et al. (2009)
PVDF/Ultem® dual-layer hollow fiber	50	15	332	71	Ong and Chung (2012)
PAAS/modified NaA zeolite dense membrane	30	10	533	436	Wei et al. (2013)
Modified PES/PVA composite membrane	60	20	444	13	Villagra Di Carlo and Habert (2013)
TFC polyamide with mPAN substrate	25	10	2425	15	Fathizadeh et al. (2013)
Thin film nanocompostie with mPAN substrate	25	10	4500	30	Fathizadeh et al. (2013)
TFC hollow fiber with PVDF substrate	50	15	1288	40	This work

Design and Fabrication of Thin-Film Composite Tri-bore Hollow Fiber (TFC TbHF) Membranes for pervaporative Isopropanol Dehydration

Dan Hua [a,b], Yee Kang Ong [b], Peng Wang [b], Tai-Shung Chung [a,b]*

[a] NUS Graduate School for Integrative Science and Engineering, National University of Singapore, Singapore 117456, Singapore

[b] Department of Chemical & Biomolecular Engineering, National University of Singapore, Singapore 117585, Singapore

* Corresponding author. Tel.: (+65) 6516 6645; fax: (+65) 6779 1936.

E-mail address: chencts@nus.edu.sg (T. S. Chung).

Abstract

Novel outer-selective thin-film composite tri-bore hollow fiber (TFC TbHF) membranes were designed and fabricated for pervaporative dehydration of isopropanol (IPA) in this work. The traditional single-bore hollow fiber (SbHF) was first used to understand the effect of interfacial polymerization conditions on pervaporation performance and to identify the most suitable procedure for fabricating the TFC TbHF membranes. Then the effects of spinning parameters (i.e. bore fluid composition, bore fluid flow rate, air gap and dope flow rate) on the geometry, morphology and pervaporation performance of TFC TbHF membranes were systematically investigated. Results show that the bore fluid composition is the most critical factor. Moreover, fine TFC TbHF membranes (outer diameter approximately 1 mm) exhibit better separation performance for IPA dehydration than thick TFC TbHF membranes (outer diameter approximately 2 mm). The optimal TFC TbHF membrane shows a satisfying separation performance with a flux of 2.65 kg m^{-2} h^{-1} and a separation factor of 246 for the dehydration of IPA/water (85/15 wt%) mixture at 50 °C. Most importantly, the newly designed TFC TbHF membrane not only shows enhanced separation performance and mechanical strengths compared with the conventional TFC SbHF, but also exhibits long-term stability, suggesting the TFC TbHF has a great potential in pervaporation applications.

Acknowledgements

The authors are grateful to National Research Foundation, Prime Minister's Office, Singapore for funding this research under its Competitive Research Program for the project entitled, "New Biotechnology for Processing Metropolitan Organic Wastes into Value-Added Products" (CRP Award No. NRF-CRP 5-2009-5 (NUS grant number R-279-000-311-281)). Special thanks are due to Dr. Shi Guimin, Dr. Zuo Jian, Dr. Zhou Zhengzhong and Mr. Tang Yupan for their valuable suggestions.

388866 Polymeric Membranes for the Filtration of Oil/Water Nanoemulsions

Abstract:

In recent years, growing environmental concerns have fueled the need for efficient oil and water separation systems. Membranes leveraging contrasting wetting properties have received significant attention for their ability to selectively permeate one component while rejecting the other. However, separation of droplets below a micron in size remains a key challenge. Examination of the wettability characteristics of a membrane gives rise to the concept of the breakthrough pressure - the maximum transmembrane pressure at which a membrane can operate retaining high selectivity. These separation principles are experimentally verified using a model system of through-pore membranes chemically engineered with desired wetting properties. To overcome the flux limitations imposed by the breakthrough pressure, we extend our findings to hierarchical polymeric membranes. A novel fabrication scheme is employed to control the skin layer thickness to optimize the permeability, and we demonstrate separation of droplets that are well below a micron in size in this system. By reducing the skin thickness layer by a factor of four, the permeability was improved significantly. These simple yet robust hierarchical membranes with engineered wetting characteristics show promise for the large-scale, efficient separation systems

Application of Numerical and Experimental Techniques for Verification and Validation in Cyclones

Evandro Balestrin, Oscar L. Sgrott Jr., Rodrigo K. Decker, Vinicyus R. Wiggers, Henry F. Meier
Regional University of Blumenau - Blumenau, Santa Catarina, Brazil
e-mail: evandrobalestrin@hotmail.com

Abstract

A verification and validation study of the gas flow in a cyclone was performed by comparing CFD (computational fluid dynamic) results with experimental data. The numerical analyses were developed by means of commercial and in-house CFD codes. The turbulent anisotropic behavior of the swirling flow in cyclones was modelled from both, the Reynolds Stress Model, and a hybrid model which combines the k-epsilon and Prandt models. The experimental data were performed with a Stereo Particle Image Velocimetry (PIV) system and the pressure drop in the cyclone was measured with a pressure probe. The measurements of the gas flow velocities were obtained in the cylindrical section, between the finder and the conical section, in terms of axial and tangential velocities. The boundary conditions for both numerical and experimental analyses were based on the inlet velocities from 8 to 16 m/s. The comparison between numerical results and experimental data showed a good agreement mainly for the axial velocity. Furthermore, the results showed that the modified cyclone presents a swirling downward flow coming from the finder duct to the cylindrical section. This behavior is not found in conventional cyclones and can be attributed to the geometrical effects of inlet and outlet. This behavior has not been discussed in the literature yet and it is presented as an alternative to increase the cyclone performance.

Keywords: Cyclones; Computational Fluid Dynamics; Verification & Validation

Introduction

Cyclones are simple devices that use centrifugal force to separate particles from gases. They are the most applied equipment in the industry for particles separation in the range from 5 to 200 μm. This preference occurs due to its high efficiency of collection and its large operating ranges for temperature and pressure, with low cost of building, operation and maintenance.

The cyclone performance is generally described in relation to pressure drop and collection efficiency. To better understand the pressure drop and particles separation in a cyclone, it is important to understand the gas flow in cyclones. An alternative way of studying the gas flow in cyclones is by solving the equations that govern it through numerical techniques, using CFD. The validation of results obtained with this technique occurs through experimental data, generally acquired with pressure probe, measurements of temperature, Stereo PIV, anemometry, among others.

However, to solve the equations that govern the flow in cyclones (Navier-Stokes equations) it is necessary to model the Reynods stress tensor using the turbulence closure model. In this study, two different turbulence closure models were applied, the Reynolds Stress Model (RSM-SSG) and a hybrid model, which combines the k-ε model, and the mixture Prandtl theory model (k-ε/Prandtl). Many of the cyclone CFD studies have been performed using these turbulence closure models, such as Peres et al. [1] and Noriler et al. [2] which studied the cyclone flow using the RSM model, and Vegini et al. [3], Meier et al. [4] and Sgrott Jr. et al. [5] which worked with k-ε/Prandtl.

Experiments

Physical experiments

Physical experiments were conducted in the test facility, illustrated in Figure 1. The physical experiments were performed in two steps, both of them working with a single phase flow. At first, pressure drop data were obtained through the pressure probe (04) where the pressure measurements were performed by ranging the cyclone inlet velocity from 8 to 16 m/s. In a second moment, by using a cyclone inlet velocity of 14 m/s and the particle image velocity system (Stereo PIV), which is composed of a PIV system (08) and a laser (03), experiments in order to obtain the velocity field were performed in the inner of the cyclone. For the PIV experiments, tracer particles with 3.89μm Sauter mean diameter and 1400kg/m³ density were loaded for the analysis of the gas phase velocity. In order to build an experimental database, the pressure drop experiments were repeated ten times, and the velocity filed experiments were performed three times.

The test facility consists of a set of tubes, where the gas phase flows until the Plexiglas cyclone (05), and then goes out from the cyclone toward to a bag filter (09). This system works with negative pressure provided by an exhauster (06) at the end of the line.

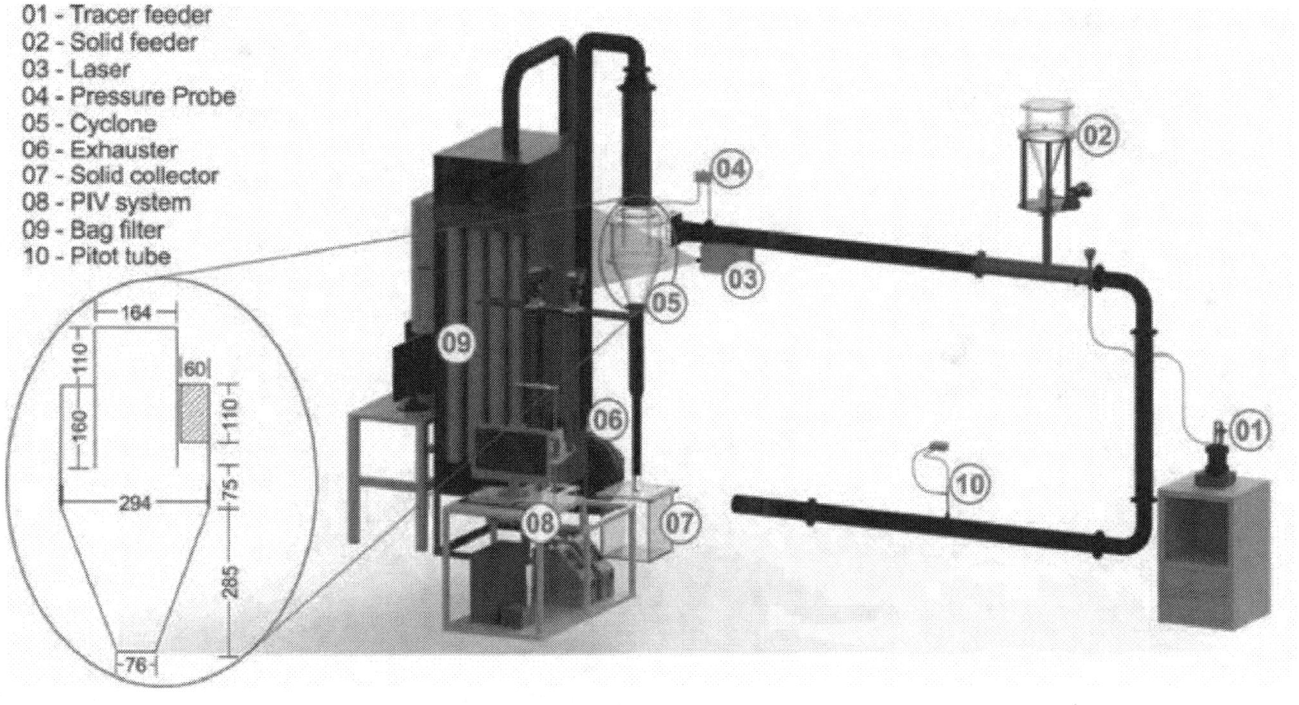

Figure 1: Test facility with cyclones dimensions in (mm).

Numerical simulations

The simulations were performed in two CFD codes, one of them an in-house code dedicated to cyclones and denominated CYCLO-EE$_5$, and another a commercial code called FLUENT provided by ANSYS.

In house code: based in an eulerian-eulerian model, with the possibility of working with up to five solid phases, each one with its own particle size, density and specific volumetric fraction. For turbulence closure, the code used an anisotropic hybrid model, which combines the k-ε model for radial and axial components of the Reynolds stress, and the mixture Prandtl theory for the tangential components. The finite volume method has been used to discretize the partial differential equations of the model, with SIMPLEC method for pressure-velocity coupling in axisymmetric domain with sttagered grid. This code was used to obtain the pressure drop of the gas phase, in order to validate de k-ε/Prandtl model by comparison with the experiments data. The simulation was performed using the cyclone dimensions, showed in Figure 1.

Commercial code: simulator with which it is possible to use various geometry configurations and fluid dynamics conditions. Therefore, two different geometry configurations were simulated, one of them with the complete experimental device (Figure 2 (a)), and another with only the Cyclone (Figure 2 (b)). In the simulations with this code, in both of the cases RSM-SSG model was used as turbulence closure and SIMPLEC method for pressure-velocity coupling. This code was used to obtain pressure drop data with the "Cyclone" geometry, and axial and tangential velocity profiles data with the "Test facility" geometry, in order to validate the RSM-SSG model making a comparison with experimental data.

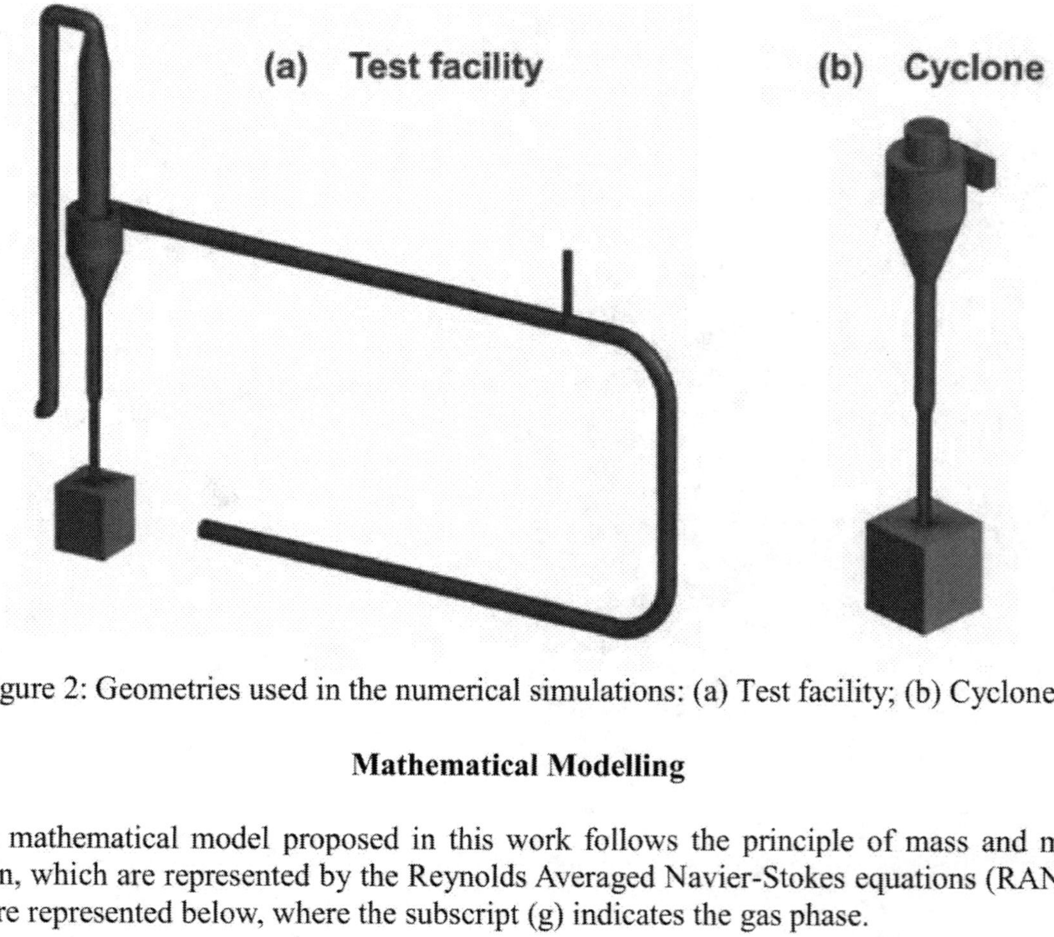

Figure 2: Geometries used in the numerical simulations: (a) Test facility; (b) Cyclone.

Mathematical Modelling

The mathematical model proposed in this work follows the principle of mass and momentum conservation, which are represented by the Reynolds Averaged Navier-Stokes equations (RANS). These equations are represented below, where the subscript (g) indicates the gas phase.

Mass conservation equation:

$$\frac{\partial}{\partial t}(\rho_g) + \nabla \cdot (\rho_g \boldsymbol{v}_g) = 0 \,. \tag{01}$$

Momentum conservation equation:

$$\frac{\partial}{\partial t}(\rho_g \boldsymbol{v}_g) + \nabla \cdot (\rho_g \boldsymbol{v}_g \boldsymbol{v}_g) = -\nabla \cdot (\boldsymbol{T}_g^{eff}) + \rho_g \boldsymbol{g} - \nabla p . \quad (02)$$

For closure of the turbulence two different models were used, the Reynolds Stress Model (RSM) and the hybrid model which combines the k- ε and Prandtl models (k-ε /Prandtl).

k-ε /Prandtl

The k-ε /Prandtl turbulence model is based on Boussinesq approximation to model the effective stresses tensor (similar the tensor for Newtonian fluid). In this case, the effective stresses tensor is directly proportional to the rate of deformation, and can be expressed as:

$$\mathbf{T}_g^{eff} = -2\mu_g^{ef}\mathbf{D}_g ; \quad (03)$$

where the effective viscosity (μ_g^{ef}), is represented by the sum of the molecular viscosity (μ_g) and the turbulent viscosity (μ_g^t):

$$\mu_g^{ef} = \mu_g + \mu_g^t . \quad (04)$$

As consequence of the anisotropy of the Reynolds stress, the axial and radial components are calculated with standard k-ε model and the tangential components with the Prandtl model. More information about this model and the multiphase model applied to cyclones flow, can be found in Meier et al. [4].

Reynolds Stress Model (RSM)

Unlike the k-ε /Prandtl model, the RSM modeling does not take into account the Boussinesq hypothesis of turbulent viscosity. This model closes the Reynolds averaged Navier-Stokes equations by solving transport equations for the Reynolds stresses, together with an equation for the dissipation rate. This means that seven additional transport equations are required to solve the tridimensional flows. Here the effective stresses tensor is expressed as:

$$\mathrm{T}_g^{eff} = \mathrm{T}^t{}_g + \mathrm{T}_g ; \quad (05)$$

where the Reynolds stress is modeled by RSM,

$$\frac{\partial}{\partial t}(\rho_g \boldsymbol{T}_g^t) + \nabla \cdot (\rho_g \mathbf{v}_g \boldsymbol{T}_g^t) = \mathbf{D_T} + \mathbf{D_M} + \mathbf{Y} + \mathbf{\Pi} - \frac{2}{3}\boldsymbol{\delta}\rho_g \varepsilon_g . \quad (06)$$

Among the various terms in the RSM equation, ($\mathbf{D_M}$) e ($\mathbf{Y}$) do not require any modelling. However ($\mathbf{D_T}$) and ($\mathbf{\Pi}$) need to be modelled to close the equations. The pressure strain ($\mathbf{\Pi}$) term can be defined in different ways. In this study, we utilized the quadratic pressure strain model proposed by Speziale, Sarkar and Gatski [6] which is referenced as RSM-SSG. More details about each term of the RSM can be found in (SHUKLA et al., [7]).

Numerical Methods

To solve the set of partial differential equations of the mathematical models, both the commercial and the in-house codes use the finite volumes method to discretize the equations. Table 1 shows the main parameters used to solve the models in both codes.

Table 1: Settings of the model conditions.

Model conditions	FLUENT	CYCLO-EE$_5$
Regime	Transient	Transient
Turbulence model	Reynolds Stress Model (RSM-SSG)	k-ε /Prandtl
Pressure-velocity coupling	SIMPLEC	SIMPLEC
Spatial discretization	- Pressure: PRESTO - Momentum: "Second Order Upwind" - Turbulent kinetic energy: "First Order Upwind" - Turbulent dissipation rate: "First Order Upwind" - Reynolds stresses: "First Order Upwind"	WUDS (Upstream Weighted Difference Scheme)

Results and Discussions

Grid test

The grid test is important to choose the more appropriated grid refinement in order to minimize the discretization errors. Two grid tests were performed, one of them with the commercial code and the other with the in-house code. Figure 3 (a) shows the grid test using the commercial code and in this case the axial velocity profile was analyzed as the function of three different grid concentrations (approximated 1, 2 and 4 million of points) with the inlet velocity of 12 m/s. Figure 3 (b) shows the grid test using the in-house code, in this case the pressure drop was analyzed as a function of five distinct grid refinements with inlet velocity of 10 m/s. It was decided to use the more refined grid in both cases.

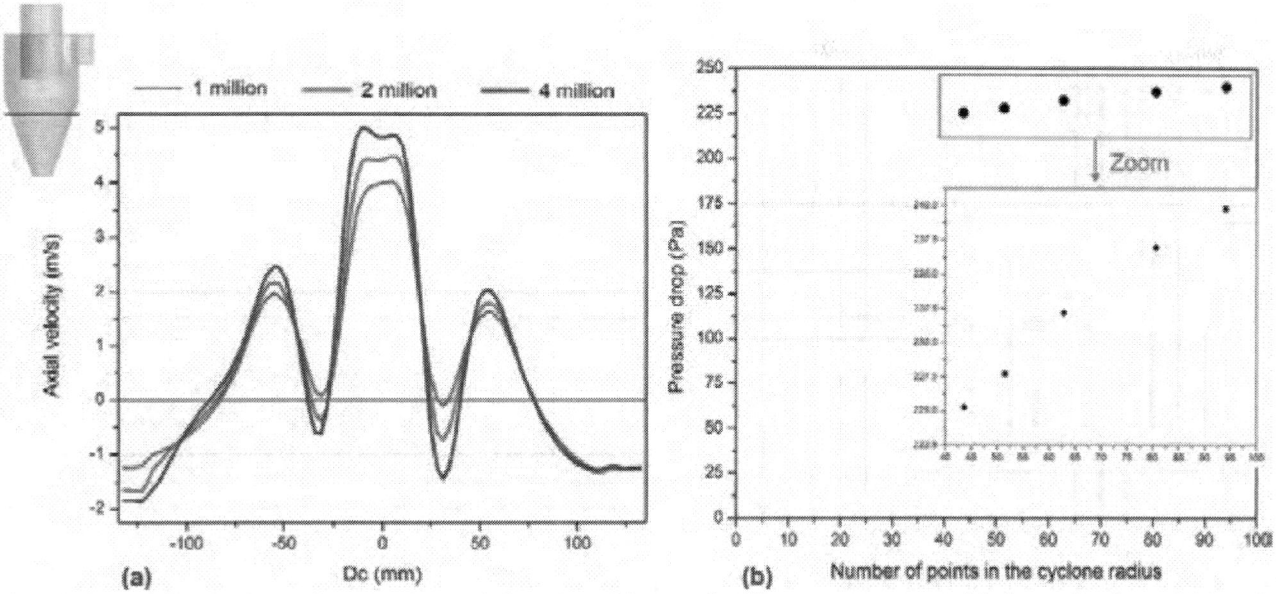

Figure 3: Grid test: (a) using commercial code with test facility geometry and inlet velocity of 12 m/s; (b) using the in-house code and inlet velocity of 10 m/s.

Validation of pressure drop prediction

Table 2 shows the average pressure drop obtained through ten experiments, the pressure drop simulated with in-house code (k-ε /Prandtl) and the pressure drop simulated with commercial code (RSM-SSG) for each velocity. This table also presents the absolute and relative errors.

Table 2: Experimental and numerical pressure drop results with the absolute and relative errors.

v(m/s)	Pressure drop (Pa)			Absolute Error (Pa)		Relative Error (%)	
	k-ε /Prandtl	RSM-SSG	Experimental	k-ε /Prandtl	RSM-SSG	k-ε /Prandtl	RSM-SSG
8	153.75	152.91	145.12	-8.63	-7.79	5.61	5.09
10	239.76	241.50	233.60	-6.16	-7.90	2.57	3.27
12	344.79	351.47	343.36	-1.43	-8.11	0.41	2.31
14	469.04	483.18	477.22	8.18	-5.96	1.74	1.23
16	612.22	636.97	632.64	20.42	-4.33	3.34	0.68

Figure 4 depicts a comparison of experimental and simulated pressure drop data, which was illustrated by two different types of graph. Therefore, based on the table and the graph results, it was concluded that the two turbulence closure models (k-ε/Prandtl and RSM-SSG) used in the study represent well the experimental pressure drop in cyclones. This conclusion can be affirmed because the relative error for velocity does not exceeded 5.61% in both models of closure turbulence, and because almost all simulated results of pressure drop was inside the error bar.

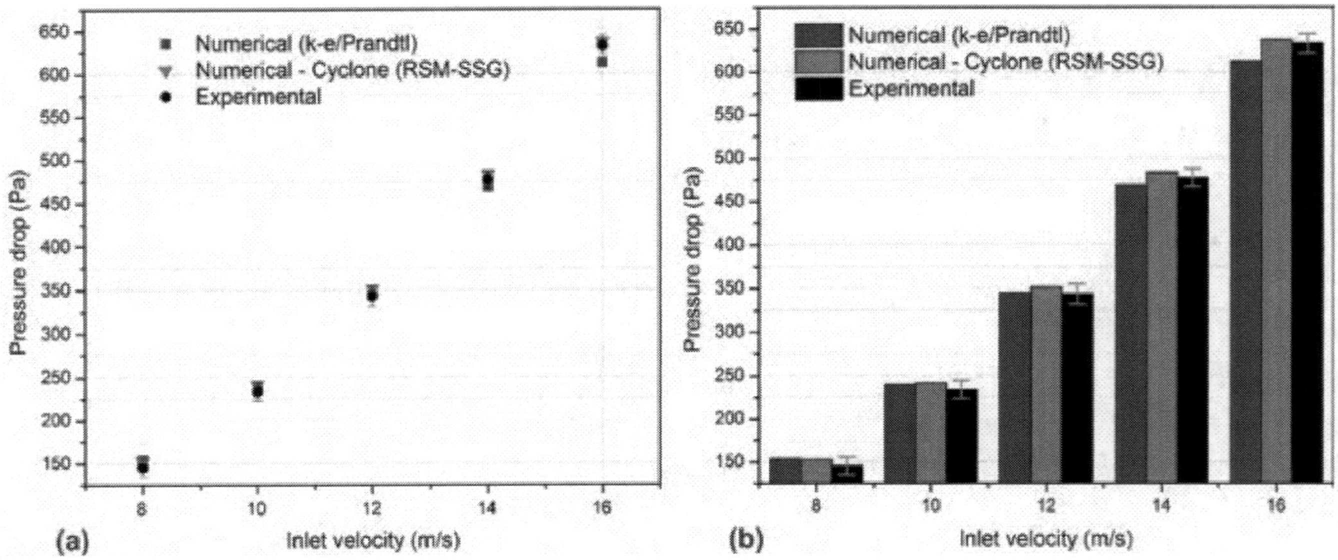

Figure 4: Comparison of experimental and numerical pressure drop results: (a) scatter plot of pressure drop; (b) bar graph of pressure drop.

Validation of the velocity profiles prediction

The experimental results of the velocity profile, acquired with the Stereo PIV system, and the numerical velocity profile, obtained through simulations performed in a commercial code with the test facility geometry are present in Figure 5, where the experimental bar represents the standard deviation. The data of velocity profile was obtained in a region located 40 mm downward of the vortex finder and with an inlet velocity of 14 m/s.

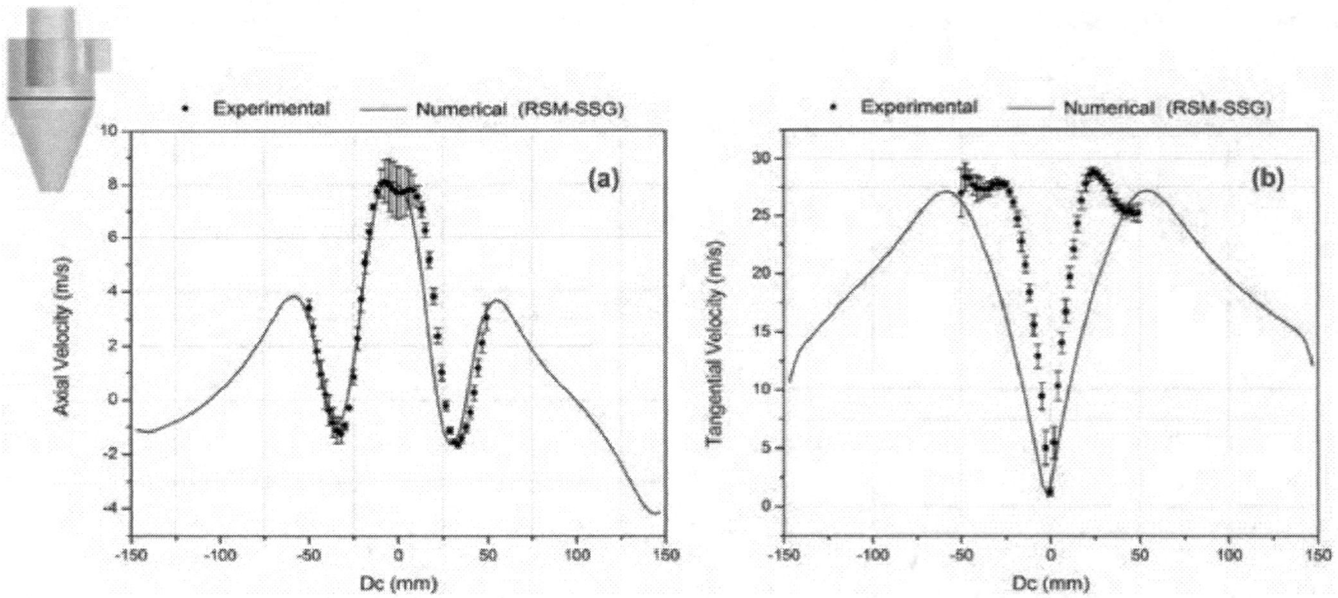

Figure 5: Comparison of experimental and numerical velocity profile with the inlet velocity in 14 m/s: (a) axial velocity; (b) tangential velocity.

Therefore, based on the comparison of experimental and numerical axial velocity profiles (Figure 5 (a)), it was concluded that the mathematical model represents well the experimental results. It was proven both qualitatively and quantitatively that the experimental and numerical results for axial velocity show the same trends. The behavior observed in the axial velocity profile of the test facility cyclone is different from commonly seen profiles in conventional cyclones. This difference was attributed to inlet and outlet geometry effects, and it can bring consequences not yet evaluated to the cyclone performance parameters.

Analyzing the tangential velocity profiles (Figure 5 (b)), the comparison of experimental and numerical profiles shows that the numerical results followed the same trends of the experimental results. However, the numerical tangential profile only represents qualitatively the experimental data. This occurred because the peaks of the numerical tangential velocity profile was dislocated toward the cyclone wall. This behavior of the numerical tangential velocity profile, perceived with the RSM-SSG model, has also been observed in other studies such as those conducted by Peres et al. (2001).

Conclusions

Through the comparison between experiments and simulations it was possible to conclude that both k-ε/Prandtl and RSM-SSG turbulence closure model predict well the pressure drop in cyclone. Also, the RSM-SSG is a good model to predict the axial velocity profile and the magnitude of the tangential velocity profile. Therefore, both study models were able to represent well the swirling flow in the inner sections of the cyclone.

It was observed that the axial velocity profile in the inner section of the cylindrical body of the test facility cyclone is different from commonly seen profiles in conventional cyclones. The test facility cyclone axial velocity profile has a third peak located in the central axis of the cyclone that does not occur in conventional cyclones. This difference was attributed to the geometry of the inlet and the outlet, and it can bring consequences not yet evaluated in the pressure drop and the collection efficiency in cyclones.

Nomenclature

ρ ≡ density;
$\mathbf{v}$ ≡ velocity ;
p ≡ pressure;
$\mathbf{g}$ ≡ gravity acceleration;
μ ≡ viscosity;
$\mathbf{D_g}$ ≡ strain rate;
$\mathbf{T}$ ≡ Reynolds stress;
$\mathbf{D_T}$ ≡ turbulent diffusion;
$\mathbf{D_M}$ ≡ molecular diffusion;
$\boldsymbol{\Upsilon}$ ≡ stress production;
$\boldsymbol{\Pi}$ ≡ pressure strain;
ε ≡ dissipation rate;
D_c ≡ cyclone diameter.

References

[1] Peres, A. P.; Meier, H. F.; Kasper, F. R. S.; Huziwara, W. K.; Mori, M., "Análise de Modelos de Turbulência e da Importância do Refinamento da Malha na Simulação 3-D do Escoamento Gasoso em Ciclone," *Congresso Brasileiro de Engenharia Mecânica,* vol. 8, 2001.

[2] Noriler, D.; Vegini, A. A.; Soares, C.; Barros, A. A. C.; Meier, H. F; Mori, M., "A New Role for Reduction in Pressure Drop in Cyclones Using Computational Fluid Dynamics Techniques," *Brazilian Journal of Chemical Engineering,* vol. 21, no. 1, pp. 93-101, 2004.

[3] Vegini A. A.; Meier, H. F.; Iess, J. J.; Mori, M., "Computational Fluid Dynamics (CFD) Analysis of Cyclone Separators Connected in Series," *Industrial & Engineering Chemistry Research,* vol. 47, pp. 192-200, 2008.

[4] Meier, H. F.; Vegini A. A.; Mori, M., "Four-phase Eulerian-Eulerian Model for Prediction of Multiphase Flow in Cyclones," *The Journal of Computational Multiphase Flows,* vol. 3, no. 2, pp. 92-105, 2011.

[5] Sgrott Jr., O. L.; Costa, K. K.; Noriler, D.; Wiggers, V. R.; Martignoni, W. P.; Meier, H. F., "Cyclones'Project Optimization by Compination of an Inequality Constrained Problem and Computational Fluid Dynamics Techniques (CFD)," *Chemical Enginnering Transactions,* vol. 32, 2013.

[6] Speziale, C. G.; Sarkar, S.; Gatski, T. B. , "Modelling the Pressure-Strain Correlation of Turbulence: An Invariant Dynamical Systems Approach," *Journal Fluid Mechanics,* vol. 227, pp. 245-272, 1991.

[7] Shukla, S. K.; Shukla, P.; Ghosh, P., "Evaluation of Numerical Schemes Using Different Simulation Methods for the Continuous Phase Modeling of Cyclone Separators," *Advanced Powder Technology,* pp. 209-219, 2011.

375056 Separation of Particles during Drop Evaporation

Abstract:

In this work, an evaporating droplet with two different sized colloidal particles is studied. During the evaporation process, the particles can be separated by their size into different rings. The finite element method was used to solve the governing equations. The fluid is assumed to be Newtonian and the droplet is assumed to be axisymmetric. The full momentum equations and continuity equation govern the fluid velocity and pressure within the droplet. The liquid evaporates and diffuses in the vapor space. The particle concentration inside the droplet is governed by the convection-diffusion equation. Using this model, particle separation and deposition onto the substrate is analyzed throughout the evaporation process. In addition to the model, experiments have been performed to confirm the ring separation using a water droplet containing 1 μm and 3 μm polystyrene spheres.

384891 Decanter Centrifuges for Kaolin Classification

Abstract:

Decanter centrifuges have long been used to classify kaolin slurry for various coating products involving coating particles below two microns.

In this paper, a model has been developed for particle classification. The paper illustrates systematically how various products can be produced accurately with the help of the model meeting the necessary specification for which different percentages of the particles should be either below 1 micron, or 2 microns, as required by the product grade. This approach is further back-up by examples.

Traditionally this has been carried out by trial-and-error by changing feed rate and bowl speed, and fine tuning by changing differential speeds and pool levels.

373525 Combined Materials Screening and Process Optimization for Large-Scale p-Xylene Separation Using Simulated Moving Beds (SMB)

Abstract:

Simulated Moving Bed (SMB) chromatography [1,2] has found significant applications in different areas including chemicals, petrochemicals, biotechnology, pharmaceuticals, and fine chemistry. The Parex process from UOP is based on SMB and is a leading technology for the recovery of *p*-xylene (1,4-dimethyl benzene) from a mixture of C_8 aromatic isomers. *p*-Xylene is used as an industrial solvent and an intermediate to produce polyethylene terephthalate (PET) for the production of films, fibers, resins and many valuable products. Because of its steady market, the demand for *p*-xylene increases 6-8% every year [3].

The SMB process for *p*-xylene uses adsorptive separation and exploits the differences in affinity of an adsorbent material for *p*-xylene relative to the other C8 isomers. The use of SMB technology enables a simulated counter-current contact between the mobile fluid phase and the stationary solid phase while avoiding the difficulties in the movement of a solid phase. This makes SMB an inherently complex process. Unlike most adsorption processes it is a continuous process and operates in a cyclic fashion. An industrial SMB process utilizes as many as 24 beds which are divided into four different zones, namely feed, desorbent, extract, and raffinate zones. The flows to the zones are maintained using a rotary valve.

Designing an SMB process with optimal bed length, diameter, and operating conditions, such as flows to each zone and switching times, is a challenging task. While significant studies [4-13] have been performed to optimize the operating conditions in order to either maximize product recovery, purity and throughput, or minimize the desorbent flow, the overall cost of *p*-xylene separation is usually not considered. Furthermore, most optimization studies for SMB-based *p*-xylene separation selects the adsorbent material *a priori*. While adsorption of xylenes on Y zeolites has been extensively studied in the literature, the potential of replacing Y zeolites with other adsorbents is often neglected or remains unexplored.

In this work, we first present a novel grey-box constrained optimization approach for the optimization of a large scale and industrial SMB process. The process consists of 24 beds and 4 zones for the separation of *p*-xylene from a typical C_8 mixture containing 23.6% p-xylene, 49.7% m-xylene, 12.7% o-xylene, and 14% ethylbenzene. The objective is to minimize the total annualized cost (the combined investment, operating and material costs) of *p*-xylene separation, while achieving the specified product (*p*-xylene) purity and recovery of 99.5% and 97%, respectively. The constrained grey-box models are developed based on the calculations of purity, recovery, and separation cost from a discretized version of a detailed nonlinear algebraic and partial differential equation (NAPDE) model [10,13] that describes the overall SMB process. The fully-discretized NAPDE model is highly nonlinear and nonconvex, and it is often difficult to solve the model using commercial solvers. In this work, we optimize the SMB process using ARGONAUT (AlgoRithms for Global Optimization of coNstrAined grey-box compUTational problems) [14], which we have developed for solving grey-box and black-box constrained complex problems. While we develop efficient strategies for solving the discretized NAPDE model as an NLP problem using a local solver such as CONOPT, the constrained grey-box optimization approach shows the capability to find better solutions and outperforms CONOPT.

We have also identified adsorbent materials for cost-effective xylene separation by combining *in silico* screening of zeolites and SMB process optimization. We not only select the most cost-effective materials, but we also attain the optimal process conditions while satisfying purity, recovery, and other process constraints. We previously demonstrated the applicability of combined *in silico* screening of materials and process optimization for post-combustion CO_2 capture [15–17] and natural gas purification [18] using zeolites. To the best of our knowledge, this work is the first to simultaneously identify cost-effective materials and optimized SMB process for *p*-xylene separation.

References

[1] Ruthven, D.M.; Ching, C.B. Counter-current and simulated counter-current adsorption separation processes. Chem. Eng. Sci. 1989, 44, 1011 - 1038.

[2] Gomes, P.S.; Minceva, M. Rodrigues, A.E. Simulated moving bed technology: old and new. Adsorption, 2006, 12, 375 - 392.

[3] Wantanachaisaeng, P.; O'Neil, K. Capturing opportunities for para-xylene production. UOP LLC., http://www.uop.com/?document=uop-capturing-opportunities-for-para-xylene-production-tech-paper&download=1.

[4] Mazzotti, M.; Storti, G.; Morbidelli, M. Optimal operation of simulated moving bed units for nonlinear chromatographic separations. J. Chromatogr. A, 1997, 769, 3 – 24.

[5] Kawajiri, Y.; Biegler, L.T. Optimization strategies for simulated moving bed and PowerFeed processes. AIChE J., 2006, 52, 1343 – 1350.

[6] Kawajiri, Y.; Biegler, L.T. A nonlinear programming superstructure for optimal dynamic operations of simulated moving bed processes. Ind. Eng. Chem. Res., 2006, 45(25), 8503 - 8513.

[7] Kawajiri, Y.; Biegler, L.T. Large scale optimization strategies for zone configuration of simulated moving beds. Compute. Chem. Eng., 2008, 32(1-2), 135 - 144.

[8] Bentley, J.; Sloan, C.; Kawajiri, Y. Simultaneous modeling and optimization of nonlinear simulated moving bed chromatography by the prediction-correction method. J. Chromatogr. A, 2013, 1280, 51 - 63.

[9] Bentley, J.; Kawajiri, Y. Prediction-correction method for optimization of simulated moving bed chromatography. AIChE J., 2013, 59(3), 736 - 746.

[10] Minceva, M.; Rodrigues, A.E. Modeling and simulation of a simulated moving bed for the separation of p-xylene. Ind. Eng. Chem. Res., 2002, 41, 3454 - 3461.

[11] Minceva, M.; Rodrigues, A.E. Two-level optimization of an existing SMB for p-xylene separation. Compute. Chem. Eng., 2005, 29(10), 2215 - 2228.

[12] Minceva, M.; Rodrigues, A.E. Understanding and revamping of industrial scale SMB units for p-xylene separation. AIChE J., 2007, 53, 138 - 149.

[13] Kurup, A.S.; Hidajat, K.; Ray, A.K. Optimal operation of an industrial-scale Parex process for the recovery of p-xylene from a mixture of C-8 aromatics. Ind. Eng. Chem. Res., 2005, 44(15), 5703 - 5714.

[14] Boukouvala, F.; Hasan, M.M.F.; Floudas, C.A. ARGONAUT: AlgoRithms for Global Optimization of coNstrAined grey-box compUTational problems, to be presented in the AIChE Annual Meeting, 2014, November 16 – 21, Atlanta, GA.

[15] Hasan, M.M.F.; First, E.L.; Floudas, C.A. Cost-effective CO_2 capture based on *in silico*screening of zeolites and process optimization. Phys. Chem. Chem. Phys., 2013, 15(40), 17601 - 17618.

[16] Hasan, M.M.F.; Boukouvala, F.; First, E.L.; Floudas, C.A. Nationwide, regional and statewide CO_2capture, utilization and sequestration supply chain network optimization. Ind. Eng. Chem. Res., 2014, 53(18), 7489 - 7506.

[17] Hasan, M.M.F.; Baliban, R.C.; Elia, J.A.; Floudas, C.A. Modeling, simulation and optimization of post-combustion CO_2capture for variable feed concentration and flow rate. 2. Pressure swing adsorption and vacuum swing adsorption processes. Ind. Eng. Chem. Res., 2012, 51(48), 15665 - 15682.

[18] First, E.L.; Hasan, M.M.F.; Floudas, C.A. Discovery of novel zeolites for natural gas purification through combined material screening and process optimization. AIChE J., 2014, 60(5), 1767 - 1785.

Evaluation of Esterification and Transesterification Reactions for Propylene Glycol Methyl Ether Acetate in Reactive Chromatography Systems

Jungmin Oh[a], Shan Tie[a], Gaurav Agrawal[a], Balamurali Sreedhar[a], Megan Donaldson[b], Alfred Schultz[b], Timothy Frank[b], Andreas Bommarius[a], Yoshiaki Kawajiri[a]

a. School of Chemical and Biomolecular Engineering, Georgia Institute of Technology, Atlanta, GA, 30332

b. The Dow Chemical Company, Midland, MI

Abstract: Reactive chromatography is a process that combines reaction and separation in a single unit that leads to a greater process performance and productivity [1]. This process is especially advantageous when the reaction is equilibrium limited, and the in-situ separation of product shifts the equilibrium in the direction of conversion increase [2].

This study focuses on the catalytic synthesis of propylene glycol methyl ether acetate (DOWANOL™ PMA glycol ether acetate) through two separate types of reactions: transesterification and esterification. Both reactions are equilibrium limited, where conversion can be increased by applying the principle of reactive chromatography. Esterification is performed with an acidic catalyst, AMBERLYST™ 15, which has high activity and stability. Transesterification is performed with a basic catalyst, AMBERLITE™ IRA-900, which shows high activity for the transesterification at low temperatures. Esterification has an advantage of an enhanced durability of the catalyst, while the transesterification has an advantage of a better separation (absence of water) and lower operating temperatures. The benefits of each reaction are discussed with comparisons throughout the study.

Case studies of process development for an ester product are discussed for the transesterification and the esterification reactions. The dynamics of batch reaction and the fixed-bed adsorptive reaction are investigated by carrying out batch reaction experiments and chromatographic pulse tests. Stirred batch reactor experiments were conducted at various temperatures and different mass transfer resistances. Reaction equilibrium and kinetic parameters together with their dependence on temperature were determined by fitting the model to the experimental data. Measurement of the adsorption equilibrium constant and reaction parameters were conducted by pulse tests using a single chromatographic column. Since resins act as both adsorbent and catalyst, experiments were performed with nonreactive mixtures to obtain adsorption parameters first, and then reactive mixtures were injected to obtain reaction parameters.

In addition to the development of models, the chromatographic reactors were optimized for each reaction and esterification and transesterification were compared for the synthesis of the ester in terms of productivity and efficiency.

References

[1] V. Gyani, S. Mahajani, Reactive Chromatography for the Synthesis of 2-Ethylhexyl Acetate. *Separation Science and Technology*, 43: 2245-2268, **2008.**

[2] A. Rodigues, C. Pereira, J. Santos. Chromatographic Reactors. *Chem. Eng. Technol.* **2012,** 35, No7, 1171-1183

386924 Predictive Screening of Column Pair Combinations for Two-Dimensional Liquid Chromatography

Abstract:

Predictive Screening of Column Pair Combinations for Two-dimensional Liquid Chromatography

Rebecca K. Lindsey,[1] Dwight R. Stoll,[2] Peter W. Carr,[1] and J. Ilja Siepmann[1,3†]

[1]Department of Chemistry and Chemical Theory Center, University of Minnesota, Minneapolis, MN 55455-0431

[2]Department of Chemistry Gustavus Adolphus College, Saint Peter, Minnesota 56082

[3]Department of Chemical Engineering and Materials Science University of Minnesota, Minneapolis, MN 55455-0431

Two-dimensional liquid chromatography (2DLC) is a powerful separation technique that employs a second column to further separate fractions taken from the first column; when utilized in an on-line fashion, 2DLC can result in much higher total peak capacities than are achievable with a single column. This computational work addresses the significant challenge of selecting an optimal column pair from the N2 possible combinations (order matters in the on-line approach) by presenting a predictive screening method for column pair selection, and provides insight into the defining characteristics of a good column pair.

[†]Corresponding Author. E-mail: siepmann@umn.edu

Capture Chromatography for Molybdenum-99 Recovery from Uranyl Solutions: Design of Stripping Processes Based on the Adsorption Mechanisms

Lei Ling*, Amanda Youker†, Dominique C. Stepinski†, George F. Vandegrift†, and Nien-Hwa Linda Wang**

* School of Chemical Engineering, Purdue University, West Lafayette, IN 47907

**Email: wangn@ecn.purdue.edu, Phone: 765-494-4081, Fax: 765-494-0805

†Argonne National Laboratory, Argonne, IL 60439

Abstract

Molybdenum-99 (Mo-99), a fission product of Uranium-235 (U-235), is the radioactive parent of the most widely used medical isotope, technetium-99 m (Tc-99 m). An efficient capture chromatography process was developed to recover Mo-99 from uranyl sulfate solutions using a TiO_2 sorbent. A graphical design method was developed to determine the minimum column volume required for the desired yield (>99%), loading time (2 hours), and pressure limit (<0.8 atm). In this study, a stripping process is developed to elute and concentrate the captured Mo-99 from the chromatography column.

The mechanisms of Mo-99 adsorption and stripping are complex, because TiO_2 has multiple adsorption sites. The capacity of each site is estimated using frontal tests with various acids and bases. Hypotheses for the mechanisms of Mo adsorption and stripping are proposed based on the capacities of the various adsorption sites. Rate model simulations based on the hypotheses agree closely with experimental Mo peaks stripped by NaOH. Simulations based on the verified models and parameters are used to identify important factors that can affect the stripping time and Mo product concentration. A small residence time can shorten the time for stripping, but it results in a broad Mo product peak because intra-particle diffusion rate is small compared to the convection rate. Since Mo peak spreading is dominated by intra-particle diffusion, extra-column dead volume is found to have negligible effects on the stripping curves if it is less than 10% of the column volume. Counter-current stripping results in a slightly sharper Mo peak than co-current stripping because the migration distance for the Mo peak is shorter. NH_4OH stripping causes tailing of the eluted Mo peaks, whereas NaOH stripping leads to sharp Mo peaks. This phenomenon can be explained by the favorable association reaction of NH_4^+ and OH^-, which reduces the effective concentration of OH^- in the solution. For this reason, NaOH is a more effective stripping agent than NH_4OH.

Solvent selection for the recovery of fermentation inhibitors; Comparison of simulated and laboratory results for liquid-liquid extraction

Mahdieh Aghazadeh[1,3], Abigail S. Engelberth[1,2,3]

1. Agricultural & Biological Engineering, 2. Environmental & Ecological Engineering, 3. Laboratory of Renewable Resources Engineering
Purdue University

Some of the degraded products from lignocellulosic biomass pretreatment have proven to inhibit fermentation process in biotechnological conversion of biomass to bioethanol. This work studies the practicality of liquid-liquid extraction with nine organic solvents in removing the most well-known fermentation inhibitors.

Aspen Plus ™ was used to simulate the extraction system. To create the most extreme case of this hypothetical fermentation broth the highest end of each concentration range from a library of reported existing inhibitors were incorporated in the Aspen Plus™ simulation.

The results indicate that the simulation outputs are not completely reliable without experimental evaluation because the property method chosen can vary the simulation outlet streams compositions greatly. NRTL, UNIF-LL, and UNIF-LBY are the property methods that were examined for their ability to predict the liquid-liquid extraction system performance. NRTL results predict that the inhibitors are extracted in the same manner for each solvent as if they all react exactly the same with the solvent. UNIF-LL and UNIF-LBY outcomes are different from NRTL. In the set of results from UNIF-LL the best solvents are ethyl acetate and butyl acetate, which have the capacity to extract all the inhibitors to varying degrees. UNIF-LBY results lead to similar general conclusion and split fraction trends as UNIF-LL; with different final numbers for split fractions. To verify which property method best suits the simulation, liquid-liquid extraction experiments were carried out with both model solution and liquid hydrolysate from dilute acid pretreatment of corn stover.

The development of industrial scale bioprocesses has come into high scientific and industrial interest over the last decades. Especially within the red biotechnology, the production of biopharmaceuticals like therapeutic antibodies shows an enormous potential for an industrial application. Within these processes, the total production costs are often dominated by up to 80% by the downstream processing based on a series of chromatographic separations. These are often used to guarantee the required high purity of the target antibodies (protein).

An interesting alternative for the separation of these highly complex mixtures is the aqueous two-phase extraction (ATPE). Using an aqueous two-phase system for extraction enables the downstream processing of proteins under mild and biocompatible conditions. Furthermore the distribution behavior of proteins within an ATPE can be influenced by the use of an additional salt as displacement agent. Effective use of this fact allows for a selective and efficient protein purification within the ATPE.

In order to develop a suitable ATPE for proteins, the interactions between protein-protein, as well as the interactions between protein-solute (phase-forming components and displacement agent) have to be known. To access these interactions, the second osmotic virial coefficient B_{22} and the cross virial coefficient B_{23} are used in this work. B_{22} characterizes the interactions between molecules of one species (protein-protein) and B_{23} considers the interactions between different molecules (protein-solute). One big advantage using this approach is that the different influence factors like pH, temperature, solute and solute concentration are captured in one measure which enables a quick characterization of the different influence factors.

Within this work, a PEG / phosphate aqueous two-phase system containing the proteins immunoglobulin G (*IgG*) and human serum albumin (*HSA*) is evaluated with respect to the partition coefficient and the precipitation affinity of the proteins. The influence of the displacement agent NaCl on the partitioning of *IgG* and *HSA* is considered. Therefore B_{22} and B_{23} values were measured for several combinations of solute (PEG, phosphate buffer and sodium chloride) and protein using static light scattering. Based on the experimentally determined B_{22} and B_{23} values a qualitative prediction of the distribution behavior and the precipitation affinity of *IgG* and *HSA* within the ATPE systems are performed. The results will be used to develop a thermodynamic model in order to select the suitability and efficiency of ATPE systems for protein purification and thus simplify the downstream processing development.

A Thermoresponsive Polymer Based on Amino Acid Ionic Liquids as Extractant for Efficient Separation of Tocopherol Homologues

Yangyang Lu,[1] Wen-Jun Wang,[1,*] Qiwei Yang,[3] Qilong Ren,[3] Bo-Geng Li,[1] and Shiping Zhu,[2,*]

[1] State Key Lab of Chemical Engineering, Department of Chemical and Biological Engineering, Zhejiang University, Hangzhou, 310027, China

[2] Department of Chemical Engineering, McMaster University, Hamilton, Ontario, Canada L8S 4L7

[3] Key Lab of Biomass Chemical Engineering of Ministry of Education, Department of Chemical and Biological Engineering, Zhejiang University, Hangzhou 310027, China

* Corresponding authors. wenjunwang@zju.edu.cn (W.-J. Wang); zhuship@mcmaster.ca (S. Zhu)

Abstracts

Products with natural bioactivity are considered to be high added-value products. However, current separation methods of these products require large solvent consumption, and have low separation efficiency and significant energy expenditure. To address these issues, we have proposed a new method of using soluble polymers as an extractantant (SPE) for the liquid-liquid extraction of α-tocopherol from tocopherol homologues. The thermoresponsive SPEs comprising an amino acid ionic liquid (AAIL) have been synthesized via reversible addition-fragmentation chain transfer (RAFT) solution copolymerization of AAIL and N-isopropylacrylamide

(NIPAM) at 65 °C using O-ethyl-S-(1-carboxy)methyl dithiocarbonate as a RAFT chain transfer agent. The SPE samples had molecular weights between 6.75-10.2 kg/mol, molecular weight distributions between 1.10-1.25, and AAIL molar fractions between 0.64-0.90. Effects of chain length and copolymer composition on separation efficiency were investigated. A distribution coefficient of 8.92 and a selectivity coefficient of δ-tocopherol to α-tocopherol of 13.8 have been achieved with the SPEs. The distribution and selectivity coefficients were found to increase with decreasing chain length of the SPE and increasing the molar ratio of NIPAM in the SPE. By lowering the temperature of the extract liquor, the SPEs were easily and fully separated from the tocopherol homologues, resulting in more than 93% recovery of tocopherols. The SPEs can be reused for multiple extraction cycles without any change to the distribution or selectivity coefficients of tocopherol. These thermoresponsive polymeric ionic liquid based SPEs turn to have high separation efficiency and low energy consumption in the liquid-liquid extraction.

This work was supported by the National Natural Science Foundation of China (Grants 21376211 and 20976153) and the Chinese State Key Laboratory of Chemical Engineering at Zhejiang University (Grants SKL-ChE-12T05 and SKL-ChE-14D01).

385856 Reactive Aqueous Two-Phase Systems for the Production and Purification of Pegylated Proteins

Abstract:

Bioprocess intensification depends heavily in the ability of combining different stages of a production process in single unitary operation. In this sense, Aqueous Two-Phase Systems (ATPS) represent a potentially adaptable stage to combine two or more operations with this purpose. As known, ATPS are a liquid-liquid fractionation technique where the product of interest partitions towards one of the phases according to its partition coefficient (K_P) which is directly related to the physicochemical properties of the compound and the phase forming chemicals. ATPS have been successfully used in the separation of different biomolecules and their use as a primary recovery stage is widely accepted. However, little work has been made in using these systems for the production and primary purification of different biomolecules.

For instance, PEGylated proteins are chemically modified proteins where at least one chain of poly (ethylene glycol) (PEG) is covalently attached to its structure without having the protein losing its properties. The advantages conferred by this modification like a lower kidney clearance, longer blood circulation times and protection from antibodies are of great importance in pharmaceutical proteins. The production and purification of PEGylated proteins however, is not a trivial task. The modified proteins have to be separated according to their number of attached PEG chains and their positional isomerism since not all of these isomers posses the same desired attributes. Previous work in our group has demonstrated that the use of PEG – phosphate salt ATPS can potentially fractionate the modified proteins from the unreacted species in a PEGylation reaction. With this line of thought, and since PEG is used in both the production and purification of these proteins a Reactive Aqueous Two-Phase System (Rx-ATPS) strategy is proposed for the *in situ* production and primary recovery of PEGylated proteins combining the reaction and primary recovery operations in a single stage.

Binodal curves were made with reactive monomethoxy-PEG (mPEG) and phosphate salts at pH 5.1 for polymers with molecular weights of 5.0 and 20.0 kDa. Rx-ATPS systems were then constructed with tie-line lengths (TLL) of 15, 25, 35 and 45% w/w and volume ratios (V_R) of 0.33, 1.0 and 3.0. Ribonuclease A (RNase A) was used as a model protein and was used at a concentration of 2 mg/mL across all of the constructed systems. After 17 hours of agitation at 4 °C samples were centrifuged and the phases were separated for their analysis by Size Exclusion Chromatography (SEC). As expected, reactions took place and the products fractionated mainly to the top PEG-rich phase while the unreacted protein fractionated towards the phosphate salt-rich phase. However it was found that Rx-ATPS design parameters have an important role in the behavior of each system. For instance, in systems where 5.0 kDa mPEG was used, the production of the PEGylated proteins reduced as TLL and V_R increased. In systems were 20.0 kDa mPEG was used, production of the conjugates increased as TLL and V_R values increased. Although some protein degradation was observed, reaction/recovery yields varied from 20% to 78% depending on the Rx-ATPS selected. It should be noted that reaction conditions in these systems vary heavily from those optimized in regular PEGylation reactions giving different results. There are still many variables to be explored but this is the first report of the use of Rx-ATPS for the production and purification of PEGylated proteins and the optimization of these systems will be addressed.

367657 Extraction of Chrysanthemum Morifolium Extractum By Ethanol Modified Supercritical CO2

Abstract:

In this study, ethanol modified supercritical CO_2 extraction of *Chrysanthemum morifolium* extractum was optimized by single factor and orthogonal array design methods with respect to the effects of extracting pressure, extracting temperature, extracting time, flow rate of CO_2, ratio of liquid to solid and ethanol concentration on the yield of *Chrysanthemum morifolium* extractum. The optimum extracting parameters were established with extracting pressure 30 MPa, extracting temperature 45°C, flow rate of CO_2 10 $kg \cdot h^{-1}$, ratio of liquid to solid 15 $mL \cdot g^{-1}$, ethanol concentration 95% and extracting time 2 h. Under such conditions, the yield of *Chrysanthemum morifolium* extractum was 29.34%. The content of flavonoids in *Chrysanthemum morifolium* was 12.46 $mg \cdot g^{-1}$.

371962 Extraction of Essential Oils from Lemon Peels By Supercritical Carbon Dioxide

Abstract:

Lemon flavor additives are widely used in foods, drinks, cosmetics and detergents. Currently, there are many methods to extract essential oils from lemon peels. However, these methods are mostly time consuming and of low extraction efficiency. Even more, some extraction methods involve the use of organic solvents which could cause safety and health concerns as well as environmental pollution. Therefore, the purpose of this study is to apply the Supercritical Fluid Extraction using Carbon Dioxide (SFE-CO_2) to extract lemon essential oils. As supercritical carbon dioxide is non-polar, ethanol and nonionic surfactants were used as cosolvents to increase its polarity and the extraction efficiency.

About 90 % of the lemon essential oils are monoterpenes, in which the main ingredients are *d*-limonene, β-pinene and γ-terpinene. Our results showed that the wet and dry yields increased with a lower extraction temperature and a higher CO_2 pressure, probably owing to the increasing density of supercritical fluid. Besides, β-pinene has a higher polarity. Therefore, its extractability by SFE-CO_2 is relatively poorer. The extraction yield increased with more β-pinene extracted out with addition of ethanol to SFE-CO_2. However, the effect of added nonionic surfactants as cosolvents to the SFE-CO_2 seems insignificant.

387915 Membrane Surface Modification with Polyelectrolyte Multilayers for Ionic Liquid Recovery for Biomass Processing

Abstract:

Ionic liquids (ILs) have often been advertised as versatile solvents with beneficial properties such as low vapor pressure, non-toxicity and excellent solubility for otherwise recalcitrant bio-polymers. As a result of these unique properties, numerous studies have been carried out to use ILs for biomass processing. High sugar yields during biomass hydrolysis have been reported. However, the high cost of ILs is inhibitory for their application in lignocellulosic biomass processing. Recycle and reuse of ILs are essential to be economically viable. Here we develop membrane reactors that are capable of retaining IL for reuse yet allowing permeation of small sugar molecules.

Ceramic flat sheet ultrafiltration membranes have been modified using layer-by-layer deposition of polyelectrolytes (PEL) involving poly(styrene sulfonate) / poly(allylamine hydrochloride) (PSS/PAH) or poly(styrene sulfonate) / poly(diallyldimethylammonium chloride) (PSS/PDADMAC) to tune the selectivity of these ultrafiltration membranes for rejecting charged species. Additional selectivity for rejecting of IL and allowing permeation of sugar molecules is achieved by varying the number of deposited PEL layers and by selecting an appropriate operating condition.

385906 Perfluorinated Membranes for the Dehydration of Ionic Liquids for Processing Biomass

Abstract:

This presentation will discuss membrane dehydration of ionic liquids (ILs) with specific focus on biomass conversion to sugars. Ionic liquids can be used in the production of sugars from cellulosic biomass. Processing of the biomass requires mixing the IL with water. In many other potential applications, control of the water level in ILs is highly desirable since ILs are highly hygroscopic. The solvent power of ILs for processing biomass is highly dependent on their water content. Given the high cost of ILs, the wet ionic liquid must be dehydrated so it can be reused multiple times in the processing of biomass. CMS has developed chemically and thermally resistant pervaporation membranes that are effective in removing water from ionic liquids. Dehydration of ionic liquids by pervaporation using a hollow fiber composite perfluorinated membrane has been demonstrated. Efficient water removal is observed despite the low driving force for water permeation, e.g., low water activity coefficient. The water permeance is found to be approximately constant over a wide range of water concentration and temperatures. The permeation of ionic liquid through the membrane is negligible, which leads to extremely high water/IL separation factors and, therefore, minimal loss of IL. Results from both laboratory and pilot membrane dehydrators will be presented. The advantage of ionic liquids dehydration by pervaporation are discussed.

Enhancement of esterification conversion using novel composite catalytically active pervaporation membranes

Weidong Zhang*, Weihua Qing, Ning Chen, Wei Li, Jiangrong Chen

State Key Laboratory of Chemical Resource Engineering,

Beijing Key Laboratory of Membrane Science and Technology, Beijing University of Chemical Technology, Beijing 100029, People's Republic of China

(*Corresponding Author at: PO Box 1#, Beijing University of Chemical Technology, Beijing 100029, People's Republic of China. Tel.: +86-10-6442-3628; Email address: zhangwd@mail.buct.edu.cn)

Email addresses list:

Weidong Zhang: zhangwd@mail.buct.edu.cn; Weihua Qing: qingweihua@gmail.com; Ning Chen: 405812375@qq.com; Wei Li: 550460472@qq.com; Jiangrong Chen: 2011200263@grad.buct.edu.cn

Esterifications are typical examples of reactions controlled by thermodynamic equilibrium in which water plays a very important role, water accumulation at the reaction zone would facilitate the hydrolysis reaction and thus lower the final conversion. Liquid mineral acid were usually used as catalyst in batch reactors, these reaction usually suffer from acid corrosion to equipment, and the difficulty of catalyst separation.

In order to achieve higher conversion, catalytic pervaporation membrane reactor (PVMR) has drawn an increasing attention recently because of its ability to integrate the functions of catalytic reaction and pervaporation into one single unit, in which the product can be selectively removed right after it is formed at the vicinity of catalyst, and therefore, effectively driving the reaction to the product side. In contrast to the traditional reactive distillation process, where the preferred temperatures for reaction process and the separation process are usually different, the separation process in catalytic PVMR can be carried out at one optimal reaction temperature, and its separation efficiency is not limited by relative volatility. Additionally, pervaporation

(PV) is especially attractive in the temperature-sensitive process because it can be operated at moderate temperatures.

Up to date, most of the membranes used in traditional PVMRs only play the role of separation without any catalytic activity. The catalysts were generally suspended in the reaction phase, the byproduct water has to diffuse a relative long distance from the bulk solution to membrane surface for separation, this bulk diffusion contributes the largest mass transfer resistance in the overall mass transfer resistances of the process, as a result, it may greatly influence the efficiency of the water removal in PVMR.

Several studies demonstrate that the equilibrium shifting could be enhanced if a catalytically active membrane is used. However, in those literatures, separation and catalytic function are integrated into a single layer membrane with dense structure, which will lead to a low catalytic activity, as most catalysts sites are wrapped in the membrane with dense structure. Therefore, it has been suggested to use composite multi-layer catalytically active membranes to optimize the separation and the catalytic performance independently.

In the present study, a novel technique is developed to prepare composite catalytically active pervaporation membranes by coating a porous catalytic layer containing polyvinyl alcohol and catalysts on a dense polyvinyl alcohol (PVA)/ polyethersulfone (PES) separation layer using immersion phase inversion. Benefited from much closer integration of reaction and separation zone, this new membrane structure may offer many potential advantages: the conversion of reactants to products can take place within the catalytic layer of membrane instead of bulk solution, thus reducing the mass transfer resistance of bulk diffusion existing in the traditional PVMRs, the water would be *in situ* removed from the immediate vicinity of catalysts, preventing subsequent unfavorable hydrolysis caused from water accumulation. As a result, the equilibrium displacement could be further enhanced. Other potential advantages include the immobilization of enzyme, higher mass transfer driving force for water resulted from higher water concentration in the catalytic layer of membrane.

In the first application, a catalytically active membrane was prepared by coating a porous catalytic layer containing polyvinyl alcohol and ion-exchange resins on a dense polyvinyl alcohol (PVA)/ polyethersulfone (PES) separation membrane. The esterification of acetic acid and n-butanol was used as a model reaction to investigate its performance. SEM images showed its structure containing three layers: the top layer was a porous catalytic layer; the middle layer was the dense PVA selective layer and the bottom layer was the commercially available PES support layer (as shown in Figure 1). Batch reaction results showed that the activity of the membrane was comparable to that of catalyst beads and the optimal preparing condition for catalytic layer has been determined as: resin particle size between 13-25μm, catalyst loading 200g/m^2, and PVA concentration of casting suspension 3 wt.%.

The result of esterification reactions in PVMRs showed that the acetic acid conversion reaches 91.4% in 20h at 85℃ by using the catalytically active membrane, while the equilibrium conversion was around 71.9% under same condition. The increase of temperature also facilitates the enhancement of acetic acid conversion. The catalytic activity and structure stability of the catalytically active membrane were further confirmed by consecutive experiments and SEM analyses.

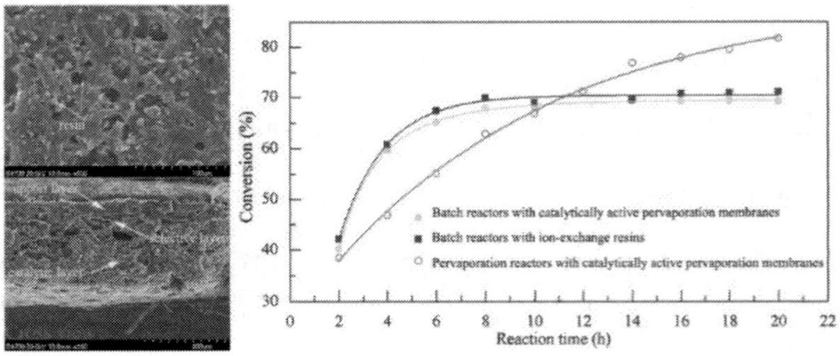

Fig.1. The conversion rate on the acetic acid conversion of esterification in different reactors, resin particle size 13-25μm, PVA concentration of casting suspension 3 wt.%, catalyst loading 200 g/m^2 (corresponding to 0.4 g resin beads),reaction temperature 75℃, reactant mole ratio 1:1.

Recently, we have also investigated the viability of this membrane preparation approach for enzymatic process. A composite catalytically active membrane immobilized with *Candida rugosa* lipase has also been prepared by immersion phase inversion technique for enzymatic synthesis of lauryl stearate in a pervaporation membrane reactor. SEM images showed that a "sandwich-like" membrane structure with a ~250 μm thick porous lipase-PVA catalytic layer uniformly coated on the PVA/PES separation layer was obtained (as shown in figure 2).

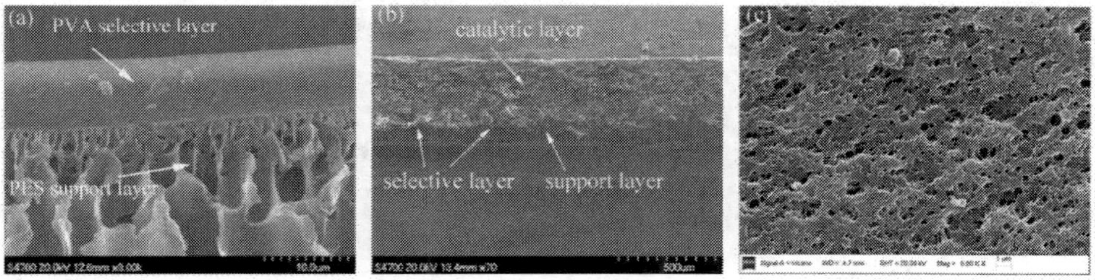

Fig. 2. SEM pictures of the catalytically active membrane: (a) Cross view of PVA/PES separation bilayer (×3000); (b) Cross view of catalytically active membrane (×70); (c) Top view of catalytically active membrane (×5000).

The optimum conditions for lipase immobilization were determined as: PVA concentration of the catalytic layer 5 wt.%, lipase concentration 2 wt.%, glutaraldehyde concentration 0.1 wt.%. The resulting lipase immobilized catalytically active membrane was proved to exhibit far superior thermal stability, pH stability and reusability than the free lipase, showing that immersion phase inversion is a viable approach to entrap lipase into polymer membranes while capable of maintaining their biological functions.

For synthesis aided with PV, the catalytically active membrane was able to overcome the equilibrium limitation by *in situ* removing the byproduct water from the system, the conversion reaches 83 % at 16 hours. An enhancement factor of 41% for the stearic acid conversion was achieved in comparison to the equilibrium conversion (as shown in figure 3). Aside from conversion enhancement, it was also found that excess water removal by the catalytically active membrane appears to improve activity of the lipase immobilized.

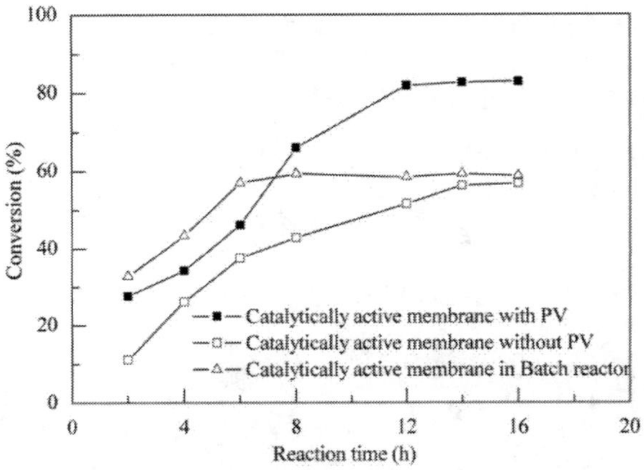

Fig. 3 Effect of reaction time on conversion of stearic acid in different reactors. Catalytically active membrane were prepared with optimum conditions determined in section 3.2.: PVA concentration of the catalytic layer 5 wt.%, lipase concentration 2 wt.% (corresponding to 45 U), glutaraldehyde concentration 0.1 wt.% . Reaction temperature 40□, reactant mole ratio 1:1.

374184 Biphasic Biodegradation of Phenol in an Osmotic Membrane Bioreactor Using Trioctylphosphine Oxide Impregnated Membranes As the Partitioning Phase

Abstract:

A membrane bioreactor (MBR) refers to the integration of a membrane filtration process into a bioreactor operating with suspended microorganisms. These bioreactors combine biodegradation and clarification, two stages in conventional activated sludge wastewater treatment process, into one single step. MBRs offer several advantages over conventional bioreactors such as high biomass concentration in the bioreactor, elimination of cell washout concerns even at high influent flow rate, high effluent quality and a low bioreactor footprint. Although MBRs are typically operated using microfiltration or ultrafiltration membranes, a new bioreactor configuration has recently been developed based on the use of forward osmosis (FO) membranes. This configuration, which is known as an osmotic membrane bioreactor (OMBR), offers several advantages over conventional MBRs: higher rejection, better effluent quality, low or no hydraulic pressure requirement, low energy consumption and low biofouling tendency.

OMBRs have been used successfully in the treatment of municipal wastewater. However, the application of OMBR in industrial wastewater treatment, especially those containing toxic and recalcitrant pollutants such as aromatic compounds, has been very limited. The primary challenge in the biodegradation of these aromatic compounds arises from their high concentration in the wastewater, which exerts substrate inhibition on the biodegrading microorganisms resulting in low or no cell growth and metabolism. This problem is aggravated in the OMBR due to the excellent rejection properties of the FO membranes, which results in high accumulation of the pollutants and an increase in the severity of their inhibitory effects. Therefore, in order to facilitate biodegradation of toxic industrial wastewater in an OMBR, the microorganisms must be protected against the detrimental effects of substrate inhibition.
In this research, biodegradation of phenol by *Pseudomonas putida*was investigated in an OMBR at inhibitory phenol concentrations of 600-2500 mg/L. Trioctylphosphine oxide (TOPO) was impregnated into polypropylene membranes and added into the OMBR to alleviate substrate inhibition. When phenol concentration in the OMBR was high, the TOPO impregnated membranes absorbed phenol to maintain a sub-inhibitory phenol concentration in the bioreactor. When phenol was metabolized and its concentration in the liquid medium decreased, it was desorbed from the absorbents and was degraded by the microorganisms. Thus, the TOPO-impregnated membranes acted as a partitioning phase for phenol sequestration and prevented the microorganisms from substrate inhibition in the beginning of the bioreactor operation, and also during bioreactor operation when phenol concentration was spiked.

In the one month of continuous operation using synthetic wastewater, the OMBR was fed phenolic wastewater at 600, 1000, 2000 and 2500 mg/L concentrations and phenol concentrations up to 2000 mg/L were completely metabolized by P. putida. Biomass concentration, pH, dissolved oxygen and salinity in the bioreactor were regularly measured. While biomass concentration in the OMBR changed with the feed phenol concentration, pH and DO values were relatively stable, and the salinity of the medium increased steadily throughout the operating period. Phenol concentrations in the bioreactor and in the effluent were regularly monitored and phenol was not detected in the effluent once the OMBR achieved steady state. The permeate flux through the FO membranes was initially quite high but it gradually decreased due to an increase in the bioreactor salinity as well as due to membrane biofouling. Cell attachment on the FO membranes was characterized and a washing cycle was developed to recover membrane performance by removing the attached cells from the membranes through osmotic backwashing. The results from the OMBR operation suggest that the OMBR coupled with a partitioning phase such as TOPO-impregnated membranes can be used for rapid and effective biodegradation of toxic and recalcitrant pollutants.

392203 Micro-Structured Ceramic Hollow Fibres for Heterogeneous Catalytic Gas Phase Reactions

Abstract:

A catalytic membrane reactor combines the processes of reaction and separation and is consequently a compact and efficient system for carrying out separations and catalytic reactions. Efficiency can be further increased by the use of a hollow fibre configuration due to the extremely large surface area per unit volume that can be achieved in this way. However, as the support structure is an integral part of the system, its properties greatly affect the performance of the reactor as a whole so it is essential to be able to control both the macro and microstructure of the support to achieve high levels of efficiency. Therefore, new techniques for manipulating ceramic hollow fibre morphology must be developed.

The formation process and morphology of asymmetric ceramic hollow fibre membranes have been studied. Experimental results obtained so far indicate that it is possible to prepare a thin membrane (pore size at ~0.1 mm) supported by a plurality of radial micro-channels of the same material in one-step. The formation of this structure is due to hydrodynamically unstable viscous fingering developed when a less viscous fluid (non-solvent) is in contact with a higher viscosity fluid (ceramic suspension containing invertible polymer binder). To better understand the relationship between non-solvent concentration in the spinning suspension and viscosity, with regard to membrane morphologies, the two parameters have been varied systematically so as to determine the dominating factor, while dynamic video-microscopy has been employed to visualize the formation of micro-channels. Finally, applications of the developed ceramic hollow fibre membranes in catalytic gas phase reactions will also be presented.

386522 Modified Ultrafiltration Membranes for Fractionation of Sugars

Abstract:

Today, production of 1st generation biofuels such as bioethanol from sugar cane and corn starch is well established[1]. Manufacturing processes that include the use of membrane-based unit operations have been described[2]. However increasing competition between food and energy production has led to significant efforts to convert lignocellulosic biomass into 2nd generation biofuels. Unlike 1st generation biofuels, production of 2nd generation biofuels is far more complex. Development of efficient separation and purification operations are essential for production of competitive 2nd generation drop-in biofuels. Membrane based separation processes are attractive as they could lead to significant process intensification and hence reduced operating costs[3].

Fractionation of sugars prior to fermentation is attractive as one can optimize the fermentation conditions for individual sugars. Here we have modified commercially available ultrafiltration membranes using layer-by-layer polyelectrolyte deposition. We investigated modification with poly(sodium 4-styrenesulfonate)/poly(diallyldimethylammonium chloride) and poly(acrylic acid)/poly(diallyldimethylammonium) chloride bilayers. Modified membranes have been characterized using infrared spectroscopy, atomic force microscopy and scanning electron microscopy. In addition we have measured surface contact angle in order to verify successful surface modification.

Dead end filtration tests have been conducted in order to determine membrane permeate flux and separation factors for model feed streams consisting of glucose, sucrose and xylose. Our results indicate that 50 kDa base membranes with 7.5 bilayers of poly(sodium 4-styrenesulfonate)/poly(diallyldimethylammonium chloride) showed the best performance with glucose to sucrose selectivity of more than 11.

Reference:

(1) Abels, C.; Carstensen, F.; Wessling, M. *J. Memb. Sci.* **2013**, *444*, 285–317.

(2) Qiu, Z.; Zhao, L.; Weatherley, L. *Chem. Eng. Process. Process Intensif.* **2010**, *49*, 323–330.

(3) Drioli, E.; Brunetti, A.; Di Profio, G.; Barbieri, G. *Green Chem.* **2012**, *14*, 1561.

386294 Biofuels Production By Fermentation/Pervaporation Coupled Process with Novel Silane Modified Silicalite/PDMS Mixed Matrix Membrane: Effects of Fermentation Components and By-Products

Abstract:

Organic permselective pervaporation membrane technology has been considered to be one of the best among various methods for the recovery of biofuels from fermentation broth. However, in addition to the desired product, fermentative organisms will produce a range of compounds that contain a variety of by-products, which could be potentially problematic for pervaporative separation. For example, during bioethanol fermentation process, the yeast will also produce longer straight-chained and branched alcohols and organic acids. Additionally, biomass pretreatment process can introduce other inhibitors, such as formic acid, acetic acid, and furfural, into the sugar feed stream to the fermentation broth.

In this presentation, the influence of different components in actual bio-ethanol fermentation broths (fermentation medium components, fermentation by-products, and the common inhibitory compounds present in lignocellulosic hydrolysates), and yeast cells on the pervaporation performance of the thin-film VTES modified silicalite/PDMS/PAN composite membrane was investigated in detail. The results showed that the composite membrane can effectively remove formic acid, acetic acid, and furfural present in lignocellulosic hydrolysates. When the model solution with dry yeast cells and the real fermentation broth containing active yeast cells were used, a slightly higher pervaporation flux was found, but the yeast cells had little effect on membrane selectivity. Based on the above results, a membrane bioreactor system comprised of a fermenter and a frame and sheet pervaporation module was developed for continuous ethanol fermentation. Owing to the efficient removal of ethanol from the fermentation broth by VTES modified silicalite/PDMS/PAN composite membrane to alleviate the inhibition to yeast cells, a continuous fermentation/pervaporation coupled process was successfully performed. The composite membrane showed an excellent stability of both flux and selectivity in the coupled process.

381436 Design of Zwitterionic Reverse Osmosis Membranes with Excellent Anti-Fouling Performance By Michael Addition

Abstract:

Membrane fouling is one of the most severe challenges to the application of reverse osmosis (RO) membrane. Zwitterion is a neutral molecule with a positive and a negative electrical charge, and proved to be a kind of non-fouling material. However, strict preparation conditions and high cost limit the application of zwitterionic materials.

In this work, an efficient and simple method to prepare zwitterionic polyamide (PA) RO membranes with excellent anti-fouling performance by Michael addition was proposed. The zwitterionic structure was designed by using common amino compounds as the positively charged section, using carboxylic and sulfonic compounds as the negatively charged section, respectively. Six different kinds of zwitterionic RO membranes were prepared based on positively charged materials, including hyperbranched polyethylenimine (PEI), tetraethylenepentamine, 1,3-diaminoguanidine monohydrochloride, and negatively charged materials, including acrylic acid (AA), 2-methyl-2-butenoic acid, 3-sulfopropyl methacrylate, 6-maleimidohexanoic acid.

Enzyme-linked immunosorbent assay (ELISA) and fluorescent protein adsorption test at different pH values were adopted to evaluate the static anti-fouling property of optimized zwitterionic membranes. Dynamic membrane fouling experiment was conducted to investigate the anti-fouling property of resultant membranes during long-term operation with lysozyme aqueous solution. Separation performance for NaCl solution was also studied.

ELISA result demonstrated that the protein adsorption quantity of resultant six zwitterionic membranes were only about 3%~11% of that of conventional PA membrane. In particular, the protein adsorption quantity of PEI-AA zwitterionic membrane was only 3%~5% in the pH range from 5.0 to 9.0, showing excellent anti-fouling property in acid, neutral, and alkaline conditions. Almost no fluorescent protein was observed on the surface of zwitterionic membranes after immersed in Cy3-BSA solution and shaking for 3 hours. On the contrary, many fluorescent points could obviously be observed on conventional PA membrane surface. In dynamic membrane fouling experiment, the permeation flux of conventional PA membrane declined rapidly to about 36% of initial flux after operation for 10 hours. However, the permeation flux of PEI-AA zwitterionic membrane declined slightly and still reached up to 90% of initial flux. Zwitterionic membranes also owned good separation performance for NaCl solution at operation pressure of 1.6 MPa. The water flux of zwitterionic membranes were even slightly higher than conventional PA membrane.

Therefore, an efficient and feasible strategy was proposed to tremendously improve the anti-fouling property of PA RO membrane, showing significant effect on the research and application of anti-fouling membranes.

398759 Polymeric ligand Affinity Membranes for Protein Separations: Experiments and Simulations

Abstract:

Bisphosphonate drived ligand has a remarkable affinity for arginine (Arg) residues on protein surface. Here, a bisphosphonate derived ligand was successfully synthesized and grafted from the surface of regenerated cellulose membrane for protein separations. Atom transfer radical polymerization (ATRP) was used for surface modification of this polymeric ligand. Two types of monomers were copolymerized with different polymerization time (ATRP 1 h, 3 h and 5 h). Besides the bisphosphonate drived monomer, which was used as the functional monomer, another hydrophilic monomer N-(2-hydroxypropyl)methacrylamide(HPMA) was used for copolymerization to enhance the flexibility of the copolymer ligand and further improve specific protein adsorption. The copolymerization of this type of bisphosphonate copolymer derivatives was successful for the first time using ATRP. Static and dynamic binding capacities were determined for binding and elution of Arg rich lysozyme. The interaction mechanism between the copolymer ligand and lysozyme was elucidated using classical molecular dynamics (MD) simulations.

386173 Hydrophobic Membrane Surface Energy Varied By Plasma Enhanced CVD: Effect on Oil/Water Emulsion Separation

Abstract:

While attention and membrane development has focused on moving the aqueous phase across a membrane, recent materials research has focused on removing the organic phase from water[1] in a method somewhat analogous to the natural geological process whereby streams and rivers are cleaned by rock and soil. However, the mechanism of oil separation from water at a hydrophobic membrane surface is not clearly understood. Tirmizi et al. studied the effect of pore size, surfactant concentration and applied pressure on separation efficiency and flux[2]. They found the separation efficiency to be independent of pore size for polypropylene membranes containing 0.02 to 0.2 μm pores and observed a linear to limiting flux behavior with pressure as they increased the surfactant concentration from zero to 30 kg/m^3. Tirmizi et al. did not consider the effect of surface energy. Ochoa et al. considered the effect of material surface energy on membrane fouling for membrane surfaces having intrinsic water contact angles of 50 to 84 degrees[3] but lower energy surfaces were not considered.

To better understand the interactions between hydrocarbon in water emulsion and the surface of a hydrophobic membrane, we vary the surface energy of hydrophobic polypropylene membranes using plasma enhanced chemical vapor deposition[4]. Trans-membrane flux and concentrations of hydrocarbon (tetradecane) and water in the permeate and retentate are monitored to determine the rate and efficiency of separation as a function of surface energy.

[1] Zhongxin Xue, Yingze Cao, Na Liu, Lin Feng, and Lei Jiang. *J. Mater. Chem. A*, 2:2445-2460, 2014.

[2] Neena Pradhan Tirmizi, Bhavani Raghuraman, and John Wiencek. *AIChE Journal,* 42(5):1263-1276, 1996.

[3] N.A. Ochoa, M. Masuelli, and J. Marchese. *Journal of Membrane Science*, 226(1{2):203-211, 2003.

[4] Ila Badge, Sarang P. Bhawalkar, Li Jia, and Ali Dhinojwala. *Soft Matter*, 9:3032-3040, 2013.

A new method of decreasing fouling of polyethersulfone hollow-fiber membranes using peptoids

Neda Mahmoudi, Maryam Asgharpour, Jamie Hestekin, Shannon Servoss

Ralph E. Martin Department of Chemical Engineering, University of Arkansas, Fayetteville, AR 72701, United States

The bio-fouling of hollow fiber membrane often presents severe problems since proteins and cells can accumulate and create a layer on the surface of hollow fibers. This layer decreases the gas exchange of hollow fibers, affects the surface structure, alters their performance, and increases the cost of health care. Due to these issues with high fouling membranes, it is necessary to create low-fouling membranes that maintain the quality of production. We used peptoid with methoxyethylamine (NMEG) side chain since NMEG which is a valuable protein that exhibit protein resistance behavior. Here, we investigate the use of peptoids to prevent fouling of gas exchange membranes.

Peptoids have been attached to the surface of polysulfone hollow fiber membranes using dopamine. The attachment of peptoid was evaluated by X-ray photoelectron spectroscopy (XPS). The efficacy of the membranes is tested by adsorbing protein on peptoid-modified and unmodified membranes using fluorescent microscope.

Keywords

Bio-fouling, Hollow-fiber membranes (HFMs), Peptoid, Polysulfone, X-ray photoelectron spectroscopy (XPS).

Rejection of Charged Spherical Particles by Charged Porous Track-etched Membranes: Implications for Enhanced Virus and Turbidity Control

The effect of surface charge on particle rejection from UF and MF membranes has been examined from both theoretical and experimental perspectives. In this work, a computational model was developed to predict the combined effects of steric, hydrodynamic and electrostatic particle - pore wall interactions on the rejection coefficient of spherical particles in cylindrical pores. The non-linear Poisson-Boltzmann equation was used to determine the electrostatic particle - wall interaction energy, and a computational fluid dynamic model was developed to determine the hydrodynamic particle-pore wall interactions. These were then incorporated in a model of particle transport to calculate rejection. Model results show that particle rejection can be increased significantly when particle and membrane carry like negative charges.

To validate the numerical model, experiments were conducted with four different spherical colloidal silica particles and two spherical viruses (MS2 and PRD1). Track-etch porous membranes were used, providing us with an easily characterized nearly cylindrical pore geometry. Experiments were conducted under conditions where both particle and membrane carry a negative charge as well as under conditions where surface charges have minimal impact. Surface charge densities on the particles were determined from zeta potential measurements. Literature values were used for the surface charge density on the pore surface in the membranes. Experiments were designed to cover a broad range of dimensionless particle sizes by changing particle and pore sizes.

Particle rejections for the colloidal silica and viruses measured in relatively high ionic strength solutions (zeta potential $\rightarrow 0$) agreed very well with model predictions for uncharged systems (Figure 1a). This confirms the validity of the pore size values, the experimental procedures used for the rejection measurements, and the accuracy of the theoretical model incorporating steric and hydrodynamic interactions. In low ionic strength solutions, the zeta potential of silica particles was near -48mV and that of the viruses was near -20mV. With no fitting parameters involved in the comparison, good agreement between experimentally determined rejection coefficients and model predictions was observed for these systems. Results for one of the silica colloids are shown in Figure 1b. Similar results were obtained for other silica particles of different size and for two different viruses.

In all cases, rejection of charged particles was greater than that of uncharged particles showing that one can potentially take advantage of electrostatic repulsion to get complete rejection for particles that are smaller than the pore size.

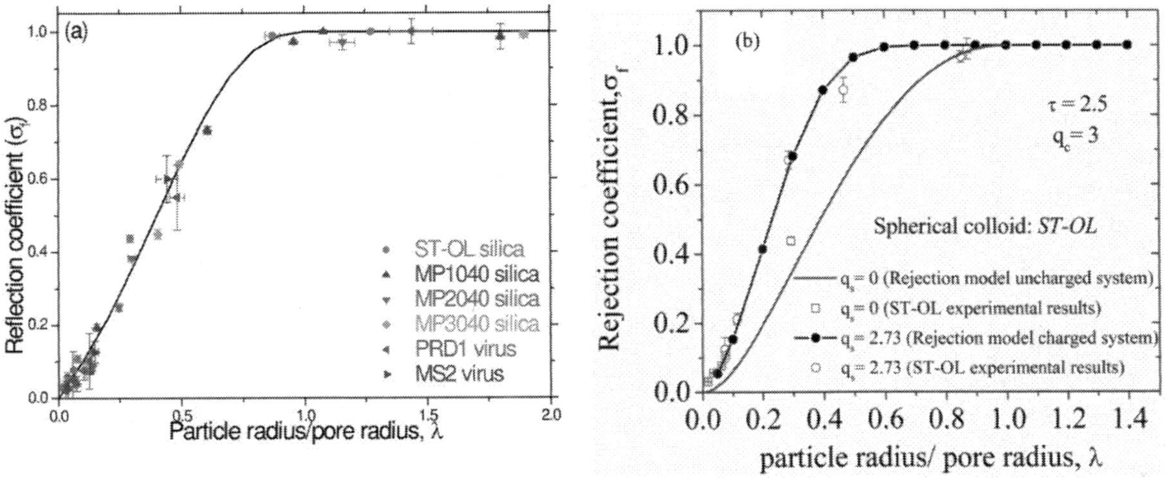

Figure 1: Quantitative comparison of experimental results for spherical particles (a) uncharged and (b) charged with theoretical predictions of the rejection coefficient.

377092 Membrane-Encapsulated Solar Fuel Generators

Abstract:

Membrane-encapsulated devices are attractive for solar-driven hydrogen production, as their design mitigates problems with concentration overpotentials, gas bubbles, and safety inherent in liquid electrolyte water-splitting devices.[1-4] However, the competing processes such as heat, water, gas, and ion transport make it nontrivial to design such devices. Here, we show analytical expressions and associated design spaces for critical membrane dimensions and material properties required for stable and efficient gas, heat, proton, and water transport. The best condition for stable gas transport in thin film or gas channels is given by the critical Damköhler number. The maximum heating of the device is governed by the Heating and Nusselt numbers, which can be tuned to the desired level. The optimal dimensions of the device, to operate under tolerable ohmic losses, correspond to the maximum value of scaled Power-loss factor. An optimal device architecture is proposed for stable and efficient operation.

References:

(1) Spurgeon, J. M.; Lewis, N. S., Proton exchange membrane electrolysis sustained by water vapor. *Energy & Environmental Science* 2011, 4, (8), 2993-2998.

(2) Xiang, C.; Chen, Y.; Lewis, N. S., Modeling an integrated photoelectrolysis system sustained by water vapor. *Energy & Environmental Science* 2013, 6, (12), 3713-3721.

(3) Haussener, S.; Xiang, C.; Spurgeon, J. M.; Ardo, S.; Lewis, N. S.; Weber, A. Z., Modeling, simulation, and design criteria for photoelectrochemical water-splitting systems. *Energy & Environmental Science* 2012, 5, (12), 9922-9935.

(4) Singh, M. R.; Stevens, J. C.; Weber, A. Z., Design of Membrane-Encapsulated Wireless Photoelectrochemical Cells for Hydrogen Production. *Journal of The Electrochemical Society* 2014, 161, (8).

Experimental and mathematical study of multicomponent ion transport in a Nafion membrane at high current densities

S. Moshtarikhah[1], N.A.W. Oppers[1], M.T. de Groot[2], J.T.F. Keurentjes[1], J.C. Schouten[1] and J. van der Schaaf[1*]

[1]*Eindhoven University of Technology, P.O.Box 513, 5600 MB Eindhoven, The Netherlands*
[2]*Akzo Nobel industrial Chemicals B.V., P.O.Box 247, 3800 AE Amersfoort, The Netherlands*
**Corresponding author: j.vanderschaaf@tue.nl*

The productivity of an electrochemical cell is determined by the current density and the selectivity of the ion-selective membrane that separates the anode and cathode compartments. For industrial processes, the productivity is limited by mass transfer to the electrodes. At current densities that exceed the mass transfer rate limit, undesired reactions occur. The mass transfer rate at the electrodes can be augmented with spinning disc electrodes and by using highly concentrated systems. This way, much higher current densities can be attained. However, virtually no data exists on the performance of ion-selective membranes at high current densities. The selectivity and conductivity of the membrane at high current densities are vital to assess the performance of the electrochemical cell and the process economics. Therefore, this paper presents data for membrane performance at high current densities up to 16 $kA.m^{-2}$. Furthermore, the ion transport in the membrane is modeled with the Nernst-Planck equations and with the Maxwell-Stefan equations [1]. The Nernst-plank equation neglects the interactions of ions and according to literature it is valid for dilute concentrations [2]. The Maxwell-Stefan equation takes the interaction of different species into account. However, it requires reliable data for diffusion coefficients [3]. The non-ideal behavior of the highly concentrated sodium hydroxide solution is modeled with the Pitzer activity model. The boundary conditions are determined with the Donnan equilibrium [4,5].

The model is validated with experiments performed in a temperature controlled electrolysis cell, using a perfluorinated sulfonate membrane, Nafion 1110, to separating the anode and cathode compartments. Sodium hydroxide is used as an electrolyte solution in order to monitor the transport of sodium ions and back-migration of hydroxyl ions to the cathode and the anode compartments respectively. The potential drop over the membrane is measured using two Haber-Luggin capillaries with reference electrodes and the tips of the capillary tubes close to either side of the membrane [4]. The selectivity of the membrane to sodium ions and hydroxyl back-migration is studied by measuring the concentration change in anode and cathode compartments and the total current passed through the membrane in a certain amount of time. The potential drop over the membrane and the membrane selectivity are determined for current densities up to 16 $kA.m^{-2}$ for a sodium hydroxide system with concentrations of 10 wt% to 32 wt%.

The Nernst-Planck model predicts that an increase in current density from 2 $kA.m^{-2}$ to 10 $kA.m^{-2}$ increases the positive ion flux from 0.02 $mol.m^{-2}s^{-2}$ to 0.1 $mol.m^{-2}s^{-2}$. The experiments show a similar trend in increase of positive ion flux from 0.02 $mol.m^{-2}s^{-2}$ to 0.07 $mol.m^{-2}s^{-2}$, however, with a clear reduction in selectivity. The model predicts a significant drop in membrane resistance from 0.045 $ohm.cm^2$ to 0.04 $ohm.cm^2$. The experiments show a bigger drop in membrane resistance from 1.3 $ohm.cm^2$ to 0.53 $ohm.cm^2$. Possibly, the decrease in selectivity measured experimentally, results in a lower voltage drop over the membrane. The transport of hydroxyl ions is much easier through the membrane than the transport of sodium ions.

Keywords: multicomponent ion transport, Nernst-Planck Maxwell-Stefan, membrane resistivity, membrane selectivity, high current density

References

[1] R. Krishna, Multicomponent mass transfer, John Wiley and Sons, 1993.
[2] S.T.P. Psaltis, T.W. Farrell, Comparing charge transport predictions for a ternary electrolyte using the Maxwell-Stefan and Nernst-Planck equations, J. Electrochem. Soc. 158 (2010) A33-A42.
[3] J.H.G. Van der Stegen, The state of the art of modern chlor alkali electrolysis with membrane cells, PhD thesis, Twente University, 2000.
[4] H. Strathmann, Ion-exchange membrane separation processes, first ed., Elsevier, 2004.
[5] M. Higa, A. Tanioka, K. Miyasaka, Simulation of the transport of ions against their concentration gradient across charged membrane, J. Mem. Sci. 37 (1987) 251-266.

381404 CO2 Separation and Plasticization Properties of Copolyimides Containing Sulfone Groups

Abstract:

Polyimide membranes have been widely applied for gas separations due to their attractive permeability, selectivity, and processing characteristics. Their use could be further expanded for CO_2 capture in natural gas purification, coal gasification, and flue gas treatment processes. The principle issue in these applications is the plasticization of the polyimide membranes at high partial pressures of CO_2, which can lead to reduced membrane selectivity and unpredictable membrane properties.

To investigate structure-plasticization relationships, three sulfonated copolyimides were selected: 6FDA/BTDA-pBAPS, 6FDA-pBAPS/DABA, and 6FDA-pBAPS/mPDA. The selection of the copolyimide structures was based on our previous study [1] in which the permeability coefficients of H_2, O_2, He, CO_2, N_2, and CH_4 gases for more than 2200 possible copolyimide structures were estimated by the group contribution method of Alentiev et al. [2]. The predicted permselectivities of the above mentioned copolyimides were located well above the upper bound of Robeson diagrams for CO_2/CH_4 and O_2/N_2separation applications. Proceeding, in another previous study of us [3] the copolyimides were synthesized and characteristics of membranes were determined both experimentally and via molecular simulation methods. The physical characteristics of simulated membranes matched well with the experimental ones and also sorption characteristics were reported.

In this work Molecular Dynamic (MD) and Monte Carlo (MC) simulations were carried out to investigate the relationship between structural properties and CO_2-induced plasticization behavior of copolyimides and CO_2/CH_4separation. Polymer Consistent Force Field (PCFF) was used to simulations. Simulated sorption coefficients agreed well with the experimental results obtained from gravimetric sorption (IGA) measurements. Diffusion coefficients of copolyimides were determined using MD simulation runs and permeabilities and perm-selectivities of each copolyimide membrane were evaluated up to 30 bars to investigate the plasticization resistance of each copolyimide.

Among the three copolyimides, 6FDA/BTDA-pBAPS provided the highest CO_2 and CH_4 gas sorption coefficients indicating that fractional free volume of 6FDA/BTDA-pBAPS is highest in the swollen state. FFV calculations confirm that CO_2 sorption-induced swelling is the highest in 6FDA/BTDA-pBAPS, while the FFV of the three copolyimides were similar before sorption. The sorption-induced increase of the FFV of 6FDA-pBAPS/DABA is the smallest which also agrees with the FAV analysis. This may be due to its rigid backbone indicated by its high Tg. Sulfone group (S) in pBAPS diamine is the prefential sorption site in all three copolyimides, but its affinity to CO_2 is highest in 6FDA/BTDA-pBAPS and lowest in 6FDA-pBAPS/DABA. Certain groups and regions in copolyimides where CO_2affinity is higher is located.

Abbreviations: 6FDA: 4,4-hexafluoro isopropylidene diphthalicanhydride; DABA: 3,5-diamino benzoic acid; BTDA: 3,3-4,4-benzophenone tetracarboxyclic dianhydride; pBAPS: bis [4-(4-aminophenoxy) phenyl] sulfone; mPDA: 1,3-phenylenediamine.

Acknowledgment: This work was supported by TUBITAK trough grant no. 113M336.

1. **Halitoglu, S.; Tantekin Ersolmaz, S.B.,** "Prediction of gas permeability coefficients of copolyimides by group contribution methods", presented at the NAMS meeting, May 12-16, 2007, Orlando, Florida.

2. **Alentiev, A. Y.; Loza, K. A.; Yampolski, Y.** "Development of the methods for prediction of gas permeation parameters of glassy polymers: polyimides as alternating co-polymers", *J. Memb. Sci.*, **167**, 91-106 (2000)

385876 Numerical and Analytical Ion Transport Model to Investigate the Influence of Convection in Electrodialysis Reversal

Abstract:

Numerical and Analytical Ion Transport Model to Investigate the Influence of Convection in Electrodialysis Reversal

Azadeh Ghorbani *, Abbas Ghassemi

Institute for Energy and the Environment New Mexico State University, Las Cruces, NM, USA.
***azadeh12@nmsu.edu**

There is no doubt that demand for fresh water will continue to increase as the world's population grows and as the global demand for consumer goods increases. Desalination processes, which remove salts from saline water to create freshwater, are a group of technologies with the potential to help meet this demand.

One of the main technologies for water desalination is Electrodialysis reversal (EDR), which uses electrical potential difference to drive ions through a porous media called an ion exchange membrane. Despite this technology's widespread use, there is not currently a complete theoretical model for how ions transfer through the ion exchange membrane under the driving force of electrical potential difference: while current models account for the driving forces of migration and diffusion, convection has not been addressed.

A model which includes all three forces could create more accurate data and help design more efficient EDR technologies. For this reason a complete model is highly desirable, and the goal of this study is to create such a model using computational fluid dynamics to include convection forces.

390531 Solvation of Vanadium Cations: Effects of Sulfuric and Triflic Acids

Abstract:

The desire, in recent decades, to replace fossil fuel with renewable green energy sources such as solar and wind has significantly increasing as a result of environmental issues. One of the limiting factors for the utilization of renewable energy sources in the electrical supply grid is the intermittent nature of these sources which requires the employment of megawatt scale electrical energy storage devices. Redox flow batteries (RFBs) are among the promising large-scale electrical energy storage systems and have received significant attention due to their superior properties. RFBs provide long life-span, simple installation, the possibility of instant charging by replacing the electrolytes, and the ability to repeatedly store and convert electrical energy into chemical energy and vice versa. In addition, RFBs have the unique advantage of having power and energy decoupled due to the separation of the electrolyte reservoirs and the battery stacks.

Of the various RFBs currently under investigation, the vanadium redox flow battery (VRFB) demonstrates design flexibility, good electrochemical activity, reversibility, and low maintenance cost. A VRFB consists of a V^{2+}/V^{3+} sulfate solution at the negative electrolyte and a VO_2^+/VO^{2+} sulfate solution at the positive electrolyte separated by a proton exchange membrane (PEM). The wide span between the standard reduction potentials of the two half-cells in these batteries produces a standard voltage of 1.25 V, which is comparable to other types of RFBs. Despite the numerous advantages of VRFBs, there still exist a number of issues that hinder the commercialization of these storage devices. For example, the poor stability of the VO_2^+ solution at high temperatures and high concentrations, results in precipitation of hydrated V_2O_5 and energy loss [6]. Although, the stability of the VO_2^+ ion may be increased by the addition of hydrochloric acid, eventually it will form a gel-type precipitate involving the V^{3+} cation. It is therefore important to understand the chemistry and thermodynamics of solvation of vanadium cations in the electrolyte for design improvements. The other challenging area arises from the membranes used in these devices. The most widely used PEM in VRFBs is Nafion which degrades as a result of exposure to the highly acidic electrolytes. It has also been found that the low ion selectivity of PEMs allows permeation of vanadium cations and cross contamination of the electrolytes. Hence, knowing the nature of the solvated cation in electrolyte can help to understand the stability of the electrolyte solution as well as the formation of other ionic complexes. Also, the study of the diffused cations within PFSA can help in designing membranes with higher ion selectivity, distinguishing the proton from vanadium cations.

Recently, we determined the structure and hydration thermodynamics of the four cations in bulk water with first principles based electronic structure calculations and a quasi-chemical theory. Here we examine the effects of the highly acidic environment in a VRFB, on the hydration of the four vanadium cations in an effort to understand the mechanism of migration of the species in the ion exchange membrane. The effects of both sulfuric and triflic (CF_3SO_3H) acid are examined as they mimic the electrolyte and membrane environment of a VRFB. Hybrid density functional theory in conjunction with a continuum solvation model was utilized to obtain the local structures of the hydrated vanadium cations in proximity of sulfuric acid, triflic acid and their conjugate anions.

The results indicate that the acids/anions do not form strong covalent bonds with the vanadium cations. It is found that the acid molecules interact directly to the V^{2+} cation while their conjugate anions (HSO_4^- and $CF_3SO_3^-$) interact with cation from outside of the V^{2+} hydration shell. Also, the direct interaction of acid with the V^{2+} cation results in the proton dissociation of only the stronger acid, triflic, and not the sulfuric acid. However, in the complexes of V^{3+} both the triflic and sulfuric acid protons fully dissociate. For VO^{2+} cation, the axial water in the first hydration shell is replaced by acids or anions. Also, the results indicate that there are no interactions between the oxo-oxygen of VO^{2+} and acidic groups, regardless of the negative atomic charge on the oxo-oxygen. In contrast, the oxo-oxygen of VO_2^+ interacts with the acidic hydrogen and protonates by triflic acid or sulfuric acid presence.

Combining Solid-State and Solution Calorimetry for the Calculation of the Gibbs Free Energy of Polymorphs to Determine the Relative Stability and Solubility

Daniel Hsieh[1], Daniel Roberts[1], Tamar Rosenbaum[1], Chiajen Lai[1], Qi Gao[1], Thorsten Rosner[2]

[1]*Drug Product Science and Technology, Bristol-Myers Squibb Co., New Brunswick, NJ, 08903, USA*
[2]*Chemical Development, Bristol-Myers Squibb Co., New Brunswick, NJ 08903, USA*

Abstract
Determining the relative stability of polymorphs for a pharmaceutical compound is very important to chemical processes, product development, and drug performance. A method for such determination using a novel pure component free energy calculation has been successfully developed [1]. This method incorporates both experimental and theoretical approaches. Perkin Elmer HyperDSC™ was used for the measurement of thermal data including the heat capacity (c_p), melting temperature (T_m), glass transition temperature (T_g) and the heat of fusion (ΔH_f). With conventional DSC – which uses a relatively low heating rate (10-20°C/min) – recrystallization can occur during the melt, causing inaccurate data due to the convolution of multiple melt/recrystallization events. However, due to the high heating rate of HyperDSC™ (up to 750°C/min), recrystallization can be avoided and heat of fusion for polymorphs can thus be reliably measured. In addition, the heat capacity for crystalline polymorphs and amorphous materials can be accurately measured via HyperDSC™ over a wide temperature range. The accurate thermal data for the pure components, combined with the derived thermodynamic model [1], provide the basis for the determination of the relative stability of the polymorphs over the desired temperature range, greatly aiding efforts in process and product development.

Although HyperDSC™ offers an innovative approach for the calculation of the Gibbs free energy [1], it requires a highly accurate heat of fusion measurement. For most pharmaceutical products, the high heating rate of HyperDSC™ is able to suppress recrystallization and generate accurate heat of fusion values. However, two extreme cases have been encountered during product development in which the accurate heat of fusion measurement becomes untenable. One case involves the rapid chemical decomposition of the drug substance upon melting even at a high heating rate of 500°C/min. The other case entails the instantaneous recrystallization of the drug substance upon melting at the maximum heating rate of HyperDSC™, 750 °C/min. The objective of this study is to develop a method that can address these limitations by determining the heat of fusion from the enthalpy data at lower temperatures by using solution calorimetry such that polymorphic form change, recrystallization and thermal decomposition are far less likely to occur. Both theoretical approaches and experimental approaches are studied and evaluated.

The theoretical approach includes the derivation of the thermodynamic equations to relate the heat of fusion to the enthalpy data at lower temperature. On the basis of this derivation the thermal data required to determine the heat of fusion values are identified. The experimental approach includes the setup and validation of the aforementioned data collection strategies.

References:

[1] Daniel Hsieh, Jun Huang, Daniel Roberts, Qi Gao, Alicia Ng, Deniz Erdemir, David Leahy, Jun Li, Ming Huang, Chiajen Lai, "Determination of Relative Stability of a Multi-Polymorph System via a Novel Pure Component Free Energy Calculation", Crystal Growth & Design 2012, 12, 5481-5490

379700 Predicting the Neat Polymorphs of Axitinib

Abstract:

In recent years, there has been significant progress in crystal structure prediction methodologies, as demonstrated by recent blind tests and other work on the prediction of the crystal structure of small pharmaceutical molecules[1]. In this paper, we discuss the application of one particular approach that has proved successful to the challenging case of axitinib. Axitinib is well-studied molecule developed by Pfizer, which has been shown to have 71 polymorphs[2] of which 5 are neat (i.e., they contain only the axitinib molecule). The focus of the present study on neat polymorphs with one molecule in the asymmetric unit, of which 4 are known experimentally (Forms I, VI, XXV and XLI).

The approach to crystal structure prediction adopted here is based on the lattice energy minimization of a large number of trial structures at 0 K in order to generate an energy landscape, in which low-energy minima are identified based on the computational model of choice. Success is dependent upon the accuracy of the lattice energy calculations, especially in terms of electrostatics and the impact of molecular flexibility, and the ability to optimise a hundreds of thousands to millions of putative structures. Such approaches have been proven to be successful in predicting the polymorphs of compounds with one or two polymorphs in many cases3.

We use a multi-stage methodology. A model is first chosen based on analysis of the conformational flexibility of the molecule. Then 4,800,000 structures are generated and locally minimised with a computationally inexpensive model, using the CrystalPredictor4,5 algorithm. Then the most promising structures (lowest in energy) are further minimised in a two-step approach with CrystalOptimizer6,7, a program performing local minimisation, adopting a highly accurate though computationally expensive model. The axitinib molecule is found to have a richly populated energy landscape. The four known structures are predicted as lattice energy minima with the chosen model. Although Form XLI is not found to be the most stable at 0 K, contrary to expectations based on data at higher temperatures, the relative stability of other forms is predicted correctly. The relative importance of the different contributions to the energy model (intramolecular energy, electrostatics, repulsion/dispersion) is analysed and provides insights into the need for future development of predictive methodologies.

Bibliography

1. C. C. Pantelides, C. S. Adjiman, A. V. Kazantsev. "General computational algorithms for ab initio crystal structure prediction for organic molecules". In S. Atahan-Evrenk and A. Aspuru-Guzik, editors, Prediction and Calculation of Crystal Structures, volume 345 of Topics in Current Chemistry, pages 25-58. Springer International Publishing, 2014.

2. A. M. Campeta, B. P. Chekal, Y. A. Abramov, P. A. Meenan, M. J. Henson, B. Shi, R. A. Singer, K. R. Horspool, Development of a targeted polymorph screening approach for a complex polymorphic and highly solvating API, Journal of Pharmaceutical Sciences, 99(9):3874-3886, 2010.

3. D. A. Bardwell, C. S. Adjiman, Y. A. Arnautova, E. Bartashevich, S. X. M. Boerrigter, D. E. Braun, A. J. Cruz-Cabeza, G. M. Day, R. G. Della Valle, G. R. Desiraju, B. P. van Eijck, J. C. Facelli, M. B. Ferraro, D. Grillo, M. Habgood, D. W. M. Hofmann, F. Hofmann, K. V. Jovan Jose, P. G. Karamertzanis, A. V. Kazantsev, J. Kendrick, L. N. Kuleshova, F. J. J. Leusen, A. V. Maleev, A. J. Misquitta, S. Mohamed, R. J. Needs, M. A. Neumann, D. Nikylov, A. M.Orendt, R. Pal, C. C. Pantelides, C. J. Pickard, L. S.Price, S. L. Price, H. A. Scheraga, J. van de Streek, T. S.Thakur, S. Tiwari, E. Venuti, and I. K. Zhitkov. Towards crystal structure prediction of complex organic compounds -a report on thefifth blind test. Acta Crystallographica Section B, 67(6):535-551, 2011.

4. P.G. Karamertzanis, C.C. Pantelides, "Ab initio-Crystal Structure Prediction – I. Rigid molecules", J. Comput. Chem., 26:304-324 (2005)

5. P.G. Karamertzanis, C.C. Pantelides, "Ab initio-Crystal Structure Prediction – II. Flexible molecules", Molecular Physics, Vol. 105 , Nos. 2-3, 273-291 (2007)

6. A.V. Kazantsev, P.G. Karamertzanis, C. S. Adjiman, C C. Pantelides, J. Chem. Theory Comput., Efficient Handling of Molecular Flexibility in Lattice Energy Minimization of Organic Crystals, 7(6):1998-2016, 2011.

A Kinetic Approach to Polymorph Prediction: Umbrella Sampling and Molecular Dynamics

Conor Parks, and Doraiswami Ramkrishna

Abstract

Crystallization has remained one of the main separation techniques employed by the pharmaceutical industries throughout its history. However, despite the long-standing tradition of this separation process, little is known about the physics concerning the nucleation event itself. This is most reflected in the stubborn problem of polymorphism: a molecule's ability to exist in more than one distinct crystalline form. Due to the strong correlation between a molecule's three dimensional crystal structure and the physical properties which it exhibits, the phenomenon of polymorphism leads to variations in a potential drug molecule's most important characteristics: solid-state stability and bioavailability. Inability to predict a molecules crystal structure from first principles has made solvent form screening the de facto industrial standard for selecting operating conditions that lead to the synthesis of the most favorable crystal structure. However, this most notably failed in 1998 for Abbott Laboratories. Two years after the release of the anti-retroviral drug Ritonavir to market, a more stable, less soluble crystalline phase appeared that resulted in dissolution failures of the soft gelatin capsules. This forced the withdrawal of the drug from the market, creating a market crisis for Abbott Laboratories. From a first principles perspective, the area of crystal structure prediction is dominated by the lattice energy minimization approach. However, this suffers the major drawback of not only assuming that the observed structure will correspond to the global minimum in free energy, but also the difficulty of including temperature and entropy effects frequently leads to these calculations being conducted at 0 K. Hence, no reference to operating conditions such as temperature, solvent, or supersaturation is accounted for in these models.

We show the first ever crystal structure prediction of a general organic molecule, tetrolic acid, based off the kinetics of nucleation. Tetrolic acid is known to crystallize into two different space groups, dependent on solvent. We use the rare event simulation technique, umbrella sampling coupled with molecular dynamics to calculate the rate of nucleation of the critical nuclei cluster in each solvent. We show the first ever demonstration that the correlation between the polarity of the solvent and the resulting crystal structure, known as the link hypothesis, is due to a kinetic predisposition imposed by the solvent. We also demonstrate the prediction of the crystal structure of tetrolic acid based off of a kinetic framework.

Prediction of energetic - energetic cocrystal forms and their competition with solvate formation

David am Ende, Andreas Klamt, Christoph Loschen

Cocrystallization has attracted increased interest as an alternative pathway for the reliable and reproducible formulation of pharmaceutical actives, with the additional advantage of having influence on the final bioavailability by the right choice of the coformer.[1] The COSMO-RS method in its COSMO*therm* and COSMO*quick* implementations has been demonstrated to be able to guide the choice of promising coformers by the prediction of the relative probability that a coformer forms a cocrystal with a given active. In addition, it was shown that a quantitative estimate of the solubility of the resulting cocrystal can be obtained.[2] .

In this presentation we will demonstrate that the same approach also works as a useful guide for the development of cocrystals of energetic compounds. Furthermore it will be demonstrated that COSMO*therm* can be used to identify preferred solvents and solvent mixtures suitable for isolating the cocrystal while avoiding strong solvate formation.

[1] Blagden, N.; de Matas, M.; Gavan, P. and York, P. crystal engineering of active pharmaceutical ingredients to improve solubility and dissolution rates. *Advanced Drug Delivery Reviews* **2007**, *59*, 617.
[2] Abramov, Y. A.; Loschen, C. and Klamt, A. rational coformer or solvent selection for pharmaceutical cocrystallization or desolvation. *J. Pharm. Sci.* **2012**, *101*, 3687.

376929 Nucleation and Growth of Organic Semiconductor Polymorphs

Abstract:

Molecular packing has a profound impact on the solid-state properties of materials. Since many organic compounds can pack into distinct crystal structures (polymorphism), controlled polymorph formation is vital to a wide range of areas, e.g., pharmaceuticals, pigments, food, explosives, etc. Recently, crystal polymorphism is emerging as an important subject in the area of organic semiconductors (OSC), given that substantial change in the electronic properties can be induced by even small changes in the crystal packing. Many of the benchmark OSCs are found to be polymorphic. However, the discovery of polymorphs in this field has been mostly by serendipity. With higher performing molecules being designed at an ever-faster pace, controlled nucleation and growth of polymorphs is becoming more and more important not only for tuning the charge transport properties, but also for producing 'organic' circuits with consistent performance and high stability.

In this study, we take a comprehensive approach to unravel the nucleation and growth of organic semiconductor polymorphs, combining advanced synchrotron X-ray techniques with molecular simulations, and charge transport characterizations with quantum chemical calculations. Using in situ X-ray diffraction techniques, new polymorphs are identified for even extensively studied molecules. The relative stability and unit cell parameters of polymorphs predicted independently by molecular simulations closely match with the experimental observations. Using quantum chemical calculations, we found significant change in the charge transfer integral across different polymorphs, which is supported by orders of magnitude difference in the measured charge carrier mobility. In addition, we infer the molecular origin of polymorphism based on structural refinement and molecular simulations.

383371 Metal-Impregnated Carbon Silica Composites for NH3 and SO2 Adsorption

Abstract:

A biphasic material capable of effectively adsorbing a variety of light gases has promising applications in technologies such as civilian and military respiratory filters. Such a material is advantageous over traditional single-phase materials that are limited to a niche of target gases. Consisting of a polar siliceous phase and a non-polar carbonaceous phase, this carbon-silica composite (CSC) previously synthesized by our group targets dilute concentrations of both acid-forming gases such as SO_2 and basic gases such as NH_3 [1,2]. Through physical intermolecular attractions and varying polar contributions from each phase, the acidity/basicity of the surface of the adsorbent can capture NH_3 and SO_2 [1,2]. Pre-adsorbed water molecules under simulated humid environments enhance the adsorption of molecules such as SO_2 by forming acids that have increased reactivity [2].

In this study, to further promote light gas adsorption, a catalytic phase consisting of impregnated transition metals has been incorporated onto the surface of CSCs. Various metal salts including $Cu(NO_3)$, $Fe_2(SO_4)_3$, and $ZnCl_2$ have been impregnated onto CSCs through solution-condensation and wet-capillary impregnation methods. [Metal]-CSCs have been characterized using crystallography, porosimetry, thermogravimetric analysis, and electron microscopy. Results show that approximately 10-30 wt% of metal salt impregnant is able to increase the capacity of the non-impregnated sample by up to ~150%.

[1] T. G. Glover, K. I. Dunne, R. J. Davis, M. D. LeVan, Microporous and Mesoporous Materials, 111, 1-11 (2008).

[2] A. M. B. Furtado, Y. Wang, M. D. LeVan, Microporous and Mesoporous Materials, 165, 48-54 (2013).

389600 Synthesis of Ionic Liquid Silica Composite Adsorbents

Abstract:

A conventional ionic liquid, 1-methyl-3-butylimidazolium bis(trifluoromethylsulfonyl)imide ([Bmim][Tf2N]), has been successfully impregnated by physical adsorption into the ordered mesoporous silicas MCM-41 and SBA-15. For MCM-41, mixed solutions of polar protic and polar aprotic solvents provide selective solvation of the IL and can overcome pore diffusion limitations caused by self-association of the IL and steric hindrance. Proper selection of the mixed solvents allows for the control of the amount of ionic liquid confined within the pores. Materials were prepared in dry environments using a Schlenk line and in ambient conditions using incipient wetness. The materials have been characterized with X-ray diffraction, nitrogen isotherms, and thermal gravimetric analysis. Carbon dioxide, hexane, and water adsorption isotherms of MCM-41, the bulk ionic liquid, and the composite material have been measured to determine composite sorption effects. The materials are complex and the gas adsorption isotherms reflect both adsorption on the silica support and the solubility of the gas in the ionic liquid.

Novel Iron Carbide-derived Carbons with Embedded Iron Nanoparticles for Ammonia Filtration

*Michael C. Mangarella, Jessica L. Ewbank, Michael R. Dutzer, Faisal M. Alamgir, Krista S. Walton**

Carbide-derived carbons (CDCs) are highly tunable porous carbons created by the extraction of a metal from a carbide matrix, typically by chlorination. CDCs have many attractive physical properties, including high bulk porosity, high microporosity, and tunable pore size. In this work, Fe_3C was chlorinated for varying chlorination times and prepared samples were investigated to fundamentally understand the extraction process of Fe. The objective of this study was to understand if Fe from the carbide precursor could be retained as an active site for adsorption. Interestingly, the residual Fe studied by X-ray absorption near edge spectroscopy (XANES), scanning electron microscopy (SEM), and transmission electron microscopy (TEM) shows the removal of Fe does not follow a layer-by-layer extraction, as observed in the chlorination of other carbides. Instead, iron carbide is shown to convert to iron (II) and iron (III) chloride nanoparticles, which remain trapped within the resulting nanoporous carbon framework. It is hypothesized that after conversion to $FeCl_3$ and $FeCl_2$, a significant percentage of iron chloride species catalyzes the formation of graphitic nanostructures that prevent their extraction through encapsulation. Depending on the synthesis conditions and post-treatment methods, the complex extraction method of Fe from Fe_3C at moderate chlorination temperatures allows for the tailored inclusion of $FeCl_2$, $FeCl_3$, or Fe nanoparticles in the resulting CDC. Tailored Fe_3CDC adsorbents synthesized through this method show exceptional results in ammonia breakthrough experiments, likely due to their high microporosity and well-dispersed active Fe nanoparticle species.

Evaluation of Molybdenum-99 Sorbents for Medical Isotope Applications

Lee-Wei Kao*, Lei Ling*, Pei-lun Chung*, Amanda Youker[†], Dominique C. Stepinski[†], George F. Vandegrift[†], and Nien-Hwa Linda Wang**

*School of Chemical Engineering, Purdue University, West Lafayette, IN 47907

**Email: wangn@ecn.purdue.edu, Phone: 765-494-4081, Fax: 765-494-0805

[†]Argonne National Laboratory, Argonne, IL 60439

Abstract

Technetium-99m is used in about 30 million diagnostic procedures annually throughout the world. Its precursor, Molybdenum-99 (^{99}Mo), is produced from the fission of highly enriched uranium (HEU). However, the four major reactors for ^{99}Mo production were built in the 1950s and 1960s, and the two largest reactors are scheduled to shut down in 2016 and 2022, respectively. A new process is being developed to produce ^{99}Mo from the fission of low-enriched uranium (LEU) salt solutions. The ^{99}Mo is captured and separated from the LEU, which is then recycled back to the reactor. The capture process must meet several constraints, including a short capture time (< 2 hours), a high capture yield (99%), and a low pressure drop (<0.8 atm). The objective of this study is to identify the sorbent which requires the smallest column volume for capture while meeting the constraints. Four sorbents (S80, T52, T5M, and Al_2O_3) are tested for potential use in the ^{99}Mo capture process. A systematic procedure for evaluating the sorbent candidates is developed. This method is based on the intrinsic adsorption and diffusion parameters. Batch tests are used to determine the Langmuir isotherm parameters of ^{99}Mo for each sorbent, and frontal tests are used to estimate the effective intra-particle diffusivities of ^{99}Mo. A general design method based on the intrinsic parameters and dimensionless groups is developed to estimate, for each sorbent, the minimum column volume which is required to meet the design constraints. This method quickly identifies S80 as the sorbent which requires the smallest column volume for capture. The pressure limit is found to have no effect on the minimum column volume for capture, but it affects the column length and diameter. Furthermore, the method reveals the sorbent properties which can reduce the column volume. The column capacity and the ratio of diffusion time to loading time control the minimum column volume. In general, the minimum column volume is found to decrease with increasing loading time, decreasing particle size, increasing diffusivity, and increasing sorbent capacity.

386405 Copper Ion Adsorption By Chitosan and Alginate Gel Nanoparticles and Microparticles for Water Purification Applications

Abstract:

Chitosan and alginate are both low-cost natural carbohydrate materials used for the removal of metal ions from aqueous solutions. This study compared the equilibrium adsorption capacity and adsorption kinetics of chitosan gel nanoparticles, calcium-alginate gel microparticles, chitosan/alginate combination particles and large alginate gel beads for copper (Cu (II)) ions in a batch system at fixed pH. Moreover, the performance of large alginate beads in a batch system compared to that in a continuous flow system was examined.

Results show that the equilibrium adsorption behavior of chitosan and alginate in the low concentration region follows the Langmuir isotherm. Alginate has significantly higher capacity than chitosan, which can be attributed to a comparatively higher density of adsorption sites. Combination particles consisting of alginate microbeads coated with chitosan nanoparticles possess an intermediate maximum adsorption capacity, corresponding to the weight ratio of the alginate and chitosan. Meanwhile, it was observed that the adsorption kinetics of large alginate gel beads were much faster than that of alginate microbeads and combination particles.

For comparison with batch studies, the adsorption behavior of fixed-bed columns packed with large alginate gel beads was studied by varying the column size and volumetric flow rate. The column had shorter active life at a higher flow rate, or at a smaller size. Conventional fixed bed models successfully fit experimental data, allowing prediction of the breakthrough time and estimation of treatment capacity relative to typical commercial ion exchange resins.

376813 Tailoring Resistivity of Engelhard Titanosilicate (ETS-10) Molecular Sieve By Carbon Addition

Abstract:

Resistive heating is a technique for regenerating adsorbents whereby electric current is passed through a material with sufficiently low resistivity (typically, 0.1 – 10 Ω·m) and heat is generated by the Joule effect. This allows for rapid and low-cost adsorbent regeneration that is decoupled from the purge gas flow. Molecular sieves are effective adsorbents because they have tailored and highly structured pore size distributions. Unlike carbonaceous materials, however, molecular sieves have high electrical resistivity that prevents their regeneration by resistive heating. In this study, different methods were investigated for adding carbon (low resistivity) to Engelhard Titanosilicate (ETS-10) molecular sieve (high resistivity, ~5E+6 Ω·m) in order to decrease its resistivity without compromising its unique physical structure. Carbon was added to ETS-10 using chemical vapor deposition (CVD) of ethanol, thermal degradation of polyvinyl alcohol (PVA), and mixing with powdered carbon. Modified samples were characterized using micropore surface analysis, X-ray photoelectron spectroscopy (XPS), X-ray diffraction (XRD), scanning electron microscopy (SEM), and resistivity measurements. XPS, SEM, and a color change from white to grey confirmed the addition of carbon onto ETS-10 using all tested methods. Results from resistivity measurements showed decreases in ETS-10 resistivity, by up to 6 orders of magnitude, for the modified samples. The BET surface area of samples prepared by CVD was between 55 and 99% lower than untreated ETS-10, suggesting that the carbon coating caused pore blockage. Conversely, using PVA degradation, decreases in BET surface area were less significant, especially at 400 °C (5% decrease). Mixing of ETS-10 with carbon significantly decreased the resistivity, but the resulting mixtures required large amounts of binder to maintain uniformity and mechanical stability, drastically decreasing the ETS-10 weight ratio. For samples with the lowest resistivity (e.g., mixed with carbon), heating rates were measured during resistive heating and compared with more traditional heating methods, with results suggesting that faster and more energy efficient heating is possible.

MOF crystal chemistry: The road to made-to-order materials for low concentration CO_2 removal and air capture

Osama Shekhah , Youssef Belmabkhout, Zhijie Chen, Vincent Guillerm, Amy Cairns, Karim Adil and Mohamed Eddaoudi
Functional Materials Design, Discovery & Development (FMD3), Advanced Membranes and Porous Materials Center, Division of Physical Sciences and Engineering, King Abdullah University of Science and Technology, 4700 King Abdullah University of Science and Technology (KAUST), Thuwal 23955-6900, Kingdom of Saudi Arabia.

Recently, MOFs were intensively investigated for intermediate and high CO_2 concentration removal applications such as post-combustion, pre-combustion capture, natural gas and biogas upgrading.[1-4] Nevertheless, the potential of MOFs to remove traces and low CO_2 concentration from gas streams was rarely debated.[4-9] The main reason for this lack of studies is that most of MOFs reported so far, with or without unsaturated metal sites (UMC) or/and functionalized ligands, exhibit relatively low CO_2 selectivity and uptake particularly at relatively low CO_2 partial pressure. To overcome this downfall, various research groups have adopted the amine grafting chemistry and the acquired knowledge from amine-supported silica,[7,8] as a prospective pathway to enhance the CO_2 adsorption energetics and uptake in MOFs and covalent organic frameworks (COFs).[9-11] Here we report how MOF crystal chemistry permits the construction of an iso-structural MOF (SIFSIX-3-Cu) based on pyrazine/Cu(II) 2-D periodic 4^4 square grids pillared by $(SiF_6)^{2-}$ anions and thus (i) allows further contraction of the pore system to 3.5Å vs 3.84Å for the parent Zn(II) derivative, (ii) enhance physical adsorption energetics and (iii) shows subsequently unprecedented CO_2 uptake and selectivity at very low partial pressures relevant to air capture and traces CO_2 removal. The resultant SIFSIX-3-Cu exhibits uniformly distributed and non-reactive CO_2 adsorption energetics and offers for the first time remarkable CO_2 adsorption properties, uptake and selectivity in highly diluted gas streams, a performance unachievable with other class of porous materials.[12]

References

[1]. Caskey, S. R., Wong-Foy, A. G. & Matzger, A. J. Dramatic tuning of carbon dioxide uptake via metal substitution in a coordination polymer with cylindrical pores. J. Am. Chem. Soc. 130, 10870-10871 (2008).

[2]. Sumida, K. et al. Carbon dioxide capture in metal-organic frameworks. Chem. Rev. 112, 724-781 (2012).

[3]. Nugent, P. et al. Porous materials with optimal adsorption thermodynamics and kinetics for CO_2 separation. Nature 495, 80-84 (2013).

[4]. Xue, D. X. et al. Tunable rare-earth fcu-MOFs: A platform for systematic enhancement of CO_2 adsorption energetics and uptake. J. Am. Chem. Soc. 135, 7660-7667 (2013).

[5]. Choi, S., Drese, J. H., Eisenberger, P. M. & Jones, C. W. Application of Amine-Tethered Solid Sorbents for Direct CO_2 Capture from the Ambient Air. Environ. Sci. Technol. 45, 2420-2427 (2011).

[6]. Goeppert, A., Czaun, M., Surya Prakash, G. K. & Olah, G. A. Air as the renewable carbon source of the future: an overview of CO_2 capture from the atmosphere. Energy Environ. Sci., 5, 7833-7853 (2012).

[7]. Sayari, A. & Belmabkhout, Y. Stabilization of amine-containing CO_2 adsorbents: Dramatic effect of water vapor. J. Am. Chem. Soc. 132, 6312-6313 (2010).

[8]. Sayari, A., Belmabkhout, Y. & Serna-Guerrero, R. Flue gas treatment via CO2 adsorption. Chem. Eng. J. 171, 760-774 (2011).

[9] McDonald, T. M. et al. Capture of carbon dioxide from air and flue gas in the alkylamine-appended metal-organic framework mmen-Mg2(dobpdc). J. Am. Chem. Soc. 134, 7056-7065 (2012).

[10] Choi, S., Watanabe, T., Bae, T-H., Sholl, D. S. & Jones, C.W. Modification of the Mg-DOBDC MOF with amines to enhance CO_2 adsorption from ultradilute Gases. J. Phys. Chem. Lett 3, 1136–1141 (2012).

[11] Guillerm, V. et al. Porous organic polymers with anchored aldehydes: A new platform for post-synthetic amine functionalization en route for enhanced CO_2 adsorption properties. Chem. Commun., **50**, 1923-1923 (2013).

[12] O. Shekhah*, Y. Belmabkhout*, Z. Chen, V. Guillerm, A. Cairns, K. Adil, and M. Eddaoudi. Nature Communication, accepted (2014).

380545 Modeling of Multicomponent Adsorption Isotherms for Bio-Butanol Separation from ABE Model Solutions

Abstract:

Depletion of fossil fuel resources, political instability in oil-producing countries and environmental challenges has attractedsignificantattention to alternative biofuel production from renewable and sustainable resources. Butanol is considered as one of the best renewable alternatives for gasoline since it has advantages over other biofuels such as ethanol. However, there are some challenges in biobutanol production such as the very low concentration of butanol and the presence of other components in the acetone-butanol-ethanol (ABE) fermentation broths. The major challenge remains to develop a cost-effective technique to separate and recover butanol as the final product.

Distillation, as a traditional recovery technique,which is still used for butanol separation, is not economically viable due to its high energy requirement. Liquid-liquid extraction, gas stripping, perstraction, pervaporation and adsorption are other techniques being investigated for biobutanol separation from dilute fermentation broths. According to different investigations, the energy requirement for adsorption is the lowest compared to other techniques.

In our previous study activated carbon F-400 was selected as the best adsorbent amongst several adsorbents that were tested. Indeed, AC F-400 has high adsorption rate, capacity and selectivity toward butanol. The breakthrough experimental results for ABE model solution (butanol 12.7, ethanol 2.2, acetone 6.1, acetic acid 4.8, butyric acid 4.9, glucose 5.3 and xylose 4.2 g/L) showed that butanol is preferentially adsorbed and the othercomponents that were also adsorbed downstream of the butanol adsorption zone werelater displaced by butanol and butyric acid during the breakthrough experiments.

In this study, different isotherm models (multicomponent Langmuir,modified Langmuir and a model based on artificial neural network (ANN)) were tested to find the most suitable model to predict a wide array of standard adsorption isotherm behavior for butanol adsorption from multicomponent solutions. Results showed that the neural network most adequately described the multicomponent equilibrium data. A neural network with a hidden layer having four neurons, including the bias neuron, was able to represent very accurately the adsorption isotherm data. A parity plot comparing the neural network model with the experimental data was generated with a regression coefficient of 0.98. Isotherms of the other species present in the fermentation broth were also derived using neural networks.

Separation of CO_2 and CO gases is a very important one for the increasing conversion of CO_2 to CO through the Reverse Water Gas Shift (RWGS) Reaction as follows:

$$CO_2 + H_2 \leftrightarrow CO + H_2O$$

Since this is a reversible reaction that does not go to completion, the separation of the reactants and their recycle back into the reactor is an important factor for the improved overall conversion of this reaction. RWGS is an important first step for the production of renewable liquid fuels, since produced CO can be reacted with more H_2 through a Fischer Tropsch (F-T) reaction to produce liquid hydrocarbons.

The present study looks at the adsorption separation of CO from CO_2 to increase the conversion of the RWGS reaction after the removal of H_2O, assuming that H_2 gas does not get adsorbed. This presentation will include the results of the screening studies performed for different types of adsorbents including activated carbons, activated aluminas, zeolites, MOFs, and SAPOs in order to determine which adsorbent shows the most promising future for this separation process. This presentation also includes break through curves as well as the economic analysis carried out with the best adsorbent chosen for this separation looking into the technical and economic viability of this technology.

383674 A Thermodynamic Model of an Adsorbed Natural Gas Tank Containing a Metal-Organic Framework Adsorbent

Abstract:

A Thermodynamic Model of an Adsorbed Natural Gas Tank

Containing a Metal-Organic Framework Adsorbent

Hongda Zhang and Randall Q. Snurr

Department of Chemical & Biological Engineering

Northwestern University, Evanston, IL 60208 USA

Adsorbed natural gas (ANG) has many advantages, including higher safety and lower cost, compared with traditional compressed natural gas storage for vehicular applications. Metal-organic frameworks (MOFs) are a promising class of adsorbents for ANG. However, in addition to methane, commercial natural gas always contains small amount of impurities including ethane and propane. These higher hydrocarbons are more easily adsorbed by the adsorbents due to their stronger interactions with the adsorbent framework atoms. In order to study the effect of these impurities on the performance of an ANG tank, we combined GCMC molecular simulations and macroscopic thermodynamics to develop a model for an ANG tank. Central to our approach are the use of Ideal Adsorbed Solution Theory (IAST) to predict mixture isotherms from single-component isotherms and the Peng-Robinson equation of state to calculate fugacity coefficients at high pressure. With this model, the performance, especially the deliverable energy, of the natural gas storage system with different MOFs can be tested over many operation (adsorption/desorption) cycles. Furthermore, screening of a small MOF database containing 120 structures has been carried out. Based on the screening result, good MOF candidates have been identified, and some interesting trends have been observed. Some MOFs that perform well with pure methane are not among the top materials when considering natural gas that includes ethane and propane. The observed trends can be used to develop guidelines for us to search for and design better MOF materials.

364687 Adsorption of NORM from Produce Water in Natural Gas Hydraulic Fracturing

Abstract:

Natural gas has become an essential energy resource in the U.S due to the increasing demand of energy, the high oil prices, and the need of foreign oil independency. The improvement in the drilling technology has allowed the rapid expansion in gas production allowing the extraction of unconventional gas such as shale gas. Hydraulic fracturing is used to extract this natural gas from shale formations. Although natural gas is cleaner-burning than coal or oil, there is a lot of controversy due to the environmental impact related to the water consumption and its treatment. For instance, hydraulic fracturing generates significant volumes of wastewater which contain dissolved chemicals, high content of salts, and significant levels of Natural Occurring Radioactive Materials (NORM) in the form of radium isotopes. Therefore, one of the biggest challenges of this industry is to develop techniques for the prediction, prevention, remediation, and appropriate disposal of oil-field NORM. In this work, we will emphasize in NORM removal from produced waters generated by hydraulic fracturing. We will use computer aided molecular design (CAMD) methods for developing new environmentally benign adsorbents for NORM removal. CAMD is generally the reverse use of the group contribution methods (GCMs) used to generate molecules having desirable properties. Thus, we will develop new group contribution methods for important properties of adsorbents based on the thermodynamics. As a result, the GCM along with CAMD based optimization methods will identify potential groups of environmentally benign adsorbents.

Adsorption of Uranium from Seawater

C. Tsouris, S. Das, J. Kim, R.T. Mayes, C.J. Janke, T. Saito, S. Dai
Oak Ridge National Laboratory, Oak Ridge Tennessee, 37831-6181, tsourisc@ornl.gov

Polyethylene fibers functionalized with amidoxime ligands have been tested in laboratory and field experiments for uranium adsorption. Uranium adsorption was scaled up from 5-gallon batch to 110-gallon flow-through columns with recycling prior to continuous field tests with seawater. Kinetic data and mechanistic modeling were employed to determine the rate-limiting mechanism of the uranium uptake process. It was revealed that the rate-limiting mechanism of the process is the complexation reaction between the amidoxime ligand and the uranyl ion in seawater. The hypothesis is that this reaction is slow in the presence of bicarbonate ions in seawater, which form uranyl tricarbonate. This species is very stable, so that the amidoxime ligand competes with bicarbonate ions in reacting with uranyl ions, and this competition slows down the uranium uptake rate by the adsorbent. Controlled kinetic experiments with simulated uranium solutions have demonstrated the effect of bicarbonate and sodium ions on the uranium uptake process. The maximum amount of uranium uptake from seawater tests was 3.3 mg U/g adsorbent after eight weeks of contact of the adsorbent with seawater. This amount was three times higher than the maximum adsorption capacity achieved in this study by a leading adsorbent developed by the Japan Atomic Energy Agency (JAEA), which was 1.1 mg U/g adsorbent at equilibrium. The initial uranium uptake rate of the ORNL adsorbent was 2.6 times higher than that of the JAEA adsorbent under similar conditions. These results are very encouraging in efforts to secure new uranium resources for future nuclear energy needs.

Fabrication of cuprous nanoparticles in MIL-101: an efficient adsorbent for the separation of ethylene and ethane mixtures

Ganggang Chang†, Zongbi Bao†, Qilong Ren†, Shuguang Deng†‡, Zhiguo Zhang†, Baogen Su†, Huabin Xing†, and Yiwen Yang†*

Key Laboratory of Biomass Chemical Engineering of Ministry of Education, Department of Chemical and Biological Engineering, Zhejiang University, Hangzhou 310027, China

Abstract

Separations of olefin/paraffin mixtures are one of the most energy-intensive processes in the chemical industry due to their very small relative volatilities and close molecular sizes. Their separations performed at a large scale are currently achieved by cryogenic distillations, which are operated at low temperature and high pressure. An alternative olefin/paraffin separation process with low energy consumption would significantly reduce operating expenses. Among all competing techniques toward this end, the adsorptive separation seems to be one of the most promising and energy-efficient processes. Recently, an emerging class of solid materials known as metal-organic frameworks (MOFs), which offer a high surface area, adjustable pore dimensions, and chemical tenability have also been examined for olefin/paraffin separation.

Over the past decades, research efforts have been mostly aimed at preparing new MOF structures and studying their applications in molecule storage and separation. Nowadays, loading of metal ions inside the porous matrices of MOFs is of current interest. Among the numerous MOFs reported so far, MIL-101 is one of the excellent MOF materials due to its extra high specific surface area, pore volume, and high thermal and chemical stability to water as well as common organic solvents. The combination of these outstanding features makes it an interesting candidate for adsorption and catalysis. Moreover, the mesoporous cages can be used to confine the metal nanoparticles and to restrict their growth, which are the promising features for their applications in adsorption. Herein, we reported the synthesis of a new π-complexation adsorbent by using MIL-101 as the support and cuprous chloride as the active component, of which CuCl was introduced to MIL-101 by the "double solvents" method. An optimized cuprous loaded MIL-101 was shown to have an enhanced ethylene adsorption capacity and higher ethylene/ethane selectivity (14.0) compared to the pure MIL-101 (1.6). The great improvement of selectivity can be attributed to the newly generated nano-sized cuprous chloride particles that can selectively interact with the carbon-carbon double bond in ethylene through the π-complexation.

The following pictures are some relative results:

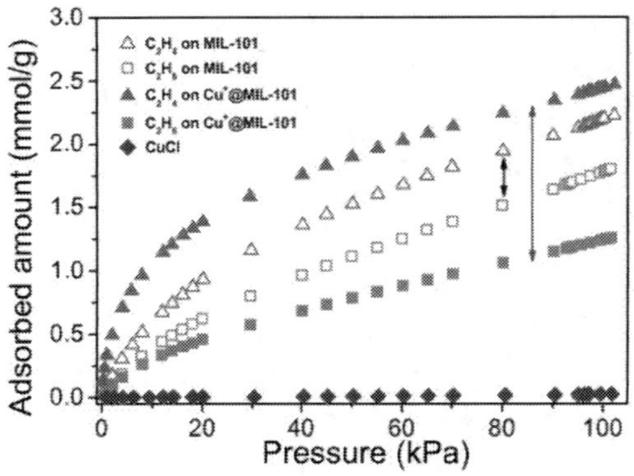

Figure 1. Ethylene and ethane isotherms for the pure, 40wt%Cu^+ loaded MIL-101 and pure CuCl at 30°C and 1bar.

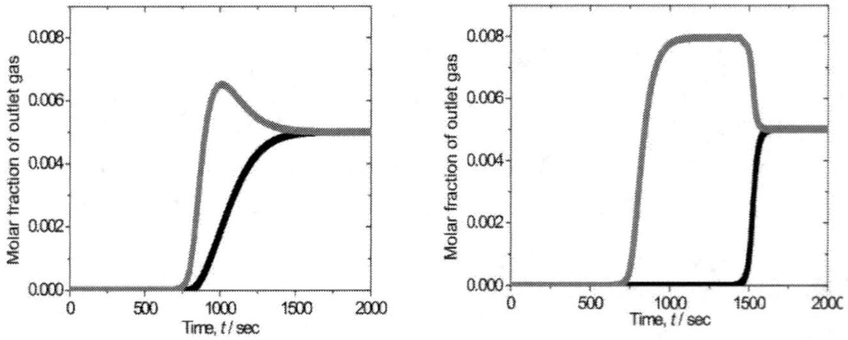

Figure 2. Breakthrough curves of equimolar mixtures of ethane/ethylene in columns of MIL-101(a) and Cu^+@MIL-101 (b) at 303.15 K and 1 bar calculated using Aspen™.(black: ethylene, red: ethane)

Reference

1. Chang, G. G.; Bao, Z. B.; Ren, Q. L.; Deng, S. G.; Zhang, Z. G.; Su, B. G.; Xing, H. B.; Yang, Y. W. Fabrication of cuprous nanoparticles in MIL-101: an efficient adsorbent for the separation of ethylene and ethane mixtures. *RSC Adv.* **2014**, DOI: 10.1039/c4ra02125h.
2. Wu, X. F.; Bao, Z. B.; Yuan, B.; Wang, J.; Sun, Y. Q.; Luo, H. M.; Deng, S. G. Microwave synthesis and characterization of MOF-74 (M = Ni, Mg) for gas separation. *Micropor. Mesopor. Mater*. **2013**, *180*, 114-122
3. Bao, Z. B.; Alnemrat, S.; Yu, L.; Vasiliev, I.; Ren, Q. L.; Lu, X. Y.; Deng, S. G. Adsorption of ethane, ethylene, propane, and propylene on a magnesium-based metal-organic framework. *Langmuir* **2011**, *27*,13554-13562
4. Bao, Z. B.; Alnemrat, S.; Yu, L.; Vasiliev, I.; Ren, Q. L.; Lu, X. Y.; Deng, S. G. Kinetic separation of carbon dioxide and methane on a copper metal-organic framework. *J. Colloid Interface Sci.* **2011**, *357*, 504.
5. Bao, Z. B.; Yu, L.; Ren, Q. L.; Lu, X. Y.; Deng, S. G. Adsorption of CO_2 and CH_4 on a magnesium-based metal organic framework. *J. Colloid Interface Sci.* **2011**, *353*, 549.
6. Saha, D.; Bao, Z. B.; Jia, F.; Deng, S. G. Adsorption of CO_2, CH_4, N_2O, and N_2 on

MOF-5, MOF-177, and Zeolite 5A, *Environ. Sci. Technol.* **2010**, *44*, 1820-1826.

ADSORPTION OF METHYLENE BLUE & CONGO RED DYE USING EGG SHELL AS ADSORBATE!!!

Olusegun A. SODEINDE and Olufemi Fadayini
Department of Chemical Engineering
Lagos State Polytechnic, Ikorodu, Lagos, Nigeria

INTRODUCTION

ENVIRONMENTAL CONTAMINATION AS A MAJOR PROBLEM IN THE SOCIETY (Vandevivere et al, 1998)

INCREASING DEMAND FOR SYNTHETIC DYE AS A MAJOR SOURCE OF TEXTILE WASTEWATER (PEARCE, 2003; DASH,2010 & KOBIRAY, 2010)

CONVENTIONAL TECHNOLOGIES EMPLOYED IN DYE REMOVAL (ONG et al, 2010; OLADOJA et al, 2008; ONG et al, 2009 & HAMEED et al, 2009).

THE STUDY: EFFECTIVENESS OF (ES) TO REMOVE CR & MB UNDER EQUILIBRIUM CONDITIONS

COMPARISM BETWEEN PHYSIOSORPTION & CHEMISORPTION (WALKER, 1998)

EXPERIMENTAL PROCEDURE

-ES COLLECTED FROM LOCAL POULTRY
-DRYING IN MICROWAVE FOR 60min AT 10°c
-COOLED & GROUND IN A MECHANICAL GRINDER
-SIEVED THROUGH A (10-50)μm SIEVES
-STORED IN AN AIR-TIGHT CONTAINER
-ADSORBATES (MB & CR) WAS COLLECTED
-ADSORPTION CARRIED OUT IN A MECHANICAL STIRRER AT 150rpm WITH 100ml SHAKING FLASK CONTAINING 50ml OF DYE SOLN

EFFECTS OF PROCESS CONDITIONS ON ADSORPTION

EFFECT OF CONTACT TIME

DOSAGE OF [ES] 'R' MIXES WITH 50ml VOL. OF 100 mg/L MB & CR IN 100ml SHAKING FLASK AT CONSTANT pH & CONSTANT TEMPERATURE AT VARYING TIME INTERVALS OF 5,10,15,30,60min for 150rpm.

CENTRIFUGING AT 10,000rpm for 2min

READING TAKING FOR DYE CONCENTRATION AT λ_{MAX} FOR MB & CR RESPECTIVELY

EFFECTS OF PROCESS CONDITIONS ON ADSORPTION

EFFECT OF ADSORBENT DOSE

-MASSES OF ES MEASURED

-MIXING WITH MB, CR & DYE SOLN AT CONSTANT pH, CONSTANT TIME & TEMPERATURE WITH 150rpm stirring

-ADSORPTIVITY AT WAVELENGTH (λ_{MAX}) OF 526nm AND 497nm FOR MB AND CR RESPECTIVELY IS MEASURED WITH SPECTROPHOTOMETER

EFFECTS OF PROCESS CONDITIONS ON ADSORPTION

EFFECT OF INTIAL DYE CONCENTRATION

-Determined By The Increase In Dye Concentration From 100 mg/L to 200 mg/L at Maximum pH, Equilibrium Contact Time & Adsorbate Dose.

EFFECTS OF PROCESS CONDITIONS ON ADSORPTION

ADSORPTION KINETIC STUDY

THE AMOUNT OF DYE ABSORBED DURING THE EXPERIMENT AND THE CALCULATED AMOUNT OF ABSORBED MB & CR WERE ANALYZED USING HO'S PSEUDO-FIRST ORDER & SECOND ORDER KINETIC MODEL

EFFECTS OF PROCESS CONDITIONS ON ADSORPTION

THERMODYNAMIC STUDY

THE EFFECT OF DIFFERENT TEMPERATURES ON THE ADSORPTION OF MB & CR DYES UNTO ES WAS INVESTIGATED AT VALUES OF 303, 313 & 323 K. THIS WAS DETERMINED BY USING THE BASIC THERMODYNAMIC RELATIONS (EQ. 12 & 13).

RESULTS & DISCUSSION

EFFECT OF pH ON MB

pH	2	4	6	7	8	10
Ce (mg/L)	44.5	31.0	25.5	22.0	15.0	5.3
q_e (mg/g)	2.775	3.45	3.725	3.9	4.25	4.735
E (%)	55.5	69	74.5	78	85	94.7

RESULTS & DISCUSSION

EFFECT OF pH ON CR

pH	2	4	6	7	8	10
Ce (mg/L)	26.7	22.53	16.5	13.36	9.05	12.9
q_e (mg/g)	3.66	3.87	54.17	54.33	24.55	4.35
E (%)	73.3	77.47	83.5	86.64	90.95	87.1

RESULTS & DISCUSSION

EFFECT OF pH ON MB & CR (DISCUSSION)

Tables show the effect of initial pH on the adsorption of MB and CR by egg shell powder at C_i of 0.1 g / L, V of 0.05 L, m of 1 g, contact time of 30 mins, at room temperature.

From the results generated, it was observed that for MB Dye, an increase in the pH of the aqueous solution favours the adsorption of the dye by the egg shell powder. As the pH value increased from 2 to 10, the dye adsorption increased from 2.775 mg / g to 4.735 mg / g, similarly, a higher removal percentage of dyes were observed, 55.5 % to 94.7 %. These results can be attributed to the effect of the solution pH on the charge of reactive group within the eggshell which, in turn makes it more effective to adsorb the dye in alkaline pH and increase the ionized able sites.

For the CR Dye, there was significant increase in the amount of dye adsorbed from pH of 2 to 8 which gave 3.665 mg / g to 4.548 mg / g and dye removal percentage of 73.3 % to 90.95 %. At pH 10, there was a significant drop in the quantity of dye adsorbed 4.355 mg / g and removal percentage of 87.1 %.

Therefore, pH value of 10 and 8 were taken as the equilibrium pH for the adsorption of EBT and CR dyes by egg shell powder respectively. Further experiments were carried out at these pH values.

RESULTS & DISCUSSION

EFFECT OF CONTACT TIME ON CR

Contact time (min)	5	10	15	30	60
Ce (mg / L)	27	18.4	11.37	7.05	3.4
q_e (mg / g)	3.65	4.08	4.4315	4.6475	4.83
E (%)	73	81.6	88.63	92.95	96.6

RESULTS & DISCUSSION

EFFECT OF CONTACT TIME ON MB

Contact time (min)	5	10	15	30	60
Ce (mg / L)	32.2	22.8	13.4	6.4	2.8
q_e (mg / g)	3.39	3.86	4.33	4.68	4.86
E (%)	67.8	77.2	86.6	93.6	97.2

RESULTS & DISCUSSION

EFFECT OF CONTACT TIME ON MB & CR

There was significant increase in the quantity of dyes adsorbed with elevation of the contact time, with values of 3.39 mg / g, 3.86 mg / g, 4.33 mg / g, 4.68 mg / g, and 4.86 mg / g for 5, 10, 15, 30 and 60 mins for the MB Dye, with percentage removals of 67.8 %, 77.2 %, 86.6 %, 93.6 % and 97.2 %.

Similarly, 3.65 mg / g, 4.08 mg / g, 4.4315 mg / g, 4.6475 mg / g and 4.83 mg / g for the same time intervals for the CR Dye, with percentage removals of 73 %, 81.6 %, 88.63 %, 92.95 %, and 96.6 %.

Therefore, since the highest amount of adsorption and percentage removal occurred at 60 mins for both dyes, the time was taken as the equilibrium time and further experiments were carried out at this time.

RESULTS & DISCUSSION

EFFECT OF ADSORBENT DOSE ON MB

Adsorbent Dose (g)	0.5	1	2	3	4	5
Ce (mg / L)	33.4	26.7	14.7	9.6	6.2	4
q_e (mg / g)	3.33	3.665	4.265	4.52	4.69	4.8
E (%)	66.6	73.3	85.3	90.4	93.8	96

RESULTS & DISCUSSION

EFFECT OF ADSORBENT DOSE ON CR

Adsorbent Dose (g)	0.5	1	2	3	4	5
Ce (mg / L)	24.6	21	18.17	12.8	8	5
q_e (mg / g)	3.77	3.95	4.09	4.36	4.6	4.75
E (%)	75.4	79	81.83	83.7	92	95

RESULTS & DISCUSSION

EFFECT OF ADSORBENT DOSE ON CR & MB

The effect of adsorbent dose on adsorption indicates that increasing adsorbent dose provides sufficient surface area to interact with the dye as witnessed by the proportional relationship between adsorbent dose and percentage of dye removal. As adsorbent dose increase 0.5, 1, 2, 3, 4 and 5 g, the percentages increase 66.6 %, 73.3 %, 85.3 %, 90.4 %, 93.8 %, and 96 % for EBT dye and 75.4 %, 79 %, 81.83 %, 83.7 %, 92 %, and 95 % for the CR dye. Adsorbent dose of 5 g is used or further experiment since the highest percentages of removal occurred at this value for both dyes.

RESULTS & DISCUSSION

EFFECT OF INITIAL DYE CONCENTRATION ON MB

Initial Dye Conc. (mg / L)	100	120	140	160	180	200
Ce (mg / L)	9.4	13	23.33	29.5	32.7	33.4
q_e (mg / g)	4.53	4.35	3.8335	3.525	3.365	4.735
E (%)	90.6	87	76.67	70.5	67.3	66.6

RESULTS & DISCUSSION

EFFECT OF INITIAL DYE CONCENTRATION ON CR

Initial Dye Conc. (mg / L)	100	120	140	160	180	200
Ce (mg / L)	8.6	11.5	16.1	22	26.7	30.5
q_e (mg / g)	4.57	4.425	4.195	3.9	3.665	3.475
E (%)	91.4	88.5	83.9	78	73.3	69.5

RESULTS & DISCUSSION

EFFECT OF INITIAL DYE CONCENTRATION ON CR & MB

The influence of initial dye concentration on the adsorption of both MB and CR dyes onto egg shell powder has been investigated between dye concentrations 100 mg / L and 200 mg / L with results obtained tabulated in the tables above. The quantity of dye adsorbed was found to decrease with an increase in the initial dye concentration. The results indicate that the percentage removal depends on the initial dye concentration. In case of low dye concentrations, the ratio of the initial number of moles of the dye ions to the available surface area of adsorbent is large. However, at higher dye concentrations, the available sites for adsorption became fewer, and hence the removal percentage which depends upon initial dye concentration decreases. Therefore, 100 mg / L is the equilibrium dye concentration.

RESULTS & DISCUSSION

LANGMUIR ISOTHERM PLOTS FOR MB DYE

Fig. :1/qe vs 1/Ce for MB at 303, 313 and 323 K
Langmuir Plots

313 K. Slope = 12.60784, Intercept = 0.22544, $r^2 = 0.9580$
303 K Slope = 15.14922, Intercept = 0.13122, $r^2 = 0.9992$
323 K Slope = 1.988407, Intercept = 0.20639, $r^2 = 0.9945$

RESULTS & DISCUSSION

FREUNDLICH ISOTHERM PLOTS FOR MB DYE

Fig.: In qe vs In Ce for MBat 303, 313, and 323 K
Freundlich Plot

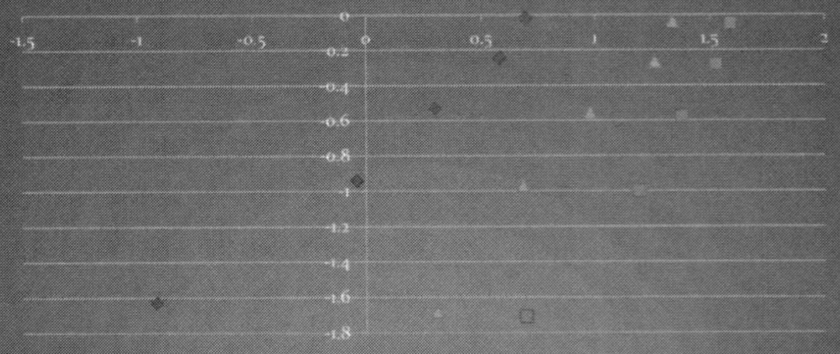

313 K. Slope = 1.957106, Intercept = 1.30382, $r^2 = 0.9855$
303 K Slope = 1.511327, Intercept = 1.37222, $r^2 = 0.9935$
323 K Slope = 0.975547, Intercept = 0.80339, $r^2 = 0.9919$

RESULTS & DISCUSSION

FREUNDLICH ISOTHERM PLOTS FOR MB DYE

Fig.: ln qe vs ln Ce for MBat 303, 313, and 323 K

Freundlich Plot

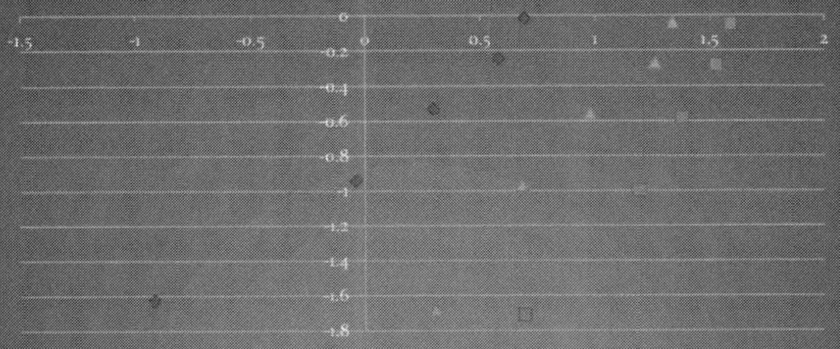

313 K. Slope = 1.957106, Intercept = 1.30382, $r^2 = 0.9855$

303 K Slope = 1.511327, Intercept = 1.37222, $r^2 = 0.9935$

323 K Slope = 0.975547, Intercept = 0.80339, $r^2 = 0.9919$

RESULTS & DISCUSSION

LANGMUIR ISOTHERM PLOTS FOR CR DYE

Fig. 11: 1/qe vs 1/Ce for CR Dye at 303, 313, and 323 K

Langmuir Plots

313 K. Slope = 5.772574, Intercept = 0.125664, $r^2 = 0.9813$

303 K Slope = 8.923824, Intercept = 0.169174, $r^2 = 0.9852$

323 K Slope = 1.765561, Intercept = 0.247697, $r^2 = 0.9978$

RESULTS & DISCUSSION

FREUNDLICH ISOTHERM PLOTS FOR CR DYE

Fig. 12: ln qe vs ln Ce for CR at 303, 313 and 323 K
Freundlich Plots

313 K. Slope = 1.221372, Intercept = 1.302627, $r^2 = 0.9794$
303 K Slope = 1.511327, Intercept = 1.37222, $r^2 = 0.9935$
323 K Slope = 0.93037, Intercept = 0.760154, $r^2 = 0.9972$

RESULTS & DISCUSSION

ADSORPTION KINETICS

Fig. 13: t/qt vs t for MB Dye at 303K, 313K, and 323K
Ho's Pseudo second-order plots

303 K. Slope = 0.4995, Intercept = 0.5402, $r^2 = 0.9987$
313 K Slope = 0.5140, Intercept = 0.6730, $r^2 = 0.9987$
323 K Slope = 0.5494, Intercept = 0.7820, $r^2 = 0.999$

RESULTS & DISCUSSION

ADSORPTION KINETICS

Fig. 14: t/qt vs t for CR Dye at 303, 313 and 323K
Ho's pseudo second-order model

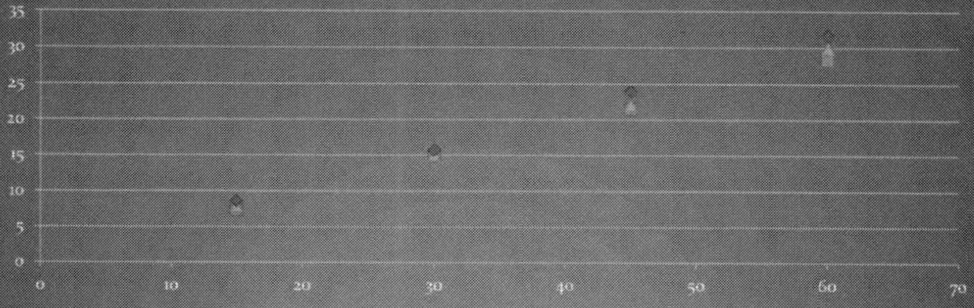

□ 303 K. Slope = 0.4615, Intercept = 0.4380, $r^2 = 0.9993$
△ 313 K Slope = 0.4877, Intercept = 0.6235, $r^2 = 0.9996$
◇ 323 K Slope = 0.5155, Intercept = 0.7002, $r^2 = 0.9997$

RESULTS & DISCUSSION

Result of Ho's pseudo-second order kinetic

		MB Dye			CR Dye		
T, K	q_e^{exp}	q_e^{cal}	K_2	r^2	q_e^{cal}	K_2	r^2
303	2.15	2.00	0.463	0.9987	2.17	0.48	0.9993
313	2.03	1.95	0.392	0.9987	2.05	0.39	0.9996
323	1.90	1.82	0.386	0.9992	1.94	0.38	0.9997

RESULTS & DISCUSSION

THERMODYNAMIC STUDY

Gibb's Free Energy Data

T, K	MB	CR
303	1.940 KJ / mol	1.850 KJ / mol
313	2.437 KJ / mol	2.450 KJ / mol
323	2.556 KJ / mol	2.598 KJ / mol

QUESTIONS?

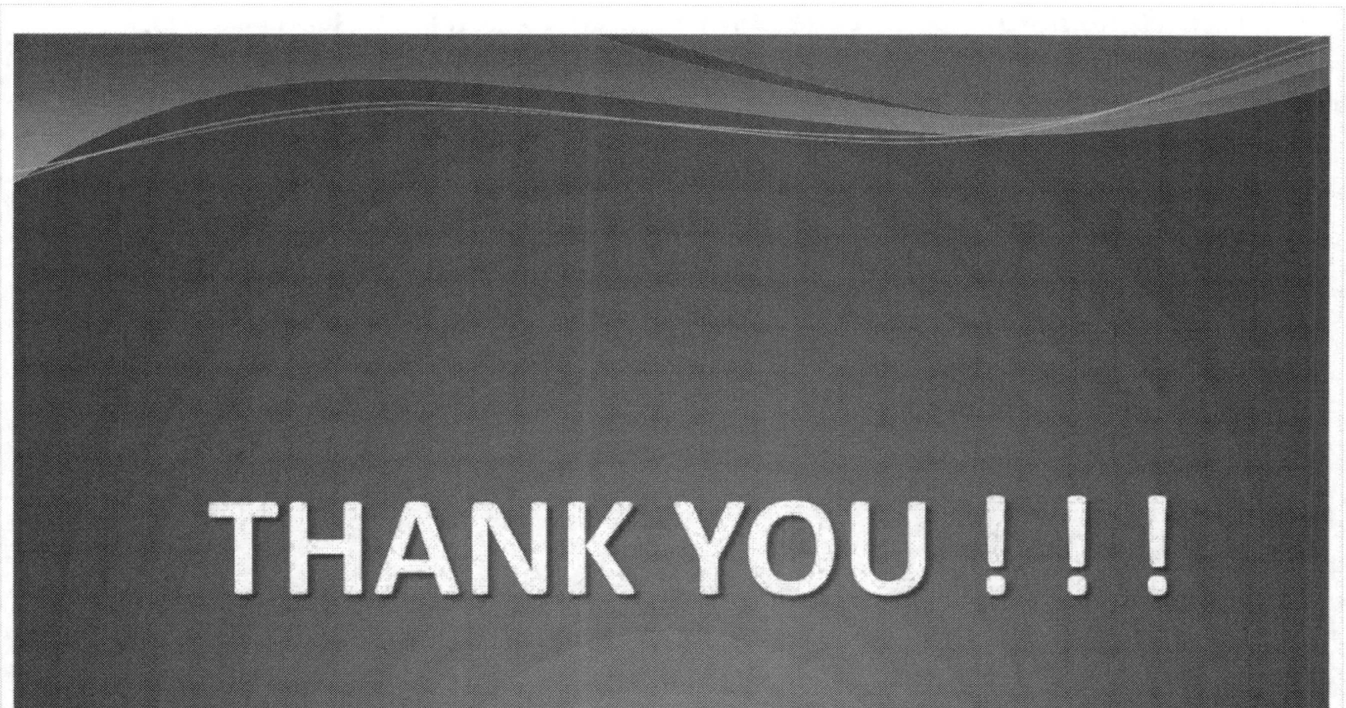
THANK YOU ! ! !

Recently, biomimetic approaches using carbonic anhydrase (CA) enzyme to capture and hydrate CO_2 offer a promising alternative to conventional solvent scrubbing for CO_2 removal. CA can catalyse the reversible hydration of CO_2, and its catalysis mechanism has been extensively studied. CA enzyme also has a CO_2/N_2 selectivity of 820 under atmosphere pressure, making the enzymatic CO_2 hydration approach especially attractive. For applications for CO_2 capture, enzyme immobilization at the gas liquid interface could potentially overcome mass transfer limitations faced with free enzymes as well as enhancing stability and recovery. In this work, membranes were developed with dual hydrophobic and hydrophilic interfaces by utilizing a combination of titania and carbon nanotube coating layers. The carbon nanotube layer was subsequently functionalized with carbonic anhydrase. The membranes were used in a liquid-gas contactor and showed enhanced CO_2 hydration rates due to the presence of the CA at the interface. Good stability and ability for regeneration was also observed.

Photosynthetic membrane protein crystals for hydrogen production

Patrick Saboe[1], Carolyn E. Lubner[2,3], Nicholas S. McCool[2], Nella M. Vargas-Barbosa[2], Hengjing Yan[4], Bryan Ferlez[3], Stanley Chan[1], Guillermo C. Bazan[4], John H. Golbeck[2,3], Manish Kumar[1,*]

1 Department of Chemical Engineering, Pennsylvania State University, University Park, PA 16802, USA
2 Department of Chemistry, Pennsylvania State University, University Park, PA 16802, USA
3 Department of Biochemistry and Molecular Biology, Pennsylvania State University, University Park, PA 16802, USA
4 Center for Polymers and Organic Solids, University of California at Santa Barbara, Santa Barbara, CA 93106, USA

*Corresponding author: 155 Fenske Laboratory, Pennsylvania State Univerity, University Park, PA, 16802, USA, (814) 865-7519, manish.kumar@psu.edu

It has recently been shown that a molecular wire can successfully tether a hydrogenase enzyme (H_2ase) and a photosynthetic protein complex, Photosystem I (PSI) to integrate hydrogen generation with effective solar energy conversion. The light harvesting component, PSI is a robust membrane-bound protein complex with the ability to provide the chemical potential energy needed to reduce protons. The H_2ase-PSI construct can generate significant amounts of hydrogen when in solution with an electron donor. Studies to increase hydrogen production rates by the attachment of the H_2ase-PSI construct to an electrode are currently focusing on single particle approaches. In order to achieve maximized electron transfer to immobilized PSI, high protein density, correct orientation and electrical connectivity to the electrode are required. An approach to design a PSI device can be borrowed from native photosynthetic membranes. When PSI is reconstituted into a lipid bilayer at high density, the protein self-assembles into 2D crystals, mimicking the dense arrays of photosynthetic complexes within membranes of purple bacteria. PSI 2D crystals have been assembled for structure determination; however, further applications of 2D protein crystals are not currently studied. A general problem with immobilizing proteins in bilayers on an electrode is that they are highly insulating to electron flow, impairing protein-electrode electron transfer events. This problem is also relevant in microbial fuel cells (MFCs). To improved current flow conjugated oligoelectrolyte polymers which insert into lipid bilayers can be added to form a conductive pathway between electrode and soluble proteins. Here we apply both 2D crystals of PSI and oligoelectrolyte polymers into a biomimetic electrochemical device to produce photocurrent. Functional reconstitution of PSI into lipid crystals was assayed by timed resolved optical spectroscopy. Incorporation of conductive polymers was confirmed by confocal microscopy. Modified gold electrons with a tethered conductive bilayer were used to directly reduce the electron donor site, P_{700} site of PSI within the crystal to produce a photocurrent. Increasing the stability, orientation and efficiency of PSI based membranes will be achieved by applying tailored amphiphilic block copolymers (BCPs) as a substitute for the lipid supporting membrane.

Title:
Development of biomimetic nanofiltration membranes using aquaporins.

Abstract:
Aquaporin is a highly selective water channel protein, which allows water molecules to pass through in single file while the transport of ions, protons and hydroxyl ions is prevented. Biomimetic membranes attempt to combine of ultra-efficient functioning of biological molecules, such as aquaporins, with the productivity of synthetic water purification membranes. The idea behind is to use or mimic the chemistry found in nature to facilitate water transport through the membrane. A successful formulation should lead to the production of water purification membranes with high selectivity along with high water flux. Aquaporins have received worldwide attention because of their potential to form biomimetic membranes with high flux and rejection for water reuse and desalination. However, chemical alterations of the aquaporins can occur during attachment to the membrane, and flattening can occur under pressure during operation. Another challenge is to design and prepare an assembly that will allow artificial biomimetic membranes with embedded aquaporins to sustain hydraulic water pressure gradients without losing their integrity and performance. Therefore, the overarching objective of this project is to form a biomimetic membrane made of unaltered aquaporin dispersed in a membrane selective layer and capable of operation under high hydraulic pressure

378634 Biomimetic Mineralization-Inspired Polymer–Inorganic Nanohybrid Membranes for CO2 Separation

Abstract:

Hybrid membranes composed of continuous polymer phase and filler phase have been extensively studied for energy-efficient gas separation, especially for CO_2 separation, gasoline desulfurization, and alcohol dehydration, etc. The multiphase characteristics of composite membranes provide more degree of freedom to manipulate multiple interactions, tailor multiscale structures and integrate multiple functionalities, compared to pristine polymer membranes. Inspired by the silica formation process mediated by silica deposition vesicle (SDV) in diatoms or sponges, silica nanoparticles (SiO_2) were in-situ formed within poly(ether-block-amide) (Pebax® MH 1657) bulk matrix through the controlled biomimetic mineralization, using either protamine or cysteamine as inducer/catalyst. The free volume of the hybrid membrane could be tailored by varying doping conditions (e.g. the concentration of silicon precursor, mineralization time and pH value). In particular, nanohybrid membrane prepared by cysteamine as inducer exhibited simultaneously high CO_2 permeability, high CO_2/CH_4 selectivity and high mechanical stability. The as-prepared membrane displayed an optimum CO_2 performance with permeation flux of 138 Barrer and the selectivity of 46 for CO_2/CH_4 mixtures.

389924 An Improved Method for Surface Modification of Porous Water Purification Membranes with Bioinspired Coatings

Abstract:

The surfaces of polysulfone and polyethersulfone ultrafiltration membranes were coated with polydopamine, a bioinspired surface modification agent that mimics the adhesive proteins found in secretions of mussels. The polydopamine coating yielded hydrophilic membranes that, under constant transmembrane pressure fouling conditions, have previously shown enhanced flux relative to unmodified membranes. When evaluated under constant permeate flux fouling, however, modified membranes exhibited higher transmembrane pressures than their unmodified analogs. This increased transmembrane pressure in the coated membranes was ascribed to the decrease in membrane permeance resulting from applying the polydopamine coating. The membrane permeance could be tuned by varying polydopamine deposition time and, even at the shortest deposition times studied here of a few minutes, a substantial increase in membrane hydrophilicity could be achieved. Therefore, polydopamine permits independent tuning of the membrane permeance and hydrophilicity. Polydopamine was deposited on a membrane of relatively high permeance until the pure water permeance of the modified membrane matched that of an unmodified membrane having lower native permeance, permitting a comparison of the fouling performance of a modified and unmodified membrane with the same pure water permeance. When unmodified and modified membranes of the same initial permeance were compared at constant flux fouling conditions, the modified membranes consistently exhibited lower transmembrane pressures and similar organic rejections to the unmodified membranes. Because many porous water purification membranes are operated at constant flux in industrial settings, an interesting methodology for membrane surface modification may be to surface modify a membrane of high permeance until the desired permeance is achieved, rather than by surface modification of a membrane that natively has the desired water transport characteristics, since the surface modification procedures almost invariably lead to lower pure water permeance.

Single Molecule Transport Characterization of a High Permeable Artificial Water Channel

Yue-xiao Shen[1], Mustafa Erbakan[2], Creedon W. Meminger[1], Junli Hou[3], Manish Kumar[1*]

[1] Department of Chemical Engineering, The Pennsylvania State University, University Park, PA, 16802, USA
[2] Department of Biomedical Engineering, The Pennsylvania State University, University Park, PA, 16802, USA
[3] Department of Chemistry, Fudan University, 220 Handan Road, Shanghai, 200433, China

*Corresponding Author, Tel.: (814)-865-7519; email address: manish.kumar@psu.edu

Abstract

Bioinspired artificial channels can combine high stability with the high permeability and selectivity found in biological channels. They have potential applications in next-generation energy-efficient and stable aqueous separation applications. This study is focused on a new architecture of artificial water channels—peptide-appended pillar[5]arenes. This study combined stopped-flow measurement with a newly developed fluorescence correlation spectroscopy technique to, for the first time, successfully determined the single channel water permeability of artificial channels. The peptide-appended pillar[5]arene channel permeability was determined to be $(2.81\pm0.11)\times10^{-17}$ cm^3/s per channel or $(9.68\pm0.38)\times10^5$ H_2O molecules/s per channel. This permeability value is only 2 magnitudes lower than that of natural water channel—aquaporins, when the low cross-sectional area (0.65 nm^2) is compared to larger cross sectional area biological water channels (~10 nm^2). This is an orders of magnitude improvement over the first-generation of the artificial water channels reported before and is the fastest one reported so far. This channel was found to have a pore size of approximately 450 Da and showed ion selectivity in the order of $NH_4^+ > Cs^+ > Rb^+ > K^+ > Na^+ > Li^+ > Cl^-$ as determined by patch clamp studies. The ability to further chemically modify the versatile chemical architecture of the pillar[5]arene channels shows promise for further improving water permeability and selectivity.

Keywords: Artificial aquaporins; artificial water channels; ion selectivity; single channel water permeability; solute rejection.

Introduction

Biological cells can conduct efficient water transport through water channel proteins known as aquaporins.[1] This membrane protein is currently being studied intensively for use in biomimetic membranes for water purification and other applications.[2-4] However, the physical and chemical stability as well as the resistance to biological attack of this protein in engineering applications is still questionable.[2] On the other hand, the determination of the atomic structures of various aquaporins and their molecular transport mechanisms provides unique strategies to engineer high permeability while maintain selectivity in artificial structures. This bioinspired strategy could open the door to a world of novel aqueous separation materials—artificial water channels.[2] The principles of the design of artificial water channels can be based on biological channels structure and self-assembly with the membrane matrix (lipid bilayer membranes). The designed channels should ideally have a tubular transmembrane structure with the thickness on the order of a few nanometers and an outer surface which can favorably interact with lipid membrane environment. They should also be able to mimic the transport properties of aquaporins while maintaining chemical stability. Research in the area of artificial water channels is still in its infancy but might herald the coming of novel energy-efficient, stable water purification materials in the near future.

Currently there are five published artificial water channels. These include: helical channels formed by self-assembled zwitterionic coordination polymers,[5] self-assembled dendritic dipeptide helical pores,[6] self-assembled imidazole channels,[7] pillar[n]arene (n=5,6)[8] and nanotubular building blocks assembled by macrocycles.[9] These channels are still far perfect compared to aquaporins when water permeability and solute selectivity is considered. This can be attributed to two reasons. First, there is a lack of available architectures that accurately mimic the structure of natural water channel proteins. Low measured permeabilities of 'first-generation' water channels also bring to fore the second challenge of accurate characterization of these low-permeability channels. Thus a critical concern is the determination of single channel permeabilities to allow for accurate comparison to natural channels. This is further complicated by the unknown insertion abilities of these channels into the membrane formed by lipids or lipid-like amphiphilic block copolymers.

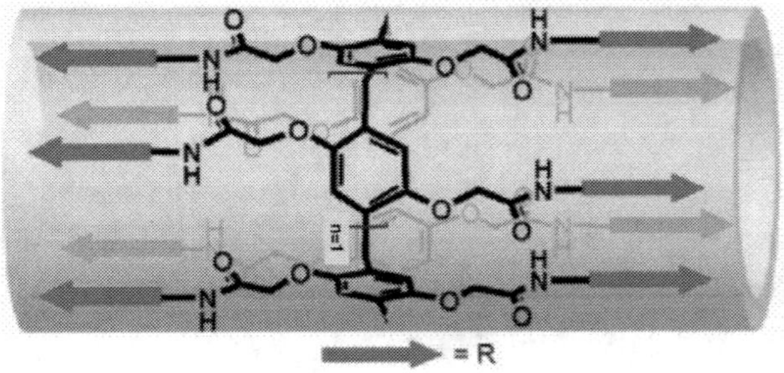

Figure 1 Structures of peptide-appended pillar[5]arene. R=D-Phe-L-Phe-D-Phe-COOH

This study focused on a new architecture of artificial water channels—peptide-appended pillar[5]arene (Figure 1). This channel is composed of peptide-bond connected amino acids— specifically phenylalanine,[10] which provides more hydrophobic regions compared to the first generation lower permeability hydrazide-appended pillar[n]arenes and is expected to improve both the membrane insertion efficiency and water permeability. The increase in water permeability is due to the increase in the hydrophobicity of inside pore surface mimicking similar surfaces in natural aquaporins and their widely studied natural analogues carbon nanotubes (CNTs). The water transport and solute rejection properties of this channel were systematically characterized in stopped-flow light scattering measurements after the channels were inserted in lipid vesicles. This technique is common in biophysical literature for aquaporin transport characterization.[11] We further labeled this channel with a specific fluorescent dye and estimated the single channel's water permeability based on a newly developed fluorescence correlation spectroscopy (FCS) technique where essentially the number of channels inserted into the wall of each vesicle can be counted. Combining the stopped-flow permeability measurements for vesicle membranes with the number of channels per unit area of membrane determined by FCS resulted in single molecule channel permeability values. Ion transport selectivity was evaluated via classic voltage-current relationship measurements using a patch clamp apparatus.[12]

Materials and methods

The channel (Figure 1) synthesis is described in detail in the previous paper.[10] Liposomes were prepared using the film rehydration method.[13] 6 mg phosphatidylcholine (PC)/phosphatidylserine (PS) (Avanti Polar Lipids, Inc., AL) mixture with a molar ratio of 4/1 was dissolved in $CHCl_3$. The peptide-appended pillar[5]arene, in DMSO, was added into the lipid solution and dried on a rotary vacuum evaporator. The resulting uniform film was dried under vacuum overnight in order to remove residual DMSO. After rehydration with 1 mL buffer containing 10 mM Hepes (pH=7), 100 mM NaCl and 0.01% NaN_3 (for biological growth control), the suspension was incubated with stirring in at 4 °C for 24 h and then subjected to extrusion through track-etched filters (3 times through 0.4 μm membranes followed by 0.2 μm membranes for 7 times, Nuclepore, Whatman, UK) to obtain monodisperse unilamellar vesicles, the size of which was characterized by dynamic light scattering (Zetasizer Nano, Malvern Instruments Ltd., UK). Permeability tests were conducted on a stopped-flow instrument (SF-300X, KinTek Corp., Snowshoe, PA). Exposure to hypertonic solutions results in the shrinkage of the vesicles due to the outwardly directed osmotic gradient created and drives water transport through the vesicles. The abrupt changes of the vesicle size lead to the increase in the light scattering at 90° according to the Rayleigh-Gans theory applied to this system.[14] The changes of light scattering caused by vesicle shrinking were recorded at an emission wavelength of 600 nm. The data were fitted to an exponential equation to calculate the exponential coefficient (k) for the initial change of the light scattering curve, and the osmotic water permeability (P_f) was calculated by using the following expression[11]

$$P_f = \frac{k}{\left(\frac{S}{V_0}\right) \times V_W \times \Delta_{osm}} \quad (1)$$

Where S is the initial surface area of the vesicles, V_0 is the initial volume of the vesicles, V_w is the molar volume of water, and Δ_{osm} is the osmolarity difference in driving the shrinking of the vesicles. The solutes in the hypertonic osmolyte included glycine (78 Da), arabinose (150 Da), glucose (180 Da), N-acetyl-D-glucosamine (221 Da), sucrose (342 Da) and dextran500 (505 Da dextran). The reflection coefficient (σ) can be used to estimate the relative rejection properties: $\sigma_{Solute}=J_{Solute}/J_{Dextran500}$,[15] where J_{Solute} and $J_{Dextran500}$ are the apparent volume flow when using solute and dextran500 in the osmolyte, respectively, and σ_{Solute} is the reflection coefficient of the solute. All the osmolarity values were measured using an osmometer (Model 3300, Advanced Instruments, Inc., MA).

The ion selectivity of the peptide-appended pillar[5]arene was studied on a bilayer workstation (Warner Instruments, LLC, CT). The diPhyPC (10 mg/mL, 20 μL) in chloroform was first re-dissolved in n-decane (5 μL) after it was dried on nitrogen gas. The resulting lipid solution (5 μL) was used to form a thin film on the aperture with a diameter of 200 μm of the Delrin cup. After the chamber (*cis* side) and the Delrin cup (*trans* side) were filled with different salt solutions, with Ag-AgCl electrodes applied directly to the two solutions and the *cis* chamber grounded, 1.0 μL lipid solution was painted on the pretreated aperture to create a planar lipid bilayer with the capacitance of 80-120 pF. The channels were then introduced into the lipid bilayer via injecting the DMSO stock (1mM, 3 μL) into the *cis* chamber followed by 5-min mixing. The membrane current was measured on a Warner BC-535D bilayer clamp amplifier and collected by PatchMaster (HEKA, Germany) with the sampling interval of 10 kHz and then filtered with a 8-pole Bessel filter at 1 kHz (HEKA, Germany). The data were analyzed by FitMaster (HEKA, Germany) with a digital filter at 100 Hz. The reversal potential (E_{rev}), obtained from the measured voltage-current data, was used to calculate the ion selectivity based on Goldman-Hodgkin-Katz equation[12]

$$E_{\mathrm{rev}} = \frac{RT}{F}\ln\left(\frac{\sum_i^N P_{M_i^+}[M_i^+]_{\mathrm{out}} + \sum_j^N P_{M_j^-}[M_j^-]_{\mathrm{in}}}{\sum_i^N P_{M_i^+}[M_i^+]_{\mathrm{in}} + \sum_j^N P_{M_j^-}[M_j^-]_{\mathrm{out}}}\right) \quad (2)$$

The single channel water permeability was determined based on a fluorescence correlation spectroscopy (FCS) technique. First the peptide-appended pillar[5]arene was labeled with 5-(and-6)-((N-(5-aminopentyl)amino)carbonyl)tetramethylrhodamine (Invitrogen™, CA) using dicyclohexylcarbodiimide (Thermo Fisher Scientific Inc., IL) as a crosslink reagent. The labeled channels were incorporated into PC/PS liposomes using the aforementioned method. The vesicles were further treated with complete dialysis followed by gel filtration chromatography to obtain the liposomes that were free of the external residual dyes. The purified liposomes were pooled for stopped-flow permeability measurement and FCS experiments; the later technique was employed to measure the average number of the labeled channels incorporated per liposome. The potential free dye encapsulated inside the liposomes were subtracted by performing the same experiment with the dye alone as control. FCS experiments were performed on a time correlated single photon counting module (Becker-Hickl GmbH, Berlin, Germany). In FCS experiments, time-resolved changes in the fluorescent intensity $F(t)$ within a small confocal volume are monitored. The fluctuations of the fluorescent intensity $\delta F(t)$ can be autocorrelated to obtained the autocorrelation function $G(\tau)$ given by[16]

$$G(\tau) = \frac{\langle \delta F(t)\rangle\langle \delta F(t+\tau)\rangle}{\langle F(t)\rangle^2} \quad (3)$$

Where t (time) and τ (lag time) vary over all times of data collection. The autocorrelation and its relation to the theoretical Gaussian confocal volume and the characteristic diffusion time τ_D is[16]

$$G(\tau) = \frac{1}{N}\left[\frac{1}{1+\tau/\tau_D}\right]\left[\frac{1}{1+(r/z)^2(\tau/\tau_D)}\right]^{1/2} \quad (4)$$

Where r and z are the radius and half height of the confocal volume, respectively; τ_D is the 2-D characteristic molecular diffusion time across a small area of radius r; N is the average number of diffusion fluorophores in the confocal volume. The number of the fluorescent molecules in the confocal volume is termed as $G(0)$. The number of labeled channels per liposomes was calculated by taking the ratio of number of particles in confocal volume before and after solubilization of the liposomes.

Results and discussion

The water permeability data are presented in Figure 2. The light scattering signal could be fitted in the form of the sum of two exponential functions due to the shrinkage of the liposomes under hypertonic conditions.[17] The fitting yielded two shrinkage rates and two permeability values: a larger value (k_1) that was almost independent of channel content and a smaller value (k_2) that increased when the channel to lipid ratio was increased. The results indicated that the contribution to the overall liposome's permeability by the channel was less than lipid background. The permeability contributed by the pillar[5]arene displayed a linear increase when the molar channel to lipid ratio increased from 0 to 0.005 (Figure 2 (b)). The channel's permeability at molar channel to lipid ratio of 0.005 reached 0.35±0.02 µm/s.

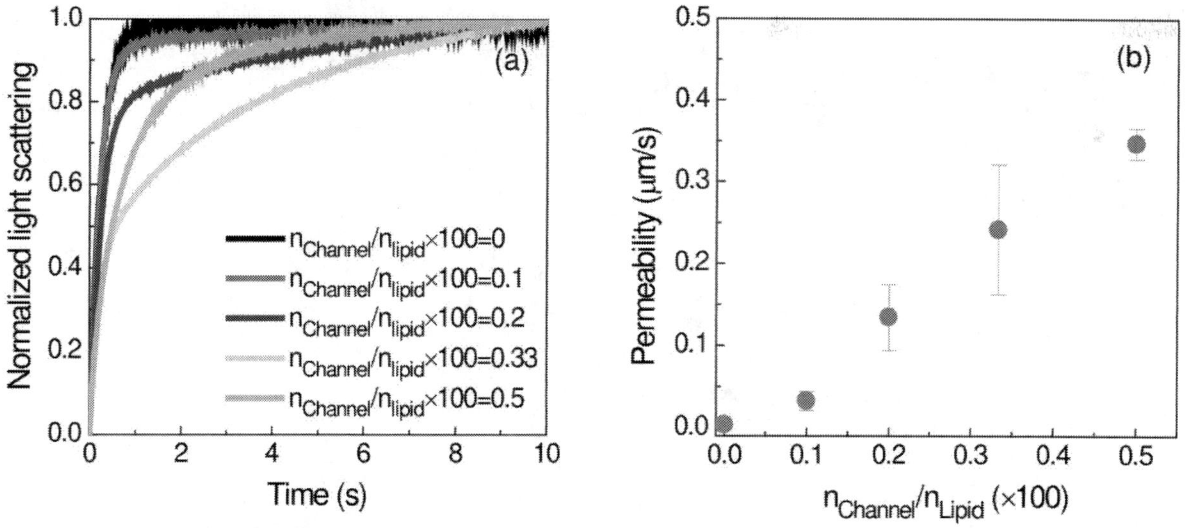

Figure 2 The permeability of lipid membranes incorporated with the channels increases with the number of channels inserted. (a) Stopped-flow experiments of liposomes with different channel to lipid ratios (molar ratio: 0, 0.001, 0.002, 0.0033 and 0.005). Liposomes were made of phosphatidylcholine and phosphatidylserine with molar ratio of 4:1. All the liposomes were abruptly exposed to a 400 mM sucrose increase in osmolarity. The data were fitted to a double-exponential equation to calculate the exponential coefficient (k) for the initial change of the light scattering curve. The smaller value (k_2) increased when the channel to lipid ratio was increased, corresponding to the channel's permeability. The osmotic water permeability (P_f) was calculated by using the following expression: $P_f = k/((S/V_0)\times V_w\times \Delta_{osm})$,where S is the initial surface area of the vesicles, V_0 is the initial volume of the vesicles, V_w is the molar volume of water, and Δ_{osm} is the osmolarity difference in driving the shrinking of the vesicles; (b) The permeabilities of the channel in the liposomes with different channel to lipid ratios (n=3).

The permeabilities induced by the solutes of different molecular size under the same osmolarity condition were found to be different. While using smaller solute such as glycine (78 Da), the light scattering signal showed a negligible k_2 (-0.0002 ± 0.0041 s^{-1}); when the molecular size of the solutes became larger, a significant increase in k_2 was shown in Figure 3 (a). The apparent permeabilities of the channel at the molar channel to lipid ratio of 0.005 increased from 0.00±0.03 μm/s to 0.75±0.06 μm/s, using glycine and dextran500 as osmolytes respectively. Since this channel has an estimated pore size of approximately 6.5 Å, slightly larger compared to that of aquaporins (~2.8 Å), this channel was expected to be partly permeable to small molecules. The resulting light scattering behavior should thus be the superposition of the outward water transport and the inward solute diffusion if the solute is not completely rejected. For glycine, the apparent permeability was close to zero, indicating the channel's ability of transporting this small amino acid as reported before.[10] The calculated reflection coefficients of each solute showed that the molecular weight cut off of this pillar[5]arene channel was around 450 Da (Figure 3 (b)). This result is similar to the previous study which showed a decrease of transport of this channel with the increasing size of the amino acids.[10]

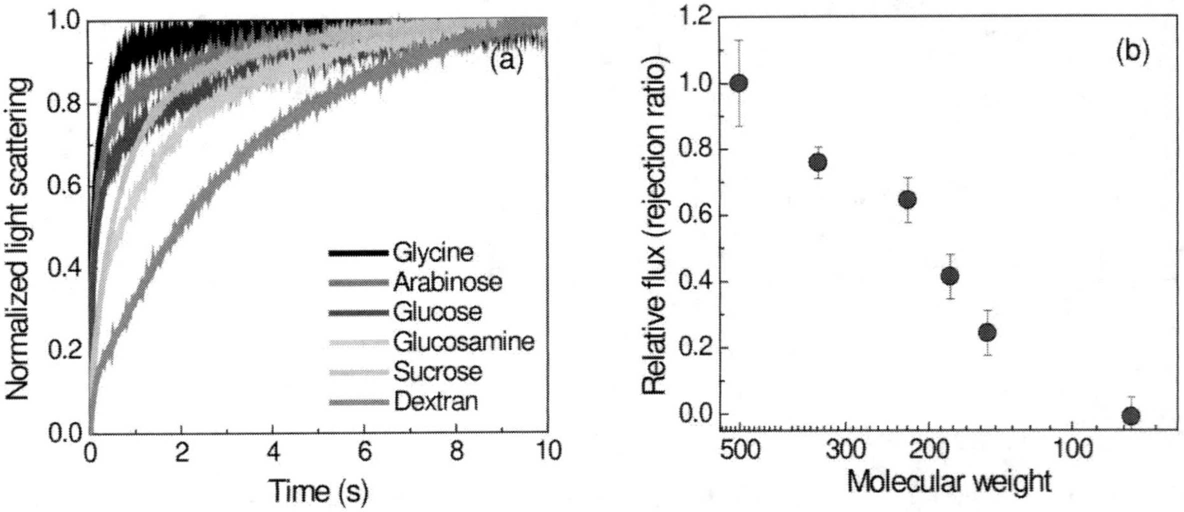

Figure 3 The permeability response of the channel incorporated liposomes to solutes in different sizes as osmolytes can be used to determine the size selectivity and thus the approximate pore size of the channel. (a) Stopped-flow experiments of liposomes (made of phosphatidylcholine (PC) and phosphatidylserine (PS) with molar ratio of 4:1, molar channel to lipid ratio is 0.005) using 400 mM glycine, 400 mM arabinose, 400 mM glucose, 400 mM glucosamine, 400 mM sucrose and 200 mM Dextran500 as different osmolytes. When the molecular size of the solutes became larger, a significant increase was shown in k_2; (b) Relative flux (or rejection ratio, based on Dextran500) of the channel. The actual osmolarity of each osmolyte was corrected by osmometer (n=3).

The ion selectivity experiment of this channel was conducted on a patch-clamp device, as shown in Figure 4 (a). The voltage-current relationship is illustrated in Figure 4 (b). The peptide-appended pillar[5]arene possesses a strong transport rate of K^+ over Cl^-; the selectivity of K^+ over Cl^- was around 10. The selectivity order was $NH_4^+ > Cs^+ > Rb^+ > K^+ > Na^+ > Li^+ > Cl^-$ (shown in Table 1).

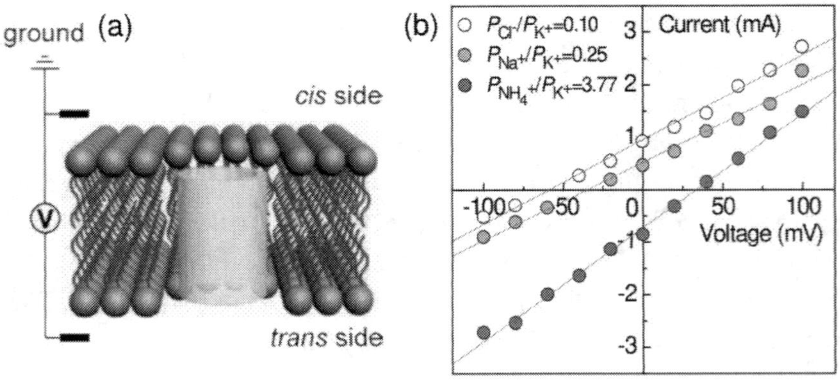

Table 1 Permeability ratio, P_X/P_{K+}

Ion	P_X/P_{K+}
Cl^-	0.10
NH_4^+	3.77
Li^+	0.15
Na^+	0.25
Rb^+	2.02
Cs^+	2.22

Figure 4 The channels show an unusual ion selectivity profile. (a) Patch clamp experiments of the peptide-appended pillar[5]arene with planar lipid bilayer; (b) Voltage-current relationship of the channel.

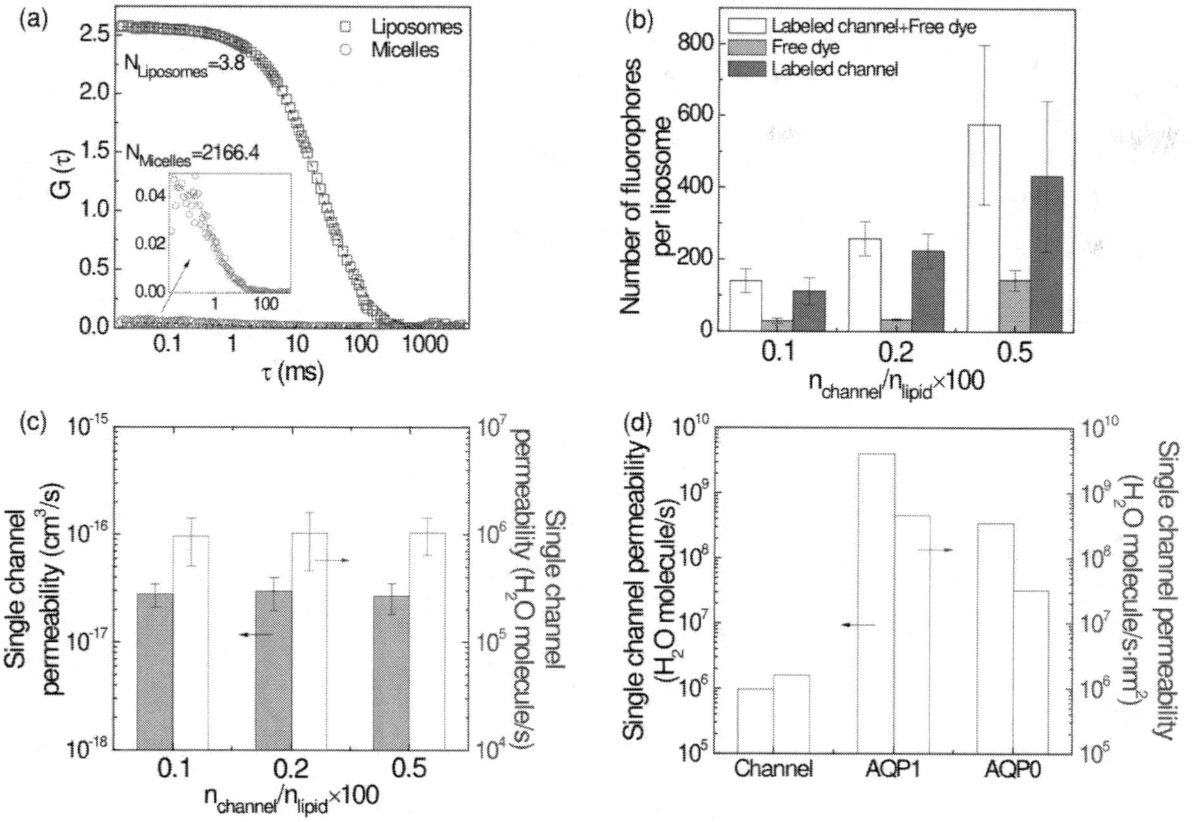

Figure 5 Fluorescence correlation spectroscopy can be used to determine the number of channels inserted per vesicle and combined with permeability measured using stopped-flow to get single channel permeabilities. (a) Autocorrelation curves of labeled channel incorporated liposomes ($n_{channel}/n_{lipid}$=0.005) before and after solubilization in 2.5% OG. A relative high correlation coefficient for liposomes indicates a low number of fluorescent particles in the confocal volume ($N_{liposomes}$). After solubilization, the number of free particles increased more than 500 times ($N_{Micelles}$); (b) Number of fluorophores (labeled channel and dye) per liposome (n=3); (c) Single channel permeability of the liposomes with different channel to lipid ratios (molar ratio: 0, 0.001, 0.002, 0.0033 and 0.005) (n=3); (d) Single channel permeability and single channel permeability per cross- sectional area compared to natural water channels—aquaporins.

The time resolved autocorrelation function $G(\tau)$ of the labeled peptide-appended pillar[5]arene incorporated liposomes, representing the time that the fluorescent particles spends within the confocal volume, is shown in Figure 5 (a). By fitting the autocorrelation curve using equation 4, we obtained the number of the fluorescent liposomes ($N_{Liposomes}$) in this confocal volume. This relatively high correlation coefficient ($G(0)$) indicates a low number of the fluorescent liposomes in the confocal volume. The liposomes were further dispersed and resolubilized in 2.5% detergent n-octyl-β-D-glucoside (OG), which releasing the labeled channels (encapsulated dyes may also be released in this step) into detergent micelles. After resolubilization, a considerable amount of fluorescent micelles ($N_{Micelles}$) contributed to a really low correlation coefficient ($G(0)$). The number of the labeled channel per liposome ($N_{Channels}$) was calculated taking the ratio of $N_{Liposomes}$ over $N_{Micelles}$. Because there might be still free dyes inside liposomes, a series of dye control experiments were performed and used to obtain the net $N_{Channels}$ per liposomes at different molar channel to lipid ratios (from 0.001 to 0.005, see Figure 5 (b)). The net $N_{Channels}$ per liposomes combined with the corresponding water permeability from stopped-flow data (not shown here) can be used to calculate the single channel water permeability at different molar channel to lipid ratios (Figure 5 (c)). Taking the average of these three samples, we estimated that the single channel permeability of the peptide-appended pillar[5]arene is $(2.81\pm0.11)\times10^{-17}$ cm^3/s, corresponding to $(9.68\pm0.38)\times10^5$ H_2O molecules/s. This value is around 3 and 2 magnitudes lower than natural water channel aquaporin 1 (AQP1)[18] and aquaporin 0 (AQP0)[17], respectively (Figure 5 (d)). If we consider the effective cross-sectional area of the artificial water channel (0.6 nm^2), AQP1 (9.0 nm^2)[19] and AQP0 (10.7 nm^2)[20], the single channel permeability of the channel per area is only 2 and 1 order of magnitudes lower compared to that of natural aquaporins and the highest permeability reported for such systems.

Conclusion

This paper systematically studied the water transport properties of a new artificial water channel—peptide-appended pillar[5]arene. The single permeability of this channel was found to be $(2.81\pm0.11)\times10^{-17}$ cm^3/s or $(9.68\pm0.38)\times10^5$ H_2O molecules/s, based on a fluorescent counting method combined with permeability test. To the best of our knowledge, this is the most efficient artificial water channel that has been reported. The permeability value is only 2~3 magnitudes lower compared to aquaporins and only 1~2 orders of magnitude lower when the cross-sectional area of the channels are taken in to account. This channel was found to have a pore size of around 450 Da and showed cations transport selectivity over anions.

Reference

1 Agre, P. Aquaporin Water Channels (Nobel Lecture). *Angewandte Chemie International Edition* **43**, 4278-4290 (2004).

2 Shen, Y.-x., Saboe, P. O., Sines, I. T., Erbakan, M. & Kumar, M. Biomimetic membranes: A review. *Journal of Membrane Science* **454**, 359-381 (2014).

3 Tang, C. Y., Zhao, Y., Wang, R., Hélix-Nielsen, C. & Fane, A. G. Desalination by biomimetic aquaporin membranes: Review of status and prospects. *Desalination* **308**, 34-40 (2013).

4 Nielsen, C. Biomimetic membranes for sensor and separation applications. *Anal Bioanal Chem* **395**, 697-718 (2009).

5 Fei, Z. *et al.* A Synthetic Zwitterionic Water Channel: Characterization in the Solid State by X-ray Crystallography and NMR Spectroscopy. *Angewandte Chemie International Edition* **44**, 5720-5725 (2005).

6 Kaucher, M. S. *et al.* Selective Transport of Water Mediated by Porous Dendritic Dipeptides. *Journal of the American Chemical Society* **129**, 11698-11699 (2007).

7 Le Duc, Y. *et al.* Imidazole-Quartet Water and Proton Dipolar Channels. *Angewandte Chemie International Edition* **50**, 11366-11372 (2011).

8 Hu, X.-B., Chen, Z., Tang, G., Hou, J.-L. & Li, Z.-T. Single-Molecular Artificial Transmembrane Water Channels. *Journal of the American Chemical Society* **134**, 8384-8387 (2012).

9 Zhou, X. *et al.* Self-assembling subnanometer pores with unusual mass-transport properties. *Nat Commun* **3**, 949 (2012).

10 Chen, L. *et al.* Chiral Selective Transmembrane Transport of Amino Acids through Artificial Channels. *Journal of the American Chemical Society* **135,** 2152-2155 (2013).

11 Borgnia, M. J., Kozono, D., Calamita, G., Maloney, P. C. & Agre, P. Functional reconstitution and characterization of AqpZ, the E-coli water channel protein. *J. Mol. Biol.* **291**, 1169-1179 (1999).

12 Hille, B. *Ionic Channels of Excitable Membranes.* (Sinauer Associates, Incorporated, 2001).

13 Kumar, M., Habel, J. E. O., Shen, Y.-x., Meier, W. P. & Walz, T. High-Density Reconstitution of Functional Water Channels into Vesicular and Planar Block Copolymer Membranes. *Journal of the American Chemical Society* **134**, 18631-18637 (2012).

14 Latimer, P. & Pyle, B. E. Light Scattering at Various Angles: Theoretical Predictions of the Effects of Particle Volume Changes. *Biophysical Journal* **12**, 764-773 (1972).

15 Meinild, A.-K., Klaerke, D. A. & Zeuthen, T. Bidirectional Water Fluxes and Specificity for Small Hydrophilic Molecules in Aquaporins 0–5. *Journal of Biological Chemistry* **273**, 32446-32451 (1998).

16 Gullapalli, R. R., Mathura, R., Dangaria, J. H., Butler, P. J. & Tabouillot, T. Integrated multimodal microscopy, time-resolved fluorescence, and optical-trap rheometry: toward single molecule mechanobiology. *BIOMEDO* **12**, 014012 (2007).

17 Tong, J., Canty, J. T., Briggs, M. M. & McIntosh, T. J. The water permeability of lens aquaporin-0 depends on its lipid bilayer environment. *Experimental Eye Research* **113**, 32-40 (2013).

18 Zeidel, M. L., Ambudkar, S. V., Smith, B. L. & Agre, P. Reconstitution of functional water channels in liposomes containing purified red cell CHIP28 protein. *Biochemistry* **31**, 7436-7440 (1992).

19 Cheng, A., van Hoek, A. N., Yeager, M., Verkman, A. S. & Mitra, A. K. Three-dimensional organization of a human water channel. *Nature* **387**, 627-630 (1997).

20 Gonen, T., Sliz, P., Kistler, J., Cheng, Y. & Walz, T. Aquaporin-0 membrane junctions reveal the structure of a closed water pore. *Nature* **429**, 193-197 (2004).

SIMULATION-AIDED CHARACTERIZATION OF BIOMIMETIC SEPARATION MEMBRANE FOR WATER PURIFICATION

Dr. Ali S. Alshami, University of North Dakota, Grand Forks, ND.

Introduction

Estimates put more than half of the world population in critical water famine and expects the number of people affected by water scarcity to grow fourfold (Sandia Report). In the developing world, where water scarcity is chronic, 80-90% of all diseases and 30% of all deaths result from poor access to clean water and sanitation (Lietner, 1998). Economic prosperity and growth is also severely hindered due to lack of available potable water. It has now been established that future peace and prosperity of a nation is intimately tied to the availability of clean, fresh potable water. Hence, developing an efficient and cost-effective water purification and desalination technology is of a paramount national strategic importance.

The current state-of-the-art of water purification and desalination technologies are membrane separation via reverse osmosis (RO) and thermal distillation (multistage flash and effect distillation) (Zhou 2005, Veerapaneni 2007). Conventional thermal desalination processes are inefficient, consume large amounts of energy and highly susceptible to corrosion and scaling. Commercial scale conventional plants use fossil fuel for energy. Even when fossil fuel is readily available at moderately low-cost, capital investment and on-going cost of operation associated with conventional thermal methods can reach prohibitive levels. Thus, we see very few plants are being built today and expectations are that their number will be declining in the years to come.

RO is a membrane process that recovers pure water from saline and brackish water systems by means of applying high pressure as the driving force. Most of the energy consumed by RO is due to the pressurization of the saline solution. The required pressure depends on the salt concentration and it is normally around 70 bars for seawater [1]. Although theoretically, the minimum required operating pressure is that needed to exceed the osmotic pressure; higher pressure, nevertheless must be used (50-80 atm) to force diffusion across the membrane cross-sectional area. Installations of RO plants are growing worldwide in efforts to keep up with the fast growing demands generated by fast world population growth. It is estimated that 42% of the world capacity of desalinated water is produced using RO systems.

In 2003, Peter Agre of John Hopkins University was awarded the Noble Prize for his discovery of aquaporins, a water-channel proteins that move water molecules across living cells walls. Since this discovery, researchers, regardless of the motive, be it financial capitalization, scientific scholorship, or humanetarian intentions, are all on a race to successfully prove the concept of successfully incorporating these biological membranes into a polymetric matix that would be capable of delivering THE required water flux, withstand operation pressure, and can be easily scaled to commercial sizes. If successful, this technology is believed to be capable of delivering the following:

1. 5-10 times more efficient water desalination plants
2. more than 70% savings on specific water consumption
3. 100 times higher water permeability coefficient compared to traditional RO membranes.
4. Superior solute retention capabilities
5. And more importantly, functional ingredients (aquaporins) are easily produced in mass quantities in a biological reactor using vectors such as the bacterium E. coli.

Research on Biomimetic membranes and their use in water purification and desalination is currently being conducted by various groups in academia, national laboratories and industry. Academic research has been primarily focusing on the fundamental physical mechanisms underlying water molecules transport across membrane walls via the protein channels. The bulk of the studies in academia have been directed towards understanding the physicochemical properties and processes controlling the water movement via molecular simulation methods. Efforts by national laboratories and industrial groups have been more experimental-base studies for purposes of either licensing and/or commercializing the technology. Regardless of the final outcome of each of these groups' efforts, each and everyone is eagerly participating in sowing the seeds of a technological revolution in water treatment and purification: a common goal to better the lives of more than one billion people around the world who are in critical need for clean drinking water.

LITERATURE SURVEY

Despite the vast literature on membrane separations, innovative chemical synthesis and molecular simulation-based design are rarely integrated. To date there appears to be no validated, simulation-aided methodology to design artificial membranes based on the critical mechanisms underlying the performance of biological membranes.

In a cell membrane, aquaporin protein channels of nanoscale dimensions allow for the passage of almost three billion water molecules per second, while blocking protons from entering the cell [2]. Aquporins constitute a large family (> 400 members) of membrane spanning proteins, for recent reviews see [3-5]. Since their initial discovery in late 1980s, until early 2000, much of the work, if not all, has been concentrating on understanding the structure and physiological roles of these proteins. Basic research conducted by various groups involved in clinical microbiology and cell biology culminated in developing a 3-dimensional structure of the proteins aquaporin [5]. Now that the structural skeleton is completely defined, explanations of water molecules movement and selectivity can be confidently provided.

The first atomic structures of aquaporins (AQP-1) were obtained by electron microscopy [6, 7]. A high-resolution structure of the *Escherichia coli* glycerol uptake facilitator (GlpF), a bacterial member of the AQP family, was solved by x-ray crystallography at about the same time [8]. In light of the successful exploration of various membrane ion channels, molecular dynamics (MD) simulations were performed on AQP water channels [9-12] soon after their structures became available. Recently, the structure of AQP1 was also solved by x-ray crystallography at high resolution [13], offering a better chance to study the dynamics and function of this water channel in atomic detail.

A few attempts to develop biomimetic membranes by incorporating aquaporin proteins into a matrix of lipids or copolyemer films have been reported in a patent and in open literature. For example, a patent was awarded in 2007 for methods of incorporating biological proteins into a copolyemer matrix to produce a biomimetic membrane with a wide variety of functionalities [14]. In another recent published study of aquaporin-based membranes, Kumar et. al. [15] described the transport characteristics of Aquaporin Z containing symmetric poly-(2-methyloxazoline)-poly-(dimethylsiloxane)-poly-(2-methyloxazoline) triblock copolymer vesicles. The authors reported water permeability coefficients that are almost one hundred times higher than commercial RO membranes.

A few corporations, mainly in the USA and Europe, are competing to bring this technology to market and capitalize on its potential. Aquaporin A/S and Danfoss AquaZ, Danish companies, and Novozymes, a USA company, to name a few are, among others, actively working on development and commercialization of this technology [15]. Industrial research and development appears to be targeting niche markets such as ultra-pure water for medical and analytical procedures. Optimistic corporations with sizable financial backing mention 2013 as a target for first commercial product to the mentioned targeted markets. The major and largest market, these corporations hope to capture, is the desalination water purification segment. Much work nonetheless remains to be done for this goal to be accomplished. Academic, national laboratories, as well as corporations, have to first resolve pertaining engineering issues a priori before a successful product can be introduced.

Due to the complex nature of the natural phenomena of biological membranes, engineering design solutions ought to be initiated at the molecular simulation level before any experimental work is to be undertaken. Engineering design parameters and performance related to water permeability, selectivity, flux rates, robustness and durability and overall structural integrity must first be investigated and validated for successful commercial production.

To accomplish the previously mentioned design specification studies, we are attempting here carrying out initial fundamental studies using Molecular Dynamics (MD) computer simulation studies. We will attempt to show that equilibrium MD simulations yield osmotic and diffusion permeability values, and result with a method to induce hydrostatic pressure differences across the membrane. We will also describe MD simulations of aquaporin based on the structure reported in [13]), in which pressure-induced water permeation was used to determine the channel's *osmotic permeability* value, which was found to agree well with experimental measurements.

METHODOLOGY

The diffusive permeability of a water channel was estimated by performing equilibrium molecular dynamics MD simulations. An approximation of the osmotic diffusivity was performed via simulating different osmotic pressures on both sides of the membrane as a mean to establish a transmembrane pressure gradient. Earlier attempts by investigators in the field were consulted where best method was adopted with needed modifications. The selected method was that of sandwiching the water layer between two simulated membrane walls. A constant force (*f*) was applied perpendicular to the membrane planes to generate the required pressure gradient, and consequently resulted with the targeted net water movement.

Another attempted method was the so-called "two-chamber method" where the unit cell consisted of two membranes and two water layers containing different concentrations of solutes. This method required large chemical potential difference which added complexity to the system and increased the possibilities of error occurrence due to departure from the much diluted systems assumed in most MD simulation in ns time scale. Additionally, previous work by other groups have also been reported where water flux from this method decreased as time lapsed and eventually stopped (ref).

The aquaporin protein channel was embedded into the lipid bilayer and solvated by adding layers of water molecules on both sides of the membrane. The system was equilibrated for 500-900 ps with the protein fixed, under standard room temperature and pressure. All simulation runs were performed using the Materials Studio simulation package. Full electrostatics was employed using the particle mesh Ewald (PME) method.

EXPERIMENTAL SETUP AND PROCEDURE

Molecular dynamics simulations are fully atomistic and were carried out using the software package Materials Studio (Accelrys Software Inc.).

Aquaporin-1. The molecular structure of AQP-1 was retrieved from the Protein Data Bank (PDB entry 1J4N, resolution 2.2). AQP-1 consists of six parallel channels. The protein was embedded in a palmitoyl-oleoyl-phophated-choline (POPC) lipid bilayer consisting of 180 lipids equilibrated separately. The whole system was packed into an amorphous cell sandwiched by two thick slabs of water. The amorphous cell module builds the molecular system in a Monte Carlo fashion, and provides a comprehensive set of tools capable of minimizing close contacts between atoms while ensuring a realistic distribution of torsion angles for used force field. The total number of atoms contained in the cell amounted to 83007. In order to minimize the finite size effect, periodic boundary conditions were enabled.

Simulation runs were performed mostly in two stages: (i) the water with the lipid bilayer were subjected to a long equilibration runs at 298 K and atmospheric pressure for 900 ns, and (ii) the protein atoms were fixed to remove bad contacts between the water and lipids, then minimization and equilibration steps were repeated for the entire cell for 500 ps each run.

RESULTS AND DISCUSSION

Membrane capability to deliver required water fluxes: simulator calculation iterations resulted with an approximate average molecular flux of about 2.2 H_2O molecules per nanosecond. This is the mean flux out of four AQP1 water channels (water conducting monomers) ran for 4 ns each. Sample results are listed in Table 1 below. Molecular water counting was performed by calculating the RMS of the molecule displacement from one end of the channel to the other.

Table 1: Unidirectional water flux through AQP1

	Number of Molecules Crossing				Molecular Flux (Molecules/ns)
	C 1	C 2	C 3	C 4	Average
Run 1	14	15	12	16	3.6
Run 2	9	8	3	11	1.9
Run 3	10	8	5	9	2.0
Run 4	12	9	9	8	2.4

The calculated flux compares well to reported values in the literature for similar simulation and experimental sittings (Kumar, et. al., 2007; Zhu, et. al., 2003; deGroot, et. al., 2001). The molecular flux translates into 1.3 x 10^{-11} moles per hour per protein which in turn results with a flux of about 150 liters/ m^2.hr. This figure compares favorably to results obtained by Sandia National Laboratories experimental runs in which they obtained a flux of about 130 liters/ m^2.hr (Rempe, et. al., 2010). These flux figures roughly translate into around 16X higher fluxes than those obtained through traditional RO membranes (DOW SW30ULE runs around 9.4 liters/ m^2.hr). Additionally, it ought to be noted that flux values reported here are only diffusional fluxes and do not include permeation due to osmotic forces. When osmotic pressures are included, fluxes would naturally increase significantly and may reach values close to those obtained when a vesicle is placed in a saline solution and resulted with a flux of about 266.4 liters/ m^2.hr (ie., 28X traditional commercial RO membranes) (see Kumar, et. al., 2007) .

Working pressure requirements and osmotic pressures: significant efforts and trials were directed at inducing a constant hydrostatic force across the unit cell, but moderate success was achieved. Difficulties centred around methods of exerting a constant force along the normal direction of the cell and maintaining its effect on the water molecules layer(s). Additional difficulties were experienced where water molecules in the water layer, including those adjacent to the entrances of the channels, when subjected to external artificial force significantly affected the number of molecules diffusing through the protein channels. Also, keeping the entire unit cell encompassing the layered system in a fixed position after applying the simulated hydrostatic pressure proved extremely challenging.

Albeit of the mentioned challenges, approximate values were obtained for the required hydrostatic pressure. The hydrostatic pressure gradient across the membrane was calculated based on the simplified equation provided by Zhu (Zhu et at., 2002):

$$\Delta P = P_1 - P_2 = \frac{nf}{A}$$

where n is the water molecules number in the water layer, and A is the cross-sectional area of the membrane. Taking a 10 Angstrom thick layer by a unit width, it would contain roughly 3000 molecules (*n*) spread over a cross-sectional area (A) of the membrane of about 1.0×10^{-16} m^2. Applying a constant force (*f*) of about 4.3 pN resulted with an average pressure gradient around 1000 Bars. This number agrees with earlier attempted efforts by researchers (Schulten, et al., 2004), but as of date we are unaware of any experimental values verifying these findings.

Using the previously obtained values for the hydrostatic pressure gradient and taking into account that the "osmotic pressure", is the hydrostatic pressure capable of producing the same "chemical potential" difference between to solutions, the diffusional and osmotic permeability were calculated. The osmotic permeability was calculated using

$$p_f = \frac{water\ flux}{concentration\ gradient} = \frac{water\ flux}{pressure\ gradient/RT}$$

And the diffusional permeability was calculated using

$$p_f = \frac{solute\ flux}{solute\ concentration\ gradient} = \frac{solute\ flux}{pressure\ gradient/RT}$$

Chemical potential difference of water in terms of pressure difference was established across the AQP1 channels resulted with a p_f of 6.3×10^{-14} cm^3/s. Equilibrium MD simulation runs resulted with a calculated p_d of 5.7×10^{-15} cm^3/s. Reported values for p_d and p_f, both are in close agreement with observation, and are of prime importance for future membrane design and scale-up for water desalination purposes.

Proteins physical and physicochemical properties and their influence on water transport: simulation runs with aquaporin 1 and also aquaporin z (most commonly used proteins in similar studies) showed no significant difference on water transport due to their physiochemical properties.

Effects and relations governing the protein-support material interactions: physical support material (polymers) was not investigated due to the added complexity of the system, large atomic volume, and consequently the much larger computational capabilities. Instead, to constrain the system and prevent it from movement, harmonic constraints were applied to the C-alpha carbon atoms of the protein and the phosphorus atoms of the lipid molecules. This was done by choosing spring constants that would fully balance the external forces.

Effects of proteins orientation and transport directionality: conducted simulation studies did not indicate any significant effects on protein orientation (perpendicular to the matrix plane only) and transport is bi-directional.

APPLICATION

The intended application(s) of this study was to provide quantitative analysis of parameters needed for design, evaluation and development of a much superior RO membrane for water purification purposes. Results from this study should prove valuable in guiding the experimental phase(s) of developing biomimetic RO membranes. It is hoped that the outcome of this research will lead to improved membrane performance and efficiency and reduce financial and environmental costs of producing ultrapure safe drinking water.

CONCLUSION AND RECOMMENDATIONS

The question this study attempted to answer was 'can a biomimetic membrane composed of water-selective aquaporin proteins embedded in lipid bilayer that is supported by a synthetic polymeric layer be ultimately utilized for water desalination purposes? They word here is "desalination" and with it, unfortunately the answer to the stated question based on efforts carried out in this study is 'may be!' The reason being is that, although the critical parameters need for design of biomimetic RO membranes and scaling up to commercial sizes were successfully obtained in this study and conferred to published values, they are , however, remain limited to 'non-saline' water. Salt concentrations and their overall influence on separation and its rejection remains an issue to be resolved only by experimental methods

Future efforts via MD simulations and experimentation will need to be carried out to obtain fundamental understanding of the physico-chemical principles that are responsible for their biomimetic membranes superior performance. These principles will then guide the design of artificial membranes

REFERENCES

1. http://www.bio.miami.edu/~cmallery/150/memb/water.channels.htm. Accessed Feb. 05, 2011.
2. Gonen, T. and T. Walz, The structure of aquaporins. Q Rev Biophys, 2006. 39(4): p. 361-96.
3. Fu, D. and M. Lu, The structural basis of water permeation and proton exclusion in aquaporins. Mol Membr Biol, 2007. 24(5-6): p. 366-74.
4. Ludewig, U. and M. Dynowski, Plant aquaporin selectivity: where transport assays, computer simulations and physiology meet. Cell Mol Life Sci, 2009.
5. Gonen, T., Sliz, P., Kistler, J., Cheng, Y. & Walz, T. Aquaporin-0 membrane junctions reveal the structure of a closed water pore. *Nature* **429**, 193–197 (2004).
6. Murata, K., K. Mitsuoka, T. Hirai, T. Walz, P. Agre, J. B. Heymann, A. Engel, and Y. Fujiyoshi. 2000. Structural determinants of water permeation through aquaporin-1. *Nature.* 407:599–605.
7. Ren, G., V. S. Reddy, A. Cheng, P. Melnyk, and A. K. Mitra. 2001. Visualization of a water-selective pore by electron crystallography in vitreous ice. *Proc. Natl. Acad. Sci. USA.* 98:1398–1403.

8. Fu, D., A. Libson, L. J. W. Miercke, C. Weitzman, P. Nollert, J. Krucinski, and R. M. Stroud. 2000. Structure of a glycerol conducting channel and the basis for its selectivity. *Science.* 290:481–486.
9. de Groot, B. L., and H. Grubmüller. 2001. Water permeation across biological membranes: mechanism and dynamics of aquaporin-1 and GlpF. *Science.* 294:2353–2357.
10. Jensen, M., S. Park, E. Tajkhorshid, and K. Schulten. 2002. Energetics of glycerol conduction through aquaglyceroporin GlpF. *Proc. Natl. Acad. Sci. USA.* 99:6731–6736.
11. Tajkhorshid, E., P. Nollert, M. Jensen, L. J. W. Miercke, J. O'Connell, R. M. Stroud, and K. Schulten. 2002. Control of the selectivity of the aquaporin water channel family by global orientational tuning. *Science.* 296:525–530.
12. Zhu, F., and K. Schulten. 2003. Water and proton conduction through carbon nanotubes as models for biological channels. *Biophys. J.* 85:236–244.
13. Sui, H., B.-G. Han, J. K. Lee, P. Walian, and B. K. Jap. 2001. Structural basis of water-specific transport through the AQP1 water channel. *Nature.* 414:872–878
14. Montemagno, C.D., J.J. Schmidt, S.P. Tozzi, Biomimetic membranes, US Patent No. 7208089, 2007.
15. Kumar, M., M. Grzelakowski, J. Ziles, M. Clark, W. Meier. 2007**.** Highly permeable polymeric membranes based on the incorporation of the functional water channel protein Aquaporin Z. P.N.A.S., 105(52), 20723-20728.
16. http://www.greentechmedia.com/articles/read/water-companies-taking-cue-from-nature-1388/. Accessed February 02, 2011.
17. Rempe, SB, Rogers, DM, Jiang, YB, Yang, S, Leung, K, Brinker, CJ, et al. 2010. Computational and experimental platform for understanding and optimizing water flux and salt rejection in nanoporous membranes. SANDIA Technical Report SAND2010-6735

365441 Influence of Controlled Fluid Shear on Nucleation Kinetics in Glycine Aqueous Solutions

Abstract:

Understanding crystal nucleation from solution is important for rationally designing and controlling manufacturing processes that involve crystallization. Both batch and continuous industrial crystallization processes typically involve fluid flow where supersaturated solutions and/or suspensions of growing crystals are pumped or agitated. It is well known that fluid shear can induce secondary nucleation and it has been experimentally demonstrated that primary nucleation can be affected by fluid shear as well (1,2). In order to gain further experimental evidence to facilitate improved understanding of effects of fluid shear on primary nucleation, we investigated nucleation in aqueous glycine solutions under well-defined flow conditions at a constant temperature.

Supersaturated glycine solutions were prepared by nanofiltering and slow cooling, and they were then subjected to controlled fluid shear in Couette cells under isothermal conditions. Average shear rates between $25s^{-1}$ and $1000s^{-1}$ were studied. Crystallization in the Couette cell was monitored using a custom built instrument that allowed transmission and light scattering measurements to be taken simultaneously in situ; the onset of nucleation was associated with a rapid increase in scattering and decrease in transmission, so induction times were readily obtained from these measurements. The role of surface was also studied by changing the diameter of the inner cylinder of the Couette cell; this allowed surfaces of $2.5cm^2/cm^3$ solution, $5\ cm^2/cm^3$ solution and $10\ cm^2/cm^3$ solution to be investigated. Due to the stochastic nature of nucleation, experiments were repeated multiple times and a statistical analysis was performed to show that the number of repetitions was sufficient for accurate trends to be deduced.

Nucleation induction times in sheared solutions were found to be considerably lower than those in unsheared solutions, while a great care was taken to avoid seeding either in solution or at solid-liquid interfaces. This suggested that fluid shear had a profound influence on primary nucleation glycine from aqueous solutions. In sheared solutions, a large number of small crystals formed in close succession, causing solutions to rapidly go turbid, so it is proposed that shear enhanced the formation of primary nuclei which then initiated extensive secondary nucleation (3). Increasing the average shear rate was found to reduce the induction time in a power law relationship with the exponent of -0.98 ± 0.1. This indicated that the total strain applied had an important effect on nucleation kinetics. Increasing the surface area was also found to notably decrease the induction times, with a power law relationship with the exponent of -0.88 ± 0.2, which indicated that the surface played an important role in nucleation kinetics as well. In situ dynamic light scattering measurements showed that colloidal scale liquid-like clusters (nanodroplets) observed in mesostructured amino acid aqueous solutions (4) had their mean size increasing upon continued shearing of solutions. This is consistent with these mesoscale clusters (in the bulk liquid phase and/or at the solid-liquid interface) playing a direct role in the primary nucleation pathway in this system as proposed recently (1), where larger clusters correlated with shorter nucleation induction times.

(1) Jawor-Baczynska, Sefcik and Moore, Crystal Growth & Design, 13, 470 (2013)

(2) Liu and Rasmuson, Crystal Growth & Design, 13, 4385 (2013)

(3) Kadam, Kramer and ter Horst, Crystal Growth & Design, 11, 1271 (2011)

(4) Jawor-Baczynska, Moore, Lee, McCormick and Sefcik, Faraday Discussions, 167, 425 (2013)

377543 Analysis of Crystallization Kinetics Based on CSD Estimates from PAT Tools

Abstract:

Nucleation and growth are the two dominating phenomena in crystallization processes. In our study, a partial crystal size distribution (CSD) was estimated for the size range from 50 μm to 500 μm by combining in situ measurements from both FBRM and ATR-FTIR. Experimental results show that the evolution of the CSD was better tracked by this framework, compared to a qualitative interpretation using chord length distribution (CLD).

To explore the process and the corresponding kinetics, seeded and unseeded batch cooling crystallization experiments were designed to start with different initial conditions and proceed through different cooling rates. The general procedure is: (1) a concentrated ethanolic solution of paracetamol is cooled to supersaturated conditions, followed by a constant-temperature plateau; (2) either seeds are added or primary nucleation occurs on the constant-temperature plateau; (3) after desupersaturation of the solution is complete, the solution is further cooled to the final temperature. The estimated CSDs from CLDs were consistent with our expectations of the design of the crystallization runs, such as initial concentrations and cooling rates. These CSDs were also verified by sieving.

A full population-balance model was solved in order to fit the experimental data and estimate kinetic parameters. A numerical method (conservation/solution element scheme) was applied to the population balance to simulate the experiments efficiently and was used in parameter estimation. Classical nucleation theory was used for primary nucleation, secondary nucleation was related to relative supersaturation and mass of crystals, and the growth rate was assumed to be size-independent.

Using the estimated parameters, the simulated runs for unseeded crystallizations suggest that the final CSD was strongly influenced by the relative importance of primary versus secondary nucleation when they occur simultaneously in the clear supersaturated solution; if secondary nucleation is dominant at that moment, then most of the crystal mass is concentrated in small size ranges. An optimal two-step cooling profile was designed to produce a final size distribution with more large crystals and it was validated experimentally.

Heterogeneous Nucleation of APIs on Engineered Biocompatible Polymer Surfaces

Li Tan, Rachel M. Davis, Prof. Allan S. Myerson, Prof. Bernhardt L. Trout

Date: May 12th, 2014

Continuous manufacturing of pharmaceuticals have significant advantages over batch processing in terms of economics, reliability, safety, and sustainability. Directly crystallize active pharmaceutical ingredients (APIs) on polymer excipient surfaces can achieve impurity separation and drug-excipient composite formation in a single step, which eliminates many solid handling steps traditionally needed in downstream processing. In order to develop such a process, we must understand the kinetics of heterogeneous nucleation of small molecule drugs on biocompatible polymer surfaces, and study how heterogeneous nucleation affects polymorph formation. The chemical interactions between APIs and polymer substrates plus the effect of surface geometries are the focuses of the present investigation.

We engineered surfaces by first manufacturing a rigid silicon mold using interference lithography to create nanometer-sized and mono-dispersed pillars of varying angles on a flat surface. Negatives of the mold were transferred to polyvinyl alcohol (PVA) films via nano-imprint lithography to create films with nano-pores. The films were submerged in solution as substrates for cooling crystallization of two API compounds, aspirin and indomethacin. A high throughput system was used to collect heterogeneous nucleation kinetics data on the patterned films at low supersaturation settings. Experimental results show that the induction time of aspirin on these surfaces were reduced by more than an order of magnitude with favorable surface chemistry, with further reduction achieved by matching surface geometry. Matching angles on the substrate surface and angles between dominant faces resulted in the greatest reduction in induction time. X-ray diffraction studies revealed favorable interactions between dominant habit faces of aspirin to PVA via hydrogen bonding. Crystallization experiments with indomethacin as the model compound show that in addition to enhancing nucleation rates, polymer chemistry and surface geometry also alters the distribution of polymorphs obtained in the experiments. Correlation between final polymorph distribution on film surface and initial nucleation events were established using Raman spectroscopy. Computational analyses were also used to corroborate experimental results and hypothesis presented.

Impact of surface chemistry on crystallization of acetaminophen

Using polymeric surfaces is a novel way to control crystallization however, the mechanism is still ambiguous. In the present study, chitosan (CS) films that were doped with different molecules were employed to investigate the impact of surface chemistry on the heterogeneous crystallization behavior of acetaminophen (APAP).Crystal growth rate and crystallization kinetic measurements of APAP on various substrates were performed. Fourier transform infrared spectroscopy was used to probe intermolecular interactions between APAP and CS films. The results were interpreted with the aid of molecular simulations. It was found that substrates with more hydrogen bonding groups that could interact with APAP promoted its crystallization.

Development of a robust API crystallization in a multi-component solvent mixture: Using high throughput automation as an enabling technology to develop comprehensive solubility maps

Authors: **Ben Cohen[1]**, Michelle Mahoney[1], Brenda Remy[1], Jun Qiu[1], Chris Sfouggatakis[1], Zhongping Shi[1], Chenchi Wang[2]

[1] Chemical Development, Bristol-Myers Squibb Co.,
[2]Drug Product Science & Technology, Bristol-Myers Squibb Co.,

Session: 2B07 Crystallization of Pharmaceutical and Biological Molecules

Knowledge of solubility is essential to process design of API crystallizations. Detailed solubility enables a basis for defining robust and optimized process conditions, such as dissolution, seeding and isolation conditions. Determination of solubility maps is often performed by manual methods. However, as the number of components grows, and hence the number of possible process combinations grows, obtaining a comprehensive solubility map becomes more time-intensive. An example of this complexity was observed in the final crystallization of a BMS API currently in development. API solubility was found to be affected by numerous factors in the crystallization: the relative amounts of the three solvents, the amount of base, and the amount of reaction byproduct, as well as temperature.

To streamline the construction of the solubility map, a high-throughput workflow was developed to measure API solubility using automation equipment. The new workflow generated accurate solubility, with units of mass fraction through the use of an internal standard, over a wide temperature range. The data set enabled construction of a reliable solubility map which led to improved understanding of the crystallization. It helped establish process ranges around important operations such as dissolution, seeding, anti-solvent addition and isolation, as well as acceptance criteria for the upstream reaction and workup. The detailed solubility knowledge provided a basis for defining optimized process conditions within a robust design space.

389891 Effect of Functional Groups on Separating CO_2 from CO_2/N_2 Mixtures Using Edge Functionalized Graphene Nanoribbons

Abstract:

Development of microporous adsorbents for separation and sequestration of carbon dioxide from flue gas streams is an area of active research. In this study, we assess the influence of specific functional groups on the adsorption selectivity of CO_2/N_2 mixtures through Grand-Canonical Monte Carlo (GCMC) simulations. Our model system consists of a bilayer graphene nanoribbon that has been edge functionalized with OH, NH_2, NO_2, CH_3 and COOH. Ab initio Møller-Plesset (MP2) calculations with functionalized benzenes are used to obtain binding energies and optimized geometries for CO_2 and N_2. These information in turn are used to validate the choice of the classical forcefield in the GCMC simulations. In addition to simulations of adsorption from binary mixtures of CO_2 and N_2, the ideal adsorbed solution theory (IAST) is used to predict mixture isotherms. Our study reveals that functionalization always leads to an increase in the adsorption of both CO_2 and N_2 with the highest for COOH. However, significant enhancement in the selectivity for $CO_2$is only seen with COOH functionalized nanoribbons. The COOH functionalization gives a 28% increase in selectivity compared to H terminated nanoribbons, whereas the improvement in the selectivity for other functional groups are much more modest. Our study suggests that specific functionalization with COOH groups can provide a materials design strategy to improve CO_2 selectivity in microporous adsorbents. Synthesis of graphene nanoplatelets with edge functionalized COOH, which has the potential for large scale produciton has recently been reported.

388406 Polymeric Moisture Swing Membranes for Carbon Dioxide Capture from Gas Mixtures

Abstract:

Processes permitting efficient carbon dioxide capture from gas mixtures could find application in large-scale industrial gas separations, production of synthetic carbon-containing fuels, and greenhouse gas removal from the atmosphere. Membrane-based processes provide several advantages over traditional carbon dioxide removal technologies, including mechanical simplicity, a relatively small footprint, and energy efficiency. Here, we describe the synthesis, characterization, and analysis of polymeric anion exchange materials for carbon dioxide concentration from gas mixtures. Transport of carbon dioxide through the membrane is promoted by an opposing flux of water, which reacts with carbon dioxide through several equilibrium reactions to form charged species (bicarbonate, carbonate, and hydroxide) within the membrane matrix. Transport of carbon dioxide through the membrane will be discussed as a function of material characteristics, including charge density, and process characteristics, such as feed stream relative humidity, crossflow velocity over the membrane surface, and carbon dioxide concentration on both sides of the membrane. The development of several membrane materials, including those based on poly(phenylene oxide) and imidazolium chemistries, will be discussed. Results of experimental gas transport studies in a crossflow test cell will be presented. Optimal operating conditions for maximum carbon capture will be proposed. Of particular interest is the integration of a carbon dioxide capture membrane into an artificial photosynthesis system, which would permit the combination of hydrogen and oxygen from solar water splitting with captured carbon dioxide from the atmosphere to create synthetic, carbon-rich fuels.

377109 Thermal Swing Sorption Process for CO2 Capture from Flue Gas

Abstract:

A variety of sorbents are being used in packed beds for CO_2 capture from flue gas. The bed is regenerated by thermally stripping CO_2 in a cyclic process. The sorbent capacity is highly relevant. The higher the sorption capacity, the lower is the energy requirement for CO_2 capture and recovery. The highest values of the CO_2 sorption capacity in published literature are between 2.5 and 3 mmol/g of the sorbent. We have developed a novel sorbent having a capacity of as much as 6.3-7 mmol/g of the sorbent and determined its selectivity with respect to N_2. We have studied the sorption and desorption behavior of the packed bed with flue gas in a cyclic process. Experimental results will be presented and discussed. Membranes are of special use in such a process.

Adsorption of CO_2, CH_4 and N_2 on N-doped Highly Porous Activated Carbons: Approach for Selective Adsorption of Flue Gas and Natural Gas Upgrading

Jun Wang[a], Rajamani Krishna[b], Jiangfeng Yang[c], Shugaung Deng[a *]

[a.] Chemical Engineering Department,

New Mexico State University,

Las Cruces, New Mexico,88003, USA

[b.] Van't Hoff Institute for Molecular Sciences,

University of Amsterdam,

Science Park 904, 1098 XH Amsterdam, The Netherlands

[c.] Research Institute of Special Chemicals,

Taiyuan University of Technology,

Taiyuan 030024, Shanxi, PR China

Abstract

Three N-doped microporous activated carbon adsorbents were synthesized with KOH as the porogen agent at pyrolysis temperatures of 600, 700, and 800°C. The adsorbent samples were characterized by X-ray diffraction, scanning electron microscopy, thermal gravimetric analysis, Raman spectra, Fourier transformed infrared (FT-IR) spectra, Energy-dispersive X-ray Spectroscopy (EDS) mapping, and volumetric nitrogen adsorption. CO_2, CH_4, and N_2 adsorption isotherms were measured at 100 kPa and three different temperatures after activation. All isotherms were correlated with the Langmuir model. The heat of adsorption was calculated using the Clausius-Clapeyron equation based on single component adsorption isotherms. These data were used to estimate the separation selectivities for CO_2/CH_4, CO_2/N_2, and CH_4/N_2 binary mixtures at 298K and 100 kPa with the ideal adsorbed solution theory (IAST) model. Extremely good selectivites were achieved due to their N-containing groups, narrow pore size distribution, and large specific surface area. In addition, breakthrough simulations for $CO_2/CH_4/N_2$ binary mixtures were conducted. The result is superior to ZIF-7 under the same condition. Based on these advantages, we infer N-ACs could be suitable and promising adsorbents for flue gas selective adsorption and natural gas upgrading.

Preliminary Results

The CO_2, N_2 and CH_4 adsorption isotherms are summarized in figure 1. The highest CO_2 capacity at 100 kPa is obtained by N-AC 600, which is 6.36 and 4.11 mmol g^{-1} at 273 and 298K, respectively.

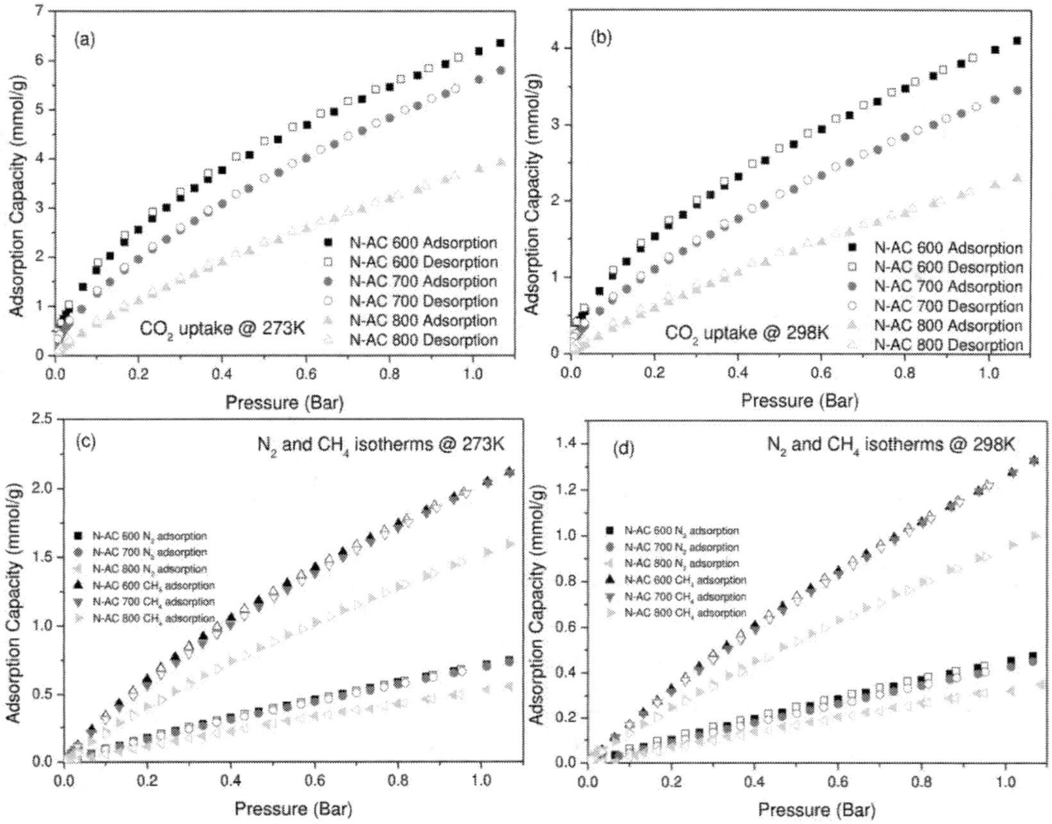

Figure 1. (a) CO2 adsorption 273K and (b) CO_2 adsorption 298K (c) N_2 and CH_4 adsorption 273K and (d) N_2 and CH_4 adsorption 298K of N-AC 600, N-AC 700, and N-AC 800.

Figure 2 presents the values of the adsorption selectivity for equimolar binary CO_2/CH_4, CO_2/N_2, and CH_4/N_2 gas mixtures maintained at isothermal conditions at 298 K in N-AC 600, N-AC 700, and N-AC 800. For CO2/CH4 and CO2/N2 mixtures, the highest adsorption selectivities are obtained with N-AC 600 of 9.2 and 47.3 at 100 kPa, respectively.

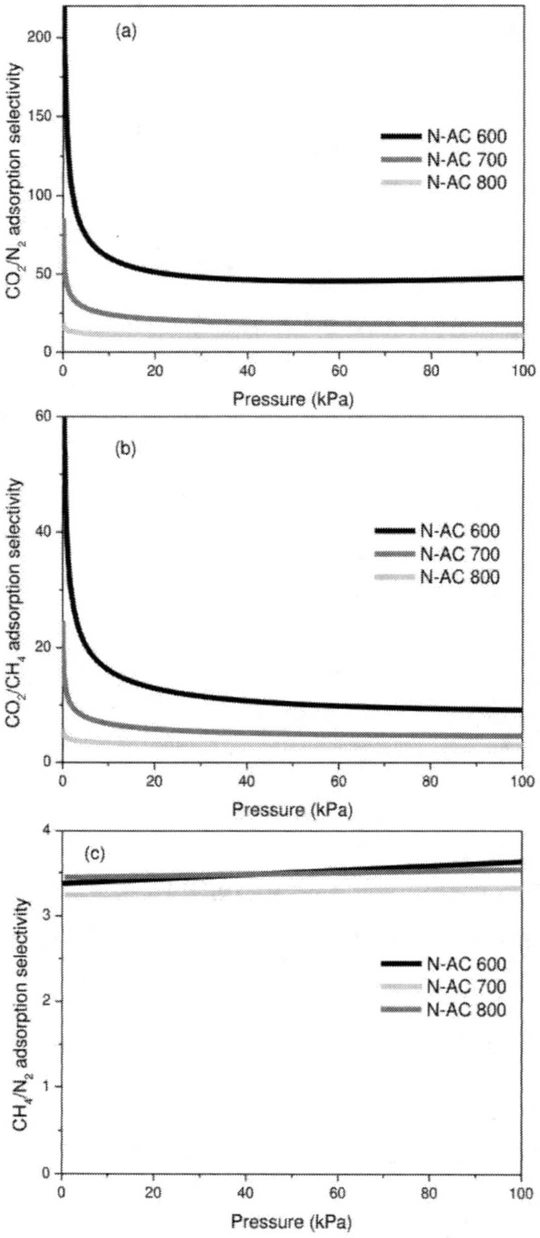

Figure 2. IAST predicted adsorption selectivities for equimolar binary mixtures of (a)CO_2/CH_4, (b)CH_4/N_2, and (c) CO_2/N_2.

For all three N-ACs the sequence of breakthroughs is N_2, CH_4, and CO_2, this is dictated by the adsorption strength. CO_2 molecules are strongly adsorbed by N-ACs, which elutes last in the sequence. The longest retention time obtained on N-AC 600 and this material has the best separation performance due to its highest CO_2 capacity and highest CO_2/N_2 and CO_2/CH_4 adsorption selectivities. We noted that there is a time interval that we can get pure CH_4 or N_2 using N-ACs.

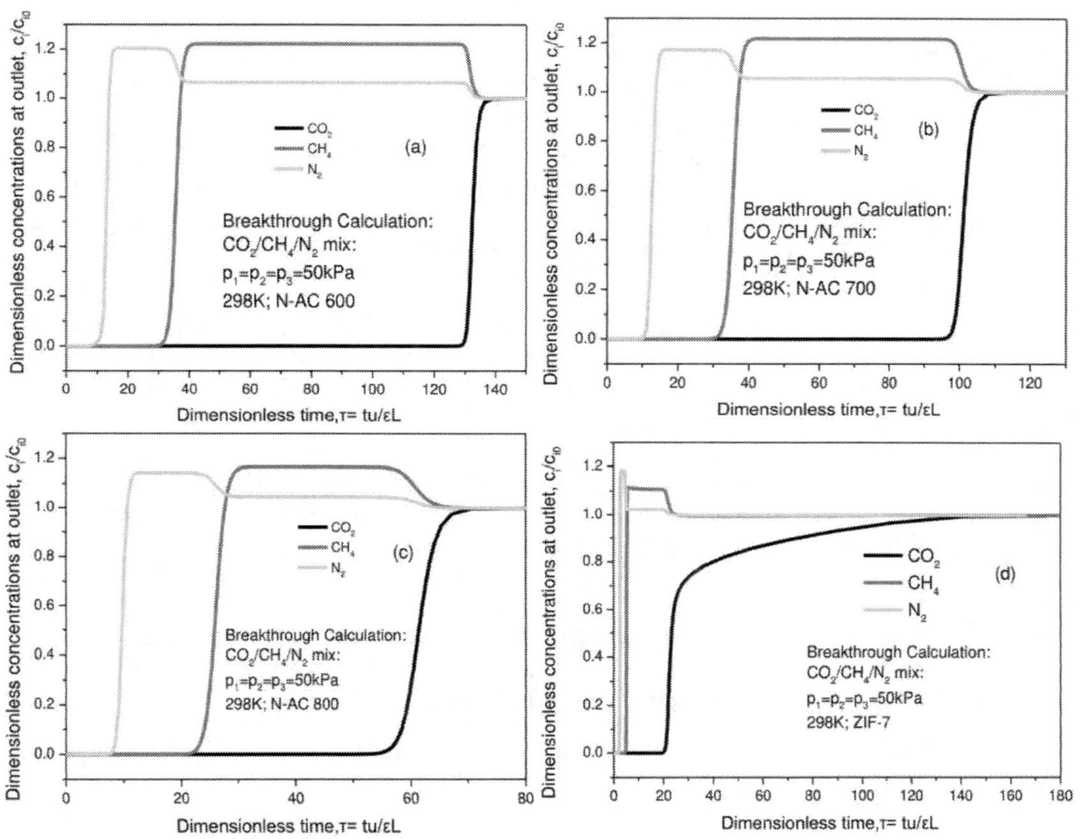

Figure 3. Breakthrough simulation data of (a)N-AC 600, (b) N-AC 700, (c) N-AC 800 at 298 K and 1 bar for an equimolar $CO_2/CH_4/N_2$ mixture, and (d) ZiF-7 at 298 K and 1 bar for an equimolar CH_4/N_2 mixture (ci, the content of outlet gas; c0, the content of raw gas).

381966 Investigation of MOF/Polyimide Interactions in Mixed Matrix Membranes for CO2 Separation

Abstract:

The separation of CO_2 from industrially important gas mixtures like natural gas, syngas and flue gas is an enormously crucial issue in chemical industry. An attractive class of material is mixed-matrix membranes (MMMs) where metal organic frameworks (MOFs) are incorporated into polymer matrices. The principle issue in these applications is the plasticization of the polymers at high partial pressures of CO_2, which can lead to reduced membrane selectivity and unpredictable transport properties at the polymer/MOF interfaces.

In this molecular simulation study, we focused on the interfacial interactions of ZIF-8/Matrimid® 5218 and ZIF-8/Ultem® 1000 MMMs for CO_2/CH_4 separation. Polyimides were constructed using Accelrys Materials Studio 5.0® software and equilibrated by a series of Molecular Dynamics (MD) run using LAMMPS package. Polymer Consistent Force Field (PCFF) was used for simulations and density, Fractional Free Volume (FFV) and Glass Trasition Temperature (Tg) of the polymers were compared with experimental values and the force-field was validated considering the agreement within these values. Monte Carlo (MC) and MD simulations were applied for estimating CO_2 and CH_4 sorption and diffusion in the polymers to validate the models.

In order to build model MMMs, a ZIF-8 cage and two polymer chains with 40 monomer units were packed in a simulation box by using Amorphous Cell Module in Materials Studio 5.0®. This corresponded to 15% per weight ZIF-8 content in the mixed matrix. The same equilibration steps were applied to the mixed matrixes, and their sorption and diffusion characteristics were investigated via MC and MD simulations.

Gas permeabilities up to 30 bar was estimated in order to investigate the plasticization of MMM's and subsequent changes at the polymer/ZIF-8 interfaces. Simulation results indicated that the interactions between the functional groups of the polymer and dangling ligands at the MOF surface may have significant contribution on the gas transport through the interface.

Acknowledgment: This work was supported by TUBITAK trough grant no. 113M336.

379522 Computational Analysis and Screening of Materials for Capture of Toxic Compounds

Abstract:

Porous materials play an important role in the capture of toxic compounds. Advances in nanotechnology have enabled the development of novel materials by a targeted molecular design approach. This has led to the emergence of numerous functional materials over the past decade. Among them, metal-organic frameworks (MOFs) are promising materials, as they allow for design of new structures from a building-block approach permitting the design of a wide variety of adsorbents tailored for specific adsorption separations.

Development of selective adsorbents for target molecules would be greatly accelerated if we had a better understanding of how target sorbates interact with surface functional groups. Molecular simulations can provide valuable information to predict interaction energies with sufficient accuracy for screening and to provide insights about the detailed nature of molecular-level interactions. The objective of this work has been to computationally discover the best functional groups and surface chemistries for the removal of toxic chemicals from air.

An inherent challenge in the capture of toxic compounds in humid conditions is the competitive adsorption of water. We have explored two approaches for overcoming this problem. In the first approach, we have used quantum chemical methods to search for functional groups or surfaces that could bind toxic molecules more strongly than water. In the second approach, we have targeted hydrophobic MOFs. Using classical Monte Carlo simulations, we have investigated the effect of perfluoroalkane chains on ammonia and water adsorption in MOFs and performed high-throughput computational screening of hydrophobic MOFs. The results suggest that if materials are too hydrophilic, they adsorb too much water. On the other hand, if they are too hydrophobic, they adsorb too little ammonia.

Adsorption of Aromatic and Nitrogen Aromatic Compounds on the MOF by the Spectroscopic, Kinetic and Computational Studies

Jun Dai[1], *Michael McKee*[2], *Jing Li*[3] *and* ***Alexander Samokhvalov***[1*], *(1)Chemistry Department, Rutgers University, Camden, NJ, (2)Chemistry and Biochemistry Department, Auburn University, Auburn, AL, (3)Department of Chemistry and Chemical Biology, Rutgers University, Piscataway, NJ*

Introduction

Nitrogen-containing heterocyclic aromatic compounds are present in petroleum, other fossils, refinery streams and commercial fuels; upon combustion of fuels, toxic nitrogen oxides are released. Metal-Organic Frameworks (MOFs) form a new class of sorbents with very high surface area and pore volume. Adsorption on the MOFs in the liquid phase was studied much less compared to adsorption in the gas phase. Naphthalene (NAP) is a prototype fused-ring aromatic hydrocarbon and indole (IND) is a prototype nitrogen-containing heterocyclic aromatic compound with fused rings that are both present in petroleum and fuels. Mechanism of adsorption of aromatic compounds *vs.* nitrogen-containing heterocyclic aromatic compounds on F300 Basolite MOF has not been investigated. We report the study of adsorption of the prototype fused-ring aromatic and nitrogen-containing heterocyclic aromatic compounds NAP and IND on F300 MOF *vs.* MIL-100(Fe) by the complementary fluorescence spectroscopy and near-UV/Vis DRS, and by density functional theory (DFT) computations.

Experimental

Chemicals. Naphthalene (NAP), indole (IND), eicosane n-$C_{20}H_{42}$, F300 Basolite were from Sigma Aldrich, spectroscopic n-heptane was from Fisher, and MIL-100(Fe) was provided by Dr. Jing Li. F300 and MIL-100(Fe) were activated as in Ref. [1].

Solid stoichiometric adsorption complex of F300 and NAP in eicosane matrix. Eicosane was melted under argon, a known amount of NAP was added, heated at 40 °C on a sonic bath until dissolution, and cooled to room temperature. To prepare the stoichiometric adsorption complex of NAP with F300 in eicosane matrix, 0.085 g F300 was activated in 1.5 cc quartz fluorescence cuvette, 1.4 cc of melted 0.08 M solution of NAP in eicosane was added, closed, allowed to solidify at room temperature at continuous shaking; this sample contained one molecule of NAP per three Fe cations in F300.

Solid non-stoichiometric mixtures of NAP and IND with F300 in eicosane matrix. A calculated amount of activated F300 (a-F300) or non-activated (n-a-F300) in a 0.5 cc quartz fluorescence cuvette under argon was mixed with melted NAP or IND, closed and allowed to solidify at room temperature while being shaken. The first set of samples had molar ratios R_1 = n(NAP)/n(F300) = 50 and R_1 = n(IND)/n(F300) = 50. The second set of samples had molar ratios R_2 = n(NAP)/n(F300) = 15 and R_2 = n(IND)/n(F300) = 15.

Solid "pure" stoichiometric adsorption complexes of F300 and MIL-100(Fe) with IND and NAP. The sample of 28.6 mg F300 or 34.0 mg MIL-100(Fe) was activated in the 0.5 cc quartz fluorescence cuvette, ca. 0.4 cc of 0.08 M solution of NAP or IND in n-heptane was added under argon (added amounts correspond to one IND or NAP molecule per one Fe in each MOF). The cuvette with suspension was closed and shaken for 1 hour to reach adsorption equilibrium. Then, the solvent in the cuvette was allowed to evaporate at $<1 \times 10^{-1}$ Torr and further at $<1 \times 10^{-3}$ Torr to remove any non-adsorbed IND or NAP. The vacuum system was vented with argon, and the "pure" adsorption complex was obtained as dry powder in the cuvette that was immediately sealed.

Spectroscopic measurements. Fluorescence spectroscopy measurements were conducted with Cary Eclipse spectrometer. Angular fluorescence accessory [2] was equipped with mini-stirrer and temperature control. Excitation wavelength λ_{exc} was 260, 280, 290, 300, or 310 nm. Numeric fitting of the spectra was conducted by Microcal Origin. Near-UV/Vis DRS spectra were measured using the spectrometer, xenon LS-1 lamp, fiber-optical reflection probe, and WS-1 diffuse reflectance standard from Ocean Optics.

Kinetics of IND adsorption on a-F300. Magnetic mini stir bar was added to 3.5 cc quartz cuvette, solid mixture of IND with a-F300 was prepared in the cuvette at $R_2 = n(IND)/n(F300) = 15$, cuvette was sealed, placed into the angular fluorescence accessory at 60 °C, and its content was continuously stirred once melted. Time-dependent fluorescence spectra were collected at $\lambda_{exc} = 310$ nm at 6.6 min intervals.

Quantum chemical calculations. All calculations were made with the Gaussian 09 program [3]. Geometries and frequencies were calculated at the B3LYP/LANL2DZ level [4-8], while single-point energy calculations were made at the B3LYP/6-31+G(2d,p)/(Fe=SDD)+GD3. Here, SDD indicates that Stuttgart effective core pseudopotenials were used in place of the core electrons on Fe atoms [9] and GD3 indicates that empirical dispersion corrections were added to the calculated energies [10].

Results and Discussion

1. Fluorescence spectra of 0.08 M solution of NAP in n-heptane and pure NAP.

Figure 1A shows the fluorescence spectrum of 0.08 M solution of NAP in n-heptane at $\lambda_{exc} = 280$ nm. The fluorescence maxima are at 312 nm (shoulder), 323 nm (strong peak), 335 nm (strong peak), 350 nm (shoulder), and 368 nm (shoulder).

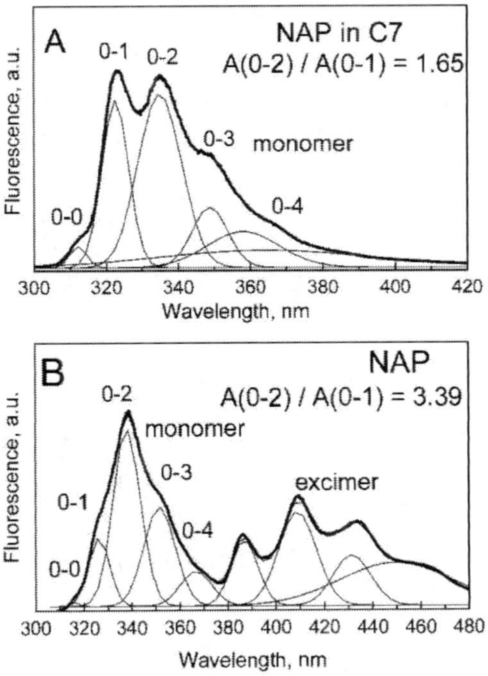

Figure 1. A) Fluorescence spectrum of 0.08 M liquid solution of naphthalene (NAP) in n-heptane (C7) at λ_{exc}=280 nm and multi-Gaussian curve fitting of vibronic subbands. B) Fluorescence spectrum of pure NAP at λ_{exc}=280 nm and multi-Gaussian curve fitting of vibronic subbands.

The 0.08 M solution is "dilute", with isolated NAP molecules being surrounded by molecules of n-alkane. The peaks in the spectrum in Fig. 1A are consistent with reported spectra of a diluted solution of NAP in n-hexane [11]. In Fig. 1A, the shoulder at 312 nm is due to the 0-0 fluorescence transition in isolated NAP molecule [11], the peak at 323 nm is due to the 0-1 vibronic transition, peak at 335 nm is due to the 0-2 transition [11], and the shoulder at 350 nm is due to the 0-3 transition coupled with collective vibrations [12]. The shoulder at 368 nm is the next member of vibronic multiplet that was detected in the gas-phase spectra of naphthalene [12]. Figure 1A also shows numeric curve fitting with the six-Gaussian function [13]. The peak areas correspond to the intensities of the fundamental 0-0 and vibronic transitions. The ratio of areas (A) of two major 0-2 and 0-1 transitions is A(0-2)/A(0-1) = 1.65. Figure 1B shows the fluorescence spectrum of pure NAP at λ_{exc} = 280 nm. Several fluorescence bands are seen at 380-480 nm that were not observed in the spectrum of solution of NAP in n-heptane (Fig. 1A): at 386 nm, 409 nm, 434 nm and 462 nm. In pure solid NAP, each molecule is surrounded with the same molecules, unlike that in 0.08 M liquid solution of NAP in n-heptane. Thus, the new bands at 380-480 nm are due to molecular associates of NAP. In Fig 1B, the peak at 386 nm is the 0-0 transition in the NAP excimer [14], while the other peaks are vibronic components of emission from the NAP excimer [15]. The ratio of areas of 0-2 and 0-1 vibronic peaks for pure NAP is A(0-2)/A(0-1) = 3.39. The difference in the A(0-2)/A(0-1) ratio of NAP monomer in pure NAP (Fig. 1B) and NAP monomer in solution (Fig. 1A) indicates coupling of the vibration at 1380 cm^{-1} in NAP molecule with vibronic states of the neighbour molecules. Thus, we can use the ratio A(0-2)/A(0-1) to learn about molecular environment of *adsorbed* NAP.

2. Fluorescence spectra of solid 0.08 M solution of NAP in eicosane and stoichiometric adsorption complex of F300 MOF and NAP in eicosane.

Figure 2A shows the fluorescence spectrum of 0.08 M solid solution of NAP in eicosane at λ_{exc} = 280 nm. This spectrum resembles that of pure NAP (Fig. 1B), therefore in solid solution of NAP in eicosane, there is a significant phase separation of NAP due to inter-molecular interactions between NAP molecules that form excimers, despite the low concentration of NAP. The ratio A(0-2)/A(0-1) is equal to 3.35 that is about the same as the one obtained with pure NAP (Fig. 1B). Thus, the ratio A(0-2)/A(0-1) is indicative of molecular environment of NAP molecules.

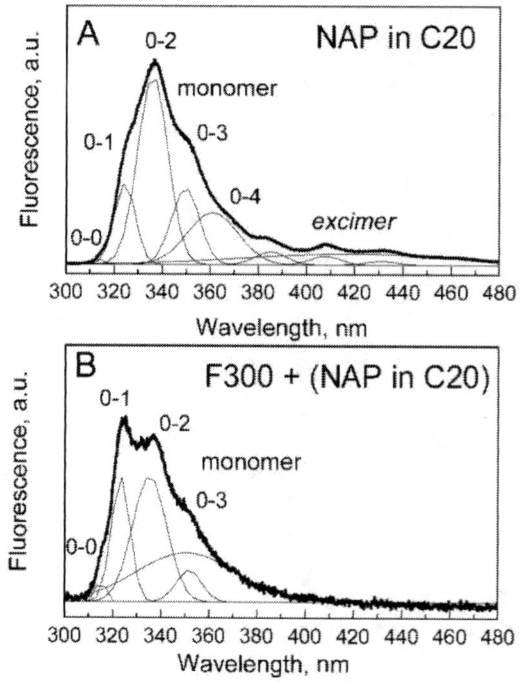

Figure 2. A) Fluorescence spectrum of 0.08 M solid solution of naphthalene (NAP) in eicosane (C20) at λ_{exc}=280 nm with multi-Gaussian curve fitting. B) Fluorescence spectrum of stoichiometric mixture of NAP with F300 MOF in eicosane matrix at λ_{exc}=280 nm with multi-Gaussian curve fitting.

Figure 2B shows the fluorescence spectrum of the stoichiometric complex of a-F300 and NAP in eicosane matrix. First, the spectrum of the stoichiometric complex (Fig. 2B) resembles that of *liquid* solution of NAP in n-heptane (Fig. 1A) where NAP molecules do not interact with each other. The absence of spectral red shift indicates a weak electronic interaction of NAP molecule with both adsorption sites of F300: Fe(III) and BTC linker. Second, the ratio A(0-2)/A(0-1)=1.86 is close to the one A(0-2)/A(0-1)=1.65 obtained with *liquid* 0.08 M solution of NAP in n-heptane containing NAP monomers. Third, there is no emission from NAP excimers in the complex with F300 (Fig. 2B), thus the dispersion of NAP molecules in the stoichiometric complex with F300 occurs. This finding indicates a significant change in molecular environment of adsorbed NAP when F300 sorbent is present, which affects the fluorescence coupled vibrations of adsorbed NAP molecule. Confinement of aromatic adsorbate molecules within the nanopore of cellulose sorbent was detected via the change in the fluorescence spectra [16]. When we use NAP as a fluorescent "self-probe" to learn about its molecular surroundings in the adsorption complex with F300, we find: 1) the absence of excimer emission in the stoichiometric complex of NAP with F300, and 2) the A(0-2)/A(0-1) ratio=1.86 that is close to that of NAP monomers. Therefore, in meso-cavity of F300, adsorbed NAP molecules are present as *monomers*, similarly to their solvated state in diluted liquid solution. This also indicates the non-polar environment of NAP molecules in the adsorption complex with F300, likely due to surrounding with BTC linkers (zero total dipole moment). Our findings represent the first direct spectroscopic evidence of an effect of molecular environment on aromatic adsorbate within meso-cavities of the MOF. We conducted the near-UV/Vis DRS study (Section 3) and quantum chemical DFT and TD-DFT simulations (Section 4).

3. Near-UV/Vis DRS of F300 MOF with NAP and IND

Figure 3A shows the near-UV/Vis DRS spectra of mixtures of a-F300 with NAP or IND at molar ratio R_1 = n(NAP)/n(F300) = 50. The new broad band at 460-600 nm manifests electronic interactions of adsorbed IND with Fe(III) CUS in a-F300, i.e. formation of coordination bonds.

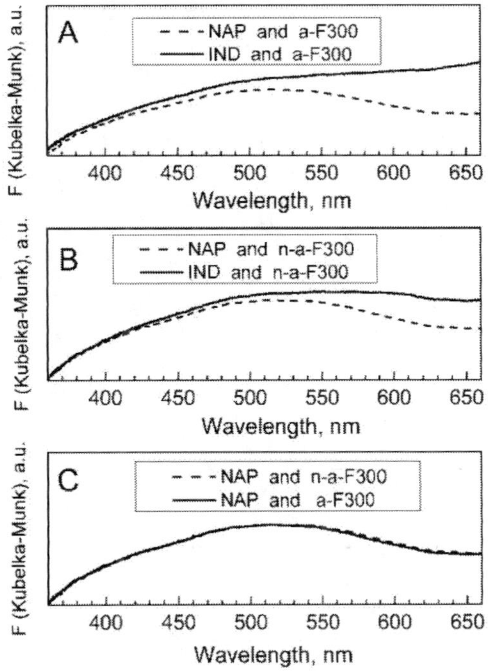

Figure 3. A) Near-UV/Vis DRS of a-F300 with NAP and IND. B) Near-UV/Vis DRS of n-a-F300 with NAP and IND. C) Near-UV/Vis DRS of NAP with n-a-F300 and a-F300.

Figure 3B shows the near-UV/Vis DRS spectra of mixtures of non-activated F300 (*n-a*-F300) with NAP or IND at molar ratio R_1 = n(NAP)/n(F300) = 50. The sample with IND still shows an enhanced optical absorbance at 460-660 nm *vs*. that with NAP that indicates the change in coordination environment of Fe(III) in n-a-F300 after the interaction with IND. The near-UV-Vis DRS spectra of "pure adsorption complexes" (see Experimental) of IND with a-F300 or activated MIL-100(Fe) are similar to those of a-F300 and activated MIL-100(Fe). Therefore, "pure" adsorption complexes are unstable without an excess of IND ligand, and cannot be spectroscopically identified. Figure 3C shows the near-UV/Vis DRS spectra of the mixture of NAP with n-a-F300 *vs*. the mixture with a-F300 at molar ratio R_1 = n(NAP)/n(F300) = 50. There are no differences between two spectra, despite a major difference in coordination environment of Fe(III) in a-F300 where adsorbed water is coordinated to Fe(III), and n-a-F300 where Fe(III) is a coordinatively unsaturated site (CUS). Therefore, no coordination bond is formed between adsorbed NAP and Fe(III) in neither a-F300 nor n-a-F300.

4. DFT computations of adsorption complexes of NAP and IND with F300

We have chosen the $Fe_3O(OH)L_6$ cluster denoted $Fe_3O(OH)L_6$ to model adsorption of NAP and IND. In $Fe_3O(OH)L_6$ cluster, each pair of Fe(III) cations in the $[Fe_3O]^{7+}$ core were connected by six bridging $C_6H_5CO_2^-$ anions (L) above and below the Fe_3O plane, and OH anion was coordinated to one of the Fe centers to form a neutral $Fe_3O(OH)L_6$ complex. Table 1 shows calculated ionization energy and reaction energies for the loss of H_2O and OH. Loss of three molecules of water from $[Fe_3OL_6(H_2O)_3]^+$ and two from $Fe_3O(OH)L_6(H_2O)_2$ is unfavourable by 17.8 and 13.6 kcal/mol per water molecule, while the OH ligand has a much larger binding energy as expected.

Table 1. Calculated Ionization Energy, and Reaction Energies for Loss of H_2O and OH, and Binding Energies (in kcal/mol) of IND and NAP to High-Spin $Fe_3O(OH)L_6$, L=$C_6H_5CO_2^-$

	Symm	B3LYP/ LANL2DZ	+GD3	B3LYP/6-31+G(2d,p)/ (Fe=SDD)+GD3	+ZPC
Fe_3OL_6	C_{2v}	0.0	0.0	0.0	0.0
$[Fe_3OL_6]^+ + e^-$	D_{3h}	149.4	149.4	139.9	140.6
$[Fe_3OL_6(H_2O)_3]^+$	D_3	0.0	0.0	0.0	0.0
$[Fe_3OL_6]^+ + 3H_2O$	D_{3h}	78.0	92.1	59.7	53.3
$[Fe_3OL_6(H_2O)_3]^+$	D_3	0.0	0.0	0.0	0.0
$Fe_3OL_6 + H_2O$	C_2	23.8	28.7	17.9	16.1
$Fe_3O(OH)L_6$	C_s	0.0	0.0	0.0	0.0
$Fe_3OL_6 + {}^{\cdot}OH$	C_{2v}	46.7	55.3	60.6	57.9
$Fe_3O(OH)L_6(H_2O)_2$	C_s	0.0	0.0	0.0	0.0
$Fe_3O(OH)L_6 + 2H_2O$	C_s	40.3	49.6	32.4	27.2
$Fe_3O(OH)L_6$ + substrate	C_s	0.0	0.0	0.0	0.0
$Fe_3O(OH)L_6$-IND	C_1	-7.6	-13.4	-18.2	-17.0
$Fe_3O(OH)L_6$-naph1 [e]	C_1	-2.8	-9.9	-11.9	-11.4
$Fe_3O(OH)L_6$-naph2 [e]	C_1	-3.9	-11.0	-13.8	-13.1

Table 1 also shows the results of calculation of binding energy of NAP or IND to $Fe_3O(OH)[(C_6H_5CO_2)_3]_2$ cluster. The binding of NAP at <4 kcal/mol indicates rather weak donor-acceptor interactions between the π system and Fe in the MOF cluster. With empirical dispersion corrections, the binding energy increases three-fold to 11.0 kcal/mol. For the complex of NAP with $Fe_3O(OH)L_6$ cluster, all Fe-C distances were longer than 3.0 Å this indicates very long and weak bonds. The binding energy of IND is larger than that of NAP by 3.9 kcal/mol: 17.0 vs. 13.1 kcal/mol. In the complex with IND, calculated Fe-N distance is 2.453 Å. In the model compound - trinuclear iron(III) complex with $[Fe_3(\mu3\text{-}O)]^{7+}$ unit [17], Fe-N bond lengths are between 2.11 and 2.203 Å. Therefore, Fe-N bond in the complex of $Fe_3O(OH)L_6$ with IND (Table 1) is somewhat longer and weaker than a Fe-N coordination bond in the model coordination compound [17]. The UV-Vis absorption spectra calculated by the time-dependent DFT (TD-DFT) are consistent with experimentally determined UV-Vis DRS spectra of adsorption complexes of F300 MOF with IND (Fig. 3), but experimentally measured UV-Vis DRS spectra are much broader than the TD-DFT calculated spectra, as expected.

5. Fluorescence of IND in the adsorption complex with a-F300

Figure 4A shows the excitation wavelength-dependent fluorescence spectra of pure IND, and Figure 4B shows the fluorescence spectra of the mixture of IND and a-F300 at R_2 = n(IND)/n(a-F300) = 15 at λ_{exc} = 280, 290, 300, and 310 nm.

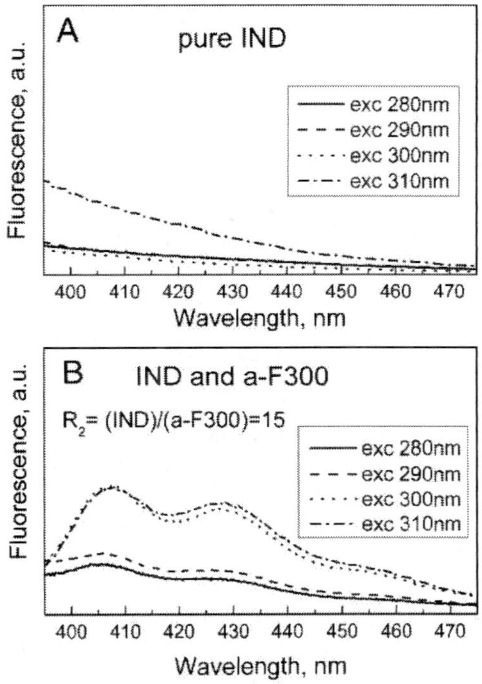

Figure 4. The wavelength-dependent fluorescence spectra at λ_{exc} = 280, 290, 300 and 310 nm for A) pure IND, and B) activated a-F300 with IND at molar ratio R_2 = n(IND)/n(F300) = 15.

For all λ_{exc}, there is a well-structured fluorescence from the mixture of IND and a-F300, with spectral maxima at 407 nm and 430 nm. The mixture of IND and a-F300 contains IND as its majority component. We assume the formation of the adsorption complex:

F300 (s) + IND (l) → Complex (s) (Equation 1)

where (s) indicates the solid substance in suspension, and (l) indicates the liquid substance in suspension. Therefore, the wavelength-dependent fluorescence spectroscopy allows spectroscopic detection of the formation of adsorption complex of a-F300 and IND, even when IND ligand is present in a large molar excess.

6. Dynamics of the formation of the adsorption complex of IND with a-F300

We measured the time-dependent fluorescence spectra during the adsorption of *liquid* IND on a-F300 at 60 °C using λ_{exc} = 310 nm, at continuous stirring, under argon. The intensity of the fluorescence from the adsorption complex (spectra as in Fig. 4B) steadily increased for about 80 min. of adsorption time. At 100-180 min., the fluorescence spectra grew more slowly, and at adsorption time >180 min., the fluorescence spectra did not change that indicates adsorption reaching an equilibrium. To build the kinetic model of adsorption of IND on a-F300 MOF in suspension, we assume Equation 1. The reactant F300 shows no fluorescence, thus no information can be obtained about the kinetic order of adsorption by this reactant. The reactant IND is a solvent present at constant concentration, thus the kinetic rate by IND as reactant cannot also be determined. The product "Complex" is present in the small amounts at the early adsorption stages, thus its concentration can be analyzed by the chemical kinetics formalism. An important assumption is made that at the small concentration of the adsorption complex, the intensity of the fluorescence is proportional to the concentration of the adsorption complex in suspension. We analyzed the dependence of the integrated fluorescence spectra of the adsorption complex on the adsorption time at constant temperature of 60 °C. At the early adsorption time, the integrated fluorescence spectrum was fitted with the linear function, and the time constant k was found to be equal to 189.3 ± 2.8 min^{-1}. The coefficient of determination R^2 (correlation coefficient) was equal to 0.998, and the error of determination was 1.5 %. Therefore, the zeroth order kinetic rate law describes the adsorption of IND on a-F300 MOF. The zeroth kinetic rate order was found in the studies of desorption of nitrogen-containing heterocyclic small molecule drugs from the adsorption complexes with MIL-100 and MIL-101 MOFs [18].

Conclusions

We report a direct spectroscopic proof of confinement of adsorbed molecule of aromatic hydrocarbon naphthalene (NAP) in meso-cavity of F300 MOF sorbent using vibronically-resolved fluorescence spectroscopy. We also report a direct spectroscopic proof of the formation of the adsorption complex with coordination bonds between Fe(III) CUS of F300 and adsorbed IND molecules, using the complementary near-UV/VIS DRS spectroscopy and the wavelength-dependent fluorescence spectroscopy. Combined spectroscopic data and the DFT and TD-DFT calculations suggest that NAP molecule in the adsorption complex with F300 MOF is confined in the mesocavity of a-F300, very weakly electronically bound to Fe(III) CUS, and strongly dispersively stabilized by the side interactions with benzene rings of the BTC linker. The IND molecule in the adsorption complex with a-F300 is electronically bound to Fe(III) CUS forming coordination bond and dispersively stabilized within meso-cavity by the side interactions with benzene rings of the BTC linker.

Acknowledgement

A.S. thanks Rutgers University for his Research Council Grant #202221. M. M. acknowledges the Alabama Supercomputer Center for a generous allocation of computer time.

References

1. Blanco-Brieva, G., et al., *Effectiveness of metal-organic frameworks for removal of refractory organo-sulfur compound present in liquid fuels.* Fuel, 2011. **90**(1): p. 190-197.
2. Lakowicz, J.R., *Principles of Fluorescence Spectroscopy.* 2006, New York: Springer.
3. Gaussian 09, R.D., M. J. Frisch, G. W. Trucks, H. B. Schlegel, G. E. Scuseria, M. A. Robb, J. R. Cheeseman, G. Scalmani, V. Barone, B. Mennucci, G. A. Petersson, H. Nakatsuji, M. Caricato, X. Li, H. P. Hratchian, A. F. Izmaylov, J. Bloino, G. Zheng, J. L. Sonnenberg, M. Hada, M. Ehara, K. Toyota, R. Fukuda, J. Hasegawa, M. Ishida, T. Nakajima, Y. Honda, O. Kitao, H. Nakai, T. Vreven, J. A. Montgomery, Jr., J. E. Peralta, F. Ogliaro, M. Bearpark, J. J. Heyd, E. Brothers, K. N. Kudin, V. N. Staroverov, R. Kobayashi, J. Normand, K. Raghavachari, A. Rendell, J. C. Burant, S. S. Iyengar, J. Tomasi, M. Cossi, N. Rega, J. M. Millam, M. Klene, J. E. Knox, J. B. Cross, V. Bakken, C. Adamo, J. Jaramillo, R. Gomperts, R. E. Stratmann, O. Yazyev, A. J. Austin, R. Cammi, C. Pomelli, J. W. Ochterski, R. L. Martin, K. Morokuma, V. G. Zakrzewski, G. A. Voth, P. Salvador, J. J. Dannenberg, S. Dapprich, A. D. Daniels, Ö. Farkas, J. B. Foresman, J. V. Ortiz, J. Cioslowski, and D. J. Fox, *Gaussian, Inc.* Wallingford CT, 2009.
4. Koch, W. and M.C. Holthausen, *A Chemist's Guide to Density Functional Theory.* 2nd Edition. 2001, Weinheim, Germany: Wiley-VCH.

5. Cramer, C.J., *Essentials of Computational Chemistry: Theories and Models* 2004, Chichester, UK: Wiley.
6. Hay, P.J. and W.R. Wadt, *Ab initio effective core potentials for molecular calculations. Potentials for the transition metal atoms Sc to Hg.* The Journal of Chemical Physics, 1985. **82**(1): p. 270-283.
7. Wadt, W.R. and P.J. Hay, *Ab initio effective core potentials for molecular calculations. Potentials for main group elements Na to Bi.* The Journal of Chemical Physics, 1985. **82**(1): p. 284-298.
8. Hay, P.J. and W.R. Wadt, *Ab initio effective core potentials for molecular calculations. Potentials for K to Au including the outermost core orbitals.* The Journal of Chemical Physics, 1985. **82**(1): p. 299-310.
9. Dolg, M., et al., *Energy-adjusted abinitio pseudopotentials for the first row transition elements.* The Journal of Chemical Physics, 1987. **86**(2): p. 866-872.
10. Grimme, S., et al., *A consistent and accurate ab initio parametrization of density functional dispersion correction (DFT-D) for the 94 elements H-Pu.* The Journal of Chemical Physics, 2010. **132**(15).
11. Privalova, A.I., et al., *Spectral and Luminescent Properties of 1-Substituted Naphthalenes.* Journal of Applied Spectroscopy, 2011. **78**(3): p. 309-317.
12. Schnepp, O. and D.S. McClure, *A Vibrational Analysis of the Fluorescence of Naphthalene Vapor.* The Journal of Chemical Physics, 1952. **20**(9): p. 1375-1383.
13. Zhang, Z.Z., et al., *Probing the Electronic Structure and Photoactivation Process of Nitrogen-Doped TiO2 Using DRS, PL, and EPR.* ChemPhysChem, 2012. **13**(6): p. 1542-1550.
14. Aladekomo, J.B. and J.B. Birks, *'Excimer' Fluorescence. VII. Spectral Studies of Naphthalene and Its Derivatives.* Proceedings of the Royal Society of London. Series A. Mathematical and Physical Sciences, 1965. **284**(1399): p. 551-565.
15. Patil, S.R. and S.B. Patwari, *Red shift in fluorescence of naphthalene doped by anthracene and perylene.* Journal of Luminescence, 1999. **82**(2): p. 115-119.
16. Tozuka, Y., et al., *Solid-state fluorescence study of naphthalene adsorption on porous material.* Journal of Colloid and Interface Science, 2002. **248**(2): p. 239-243.
17. Lalia-Kantouri, M., et al., *A Trinuclear Iron(III) Complex Containing the Semi-Cubane Fe-3(mu(3)-O) (7+) Core: Structural, Spectroscopic, Magnetic and Electrochemical Study.* Zeitschrift Fur Anorganische Und Allgemeine Chemie, 2010. **636**(3-4): p. 531-538.
18. Horcajada, P., et al., *Metal–Organic Frameworks as Efficient Materials for Drug Delivery.* Angewandte Chemie International Edition, 2006. **45**(36): p. 5974-5978.

367694 Water-Stable Uio-66-Derived Metal-Organic Frameworks for Highly Efficient CO2 Separation

Abstract:

A library of 20 UiO-66-derived metal-organic frameworks (MOFs) was synthesized in a combinatorial approach involving mixed ligand copolymerization and two post-synthetic modifications (PSMs) in tandem. These MOFs exhibit excellent water stabilities in a pH range of 1 to 12, together with high CO_2 uptake capacities and selectivities revealed by the analysis of 147 isotherms, which make them promising adsorbents in adsorption-based CO_2 separations such as post-combustion CO_2 capture and natural gas upgrading.

Annual Meeting, Atlanta, GA

Efficient Adsorptive Desulfurization over Heteroatom-doped Porous Carbons by Carbonization of an Organic Salt

Yawei Shi, Guozhu Liu*, Li Wang, Xiangwen Zhang

Key Laboratory of Green Chemical Technology of Ministry of Education, School of Chemical Engineering and Technology, Collaborative Innovation Center of Chemical Science and Engineering (Tianjin), Tianjin University, Tianjin 300072, China.

*Email: gliu@tju.edu.cn

Annual Meeting, Atlanta, GA

Content

- Introduction
- Preparation and Characterization
- Adsorption isotherms
- Effect of Porosity
- Effect of heteroatoms
- Regeneration
- Summary

2

Introduction

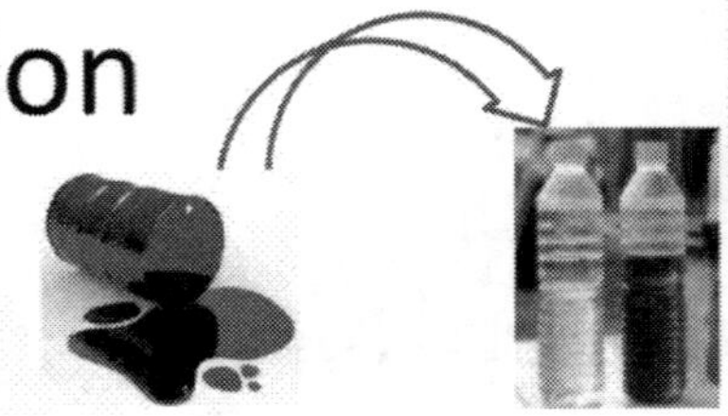

Sulfur: 12700 ppmw

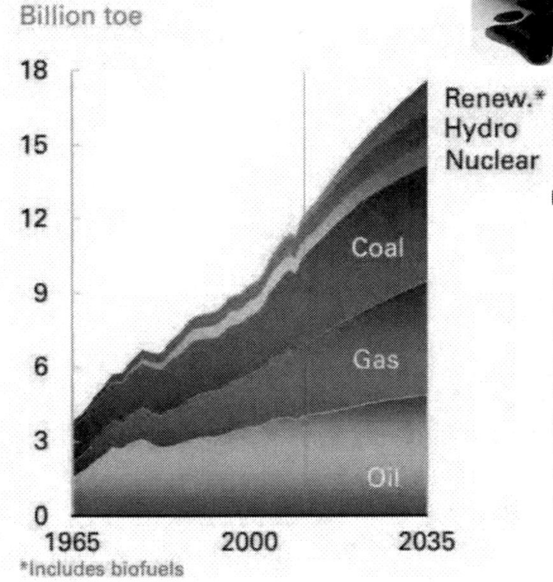

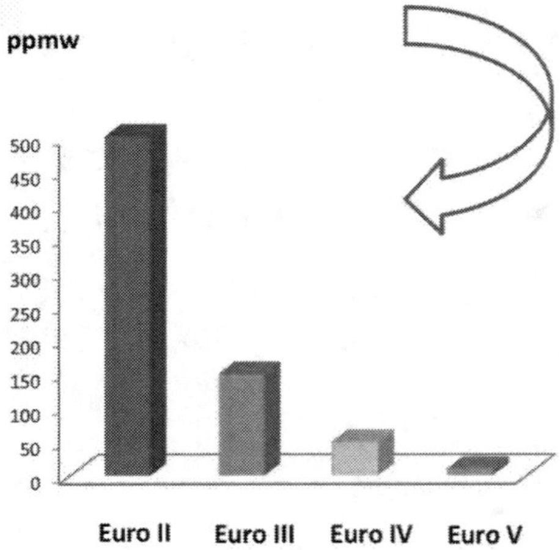

BP Energy Outlook 2035
RSC Adv 2012;2:759-83

3

Introduction

14AIChE
Annual Meeting, Atlanta, GA

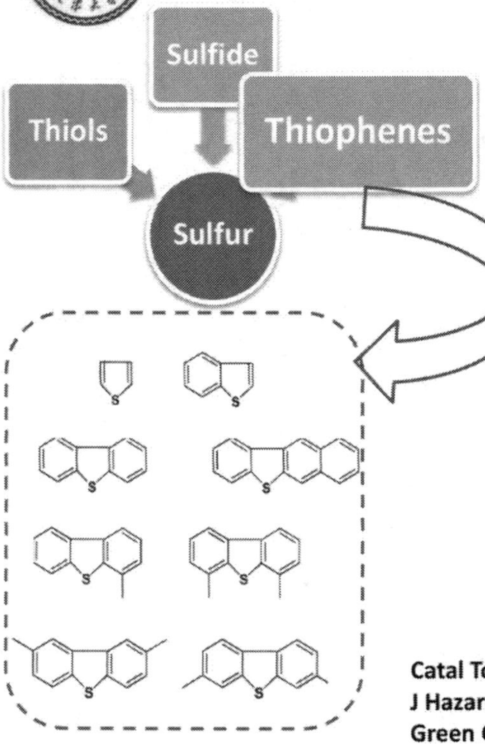

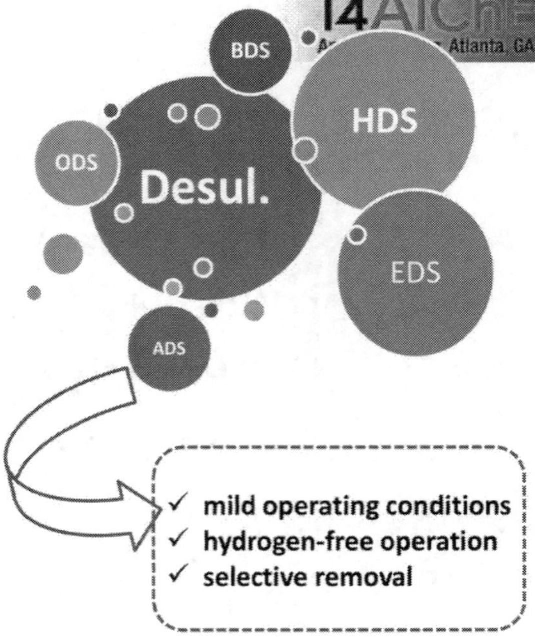

Catal Today 2010;153:1-68.
J Hazard Mater 2013, 244-245, 444.
Green Chemistry 2010, 12, 1139.

4

Introduction

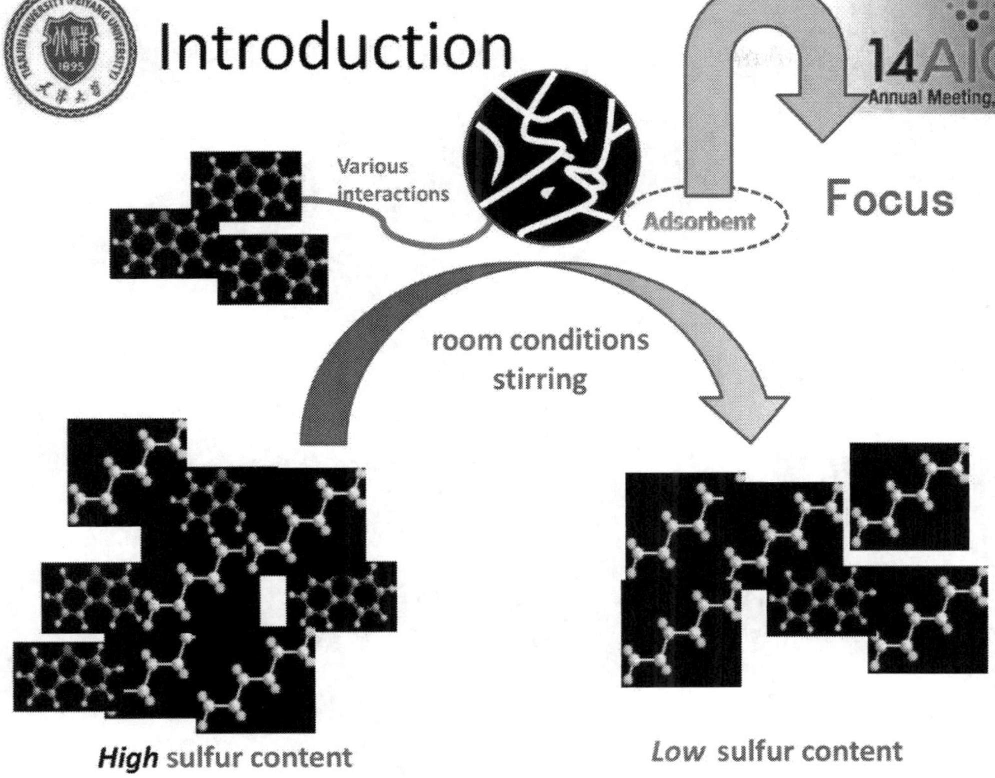

5

Preparation and Characterization

14AIChE Annual Meeting, Atlanta, GA

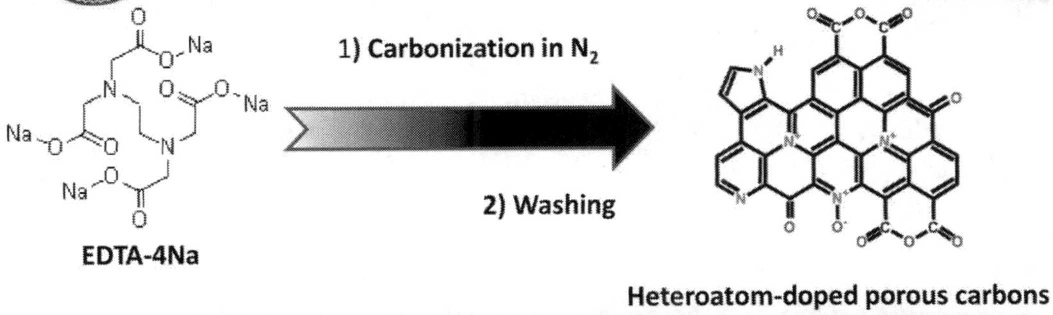

Sample	S_{BET} (m^2g^{-1})	V_{total}(cm^3g^{-1})	$V_{<1nm}$(cm^3g^{-1})	$V_{<1nm}$/ V_{total}
C-600	382	0.260	0.135	0.52
C-700	932	0.594	0.260	0.44
C-800	1518	1.056	0.244	0.23

	CHN Elemental Analysis				XPS		
Sample	C	H	N	O	C	N	O
C-600	76.6	0.3	7.0	16.1	81.2	6.0	12.8
C-700	84.1	1.0	3.1	11.8	89.0	2.9	8.1
C-800	90.0	0.7	2.6	6.7	89.8	1.8	8.4

6

Adsorption isotherms

14AIChE Annual Meeting, Atlanta, GA

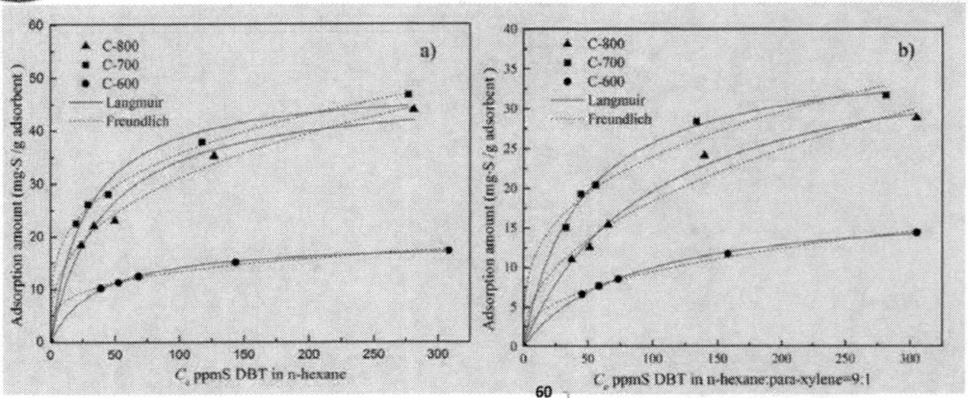

Experimental conditions:
25 °C, dosage=1g/L, C_0<358ppmwS

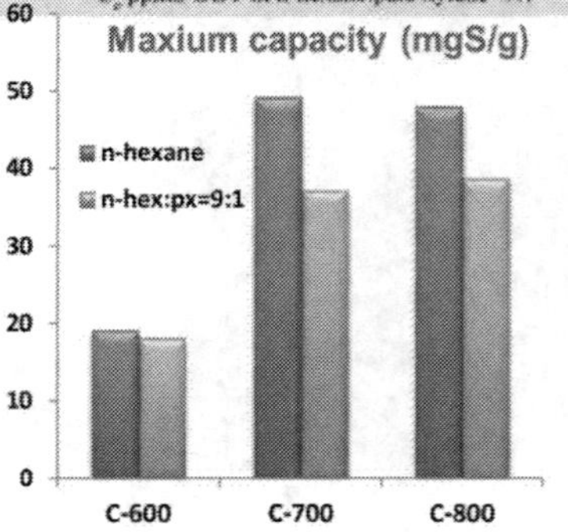

7

Adsorption isotherms

14AIChE Annual Meeting, Atlanta, GA

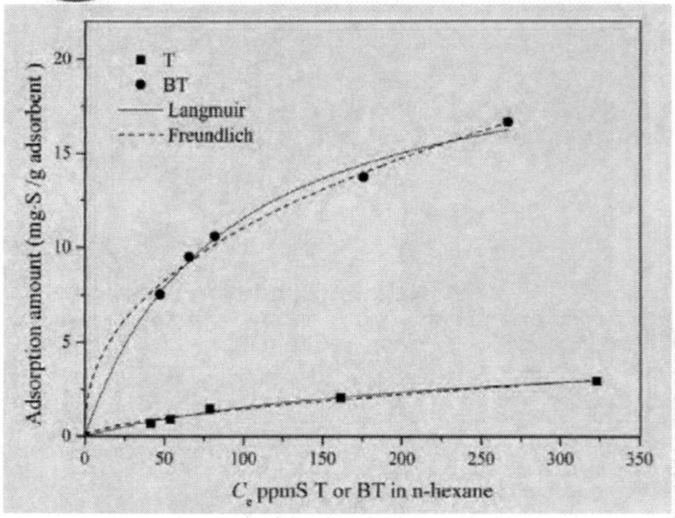

Experimental conditions:
25 °C, dosage=1g/L, C_0<358ppmwS

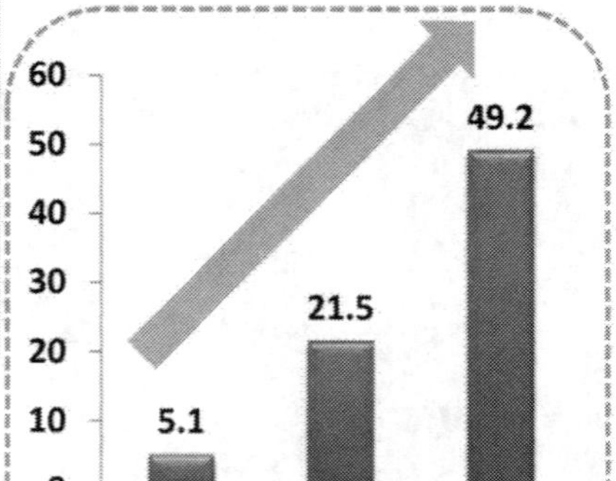

8

Effect of Porosity

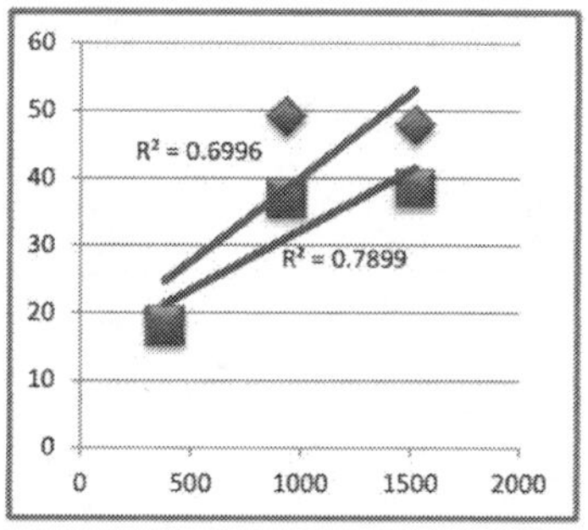

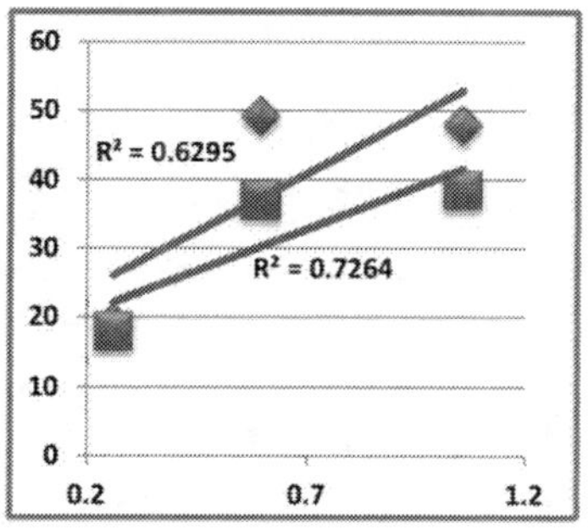

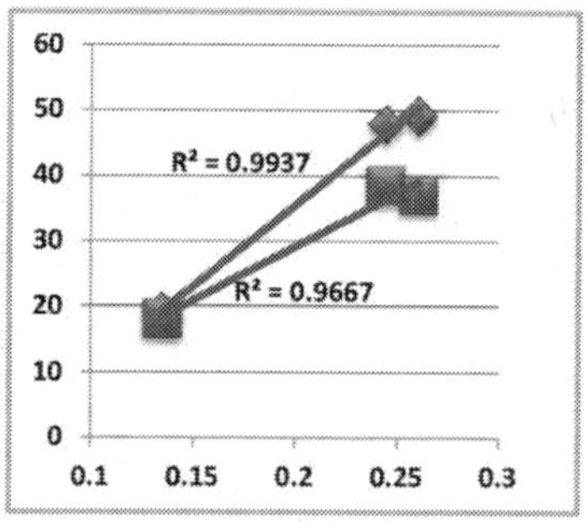

S_{BET} V_t $V_{<1nm}$

9

Effect of heteroatoms

- Oxidation modification *changed* the capacity significantly through an increase of the oxygen-containing functional groups.
- The *low* amount of carboxylic acid groups in our work.
- The adsorption capacity may be further *enhanced.*

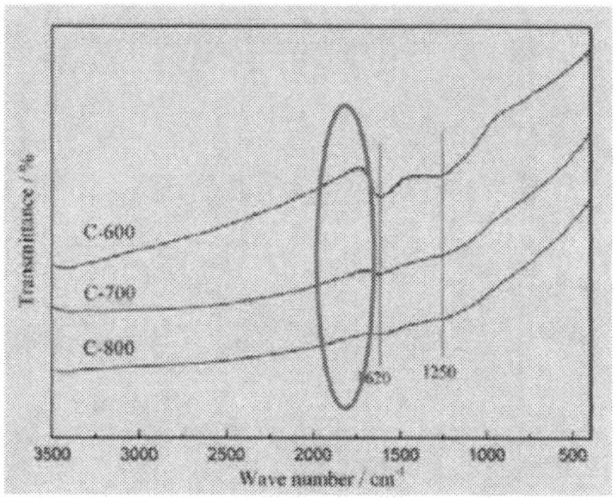

	Nitrogen content (at%)					Oxygen content (at%)			
Sample	Total	N-6	N-5	N-Q	N-X	Total	O-1	O-2	O-3
C-600	5.36	1.25	3.28	0.38	0.45	9.97	3.00	6.55	0.42
C-700	2.59	0.12	1.51	0.66	0.31	6.20	1.38	4.53	0.30
C-800	1.54	0.05	0.73	0.34	0.43	6.43	0.00	5.96	0.47

Appl.Catal B-Environ. 87 (2009) 190-199; Chem. Eng. J. 242 (2014) 211-219; Catal. Today 212 (2013) 31-37; Carbon, 45 (2007) 3042-3044.

10

Effect of heteroatoms

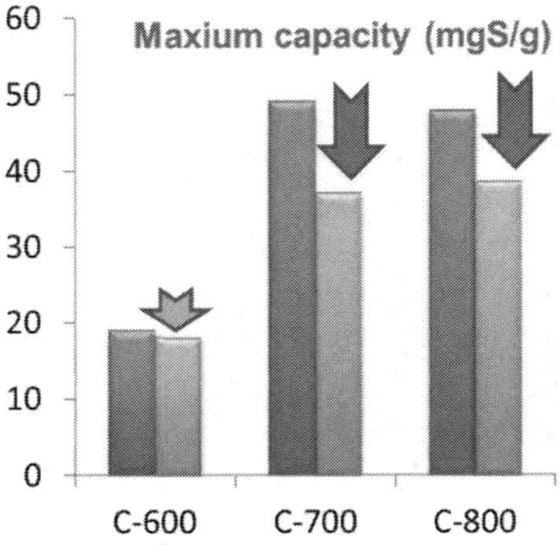

	CHN EA		XPS	
Sample	N	O	N	O
C-600	7.0	16.1	6.0	12.8
C-700	3.1	11.8	2.9	8.1
C-800	2.6	6.7	1.8	8.4

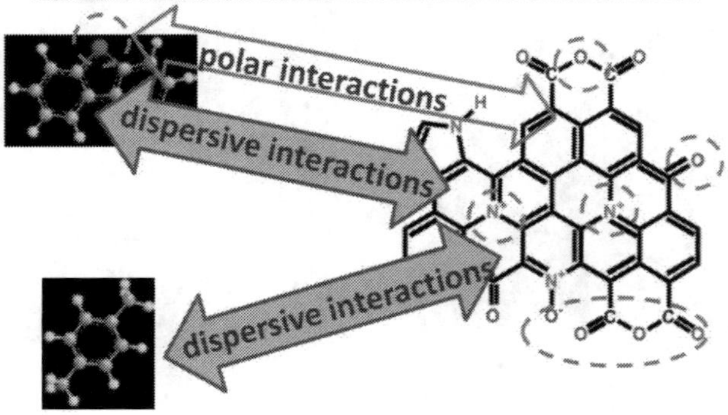

Regeneration

- C-700 and C-800 were tested for regeneration.
- The used adsorbents were separated, washed with *para*-xylene, and dried overnight for the next adsorption run.

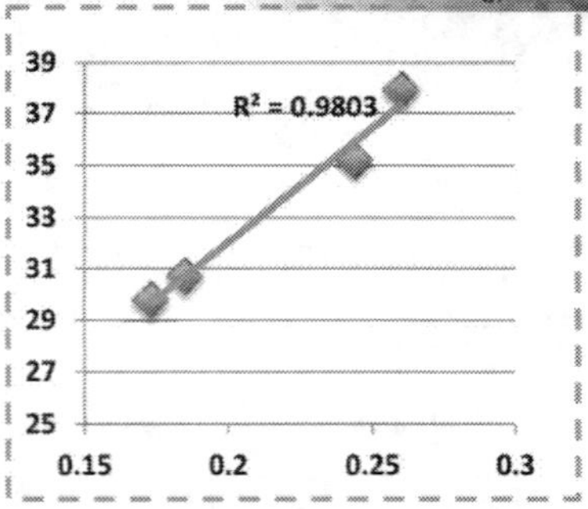

	S_{BET} (m^2g^{-1})	$V_{<1nm}$(cm^3g^{-1})	Retention in $V_{<1nm}$	q_e[a]	Retention in q_e
C-700	932	0.260	---	37.9	---
C-700-R	728	0.185	71%	30.7	81%
C-800	1518	0.244	---	35.2	---
C-800-R	1105	0.173	71%	29.8	85%

[a] mg·S/g; Experimental conditions: C_0=178ppmwS, adsorbent dosage=1g/L, t=25° C, T=1h

Summary

14AIChE
Annual Meeting, Atlanta, GA

✓ Heteroatom-doped carbons were efficient adsorbents for adsorptive desulfurization.

✓ Volume of pores smaller than 1 nm played a determining role on the adsorption capacity.

✓ Heteroatoms favored the polar interaction between DBT and the carbons.

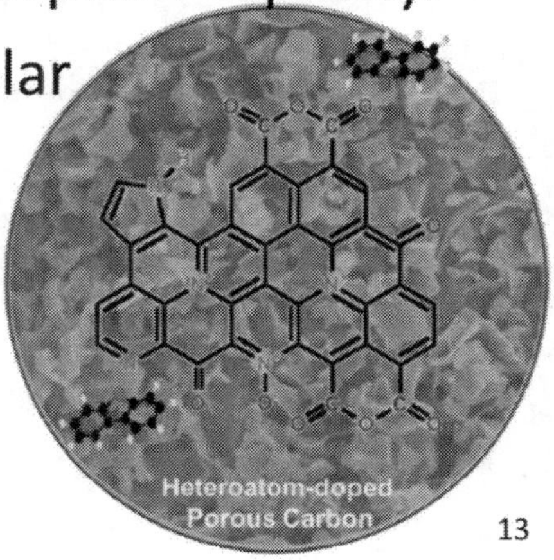

13

14AIChE
Annual Meeting, Atlanta, GA

THANK YOU!

14

388067 Novel Adsorbent from Agricultural Waste-Cashew Nut Shell for Methylene Blue Dye Removal: Optimization By Response Surface Methodology

Abstract:

Activated carbon, Cashew Nut Shell (CNS), prepared from an agricultural waste has been utilized as an adsorbent for the removal of Methylene Blue (MB) dye from an aqueous solution. A 2^4 full factorial central composite design is successfully employed for experimental design and analysis of the results. Batch adsorption study is carried out with the variables pH, adsorbent dose, initial dye concentration and time. The Response Surface Methodology (RSM) is applied to model the process and optimize the variable. The experimental values were in good agreement with predicted values. The optimum values of pH, adsorbent dose, initial dye concentration and time are found to be 10, 2.1846 g/L, 50 mg/L and 63 minutes for complete removal of MB dye respectively.

362626 Efficient Removal of Fluoride By Lanthanum Modified Carbon Nanoparticles (LMCN)

Abstract:

Excessive fluoride existing in water could result in serious environmental concerns as well as adverse health impact on human being. In the current study, a novel lanthanum-modified carbon nanoparticle (LMCN) was prepared as adsorbent and applied for fluoride removal from aqueous solution. Excellent removal efficiency was observed under a wide range of solution pH, and the maximum adsorption capacity at pH 7 reached up to 48.70mg/g. Further, it was found that 90% removal could be accomplished within the first 1 hour. The adsorption performance can be modeled by Freundlich isotherm at lower equilibrium solute concentration, while Langmuir isotherm better fits the adsorption process when solute concentration is higher. Moreover, the influence of co-existing anions such as PO_4^{3-}, HCO_3^-, SO_4^{2-}, SiO_3^{2-}, AsO_4^{3-} and humid acid (HA) on the adsorption process was studied. It was found that the fluoride removal efficiency decreased with the increase of ionic strength in the aqueous solution, while no significant impact was identified with the presence of HA. To conclude, LMCN as a low-cost adsorbent can provide fast fluoride removal from aqueous solution with considerably high adsorption efficiency, which leads to potentially wide applications in water treatment.

Maxwell-Stefan Kinetics Applied to Net Adsorption

Ali M. Farran and Orhan Talu
Cleveland State University
Cleveland, Ohio 44115, USA
(o.talu@csuohio.edu)

The ***net*** adsorption thermodynamic framework was introduced by Gumma and Talu (*Langmuir* 2010, 26(22), 17013-17023). Net adsorption fundamentally differs from ***excess*** and ***absolute*** adsorption by the placement of Gibbs dividing surface which is necessary to quantify adsorption. Net adsorption is shown to be the natural choice for microporous solids. It removes any ambiguity by circumventing questionable measurements and quantities such as; 1) solid volume necessary for excess adsorption which is typically based on "non-adsorbing" helium assumption, and 2) pore volume necessary for absolute adsorption which is typically based on liquid density assumption in micropores.

Among various approaches to describe diffusion in microporous solids, the Maxwell-Stefan (MS) formalism has been extensively developed by R.Krishna and collaborators using excess (or absolute) adsorption with some success. In this paper, we will show 1) the MS formulations for net adsorption, and 2) contrast the results to those with excess adsorption. Previously published data collected with Single-Crystal-Membrane technique will be used to demonstrate the similarity and differences between different formulations wherever available.

391704 On the Use of Volumetric Frequency Response for Understanding the Transport of Gases in Commercial Adsorbents

Abstract:

Frequency response methods have been developed as an alternative technique for the determination of mass transfer mechanisms controlling the transport of gases into and out of microporous adsorbent materials. Recent results in the literature are showing that this technique seems to be the best for discriminating the controlling mechanism. It is also possible to reveal two or more different controlling mechanisms that may dominate in different frequency ranges.

The present talk will focus on the use of the volumetric frequency response (VFR) approach for the identification on the main mechanism governing the transport of gases in commercial zeolite and activated carbon adsorbents. In a recently commissioned VFR apparatus, experiments were carried out with pure gases at pressures of 200, 400 and 750 Torr, temperatures of 25, 40 and 55 ^{o}C and frequencies between 0.7×10^{-5} and 10 Hz. The results are showing that the transport mechanism can be unequivocally identified when fitting the frequency response curves to various transport models. The results are also revealing that it is important to carry out experiments at different temperatures and pressures to ensure the correct transport mechanism has been identified. During this presentation the latest results obtained from the system will be presented and interpreted.

Experimental and modeling study of adsorptive water removal from dichloromethane in a molecular sieve packed bed

Slavisa Jovic[1], Yashasvi Laxminarayan[1], Jos Keurentjes[1], Jaap Schouten[1] and John van der Schaaf[*1]

[1]*Laboratory of Chemical Reactor Engineering, Department of Chemical Engineering and Chemistry, Eindhoven University of Technology, P.O. Box 513, 5600 MB Eindhoven, The Netherlands*

[*]*Corresponding author: j.vanderschaaf@tue.nl*

Removal of impurities from chemicals and solvents is often crucial for selective production of pharmaceuticals. Dichloromethane (methylene chloride; DCM) is one of the most widely used chlorinated solvents, because of its ability to dissolve many organic and organometallic compounds [1]. For specialty pharmaceuticals, DCM should be exceedingly pure and have low water content. The removal of water by adsorption on 3A molecular sieves appears promising [2]. In this study, adsorption equilibrium of water in DCM on molecular sieves 3A at room temperature are determined. Also, adsorption breakthrough curves are measured to verify a mathematical packed bed model. The adsorption breakthrough curves were determined at room temperature using a column with a diameter of 31 mm and a height of 700 mm. DCM saturated with water was pumped upwards through the column, with flow rates of 2 to 7 g/s. The water content of the DCM stream was measured inlet and outlet of the column containing 385 g of adsorbent. In the packed bed model, liquid flow was described as plug flow with axial dispersion. The particle mass balance included diffusion in macropores, surface diffusion, and pseudo first-order adsorption kinetics. In order to solve the set of partial differential equations simultaneously, packed bed model is discretized in space. Macropore diffusion, external mass transfer and axial diffusion coefficients are estimated with correlations from literature. The activity of water in DCM is estimated with a modified NRTL equation and validated with mutual solubility data.

The measured values of adsorption equilibrium are slightly lower than the literature data [3], which is explained by the presence of binding material in the industrial-grade molecular sieve particles [4]. The molecular size of DCM (9,2 Å) is too large to enter the 3 Å micropores. Therefore, it can be concluded that DCM does not affect the adsorption. A modified NRTL model gives good prediction of mutual solubility, within 10 % error.

Adsorption breakthrough curves were measured for five different flow rates: 2 g/s, 3 g/s, 4 g/s, 5 g/s and 7 g/s. Packed bed model fitted experimental data within 86 % confidence level, using a surface diffusion coefficient as a fitting parameter. Sensitivity analysis, together with experimental results, showed that diffusional resistances play dominant role for this system. Thus, use of smaller particles is recommended, in respect to the pressure drop over the packed bed. The external mass transfer and the axial diffusion coefficients do not play a significant role under the investigated experimental conditions. Since diffusional resistances are limiting, bed is more utilized at lower flow rates. Presented model can be used to determine the performance of a large scale packed bed for adsorptive drying of DCM.

References

[1] M. Rossber, W. Lendle, G. Pfleiderer, A. Togel, T. R. Torkelson, K. K. Beutel, Chloromethanes, in: Ullmann's Encyclopedia of Industrial Chemistry, Electronic Release, Wiley-VCH, Weinheim 2013

[2] C. Pahl, C. Pasel, M. Luckas, D. Bathen, Chem. Ing. Tech. 83 (2011) 177

[3] M. Llano-Restrepo, M. A. Mosquera, Fluid Phase Equilib. 283 (2009) 73

[4] R. Gopal, B. R. Hollebone, C. H. Langford, R. A. Shigeishij, Sol. Energy 28 (1982) 421

364303 New Diffusive Intermediates Under Polar Condition for CO2 Adsorption in Silica-Supported Amine Sorbents

Abstract:

The adsorption process of CO_2 on highly loaded, silica-supported, polyethylenimine-impregnated sorbents has been investigated based on a reaction-diffusion model that takes zwitterions as the diffusive intermediate. Simulation results qualitatively replicate the experimental results under dry conditions. An associated parameter sensitivity and quantum chemistry study shows that the diffusive intermediates control the capacity of CO_2 uptake in sorbents but that zwitterions are likely unstable under anhydrous conditions. Based on quantum chemical results, a new model proposes physically-bonded moieties involving water, amines and CO_2 as new diffusive intermediates. The new model will be analyzed and compared with experimental TGA results.

Abstract: AIChE 2014

Modeling Uptake of an Emerging Contaminant from Water & Wastewater Using Zeolite Beta

Scott A. Grieco, Ph.D., P.E.
Vice President & Practice Area Leader – O'Brien & Gere Engineers, Inc.

Bandaru V. Ramarao, Ph.D.
Professor – State University of New York Environmental Science & Forestry

Emerging contaminants are becoming a global environmental problem as there is increasing information linking them to environmental and human health hazards. Identification and mitigation of these constituents at the source before entering the environment is crucial if impacts to the ecosystem and humans are to be eliminated. One emerging contaminant of interest is tris-2-chloroethyl phosphate (TCEP), an organo-phosphate ester used as a plasticizer and flame retardant. TCEP is of special interest due to potential toxicity, prevalence in the environment, and difficult to remove through conventional wastewater treatment operations. In our previous published work, TCEP was shown to be removed efficiently (>97%) using highly siliceous Zeolite β.

Kinetics of TCEP adsorption were evaluated in surrogate background (natural organic material and salts) and in a competitive adsorption environment with methyl tert-butyl ether (MTBE). Fast uptake of TCEP (near equilibrium conditions were reached in <60 minutes) was observed. The results show that Zeolite β is a successful approach to remove TCEP from water and wastewater matrices and could be a useful technology to treat other hydrophilic persistent ECs.

For many zeolite sorption applications, and specifically for the TCEP/Zeolite β system, the molecular size of the adsorbate is comparable to the diameter of the zeolite channel. As such, the molecules may never completely escape the potential field of the zeolite lattice. Diffusion in this case is considered configurational diffusion (CD) and may preclude a clear separation between diffusion coefficients for pore & surface diffusion. In support of this premise, existing uptake models using single or constant diffusivities did not provide a good representation of the uptake data. It was theorized that, a new model incorporating variable diffusivity with a loading-dependent factor would be more representative for CD in liquid phase applications. Therefore, a CD model is proposed to simulate variable diffusivity in the uptake of TCEP in zeolite.

The CD model represented the diffusion of TCEP in different Zeolite β materials under varying conditions of background organics, salts, competing adsorbates, and initial concentrations. The CD model describes the decreased diffusional mobility upon the addition of salts and completing organics. The new CD model showed flexibility to describe configurational diffusion better than existing models using constant or single diffusivity values.

386584 Understanding Dominating Mass Transfer Resistance Using Frequency Response Method

Abstract:

Knowledge for mass transfer is very important for developing effective adsorbents and catalysts that utilize differences in mass transfer rates for different molecules. Mass transfer limitations often play an important role in the overall rate of reaction; the rate of conversion and product formation in the catalytic systems. Various mass transfer steps can be encountered in porous adsorbent or catalyst, such as molecules diffusing through the various systems of pores and surface barrier through pore openings. The slowest mass transfer step (giving longest resistance time) is quite possibly the overall rate limiting step for adsorption and catalytic reaction. It is very important to understand the dominating mass transfer mechanisms and measure accurate mass transfer rate constants in order to develop the suitable adsorbents/catalyst and process design.

Frequency response is a unique method to discriminate among different mass transfer mechanisms by introducing harmonic perturbations at various frequencies of oscillation. A novel pressure-swing frequency response (PSFR) has been developed to have capability to measure mass transfer at operating conditions close to practical interest. Examples are given to demonstrate how the dominating resistance can be determined for various systems. Micropore diffusion is most common for mass transfer in zeolites. Heat effects can also be very important when heat dissipation is much slower compared to mass transfer (for example for CO_2 diffusion). Particle size difference or/and adsorbent structure can relate to parallel micropore diffusion steps. Mass transfer study at various pressures and temperatures can also provide a fundamental understanding of loading dependence for diffusion rates.

In this work we will present some examples of application of PSFR technique to study mass transfer in several classes of adsorbents.

SAPO-34, a member of the silico-alumino-phosphate zeolite minerals, exhibits the same framework structure as the naturally occurring chabazite. The chabazite topology consists of a three-dimensional pore system of ellipsoidal cages (6.8 x 10$Å^2$) interconnected by eight membered windows (3.8 x 3.8$Å^2$), where each cage possesses six windows. Due to this small pore system and its catalytic properties, SAPO-34 is a very interesting material for molecular sieve separations and selective catalysis. It's well known as the catalyst of excellence for the Methanol-To-Olefins (MTO) process (Stocker, 1999; Vora et al., 2009). A lot of attention has been given to the potential of SAPO-34 to separate small gaseous molecules, like nitrogen, carbon dioxide and methane (Ackley et al., 2003; Ruthven & Reyes, 2007).

Recently, our group investigated the adsorption and separation of larger molecules, C1-C8 1-alcohols, on SAPO-34 and concluded smaller alcohols are preferentially adsorbed over longer alcohols, with an almost full exclusion of alcohols larger than propanol (Remy et al., 2011). This was proven to be very useful for the recovery and purification of biobutanol, where 1-butanol needs to be separated from its side-product ethanol (Cousin Saint Remi et al., 2012). Preliminary results indicated that this full exclusion of the larger molecule could be related to transport properties inside the SAPO-34 crystals. Later, liquid phase batch uptake measurements confirmed this behavior. They highlighted a non-uniform chain length dependent uptake and a complex mixture mechanism of 1-alcohols in SAPO-34 (Cousin Saint Remi et al., 2013). The further exploration of these phenomena is of great importance in the efficient design of new separations process, like for the recovery and purification of biobutanol by adsorption.

In this study, we present a detailed investigation of diffusion properties of single compound and mixtures of 1-alcohols in SAPO-34 crystals, by means of microscopic and macroscopic techniques, mainly the zero-length column method and interference microscopy. The experimental approaches, the uptake and desorption curves, the diffusion profiles inside the crystals, as well as the diffusion coefficients will be presented, and interpreted in relation to the separation behavior of the SAPO-34 molecular sieve.

Acknowledgements

J. Cousin Saint Remi is grateful to the Agency for Innovation by Science and Technology in Flanders (IWT) for the financial support. C.L. Cavalcante Jr. is grateful to Conselho Nacional de Apoio ao Desenvolvimento Cientifico e Tecnologico (CNPq, Brazil) for financial support in a sabbatical stay at VUB.

References

Ackley, M.; Rege, S.; Saxena, H. Application of Natural Zeolites in the Purification and Separation of Gases. *Micropor. Mesopor. Mat.* **2003**, *61*, 25-42.

Cousin Saint Remi, J.; Baron, G.; Denayer, J. Adsorptive Separations for the Recovery and Purification of Biobutanol. *Adsorption* **2012**, *18*, 367-373

Cousin Saint Remi, J.; Baron, G.; Denayer, J. Nonuniform Chain-Length-Dependent Diffusion of Short 1-Alcohols in SAPO-34 in Liquid Phase. *J. Phys. Chem. C*, **2013**, *117*, 9758–9765.

Remy, T.; Cousin Saint Remi, J.; Singh, R.; Webley, P.; Baron, G.; Denayer, J. Adsorption and Separation of C1-C8 Alcohols on SAPO-34. *J. Phys. Chem. C* **2011**, *115*, 8117-8125.

Ruthven, D.; Reyes, S. Adsorptive Separation of Light Olefins from Paraffins. *Micropor. Mesopor. Mat.* **2007**, *104*, 59-66.

Stocker, M. Methanol-to-Hydrocarbons: Catalytic Materials and Their Behavior. *Micropor. Mesopor. Mat.* **1999**, *29*, 3-48.

Vora, B.; Chen, J.; Bozzano, A.; Glover, B.; Barger, P. Various Routes to Methane Utilization - SAPO-34 Catalysis Offers the Best Option. *Catal. Today* **2009**, *141*, 77-83.

The vegetative water (10's -100's m^3/d) generated in two or three phase olive mill oil production is heavily loaded with COD and polyphenols. The polyphenols are phyto-toxic and can also severely impact standard municipal wastewater treatment plants. At the same time these polyphenols in concentrated isolated form can be quite valuable as byproducts as nutriceuticals, antibacterial agents and as pharmaceutical precursors with values approaching $100/kg or more. In this talk we present our experience applying a combined membrane process involving pH adjustment, submerged UF, nanofiltration and membrane contactors to isolate highly valued components such as tyrosol and hydroxytyrosol.

Fluxes in the UF step could be kept at 20 L/m2-h with periodic backwash, but large agglomerations of particulates had to be previously separated to prevent blinding of the fibers. Fluxes in the NF step were kept at 20-30 L/m2-h. Polyphenols were reduced from 2800-4000 mg/L in the original OMWW to ~200-400 mg/L in some of the NF permeates. Hydroxytyrosol fraction in polyphenols could be increased to greater than 90-95% in some of the downstream process streams. Further refinement is ongoing on each of the process steps to maximize polyphenol recovery.

390039 Design & Construction of a Pilot-Scale Osmotic Heat Engine

Abstract:

Pressure retarded osmosis (PRO) is a method used to capture the energy stored in salinity gradients. While open-loop PRO using freshwater and seawater has been considered, using a dilute feed solution and highly concentrated engineered draw solution, a closed loop has been shown to yield higher power densities. This closed-loop process, known as the osmotic heat engine (OHE), can produce electricity while using low-grade or waste heat to recover the draw solution concentration. Applications of such a technology include waste heat recovery and grid storage for intermittent renewable energy sources.

The majority of previous work on the OHE has been largely theoretical, but in this study a pilot-scale OHE was constructed using an ammonia-carbon dioxide draw solution. A previous study was conducted on the bench-scale to understand the power density of membranes at elevated temperature similar to OHE conditions. A spiral-wound membrane element was used for the membrane step, with draw solution pressures ranging from 0 to 20 bar. Diluted draw solution was sent to a bank of falling-film ammonia strippers and condensers to recover the concentration of the draw. Both streams used a nitrogen sweep gas to remove ammonia in the stripping section while providing ammonia to the condenser. Altering the temperatures of the heating and cooling fluids could also dictate the overall temperature of the OHE. Actual stripping effectiveness will be compared to model-based predictions. This platform allows for future optimization of the OHE for overall energy efficiency.

389665 Self-Assembly of Graphene Oxide and Polymer Nanohybrid Membranes for Separation Application

Abstract:

Organic-inorganic nanohybrids are believed to be one of the most promising membrane materials for separation applications. Graphenen oxide could be integrated with other functional materials at a molecular/nanometer scale to create multifunctional GO-based composites because of its numerous oxygen-containing functional groups. In this study, nanohybrid membranes were prepared by incorporating graphene oxide (GO) into polymer for nanofiltration and pervaporation. Firstly, the PEI-modified GO and polyacrylic acid was sequentially assembled onto hydrolyzed polyacrylonitrile (PAN) ultrafiltration supporting membrane. It was found that the resulting membranes had excellent dye removal capacity and separation performance of monovalent and divalent ions. Additionally, it was demonstrated that this GO/PECs nanohybrid membrane is also a good candidate for the pervaporation dehydration of different solvent-water mixtures. In the second part, GO could be doped into poly(vinyl alcohol) (PVA) solutions to form nanohybrids. The PVA-GO nanohybrid could be assembled onto PAN membrane to form "pore-filling" membrane by dynamic pressure-driven assembly. The composite membrane could be used for separating toluene/n-heptane mixtures. With respect to the separation factor of pure PVA membrane, an improvement in the separation factor in the pervaporation of toluene/n-heptane mixture was observed after doping of PVA with GO. The "pore-filling" structure could effectively reduce swelling of the nanohybrid membrane. Furthermore, with the doping of GO, the thermal and mechanical properties of organic-inorganic nanohybrid membrane significantly improved. These results strongly suggest that GO has much potential in applications for preparing organic-inorganic nanohybrid membrane.

389994 Simultaneous Spray Self-Assembly of Highly Loaded ZIF-8–PDMS Nanohybrid Membranes Exhibiting Exceptionally High Biobutanol-Permselective Pervaporation

Abstract:

The ability to obtain a maximum loading of inorganic nanoparticles while maintaining uniform dispersion in the polymer is the key to the fabrication of mixed-matrix membranes with high pervaporation performance in bioalcohol recovery from aqueous solution. Herein, we report the simultaneous spray self-assembly of a zeolitic imidazolate framework (ZIF)–polymer suspension and a cross-linker/catalyst solution as a method for the fabrication of a well dispersed ZIF-8–PDMS nanohybrid membrane with an extremely high loading. The dispersion uniformity of the fillers and their compatibility always had an important effect on the pervaporation performance of the membranes. The ZIF-8–PDMS membrane showed excellent biobutanol-permselective pervaporation performance. When the ZIF-8 loading was increased to 40 wt%, the total flux and separation factor could reach 4846.2 g/m^{2}h and 81.6, respectively, in the recovery of n-butanol from 1.0 wt% aqueous solution (80 ^{0}C). It exhibited an optimal overall pervaporation performance as compared with that of other reported membranes. We found that butanol molecules mainly pass through the inner channel of ZIF-8, whereas water molecules mainly pass through the interface between PDMS and ZIF-8 particles in such ZIF-8–PDMS membranes with an extremely high loading. In view of the versatile architectures and customizable chemical functionalities of MOF materials, as well as the simplicity, speed, and reproducibility of the spray self-assembly technique, this technology may contribute to the preparation of various MOF hybrid membranes and extend the use of these membranes to industrial applications.

382947 Novel Perfluorinated Membrane for Drying Solvents

Abstract:

Data are presented on the dehydration of EtOH, IPA, MEK, THF and [C_2mim][OAc] by pervaporation. The tests were done on a pilot scale using a chemically and thermally resistant membrane system. The membrane consists of a perfluorinated polymer backed by a chemically and thermally resistant hollow-fiber porous support. The membrane is packed in cartridges that have several square feet of active area. The "wet" solvent is fed to the pervaporation system in the liquid phase. The feed to the membrane is separated into two streams: the retentate, which is solvent rich, and the permeate, which is water rich and is in the vapor phase. The data show that high levels of drying can be achieved with high solvent recoveries and rapid rates of water removal. The pervaporation process is especially advantageous for drying organics that are difficult to dry by conventional technologies (e.g., azeotropes or solvents that have boiling points close to water), heat sensitive chemicals, reacting mixtures in which the chemical reaction is inhibited by the presence and/or generation of water. This technology is been used commercially for the successful dehydration of lubricating oils and phosphate esters. The pervaporation process can be combined with conventional distillation, e. g., for recovering the solvent that goes to the water-rich stream from the pervaporation process. Examples of this hybrid membrane-distillation drying process will be presented.

363073 Sustainable Water Recovery from Oily Wastewater Via Forward Osmosis – Membrane Distillation (FO-MD)

Abstract:

Oily wastewater from industry and domestic sewage is one of the main pollutants to the environment in the world. With the production of crude oil and natural gas, an aqueous stream named "produced water" is normally accompanied due to the hydraulic fracturing process. The produced water, which contains dispersed oils, suspended particles and dissolved solutes, constitutes the largest waste stream from oil and gas manufacturing industries. Increasing public interests have been accumulated recently in reducing the quantity of contaminants that are released to the environment during the treatment of the produced water.

This study proposed and investigated a hybrid forward osmosis – membrane distillation (FO-MD) system for sustainable water recovery from oily wastewater by employing lab-fabricated FO and MD hollow fiber membranes. Stable oil-in-water emulsions of different concentrations with small droplet sizes (< 1 μm) were firstly prepared and applied as the feed solution in the FO process. Fouling was immediately observed in the FO mode and was quite low on the cellulose triacetate (CTA) –based thin film composite (TFC) membranes. Moreover, slight increment of fouling was observed in the first few hours and the water flux was then stabilized over 24 hours. The characterizations of water flux and solute rejection in separate FO and MD processes revealed that a high water flux, good NaCl rejection, impressively high retention of oil droplets and partial permeation of acetic acid could be achieved. Finally, an integrated FO-MD system was developed to treat the oily wastewater containing petroleum, surfactant, NaCl and acetic acid at 60°C in the batch mode. The water flux in FO undergoes three-stage decline due to fouling and reduction in osmotic driving force, but is quite stable in MD regardless of salt concentration. Oily wastewater with relatively high salinity could be effectively recovered by the FO-MD hybrid system while maintaining large water flux, at least 90% feed water recovery could be readily attained with only trace amounts of oil and salts, and the draw solution was re-generated for the next rounds of FO-MD run. Interestingly, significant amount of acetic acid was also retained in the permeate for further reuse as a chemical additive during the production of crude oil. The work has demonstrated that not only water but also organic additives in the wastewater could be effectively recovered by FO-MD systems for reuse or other utilizations.

375208 Water Vapor Permeation through Amine-Containing CO2-Selective Facilitated Transport Membranes for Hydrogen Purification

Abstract:

Abstract

We have synthesized carbon dioxide-selective membranes by incorporating amino groups into polymer networks. The membranes have shown high carbon dioxide permeability and selectivity vs. hydrogen, carbon monoxide and nitrogen up to 180°C. Hydrogen sulfide permeates through the membrane much faster than carbon dioxide due to much higher reactivity as a result of small-size proton transfer, allowing nearly complete removal of hydrogen sulfide to about 10 ppb in the treated synthesis gas before water-gas-shift (WGS) reaction. In some applications, it is also required to selectively separate CO_2 by restricting the amount of water permeated. The amine-containing membranes are highly hydrophilic, and they exhibit not only high CO_2 permeance but also high water vapor permeance. This work demonstrates various water and CO_2 transport results exhibited by these thin-film composite amine membranes. Water vapor permeance higher than 3500 GPU has been experimentally demonstrated with >3000 Barrers of CO_2 permeability and >150 CO_2/H_2 ideal selectivity. Gas permeation experiments showed water vapor permeance to be not sensitive to the amine-containing selective layer thickness. The water transport resistance offered by the polymeric substrate appeared to be the controlling mass transfer resistance, independent of the resistance offered by the selective amine layer. Water vapor permeance was experimentally demonstrated to reduce on increasing the polymeric substrate resistance via increasing the number of substrate layers used to synthesize the membrane. The hydrophilic amine-containing selective layer was expected to retain moisture and served as the governing layer for CO_2 and H_2 permeation. Effect of the extent of water retention by the polymer matrix on the overall transport result was also investigated by use of different fixed-site amine carriers. The use of commercial polyvinylamine (containing >60% salts) as a fixed carrier was shown to retain more moisture and aid the CO_2 permeance as compared to the laboratory synthesized pure polyvinylamine (without containing salts). Water vapor permeance was shown to be related to the moisture retention capacity of the matrix. Furthermore, the use of a relatively hydrophobic PDMS as an additional barrier/support layer was shown to reduce the water permeance to as low as 600 GPU.

375298 Palladium Membranes for Portable Power Applications

Abstract:

Hydrogen selective membrane technology offers an opportunity to utilize reformed hydrocarbon feed stocks as hydrogen carriers for low temperature fuel cells. However, existing membrane technology struggles to meet cost and reliability targets, particularly when considering the challenging portable marketplace where thermal cycling durability and high flux become paramount. In this work, the latest efforts in engineering a palladium composite membrane optimized for portable power applications is discussed. The rate limiting mechanism of transport is identified, and membranes are tested under various synthetic reformed fuel atmospheres – such as reformed methanol and water gas shift atmospheres. Operating under moderate pressures (20psi), the membranes produce 'fuel cell quality' hydrogen from the aforementioned reformed fuel atmospheres. Thermal cycling stability is also discussed.

386223 The Characterization of Supported Molten Metal Membranes for Hydrogen Separation

Abstract:

Hydrogen separation membranes such as Pd, Pd alloys are being applied in the reformers as they prompt the reaction conversion and lower the working temperature which makes it more energy efficient, compact and highly selective towards hydrogen as compared to the traditional technologies. However, they are expensive, sensitive to poisons, and lack adequate durability. Therefore, the search for novel and cheaper materials to replace palladium is ongoing. We are, thus, developing entirely novel supported molten metal membranes (SMMMs) based on low melting metals and their alloys for their inherent ability to avoid sintering, hydrogen embrittlement, and thermal mismatch between the membrane and the support, and potentially their higher permeability. Since this molten metal as hydrogen separation was never investigated before, therefore, the attempt of hydrogen permeability characterization was performed via experiments and theoretical calculations. Also in this presentation, we will describe the promise of SMMM and the pitfalls we have so far encountered in this quest.

Before membrane characterizing, an understanding of developing thin and stable SMMMs is required. To build SMMMs, the wettability and reactivity between porous support and the molten metals are two major issues. This is a fine balance since some chemical interaction between the liquid metal and the support is necessary for adequate wetting, while too much of it dooms the membrane. Therefore, one of the goals of this work was to find a porous support material with both good wettability and stability. Dense Ga molten metal membrane was made using previously screened materials and showed that the hydrogen permeance of molten Ga is comparable to Pd membrane at 500-550. Since there is the lacking of Ga hydrogen permeance in the literature as well as its hydrogen solubility which makes it difficult to verify the credibility of experimental measurements. The attempt of calculating Ga hydrogen permeability using Margules type equations was also made in this work for comparison.

Abstract for AIChE 2014 Annual Meeting
Atlanta, GA, USA
November 16 – 21, 2014

Session: 02D09 Membranes for Hydrogen Purification

Material Investigation for Hydrogen Purification and Storage Applications: Hydrogen Solubility in Transition Metals Predicted by Density Functional Theory Combined with Thermodynamics Principles

Kyoungin Lee, Mengyao Yuan, Jennifer Wilcox
Department of Energy Resources Engineering, Stanford University, Stanford, CA

Hydrogen production and purification are often achieved by steam-methane reforming (SMR) of natural gas and subsequent hydrogen separation techniques. Metallic membranes for such separation techniques have been widely developed with palladium (Pd)-based materials. Due to the high and unstable price of Pd as well as its limited membrane permeability, non-Pd-based metallic membranes have been investigated. Materials modeling of such membranes may provide an efficient way to predict membrane properties and may lead to the design of new materials with optimal properties thereby saving time and effort associated with carrying out a large number of exploratory experiments.

Hydrogen solubility is an important parameter that affect hydrogen-selective membrane properties because the hydrogen permeability through a metallic membrane is half the product of the solubility and diffusivity of atomic hydrogen in the bulk metal. Accurate solubility predictions is crucial for the prediction of overall membrane properties. Our study may provide useful information in the search for new materials for hydrogen-selective membranes and hydrogen storage by identifying a proper alloy for the Pd-based and non-Pd-based membranes that will enhance hydrogen solubility.

We applied density functional theory (DFT) calculations combined with statistical thermodynamics principles to predict solubility in various transition metals including group V metals. The metal systems included in this study are: vanadium (V), niobium (Nb), tantalum (Ta), tungsten (W), nickel (Ni), palladium (Pd), platinum (Pt), copper (Cu), silver (Ag), and gold (Au). There are two interstitial binding sites for hydrogen in these metals, which are octahedral (O-) and tetrahedral (T-) sites. Assuming that binding in either the O- or the T-site dominates the entire solubility, we made solubility predictions for each case.

The hydrogen solubility is predicted to be within an order of magnitude from known experimental results for most of the cases. As examples, solubility plots predicted for V and W and their corresponding experimental values reported in the literature are shown in Figure 1. Solubility has no unit in this study, but is the ratio of the number of hydrogen atoms to the number of metal atoms (H/M). The solubility trend of T-site binding describes the experimental measurements more closely than that of O-site binding. Hydrogen solubility in V decreases with increasing temperature, whereas the opposite is observed for hydrogen in W. This trend is related to the sign of the binding energy predicted. The binding energies of atomic hydrogen in the T-site are -0.33 and 0.93 eV for V and W, respectively. The negative binding energy of H in V indicates a stronger interaction between H and the metal, which leads to an exothermic absorption process and a negative heat of absorption. On the other hand, the positive binding energy of H in W indicates that absorption is unfavorable but will be promoted with the addition of heat, indicating an endothermic process.

Based on the predicted solubility trend, hydrogen binding in the metals with the body-centered cubic (bcc) structure is observed to be favored in the T-site over the O-site, whereas the opposite is true in the metals with the face-centered cubic (fcc) structure. We compared the size of the metal lattice and interstitial sites with the size of the hydrogen atom using a simple hard-sphere model to explain the solubility trend. It is known that large interstitial atoms such as carbon or nitrogen can be inserted with lattice expansion, which involves an energy barrier, but the small hydrogen atom would rather bind at the site that offers enough space than expanding the lattice to bind at the smaller site.

This structural effect partially, but not entirely, explains the solubility trend. In the case where metals with similar lattice sizes have completely different solubility trends, vibrational frequencies were predicted to be similar, but the binding energies of hydrogen were significantly different. For example, V has a lattice constant close to that of W, and the highest hydrogen vibrational frequency in the tetrahedral interstitial site is 1520 and 1530 cm^{-1} for V and W, respectively. As previously mentioned, the binding energies of atomic hydrogen in the tetrahedral interstitial site are -0.33 and 0.93 eV for V and W, respectively. The frequency is sensitive to the geometry of the confined interstitial site, but the significant difference in binding energy may be induced from the differences in the electronic structures. Charge and density of states analyses are being performed to further explain the trends in solubility predictions between metals with a similar lattice constant.

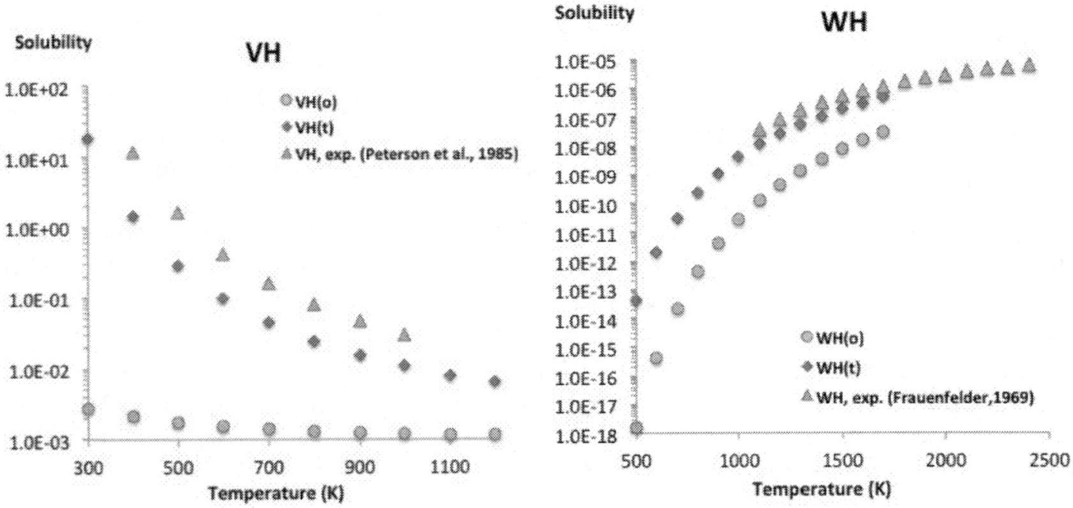

Figure 1. Solubility of hydrogen in V (left) and in W (right) predicted in this study and from experiment. [1]

References

[1] Peterson, D. T.; Nelson, S. O., Isopiestic Solubility of Hydrogen in Vanadium Alloys at Low Temperatures. *Metal. Trans. A* **1985**, 367-374.; Frauenfelder, R., Solution and diffusion of hydrogen in tungsten. *J. of Vacuum Sci. and Tech.* **1969,** *6* (3), 388-397.

386772 Performance of Pd-Ag Membranes Fabricated By a Novel Electroless Plating Method for Hydrogen Separation

Abstract:

Uses of fossil fuels in transportation and power generation release greenhouse gases. Hydrogen has the potential to be an alternative clean energy carrier. Hydrogen fueled proton exchange membrane fuel cell (PEMFC) can be used in transportation and stationary power generation to reduce greenhouse gases emission. In this context, high purity of hydrogen is the fuel of choice for PEMFC. Membrane-based process is a key technology in production and separation of high purity hydrogen, whether it is from biomass gasification, coal gasification or steam reforming of hydrocarbons and/or alcohols. In this work, dense Pd-Ag composite membranes on microporous stainless steel substrate (MPSS) were fabricated by a novel surfactant induced electroless plating (SIEP) process. A cationic surfactant, dodecyl trimethyl ammonium bromide (DTAB) was used in Pd and Ag baths for the improved deposition of metals on MPSS substrates. Fabricated membranes are characterized and tested for thermal and mechanical stability, structural integrity and H_2-permselectivity. Morphological features of the membranes at pre- and post-annealing conditions are studied by using SEM, XRD, EDS, and AFM techniques. Permeability tests and long-term performance under thermal cycling are carried out with hydrogen at temperatures and transmembrane pressures in the range of 250 - 550 °C and 20 - 100 psig, respectively. The performance of the SIEP process developed Pd-Ag membrane will be compared with the membranes fabricated using conventional electroless plating method.

386416 Additive Effects on the Growth and Polymorph Transformations of Calcium Pyrophosphate Crystals

Abstract:

Calcium pyrophosphate deposition disease (pseudogout) is a rheumatologic condition caused by the formation of calcium pyrophosphate (CaPP) crystals in joints. Nearly half of people over the age of 80 may develop CaPP crystals in their joints.[1] Yet, the mechanism by which CaPP crystals form in the joints is not well established. Furthermore, there is currently no known treatment to inhibit the growth of CaPP crystals. Mechanistic understanding of how additives effect nucleation, growth, and polymorph selection of CaPP crystals is necessary to develop additives that inhibit their formation or growth.

We use atomistic modeling to study additive effects on the CaPP monoclinic dihydrate crystal structure found during in vivo studies of human joints.[2] To properly model the effects of additives, we developed a new force field using the CHARMM force field potential as a starting point. The force field accurately reproduces the crystal structures of the monoclinic dihydrate, and a tetrahydrate form of CaPP.

Simple metal salts have been shown experimentally to influence crystal growth and polymorph transformations of CaPP crystals in aqueous solution. In order to better understand the effects of salts on CaPP crystal growth and polymorph transformations, we calculate binding free energies between metal cations and CaPP crystal faces.

References

[1] Guerne, P. A., and R. Terkeltaub, "Calcium pyrophosphate dihydrate crystal deposition: epidemiology, clinical features, diagnosis, and treatment," Chapter 21 in Terkeltaub, R., ed., "Gout and Other Crystal Deposition Arthropathies", Elsevier Heath Sciences, Philadelphia, PA (2012).

[2] McCarthy, D. J., N. N. Kohn, and J. S. Faires, "The Significance of Calcium Pyrophosphate Crystals in the Synovial Fluid of Arthritic Patients: The "Pseudogout Syndrome". II. Identification of crystals," Ann. Intern. Med., 56, 838-845, (1962).

382726 Computational Studies of Homogeneous Nucleation of Ice in Molecular Models of Water

Abstract:

Water is undoubtedly one of the most important molecules on earth, and is present in many environments on our planet. In many of these environments, freezing of water- or lack thereof- is consequential. However, the freezing of water can be successfully suppressed for temperatures as low as 40 K below its melting point, which is due to large free energy barriers that separate the metastable liquid basin from the stable crystalline basin. It is therefore impossible to fully understand and predict the behavior of systems containing supercooled water without having any knowledge about the rate and the mechanisms of the freezing in those systems. Molecular simulations are invaluable tools in this endeavor, since the existing experimental techniques lack the spatiotemporal resolutions necessary for identifying the molecular-level events that lead to crystallization. However, the problem of calculating the rate of homogeneous nucleation of ice is one of the most challenging open problems in computational statistical physics, besting the efforts of generations of computational scientists [1]. Indeed, direct calculations of nucleation rates have only been done for mW, a monoatomic coarse-grained model of water, [2,3] and indirect estimates have been obtained for more sophisticated models of water using certain a priori assumptions [4] (e.g. the validity of the classical nucleation theory). To this date, no direct calculation has been reported for fully atomistic models of water, like the TIP4P family of models. In this work, we employ a coarse-grained version of the forward-flux sampling (FFS) algorithm [5] to perform the first direct calculation of the homogeneous nucleation rate in a fully atomistic model of water. We also elaborate on the mechanism of freezing in the corresponding system.

[1] Brukhno, Anwar, et al., *J. Phys.: Condens. Matter* **20**, 494243 (**2008**).

[2] Li, Donadio, et al., *Phys. Chem. Chem. Phys.* **13**, 19807 (**2011**).

[3] Li, Donadio, et al., *Nat. Comm.* **4**, 1887 (**2013**).

[4] Sanz, Vega, et al. *J Am. Chem. Soc.* **135**, 15008 (**2013**).

[5] Allen, Frenkel, et al., *J. Chem. Phys*. **124**, 194111 (**2006**).

Layered crystal growth occurs via steps flowing across the surface of a face and completing successive layers. These steps appear as a result of screw dislocations (spiral growth), or two-dimensional (2D) nuclei forming on the surface. The mechanism with the largest normal growth rate for that face will dominate, which is spiral growth at low supersaturation and 2D nucleation & growth at higher supersaturation. In addition, incorporation into kink sites in either mechanism can be limited by attachment kinetics or mass transport.

A mechanistic spiral growth model has been developed that performs well at predicting the crystal habit of organic molecules grown from solution at low supersaturation[1] and has also been adapted to consider 2D nucleation and growth[2]. In the present work, this 2D nucleation & growth model has been generalized and divided into two regimes: birth and spread, with negligible initial nucleus area, and polynuclear, with negligible nucleus growth due to rapid step annihilation. A variety of crossovers exist: 2D birth and spread to 2D polynuclear, spiral to 2D, and the crossover to mass transport-limited, which can happen from any of these regimes. These crossovers are calculated by comparing growth rates and solving for the supersaturation at which they are equal; knowledge of them allows relative growth rates to be calculated at any supersaturation for a given system, and therefore the crystal morphology can be predicted in each regime.

In calculating crossovers or relative growth rates, the rate constants for kink attachment/detachment (k_+ & k_-) and 2D nucleation rate prefactor (κ_{2D}) appear, which are expensive to determine from molecular simulations. Instead, the ratio of κ_{2D} / k_- can be evaluated, adapting theory for the stationary nucleation rate[3] to develop an expression for κ_{2D} using a system-specific 2D nucleus geometry. This technique allows crossovers and relative growth rates to be calculated using parameters we can obtain through the mechanistic model, with the exception of mass transport effects that still require estimates for these rate constants. This has been applied to small centrosymmetric molecules such as naphthalene to predict crossovers and supersaturation dependent morphologies.

References

1. Snyder, R. C., & Doherty, M. F. (2009). Predicting crystal growth by spiral motion. *Proceedings of the Royal Society A: Mathematical, Physical and Engineering Science*, *465*(2104), 1145-1171.
2. Lovette, M. A., & Doherty, M. F. (2012). Predictive Modeling of Supersaturation-Dependent Crystal Shapes. *Crystal Growth & Design*, *12*(2), 656-669.
3. Kashchiev, D. (2000). Nucleation: Basic Theory with Applications. Butterworth-Heinemann, Oxford.

385206 Identification of Polar Plots of Crystal Dissolution Rates Using Hot-Stage Microscopy

Abstract:

Polar plots are the representations of the anisotropic surface properties such as surface energies and growth rates of crystalline materials. The steady-state morphologies of growing crystals are usually obtained from Wulff constructions on the polar plots of growth rates,[1] whereas the morphologies of dissolving crystals are known to have no steady-states.[2]Here we show that the dissolving crystal can attain steady-state morphologies under certain conditions. The Wulff construction on the polar plots of dissolution time (or slowness) can be used to identify such steady-state morphologies. It is shown that the dissolving crystal can attain faceted morphology composed of fast dissolving faces. The evolution of dissolving crystal towards faceted morphology involves disappearance of slow-dissolving faces, which also causes vanishing of curvatures from crystal surface. This article presents a method to experimentally determine polar plots from the dynamic images of crystals obtained from hot-stage microscopy. The method relies on the solution of the characteristics for crystal dissolution. The methodology is demonstrated to obtain polar plots of Succinic acid at different sub-saturations.

References:

(1) Singh, M. R.; Verma, P.; Tung, H.-H.; Bordawekar, S.; Ramkrishna, D., Screening Crystal Morphologies from Crystal Structure. *Crystal Growth & Design* 2012, 13, (4), 1390-1396.

(2) Snyder, R. C.; Doherty, M. F., Faceted crystal shape evolution during dissolution or growth. *AIChE journal* 2007, 53, (5), 1337-1348.

Multi-objective optimization of an unseeded batch cooling crystallization system for shape and size manipulation

David Acevedo[1], Zoltak K. Nagy[1]

[1]School of Chemical Engineering, Purdue University, West Lafayette, IN, 47906, USA

Crystallization processes commonly have multiple performance objectives such as size, purity, yield, and shape. These properties can affect the final product properties such as bioavailability, bulk density, and flowability. The optimization and control of final crystal size and shape distribution have been the common objectives for this type of studies. Recent work developed a classification method based on the effect of temperature profiles on the nucleation and growth rates of two different characteristic crystal dimensions [1]. This provides an indication of the achievable shape and size of final crystals by manipulating the supersaturation. Furthermore, recent studies in the literature have been focused on the optimization and control of the evolution and final target shape for seeded crystallization systems [2]. In this work the optimization of unseeded batch cooling crystallization systems is studied in a multi-objective framework by implementing a multidimensional population balance model (PBM). The length mean size and a target aspect ratio (AR) of final crystals were considered as objectives which gives rise to a set of optimal solutions known as *Pareto-optimal* solutions.

The multi-objective optimization framework was studied for two case scenarios of different crystallization kinetics behavior: nucleation or growth difference dominated systems. The two model compounds were paracetamol and potassium dyhidrogen phosphate (KDP), the former being nucleation dominated whereas the latter growth dominated systems, respectively. The optimization of KDP demonstrated that the optimal cooling profile varied from a linear to programmed cooling profile depending of the targeted objectives weights. However, a decrease of only ten percent was achieved for the final AR between the single optimization formulations, maximization of mean size and minimization of AR. This shows that most crystallization systems tend to achieve an equilibrium shape which limits its variability due to manipulations in the supersaturation. The effect of various optimal cooling profiles were validated experimentally. Furthermore, the evolution of the Pareto-optimal front due to variations in batch time was studied using KDP as case study. Longer batch time has a greater impact in the optimal AR that could be achieved since the width characteristic dimension can achieve a more linear supersaturation dependence. Moreover, the results of the multi-objective optimization demonstrated a relation between the classification of crystallization kinetics and the set of optimal cooling profiles.

References

[1] Acevedo, D.; Nagy, Z.K. Systematic classification of unseeded batch crystallization systems for achievable shape and size analysis. *Journal of Crystal Growth.* **2014**, 394, 97-105.

[2] Liu, J.J.; Hu, Y.D.; Wang, X.Z. Optimization and control of crystal shape and size in protein crystallization process. *Computers and Chemical Engineering.* **2013**, 57,133-140.

Formation Kinetics of Binary Hydrates in Brine Solution

Lichao Cai, Brian A. Pethica, Pablo G. Debenedetti and Sankaran Sundaresan
Chemical and Biological Engineering, Princeton University, Princeton, NJ 08540

Clathrate hydrates are ice-like, nonstoichiometric crystalline solids formed by water and guest molecules encapsulated in the various-sized water cavities. While the early attention they attracted was due to hydrate blockage of oil pipelines, new applications have been identified, such as natural gas storage, carbon dioxide capture and desalination. However, earlier attempts at desalination based on single-component hydrates suffered from the unfavorable cost of refrigeration and/or pressurization because either high pressure (e.g. methane hydrate) or relatively low temperatures (e.g. refrigerant-based hydrate) are necessary for such hydrates to form. These considerations have motivated research to identify hydrates with extended ranges of thermodynamic stability, and the corresponding kinetics for process design and optimization.

A potential solution is the use of binary clathrate hydrates with superior thermodynamic stability. In light of the fact that hydrates have cavities of different sizes, and occupation of each type contributes to stabilizing the structure, no single hydrate former can fit call cavities effectively at the same time. However, by involving two guest species simultaneously – fitting the large and small cavities, respectively – clathration is boosted and yields a binary hydrate that is stable under conditions where neither of the corresponding single-component hydrates can form. Usually, this requires a large liquid hydrate former such as cyclopentane to occupy the large cavities, and a small helper gas such as methane to occupy the smaller cavities. This hydrate formation involves four phases – hydrate, cyclopentane-rich liquid, water-rich liquid and the vapor phase.

Given the thermodynamic advantage, the formation kinetics of these binary clathrate hydrates are also important when it comes to applications. The corresponding hydrate-based processes could be impractical should these hydrate have unfavorably slow formation rates. Also, the understanding of typical formation rates and the rate-limiting factors is necessary for process evaluation and optimization. However, their formation kinetics are not well understood compared to single-component hydrates.

In a recent study on cyclopentane-methane hydrate formation in a stirred reactor, we have found the kinetics to be largely controlled by mass transfer, especially the delivery of the gas species across the hydrate layer surrounding cyclopentane droplets. However, this initial work was done with fresh water. Since many hydrate-based processes involve hydrate formation in brine solutions (with desalination as a prominent example), it is necessary to investigate how salt affects the formation kinetics of binary clathrate hydrates. As salt-free hydrate grows and expels salt to the surrounding aqueous phase, the local salt concentration is elevated, and this may raise an additional mass transfer barrier for hydrate formation. In this follow-up experimental study, the formation kinetics of cyclopentane-methane hydrate in brine solution is studied in a stirred reactor to address the mass transfer of salt. By analyzing the growth rates under various operating conditions compared to those established with fresh water, the role played by salt is discussed.

382800 Effects of Surface Chemistry on Heterogeneous Methane Hydrate Nucleation

Abstract:

Gas hydrates represent an important class of crystalline solids that have applications in wide ranging fields such as petrochemical industry, environment and energy. Gas hydrates comprise cages formed through hydrogen bonding between water molecules entrapping gas (e.g. methane, ethane) in them. Naturally occurring methane hydrates are estimated to be significant sources of energy. The fact that the solubility of methane in water can increase by few orders of magnitude when transitioning from liquid solution to crystalline solids is remarkable. However, the mechanism(s) through which this transition occurs is not well understood. Recent simulation studies, including microsecond long simulations, have provided valuable insights into developing the molecular picture of homogeneous methane hydrate nucleation. It has also been shown that the probability of observing homogeneous nucleation is considerable lower than heterogeneous nucleation. Therefore, it is important to investigate the effect of surfaces on methane hydrate nucleation. Water behavior is strongly influenced by the surface and therefore, it is expected that hydrate nucleation will also be affected by the presence of surface. In our work, we use microsecond long molecular dynamics simulations to investigate the effects of surface chemistry on hydrate nucleation. Specifically, we have performed simulations of methane-water solutions in the presence of hydrophobic and hydrophilic self-assembled monolayers at hydrate forming conditions (i.e. low temperature and high pressure). We observe methane hydrate nucleation in the presence of both surfaces. Interestingly, aspects of hydrate nucleation such as number of cages observed and location of these cages relative to the surface are affected by the surface chemistry. In our presentation, we will discuss hydrate nucleation near these surfaces and the origins of these effects. Our results will provide insights into using surface chemistry to engineer hydrate nucleation.

381662 A Scalable Method for Producing Graphene Membranes with Nanofiltration Properties

Abstract:

We have developed a novel and mass-producible method for fabricating supported Graphene-based membranes. Rheologically-tailored Graphene Oxide (GO) dispersions are cast on microfiltration Nylon (pore size ~ 0.2 μm) support membrane by knife coating – a very simple and rapid method to produce ~10 cm×15 cm membranes in less than a minute. Scanning electron and polarized light microscopy reveal well-packed and ordered structure. The microstructure can be tuned by shear rate and rheological characteristics of the coating fluid. Preliminary filtration studies have shown large water flux (25 litres/m^2-hour-bar) and 100% rejection of Direct Yellow 50 (956.82 g/mol), > 99% for Methylene blue (319.85 g/mol), > 92% for Rhodamine B (479.02), ~ 86% for Methyl Viologen (257.16 g/mol) indicating the suitability as a novel nanofiltration membrane. Rejection mechanisms are reliant on physical sieving and electrostatic interactions with minimal adsorption. This new process of ideal asymmetric membranes on macroporous supports is scalable and has the potential to be translated to large areas easily.

383681 Preparation of Amorphous Carbon Molecular Sieve Membranes Via Plasma-Enhanced Chemical Vapor Deposition

Abstract:

Amorphous carbon membranes were fabricated on porous ceramic substrates by means of plasma-enhanced chemical vapor deposition (PECVD) using propane as the carbon precursor. The gas permeation measurement revealed that PECVD-derived amorphous carbon membranes possessed molecular sieving properties. Although the amorphous carbon layer was deposited at room temperature, the membrane showed a good thermal stability.

391706 Single Bed Apparatus Designed for Studying Rapid Pressure Swing Adsorption Concepts

Abstract:

Pressure swing adsorption (PSA) is a cyclic gas separation technique that is just over 50 years old. It has many successful commercial applications including gas dehydration, air separation, hydrogen purification, and solvent vapor recovery. More applications are also on the horizon, including CO_2 capture from flue gas

Traditional, commercial PSA systems utilize cycle times on the order of minutes to tens of minutes, with individual step times no shorter than 0.5 min (30 s) or so. However, since the mid 80's there has been a great deal of interest in trying to reduce the cycle time to make the columns smaller and thus increase the feed throughput, which is the amount of gas processed per unit time per unit mass of adsorbent. These so called rapid PSA processes have met with limited success, however, until only recently.

Within the past 10 years, SeQual Technologies in San Diego and QuestAir in Vancouver (among less notable others) developed the rapid PSA concept into viable, commercial processes using rotary valve and rotary bed technology, respectively. They both have commercial processes that operate with cycle times on the order of a few seconds, with individual step times of less than 1 s. SeQual has been focusing on medical oxygen technology, while QuestAir has been focusing on hydrogen technology.

However, a paucity of work has been done on other rapid PSA applications; and overall, the understanding of the rapid PSA concept is quite limited with little fundamental information available. This leaves considerable room not only for the development of new theories and mechanisms underlying the rapid PSA concept, but also for the development of new applications for rapid PSA processes. The goal of this project has been to shed some light on both of these notions.

One of the shortcomings that have apparently limited the development of rapid PSA processes is a fundamental understanding of the controlling mass transfer mechanism that governs the performance. It has become clear to Ritter and co-workers over the past few years that the kind of information needed cannot be obtained by conventional measurement techniques, such as analyzing a column breakthrough curve. However, it can be obtained by building a prototypic rapid PSA process and then calibrating a rapid PSA process simulator to the performance data by varying the mass transfer coefficient. A 1-bed, rapid PSA system has been constructed for this purpose. So far, this rapid PSA system has been characterized by carrying out dynamic cycling experiments with single gases such as He, CO_2 and N_2, both with an empty bed and with the bed filled with beaded 13X zeolite over a wide range of cycle times and process conditions. Pressurization and depressurization step times as short as 0.5 s are possible in this system under both positive and negative gauge pressures and at different temperatures. The separation of a model binary gas mixture will be studied next. This presentation will discuss the latest results obtained with this unique 1-bed rapid PSA apparatus.

390670 Direct Measurement of Adsorption Thermodynamics of CO2 on Type I Adsorbents

Abstract:

The combination of an automated volumetric adsorption apparatus (Micromeritics ASAP 2020) and a commercial differential scanning calorimetry (DSC) is versatile apparatus to study adsorption energies of CO2 on solid aminosilica adsorbents. Since volumetric adsorption instrument controls the gas dosing precisely, high energy sites can be investigate at low coverages while interfaced with the DSC, which delivers high resolution data on the heat of adsorption. Another capability of this combination is to measure differential and integral heat of adsorption separately for each equilibrated pressure point while the adsorption isotherm is collecting via the volumetric adsorption apparatus. Therefore, one can investigate the energetic heterogeneity of adsorption sites and lateral interaction of adsorbed gas at various gas coverage. This combination has been used to measure the complete thermodynamic profile such as heat of adsorption, Gibbs free energy, and entropy at various temperatures on polyethyleneamine (PEI), which was impregnated on silica material for CO_2 adsorption.

A unique behavior has been observed for the adsorption of CO_2 over 40wt% PEI/SiO_2 at low CO_2 partial pressures (<0.4 Torr). In general, it is expected that the highest heat of adsorption will be measured at the lowest coverage, but we have determined that the heat of adsorption actually increases with inceasing coverage of CO2, which suggest a minimum requirement of CO_2 partial pressures to chemically adsorb the CO_2 gas on the amine groups.

AIChE Annual meeting, Atlanta, November 16-21 2014

Pure and Binary Net Adsorption of CH4, CO2 and C2H4

Orhan Talu, Prahar Vaidya and Aaron Moran
Cleveland State University
Cleveland, Ohio 44115, USA
(o.talu@csuohio.edu)

The ***net*** adsorption thermodynamic framework was introduced by Gumma and Talu (*Langmuir* 2010, 26(22), 17013-17023). Net adsorption fundamentally differs from ***excess*** and ***absolute*** adsorption by the placement of Gibbs dividing surface which is necessary to quantify any adsorption process. Net adsorption is a natural choice for microporous solids. It removes any ambiguity in adsorption phenomena by circumventing questionable measurements and quantities such as; 1) solid volume necessary for excess adsorption (typically based on non-adsorbing helium assumption), and 2) pore volume necessary for absolute adsorption (typically based on liquid density assumption in micropores).

This paper presents the first set of net adsorption data collected for pure and binary mixtures of methane, carbon dioxide and ethylene in silicalite at several temperatures. The measurement techniques are discussed and contrasted with primarily excess adsorption. Advantages and disadvantages are discussed. Data is examined for thermodynamic consistency and compared to ideal adsorbed solution predictions for both net and excess adsorption.

381481 Adsorption Kinetics Measurement Using a Volumetric System

Abstract:

Examples of adsorption kinetics with powders and pellets are presented in order to investigate in detail the use of the Quantachrome Autosorb iQ for the determination of mass transfer kinetics in nano-, meso- and macro-porous adsorbents.

The interpretation of volumetric dynamic experiments is extended to the uptake of argon at liquid argon temperatures in Y-type zeolites and the isothermal solution to the dynamic response is adapted for the case where the dosing volume is at a higher temperature. Differential pressure increments as well as large pressure steps are considered.

Carbon dioxide and nitrogen adsorption in pellets of 5A and 13X zeolites is used to discuss the interpretation of experiments where mild and significant heat effects are present. A description of how to plot the measured dosing cell pressure vs time is used to identify the controlling regimes at various stages of the adsorption dynamics - valve effect; near-isothermal uptake; and heat limited approach to final equilibrium.

Adsorbent Screening for Post-Combustion Carbon Capture Based on Optimum VSA Process Performance: A Simple Method Based on Equilibrium Isotherm Properties

Maninder Khurana and Shamsuzzaman Farooq
Department of Chemical and Biomolecular Engineering, National University of Singapore, 4 Engineering Drive 4, Singapore 117585

ABSTRACT

New adsorbents are being synthesized in large numbers claiming suitability for post-combustion carbon capture and concentration (CCC). However, no effective method has yet been developed to evaluate their performance in an actual CCC process. Proper screening of an adsorbent for CCC must either be based on extensive process experimentation or process simulation and optimization using realistic inputs for adsorption equilibrium and rate kinetics of fluid transport into the adsorbent pores. Experimental evaluation of adsorbents synthesized is too resource and time intensive to be practically feasible for routine screening. On the other hand, process simulation and optimization is computationally demanding and generally not available to all material researchers.

Since most of the adsorbents for post-combustion capture are equilibrium controlled, their equilibrium isotherm properties strongly influence the process performance. A proper relationship between easily quantifiable isotherm properties on process performance has not been established yet. Performance indicators such as breakthrough times, high loading capacity, etc., for CO_2 have not been proven to be directly correlated to process level performance of the adsorbent.

In the current work, a predictive correlation has been developed for estimating minimum energy penalty for post-combustion CCC yielding 95% purity and 90% recovery for an adsorbent as a function of some easily quantifiable CO_2 and N_2 equilibrium isotherm properties using a 4-step VSA process with light product pressurization (LPP). The 4-step VSA process with LPP has been reported in the literature to be the best cycle for post-combustion capture from dry gas using 13X zeolite. A closer physical insight of this process gives further assurance that its effectiveness should in general apply to adsorbents with favorable CO_2 isotherms. The process simulation is based on a non-isothermal, non-isobaric model, which uses extended dual site Langmuir isotherm to represent binary equilibrium and assumes macropore molecular diffusion controlled transport kinetics. The equilibrium isotherm model has been shown in the literature to reliably capture experimental results for a wide variety of systems, while the transport model is justified for an equilibrium controlled process. Thus, the chosen cycle and features of the simulation model are most appropriate for developing a versatile adsorbent screening tool for CCC application.

A parametric study has been conducted using design of experiments for estimating the effects of the following five CO_2 equilibrium isotherm properties: binary mixture selectivity over N_2, Henry's constant, loading, non-linearity and local slope at CO_2 partial pressure in the feed gas on the minimum energy penalty of the VSA process. Each combination of the aforementioned CO_2 isotherm properties, representing a possible adsorbent, has been optimized for the operating process variables using genetic algorithm based optimization. The significance of the independent variables and their interactions were tested using the analysis of variance (ANOVA) with 95% confidence limits (α= 0.05). The correlation has been validated by comparing with the minimum energy predictions from full process optimization for several known and promising hypothetical MOF adsorbents reported in the literature.

It has been found that the optimized process performance of any adsorbent is most sensitive its binary mixture selectivity for CO_2 over N_2. An important non-intuitive observation is that for the same binary selectivity, higher CO_2 loading, which also means increased N_2 loading, increases energy penalty for the system. The meta-model provides the much needed easy-to-use tool for material researchers for evaluating new adsorbents.

REMOVAL OF TOTAL ORGANIC ACID ANIONS AND HEAVY METAL IONS FROM INDUSTRIAL LEAN AMINE SOLVENT USING BIOPOLYMERIC CALCIUM ALGINATE

*Priyabrata Pal, Anjali A. Edathil and Fawzi Banat**

The Petroleum Institute, Abu Dhabi, United Arab Emirates

Email: fbanat@pi.ac.ae

Introduction

Treating of natural gas with Methyldiethanolamine (MDEA, 50 weight% used by Gasco Habshan, Abu Dhabi) is the best industrial method used for effective removal of H_2S and CO_2 in natural gas sweetening unit. The removal of heavy metal ions and heat stable salts (HSS) contaminants from lean MDEA is always a challenge to the gas industry. Natural biopolymers such as alginate are considering increasing attention due to their extraordinary affinity towards adsorption. Alginates (Alg) are hydrophilic natural polysaccharides obtained from brown seaweed algae, consisting of different proportions and sequential arrangements of α-L-guluronic acid (G-block) and β-D-mannuronic acid (M-block) units in linear chain. Researchers have demonstrated that calcium-alginate gel beads can remove heavy metals from industrial pollutants [1, 2].

In the present work, bio-polymeric adsorbent calcium alginate (Ca-Alg) was used for the first time to remove total contaminants present in lean MDEA [3]. The main aim of this work was to optimize the preparation procedure of calcium alginate hydrogel for removing both total organic acid anions as HSS and heavy metal ions like iron and chromium present in the industrial lean MDEA at varying temperature conditions observed in middle-east (23°C-53°C). The kinetic studies, two parameter equilibrium isotherm models like Langmuir and Freundlich and thermodynamic parameters like Gibb's free energy, enthalpy of adsorption were determined to check the feasibility of adsorption process.

Experimental

Materials

Alginic acid sodium salt of 91% food grade was obtained from Loba Chemie Pvt. Ltd, India and calcium chloride dihydrate was obtained from Merck KGaA, Germany. Lean amine containing 45 weight % MDEA was obtained from Gasco, Habshan, Abu Dhabi for the analysis.

Methods

The Ca-Alg gel bead preparation procedure was optimized using various conditions by dropping different weight % sodium alginate into calcium chloride solution. The concentration of sodium alginate and calcium chloride was optimized in terms of removal efficiency of total contaminants present in lean MDEA. The optimized Ca-Alg beads were prepared by drop wise addition of 1.0 weight % sodium alginate (Na-Alg) into 1.0 (M) $CaCl_2$ solution under constant stirring to prevent agglomeration. The beads were kept in $CaCl_2$ solution for 24 hours for complete polymerization and finally, washed with water at 60^0C under constant stirring to remove excess $CaCl_2$. The produced beads were further dried into a hard mass by keeping the beads in an oven at 60°C overnight [2]. The dry beads were used in the batch experimental studies as it can be easily handled and stored; making it as a good commercial adsorbent.

Instrumentation

The thermal degradation property of the calcium alginate beads were measured using a TGA, NetzschSta 409 PC/PG, Germany. The SEM and FTIR images were recorded and analyzed using FEI Quanta 200, Netherland and FTIR, Nicolet iS10, Thermo Scientific instrument with OMNIC Software respectively. The concentration of total organic acid anions in the lean amine solution was determined using UV-VIS Spectrophotometer (DR5000, Hach Lange; Test kit 365). Inductively coupled Plasma Optical spectroscopy (ICP-OES, Optima 8000; Perkin Elmer) was used for the elemental analysis of metal ions.

Batch Adsorption Studies

The kinetics and equilibrium parameters for the removal of total organic acid anions and heavy metal ions were studied from batch adsorption experiments using mixed batch reactor technique. Varying amount of Ca-Alg beads (0.2 – 5 gm) were added into 10 ml lean MDEA solution taken in a 150 ml conical flask and allowed to equilibrate in a water bath shaker (Dihan, Kora) at 110 rpm at different temperatures ranging from 23°C-53°C for 4 hours. Table 1 shows the equations used for the adsorption studies.

Table 1 Equations used for adsorption studies

Name	Equations	Parameters
Pseudo-second-order kinetics	$\frac{1}{(q_e - q_t)} = k_2 t$	q_t and q_e = The amounts of adsorbent adsorbed t = time, minute k_2 = Pseudo-second-order rate constant
Langmuir Isotherm	$q_e = \frac{q_{max} k_L C_e}{1 + k_L C_e}$	q_{max} = Maximum sorption capacity C_e = Equilibrium adsorbate concentration k_L =Langmuir isotherm constant

Freundlich Isotherm	$q_e = k_F C_e^{1/n}$	k_F = Freundlich constants n = constant of adsorption
Thermodynamic parameters	$\Delta G = -RT ln k_L$ $\Delta G = \Delta H - T\Delta S$ $ln k_L = -\left(\frac{\Delta H}{RT}\right) + \left(\frac{\Delta S}{R}\right)$	ΔG = Change in Gibbs free energy ΔH = Change in Enthalpy ΔS = Entropy change

Results and Discussions

Thermo Gravimetric Analysis (TGA)

Figure1 shows the thermal degradation of calcium alginate using TGA. The weight loss observed up to 200°C was caused by the dehydration of calcium alginate beads. In order to explain the thermal degradation process, it was observed that calcium carbonate was formed at 150°C and as temperature was increased calcium oxide and calcium hydroxide were found when heated to 700°C [4].

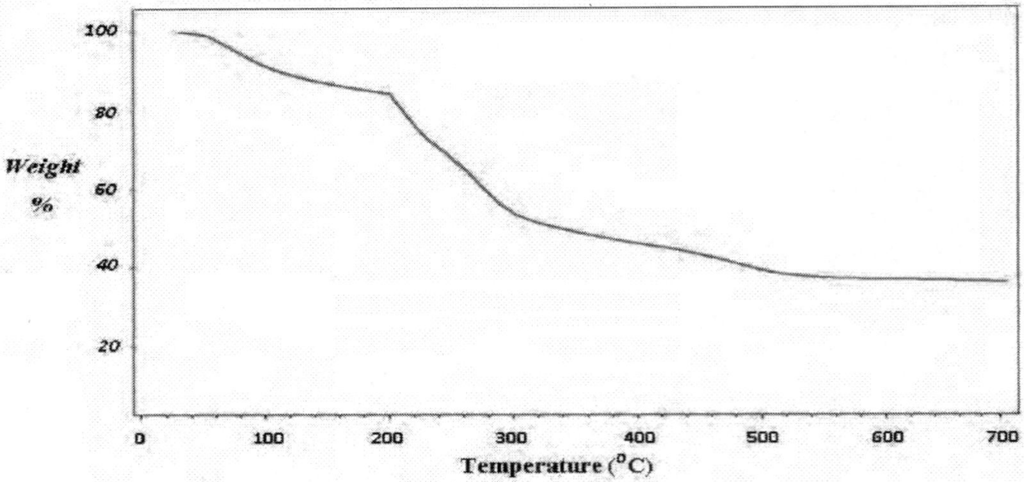

Figure1. Thermal degradation curve of calcium alginate beads

Fourier Transform Infrared (FTIR) Spectroscopy

The difference between the pure Ca-Alg hydrogel beads and adsorbed with MDEA were studied using FTIR analysis and shown in Figure2. A consistent shift in % of transmittance for Ca-Alg with pure and adsorbed with MDEA was due to adsorption of organic acid anions on the adsorbent surface [5].

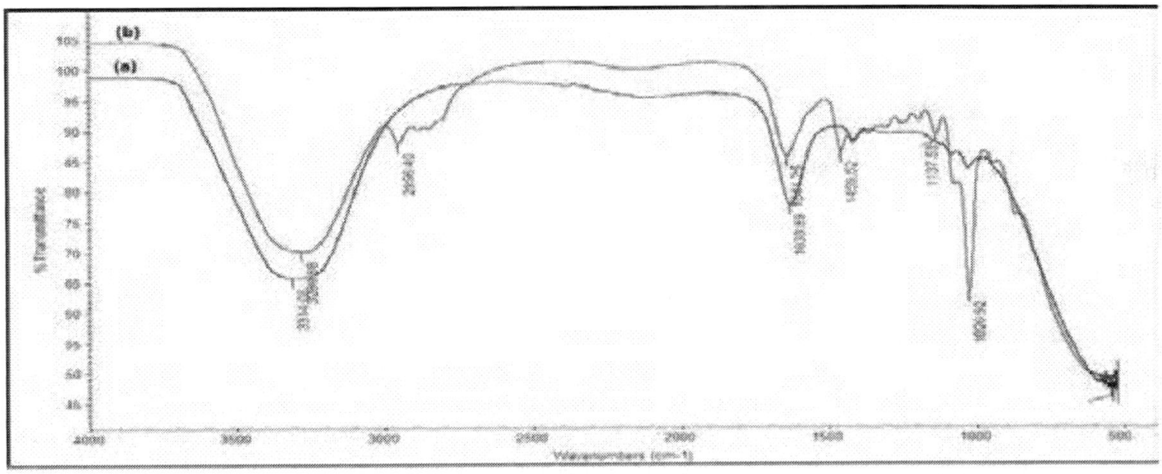

Figure2. FTIR Spectra of Ca-Alg beads a) before and b) after adsorption studies

Scanning Electron Microscope (SEM)

The SEM images (Figure3) obtained for Ca-Alg beads before (a) and after adsorption (b) were showed that the change in structure of the Ca-Alg beads due to adsorption reaction occurring between functional groups of alginate and the total contaminants present in lean MDEA as organic acid anions and heavy metal ions.

Figure3. SEM analysis of Ca-Alg beads a) before and b) after adsorption studies

Analysis of total organic acid anions (HSS) and heavy metal ions

The concentration of total organic acid anions (e.g. glycolate, formate, acetate, propionate, butyrate, valerate) in the lean MDEA was found to be 5685 ppm and the major metal ions were found to be chromium (615.7 ppb) and iron (1114 ppb) respectively [3].

Kinetic studies

To observe the kinetic behaviour of adsorption studies, 2.0 gm of Ca-Alg was taken in 10 ml lean MDEA and kept in the water bath shaker at 23°C. Sample was collected every half an hour to check the kinetic behavior. It was observed that the amount of organic acid anions adsorbed followed pseudo-second-order kinetics having rate constant 0.0124 g/mg. minute.

Adsorption Equilibrium Isotherm

The intake capacity of organic acid anions and heavy metal ions on Ca-Alg adsorbent at different equilibrium concentration was shown in Figure3. The isotherms were fitted two parameter isotherm models (Langmuir and Freundlich) to identify the interaction with the bio-polymeric sorbent. Based on the data fit, total organic acid anions follow a Langmuir type monolayer adsorption. The maximum adsorption capacity (q_{max}) values for total organic acid anions predicted from the Langmuir isothermal model were 909.09, 769.23, 625, 526.315 mg/g at 23°C to 53°C respectively. Freundlich isotherm also indicated a favorable adsorption. The negative values of Gibbs free energy (-12.948KJ/mol at 23°C) explained the spontaneous nature of the adsorption reaction and the positive values of enthalpy (2.596 KJ/mol) explained the endothermic nature of adsorption of total organic acid anions on Ca-Alg adsorbent.

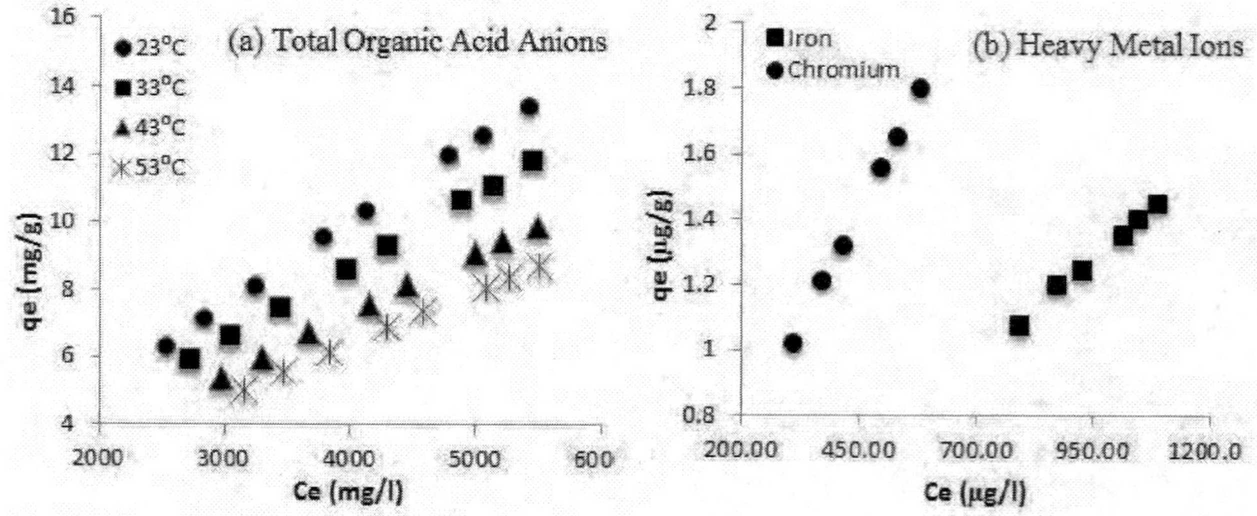

Figure4. Intake capacity of (a) organic acid anions and (b) heavy metal ions at room temperature

Regeneration of adsorbent

The organic acid anions and metal ions adsorbed on the adsorbent were desorbed by washing thrice with distilled water. Following the recovery of the adsorbent, it was thoroughly dried back to hard mass by keeping in oven at 60°C and was reused for further adsorption-desorption cycles. It was observed that even after 7 cycles of adsorption and desorption, the variation of residual concentration were less than 10%, demonstrating the extended usefulness of Ca-Alg adsorbent.

Conclusions

The batch adsorption studies using calcium alginate beads were tested to remove both total organic acid anions (55.58%) and heavy metal ions like chromium (52.82%) and iron (35.13%) from the Industrial lean amine at 23°C using 5.0 gm of Ca-Alg beads and it was observed that Langmuir isotherm best explained the adsorption process. The predicted maximum uptake capacity (q_{max}) for the removal of organic acid anions was 909.09 ppm at 23°C and showed spontaneity of the adsorption process (ΔG = -12.9484kJ/mol). The adsorbent could be regenerated using water and reused without considerable loss of its adsorption capacity through10 subsequent adsorption-desorption cycles, making it the best natural commercial sorbent.

Acknowledgement

The authors are grateful to Gas Research Centre, The Petroleum Institute, Abu Dhabi for funding the project. Sincere thanks to The GASCO, Habshan unit for their continued co-operation and support.

References

[1] M. M. Araújo, J.A. Teixeira, Trivalent chromium sorption on alginate beads, *J. Inter. Biodeter. & Biodeg.*, 40-1(1997) 63–74.

[2] Hyun Gyu Park, Tae Won Kim, Activated carbon containing alginate adsorbent for the simultaneous removal of heavy metals and toxic organics, *J. Pro. Chem*, 42 (2007), 1371-1377.

[3] P. Pal, F. Banat, Contaminants in Industrial Lean Amine Solvent and their removal using biopolymers–A new aspect. *Journal of Physical Chemistry & Biophys*ics, 2014, 4:1.

[4] Qing-shan Kong, Bing-bing Wan, Thermal Degradation and Flame Retardancy of Calcium Alginate Fibers, *Chinese Journal of Polymer Science,* 27-6 (2009) 807–812.

[5] M. Ruiz, C. Tobalina, H. Demey-Cedeñ, Sorption of boron on calcium alginate gel beads, *J. Rea. Fun. Poly.*, 73-4 (2013) 653–657.

365763 Microscopic View on Different Types of Transport of Light Gases in Mixed-Matrix Membranes: A Diffusion NMR Study

Abstract:

Mixed-matrix membranes (MMMs) are promising gas-separation membranes which consist of molecular sieve particles dispersed throughout a polymer matrix. Recently, reports in the literature have communicated examples of MMMs with remarkable sorption and diffusion selectivities which exceed the intrinsic properties of the polymer matrix. While macroscopic transport measurements are important for industrial application, gaining a complete microscopic understanding of the transport properties of MMMs from only macroscopic observations is not feasible due to the complexity of MMM structure.

Here we report an application of a pulsed field gradient (PFG) NMR technique to resolve different types of gas diffusion in MMMs. Such resolution was made possible by employing PFG NMR with a high field (17.6 T) and high field gradients (30 T/m). The diffusion types that were resolved, viz. studied separately, by this microscopic technique include gas diffusion inside the molecular sieve particles, transport in the polymer matrix between the particles and exchange of gas molecules between the molecular sieve particles and the polymer matrix. As an example, we present microscopic diffusion results for ethylene sorbate in a MMM composed of 6FDA/DAM polymer and ZIF-8 particles. It will be shown that a detailed knowledge of the different types of microscopic diffusion revealed by PFGNMR enhances understanding of long-range gas transport, i.e. transport on the length scales of displacements comparable with the thickness of MMMs.

374875 CO_2-Selective Membranes Containing Amino or Ethylene Oxide Groups for Carbon Capture from Flue Gas

Abstract:

This presentation covers two types of CO_2-selective membranes for carbon capture from flue gas: (1) amine-based membranes and (2) ethylene oxide (EO)-based membranes. The former is based on the facilitated transport mechanism, while the latter is based on the solution-diffusion transport mechanism. For the membranes containing amino groups, a high-molecular-weight of polyamine was successfully synthesized. Different selected aminoacid salts were incorporated into the polyamine solution for membrane preparation. The polyamine served as a fixed carrier, and the amino acid salts acted as mobile carriers. The amine-containing membranes showed a CO_2 permeance of more than 1100 GPU and a CO_2/N_2 selectivity of at least 200 at 57°C. For the membranes containing EO groups, we have synthesized membranes with polyamide-polyethylene oxide (PA-PEO) copolymer blended with polyethylene glycol dimethyl ether (PEG-DME) or polyethylene glycol (PEG). We have also synthesized a polydimethylsiloxane (PDMS) membrane reproducibly with a very high CO_2 permeance, which can be used as a caulking layer for our thin membranes. The membrane containing EO groups we synthesized was SO_2-insensitive but had a low CO_2/N_2 selectivity. The resulting membranes showed a CO_2 permeance of 940 GPU and a CO_2/N_2 selectivity of 25 at 57°C. The syntheses and transport properties of the membranes will be presented. The effects of different aminoacid salt mobile carriers, EO-containing small molecules and membrane thicknesses on the membrane transport performance will be discussed. We are continuing to improve the transport performance of both types of CO_2-selective membranes.

Zeolite like MOF membranes and their related gas separation properties

Osama Shekhah, Raja swaidan, Youssef Belmabkhout, Ingo Pinnau and Mohamed Eddaoudi
Functional Materials Design, Discovery & Development (FMD3), Advanced Membranes and Porous Materials Center, Division of Physical Sciences and Engineering, King Abdullah University of Science and Technology, 4700 King Abdullah University of Science and Technology (KAUST), Thuwal 23955-6900, Kingdom of Saudi Arabia.

Recently, we introduced successfully a new approach based on liquid phase epitaxy (LPE) to construct continuous ZIF-8 thin film on gold support.[1] Consequently, the achieved control over film growth thickness and continuity offer potential to effectively implement the LPE technique for the fabrication of continuous ZIF-8 membranes on porous supports.[2-5] Here we report the use of the LPE method for the successful controlled growth of ZIF-8 film on a porous alumina support. The range of gas permeance measured on the ultrathin ZIF-8 membrane fabricated by the LPE growth method was at least 100 times lower than the corresponding results reported in the literature although its thickness is one of the thinnest ever reported (0.5 and 1.6 μm). This, in conjunction with clear time-lag behavior and high O_2/N_2 selectivity (2.5) indicates high quality, reproducible membranes with negligible defects and gaps. The resulting ZIF-8 membrane was thoroughly characterized using state-of-the-art adsorption and permeation techniques. Single-gas permeation studies were performed for relevant gases (H_2, CO_2, O_2, N_2, CH_4, C_2H_4, C_2H_6, C_3H_6,C_3H_8 and n-C_4H_{10}) and gas mixtures (C_3H_6/C_3H_8 and CH_4/n-C_4H_{10}) using the time-lag method.[25] Particularly, we report for the first time unprecedented and unique findings pertaining to CH_4/n-C_4H_{10} mixture permeation through the ZIF-8 membrane.

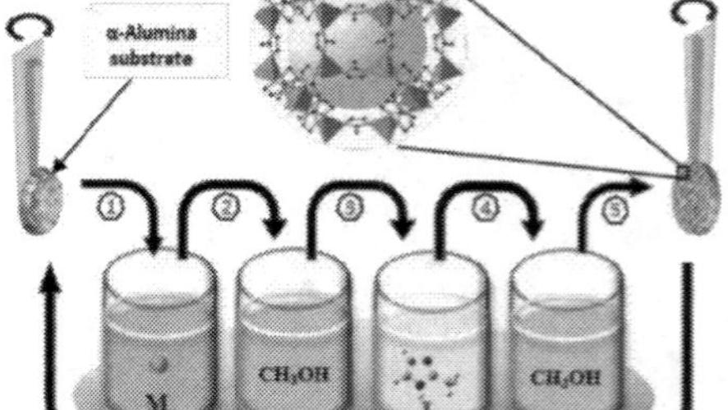

Figure 1. A schematic representation of the LPE method used for the growth of ZIF-8 membrane. The substrate is immersed in metal ions solution (1), washed with solvent (2), immersed in linker solution (3) washed with solvent (4). The process is repeated to grow more cycles (5).

References

[1] O. Shekhah, M. Eddaoudi, *Chem. Comm.*, 2013, **49**, 10079.
[2] O. Shekhah, *Materials*, 2010, **3**, 1302.

[3] O. Shekhah, H. Wang, S. Kowarik, F. Schreiber, M. Paulus, M. Tolan, C. Sternemann, F. Evers, D. Zacher, R. A. Fischer, C. Wöll, *J. Am. Chem. Soc,*. 2007, **129**, 15118.
[4] S. W. Rutherford, D. D. Do, *Adsorption*, 1997, **3**, 283.
[5] O. Shekhah et al. Chem. Commun., 2014, 50, 2089--209

382019 Sorption of CO2/CH4 Mixtures in a Polymer of Intrinsic Microporosity (PIM) : Effect of Temperature, Presence of a Tetrazole Unit (TZ PIM) and Thermal Treatment

Abstract:

The aim of this work is the study of solubility selectivity with respect to CO_2/CH_4 mixture of a polymer of intrinsic microporosity (PIM-1) and a modified PIM containing tetrazole unit (TZ-PIM). Sorption experiments were carried out at 25°C, 35°C and 50°C up to a pressure of 35 atm at three molar fractions of CO_2 (10.9, 31.3, 50.9%). Pressure decay equipped with gas chromatograph was used for the experiments.

The solubility coefficient of CO_2 is lower for mixed gas experiment than for pure gas experiment, as well as the solubility coefficient of CH_4, however the decrease of solubility coefficient of CH_4 is higher than that of CO_2. This behaviour gives rise to mixed gas solubility selectivity values higher than corresponding pure gas solubility selectivity. Sorption enthalpies were calculated at different gas mixture compositions.

The same mixed gas sorption experiments were also performed in TZ-PIM (PIM containing tetrazole unit) [3] at 25°C, 35°C and 50°C and molar fraction of CO_2, 11.3% CO_2, 30.7% CO_2 and 50.3% CO_2 up to a pressure of 35 atm. Solubility selectivity of TZ-PIM for CO_2/CH_4 mixture is higher than solubility selectivity of PIM-1. This is due to the tetrazole group which improves the solubility of CO_2 while leaving unaffected the solubility of CH_4. Solubility coefficient for both gases decreases with increasing temperature.

Finally, in order to study the physical ageing, we performed a series of tests in which the ageing of films of PIM-1 and TZ-PIM is accelerated with high temperature treatments under vacuum at 120, 140 and 160°C. The effect of thermal treatment is more evident on the diffusivity than on the solubility coefficient, in both types of membranes. In particular the diffusivity of both gases decreases with increasing heat treatment temperature. Trends of diffusivity with concentration are also thoroughly inspected.

[1]O. Vopicka, M. G. De Angelis, N. Du, N. Li, M. D. Guiver, G. C. Sarti, *Mixed gas sorption in glassy polymeric membranes: II. CO2/CH4 mixtures in a polymer of intrinsic microporosity (PIM-1)*, J. Membrane Sci. 2014, 459, pp. 264 - 276

[2] O. Vopicka, M. G. De Angelis, G. C. Sarti, *Mixed gas sorption in glassy polymeric membranes: I. CO2/CH4 and n-C4/CH4 mixtures sorption in poly(1-trimethylsilyl-1-propyne) (PTMSP)*, J. Membrane Sci. 2014, 449, pp. 97 - 108

[3] N. Du, G. P. Robertson, M. M. Dal-Cin, L. Scoles, M. D. Guiver Polymers of intrinsic microporosity (PIMs) substituted with methyl tetrazol, Polymer 53 (2012) 4367-4372

Fabrication and characterization of PBI-Matrimid blend gas separation membranes

Leslie Schulte, Shuzhen Qiu, Michael Wales, Dr. Mary Rezac

Membranes act as barriers between two phases. One way that membranes are used is to produce streams that are enriched in one or more of the components of the feed. The use of membranes in industrial settings requires the membrane to be stable and inexpensive with large fluxes and selectivities. In order to achieve large fluxes, the separating layer of the membrane must be thin but in order to separate components, this layer must also be defect-free. Currently, membranes are used in reverse osmosis, gas separations and as membrane reactors. To expand the use of membranes, we intend to make a more thermally stable asymmetric polymeric membrane by blending two polymers; PBI (polybenzimidazole) and Matrimid 5218. PBI has significant thermal stability (T_g = 435°C) but low permeabilities. Matrimid has some thermal stability (T_g = 320°C) but larger permeabilities. By combining these two polymers, we hope to meet all four requirements for a gas separation membrane.

Membranes were fabricated using phase inversion to produce a thin, defect-free skin layer on top of a porous substructure. This presentation will focus on factors in the fabrication process that affect the gas transport properties of blend membrane and characterization of the blend membrane. We've developed a process that produces a membrane with equal parts of PBI and Matrimid that have H_2 fluxes above 100 GPU at room temperature. Selectivities are close to a dense film of either material suggesting the skin layer is defect-free. Gas fluxes, including N_2, CH_4, O_2, CO_2 and H_2, have been measured from room temperature to 300°C. Blend membranes demonstrate better thermal and chemical resistance than pure Matrimid membrane. Thermal post-treatments have been investigated to improve gas transport properties. Thermal treatments result in a membrane with better selectivities but lower fluxes at all temperatures. The blend membrane demonstrated consistent gas flux at 200°C for 25 days, with nitrogen fluxes rising from 15.3 to 16.7 over this time and $\alpha_{H2/N2}$ decreasing from 7.5 to 5.7. It is our hope by developing a procedure for fabricating these blend membranes and characterizing them, our work will enable the use of membranes in new applications.

Development of Polyimide Membranes with Dual Selectivity for CO_2 Separations

Shuagjiang Luo (sluo2@nd.edu), Gregory Kline (gkline@nd.edu), Ruilan Guo* (rguo@nd.edu)

Department of Chemical and Biomolecular Engineering, University of Notre Dame
Notre Dame, IN 46556-5637

Separations involving CO_2 are very important in many industrial processes such as syngas ratio adjustment, natural gas sweetening for clean energy supply or carbon capture from flue gas for environmental remediation. Conventional glassy polymer membranes have demonstrated insufficient capability for CO_2 removal in that the performance of these membranes is primarily controlled by the diffusivity selectivity, i.e., size discrimination between various gas molecules. In addition membrane plasticization due to sorption of condensable CO_2 further limits their separation performance. Polymeric membranes based on poly(ethylene oxide) (PEO) have recently attracted growing interests in CO_2-related separations in that the ethylene oxide (EO) units have high affinity towards CO_2 promoting high separation efficiency due to high solubility selectivity. In this paper, we report the synthesis and characterizations of a series of new PEO-containing segmented copolymer membranes that are designed to have dual selectivity for CO_2 separations, i.e., diffusivity selectivity and solubility selectivity. These newly designed and synthesized copolymers consist of alternating PEO sequences and rigid iptycene-containing polyimide segments, wherein the bulky iptycene-containing imide sequences provide high gas permeability and diffusivity selectivity, while the CO_2-philic characteristic of PEO sequences provides high solubility selectivity. The copolymers are systematically varied in their chemical structure in terms of PEO sequence length, PEO weight fraction, and iptycene structure. FTIR and ^{1}H NMR confirm the desired macromolecular structure of the copolymers. Physical properties such as glass transition temperature, crystallinity, and thermal stability have been shown to strongly depend on the chemical structure/composition of the copolymers. Depending on the copolymer composition and the intra- and inter-molecular interactions between the soft PEO sequence and hard iptycene units, these membranes show varied transport properties that have promise for efficient CO_2 separation. Synthesis and characterization of these new PEO-containing copolyimides will be presented. The dependence of macroscopic transport properties on microscopic structures for these new membranes will be discussed.

Model-Free Control of Fractional Crystallization

The Department of Energy must extract, vitrify, and encase the hazardous constituents contained in 56 million gallons of nuclear waste currently stored at the Hanford site in the state of Washington. Separating non-radioactive components from the waste can reduce costs and expedite cleanup efforts (Herting, 1996; Nassif et al., 2008). In particular, sodium salts—which are non-radioactive and innocuous by themselves—make up a significant portion of the unprocessed waste and potentially can be removed with a crystallization-separation operation and treated as low-activity waste. However, as nuclear waste solutions contain many solutes that can crystallize, methods for reliably monitoring and controlling *multi-solute* crystallizations are needed. The operation in such a circumstance is referred to as fractional crystallization.

In this work we explore the use of process analytical measurement techniques to implement model-free methods for controlling fractional crystallizations of a model two-solute system (sodium nitrate and sodium sulfate in water) with the objective of improving the crystal size distribution to facilitate separation. Model-free control strategies—such as supersaturation or concentration control (Fujiwara et al., 2005)—have been successfully applied to control single-solute crystallization (Nagy and Braatz, 2012), but have not been extended to multicomponent crystallizations. The presence of multiple solutes affects solution behavior and can convolute measurement signals. This introduces complications in the supersaturation control strategy, which relies on solubility information and concentration measurement feedback (via infrared or other spectroscopic techniques). In considering these complications for the model two-solute solution we developed methodologies for quickly gathering the solubility of a multicomponent solution and also extracting the concentration of multiple components from the complex solution infrared absorbance spectra. Applying these techniques we are able to apply supersaturation control during fractional crystallization.

Fujiwara, M., Nagy, Z. K., Chew, J. W. and Braatz, R. D. 2005. First-Principles and Direct Design Approaches for the Control of Pharmaceutical Crystallization. *Journal of Process Control* 15, 493-504

Herting, D. L. 1996. Clean Salt Process - Final Report. Environmental Management

Nagy, Z. K. and Braatz, R. D. 2012. Advances and New Directions in Crystallization Control. *Annual Review of Chemical and Biomolecular Engineering, Vol 3* 3

Nassif, L., Dumont, G., Alysouri, H. and Rousseau, R. W. 2008. Pretreatment of Hanford Medium-Curie Wastes by Fractional Crystallization. *Environmental Science & Technology* 42, 4940-4945

Cooling Crystallization within Dual-Impinging-Jet Mixers: Mathematical Modeling, Theory Analysis, and Experimental Validation

Mo Jiang and Richard D. Braatz
Department of Chemical Engineering, Massachusetts Institute of Technology

An objective in pharmaceutical crystallization is to control crystal size distribution (CSD), to improve process efficiency and product quality. An effective way to improve the control of CSD is through combining continuous seeding and growth with concentration control [1]. One well-studied approach for continuous seeding is to combine solution and anti-solvent streams in a dual-impinging-jet (DIJ) mixer (e.g., [1]–[3]). Rather than anti-solvent crystallization as commonly used, this DIJ mixer combines hot and cold saturated solutions to generate small uniform seed crystals [4], which appears to be the first time a cooling DIJ nucleation has been reported. The product crystals were less than 10 microns in length, placing them in the correct size range for direct application in inhalers [5].

Theoretically, if the mixing was perfect (that is, concentration and temperature were completely mixed at the molecular scale in the mixer), the average supersaturation level would be too low to nucleate crystals in a cooling DIJ mixer. The surprising result of nucleating crystals by combining hot and cold saturated solutions in a DIJ mixer motivated the theoretical analysis of the system, which showed that nucleation was enabled by the much faster energy transfer than mass transfer rates near the interface. This difference in rates enabled the temperature of the hot solution to drop to approximately the average temperature of the two solutions before its solution concentration had significantly changed, resulting in a supersaturation sufficiently high to nucleate crystals.

The theoretical analysis is supported with process simulations. The two-dimensional energy and molar balances are solved using COMSOL input with an analytical solution for the velocity field, to generate spatial distributions of temperature, concentration, and supersaturation near the hot-cold interface within a cooling DIJ mixer. In the most important spatial region of importance for characterization of nucleation, the two-dimensional fields are shown to be very close to analytical solutions derived from a one-dimensional approximation of the energy and molar balances. This simplification applied the derivation of design criteria to facilitate quick assessment of whether any particular pharmaceutical-solute combination will nucleate crystals in a cooling DIJ mixer, based on the physicochemical properties. These criteria could save time and material by avoiding or reducing trial-and-error experiments, which is helpful at the early stage of pharmaceutical process development. The process simulations are then expanded to include the full solution of the spatially-varying population balance model [1] to calculate the crystal size distributions throughout the spatial domain. The mathematical models and methodologies are validated for the crystallization of L-asparagine monohydrate (LAM) in aqueous solution.

References

[1] X. Y. Woo, R. B. H. Tan, and R. D. Braatz, "Precise tailoring of the crystal size distribution by controlled growth and continuous seeding from impinging jet crystallizers," *CrystEngComm*, vol. 13, no. 6, pp. 2006–2014, 2011.

[2] B. K. Johnson and R. K. Prud'homme, "Chemical Processing and Micromixing in Confined Impinging Jets," *AIChE J.*, vol. 49, no. 9, pp. 2264–2282, 2003.

[3] M. Midler Jr., E. L. Paul, E. F. Whittington, M. Futran, P. D. Liu, J. Hsu, and S. H. Pan, "Crystallization Method to Improve Crystal Structure and Size," U.S. Patent 5,314,506A, May 24, 1994.

[4] M. Jiang, M. H. Wong, Z. Zhu, J. Zhang, L. Zhou, K. Wang, A. N. Ford Versypt, T. Si, L. M. Hasenberg, Y. E. Li, and R. D. Braatz, "Towards achieving a flattop crystal size distribution by continuous seeding and controlled growth," *Chem. Eng. Sci.*, vol. 77, pp. 2–9, 2012.

[5] T. G. D. Capstick and I. J. Clifton, "Inhaler technique and training in people with chronic obstructive pulmonary disease and asthma," *Expert Rev. Respir. Med.*, vol. 6, no. 1, pp. 91–103, Feb. 2012.

383475 Crystallization Kinetics in an Airlift Crystallizer: A Parameter Estimation Study

Abstract:

Batch solution crystallization is one of the oldest downstream processes, which involves the contribution of several phenomena such as growth, nucleation, agglomeration, dissolution etc. Among those kinetic processes the growth of seeds is often the only desirable one, whereas primary and secondary nucleation and agglomeration give rise to problems in both process and product control leading to variability in the product quality and unwanted crystal size distributions (CSD), shapes and impurity uptakes.

Crystallization in traditional stirred-vessel crystallizers, in which especially secondary nucleation is difficult to avoid, often leads to a wide multimodal CSD. Recently it has been shown [1,2] however that when crystallization is performed in a recently developed air lift crystallizer a strong suppression in the amount of fine crystals can be obtained compared to a stirred crystallizer; probably due to the suppression of the rate of secondary nucleation by attrition.

To prove this suppression in the rate of secondary nucleation and to get more insight in the kinetics and thus in the performance in the airlift crystallizer a more quantitative study is necessary in which the nucleation and growth phenomena are modelled and analyzed in both type of crystallizers.

The mixing strategy is the major distinction between an air lift and a stirred vessel crystallizer. In an airlift crystallizer, injected air bubbles are responsible for the mixing and the suspension of the crystals in an airlift, whereas the impeller fulfils these tasks in a stirred vessel crystallizer. It is most likely that the differences in hydrodynamics between the two types of crystallizer could have a strong impact on the crystallization kinetics. Especially secondary nucleation might be affected due to the absence of the high energetic collisions between the impeller and the crystallizer which are dominant in a stirred vessel crystallizer. [3-6].

In this contribution, the growth and nucleation behavior during batch operation in an air lift crystallizer for L-ascorbic acid/water system has been analyzed. Model verification and parameter estimation was performed based on experiments under a range of process conditions. A comparison is made between the results of the eighteen liter air lift crystallizer and that of a five liter stirred tank crystallizer to get more insight in dominating crystallization mechanisms in the airlift; allowing a further optimization of this novel type of crystallizer.

Crystallization experiments, described earlier [1], have been used with a number of different supersaturation profiles, batch times and seed loadings to estimate the parameters of the crystallization models in a model verification and parameter estimation procedure. The measured supersaturation profile and CSD of seeds and final product from these experiments were used to select the most appropriate models for nucleation and growth and to estimate the relevant kinetic parameters using a variety of models implemented in gPROMS Model Builder.

Both size dependent and size independent growth models are used along with different models for primary and secondary nucleation. Different secondary nucleation mechanisms were taken into account ranging from attrition based to activated nucleation mechanism [2,3,5,6,7].

The results show indeed a strong suppression of the secondary nucleation by attrition in the airlift crystallizer compared to stirred vessel crystallizer under all studied process conditions, leading to larger crystals and a reduced number of fines. However, the supersaturation profile and CSD in the airlift crystallizer cannot be simply described by a reduction in attrition rate. The parameter estimation study revealed that a different, activated nucleation model is needed to describe the experimental results properly. Apparently in the absence of the impeller in the airlift crystallizer, which avoids the attrition caused by the crystal-impeller and crystal-baffle collisions, an alternative secondary nucleation mechanism becomes dominant. An activated surface nucleation model, in which it is assumed that preordered clusters or pre-nuclei are formed in the immediate vicinity of the crystal surface or on the crystal surface [7] gives the best fit for the experimental data of the air lift crystallizer.

This means that although secondary nucleation by attrition is suppressed in airlift crystallizer, secondary nucleation is not completely absent in this type of crystallizer for the studied process conditions. The indication of the presence of an alternative nucleation mechanism is important as it opens possibilities to optimize the production of high quality crystals in this type of crystallizer.

Acknowledgments: *This work was supported by the European Commissions Framework 7 program through the OPTICO consortium*

References:

1. *R. Lakerveld, J.J.H van Krochten, H.J.M.. Kramer, An airlift crystallizer can suppress secondary nucleation at a higher supersaturation compared to a stirred crystallizer, submitted (2013)*
2. *Soare, A., Lakerveld, R., van Royen, J., Zocchi, G., Stankiewicz, A.I., Kramer, H.J.M., 2012. Minimization of Attrition and Breakage in an Airlift Crystallizer. Ind Eng Chem Res 51, 10895-10909.*
3. *J. W. Mullin, M. Chakraborty, K. Mehta, Nucleation and growth of ammonium sulphate crystals from aqueous solution, Journal of Applied Chemistry, 20, 12 (1970)*
4. *A. Soare, S. A. Pérez Escobar, A. I. Stankiewicz , M. Rodriguez Pascual , H. J. M. Kramer, 2-D Flow and Temperature Measurements in a Multiphase Airlift Crystallizer,* Ind. Eng. Chem. Res., *52, 34 (2013)*
5. *S.K. Bermingham, H.J.M. Kramer, and G.M. van Rosmalen, Towards on-scale crystalliser design using compartmental models. Computers & Chemical Engineering, 22 (1998)*
6. *Kramer, H.J.M., S.K. Bermingham, and G.M. van Rosmalen, Design of industrial crystallisers for a given product quality. Journal of Crystal Growth,. 198 (1999)*
7. *A. Mersmann, B. Braun, M. Loffelmann, Prediction of crystallization coefficients of the population balance, Chemical Engineering Science, 57, 20 (2002)*

376312 Protein Crystal Shape and Size Control Using a Plug Flow Crystallization Configuration

Abstract:

Crystallization plays a key role in the context of separation and purification methods for the production of therapeutic drugs. Considering the fact that crystal size and shape distributions have significant influence on the bioavailability of drugs such as the dissolution rate, filterability in the subsequent downstream process, and stability as a carrier to the target site, it becomes of particular interest to the pharmaceutical industry to produce crystals with desired size and shape distributions [1]-[2]. Traditionally, batch crystallization processes have been widely used in the pharmaceutical industry. However, owing to a few well-known potential drawbacks OF batch processes such as batch-to-batch variability, the difficulty in the scale-up, mixed suspension mixed product removal (MSMPR) crystallization process has received growing attention, and many efforts have been made in order to produce crystals from the MSMPR process with a higher production rate and desired product quality [3]-[4]. However, due to the presence of back-mixing, those crystals nucleated at a later stage during the crystallization process will reside a relatively short amount of time in the crystallizer and thus they will end up leaving the crystallizer with undesired size and shape distributions. To this end, plug flow crystallizer (PFC) has been proposed to produce crystals with desired size and shape distributions [5]-[7].

Motivated by the above considerations, the present work focuses on modeling of a PFC with 5 segments for the production of tetragonal HEW lysozyme crystals through kinetic Monte Carlo (kMC) simulation methods in the way described in [8] using the rate equations originally developed by [9]. A seeding strategy is used to decouple the nucleation from the crystal growth processes [10]. Furthermore, a constraint on the supersaturation level is imposed so that the system is enforced to stay in the metastable regime where the degree of primary nucleation is negligible [11]. Then, a population balance equation (PBE) is presented to describe the spatio-temporal evolution of the crystal volume distribution, and by applying the method of moments (MOM) to the PBE, a reduced-order moment model is derived because kMC models are not readily available in a closed form. Along with the mass and energy balance equations, the leading moments are used for the estimation of the spatio-temporal evolution of the crystal size and shape distributions in an optimization problem. More specifically, the crystallizer jacket temperatures at each segment and the superficial flow velocity are chosen as the decision variables in the optimization problem and the objective function is to minimize the squared deviation of the average size and shape of crystals produced throughout the PFC. Besides, dynamic models are developed and used for the design of a feed-forward control (FFC) strategy for the production of crystals with desired size and shape distribution properly suppressing the undesired effect caused by disturbances in the inflow solute concentration and crystal fines distribution.

[1] Patience DB, Rawlings JB. Particle shape monitoring and control in crystallization process. AIChE J. 2001;47:2125-2130.

[2] Wang L, Lee MH, Barton J, Hughes L, Odom TW. Shape-control of protein crystals in patterned microwells. J. Am. Chem. Soc. 2008;130:2142-2143.

[3] Griffin DW, Mellichamp DA, Doherty MF. Reducing the mean size of API crystals by continuous manufacturing with product classification and recycle. Chem. Eng. Sci. 2010;65;5770-5780.

[4] Alvarez A, Singh A, Myerson AS. Crystallization of cyclosporine in a multistage continuous MSMPR crystallizer. Crysal Growth & Design. 2011;11:4392-4400.

[5] Eder R, Schmitt E, Grill J, Radl S, Gruber-Woelfler H, Khinast J. Seed loading effects on the mean crystal size of acetylsalicylic acid in a continuous-flow crystallization device. Cryst. Res. Technol. 2011;3:227-237.

[6] Ridder B, Majumder A, Nagy Z. Population balance model based multi-objective optimization of a Multi-Segment Multi-Addition (MSMA) continuous plug flow antisolvent crystallizer. Ind. & Eng. Chem. Res. 2014;53:4387-4397.

[7] Alvarez A, Myerson A. Continuous plug flow crystallization of pharmaceutical compounds. Crysal Growth & Design. 2010;10:2219-2228.

[8] Kwon JS, Nayhouse M, Christofides PD, Orkoulas G. Modeling and control of crystal shape in continuous protein crystallization. Chem. Eng. Sci. 2014;107:47-57.

[9] Durbin SD, Feher G. Simulation of lysozyme crystal growth by the Monte Carlo method. J. Cryst. Growth. 1991;110:41-51.

[10] Liu JJ, Ma CY, Hu YD, Wang XZ. Effect of seed loading and cooling rate on crystal size and shape distributions in protein crystallization - a study using morphological population balance simulation. Comp. & Chem. Eng. 2010;34:1945-1952.

[11] Shi D, Mhaskar P, El-Farra NH, Christofides PD. Predictive control of crystal size distribution in protein crystallization. Nanotechnology. 2005;16:S562-S574.

384909 Crystallization Process Design: Accelerating Process Design through Modelling and Control

Abstract:

Pressure to deliver robust, high-performing processes at scale with compressed timelines and greater process understanding is increasing. Advanced modelling and process control represent one possible approach to accelerating process development while delivering on process understanding and performance. These approaches compliment and support the experimental programs undertaken during process development.

In this talk, the application of novel crystallization process models, continuous crystallization operation and advanced process control to the design of batch crystallization processes will be discussed. The deployment of these tools allows for the systematic mapping and interrogation of the design space, streamlining the development of a robust process at scale. A number of case studies applying this approach to process design will be presented.

Title

Steady-state and Startup Optimization for Combined Cooling and Antisolvent Crystallization (CCAC) in Continuous Mixed Suspension Mixed Product Removal (MSMPR) Cascade Crystallizers

Authors

Yang Yang and Zoltan K. Nagy, Chemical Engineering, Purdue University, West Lafayette, IN

Abstract

Crystallization is widely used as an important separation and purification technique in pharmaceutical industry [1]. Traditionally, most crystallization processes are performed in batch systems due to the high operating flexibility and short development time. However, in recent decades continuous crystallization is of great interest due to its enhanced reproducibility, reduced operating cost, simplified scale up and improved efficiency of process and equipment as compared with the traditionally used batch crystallization [2]. Similarly to the batch crystallization, either cooling or antisolvent addition could be used to generate supersaturation in a continuous crystallization process. Alternatively, combined cooling and antisolvent crystallization (CCAC) is also possible in continuous crystallization. Because in batch systems the CCAC approach could result in higher yield and larger control freedom on the multiple crystal properties like crystal size, shape and crystalline form as compared with cooling or antisolvent addition only crystallization [3].

Mixed suspension mixed product removal (MSMPR) crystallizer is one of the most important continuous crystallizers with its benefits demonstrated in the literature [2][4-7].

In continuous manufacturing, the control of startup is always a very crucial problem, due to the considerable amount of time product that can be wasted during this time. Methods to shorten the startup duration time and reduce the amount of waste in continuous crystallization are desirable.

This work presents a comprehensive study on the various control and optimization approaches for both the steady-state and startup behaviors of a continuous MSMPR cascade crystallizer through a model-based framework. Both cooling and antisolvent addition are applied to generate supersaturation at all cascade stages in order to provide the largest degree of freedom that can lead to the maximum steady-state crystal size achievable through an offline open-loop control approach. The optimal operating trajectory is obtained and compared with the optimal batch trajectory under similar scenario. Such comparison proves that the batch system can achieve larger crystals as compared with the cascade system. However, the optimal trajectories of the two systems show great similarities on both temperature and antisolvent dimensions.

In addition, different methods to shorten the startup duration time and reduce the waste amount have been discussed and proved to be feasible using a same open-loop optimal control framework. These methods include changing the initial solution compositions, applying dynamic antisolvent profiles, applying dynamic temperature profiles, and applying dynamic combined antisolvent/temperature profiles. The most efficient and practical way to achieve the goal turns out to be using dynamic antisolvent profiles only during startup. The proposed approach could reduce with approximately 50% of the startup duration and the waste amount while having no impact on the final steady-state product qualities.

References:

[1] Myerson, A. S. Handbook of Industrial Crystallization, 2nd ed.; Butterworth-Heinemann: Boston, 2002.

[2] Alvarez, A. J.; Singh, A.; Myerson, A. S. Crystallization of cyclosporine in a multistage continuous MSMPR crystallizer. *Cryst. Growth Des.* **2011**, 11, 4392-4400.

[3] Yang, Y.; Nagy, Z. K. Model-based systematic design and analysis approach for unseeded combined cooling and antisolvent crystallization (CCAC) systems. *Cryst. Growth Des.* **2014**, 14, 687-698.

[4] Wong, S. Y.; Tatusko, A. P.; Trout, B. L.; Myerson, A. S. Development of continuous crystallization processes using a single-stage mixed-suspension, mixed-product removal crystallizer with recycle. *Cryst. Growth Des.* **2012**, 12, 5701-5707.

[5] Quon, J. L.; Zhang, H.; Alvarez, A.; Evans, J.; Myerson, A. S.; Trout, B. L. Continuous crystallization of aliskiren hemifumarate. *Cryst. Growth Des.* **2012**, 12, 3036-3044.

[6] Zhang, H.; Quon, J.; Alvarez, A. J.; Evans, J.; Myerson, A. S.; Trout, B. Development of continuous anti-solvent/cooling crystallization process using cascaded mixed suspension, mixed product removal crystallizers. *Org. Process Res. Dev.* **2012**, 16, 915-924.

[7] Vetter, T.; Burcham, C. L.; Doherty, M. F. Regions of attainable particle size in continuous and batch crystallization processes. *Chem. Eng. Sci.* **2014**, 106, 167-180.

AUTHOR INDEX

AUTHOR INDEX

AUTHOR INDEX

AUTHOR INDEX

AUTHOR INDEX

AUTHOR INDEX

AUTHOR INDEX

AUTHOR INDEX

AUTHOR INDEX

AUTHOR INDEX

AUTHOR INDEX

AUTHOR INDEX

AUTHOR INDEX

AIChE
120 Wall Street, FL 23
New York, NY 10005-4020

ISBN 978-1-5108-1272-7

9 781510 812727

CURRAN ASSOCIATES INC.
proceedings.com